ACID REIGN '95?

This publication
is printed on
totally chlorine free paper
from Håfreströms AB,
based on
totally chlorine free Z-pulp
from Södra Cell AB.

Håfreströms AB, S-464 82 Åsensbruk, Sweden. Tel. +46 530 365 00, Fax +46 530 309 97

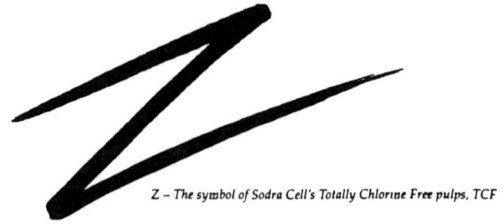

Z – The symbol of Sodra Cell's Totally Chlorine Free pulps, TCF

Södra Cell AB, S-351 89 Växjö, Sweden. Tel. +46 470 89 000, Fax +46 470 479 53

Acid Reign '95?

*Proceedings from the 5th International Conference
on Acidic Deposition: Science & Policy
Göteborg, Sweden, 26-30 June 1995*

Volume 4
*Emissions, Atmospheric and Deposition Processes,
Acidification outside Europe and North America, Critical Loads,
Regional Air Pollution Control Strategies, Effects on Materials*

Edited by

PERINGE GRENNFELT

*Swedish Environmental Research Institute,
Göteborg, Sweden*

HENNING RODHE

*Department of Meteorology, Stockholm University,
Stockholm, Sweden*

EVA THÖRNELÖF

*Swedish Environmental Protection Agency,
Stockholm, Sweden*

and

JOE WISNIEWSKI

*Wisniewski and Associates Inc.,
Falls Church, Virginia, USA*

Reprinted from *Water, Air, and Soil Pollution* 85(1–4), 1995

Springer Science+Business Media, LLC

A C.I.P. Catalogue record for this book is available from the Library of Congress.

ISBN 978-94-010-3745-7   ISBN 978-94-007-0864-8 (eBook)
DOI 10.1007/978-94-007-0864-8

Printed on acid-free paper

All Rights Reserved
© 1996 Springer Science+Business Media New York
Originally published by Kluwer Academic Publishers in 1996
Softcover reprint of the hardcover 1st edition 1996
No part of the material protected by this copyright notice may be reproduced or
utilized in any form or by any means, electronic or mechanical,
including photocopying, recording or by any information storage and
retrieval system, without written permission from the copyright owner.

Foreword	xiii–xiv
Preface	xv–xvi

PART I
EMISSIONS AND THEIR CONTROL

LING QI, JIMING HAO and MINGMING LU / SO2 Emission Scenarios of Eastern China	1873–1878
P.D. BAILEY / Modelling Future Vehicle Exhaust Emissions in Europe	1879–1884
BERND SCHÄRER / Recent Developments in Technologies and Policies in Germany to Control Acid Deposition	1885–1890
H.M. ApSIMON, S. COULING, D. COWELL and R.F. WARREN / Reducing the Contribution of Ammonia to Nitrogen Deposition across Europe	1891–1896
L. LENHART and R. FRIEDRICH / European Emission Data with High Temporal and Spatial Resolution	1897–1902
V. LIBLIK, A. RÄTSEP and H. KUNDEL / Pollution Sources and Spreading of Sulphur Dioxide in the North-Eastern Estonia	1903–1908
C.W.A. EVERS, J.J.M. BERDOWSKI and P.F.J. VAN DER MOST / Acidifying Emissions in the Netherlands	1909–1914
KENTARO MURANO, SHIRO HATAKEYAMA, TSUGUO MIZOGUCHI and NAOMI KUBA / Gridded Ammonia Emission Fluxes in Japan	1915–1920

PART II
ATMOSPHERIC PROCESSES AND AIR POLLUTION CLIMATE

HUANG MEIYUAN, WANG ZIFA, HE DONGYANG, XU HUAYING and ZHOU LING / Modeling Studies on Sulfur Deposition and Transport in East Asia	1921–1926
Y. ICHIKAWA and S. FUJITA / An Analysis of Wet Deposition of Sulfate Using a Trajectory Model for East Asia	1927–1932
A. SEMB, J.E. HANSSEN, F. FRANCOIS, W. MAENHAUT and J.M. PACYNA / Long Range Transport and Deposition of Mineral Matter as a Source for Base Cations	1933–1940
N.C. MCARDLE, G.W. CAMPBELL and J.R. STEDMAN / Wet Deposition of Non Sea-Salt Sulphate in the United Kingdom: The Influence of Natural Sources	1941–1947
G. DOLLARD, D. FOWLER, R.I. SMITH, A.-G. HJELLBREKKE, K. UHSE and M. WALLASCH / Ozone Measurements in Europe	1949–1954
A. BAJIĆ and V. ĐURIČIĆ / Precipitation Chemistry and Atmospheric Processes in the Forested Part of Croatia	1955–1960
J.D. WHYATT, J.R. STEDMAN, S.E. METCALFE and G.W. CAMPBELL / Measurements of Precipitation Composition at UK EMEP Sites 1987–1992 and Comparison with the HARM Model	1961–1966
Y. DOKIYA, K. TSUBOI, H. SEKINO, T. HOSOMI, Y. IGARASHI and S. TANAKA / Acid Deposition at the Summit of Mt. Fuji: Observations of Gases, Aerosols and Precipitation in Summer, 1993 and 1994	1967–1972
W. VAUTZ, M. SCHILLING, F.L.T. GONCALVES, M.C. SOLCI, O. MASSAMBANI and D. KLOCKOW / Preliminary Analysis of Atmospheric Scavenging Processes in the Industrial Region of Cubatao, Southeastern Brazil	1973–1978
K. ACKER, D. MÖLLER, W. WIEPRECHT and ST. NAUMANN / Mt. Brocken, a Site for a Cloud Chemistry Measurement Programme in Central Europe	1979–1984
M. RADOJEVIC, B.J. TYLER, S. HALL and N. PENDERGHEST / Air Oxidation of S(IV) in Cloud-Water Samples	1985–1990
S. ARMALIS / Measurements of Atmospheric Particulate Carbon in Lithuania	1991–1996
A. VIRKKULA, M. MÄKINEN, R. HILLAMO and A. STOHL / Atmospheric Aerosol in the Finnish Arctic: Particle Number Concentrations, Chemical Characteristics, and Source Analysis	1997–2002
A. MILUKAITE, V. JUOZEFAITE, A. MIKELINSKIENE, B. GIEDRAITIS, A. GALVONAITE, P. JURGUTIS / Evaluation of Extreme Pollution Episodes over Lithuania in 1980–1994	2003–2008

J. ZWOZDZIAK, A. ZWOZDZIAK, G. KMIEC AND K. KACPERCZYK / Some Observations of Pollutant Fluxes over the Sudeten, South-Western Poland   2009–2013

H. HAYAMI and Y. ICHIKAWA / Development of Hybrid LRT Model to Estimate Sulfur Deposition in Japan   2015–2020

J. LANGNER, C. PERSSON and L. ROBERTSON / Concentration and Deposition of Acidifying Air Pollutants over Sweden: Estimates for 1991 Based on the MATCH Model and Observations   2021–2026

A. LINDSKOG, S. SOLBERG, M. ROEMER, D. KLEMP, R. SLADKOVIC, H. BOUDRIES, A. DUTOT, R.H. HAKOLA, R. SCHMITT and H. ARESKOUG / The Distribution of NMHC in Europe: Results from the EUROTRAC TOR Project   2027–2032

H. PLEIJEL, G. WALLIN, P. KARLSSON, L. SKÄRBY and G. SELLDÉN / Gradients of Ozone at a Forest Site and over a Field Crop – Consequences for the AOT40 Concept of Critical Level   2033–2038

J. ISAKSON, E.A. SELIN LINDGREN, V.L. FOLTESCU, J.M. PACYNA and K. TØRSETH / Behaviour of Sulphur and Nitrogen Compounds Measured at Marine Stations Lista and Säby in Scandinavia   2039–2044

Y. ANDERSSON-SKÖLD and S. JANHÄLL / Simulations of Ozone Formation from Different Emission Sources in Sweden   2045–2050

M. GALPERIN, M. SOFIEV and O. AFINOGENOVA / Long-Term Modelling of Airborne Pollution Within the Northern Hemisphere   2051–2056

PART III
DEPOSITION PROCESSES

M.A. SUTTON, D. FOWLER, J.K. BURKHARDT and C. MILFORD / Vegetation Atmosphere Exchange of Ammonia: Canopy Cycling and the Impacts of Elevated Nitrogen Inputs   2057–2063

JAN DUYZER, HILBRAND WESTSTRATE and SAM WALTON / Exchange of Ozone and Nitrogen Oxides between the Atmosphere and Coniferous Forest   2065–2070

R.T. KARABAN and M.L. GYTARSKY / Studies of Precipitation Contamination Levels over the North-Western Forests of Russia Subject to Emissions from the Two Nickel Smelters   2071–2076

S. GUERZONI, A. CRISTINI, R. CABOI, O. LE BOLLOCH, I. MARRAS and L. RUNDEDDU / Ionic Composition of Rainwater and Atmospheric Aerosols in Sardinia, Southern Mediterranean   2077–2082

G.P. AYERS, R.W. GILLETT, N. GINTING, M. HOOPER, P.W. SELLECK and N. TAPPER / Atmospheric Sulfur and Nitrogen in West Java   2083–2088

G.P. AYERS, H. MALFROY, R.W. GILLETT, D. HIGGINS, P.W. SELLECK and J.C. MARSHALL / Deposition of Acidic Species at a Rural Location in New South Wales, Australia   2089–2094

D.E. CONLAN, S.J. LINDLEY and J.W.S. LONGHURST / Spatial and Temporal Variability in Precipitation Chemistry in the Urban Area of Greater Manchester   2095–2100

J.W. ERISMAN, C. POTMA, G. DRAAIJERS, E. VAN LEEUWEN and A. VAN PUL / A Generalised Description of the Deposition of Acidifying Pollutants on a Small Scale in Europe   2101–2106

D. FOWLER, I.D. LEITH, J. BINNIE, A. CROSSLEY, D.W.F. INGLIS, T.W. CHOULARTON, M. GAY, J.W.S. LONGHURST, and D.E. CONLAND / Orographic Enhancement of Wet Deposition in the United Kingdom: Continuous Monitoring   2107–2112

D. FOWLER, R. MOURNE and D. BRANFORD / The Application of ^{210}Pb Inventories in Soil to Measure Long-Term Average Wet Deposition of Pollutants in Complex Terrain   2113–2118

D.W.F. INGLIS, T.W. CHOULARTON, A.J. WICKS, D. FOWLER, I.D. LEITH, B. WERKMAN and J. BINNIE / Orographic Enhancement of Wet Deposition in the United Kingdom: Case Studies and Modelling   2119–2124

K. JYLHÄ / Depositon around a Coal-Fired Power Station during a Wintertime Precipitation Event	2125–2130
G. KMIEC, K. KACPERCZYK, A. ZWOZDZIAK and J. ZWOZDZIAK / Acid Pollutants in Air and Precipitation/Deposition at the Sudeten Mountains, Poland	2131–2136
U.C. KULSHRESTHA, A.K. SARKAR, S.S. SRIVASTAVA and D.C. PARASHAR / Wet-Only and Bulk Deposition Studies at New Delhi (India)	2137–2142
U.C. KULSHRESTHA, A.K. SARKAR, S.S. SRIVASTAVA and D.C. PARASHAR / A Study on Short-Time Sampling of Individual Rain Events at New Delhi during Monsoon, 1994	2143–2148
R. LAVRINENKO / Nitrogen Compounds in Atmospheric Precipitation	2149–2154
O. LE BOLLOCH and S. GUERZONI / Acid and Alkaline Deposition in Precipitation on the Western Coast of Sardinia, Central Mediterranean (40°N, 8°E)	2155–2160
MEDHA S. NAIK, G.A. MOMIN, A.G. PILLAI, P.D. SAFAI, P.S.P. RAO and L.T. KHEMANI / Precipitation Chemistry at Sinhagad – a Hill Station in India	2161–2166
S. SETO, M. KITAMURA, A. MORI, I. NOGUCHI, T. OHIZUMI, T. TAKEUCHI, T. DEGUCHI, and H. HARA / Relationship between Wet Deposition of Sulfate and Nitrate and Rainfall Amount in Japan	2167–2172
E.P. VAN LEEUWEN, G.P.M. DRAAIJERS, C.J.M. POTMA, W.A.J. VAN PUL and J.W. ERISMAN / The Compilation of Measurement Based European Wet Deposition Maps of Acidifying Components and Base Cations	2173–2178
S. VIDIČ / Deposition of Sulphur and Nitrogen Compounds in Croatia	2179–2184
A.-J. LINDROOS, J. DEROME and K. NISKA / Snowpack Quality as an Indicator of Air Pollution in Finnish Lapland and the Kola Peninsula, NW Russia	2185–2190
J. SOVERI and K. PELTONEN / Evaluation of the Changes in Regional Wintertime Deposition in Finland during 1976–1993	2191–2197
A. ALEBIC-JURETIC / Trends in Sulphur Dioxide Concentrations and Sulphur Deposition in the Urban Atmosphere of Rijeka (Croatia), 1984–1993	2199–2204
M.F. HOVMAND and H.V. ANDERSEN / Nine Years of Measurements of Atmospheric Nitrogen and Sulphur Deposition to Danish Forest	2205–2210
H.V. ANDERSEN and M.F. HOVMAND / Ammonia and Nitric Acid Dry Deposition and Throughfall	2211–2216
K. PETERS and G. BRUCKNER-SCHATT / The Dry Deposition of Gaseous and Particulate Nitrogen Compounds to a Spruce Stand	2217–2222
K. PILEGAARD, N.O. JENSEN and P. HUMMELSHØJ / Seasonal and Diurnal Variation in the Deposition Velocity of Ozone Over a Spruce Forest in Denmark	2223–2228
M. FERM and H. HULTBERG / Method to Estimate Atmospheric Deposition of Base Cations in Coniferous Throughfall	2229–2234
H. HULTBERG and M. FERM / Measurements of Atmospheric Deposition and Internal Circulation of Base Cations to a Forested Catchment Area	2235–2240
H. LIPPO, J. POIKOLAINEN and E. KUBIN / The Use of Moss, Lichen and Pine Bark in the Nationwide Monitoring of Atmospheric Heavy Metal Deposition in Finland	2241–2246
J.N. CAPE, L.J. SHEPPARD, J. BINNIE, P. ARKLE and C. WOODS / Throughfall Deposition of Ammonium and Sulphate during Ammonia Fumigation of a Scots Pine Forest	2247–2252
G.P.J. DRAAIJERS and J.W. ERISMAN / A Canopy Budget Model to Assess Atmospheric Deposition from Throughfall Measurements	2253–2258
K. HANSEN / In-Canopy Throughfall Measurements in Norway Spruce: Water Flow and Consequences for Ion Fluxes	2259–2264
J. SHUBZDA, S.E. LINDBERG, C.T. GARTEN and S.C. NODVIN / Elevational Trends in the Fluxes of Sulphur and Nitrogen in Throughfall in the Southern Appalachian Mountains: Some Surprising Results	2265–2270
E. HALLGREN LARSSON, J. KNULST, G. MALM and O. WESTLING / Deposition of Acidifying Compounds in Sweden	2271–2276

PART IV
ACIDIFICATION OUTSIDE EUROPE AND NORTH AMERICA

W. FOELL, C. GREEN, M. AMANN, S. BHATTACHARYA, G. CARMICHAEL, M. CHADWICK, S. CINDERBY, T. HAUGLAND, J.-P. HETTELINGH, L. HORDIJK, J. KUYLENSTIERNA, J. SHAH, R. SHRESTHA, D. STREETS and D. ZHAO / Energy Use, Emissions, and Air Pollution Reduction Strategies in Asia 2277–2282

R.L. ARNDT and G.R. CARMICHAEL / Long-Range Transport and Deposition of Sulfur in Asia 2283–2288

G.R. CARMICHAEL, M. FERM, S. ADIKARY, J. AHMAD, M. MOHAN, M.-S. HONG, L. CHEN, L. FOOK, C.M. LIU, M. SOEDOMO, G. TRAN, K. SUKSOMSANK, D. ZHAO, R. ARNDT and L.L. CHEN / Observed Regional Distribution of Sulfur Dioxide in Asia 2289–2294

WENXING WANG and TAO WANG / On the Origin and the Trend of Acid Precipitation in China 2295–2300

HANS M. SEIP, ZHAO DIANWU, XIONG JILING, ZHAO DAWEI, THORJØRN LARSSEN, LIAO BOHAN and ROLF D. VOGT / Acidic Deposition and Its Effects in Southwestern China 2301–2306

HIROSHI HARA, MORITSUGU KITAMURA, ATSUKO MORI, IZUMI NOGUCHI, TSUYOSHI OHIZUMI, SINYA SETO, TADASHI TAKEUCHI and TERUYUKI DEGUCHI / Precipitation Chemistry in Japan 1989–1993 2307–2312

G.P. AYERS, R.W. GILLETT, P.W. SELLECK and S.T. BENTLEY / Rainwater Composition and Acid Deposition in the Vicinity of Fossil Fuel-Fired Power Plants in Southern Australia 2313–2318

J.C.I. KUYLENSTIERNA, H. CAMBRIDGE, S. CINDERBY and M.J. CHADWICK / Terrestrial Ecosystem Sensitivity to Acidic Deposition in Developing Countries 2319–2324

J.A. MORALES, C. BIFANO and A. ESCALONA / Rainwater Chemistry at the Western Savannah Region of the Lake Maracaibo Basin, Venezuela 2325–2330

R.W. SKOROSZEWSKI / Sulphate Deposition to a Small Upland Catchment at Suikerbosrand, South Africa 2331–2336

LENNART ROBERTSON, HENNING RODHE and LENNART GRANAT / Modelling of Sulfur Deposition in the Southern Asian Region 2337–2343

M. INAGAKI, M. SAKAI and Y. OHNUKI / The Effects of Organic Carbon on Acid Rain in a Temperate Forest in Japan 2345–2350

O. NAGAFUCHI, R. SUDA, H. MUKAI, M. KOGA and Y. KODAMA / Analysis of Long-Range Transported Acid Aerosol in Rime Found at Kyushu Mountainous Regions, Japan 2351–2356

I. NOGUCHI, T. KATO, M. AKIYAMA, H. OTSUKA and Y. MATSUMOTO / The Effect of Alkaline Dust Decline on the Precipitation Chemistry in Northern Japan 2357–2362

M. RADOJEVIC and L.H. LIM / Short-Term Variation in the Concentration of Selected Ions within Individual Tropical Rainstorms 2363–2368

M. RADOJEVIC and L.H. LIM / A Rain Acidity Study in Brunei Darussalam 2369–2374

PART V
CRITICAL LOADS

HARALD SVERDRUP, PER WARFVINGE and KAJ ROSÉN / Critical Loads of Acidity and Nitrogen, Based on Multiple Criteria for Different Swedish Ecosystems 2375–2380

J.-P. HETTELINGH, M. POSCH, P.A.M. DE SMET and R.J. DOWNING / The Use of Critical Loads in Emission Reduction Agreements in Europe 2381–2388

G.P.J. DRAAIJERS, E.P. VAN LEEUWEN, C. POTMA, W.A.J. VAN PUL and J.W. ERISMAN / Mapping Base Cation Deposition in Europe on a 10 × 20 km Grid 2389–2394

V.N. BASHKIN, M.YA. KOZLOV, I.V. PRIPUTINA, A.YU ABRAMYCHEV and I.S. DEDKOVA / Calculation and Mapping of Critical Loads of S, N and Acidity on Ecosystems of the Northern Asia 2395–2400

SHAODONG XIE, JIMING HAO, ZHONGPING ZHOU, LING QI and HANHUI YIN / Assessment of Critical Loads in Liuzhou, China Using Static and Dynamic Models — 2401–2406

J.P. PARTY, A. PROBST, E. DAMBRINE and A.L. THOMAS / Critical Loads of Acidity to Surface Waters in the Vosges Massif (North-East of France) — 2407–2412

ROLAND BOBBINK and JAN G.M. ROELOFS / Nitrogen Critical Loads for Natural and Semi-Natural Ecosystems: The Empirical Approach — 2413–2418

A. HENRIKSEN, M. POSCH, H. HULTBERG and L. LIEN / Critical Loads of Acidity for Surface Waters – Can the ANC_{limit} Be Considered Variable? — 2419–2424

T.E.H. ALLOTT, P.N.E. GOLDING and R. HARRIMAN / A Palaeolimnological Assessment of the Impact of Acid Deposition on Surface Waters in North-West Scotland, a Region of High Sea-Salt Inputs — 2425–2430

GUN LÖVBLAD, PERINGE GRENNFELT, OLLE WESTLING, HARALD SVERDRUP and PER WARFVINGE / The Use of Critical Load Exceedances in Abatement Strategy Planning — 2431–2436

A. HENRIKSEN / Critical Loads of Acidity to Surface Waters – How Important is the F-Factor in the SSWC-Model? — 2437–2441

J.R. HALL, S.M. WRIGHT, T.H. SPARKS, J. ULLYETT, T.E.H. ALLOTT and M. HORNUNG / Predicting Freshwater Critical Loads from National Data on Geology, Soils and Land Use — 2443–2448

L. RAPP and K. BISHOP / Sulphur Deposition and Changes in Swedish Lake Chemistry 1988–1993 — 2449–2454

R. HARRIMAN, E.E. BRIDCUT and H. ANDERSON / The Relationship between Salmonid Fish Densities and Critical ANC at Exceeded and Non-Exceeded Stream Sites in Scotland — 2455–2460

D. TURNBULL, C. SOULSBY, S. LANGAN, R. OWEN and D. HIRST / Macroinvertebrate Status in Relation to Critical Loads for Freshwaters: A Case Study from N.E. Scotland — 2461–2466

C.J. CURTIS, T.E.H. ALLOTT, R.W. BATTARBEE and R. HARRIMAN / Validation of the UK Critical Loads for Freshwaters: Site Selection and Sensitivity — 2467–2472

K.P. MACPHEE, S.J. LANGAN and M.F. BILLET / Critical Loads for Soils and Waters in a Selected Scottish Catchment — 2473–2478

M. KERNAN / The Use of Catchment Attributes to Predict Surface Water Critical Loads: A Preliminary Analysis — 2479–2484

D.J. TERVET, D.A. RENDALL and A.B. STEPHEN / Critical Loads – A Valuable Catchment Management Tool? — 2485–2490

E.J. WILSON, R.A. SKEFFINGTON, E. MALTBY, P. IMMIRZI, C. SWANSON and M. PROCTOR / Towards a New Method of Setting a Critical Load of Acidity for Ombrotrophic Peat — 2491–2496

SIMON J. LANGAN, HARALD U. SVERDRUP and MALCOLM COULL / The Calculation of Base Cation Release from the Chemical Weathering Of Scottish Soils Using the Profile Model — 2497–2502

R.I. SMITH, J.R. HALL and D.C. HOWARD / Estimating Uncertainty in the Current Critical Loads Exceedance Models — 2503–2508

MATTIAS ALVETEG, HARALD SVERDRUP and PER WARFVINGE / Regional Assessment of the Temporal Trends in Soil Acidification in Southern Sweden, Using the SAFE Model — 2509–2514

ANDREAS BARKMAN, PER WARFVINGE and HARALD SVERDRUP / Regionalization of Critical Loads under Uncertainty — 2515–2520

J. HALL, K. BULL, M. BROWN, H. DYKE, J. ULLYETT and M. HORNUNG / The Effects of Scale and Resolution in Developing Percentile Maps of Critical Loads for the UK — 2521–2526

K.R. BULL, M.J. BROWN, H. DYKE, B.C. EVERSHAM, R.M. FULLER, M. HORNUNG, D.C. HOWARD, J. RODWELL and D.B. ROY / Critical Loads for Nitrogen Deposition for Great Britain — 2527–2532

D. KURZ, U. EGGENBERGER and B. RIHM / Evaluating Critical Loads of Acidity for Swiss Forest Soils: Comparison of Two Calculation Methods 2533–2538

W. MIKUŁA / Sulphate Sulphur Concentration in Vegetable Crops, Soil and Ground Water in the Region Affected by the Sulphur Dioxide Emission from Płock Oil Refinery (Central Poland) 2539–2546

W. MILL / Critical Loads Mapping in Poland: Lessons Learned 2547–2552

G. KOPTSIK and S. KOPTSIK / Critical Loads of Acid Deposition for Forest Ecosystems in the Kola Peninsula 2553–2558

M.YA. KOZLOV, V.N. BASHKIN and O.M. GOLINETS / Uncertainty Analysis of Critical Loads for Terrestrial Ecosystems in Russia 2559–2564

JEAN-PAUL HETTELINGH, HARALD SVERDRUP and DIANWU ZHAO / Deriving Critical Loads for Asia 2565–2570

J. SHINDO, A.K. BREGT and T. HAKAMATA / Evaluation of Estimation Methods and Base Data Uncertainties for Critical Loads of Acid Deposition in Japan 2571–2576

A.M. VAN TIENHOVEN, K.A. OLBRICH, R. SKOROSZEWSKI, J. TALJAARD and M. ZUNCKEL / Application of the Critical Loads Approach in South Africa 2577–2582

PART VI
STRATEGIES FOR REGIONAL AIR POLLUTION CONTROL INCLUDING ECONOMIC ASPECTS

M.R. HOLLAND / Assessment of the Economic Costs of Damage Caused by Air-Pollution 2583–2588

M. BROWN, H. DYKE, S.M. WRIGHT, R.A. WADSWORTH, K.R. BULL, A. FARMER, S. BAREHAM, S.E. METCALFE, D. WHYATT and C. POWLESLAND / Estimating the Impact of Air Pollution on Environmentally Valuable Sites 2589–2594

M. AMANN, M. BALDI, C. HEYES, Z. KLIMONT and W. SCHÖPP / Integrated Assessment of Emission Control Scenarios, Including the Impact of Tropospheric Ozone 2595–2600

C.A. GOUGH, M.J. CHADWICK, B. BIEWALD, J.C.I. KUYLENSTIERNA, P.D. BAILY and S. CINDERBY / Developing Optimal Abatement Strategies for the Effects of Sulphur and Nitrogen Deposition at European Scale 2601–2606

T.J. SULLIVAN and B.J. COSBY / Testing, Improvement, and Confirmation of a Watershed Model of Acid-Base Chemistry 2607–2612

R.A. WADSWORTH, M.J. BROWN, K.R. BULL, S.E. METCALFE, D. WHYATT and C. POWLESLAND / A Comparison of Different Ranking Schemes for Assessing the Effect of Large Point Sources of Pollution in the UK 2613–2618

S.E. METCALFE, J.D. WHYATT, R.G. DERWENT, K. BULL and H. DYKE / Spatial Variability in Emissions Reduction Strategies for Sulphur and Nitrogen in the UK 2619–2624

J.W.S. LONGHURST, S.J. LINDLEY and D.E. CONLAN / Emissions of Acidifying Air Pollutants in the North West Region of England 2625–2630

J.W.S. LONGHURST, J. BANTOCK, S.E. HARE and D.E. CONLAN / Changing Public Interest in, and Awareness of, Acid Deposition: Some Evidence from the UK 2631–2636

Z.M. KARACZUN / Policy of Air Protection in Poland 2637–2642

PRAMOD K. JHA / Pollution Preventing Efforts and Strategies for the Kathmandu Valley 2643–2648

R.A. WADSWORTH an M.J. BROWN / A Spatial Decision Support System to Allow the Investigation of the Impact of Emissions from Major Point Sources under Different Operating Policies 2649–2654

PART VII
AIR POLLUTION EFFECTS ON MATERIALS

R.N. BUTLIN, T.J.S. YATES, M. MURRAY and G. ASHALL / The United Kingdom National Materials Exposure Programme	2655–2660
D. KNOTKOVA, P. BOSCHEK and K. KREISLOVA / Effect of Acidification on Atmospheric Corrosion of Structural Metals in Europe	2661–2666
S.E. HAAGENRUD, J.F. HENRIKSEN and T. SKANCKE / Mapping of Urban Material Degradation from Available Data	2667–2672
A.A. MIKHAILOV, M.N. SULOEVA and E.G. VASILIEVA / Environmental Aspects of Atmospheric Corrosion	2673–2678
E.C. SPIKER, R.P. HOSKER Jr., V.C. WEINTRAUB and S.I. SHERWOOD / Laboratory Study of SO_2 Dry Deposition on Limestone and Marble: Effects of Humidity and Surface Variables	2679–2685
P. MAYERHOFER, M. WELTSCHEV, A. TRUKENMÜLLER and R. FRIEDRICH / A Methodology for the Economic Assessment of Material Damage Caused by SO_2 and NO_x Emissions in Europe	2687–2692
R.N. BUTLIN, T.J.S. YATES and B. CHAKRABARTI / Mapping of Critical Loads and Levels for Pollution Damage to Building Materials in the United Kingdom	2693–2699
A. REISENER, B. STÖCKLE and R. SNETHLAGE / Deterioration of Copper and Bronze Caused by Acidifying Air Pollutants	2701–2706
J.F. HENRIKSEN / Reactions of Gases on Calcareous Stones under Dry Conditions in Field and Laboratory Studies	2707–2712
C.M. GROSSI, M. MURRAY and R.N. BUTLIN / Response of Porous Building Stones to Acid Deposition	2713–2718
A.G. NORD and K. TRONNER / Effect of Acid Rain on Sandstone: The Royal Palace and the Riddarholm Church, Stockholm	2719–2724
K. TRONNER, A.G. NORD and G.CH. BORG / Corrosion of Archaeological Bronze Artefacts in Acidic Soil	2725–2730

PART VIII
EPILOGUE

Conference Songs

Foreword

The acidification problem is still an area of great concern. Many areas in the world are subject to large, in some cases increased, deposition of acidifying substances. Scientific research has played a crucial role in the discovery and exploration of the problems as well as a basis for the development of control strategies for the more than 25 years that have passed since Svante Odén first presented his results. Even today scientific research is important as a tool for policy, most clearly observed in the effect-oriented second sulphur protocol under the Convention on Long-range Transboundary Air Pollution signed in Oslo 1994. Without a close international scientific cooperation this protocol would have never been able to develop.

The 5th Conference on Acidic Deposition *Acid Reign '95?* should be seen in the context of the ongoing process to strengthen the scientific background for policy. It was therefore a great pleasure for Sweden to host the conference. It became a success in many respects. First the number of scientists and presentations made this conference the most comprehensive ever on acidic deposition. From the organizers, we were extremely pleased to see the number of participants and contributions from countries outside Western Europe and North America. These participants turned the focus to areas showing signs of an increased acidification problem and areas still under heavy pressure from acidic deposition. An excursion to the Czech republic prior to the conference underpinned this interest.

We were also very pleased with the quality of and interest in the scientific presentations. The approximate 600 posters together with the more than 150 oral presentations gave the participants a deeper understanding of all aspects of the acidification phenomenon.

The title of the conference is a play on words. It raises the question if acid rain reigns in nature or if nature and the natural processes can proceed without being severely affected by the deposition of acidifying pollutants. Conclusions from this conference indicated that the reductions already achieved show signs of improvement in the ecosystems. Forthcoming reductions will lead to further improvements. The road to success, however, involves - as a necessity - international cooperation on a variety of levels. Let us hope that it by next conference is realistic to state: Acid rain does not reign any more!

To handle our heritage responsibly means an environment awareness directly related to quality of life. A culture can never select its way over nature. Albert Einstein states that respect for nature and mankind should always be the target for all technical progress. The famous botanist Carl von Linné writes more than 200 years ago in his "The Wonder of Nature" that the nature of matters does not at once reveal its secret; we consider us informed but we are only in the fore-court.

By means of increased knowledge through research efforts and international cooperation we have to realize that continued economic growth is possible to achieve in harmony with the cycle of nature. The driving force behind economic growth is new technology. Possibilities to reduce consumption of limited resources are thereby created. At the same time, however, there are risks for change in the process of evolution. But, the new concept considers an awareness that the threats in the development of the society tie the global economy and global ecology together.

Air pollution and acidification constitute a threat to all of us. Slowly, we have to adjust ourselves to an accelerating ecological dependence between countries. As every other organism, man is part of his own surroundings. Disturbances in the external environment no doubt signify changes also of the internal nature. To my mind, therefore, necessity, even duty, to protect nature and the environment is based upon a variety of reasons - moral, ethical, commercial, cultural, aesthetic and scientific.

From these aspects, it is obvious that in a wider and more intensive sense we have to make our voice heard in the international arena, in debates and influencing public opinion. The rate of change and global improvements of environmental conditions should increase. Harmonizing environment laws in different countries might constitute a highway to such a target. It should be recognized that environmental matters have changed character within industry. It is not any longer just to follow laws and restrictions. Environmental matters have become a means of competition, a qualification of success, thereby a matter of strategy selection on the highest level.

The road to success implies confidence and cooperation between individuals and nations. This international conference provided an excellent opportunity for scientists to evaluate facts and new results within acidification perspectives and to meet with representatives of industry and the community as a whole.

I am convinced that major scientific challenges will be formulated to result in cost-effective strategies to meet future demands for reducing or eliminating present acidification threats. This conference has contributed to increased knowledge and scientific understanding and the Proceedings of the conference can act as a new basis for formulating control measures and further research.

Lennart Schotte
Professor
Chairman of the Organizing Committee

Preface

The "Acid Reign '95?" conference in Göteborg, Sweden, 26-30 June 1995, was the 5th conference in a series starting in Columbus, Ohio in 1975 and including Sandefjord, Norway in 1980, Muskoka, Canada in 1985 and Glasgow, Scotland in 1990. For an issue that seemed to have taken a secondary role relative to other environmental issues (e.g. climate change, ozone depletion, biodiversity), the great scientific and political interest in this conference (more than 750 scientific contributions were presented) shows, unquestionably, that acid deposition remains a very important international environmental issue. The sulfur problem in parts of Eastern Europe is still critical and there is a continuing urgency to find the most appropriate methods for revitalizing and/or restoring the soils and waters which have been severely damaged by decades of acid deposition. Acidification problems continue to surface in new areas of the globe. Sulfur emissions in South and East Asia continue to increase at a fast rate with increasing risks for severe impacts on humans and their ecosystems. Europe and North America are facing large costs to meet requirements stipulated by air quality standards and critical loads. The use of cost effective control and mitigation strategies for meeting these demands remains a scientific, political, engineering, economic and social challenge.

The scientific contributions to the conference have resulted in more than 400 peer-reviewed scientific papers collected in this Special Edition of The Journal of Water, Air, and Soil Pollution. It is the largest condensed volume of acid deposition papers ever assembled and it is likely to serve as a reference base for years to come.

The papers in this volume are divided into 20 different topics, each topic representing an important aspect of the acidification phenomenon. The papers appear in four hard-bound books, the first containing the plenary talks and the other three the remaining scientific papers.

Acknowledgements

We acknowledge the input of all of the authors who contributed papers for the Conference Proceedings. We also acknowledge all who have helped us in the scientific review of the papers as well as those who advised us in the preparation of the conference programme and helped in many other ways.

We acknowledge the support of the organisers of the Conference namely, the Swedish Environmental Protection Agency, The Swedish Environmental Research Institute and the Provincial Government of Göteborgs and Bohus län. We also appreciate additional funding support from a variety of organizations and companies all listed on the back page.

We acknowledge the input and guidance both from the Steering and the Scientific Committees. We specifically cite the enormous and very much appreciated efforts of Gunilla Pihl Karlsson, Gun Olsson and Elizabeth Ritter, who continuously persisted in tracking down and keeping timetables among the editors and authors.

We acknowledge Billy McCormac, Editor of Water, Air, and Soil Pollution who always provided useful guidance and Martinus Ader, Kluwer Academic Publishers, whose copy editing is simply the best.

Co-Editors:
Peringe Grennfelt
Henning Rodhe
Eva Thörnelöf
Joe Wisniewski

SO2 EMISSION SCENARIOS OF EASTERN CHINA

LING QI, JIMING HAO, MINGMING LU

Department of Environmental Engineering, Tsinghua University, Beijing 100084, P. R. China

Abstract. Under the National Key Project in Eighth Five-year Plan, a study was carried out on forecasting SO2 emission from coal combustion in China, with a special emphasis on eastern area. 3 scenarios, i.e. "Optimistic", "Pessimistic" and "Business as Usual" scenario were developed trying to cover changing scale of coal consumption and SO2 emission from 1990 to 2020. "Top-down" approach was employed, and coal consumption elasticity was defined to project future economic growth and coal consumption. SO2 emission scenarios were outlined, based on coal consumption, estimated sulfur content level and prospective SO2 control situation. Emission level for each 1° longitude × 1° latitude grid cell within eastern China was also estimated to show geographical distribution of SO2 sources. The results show that SO2 emission of China will increase rapidly, if current situation for energy saving and SO2 control is maintained without improvement; measures enhanced reasonably with economic growth could stop further increase of emission by 2010; realization of more encouraging objective to keep emission at even below 1990 level needs, however, more stringent options. Share of eastern China in country's total emission would increase until 2000, while general changing tendency would principally follow the scenarios of whole country.

1. Introduction

With the rapid economic development, energy consumption of China has increased significantly in the last decade. Coal combustion, as the main process of primary energy consumption, has generated large amount of acid precursors such as sulfur dioxide (SO2), nitrogen oxides (NOx) *et al.* and emitted to atmosphere. In 1990, SO2 emission of China was about 15 million tons, and considerable areas in southern and eastern China have been polluted by acid rain.

Eastern China is a relatively developed area with multi-sectoral industry, dense population and long coastal line. With China's economic reform and opening, economic boom in eastern area has imposed stresses on environment, in which urban SO2 pollution and regional acid deposition are especially outstanding. According to the data from national monitoring network, some regions in eastern China have highest frequency of acid rain in China. Since continuous economic growth in next few decades would have large potential of acid precursor emission, and the flat territory in this area may facilitate long-range transport of pollutants, thus contribute to acid deposition in surrounding area, it is very meaningful to outline and analyze potential emission scenarios, and provide information for decision-makers to formulate emission control and pollution mitigation strategies on acid deposition in eastern China and its surrounding area.

This study was carried out as part of *National Key Project in Eighth Five-year Plan*. While former study showed SO2 was the main acid precursor in China, this study was focused on SO2 emission only. The studied area involved 10 eastern provinces and 7 neighboring provinces (Fig.1). The incorporation of neighboring provinces was to facilitate the succeeding study on sulfur deposition scenarios of eastern China, therefore the provinces selected were those have long SO2 retention time (Sheng et. al. 1990), thus could make inneglegible contributions to sulfur deposition at eastern provinces.

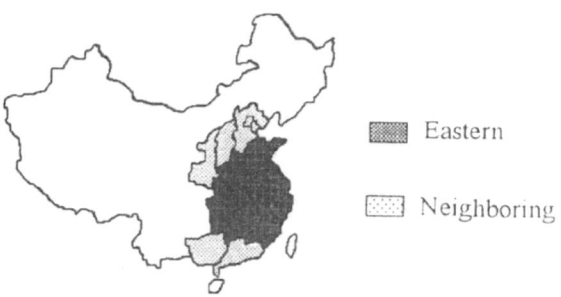

Fig.1. Studied provinces and their location

2. Methodology

In the study, a general procedure was followed to project SO2 emission from the base year 1990 up to 2020. The procedure can be simply described as:

economic development → energy consumption → pollutant emission

The "Scenario analysis" approach was employed all through the study. Scenarios are a group of changing tendencies which simulate alternative energy, economy and environment future under different assumptions. By scenario analysis, potential future conditions can be foreseen in order to compare various options reflected by corresponding scenarios, and recommend the options to realize most desirable future. In this study, 3 scenarios were designed trying to cover changing scale of coal consumption and SO2 emission from 1990 to 2020. These scenarios were named: "Optimistic" (possible lowest SO2 emission, corresponding to most intensive measures could China afford to take on emission control and energy saving, and lowest values assumed for economy growth rate and sulfur content of coal), "Pessimistic" (possible highest SO2 emission, corresponding to do-nothing situation on emission control and energy saving since the base year, and highest values assumed for economy growth rate and sulfur content of coal), and "Business as Usual" (BAU, corresponding to the most probable situation with moderate measures on emission control and energy saving, and medium parameter values).

The "Top-down" approach was adopted to project economic growth and coal consumption. Differing with "Bottom-up" approach which studies energy system from the base level, "Top-down" approach attempts to capture aggregated behavior of energy system, and relies more upon macroeconomic parameters reflecting overall change of the system. For China, which just started transition from centrally planned economy to market economy, the detailed information of base-level energy system was not adequate, while official plans on macroeconomic, energy production and consumption sectors were still very comprehensive, and would be effective in near future. Thereafter, "Top-down" approach seemed more suitable for this study.

3. Coal consumption

Coal has always played a dominant role in primary energy system of China. Since 1960, share of coal in primary energy consumption has maintained a percentage over 70 (National Statistics Agency of PRC, 1992), consequently coal combustion has made great contribution to total anthropogenic SO2 emission. Under this situation, coal consumption, instead of energy consumption, was projected in this study, and an adjustment was considered while estimating total SO2 emission. Although this simplifying approach seemed not as accurate as projection on the basis of total primary energy consumption, it was convenient for use under Chinese situation that average sulfur content and combustor emission factor for other energy resources were insufficient.

3.1 METHOD AND PARAMETER VALUE

3 scenarios for coal consumption as "High", "Medium", and "Low" were outlined, corresponding to prescribed "Pessimistic", "BAU" and "Optimistic" scenarios, respectively. Growth rate method was employed to project future economic development, and elasticity method was applied to relate coal consumption with GNP increase.

Elasticity of coal consumption (CCE) was defined referring to that of energy consumption (ECE), as ratio of the changing rate of coal consumption to that of GNP. Initial CCEs for the country and studied provinces were regressed using data from 1985 to 1990. CCE value would be lower than ECE value, and changing tendencies differ within next 30 years: ECE value of China will firstly decline then recover, since energy saving resorting to enhanced administration in the first decade has more potential than resorting to new technology and industrial structure adjustment in next 2 decades; while CCE will keep declining because share of coal in Chinese energy system would decrease despite its dominant role maintain, and popularization of technological energy saving and shifting to "high product value, low energy consumption" industry will indubitably reduce coal consumption per unit product value. Based on this principle, CCE values were determined in light of their initial values and ECE values assumed.

3.2. RESULTS

As shown in Fig.2, coal consumption for "High" scenario can be 30% higher than "Low" scenario in 2020, thus different economic development rate and energy saving intensity (reflected by CCE) could lead to significant difference on total coal consumption. Comparing with projected China's coal production, future coal demand could be met domestically for "Medium" and "Low" scenarios, and not for "High" scenario. Within the projecting period, China seems unlikely to be a coal importer, therefore "High" scenario reflected a situation with a deficit between coal supply and demand, hence shall be avoided by enhancing energy conservancy, adjustment of industry and energy structure.

Aggregated coal consumption of eastern, neighboring and other provinces would change following general tendency of whole country. Fig.3 presents coal consumption of each under "Medium" scenario. It clearly shows that coal consumption of eastern China would increase faster than other areas, and has a bigger and bigger share in that of the country. This situation mainly results from faster economic growth projected for eastern area.

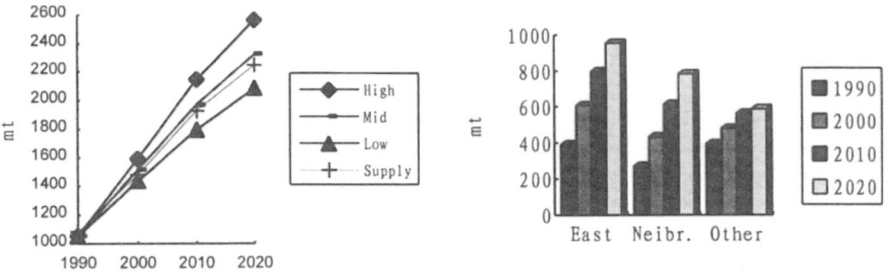

Fig.2. Scenarios for coal consumption of China   Fig.3. Coal consumption of different areas in China

4. SO2 emission

4.1. PARAMETER VALUE

4.1.1. Average Sulfur Content of Coal

Sulfur content of coal is a key factor in calculation of SO2 emission. Before SO2 pollution drew wide public attention, sulfur content was not emphasized in coal quality analysis in China. Therefore these data were usually not available for coal mines in China then. In recent years, several studies have been conducted attempting to provide systematic data, yet no data obtained unanimity. In this study, data from a research on SO2 emission charge policy were adopted (Hao J. et al.). Being calculated based on sulfur contents of coal from most national and local mines and their share in total coal consumption of each province, these data showed that the average sulfur content in coal of China was about 1.2%, while that of eastern provinces ranged from 0.8% to 2.9%.

4.1.2. SO2 Control Situation

China has just started to control SO2 emission. While power plant boilers contribute to most SO2 emission in developed countries, industrial boilers and household coal combustors contribute at similar extent with power plant boilers in China. Considering large amount of separated industrial and household coal combustors, 5 practical SO2 control technologies were selected (Table I), from the promising and suitable technologies for China recommended by a special study (China Energy Research Society, 1991).

The share of coal consumption toward which a special SO2 emission abatement technology would be applied was another important factor for estimating SO2 emission. To the "Pessimistic" scenario assumed as "Do-nothing" scenario, the situation would maintain as that of 1990. To the "Optimistic" and BAU scenarios, the situation had to be assumed according to available planning of relevant industrial sectors (Table I). For example, the share of coal consumption for FGD usage was estimated based on the percentage of installation capacity with FGD projected by power industry sector, and contribution of power industry to total coal consumption.

TABLE I. SO2 control situation of China for 3 scenarios

		Industrial boiler (including power plant boiler)			Household combustor	
Technology		Coal pretreatment	Briquet	FGD	Briquet	Fuel switch *
SO2 removal (%)		40	40	90	50	100
Share in coal consumption(%)						
	1990	18	0**	0	0**	0.9
BAU:	2000	29.6	1	1	0	2.8
	2010	40.6	2	2	1	4.5
	2020	51.6	5	3	2	6.5
Optm:	2000	33.0	5	5	2	5
	2010	48.5	7	10	3	10
	2020	60.0	8	15	4	15

* Involving coal gasification and district heating;   ** No desulphurizer was added to briquet used then.

4.2. RESULTS AND DISCUSSION

SO2 emission scenarios for the country are shown in Fig.4. The "Pessimistic" scenario predicts a rather serious future: total SO2 emission would continuously increase, and triple 1990 level in 2020, if no further SO2 control measure would be undertaken since 1990. The "BAU" scenario is less serious but still uneasy: SO2 emission would continue to increase conspicuously until 2010, and more slightly thenceforth. This scenario seems coinciding well with typical module of economic development: grow rapidly initially, and more moderate later, as entering "post-industrialization" period. The "Optimistic" scenario draws an encouraging picture: total SO2 emission would increase slightly until 2000, then start to decrease and attain an level below 1990 level in 2020. Nevertheless, realization of this challenging future requires stringent measures for energy saving and SO2 control.

As coal consumption, aggregated SO2 emission of different areas would change with the general tendency of the country, and would not differ obviously for each scenario either. Fig.5 shows the share of each area in China's total emission for BAU scenario. It could be found that share of eastern and neighboring provinces would increase conspicuously until 2000, then keep stable with slight fluctuation and increase, respectively; while the share of other provinces would decline considerably until 2000, then keep declining at more moderate rate. It may be concluded that the contribution of eastern and neighboring provinces would increase as their economies grow more rapidly than other provinces, and the share of other provinces would decrease but still maintain an inneglegible level due to initial large share and relatively poorer control situation in future.

SO2 emission for each 1° longitude × 1° latitude grid cell within studied provinces was calculated based on emission of relevant provinces and cities, and graded emission level for each cell was mapped. The results show that emission of many grid cells would upgrade within next 30 years, while geographical distribution of those cells with high emission level would not change much. An interesting result was that most cells with high emission level were locating in northeastern China where acid rain was seldom observed, while most cells in southeastern China with high frequency of acid rain have relatively lower emission intensity, thus a meaningful conclusion could be inferred that acid deposition in eastern China results from regional emission sources as well as local sources.

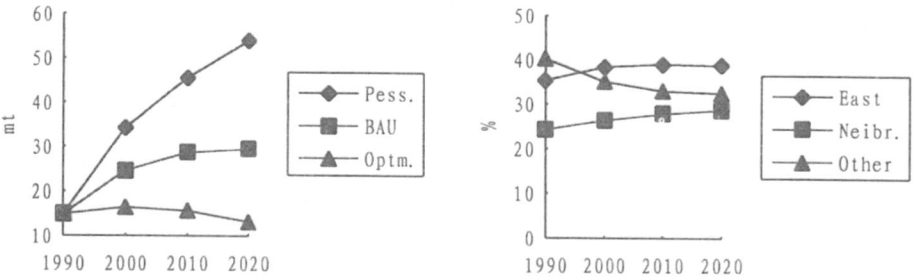

Fig.4. Scenarios for SO2 emission of China   Fig.5. Shares of SO2 emission for different areas of China

5. Conclusion

For China, if current situation for energy saving and SO2 control is maintained without enhancement, coal consumption and SO2 emission would increase very rapidly within next 30 years: coal demand would exceed domestic supply capability and SO2 emission triples 1990 level at 2020. If measures are enhanced reasonably with economic growth, domestic coal demand and supply could be balanced principally, and SO2 emission stabilized after 2010. However, an encouraging objective to maintain emission at 1990 level (even below that later), could also be realized, once more stringent options were carried out. For eastern China, due to relatively faster economic growth, its share in country's total emission would increase until 2000, while general changing tendency would principally follow the scenarios of whole country.

Acknowledgments

This study was supported by State Science and Technology Commission and National Environmental Protection Agency, as the *National Key Project in Eighth Five-year Plan*.

References

China Energy Research Society: 1991, Research on Reasonable and Effective Utilization of 1.4 gt Coal, Project report of State Science and Technology Commission, p.72-79.
Hao J., Xi D. *et al.*: 1993, Study on SO2 Emission Charge Policy toward Industrial Coal Combustion, Project report of National Environmental Protection Agency, p.11-28.
National Statistics Agency of PRC: 1992, China Energy Yearbook of 1991, China Statistics Press, p.135.
Sheng P., Mao J. *et al.* : 1988, Estimation on Long-range Transport of Air Pollutant in China, Project report of State Education Commission, p.10.

MODELLING FUTURE VEHICLE EXHAUST EMISSIONS IN EUROPE

P.D. BAILEY

Stockholm Environment Institute at York, University of York, PO Box No. 373, York YO1 5YW, UK

Abstract. Vehicle exhaust emissions are significant sources of air pollution in many European countries. Emissions of oxides of nitrogen and volatile organic compounds contribute to transboundary air pollution, and the environmental problems of acidification, eutrophication and tropospheric ozone formation. This paper presents estimates of future exhaust emissions from road vehicles within countries of the United Nations Economic Commission for Europe region for two control scenarios. Alternative regulatory standards are examined including a Current Plans scenario and a Best Available Technology scenario. Calculations presented in this paper suggest that Current Plans could reduce emissions of oxides of nitrogen from road vehicles by about 70 per cent and that the application of Best Available Technology to gasoline and diesel vehicles could reduce emissions by 80 per cent. The paper ends with a discussion of the many interactions between transport and the environment and it is suggested that combinations of regulatory approaches will be needed if the environmental impact of road transportation is to be reduced in the future.

Key words: emission standards, best available technology, oxides of nitrogen.

1. Introduction

Air pollution from road transport is one of the most significant environmental problems in Europe with present levels of road transport emissions causing widespread damage to human health and the environment. The three pollutants examined in this paper, carbon monoxide (CO), volatile organic compounds (VOC) and oxides of nitrogen (NO_x) contribute to a wide range of environmental problems, including urban health problems, materials damage, the formation of ground-level ozone, acidification and eutrophication of natural ecosystems (Royal Commission on Environmental Pollution, 1994). These problems range from local to the international transboundary scale and policies to reduce the impact of road transport emissions are formulated at the local, national and international level.

This paper examines national and international policies to control road vehicle emissions and concentrates on the application of vehicle exhaust emission standards. Emission standards have been the most widely used policy instrument for controlling emissions in the past (Dunne and Greening, 1993). Standards are not the only instrument for controlling road vehicle emissions and a wide range of policies are probably required if air pollution from road transport is to be controlled and, perhaps more importantly, if the other environmental impacts of road vehicle use are to be limited. Switching mode from road to rail, cycling or walking can bring a wide range of environmental benefits, however, the future level of road transport activity is likely to grow in the absence of significant incentives for mode switching and emission standards do have the potential for reducing some of the effects of road vehicles on the environment.

2. Emission control scenarios

Four countries are analysed in this paper. Two countries, the Netherlands and UK, have been members of the European Union (EU) for several years, whilst another, Sweden, has recently joined. The other country, Poland, has an economy and transport sector experiencing dramatic changes. National forecasts or scenarios of road transport growth are used to project vehicle numbers and activity levels to the year 2010 for the analysis in this section. It is recognised that many countries now present more than one forecast or scenario of road transport growth because of the inherent uncertainty in such an exercise. In the analysis, however, this has been simplified into one projection for each country to avoid a vast number of scenarios being examined. The UK projection is based upon Department of Transport data (Department of Transport, 1992); the mid value of the lower and upper forecasts is used. The projection from the Netherlands is obtained from the National Institute of Public Health and Environmental Protection analysis of the environmental outlook in the Netherlands (RIVM, 1992; RIVM, 1994). The projected increase in vehicle activity in Poland is adapted from Anderson (1994). Finally, growth in road transport activity in Sweden was inferred from work by the Swedish Environmental Protection Agency (Naturvårdsverket, 1993). Table 1 shows the projected total vehicle-kilometres (veh-km) for all road vehicles for the period 1995 to 2010 assumed in this paper.

TABLE 1

Projection total annual road vehicle distances in the Netherlands, Poland, Sweden and the UK to the year 2010 (includes mopeds, motorcycles, cars, vans, trucks and buses)

Country	1995 (billion veh-km)	2000 (billion veh-km)	2005 (billion veh-km)	2010 (billion veh-km)
Netherlands	107	114	122	131
Poland	73	97	122	149
Sweden	69	74	78	83
UK	480	539	596	653

During the 1970s, 1980s and early 1990s vehicle exhaust emission standards were tightened for both passenger cars and heavy duty vehicles (Dunne and Greening, 1993). Typically the standards in Europe lagged behind the United States standards, particularly California, but this position is beginning to change. Many of the European

Union proposals for later in this decade are likely to be as strict as the United States' standards (Needham *et al.*, 1993; Hadded *et al.*, 1994). Furthermore, many of the proposed standards being examined in Europe can be considered as forcing technological change rather than following developments in technology as was previously tending to happen. Reflecting this emphasis, two emission control scenarios are considered in this paper.

2.1 CURRENT PLANS SCENARIO

The first scenario examines present policies and standards and is called "Current Plans". This scenario assumes that existing policies continue until the end of the scenario period, the year 2010. In the Netherlands and UK, the EU "Stage II" standards agreed for implementation in 1995/1996 define the emission rates for both light-duty and heavy-duty vehicles in the Current Plans scenario (Dunne and Greening, 1993; Needham *et al.*, 1993; Hadded *et al.*, 1994). The situation in Sweden is more complicated as three emission classes of vehicles are available and economic incentives have been introduced to encourage the purchase of lower emission vehicles (Swedish Environmental Protection Agency, 1993). It is assumed that for Sweden the Current Plans scenario implies that all light-duty vehicles will meet Californian "TLEV" standards after the year 2000 and heavy-duty vehicles meet the Europe Stage II standard, as in the other EU countries. Poland has adopted many of the EU standards for its recent transport sector environmental regulations (Anderson, 1994). It is assumed in the Current Plans scenario that all vehicles meet the present EU Stage I standards, but not the Stage II regulations, as no further commitments have been agreed.

2.2 BEST AVAILABLE TECHNOLOGY SCENARIO

The second scenario, is a Best Available Technology (BAT) scenario which quantifies the impact upon emissions of control technologies that are likely to be available in the near future. The EU Stage III standards, planned for introduction in 1999/2000, should be achievable in the next few years by a range of technological options for both light-duty and heavy-duty vehicles, for example cold start emissions from gasoline vehicles could be reduced by electrically heated catalysts. Combinations of techniques, such as the introduction of low sulphur fuels and the application of exhaust gas recirculation, should enable the Stage III standard to be met by heavy-duty diesel vehicles. It is assumed in the BAT scenario that light-duty and heavy-duty vehicles meet the EU Stage III standards in all four countries.

2.3 SCENARIO RESULTS

This section presents the results of the scenarios for each of the countries in 2010. This target year has been chosen as it is sufficiently far into the future to assume that the existing vehicle fleet has been replaced by vehicles that meet the scenario standard. It is possible to calculate intermediate years if assumptions are made about the proportion of new and old vehicles in any year of interest. The calculations are performed in a spreadsheet model. Twelve types of vehicles are modelled, categorised according to the size and engine type of vehicle. The vehicle types modelled are two-stroke mopeds and motorcycles, four-stroke motorcycles, three sizes of gasoline cars, diesel cars, gasoline vans, diesel vans, gasoline trucks and buses, medium-sized diesel trucks, heavy diesel trucks and diesel buses. The number of vehicles of each type, the average annual distance driven per vehicle, and the emission factors per kilometre are entered into the model (Bailey and Gough, 1994). Emission factors were based upon values from several sources (CORINAIR, 1992; Gillham et al., 1994; Needham et al., 1993; Hadded et al., 1994; UNECE, 1991). The model was checked and calibrated against the 1990 emission inventory produced by Warren Spring Laboratory (Gillham et al, 1994) and the results are shown in Table 2. Although the spreadsheet model results are all lower, the values are in reasonable agreement considering the relatively large uncertainties involved, for example, emission factors are difficult to measure and it is often hard to find close agreement amongst different sources of these values. It is felt that the spreadsheet model is sufficiently accurate for the type of calculations made in this paper.

TABLE 2

Comparison of Warren Spring inventory and spreadsheet model results for the UK in 1990

Exhaust emissions	Warren Spring Inventory	Spreadsheet model
CO (kt)	6023	5062
VOC (kt)	829	680
NO_x (kt)	1383	1303

The results for the two scenarios for the four countries are displayed in Table 3. The calculations assume that all vehicles meet the scenario standard and remain in compliance in the year 2010. Despite the projected increases in total road transport activity, the Current Plans scenario effectively reduces the three pollutants compared to current levels in all countries, for example, emissions of NO_x could be reduced by about 70 per cent in the year 2010 compared to 1990 levels in the EU countries and by over 60 per cent in Poland. The BAT scenario reduces emissions further in all countries, in particular, VOC emissions are considerably reduced through the application of control technologies capable of meeting the EU Stage III regulations. Emissions of NO_x could be reduced by around 80 per cent through the application of Best Available Technology to diesel and gasoline vehicles assuming that the new vehicle standards are maintained in service.

TABLE 3

Emissions from road vehicles in the year 2010 for the two scenarios

Country and scenario	CO (kt)	VOC (kt)	NO_x (kt)	Approximate NO_x reductions from 1990
The Netherlands				
Current Plans	295	74	113	71%
BAT	243	29	86	78%
Poland				
Current Plans	431	123	161	64%
BAT	291	27	80	82%
Sweden				
Current Plans	180	20	54	70%
BAT	170	12	40	78%
UK				
Current Plans	1480	258	403	71%
BAT	1307	94	288	79%

3. Discussion

Estimating future emissions from road transport is a difficult activity. Many of the parameters required are highly uncertain, such as emission factors or the future vehicle stock. The scenario approach has the advantage that the assumptions define the scenario and different scenarios can be used to explore possible future outcomes. The two scenarios analysed here show that technological emission controls do have the potential to reduce road vehicle exhaust emissions significantly, typically by between 70 and 80 per cent compared to 1990 levels. However, this conclusion requires careful qualification. The scenarios assume that all vehicles remain in compliance with the standards over their operating lives, but this will only occur with strict in-service testing of vehicles. The total number of vehicles, the ratio of gasoline to diesel vehicles and the distance travelled will all have an impact upon future emission levels, independently of the emission standards analysed in the scenarios. The results in this paper should be seen as illustrative of what may be achievable with emission control technology in the future, rather than precise forecasts of future events. These results, in combination with estimates of NO_x and ammonia emissions from other pollution sources, can be used to estimate future deposition levels of nitrogen. In this way it is possible to relate emissions control strategies for the transport sector to future nitrogen deposition levels and the implied exceedances of critical loads for acidity and eutrophication.

In many respects, this analysis only considers a small aspect of the transport and environment issue. The three pollutants analysed are significant local and regional pollutants, but many other impacts of vehicles should be considered when analysing the environmental impacts of road transport. The health impact of particulate emissions from diesels is the cause of widespread concern and these emissions should be reduced as a result of the European Stage II and III standards. However, without improvements in vehicle fuel efficiency, carbon dioxide emissions will increase from the road transport sector in both scenarios, and noise levels could increase. The development of road transport infrastructure, that may result from the increased activity levels in the scenarios, could impose pressures on land use because of the need for new roads and the mining of minerals to build them. It is for these reasons that the switch to other transportation modes, such as rail or cycling, are often advocated as a broader solution to the problem of transport and the environment. It is likely that a combination of policy instruments, which includes the application of tighter emission standards such as those analysed in this paper, will be required to reduce the impact of road vehicles upon the environment in many European countries.

References

Anderson, G. (1994). Addressing mobile sources of air pollution in Poland: emerging trends and policy options. *European Environment* 4(6), 9-14.

Bailey, P.D. and Gough, C.A. (1994). *Mobile Sources' Emissions and Abatement Costs*. Paper presented to the December 1994 meeting of the UNECE Task Force on Integrated Assessment Modelling, Prague.

CORINAIR (1992). *Default Emission Factors Handbook (Second Edition)*. CORINAIR Inventory, Commission of the European Communities, CEC-DG XI, Brussels.

Department of Transport (1992). *Transport Statistics Great Britain 1992*. HMSO, London.

Dunne, J.M. and Greening, P.J. (1993). European emission standards to the year 2000. In *Worldwide Engine Emission Standards and How to Meet Them*. The Institute of Mechanical Engineers, London.

Gillham, C.A., Couling, S., Leech, P.K., Eggleston, H.S., Irwin, J.G. (1994). *UK Emissions of Air Pollutants 1970-1991 (Including Methodology Update)*. Warren Spring Laboratory, Stevenage.

Hadded, O., Stokes, J., Grigg, D.W. (1994). Low emission vehicle technology for ULEV and European Stage 3 emission standards. In *Automobile Emissions and Combustion*. The Institute of Mechanical Engineers, London.

Naturvårdsverket (1993). *Trafik och miljö*. Rapport 4205, Naturvårdsverket, Sweden.

Needham, J.R., Such, C.H., and Nicol, A.J. (1993). Fuel efficient and green - the future heavy-duty diesel. In *Worldwide Engine Emission Standards and How to Meet Them*. The Institute of Mechanical Engineers, London.

RIVM (1992). *National Environmental Outlook 1990-2010: 2*. National Institute of Public Health and Environmental Protection, Bilthoven.

RIVM (1994). *National Environmental Outlook 1993-2015: 3, Summary*. National Institute of Public Health and Environmental Protection, Bilthoven.

Royal Commission on Environmental Pollution (1994). *Transport and the Environment*. Royal Commission on Environmental Pollution, Eighteenth Report. HMSO, London.

Swedish Environmental Protection Agency (1993). *Motor Vehicle Air Pollution Control Regulations in Sweden: A Guide*. Swedish Environmental Protection Agency, Solna, Sweden.

UNECE (1991). *Protocol to the 1979 Convention on Long-range Transboundary Air Pollution concerning the Control of Emissions of Volatile Organic Compounds or their Transboundary Fluxes*. ECE/EB.AIR/30. United Nations Economic Commission for Europe, Geneva.

RECENT DEVELOPMENTS IN TECHNOLOGIES AND POLICIES IN GERMANY TO CONTROL ACID DEPOSITION

BERND SCHÄRER

Federal Environmental Agency, Postfach 33 00 22, D-14191 Berlin

Abstract. In Germany, acidifying emissions have decreased since the mid-eighties, but are still at levels that cause environmentally harmful acid deposition and thus make further action necessary. The driving force behind such actions is the precautionary principle laid down in pollution control legislation. It is implemented as a requirement to minimize emissions and mandates the parties concerned to formulate and implement emission control requirements based on the state of the art, and to update them as technological advances are made. As the scope for restructuring energy supply and switching to environmentally friendlier sources is very limited in Germany, strategies for controlling inevitable NO_x and SO_2 emissions will continue to be directed at further improving the technical systems (besides necessary changes in lifestyle). Since the large-scale retrofit programmes were initiated in the eighties, technological advances have provided some scope for a further tightening of emission reduction requirements for SO_2 and NO_x. In addition, there is some potential for further reducing these emissions through more energy-efficient demand- and supply-side technology.

Key words: acidifying emissions, control technologies, energy efficiency, policy instruments

1. Introduction

The most important air pollutants and sources causing acidification are sulphur dioxides (SO_2) from combustion plants, ammonia (NH_3) from agriculture and nitrogen oxides (NO_x) from motor vehicles and combustion plants. Acidification problems can be dealt with at the various levels at which they arise:

Compensatory measures - including liming and fertilization but also payment of compensation - are usually actions taken as a last resort to avert or offset damage. The applicability of such measures is limited. Environmental quality standards, i. e. maximum-permissible concentrations of pollutants in air and maximum-permissible deposition rates, constitute thresholds derived on the basis of knowledge about the effects of pollutants. Comprehensive concepts for environmental quality standards are, however, difficult to develop because the underlying knowledge about effects is often incomplete and their implementation requires emission and product standards directly aimed at the emission sources, involving complicated dispersion modelling. Nevertheless, effect-based environmental quality standards, as a measure for the carrying capacity of the ecosystem, constitute the right methodological approach because they place the assets to be protected and the carrying capacity of the ecosystem at the heart of the matter.

A point to be noted for the German air pollution control concept is the independent function of emission and product requirements which mainly reflect technical feasibility of avoiding emissions at source. All the measures mentioned are used in Germany to combat acidification problems, although their practical relevance differs greatly.

2. Technology based air pollution control - concept and regulations

When the subject of acidification with its catchwords "acid rain" and "forest die-off" was taken up by policymakers in the late 1970s, effective measures could be initiated relatively quickly, as there was knowledge about both the causes and sources as well as about remediation technologies (Schärer *et al.*, 1980). What was also of crucial importance, however, was that the regulatory framework for combating acidification processes and damage to forests did not have to be created but was already in place. The basis for the measures was the principle of "precautionary action", which was incorporated in the Federal Immission Control Act (Bundes-Immissionsschutzgesetz - BImSchG) as early as 1974 and requires technology-based emission control requirements to be formulated for all sources of air pollution. It mandates decision-makers to take actions which go beyond an averting of concrete environmental hazards and do not necessarily require scientific evidence of the existence of such hazards as a prerequisite. Rather, measures are to be taken in an anticipatory manner, so as to avoid and reduce environmental risks. Consequently, in Germany, technological development is a main driving force behind decisions taken in environmental policy and behind actions taken to enforce legislation regulating high-emission-volume air pollutants.

For the purposes of air pollution control, this abstract imperative is implemented by requiring all relevant emission sources to minimize emissions on the basis of the state of the art. State of the art is taken to mean in this context the state of development of advanced processes, of facilities or of modes of operation which is deemed to indicate the practical suitability of a particular technique for restricting emission levels. Within the framework of the treaty on the unification of the two German states it was agreed that the requirements existing in the original (western) Federal Republic should be applied identically in the new federal states (former GDR). Especially power plants and industrial plants should comply with the same emission standards within a certain time frame, but by the year 1999 at the latest (Bundesumweltministerium, 1992).

The single most important regulation for the control of acidifying emissions is the Ordinance on Large Combustion Plants of 1983, which consistently translates the technology-based air pollution control concept into emission control requirements. Applying to all combustion plants with a thermal rating of 50 MW and more, it lays down emission control requirements, provisions concerning the release of flue gas, emission measurement and monitoring, as well as a binding timeframe within which existing plants have to be retrofitted or shut down. As of 1 April 1993 at the latest, all existing plants had to comply with the same emission limits as new plants.. A comprehensive overview of technologies to clean up power plants is provided by Schärer, 1993. In the new federal states, the deadline for the retrofitting of major power plants is 1 June 1996.

Next followed, in 1986, the amendment to the Technical Instructions on Air Quality Control (TA Luft). The tightened TA Luft emission requirements for sulphur and nitrogen compounds covered all other combustion plants and industrial processes subject to licensing. In the original federal states, these were to be retrofitted or shut down by 1994 according to a graded time schedule oriented towards the quantity and noxiousness of the emissions. The time periods applicable in the new federal states will run out by 1999 at the latest. Due to their particular relevance, waste incinerators were regulated separately by way of the 17th Ordinance under the BImSchG (17.BImSchV).

Finally, the Ordinance on Small Combustion Plants, amended in 1993/94, covered smaller plants not subject to licensing (< 1 MW$_{th}$). In addition to improved combustion techniques, this Ordinance prescribes minimum efficiencies for hot-water boilers, maximum-permissible values for heat losses caused by the waste gas as well as requirements for the sulphur content of the fuels.

A significant portion of NO$_x$ emissions (more than 50%) originates from the transport sector. The most effective control technology for Otto engine cars is the three-way catalyst. The implementation of the catalyst was difficult as it touched upon the responsibility of the European Community. Lengthy negotiations finally resulted in a stepwise introduction of the catalyst throughout the EC, for larger-engine new cars ($>$ 1400 ccm) beginning from 1991 and for all new cars beginning from 1993. Some further reduction in emissions can be expected for the years to come as the share of catalyst-equipped cars in the fleet increases. Further improvements will result from the optimization and preheating of catalysts as well as improved surveillance of the car fleet. Technical improvements are also possible to reduce the emissions of heavy-duty vehicles, for which European limit values (EURO I to IV) have already been adopted or under discussion. As for SO$_2$ emissions, the EC Directive on gas oil will make diesel fuel with a maximum sulphur content of 0.05 wt.% mandatory throughout the European Union as of 1 October 1996. In Germany, fuel labelling provisions are to ensure that low-sulphur fuel gains a large share of the market ahead of time. In conclusion it should be stated that there is a general need for an environmentally friendly optimization of the traffic structure and traffic flows, which arises in particular from the need for reducing CO$_2$ emissions. In addition to technical measures, measures must therefore be taken to avoid traffic and shift it to environmentally more compatible modes of transportation.

The sector which from environmntal point of view evidently is most difficult to regulate in Germany is agriculture. Livestocks are the most important source of ammonia emissions, which are released during the storage and application of slurry as well as in animal housing. Emission control requirements are laid down in the TA Luft, and provisions regulating the spreading of slurry are included in the Fertilization Ordinance, which is currently being prepared, and in the Fertilizer Act.

3. Trend in acidifying air pollutants

In Germany, acidifying emissions as a whole have been on a downward trend since the early 1980s. However, in terms of the reductions achieved, marked differences exist between the various pollutants and, above all, between the trends in western and eastern Germany. Drastic reductions, of about 75%, were achieved for SO$_2$ in western Germany in the 1980s, with a 90% reduction in the emissions from power plants, the main emission source. By contrast, emissions in the former GDR increased considerably yet again in the 1980s, since lignite rich in sulphur was used for energy supply and the plants were not equipped with emission control technology (see table I).

TABLE I
Trends in Emissions in Germany, 1980/1990

		1980, kt			1990, kt		
		East	West	Σ	East	West	Σ
NH_3	Fertilizer applications	30	62	92	20	55	75
	Animal husbandry	218	496	714	173	481	654
	Industry	13	9	22	10	6	16
	Other sources	2	4	6	2	12	14
	Total	**263**	**571**	**834**	**205**	**554**	**759**
NO_x	Other Transport	93	223	316	80	250	330
	Road transport	100	1364	1464	137	1519	1656
	Households	3	87	90	4	73	77
	Small-consumers	7	56	63	6	36	42
	Industry	74	396	470	72	247	319
	Power plants	237	800	1037	274	335	609
	Total	**514**	**2926**	**3440**	**573**	**2460**	**3033**
SO_2	Other Transport	34	22	56	18	15	33
	Raod transport	24	67	91	26	40	66
	Households	428	196	624	369	83	452
	Small-consumers	141	142	283	90	51	141
	Industry	557	860	1417	487	394	881
	Power plants	3136	1879	5015	3765	295	4060
	Total	**4320**	**3166**	**7486**	**4755**	**878**	**5633**

Since German unification in 1990, the restructuring of the energy industry in the new federal states has already caused a drastic reduction in emissions there. The shut-down of obsolete power plants, the retrofitting of existing and the construction of modern, efficient power plants are factors contributing to this trend, as are fuel switches in industry and private households where space heating still is a major source for locally high SO_2 concentrations. SO_2 emissions from power plants are expected to decrease in eastern Germany by 80% by mid-1996, which is the deadline by which major power plants have to be retrofitted with flue-gas desulphurization and $DeNO_x$ systems. The complete implementation of present German regulations will cause SO_2 emissions in eastern Germany to decrease by 95% and also for Germany as a whole, the target of an 87% reduction laid down in the 2nd ECE SO_2 Protocol will be reached (see figure 1, Bundesumweltministerium, 1995). A comparable positve trend cannot be seen for NO_x. The reductions achieved by controlling emissions from combustion plants and passenger cars with catalytic converters will be cancelled out in part by additional NO_x emissions caused by the growth in traffic. However, present regulations will result in a 37% reduction in NO_x emissions from 1990 to 2005. As relevant regulations are lacking to a large extent, NH_3 emissions are not expected to decrease markedly, either.

The decrease in SO_2 emissions has had a positive impact on pollution levels. Since 1988, the Umweltbundesamt's air pollution monitoring stations have recorded ambient air concentrations of SO_2 markedly lower than those of previous years, with a decrease in eastern Germany of about 30% and in western Germany of about 70%. Relatively mild winters had also contributed to this trend. The air pollution measurement data also

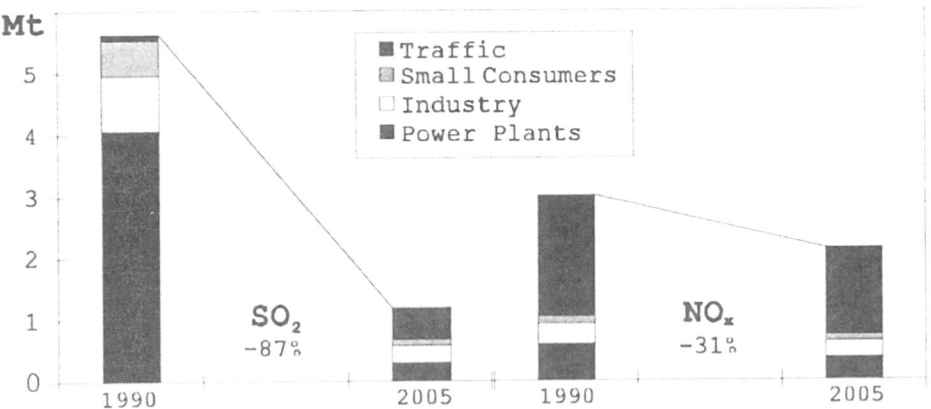

Fig. 1: Emissions in Germany: Trend scenario 2005

reflected the drastic emission reductions caused by the economic collapse in the new federal states after unification. Concerning depositions, the new ECE Sulphur Protocol is based on so-called critical loads, i.e. deposition values for sulphur compounds below which, according to present knowledge, the protection of ecosystems and the groundwater is ensured for the long term. According to calculations performed using the IIASA RAINS model, the use of measures considered at present to reflect the state of the art will not be sufficient to meet critical loads throughout Germany. This holds true even if all countries which "export" SO_2 to Germany likewise take measures reflecting the state of the art. In view of this, it should already be clear that further advances in the state of the art will have be made and that, in addition, consumption patterns and lifestyles will have to undergo considerable changes in order to achieve further progress.

4. Perspectives for the technology-based regulatory concept

A renewed discussion over emission limits for large combustion plants has started among policymakers in Germany, also because an amendment to the EC Directive on large combustion plants is planned for 1996. More stringent emission limits have been proposed, as follows:
- for SO_2, 200 mg/m^3 and 95 - 98% removal efficiency instead of 400 mg/m^3 and 85% removal efficiency;
- for NO_x from coal-fired dry bottom furnaces, 100 mg/m^3 instead of 200 mg/m^3.

More stringent limit values are also being discussed for plants using liquid and gaseous fuels as well as for smaller plants.

The technological advances made since the 1980s are cited as the reason for these initiatives. Desulphurization efficiencies in excess of 95% have in fact already been required in permits for power stations in eastern Germany. These developments are mainly based on the optimization of lime scrubbing systems. Possibilities for reducing NO_x emissions through technical means have not yet been fully exploited, either. Improving air staging, increasing pressure and the use of more finely ground coal are primary measures with which NO_x emissions from combustion plants can be further reduced. In addition, SCR (Selective Catalytic Reduction) technology has been improved. SCR technology is already being employed at coal-, some lignite- and oil-fired boilers, waste incinerators, in the glass industry and in the manufacture of nitric acid. Further potentials for SCR are in the cement industry, one of the main industrial emitters of NO_x, and in lignite-fueled power plants. For smaller plants such as fluidized bed furnaces, however, the cost/benefit relationship in using SCR would still have to be examined.

Now that the possibilities for reducing the emissions of SO_2 and NO_x from stationary sources through add-on technologies and primary measures have largely been exhausted, energy conservation measures are the main area for achieving for further reductions. An additional motive for measures to improve energy efficiency is the need to reduce CO_2 emissions in order to protect the global climate. For this, the Federal Government has formulated an ambitious objective which Chancellor Kohl, at the World Climate Summit in Berlin, specified further in the form of a pledge to reduce CO_2 emissions by 25% by the year 2005 versus 1990 levels.

In order to meet this target, energy efficiency standards based on the state of the art could, in principle, be formulated for the entire spectrum of energy supply and energy use, although the use of other instruments to this end is also conceivable. Energy efficiency standards have so far mainly been used for electrical appliances, boilers and the energy demand of buildings, but also to regulate the specific energy consumption of motor vehicles. A question that has been discussed for some time now in Germany is whether the complex energy structure of power plants and large-scale industrial processes should also be regulated on the basis of the state of the art of energy efficiency. With the Heat Management Ordinance (draft), the Umweltbundesamt has developed a new set of instruments, covering all energy-relevant installations subject to licensing (Umweltbundesamt, 1992).

5. Conclusion

On the whole, it appears that regulation with a view to controlling acidifying pollutants has exhausted technical ways and means (best available technology) in Germany. The technology-based air pollution control concept as the driving force behind environmental progress is approaching its limits. To achieve the ecological improvements that still need to be made, environmental policy will have to draw more heavily on planning-related measures and instruments providing economic incentives.

References

Bundesumweltministerium: 1991, Fünfter Immissionsschutzbericht, 13 - 18
Schärer, B. et al.: 1980, Umweltbundesamt - Texte, Luftverschmutzung durch Schwefeldioxid, 123 pages
Schärer, B.: 1993, Staub-Reinhaltung der Luft, Technologies to clean up power plants, 27 - 92 and 157 - 160
Umweltbundesamt: 1992, Jahresbericht, 170 - 173
Bundesumweltministerium: 1995, Sechster Immissionsschutzbericht

REDUCING THE CONTRIBUTION OF AMMONIA TO NITROGEN DEPOSITION ACROSS EUROPE

H.M.ApSIMON, S.COULING, D.COWELL and R.F.WARREN

Imperial College Centre for Environmental Technology, London SW7 2PE, U.K.

Abstract. European emissions of reduced nitrogen, arising principally from agriculture, are comparable with those of oxidised nitrogen from mobile and stationary combustion sources. It is therefore important to include ammonia emissions in working towards a new protocol on nitrogen under the programme of the UN Economic Commission for Europe on the control of transboundary air pollution. However the nature of the sources and the subsequent atmospheric transport and chemistry are very different from other acidifying pollutants. This paper describes work in hand under the MARACCAS project to compare agricultural activities in different European countries and to assess the applicability and efficacy of potential abatement measures. The aim is to derive abatement costs for each country relating successive emission reductions to the costs of achieving them, to be used by the UN ECE Task Force on Integrated Assessment Modelling (TFIAM) - in particular with our Abatement Strategies Assessment Model, ASAM. The paper will also address the large uncertainties involved in integrated assessment modelling with respect to ammonia, and suggest how these may be allowed for in deriving cost-effective abatement strategies.

Key words: emissions, nitrogen, ammonia, agriculture, integrated assessment modelling, abatement strategies

1. Introduction

In developing the second protocol for sulphur, the Oslo protocol, maps of critical loads across Europe were used as an indication of sustainable levels of deposition of sulphur. Integrated assessment models including the RAINS model of IIASA (Alcamo et al.,1987), the CASM model of the Stockholm Environment Institute (SEI,1991), and our own model ASAM (Abatement Strategies Assessment Model- ApSimon et al.,1994) were used to indicate cost-effective strategies for reducing the exceedance of these critical loads. Under the Task Force on Integrated Assessment Modelling (TFIAM) of the UN Economic Commission for Europe (UN ECE) a similar procedure is now being applied to nitrogen deposition in the context of development of a new protocol for NOx. This is far more complicated because NOx is a contributor to problems of tropospheric ozone and eutrophication as well as acidification, and for the latter two effects nitrogen deposition from ammonia is also important. Accordingly all three integrated assessment models are being adapted to treat a combination of sulphur and nitrogen species in relation to acidification and eutrophication. Thus the ASAM model now includes SO_2, NOx from stationary combustion, and from the transport sector, and ammonia, with the potential to add VOCs in relation to ozone formation in due course.

In order to include ammonia in these models it is necessary to have information on the emissions from each country, the transport through the atmosphere from the source areas to deposition across Europe, the sensitivity of the ecosystems affected by this deposition, and on the measures which can be taken to abate the emissions and hence reduce the effects. To derive cost-effective strategies it is also necessary to know not only the abatement techniques, but also their costs. In addition the variability in agricultural systems and climates in different countries greatly affects both the emissions, and the extent to which they can be controlled.

This paper therefore begins by discussing the contribution of ammonia to transboundary air pollution as compared with NOx. A brief outline of ASAM is then given as an example of an integrated assessment model, and the way in which emission reduction scenarios for Europe may be derived for a combination of pollutants and effects to suggest the levels of abatement needed in each country. The paper concludes with the measures applicable to reducing ammonia emissions, and the work in progress in the MARACCAS project to assemble the relevant information on agricultural activities in each country in order to include ammonia in such integrated assessment.

2. The role of ammonia in strategies to reduce transboundary air pollution.

Emissions of nitrogen as ammonia in Europe come mainly from agriculture (>90% for most countries), particularly from livestock farming. Such emissions of nitrogen in reduced form are comparable with those as oxidised nitrogen from the combination of stationary combustion sources and mobile sources. This is illustrated in table 1 taken from the work under the European Monitoring and Evaluation Programme (EMEP,1994) for those countries who had submitted official estimates of their ammonia emissions. It should be noted that many countries still have only very preliminary estimates of such emissions, and that even in the new CORINAIR inventory for 1990 default values for emission factors have been used largely based on experience in the Netherlands.

TABLE 1: Emissions of ammonia nitrogen and oxidised nitrogen and the respective portions deposited a) within the country of origin and b)summed over other countries (all in units of 100 tonnes of N per year)

Country	Emission NH4-N in country	Portion deposited in country	Portion dep. on other countries	Emission NOx in country	Portion deposited in country	Portion dep. on other countries
Austria	815	403	219	676	46	265
Belgium	642	261	265	1017	28	546
Denmark	1153	491	323	861	20	392
Finland	354	195	114	883	116	314
France	6522	3704	1157	5326	843	1170
Germany	6259	3458	1717	9709	1556	4392
Hungary	1400	620	467	724	61	354
Ireland	1038	516	158	350	18	90
N'lands	1927	832	771	1680	55	860
Norway	313	168	47	700	58	159
Poland	4529	2387	1266	3896	517	1769
Portugal	766	335	105	642	50	123
Sweden	428	226	88	1217	155	388
Switz'land	502	269	149	560	40	208
U.K.	2306	1140	333	8458	629	2450
Europe-total	65421			70675		

It can be seen from table 1 that per ton of nitrogen emitted, the ratio of the nitrogen deposited within the country of origin to that exported to other countries is far greater in the case of ammonia nitrogen than for oxidised nitrogen. That is, a country will improve its own situation more by reducing transboundary contributions of ammonia nitrogen than of oxidised nitrogen. There may also be an extra incentive to reduce nitrogen deposition by controlling ammonia emissions in those countries where reductions of NOx can lead to increased concentrations of ozone unless accompanied by reductions in VOCs.

Ammonia is a reactive gas which is emitted as a result of volatilisation from ground-level or low level sources. As such there is competition between upward diffusion into the atmosphere and local re-deposition close to the source which depends on surface vegetation. The portion which disperses up through the atmosphere reacts with acidic gases and aerosols forming ammonium particulates which are not readily dry deposited like the gaseous ammonia, and hence give rise to longer range transport until deposited in rain. The presence of ammonia also influences the cloud chemistry, and the uptake and oxidation of sulphur dioxide in clouds. The atmospheric transport of ammonia, and its interaction with other species is described in more detail in ApSimon et al. (1994). The point to note here is that deposition is extremely patchy, and highly correlated with the spatial distribution of emissions. There may therefore be high priorities for reducing ammonia emissions in the vicinity of sensitive ecosystems with excess nitrogen deposition.

3. Inclusion of ammonia in the integrated assessment model, ASAM

For the reasons given above, ammonia has been included in the new version of our integrated assessment model, ASAM. This model takes as a starting point projected future emissions, and uses source-receptor relationships based on atmospheric transport modelling of EMEP (1994) to derive the associated deposition across Europe. Critical load maps assembled from national data by the Co-ordinating Centre on Effects at RIVM in the Netherlands (CCE, 1994), are used to indicate the sustainable levels of deposition and derive the excess deposition or "exceedance". In the new model there is provision for independent specification of critical loads with respect to acidification (due to the combined effect of sulphur and nitrogen) and eutrophication (due to nitrogen deposition alone). Different weightings may be attached to the two effects.

Measures which may be taken to reduce emissions are incorporated as tables for ammonia and each of the other pollutants, specifying the emission reductions which may be achieved, listed in order of increasing cost per unit of emission abated. The ASAM model has a rather different approach from RAINS and CASM in that it derives a prioritised sequence of steps across the different pollutants and countries in order to reduce deposition towards the desired goals as fast as possible as a function of cost. Thus each step is selected by scanning through the different sources and pollutants in all the countries, and computing the benefits that will accrue in the form of reduced exceedances by taking the next abatement measure not yet implemented for each source and pollutant. That combination of source and pollutant which has the highest ratio of benefit to cost is then chosen for implementation, and the process repeated with the improved environmental situation. An advantage of this approach is that it does not matter if the target deposition maps can be attained or not; and grid-squares in the map which are driving the abatement strategy, or are difficult to attain, are clearly identified. Improved ecosystem protection can be shown as a graph against cumulative expenditure, indicating how the marginal cost per hectare brought under protection gets increasingly expensive. Tables of the emission reductions assigned to each country for each pollutant, including ammonia, can be produced at specified intervals of total expenditure, or when deposition has been reduced to the desired levels.

4. The MARACCAS project on the potential for abating ammonia emissions in Europe

In order to include ammonia in integrated assessment it is necessary to have the information on the abatement techniques, their applicability and efficiency, and their costs. Accordingly the MARACCAS project (Model for the Assessment of Regional Ammonia Cost Curves for Abatement Strategies) has been established to gather agricultural information from each country and use it in a computer model to analyse the potential for reduction of ammonia emissions as a function of cost for that country. The basic methodology is described below, and was approved at a workshop held at Culham in the UK in October 1994, together with a list of possible abatement measures. These have been further refined and discussed at a workshop under the UN ECE Working Group on Technology in the Netherlands in June 1995. Meanwhile a questionnaire has been sent out to each country to gather the relevant data, which are being fed into the MARACCAS program (written in EXCEL, a spreadsheet software package).

The major part of the ammonia emissions originate from livestock wastes, and it is on these that most effort has been concentrated in MARACCAS, although fertiliser use on crops and reduced use of urea is also included. The approach used is illustrated in figure 1. It starts with the intake of nitrogen per animal in the diet (this varies considerably from country to country in some cases), and the ammonium nitrogen generated in wastes. This is then followed to the pasture, or through the pathway of animal housing to storage and thence to spreading or disposal. A distinction is made between liquid and solid wastes. Some countries, such as the Netherlands, use predominantly slurry based systems, whereas other countries use far more bedding and straw, generating solid wastes. This makes a large difference to the degree of abatement which may be attained.

Figure 4.1. Scheme for assessing livestock sectors in MARACCAS

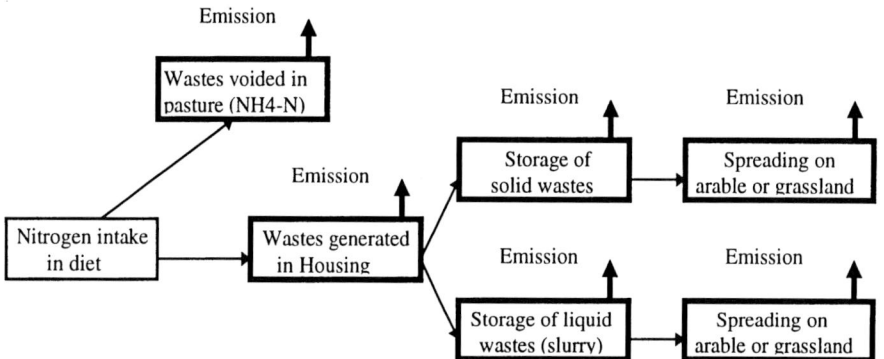

The scheme is applied separately to dairy cows, other cattle, pigs, sheep and poultry. Emissions to the atmosphere, indicated in the figure by the thick arrows, are specified by volatilisation rates as percentages of the NH4-nitrogen in the compartment generating the emissions. They will depend on the system used (e.g. storage of slurry in tanks, lagoons or weeping walls) and on other parameters such as the duration of storage, or grazing period By following the flow of ammonium nitrogen from one compartment of the model to another the interaction between control measures is explicitly included. This is important as the introduction of one measure may affect the overall reduction in emissions achieved by another. The abatement measures simulated in the model are summarised in table 2. Practical experience with these techniques is limited to a few countries, especially the Netherlands. The efficiencies assigned to the measures reflect the fact that in other countries they may well be less effective- for example injection of slurries may reduce losses of ammonia to the atmosphere by 95% in sandy soils in the Netherlands, but have far less effect on other soils and in wetter climates. It is important to recognise the large uncertainties, and a range has been assigned as well as a best estimate for the average efficiency of each measure.

In addition the applicability of measures may be limited. For example it is not possible to inject slurries on stony soils or steep slopes, and solid wastes can not be ploughed in where they can only be spread on grassland. It is also important to recognise

Table 2: Efficiencies of abatement measures

Abatement measure	efficiency range %	best estimate %	applicability	other comments
spreading of wastes a) liquids/slurries i) high efficiency measures e.g. injection	20-95	70%	not on stony or water-logged soils or steep slopes >12%	may increase nitrate leaching, and need extra storage
ii) medium efficiency e.g. slit injection, trailing shoe	20-80	50%		
iii) low efficiency e.g. band spreading, sprinkler boom	0 - 50	30%		
b) solid wastes/FYM rapid ploughing (< 4hrs)	20-90	60%	not on grassland except when re-seeding	
storage of wastes :- a) liquids/slurries covering storage tanks:- i) high efficiency - rigid lids	70-90	80 %	not applicable to lagoons	
ii) medium efficiency - e.g. LECA,foil	50-80	60%		
iii) low efficiency - e.g. crust, straw, + filling tanks from bottom	0-50	30%		
b) solid wastes - no measures identified				
animal housing :- a) cattle scraper-sprinkler-flushing systems	0-50	30%	size of unit building restriction	need careful op./maintenance
b) pigs (fattening, sows) - successive measures i) change floor grid ii) gutters under floor iii) flushing with aerated slurry	0-20 0-30 0-60	10 20. 30%	solid floors need rebuilding	
ultra low emission housing (e.g. using acid)				very expensive
filters/biofilters	60-90	80%	only on mechanic-ally ventilated buildings	
poultry: a) broilers deep litter-forced ventilation to dry b) egg laying: change housing system +introduce belt system+drying of wastes +effect of watering system e.g. nipple	(10-80)			
dietary change pigs (2 steps:-phase feeding +lysine) poultry	0-30	(15) (10)		depends on N in diet

possible side-effects such as increased nitrate leaching, and take steps to minimise them. This may entail extra costs such as more storage so that wastes can be spread at the optimum time when the nitrogen will be taken up by crops. The costs of implementation can also vary considerably from one situation to another. The most cost-effective steps are usually found to be those concerned with spreading and storage of wastes, and losses from housing are usually more expensive to reduce per kilogram of nitrogen saved. The costs will also depend on whether existing facilities can be modified, or whether they would need to be replaced. Costs also tend to be higher for small units- for example in Switzerland where farms are generally very small.

The reductions which can be achieved also depend on the underlying unabated emissions. Thus average emissions per animal may vary by a factor of up to 2 between an intensive system with high fertiliser use and productivity, and low intensity farming. Reductions in the intensity of farming and the density of fertiliser application, plus lower animal numbers, would complement other measures- but need to be viewed in a wider context than abatement of ammonia emissions alone.

5. Results and conclusions

The information gathered from countries is still being processed through the MARACCAS program to derive tables of potential emission reductions and their costs (to be referred back to the countries for comment before publication). Whereas a few countries such as the Netherlands hope to cut back ammonia emissions from agriculture by at least 50% and possibly up to 70%, in many countries the potential reduction will be less than 30%. Overall the maximum feasible reduction in Europe without radical agricultural reform is likely to be of the order of 30%. Note also the large uncertainties in the future of agriculture, especially in central and eastern Europe where production has dropped by 30% in some cases as large collective farms have been divided among private ownership. Even in western Europe projected livestock numbers and future agricultural policy in the European Union are uncertain. For some agricultural sectors costs of high levels of abatement could exceed their profit margins, so that the economic impact would be very significant. Overall it is important to appreciate the much greater uncertainties in addressing ammonia as compared with sulphur. In integrated assessment use of a single cost curve based on a table of best estimates for each country must be extended to encompass the likely range of uncertainty (e.g. see table 2). Having identified those countries where abatement of ammonia emissions is likely to be important, more detailed attention can then be given to their specific circumstances.

Acknowledgements
We thank all those who have contributed to the MARACCAS project, and the UK Ministry of Agriculture and Department of the Environment for their financial help and support.

References

Alcamo J et al.: 1987, Acidification in Europe: a simulation model for evaluating control strategies. Ambio 16, 232-245

ApSimon H.M., Barker B.M. and Kayin S: 1994, Modelling studies of the atmospheric release and transport of ammonia in anticyclonic episodes. Atmos.Env.28, 665-678

ApSimon H.M., Warren R.F. and Wilson J.J.N: 1994, The Abatement Strategies Assessment Model-ASAM; applications to reductions of sulphur dioxide emissions across Europe. Atmos.Env.28, 649-663.

EMEP: 1994, Transboundary acidifying pollution in Europe: calculated fields and budgets 1985-1993 EMEP/MSC-W report 1/94.

SEI: 1991, An outline of the Stockholm Environment Institute's Co-ordinated Abatement Strategy Model, CASM. SEI report November 1991, University of York, Heslington, York, UK.

CCE: 1994, Calculation and mapping of critical loads in Europe. RIVM report No259101003

EUROPEAN EMISSION DATA WITH HIGH TEMPORAL AND SPATIAL RESOLUTION

L. LENHART, R. FRIEDRICH

Institute of Energy Economics and the Rational Use of Energy
University of Stuttgart, Heßbrühlstraße 49a, 70565 Stuttgart, Germany

Abstract. The estimation of acidic deposition strongly depends on the availability of accurate emission data. The atmospheric models, that calculate concentrations and depositions of pollutants, need data in a high temporal resolution (e.g. daily, 6-hourly or even hourly data). However, although some progress has been achieved concerning annual emission data for Europe (e.g. CORINAIR 90), only very little information is available about the temporal variation of these emissions during a year. Therefore, within the EUROTRAC-GENEMIS project special emphasis was laid on the development of methods to generate emissions with a high temporal and spatial resolution. As results the temporal and spatial distribution of SO_2-and NO_x-emissions are shown. The results indicate, that the emissions vary considerably over time and that the use of simple patterns for the temporal disaggregation is not sufficient for modelling and assessment of acidic deposition.

Keywords: European emission data, temporal disaggregation of emissions, spatial disaggregation of emissions

1 Introduction

For the investigation of phenomena which are highly dependent on short and medium term conditions (e.g. photo-oxidant formation) emission data with high temporal resolution are required. These data must be as accurate as possible, since they are regarded as the major source of uncertainty for the results of atmospheric transport and chemistry models (Heymann, *et al.*, 1993). While annual emission inventories experienced improvements, not much information has been available about the temporal distribution of those emissions. The GENEMIS project has been established to work out methods for the temporal disaggregation of annual emission data, improvements of emission inventories esp. for Eastern Europe, validation of emission data and for the development of methods for update. The results, if being implemented by modelers, can help to build the background for political answers to the pollution problems in Europe.

2 Methods and results

First approaches applied in the past were based on plausible assumptions and yield simple patterns of relative emissions in time (e. g. PHOXA and LOTOS inventory, EMEP, EUMAC). This patterns can be multiplied by annual emissions and give first order estimations of seasonal and daily variations of the emissions (Axenfeld *et al.*, 1987, Meinl *et al.*, 1989, Veldt , 1992). National or regional differences of the temporal behavior of emission sources have been neglected. Also, with the exception of biogenic emissions and VOC evaporation, the strong influence of the temperature on emissions has not been taken into consideration. These methods were seen to be too rough for a reliable estimation of the emissions. Therefore, methods to estimate hourly emission data with higher accuracy and reliability have been developed by analyzing results from a large number of research projects. These methods use various available sources like e.g. fuel use, temperature, degree days, working time etc. as indicators for the temporal resolution of activities causing the emissions. Table 1 gives an overview of the most important indicators for nine emission source sectors (see [6]).

TABLE I
GENEMIS indicator data for the estimation of emissions

Sector	Indicator data monthly resolution	Indicator data daily resolution	Indicator data hourly resolution
Power plants	fuel use	load curves	load curves
Industrial combustion	fuel use, regression analysis (using temperature, degree days, production)	working time, holidays, regression analysis (using temperature, degree days, production)	working time
Small consumer combustion	fuel use, regression analysis (using temperature, degree days, production)	user behavior, regression analysis (using temperature, degree days, production)	user behavior
Combustion in households	degree days, temperature	degree days, temperature	user behaviour, measurements and model calculations of fuel consumption
Refineries	oil throughput, fuel use	working times, holidays	working times, shift times
Industrial processes	production	working times, holidays	working times, shift times
Solvent use	production	working times, holidays	working times, shift times, user behavior
Road traffic	traffic counts	traffic counts	hourly traffic counts
gasoline evaporation	temperature	temperature	temperature
Air traffic	LTO cycles, number of passengers and freight	LTO cycles, number of passengers and freight	LTO cycles, number of passengers and freight
Biogenic emissions	temperature, radiation	temperature, radiation	temperature, radiation

Qualitative expert judgements, plausibility tests and some limited comparisons between measurements and calculations (e.g. for road traffic) show that the accuracy of the temporal resolution has improved significantly with this methods. As an example for GENEMIS indicator-data the monthly fuel consumption of public power plants in France and Spain is displayed in fig. 1. Emissions of combustion sources are related to their actual fuel use, which is typically higher in winter than in summer. Fig. 1 indicates, that this seasonal variation is much higher in France than in other European countries. The higher share of nuclear power reduces the contribution of other combustion plants in summer. The disaggregation from day to hour for power plant emissions can be derived from load curves. First assumptions of earlier patterns imply seasonal under- or overestimation (see also fig. 1).

The emissions of industrial combustion depend on production rates controlling the energy consumption for production processes, outside temperature controlling energy consumption for process heat, degree days as a measure of energy demand for space heating and working times. Small consumer combustion includes institutional and commercial fuel consumers, farms, etc. and is related to fuel consumption, outside temperature, degree days, working time and regional user behavior. A regression analysis leads to equations, which describe the time variation as a function of these parameters.

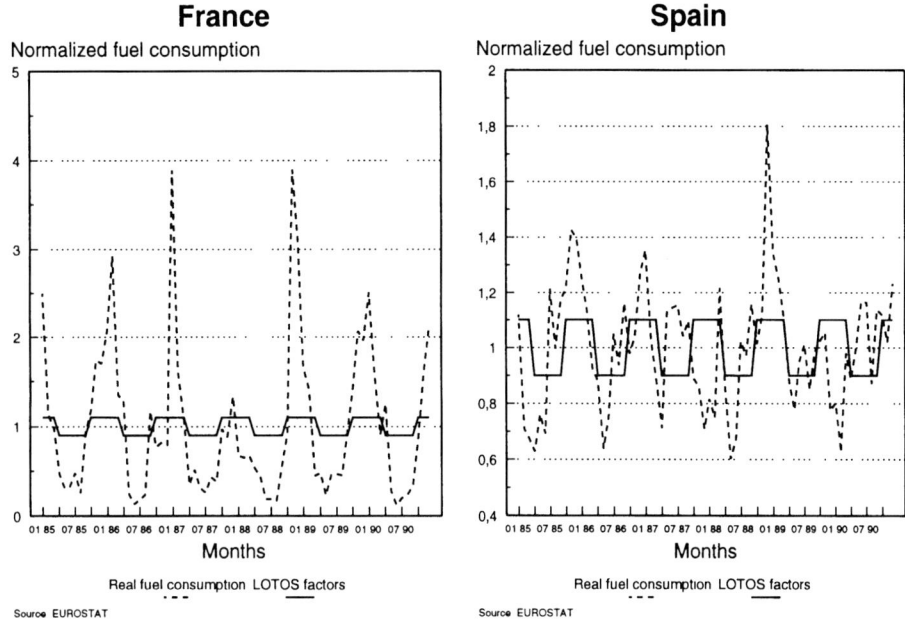

Fig. 1. Monthly total fuel use of public power plants 1985-1990 for France and Spain in relative units.

The sector solvent use poses some problems for emission inventorying, since it is characterized by a huge number of small and heterogeneous emission sources. A correlation of emissions to production, working times, holidays and user behavior is the most reasonable assumption. For private solvent use higher emissions are assumed in summer and on working days. Hourly emissions of this sector can be estimated by taking into account working times.

Road traffic belongs to the most important emission sources for NOx, CO and VOC. Fig. 2 illustrates the total hourly NOx-emissions of all main sources for two urban and two rural regions. Traffic rush hours in the morning and in the evening can be identified. The contributions of the other sectors change from region to region, depending on the location of power plants or industrial facilities. Differences between the total day and night emissions for example range from factor 4 in Northern Jutland up to factor 12 in Alto Alentejo.

The application of the GENEMIS method proofs a considerable spatial and temporal variation of most emissions. The impact of the deviations from patterns which are formerly used (up to 600% for public power plants, 30% for industrial combustion and solvent use, 50% for road traffic, 100% for small consumers) on the results of modellers must be regarded as high and require more detailed investigations in the future.

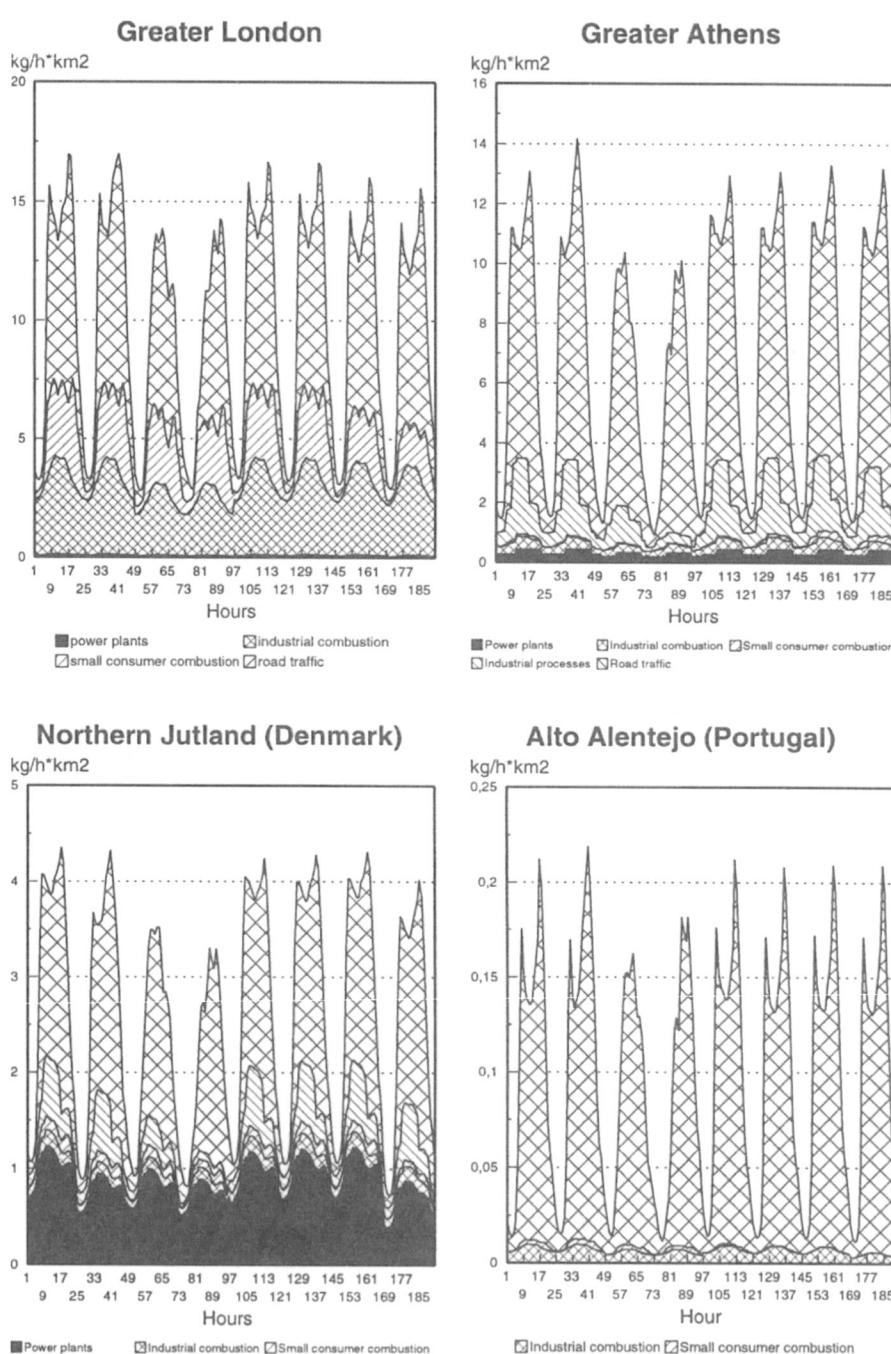

Fig. 2. Total hourly NOx-emissions for urban regions (London, Athens) and rural regions (Northern Jutland in Denmark and Alto Alentejo in the East of Portugal)

Other examples for the numerous results are shown in fig. 3 (hourly European NO$_x$-emissions for Sunday, October 14th, 1990 at 4 am and 4 pm) and fig. 4 (daily SOx-emissions of Germany, France and Spain in 1990).

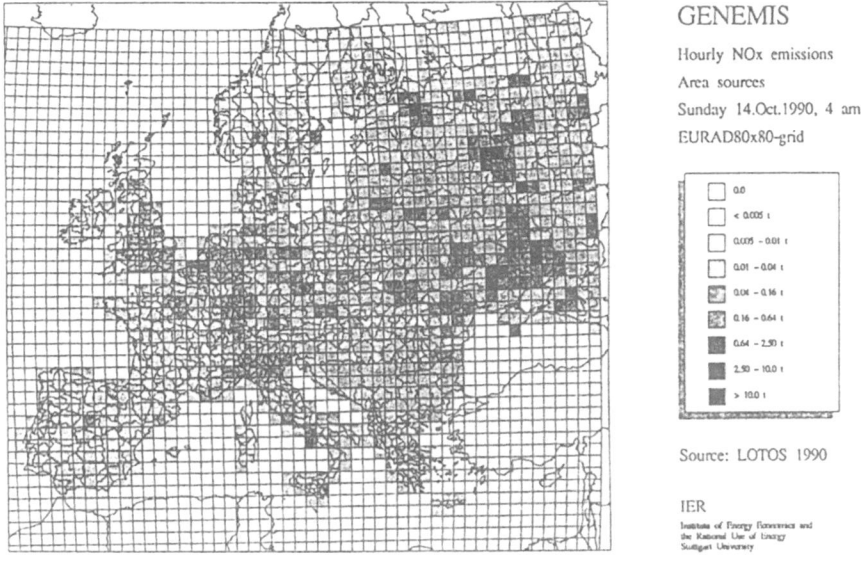

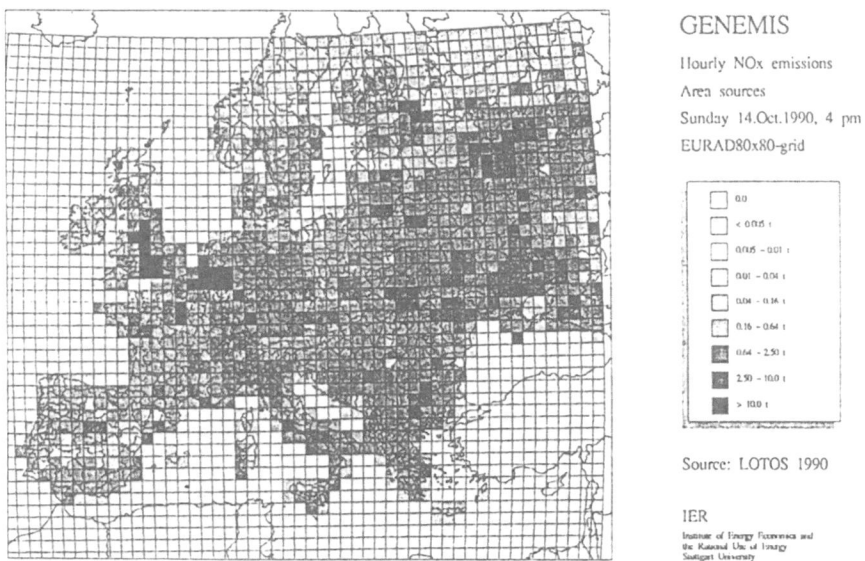

Fig. 3. Hourly European NOx-emissions for Sunday, October 14th, 1990 at 4 am and 4 pm

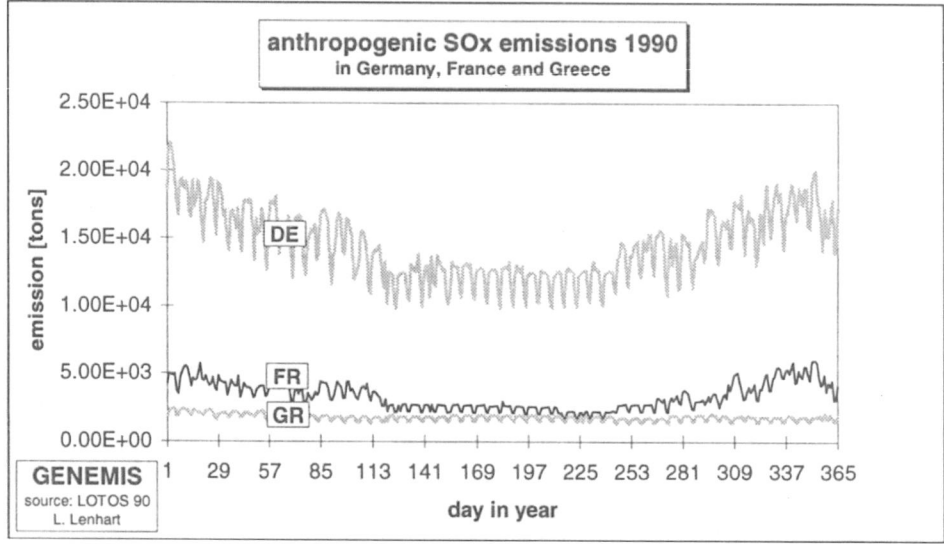

Fig. 4. Daily anthropogenic SO_x emissions of Germany, France and Spain in 1990

3. Conclusion

The use of indicator data yields the most plausible basis available for the description of the temporal distribution of emissions. Based on statistical and meteorological data, a strong variation in time can be detected for most of the source sectors. This dependency differs from country to country and from sector to sector. The GENEMIS inventory provides reasonable estimations of disaggregated emissions, which can be considered more reliable and nearer to reality than simple patterns as used by earlier projects. The use of the GENEMIS method can help to improve the quality of atmospheric modeling results.

References

Axenfeld F., J. Münch, H. Effinger, L. Janicke, PHOXA emission data base - Federal Republic of Germany 1980/1982, Dornier System i.A. des Umweltbundesamtes, 1987.

Friedrich R., M. Heymann, Y. Kasas, The calculation of temporal resolution and updates of emission data, in: EUROTRAC Annual Report, part 5, 19-58.

Meinl H., Münch J., Schubert S., Stern R., PHOXA Emissions data base -within the framework of control strategy development, PHOXA-report no. 1, MT-TNO Report 88-120, Netherlands Organisation for Applied Scientific Research, Apeldoorn 1989.

Veldt C., Updating and upgrading the PHOXA emission data base to 1990, TNO report, TNO Institute of Environmental and Energy Technology, Apeldoorn, March 1992.

Heymann M., A. Trukenmüller, R. Friedrich, Development Prospects for Emission Inventories and Atmospherical Transport and Chemistry Models (DEMO), Report to the Institute for Prospective Technological Studies, Joint Research Centre Ispra, Novembre 1993.

Friedrich R., Generation of Time-dependant Emission Data, EUROTRAC Symposium, 23-27 March 1992, Garmisch-Partenkirchen, SPB Academic Publishing, Den Haag, 1993, 255-268.

Builtjes P., The LOTOS-Long Term Ozone Simulation-project, Summary report, TNO-report, IMW-R 92/240, Order no.51555, 1992.

Bouscaren R., Veldt C., Zierock K.H., CORINE: Emission inventory feasibility study, Final Report, CITEPA, Paris 1986.

POLLUTION SOURCES AND SPREADING OF SULPHUR DIOXIDE IN THE NORTH-EASTERN ESTONIA

V. LIBLIK, A. RÄTSEP and H. KUNDEL

North-East Estonian Department, Institute of Ecology, 15 Pargi Str., Jôhvi, Estonia

Abstract. The greatest sources of atmospheric emissions of SO_2 in Estonia are caused by power plants (TP) which use oil shale. Since 1990 the amount of SO_2 discharges has continuously decreased due to fall in production of electric energy, and it was from TP as follows: in 1990-1991 about 180-200 thousand tons, in 1992 about 140 and in 1993-1994 about 100 thousand tons. In 1990 the annual mean emission intensity of SO_2 from all North-East (NE) Estonian pollution sources was fixed to be about 6.kg/s, with a maximum of 9.5-11 kg/s in winter period. In 1992-1993 the corresponding values were 3.5-4.6 and 5.1-6.8 kg/s. The single maximum concentrations (MC, per 30 min.) of SO_2 in the overground air layer would be in the ranges 25-450 $\mu g/m^3$ depending on emission intensity and wind parameters. The annual mean concentrations are below 25 $\mu g/m^3$ on the main territory, but may be up to 50-75 $\mu g/m^3$ near the power plants. In Kohtla-Järve town the annual mean values of 15.8-19.1 $\mu g/m^3$ and MC values of 271-442 $\mu g/m^3$ were fixed during 1991-1994 by automatic air monitoring system. Many arable lands, forest areas and wild-life preserves are subjected to relatively high sulphur precipitation loads, exceeding 0.5 g S/m^2 per year, of which the role of emissions from local sources is about 60-95%. On the basis of air pollution concentration maps, the landscape of NE Estonia is classified into zones of high, moderate and low pollution level.

1. Introduction

The atmospheric influx of sulphur dioxide is one of the factors affecting most seriously the ecosystems in North-East (NE) Estonia (Liblik and Rätsep, 1994). Sulphur dioxide is mainly emitted as a result of the combustion of oil shale in the thermal power plants (TP), but also when other solid, liquid and gaseous fuels such as crude oil, shale oil, petroleum, gas from semi-coking of oil shale etc. are used in the boiler houses and furnaces. Of five TP using oil shale as fuel (the sulphur content of wet matter is about 1.4-1.6 %), two gigantic ones, the Baltic (1624 MW) and Estonian (1610 MW) plants are near the town of Narva, while those of lower capacity are located at Kohtla-Järve, Ahtme and Kiviõli (Fig. 1).

The average SO_2 concentrations in flue gases from Baltic and Estonian TP, 1500-2600 mg/m^3 (Aunela *et al.*, 1995), are high compared with 400 mg/m^3 according to the European Community guideline (Umweltpolitik, 1992). The effect of the SO_2 emission is not limited to NE Estonia only, but reaches also, on a noticeable scale, as far as the neighbouring territories within Estonia and even the nearby states. The sulphur deposition in southern Finland due to emissions from Estonia is about 100 $mg/m^2/year$ which is about 10-12% of the total sulphur deposition, and in central part 30 $mg/m^2/year$ (Ministry of the Environment of Finland, 1991). In Estonia the annual intensity of sulphur precipitation is measured in the ranges of 0.54-1.7 g S/m^2 on the northern coast of Lake Peipsi up to 1.8-19.7 g S/m^2 near the Baltic TP (Voll *et al.*, 1988), which coincides with calculation data (Klimova, 1993; Kaasik, 1995).

In accordance with atmospheric air quality normatives valid in Estonia, concentration of SO_2 in the ambient air, beyond the sweep of industrial territories and sanitary zones, is allowed to exceed neither the Maximum Permissible Limit Concentration (PLC$_m$, measured per 30 minutes) nor the mean limit concentration level per day (PLC_{mean}), which are 500 µg/m³ and 50 µg/m³, respectively. In Nordic states the recommended permissible mean concentration of SO_2 is declared in the atmosphere of arable lands and forested areas at the level of 25 µg/m³, and the annual sulphur precipitation load must not exceed 0.5 g S/m² (Brodin and Kuylenstierna, 1992).

The study about spreading and distribution of sulphur dioxide dependent on emission intensity and wind parameters on NE Estonian natural and dwelling areas is carried out here, with the purpose to divide the landscape of NE Estonia into zones according to air pollution concentration levels.

2. Methods

The study was carried out by a special imitation system for detailed assessment of atmospheric air quality and for investigation of dynamics of formation of multicomponential concentration fields (Liblik et al., 1995). All location parameters and technical characteristics of sulphur dioxide pollution sources were collected and systematized in the Data Bank (DB) of NE Estonian Air Pollution Sources.

For calculations of maximum and mean concentration fields on the overground air layer over the landscapes the semi-empirical modelling method (OND-86, 1987) of Main Geophysical Observatory (St.Petersburg, Russia) had been used. The model is fit for territories with area up to 100*100 km with a relatively long life-time components, as calculations were applied to dry deposition conditions, while interactions between components in the atmosphere and washout by precipitations are not taken into account. Meteorological data required to describe dispersion consist of regional coefficient of atmospheric temperature stratification, wind direction, wind velocity, ambient air temperature etc.

For every network point in a regional map with a space of 2-4 km, the maximum concentrations (MC) of the SO_2 and the critical wind direction and velocity were calculated at which the MC values might be formed. The mean concentrations of SO_2 in the given point of landscape for periods longer than a month, a quarter, a season and a year, were calculated on the basis of average emission intensity data, wind velocity and repetition of wind direction for observation period. The results of calculation were corrected according to monitoring data. The correction coefficient F_k was developed for that purpose, which is specific for regional air pollution situation. F_k takes into consideration both irregular dispersion of pollutant, which is caused by unstable character of wind directions and by non-linear spreading of pollutant on the distance from source to given point, and also so-called "local background level" which forms due to existence of unknown and disregarded sources. The factual mean concentration $C_{kr,f}$ for a wind rhumb (N, NE, E, SE, S, SW, W or NW) was calculated by the following formula (Liblik et al., 1995):

$$C_{kr,f} = (100\ C_{kr} + C_{kp} F_k)(100 + F_k)^{-1}$$

where C_{kr} is the mean concentration for wind rhumb at concrete wind velocity (without accounting of F_k) and C_{kp} is the mean concentration in the given point of the landscape (the sum of products $C_{kr}*S_r*10^{-2}$, where S_r is repetition of the wind rhumb per given period, %).

In calculations of C_{kr} all NE Estonian individual pollution sources affecting the given point and nine wind directions (after each 5 degrees) in every rhumb were taken into account. F_k was determined by comparison of calculated concentration fields with air monitoring data during 3-4 years, using factor and multiregression analyses. In NE Estonia in the region of TP and boiler houses the value of F_k for SO_2 discharges ranges between 20-30%. To compare calculated data of MC and $C_{kr,f}$ with factual pollution level the results obtained by "OPSIS" monitoring system (Sweden) were used, which was installed in Kohtla-Järve in 1991.

3. Results and Discussion

3.1. EMISSION SOURCES AND EMISSIONS OF SO_2

The total quantity of registered stationary point sources is about 150 (1993), together with boiler-houses, power generating facilities of chemical plants etc. The sulphur dioxide emissions of TP are responsible for 70-78% of the total SO_2 discharges in Estonia and about 92-93 % of the annual quantity emitted from all NE Estonian sources. Table 1 shows that since 1990 the amounts of SO_2 emissions have continuously decreased, first of all due to fall in production of electric energy. In calculations of SO_2 amounts the actual reaction rates of oil shale sulphur with calcium (Ca-S reaction) in the TP-17 and TP-67 type boilers (Baltic TP) at the level of 65-75% and TP-101 boilers (Estonian TP) about 81% were taken into account (Ministry of the Environment of Finland, 1991; Liblik and Rätsep, 1994). Obtained emission data (Table 1) coincide with measurements (Aunela et al., 1995). The annual discharges of SO_2 amounting 7600-14500 t (1990-1994) emitted by the other sources cause mainly local air pollution effect.

The annual average discharge intensity of sulphur dioxide affecting NE Estonian area varied between 3.5-4.6 kg/s in 1992-1994 up to 6 kg/s in 1990-1991. Really, the maximum values of concentration fields take place in cold season (from October to March) when discharge intensity of SO_2 may be higher by 40-60% or more. In 1990 the emission intensity from all NE Estonian pollution sources was about 9.5-11 kg/s as maximum, and in 1992-1993 in the ranges of 5.1-6.6 kg/s, as maximum.

TABLE 1

Annual emission of sulphur dioxide into atmosphere in NE Estonia, thousand tons

Names of power plants	1990	1991	1992	1993	1994
Baltic	120.0	106.6	74.9	53.6	51.1
Estonian	71.1	65.7	53.9	43.4	46.7
Kohtla-Järve	5.6	5.5	4.7	3.7	3.3
Ahtme	3.6	3.8	3.3	2.8	2.9
Kiviõli	3.0	2.9	2.7	2.6	2.8
Total from TP	203.3	184.5	139.5	106.1	106.8
Total in NE Estonia	217.8	198.3	150.4	115.7	114.4

3.2. FIELDS OF MAXIMUM AND MEAN CONCENTRATIONS

Investigations by calculation methods show that the areas affected by TP involve the whole territory of NE Estonia, especially in case of east (8 %) and north-east (10 %) winds as the result of merging of the pollutants plumes (Fig. 1, Table 2).

In the overground air layer of landscapes and dwelling regions (area about 3400 km^2) the MC values of SO_2 would be in the ranges of 25-450 µg/m^3, depending on emission intensity and wind parameters. The maximum pollution level of SO_2 does not exceed the PLC_m value, i.e. 500 µg/m^3 (Figure 1). The regions most heavily polluted with sulphur dioxide are situated near the Baltic TP (300-450 µg/m^3), in the Kohtla-Järve-Ahtme area (100-350 µg/m^3) and in the town of Kiviõli (about 150 µg/m^3). On the bulk of the NE Estonian territory the MC are forming at a wind velocity of 2-4 m/s (on the height of 10 m from surface), and at 5-9 m/s in the surroundings within a radius of 5-15 km from the stacks of the TP.

The mean concentrations of SO_2 per year, season, month or another period in the given locality of landscape are different from MC values, as they essentially depend on factual distribution of wind directions and average wind velocity during observable period.

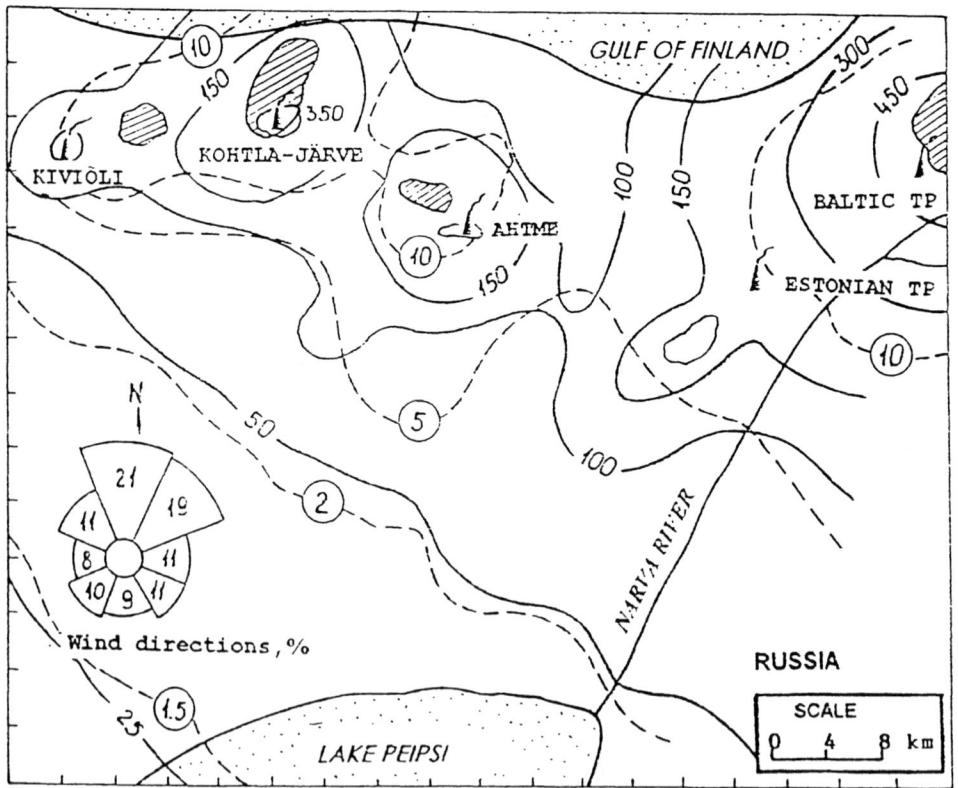

Fig. 1. Isolines of the short-time maximum (continued lines) and annual mean (broken lines) concentrations in µg/m^3 of SO_2 in the overground air layer in NE Estonia. Striped areas show the regions where the annual mean concentrations will be more than 25 µg/m^3 up to 75 µg/m^3.

The highest mean concentrations (30-80 µg/m^3) take place for periods of eastern winds (about 8 %): for most northern part of NE Estonia (coast of Gulf of Finland) the level of 25 µg/m^3 is exceeded. Fig. 1 shows that annual mean concentrations of SO$_2$ are below 25 µg/m^3 on the main territory, but may be up to 50-75 µg/m^3 near the TP on the leeward direction of the south and south-west winds (in all 40 %).

In the town of Kohtla-Järve the annual means of 15.8-19.1 µg/m^3 and MC values 271-442 µg/m^3 in 1991-1994 are fixed by automatic monitoring station "OPSIS", which gives confidence to the calculation data. The divergency was obtained in the ranges of 15-25% for SO$_2$ monthly, yearly and wind rhumbs mean concentrations at the distance of 4-50 km from stacks of TP (the Baltic, Estonian, Kohtla-Järve, Ahtme). At the same period by wind direction from relatively "fresh air sides" (from north and north-east) with respect to local sources the annual mean concentrations of SO$_2$ are lying on the interval of 2.5-5.6 µg/m^3, which may be caused by long-distance transference from Finland and Karelia, first of all.

TABLE 2

Scopes of areas affected with SO$_2$ around the power plants

Names of TP	Average height of stacks (m)	Distance from the stacks (km) dependent on SO$_2$ overground concentration value			
		maximum *)	200 µg/m^3 (40% PLC$_m$)	25 µg/m^3 (5% PLC$_m$)	5 µg/m^3 (1% PLC$_m$)
Baltic	160	4.0 (400)	8-12	34-50	more 100
Estonian	250	7.1 (94)	-	20-35	59-86
Kohtla-Järve	150	2.4 (135)	-	10-15	26-36
Ahtme	95	1.7 (275)	2-3	9-16	22-24
Kiviõli	96	1.2 (160)	-	6-9	16-21

*) the maximum overground concentration is shown in the brackets, µg/m^3

TABLE 3

Distribution of NE Estonian landscape into zones by SO$_2$ pollution level

PLC$_m$		PLC$_{mean}$		Sulphur deposition		Location of areas
µg/m^3	Pollution level	µg/m^3	Pollution level	g S/m^2 per year	Pollution level	
300-450	Moderate	10-25 and more	High	8-15	Very high	Near Baltic TP, area to north-west
100-300	Relatively low	5-10	Moderate	3-8	High	From Estonian TP to west; area from Ahtme to Kiviõli
25-100	Low	1.5-5	Relatively low	1-3	Moderate	The Central part to northern coast of Lake Peipsi
below 25	Very low	below 1.5	Low	0.4-1	Relatively low	The south-western area

3.3. POLLUTION ZONES

On the basis of obtained air pollution concentration maps the landscape of NE Estonia may be classified into zones of high, moderate and low pollution level in respect to PLC_m and PLC_{mean} values, and also to sulphur deposition load (Table 3).

Analysis of the data shows that sulphur precipitation per surface unit depend on SO_2 content in the air. It must be emphasized here that about half of sulphur deposition in Estonia is due to long-distance transfer from other countries (Kallaste et al., 1991). In NE Estonia due to powerful pollution sources, the role of SO_2 emissions from local sources is at the first place in forming of sulphur deposition load and it will be 60-95%. In this result many forest areas, wild-life preserves and other unique landscapes are subjected to relatively high sulphur precipitation loads, exceeding 0.5 g S/m^2 per year.

4. Conclusion

It seems clearly that sulphur dioxide influxes are affecting essentially air pollution situation in all North-East Estonian territory, especially in the northern part, near power plants and with east and norht-eastern winds. The zonation of polluted areas helps more radically to understand the changes in the ecosystems and nature, also to select areas for testing and investigation the state of ecosystems in the future.

Acknowledgments

This study was partially supported by the Estonian Science Foundation (Grant No. 1200).

References

Aunela, L., Häsänen, E., Kinnunen,V. et al.: 1995, Atmospheric emissions from two Estonian power plants using oil shale. *Proceedings of the 10th World Clean Air Congress (Espoo-Finland, May 28-June 2, 1995)*, Vol.1, No. 052.
Brodin Y.-W. and Kuylenstierna, J.C.: 1992, *Ambio*, 21, 332-338
Kaasik, M.: 1995, Air pollution dispersion and deposition model Aeropol. *Proceedings of the 10th World Clean Air Congress (Espoo-Finland, May 28-June 2, 1995)*, Vol.2, No. 259.
Kallaste, T., Roots, O., Saar,J. and Saare, L.: 1992, Air pollution in Estonia 1985-1990, Helsinki, 58 pp.
Klimova, E.: 1993, *Oil shale*, 10/1, 67-78.
Liblik, V., Kundel, H. and Rätsep, A.:1995, Air quality management on the areas of North-East Estonia, *Proceedings of the 10th World Clean Air Congress (Espoo, Finland, May 28-June 2, 1995)*, Vol. 3, No. 511.
Liblik,V. and Rätsep, A.: 1994, Pollution sources and distribution of pollutants. - *The influence of natural and anthropogenic factors on the development of landscapes. Inst.of Ecol., Est.Acad. Sci.*, Publ. 2.. Tallinn, 70-93.
Ministry of the Environment of Finland: 1991, *Synthesis Report*. Plancenter LTD, pp. 44-45.
OND-86: 1987, The method for calculation of concentrations of harmful substances, contained in the releases of enterprises, into the atmospheric air, Leningrad. 48 pp. (in Russian).
Umweltpolitik: 1992, Bericht der Bundesregierung an den Deutschen Bundestag Fünfter Immissionsschutzbericht der Bundesregierung. *Eine Information des Bundesumweltministeriums, Drucksache 12/4006*, p. 123.
Voll, M., Trapido,M., Luiga, P., Haldna,J., Palvadre,R. and Johannes,H.:1988, Emissions of energetical facilities and oil shale processing industry into atmosphere. - *In.: Ecological researches. Inst. of Chemistry, Est. Acad. Sci.* Tallinn, pp.80-97 (in Russian).

ACIDIFYING EMISSIONS IN THE NETHERLANDS

C.W.A. Evers[1], J.J.M. Berdowski[2] and P.F.J. van der Most[2]

[1] *Inspectorate for Environmental Protection, Ministry of Housing, Spatial Planning and the Environment,*
's-Gravenhage,
The Netherlands, [2] *TNO Institute for Environmental Sciences, Delft, The Netherlands*

Abstract. The emission of acidifying compounds to air in the Netherlands, expressed as acidifying equivalents, consisted in 1992 mainly of NO_x (45%), NH_3 (35%) and SO_2 (20%). Transportation, agriculture and large combustion plants each contributed about 30% to the national total emission of acidifying compounds. The emissions from transportation activities mainly consisted of NO_x, while in agriculture NH_3 emission strongly dominated. Combustion processes in large combustion plants resulted both in SO_2 emissions (especially from refineries) and NO_x emissions (especially from public power plants). The total emission of acidifying substances decreases steadily in the Netherlands. The emission in 1992 was 24% lower than in 1985. It is expected to decrease further in future. The emission levels in 1992 and 1993 still are more than twice as high as the emission objective for the year 2000, set by Dutch environmental policy.

Keywords: Emissions, acidification, SO_2, NO_x, NH_3, Netherlands.

1. Introduction

The Dutch emission inventory system enables the registration, analysis and localization of emission data of both industrial and non-industrial sources in the Netherlands. The results can be used to test the effectiveness of governmental environmental policy. These activities are part of the policy evaluation tasks of the Inspectorate General for Environmental Protection (IGEP) and of the Ministry of Transport, Public Works and Water Management.

The emission inventory takes place in cycles of one year. Recently, the most relevant results of the Dutch emission inventory for 1992 have been published (Berdowski *et al.*, 1994; Pulles, 1994). In that cycle the emissions in 1992 to air and water from about 800 major companies have been registered. These 800 companies are the most important contributors to the total industrial emissions in the Netherlands. The emissions of these companies are stored within the individual inventory system. The emissions from the smaller enterprises and from diffuse non-industrial sources are stored in the collective emission inventory system.

The goal of this paper is the presentation of the objectives and structure of the Dutch emission inventory and the analysis of the data on acidification with respect to environmental policy in the Netherlands.

2. Objectives and structure of the emission inventory

2.1. OBJECTIVES

The Dutch emission inventory system comprises the registration, analysis and localization of emission data of both industrial and non-industrial sources in the Netherlands. The objective of the emission inventory is to monitor the emissions from sources of air and water pollution on a national scale. This information is used to evaluate the progress of environmental policy and to provide national and international bodies with official data on emissions within the country.

TNO is commissioned by the Department for Emission Inventory and Information Management of the IGEP to perform the inventory of emission data. The Inspectorate on its turn acts as a commissioner on behalf of both the Ministry of Housing, Spatial Planning and the Environment, and the Ministry of Transport, Public Works and Water Management.

Based upon the above objectives the following tasks of the emission inventory are defined:
a) to determine and collect emissions of industrial and non-industrial sources in the Netherlands and store them in the databases of the inventory system;
b) to analyse emission data with respect to compound, target group and industrial branches, environmental theme and to the location of origin of emissions;
c) to assess the effects of environmental policy and to evaluate to what extent policy goals are achieved;
d) to ascertain trends by comparing the results for the subsequent inventory years;
e) to supply emission data to local (e.g. provinces) and international (e.g. EC) authorities and other interested parties (e.g. for modelling studies).

The emission inventory system contains emissions to air, water and soil both from industrial and non-industrial sources. For the most relevant industries, the emissions are registered individually based upon information of each individual plant, installation and appliance. The emissions of the other, the smaller industrial as well as non-industrial source are calculated collectively with statistical data as population density, number of employees etc.

The emission inventory comprises two coupled information systems:
a) the Individual Emission Inventory system (IEI), containing emissions to air and water for major individual industries;
b) the Collective Emission Inventory system (CEI), storing spatial resolved activity data and emission factors, from which spatial resolved emission estimates are calculated and mapped.

2.2. INDIVIDUAL EMISSION INVENTORY SYSTEM

For the most relevant industries, the Individual Emission Inventory system (IEI) contains a large number of specific compounds emitted to air and water. Each individual plant, installation and, in most plants, each device is examined and significant data at the different levels are stored in the inventory. Together with information on the sources (points of release) within the company each emission is categorized as to the origin, the chemical nature and the location.

The companies concerned provide the data required to estimate emissions on a voluntary basis. This information is treated as confidential and is administered by the Department for Emission Inventory and Information Management of the IGEP. Only a restricted number of persons, the primary environmental authorities included, have access to the data on individual industries. Others may obtain data at an aggregated level only.

However, with respect to the year of emission, from 1990 onwards all emission data are on request available for public, under the restriction of total annual emissions per compound for the whole plant.

Nowadays, the IEI takes place in cycles of one year. The first inventory cycle started in 1974 and was completed in 1981. During this period about 6,300 companies with approximately 20,000 installations were registered. One important result was that about 10% of the companies caused 97% of the air pollution. Hence it was decided to use a more efficient selection of companies for the following cycles. This selection is based both on the (expected) contribution of companies to the emission within the country, and on the environmental impact of the emitted compounds. The selection for the base year 1992 was done similarly.

The data set for the year 1992 of the IEI of which the results are presented in this paper, essentially, is a matrix of approximately 900 emitted substances and about 800 companies. Each matrix cell contains a new matrix, in which the emission of a substance is attributed to an installation and an emission point. Because of this, a link has been established between an emission and its underlying processes. The inventory of emissions to water is more simple as the link with processes is disconnected in many occasions. In most cases, the total, aggregated emission of a company to water has been recorded.

2.3. COLLECTIVE EMISSION INVENTORY SYSTEM

The Collective Emission Inventory system (CEI) does not take place in cycles. The underlying data are, however, also yearly updated. The Collective system stores the emission data of other (smaller) companies and of diffuse emissions from road traffic and other mobile sources, from households and from land use related sources as agriculture and nature. The emissions are estimated with statistical data such as number of inhabitants, houses, cars, jobs etc. and by use of emission factors.

Furthermore, the collective emission inventory system contains all kind of basic data related to the infrastructure of the Netherlands, such as geographical information about houses, traffic roads, railway roads, airports, shipping routes and smaller companies. Also general data are incorporated such as the type of soil, the nationwide sewer system and the drainage system for waste water.

No restrictions are imposed on use of data from the CEI. The data bases are administered by the Department of Emission Inventory and Information Management and requests concerning use of the data are to be addressed to this Department.

2.4. SELECTED COMPOUNDS FOR PRESENTATION

From the circa 900 registered substances about 60 are selected for presentation and discussion in annual reports. Among them are the so-called 'priority' compounds, which the Dutch environmental policy is primarily aimed at. Furthermore the results of the analysis of emission data are presented with respect to both environmental themes and target groups, i.e. source categories, as distinguished in the National Environmental Policy Plan (VROM, 1994).

3. Analysis of the results with respect to environmental policy

3.1. INTRODUCTION

The Dutch environmental policy is arranged along two different dimensions:
- Firstly, the target groups or source categories, representing important groups of polluters, which have a certain level of homogeneity in common with regard to environmental

problems. Many target groups have been formed on the basis of their emissions, but the type of feedstock used can be an important determinant as well.

Important emitting target groups are industry, traffic, agriculture and consumers. The industrial target groups are subdivided into some 30 industrial branches. Each of them is treated separately in the frame of national policy on target groups. The contributions of the various target groups to the emissions of the major pollutants to air and water are given. Also geographical patterns of emission densities of some pollutants are presented.

Secondly, the environmental themes, representing the major environmental problems. The themes relevant to the annual reports are:
- Climate change
- Ozone depletion
- Acidification
- Eutrophication
- Dispersion of toxic substances

3.2. ACIDIFICATION

The environmental theme acidification is related to the emission of the acid substances SO_2, NO_x and NH_3. The emissions of SO_2, NO_x and NH_3 are expressed in acidification equivalents (Adriaanse, 1993). The emission data in Gg/year are multiplied by a correction factor, shown in Table I. Dutch environmental policy employs the unity of Aeq, being equal to 1 Mg emitted pollutant (Adriaanse, 1993). In that way the relative contributions of the different target groups can be shown for the total emissions relevant for the environmental themes.

For 1992 the total emission of acidifying substances in the Netherlands amounted to 28,100 10^{-3} Aeq (Figure 1), indicating a decrease with nearly 10% compared with the 1990 emission data. This reduction partly is caused by the decrease of NH_3 emissions from agriculture and the decrease of industrial SO_2 emissions. NO_x emissions do not decrease, mainly due to the increasing traffic density which counterbalances the decreases in emission factors and emissions in other target groups.

The total Aeq emission for 1992 was about 2.2 times higher than the policy objective for this environmental theme for the year 2000. This factor does not differ very much for the three individual substances involved. The contribution of the three substances to total Aeq emission in 1990 was 19% SO_2, 45% NO_x and 36% NH_3.

The spatial distribution of several target groups to Aeq emissions is graphically presented in Figure 2. The largest contributions are caused by the SO_2 emissions of the refineries and power plants, by the NH_3 emission from agriculture, and by the NO_x emissions from traffic.

The spatial distribution of the emission of acidifying substances is presented in Figure 2A. The important industrial areas can be recognized, as well as the areas of factory farming.

SO_2 emissions contribute about 19% to total acidifying substances. Important contributors to these emissions are the oil refineries and the power plants. Figure 2B presents the spatial distribution of SO_2 emissions within the target groups. Emissions are concentrated within the most important harbour regions within the country.

TABLE I

Contribution of the various target groups to emissions of acidifying substances to air in the Netherlands in 1992 (10^{-3} Aeq/year). 1 Aeq equals the emission of 1 Gg SO_2, NO_x or NH_3 times the factor of a substance as shown in the table.

Substance	SO_2	NO_x	NH_3	Total Aeq	Contribution per target group (%)
Factor	0.0313	0.0217	0.0588		
Refineries	1,860	391	0.412	2,250	8.0
Energy sector	904	1,480	-	2,390	8.5
Waste incineration	131	91.4	-	222	0.8
Chemical industry	484	672	187	1,340	4.8
Other industry	912	1,400	131	2,440	8.7
Agriculture	28.7	423	8,580	9,030	32.2
Transport and traffic	938	7,380	30.7	8,350	29.8
Residential consumption[1]	39.7	460	623	1,120	4.0
Nature	-	354	553	907	3.2
Estimate 1993	5,250	12,200	10,100	27,500	98.2
Total 1992	**5,300**	**12,700**	**10,100**	**28,100**	**100**
1990	6,290	12,500	12,000	30,800	109
1988	7,400	12,600	13,700	33,700	120
1985	7,940	12,300	14,600	34,900	124
Emission objective for the year 2000	2,340–2,810	5,170–5,280	4,820	12,300–12,900	

1   Including target groups: Building industry, Water companies, Sewage works and Retail trade.

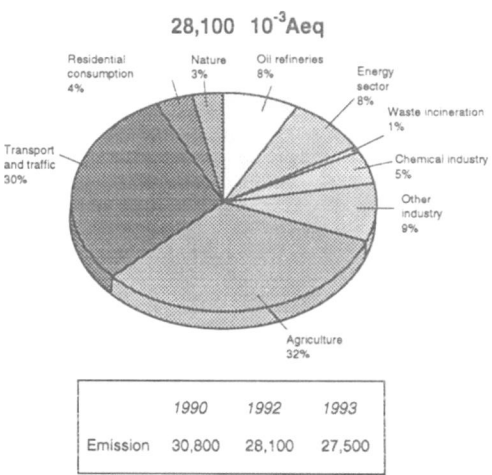

Fig. 1. Relative contribution of target groups to emission of acidifying substances in the Netherlands in 1992 (10^{-3} Aeq/year).

NH_3 emissions in the Netherlands for about 85% are the result of agricultural activities, namely, production and utilization of animal manure. Figure 2C illustrates the NH_3 emission pattern due to agriculture in relation with the location of forests. It shows that the most important emissions are on sandy soils close to forests. Figure 2D shows the NO_x emissions related to road traffic. The most densely populated areas and the national highway system can be recognized on the map.

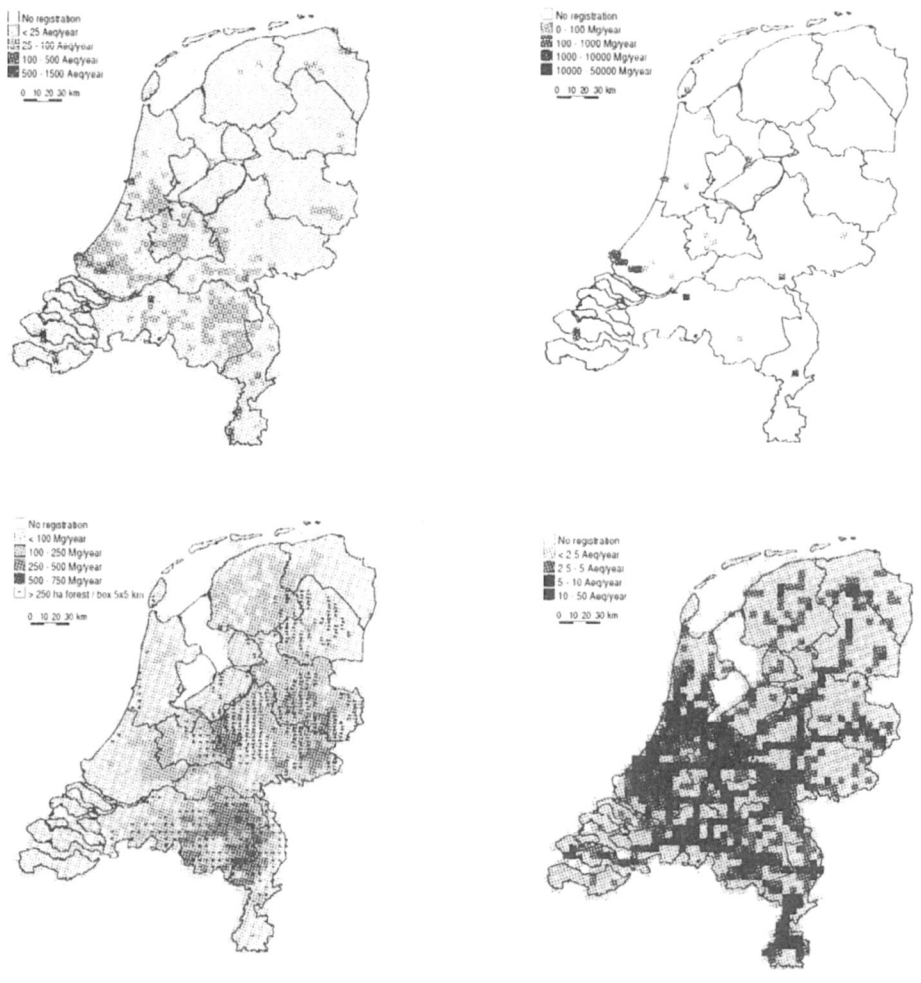

Fig. 2. Emission densities of acidifying compounds to air in the Netherlands in 1992 (5 x 5 km squares, sources and compounds as indicated) in Aeq/year.

4. References

1. Adriaanse, A.: 1993, *Environmental Policy Performance Indicators*, SDU, 's-Gravenhage, The Netherlands. ISNB 90 12 08099 1, 175 pp.
2. Berdowski, J.J.M., Van der Auweraert, R.J.K., Van der Most, P.F.J., Thomas, R. and Zonneveld, E.A.: 1994, *Emissions in The Netherlands – 1992. Trends, themes and target groups* (in Dutch), Publikatiereeks Emissieregistratie 20, Ministry of Housing, Spatial Planning and the Environment, 's-Gravenhage, The Netherlands, 212 pp.
3. Pulles, M.P.J.: 1994, *Emission Inventory in The Netherlands, Emissions to air and water in 1992*, Publikatiereeks Emissieregistratie 22, Ministry of Housing, Spatial Planning and the Environment, 's-Gravenhage, The Netherlands, 56 pp.
4. Ministry of Housing, Spatial Planning and the Environment (VROM): 1994, *National Environmental Policy Plan 2, Tweede Kamer, vergaderjaar 1993–1994*, 23 560, nrs. 1–2.

Gridded Ammonia Emission Fluxes in Japan

Kentaro Murano[1], Shiro Hatakeyama[1], Tsuguo Mizoguchi[2] and Naomi Kuba[3]

[1]. *National Institute for Environmental Studies, 16-2 Onogawa, Tsukuba City, Ibaraki 305, Japan.* [2]. *The Institute of Public Health, Shirokanedai 4-chome, Minatoku, Tokyo, 108, Japan.* [3]. *CRC Research Institute Inc., 2-7-5 Minamisuna, Kotoku, Tokyo, 136, Japan*

Abstract. In order to fully understand the acidification of precipitation, it is essential to determine ammonia emissions. Detailed gridded emission fluxes of NH_3 have been compiled in Europe. In East Asia they have been determined on a national basis(Zhao and Wang, 1994). In Japan we have calculated NH_3 emission fluxes on a 1° latitude x 1° longitude basis for livestock and the application of fertilizer. Livestock emission factors developed by W.A.H. Asman(Asman, 1992) for Europe were used 23.04 and 5.36 kg NH_3/animal/yr. for cattle (dairy cows and beef), and pigs, respectively. Domestic animal population data was collected by prefecture and apportioned to grid cells based on the prefectural area in each grid cell. For fertilizer emissions, NH_3 emission were calculated assuming a 10% ammonium nitrogen evaporation rate for ammonium sulfate, urea, and other nitrogen-containing fertilizers. Since prefectural fertilizer data were not available, total fertilizer usage for Japan was distributed to prefectures based on cultivated area. The maximum calculated NH_3 emission fluxes for each of the three animal categories were as follows: Dairy cows, 4730 (Hokkaido), beef cattle, 4540 (Kyushu) and pigs, 3480 (Kanto) tonnes NH_3/grid/yr. The total NH_3 emissions due to livestock in Japan were 4.6, 6.0 and 4.4 x 10^4 tonnes NH_3/yr. from dairy cows, beef cattle and pigs, respectively. The overall total NH_3 emission from livestock and the application of fertilizer was 2.0 x 10^5 tonnes NH_3/yr. The NH_3 emission by Japan is small compared to those of most European countries.

Key Words: Emission inventory, ammonia, Japan, livestock, application of fertilizers.

1. Introduction

In order to fully understand the acidification of precipitation, it is essential to determine ammonia (NH_3) emissions, because NH_3 is one of the most important neutralizing components of the atmosphere. When deposited to the ground, it may be oxidized to nitrate ion with the concurrent release of two hydrogen ions to the soil. Hence NH_3 is considered to be an air pollutant which is harmful to the soil. Detailed gridded emission fluxes of NH_3 from livestock and the application of fertilizer have been compiled for Europe (Buijsman et al., 1987, Asman et al., 1988, Asman, 1992).

When we evaluate the total deposition of chemicals to the environment in Europe, North America and Japan, the amounts of SO_4^{2-} and NO_3^- deposition are similar, however, the NH_4^+ deposition in Japan is larger than that in the other regions (Interim Assessment, The Causes and Effects of Acidic Deposition, Tamaki et al., 1991). In order to understand the characteristics of acid deposition in Japan, NH_3 emission fluxes must be determined. In the near future, the transboundary air pollution among east Asian countries will be evaluated with an emission, transport, conversion and deposition model. The detailed gridded emission flux map of SO_2 and NOx has been compiled for Asia (Akimoto and Narita, 1994) as the basic data on model run. It is essential to establish an NH_3 emission flux map to conduct a simulation model of air pollutants among east Asian countries. The East Asian NH_3 emission fluxes have been determined by Zhao(Zhao and Wang, 1994), however, they are on a national basis. We have calculated NH_3 emissions from livestock and the application of fertilizer for Japan on a 1° latitude x 1° longitude grid.

2. Methods

Detailed NH_3 emission fluxes for Europe have been compiled in order to run air pollution simulation models among European countries. Only livestock and the application of fertilizer have been taken into account to obtain NH_3 emission fluxes. We used the animal emission factors developed by W.A.H. Asman(Asman, 1992) for Europe 23.04 and 5.36 kg NH_3/animal/yr. for cattle(dairy cows and beef) and pigs, respectively. In Japan we generally have much more rainfall and the average ambient temperature is higher compared to European countries. Thus the former may actually make emission factors be lower in Japan and the latter may make emission factors be higher than those factors for Europe. Perhaps these differences cancel each other. In Europe, NH_3 emissions from cattle and pigs cover 84% of the NH_3 emission from livestock. Only NH_3 emissions from cattle and pigs were considered in this report. Domestic animal population data was collected by prefecture and apportioned to grid cells based on the prefectural area in each grid cell (1° latitude x 1° longitude). We used Information Related to Livestock Improvement (pub. Ministry of Agriculture, Forestry and Fisheries, Livestock Industry Bureau Animal Production Division, 1992) and the Data Book on Livestock of Hokkaido '92 (pub. Hokkaido 1992). For fertilizer NH_3 emissions, we assumed a 10% ammonium nitrogen evaporation rate for ammonium sulfate, urea, and other nitrogen-containing fertilizers. Since prefectural fertilizer data were not available, total fertilizer usage for Japan was apportioned to prefectures based on cultivated area. We used the Yearbook of Fertilizer (pub. The Association of Fertilizer, 1992) and the Statistics Table on Agriculture of Hokkaido (pub. Hokkaido, 1992).

3. Results and Discussion

The NH_3 emission fluxes from pigs are shown in Fig. 1. The NH_3 emission fluxes calculated for Northern Kanto were between 2,500 and 3,500 tonnes NH_3/grid/yr. Those for Southern Kanto were greater than 2,000 tonnes NH_3/grid/yr. In Kyushu there is one grid showing between 3,000 and 2,500 tonnes NH_3/grid/yr. The maximum calculated NH_3 emission fluxes for each of the three animal categories were as follows: Dairy cows, 4730 (Hokkaido), beef cattle, 4540 (Kyushu) and pigs, 3480 (Kanto) tonnes NH_3/grid/yr. The five largest NH_3 emission fluxes by dairy cows ranged from 4,730 to 1,920 tonnes NH_3/grid/yr. and were situated in Hokkaido (4 grids) and in northern Kanto (1 grid). The five largest NH_3 emission fluxes by beef cattle ranged from 4,540 to 2,350 tonnes NH_3/grid/yr. and were located in Kyushu (3 grids), in northern Kanto (1 grid) and in Tohoku (1 grid). The highest total emission flux from livestock including dairy cows, beef cattle and pigs of 8,490 tonnes NH_3/grid/yr. was observed for northern Kanto and the second highest value of 7,880 tonnes NH_3/grid/yr. was observed for southern Kyushu (Fig. 2). Grids with emission fluxes larger than 5,000 tonnes NH_3/grid/yr. exist in Hokkaido (2), Kanto (2) and southern Kyushu (3), respectively.

There are no grids with fertilizer emission fluxes larger than 2,500 tonnes NH_3/grid/yr. (Fig. 3) and differences between grids are small. However, relatively high emission fluxes were calculated for Hokkaido, Kanto and northern Kyushu. The maximum combined emission flux from both livestock and the application of fertilizer calculated for the northern part of Kanto was 10,600 tonnes NH_3/grid/yr. (Fig. 4). High emission fluxes of 7,000 to 9,000 tonnes NH_3/grid/yr. were calculated for eastern part of Hokkaido, the Kanto plain and Kyushu.

Total livestock NH_3 emissions in Japan were 4.6, 6.0 and 4.4 x 10^4 tonnes NH_3/yr. from dairy

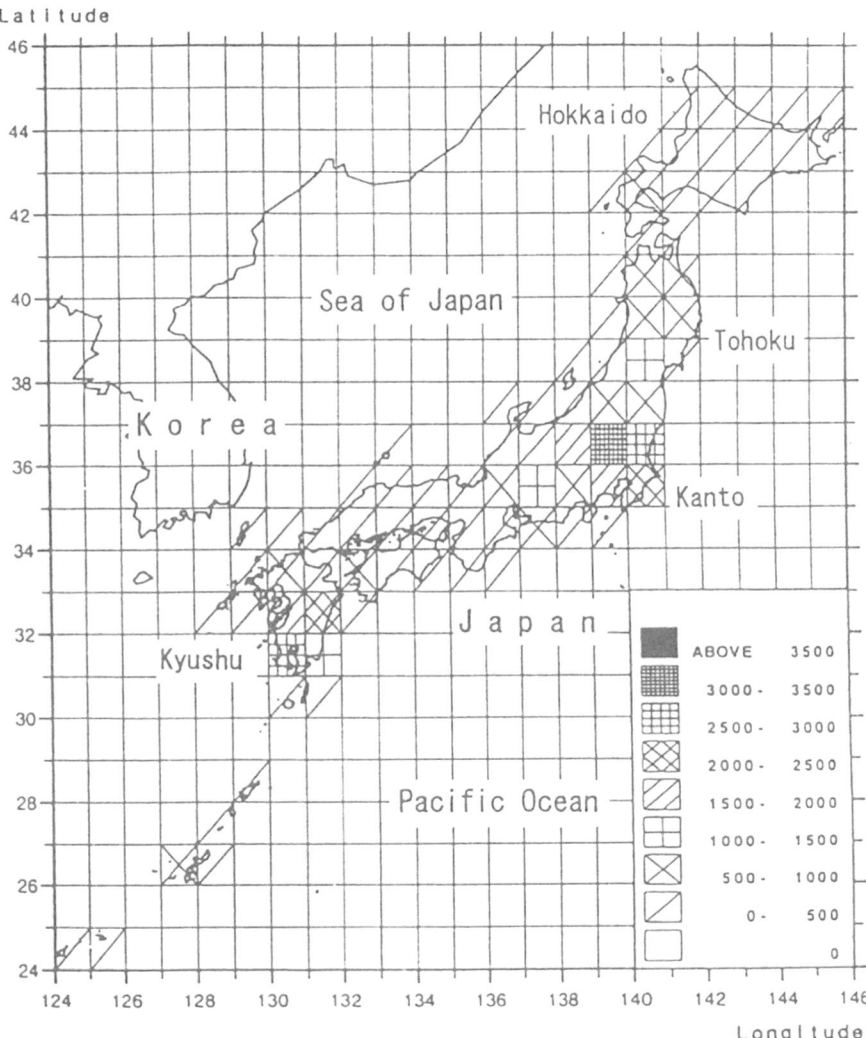

Fig.1. Gridded NH_3 emission fluxes from pigs in Japan(tonnes NH_3/grid/yr.)

cows, beef cattle and pigs, respectively. The total domestic emission from the application of fertilizer was 5.0×10^4 tonnes NH_3/yr. The overall total NH_3 emission from livestock (cattle and pigs) and the application of fertilizer was 2.0×10^5 tonnes NH_3/yr. The Japanese NH_3 emission due to livestock (without poultry) and the application of fertilizer were estimated to be 14 and 9.7×10^4 tonnes NH_3/yr.(Zhao and Wang, 1994), respectively. Those estimates are within 10% difference of our calculation for livestock, however, our calculation of the emission due to the application of fertilizer is approximately 50% smaller. A comparison of our data to those for other countries indicates that the emission flux due to the application of fertilizer in Japan is similar to those in the former West Germany and the United Kingdom, however, the NH_3 emission flux from cattle is larger in the former West Germany and in the UK compared to that in Japan. The total national NH_3

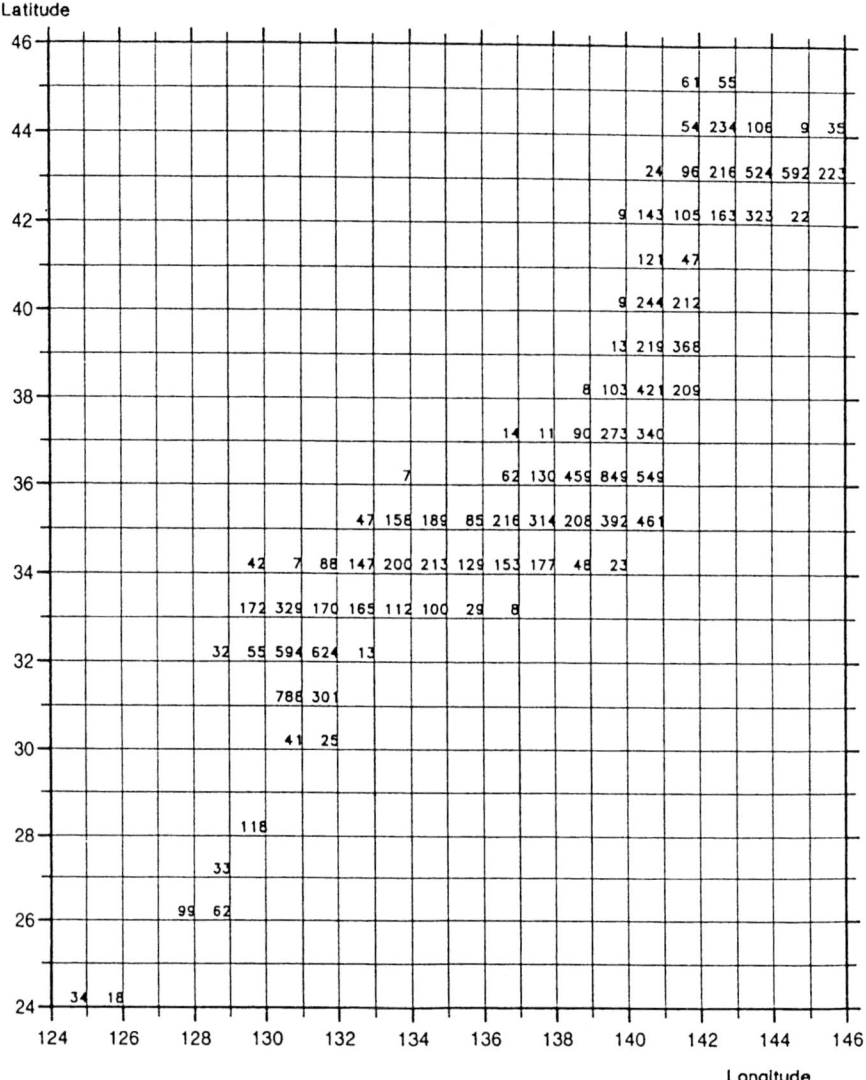

Fig. 2. Gridded NH_3 emission fluxes from cattle and pigs in Japan(10 tonnes NH_3/grid/yr.)

emission by Japan is small compared to those of most European countries. .

We compared the total NH_3 emission amount to the total NH_4^+ deposition in Japan. Acid deposition monitoring throughout the year was conducted in Japan from 1983 to 1987 with bulk samplers (Tamaki et al., 1991). The results indicate that the minimum, average and maximum deposition rates are 11, 24 and 61 x 10^4 tonnes NH_3/yr. Hence the average deposition is about 20% larger than the total annual NH_3 emission.

4. Conclusion

In Asia, gridded emission fluxes of SO_2 and NOx were compiled, however, there was no gridded emission fluxes of NH_3. We have calculated NH_3 emission fluxes on a 1° latitude x 1° longitude basis for livestock and the application of fertilizer in Japan using emission factors developed in

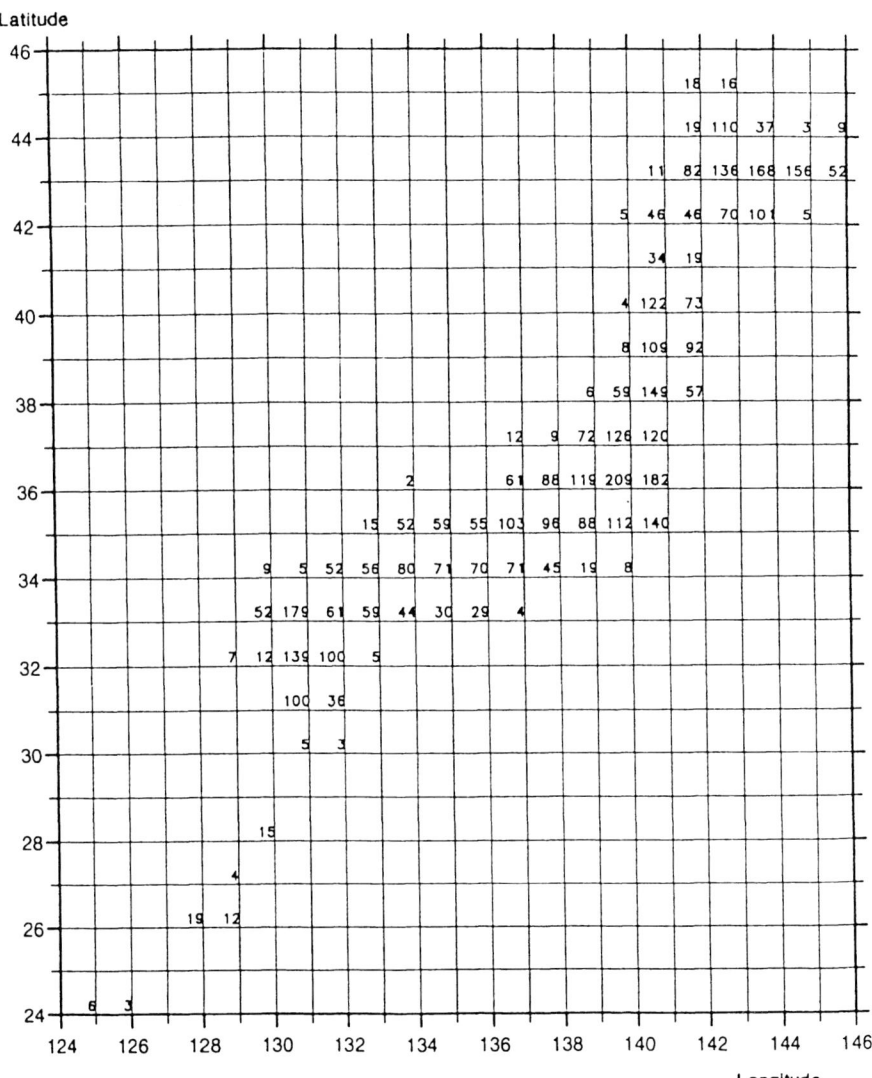

Fig. 3. Gridded NH_3 emission fluxes from the application of fertilizer in Japan (10 tonnes NH_3/grid/yr.)

Europe. The total NH_3 emissions due to livestock in Japan were 4.6, 6.0 and 4.4 x 10^4 tonnes NH_3/yr. from dairy cows, beef cattle and pigs, respectively. The overall total NH_3 emission from domestic animals and the application of fertilizers was 2.0 x 10^5 tonnes NH_3/yr. The larger emission fluxes from domestic animals and the application of fertilizers were observed in Hokkaido, Kanto and Kyushu districts The most dominant contributor in each category is beef cattle. The NH_3 emissions in Japan are small compared to those of most European countries.

Acknowledgments

This research was funded by the Japan Environment Agency's Global Research Fund. This work is also supported by the IGAC/GEIA activity convened by Dr. T.E. Graedel.

References

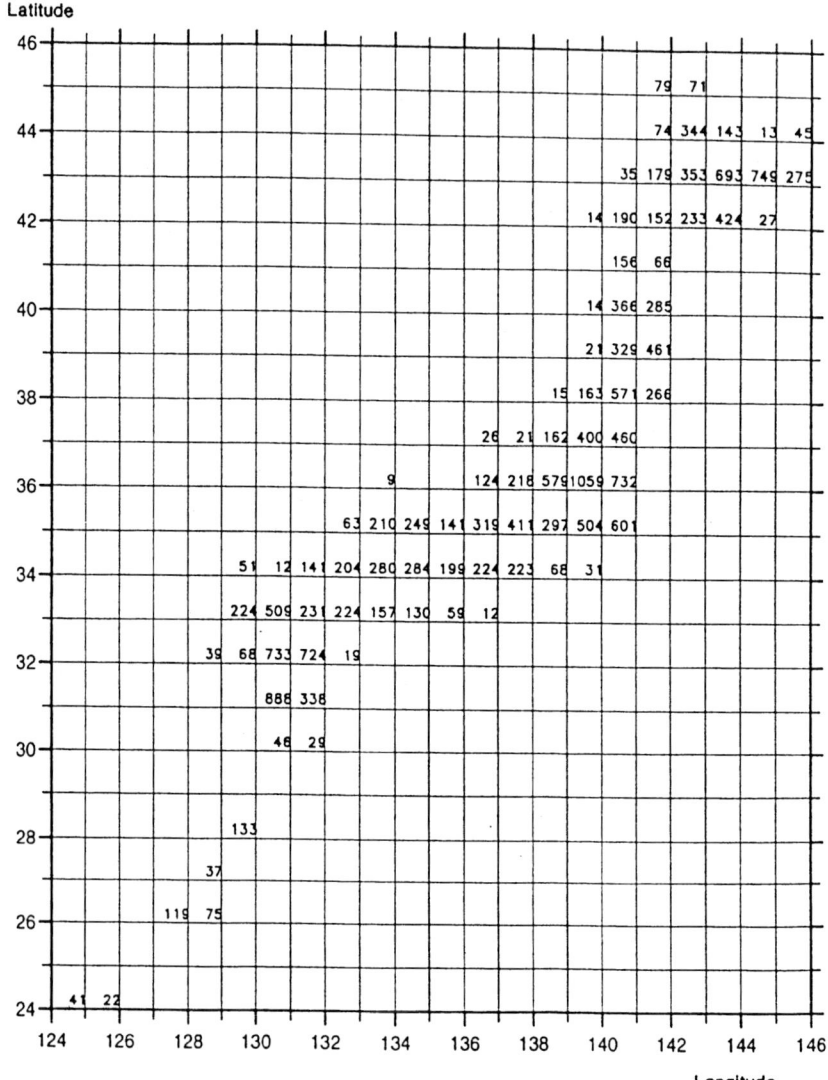

Fig. 4. Gridded NH$_3$ combined emission fluxes from livestock and the application of fertilizer in Japan (10 tonnes NH$_3$/grid/yr.)

Akimoto, H. and Narita, H.: 1994, *Atmos. Environ.* **28**, 213-225.
Asman, W. A. H.: 1992, Ammonia emission in Europe: Updated emission and emission variations, Report 481507002, Bilthoven, The Netherlands.
Asman, W. A. H., Drukker, B. and Janssen, A. J.:1988, *Atmos. Environ.* **22**, 725-735.
Buijsman, E., Maas, H. F. M. and Asman, W. A. H.:1987, *Atmos. Environ.* **21**, 1009-1021.
Interim Assessment, The Causes and Effects of Acidic Deposition, III., U.S.A.
Tamaki, M., Katou, T., Sekiguchi, K., Kitamura, M., Taguchi, K., Oohara, M., Mori, A., Wakamatsu, S., Murano, K., Okita, T., Yamanaka, Y. and Hara, H.: 1991, *Nippon Kagaku Kaishi* (in Japanese). 667-674..
Zhao, D and Wang, A.:1994, *Atmos. Environ.* **28**, 689-694.

MODELING STUDIES ON SULFUR DEPOSITION AND TRANSPORT IN EAST ASIA

HUANG Meiyuan, WANG Zifa, HE Dongyang, XU Huaying, ZHOU Ling
Institute of Atmospheric Physics, Chinese Academy of Science
Beijing 100029, China

Abstract

A three dimensional regional Eulerian model of sulfur deposition and tranport has been developed. It includes emission, transport, diffusion, gas-phase and aqeous-phase chemical process, dry depostion, rainout and washout process. A "looking up table" method is provided to deal with the gas-phase chemical process including sulfur transfer. Calculated values have reasonable agreement with observations. Distribution of sulfur deposition and transport in East Asia are also analyzed in the paper. Simulation shows that sulfate (SO_4^{2-}) is the main substance to transport in long range transport. Some amount of sulfur emission of different countries transport across boundaries, but the main origin of sulfur deposition in each country in East Asia is from herself. Furthermore, some transport paths on different layers and outlet or inlet zones are found.

Keywords: Sulfur deposition, modeling, East Asia, transport.

1. Introduction

Economy of East Asia has been made great progress in recent years, but it also leads to some environmental problems such as atmospheric pollution, acid rain etc.. These problems have been increasing in recent ten years. Pollutants can stay in air for long time, so their long range transport even the so-called transboundary pollution problems are followed with great interest by the governments and scientists.

Model calculations using numerical models are useful not only to investigate the mechanism and distribution of pollutant deposition and transport but also to predict the changing trend in the future. So many models have been already presented to predict long range transport and acid deposition, however, most of them have been applied to Europe or North America (Eliassen and Saltbones (1983), Chang et al.(1987), Iversen et al.(1989), Langner et al.(1991), Zlatev and Rodhe(1992)). Few examples concerning the modeling studies on these problems in East Asia could be found. Kotamarthi and Carmichael(1990) pointed out that the overseas transport of pollutions are of much importance to Japan. Murao et al.(1993) showed that sulfur dioxide(SO_2) emission from volcanoes has a large contribution to the deposition of sulfur around Japan.

It is the purpose of the paper to get some views about the deposition distribution and transport in East Asia. Section 2 describes the model with particular focus on the treatment of gas-phase chemical reaction process and comparison with observations. Sulfur deposition distribution and transport are given in Section 3.

2. Model Description and Comparison with Observation

2.1 Model Description

In a spherical and sigma(σ) coordinate system, the model is described mathematically by a system of partial differential equation of the form:

$$\frac{\partial}{\partial t}(\Delta H \cdot C_t) + \frac{\partial}{R \partial \varphi}(u \cdot \Delta H \cdot C_t) + \frac{\partial}{R \cos\theta \partial \theta}(v \cos\theta \cdot \Delta H \cdot C_t) + \frac{\partial}{\partial \sigma}(W \cdot C_t) =$$

$$\frac{K_\varphi \partial}{R^2 \cos^2 \theta \partial \varphi}\left(\Delta H \cdot \frac{\partial C_t}{\partial \varphi}\right) + \frac{K_\theta \partial}{R^2 \cos\theta \partial \theta}\left(\Delta H \cdot \cos\theta \frac{\partial C_t}{\partial \theta}\right) + \frac{\partial}{\partial \sigma}\left(\frac{K_\sigma \partial C_t}{\Delta H \partial \sigma}\right) \quad (1)$$

$$+ S \cdot \Delta H + P \cdot \Delta H - R_d \cdot \Delta H - M \cdot \Delta H$$

where C_t is the concentration of pollutants, t is the time, θ, φ is latitude and longitude, R is Earth radius, $K_\theta, K_\varphi, K_\sigma$ are diffusion coefficients in different directions, u,v are horizontal wind velocity, S is emission rate, R_d is dry deposition, W is scavenging term, P is chemical transfering term, the terrain-following coordinate σ(θ,φ) can be written by

$$\sigma(\theta,\varphi) = \frac{z - h(\theta,\varphi)}{H(\theta,\varphi) - h(\theta,\varphi)} = \frac{z - h}{\Delta H} \quad (2)$$

where H(θ,φ) is the height of tropopause layer, h(θ,φ) is the height of relief, z is geopotential height, the equivalent vertical velocity W is computed using the mass continuity equation in sigma (σ) coordinate:

$$W = \omega - \frac{u}{R\cos\theta}\left(\frac{\partial h}{\partial \varphi} + \sigma \frac{\partial \Delta H}{\partial \varphi}\right) - \frac{v}{R}\left(\frac{\partial h}{\partial \theta} + \sigma \frac{\partial \Delta H}{\partial \theta}\right) \quad (3)$$

Considering the fractional area of cloud coverage in grid, chemical term in equation (1) is written as

$$P = (1 - \alpha) \cdot \mu \cdot C_{SO2} + \alpha \cdot \mu_q \cdot C_{SO2} \quad (4)$$

where the fractional area (α) calculated with equation (10), μ_q is chemical reaction rate of SO_2 in clouds and less than 3.0×10^{-5}, μ is oxidation rate of SO_2 in air and can be looked up from a table of oxidant rate which calculated by a chemical model. We don't couple the chemical model into 3D transport model in real-time because of much more CPU time. As a substitute, we choose five main factors such as temperature, humidity, light intensity, cloud and fog coverage and pollutant concentration, which have most effects on the oxidant rate of SO_2, then calculated the oxidant rate under different conditions of these five factors by the gas-phase chemical model and get a big table of oxidant rate of SO_2. At last, in the transport model it can look up the value of μ(in equation (4)) from table under varying conditions.

There are 31 species, 52 chemical reactions in the gas-phase chemical model (He and Huang, 1992). Photolysis rate(j) in the grid column containing clouds can be written as

$$J = J_{clear}[1 + \alpha(F_{cld} - 1)] \quad (6)$$

where j_{clear} is the coefficient of photolysis rate in clear sky, α is the fractional area of cloud coverage, F_{cld} is correction function of photolysis rate depending on solar senior angle, cloud layer, cloud droplet radius etc.(Chang et al. (1987)).

There are many factors such as atmospheric degree of stability, land type, surface wetness and roughness influencing on the dry deposition rate of sulfur. A sub model considering these facts is constructed to calculate dry deposition rate. Dry deposition rate V_g is calculated by.

$$V_g = \frac{K_z \partial C}{C \cdot \partial z} = (r_a + r_s + r_c)^{-1} \qquad (7)$$

The values of V_g for SO_2 are from 0.2 to 2.2cms^{-1}, for SO_4^{2-} from 0.05 to 0.5cms^{-1}, which vary both on surface and diurnal bases.

The in-cloud and below-cloud scavenging of sulfur is very complex. We cite scavenging coefficient of rain, which is effected by the rain intensity, drop spectrum and pollutants concentration. Scavenging term is written as
$$M = W_a \cdot C_i$$
According to Xu (1992) and Peng(1992), we assumed that
$$W_{aSO2} = (4.3 + 0.78 \ln C_{so2}) + (0.14 - 0.019 \ln C_{SO2}) \cdot I$$
$$W_{aSO4} = 0.33 I^{0.83} \qquad (8)$$

where I(mms^{-1}) is the rain intensity and C(μgm^3) is the concentration of pollutants.

Datasets are obstained from ECMWF(Europe Center for Middle-range Weather Forecasts). The contents of these data include horizontal wind(u,v), σ-wind, geopotential height, temperature(T), relative humidity(RH). The seven levels are following the pressure levels:1000, 850, 700, 500, 400, 300, 100 (the units are hpa). The period between inputs is 12 hours. The necessary for the model meteorological data are obtained from the input meteorological fields by appropriate interpolation. Rainfall datasets are interposed by observed data from Meteorological Agency of China. The fraction of cloud coverage in grid is calculated from relative humidity:

$$\alpha = \frac{RH - RHC}{1 - RHC} \qquad (RH > RHC) \qquad (10)$$

Where RHC is critical relative humidity, which changed with seasons and altitude.

The current model uses a horizontal grid with 1°×1° resolution covering the East Asia from 16°N to 50°N, from 98°E to 146°E (See Fig 1). The vertical extending consists of eight layers from surface to tropopause along a sigma (σ) coordinate. The boundary conditions used to solve equation (1) are set as these: pollutants can only be promised to transport out of and not into boundary. It is assumed that surface is seen as absorbed boundary while top boundary as closed one. Numerical advection over 48×34 grid is based on the upwinding scheme and the filtering routine. Greater stability of the numerical soloution is obstained while the Courant number (uΔt/Δx) is less than 1.

The emission inventory for SO_2 (See Fig 1) used in the model is taken from three origins. Emissions in mainland of China are obtained from China Environment Agency, those in other places are taken from Akimoto(1994) and volcano emission is added in Japan according to the data of Fujita. It is assumed that of the total sulfur emissions in each grid square, 80% is emitted as SO_2, 5% as SO_4^{2-}, and remaining 15% is deposited within the grid square (Eliassen and Saltbones (1983) ,Zlatev et al. (1992)).

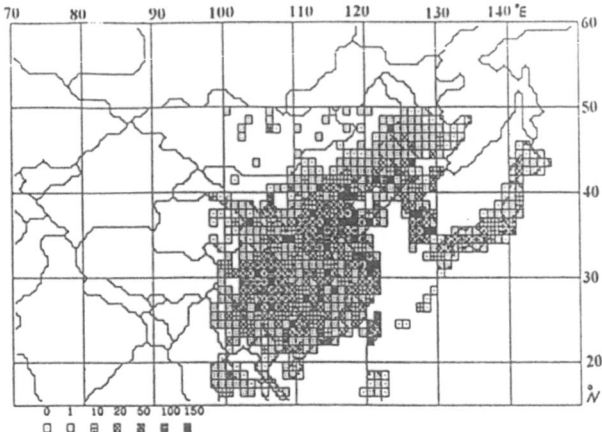

Fig 1 Emissions in East Asia (1991) and Model Resolution (unit:GgS/year)

2.2 Comparison Results

The comparisons were carried out by using measurements of the concentrations for SO_4^{2-} in precipitation. SO_4^{2-} Concentrations in precipitation were measured in July at about 100 mearsuement stations in China and Japan. And comparisons between calculated and measured SO_4^{2-} have been carried out for these stations. A scatterplot diagram for the calculated and observed SO_4^{2-} concentration in precipitation is shown in Fig 2. Considering some observation stations (about 60) stated in cities and represented limited areas, it is natural that the calculated values in these stations are lower than observed. The comparison of the calculated and observed has reasonable agreement with a correlation coefficient of 0.80, but with a tendency to nderestimate SO_4^{2-} concentration in precipitation and average error of 23%.

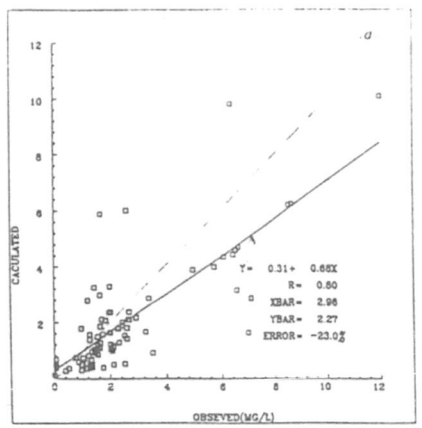

Fig 2 Scatterplot diagram for the calculated and observed SO_4^{2-} concentration in precipitation (mg/l)

3. Results and Discussion

The distribution of SO_2 and SO_4^{2-} deposition in East Asia in January is shown in Fig 3. SO_2 deposition distribution is very sensible to the distribution of emissions' intensity. Its high value areas locate in Southwest of China (Sichuan and Guizhou Province), South of Taiwan, Shandong peninsular, Korea peninsular and center of Japan. Deposition of SO_4^{2-} is less sensible to emission than that of SO_2. Its fewer high value centers and wider scale show that transport scale and distance of SO_4^{2-} are wider and longer than that of SO_2.

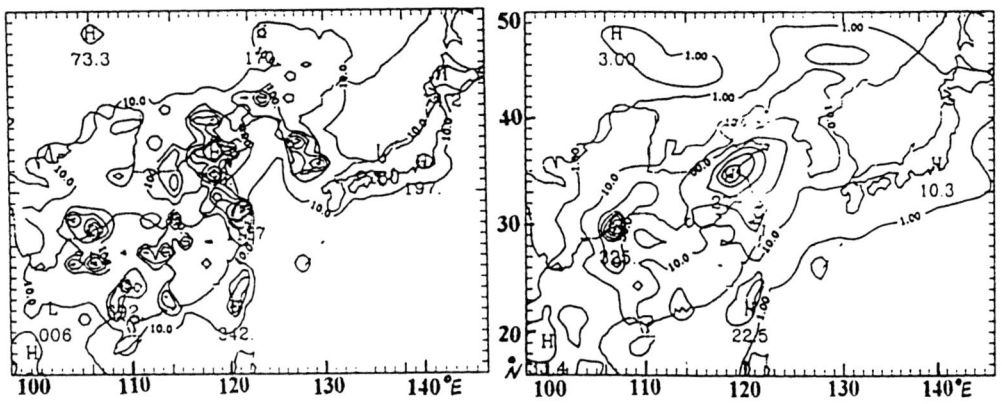

Fig 3 The Distribution of Deposition for SO_2 and SO_4^{2-} in East Asia (January ,1989)
Units: mgm^{-2} (a , SO_2 b , SO_4^{2-})

Table 1 shows the exports of sulfur emission of countries in East Asia in 1989. Parts of sulfur emission emitted by these countries is transported across boundary. The proportion of outlet is different with sulfur types. For total sulfur deposition and SO_2 deposition ,58-95% of emissions are deposited on their own countries, their exports are less than 45 percent. But for SO_4^{2-} deposition ,except China, their exports are more than 60 percent, some countries such as Japan and South Korea more than 80 percent. Seas and the Pacific Ocean are accepted a great amount of sulfur deposition, e.g. 30% of South Korea's emission, 28% of Japan's, 20% of China's, moreover, the percentage deposition for SO_4^{2-} in seas and the Pacific Ocean is higher than that for SO_2.

Table 1 Percentage Contribution of Deposition of Different Countries' Sulfur Emission in East Asia in 1989(note:Others includes Seas,in air and out of boudaries)

	Emission	receptor				
Source	tonS	Mongolia	China	S. Korea	Japan	Others
Mongolia	43700	57.8	11.4	0.3	0.3	30.2
China	9134100	0.0	71.2	0.3	0.3	28.2
S.Korea	506700	0.0	<0.1	62.7	3.0	34.3
Japan	986300	0.0	<0.1	<0.1	71.0	28.9

Table 2 Percentage Contribution of Total Sulfur Deposition from One Country to Others in East Asia in 1989

	receptor			
source	Mongolia	China	S. Korea	Japan
Mongolia	99.7	0.1	<0.1	<0.1
China	0.3	99.5	7.9	3.5
S. Korea	0.0	0.0	90.9	2.1
Japan	0.0	0.0	<0.1	93.7
Others	0.0	0.4	1.2	0.7

The origins of sulfur deposition in different countries in East Asia are given in table 2. It can be concluded that the main origin of sulfur deposition in each country in East Asia is from himself. Because of different areas and wind fields, the ratio of different origin is not same with each other.The pollutants transport usually from west to east, and in eastern countries or places there are more pollutants flown into the Pacific Ocean.

Reference

Akimoto H.: 1994, "Distribution of SO_2,NOx,and CO2 Emissions from Fuel Combustion and Industrial Activities in Asai with $1°\times1°$ Resolution",Atomosphric Enviroment,28,213-225.

Chang, J.S., Brost, R.A., Isaksen, I.S., Madronich, S., Middleton, P., Stockwell, W.R., and Walcek, C.J.: 1987, "A Three Dimensional Eulerian Acid Deposition Model: Physical Concepts and Formulation",Journal of Geophysical Research,92,No.D12, 14,681-14,700.

Eliassen, A. and Saltbones, J.: 1983, "Modeling of Long-range Transport of Sulfur over Europe:A two-year model run and model experiment ",Atmos.Environ.,17,1457-1473.

He,Dongyang and Huang,Meiyuan: 1992, "A Photochemical Model for Regional Air Quality Simulation" ,Journal of Enviromental Science ,12(2),192-198.

Iversen ,T., Saltbones,J., Sandnes,N., Eliassen,A. and Hov,Ø.: 1989, "Airborne Transboundary Transport of Sulfur and Nitrogen over Europe---Model Description and Caculatons",EMEP MSCW Report 2/89, Det Norske Meteorological Institute Technical Report , No. 80.

Kotamarthi, V.R. and Carmichael G.R. :1990, "The long rang transport of pollutants in the Pacific Rim Region. ", Atmos. Environ., 24A,1521--1524.

Langner, J. and Rodhe, H.: 1991, "A Global Three-Dimensional Model Of the Tropospheric Sulfur Cycle", Journal of Atmospheric Chemistry,13,225-263.

Murao, N., Katatani, N., Sasaki, Y.,Okamoto, S., Koobayashhi, K. : 1993, "A Modeling Study on Acid Rain in East Asia", International Conference on Regional Environment and Climate Changes in East Asia, 305-309.

Peng, H. and Qin, Y.:1992, "Parameterization of rain scavenging, for aerosol", Atoms. Science(in Chinese),Vol.16, No.5, 622-631.

Xu, L. and Qin, Y.: 1992, "Parameterization of scaveging process for gas in below-cloud layer", Environment Chemistry(in Chinese), Vol. 11, No.1, 1-10.

Zlatev,Z., Christensen, J.and Hov:, Ø.: 1992, "A Eulerian Air Pollution Model for Europe with Nonliear Chemistry" ,Journal of Atmospheric Chemistry,15,1-37.

AN ANALYSIS OF WET DEPOSITION OF SULFATE USING A TRAJECTORY MODEL FOR EAST ASIA

Y. ICHIKAWA and S. FUJITA

Central Research Institute of Electric Power Industry
11-1 Iwato Kita 2-Chome, Komae-shi, Tokyo, 201 Japan

Abstract. A long-range transport model for East Asia was developed to estimate the wet deposition of sulfate. The model is a trajectory type which is appropriate for long-term analysis. Trajectories of air masses are calculated by tracing the wind field which changes spatially and temporally. The processes of reactions, rainout removal, intake of sulfate in cloud water into rain water, and dry and wet depositions are considered. It is possible to calculate the concentration of sulfate in precipitation at a receptor by performing material balance in a grid box containing the receptor.

The results obtained by the long-range transport model were evaluated through comparison with observation data of acidic deposition. The observation was conducted at 21 stations throughout Japan for one year. The calculated amount of wet deposition of sulfate in Japan was 0.22Tg/y in S equivalent, while the observed amount was 0.29Tg/y. The long-range transport model can predict almost 80% of observed wet deposition. The contributions of domestic anthropogenic sources and volcanic eruption to wet deposition of sulfate in Japan were estimated using the long-range transport model. The ratio of the deposition of sulfate due to Japanese anthropogenic sources to that due to the Asian continental sources was about 1 to 2. Since air stream from the direction of the Asian continent dominates during winter, the contribution of Japan to wet deposition in the region which faces the Sea of Japan amounted to less than 15%. The contribution of the sulfur oxides from volcanoes was about 20%.

Key words: acidic deposition, wet deposition, sulfur oxides, East Asia, long-range transport model

1. Introduction

Acidic deposition began to attract attention as a regional-scale environmental problem in the beginning of the 1970's in Europe and North America. Recently, acid rain problems have also extended to Asia, because of a significant increase in atmospheric emissions resulting from high economic and population growth rate. We established a nationwide network in Japan for the observation of acid rain in 1987 and have been conducting observations since then. In this paper, a long-range transport model for East Asia was developed to analyze wet deposition of sulfate and the validity of the model was examined on the basis of the data obtained from the above observation. A large number of long-range transport models of acidic substances for Europe and North America were developed in the 1970's and the 1980's (e.g., Eliassen et al., 1982; Carmichael et al., 1986; Chang et al., 1987). In the 1990's, some analyses of long-range transport for Asia were begun (e.g., Katatani et al., 1991; Carmichael, 1991; Kitada and Tanaka, 1992) but there is hardly any case which predicts acidic deposition and examines the accuracy of the prediction all year round for various places in Japan. The results of our long-range transport model were evaluated through comparison with the data of acidic deposition observed at 21 stations throughout Japan for one year. This paper also describes the contributions of domestic anthropogenic sources and volcanic eruption to wet deposition of sulfate in Japan.

2. Long-range transport model

There are two basic approaches to the analysis of long-range transport of acidic substances: use of a trajectory model and an Eulerian model. The trajectory model treats physical and chemical processes involved in acidic deposition simply and uses routinely obtained meteorological data. The model is suitable for long-term analysis, covering a year, for example. The Eulerian model enables detailed analysis of physical and chemical processes, but cannot provide good results without sufficient data on emission, meteorology, climate and geography, and an excellent computational environment. The model is suitable for episodal analysis within a limited period. Analyses for the period extending over at least one year are required to estimate the contributions of domestic anthropogenic sources and volcanic eruption to wet deposition of sulfate. A trajectory model was therefore developed in this paper.

Our trajectory model provides estimates of sulfur oxides (SO_x) concentration in air and sulfate concentration in rain. Trajectories of air mass are calculated by tracing the wind field, which changes spatially and temporally. Trajectories are traced every 3 hours for 10 days and calculated on an 850 hPa isobaric surface. The concentration distribution of pollutants is assumed to be normal in the horizontal direction and uniform in the vertical direction within the mixing layer. The height of the mixing layer is assumed to be 1000 m. Physical and chemical processes considered in this model are shown schematically in Figure 1. The values of parameters vary widely depending on meteorological, geographical, and climatic conditions. In our model, the values are constant or simple functions of the precipitation intensity. The ratio of reactions, intake of cloud and rain water, and dry and wet deposition processes are assumed to be linearly proportional to concentration. Then, analytical solutions are obtained for sulfur dioxide (SO_2) and particulate sulfate concentrations in air, and sulfate concentration in cloud water. It is possible to calculate the concentration of sulfate in rain in a receptor by performing material balance in the grid box containing the receptor.

Anthropogenic emissions of SO_2 per grid square in East Asia were obtained by revising Fujita et al.'s data (Fujita et al., 1991). The region of the study includes Japan, China, Taiwan, South Korea and North Korea and was divided into 54 by 54 grids, with grid size equivalent to 80km by 80km in the center of the region. Anthropogenic SO_2 emissions from East Asia for 1986 were about 21×10^6 t/y. The breakdown of this figure is as follows: 0.9×10^6 t/y for Japan; 18×10^6 t/y for China; 0.3×10^6 t/y for

Taiwan; 1.0×10^6 t/y for South Korea; and 0.9×10^6 t/y for North Korea. There were 12 active volcanoes in Japan at the end of the 1980's. The SO_2 emissions from volcanoes were estimated to be approximately 1.5×10^6 t/y. The acidic deposition from volcanoes was also calculated.

High-altitude wind data were obtained from the Aerological Data of Japan and weather charts, edited by the Japan Meteorological Agency (JMA) and published by the Japan Weather Association (JWA). There are 54 observation points in East Asia. Winds are observed at 0 and 12 Greenwich mean time. Data of precipitation for each day were obtained from SDP (JMA weather station) data and World data. Both data are edited by JMA. There are 345 observation points in East Asia.

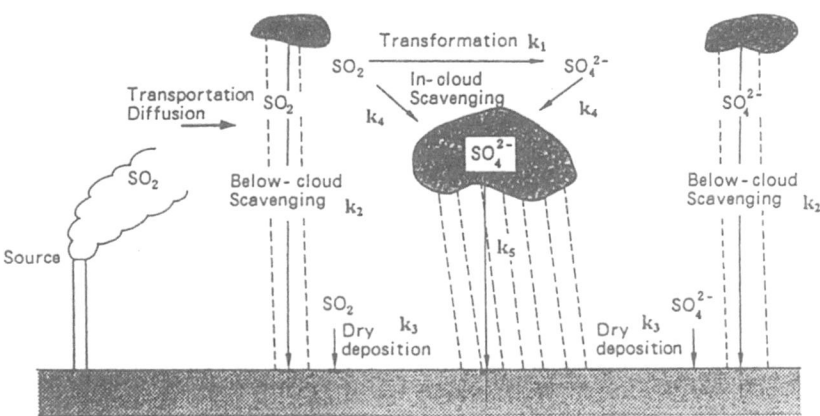

Fig.1. Schematic of the long-range transport model. k_1: 0.01 1/h for winter, 0.03 1/h for summer, k_2: 0.1I 1/h, k_3= Vd/H, where the values of Vd vary from 0.0002 to 0.006 m/s depending on substance, season and receptor location., k_4: 0.02 1/h, k_5: 0.1I 1/h. I: precipitation intensity(mm/h), Vd: deposition velocity (m/s), H: mixing height(m).

3. Prediction and analysis of wet deposition of sulfate

3.1. Observation of wet deposition

The long-range transport model was evaluated using observation data of acidic deposition. The location of observation stations is illustrated in Figure 2. The observation stations indicated by white circles (○) represent the areas divided by solid lines on the main Japan Islands according to climatic and geographical conditions. Black circles (●) and white circle with plus sign (⊕) show the observation stations at small islands and our research institute, respectively.

A wet-only sampler with an aperture of 190 cm² was used to collect precipitation samples. The precipitation sensor responds to rain droplets with a diameter larger than 0.5mm. Precipitation samples were collected in a five-liter polyethylene bottle installed in the sampler. Samples were collected at 10-day intervals. Sulfate and sodium ion in precipitation were analyzed by ion chromatography and atomic absorption spectrometry, respectively. Sulfate concentration originated from non-seasalt sources was estimated using sodium ion concentration.

3.2. Prediction of wet deposition

Calculations using the long-range transport model were carried out for the period from October 1988 to September 1989. This period corresponds to the second year of our observation. The selection of this period is based on the facts that insufficient observation data were obtained in the first year and that the data on volcanic emissions were not reliable after the third year.

Figure 2 compares the annual wet depositions of sulfate predicted by the long-range transport model with those observed. Numbers accompanying the squares correspond to the observation points indicated on the map of Japan. Calculation was carried out for the grid with size equivalent to about 80 km^2. The predicted total wet deposition in Japan was 0.22 Tg/y in S equivalent, while the observed one was 0.29 Tg/y in S equivalent. The model tends to give slightly lower values than the observed ones. This is probably due to the facts that the influence of Russian Far East is not taken into consideration and that the emission data for 1986 were used for the prediction of wet deposition in 1988 and 1989. When the SO_2 emission from Russian Far East is assumed to be 1.0 g/m^2/y according to Spiro et al. (1992), the predicted values for the evaluation points in the northern part of Japan, Nos.1~5 in Figure 2, increase by 0.1~0.2 g/m^2/y. In addition, it is said that there has been an increase of 15% in emissions in China and Korea between 1986 and 1989. Considering this increase, the predicted values are certain to increase. It is concluded from these considerations that the long-range transport model can predict wet deposition of sulfate with high accuracy.

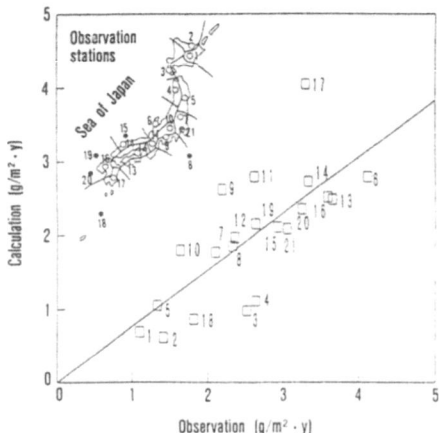

Fig.2. Comparison of wet depositions predicted by the long-range transport model with those observed at 21 points of the sampling network for acidic depositions.

y=0.76x+0.00 (x: observation, y: calculation), correlation coefficient: 0.73.

3.3. Estimation of sources contributing to wet deposition in Japan

The contributions of domestic anthropogenic and volcanic sources to wet deposition of sulfate in Japan were estimated using the long-range transport model. Values in percent under the circle graph in Figure 3 indicate the contribution of volcanoes. The contribution amounted to 20% for the whole area of Japan. The circle graph in Figure 3 shows the contributions of East Asian countries to the total wet deposition in Japan excluding the wet deposition arising from volcanic sources. For the whole of Japan, the contributions of China, Japan, and Korea are approximately one-half, one-third, and one-sixth of the total contribution, respectively. Less than 1 % is due to anthropogenic emissions from Taiwan. Emissions from both volcanic and anthropogenic sources in Japan account for about 50 % of the total wet deposition of sulfate in Japan.

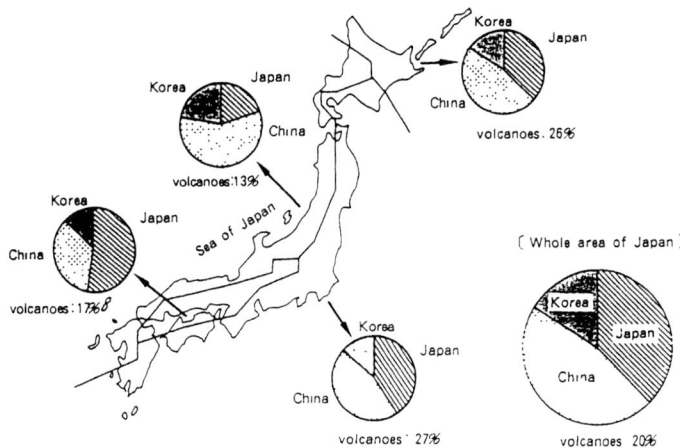

Fig.3. Sources contributing to wet deposition of sulfate in Japan.
  Circle graphs show the contributions of anthropogenic emissions excluding those of volcanic emission.

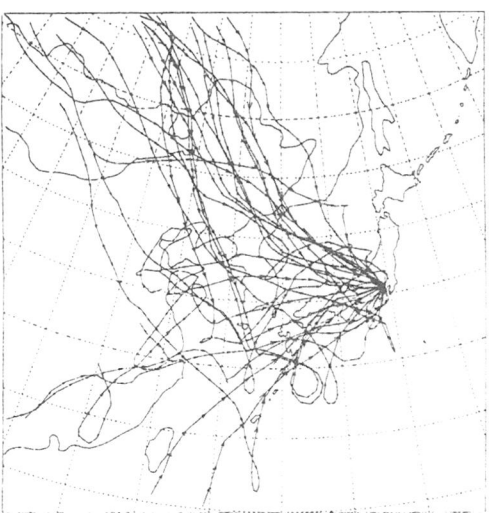

Fig.4. Back trajectories arriving at Tokyo (February, 1989).

Figure 4 shows the back trajectories arriving at Tokyo in February 1989. The trajectories were calculated on an 850 hPa isobaric surface. Air stream from the direction of the Asian continent dominates from October through May, as shown in this figure. The contribution of domestic sources to the region which faces the Sea of Japan is therefore low as compared with that to other regions. The contribution of Asian continental sources to the coastal region of the Japan Sea amounted to more than 85 % in winter. In summer, air flows from various directions to Japan. The contribution of Asian continental sources to wet deposition in the whole area of Japan in summer is about 40 %.

4. Conclusion

A trajectory-type long-range transport model was developed to estimate the wet deposition of sulfate. The results calculated by the model were evaluated through comparison with observation data of acidic deposition. The observation was conducted at 21 stations throughout Japan for one year. The model showed relatively high accuracy of prediction. The contributions of domestic anthropogenic and volcanic sources to the total wet deposition of sulfate in Japan were estimated using the long-range transport model. The contribution of the emissions from both domestic anthropogenic and volcanic sources was approximately 50 %.

Dry deposition strongly depends on the ground surface concentrations of SO_x. When a large source exists in the vicinity of the receptor and the model does not take vertical concentration distribution into account, SO_x concentration near the ground surface may be underestimated by a one-layer model such as our model. In that case, it is important that the model be expanded to include the vertical distribution of emissions and concentration.

References

Carmichael, G.R., Peters, L.K., Kitada, T.: 1986, *Atmos. Environ.* **20.1**, 173-188.

Carmichael, G.R.: 1991, Third Annual Conference on Acid Rain and Emissions in Asia, Bangkok, Thailand, Resource Management Associates, Argonne National Laboratory and Asian Institute of Technology, 36-45.

Chang, J.S., Brost, R.A, Isaksen, I.S.A., Madronich, S., Middleton, P., Stockwell, W.R., Walcek, C.J.: 1987, *J.Geophys. Rev.* **92.D12**, 14681-14700.

Eliassen, A., Hov, O., Isaksen, I.S.A., Saltbones, J., Stordal, F.: 1982, *J.Appl. Met.* **21**, 1645-1661.

Fujita, S., Ichikawa, Y., Kawaratani, R., Tonooka, Y.:1991, *Atmos. Environ.* **25A.7**, 1409-1411.

Katatani, N., Murao, N., Okamoto, S.: 1991, Emerging Issues in Asia, Proceedings of the 2nd IUAPPA Regional Conference on Air Pollution, II, Seoul, Korea, Korean Air Pollution Research Association, 59-64.

Kitada, T., Tanaka, K.: 1992, Air Pollution Modeling and its Application IX, van Dop, H. and Kallos, G. ed., Plenum Press, New York, 445-454.

Spiro, P. A., Jacob, D. J., Logan, J. A.: 1992, *J.Geophys. Rev.* **97.D5**, 6023-6036.

LONG RANGE TRANSPORT AND DEPOSITION OF MINERAL MATTER AS A SOURCE FOR BASE CATIONS.

A. SEMB[1], J.E. HANSSEN[1], F. FRANCOIS[2], W. MAENHAUT[2] AND J.M. PACYNA[1].

[1] *Norwegian Institute for Air Research, N-2007 Kjeller, Norway,* [2] *Institute for Nuclear Sciences, Proeftuinstraat 86, B-9000 Gent, Belgium.*

Abstract. Mineral dust in the atmosphere is generally alkaline, and is a source of base cations in precipitation. Annual emissions of particles from large combustion plants and industrial processes in Europe is of the order of 24 million tonnes, and the calcium content may be as high as 1.4 million tonnes. Emissions from diffuse sources such as agricultural activities, construction and quarrying are much less well known. Emissions of dust from the Sahara have been estimated to more than 200 tonnes, with a calcium content of 3-5%.

In northern Europe, airborne concentrations of calcium and the concentration of calcium in precipitation are generally consistent with the anthropogenic emissions and their regional distribution. Transport of dust from Sahara is a major source of base cations in precipitation around the Mediterranean Sea, but the influence diminishes further north. The concentration of calcium in precipitation decreases from south to north and from east to west in Europe. Dry deposition of alkaline particles is not well documented, but may be at least as high as the input by precipitation in regions where there are large emissions. There are still large uncertainty gaps with respect to emissions, transport and deposition of calcium-containing particles.

Key words: acid precipitation, base cations, emissions, airborne concentrations, deposition in Europe

1. Introduction.

In addition to the acid rain components, *viz.* sulphate, nitrate, ammonium and hydrogen ions, and sea-salt components from sea-spray, precipitation also contains variable amounts of base cations such as calcium, potassium and magnesium from terrestrial sources. These base cations reduce to some extent the acid deposition critical load, and their deposition have been demonstrated to be comparable to the release of base cations from soil minerals (e.g. Åberg *et al.*, 1989).

Very little has been done to quantify the deposition of base cations in relation to acid deposition, although some concern has been expressed about the fact that pollution control may have decreased this deposition of base cations in Europe (Lövblad, 1987). A strong decrease in the concentration of base cations in precipitation samples for the period 1983-1990, has been documented by Hedin *et al.* (1994). In addition to possible industrial emissions, however, base cations may also be derived from aeolian dust, either from agricultural areas within Europe, or from desert area outside Europe. Episodes with red- or ochre-stained deposition of Saharan dust have been observed even in Northern Europe, and are well documented from the glaciers of Switzerland (e.g. Glawion, 1939, and Haeberli, 1978). Soil dust transport from agricultural areas in Europe is less well documented, but should also be of some importance in Europe.

It is the purpose of this paper to examine some of the available evidence with respect to acid-neutralizing airborne particles, and base cations in precipitation.

2. Emissions.

Recent assessments of particle fluxes into the atmosphere, prepared for the Intergovernmental Panel on Climate Change by Jonas *et al.* (1995), indicates that anthropogenic emissions estimated at 126 million tonnes with and uncertainty range of between 50 and 160 million tonnes contribute only about 10% of the total budget of particles.

The European emissions of primary anthropogenic particles are estimated at 24.1 million tonnes per year as an average for the period 1987-1989 (Nowicki, 1993). A major part of these emissions; estimated at 81 %, are from sources in Eastern Europe.

A summary of the estimated emissions is given in Table I.

TABLE I.
Emissions of particles from anthropogenic primary sources in Europe(in million tonnes/year), with estimated calcium content (in kilotonnes/year).

Country/ region	Power production	Cement	Ferrous metallurgy	Other industries, municipal, agriculture	Sum	Calcium equivalents
Former CSFR	0,5	0,3	0,3	0,6	1,7	145
Poland	0,8	0,5	0,4	1.3	3,0	238
Former DDR	0,7	0,5	0,2	0,8	2,2	208
Former USSR	2,8	0,8	0,8	4.0	8,4	464
Rest of Eastern Europe	1,4	0,5	0,4	1,9	4,2	262
Scandinavia, Austria, Switzerland	0,1	0,07	0,07	0,26	0,5	35
EEC countries	1,3	0,5	0,4	1,9	4,1	260
Sum particles, million tonnes	7,6	3,17	2,57	10,76	24,1	-
Sum calcium equivalents, kilotonnes	152	1014	231	205	1602	-

Based on measurements in Polish power plants (e.g. Pacyna 1980), the calcium content of particles from power plants and other solid fuel combustion sources may be estimated at 2%. In addition to power plants, cement production and ferrous metallurgy are important emission sources of airborne particles. The numbers in Table I are taken from national emission surveys in Poland (GUS, 1981) and in the former Soviet Union (MEPNR, 1994), and have been extrapolated to other countries in Europe using data for the production of cement, iron and steel in the respective countries.

The CaO content of fine particles emitted from cement plants ranges from 40 to 50%(MRI, 1971). CaO contents for fine particles from the iron and steel industry employing open hearth and basic oxygen furnaces range from 1 to 5%, while the CaO content of fine particles from electric arc furnaces is from 5 to 20%. The latter type of furnace is more common in Europe today.

Emissions from municipal and agricultural activities, and from other industries, are highly uncertain, as are the the acid-neutralizing capacities of these emissions. A calcium content of 2% has been used in Table 1.

This gives an annual emission of 1.6 M tonnes, which may be compared to an annual emission of sulphur dioxide from the same countries of 35 M tonnes. The emissions of alkaline materials are, in other words, sufficient to neutralize only about 10% of the sulphur dioxide emissions.

Very large amounts of airborne particulate matter are produced by wind erosion in desert and other arid areas, such as the Sahara desert. Junge (1979) estimated the annual emissions from Sahara to be from 60 to 200 million tonnes, and more recent investigations indicate that this number is an underestimate. Arimoto et al. (1995) have found the average calcium content to be around 3-5%. A large part of this dust is transported westward over the Atlantic ocean, but transport northwards across the Mediterranean sea (Loye-Pilot et al., 1986, Chester et al, 1984) and eastwards to the Middle East (Ganor and Mamane, 1982) is also important.

3. Airborne concentrations of particulate matter and base cations.

There is not much information on the base cation contents of airborne particulate matter. Studies of the elemental composition have largely focussed on heavy metals and trace elements enriched in aerosols relatively to crustal material(e.g. Rahn, 1976).

We have chosen to look in more detail on the airborne concentrations at two sites in southern Norway, largely because of the availability of comparatively recent data.

Table II gives average concentration values for the period February 1991- December 1992. Samples are collected with a two-filter sampler, which excludes particles larger than ≈10 µm a.e.d., and with a size separation at about 2 µm. The sampling method and the chemical analysis methods are described by Maenhaut et al. (1993).

TABLE II.

Average airborne concentration in ng/m^3 of some elements at Birkenes.

		Na	Al	Si	S	K	Ca	nss Ca	Pb
1991	Fine	113	15	70	773	38	12	7	4
	Coarse	273	54	114	131	37	47	37	1
1992	Fine	109	14	44	576	23	12	7	3
	Coarse	305	60	125	112	37	45	41	1

Since these sites are relatively close to the North Sea, concentrations of sea-salt aerosol contribute to the observed concentrations of calcium. Therefore the concentration of non-marine calcium has also been calculated, using sodium as a measure of the sea-salt contribution. The seasalt contribution to the potassium concentrations is less significant.

The concentration levels are relatively similar for these two years, but examination of the concentration records show some periods with clearly increased concentrations of calcium and potassium, as well as other typical mineral dust elements (Figure 1).

Back-trajectories for the 3-week period in May-June show that the air had passed across Sweden from the Baltic countries and Russia.

Even when this period is excluded, calcium is generally highly correlated with crustal elements such as Si ($r=0.91$), Al ($r=0.87$), K ($r=0.82$), not correlated with seasalt, and only weakly correlated with pollution components such as sulphate or Pb ($r=0.5-0.6$).

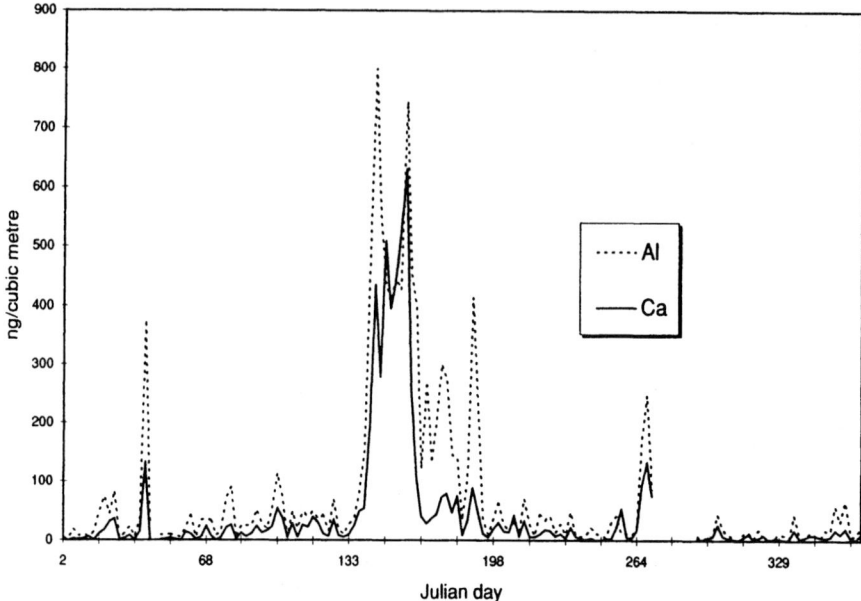

Fig.1.Measured airborne concentrations of calcium and aluminium at Birkenes, Norway during 1992

Earlier studies of the elemental composition of aerosols at Birkenes have also shown that high concentrations of crustal elements occur in connection with transport from the east or southeast (Amundsen et al., 1992).

It is interesting to note, however, that, in addition to reports of red rain caused by Saharan dust, there are also historical reports of dustfall in Scandinavia which appeared to have been caused by dust storms near the Black Sea.

The ratio of potassium to calcium in igneous rocks is typically 2:3, but the amounts of potassium collected on the filter samples is actually higher than the amounts of calcium. A relatively large fraction of the potassium in airborne particles is associated with particles less than 2μm a.e.d. Both clay minerals and plant material may contain relatively more potassium than calcium. It cannot be explained by local sources cuch as wood combustion.

4. Base cations in precipitation.

Contamination of precipitation samples by mineral dust at the precipitation collection site is a recognized problem in precipitation chemistry studies, and careful selection of monitoring sites as well as use of wet-only samplers and rigorous sampling procedures may be necessary to avoid this problem. Within the EMEP network, precipitation is sampled on a daily basis and wet-only samplers are generally used. If not, samplers are rinsed every day, even if there is no precipitation.

Table III gives some examples of precipitation-weighted annual average concentrations at a selection of EMEP sites. Calcium is the dominant base cation in precipitation samples, followed by potassium. There are clear regional gradients in the observed concentrations, particularly for calcium, and also some significant differences for sites which are close to each other. Concentrations decrease from south to north, and from east to west in Europe. Apart from Illmitz in Austria, which is quite close to the dry Hungarian plain, and from the sites in the Mediterranean area, Eastern Europe has generally higher concentrations than North-Western Europe. This is in general accordance with the emission estimates above, and is even more convincing when the precipitation chemistry data are interpolated by kriging to give maps of Ca concentrations in precipitation or deposition.(Figure 2)

TABLE III.

Precipitation-weighted mean concentrations in microequivalents per litre in precipitation at some EMEP sites in 1992. (Schaug et al., 1994).

	SO_4--S	nss Mg	nss Ca	nss K
Toledo, Spain	31	2	10	1
Arabba, Italy	39	5	16	2
Donon, France	31	0	4	1
Schauinsland, Germany	30	3	5	2
Deuselbach, Germany	44	3	13	2
Illmitz, Austria	96	43	97	4
Svratouch, Czech Republic	79	4	14	3
Kosetice, Czech Republic	62	3	9	3
Chopok, Slovakia	93	7	20	5
Langenbrügge, Germany	62	6	14	4
Jarczew, Poland	81	3	9	2
Tange, Denmark	37	-1	2	3
Witteven, Netherlands	66	0	4	-1
High Muffles, England	56	1	4	0
Eskdalemuir, Scotland	30	2	4	1
Rörvik, Sweden	44	1	2	1
Ähtäri, Finland	47	1	2	1
Virolahti, Finland	71	4	18	8

Examination of a 13-year record of precipitation chemistry at Birkenes show that there is not a very large difference in the ratio of mean monthly deposition of nss K or Ca to the mean monthly precipitation amount over the year, but there tends to be a little higher depostion of Ca in the spring months (March-May). This could be interpreted as a wind erosion or soil tillage effect.

In comparison with the air samples, precipitation samples contain much more calcium in relation to potassium (Table IV). This is at least partly a solubility phenomenon, calcium silicates are more easily weathered than potassium silicates, and potassium is strongly adsorbed to clay minerals.

The comparison also shows that the concentration of calcium in the precipitation relative to calcium in air samples is higher than the corresponding ratio for elements that occur mainly in submicron size particles. The ratio is also very close to the ratio of the concentration of calcium in air to the concentration of calcium in precipitation reported from the APIOS data in Canada by Eder and Dennis (1990).

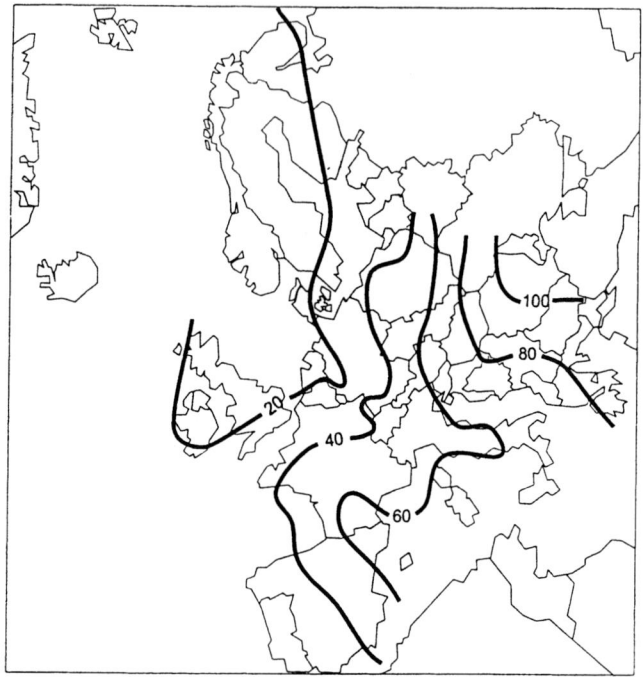

Fig. 2. Concentration of excess Ca in precipitation, interpolated from weighted annual mean concentration at EMEP sites for the period 1983-1987. The unit is µeq./litre.

TABLE 4.

Annual average concentrations in air and in precipitation, and calculated average scavenging coefficients at Birkenes

		Na	nss S	nss Ca	nss K	Pb
Air, ng/m3	1991	384	904	44	75	5
	1992	414	688	48	60	4
Precipitation, µg/l	1991	1520	750	82	85	4
	1992	1170	740	76	28	3
Scavenging	1991	5,0	1,04	2.3	1.4	1.0
ratio*(x 10^{-3})	1992	3.5	1.3	1.9	0.5	0.9

(*ng/kg of precipitation/ng/kg air)

Although there are many observations of deposition of Saharan dust in Northern Europe, for example in March 1991 (Franzén et al, 1994), it is difficult to use these observations to estimate the input of calcium from this source. In the Mediterranean countries the effects of episodic Saharan dust deposition on the precipitation chemistry is much more conspicuous, and may be interpreted to determine the influence of Saharan dust on the depositon of alkaline material. Rodá et al. (1991) found that the calcium concentrations in red or ochre-stained rain samples collected in north-eastern Spain was typically 20 times higher than in rain which did not contain Saharan dust. Of a total annual Ca deposition of 500 eq/ha, about 50% was associated Saharan dust episodes. De Angelis and Gaudichet

(1991) found that Saharan dust contributed by more than 22% of the average Ca deposition of 0.24 g/m2 over 30 years to the glaciers on Mont Blanc, but that the inputs vere very variable from year to year.

5. Dry deposition.

Large particles should be efficiently captured by surface roughness elements, but quantitative estimates of large particle dry deposition based on airborne concentrations and deposition rates are not available. However, inputs have been estimated in connection with forest stand and catchment mass balances, these estimates are partly based on canopy throughfall sampling and analogy with sodium. Two recent studies report Ca "dry deposition"to forest stands of 0.17 and 1.3 times the input by precipitation, in the Black Forest and Solling area in Germany, respectively (Feger, 1995 and Manderscheid et al., 1995). However, these report bulk precipitation inputs which are significantly higher than the precipitation deposition from the EMEP data. There is clearly need for more direct and accurate information with respect to base cation dry deposition.

6. Conclusions.

Although the information of both emissions, airborne concentrations and deposition is scattered and fragmentary, deposition of base cations by precipitation in Northern Europe seem to be largely caused by anthropogenic emissions of mineral particles with an excess of calcium silicates and calcium carbonates. On an equivalent basis, this input varies from 5-20% of the sulphate deposition. The emissions are being reduced as part of emission control programmes, however. In the Mediterranean area, the input of calcium and minerals is dominated by Saharan dust, which typically contain 3-5% of calcium as calcite. This input is not significant in Northern Europe.
Much less is known about the possible influxes from arid and desert areas in Central Asia to Southeastern Europe.
Dry deposition of alkaline mineral matter is potentially also an important source of base cations. Better information on emission rates, size distribution and chemical composition of mineral dust emissions is required to carry out quantitative estimates of dry and wet deposition on a regional basis.

Aknowledgement.

The map of Ca concentrations in precipitation over Europe was prepared by our late colleague Ulf Pedersen. We are grateful to J.Schaug and our co-workers in EMEP for access to the EMEP precipitation chemistry data, and to Sverre Solberg, NILU, and the Norwegian Meteorological Institute for air-mass trajectories.

References

Åberg, G.I., Jacks, G and.Hamilton, P.J.(1989). Calcium budgets for catchments as interpreted by strontium isotopes. *Nord. Hydrol.*, **20**, 85-96.

Amundsen, C.E., Hanssen, J.E:, Semb, A., and Steinnes, E.(1992). Long-range transport of trace elements to southern Norway. *Atmospheric Environment*, **26A,** 1309-1324.

Arimoto, R., Duce, R.A., Ray, B.J., Ellis, W.G., Cullen, J.D., and Merrill, J.T.(1995) Trace elements in the atmosphere over the North Atlantic. *J. Geophys.Res.*, **100**, 1199-1213.

Chester, R., Sharples, E.J., Sanders, G.S., and Saydam, A.C.(1984). Saharan dust incursions over the Tyrrhenian sea. *Atmospheric Environment*, **18**, 929-935.

De Angelis, M. and Gaudichet, A.(1991). Saharan dust deposition over Mont Blanc(French Alps) during the last 30 years. *Tellus*, **43B**, 61-75.

Eder, B.K., and Dennis, R.L.(1990) On the use of scavenging ratios for the inference of surface-level concentrations and subsequent dry deposition of Ca^{2+}, Mg^{2+}, Na^+ and K^+. *Water, Air, Soil Pollut.*, **52**, 197-216.

Feger, K.H.(1995) Solute fluxes and sulfur cycling in forested catchments in SW Germany as influenced by experimental (NH4)2SO4 treatments. *Water, Air, Soil Pollut.*, **79**, 109-130.

Franzén, L.G., Hjelmroos, M., Kållberg, P, Brorström-Lundén, E., Juntto, S., and Savolainen, A-L.(1994). The "yellow snow" episode of northern fennoscandia, March 1991. - A case study of long-range transport of soil, pollen and stable organic compounds. *Atmospheric Environment*, **28**, 3587-3604.

Ganor, E. and Mamane, Y.(1982). Transport of Saharan dust across the Eastern Mediterrranean, *Atmospheric Environment*, **16**, 581-587.

Glawion, H.(1939). Staub und Staubfälle in Arosa, *Beitra. Phys. Frei. Atmos.*, **25,**, 1-43.

GUS, (1981) Przemyslowe zanieczyszczenia i ochrona powiertza atmosferycznego 1975-1980. Glowny Urzad Statystyczny, warzawa, Poland(in Polish).

Haberle, W.(1977). Sahara dust in the Alps-a short review. *Z. f. Gletscherkunde u. Glazialgeol.*, **13**, 206-208.

Hedin, L.O., Granat, L., Likens, G.E., Buishand, T.A., Galloway, J.N., Butler, T.J., and Rodhe, H.(1994). Steep declines in atmospheric base cations in regions of Europe and North America. *Nature*, **367**, 351-354.

Jonas, P.R., Charlson, R.J. and Rodhe, H.(1995). IPCC Report. Chapter 3: Aerosols. Cambridge University Press, Cambridge.

Junge, C.(1979). The importance of mineral dust as an atmospheric constituent. In: Morales, C. (Ed.): Saharan Dust, pp 49-60. Wiley, Chichester.(SCOPE report 14).

Lövblad, G.(1987). Utsläpp til luft av alkali(Emissions of alkaline material to the air). Gothenburg, Institutet för Vatten- och Luftvårdsforskning.

Loye-Pilot, M.D., Martin, J.M: and Morelli, J.(1986). Influence of Saharan dust on the rain acidity and atmospheric input to the Mediterranean. *Nature*, **321**, 427-428.

Maenhaut, W., Ducastel, G., Hillamo, R., Pakkanen, T., and Pacyna, J.M.(1993). Atmospheric aerosol studies in southern Norway using size-fractionating sampling devices and nuclear analytical techniques. *J. Radioanal. Nucl. Chem.*, **167**, 271-281.

Manderscheid, B., Matzner, E., Meiwes, K-J., and Y.Xu.(1995). Long-term development of element budgets in a Norway spruce(*Picea abies*(L.) Karst.) forest of the German Solling area. *Water, Air, Soil Pollut.*, **79**, 3-18.

MEPNR, (1994) State of the Environment of the Russian Federation: 1993. National Report. Ministry for Environment Protection and Natural Resources of the Russian Federation, Moscow, Russia.

MRI(1971). Particulate Pollutant System Study. Vol. III - Handbook of Emission Properties. Midwest Research Institute, MRI Project No. 3326-C. Durham, N:C:

Nowicki, M.(1993). State of the Environment in Poland. Chapter 8: Air. State Inspectorate for Environmental Protection, Warzawa, Poland. 1993

Pacyna, J.M.(1980). Coal-fired power plants as a source of environmental contamination by trace metals and radionucleides. Habilitation Thesis, Technical University of Wroclaw, Poland.

Rahn, K. A.(1976). The chemical composition of the atmospheric aerosol. Tech. rep., Graduate School of Oceanography, University of Rhode Island, Kingston, R.I.

Rodá, F., Bellot, J., Avila, A., Escarré, A. Piñol, J. and Terradas, J.(1991). Saharan dust and the atmospheric input of elements and alkalinity to Mediterranean ecosystems. *Water, Air, Soil Pollut.*, **66**, 277-288.

Schaug, J., Pedersen, U., Skjelmoen, J.E., and Arnesen, K.(1994). EMEP Data Report 1992. Norwegian Institute for Air Research, Kjeller

WET DEPOSITION OF NON SEA-SALT SULPHATE IN THE UNITED KINGDOM: THE INFLUENCE OF NATURAL SOURCES.

N.C. MCARDLE[1], G.W. CAMPBELL[2] AND J.R. STEDMAN[2]

[1] *School of Environmental Sciences, University of East Anglia, Norwich, UK*
[2] *AEA Technology, National Environmental Technology Centre, Culham, Abingdon, UK*

Abstract Because of its position to the west of Europe, much of the wet sulphur deposition in the west of the UK is background in the sense that it is not attributable to pollutants emitted within Europe less than four days previously. There are both natural and anthropogenic sources of this sulphur. An important natural source, especially during the summer, is dimethylsulphide (DMS) produced by marine phytoplankton. To identify the contribution of marine biogenic sulphur we have measured stable sulphur isotope ratios in precipitation. We show that biogenic sulphur is significant in summer but contributes little in winter and that around 5-10% of the annual background wet sulphur deposition is due to biogenic sources. During July and December 1993, airflow across the UK was predominantly westerly. The measured biogenic component of precipitation sulphate accounted for around 30 % of background sulphate in July but was negligible in December. Investigation of five day back-trajectories for the period indicated little opportunity for re-circulation of European emissions, suggesting that other (non-DMS) natural sources and non-European anthropogenic emissions were responsible for most of the background sulphur.

Key words: United Kingdom, Europe, background, biogenic sulphur, sulphur isotope ratio

1. Introduction

Using a simple model of sulphur wet deposition (Fisher, 1978) we have estimated that ~70 % of the sulphur deposited in the UK is from UK sources and ~16 % is directly attributable to European sources. The remaining 13 % is due to background sulphur and is a combination of natural and long-range transported anthropogenic sulphur. However, in many areas that are particularly sensitive to acidification in the upland areas of North and West UK the average contribution of background sulphur is estimated to be ~30 %.

Marine biogenic and volcanic emissions are the dominant natural sources of non sea-salt sulphur to the atmosphere, and in the Northern hemisphere these two sources are of aproximately equal magnitude (Bates *et al,* 1992). Although a variety of volatile sulphur gases are produced biologically and/or photochemically in the surface waters of the ocean, dimethylsulphide (DMS) is considered to be the dominant contributor to atmospheric non sea-salt sulphate (Andreae, 1985, Bates *et al,* 1987, Berresheim, 1987). The oxidation of DMS released into the marine atmosphere results in a range of compounds with sulphate and methane sulphonic acid (MSA) the major stable end products (Turnipseed and Ravishankara 1993 and refs. therein).

In this paper, we attempt to quantify the contribution of marine biogenic sources to background NSS sulphate in the UK, by bringing together the results of a detailed study carried out by the University of East Anglia (UEA) of the isotopic composition of sulphate in rain and particulate and results from routine monitoring work carried out by AEA Technology (AEA) as part of the UK contribution to EMEP.

2. Methods

Daily measurements of rainfall composition, sulphur dioxide and particulate sulphate have been made at Lough Navar (54 26N 7 54W), Strathvaich Dam (57 44N 4 46W) and Yarner Wood (50 36N 3 43W) for the seven year period from 1987 to 1993 (see Figure 3 for locations) as part of the UK contribution to EMEP. Rainfall was collected using wet-only collectors with subsequent analysis by ion chromatography. Sulphur dioxide concentrations were monitored using hydrogen peroxide bubblers with ion chromatographic analysis of sulphate. Particulate sulphate was collected on Whatman 40 filters and sulphate was determined by X-ray fluorescence.

For the isotope study, open samplers were used to collect weekly precipitation samples at three sites in Wales (see Figure 3): Beddgelert ($53°3'N$, $4°9'W$); Plynlimon ($52°26'N$, $3°44'W$) and Llyn Brianne ($52°8'N$, $3°43'W$), with high volume aerosol samples also collected at Plynlimon. Sulphate and MSA were determined by ion chromatography and the isotopic ratios by stable isotope mass spectrometry. Sodium was used as a sea-salt tracer in both studies.

3. Results

3.1 ROUTINE MONITORING MEASUREMENTS

Lough Navar and Strathvaich Dam are both sites for which there are few sources to the west whereas Yarner Wood, in south-western England, does have some hundreds of thousands of people to the west and, although there is little large industry, there are significant emissions of SO_2 of around 10ktonnes per year.

TABLE 1
Statistical summary of S concentrations at UK EMEP sites using all data and a subset selected on the basis of EMEP back-trajectories which do not pass east of the measurement site.

	sulphur dioxide (μgSm^{-3})				particulate sulphate (μgSm^{-3})				NSS sulphate in rain (μM)			
	25%	75%	mean	N	25%	75%	mean	N	25%	75%	mean	N
Lough Navar												
all	.3	.9	.9	2503	.4	1.4	1.0	2460	2.7	7.9	8.2	1446
west	.3	.7	.6	1467	.4	.8	.7	1458	2.3	6.2	5.2	820
Strathviach Dam												
all	.2	.7	.6	2167	.4	.9	.8	2210	2.0	7.2	6.8	1124
west	.2	.4	.4	897	.3	.6	.5	928	1.7	4.5	4.1	520
Yarner Wood												
all	.4	2.3	2.1	2268	.5	1.7	1.3	2237	3.7	13.9	12.9	910
west	.2	.6	.6	539	.4	.9	.7	541	2.9	7.9	6.7	365

To assess mean background levels at these sites, we have picked out days on which EMEP 96-hour back trajectories arriving at the site did not stray to the east of an EMEP grid line passing through the site. A summary of the data for these days and all days is given in Table 1. The mean NSS sulphate concentrations in westerlies at Lough Navar and Strathvaich are 5 and 4μM respectively. The 7μM observed at Yarner Wood demonstrates the relatively small contribution of local sources relative to background.

Figure 1 shows the contribution that these westerly days make to the total deposition of NSS sulphate at Lough Navar and Strathvaich Dam. The annual deposition in westerlies is around $0.1gSm^{-2}$ at Strathvaich and $0.13gSm^{-2}$ at Lough Navar. The total deposition from background sources will be larger than these values due to the unquantifiable contribution from background sources to sulphur deposition on the non Westerly days.

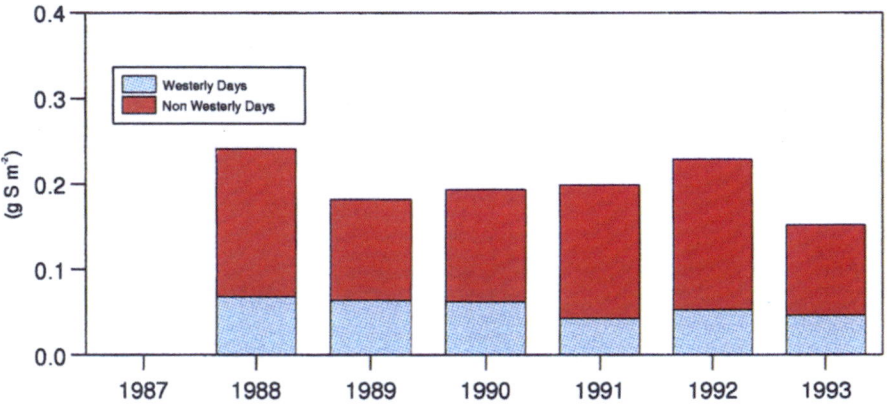

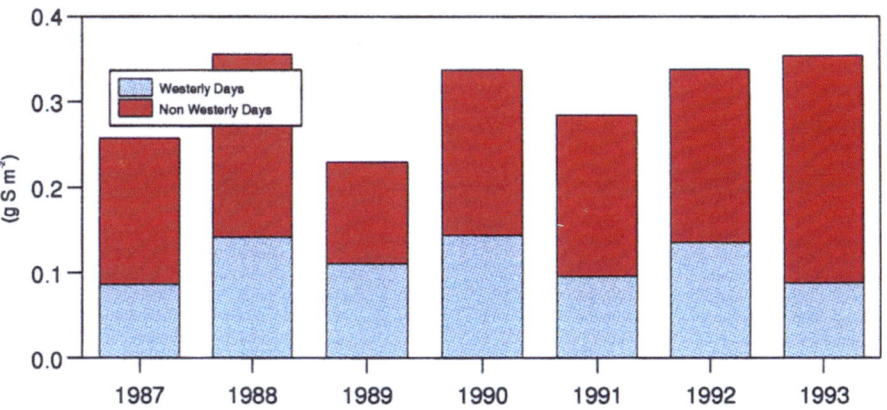

Figure 1. Annual deposition of NSS sulphate at Strathvaich Dam and Lough Navar measured by wet-only collector, showing the contribution from precipitation on days when EMEP back-trajectories did not pass east of the monitoring site.

3.2 STABLE ISOTOPE STUDY

The ratio of the stable isotopes $^{34}S/^{32}S$ in atmospheric samples offer a means of discriminating between anthropogenic and biogenic (from DMS oxidation) sulphur because the $\delta^{34}S$ values are different for the two sources, where

$$\delta^{34}S = \left[\frac{(^{34}S/^{32}S)_{sample}}{(^{34}S/^{32}S)_{standard}} - 1 \right] \times 1000$$

The standard is the Canyon Diablo Troilite and the units ‰.

Anthropogenic sulphur tends to have $\delta^{34}S$ values in the range 0-5 ‰ (Newman & Forrest, 1991, McArdle, 1993). The $\delta^{34}S$ of sea-water sulphate is consistent worldwide at +20 ‰ (Sasaki, 1972). Theoretical estimates of the $\delta^{34}S$ of DMS derived sulphur predict values close to that of sea-water (Calhoun & Bates 1989, Wadleigh 1989, McArdle 1993). A $\delta^{34}S$ value of ~22 ‰ was derived for marine biogenic sulphur from a study of aerosol samples collected at Mace Head, Eire and Ny Ålesund, Spitsbergen (McArdle 1993, McArdle & Liss in press). Although the $\delta^{34}S$ values for sea-salt and biogenic sulphate are close, a sea-salt correction can be made. Unfortunately isotopes cannot be used to determine the magnitude of a volcanic component as $\delta^{34}S$ values are likely to coincide with those of anthropogenic sulphur (Lein, 1991). Once end member values have been defined for a location a mass balance equation can be used to calculate the contribution of each source.

As MSA is derived solely from the oxidation of DMS it can be used as a tracer for DMS production. The aerosol concentrations found in Wales are shown in Figure 2, and the distribution and concentrations are very similar to those at Mace Head (McArdle 1993, Savoie et al. in press).

Concentrations of sulphate from DMS oxidation were calculated using the biogenic end member value (McArdle & Liss in press) of +22 ‰. An anthropogenic end member value of ~3 ‰ was derived from winter samples, assuming that, as DMS production at these latitudes has been found to be strongly seasonal (Leck et al. 1990, Turner et al. 1988 and Figure 2), most of these contain insignificant amounts of biogenic sulphur.

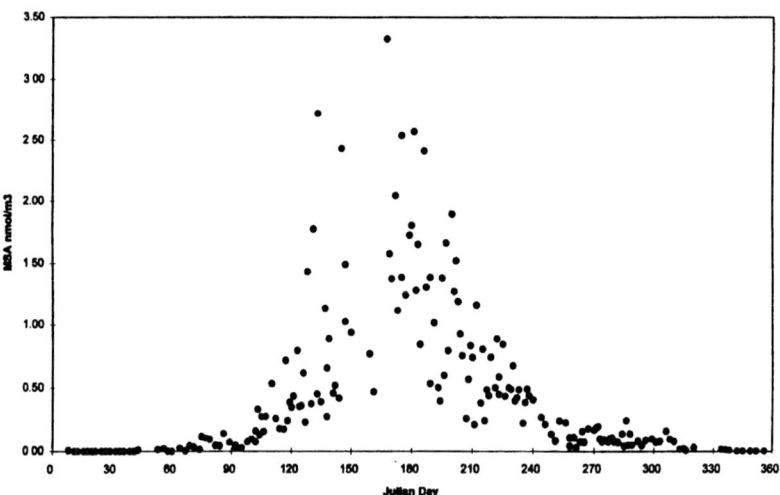

Figure 2. MSA concentrations in aerosol samples collected at Plynlimon from March 1993 to March 1994. The maximum value of ~3.5 nmol/m³ is close to the maximum of 4 nmol/m³ found at Mace Head for the period March 1991 to August 1992 (McArdle & Liss in press).

In the precipitation samples for the year February 1993-94 ~5% of the NSS sulphate at Beddgelert was biogenic with <5 % at Plynlimon and Llyn Brianne. The annual mean concentrations of NSS sulphate at these sites are 11-14µM (Campbell *et al*, 1994), i.e. around three times the background level of 4-5µM (see Section 3 earlier). Assuming one third of the wet deposition at these Welsh sites is due to background, around 5-15% of this can be attributed to biogenic sources.

3.3. CASE STUDIES: JULY AND DECEMBER 1993

These two months were selected for closer investigation as air flow across the UK was predominantly westerly throughout. Inspection of five-day isobaric back-trajectories at pressures between 950mb and 700mb for the two periods indicated that sites to the west of the UK would not have been influenced by European sources on most days of both months. Trajectories at one or more pressure levels passed over Europe on less than five days each month.

Figure 3 summarises wet deposition data from the UEA and AEA measurements. In July, the mean concentration of NSS sulphate varied from 4 µM at Strathvaich to 12µM at Beddgelert. Concentrations were larger at the sites in Wales and at Yarner Wood probably due to UK and Eire sources. The biogenic sulphate concentration was variable between the Welsh sites at ~1 to 2µM with a mean of 1.4µM. If it is assumed that the biogenic sulphate concentrations were the same at Strathvaich and Lough Navar as at the Welsh sites, biogenic sulphate made up around 30 % of the background NSS sulphate.

In December, the mean concentration of NSS sulphate varied from 2.5µM at Lough Navar to 5µM at Plynlimon and Llyn Brianne. Concentrations of NSS sulphate were smaller than in July at all sites, especially those with significant UK/Eire contributions, presumably because of the smaller oxidation rates, larger wind speeds and larger rainfall volumes. The mean NSS sulphate concentrations for the whole UK monitoring network were unusually small in December 1993 (Campbell *et al*, 1994). Biogenic sulphate concentrations were negligible, as might be expected in winter.

However, the mean concentrations of sulphate in rain at the Welsh sites were around twice those at Lough Navar and Strathvaich, indicating that half of the NSS sulphate in the rain was derived from UK and Eire sources.

It is clear that while biogenic sources account for a significant proportion of NSS sulphate in summer, most of the annual background deposition is derived from other sources. The only potential source of volcanic emissions in the North Atlantic is Iceland. Although Spiro *et al* (1992) show significant emissions around Iceland in their global sulphur emissions inventory for 1980, Hekla was erupting at this time. During non-eruptive periods most of the SO_2 released by magma bodies is trapped by overlying hydrothermal systems (e.g. Ágústsdóttir & Brantley 1994). It is therefore unlikely that Iceland forms a consistent source of volcanic sulphur to the atmosphere. The remainder of the background sulphur is probably transported across the Atlantic. It is interesting to note that Tarrason *et al.* (1995), in a modelling study of trans-Atlantic transport, found that the contribution of North American sources to sulphur deposition in Ireland was largest in summer.

4. Summary and Conclusions

This paper addresses the question of the contribution of background sulphate to acidic deposition in the United Kingdom. From the measurement point of view, this can be estimated by back-trajectory analysis at west coast sites and a mean NSS sulphate

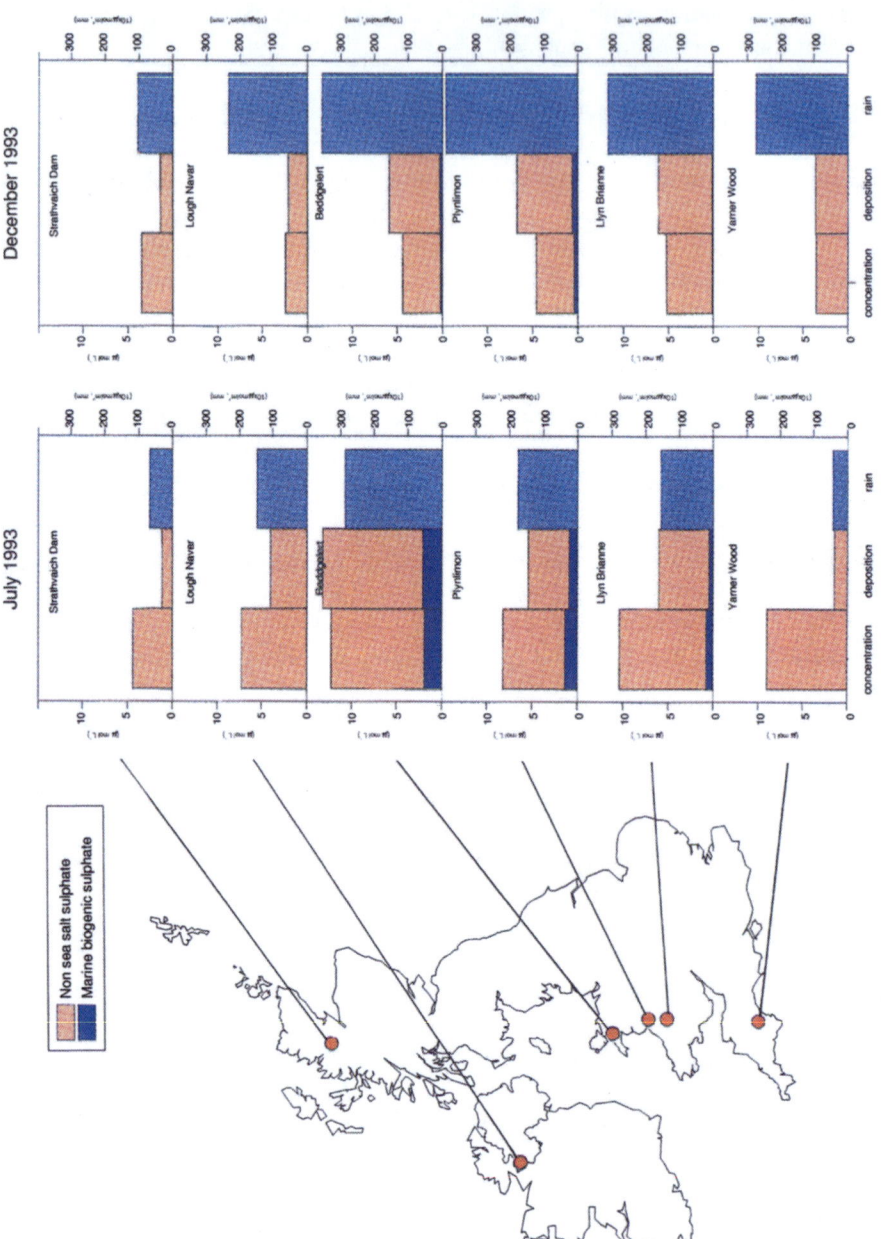

Figure 3. The mean concentrations of NSS sulphate in rainfall (mM) during July and December 1993. The biogenic component of NSS sulphate is shown at the three sites where it was measured.

concentration of 4-5 μM can be derived. If it is assumed that all rainfall includes a background sulphur component then this mean concentration would contribute $0.16 gSm^{-2}$ for each 1000 mm of rain. However, if days for which 4 day back-trajectories do not pass over European sources at all are selected, around $0.1 gSm^{-2}$ per 1000 mm rain in western Britain can be unambiguously assigned to background sources.

We have used stable sulphur isotope ratios of NSS sulphate in rain to quantify the contribution of marine biogenic sources and have shown they account for around 30% of background wet S deposition in summer months but only around 5-15% for the year as a whole. While we have not been able to quantify the other contributions to background S, we have used two months in 1993 during which there was little indication of recirculation of air masses from Europe. It seems likely that trans-Atlantic transport contributes significantly to background deposition in the west of the UK.

In conclusion, we have shown that the current contribution of biogenic sources to background deposition in the UK is quite small. There may be a natural component from volcanic sources but we cannot quantify this. Background deposition in the UK now has a large anthropogenic component which will decrease as emissions in Europe and North America decrease after implementation of the Oslo protocol. By 2010 UK emissions of sulphur dioxide will be less than 1 Mtonnes. However, this, together with natural background deposition would mean that present critical loads for sulphur deposition in the most sensitive areas are likely to be exceeded in the UK.

Acknowledgments

The UK precipitation monitoring network forms part of the research programme of the UK Department of the Environment, contract PECD 7/12/116. The work at UEA was funded by the UK Department of Trade and Industry and the Countryside Council for Wales. Thanks are also due to the Norwegian Meteorological Institute and to Dr Steve Dorling of UEA who provided us with back trajectory data. The views espressed in this paper are those of the authors, not those of the funding organisations.

References

Ágústsdóttir, A.M. & Brantley, S.L. (1994) *Journal of Geophys. Res.* B99, 9505-9522.
Andreae, M.O. (1985) The emission of sulfur to the remote atmosphere. In *The Biogeochemical Cycling of Sulfur and Nitrogen in the Remote Atmosphere*; Eds. Galloway, J.; Charlson, R.; Andreae, M.; Rohde, M.; Reidel: Dordrecht, 1985; 5-25.
Bates, T.S.; Charlson, R.J.; Gammon, R.H. (1987) *Nature* 329, 319-21.
Bates, T.S., Lamb, B.K., Guenther, A., Dignon, J., Stoiber, R.E. (1992) *J. Atmos. Chem.* 14 315-337.
Berresheim, H. (1987) *J.Geophys.Res.* D92, 13245-62.
Calhoun, J.A.; Bates, T.S. (1989) Sulfur isotope ratios tracers of non-sea salt sulfate in the remote atmosphere. In *Biogenic sulfur in the environment*; Saltzman, E.S.; Cooper, W.J., Eds; ACS Symposium Series 393, Chapter 22; American Chemical Society: Washington, DC, 1989.
Campbell, G.W., Stedman, J.R., Downing, C.E.H, Vincent, K, Hasler, S & Davies, M. (1994) Acid Deposition in the United Kingdom 1993. AEA report AEA/CS/16419029/001. AEA Technology, Abingdon, UK.
Fisher (1978) *Atmospheric Environment*, 12, 489-501.
Leck, C., Larsson, U., Bagander, L.E., Johansson, S., Hajdu, S. (1990) *J. Geophys. Res.* 95 C3, 3353-3363.
Lein, A.Y. (1991) Flux of volcanogenic sulphur to the atmosphere and isotopic composition of total sulphur. In *Stable Isotopes: Natural and Anthropogenic Sulphur in the Environment*. SCOPE 43. Eds. Krouse, H.R. & Grinenko, V.A. Wiley.

McArdle, N.C. (1993) The use of stable sulphur isotopes to distinguish between natural and anthropogenic sulphur in the atmosphere. *Ph.D. University of East Anglia.*

McArdle, N.C. & Liss, P.S. (1995) Isotopes and atmospheric sulphur. *Atmos. Environ.* in press.

Newman, L. & Forrest, J. (1991) Sulphur isotope measurements relevant to power plant emissions in the northeastern United States. In *Stable Isotopes: Natural and Anthropogenic Sulphur in the Environment.* SCOPE 43. Eds Krouse, H.R. & Grinenko, V.A. Wiley.

Sasaki, A. (1972) Variations in sulphur isotopic composition of oceanic sulphate. *24th I.G.C., 1972, Section 10, pp342-45.*

Savoie, D.L., Arimoto, R., Prospero, J.M., Duce, R.A., Graustein, W.C., Turekian, K.K., Galloway, J.N., Keene, W.C. (1995) *J. Geopys. Res.* in press.

Spiro, P.A., Jacob, D.J. & Logan, J.A. (1992) *J. Geophy. Res.*, 97(D5), 6023-6036.

Tarrason, L, Turner, S & Fløisand, I (1995) An estimation of seasonal DMS fluxes over the north Atlantic Ocean and their contribution to European pollution levels. Submitted to *J. Geophys. Res.*

Turner, S.M.; Malin, G.; Liss, P.S.; Holligan, P.M. (1988) *Limnol.Oceanogr.* 33, 364-75.

Turnipseed, A.A. & Ravishankara, A.R (1993) The atmospheric oxidation of dimethyl sulfide: elemantary steps in a complex mechanism. In *Dimethylsulphide: Oceans, Atmospher, and Climate.* 185-195. Eds Restelli, G. & Angeletti, G.ECSC, EEC, EAEC, Brussels and Luxembourg. Netherlands.

Wadleigh, M.A. (1989) Geochemical characterization of coastal precipitation, natural versus anthropogenic sources. *Ph.D. McMaster University.*

OZONE MEASUREMENTS IN EUROPE

[1]G. DOLLARD, [2]D. FOWLER, [2]R. I. SMITH, [3]A.-G. HJELLBREKKE, [4]K. UHSE & [4]M. WALLASCH

[1]AEA Technology, National Environmental Technology Centre, UK [2]Institute of Terrestrial Ecology, Edinburgh, UK
[3]Norwegian Institute for Air Research, Norway [4]Umweltnundesamt Offenbach, Pilostation Frankfurt, Germany

Abstract. Ozone measurements have been a part of EMEP since its third phase in 1984-1986 and since 1988 data have been collected systematically. By 1992 data for 76 sites were being collected by the Chemical Co-ordinating Centre in NILU. The mean ozone concentration increases from 20-25ppb in the western and northern fringes to 30-35 in central areas of Europe. There is also evidence from the last decade of an upward trend of up to 0.5ppb y^{-1} at rural sites in the UK. The data have been analysed to estimate the spatial patterns in AOT 40 for ozone effects on crops and forests. The data show that the critical level for cereal crops of 5300 ppb.h above a threshold of 40 ppb is exceeded over almost all of continental Europe south of 65°N and over most of S.Britain. A similar exercise for the AOT 40 for the forest again shows exceedances of the critical load of 10^4 ppb.h across all the mapped area of Continental Europe south of 65°N including S.Britian. As land use for forestry and ozone dose both increase with altitude, and these effects have not so far been incorporated in the AOT 40 assessment for forests, the degree of exceedence for forests may have been significantly under-estimated.

1. Introduction

Recent reviews of surface ozone concentrations have drawn attention to the changes in mean annual ozone concentrations since the beginning of the century (Cartalis and Varotsos, 1994; Voltlz and Kley, 1988) and to seasonal regional variability in concentration (Oltmans and Levy, 1994; Beck and Grennfelt 1994). The measurements show the mean ozone concentrations have increased fro about 10 ppb in the early years of this century to between 20 and 30 ppb during recent decades. The current mean ozone concentrations are close to the lower threshold for chronic injury to sensitive vegetation. In most biological responses to environmental stress, including ozone, the relationship between biological response and O_3 concentration is sigmoidal. This has been simplified as a step function to define an arbitrary threshold above which ozone concentrations are considered damaging to vegetation (Ashmore, 1994). In analysing a series of detailed studies a crop yield - ozone dose relationships, Fuhrer (1994) demonstrated that the relationship between relative yield of several crops and accumulated exposure over a threshold of 40 ppb (AOT 40) was approximately linear. Using the wheat yield AOT 40 relationship from well replicated field studies, a critical level of 5300 ppb.h was established as the value equivalent to a 10% reductions in yield relative to crop yield in clean air. The data used to define an equivalent AOT 40 for forest relies more on physiological responses than growth and yield data, and indicates critical levels of 10^4 ppb.h above 40 ppb (Skarby, 1994). In this paper we quantify the geographical variability in the AOT 40 in Europe for cereal crops and forest using the EMEP monitoring network

2. Ozone monitoring

Measurements of ozone throughout Europe are provided largely by combining the National network data for many of the countries in Western Europe. These data are collected and analysed as a part of the EMEP activities of the Chemical Co-ordinating Centre at NILU in Norway. The data base provides measurements at 96 monitoring stations stretching from 79°N in Norway to 45°S in Italy and from 9°W in Ireland to 29°E in Finland. The area covered is much smaller than the total area of Europe and the Mediterranean area in particular is not covered by this data set. However, a broad geographical range in sites is included and these may be considered to represent a N-S transect through Europe from the Artic to Northern Italy. In general the sites are rural and not influenced strongly by local sources of NO_x and most are suitable for characterizing the regional O_3 concentration field.

Much of the recent literature has focused on changes in O_3 concentration, and while evidence of a long-term change is clear, current trends are more complex with some peripheral regions of Europe reporting clear upward trends (as illustrated by data for rural sites in the UK, 1978 to 1994, Figure 1) while in southern Germany and Alpine countries the major increases were during the 1980s (WMO, 1995).

2.1 MEAN CONCENTRATIONS

The concentrations vary between the remote northern sites in Finland and Norway at which annual mean concentrations are generally in the range of 20 to 25 ppb (2 µg m^{-3} ≡ 1 ppb) to the Austrian and Italian sites with values between 30 and 40 ppb (Figure 2). There are several years of data within the network at which the annual mean exceeds the threshold above which some sensitive plants may be damaged (40 ppb). The annual mean conceals great temporal variability such that all sites with mean annual O_3 concentrations of 30ppb or more, experience summer monthly mean values in excess of 40 ppb and almost all sites in the network experience

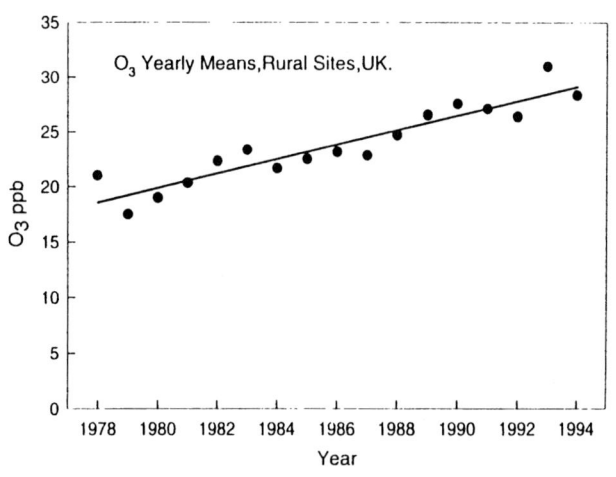

Fig. 1. Trend in annual mean O_3 concentration averaged across all rural stations in the UK 1978-1994

significant periods of exposure to concentrations in excess of 40 ppb. There is also considerable variability within the broad spatial pattern of increasing concentrations from the northern and western boundaries of Europe to the more continental and Mediterranean area. Some of the variability results from local contamination at sites in industrial regions but a larger effect is that of altitude above the mean terrain which results in larger mean windspeeds and a shorter duration of period when sites become decoupled from the majority of the planetary boundary layer by nocturnal stability and stratification of the air close to the ground (Garland and Derwent, 1979). The mean concentrations are not only of interest in the trend and analysis, as some sites have mean concentrations large enough to cause chronic injury to sensitive vegetation (Figure 3). The mean concentrations are not however, of great relevance for an assessment of effects on vegetation. For this assessment it is necessary to quantify the exposure to O_3 above the threshold for damage. For this approach, various indices have been developed to quantify exposure, generally as a product of time and O_3 concentration above a threshold (Lefohn et al., 1988). There is also a need to develop tools which permit the interpolation of spatial O_3 exposure across the landscape from the monitoring stations. The effect of altitude on the diurnal cycle in ozone concentration provides a method for high resolution mapping of

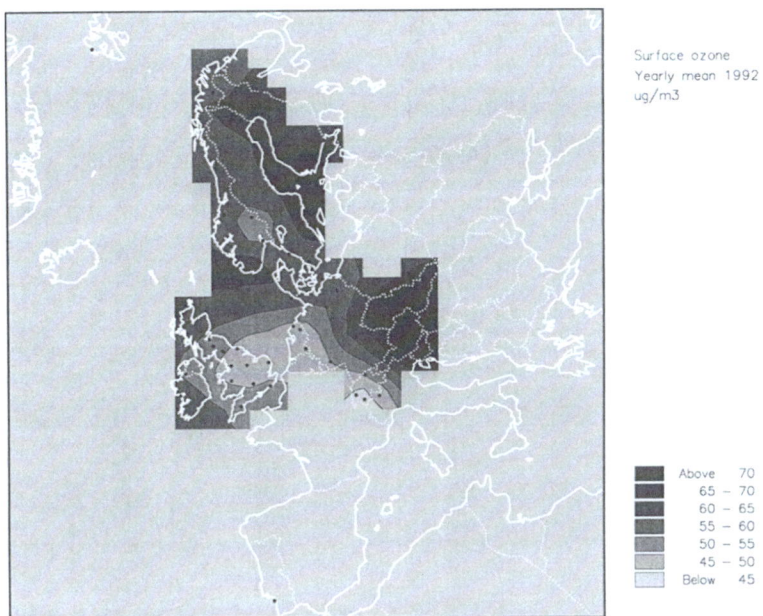

Fig. 2. Mean annual surface ozone concentrations from EMEP measurement stations in Europe 1992.

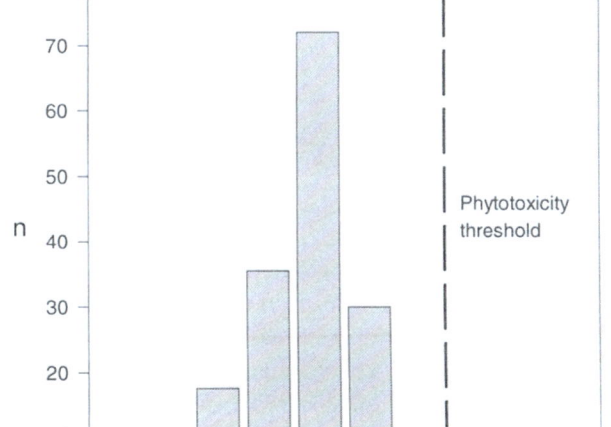

Fig. 3. Frequency distribution of annual mean O_3 concentrations at 96 EMEP stations 1988-1992.

concentration or dose, AOT 40 for example. The period of the day during which ozone concentrations on the surface are representative of those averaged through the planetary boundary layer varies with season and meteorology but during the summer it is generally between about 1200 and 1800 solar time. This, as shown by Fowler, et al. (1995) may be used to provide a regional map of ozone concentration for the period of the day with largest values and is similar to the approach of Beck and Grennfeld (1994) in mapping the diurnal maximum concentration. The degree to which the different parts of the countryside become decoupled is strongly influenced by windspeed (and hence altitude) so that the ratio of mean concentration for the period 1200-1800 to that for the whole day is a function of altitude and serves as a basis for high resolution mapping of mean values or AOT 40 (Fowler et al., 1995).

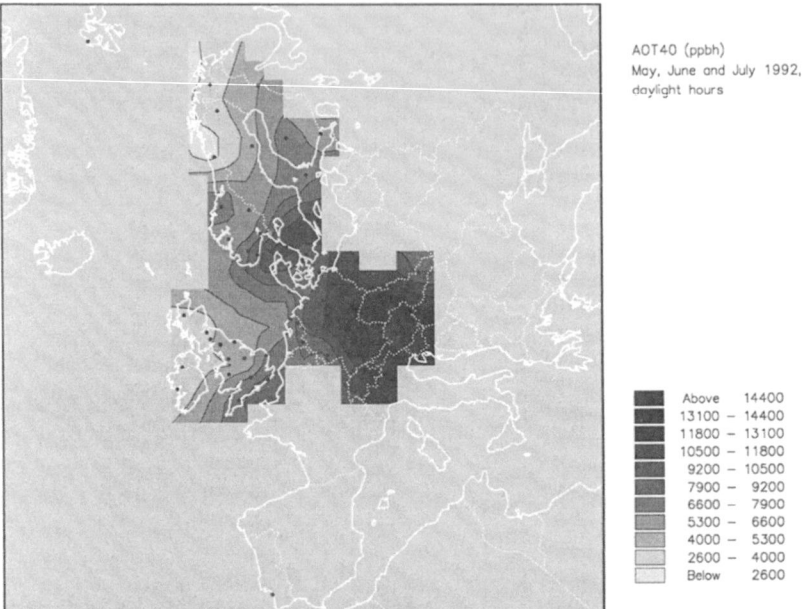

Fig. 4. An interpolated map of AOT 40 (ppb.h) for wheat, May-June and July 1992 daylight hours. (CL5300 ppb.h)

2.2 AOT40 MAPS FOR EUROPE

The goal of extending the high resolution mapping of ozone from the work done in the UK, Switzerland and Austria to Europe has not been attempted, largely due to a lack of suitable data. However, the interpretation of AOT 40's for wheat and forest directly from the monitoring data provide a revealing illustration of the geographical extent of the region in which the critical levels of ozone for crop yield and forest effects are exceeded.

Wheat
Figure 4 shows the interpolated map of AOT 40 for wheat, based on the three months May, June and July daylight hours. The figure shows most of the continental Europe south of 65° and a large fraction of southern Britain in which the critical level of O_3 for a 10% yield loss is exceeded. In fact there are parts of southern Germany, Austria and Italy where the exceedances would be expected to lead to 20-30% yield reduction for spring wheat. Overall for Europe the precise magnitude of yield loss remains uncertain as a result of uncertainties in both the inter-variety and crop responses relative to those used in the experimental work and the fine spatial resolution of O_3 dose. However, it is clear that the O_3 represents the major phytotoxic pollutant, and may be causing yield losses of the order of 10 to 20% across Europe. Furthermore, the data available for countries on the northern boundary of the Mediterranean show a much larger frequency and duration of potentially phytotoxic ozone concentrations, so that for the countries of southern Europe, much larger effects may be expected.

Forest
The AOT 40 interpolated map for forest (Figure 5) shows that the critical level of 10^4 ppb.h is exceeded over most of the mapped area, with the exception of the coastal fringe of north west Europe including parts of north west Britain and northern Scandinavia. The magnitude of the exceedence is large with many sites in the forest region of southern Germany and Alpine countries exceeding the critical level by a factor of 2 or more. These simple interpolations

contain little of the altitude enhancement in ozone dose which, for forests is an important feature.

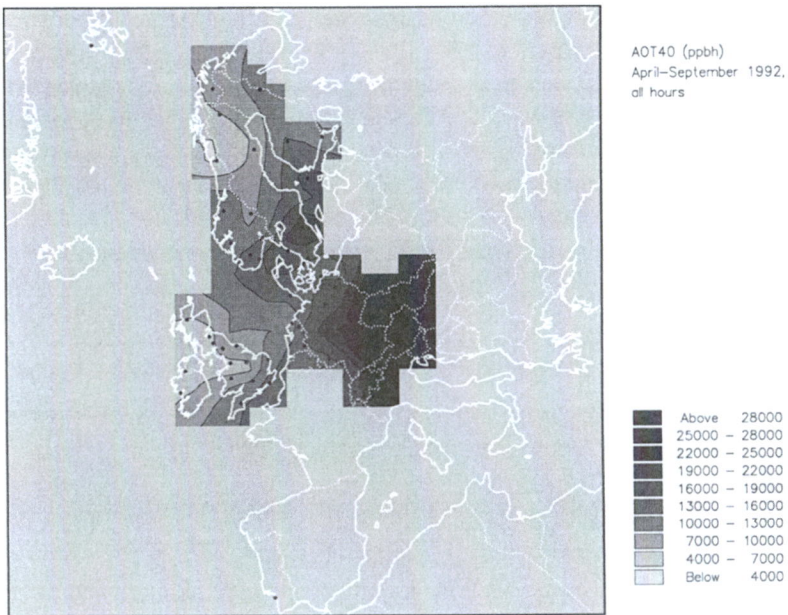

Fig. 5. An interpolated map of AOT 40 (ppb.h) for forest April-September 1992 all hours. (CL 10⁴ pp.h).

In many European countries forests become a larger fraction of land use with increasing altitude and the EMEP O_3 data set show that ozone dose increased with altitude. The data in Figure 6 show the relationships between the ratio 24 hour AOT40/ 1200-1800 AOT40 and altitude. The slope of the relationship is significant and at $9.6 \times 10^{-4} \times alt(m)$ is of the same order as that obtained for UK sites to enable high resolution mapping of O_3 exposure. The effect of incorporating the effects of altitude with a high resolution land use map in Europe would increase the magnitude of exceedence generally, but especially in Alpine countries where large areas show very large exceedences already.

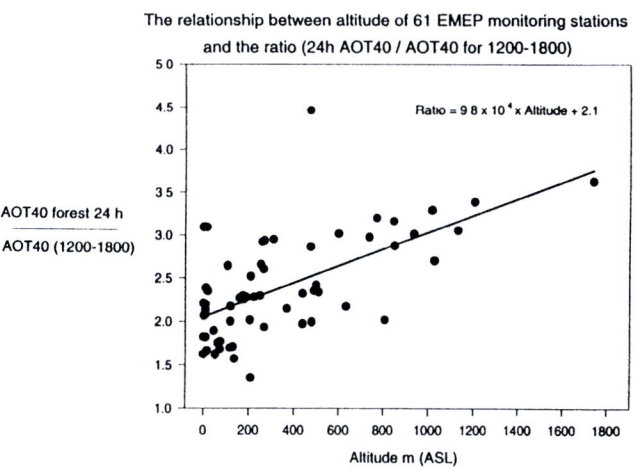

Fig. 6. The relationship between altitude of 61 EMEP monitoring stations and the ratio (24h AOT40/AOT40 for 1200-1800)

3. Conclusions

The application of AOT 40 criteria to regional O_3 concentration measurements in Europe allow estimates to be made of the area of exceedence for agricultural crops and forest. The analysis shows most of the mapped area to exceed critical levels for wheat and forest. In general the exceedence for arable crops lies in the range showing 10 to 20% reductions in yield. For forests, again the majority of the mapped area of Europe exceed the critical level, and in the Alpine countries the exceedance is by a factor of between 2 and 3. The dose and hence exceedance increase with altitude and these effects have yet to be incorporated in the European scale mapping, the current exceedences should be treated as underestimates.

The EMEP data set is restricted to the most recent decade of measurements, which is quite short for an analysis of trends. However, taken with recently published data for a much longer time series, analysis indicates increases in actual mean O_3 concentrations in part of NW Europe of the order 0.2 to 0.7 ppb y^{-1}

References

Ashmore, M.R.:1994, "Critical Levels and Agriculture in Europe." In: *Critical Levels for ozone a UN-ECE workshop report.* (J.Fuhrer and B.Achermann Eds.), p.22-4.

Beck, J.P. and Grenfelt, P.: 1994, "Estimate of ozone production and destruction over northwestern Europe." *Atmos. Environ.,* **28**, 129-140.

Cartalis, C. and Varotsos, C.: 1994, "Surface ozone in Athens, Greece, at the beginning and the end of the twentieth century." *Atmos. Environ.,* **28**, 3-8.

Fowler, D., Smith, R.I., Coyle, M., Weston, K.J., Davies, T.D., Ashmore, M.A. and Brown, M.: 1995, "Quantifying the fine scale (1km x 1km) exposure and effects of ozone. Part 1. Methodology and application of effects on forests". *Water, Air, Soil Pollut.* (In press).

Fuhrer, J.:1994, "The critical level for ozone to protect agricultural crops - An assessment of data from European open-top chamber experiments." In: *Critical Levels for ozone a UN-ECE workshop report.* (J.Fuhrer and B.Achermann Eds.), p.42-57.

Garland, J.A. and Derwent, R.G.: 1979, "Destruction at the ground and the diurnal cycle of concentration of ozone and other gases" *Quart J.R. Met Soc. **105**, 169-183.*

Lefohn, A.S., Laurence, J.A. and Kohut, R.J.: 1988, "A comparison of indices that describe the relationship between exposure to ozone and reduction in the yield of agricultural crops." *Atmos Environ.,* **22**, 1229-1240.

Oltmans, S.J. and Levy II, H.: 1994, "Surface ozone measurements from a global network" *Atmos Environ.,* **28**, 9-24.

Skarby, L.: 1994, "Critical levels for ozone to protect forest trees. "In *Critical Levels for ozone a UN-ECE workshop report.* (J. Fuhrer and B. Achermann Eds.), p.74-87.

Voltz, A. and Kley, D.: 1988, "Evaluation of the Montsouris series of ozone measurements made in the nineteenth century. *Nature,* **332**, 240-242.

WMO: 1995, "Scientific Assessment of Ozone Depletion: 1994". World Meteorological Organization Global Ozone Research and Monitoring Project - Report No 37.

PRECIPITATION CHEMISTRY AND ATMOSPHERIC PROCESSES IN THE FORESTED PART OF CROATIA

A. BAJIĆ and V. ĐURIČIĆ

Meteorological and Hydrological Service of Croatia, Grič 3, 10000 Zagreb, Croatia

Abstract. The influence of mesoscale weather patterns on the chemical composition of daily precipitation samples is analysed. The data of pH, sulphur from sulphates and total nitrogen are analysed for two rural sites: Plitvice station in forested part of Central Croatia (1981 to 1990) and Puntijarka suburban station on the mountain near Zagreb, the capital of Croatia (1982-1991).

The two prevailing weather types in precipitation days are selected and the comparison of chemical composition of precipitation is made for each of them. The frequency distributions of pH, sulphur and nitrogen show that concentration of major ions in precipitation apparently depends on the regional scale weather type.

It is shown that the seasonal variation of deposition is related to the seasonal variation in precipitation amount. In both weather types Plitvice receives more pollution than Puntijarka that is closer to urban and industrial pollution sources. Both locations are under the prevailing influence of regional pollution sources.

Key words: synoptic weather type, precipitation chemistry, forested part of Croatia

1. Introduction

There is considerable public concern about possible changes in atmospheric composition due to anthropogenic influence. The chemistry of precipitation at remote sites is of interest for several reasons. Chemical composition of precipitation measured at sampling points far from industrial and urban areas is useful as an indicator for the geographical area and it allows us to examine the extent of anthropogenic contamination on meso and larger scales. Precipitation chemistry may change with time in response to changes in emission, meteorological factors, physical and chemical transformation, etc. In this paper meteorological conditions classified in different weather types are used.

The purpose of the paper is to relate chemical composition of daily precipitation samples to prevailing synoptic weather patterns. The analysis is performed for two rural sampling sites: Plitvice, in a forested part of Central Croatia and Puntijarka near Zagreb. The data of pH, sulphur from sulphates and total nitrogen are analysed for the period of ten years (1981-1990 for Plitvice and 1982-1991 for Puntijarka).

2. Data and area of study

Systematic monitoring of forest condition has been carried out in Croatia since 1987. Results obtained by this monitoring showed that the most susceptible area is in the central mountainous part of Croatia, which is the area of exceptional natural value. In this area Plitvice Lakes National Park with area of 195 km^2, established in 1949 and registered as a UNESCO world heritage site in 1979, is under the stress of high acidic deposition (Figure 1). Measurements of chemical composition of precipitation were established in May 1980, and since then, bulk daily precipitation samples have been collected (λ= 15°37.5', ϕ=44°53', h=595 m).

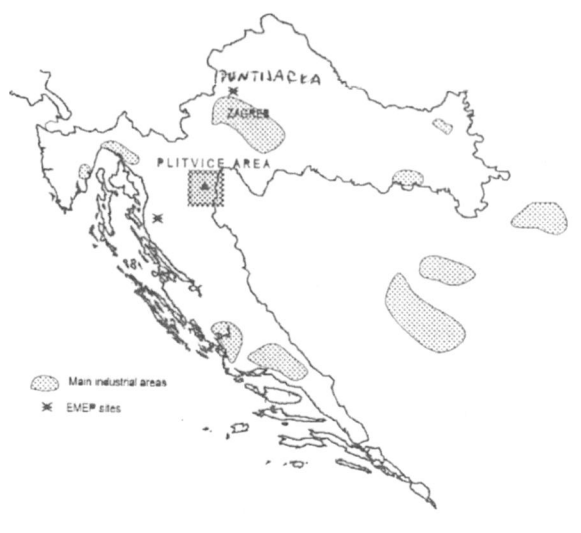

Figure 1. Geographical situation of the area of study and sampling site locations.

Chemical analyses are performed in the central laboratory at Meteorological Service in Zagreb by standard analytical methods.

Puntijarka ($\lambda=15°58'$, $\phi=45°55'$, h=988 m) is the rural station situated on the top of Medvednica mountain, about 10 km air length to the north from the centre of Zagreb, the greatest urban and industrial city in Croatia. A ten-year period of data (1981-1990, with exception of 1983 for Plitvice and 1982-1991 for Puntijarka) is used for study of some characteristic features in sulphur (SO_4^{2-}-S) and nitrogen (NO_3^{-}-N + NH_4^{+}-N) concentrations and pH value of precipitation. In order to relate synoptic scale flow patterns to chemical composition of precipitation the day-to-day weather types have been examined.

3. Precipitation favourable synoptic weather patterns

Since weather is considered to be the ultimate forcing function for many, if not most, environmental processes, we analysed data on precipitation chemical composition obtained in days with different weather patterns. The day-to-day weather at the continental part of Croatia is organised on the basis of mesoscale atmospheric circulation patterns into relatively few types that can provide an environmental baseline inventory specially for the considered region. Daily sampling allowed us to assume a good correspondence between the precipitation sampled and the meteorological situation recorded in the chart for a particular day.

Precipitation at the considered sampling sites occurs mainly with two different synoptic patterns. They receive most precipitation (40-45% of total precipitation amount) in connection with frontal passages and orographic lifting of moist air connected with so called precipitation weather type (PWT) (Lončar and Bajić, 1994). This weather type is characterised by strong winds and advection of warm and moist air dominantly from NW, convergence of horizontal air flow and air lifting along the axis or in the centre of cyclonic activity over the western and north-western Europe. Thus, such situations could be considered as favourable for the regional and long-range transport of pollutants.

Radiation weather types (RWT) are characterised by zero pressure gradient field dominantly accompanied by weak winds of variable direction. In such situations local effects prevail and precipitation is mainly convective. Precipitation amount in days with

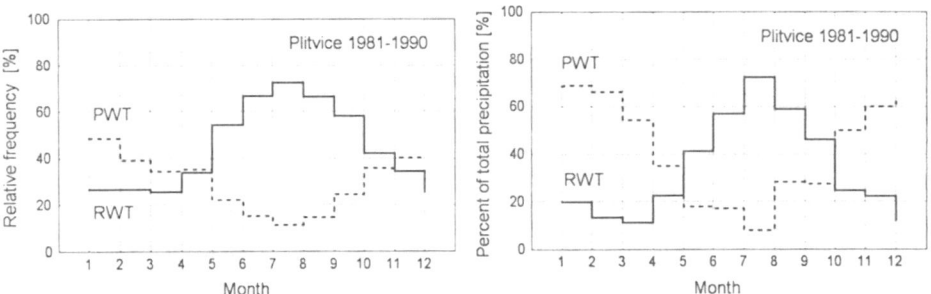

Figure 2. Annual courses of: left - the precipitation (PWT) and radiation (RWT) weather types relative frequencies; right - the percentage of total precipitation amount registered during the PWT and RWT.

radiation weather type is comparable to the amount of precipitation weather type (33-34% of total precipitation).

During winter and spring precipitation preferentially accompanied PWT (Figure 2). More than 50% of total precipitation amount was measured during that weather type. The situation is quite different in summer months when synoptic patterns were classified as radiation weather type with 50-75% of total precipitation amount (in 60% of all days with precipitation). Such seasonal differences in mesoscale weather cause differences in the chemical composition of precipitation.

4. Chemical composition of precipitation

The absolute frequency distribution of pH, sulphate and nitrogen (Figures 3 and 4) for radiation and precipitation weather types shows that chemical composition of precipitation apparently depends on the local weather type. There are evident differences between two stations. It appears that precipitation is more acidic for RWT than for PWT days in the Plitvice area, while at Puntijarka the opposite. The acid precipitation frequency (pH<5.0 according to Charlson and Rhode, 1982) over the period of 10 years is 38% for RWT and 22% for PWT at Plitvice, while 12% in RWT and 11% in PWT at Puntijarka.

Concentrations of sulphur and nitrogen in precipitation are in the high range of values. The fitted log-normal distributions are shifted to the greater concentrations for radiation than for precipitation weather type. Nevertheless, the precipitation volume weighted averages of sulphate concentrations do not differ significantly for those two categories.

The differences between sulphur content in precipitation in two considered weather types are a little bit greater at Puntijarka, but the volume weighted averages of SO_4^{2-}-S and total nitrogen concentrations are smaller at Puntijarka than at Plitvice.

To assess and study the effects on the soil, vegetation and human environment deposition rates of sulphur and nitrogen should to be analysed. In Figure 5 relations between seasonal precipitation amounts and deposition of sulphur and total nitrogen for two considered weather types are presented. Deposition increases with precipitation but not linearly. At Plitvice the differences between radiation and precipitation weather types increase with precipitation amount for sulphur deposition while for nitrogen deposition it is the opposite. Although the deposition reveals a strong relation with precipitation amount, there is a considerable scatter within each subset.

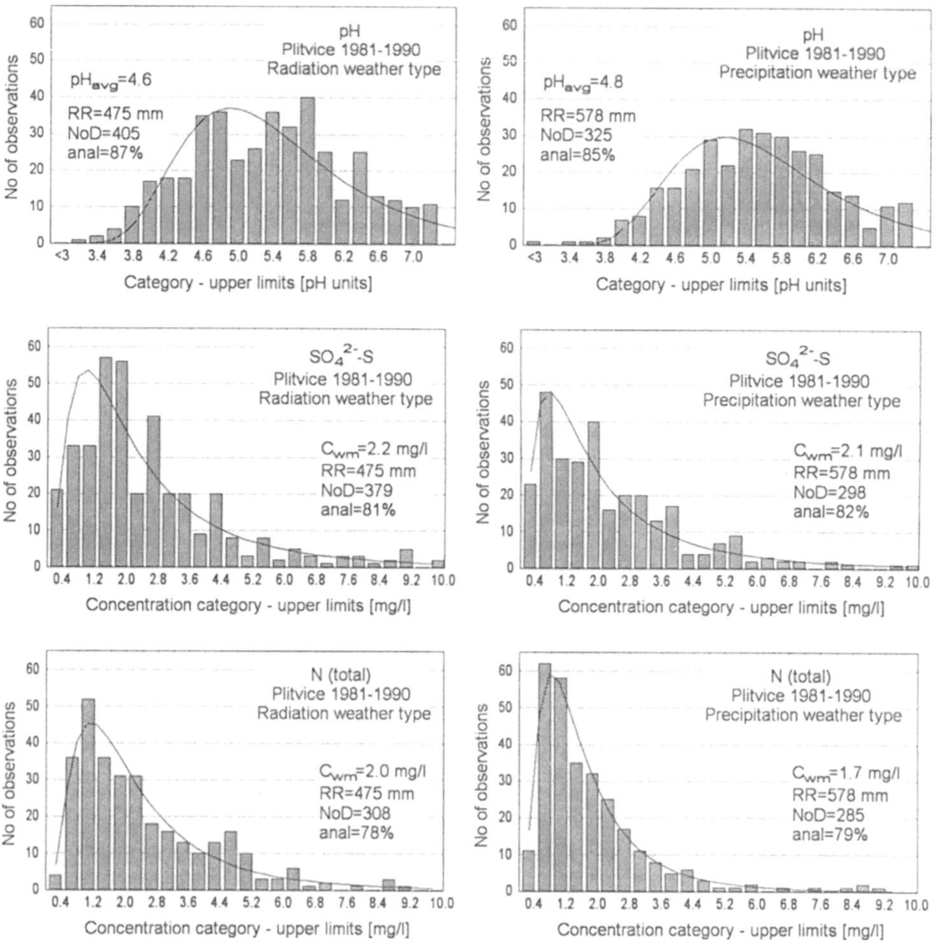

Figure 3. Absolute frequency histograms of pH values, SO_4^{2-}-S and N (total) concentrations with fitted log-normal distribution curve in Plitvice (1981-1990) for radiation and precipitation weather type days. pH_{avg} - the average pH value obtained from precipitation volume weighted H^+ concentration, C_{wm} - the precipitation weighted arithmetic mean concentration, RR - the mean annual precipitation amount, NoD - the number of days with measurements for specific component, anal - the percent of the total precipitation reported analysed for specific component.

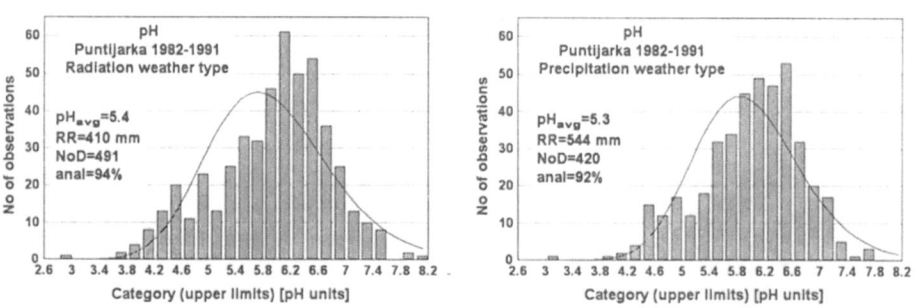

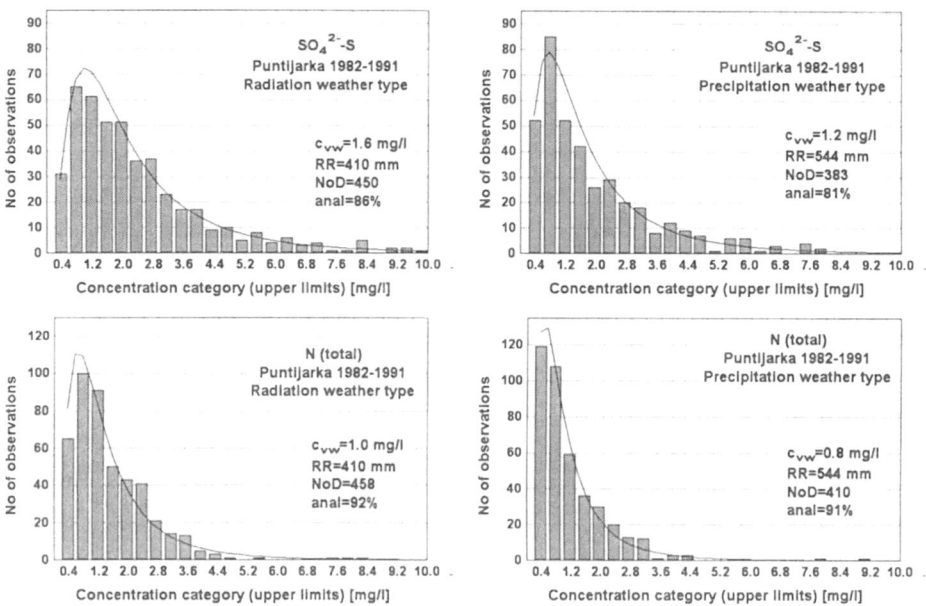

Figure 4. Same as Figure 3 for Puntijarka.

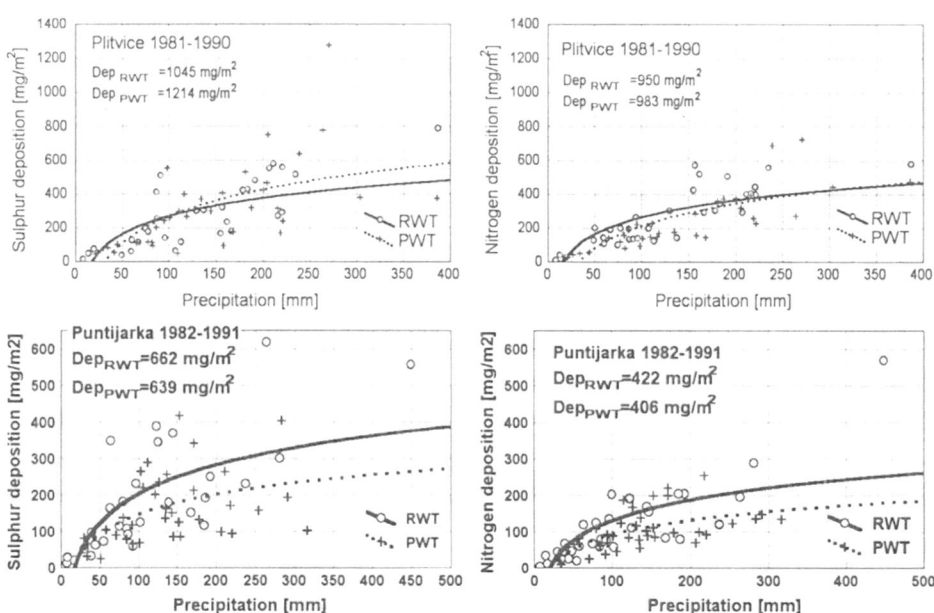

Figure 5. Seasonal sulphur (left) and nitrogen (right) deposition as a function of seasonal precipitation for radiation (RWT) and precipitation (PWT) weather types at Plitvice (1981-1990) and Puntijarka (1982-1991).

Quite different behaviour shows the data from Puntijarka. Both sulphur and nitrogen deposition increase with increasing precipitation amount more in RWT then in PWT. The scatter in seasonal deposition rates is much greater for sulphate than for nitrogen in both weather types. Notice that the deposition of both compounds is at Plitvice almost as twice as at Puntijarka, while the mean total annual precipitation amount is only 20% greater.

5. Concluding remarks

Using the ten year period of data we have analysed the influence of mesoscale weather patterns on the chemical composition of precipitation at two stations in the forested regions: one in central part of the country rather far from local pollution sources, the other in the vicinity of Zagreb, the greatest town in Croatia. The main results may be summarized as follows:
- two precipitation favourable weather types could be separated: *precipitation weather type*, characterised by meteorological conditions favourable for the regional and long-range transport of pollutants and *radiation weather type*, with domination of local effects and convective precipitation;
- prominent relation exists between the chemical composition of precipitation and mesoscale weather patterns;
- the sulphur deposition is more influenced by meteorological conditions than the deposition of nitrogen.

It has been shown that National Park Plitvice is under the stress of acidification even more than rural station close to Zagreb. More than 60% of total precipitation is found to be acidic at Plitvice, while only 31% at Puntijarka. The presented paper indicates that it could be result of regional transport. This should be confirm by further analyses of precipitation chemistry data at other stations. Trajectory analysis and a few individual case studies from regional network stations during well-defined synoptic conditions would undoubtedly be rewarding. Under certain circumstances in radiation weather type the effects of local urban sources become important at Puntijarka. On the contrary, in some situations, specially in winter, mixing height in Zagreb is rather small so urban pollution does not reach Puntijarka (988 m asl) at all (Cvitan, Lončar, 1992).

References

Charlson, R.J. and Rodhe, H: 1982. "Factors controlling the acidity of natural rainwater", Nature, **295**, 683-685.

Cvitan, L. and Lončar, E.:1992. "Natural Potential of Air Ventilation in Zagreb", Proceedings of International Congress Energy and Environment, 28-30 Oct. 1992, Opatija, Croatia, 179-185 (in Croatian).

Đuričić, V. and Vidič, S.:1992. "Acid precipitation in the Northern Adriatic", Airborne pollution of the Mediterranean Sea, UNEP/WMO, MAP Technical Report Series No. **64**, 137-153.

Đuričić, V.: 1992. "Mokro taloženje na području Zagreba", (Acid deposition at Zagreb). International Congress Energy and Environment, Opatija, 187-194.

Lončar, E. and Bajić, A.: 1994. "Tipovi vremena u Hrvatskoj", (Weather types in Croatia). Hrv. Meteor. Časopis **29**, 31-42.

MEASUREMENTS OF PRECIPITATION COMPOSITION AT UK EMEP SITES 1987 - 1992 AND COMPARISON WITH THE HARM MODEL

J.D. WHYATT [1], J.R. STEDMAN [2], S.E. METCALFE [3] and G.W. CAMPBELL [2]

[1] School of Geography and Earth Resources, University of Hull, Hull HU6 7RX, UK,
[2] AEA Technology, National Environmental Technology Centre, Culham Abingdon, UK,
[3] Department of Geography, University of Edinburgh EH8 9XP, UK

Abstract. Precipitation composition has been measured daily at five UK EMEP sites since 1987. Sulphur dioxide and sulphate aerosol concentrations are also measured daily at the sites. Back trajectories and wind sectors calculated by the Norwegian Meteorological Institute have been used to characterise the variation in wet deposition in terms of air mass source. Contributions to wet deposition from various source regions have been estimated for Eskdalemuir. Observations from the EMEP sites have been compared with output from the Hull Acid Rain Model (HARM). HARM is a Lagrangian model using simplified meteorology but straight-line trajectories. Results are compared on a site-by-site and sector-by-sector basis and the model reproduces the general features of pollutant concentration and wet deposition indicated by the measurements. The possible effects of future reductions in emissions of SO_2 and NOx on precipitation concentrations by wind sector are described.

Key words: UK EMEP sites; precipitation concentration; wind sectors; modelling

1. Introduction

Daily measurements of precipitation composition, SO_2 and sulphate particle (SA) concentrations have been made at the five UK EMEP sites since 1987 as part of the UK National Acid Rain Monitoring Network (Campbell, Stedman et al, 1994). Using back trajectories and wind sectors calculated by the Norwegian Meteorological Institute it is possible to derive the contributions of the measured pollutants, as well as calculated wet deposition, from different source regions. Such source attribution plays an important part in modelling studies used to assess the likely impact of future emissions controls. In this paper we compare data from the EMEP sites with output from the Hull Acid Rain Model (HARM) to provide an indication of model performance with respect to the relative contributions from different sectors to total pollution concentrations and depositions. The locations of the EMEP sites and the distribution of SO_2 emissions across the UK are shown in Figure 1.

2. The Hull Acid Rain Model

HARM is a Lagrangian statistical model used to estimate annual pollutant levels and loads of species of S, N and HCl across the UK. The coupled chemistry of the model has been described elsewhere (Metcalfe et al, 1995a) and is mass conservative. Although the chemistry in HARM is quite complex, the model employs a highly simplified meteorology assuming: constant boundary layer height (800 m); instantaneous mixing; constant wind speed (10.4 m s^{-1}) (Jones, 1981) and the use of a single representative wind rose for the whole of the UK. Wet scavenging coefficients are based on the assumption of constant drizzle. Emissions, transformations and removal of each pollutant is calculated along 24 trajectories (every 15°) running towards the designated receptor site. Trajectories have a maximum travel time of 96 hours crossing both EMEP area and UK emission sources. At the receptor site pollutant contributions are aggregated up in to 8 x 45° sectors and then weighted using the wind rose. There is no unattributed element. Given this highly simplified representation of meteorology and as HARM is used for source attribution in the context of policy development, it is important to test the model against data from the UK's EMEP sites for which it is possible to allocate measured pollutant concentrations.

3. Measurements

Measurements at the UK EMEP sites (Figure 1) follow EMEP protocols and results have been reported in full elsewhere (Schaug *et al*, 1992). Rainfall was collected using a wet-only collector and analysed by ion chromatography. Using back trajectories calculated by the Norwegian Meteorological Institute, each days measurements at the sites have been assigned to 45° wind sectors (1 - 8). If less than half of the position markers for the calculated trajectories lay within one sector then the day was assigned to a ninth 'indeterminate' sector. The number of days assigned to sector 9 depends upon the prevailing meteorological conditions and is common under cyclonic conditions. Sector arithmetic mean concentrations for the period 1987 - 92 have been calculated for SO_2 and SA, as have precipitation weighted mean concentrations for non-marine sulphate (pSO_4-S) and nitrate (pNO_3-N). The percentage of total deposition of S from each sector has also been calculated.

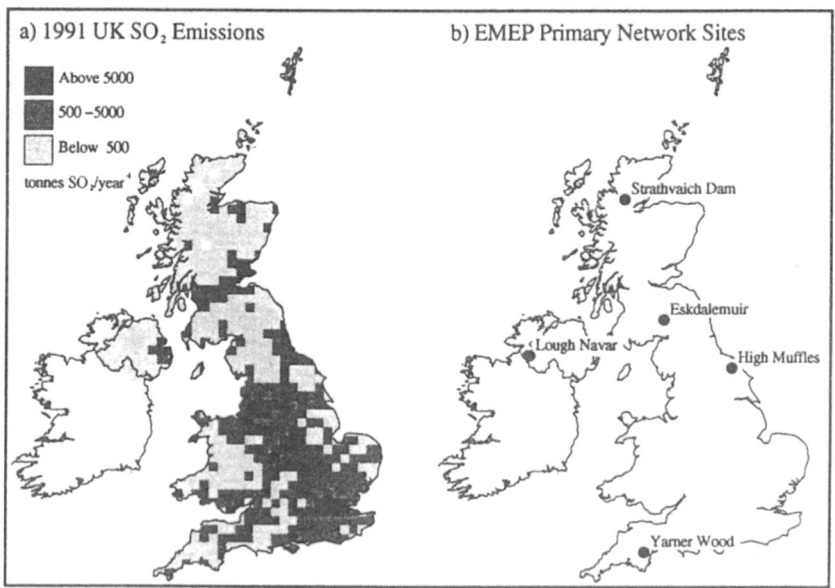

Fig. 1. Distribution of SO_2 emissions across the UK and locations of EMEP monitoring sites.

4. Data-Model Comparison

The model estimates values close to those observed for precipitation concentrations. Sectoral contributions for precipitation concentrations of pSO_4-S and pNO_3-N are shown in Figure 2. Given the highly simplified meteorology in the model the correspondence is surprisingly good. The sector with the highest concentration (2, 3 or 4) is generally correctly identified by HARM, clearly reflecting the importance of both European sources and the areas of highest emissions in the UK (Figure 1). The modelled estimates of concentrations in the westerly sectors as at Lough Navar and Strathvaich Dam are rather smaller than the measured values.

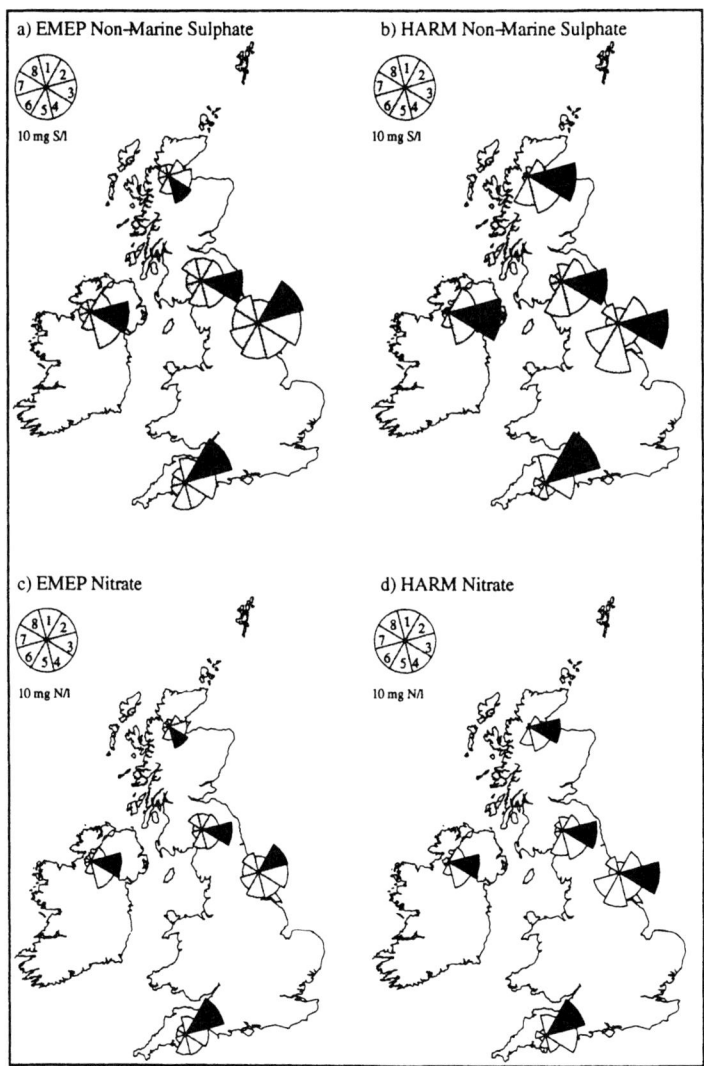

Fig. 2. Comparison of measured and modelled precipitation concentrations of SO_4-S and NO_3-N by sector (most heavily polluted sector shown in black)

In Figure 3 we compare modelled and measured estimates of the percentage of non-marine wet S deposition by sector for the site at Eskdalemuir. In this figure the measured wet deposition from sector 9 has been ignored, only the contributions from sectors 1 - 8 are presented. The model correctly predicts that at this site most deposition is associated with westerly sectors, rather than the high concentration easterlies (cf. Figure 2). There is, however, considerable model underestimation of the contributions from sector 6. The attribution of deposition at sites such as Eskdalemuir is due to the very large rainfall amounts brought by westerly winds, even though concentrations are relatively low. At Eskdalemuir 28% of total wet deposition is assigned to sector 9. Concentrations in sector 9 are generally high. If it were possible to distribute the sector 9 deposition between all of the sectors 1 to 8, it is likely that the percentage of deposition from easterly sectors would increase to be more in line with modelled values.

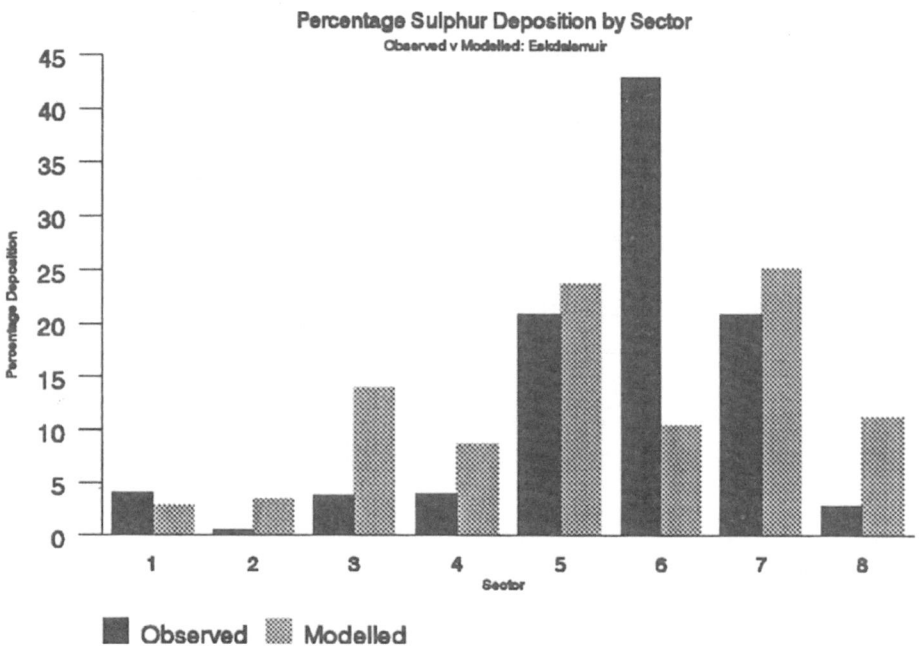

Fig. 3. Percentage of total non-marine S deposition by sector, observed and modelled for Eskdalemuir

Sector 9 days at Lough Navar (Figure 1) were further sub-divided into 'westerly days' (days when none of the trajectory markers strayed to the east of the EMEP grid line passing through the site) and non-westerly days. Eighteen percent of the sector 9 days were classified as westerly, this contributed 27% of the total sector 9 rainfall and 10% of the sector 9 total non-marine sulphate deposition. The westerly sector 9 days were generally wetter than the non-westerly sector 9 days, but non-marine sulphate concentrations were typical of westerly sectors such as sectors 6, 7 and 8; the non-westerly sector 9 days concentration being more consistent with that of sector 5.

5. Modelling Future Concentrations

As HARM appears to provide a good representation of the source direction of pollutants in rain, a sectoral analysis has been carried out on modelled concentrations after the implementation of emissions controls. Figure 4 shows pSO_4-S and pNO_3-N concentrations assuming the implementation of the 1994 S protocol and an N reduction scenario (for further details of the emissions scenario see Metcalfe *et al*, 1995b). As might be expected, the concentrations of S in easterly and southerly sectors shows a marked decline reflecting reduced emissions across mainland Europe and the major source areas of the UK (Figure 1). The dominant sector is unchanged (Figure 2). For N the trend is similar, but less pronounced, which is unsurprising given the smaller reduction in NOx emissions included in the model run.

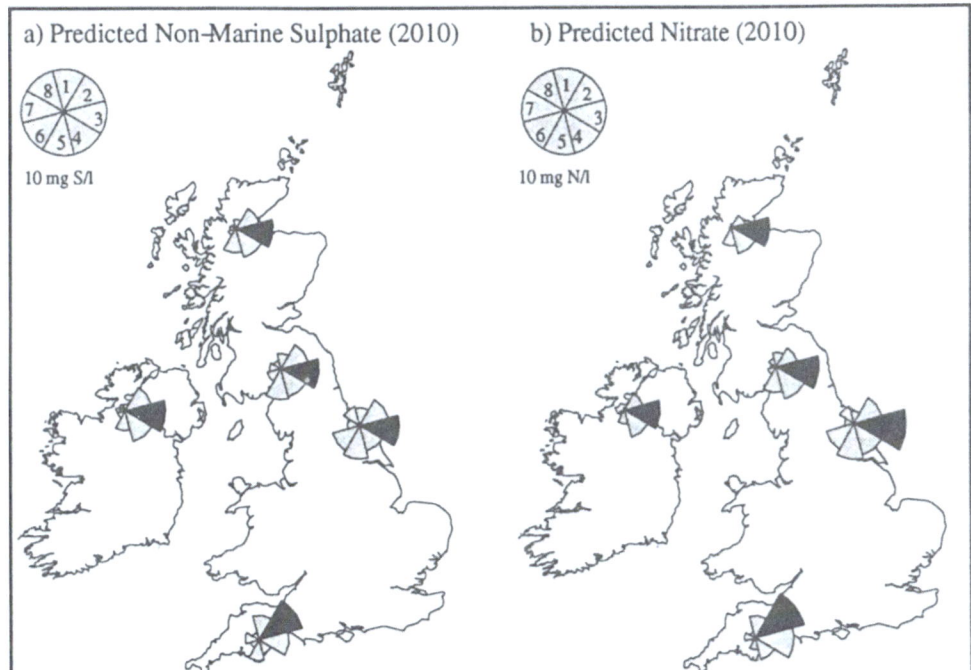

Fig. 4. Modelled precipitation concentrations of SO_4-S and NO_3-N by sector following the implementation of the 1994 S protocol and reductions in NOx emissions (most heavily polluted sector shown in black)

6. Conclusions

A comparison of sectoral data for pollutant concentrations and depositions from the UK's EMEP monitoring sites with output from the HARM model shows generally good agreement in spite of the highly simplified meteorology of the model. The role of days attributed to EMEP sector 9 has been considered with respect to Eskdalemuir and Lough Navar and it seems that pollution in this sector is more likely to be of easterly, rather than westerly, origin which would further improve the match between modelled and measured deposition. Based on model output from emissions reduction scenarios it appears that precipitation concentrations of S and oxidised N would continue to be greatest in sectors coming from the east.

Acknowledgements

This work was supported by Air Quality Division, DoE under contract number PECD 7-12-64 (SEM and JDW) and PECD 7-12-116 (JRS and GWC).

References

Campbell, G.C., Stedman, J.R. et al.: 1994, *Acid Deposition in the UK: Data Report 1993*, Culham AEA Technology Report AEA/CS/16419029/001.
Jones, A.: 1981, *The estimation of long range dispersion and deposition of continuous releases of radionuclides to atmosphere*, National Radiological Protection Board Report NRPB R-123.
Metcalfe, S.E., Whyatt, J.D. and Derwent, R.G.: 1995a, *Q.J.R. Meteorol. Soc.*, 121, 1387-1411
Metcalfe, S.E., Whyatt, J.D., Derwent, R.G., Bull, K.R. and Dyke, H. : 1995b, Water, Air and Soil Pollut., in press.
Schaug, J., Pedersen, U., Skjelmoen, J.E. and Arnesen, K.: 1992, *EMEP Data Report 1992*, EMEP/CCC/Report 5/94.

ACID DEPOSITION AT THE SUMMIT OF MT. FUJI: OBSERVATIONS OF GASES, AEROSOLS AND PRECIPITATION IN SUMMER, 1993 AND 1994

Y. DOKIYA[1], K. TSUBOI[1,*] H. SEKINO[1], T. HOSOMI[1], Y. IGARASHI[2] and S. TANAKA[3]

[1] *Meteorological College, 7-4-81 Asahicho Kashiwa, 277, Japan,* [2] *Meteorological Research Institute, Nagamine, Tsukuba, 305, Japan,* [3] *Faculty of Science and Technology, Keio University, 3-14-1 Hiyoshi, Yokohama, 223, Japan,* * *present adress:Tokushima Meteorological Observatory, Yamatocho, Tokushima, 770, Japan*

Abstract. Intensive observations of chemical species in aerosols, gases and other samples at the summit of Mt. Fuji and at Tarobo (at 1300m on the mountain'ts southern slope) was performed from July 28 to Aug. 3, 1993 and from July 25 to 30,1994. The most interesting observation was the abrupt increase in the sulfate concentration in aerosol collected in July, 1993 just after the typhoon (number 9306) passed the Japanese archipelago and the wind direction shifted from south to west. Chemical analysis indicated this aerosol was acidic. In contrast, the summit aerosol observed in 1994 was not acidic following a less dramatic rise in sulfate content. Back trajectory analyses were used to extrapolate from these measurement to an inventory of polluted air over the Asian Continent. The concentrations of gaseous SO_2 and HCl remained low during both observation periods, with some higher concentrations of NH_3.

Key Words: SO_4^{2-} in aerosol, NH_4^+ in aerosol, summit of Mt. Fuji, Back trajectory

1. Introduction

The summit of Mt. Fuji, the highest mountain in Japan, is a solitary 3776 m peak, which is considered to be in the free atmosphere and presumably free from local pollution. The Japan Meteorological Agency (JMA) has maintained a weather station at the summit since 1932. However, because of the severe meteorological conditions, no data has been available on the amount of precipitation (Fujimura,1971). The authors have measured chemical species in the precipitation since August, 1990 in order to evaluate the site as a background station and to obtain information on the long range transport of chemical species. The concentrations of chemical species in samples from Mt. Fuji were generally found to be much lower compared with those sampled at lower elevation sites (Dokiya et al.,1993). However, during some of the precipitation events caused by the Baiu front or cold fronts, the concentration of sulfate was found to be higher even at the summit (Maruta et al., 1993).

Elucidation of the long range transport of chemical species using only precipitation data is difficult. We intensively sampled trace gases and aerosols as well as precipitation and fog samples during the periods from July 27 to Aug. 3, 1993 and from July 25 to 30, 1994, to provide more detailed information relevant to long range pollutant transport studies. This paper focuses primarily on the sulfate and ammonium in aerosols.

2. Materials and methods

The location of the sampling sites are shown in Fig. 2 in the session of results. A variety of sampling and analytical techniques were employed during the summer 1993 and 1994 campaigns, however, this paper deals solely with gases and chemical species in aerosol.

Alkali impregnated filters used for SO_2 and HCl sampling were prepared by washing 47 mm diameter ADVANTEC 5A filters three times with purified water and then soaking them for 30 minutes in 1% sodium carbonate and 1% glycerin solution. Phosphate impregnated filters for NH_3 sampling were prepared in the same manner using 1 % phosphoric acid in place of the 1% sodium carbonate. Nuclepore membrane filters (47 mm diameter) were used for aerosol samples in the front of the sampling line.

Tanaka et al. (1987) have developed a sequential technique of sampling aerosol, sulfur dioxide and ammonia, modifying other pre-existing impregnated filter techniques. The collection efficiency of this system for sulfur dioxide and ammonia was reported to be more than 96%. The detection limit for both these gases was 2 ppb when 0.3 m^3 air was sampled. Six sets of this system are installed in a small box with shelves, with a timer and electrical valves for automatic sample changing, making it is easy to collect samples even at remote sites where available laboratory equipment is poor.

Low volume Andersen samplers were also used to collect samples for evaluation of the size distribution of chemical species in the aerosol, simultaneously at the summit and at Tarobo (1300 m on the southern slope) in 1993 but only at the summit in 1994.

The filters with sample were carefully sealed and carried back to the analytical laboratory, where they were processed by extraction with distilled water for 15 min. under sonication. The concentrations of chloride, nitrate, sulfate, ammonium, sodium, potassium, calcium and magnesium ions were determined by ion chromatography (YEW IC-7000D).

3. Results and Discussion

3.1 CONCENTRATIONS OF GASES

The concentrations of HCl and SO_2 (Table 1) were very low, compared to those at remote mountainous sites such as the Happo mountains. The concentration of NH_3, however, was found to be higher, showing a diurnal change, high [NH_3] during days when atmospheric temperatures were high and low at night. Since the better weather in 1994 brought more mountain climbers than came during 1993 and higher ammonia concentrations were observed in 1994 (Table 1), we suspect that microbial activity fueled by human excrement left by the climbers was the source of ammonia gas at the summit during summer.

TABLE I

Trace gas concentrations, average (range) in nmol m^{-3} at Mt. Fuji'as summit

Sampling year	HCl	SO$_2$	NH$_3$
1993	4.0 (<1.0-8.0)	1.9 (<0.4-4.8)	64.6 (32.1-146.7)
1994	4.1 (<0.4-12.6)	0.7 (<0.4-4.5)	205.3 (<10-320.0)

It is known that this filtration method is not an ideal method to sample gases and aerosols because of some undesirable chemical reactions which occur on the filters and cannot be completely eliminated, especially in the case of ammonia. However, the relative simplicity and light weight of the equipment has enabled us to do airborne sampling even at difficult sites such as the summit of Mt. Fuji. For the data shown above, the artifact of concern should be small if it occurred at all, because the amount of the air sampled in this work was around 5 m^3, which is in the range 4-6 m^3, which Ferek et al (1991) recommended to avoid interactions with aerosols on the prefilter in ambient air.

3.2 CONCENTRATION OF CHEMICAL SPECIES IN AEROSOL

Concentrations of chloride, nitrate, sulfate, ammonium, sodium and calcium ions in Mt. Fuji aerosols were also be very low compared with those from the plain. Aerosol sulfate concentrations rose abruptly on July 31, 1993, and on July 27, 1994 (Figure 1 a). The sulfate concentration increase was greater in 1993, reaching almost 50 nmol m^{-3}, than that in 1994. In contrast, the ammonium concentration in aerosols collected during summer 1993 remained low throughout the observation period. The ammonium content of aerosols in summer 1994 increased at the same time as the sulfate concentration rose but the amount of the ammonium increase was greater than that expected if all of it had been in the form of ammonium sulfate (Figure 1 b). Significant amounts of nitrate (0.3-20 nmol m^{-3}) were observed in 1994 aerosol, in contrast to 1993 aerosol in which the nitrate concentration remained very low (<0.3 nmol m^{-3}).

The balance of cations and anions was calculated to determine if the aerosol was acidic or neutral. The contribution of calcium ion to the neutralization of sulfate acidity was small, amounting to 5-20 % of the total sulfate acidity in 1993 aerosol and even less in 1994 aerosol. The aerosols sampled during the latter half of the observation period in 1993 (when sulfate concentrations were high) were acidic, in other words the bulk of them might have existed in the form of sulfuric acid. In contrast, the sulfate acidity in the 1994 aerosol was neutralized by ammonium throughout the observation period.

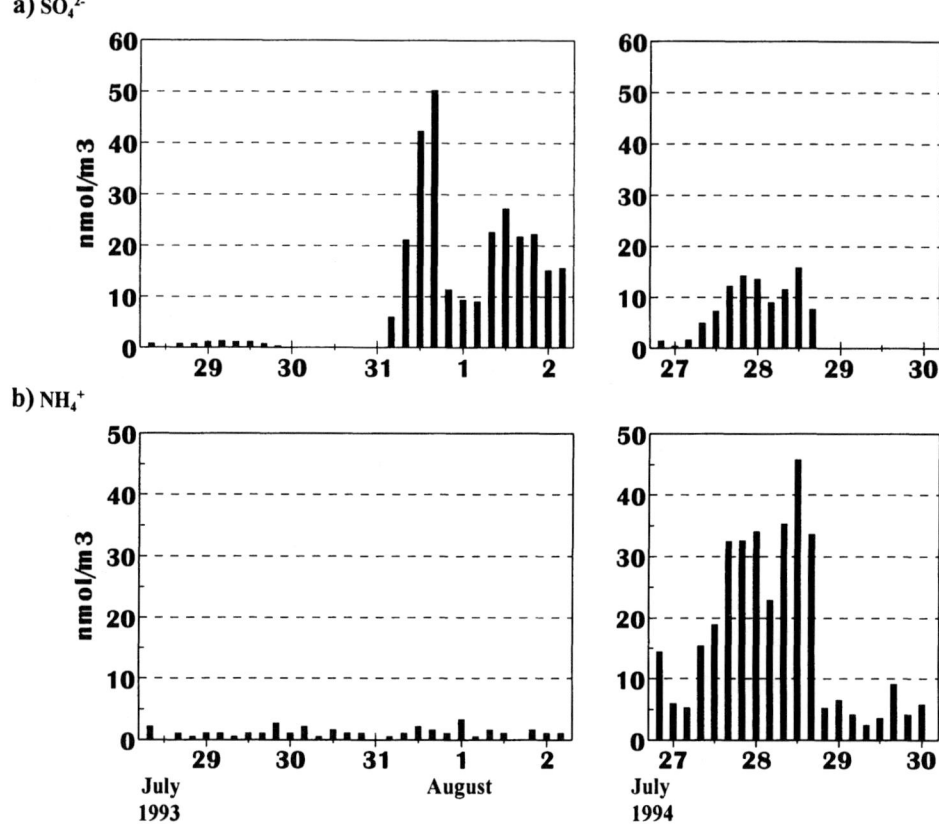

Fig. 1. Concentration of sulfate and ammonium in aerosols a) SO_4^{2-}, b) NH_4^+

3.3. BACK TRAJECTORY STUDY

To identify the reason for the different behavior of sulfate ions in the aerosols, back trajectories were calculated from Global Analysis (GAnal) data of JMA for every 12 hour period (Fig. 3 a & b). Clearly the air mass at the summit of Mt. Fuji in 1993 was influenced by southern marine air during the first half of the observation period and then the air mass changed, with greater influence by westerlies. Typhoon 9306, which reached Nagasaki on July 30, influenced precipitation on Honshu (the largest Japanese island upon which Mt. Fuji is located). After the typhoon, the air mass from the western direction carried sulfate-rich aerosols which had not been exposed to ammonium ion nor calcium ion during transport.

Back trajectory analysis shows that the air mass at the summit of Mt. Fuji during the 1994 observation period came predominantly from the south but sometimes passed slowly over the Kanto Plain, which includes the Tokyo metropolitan area. This finding suggests that the source of the higher sulfate concentrations observed during the first half of the observation period (July 28-29) could be polluted air from the Tokyo metropolitan area. This explanation is consistent with the observation that during this time the sulfate acidity was fully neutralized by ammonium.

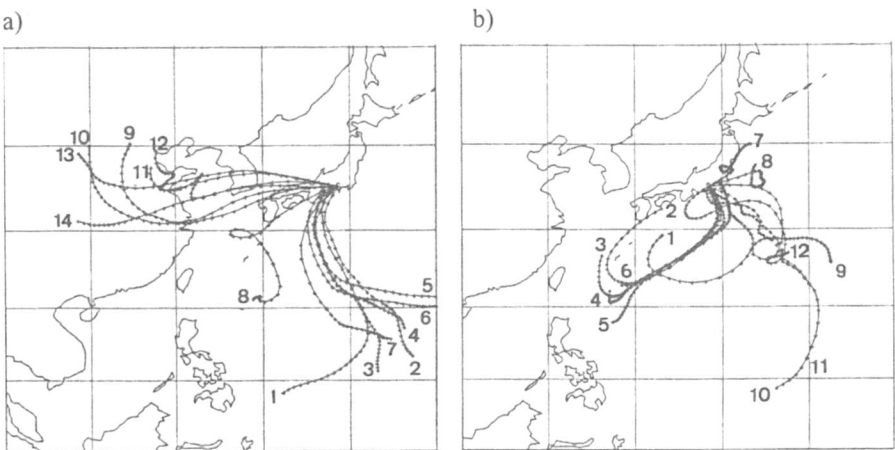

Fig.3. Back trajectory analyses for air sampled on Mt. Fuji
a) from 27 July through 2 August, 1993: 1:July 27 9:00(local time), 2: July 27 21:00, 3: July 28 9:00, 4: July 28 21:00, 5: July 29 9:00, 6: July 29 21:00, 7: July 30 9:00, 8: July 30 21:00, 9: July 31 9:00, 10: July 31 21:00, 11: August 1 9:00, 12: August 1 21:00, 13: August 2 9:00, 14: August 2 21:00, b) from 25 through 30 July, 1994: 1: July 25 9:00, 2: July 25 21:00, 3: July 26 9:00, 4: July 26 21:00, 5: July 27 9:00, 6: July 27 21:00, 7: July 28 9:00, 8: July 28 21:00, 9: July 29 9:00, 10: July 29 21:00, 11: July 30 9:00, 12: July 30 21:00, Each small tick on the line indicates 6 hours.

3.4 PARTICULATE SIZE AND THE CHEMICAL SPECIES

The concentrations of sulfate and ammonium in the particles of various sizes were determined. Both species were distributed mainly in small (< 1 μ m) particles during both sampling periods. One peculiar feature is that the range of sulfate concentration observations for 1993 aerosol samples was much higher than that for ammonium, presumably reflecting the fact that the highly concentrated sulfate in the aerosol during July 30-Aug. 2 (after the typhoon passed) was in the form of sulfuric acid. The distribution patterns of these species in Tarobo aerosol in 1993 were different from those at the summit, showing less pronouced concentration difference between them. The concentration ranges of these two species for 1994 aerosol samples also similar.

Conclusion

Intensive observations of gases and aerosols at the summit of Mt. Fuji were performed during the summers of 1993 and 1994 to evaluate the sources of chemical species in the precipitation there. From July 26 to Aug. 3,1993, the concentrations of gaseous HCl and SO_2 were low and comparable to those at remote sites. However, the concentration of NH_3 during this period was higher with diurnal variation suggesting some influence of mountain climbers. Similar tendencies were found for gases from July 26 to 30, 1994.

The concentration of sulfate in the aerosol increased abruptly after a typhoon passed on July 30, 1993. These aerosols with high sulfate concentrations were found to be acidic and back trajectory analysis indicated that they advected to Mt. Fuji from the west. In contrast, the higher sulfate concentrations sampled during the first half of the July 1994 observation period seemed to originate from local sources, presumably in the Kanto area.

Acknowledgment

This study was funded in part by the Sumitomo Foundation. The authors are grateful to two anonymous reviewers whose comments lead to the improvement of this manuscript.

References

Dokiya, Y., Tsuboi, K., Maruta, E.: 1993, *Tenki.* **40**, 539-542.
Dokiya, Y., Tsuboi, K., Sekino, H., Maruta, E., Tanaka S.:1994, Proceedings of 7th IUAPPA Regional Conference on Air Pollution and Waste Issues, Nov. 2-4, Taipei, Vol. **II**, pp173-177
Ferek, R.J., Hegg, D.A., Herring, J.A., Hobbs P.V.: 1991, *J. Geophys. Res.* **96** 22373-22378
Fujimura, I. :1971, General Research report of Mt. Fuji, Fujikyukou Pub. pp220-296
Maruta, E., Tsuboi, K., Dokiya, Y.: 1993, *Environ. Sci*, **6**, 311-320
Tanaka, S., Komazaki,Y., Ikeuchi, Y. Hahsimoto, Y.: 1987, *Bunseki Kagaku*, **36**, 164-168

Preliminary Analysis of Atmospheric Scavenging Processes in the Industrial Region of Cubatao, Southeastern Brazil

W. Vautz[1], M. Schilling[1], F.L.T. Goncalves[2], M.C. Solci[3], O. Massambani[2], D. Klockow[1]

[1] Institut für Spektrochemie und Angewandte Spektroskopie - Dortmund - Germany
[2] Departamento de Ciencias Atmosfericas / Universidade de Sao Paulo - Brazil
[3] Departamento de Quimica / Universidade de Londrina - Brazil

Abstract. During a German-Brazilian research project on vegetation damage, atmospheric scavenging processes were investigated, considering transport and local emissions. For this goal, radar precipitation data was evaluated together with analyses of fractionated rain samples of one stratiform and one convective rain event. The temporal variation of pollutant concentration in rain water and the meteorological conditions were compared for both rain events at two stations: one at the top of the Serra do Mar mountain ridge (~ 900 m) with minor influence of local emissions and one at sea level, close to the high indutrialized area of Cubatao. At the station close to the industrial plants, major influence of the local sources on wet deposition was determined. During the stratiform rain event, an additional influence of transport could be observed. On the top of the Serra do Mar, major influence of the local sources could only be observed during the convective rain event, while transport is mainly responsible for wet deposition during the stratiform event.

Keywords: wet deposition, transport, scavenging, rain-out, wash-out

1. Introduction

Chemical constituents of ambient air influence the atmospheric water cycle, and the material deposited by rain may affect soil as well as vegetation. The amount of trace substances deposited by rain is influenced by local emissions (accumulation, wash-out) as well as by transport of pollutants (rain-out), the type of rain (the drop size distribution influences rain-out and wash-out efficiency), and the height above sea level (Baron et al. 1993). During a rain event with stable meteorological conditions and without local emissions, the concentration of trace substances in rain water is high in the beginning followed by a steep decrease (Durana 1992). A comparison of rain events with different meteorological conditions can be helpful to investigate the influence of transport and/or local emissions. For frontal rain (transport, rain-out, wash-out), completely different behavior of the mean values and the variation of species concentration in rain water can be expected compared to convective precipitation (no transport, mainly wash-out, see Lacaux et al. 1992). In general, higher concentrations can be expected from frontal rain because of a drop size distribution which is more efficient for scavenging (Lacaux et al. 1992).

In order to promote such investigations, a joint bilateral German-Brazilian research project has been designed in the frame of different modules (atmospheric circulation and mass transport, chemistry of the local atmosphere, vegetation damage, effects of air pollution on soil). The location chosen for this study is the industrial region of Cubatao, located in the state of Sao Paulo in southeastern Brazil, close to the Serra do Mar mountain ridge with native tropical forests. During intensive measuring phases, two different rain events were investigated. Meteorological data were available from routine monitoring stations and radar observations. Additionally fractionated rain samples have been analysed.

The influence of transport of pollutants to the investigated area should be investigated in comparison to the influence of local emissions and their scavenging on wet deposition. Therefore two sources were chosen with significantly different emissions and locations relative to a monitoring station. The sources are a fertilizer plant from which emissions of sulphate and ammonia can be expected and a refinery with emissions of formaldehyde and acetaldehyde. Measuring wet deposition of those species during different meteorological conditions and at different places should enable conclusions on the contribution of transport processes to wet deposition.

2. The location of the experiment

The region of Cubatao is located at 23° S 46° W at sea level close to the coast. About 23 industrial plants - mainly chemical, metallurgical and fertilizer industry - are located in about 40 km², close to the steep terrain of the Serra do Mar mountain ridge (about 900 m, see Figure 1). The horizontal distance from the bottom of the basin to the top of the mountains is about 3 km (Figure 2). Meteorological situations include periods of stagnating air masses, land-sea breeze circulation, frontal and isolated convective systems. The topography of the region causes high pollutant concentrations for many meteorological situations. As a result, soil and vegetation in the region and in the neighbourhood have suffered from pollutant deposition by rain, fog or particulate matter.

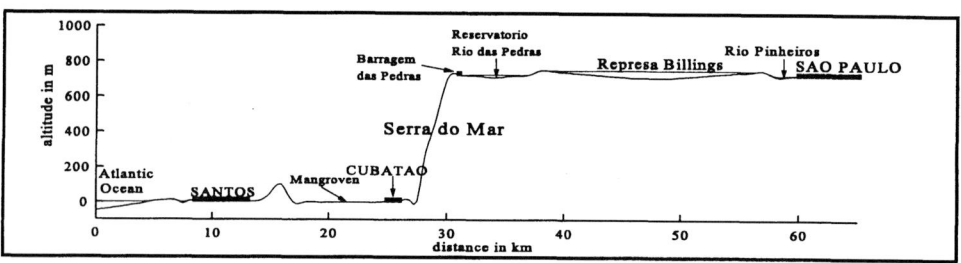

Fig. 1. Profile of the investigated area (Dippell 1995).

Two monitoring stations, including meteorological instruments as well as rain collectors, are located in the investigated area: one in the Mogi Valley (MS) at about sea level, close to a fertilizer plant, about 10 km NE of a refinery and one close to the village of Paranapiacaba (PS) at the end of the Mogi Valley on the top of the Serra do Mar about 800 m above sea level and 8 km NE of MS. A further meteorological station, operated by the Institute of Astronomy and Geophysics (IAG), University of Sao Paulo, is located about 30 km WNW of PS on the plateau of the Serra do Mar. Also on the top of the Serra, about 20 km NE of PS, the weather radar *Ponte Nova* is installed.

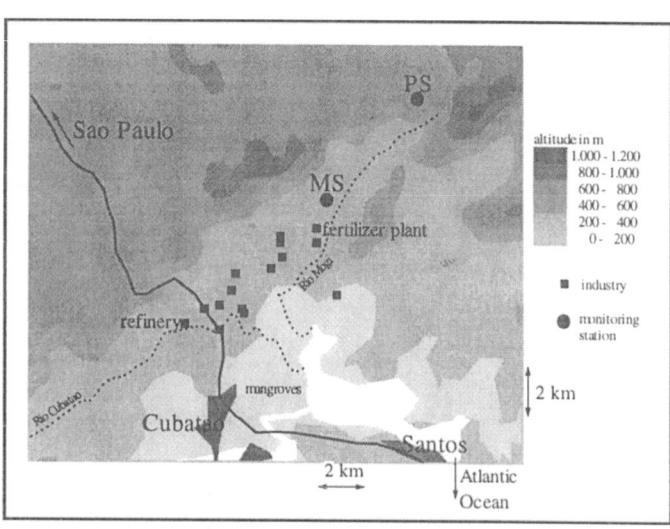

Fig. 2. The investigated industrial area of Cubatao, Sao Paulo State, Southeastern Brazil. Major industries and monitoring stations are indicated (Dippell 1995).

3. Data Acquisition

3.1. Analyses of fractionated rain samples

Fractionated rain samples were taken in polyethylene tubes with a wet only collector, equipped with a polyethylene funnel of 500 mm diameter. Sample tubes were changed manually when at least 10 mL of rain - the minimum amount for chemical analyses - were reached. After collection, the sample was devided into two parts and stored in polyethylene tubes: One part for analyses of NH_4^+ and SO_4^{2-} without treatment and another part for the determination of form- and acetaldehydes, treated with a small amount of HgI_2 to avoid decomposition of organic species.

The analyses of SO_4^{2-} were carried out by ion chromatography with suppressed conductivity (SYKAM LCA09 column, 5 mM Na_2CO_3 + 50 mg 4-Hydroxybenzonitrile + 2 % CH_3CN as eluent). NH_4^+ was determined by using the indophenol blue procedure and photometric measurements at 546 nm. For aldehyde measurements, the rain sample was treated with a 2,4-dinitrophenylhydrazine as derivatization reagent, and the corresponding hydrazones of form- and acetaldehyde were determined by HPLC separation (column: Nucleosil 300C18 / 5 µm / 250*4 mm; eluent: CH_3CN-H_2O 60-40 (v/v) und UV detection at 365 nm).

3.2. Weather radar

The Sao Paulo Weather Radar *Ponte Nova* (S-Band microwave remote sensor) is monitoring an area of 180 km radius, operated by the Sao Paulo Water Authority (DAEE/FCTH). Good agreement of rain fall rates from radar and rain gauges have been found in many investigations (e.g. Smirnov et al. 1994). The radar system uses a 10 cm electromagnetic wavelength, generated in pulses of 2 microseconds with a pulse repetition frequency of about 250 pps. 360 degrees are scanned in azimuth and 30 degrees in elevation. From the raw data the system generates CAPPI maps (**C**onstant **A**ltitude **P**lan **P**osition **I**ndicator), which display the precipitation intensity observed at a 2x2 km grid at 3 km height as well as ECHOTOP maps, which display the maximum height reached by precipitation echoes. From these maps, which are available every ten minutes, intensity and size of the precipitation system can be determined.

4. Results and Discussion

4.1. Meteorological situation on March 17th, 1992

The period from 16th to 20th of March 1992 was influenced by a cold front, located in the southeast of Brazil as could be seen from the pressure charts. During a fast development, the front reached the eastern border of the state of Sao Paulo on March 17th. Stability of the air on March 17th and instability on March 19th in the morning of both days was determined by radiosonde data (vertical temperature and wind profiles).

During the passage of the front, a large area with rain intensities of about 5 mm/h moved over the Cubatao area, including smaller areas with intensities up to 25 mm/h. The precipitation system was moving with ~15 km/h from NW to SE as observed by radar and radiosonde data for the 600-700 hPa level. The investigated rain cloud was idendified as Nimbostratus with shallow raincloud tops and hydrometeors up to 7 km height. The precipitation structure is typically a result of a slow lift of air masses in a shallow frontal zone. A change in surface wind direction from ENE to WNW was observed at IAG station at 12^{00}. Because of the location of PS, the same change in wind direction can be expect for this station, but not for MS, which is influenced strongly by local circulation in the Mogi valley.

Precipitation was observed at MS from 10^{45} to 12^{00} and at PS from 11^{11} to 12^{17} (Figure 3). This is in good agreement with the radar observations. At both stations fractionated rain samples were taken during this period.

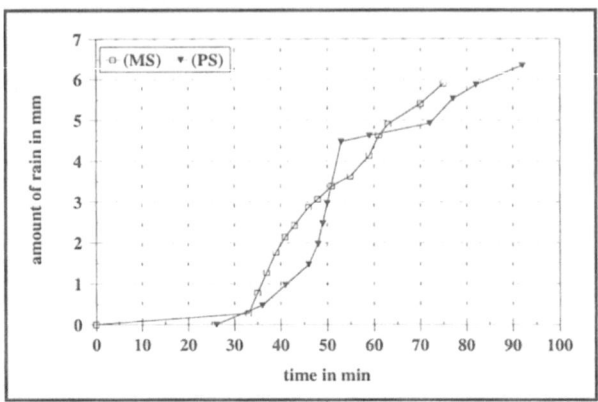

Fig. 3 The amount of rain on March 17th, 1992 (t=0 at $10^{5)}$). A very similar development at both stations can be seen.

4.2. Meteorological Situation on March 19th, 1992

After the passage of the front, on March 19th the large instability of the post frontal air mass together with surface heating caused the development of deep convection clouds (Altocumulus) in the area. From 13^{00} to 14^{00} the dominant wind at IAG was WNW, changing to weaker winds from E after the passage of a large isolated rain cloud. This Cumulonimbus cloud moved slowly from WNW to the Cubatao area.

The radar image at 14^{13} showed a large rain cell in the neighbourhood of MS and PS. At 14^{20} the rain reached MS, where precipitation was observed from 14^{20} until 16^{00}. At that time, the size of the rain cell was about of 80 km x 60 km 12 km height. At PS, rain was observed from 14^{30} until 16^{5}. For a short period with high rain inten-sities, MS was in the centre of the rain cell (Figure 4). At PS only light rain was observed, lasting about one hour longer than at MS.

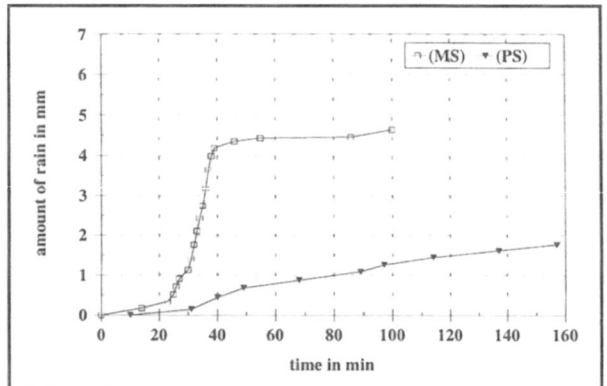

Fig. 4 The amount of rain on March 19th, 1992 (t=0 at 14^{20}). Significantly different development at both stations can be seen.

4.3. Ammonia and sulphate concentration

Considering a time shift caused by the different arrival time of the precipitation system at both stations on March 17th (see Figure 3), rain intensity, duration and amount is in good agreement at MS and PS. Therefore in Figure 5 the concentrations are displayed versus the amount of rain. The temporal variation of NH_4^+ and SO_4^{2-} concentration also is very similar at both stations but the values are significantly higher at MS than at PS (Figure 5). This is due to the strong influence of the local emissions at MS and to the neglectable contribution of transport. The highly polluted air - mainly a result of emissions of the fertilizer plant close to MS - is responsible for the high wet deposition of NH_4^+ and SO_4^{2-} at that location. Both substances are removed by wash-out processes.

On March 19th MS and PS are affected by the same rain cell. Therefore the concentration variation is displayed versus time in Figure 6. At MS, NH_4^+ and SO_4^{2-} concentrations are lower than on March 17th. This can be explained by the different type of rain: the drop size distribution of stratiform rain (many little drops) is more efficient for wash-out. The concentration of both substances are in the same range at MS and PS on March 19th, with only little higher concentrations at MS. Because of the convective dynamics and the instability of the air mass, the emissions of NH_4^+ and SO_4^{2-} from the fertilizer plants nearby MS are well distributed in the area and influence even PS.

It can be concluded, that wet deposition of NH_4^+ and SO_4^{2-} is mainly a result of the high emission rates from the industrial area of Cubatao. This causes high deposition rates close to the source, for instability of the air masses also at the top of the Serra do Mar mountain ridge. Even if air masses are transported towards the investigated area, no transport of pollutants can be observed.

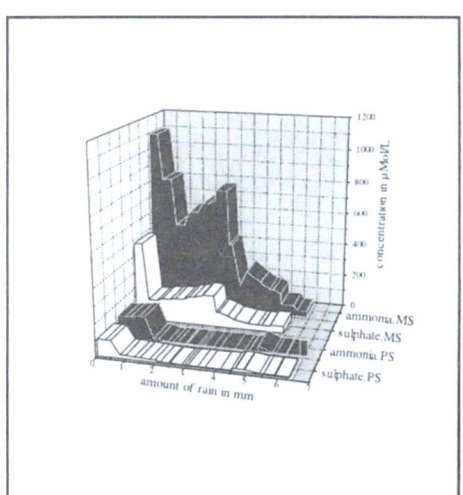

Fig. 5. Ammonia and sulphate concentration versus amount of rain at MS and PS on March 17th.

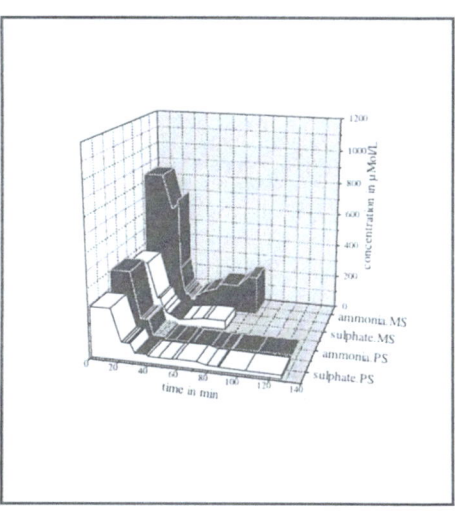

Fig. 6. Ammonia and sulphate concentration versus time at MS and PS on March 19th.

4.4. Formaldehyde and Acetaldyde concentration

Precipitation weighted mean values of aldehyde concentrations in rain water are shown in table 1 for both rain events. On March 19th, the concentrations were significantly higher than on March 17th as a result of the stagnating air masses over the Cubatao area: the air mass accumulated emissions from local sources. On March 17th, with wind coming from ENE at PS the concentrations do not originate from the Cubatao area - they must be a result of transport, possibly from Sao Paulo. The reason for the lower concentrations at MS at that day may be different source regions for both stations. A strong local influence - similar to NH_4^+ - cannot be observed.

The acetaldehyde concentration is about 4 times lower than the formaldehyde concentration for both events and of both stations. This relation cannot be transferred automatically to the gas phase, because the solubility of acetaldehyde in water is significantly lower compared to formaldehyde.

It can be concluded, that wet deposition of aldehydes is higher for the convective rain event, caused by local emissions. During the stratiform rain event a significant partition of transport was observed.

	Formaldehyde in µMol/L		Acetaldehyde in µMol/L	
	MS	PS	MS	PS
March 17th, stratiform	1,97	4,12	0,55	1,05
March 19th, convective	16,23	16,82	2,18	3,23

Table 1. Aldehyde mean concentrations in rain water at MS and PS during the rain events on March 17th and 19th.

5. Conclusions

During the investigated rain events, local emissions of atmospheric trace substances and their removal by rain have strong influence on wet deposition, especially close to sources as could be seen from ammonia and sulphate wet deposition. Because of the high amount of substances emitted in the Cubatao region, this leads to high deposition rates in this area, but also in the neighbourhood when the boundary layer is mixed well (e.g. for convective rain events). For stratiform rain events the influence of transport of pollutants (e.g. traffic emissions from Sao Paulo area) can exceed local influences in the neighbourhood of the Cubatao area (e.g. at PS). Even at MS an influence of transport on wet deposition can be observed during stratiform rain events, but it is exeeded by the strong influence of the local sources. In general, wet deposition of pollutants emitted by local sources can be expected to be much higher than the partition of transport.

Acknowledgement

Financial support of the German Ministry for Education and Research, the Ministry of Science and Research of Northrhine-Westfalia , the *Centro Tecnologico e Hidraulico,* the *Secretaria do Meio Ambiente* and the *Conselho Nacional de Desenvolvimento Cientifico e Tecnologica*, Brazil is greatfully acknowledged.

References

Baron, J., Denning, A.S. 1993. The Influence of Mountain Meteorology on Precipitation Chemistry at low and high Elevations of the Colorada Front Range, U.S.A. Atmospheric Environment, Vol. 27A, No. 15, pp. 2337-2349

Chaumerliac, M., Rosset, R., Renard, M., Nickerson, E.C. 1992. The Transport and Redistribution of Atmospheric Gases in Regions of Frontal Rain. Journal of Atmospheric Chemistry, Vol. 14, pp. 43-51

Dippell, J., 1995. Vergleich natürlicher und anthropogener Emissionen der reduzierten Schwefelverbindungen H_2S, COS und CS_2 in einer tropischen Industrieregion. Theses, Centre for Environmental Research, University of Frankfurt

Durana, N., Casado, H., Ezcurra, A. 1992. Experimental study of the Scavenging Process by means of a Sequential Precipitation Collector, Preliminary Results. Atmospheric Environment, Vol. 26A, No. 13, pp 2437-2443

Lacaux, J.P., Delmas, R., Kouaido, G., Cros, B., Andreae, M.O. 1992. Precipitation Chemistry in the Mayombé Forest of Equatorial Africa. Journal of Geophysical Research, Vol. 97, No. D6, pp. 6195-6206

Sakugaea, H., Kaplan, I.R. 1993. Measurements of H_2O_2, Aldehydes and Organic Acids in Los Angeles Rainwater: Their Sources and Deposition Rates. Atmospheric Environment, Vol. 27B, No. 2, pp. 203-219

Smirnov, M.T., Hagen, M., Evtushenko, A.V., Kutuza, B.G., Meischner, P.F., Petrenko, B.Z. 1994. Estimation of Rain rate by Microwave Radiometry and Active Radar during CLEOPATRA '92. Contributions to Atmospheric Physics, Vol. 67, No. 4, pp. 303-312

MT. BROCKEN, A SITE FOR A CLOUD CHEMISTRY MEASUREMENT PROGRAMME IN CENTRAL EUROPE

K. Acker, D. Möller, W. Wieprecht and St. Naumann

Brandenburg Techn. Univ. Cottbus, Department of Air Chemistry, Rudower Chaussee 5, D-12484 Berlin

Abstract. We present results from the Brocken Cloud Chemistry Measurement Project (BROCCMON) which started in 1991. Since 1992 the full programme is running, based on continuous measurements (e.g. trace gases, meteorology, liquid water content), cloud water sampling and analysis and intensive measurement campaigns. The observed high variability of cloud water composition we explain with cloud dynamic and microphysical behaviour of clouds and differences in the air mass characteristics. During the measurement period 1992-1994 we observed an increase in cloud water acidity (by a factor of 3) and we found photochemical conditions typically for summersmog situations. Our preliminary data also show that an understanding of tropospheric ozone balance would be incomplete without consideration of chemical processes within clouds. A long-term goal of our programme is to establish a cloud chemistry climatology which is representative for the region.

Key words: atmospheric pollution, cloud chemistry, cloud water acidity, ozone, trajectory analysis

1. Introduction

The emission situation e.g. in Eastern Germany changed rapidly in the past 5 years. Several emitted pollutants like sulphur dioxide (SO_2), ammonia (NH_3), hydrochloric acid (HCl) and dust decreased in the last five years with yearly rates of about 10%, mainly caused by decreasing coal consumption and drastic structural changes in agriculture. On the other hand, an increase of photooxidant precursors nitrogen oxides (NO, NO_2) and volatile organic compounds (VOC) are observed due to the increasing traffic.

Chemical processes in cloud water are often more effective than in gas phase and influence the budget of acid components, photooxidants and aerosols within the troposphere. From the beginning of our programme BROCCMON until now we analyzed more than thousand cloud water samples with related information on atmospheric trace gases (ozone (O_3), SO_2, NO, NO_2 and hydrogen peroxide (H_2O_2)) and on meteorological and cloud physical parameters. Here we will show some results characterizing the relation between chemical composition and microphysical parameters of clouds and the air pollution situation. Within our scientific programme we investigate the influence of changing emissions on acid formation, on the distribution of trace species between the gas and liquid phase as well as we study the role of clouds in the oxidant chemistry.

2. Measurement Programme

Our investigations were carried out at the highest elevation in the northern part of Central Europe, Mt. Brocken/Harz (51.80° N, 10.67° E; 1142 m a.s.l.). Due to the high occurrence of clouds (30-50% of time from April to October) the measurement programme is focused on cloud water chemistry. The station is equipped with an automatic weather station. The measurements of trace gases are made using commercial automatic analyzers. Cloud water is collected using a passive string collector of Mohnen-type (Mohnen and Kadlecek, 1989) in combination with an automatic sampling unit and analyzed by ion chromatography. Since June 1993 we collected cloud water with 1-h time resolution to be in coincidence with the time scale of synoptic observations (cloud type, cloud frequency, cloud base) and the typical time scale of weather changes. Liquid water content (LWC) of clouds is measured continuously by a laser diffraction technique using a Gerber Particulate Volume Monitor. The yearly mean of LWC lies between 250 and 300 mg/m^3. The whole measurement and analytical procedure are under QA/QC control based on adopted US-EPA project (Mohnen and Vong, 1993). More details about the experimental programme are described in Möller et al., 1994, 1995. Back trajectories are computed by an isentropic model using a mixed dynamic/kinematic procedure (Reimer and Scherer, 1991). Further transport parameters (e.g. time of transport in clouds and in the mixing layer) were analyzed by our programme TRAP.

3. Clouds and cloud water chemistry

We found large differences in the chemical composition of collected cloud water from event to event (Table I), much larger than what is known from precipitation chemistry studies (Hansen et al., 1994). The LWC of clouds predominantly determines the total ionic content (TIC, TIC = sum of liquid phase concentrations of chloride (Cl$^-$), nitrate (NO$_3^-$), sulphate (SO$_4^{2-}$), sodium (Na$^+$), ammonium (NH$_4^+$), potassium (K$^+$), calcium (Ca^{2+}), magnesium (Mg^{2+})). Based on all 570 1-hour samples of cloud water collected in 1993 from nonprecipitating clouds, we found that the relationship between LWC and TIC is best approximated by a power function (Möller et al., 1995), see Figure 1. Other cloud physical parameters (droplet size, cloud type, level of cloud base and top) also influence the chemical composition.

Studies from literature (Mazin and Khrgian, 1989) and our own results showed for most cloud types an approximately linear increase of LWC with altitude above cloud base. From the maximum at 80-90% of the total cloud depth to the cloud top the LWC decreases sharply. These studies indicate that variations of LWC from event to event (and, also within events) could, at least partly, be attributed to different altitudes of the cloud base. On the other hand, an increase of the ionic content near the cloud base could not be explained only by droplet evaporation (decrease of LWC) but also by scavenging of trace gases and aerosols due to mixing processes with subsaturated air (Arends et al., 1994). Differences in air mass history is another reason for the deviations of data points from the power function between LWC and ionic content (Figure 1) mentioned above. Using 72-h back trajectory calculations we characterized selected cloud events being typical for air pollution influence by different European regions (Figure 2). The differences in cloud water composition for that examples are documented in Table I.

The cloud water samples collected from September, 30, to October, 01, 1993 were

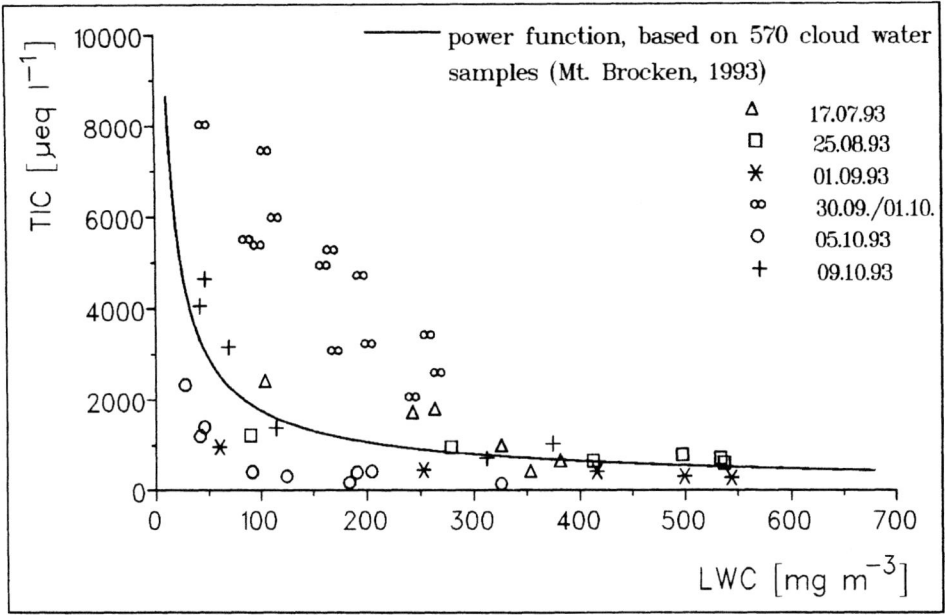

Fig. 1 Relationship between liquid water content (LWC) and total ionic content (TIC) of cloud water samples

Fig. 2 Mean 72-hour back trajectories for selected events

characterized by very high concentrations of SO_4^{2-}, NO_3^-, NH_4^+, Ca^{2+} and Mg^{2+}. The acidity was also high (e.g. pH =3.07 between 2-3 CET) but partly compensate by the high level of alkaline components. More than 36 hours before reaching the Brocken, the air masses were under extremely stagnant flow conditions in Bohemia. Due to the industrialized and high polluted areas in southeastern Central Europe, also high trace gas concentrations were observed. Air masses from the main wind direction west-southwest (Oct. 9, 1993 and June 9 to 10, 1994) crossing also industrial areas showed much lower concentrations of trace species in the gas phase (SO_2, NO_x) and in the liquid phase (SO_4^{2-} and Ca^{2+}), but partly low pH-values, too. Cloud water samples with a very low ionic content were obtained from air masses with an origin from the north (case Sept.1, 1993).

TABLE I

Mean chemical composition of all 1-hour cloud water samples collected in 1993 and for selected events (in µeq/l; LWC in mg/m³) including concentrations of gas phase components (in ppb), CET=Central European Time

	1993			selected events			
	mean	min	max	1.9. 93 6-11CET	30.9./1.10.93 15-4CET	9.10.93 9-14CET	9./10.7.94 23-6CET
Cl^-	68	0.3	2401	26	52	197	86
NO_3^-	280	0.5	5946	105	1038	400	442
SO_4^{2-}	265	4	4169	148	1508	401	336
Na^+	60	2	2444	18	37	216	91
NH_4^+	410	2	8083	163	1596	766	369
Ca^{2+}	54	3	2245	17	423	42	33
Mg^{2+}	26	5	693	9	86	49	24
H^+	83	<0.1	1500	76	289	47	470
LWC	278	25	863	355	166	182	387
O_3	37**	10**	77**	14-30*	13-29*	16-32*	21-30*
SO_2	4**	0.1**	38**	0.1-1*	2-56*	0.6-3*	0.5-4*
NO_2	4**	0.1**	27**	0.2-1.8*	0.7-24*	2-14*	3-9*

*   minima and maxima of gas phase concentration during the event
** based on daily means

Since the beginning of our continuous measurements at the Brocken station in 1992 until 1994 we observed a permanent increase in cloud water acidity by a factor of about 3. The mean hydrogen ion (H^+) concentration of all collected cloud water samples was in 1992 39 µeq/l (n = 131), in 1993 83 µeq/l (n = 1054) and 146 µeq/l in 1994 (n = 847).

This results are in agreement with data from the precipitation network in the New German States (Brüggemann and Rolle, 1995). More detailed analyses of our data sets (e.g. sector analysis) and cloud water monitoring over a longer period are necessary to clarify reasons for the observed behaviour of cloud water pH.

4. Clouds and photooxidants

In the continuous record of O_3 at Mt. Brocken we often observed that the ozone concentration decreases rapidly with passing clouds (Acker et al., 1995). Laurila et al. (1995) reported that depleted ozone concentrations (of roughly 10 ppb) are measured in cloudy conditions at Pallas station/Finland. We found indications that polluted air masses have a preference for heterogeneous ozone destruction. Cloud water samples collected during events with ozone depletion showed several times larger concentrations for NO_3^-, SO_4^{2-}, NH_4^+, Ca^{2+} and Mg^{2+} in comparison to samples of events without ozone depletion. This differences were not caused by different LWC. We also found large differences in the mean cloud duration for both cases and in the time the air masses were transported within clouds during the two days before reaching Mt. Brocken. The events without ozone depletion lasted longer compared to the events with ozone depletion. Because of the high portion of transport in clouds (43%), we believe that events without ozone depletion were associated with precipitation on their pathway and, consequently, with wet removal of trace species before they reach the Brocken. For events with ozone depletion the portion of transport within clouds is quite low (12%).

Results of aircraft measurements in 1992 (Schaller et al., 1992) and of two field measurement campaigns in the Harz region, during which high concentrations of ozone (>100 ppb), hydrogen peroxides (>1.5 ppb) and VOCs were observed in the atmospheric boundary layer, show clearly that the problem of summer smog already exists in the New German States. Ground based measurements were made in July 1993 and 1994 at the Brocken summit and at a second station (400 m a.s.l.) in the lower Harz, between Brocken and the industrial region Halle-Leipzig. At both sites a steady increase of the daily maxima and average O_3-concentration has been observed with a rate of 8-10 ppb/day.

5. Concluding Remarks

Despite decreasing emissions in East Germany and partly in Eastern Europe, responsible for acidic and alkaline substances, we observed a permanent increase in cloud water acidity between 1992 and 1994. With changing the emission pattern also an increase of summersmog periods could be assumed due to emissions of pollutant precursors (e.g. VOCs, NO_2), caused by industries and increased traffic.

We found strong indications for ozone depletion within clouds on regional scale. The regional ozone budget would be incomplete without consideration of heterogeneous processes. Our recommendation is to enforce activities for inclusion of aerosol and cloud chemistry moduls in mesoscale transport/transformation models.

The intensive cloud water monitoring programme at Mt. Brocken is aimed to be continued at least in the next few years including registration of cloud parameters to document changes in the air and cloud water pollution situation and to find out more detailed reasons of the high variations in cloud water composition.

Acknowledgement

This work was founded from the German Ministry for Science and Research within the projects SANA and EUROTRAC and by the Environmental Protection Board of the Land Saxonia-Anhalt/Germany (summersmog campaigns). We thank the Deutscher Wetterdienst for providing the cloud observation data and Dr. E. Reimer (FU Berlin) for calculating the back trajectories.

References

Acker, K., Möller, D. and Wieprecht, W.: 1995, *Naturwiss.* **82**, 86-89.
Arends, B.G., Kos, G.P.A., Maser, R., et al.: 1994, *J. Atm. Chem.* **19**, 59-85.
Brüggemann, E. and Rolle, W.: 1995, In: *Acid Rain Research: do we have enough answers?* (ed. G.J. Heij and J.W. Erisman), Elsevier Sci. Publ., Amsterdam, 403-406.
Hansen, K., Draaijers, G., Ivens, W.P., et al.: 1994, *Atmos. Environ.* **28**, 3195- 3205.
Laurila, T., Plathan, P., Hakola, T., Koskinen, T. and Lättila, H.: 1995, EUROTRAC Newsletter **15**, 2-4.
Mazin, J.P. and Khrgian, A. Kh.: 1989, *Handbook of clouds and cloudy atmosphere*, Leningrad, Gidrometeoizdat, 336-338.
Mohnen, V.A. and Kadlecek, J.A.: 1989, *Tellus* **41B**, 79-91.
Mohnen, V. A. and Vong, R. J.: 1993, *Environ. Rev.* **1**, 38-54.
Möller, D., K. Acker and Wieprecht, W.: 1994, In: *Physico-Chemical Behaviour of Air Pollutants* (eds. G. Angeletti and G. Restelli), Report EUR 15609/2 EN, Office for official publ. of the EU, L-2985 Luxembourg, 968-974.
Möller, D., Acker, K. and Wieprecht, W.: 1995, *J. Aerosol Sci.* **26**, in press.
Reimer, E, and Scherer, B.: 1991, In: *Proc. of the 19th ITM on Air Pollution Modelling and its Application 1991 in Crete*, Vol. II, 421-428.
Schaller, E., Werhahn, J. and Meyer-Wyk, M.: 1992, In: *IFU Schriftenreihe Bd.11*, Wiss.-Verlag Dr. W. Maraun, (ISBN 3-927548-47-2), Frankfurt/M., 67pp.

AIR OXIDATION OF S(IV) IN CLOUD-WATER SAMPLES

M. RADOJEVIC[1], B.J.TYLER[2], S.HALL[2] and N. PENDERGHEST[2]

[1] *Department of Chemistry, University of Brunei Darussalam, Bandar Seri Begawan, Brunei Darussalam.*
[2] *Department of Chemistry, University of Manchester Institute of Science and Technology, Manchester M60 1QD, U.K.*

Abstract. Aerobic oxidation of S(IV) was investigated in cloud-water samples collected at Great Dun Fell, U.K., as part of a wider project into cloud-water chemistry. The rate was found to be first-order in S(IV) concentration, and the reaction rate constant, $k_{S(IV)}$, was found to vary from 10^{-5} to 10^{-3} s^{-1}. The rate constant was highly correlated with H$^+$ concentration (pH 3.5 to 6.5) and Fe concentration (<0.02 to 3×10^{-6} mol dm^{-3}). The aerobic oxidation of S(IV) does not contribute significantly to SO$_2$ oxidation in clouds at Great Dun Fell, however, the reaction may be of consequence in clouds and fogs at polluted or urban sites with elevated trace metal concentrations. Also, this reaction may be responsible for the oxidation of S(IV) in cloud-water samples during storage.

Key words: cloud-water, sulphur dioxide, oxidation, catalysis, trace metals, iron, acidity, pH.

1. Introduction

Sulphur dioxide (SO$_2$) dissolves in cloud droplets to give S(IV) species that are subsequently oxidised to sulphate by a variety of aqueous-phase mechanisms. These processes are widely recognised as being fundamental to the formation of atmospheric acidity, and the oxidation of S(IV) by H$_2$O$_2$, by O$_3$ and by O$_2$ in the presence of trace metal catalysts are of particular interest. Sulphur (IV) oxidation mechanisms have been studied intensively in the laboratory using pure water solutions, and more recently using rainwater samples. These studies have been reviewed (Radojevic, 1982). In this publication we report results of a laboratory study into the aerobic oxidation of S(IV) in stored cloud-water samples collected at Great Dun Fell, a rural site in the U.K.

2. Experimental

Cloud-water samples were collected on the summit of Great Dun Fell (GDF), U.K. The field site and details of the cloud-water collector have been described elsewhere (Chandler et al, 1988). Samples were stored at 4 °C and the kinetic experiments were performed six to twelve months after collection. The experimental method for studying S(IV) oxidation in aqueous solutions has been described previously (Clarke and Radojevic, 1983) and employed for measuring S(IV) oxidation rates in rainwater samples (Clarke and Radojevic, 1987). One hundred cm^3 test solutions were employed and the experimental run was started by addition of a stock Na$_2$SO$_3$ solution to give an initial S(IV) concentration of ca. 50×10^{-6} mol dm^{-3} in aerated cloud-water. Aeration was achieved by bubbling air through the solution at 1 dm^3 min^{-1}; in some experiments pure oxygen was used instead. The pH was monitored independently in an aliquot of the test solution sampled at the beginning of each run. It was considered undesirable to place the pH electrode directly into the test solution because of possible interference in the oxidation reaction (Clarke and Radojevic, 1983). Five cm^3 aliquots of test solution were analysed for S(IV) during the course of the experimental run using a colorimetric method described previously (Clarke and Radojevic, 1983).

Kinetic experiments were performed with both filtered and unfiltered samples. Sample filtration was by means of Duropore (polyvinylidene difluoride) filter papers purchased from Millipore, and these were of a hydrophillic variety with a pore size of 0.45x10^{-6} m. All experiments reported here were carried out in the diffuse laboratory light. In experiments were the pH was intentionally altered, the pH was adjusted on samples in the reaction vessel and not in the original sampling bottles. Filtered and unfiltered cloud-water samples were analysed for 45 elements, including a number of transition metals (Fe, Mn, Zn, Cu, V), by inductively coupled plasma (ICP).

3. Results

3.1. S(IV) CONCENTRATION DEPENDENCE

Reaction was followed almost to completion (≥90% reaction) in most of the samples and plots of log[S(IV)] versus time gave straight lines as shown in Figure 1. The rate could therefore be expressed as :

$$-d[S(IV)]/dt = k_{S(IV)} [S(IV)]$$

Values of $k_{S(IV)}$ were determined from the slopes of the first-order plots. Values from 2x10^{-5} to 1.3x10^{-3} s^{-1} were determined for $k_{S(IV)}$ in ten unfiltered cloud-water samples at 25 °C and without pH adjustment (pH 3.5 to 6.5). These are within the range of values of 10^{-5} to 10^{-2} s^{-1} reported previously for $k_{S(IV)}$ in rainwater samples (Radojevic, 1982). The concentration of dissolved oxygen in aerated water at 25°C can be calculated to be 2.64x10^{-4} mol dm^{-3}, well in excess of S(IV) concentrations in the present experiments or in cloud-water (Radojevic et al, 1990). At these conditions the S(IV) oxidation is independent of the O$_2$ concentration (Radojevic, 1984). Reproducibility tests showed that there could be a ±5 to 30% variation in the quoted values of $k_{S(IV)}$. Problems of reproducibility are common in experiments of this kind and considerably higher discrepancies have been reported for pure, laboratory grade water (Radojevic,1984). Values of $k_{S(IV)}$ are summarised in Table I for nine unfiltered cloud-water samples for which pH, trace metal and sulphur measurements were also available. Adding Na$_2$SO$_3$ to cloud-water raises the initial experimental pH although this effect diminishes with increasing sample acidity. Oxidation of S(IV) results in a decrease in the pH during the course of the experiments, however, these changes were generally less than 0.3 of a pH unit and only occasionally as high as 0.5 of a pH unit (at pH of 6). The average experimental pH is quoted in Table I together with the sample pH.

3.2. pH DEPENDENCE

Variation in $k_{S(IV)}$ with pH is illustrated in Figure 2 for cloud-water samples and rates were about an order of magnitude faster than for distilled-deionised water. For cloud-water at 25 °C:

$$k_{S(IV)} = 0.097 \, [H^+]^{ca.0.5} \, s^{-1}$$

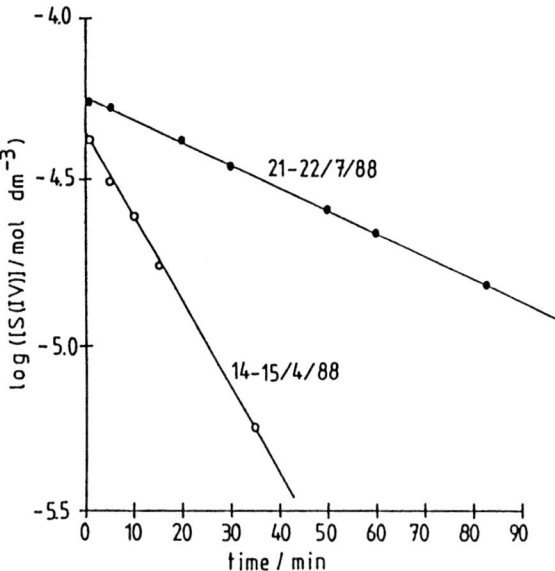

Fig. 1. First-order plots for $k_{S(IV)}$ oxidation in selected cloud-water samples (unfiltered, 25 °C)

TABLE I

Summary of results of kinetic experiments, trace metal concentrations, sulphur and pH in unfiltered cloud-water samples. All the elements, including sulphur were determined by ICP analysis. S is the concentration of sulphur in cloud-water prior to addition of sulphite. Rate constant was determined at 25 °C.

Sample	Fe	Mn	Zn	Cu	V	S	pH	pH$_{exp}$	$k_{S(IV)}$ (s^{-1})
			Concentration (10^{-6} mol dm^{-3})						
14-15/4/88	3.26	0.44	2.62	0.131	0.147	169.8	4.0	4.2	1.3x10^{-3}
19-20/4/88	1.57	0.31	0.81	0.022	0.086	>345	4.4	5.0	4.6x10^{-4}
20-21/7/88	0.21	0.18	1.92	0.047	<0.020	69.8	3.6	4.1	3.7x10^{-4}
21-22/7/88	0.53	0.075	1.14	0.046	<0.020	53.8	3.7	3.8	2.6x10^{-4}
6-7/4/88	0.19	0.17	0.68	<0.013	<0.020	42.9	4.5	5.5	2.0x10^{-4}
18-19/7/88	<0.02	0.087	0.34	<0.013	<0.020	57.9	5.0	6.2	3.7x10^{-5}
8-12/4/88	0.02	0.49	0.44	<0.013	0.025	161.0	6.1	6.5	3.0x10^{-5}
20-21/4/88	<0.02	0.027	0.11	<0.013	<0.020	32.6	5.2	6.5	3.0x10^{-5}
19/7/88	<0.02	<0.007	0.16	<0.013	<0.020	36.5	5.6	6.5	2.0x10^{-5}

Experiments were also carried out in which the pH of cloud-water samples was intentionally altered over the range 3 to 6.5 by additions of H_2SO_4 or NaOH. There was considerable variation in the pH dependence between individual samples from little or no dependence ($k_{S(IV)} \propto [H^+]^{ca.0}$) to a high of $k \propto [H^+]^{ca.0.5}$. The H^+ dependence was weakest in samples with low levels of trace metals and low values of experimentally determined $k_{S(IV)}$ while high dependencies were found in samples with high metal concentrations and high values of $k_{S(IV)}$. This was confirmed in experiments using cloud-water in which trace metal concentrations were below the ICP detection limits and the determined values of $k_{S(IV)}$ low ($k_{S(IV)} < 10^{-4}$ s^{-1}). Addition of Fe to these samples markedly increased the pH dependence of the reaction. Varying the pH was found to have little effect on trace metal concentrations in either filtered or unfiltered cloud-water samples, suggesting that dissolution was completed during storage.

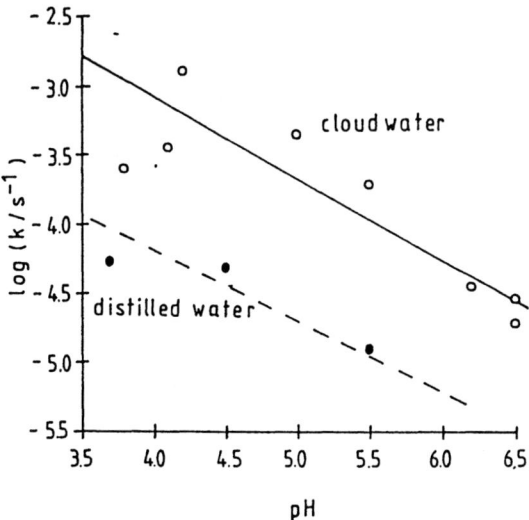

Fig. 2. Variation in the first-order reaction rate constant, $k_{S(IV)}$, with experimental pH for unfiltered cloud-water samples at 25 °C. A sample collected on 7-8/4/88 is also included in the figure but not in Table 1 as no elemental analyses were available. Single point for two samples collected on 8-12/4/88 and 20-21/4/88 (see Table I).

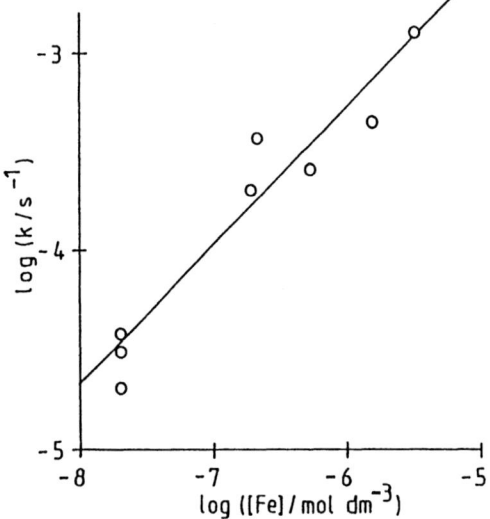

Fig. 3. Variation in the first-order reaction rate constant, $k_{S(IV)}$, with Fe concentration for unfiltered cloud-water samples at 25 °C. Single data point for samples collected on 8-12/4/88 and 20-21/4/88 (see Table I).

3.3 TRACE METAL CATALYSTS

The reaction rate constant $k_{S(IV)}$ was found to be strongly correlated with Fe concentrations as shown in Figure 3 (r=0.968). No significant relationship was noted between $k_{S(IV)}$ and the concentrations of Mn (0.615), Cu (0.836), or V (0.708); however, a significant positive correlation was found with Zn (0.909). These observations are similar to those reported previously for rainwater (Clarke and Radojevic, 1987). The rate of S(IV) oxidation was also studied in ten cloud-water samples collected in the autumn of 1989 and acidified to pH 4 before each kinetic run. Concentrations of trace metals in these samples were generally below the ICP detection limits, and values of $k_{S(IV)}$ in the range of 1.3 to 7.8×10^{-5} s^{-1} were determined, similar to those observed in distilled water (see Figure 2). Addition of Fe to these samples was found to greatly increase the S(IV) oxidation rate. Values of $k_{S(IV)}$ of 3.2×10^{-4} and 1.3×10^{-3} s^{-1} were measured respectively in samples containing 0.6 and 1.2×10^{-6} mol dm^{-3} Fe additions. This suggests an order of about two with respect to iron at constant pH, which is close to the order observed with respect to Fe in rainwater (Clarke and Radojevic, 1987). For cloud-water samples collected in the present study there was a strong inverse relationship between the pH and Fe concentrations with acidic samples having higher concentrations than high pH samples. The data in Table I show that $[Fe] \propto [H^+]^{ca.1}$. This may be an indication of greater pollution associated with more acidic events, or it could be due to a dilution effect (Clarke and Radojevic, 1987). Increased leaching of trace metals from suspended particles with increasing acidity could also result in this relationship (Williams et al, 1988). Trace metal concentrations and values of $k_{S(IV)}$ determined in filtered cloud-water samples were not significantly different from those measured in unfiltered samples. Therefore, an important influence of heterogeneous catalysis by suspended particles may be discounted and S(IV) oxidation is affected primarily by soluble species.

4. Discussion

4.1 ATMOSPHERIC OXIDATION RATES

The rate of the trace metal catalysed oxidation of SO_2 by O_2 as determined in the present work is compared to that by O_3 and by H_2O_2 in Figure 4 for conditions typical of GDF. Sulphur dioxide conversion is dominated by H_2O_2 and O_3 reactions; however, trace metal catalysed oxidation by O_2 could be competitive in acidic droplets containing low H_2O_2 concentrations. Sulphur(IV) has been observed in cloud-water at GDF on occasions of low H_2O_2 concentrations ($<10^{-6}$ mol dm^{-3}) and low pH (3 to 4) (Radojevic at al, 1990). In these samples S(IV) oxidation may have been dominated by the aerobic, trace metal catalysed mechanism. In view of the short lifetime of clouds at GDF (1 to 10 minutes) trace metal catalysed oxidation of S(IV) by O_2 will not contribute significantly to sulphate formation. In fact, we observed no significant correlation between the measured S(IV) oxidation rates and the sulphate content of cloud-water samples suggesting that sulphate was predominantly formed via other mechanisms. Aerobic oxidation may, however, contribute to the observed oxidation of S(IV) during sample storage. It may also play an important role in fogs and clouds at polluted sites. Concentrations of Fe as high as 4.2×10^{-4} mol dm^{-3} have been measured in urban fog-water (Munger et al, 1983).

4.2 REACTION MECHANISM

The mechanism of the Fe catalysed reaction has been studied extensively in laboratory studies but it still remains to be fully resolved (Radojevic, 1992). Comparisons of our results with those obtained in pure water studies of the metal catalysed aerobic oxidation of S(IV) are difficult to make because of the large discrepancies between studies reported in the literature. Furthermore, the situation in real samples containing a mixture of components is complicated by the possibility of synergistic and inhibitory effects.

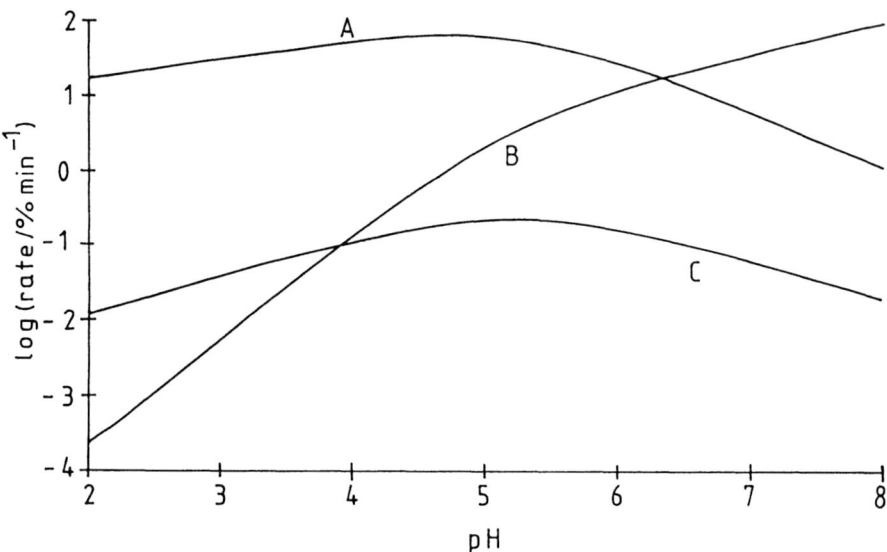

Fig. 4. SO_2 conversion rates (% min^{-1}) in a typical cloud. Curve A gives the variation of rate with pH for oxidation by H_2O_2; curve B for oxidation by O_3; and curve C for oxidation by O_2 in the presence of Fe^{3+}. Chosen conditions were a liquid water content of 0.5 g m^{-3}, 10 °C, 1 ppbv of H_2O_2, 30 ppbv of O_3. Kinetic data taken from Penkett et al (1979a) for oxidation by H_2O_2 and by O_3, and from this work for oxidation by O_2 catalysed by Fe^{3+}.

Acknowledgments

We should like to thank G.J. Dollard from the Environmental and Medical Sciences Division, UKAERE, Harwell for providing the trace metal analysis and the CEGB (National Power) for financial assistance.

References

Chandler, A.S., Choularton, T.W., Dollard, G.J., Gay, M.J., Hill, T.A., Jones, B.M.R., Morse, A.P., Penkett, S.A. and Tyler, B.J.: 1988, *Atmos. Environ.* **22**, 683-694.
Clarke, A.G. and Radojevic, M. :1983, *Atmos. Environ.* **17**, 617-624.
Clarke, A.G. and Radojevic, M. :1987, *Atmos. Environ.* **21**, 1115-1123.
Munger, J.W., Jacob, D.J., Waldman, J.M. and Hoffmann, M.R.: 1983, *J. Geophys. Res.* **88**, 5109-5121.
Penkett, S.A., Jones, B.M.R., Brice, K.A. and Eggleton, E.A.J.: 1979a, *Atmos. Environ.* **13**, 123-137.
Penkett, S.A., Jones, B.M.R., Brice, K.A. and Eggleton, E.A.J.: 1979b, *Atmos. Environ.* **13**, 139-147.
Radojevic, M.: 1984, *Environ. Technol. Lett.* **5**, 549-566.
Radojevic, M.: 1992, "SO_2 and NO_x Oxidation Mechanisms in the Atmosphere", in (eds.) Radojevic, M. and Harrison, R.M. *Atmospheric Acidity: Sources, Consequences and Abatement*, Elsevier Applied Science, London, pp.73-137.
Radojevic, M., Tyler, B.J., Wicks, A.J., Gay, M.J. and Choularton, T.W.: 1990, *Atmos. Environ.* **24A**, 323-328.
Williams, P.T., Radojevic, M. and Clarke, A.G.: 1988, *Atmos. Environ.* **22**, 1433-1442.

MEASUREMENTS OF ATMOSPHERIC PARTICULATE CARBON IN LITHUANIA

S. ARMALIS

Department of General and Inorganic Chemistry, Vilnius University, Naugarduko 24, 2006 Vilnius, Lithuania

Abstract. During 1985 - 1989 particulate carbon concentrations were determined at suburban and rural sites in Lithuania. The main part of the continuous measurements was conducted at Preila background station (Curonian spit). Daily samples of airborne particulate matter were collected on glass fiber filters or by impactors and analysed for total carbon (TC), benzene extractable organic carbon (BEOC) and elemental carbon (EC) using combustion and optical techniques. Monthly median concentrations of total particulate carbon varied in the range 2 - 22 $\mu g/m^3$ during the period of measurements. For all the forms of particulate carbon the strong seasonal dependence was observed. The very regular seasonal cycle is characteristic for the ratio EC/TC: 0.10 - 0.15 and 0.30 - 0.35 for warm and cold periods respectively. Size distribution measurements shows that the main part of TC is in the sub-micron particle fraction - about 80 - 90 % in summer and even 92 - 98 % in winter. The concentrations of TC in size fraction <1.2 μm have sharp winter peaks and summer minima. Very low concentrations and summer maxima for TC in the fraction >3.5 μm have been observed. The concentrations of EC in precipitation varied from 0.02 to 0.7 mg/l and average wet deposition flux of EC was 5 - 7 and maximum 15 - 20 mg/m^2 per month.

Key words: particulate carbon, elemental carbon, size distribution, wet deposition, Lithuania

1. Introduction

It is generally accepted that carbonaceous material is a major component of airborne particulate matter, with carbon responsible for 10-30 % of total aerosol mass (Wolff *et al.*, 1982). The main species in which carbon occurs in aerosols are organic carbon and elemental carbon (EC). The contribution of carbonate carbon to total carbon usually doesn't exceed 2-5 % (Mueller *et al.*, 1972).

Particulate carbon, especially elemental carbon, plays important roles in atmospheric chemistry and physics and is involved in a number of local and global environmental effects. The process of catalytic oxidation of sulphur dioxide to sulphate on soot particles in polluted atmospheres was known more than twenty years ago (Novakov *et al.*, 1974).

Unfortunately, despite the well-known catalytic, climatic and health effects the long-term studies on atmospheric particulate carbon are rather sparse in Europe (e.g., Brosset and Åkerström, 1972; Heintzenberg and Winkler, 1984). It should be noted also that most of studies deal only with elemental carbon although the total carbon (TC) consists at least of 70-80 % of organic carbon (Cachier *et al.*, 1989). Also, EC deposition fluxes have been reported only for Sweden (Ogren *et al.*, 1984).

The aim of this study is to present the results of particulate carbon measurements in Lithuania during 1985-1989. The measurements have been focused on the variability of the concentrations of total, benzene extractable and elemental carbon, the size distribution of carbonaceous particles and wet deposition of EC. It has to be noted that data on TC and EC daily concentrations at Preila background station are available until 1995.

2. Materials and methods

The main part of the continuous measurements was carried out at a coastal background station Preila (Curonian spit). Some minor episodes of simultaneous measurements were conducted also at Vilnius suburb and at rural site Molėtai (280 km to the east of Preila).

Daily samples of airborne particulate matter were collected on 2x 2 cm glass fiber filters (Whatman GF/B) at a flow rate of 8 l min^{-1}. Sampling duration 3-4 days and a flow rate of 17 l min^{-1} was used for home-made impactor. The filters or impaction plates (made from aluminium foil) were heated prior the sampling at 450 °C in air for an hour to remove carbonaceous contaminants. The 50 % cut-off points of the three-stage impactor were following: stage 1 - 3.5 µm, stage 2 - 1.2 µm, stage 3 - 0.5 µm (the fourth stage - filter GF/B).

The precipitation samples for elemental carbon analysis were taken on monthly basis to 0.5 l glass jars of 50 cm^2 of the cross-sectional area. This kind of the precipitation samplers was used because of easier removal of EC from the sampler. The concentrated solution of calcium chloride was added to the samplers during the winter months to protect the sample from freezing.

The total carbon determination in the samples was performed by the combustion technique. The impaction plate or 1 cm diameter disc from the exposed glass fiber filter was placed into a quartz combustion tube and the sample was oxidised in oxygen at 700 °C. After the conversion of carbon dioxide to methane on nickel catalyst the chromatographic determination with flame ionisation detector was used. The detection limit of the technique was 0.3 µg of carbon in the sample.

The 6 hour extraction in a Soxhlet apparatus was used for benzene extractable organic carbon (BEOC) determination. The BEOC concentration was calculated as a difference between the amount of TC on the filter before and after extraction.

The optical method based on the measurement of visible light reflectance was used for the determination of elemental carbon on the filters (Delumyea et al., 1980). A calibration curve was prepared by using the soot of butane flame. The detection limit was 0.2 µg of EC in the sample.

Precipitation samples for EC determination were filtered through a 0.2 µm pore polyamide membrane filter. The collector was washed with an alkaline solution of H_2O_2 for the digestion of biogenic material (Oğren et al., 1983) and the filter also added to the mixture. After the digestion overnight the mixture was neutralised with HCl, the filter was taken out and the surfactant was added. Prior the refiltration through the new membrane filter the dispersion of 30 min in an ultrasonic bath was applied. The light reflectance method and the standards from a commercial carbon black (Monarch 71) were used to determine the amount of EC on the filter. The detection limit was 2 µg l^{-1} EC for 100 ml sample volume.

3. Results and Discussion

The concentrations of TC, BEOC and EC and the ratios BEOC/TC and EC/TC as monthly medians at background station Preila and Vilnius suburb for 1985-1989 are presented in Figure 1. It can be seen that for all the forms of particulate carbon the strong seasonal dependence is observed. The TC concentrations for winter months are 2-4 times, EC and BEOC - even 4-8 times higher than during summer.

The most regular seasonality for the ratio EC/TC indicates that particulate elemental carbon originates mainly from combustion of carbon-containing fuel for heating. Of course, in some extent, the photochemical reactions can contribute to the concentrations of particulate organic carbon in summer time. The high values of the BEOC/TC ratios, showing the high content of non-polar organic compounds in aerosol, confirm the decisive role of primary emissions to the composition of particulate carbon. It is interesting also to note that EC and BEOC concentrations measured at Vilnius suburb were only 1.5 - 2 times higher than at background station (Figure 1).

The variation range of daily concentration of particulate carbon is quite large and depends on season. As shown in Table I, the highest concentrations of TC exceed the lowest ones 15 - 17 times, however, in the case of EC or BEOC this difference can be much higher - even two orders. Moreover, the minimum daily concentrations, except for EC, are very close both in winter and in summer.

TABLE I

Daily concentrations ($\mu g\, m^{-3}$) of particulate carbon at Preila station (1985-1989)

Particulate carbon	Cold period (November-April)			Warm period (May-October)		
	min	max	mean	min	max	mean
TC	2.0 - 2.5	30 - 35	9.4	1.5 - 2.0	12 - 15	5.7
EC	0.10 - 0.15	15 - 20	3.7	0.02 - 0.05	2 - 3	1.2
BEOC	0.05 - 0.07	5 - 7	3.2	0.03 - 0.06	3 - 4	1.9

Figure 2 illustrates the size distribution of particulate carbon at Preila station. The analysis of data has shown that the main part of TC is in the size fraction <1.2 μm - about 80 - 90 % in summer and even 92 - 98 % in winter. It can be seen from Figure 2 that also very distinct seasonality is characteristic for sub-micron fraction of TC. Although the concentration of TC in the size fraction >1.2 μm only slightly depends on season, the sub-fraction >3.5 μm despite its low contribution to total particulate carbon shows inverse seasonality - summer maxima and winter minima. This phenomenon can be explained by the biogenic origin of coarse carbonaceous particles. The simultaneous size-segregated measurements at coastal (Preila) and forested Eastern part of Lithuania has confirmed that mass median diameter of TC particles is higher in the latter case.

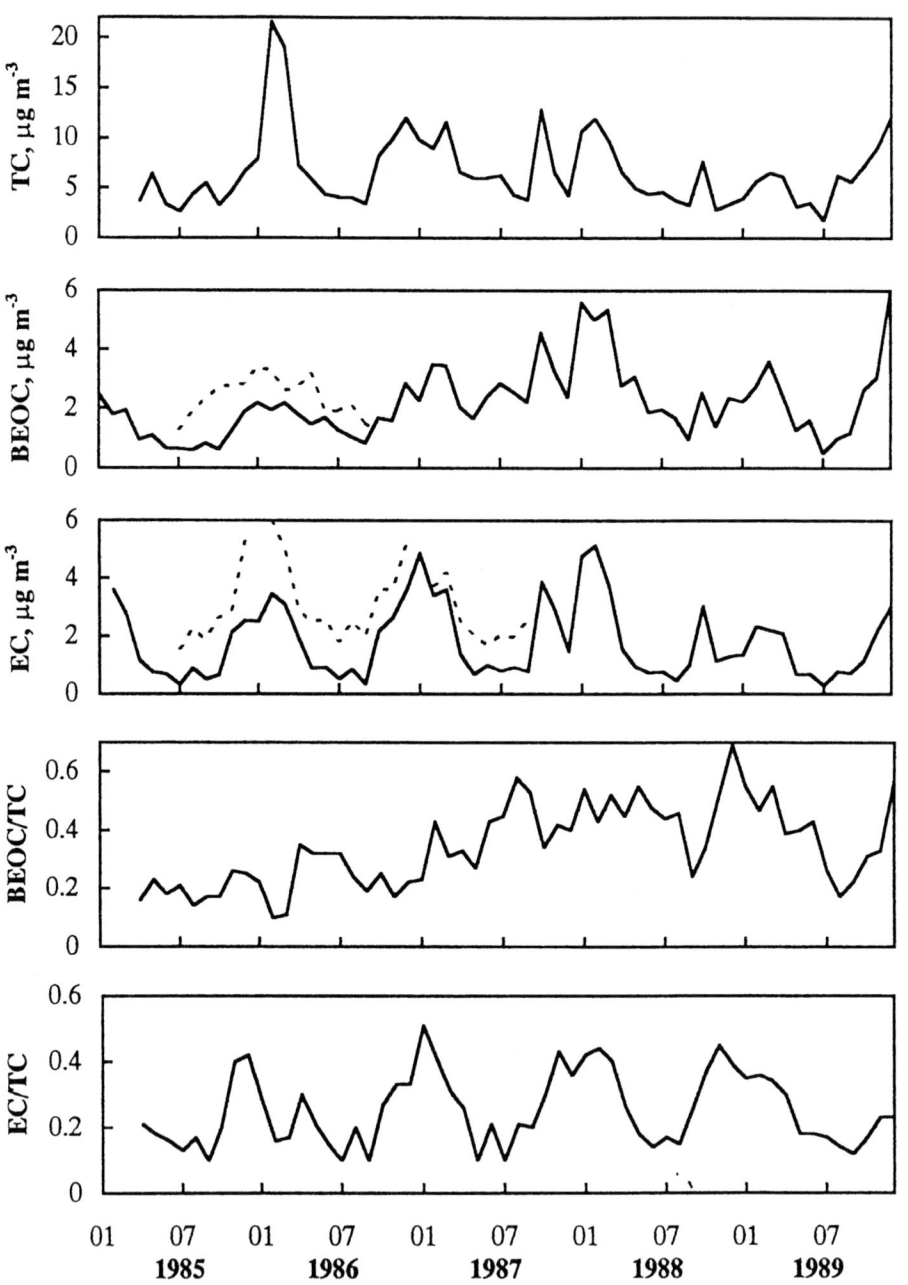

Fig. 1. Monthly median concentrations of particulate carbon and their ratios at background station Preila and at Vilnius suburb (dashed line).

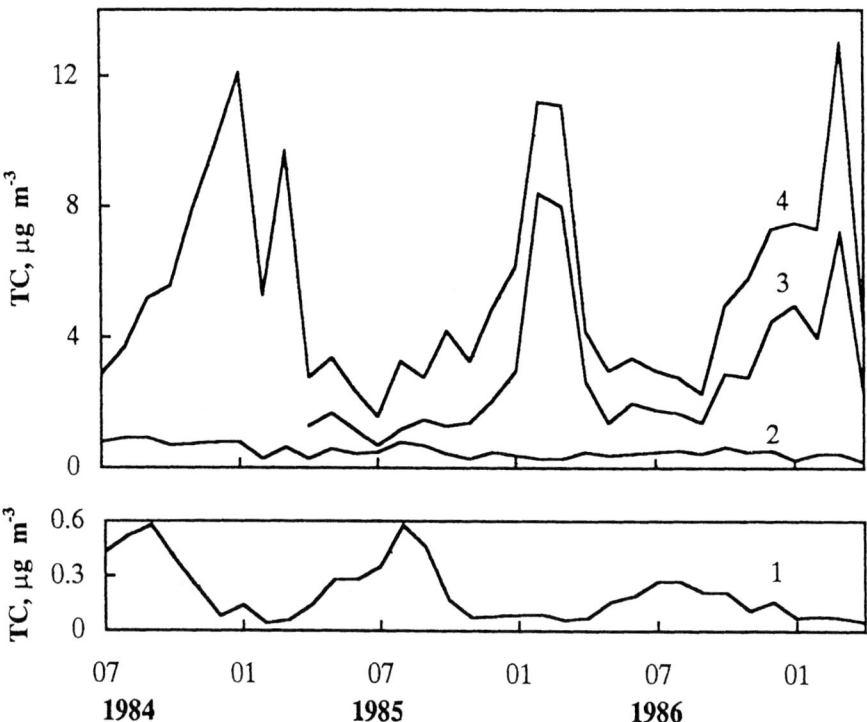

Fig. 2. Monthly mean concentrations of total particulate carbon in different size fractions at Preila station (1984.06 - 1987.04). Particle size fractions: 1 - > 3.5 μm, 2 - > 1.2 μm, 3 - 0.5 - 1.2 μm, 4 - < 0.5 μm.

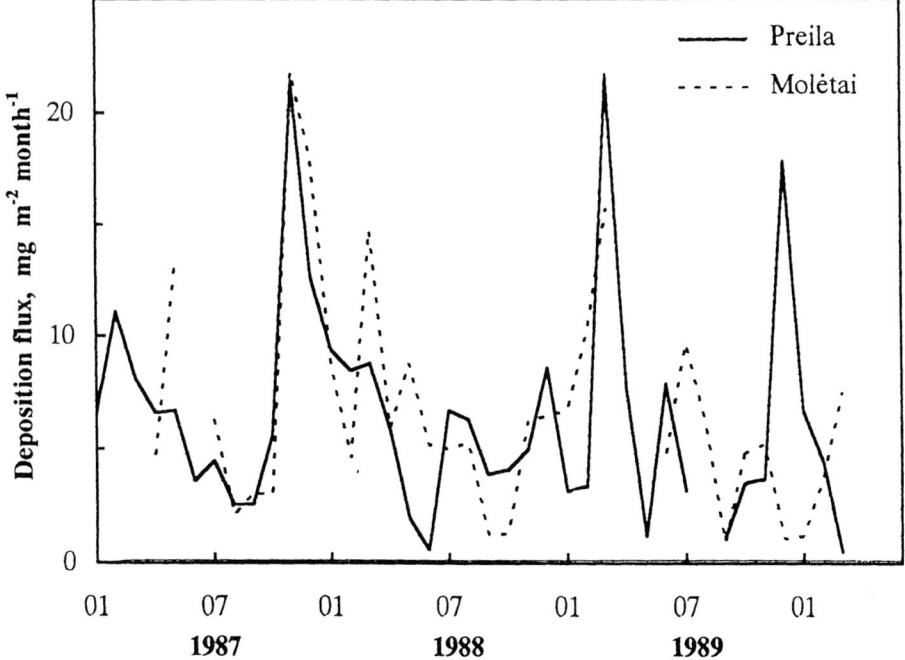

Fig. 3. Elemental carbon wet deposition flux values at Preila and Molėtai sites (1987.01-1990.03).

Elemental carbon wet deposition fluxes are presented in Figure 3. It can be seen that monthly deposition flux values varied from 1 to 20 mg m^{-2} both for Preila and Moletai sites, what corresponds approximately the range of EC concentrations in the precipitation 0.02 - 0.7 mg l^{-1}. High winter peaks of EC in precipitation are observed like in the case of aerosol carbon. Mean monthly wet deposition flux of EC can be estimated as 7 mg m^{-2} and it results in deposition of 5000 t of elemental carbon to the area of Lithuania annually. The estimated EC deposition flux value is very close to that for rural sites in Sweden - about 6 mg m^{-2} month^{-1} (Ogren et al., 1984).

4. Conclusions

The results obtained are in good agreement with the results reported for rural sites of Western Europe and the USA (Heintzenberg and Winkler, 1984; Ogren et al., 1984; Cachier et al., 1989; Wolff et al., 1982; Shah et al., 1986), what indicates that particulate carbon plays an important role in all the areas and for all the types of air pollution.

Acknowledgement

The author would like to thank all his former colleagues from the Institute of Physics, Lithuanian Academy of Sciences, without whose technical assistance this study would not have been possible.

References

Brosset, C., Åkerström, Å.: 1972, "Long Distance Transport of Air Pollutants - Measurements of Black Particulate Matter (Soot) and Particle-borne Sulphur in Sweden during the Period of September-December 1969", Atmos. Environ., **6**, 661-673.
Cachier, H., Bremond, M.-P., Buat-Ménard, P.: 1989, "Determination of Atmospheric Soot Carbon with a Simple Thermal Method", Tellus, **41B**, 379-390.
Delumyea, R.G., Chu, L.-C., Macias, E.S.: 1980, "Determination of Elemental Carbon Component of Soot in Ambient Aerosol Samples", Atmos. Environ., **14**, 647-652.
Heintzenberg, J., Winkler, P.: 1984, "Elemental Carbon in the Urban Aerosol: Results of a Seventeen Month Study in Hamburg, F.R.G.", Sci. Total Envir., **36**, 27-38.
Mueller, P.K., Mosley, R.W., Pierce, L.B.: 1972, "Chemical Composition of Pasadena Aerosol by Particle Size and Time of Day. IV. Carbonate and Noncarbonate Carbon Content", J. Coll. Interface Sci., **39**, 235-239.
Novakov, T., Chang, S.G., Harker, A.B.: 1974, Sulfates as Pollution Particulates: Catalytic Formation on Carbon (Soot) Particles", Science, **186**, 259-261.
Ogren, J.A., Charlson, R.J., Groblicki, P.J.,: 1983, "Determination of Elemental Carbon in Rainwater", Anal. Chem., **55**, 1569-1572.
Ogren, J.A., Charlson, R.J.: 1984, "Wet Deposition of Elemental Carbon and Sulfate in Sweden", Tellus. **36B**, 262-271.
Shah, J.J., Johnson, R.L., Heyerdahl, E.K., Huntzicker, J.J.: 1986, "Carbonaceous Aerosol at Urban and Rural Sites in the United States", J. Air Poll. Control Ass. **36**, 254-257.
Wolff, G.T., Groblicki, P.J., Cadle, S.H., Countess, R.J.: 1982, "Particulate Carbon at Various Locations in the United States", In: Particulate Carbon: Atmospheric Life Cycle (eds. G.T. Wolff and R.L.Klimisch), Plenum Press, New York, p. 297-315.

ATMOSPHERIC AEROSOL IN THE FINNISH ARCTIC: PARTICLE NUMBER CONCENTRATIONS, CHEMICAL CHARACTERISTICS, AND SOURCE ANALYSIS

A. VIRKKULA[1], M. MÄKINEN[1], R. HILLAMO[1], AND A. STOHL[2]

[1]Finnish Meteorological Institute, Air Quality Department, Sahaajankatu 22 E, SF-00810 Helsinki, Finland
[2]Institute for Meteorology and Geophysics, University of Vienna, Hohe Warte 38, A-1190 Vienna, Austria.

Abstract. Atmospheric aerosols have been measured in Finnish Lapland from January 1992 to June 1994. A seasonal cycle in particle number concentration, measured using a condensation particle counter, is observed. Aerosol samples have been collected by a virtual impactor (VI) in two size ranges. The filters have been changed every 48 hours. Most fine particle samples are acidic: 91% of the measured anions (neq/m^3) to measured cations (neq/m^3) ratios are above one. According to a trajectory statistical analysis, high sulphate concentrations come to Finnish Lapland from all of continental Europe. The source areas of the highest observed ammonium concentrations are in Western Europe.

1. Introduction

Aerosol measurements have been carried out in Finnish Lapland at a measurement station located at Sevettijärvi (69°35'N, 28°50'E, 130 m AMSL) (Fig. 1), 400 km north of the Arctic Circle. The closest bay of Barents Sea is 30 km NE (50°) of the station. By far the largest air pollution source area in the vicinity is the industrial area in Nikel/Zapolyarnyi, Russia, approximately 60 km SE (105°) of the measurement station. Pollution episodes from the industrial areas in Kola Peninsula typically occur two or three times monthly. The rest of the time either long-range transported pollutants from different parts of Europe or cleaner air from the Arctic Sea arrive at the measurement site. Results of the measurements at Sevettijärvi have been previously presented for example at Mäkinen (1994) and Virkkula et al. (1993 and 1994).

In this paper, particle number concentrations measured by two different monitors, chemical characteristics of aerosol and source analyses of ionic species are presented.

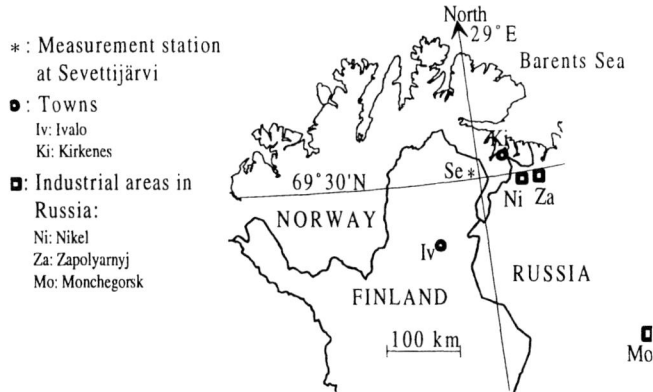

Figure 1. Location of Sevettijärvi measurement station and some population centres in Finnish, Norwegian and Russian Lapland.

2. Methods

At the station, particle number concentrations have been monitored using a condensation particle counter (CPC, TSI Model 7610) that detects particles greater than 14 nm in diameter, and using a laser particle counter (LPC, TSI Model 7430) that counts particles in two size channels (D > 0.3 µm and D > 0.5 µm). Particles with D > 2 µm are removed by the inlet tubing. The monitor data are stored as 1 h averages. Aerosol samples have been collected by a virtual impactor (VI) in two size ranges (aerodynamic diameter D_a < 2.5 µm and 2.5 - 15 µm). The filters have been changed every 48 hours. The VI samples have been analyzed for major water soluble inorganic ionic species (SO_4^{2-}, NO_3^-, Cl^-, NH_4^+, Na^+, K^+, Ca^{2+}, Mg^{2+}) by ion chromatography. Particle size distributions have been measured occasionally by 11-stage Berner low pressure impactors. After sampling the polycarbonate films have been analyzed by ion chromatography. The MICRON data inversion code (Wolfenbarger and Seinfeld, 1990) has been used to further evaluate the impactor raw data, i.e., concentration of each analyzed ion per stage. Gaseous air pollutants (SO_2, NO_2, and O_3) have been monitored by differential optical absorption spectroscopy. Meteorological parameters are monitored continuously.

Transport routes have been studied by calculating three-dimensional back-trajectories using the TRADOS model developed at the Finnish Meteorological Institute (Valkama and Rossi, 1992). The TRADOS trajectories are computed from the numerical meteorological forecasts of the Finnish version of the Nordic HIRLAM (High Resolution Limited Area Model) weather prediction model. For the year 1992 trajectories were calculated using the FLEXTRA trajectory model, which is described in detail by Stohl et al. (1995). FLEXTRA is based on analyzed model-level wind fields of the T213 L31 weather prediction model of the European Centre for Medium-Range Weather Forecasts (ECMWF). Three or four 96 hour back-trajectories arriving at ground level, 950 hPa and at 900 hPa have been calculated for each day.

The trajectories have been used both for analyzing single episodes and for statistics. A statistical method developed by Stohl (1995) has been used for identifying source areas. The method utilizes ambient air pollutant concentration measurements at a receptor site and corresponding back trajectories arriving at that site. The method is a two-step iterative procedure, which gives the source areas more accurately than previous methods.

3. Results and Discussion

Approximately 50 % of hourly averaged sulphur dioxide concentrations are very low (< 1 µg/m³). During pollution episodes from the south concentrations are typically in the range from 5 to 20 µg/m³. The highest SO_2 concentrations (>100 µg/m³) are measured during episodes from Nikel. In spring, ultrafine particle number concentrations, measured by the CPC, are higher both in cleaner air and during the Nikel episodes. This seasonal cycle can be explained by photochemical particle formation during the transport. Shaw (1991) has observed similar seasonal cycle in Central Alaska. In the accumulation mode particles (0.3 µm < D_a < 2 µm), measured by the LPC, concentrations have been lowest during the summer months. This is consistent with the fact that particulate sulphate concentrations are lowest in summer.

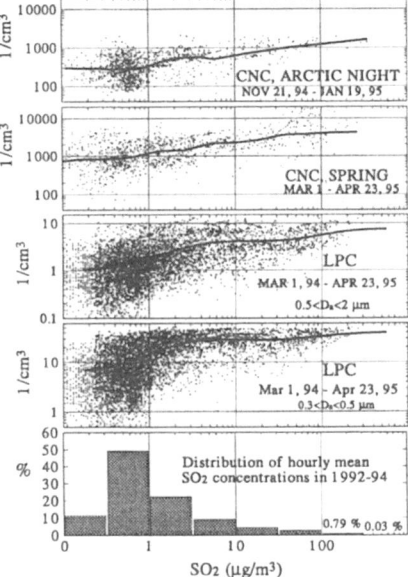

Figure 2. Simultaneously measured hourly averaged SO_2 concentrations and particle number concentrations (thick lines: weighted least square fit) and the distribution of SO_2 concentrations.

Sulfate is the most abundant inorganic ion in the fine particles (Table 1). The analyzed ions constitute most of the mass in fine particle samples. Li and Winchester (1989) have shown that contributions of organic ions like formate, acetate, propionate, and pyruvate are comparable to those of inorganic ions. It may be assumed that the missing mass in the fine particle mode is mostly composed of these ions and of carbon (Sheridan, 1989).

Table 1. Average contribution (in percents) of ion mass concentrations to the total mass concentration. Samples from July 16, 1992 to June 25, 1994.

	Cl⁻	NO_3^-	SO_4^{2-}	Na^+	NH_4^+	K^+	Mg^{2+}	Ca^{2+}	Not identified	Average total mass (µg/m³)
Fine	6.8	2.7	32.9	5.6	5.3	0.4	0.7	0.3	45.3	3.7
Coarse	20.1	2.5	4.7	12.1	1.0	0.7	1.4	0.8	56.7	1.6

Percentages are calculated from 100% × {Σ([ion X]/[total mass of filter])}/N

Most fine particle samples are acidic: 91% of the measured anions (neq/m³) to measured cations (neq/m³) ratios are above one (Fig. 3). H^+ is not analyzed, and thus it is assumed that the cation deficit can be explainded by H^+. During pollution episodes from Kola peninsula industrial areas particle samples often contain over six times more measured anions than cations. The acidity of the fine particle samples is mainly due to sulphate.

Fine particles measured during flows from the south are also acidic as shown in the size distribution of a sample collected in November 23-25, 1993 (Fig. 4), during an episode which was detected at most measurement stations in Finland. At Sevettijärvi SO_2 concentration rose to over 20 µg/m³, which is not very usual during southern flows. The source area of this episode was calculated by TRADOS to be in the region of St.Petersburg.

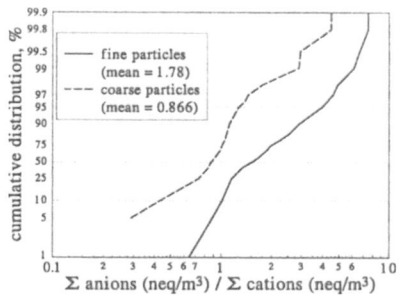
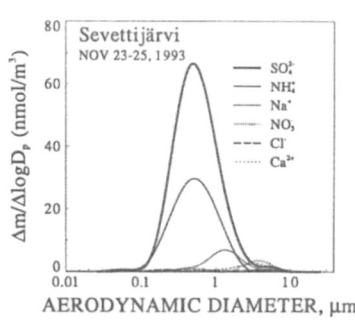

Figure 3. Cumulative distribution of ionic balance of Sevettijärvi filter samples in 1992 to 1994.

Figure 4. Size distribution of ions in aerosol in an episode, whose source was calculated to be St. Petersburg area.

The source area of high SO_2 concentrations is easy to find even by simple wind comparisons. The average SO_2 concentrations were highest in the wind direction sector which points directly to Nikel (Fig. 5) and most of the SO_2 exposure (concentration × exposure time) is accumulated during winds from the east.

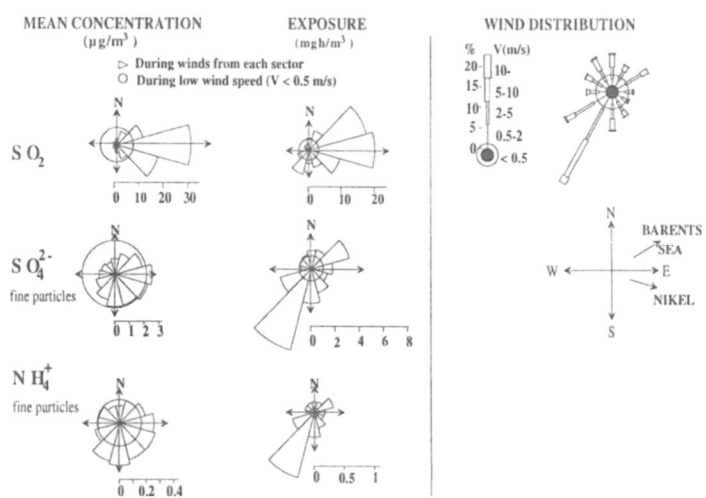

Figure 5. Mean concentration and exposure of sulphur dioxide, sulphate, and ammonium in each wind sector and wind distribution at Sevettijärvi in April 1992 - June 1994.

The source areas of sulphate and other particulate components, are much more difficult to define. Wind distribution shows only that most of the sulphate and ammonium exposure is accumulated during southwestern winds. Source areas of sulphate, ammonium, and sodium were calculated using the trajectory statistical method developed by Stohl (1995) (Fig. 6).

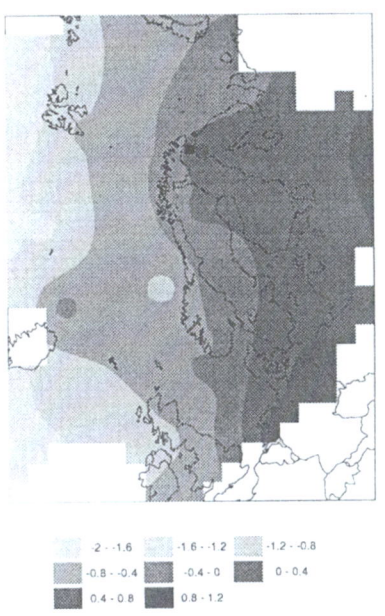 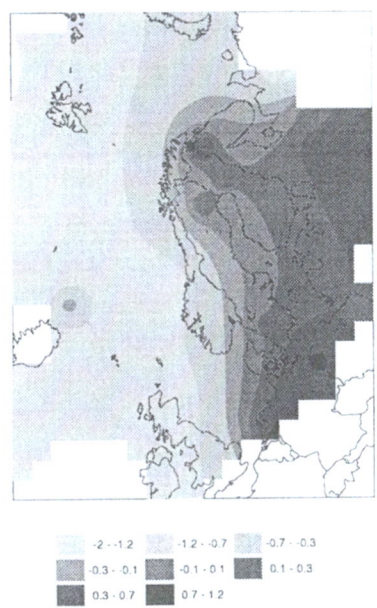

Total SO_4^{2-}, fine partices, $<C> = 1.52$ μg/m³   NSS SO_4^{2-}, fine particles, $<C> = 1.48$ μg/m³

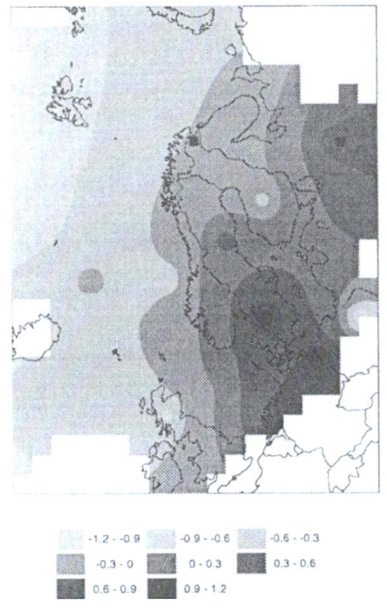 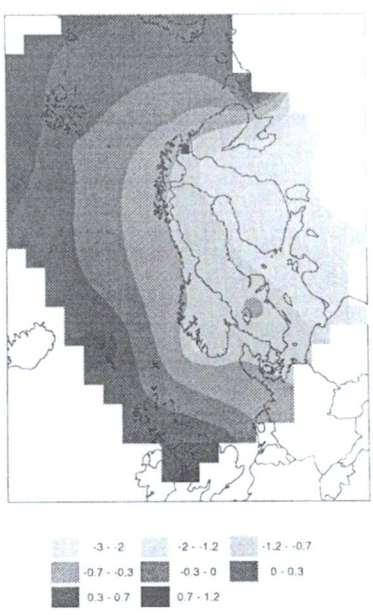

Total SO_4^{2-}, fine partices, $<C> = 1.52$ μg/m³   NSS SO_4^{2-}, fine particles, $<C> = 1.48$ μg/m³

Figure 6. The average observed concentration of total and non-sea-salt (NSS) sulphate, ammonium and sodium at Sevettijärvi, when the trajectories have passed over each grid square. Fine particle samples are from January 1992 to June 1994, coarse particle samples from July 1992 to June 1994. The 3D trajectories used for the analysis arrive at 950 hPa level. The concentrations are expressed in *lg(C/<C>)* where *<C>* is the arithmetic mean concentration of all the filters. The white areas are those grid squares, which have been crossed by less that ten trajectories.

The ability of the method to determine source areas is best shown when it is applied to sodium, whose source area is very clear: the calculated source area of sodium in coarse particles follows almost perfectly the Norwegian coast. The apparent source area in Britain is due to the short transport range of the coarse particles and to the fact that most trajectories from British Isles to Sevettijärvi come over Northern Atlantic.

According to the analysis, high sulphate concentrations come to Finnish Lapland from all of continental Europe. The analysis also shows clearly that the industrial areas around Nikel are a source of sulphate, even though they are so close to the Sevettijärvi station that SO_2 has little time to be oxidized to sulphate. The apparent source of sulphate close to Iceland may be interpreted to be the continuous volcanic activity of the island. The source areas of the highest ammonium concentrations are in Western Europe. This is also an expected result, as the Western European agriculture is a strong emitter of ammonia.

5. Summary and Conclusions

A seasonal cycle of ultrafine particle number concentrations, measured by the CPC, has been observed. It can be explained by photochemical particle formation during the transport. In the accumulation mode particles ($0.3 \: \mu m < D_a < 2 \: \mu m$), measured by the LPC, concentrations have been lowest in the summer months. The chemical analysis of the virtual impactor samples shows that most fine particles are acidic. The acidity of the fine particle samples is mainly due to sulphate. The new trajectory statistical method has proven to be a powerful method in analyzing source areas of various atmospheric components, even when measurements of only one station are used.

References

Mäkinen, M., Pakkanen, T., Hillamo,R., Virkkula,A., and Mäkelä,T. (1994) Ion balance of aerosol in Northern Finland. *Journal of Aerosol Science* **25**, May 1994, S299-300.

Li, S.-M. and Winchester, J. (1989) Geochemistry of organic and inorganic ions of late winter arctic aerosols. *Atmos. Env.* **23** 11 pp. 2401-2415.

Shaw, G., E. (1991) Physical properties and physical chemistry of Arctic aerosols. In Sturges, W.T. (Editor) *Pollution of the Arctic atmosphere*. Elsevier Science Publishers LTD, England. pp. 123-154

Sheridan, P.J. (1989) Characterization of size segregated particles. *Atm. Env.* **23** 2371-2386.

Stohl, A. (1995) Trajectory statistics - a new method to establish source-receptor relationships of air pollutants and its application to the transport of particulate sulfate in Europe. Submitted to *Atmos. Env.*

Stohl, A., Wotawa, G., Seibert, P., and Kromp-Kolb, H. (1995) Interpolation errors in wind fields as a function of spatial and temporal resolution and their impact on different types of kinematic trajectories. *J. Appl. Meteor.* In press.

Valkama, I. and Rossi, J. (1992) Description of the model TRADOS. In Klug, W., Graziani, G., Grippa, G., Pierce, D. & Tassone, C. *Evaluation of long range atmospheric transport models using environmental radioctivity data from the Chernobyl accident. The ATMES report*, Elsevier Publ.

Virkkula, A., Hillamo, R., Mäkinen, M., Mäkelä, T., and Pakkanen, T. (1994) Number concentrations of atmosheric aerosol in the Finnish Arctic. *Journal of Aerosol Science* **25**, May 1994, pp.S25-S26.

Wolfenbarger, J.K. and Seinfeld, J.H. (1990) Inversion of aerosol size distribution data. *J.Aerosol Sci.* **21**, 227-247

EVALUATION OF EXTREME POLLUTION EPISODES OVER LITHUANIA IN 1980-1994

A.MILUKAITE, V.JUOZEFAITE, A.MIKELINSKIENE,
B.GIEDRAITIS, A.GALVONAITE, P.JURGUTIS

Institute of Physics, A.Gostauto 12, 2600 Vilnius, Lithuania

Abstract. The investigation of SO_2, NO_2, benzo(a)pyrene (BP) and soot (C_{el}) has been carried out daily in the atmospheric air in the background station Preila (East shore of the Baltic Sea) since 1980. Over this period. daily concentrations of pollutants varied in the wide intervals in warm and cold period as well. From 15 years data typical episodes of highest and lowest concentrations of pollutants are chosen and analysed with respect to the air masses trajectories and meteorological conditions. The highest concentrations of SO_2, NO_2, BP, C_{el} were fixed, when the air masses passing Lithuania have been formed over Great Britain and Central Europe.

Key words: SO_2, NO_2, benz(a)pyrene, soot, concentration, air mass trajectories, Baltic sea.

1. Introduction

Observation of high concentrations of various pollutants at the cleanest sites of the world allow to admit that the reason of different air pollution depend on the long-range transport of pollutants. Black and white episodes have been detected by studying SO_2, NO_2, soot and NH_4^+, SO_4^{2-}, NO_3^- in the atmosphere over Sweden on several episodes in 1973-1975. In November 1975, during the black episode, the concentrations of SO_2, NO_2 and soot varied in intervals 1.0-15.0 ppb, 1.5-33.0 ppb and 5.0-38.0 µg/m^3, respectively (Brosset, 1976). During the winter of 1977, on several days, the concentrations of polycyclic aromatic hydrocarbons (PAH) have been observed. Among them the concentration of BP varied in the interval 0.01-2.35 ng/m^3 in the Norwegian sampling station, Birkeness, and 0.09-4.32 ng/m^3 in the Swedish sampling station, Rorvik. Variation of concentration of PAH have been related to the trajectories of the air masses (Bjorseth *et al.*, 1979). The dependency of the concentrations of PAH, SO_2, NO_2, soot and chlorinated hydrocarbons on long-range transport have been studied in 12 episodes on duration 3-4 days of sampling period in 1989-1990 (Lunden *et al.*, 1994). The first attempts to coordinate meteorological parameters and concentrations, of more than 30 pollutants, have been done on the eastern shore of the Baltic sea (Sopauskiene et al., 1989, Milukaite et al., 1994). As a continuation of this problem is the present study, purpose of which is to characterize the extreme high and low concentrations of acydifying and other accompanying pollutants, to identify regions of their sources and regularities of their transport to Lithuania over the last 15 years.

2. Materials and Methods

Since 1980 the samples of gaseous and aerosol pollutants have been collected daily, from 9^{00} a.m. to 9^{00} a.m. of the other day, at the background station Preila. The acetilcelulose fibre filters AFA-XA-20 have been used for BP sampling and glass fibre filters for sampling of soot (C_{el}). Atmospheric aerosols were collected at the velocity of drawing up to 2 m^3/h. Benzo(a)pyrene has been detected by the spectrofluorescence method at the temperature of liquid nitrogen. The sensitivity of the method is 0.02 ng/m^3, the repeatability ±20% (Milukaite, 1986). C_{el} has been determined by reflectance method calibrated by the gas chromatographic method based on the burning of aerosol samples in the atmosphere of oxygen. The sensitivity of the method is 0.01 µg/m^3 (Armalis, 1986).

SO_2 has been adsorbed in the glass tubes with an active layer of $NaHCO_3$ and determined by the turbidimetric method. The sensitivity of the method is 0.2 µg/m^3. NO_2 has been absorbed in the trietalonamine solution and the concentration of NO_2 has been determined by the colorimetric method. The sensitivity of the method is 0.3 µg/m^3 (Giedraitis, 1986).

The region from which the pollutants were supposed to be transported to the observation site was determined by the back trajectories of the air mass transport. They were calculated from the maps of the baric topography on the level of 850 mb for the preceding 48 hours (Galvonaite et al., 1986).

3. Results and Discussion

During the investigation period drastic differences in daily concentrations of pollutants have been observed in the background station Preila. Several episodes of high (in some cases extremely low) concentrations, clearly influenced the monthly concentration of BP (for the duration of 15 years) were chosen and compared with the concentrations of SO_2, NO_2 and C_{el} (Tables 1 and 2). Date of investigation presented in Tables 1 and 2 show, that during the extreme episodes concentrations of pollutants on several days varied in intervals: 0.2-43.0 µg/m^3 for SO_2, 0.3-50 µg/m^3 for NO_2, 0.02-25.08 ng/m^3 for BP and 0.01-30.05 µg/m^3 for soot. Correlation of concentrations of various pollutants with microclimatic parameters did not solve the problem of extreme concentrations (Milukaite at al., 1994). The reason of differences in concentrations have been searched in atmospheric circulation. With this purpose concentrations of pollutants on extreme episodes were analysed with respect to concrete air mass trajectories at that time. (Fig.1). Five episodes from cold (E_c) and warm (E_w) seasons have been analysed. Episode E_{1c} shows the extremely high concentrations of BP, SO_2, NO_2 that are coming to the sampling station from the "black triangle" - triangular area 25000 km^2 at the borders between the Czech, Germany and Poland, where local soft coal with a high content of sulphur has been used in power plant and industry. E_{2c} and E_{3c} reflects the major influence of England to the concentrations of BP, SO_2, NO_2 and soot. E_{4c} shows the highest concentrations of BP, SO_2 and soot when the air masses are passing "black triangle" and the moderate level of concentrations of the same pollutants when air masses are coming from the northern part of the West Europe. The high level of NO_2 is observed at this time. Recently episode E_{5c} shows that the level of BP concentration is

TABLE I

Concentration of the analysed pollutants during the extreme episodes of pollution in cold season in Preila

Episode	Number of trajectory	Date	BP ng/m^3	soot μg/m^3	SO$_2$ μ/m^3	NO$_2$ μg/m^3
E_{1c} 1984 December			$\bar{c}$ =1.49		$\bar{c}$ =12.5	$\bar{c}$ =15.8
	1	18-19	4.88	-	29.4	<0.3
	2	19-20	13.50	-	24.9	37.5
	3	20-21	15.40	-	33.9	31.6
	4	21-22	5.68	-	30.6	31.5
	5	22-23	4.80	-	0.8	3.1
	6	23-24	1.26	-	6.6	11.9
E_{2c} 1986 October			$\bar{c}$ =2.15	$\bar{c}$ =3.35	$\bar{c}$ =11.5	$\bar{c}$ =16.7
	1	10.16-17	14.80	9.40	43.0	4.5
	2	17-18	19.05	0.66	21.0	10.0
	3	18-19	16.90	2.60	22.0	32.0
E_{3c} 1986 November			$\bar{c}$ =2.15	$\bar{c}$ =3.35	$\bar{c}$ =11.5	$\bar{c}$ =16.7
	4	11. 6- 7	0.24	0.01	<0.2	5.6
	5	7- 8	21.21	1.10	1.2	42.0
	6	9-10	18.72	4.00	1.7	10.0
	7	10-11	4.49	5.00	10.0	47.0
	8	11-12	25.08	5.20	8.0	24.0
E_{4c} 1990 February			$\bar{c}$ =1.32	$\bar{c}$ =2.95	$\bar{c}$ =5.4	$\bar{c}$ =17
	1	02. 1- 2	6.40	5.77	8.3	20.0
	2	2- 3	5.40	5.92	5.4	26.0
	3	3- 4	2.86	4.84	8.6	28.0
	4	4- 5	2.07	1.23	-	-
	5	5- 6	4.32	5.13	5.1	29.0
	6	6- 7	8.09	30.05	15.4	28.0
	7	7- 8	3.58	12.25	12.9	33.0
	8	8- 9	1.11	1.89	8.2	24.0
E_{5c} 1994 December			$\bar{c}$ =1.35		$\bar{c}$ =7.2	$\bar{c}$ =13.8
	1	12. 3- 4	14.40	4.69	20.1	4.9
	2	4- 5	12.80	25.74	3.4	9.7
	3	8- 9	11.84	10.90	3.9	20.0
	4	9-10	4.80	3.57	12.7	25.3
	5	10-11	11.84	1.76	5.9	1.3
	6	11-12	0.96	0.40	3.9	17.5

the same, as 10 years before, when air masses are coming from England and "black triangle", while the concentrations of SO$_2$ and NO$_2$ are lower in recent years. E_{1w} and E_{2w} episodes show that the high concentrations of BP, SO$_2$, and NO$_2$ were observed in Preila when air masses were coming from England in warm season as well. During the episode E_{3w} different air masses carried mostly moderate concentrations of BP and SO$_2$ and in several days relatively high concentrations of NO$_2$ from the East Europe. The air masses from the East in most cases are following by anticyclone from Eurasia. This process usurps the wide territories, is long in term and often transport to Lithuania moderately high concentrations of NO$_2$. The repeatability of eastern air masses is relatively low (~20%) and the influence of anthropogenic loading is inconsiderable, while the air masses of the West and North-West directions are prevailing in Lithuania. Episode E_{4w} shows that the low concentrations of BP, SO_2 and soot but high concentrations of NO_2 are coming with the air masses from Scandinavia. In the episode E_{5w} on 15-16 of September the highest concentration of BP in warm season consisting 23.04 ng/m³ was received. High concentrations of soot were observed on 15-16 of September as well. The episode E_{5w} is mostly polluted by BP in warm season in the

TABLE II

Concentration of the analysed pollutants during the extreme episodes of pollution in warm season in Preila

Episode	Number of trajectory	Date	Concentration			
			BP ng/m³	soot µg/m³	SO_2 µg/m³	NO_2 µg/m³
			c =1.83		c =2.7	c =2.6
E_{1w}	1	06.15-16	6.35	-	-	2.2
1982	2	20-21	1.12	-	1.46	4.7
June	3	21-22	10.66	-	1.10	1.5
			c =1.83		c =2.7	c =2.6
E_{2w}	4	09.16-17	<0.02	-	6.6	7.3
1982	5	17-18	6.65	-	3.7	6.0
September	6	21-22	6.66	-	18.4	10.5
			c =0.50		c =4.1	c =21.4
E_{3w}	1	08. 4- 5	0.46	-	1.1	-
1984	2	5- 6	1.08	-	3.1	7.6
August	3	6- 7	1.80	-	3.0	2.5
	4	7- 8	0.74	-	0.5	38.6
	5	8- 9	0.96	-	10.1	1.4
	6	9-10	0.68	-	1.8	20.7
	7	10-11	0.84	-	3.5	40.6
	8	11-12	0.66	-	3.0	9.7
	9	12-13	0.90	-	5.1	9.7
	10	15-16	0.68	-	3.1	14.1
			c =0.12	c =0.96	c =1.8	c =17.4
E_{4w}	1	08. 5- 6	0.02	0.09	2.4	8.0
1989	2	6- 7	0.05	0.12	1.0	35.0
August	3	7- 8	0.02	0.34	1.2	43.0
	4	8- 9	0.02	2.41	3.2	29.0
	5	9-10	0.02	0.71	1.2	38.0
	6	10-11	0.02	0.39	2.0	47.0
	7	11-12	0.80	2.41	2.6	47.0
	8	16-17	0.71	4.22	2.6	11.0
			c =1.16	c =1.00	c =1.6	c =13.2
E_{5w}	1	09.10-11	3.02	1.18	1.1	3.7
1991	2	11-12	8.00	0.29	1.1	3.7
September	3	15-16	23.04	4.69	1.1	0.9
	4	16-17	4.03	0.63	7.9	6.1
	5	25-26	1.10	-	-	-
	6	26-27	21.60	-	3.7	3.7

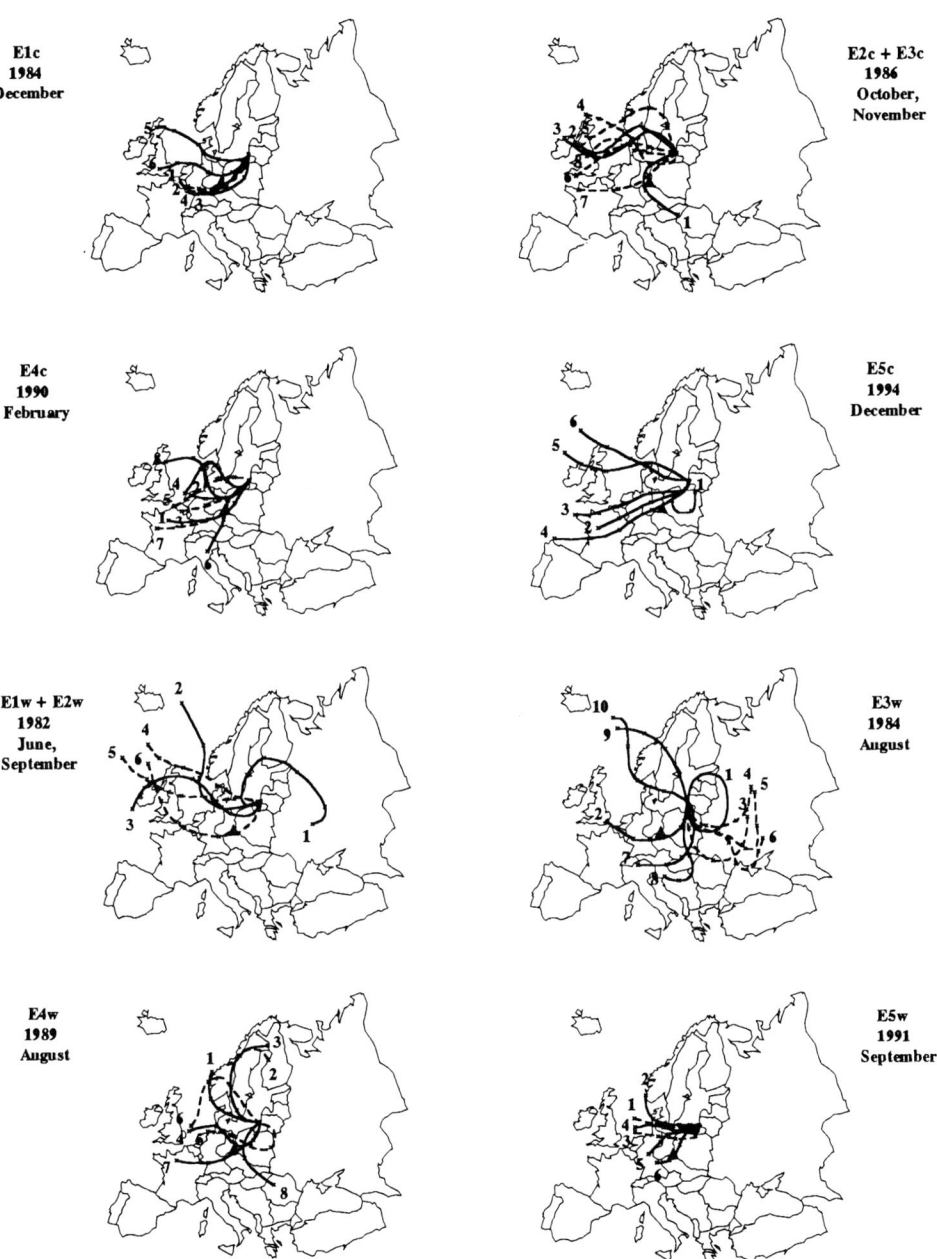

Fig.1. 48h back trajectories of air masses during the extreme episodes of atmospheric pollution in Preila in cold and warm season.

background station Preila per 15 years and is not in accordance with the SO_2 and NO_2 concentrations, while the air masses for the duration 48 hours were coming slowly from the North sea. The longer period of back air masses trajectories needs to be analysed, to identify the source of pollution. In warm season of the year the concentrations of pollutants during the period of maximum episodes are twice lower if compared its with cold season. Among of the presented episodes is evident that the low concentrations are characteric for the air masses having been formed in Atlantic ocean and passing through the Scandinavia or North sea (E_{3w} 9,10 tr.E_{4c} 8 tr., E_{5c} 6 tr.). Relatively low concentrationsof pollutants were observed when the centre of baric derivatives was over Lithuania (E_{3w} 1 tr.,). In some episodes differences in concentration of various pollutants in the same air mass may appear by, partly different, sources, different physical stage and different residence time of each pollutant.

4. Conclusions

One of the main factors defining the extreme episodes of pollution in the background station is the region of air mass formation: the highest concentrations of SO_2, BP and soot is coming to the sampling station with the air masses passing "black triangle" and England, the highest concentrations of NO_2 is coming to Lithuania with the air masses from the northern part of the West Europe, the lowest concentrations of BP, SO_2 and soot are observed in the Atlantic air masses coming via Scandinavia.

References

Armalis S., Nika A.: 1986, "Methods for organic and elemental carbon determination in atmospheric aerosols", Atmospheric Physics, Vilnius, **11**, 155-159. (in Russian)
Bjorseth A., Lunde G. and Lindskog A.: 1979, "Long range transport of polycyclic aromatic hydrocarbons", Atm. Envir., **13**, 45-53.
Brosset C.: 1976, "Air-borne particles: Black and white episodes", Ambio, **5, 4**, 157-163.
Galvonaite A. & Mikelinskiene A.: 1986, "On the influence of wind regime on the transportation of pollutants over Lithuania", Atmospheric Physics, Vilnius, **11**, 60-65 (in Russian).
Giedraitis B. 1986. Analysis of samples. In: Protection of atmosphere from pollution, Vilnius, **11**, 14-16 (in Russian).
Lunden E., Lindskog A., Mowrer J.: 1994, "Concentrations and fluxes of organic compounds in the atmosphere of the Swedish West coast", Atm. Environ., **28**, 3605-3615.
Milukaite A.1986, Fine-structural fluorimetry. In: Unificated methods of the environment background monitoring, Moscow, p. 46-53 (in Russian).
Milukaite A., Giedraitis B. Juozefaite V., Galvonaite A.: 1994, "The influence of long-range pollutant transport on the variation concentrations of gaseous and particulate admixtures in atmospheric air in Lithuania" Proceedings of EUROTRAC Symposium, 1994, Garmisch-Partenkirchen, Germany 11-15 April, 1994. SPB Academic Publishing bv. Den Haag. The Netherlands. 741-746.
Nielsen T.: 1988, "The decay of benzo(a)pyrene and cyclopentino(cd)pyrene in the atmosphere", Atm. Environ., **22**, 10, 2249-2254.
Sopauskiene D., Davidaviciene L., Galvonaite A.: 1989, "Meteorological aspects of South Baltic background aerosol formation", Atmospheric Physics, Vilnius, **13**, 140-148 (in Russian with English summary).

SOME OBSERVATIONS OF POLLUTANT FLUXES OVER THE SUDETEN, SOUTH-WESTERN POLAND

J. ZWOZDZIAK, A. ZWOZDZIAK, G. KMIEC and K. KACPERCZYK

Environment Protection Engineering Institute, Technical University of Wroclaw, Wybrzeze Wyspianskiego 27, 50-370 Wroclaw, Poland

Abstract: Field investigations conducted over the period 1988-1993 proved large variations in daily SO_2, total sulphur and trace metals concentrations over the Sudeten within the Polish part of the Black Triangle. Special attention should be paid to the concentrations of such trace metals as zinc and cadmiun. A detailed analysis of a synoptic chart for central Europe in the days with SO_2 episodes as well as in the days preceding and following them allowed us to separate two general categories of meteorological conditions that can result in the occurrence of high pollutant concentrations, a slowly moving high pressure system or a weather front.
Key words: atmospheric aerosol composition, pollutant episode, Sudeten

1. Introduction

In the years 1988/1989, there were built-up four measurement stations in the western part of the Sudeten Mountains, on the Polish-Czech border within the extremely polluted region of the Black Traingle. Since 1989 several research projects have been undertaken based on weekly or monthly intensive field measurement campaigns [Zwozdziak et al., 1990, 1993, 1995].

The aim of this paper is to provide the evidence of the great variability of pollutant levels and to present the complexity of chemical changes in the atmospheric aerosol composition in the region of interest.

2. Materials and methods

Aerosol samples were collected for 24-hour during a week (once a month from January to December, 1989) or a month (in February, May, July and October, 1990-1993) of field campaigns at four sampling sites located in the western part of the Sudeten Mts. Thus, the sites No. 1 and No. 2 were placed at the summits of Sniezne Kotly (1415 asl) and Szrenica (1315 m asl), respectively. The site No. 3 was situated at the altitude of 762 m asl (Rozdroze), 10 km from the summit of Szrenica in the NW direction and the site No. 4 was located at the RIVM site (Czerniawa, 650 m asl), approx. 15 km from the summit of Szrenica in the NW direction. The samples were analysed for sulphur dioxide (SO_2), nitrogen dioxide (NO_2), and total sulphur using conventional colorimetric techniques with the detection limits of 0.1 µg SO_2, 0.04 µg NO_2 and 0.5 µg S in 1 ml absorbing solution [Warner, 1976; Harrison and Perry,1986].

In the field campaigns during 1994 at the Szrenica site, an Andersen automatic dichotomous impactor (Series 245) was used for aerosol sampling and fractionating. The inlet cut-off point in particle diameter was 10 µm, while the size-fractionating point was

2.5 μm. Both fine and coarse aerosol samples were analysed for metal concentrations using inductively coupled plasma spectrometer (PU 7000 Philips Scientific). The filters were extracted with nitric acid (Merck, Suprapur) prior the analysis.

For the analysis of pollutants and meteorological data for episodes in 1994 the results from the automatic station for air monitoring placed in Czerniawa were used. This station is under control of the Provincial Inspectorate of Environment Protection in Jelenia Gora, Poland. A detailed analysis of a synoptic situation in central Europe in the days of episodes as well as in days preceding and following them was obtained from the Polish Weather Service.

3. Results and Discussion

The pollutant concentrations in the air over the Sudeten are characterized by a large variability. Annual avarages at the four sampling sites ranged from 6 to 35 μg SO_2 /m^3, from 3 to 6 μg NO_2 /m^3 and from 5 to 18 μg S/m^3. Based on the earlier investigations we tried to find some crucial relationships between the daily pollutant concentrations and meteorological parameters. The attempt to find a satisfactory correlation had failed [Zwozdziak et al., 1994]. Next, the results were classified according to the sectors of the inflow of air masses and similar analyses were carried out. No wonder that the attempts to average the concentrations obtained reveal that the greatest variations and maximum concentrations of sulphur compounds as well as other pollutants (e.g. iron, zinc) occurred when the winds were blowing from the northwest to south directions. Under those conditions, this area is exposed to polluted air masses originating in this part of the Black Triangle where extracting and power industries are concentrated.

In this study we applied the other method allowing the explanation of SO_2 concentrations on the basis of the hourly measurements of its concentrations, wind direction and speed in Czerniawa. The episodes of concentrations were chosen from the whole year 1994. We deal with the episode when the SO_2 concentrations in the air exceed 200 mg/m^3. The results are presented in Table I. Over the period of one year 34 episodes were recorded. In this region, a typical episode has lasted only for a few hours (from 2 to 4 hours). Their analysis allows us to draw the following general conclusions:

– The episodes occur in every month, but their frequency increases slightly during cold months. The episodes occur at any time of day.

– During episodes the correlation between the concentrations of SO_2 and NO_2 is high (corr.coef.>0.7, sign. level= 0.01), and that between the concentrations of SO_2, NO_2 and NO is satisfactory (corr.coef.= 0.4-0.6, sign. level = 0.01). During the inflow of air masses from eastern directions at night, the concentration of NO increased slightly; at the same time the concentrations of SO_2 and NO_2 were very low.

– The episodes of SO_2 concentrations are most frequently associated with the change in wind direction from westerly (SW-NW) to south-easterly and vice versa.

A detailed analysis of a synoptic situation in central Europe in the days of episodes as well as in days preceding and following them allows us to separate two general categories of meteorological conditions that can, under some circumstances, lead to the occurrence of high pollutant concentrations.

TABLE I

Date	Max. conc. [µg/m^3]	Duration	Wind direction [deg]	Category of meteorological conditions
06.02	248	9pm - 12pm	301 - 330	weather front warm
07.02	221	2pm - 3pm	301 - 330	weather front warm
09.02	332	1pm - 2pm	260 - 280	weather front occluded
19.02	232	10am - 1pm	260 - 273	weather front cold
23.02	385	9am - 12am	274 - 303	weather front warm
24/25	507	9pm - 1am	262	weather front warm, cold
25.02	248	10pm - 11pm	270	weather front stationary
05.03	290	3am - 5am	270	weather front cold
21.04	235	3pm - 4pm	317 - 320	high pressure system
29.04	676	9pm - 11pm	202 - 238	high pressure system
25.05	210	8pm - 9pm	250 - 280	weather front warm, cold
26.05	283	2am - 3am	260 - 270	weather front warm, cold
01.06	323	10pm - 11pm	260	high pressure system
07.06	442	1am - 6am	250 - 260	weather front warm
08.06	467	6am - 7am	260 - 270	high pressure system
09.08	208	10pm - 11pm	270 - 300	weather front cold
16.08	275	1pm - 2pm	295 - 300	high pressure system
21.08	295	5am - 6am	254 - 260	high pressure system
24.08	291	9pm - 10pm	236 - 300	weather front warm
26.08	303	8pm - 10pm	268	weather front occluded
28/29	365	11pm - 2am	242 - 256	weather front cold
01.09	222	10pm - 11pm	260 - 266	weather front warm
23.09	307	4am - 6am	220	high pressure system
25.09	330	8am - 10am	260 - 300	weather front cold
25/26	262	11pm - 1am	221 - 245	weather front cold
05.10	246	5pm - 6pm	260 - 270	high pressure system
10/11	861	10pm - 1am	stagnation	high pressure system
12.10	276	8pm - 9pm	260 - 268	high pressure system
13.10	680	10 hours	268 - 289	high pressure system
15/16	507	10pm - 2am	267 - 273	high pressure system
22.11	253	1am - 2am	269 - 273	high pressure system
27.11	345	0.00 - 4pm	265 - 291	weather front warm
10.12	775	3am - 4am	260 - 267	weather front cold
21.12	237	9am; 10pm	stagnation	weather front cold, occluded

The first of these is the high pressure system with a center under the Middle Europe with its light or moderate winds. The episode occurs when the center of the high pressure system moves slowly in the eastern direction which is associated with a gradual change in the wind direction. Then the highly polluted masses of air which stagnate in the Czech Valley move slowly over the Sudeten ridges. Such cases were recorded in April (21;29), June (1;8), August (16, 21), September (23), October (10-11, 13, 15) and November (22). The second category of meteorological conditions differs significantly in character from the stagnation that prevails in high pressure systems. Several examples of the occurrence of high SO2 concentrations near weather fronts were found in February (7, 9, 23, 24/25), March (5), May (25/26), August (24, 26, 28, 29), November (27) and December (10, 21). Such cases were described earlier [Zwozdziak, 1993] when analysing the concentrations of sulphur compounds in the air of the Mount Szrenica.

Table II summarizes the results of heavy metals concentrations obtained during field campaigns in July, 1994 and October, 1994 for the Szrenica site. In Table II, great temporal variation in the elemental concentrations (reflected by the standard error) and significant concentrations of many metals present in both aerosol fractions are shown. Total concentrations of cadmiun, zinc, copper, iron change generally in the range of about an order of magnitude. No significant differences in elemental concentrations of fine and coarse aerosol fractions suggest the possibility of their multiple sources as well as the effect of ambient aerosol dynamics on shaping the metal distributions.

TABLE II

Concentration of some elements in the aerosol of Szrenica in the Sudeten

Element	Coarse particles [ng/m^3]			Fine particles [ng/m^3]		
	Mean	Median	(S.E.)	Mean	Median	(S.E.)
Al	307.9	278.0	42.7	162.1	103.0	39.8
Ca	1018.0	712.0	23.3	594.6	491.0	99.6
Mg	66.4	58.0	9.2	78.7	42.2	25.2
Na	524.0	493.0	71.0	441.0	323.0	84.0
K	793.0	819.0	66.0	754.0	699.0	121.0
Fe	259.0	267.0	36.0	167.0	95.0	35.0
Mn	5.6	5.6	0.8	4.5	3.7	0.6
Pb	76.0	31.0	25.0	73.0	48.0	16.0
Zn	157.0	82.0	46.0	123.0	83.0	30.0
Cu	19.3	18.6	1.7	16.5	17.2	2.4
Cd	3.1	2.1	0.8	1.8	1.2	0.5

Note: The standard errors (S.E.) of the means are given to illustrate the fluctuations of the means. (S.E.) is calculated using n=23 values

If we compare the mean concentrations obtained in this study with those reviewed for the other rural areas in Europe [Lee et al., 1994; Chester et al., 1994], it is evident that the mean concentrations of manganese and lead found in this study are of the same order of magnitude as those reported in trace element studies carried out at the rural sites in UK and in the North Sea. Daily cadmium and zinc concentrations in this study ranged approximately from 1 to 10 ng/m^3 and from 30 to 1300 ng/m^3, respectively, and were roughly consistent with those reported for various urban sites in Europe. The concentrations of iron in the range from 190 to 540 ng/m^3 at the Szrenica site were also higher than those found at the rural sites in Europe and were comparable to the concentrations at the urban sites [Lee et al., 1994]. From all studies of metals, special attention should be paid to those dealing with the zinc and cadmium concentrations. We think that particularly zinc might have been a marker for brown coal combustion processes in the Black Triangle. Very high zinc concentrations were also determined in precipitation samples [Kmiec et al., 1994].

4. Conclusions

The important factor in day to day variability of pollutant levels observed is the synoptic-scale circulation of air masses prior to their arrival. High concentrations of pollutants are likely to be found under the following meteorological conditions: (1) slowly moving anticyclonic systems with gradual change in the wind direction (from NW or SW to SE and vice versa) ; (2) weather fronts oriented towards west - east so that the low level winds ahead the front continue to accumulate pollutants. These conditions will be studied in the future.

Concentrations of atmospheric cadmium, zinc and iron were found to be broadly comparable with those characteristic of some urban areas in Europe and several times higher than those at rural sites. The elements did not show significant variability in particle size distribution of ambient aerosol.

Acknowledgement

This study is partly supported by the European Union's Co-operation in Science and Technology with Central and Eastern European Countries (the Emission Abatement Strategies and the Environment (EASE) for the Black Triangle project).

References

Chester, R., Bradshaw, G.F., Corcoran, P.A.: 1994, *Atmos. Environ.* 28, 2873-2883.
Harrison, R.M. and Perry, R., : 1986, *Handbook of Air Pollution Analysis*, Chapman & Hall, London, p.279.
Kmiec, G., Zwozdziak, J.W., Zwozdziak, A.B.: 1995, *Environ. Prot. Engng*, in press.
Kwiatkowski, J. and Holdys, T.: 1985, *Warunki klimatyczne Karkonoszy* (in Polish) in *Karkonosze Polskie* (ed. Jahn, A.), Ossolineum Wroclaw, p. 87.
Lee, D.S., Garland, J.A., Fox, A.A.: 1994, *Atmos. Environ.* 28 , 2691-2713.
Warner, P.O.: 1976, *Analysis of Air Pollutants*, John Wiley & Sons, New York, London, p.108.
Zwozdziak, J. W. and Zwozdziak, A.B.: 1990, *Environ. Prot. Engng* 16 , 89-98.
Zwozdziak, J .W. and Zwozdziak, A.B.: 1994, in: *Pollution Control and Monitoring* (eds: Baldasano, J.M., Brebbia, C.A., Power, H., Zanetti, P.) vol.2, Computational Mechanics Publications, Southampton Boston, 261-269.
Zwozdziak, J.W.: 1993, *Arch. Ochr. Srodow.* (in Polish) , 3-4, 11-23.
Zwozdziak, J.W., Zwozdziak, A.B., Kmiec, G.: 1995, *Environ. Prot. Engng*, in press.

DEVELOPMENT OF HYBRID LRT MODEL TO ESTIMATE SULFUR DEPOSITION IN JAPAN

H. HAYAMI* AND Y. ICHIKAWA
Central Research Institute of Electric Power Industry
2-11-1 Iwado Kita, Komae-shi, Tokyo 201, Japan

Abstract. We have developed a hybrid long-range transport (LRT) model to estimate long-term sulfur deposition amounts in Japan. This model combines a trajectory model for the LRT with an Eulerian model for the short-range transport and deposition. The hybrid model shows the ability to predict concentrations influenced by large nearby sources, which the trajectory model we previously developed consistently underestimated. The hybrid model is designed as an engineering model, which allows for long-term estimation without the requirements of detailed data on meteorology, surface condition, and emission over the whole domain, and huge computer resources required for comprehensive Eulerian models.

1. Introduction

Many long-range transport(LRT) models have been developed to investigate acidic deposition. Most of these efforts have been devoted to the estimation of acidic substances in North America and Europe (e.g., Carmichael *et al.*, 1986). Recently, modeling efforts have begun in East Asia where anthropogenic emissions are increasing rapidly and the vast differences in climatology and topography pose significant challenges to LRT models (e.g., Murao *et al.*, 1993).

We previously developed a one-layer trajectory LRT model to estimate sulfur deposition in Japan (Ichikawa and Fujita, 1993). The model will be referred to the box-trajectory model. The results of the box-trajectory model agreed well with the observed data of the dry deposition in less pol-

---

*Present address: Dep. of Chemical & Biochemical Eng., The University of Iowa, Iowa City, IA 52242 USA

luted areas and the wet deposition over Japan. The box-trajectory model seems applicable for prediction in less polluted areas. For the dry deposition in polluted areas, however, the box-trajectory model consistently underestimated the observed data. This may be due to the assumed vertical uniformity of the model. The box-trajectory model assumes a uniform concentration between the ground and the mixing height. The concentrations observed in polluted areas are thought to be strongly affected by the emission intensity and height of the nearby sources. Because the box-trajectory model doesn't express the vertical profile, the dry deposition, i.e., the concentrations near the surface might be underestimated in polluted areas. An Eulerian LRT model could express the profile and bring better results, but it is difficult to prepare the necessary detailed emission and meteorological data over the Asian domain. Substances emitted from far sources are expected to be well mixed during the LRT and uniformly distributed in the vertical direction, which means the LRT could be represented with the box-trajectory model.

Based on the above consideration, we propose a new engineering LRT model, a hybrid LRT model, which combines a trajectory LRT model with an Eulerian model. This combination could bring more accurate predictions by retaining the strong points of both the trajectory and Eulerian models.

2. Model Description

The hybrid model treats advection and diffusion of gaseous sulfur dioxide SO_2, particulate sulfate $SO_4^{2-}(p)$, and sulfate in cloud water $SO_4^{2-}(c)$; dry deposition of SO_2 and $SO_4^{2-}(p)$ to the surface; conversion of SO_2 to $SO_4^{2-}(p)$; and in-cloud process of SO_2 and $SO_4^{2-}(p)$ to $SO_4^{2-}(c)$. Sub-cloud processes including SO_2, $SO_4^{2-}(p)$, and $SO_4^{2-}(c)$ to sulfate in rainwater $SO_4^{2-}(r)$ are also calculated. Each process is assumed to be driven with a linear rate proportional to the concentration of the individual substance. The trajectory model used in the hybrid model is the box-trajectory model, which is described in Ichikawa and Fujita (1995, presented in this issue), so we don't describe it in this paper.

2.1. EULERIAN MODEL

To use for the hybrid model, we developed a three-dimensional Eulerian model which treats the same processes and substances as the box-trajectory model. All the substances satisfy the advection-diffusion equation,

$$\frac{\partial c_i}{\partial t} + \frac{\partial u_j c_i}{\partial x_j} = \frac{\partial}{\partial x_j}\left[K_{jj}\frac{\partial c_i}{\partial x_j}\right] + R_i + E_i, \qquad (1)$$

where c_i is the concentration of the substance i, u_j is the wind velocity in the direction x_j, K_{jj} is the diagonal component of the eddy diffusivity tensor, R_i denotes the process term and E_i is the source term. This equation is solved using time-splitting; i.e.,

$$\frac{\partial c_i}{\partial t} + L_x c_i + L_y c_i = 0 \tag{2}$$

$$\frac{\partial c_i}{\partial t} + L_z c_i = 0 \tag{3}$$

$$\frac{\partial c_i}{\partial t} = R_i + E_i, \tag{4}$$

where L_j represents the one-dimensional transport operator:

$$L_j c_i = \frac{\partial u_j c_i}{\partial x_j} - \frac{\partial}{\partial x_j}\left(K_{jj}\frac{\partial c_i}{\partial x_j}\right).$$

Equations 2 and 3 are discretized by the Petrov-Galerkin finite element method and time-integrated by the Crank-Nicolson method. Both methods are used by Carmichael et al. (1986), but we two-dimensionally solve the horizontal equation to reduce the numerical diffusion. Because the rate of each process is linearly proportional to the concentration of the individual substance, Equation 4 is expressed as

$$\begin{bmatrix}\partial c_1/\partial t \\ \partial c_2/\partial t \\ \partial c_3/\partial t\end{bmatrix} + \begin{bmatrix} k_1+k_2+k_4 & 0 & 0 \\ -(3/2)k_1 & k_2'+k_4' & 0 \\ -(3/2)k_4 & -k_4' & k_5\end{bmatrix}\begin{bmatrix}c_1\\c_2\\c_3\end{bmatrix} = \begin{bmatrix}E_1\\E_2\\0\end{bmatrix}, \tag{5}$$

See Table 1 regarding the k's, and the c's and E's indicate the concentrations and emissions of SO_2, $SO_4^{2-}(p)$, and $SO_4^{2-}(w)$, respectively. Equation 5 has an analytical solution. The Eulerian model was evaluated by Hayami and Ichikawa (1994). The concentration of $SO_4^{2-}(r)$ is calculated by

$$I \cdot [SO_4^{2-}(r)] = \int_{z_0}^{z_t} \left(\frac{3}{2}k_2 c_1 + k_2' c_2 + k_5 c_3\right) dz + \frac{3}{2}k_3 c_1(z_0) + k_3' c_2(z_t), \tag{6}$$

where I is precipitation intensity (m/s), z_0 and z_t the height of the model top and bottom, respectively. See Table 1 regarding k_3 and k_3'.

The wind field required for the Eulerian model is produced based on the AMeDAS (Automated Meteorological Data Acquisition System) data and Aerological Data in Japan. The AMeDAS data set also includes hourly precipitation ammounts.

TABLE 1. Simulation parameters of the hybrid model

Processes	Notation	Used values*
conversion rate of SO_2 to $SO_4^{2-}[h^{-1}]$	k_1	0.01 (w)
		0.03 (s)
wet deposition velocities of SO_2, $SO_4^{2-}[h^{-1}]$	k_2, k_2'	$0.1 \times I$
dry deposition velocity of $SO_2 [cm/s]$	k_3	0.20, out of the box
		0.47 (w), in the box
		0.54 (s), in the box
dry deposition velocity of $SO_4^{2-}[cm/s]$	k_3'	0.10, out of the box
		0.13 (w), in the box
		0.15 (s), in the box
cloud scavenging rates of SO_2, $SO_4^{2-}[h^{-1}]$	k_4, k_4'	0.02
transfer rate of SO_4^{2-} of cloud water into rain water$[h^{-1}]$	k_5	$0.1 \times I$
fraction of SO_4^{2-} emitted [-]		0.03

*) Fujita et al. (1991) for dry deposition velocities and Ichikawa and Fujita (1993) for others. I indicates precipitation intensity [mm/h]. (w) is for winter and (s) for summer.

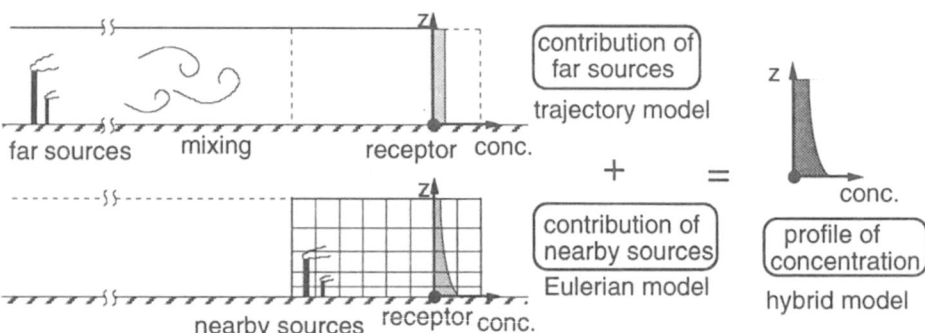

Figure 1. The concept of the hybrid model.

2.2. COMBINATION OF THE MODELS

The box-trajectory model by Ichikawa and Fujita (1993) predicts the concentrations by performing material balance in a box (80 km × 80 km × 1 km). The hybrid model has the box replaced by the Eulerian model with cells of 10 km × 10 km × 5 layers. The box-trajectory model is used to provide the concentrations in the box assuming no emission from sources within the box. On the other hand, the Eulerian model calculates the concentration in each cell in the box, considering all the sources in the box and assuming no substances are advected into the box. At the present time, the concentration predicted by the hybrid model is the combination of concentration calculated by the Eulerian model and the concentration calculated by the trajectory model. Figure 1 shows the concept of the hybrid model.

This method, which combines the two model results, holds for the linear processes. For non-linear processes, one approach would be to use the trajectory model to provide the lateral boundary conditions for the Eulerian model.

3. Model Simulations

To evaluate the ability of the hybrid model, we conducted some simulations and compared predicted results to observed concentrations of SO_2, SO_4^{2-}(p) and SO_4^{2-} in rain water during winter and summer at a city, Komae. Komae is an industrial suburb located in the metropolitan Tokyo area. The box-trajectory model had the largest underestimation of the concentrations at Komae. The parameters used are listed in Table 1. Anthropogenic source data estimated by Fujita et al. (1991) are used for emission rate of SO_2 over the whole model domain (about 4,500 km × 4,500 km).

Figure 2 shows the comparison of calculated concentrations of SO_2, SO_4^{2-}(p) and SO_4^{2-}(r) by the hybrid model with the measurements. The results calculated by the box-trajectory model are also shown. The predicted values by the hybrid model are 5.7 times for SO_2, 2.7 times for SO_4^{2-}(p), and 2.2 times for SO_4^{2-}(r) higher than those by the box-trajectory model, and better agree with the measurements, especially for SO_2. The improved predictions could be caused by the increased vertical resolution of the hybrid model which allows for the vertical expression of the emission intensity and behavior of sulfur compounds. The SO_2 concentration of the surface is most influenced by the locational relationship between a receptor and sources because SO_2 is a primary substance. In the box-trajectory model, the 80 km × 80 km box includes not only the industrial area but also the mountains and sea with few emissions and the concentrations were diluted and averaged in the box. The higher resolution in vertical and horizontal directions of the hybrid LRT model better represents the horizontal distribution of emissions and the concentrations. The prediction of SO_4^{2-}(p) and SO_4^{2-}(r) which are secondary products is also improved by this higher resolution.

4. Summary

We previously underestimated concentrations of sulfur compounds in polluted areas in Japan by using the box-trajectory model. The hybrid model

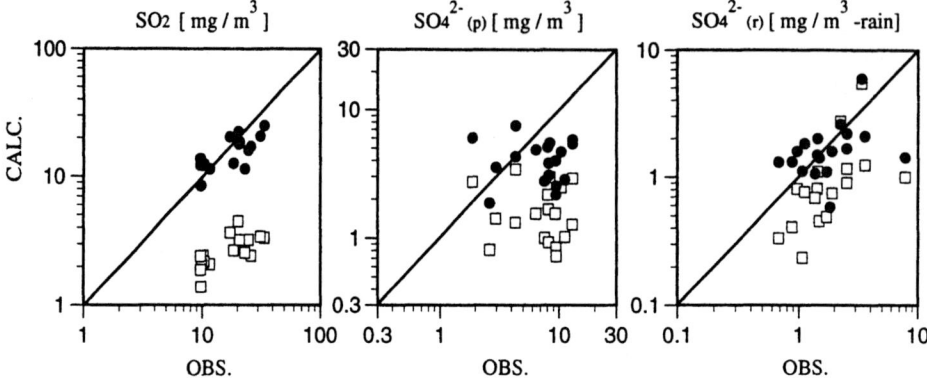

Figure 2. Comparison of calculated sulfur compounds concentrations with the observed values in Komae. The black circles indicate the results of the hybrid model and the squares by the box-trajectory model.

we have developed shows better agreement with the observed concentrations at Komae, where we greatly underestimated the concentrations which were mostly influenced by the nearby sources. By using the hybrid LRT model, the contribution of far sources to the wet and dry deposition of sulfur can be estimated. For example, it can be estimated that 11% of the dry deposited sulfur in winter comes from outside of Komae and its surrounding industrial area. By changing the method to combine both models and developing simplified schemes of NOx chemistry, the hybrid model could be applied to the estimation of NOx deposition.

References

Carmichael, G. R., Peters, L. K. and Kitada, T. (1986) A Second Generation Model for Regional Transport/Chemistry/Deposition, *Atmospheric Environment*, **20**, 173-188.

Fujita, S., Ichikawa, Y., Kawaratani, R. K. and Tonooka, Y. (1991) Preliminary inventory of sulfur dioxide emissions in East Asia, *Atmospheric Environment*, **25A**, 1409-1411.

Fujita, S., Takahashi, A., and Muraji, Y. (1991) An estimation of dry deposition of sulfur compounds in Japan, *J. Japan Soc. Air Pollut.*, **25**, 343-353.

Ichikawa, Y. and Fujita, S. (1993) An Analysis of Wet Deposition of Sulfate in East Asia, *CRIEPI report, No. T92041.* Central Research Institute of electric Power Industry, Japan.

Ichikawa, Y. and Fujita, S. (1995) An analysis of wet deposition of sulfate using a trajectory model for East Asia, presented in this meeting.

Hayami, H. and Ichikawa, Y. (1994) Development of the Method to Predict Deposition Amount of Acidic Substances. -Construction of a Framework of the Hybrid Type Long-Range Transport Model-, *CRIEPI report, No. T93098.*

Murao, N., Katatani, N., Sasaki, Y., Okamoto, S. and Kobayashi, K. (1993) A Modeling Study on Acid Deposition in East Asia, *Proc. International Conf. on Regional Environment and Climate Changes in East Asia*, Nov.,1993, Taipei

CONCENTRATION AND DEPOSITION OF ACIDIFYING AIR POLLUTANTS OVER SWEDEN: ESTIMATES FOR 1991 BASED ON THE MATCH MODEL AND OBSERVATIONS

J. LANGNER, C. PERSSON and L. ROBERTSON

Swedish Meteorological and Hydrological Institute (SMHI), S-601 76 Norrköping, Sweden

Abstract. The MATCH (Mesoscale Atmospheric Transport and CHemistry) model has been developed as a tool for air pollution assessment studies on different geographical scales. MATCH is an Eulerian atmospheric dispersion model, including physical and chemical processes governing sources, atmospheric transport and sinks of oxidized sulfur and oxidized and reduced nitrogen. Using a combination of air and precipitation chemistry measurements and the MATCH model, the national and long-range transport contributions to air pollution and deposition can be quantified in the model region. The calculations for the year 1991 show that the Swedish import was about 4.5 times larger than the export for sulfur and about six times larger for reduced nitrogen, while the Swedish import of oxidized nitrogen only exceeded the export by 10%. Using the MATCH system we estimate the long-range transport in an independent way compared to EMEP. Comparison between the EMEP and MATCH calculations for 1991 show that the total deposition of oxidized nitrogen over Sweden is similar, while the EMEP-values for total deposition of oxidized sulfur and reduced nitrogen are 25% respectively 40% smaller than what is obtained from MATCH.

1. Introduction

Deposition of acidifying pollutants in Sweden is a well known problem. Long-range transport was suggested as a major cause of increased acidification of air and precipitation in Sweden by Svante Odén in the late 1960's (Odén, 1976). Since then considerable efforts have been made on both national and European scales in order to quantify and understand the phenomenon. The work on the European scale has been coordinated within EMEP (Co-operative programme for monitorig and evaluation of the long range transmission of air pollutants in Europe, EMEP, 1980). EMEP runs a model covering Europe which is capable of allocating the deposition of oxidized sulfur and oxidized and reduced nitrogen compounds on a 150x150 km horizontal grid to the emitting countries on an annual basis (Eliassen and Saltbones, 1983). EMEP has provided key information in the negotiation of emission reduction protocols in Europe.

In addition to EMEP most countries run national acidification programmes. Sweden has a long tradition of monitoring the chemical composition of precipitation, and the first network was established in the 1950's. Since then monitoring efforts on the chemical composition of both air and precipitation have been expanded and now include measurements of sulfur and nitrogen compounds at several background locations as well as a dense network of through-fall measurements in the southern part of Sweden. Modelling work aimed at utilizing these data to provide additional information and generalizations have so far been limited in Sweden. To optimize efforts such as liming and national emission controls to limit the effects of acid deposition, information about deposition of acidifying pollutants with higher resolution than currently available from EMEP is desirable. In describing the effects of acidification on the ecosystem level, information should be available at least on the size of the ecosystems. The work presented here is an attempt to meet these requirements by combining model calculations for national emissions with careful analysis of observations linked with high resolution meteorological data.

2. Modeling system

The MATCH (Mesoscale Atmospheric Transport and CHemistry) modeling system consists of three parts: A regional atmospheric dispersion model including modules for emission, chemistry and deposition of sulfur and nitrogen compounds. An objective analysis system for air- and precipitation chemistry data, and an objective analysis system for meteorological data.

2.1. DISPERSION MODEL

The MATCH model (Persson and Robertson, 1991; Persson et al., 1994) is a three dimensional Eulerian atmospheric dispersion model. The model requires meteorological data from an external archive at regular time intervals (usually three or six hours) in order to calculate transport, chemistry and deposition. The model version used here has three layers in the vertical. The first layer has a fixed height of 75 m. The top of the second layer is taken to be the same as the mixing height, and the top of the third layer is fixed at a certain level (~1.5 km in winter, ~2.5 km in summer). It is easy to add additional layers if necessary, but for calculations over areas of the size of Sweden or smaller three layers was judged to be sufficent. The horizontal resolution for the calculations over Sweden is 20x20 km while a 5x5 km resolution has been used for subregions in Sweden.

Horizontal advection is calculated using a fourth order flux correction scheme (Bott 1989). Vertical advection is calculated using an upstream scheme. Vertical transport is also induced by turbulent vertical diffusion and the spatial and temporal variations of the mixing height.

2.1.1. Emissions
Emissions can be specified both as area and point sources. In the calculations over Sweden as a whole all emissions where treated as area sources and partioned between the different levels in the model based on point source statistics from subregions in Sweden. The emission totals used for 1991 are given in Table I.

2.1.2. Chemistry and deposition
The chemistry in the model deals with sulfur oxides and oxidized and reduced nitrogen compounds and is almost identical to that used in the EMEP model (Iversen et al., 1989), the main difference being the specification of ozone (O_3) concentrations. Here analysed O_3 distributions with three hourly time resolution are generated from observations (c.f. section 2.2). A local adjustment of the O_3 concentration with regard to local NO- and NO_2-concentration and solar radiation is also done.

Wet scavenging of the different species is proportional to the precipitation rate and a species specific scavenging coefficient. Dry deposition velocities are specified as a function of the surface characteristics (fraction forest, field etc). Scavenging coefficients and deposition velocities have values close to those used in the EMEP calculations.

Fig. 1. Background air and precipitation chemistry stations used in the analysis for 1991.

2.2. ANALYSIS OF AIR AND PRECIPITATION CHEMISTRY DATA

The dispersion model described above, combined with national emission estimates and meteorological data provides daily estimates of concentrations in air and precipitation as well as dry and wet deposition of the simulated sulfur and nitrogen compounds. These results refer to contributions from sources within the model area (in this case Sweden). To derive distributions of the contribution from sources outside Sweden the following method is employed: Model calculated daily contributions from Swedish sources are deducted from observed daily values of concentration in air and precipitation at background locations on a point by point basis (a map with the stations included in the analysis is given in Figure 1). The residual is termed long-range transport contribution. These residuals are analysed using an optimum interpolation method, where differences in observation quality can be accounted for, to give distributions of long-range transport contributions of concentrations in air and precipitation over the whole modeling domain. The basic idea behind this method is that the long-range transport contributions to the concentrations can be expected to vary more smoothly in space than the total concentrations which are affected to some extent by local sources, and should therefore be more suitable for interpolation. Long-range transport wet deposition is then calculated by multiplying with the observed precipitation field (c.f. section 2.3). Long-range transport dry deposition is calculated by running the long-range transport air concentrations through the dry deposition module of the dispersion model. Considerable efforts have been spent on quality control of both input chemical observation data and resulting analyzed concentration distributions. The objective analysis scheme is a very useful tool for identifying different kinds of errors in the observations. A similar method using total concentrations has been developed by van Pul et al. (1994). The difference here is the possibility to separate the national and long-range transport contributions. In order to gain confidence in the method comparisons have been made with independent air chemistry measurements. One such comparison taken from a study over the Swedish west coast region using the same method as described here is given in Figure 2. It shows a time series of the concentration of NO_2 in central Gothenburg for the second half of 1991. Both the overall concentration level and episodes are captured favorably.

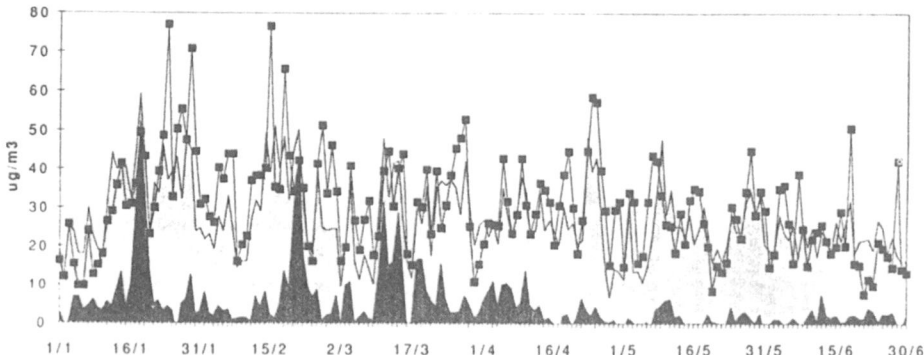

Fig. 2. Comparison, for a measuring station in central Gothenburg, between long-range transport contribution (dark shaded) plus regional contribution (light shaded) from the Swedish west-coast area and observed (black squares) total NO_2 concentrations in air for the second half of 1991 ($\mu g\ N/m^3$).

2.3. METEOROLOGICAL ANALYSIS

The dispersion model requires meteorological data to calculate transport, chemistry and deposition processes. For studies over Sweden an objective meteorological analysis system has been developed. The system makes use of routine meteorological observations to derive the wind, turbulence and precipitation fields required by the dispersion model at three hour intervals. A high resolution data base for topography and land use (fraction of forest, field, water, urban) is used in the analysis.

The precipitation analysis is given special attention: About 800 stations measuring daily precipitation is combined with precipitation and weather information from synoptic stations to give precipitation fields with three hourly time resolution and high horizontal resolution. Corrections for sampling losses and topography are also applied.

3. Results and discussion

Examples of results from the MATCH-Sweden calculations for 1991 are shown in Figure 3 which shows the annual contributions to the deposition (dry + wet) of oxidized sulfur from emissions in Sweden, long-range transport and also the total deposition of oxidized sulfur. Long-range transport dominates in south western Sweden and over the inland parts of central and northern Sweden while Swedish contributions are comparable along the east coast of central and northern Sweden. Note the pronounced maximum in deposition over the Swedish west-coast area which is due to a combination of high concentrations of sulfate in precipitation and a maximum in precipitation on the edge of the highlands in southern Sweden.

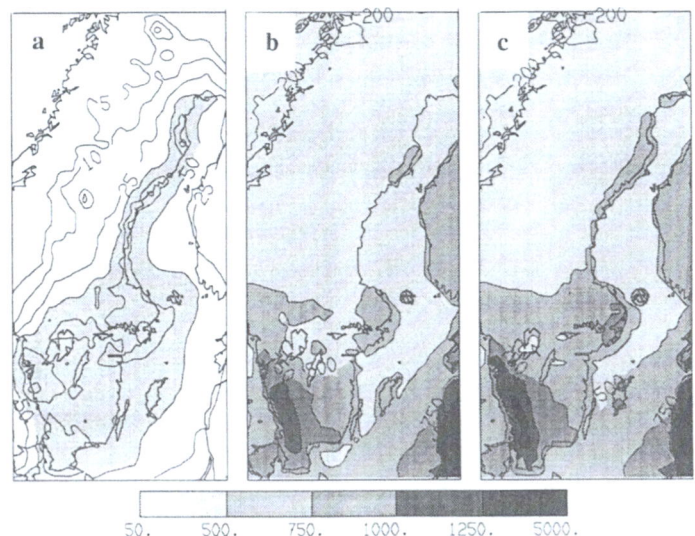

Fig. 3. Calculated annual (year 1991) dry + wet deposition of oxidized sulfur: a) contribution from Swedish sources, b) contribution from long range transport, c) total (mg S/m^2).

TABLE I

Comparison between EMEP and MATCH estimates of the Swedish and total contribution to deposition of oxidized sulfur and oxidized and reduced nitrogen over Sweden for 1991 (100 ton/year). The assumed Swedish emissions in the different models are also given.

	Sulfur	NO_x- nitrogen	NH_x - nitrogen
Swedish Emissions:			
EMEP (without shipping)	550	1184	420
MATCH (including shipping)	700	1184	440
Swedish deposition over Sweden:			
EMEP	152	148	220
MATCH	243	207	276
Total deposition over Sweden:			
EMEP	1740	1200	720
MATCH	2300	1300	1260

3.1. COMPARISON WITH EMEP

Using the MATCH system we estimate long-range transport in an independent way compared to EMEP (Tuovinen et al., 1994). Comparison between the EMEP and MATCH calculations for 1991 (Table I) show that the total deposition of oxidized nitrogen over Sweden is similar, while the EMEP-values for total deposition of oxidized sulfur and reduced nitrogen are 25 and 40% smaller respectively than what is obtained from MATCH. The Swedish contributions to deposition over Sweden are also somewhat smaller (20-38%) in the EMEP model compared to MATCH. When looking closer at different regions in Sweden even larger differences can be noted. The coarse resolution of the emission data, meteorological data and dispersion model in the EMEP calculations here obviously plays an important role. Some small changes, both in emission estimates and model parameters were made in the version of the EMEP model used by Tuovinen et al. compared to previous versions. Comparing the MATCH calculations with the previous version of the EMEP model (Sandnes, 1993) gives a closer agreement both for total and Swedish contributions to deposition over Sweden

4. Conclusion

A modeling system capable of providing high resolution estimates of national and long-range transport contributions to concentrations in air and to dry and wet deposition of sulfur and nitrogen compounds over Sweden has been developed. Comparisons with the EMEP calculations indicate substantial differences for total deposition of oxidized sulfur and reduced nitrogen With lower estimates derived from the MATCH system.

Acknowledgements

This work has received financial support from the Swedish Environmental Protection Agency. Air and precipitation chemistry data were provided by the Norwegian Institute for Air Research and by the Swedish Institute for Water and Air Research (IVL) respectively. IVL has also provided the gridded swedish emission data for 1991.

References

Bott, A.: 1989, *Mon. Wea. Rew.* **117**, 1006 - 1015.
Eliassen, A. and Saltbones, J.: 1983, *Atmos. Environ.* **17**, 1457-1473.
EMEP: 1980, *Summary report of the Western Meteorological Synthesizing Centre for the first phase of EMEP*, EMEP/MSC-W, Norwegian Meteorological Institute, Oslo, Norway.
Iversen, T., Saltbones, J., Sandnes, H., Eliassen, A. and Hov Ø.: 1989, *Airborne Transboundary Transport of Sulphur and Nitrogen over Europe - Model Description and Calculations*, EMEP/MSC-W Report 2/89, DNMI, Oslo, Norway.
Odén, S.: 1976, *Water, Air, and Soil Pollut.* **6**, 137-166.
Persson, C. and Robertson, L.: 1991, In H. van Dop and D. G. Steyn (eds.), *Air Pollution Modeling and Its Application VIII*, Plenum Press, New York, pp 649-650.
Persson, C., Langner, J. and Robertson, L.: 1994, In S-E. Gryning and M. M. Millán (eds.), *Air Pollution Modelin and Its Application X*, Plenum Press, New York, pp 9-18.
van Pul, W. A. J. Erisman, J. W., van Jaarsveld, J. A. and de Leeuw, F. A. A. M.1994, In S-E. Gryning and M. M. Millán (eds.), *Air Pollution Modelin and Its Application X*, Plenum Press, New York, pp 625-627.
Sandnes, H.: 1993, *Calculated budgets for airborne acidifying components in Europe: Calculated fields and budgets 1985-93*, EMEP/MSC-W Report 1/93, DNMI, Oslo, Norway.
Touvinen, J-H., Barret, K. and Styve, H.: 1994, *Transboundary acidifying pollution in Europe: Calculated fields and budgets 1985-93*, EMEP/MSC-W Report 1/94, DNMI, Oslo, Norway.

THE DISTRIBUTION OF NMHC IN EUROPE: RESULTS FROM THE EUROTRAC TOR PROJECT

A. LINDSKOG[1], S. SOLBERG[2], M. ROEMER[3], D. KLEMP[4], R. SLADKOVIC[5], H. BOUDRIES[6], A. DUTOT[6], R. , H. HAKOLA[8], R. SCHMITT[9] and H. ARESKOUG[10].

[1]*IVL, P.O.Box 47086, S-402 58 Göteborg, Sweden,* [2]*NILU, P.O. Box 100, N-2007 Kjeller, Norway,* [3]*TNO, P.O.Box 6011, NL-2600 JA Delft, The Netherlands,* [4]*KFA, P.O. Box 1913, D-5170 Jülich, Germany,* [5]*IFU, Kreuzeckbahnstr. 19, D-82467 Garmisch-Partenkirchen, Germany,* [6]*Université Paris XII-Creteil, LISA, 61 Ave de Général de Gaulle, F-94010 Creteil, France,* [7]*University of East Anglia, School of Environmental Science, Norwich NR4 7TJ, U.K.,* [8]*FMI, Air Quality Department, Sahaajankatu 22 E, SF-0081 0 Helsinki, Finland,* [9]*Meteorologie Consult, Auf der Platt 47, D-61 179 Glashütten, Germany,* [10]*ITM Air Pollution Laboratory, Stockholm University, S-106 91, Stockholm, Sweden.*

Abstract. Within the EUROTRAC sub project TOR, an European network of advanced monitoring stations situated at representative sites is operated, starting in 1988. Within the EMEP Co-operative Programme for Monitoring and Evaluation of the Long-range Transmission of Air Pollutants in Europe measurements of VOC started in August 1992. In this study the combined TOR-EMEP data base has been used. Sector analyses have been performed in order to distinguish between unpolluted and the polluted air masses at each site. The seasonal averages have been calculated for each sector and site, and the results are discussed. Data allocated to the unpolluted sectors represents the common European background, and in the winter similar NMHC concentrations are found at all sites north of $50^\circ N$. These sites cover the area from North Europe into the Arctic ($79^\circ N$), and the small north-south gradient indicates that NMHC builds up in the northern troposphere in winter, probably due to an efficient meridional mixing and slow photochemical reactions.

Key words. NMHC, winter averages, background air, Europe.

1. Introduction

Non-methane hydrocarbons (NMHC) are the key species in many environmental processes related to atmospheric chemistry, including ozone formation and acid deposition. Due to the growing awareness about their importance, the number of sites measuring NMHC has increased over the past few years. Tropospheric Ozone Research (TOR) is a EUROTRAC sub project studying the tropospheric ozone cycle on the European scale, in terms of the chemistry of photochemical formation and destruction and transport mechanisms. An European network of advanced monitoring stations situated at representative sites was established within the project, starting in 1988 (Cvitas and Kley, 1994). Within EMEP measurements of VOC started in August 1992.

In this study, which is a part of the TOR task #2 (Lindskog et al, 1995), the combined TOR-EMEP data base has been used to describe the distribution of a selected number of hydrocarbons over Europe. Of special interest is the increase in NMHC concentrations in unpolluted air masses during the winter months. This precursor reservoir is suggested to contribute to the spring maximum of ozone which appears in Northern Europe (Penkett and Brice, 1986; Hov et al, 1989; Penkett et al, 1993; Lindskog and Moldanová, 1994).

2. Description of the data set.

The different sites covered in this study is presented in Table 1, where also the time series used are given. The locations are shown on the map in Figure 1. The sampling frequency and the time of the day when the sampling is performed differs from site to site, thus the number of observations varies a lot among sites and years. In this evaluation we have chosen to overlook these discrepancies and use all the data available. Sector analyses have been performed in order to identify unpolluted and polluted air masses at each site. For most of the sites this classification is based on trajectories. EMEP trajectories calculated by NILU are based on meteorological data supported by the Norwegian Meteorological Institute/EMEP Meteorological Synthesizing Centre West (DNMI/EMEP-MSCW), and are 4 days' backwards isobaric trajectories at the $\sigma=0.925$ level every six hours. In some cases only a limited number of transport directions was used leading to a substantial reduction in data, which makes some of the interpretations a bit uncertain. In the case of Schauinsland and Moerdijk the selection is based on ground wind measurements.

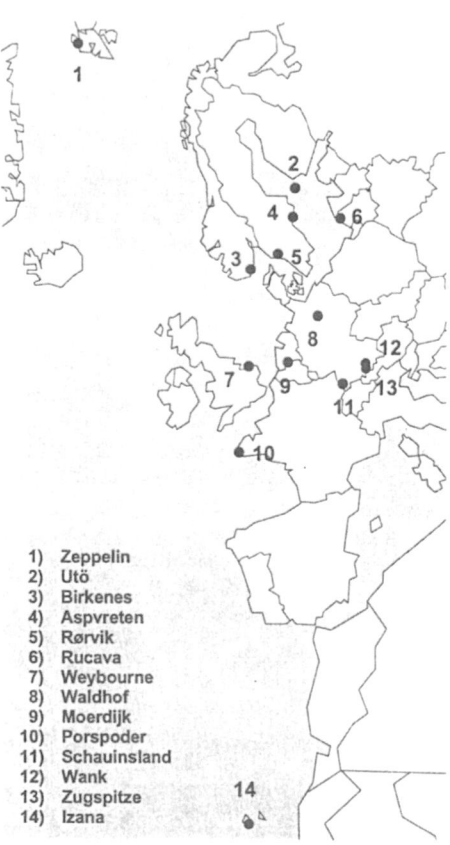

1) Zeppelin
2) Utö
3) Birkenes
4) Aspvreten
5) Rørvik
6) Rucava
7) Weybourne
8) Waldhof
9) Moerdijk
10) Porspoder
11) Schauinsland
12) Wank
13) Zugspitze
14) Izana

Figure 1. Location of the sites.

TABLE 1
The data set

Site	Location	Network	Time series
Izana, Tenerife	28°18'N;16°30'W; 2368 m asl	TOR	1990-1993
Zugspitze, Germany	47°25'N;10°59'E; 2962 m asl	TOR	1987-1990
Wank, Germany	47°31'N;11°09'E; 1776 m asl	TOR	1987-1990
Schauinsland, Germany	47°54'N;7°48'E; 1220 m asl	TOR	1989-1993
Porspoder, France	48°30'N;04°46'W; 20 m asl	TOR	1992-1993
Moerdijk, The Netherlands	51°41'N;04°32'E; 1 m asl	TNO/TOR	1989-1991
Waldhof, Germany	52°48'N;10°45'E; 74 m asl	EMEP	1992-1994
Weybourne, U.K.	52°57'N;01°07'E; 15 m asl	TOR	1993-1994
Rucava, Latvia	56°13'N; 21°13'E; 18 m asl	EMEP	1992-1994
Rörvik, Sweden	57°23'N;11°55'E; sl	TOR	1989-1993
Aspvreten, Sweden	58°48'N;17°39'E; 20 m asl	TOR	1991-1993
Birkenes, Norway	58°23'N;08°15'E; 116 m asl	TOR	1988-1994
Utö, Finland	59°47'N;21°23'E; 7 m asl	TOR	1992-1993
Zeppelin, Norway	78°55'N;11°54'E; 474 m asl	TOR	1989-1994

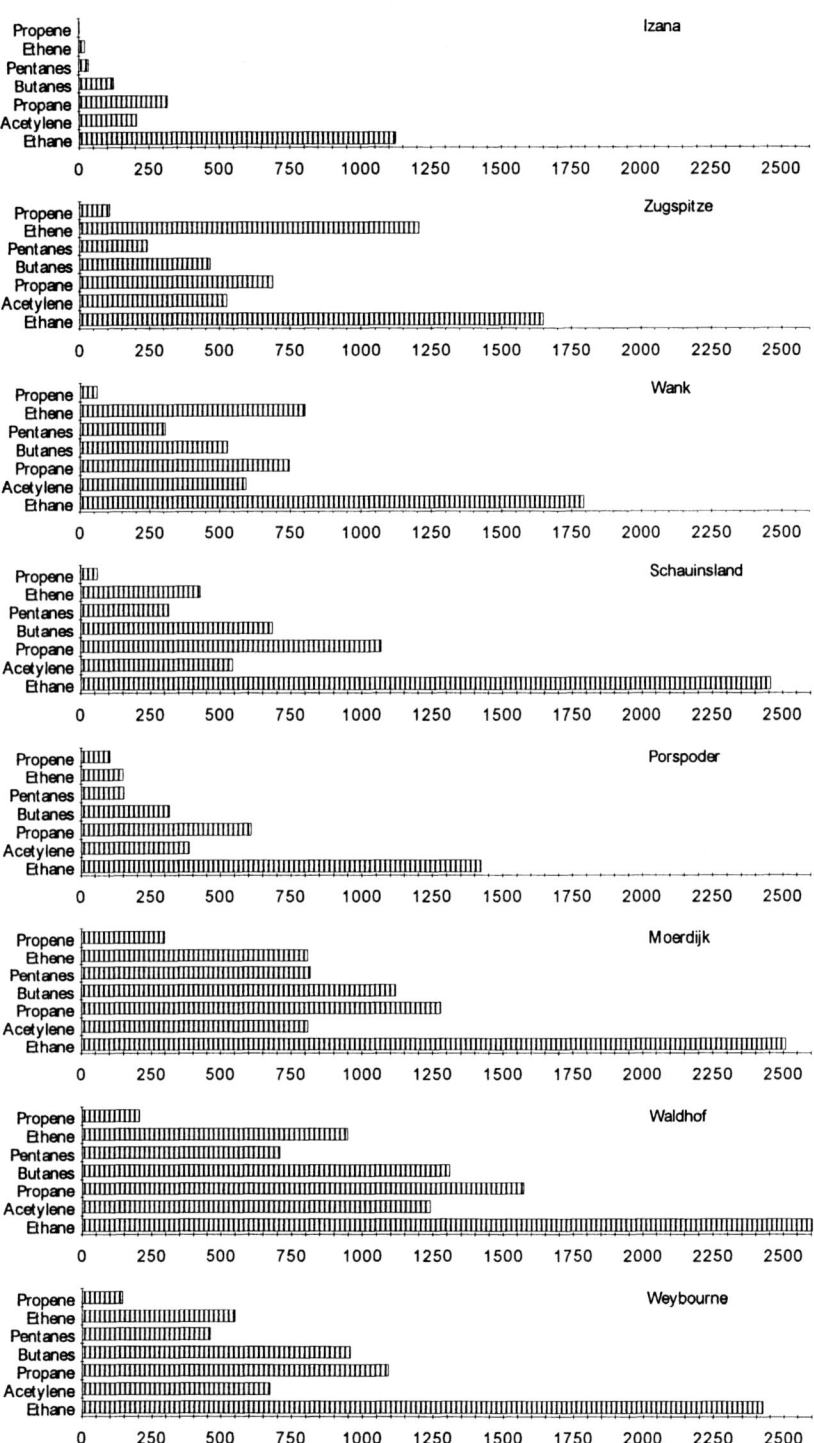

Figure 2a. The winter (Nov., Dec., Jan., Feb.) averages in ppt(v) of a selected number of NMHC in the unpolluted sector of each site. The hydrocarbons are arranged in accordance to their reactivity with the OH radical, with the most stable at the bottom.

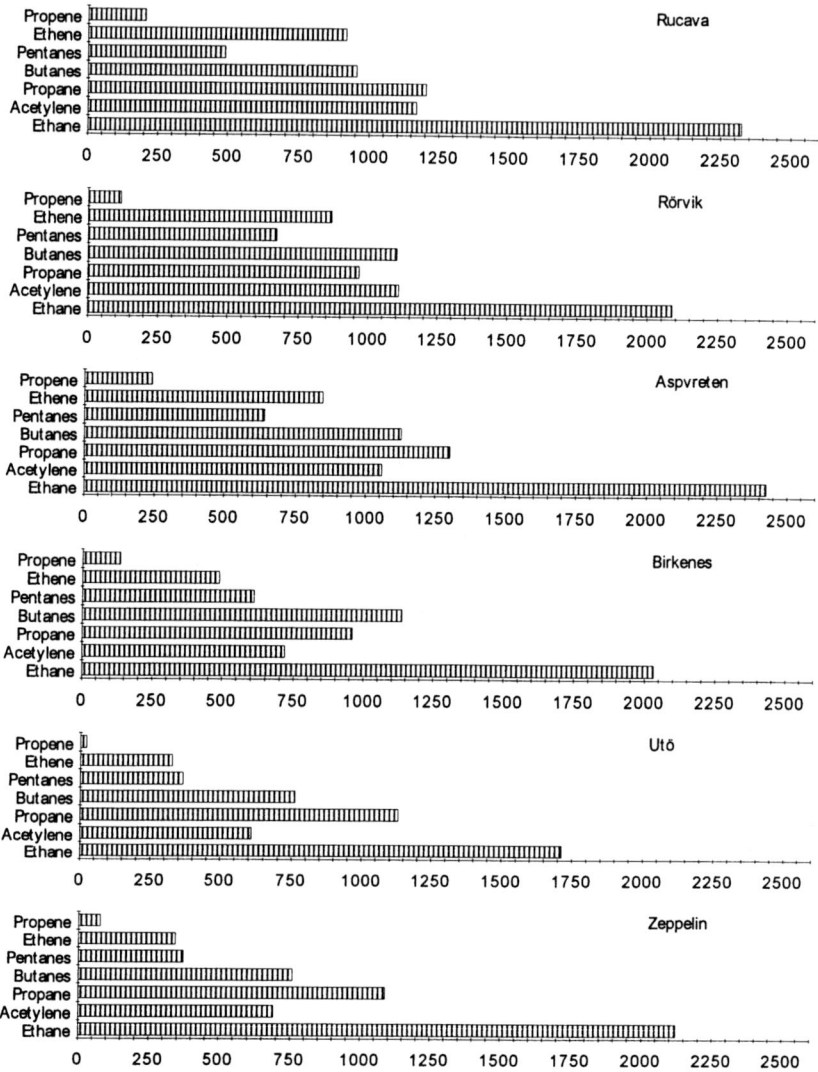

Figure 2b. The winter (Nov., Dec., Jan., Feb.) averages in ppt(v) of a selected number of NMHC in the unpolluted sector of each site. The hydrocarbons are arranged in accordance to their reactivity with the OH radical, with the most stable at the bottom.

3. Results and discussion

Between 30 and 40 individual NMHC are identified and quantified at each site. In this study only 9 are considered. These are, enumerated according to increasing reactivity with the hydroxy radical, ethane, acetylene, propane, i-butane, n-butane, i-pentane, n-pentane, ethene and propene.

The importance of the build up of ozone precursors during the winter at northern latitudes has been pointed out before. Due to the low concentration of OH radicals in the winter atmosphere the chemical lifetime of NMHC is increased. Typical chemical lifetimes of NMHC in winter according to Lightman et al (1990) are in days: ethane 265, acetylene 77, propane 48, butane 20, pentane 12, ethene 5, and propene 1. Ethene and propene also react with ozone at approximately the same time scales as given for OH.

Assuming a characteristic time scale for mixing in the lower northern troposphere north of $50°N$ of one month, one would expect ethane, acetylene and propane to be well mixed in the whole of this region during winter. The chemical lifetime of butane is somewhat less than the assumed time for mixing, but still long enough to expect a significant build-up throughout the remote northern troposphere. Ethene and propene also have longer chemical lifetimes in winter, giving rise to higher concentrations, however, because the lifetimes are much shorter than the time scale of large scale mixing, it is not fair to regard this as an accumulation of these components. Data allocated to the unpolluted sectors represents the common European background, and in the winter all sites north of $50°N$ are showing almost the same levels. The average concentrations and standard deviations (in brackets) are for ethane 2264 (290), acetylene 937 (311), propane 1228 (247) and the butanes 1074 (215). The average of the sum of the 4 components calculated as ppt carbon is 14380 (2467) for the sites north of $50°N$ compared with 8045 (2903) for the sites south of $50°N$. The concentrations are given in Figure 2. The measurement sites cover the area from North Europe into the Arctic ($79°N$), and the small north-south gradient indicates that several NMHC are being mixed throughout the northern troposphere in winter, probably due to an efficient meridional mixing and slow photochemical reactions. The southern locations will more often be located south of the polar front and thus influenced by subtropical air masses with a more efficient oxidation and presumably lower concentrations of NMHC (Lightman et al, 1990). The concentrations of NMHC during summer are much lower than during winter and the differences among sites are larger compared with the winter values (Lindskog et al, 1995).

Beside the source strength and the atmospheric dispersion the concentrations are also dependent of the chemical degradation, as discussed above. To better illustrate this, one can use the relative distribution of individual NMHC (the NMHC profile) in the comparison (Figure 3). The NMHC profile thus reflects the age of the air mass, defined as the integrated photochemical oxidation (Integral [OH] dt) since the time of the last major emissions, and not the time itself. From this comparison one can see that the more southern sites differ from the others and seem not to be comprised by the same large scale meteorology. Zugspitze and Wank are special with an unexpected high relative concentration of ethene. A larger share of the more stable compounds, ethane, acetylene and propane, implies that the clean winter sector in these cases are dominated by aged air masses. For most of the sites, these three compounds amount to about 60%. For the marine sites and Schauinsland, the share is larger, 70 - 80%. This is even more evident at Izana, where these compounds constitute about 90%. The relative amount of propene is surprisingly large at Porspoder, amounting to 15% of the reactive compounds (the sum of *i*-butane, *n*-butane, *i*-pentane, *n*-pentane, ethene and propene), thus indicating a possible influence from a more local source.

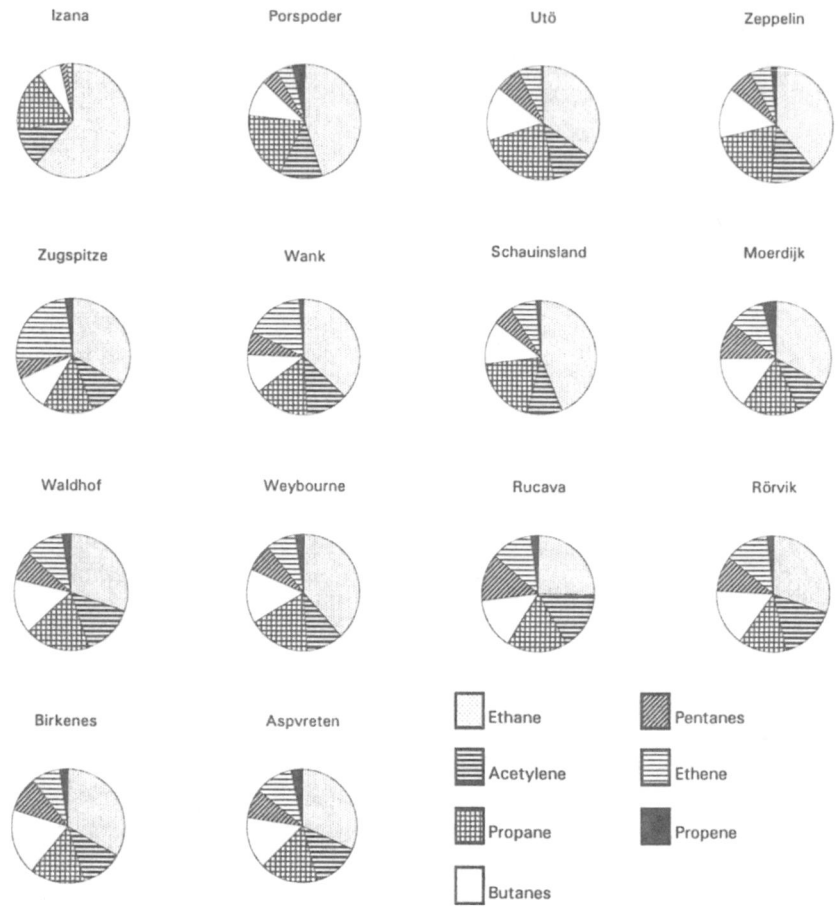

Figure 3. The relative distribution of the winter averages of individual NMHC measured in the unpolluted sector at each site.

References

Cvitas, T. and Kley, D. eds: 1994, The TOR Network, EUROTRAC Special Publications, ISS, Garmisch-Partenkirchen.
Hov, Ø., Schmidbauer, N. and Oehme, M.: 1989, *Atmos. Environ.* **23**, 2471-2482.
Lightman, P., Kallend, A.S., Marsh, R.W., Jones, B.M.R., Penkett, S.A.: 1990, *Tellus*, **42B**, 408-422.
Lindskog, A. and Moldanová, J.: 1994, *Atmos. Environ.*, **28**, 2383-2398.
Lindskog, A., Solberg, S., Roemer, M., Klemp, D., Sladkovic, R., Boudries, H., Dutot, A., Burgess, R., Hakola, H., Laurila, T., Schmitt, R., Areskoug, H., Haszpra, L., Mowrer, J., Schmidbauer, S., and Esser, P.: 1995, The emission and distribution of ozone precursors over Europe, Manuscript submitted to the EUROTRAC TOR Final Report.
Penkett, S.A. and Brice, K.A.: 1986, *Nature*, **319**, 655-657
Penkett, S.A., Blake, N.J., Lightman, P., Marsh, A.R.W., Anwyl, P., and Butcher, G.: 1993, *J. Geophys. Res.*, **98D**. 2865-2885.
Simpson, D.: 1993, *Atmos. Environ.*, **27**, 921-943.

GRADIENTS OF OZONE AT A FOREST SITE AND OVER A FIELD CROP-CONSEQUENCES FOR THE AOT40 CONCEPT OF CRITICAL LEVEL

H. PLEIJEL[1], G. WALLIN[2], P. KARLSSON[1], L. SKÄRBY[1] and G. SELLDÉN[2].

[1] Swedish Environmental Research Institute (IVL), P. O. Box 47086, S-402 58 Göteborg, Sweden,
[2] Botanical Institute, University of Göteborg, Carl Skottsbergs Gata 22, S-41319 Göteborg, Sweden

Abstract. Ozone concentrations were measured at a wind-exposed edge of a 60 year-old 15-20 m tall Norway spruce forest in south-west Sweden and simultaneously over a barley field 5 km away for 27 days. At the forest site, measurements were performed at 3 and 13 m height 15 m in front of the forest edge, at 3 m height 15 m into the forest, and at 3 and 13 m height 45 m into the forest. Measurements at 3 m were made with three replicate tubes separated by 10 m. Differences between replicates were small. At 13 m height, the concentration (24-hr-average) 45 m into the forest was 95% of that in front of the forest edge. The average concentration at 3 m height did not vary strongly with the distance into the forest, but was 86% of that at 13 m in front of the forest edge. For AOT40 (Accumulated Exposure Over Threshold 40 ppb ozone), the differences between different positions were larger. At the 13 m level the AOT40 (day and night) was 88% of that in front of the forest 45 m into the forest. The AOT40 at 3 m was 71% of that at 13 m outside the forest. At the crop site, the ozone concentration at 1.1 m (0.1 m above the canopy), was 78% of that at 9 m (06.00-22.00). The AOT40 at 1.1 m above the ground, however, was only 50% of that at 9 m, indicating that serious errors can arise if ozone monitoring data are used uncorrected in dose-response relationships based on measurements performed at plant height. The ozone concentration for the whole period differed very little between 9 m height at the crop site and 13 m height at the forest site outside the forest during daytime conditions (06.00-22.00). Night-time (22.00-06.00) values were only 21% at the crop site of those at the forest site due to the stronger night inversion development in the agricultural environment compared to the wind exposed forest edge. The results suggest that variations in topography and vegetation are important to consider when combining ozone monitoring data with dose-response functions.

Key words: AOT40, barley, ozone, ozone concentration gradient, spruce

1. Introduction

Ozone is generally considered the most important gaseous air pollutant in rural areas of western Europe and the United States of America. This is the background to the work with critical levels for ozone, which started at a conference in Bad Harzburg 1988 (Guderian, 1988) arranged by the UNECE. At that stage only average concentration values for different duration periods were considered. At a workshop in Egham 1992 it was agreed that the accumulated exposure over a threshold (AOT) would give a more relevant measure of the exposure which is important for plant response (Ashmore & Wilson, 1994). The critical level concept of AOT was further developed at a conference in Bern in 1993, where a threshold level was set tentatively to 40 ppb (nl l^{-1}) for both crops and forest trees (Fuhrer & Achermann, 1994).

Most experiments on which the critical level concept is based, have been performed using open-top chambers. In the chamber, the ozone concentration is measured relatively close to the plant surface and the air mixing is good (Pleijel et al., 1994). At monitoring sites, ozone concentrations are normally measured at least some metres above the ground or vegetation in order to represent the regional ozone situation. If dose-response relationships obtained in open-top chamber experiments are used to estimate effects of ozone based on ozone concentrations from monitoring stations an error arise due to the fact that a strong vertical ozone concentration gradient exists above a plant canopy. The AOT40 concept of critical level is likely be more sensitive to this type of error than indices based on concentration means, since a modest reduction in concentration results in a large decrease in AOT40, in the concentration range common over much of Europe (Tuovinen & Laurila, 1993). The aim of the present investigation was to quantify how much the AOT40 differs from the local background in different parts of a coniferous forest and at different heights above a field crop.

2. Materials and methods

Ozone was measured at a wind exposed edge of a 18-20 m tall, 60 year-old Norway spruce (*Picea abies* (L.) Karst.) forest in south-west Sweden. The site, called Antens kapell, is situated 10 km north west of the town Alingsås and approximately 60 km from the Swedish west coast at an altitude of 100 m. Ozone concentrations were measured at 13 m and 3 m above the ground in front of the forest, at 3 m above the ground 15 m into the forest and at 13 and 3 m above the ground 45 m into the forest. The measurements at 3 m above the ground were made with three replicate tubes 10 m apart at each distance from the forest edge. All measurement points were scanned with 15-20 min interval. At the agricultural site (Östad säteri), ozone concentrations were measured at 9, 5, 2 and 1 m above the ground of an almost 1 m tall barley (*Hordeum vulgare* L.) crop, which covered a field of 4-5 ha. The ozone concentration was measured twice per hour in all four points at the agricultural site. At both sites the ozone concentrations were monitored using Thermo Environmental 49 analysers. All PTFE tubes were 50 m long and attached to time share systems. The measurements on both sites lasted from 7 July to 2 August 1989 (27 days). The wind speed and direction was measured in front of the forest at 10 m above the ground with a Young model 12305 Microvane anemometer. Data were logged at both sites using Campbell CR10 data loggers.

3. Results

3.1. WIND SPEEDS

Daytime daily average wind speeds were always below 4 m s^{-1} and the daytime average for the period was 1.6 m s^{-1} with a standard deviation of 0.7 m s^{-1}. Night-time daily average wind speeds were always below 3 m s^{-1}. and the night-time average for the period was 0.7 m s^{-1} with a standard deviation of 0.8 m s^{-1}. Thus, wind speeds were rather low during the period of measurement.

3.2. FOREST SITE

In Table I the averages, standard deviations and coefficients of variation for the replicate measurements ozone concentrations and AOT40s are shown. The data show that neither the heterogeneity of the forest, nor differences between the measurement devices was an important source of variation.

TABLE I

The averages, standard deviation and coefficients of variation of the ozone concentration and AOT40 for replicate ozone measurements (n=3), at 3 m above the ground, in front of the forest, 15 m into the forest and 45 into the forest

	Average	Standard deviation (n = 3)	Coefficient of variation, %
Concentrations			
outside forest	28.7	0.10	0.34
15 m	28.0	0.13	0.47
45 m	28.2	0.17	0.60
AOT40			
outside forest	1746	26	1.46
15 m	1658	17	1.00
45 m	1638	26	1.58

Figure 1a shows the average ozone concentrations during the day and night (22.00-06.00, which largely corresponded to the astronomic sun hours), respectively, for the different measurement points. The ozone concentration at the 3 m level was similar at all measuring points. At the 13 m level the concentration 45 m into the forest had a 24-hr-average which was 95% of that outside the forest. The concentration at the 3 m level outside the forest was 86% of that at 13 m and the corresponding figure for the location 45 m into the forest was 90%.

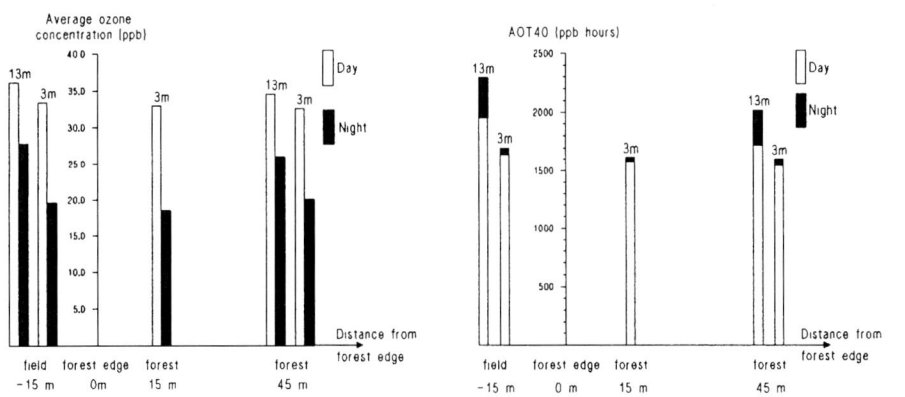

Fig. 1. a) Average ozone concentrations (ppb) for day and night (22.00-06.00) at the forest edge: outside the forest at 3 and 13 m height, 15 m into the forest at 3 m height and inside the forest at 3 and 13 m height, b) AOT40 (ppb hours) for day and night at the forest edge: outside the forest at 3 and 13 m height, in the very edge of the forest at 3 m height and inside the forest at 3 and 13 m height.

In Figure 1b the corresponding AOT40 values are shown. Differences between day and night were much more pronounced for AOT40 compared to average concentration. The 13 m level at 45 m into the forest had 87% of the AOT40 at the 13 m level outside the forest. The AOT40 at 3 m outside the forest was 75% of that at 13 m and the corresponding percentage for 45 m into the forest was 81%. Night-time ozone contributed most to the total AOT40 at the 13 m level outside the forest but only with 15% of the total AOT40. Branches with needles were present both at the 3 and 13 m levels and, thus, both levels are relevant for plant uptake of ozone. Although the average concentrations during the night were not much lower than the 24-hr-averages, the night-time contribution to the AOT40 was only 15% at 13 m outside the forest and 4% at 3 m inside the forest.

3.3. AGRICULTURAL SITE

In Figure 2a and 2b the average day and night ozone concentrations and AOT40, respectively, at the agricultural site are shown. The vertical concentration gradient is rather steep, especially during the night and for AOT40 it is obvious that the height gradient is also very important. In fact, the daytime AOT40 at 1.1 m above the ground, that is immediately above the canopy, was only half of that at 9 m. Night-time ozone contributed to AOT40 at the agricultural site only at 9 m above the ground and the contribution was less than 5%.

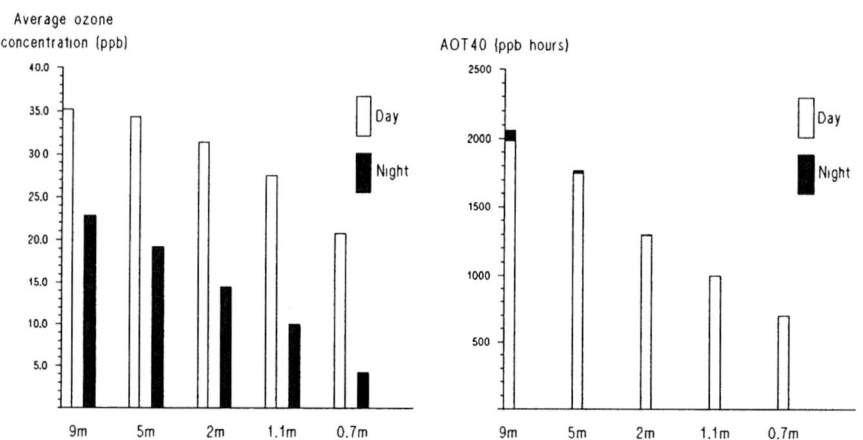

Fig. 2. a). Ozone concentrations (ppb) during day and night at different heights at a barley field, b) AOT40 (ppb hours) during day and night at different heights at a barley field.

3.4 COMPARISONS BETWEEN THE FOREST AND THE CROP SITES

In Figure 3 the daytime two-hour mean ozone concentrations at 13 m outside the forest at the forest site is plotted against the concentrations at 9 m above the ground at the crop site. The values are close to the one-to-one line. Only at the lowest ozone concentration there is a tendency towards lower concentrations at the agricultural site corresponding mainly to early morning and late evening concentrations.

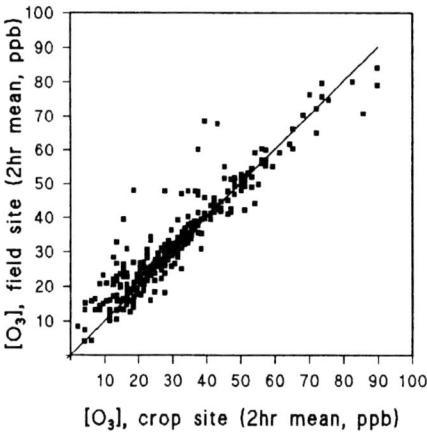

Fig 3. Daily ozone concentrations (ppb) at 13 m outside the forest site plotted against the ozone concentrations at 9 m above the ground of the barley field.

In Table II the average ozone concentrations and AOT40s for the measurement period are shown. It can be seen that during the day both the concentration and the AOT40 were virtually the same at both sites, representing the same regional background ozone concentration. During the night the differences grow larger and much more so for AOT40 than for the ozone concentration.

TABLE II

Average ozone concentration and AOT40 at 9 m above the ground at the crop site in percent of the corresponding values at the 13 m level outside the forest for the 27 day period

	concentration	AOT40
day	98%	98%
night	82%	21%
day and night	93%	90%

4. Discussion

At the forest site the ozone gradients were not very strong on average. Even the AOT40 did not vary much between different positions of measurement inside and outside the forest. This is explained by the high roughness of the spruce forest which causes air mixing by means of mechanical turbulence most of the time. The coupling between the atmosphere and the forest stand is therefore good (Jarvis & McNaughton, 1986). As a consequence the error in assuming that ozone measurements performed at monitoring sites representing the conditions of nearby spruce forest sites is rather small, although it can lead to limited overestimations of ozone effects. The dark period contributed with a small proportion of the AOT40 only. This makes the assumption (Fuhrer & Achermann, 1994) that night-time exposure should be included in the AOT40 for forests (but not for crops) uncontroversial under the conditions prevailing in the present study. This may, however, vary between times and sites.

At the agricultural site, the ozone concentration gradient turned out to be of large importance. AOT40 values obtained at different heights were very different. Simply putting ozone concentrations from monitoring sites, measured at 5 m height or higher, into dose-response relationships obtained by measuring the ozone concentrations close to the crop, like in most open-top chamber studies, would lead to over-estimations of yield loss up to 100%. This will have to be taken into consideration in future use of the AOT40 exposure index in yield loss predictions. It should be kept in mind that the wind speeds were comparatively low during the period of measurement in the present study, which may have enhanced the formation of ozone gradients. On the other hand, the period was very sunny, and strong radiation tend to prevent such gradient formation due to thermal convection.

The surface of a field crop has a lower roughness than a spruce forest. The friction exerted by the plant canopy on a moving air mass is smaller for the crop and thus the generation of mechanical turbulence. This makes the exchange between air layers slower and results in stronger gradients of for instance the ozone concentration. The conclusion is that a correction for the ozone gradient is necessary when using an exposure index of the AOT40 type in the case of crops, but for forests such a correction only means a smaller improvement of the estimations of effects. The correction for crops can be based on the logarithmic relationship between height and ozone concentration or on empirical relationships. It will have to take into account factors such as wind speeds and differences in roughness due to crop and topography.

The highest measurement positions in the present investigation (ca 10 m) resulted in very similar results for daytime conditions at both sites. Thus, during the day they both represent the same regional background of ozone. For night-time conditions, however, the concentrations tended to be lower at the crop site due to the more favourable conditions for inversion development in that type of environment.

Acknowledgements

Thanks are due to Mr. Gunnar Omstedt (SMHI) and Mr. Björn Berglind for assistance in the design and set-up of the measurements. The study was funded by the Swedish National Environment Protection Board.

References

Ashmore, M.R. & Wilson, R.B. (Eds): 1994, *Critical Levels of Air Pollutants for Europe*, Department of the Environment, UK, 209 pages.
Fuhrer, J. & Achermann, B.: 1994, Critical levels for ozone. A UN-ECE workshop report, *Schriftreihe der FAC Liebefeld* 16, 328 pages.
Guderian, R.: 1988, Critical levels for effects of ozone (O_3), In: *Final Draft Report of UNECE Critical Levels Workshop*, Bad Harzburg, 51-78.
Jarvis, P. G. and McNaughton, K. G.: 1986, *Advances in Ecological Research* 15, 1-49.
Pleijel, H., Wallin, G., Karlsson, P., Skärby, L., Selldén G.: 1994, *Atmospheric Environment* 28, 1971-1979.
Tuovinen, J.-H. & Laurila, T.: 1993, "Ozone concentration and exposures in Finland", In (Ed) Antila, P.: *Proceedings of EMEP Workshop on the Control of Photochemical Oxidants in Europe*, April 20-22, 1993, Porvoo, Finland, 15-24.

BEHAVIOUR OF SULPHUR AND NITROGEN COMPOUNDS MEASURED AT MARINE STATIONS LISTA AND SÄBY IN SCANDINAVIA

J. Isakson[1], E.A. Selin Lindgren[1], V.L. Foltescu[1], J.M. Pacyna[2] and K. Tørseth[2]

[1] *Department of Environmental Physics, Göteborgs University and Chalmers University of Technology, S-41296 Göteborg, Sweden.* [2]*Norwegian Institute for Air Research (NILU), P.O. Box 100, N-2007 Kjeller, Norway.*

Abstract. Long distance transport of various air pollutants has been studied at two Scandinavian coastal sites. The measuring period has covered one year divided into 4 campaigns. Seasonal variations for sulphur and nitrogen compounds as well as anthropogenically emitted metals are reported. Concentrations of Pb and Zn as well as the ratio of V/Ni concentrations have been used to trace sulphuric episodes. Covariation of anthropogenic pollutants for the two sites is demonstrated.

Key words: sulphur, nitrogen, Scandinavia, long distance transport, elemental tracers.

1. Introduction

Trace metals are emitted into the atmosphere by both natural and anthropogenic sources (e.g Nriagu, 1988). Various human activities contribute such as power generation from coal and oil, vehicle combustion engines, metal smelters of different kinds and waste incineration. Both crude oil and coals contain Ni and V in varying concentrations. Pb as a fuel additive is still widely used in eastern Europe. Zn has several sources like waste incineration and nonferrous smelters. Coal and oil combustion as well as metal production are the dominant sources for SO_2 and NO_x emissions into the atmosphere and therefore the accompanying metals act as tracers of these activities. The two Nordic groups at Göteborg/Chalmers University and NILU have studied fluxes of sulphur nitrogen and metal species in marine and continental air masses.

In this paper we intend to show that an episode of high concentrations in S and N compounds can often cover the whole southern part of Scandinavia and that such an episode can be characterised in origin by trace elements such as Pb, Zn, Ni and V.

2. Experimental Methods and Techniques

In order to cover different seasons the measurements were conducted in four separate campaigns, each lasting one month. The campaign dates were:
1. 93-11-14 to 93-12-12 2. 94-2-13 to 94-3-20 3. 94-5-12 to 94-6-10 and 4. 94-7-18 to 94-8-20.
Meteorological conditions during the campaigns were such that the first two campaigns represented winter season and the two latter summer season. The two stations in this work are situated at Säby on the Swedish west coast and Lista on the Norwegian south coast respectively. The approximate distance between the stations is 300 km over the Skagerrak sea. Air mass transport between the stations will take from a few to some ten hours for moderate wind speeds. The Säby station is located around 45 km northwest of the city of Göteborg, positioned some hundred metres from the sea and about 20 m above sea level. The Lista station is located on the shore of the North Sea close to the southernmost point of Norway. The instruments are positioned approximately 30 m from the sea and 13 m above sea level.

Particulate samples were taken with a PM_{10} dichotomous impactor.with a 2.5 μm cutpoint for fine particles. The filters were analysed with the EDXRF instrument at Chalmers described in an earlier work (Öblad, 1986). NO_x and SO_2 were both mesasured in real time with a Chemiluminescence NO_x analyser (det. limit 1ppb) and a Thermo Environmental Instruments model 43S (det. limit 0.1 ppb) respectively. Hourly averages from both instruments were recorded in a Campbell CR10 logger together with local meteorological data. Sampling of gaseous and particulate sulphur- and nitrogen compounds at Lista was performed using the annular denuder/filterpack method. All main components were determined using ion chromatography (IC).

Data from all campaigns were analysed using 4 day back trajectories at a pressure height of 925 hPa calculated at NILU. Back trajectories were calculated for each sixth hour and as impactor filters were changed at 12.00 GMT each measurement could be caracterized by four consecutive trajectories. In order to classify the origins of the air masses they were divided into different sectors according to the notation in the tables. Since a marine air mass reaching Säby probably has been slightly changed when passing Scotland or Denmark it is named modified marine. The same applies to continental air masses that have been modified on the route over the Baltic sea.

3. Results and Discussion

3.1 Sulphur and Nitrogen compounds

Table 1 shows mean concentrations of various S- and N-compounds measured at the two nordic stations measured in four campaigns.In the Säby table S(f,p) means fine particulate sulphur. It should be noted that both S- and N- values are much higher in winter than in summer. In the Lista table SO4-S means particulate S but this value can sometimes be disturbed by seasalt particles due to the sampling method. the values are therefore slighly higher than the Säby values.

TABLE 1 Mean concentrations of fine particulate S (f,p) compounds and gaseous S- and N oxides for different origins during the two winter (W) and summer (S) campaigns at Säby. m.mar = modified marine, m.co.east = modified continental east, m.co.south = modified continental south.

air mass	Säby								Lista			
	SO2-S		S(f,p)		NOx-N		NH3-N		SO4-S			
type	W	S	W	S	W	S	W	S	W	S		
marine							0.08	0.34	0,89	0.64		
m.mar	0.43	0.38	0.50	0.53	1.9	0.67	0.24	0.75	1.2	1.4		
m.co.east	1.9	0.33	2.0	0.58	2.8	0.79	0.20	1.1	2.3	1.3		
m.co.south	1.8	0.57	3.0	2.2	5.9	0.75						

3.2 Time series of particulate S.

Figure 1 shows the level of particulate S during a major pollution episode lasting from 29 Nov. until 1 Dec. During the period both stations received almost exactly the same air mass originating in central eastern Europe (Poland) and it is worth to note the similar pattern in concentrations for the two sites. Very high concentrations of SO_2, nitrogen species and anthropogenic trace metals were also measured at both Säby and Lista, particularly of Pb, Zn, Ni and V. The large difference in sulphate concentrations between the two sites in 26 Nov. is a consequence of completely different origin for the air masses at the two sites.

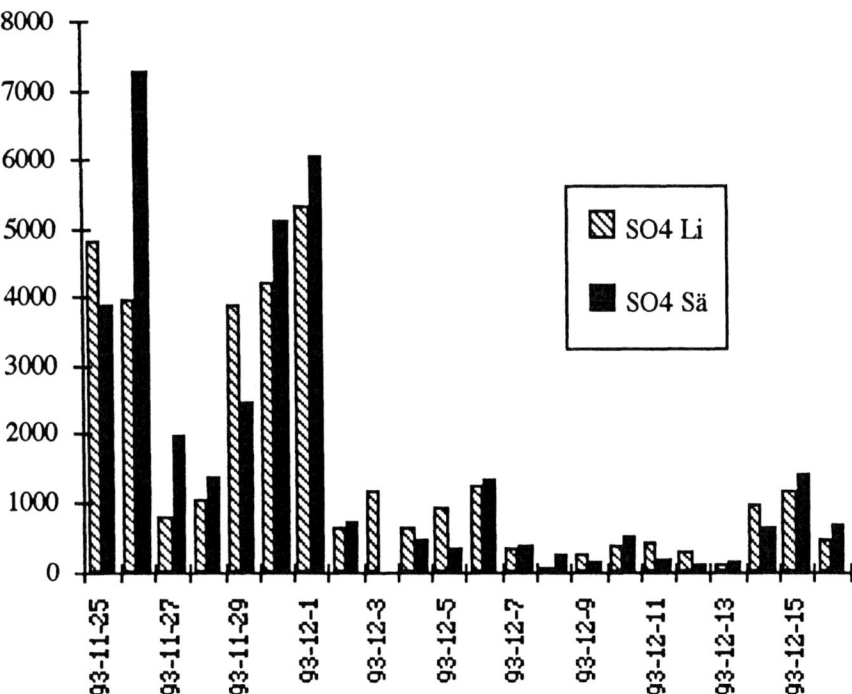

FIG. 1. Concentrations of fine particulate SO_4-S in ng/m^3 for Säby and Lista during the first campaign (winter).

Figure 2 shows the intimate covariation of Pb and fine particulate S that is sometimes found at the Säby station. Trajectory analysis shows that between 23 and 25 May the air mass originated in the S:t Petersburg area. The air passed over Estonia and later turned so that it passed over Poland. Both these areas are known to emit large quantities of SO_2. High leaded fuel is also normally in use in these countries. Between 1 and 3 June the air mass passed over southern England and Belgium. Apart from the high sulphate concentration it is interesting to note that although the EC members have agreed on a reduction on lead additives to petrol (O'Riordan, 1989) strong Pb episodes are still recorded.

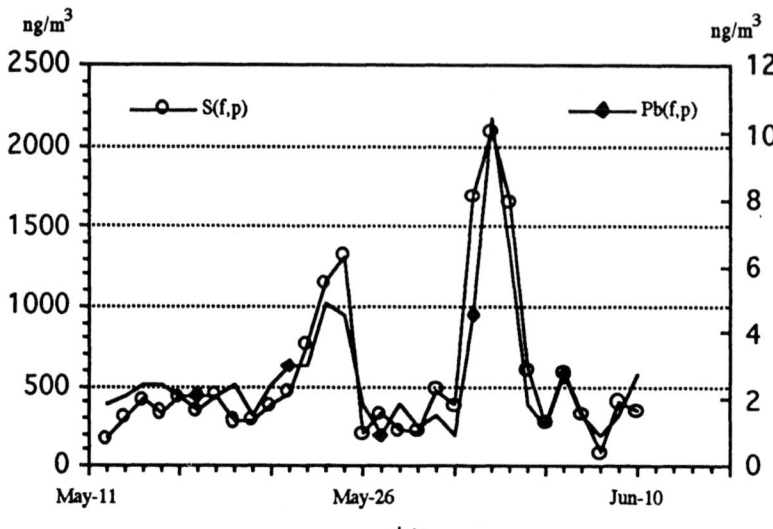

FIG. 2. Variation of Pb and fine particulate S during the third campaign.

3.3 Trace metals

Fine particles (< 1-3 µm) are the ones that are most likely to become dispersed over long distances. At a coastal station sea salt is always present and these particles tend to become larger than those in the accumulation mode. For marine and modified marine air masses the measured coarse particle (2.5 -10 µm) concentration ratios: S/Cl, K/Cl, Br/Cl and Ca/Cl are almost identical to the same ratios calculated for sea water salt. The choice of trace elements also depends on their possible natural or anthropogenic origin. In this work V, Ni, Zn and Pb have been used as tracers for anthropogenic aerosol. The relative enrichment in air compared to sea water for each of these elements is well over 1000. For Pb the enrichment factor varies between 40000 and $2 \cdot 10^6$. Enrichment relative to mean crust for Zn, Pb, Ni and V are respectively: 80-300, 1000, 30-50, 20-30. It is generally agreed that an enrichment factor of more than 10 implies that the element in question has a non-natural origin (e.g. Heidam 1985).

TABLE 2 Mean concentrations of selected trace metals in fine particles in air masses of different origins during the two winter (W) and summer (S) campaigns at Säby. See table 1 for explanations of air mass types.

air mass type	V(f) (ng/m^3)		Ni(f) (ng/m^3)		Zn(f) (ng/m^3)		Pb(f) (ng/m^3)	
	W	S	W	S	W	S	W	S
m.mar.	2.2	2.1	1.9	2.2	6.3	4.2	3.8	2.0
m.co.east	6.1	0.88	3.7	2.0	27	6.1	15	2.7
m.co.south	5.9	4.5	3.0	3.3	40	9.0	24	6.2

It is especially worth noting that the Zn and Pb values are much higher in winter than in summer in the same manner as the acidifying S- and N-compounds. V and Ni on the other hand does not show such a strong variation. The most probable reason for this is that in wintertime much oil is used for heating in the whole region.

In a similar investigation conducted in Birkenes at the southernmost Norway Amundsen et al. report concentrations of several trace metals as well as sulphur compounds (Amundsen, 1992). Their measurements add further support for the validity of our findings as they were done in the same area. A comparison of two kinds of episodes (England and eastern Europe) gave almost equal concentrations for V, Zn, Pb and SO_4-S in Birkenes and Säby.

3.4 Elemental ratios as tracers for pollution origin.

The V/Ni ratio can be used to discuss the contribution of oil combustion emissions to the air contamination. In a coal combustion region this ratio is usually 1.0 or lower while it often reaches 2.5 in a oil combustion area. (pacyna,1994). Poland is burning mostly coal while comparable amounts of coal and oil are used in the former USSR.

Figure 3 demonstrates the use of the V/Ni ratio as a tracer for sulphuric pollution arriving at the Swedish west coast. As mentioned above the episode 23-25 May originated in the S:t Petersburg area. The V/Ni emission ratio for this area is 1.8-1.9 (Pacyna, 1994) as compared to the calculated 1.5. The steep decrease between the 25:th and 26:th of May in the V/Ni ratio can be explained by the air mass moving over Poland with an emission ratio of 1.0 (Pacyna, 1994). The next sulphur episode originated in Southern England and Belgium. The V/Ni ratio points to a relatively larger influence from coal combustion. For both episodes the NO_x value was elevated. During the second episode the NO_x concentration was raised 5 fold.

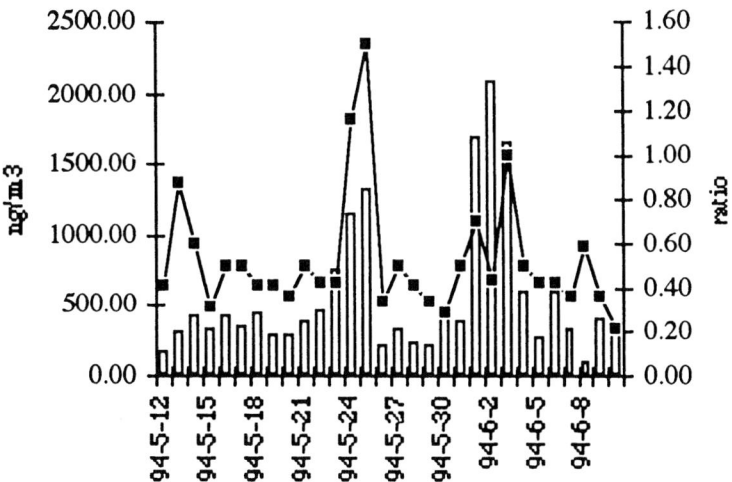

Fig. 3. Variation of fine particulate S (bars) and V/Ni ratio (line) during the third campaign.

4. Conclusions

- It has been shown (table 1 and 2) that anthropogenic species, eg trace metals in fine aerosol particles and S- and N- compound levels at the Swedish west coast in winter time are higher than summer levels, even when the air mass arrives from the sea. It is therefore relevant to question to what extent the North Sea is influenced by the European countries in such a way that there always seems to exist a "background" level of anthropogenic pollutants.
- From table 1 it is seen that sulphur in particulate form is dominant over SO_2-S in southern Scandinavia.
- Anthropogenic metals have been shown (figures 2 and 3) useful to trace the origin of acidifying pollution. Especially the V/Ni ratio varies strongly between clean and polluted air.

References

Amundsen, C.E., Hanssen, J.E. Semb, A. and Steinnes, E., (1992), Long-range atmospheric transport of trace elements to southern Norway. Atmos. Environ., **26A**, 1309-1324.

Heidam, N.Z. (1985), Crustal enrichments in the arctic aerosol. Atmos. Environ., **19**, 2083-2097.

Nriagu, J.O. and Pacyna, J.M. (1988) Quantitative assesment of worldwide contamination of air, water and soils by trace metals. Nature, vol 333 134-139.

O'Riordan, T. (1989), Air pollution legislation and regulation in the European community: A review essay. Atmos. Environ., **23**, 293-306.

Pacyna, J.M. (1994) Economic Commision for Europe. Convention on Long-range Transboundary Air Pollution. Task Force on Heavy Metals Emissions State-of-the Art Report. Prague, the Czech Republik, June 1994.

Öblad, M. and Selin, E., (1986) Measurements of elemental composition in background aerosol on the west coast of Sweden. Atmos Environ., **20**, 1419-1432.

SIMULATIONS OF OZONE FORMATION FROM DIFFERENT EMISSION SOURCES IN SWEDEN

Y. ANDERSSON-SKÖLD[1] and S. JANHÄLL[2]

[1]*Melica, Fjällgatan 3E, S-413 17 Göteborg, Sweden* [2] *IVL, P.O.Box 47086, S-402 58 Göteborg, Sweden*

Abstract. An emission inventory concerning volatile organic compounds (VOC) and their emission profile linked to their sources in Sweden has been undertaken. The inventory has been used in model simulations to predict the ozone formation from different emission source categories in Sweden. The studies have been carried out using the IVL photochemical trajectory model for two types of air masses which describes clean and polluted air. In Sweden mobile sources contribute to 45 % by mass of the total national VOC emissions, 58 % of the NOx emissions and to at least 43 % of the ozone formation from national sources. In general, the ozone formation in Sweden is more dependent and sensitive to emissions of NOx rather than VOC.

Key words. Swedish emissions, NOx, VOC, ozone formation, model simulation.

1. Introduction

In order to reduce the formation of ozone over Europe emissions of NOx and VOC have to be reduced. The most efficient means of ozone abatement varies from country to country depending on the inherent emission conditions.

During the last decade questions have been raised whether to reduce the NOx emissions, the VOC emissions or both, and whether each species in the VOC emissions shall be reduced by the same fraction, or according to the ozone forming ability? As a tool in determining the most efficient abatement strategy for each country it is of interest to determine the extent each emission source category contributes to the ozone formation. This study focuses on Swedish VOC emissions from different source categories, their VOC profiles and their contribution to the ozone formation.

The study includes an emission inventory and computer model simulations. The simulations have been conducted using the IVL photochemical trajectory model (Andersson-Sköld et al., 1992, Andersson-Sköld; 1995) which is a chemically expanded version of the Harwell photochemical trajectory model (Derwent and Hough, 1987, Derwent and Jenkin, 1992). The model describes the chemical development in a parcel of boundary layer air, including an explicit description of the atmospheric oxidation of more than 80 of the anthropogenic VOCs emitted over Europe. The air parcel is split into two layers to provide a description of the diurnal variation in the boundary layer depth and the exchange of pollutants at the ground surface. A detailed description is given in Janhäll and Andersson-Sköld (1995) and Pleijel et al., (1992).

The contribution to ozone formation from Swedish emissions of NOx and VOC have been simulated for different types of air masses which describe clean and polluted air transported to Sweden under different meteorological conditions. Simulations have also been conducted concerning the ozone formation ability specific to the VOC profiles from the source categories. These simulations have only been conducted for a high pressure situation around midsummer.

2. Methods

The total emissions of Non Methane Volatile Organic Compounds (NMVOC) in Sweden linked to their sources have been obtained from the Swedish Environmental

Table I. Emission sources of NOx[7,20] and VOC[21] in Sweden.

Emission source category	Total emission (kg /km² and year) NOx (as NO$_2$)	VOC	VOC profile references
Road Traffic	396.5	404.9	4, 5, 10, 13, 23, 24
Working Machinery	188.4	47.7	9, 11
Other Mobile sources	159.5	43.3	6, 9, 11, 17, 18, 22
Wood Combustion	4.9	331.9	1
Energy	166.5	23.6	9
Industry	24.7	186.2	13
Domestic Use, Pesticides	-	98.9	21, 23
Total	940.5	1136.5	

Protection Agency (SNV, 1992). These emissions were further divided into individual species following specific source profiles, which have been obtained according to their sources. A further detailed description of how the profiles where obtained is presented in Janhäll and Andersson-Sköld, 1995. The inventory includes 80 different VOCs. The sources were grouped in seven emission categories, i.e. road traffic, working machinery, other mobile sources in which sea traffic, air traffic and track bound traffic are included, wood combustion, energy, industry and domestic use and pesticides. The NOx emissions have been obtained from the Corinair '90 inventory.

The total Swedish emissions of NMVOC and NOx from the individual emission source categories are shown in Table I together with the references used to obtain the VOC profile for each category. The average fraction of each individual VOC species obtained for the Swedish VOC emissions is given in Table II.

Table II. The individual VOC emission species treated explicitly in the IVL photochemical trajectory model, numbers are parts of total NMVOC (% by mass).

1.21	ethane	0.58	propane	1.96	n-butane	1.33	i-butane	1.04	n-pentane
2.06	i-pentane	0.70	n-hexane	0.96	2-methylpentane	0.79	3-methylpentane	1.59	heptane
0.46	octane	2.11	2-methylheptane	0.26	n-nonane	0.36	2-methyloctane	0.88	n-decane
0.83	2-metylnonane	0.63	n-undecane	0.65	2-methyldecane	0.42	n-dodecane	2.11	methylcyclohexane
5.60	ethene	12.71	propene	0.27	1-butene	0.56	2-butene	1.80	i-butene
0.39	1-pentene	0.42	2-pentene	0.18	2-methyl-1-butene	0.22	2-methyl-2-butene	2.52	ethylene
4.94	benzene	5.36	toluene	1.93	o-xylene	2.03	m-xylene	2.01	p-xylene
1.13	ethylbenzene	0.70	1,2,3-TMB	2.50	1,2,4-TMB	0.97	1,3,5-TMB	1.24	m-ethyltoluene
1.06	p-ethyltoluene	0.71	o-ethyltoluene	0.62	n-propylbenzene	0.32	i-propylbenzene	0.24	styrene
2.48	hcho	0.37	ch3cho	0.42	c2h5cho	0.13	c3h7cho	0.07	i-propanol
0.01	c4h9cho	0.30	benzaldehyde	0.55	ch3coch3	2.48	ch3coc2h5	0.34	mibk
3.19	methanol	4.47	ethanol	0.45	propanol	3.65	n-butanol	0.03	methylacetate
0.57	ethylacetate	0.37	n-butylacetate	0.37	s-butylacetate	0.00	dimethylether	0.24	diethylether
0.15	mtbe	5.47	acetic acid	0.02	glyoxal	0.01	methylglyoxal	0.22	ch2cl2
0.02	ch3chcl2	0.02	ch2clch2cl	1.18	ch3ccl3	0.16	c2hcl3	0.02	c2cl4
0.13	acroleine	0.06	methylmercaptan	0.53	DMS	0.09	DMDS	0.07	furfural

The study has been conducted for air masses arriving to Sweden over the North Sea. The concentrations used to describe the air parcel as it reaches Sweden have been obtained from the TOR Monitoring station at Rörvik, located at the Swedish west coast (Lindskog, 1995). The simulated air masses can be described by a high pressure situation with clean air (Atlantic origin), a more common weather situation with clean air, a high pressure situation with polluted air (European continental and south UK origin) and a more common weather situation with polluted air. The initial concentrations and meteorological parameters used to describe the simulated air masses are shown in Table III.

Table III. The simulated air masses described by initial concentrations and meteorological parameters used in the simulations. The average ozone concentrations over 12 and 36 hours calculated for with and without Swedish emissions. The average contributions from the Swedish emissions to the ozone concentrations are given in per cent.

Type of air mass	Initial concentrations1		12 hours	36 hours
Clean air High pressure Clear sky[3]	O_3 60 ppb NO 0.13 ppb C_3H_8 0.142 ppb C_5H_{12} 0.040 ppb C_3H_6 0.016 ppb	NO_2 0.71 ppb C_2H_6 0.783 ppb C_4H_{10} 0.119 ppb C_2H_4 0.041 ppb C_2H_2 0.102 ppb	56-51 ppb (10%)	48-35 ppb (28%)
Clean air "common weather"[4]	O_3 43 ppb NO 0.30 ppb C_3H_8 0.285 ppb C_5H_{12} 0.139 ppb C_3H_6 0.035 ppb	NO_2 1.25 ppb C_2H_6 1.177 ppb C_4H_{10} 0.271 ppb C_2H_4 0.139 ppb C_2H_2 0.179 ppb	49-45 ppb (8%)	45-35 ppb (23%)
Polluted air High pressure Clear sky[3]	O_3 68 ppb NO 0.19 ppb C_3H_8 0.230 ppb C_5H_{12} 0.167 ppb C_3H_6 0.022 ppb	NO_2 0.49 ppb C_2H_6 0.893 ppb C_4H_{10} 0.278 ppb C_2H_4 0.116 ppb C_2H_2 0.196 ppb	61-55 ppb (9%)	51-37 ppb (27%)
Polluted air "common weather"[4]	O_3 47.5 ppb NO 0.31 ppb C_3H_8 0.356 ppb C_5H_{12} 0.347 ppb C_3H_6 0.039 ppb	NO_2 1.70 ppb C_2H_6 1.077 ppb C_4H_{10} 0.570 ppb C_2H_4 0.236 ppb C_2H_2 0.335 ppb	53-50 ppb (6%)	48-38 ppb (20%)

1. The values are obtained from the TOR monitoring station at Rörvik (Lindskog, 1995)
2. The temperature has been set to 14-22.3 °C in all simulations corresponding to the growing season average temperature, and the meteorological parameters are obtained from Pleijel, et al, 1992.
3. Wind speed 2-3.5 m/s, mixing depth 150-1000 m, date 21/6 and cloudiness 0
4. Wind speed 4-7 m/s, mixing depth 150-1400 m, date 15/7 and cloudiness 2

To study the contribution of Swedish emissions to the ozone concentration measured in Sweden this work has focused on the south of Sweden. In the south of Sweden the emissions are more pronounced and the solar intensity higher than in the north. In addition there is data available to describe the chemical composition of the air masses the Swedish west coast (Lindskog, 1995). The simulations have been carried out where emission source categories are successively removed, until there are no Swedish emissions of NOx or VOCs left. This study has been conducted for all the air masses described in Table III above.

In order to compare the relative importance of the Swedish emission source categories, the sources have been individually excluded, one at each simulation run in contrast to being successively removed. These simulations treat from each emission category the relative influence of NOx, of VOC and of both NOx and VOC. These simulations have only been conducted for the clean air mass under sunny, high pressure conditions.

The above simulations describe the relative importance of NOx and VOC on ozone formation in Sweden. They include both the mass and the VOC profile dependence. In order to study the VOC profile dependence only, simulations have been carried out where an equal amount by mass of VOC from each sector has been added, one sector at a time, to the air mass and the additional ozone formation has been calculated.

3. Results

The ozone concentrations, obtained over the first 40 hours of the simulations, where one emission source category at a time has successively been removed until there are no Swedish emissions of NOx nor VOCs left, are shown in Figure 1 for the clean air high pressure situation.

Table III shows the average ozone concentration obtained over the first 12 and 36 hours for the simulations where all south Swedish emissions are considered and where no Swedish emissions are considered are shown together with the obtained ozone reduction for some emission sectors. As can be seen in Figure 1a and Table III over the first 12 hours after the air mass has reached Sweden, the national ozone contribution is less than 10 %. The national contribution increases, as expected, with the time of the simulation. If the air mass stays over Sweden for 36 hours the national contribution increases to 30 % in average over that period.

According to the results presented in Figure 1 and Table III the significance of Swedish emissions on ozone concentration increases with time especially after night. In the simulations presented here the importance may be overestimated as a consequence of a to high value used for the deposition of ozone over night. The treatment of deposition and other heterogeneous parameters are to be improved in the IVL model but has here been treated in the standard IVL way (Derwent and Hough, 1987; Derwent and Jenkin, 1992; Andersson-Sköld *et al.*, 1992; Pleijel *et al.*, 1992; Andersson-Sköld, 1995; Janhäll and Andersson-Sköld, 1995). It has been shown from measurements, however, that the ozone concentrations in Sweden may increase in the same way as calculated for the "today" and "wood combustion removed" cases due to Swedish emissions in the south of Sweden (Kindbom *et al.*, 1995).

In Figure 1b the corresponding ozone concentrations are shown for simulations when only the VOC emissions from each source category are taken away. The ozone reduction is much less pronounced than when both NOx and VOC are reduced. Over the first 12 hours the ozone reduction is insignificant to the VOC reduction and the reduction over 36 hours only corresponds to 5 ppb or 10 % of the total ozone concentration obtained for all Swedish emissions. This illustrates that the ozone formation in Sweden is more sensitive to NOx emissions than VOC emissions.

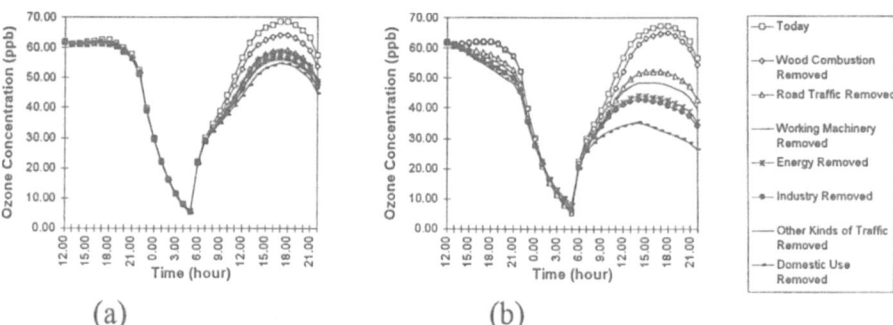

Fig 1a and b. Simulated ozone concentrations of a clean air mass arriving to Sweden under a high pressure situation, when average Swedish emissions are assumed (□) and (a) when one emission source category after another has been removed until there are no Swedish emissions of NOx or VOCs left (--) and (b) the same simulation when only VOC emissions are affected.

The importance of each individual emission category is not possible to calculate by this kind of simulation as the order in which the sectors are removed affects the calculated ozone concentration. To obtain the relative contribution from each emission source category the categories have to be removed one at the time. The results from these simulations are given in Figure 2 where the relative contribution to the ozone concentration from each emission category in Sweden is shown.

From Figure 2 it can be seen that the mobile sources give the greatest contribution to the Swedish ozone production and road traffic is the greatest single source, 30 %. According to Figure 2, wood combustion also represents a significant contribution. The contribution is, however, overestimated in this study since the emissions in reality are limited almost only to the winter half year, but in this study they are treated as constant over the whole year. The influence may however be valid in spring, when it is still cold but the solar activity becomes important for Scandinavian ozone formation (Andersson-Sköld, 1993).

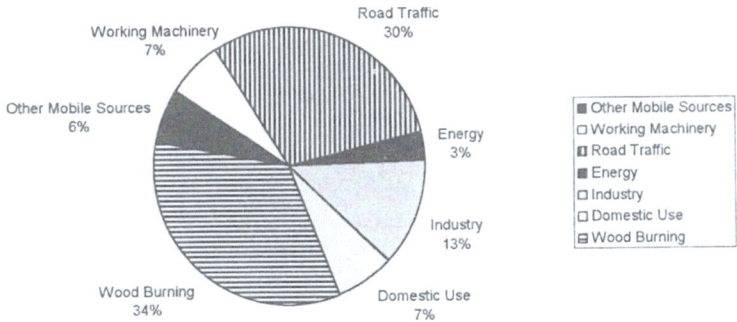

Fig 2. Relative contribution from emission source categories in Sweden to ozone.

The ability for each emission category to produce ozone with respect to its VOC profile has been calculated in clean air under a high pressure situation and in polluted air under a situation with more common weather. The ozone production ability due to a similar added amount by mass of VOC is shown in Figure 3. According to Figure 3 there is a difference in ozone formation ability between the emission categories studied. The mobile sources have an average ozone formation ability whereas the domestic use has the lowest and the industrial sector has the highest ozone formation ability according to their VOC profiles.

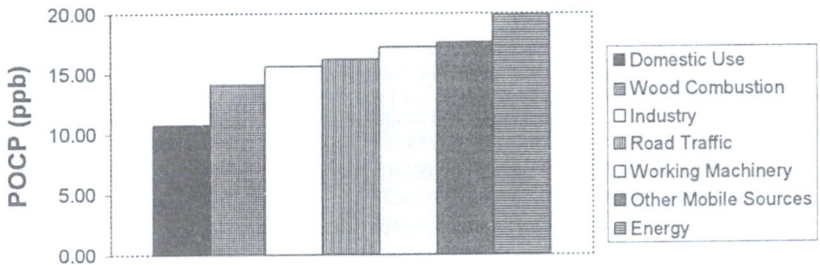

Fig 3. Ozone formation ability for the emission categories in Sweden.

4. Conclusions

The ozone production in Sweden is mainly sensitive to NOx emissions. For air masses passing over Sweden in less than 12 hours, the Swedish ozone forming contribution is not affected at all by the VOC emissions and only by 10 % by the Swedish NOx emissions.

The greatest national source to ozone in Sweden is represented by the mobile sources which are dominated by road traffic. The high ozone production in this case results from relatively high emissions of NOx and VOC. The VOC species profiles of mobile sources have an average ozone formation ability compared to other VOC source categories in Sweden.

Acknowledgement

The Swedish Environmental Protection Agency (SNV) and Swedish Environmental Research Institute (IVL) are gratefully acknowledged for their financial contributions. Dave Cooper is greatfully acknowledged for the language check.

References

Andersson-Sköld, Y.: 1993, Regional oxidantbildning - Till följd av utsläpp från vedeldning. SNV 4231, Solna, Sweden

Andersson-Sköld, Y.: 1995, Updating the Chemical Scheme for the IVL Photochmical trajectory model, IVL B-1151, Göteborg, Sweden

Andersson-Sköld, Y., Pleijel, K., Grennfelt, P., Rohndahl, L.: 1992, Photochemical ozone creation potentials: a study of different concepts. *J. Air Waste Manage. Assoc.* **42:9**, pp.1152-1158

Berglund P. M., Petersson G.: 1990, Hazardous petrol hydrocarbons from refuelling with and without vapour recovery. *The Science of the Total Environment*, **91**, pp 49-57

Bilindustriföreningen: 1994, Bilismen i Sverige1994, Stockholm, Sweden

Cooper, D. A., Peterson, K. P. and Simpson, D.: "Hydrocarbon, PAH and PCB emissions from ferries: a case study in the Skagerak-Kattegatt-Öresund region", submitted to Atmospheric Environment 1995

Corinair Emission Inventory: 1990

Derwent, R.G., Jenkin, M.E.: 1991, Hydrocarbons and the long range transport of ozone and PAN across Europe, *Atmos. Environ.*, **25a**, p 1661

Quality of Urban Air Group: 1993, Urban Air Quality in the United Kingdom-First report of the Quality of Urban Air Group., January 1993

Grennfelt, P., Hov, Ö., Derwent, R. G.: 1993, Second Generation Abatement Strategies for NOx, NH$_3$, SO$_2$ and VOC, IVL B-1098, Göteborg, Sweden

Hoekman, S. K.: 1992, Spediated Measurements and Calcula... *Environ. Sci. Technol.* **26**, pp 1206-1216

Hough, A.M., Derwent, R.G.: 1987, Computer modelling studies of the distribution of photochemical ozone production between different hydrocarbons. *Atmos. Environ.* **21**, pp 2015-2033.

Janhäll, S., Andersson-Sköld, Y.: 1995, Emission Inventory of NMVOC species in Sweden, IVL B-1193

Kindbom, K., Lövblad, G., Peterson, K., Grennfelt, P.: 1995, Concentrations of tropospheric ozone in Sweden, *Ecological Bulletins* **44**: pp 35-42

Lindskog, A.: 1995, private communication

Pleijel, K., Andersson-Sköld, Y., Ohmstedt, G.: 1992, Atmosfärkemiska och meteorologiska processers betydelse för mesoskalig ozonbildning, IVL B-1056, Göteborg, Sweden

Pleijel, K., Moldanova, J., Andersson-Sköld, Y.: 1993, Kemisk modellering av en flygplansplym i fria troposfären, IVL- B-1104, Göteborg, Sweden

Sjöfartsverket: 1994, Åtgärder - luftföroreningar från den marina sektorn, rapport regeringsuppdrag 45-9371263, Sjöfartsverket, 601 78 NORRKÖPING, Sweden

SNV: 1990:1, Strategi för flyktiga organiska ämnen (VOC) SNV 3763, Solna, Sweden

SNV: 1993, Ett miljöanpassat samhälle Naturvårdsverkets aktionsprogram Miljö 93, Solna, Sweden

SNV: 1994:1, Utsläpp till luft av flyktiga organiska ämnen, 1992 SNV 4312, Solna, Sweden

Östermark U.: 1993, Volatile Hydrocarbons in Vapour and Exhaust from Alkylate-based and conventional petrol. Kemisk miljövetenskap, Göteborgs Universitet, Sweden

Östman, M.: 1994, *Kemikalieinspektionen*, Solna,Sweden, private communication

LONG-TERM MODELLING OF AIRBORNE POLLUTION WITHIN THE NORTHERN HEMISPHERE

M.GALPERIN, M.SOFIEV, O.AFINOGENOVA
Meteorological Synthesizing Centre-East of EMEP, Russia, 117292, Moscow, Str.Kedrova, 8, k.1

Abstract. A long-term modelling (1991-1994) of oxidised sulphur, bound nitrogen and some heavy metals has been carried out by MSC-E/EMEP for the Northern Hemisphere. The transport unit of the model is an Eulerian scheme which could be classified as Pseudo-Lagrangian one. Vertical distribution described by means of Gaussian approximation and the exchange with the free troposphere are taken into account. Vertical movement is calculated proceeding from local mixing conditions, state of the surface, its height (topography) etc. The chemical unit for acid compounds contains 25 reactions and 14 compounds including sulphur and nitrogen compounds peroxyacetylnitrate, tropospheric ozone, volatile hydrocarbons (but methane) are considered as a whole via ozone creation potential. The model time step is 1 hour, meteorological data (winds, temperature, precipitation etc.) cover 6-hour intervals. The model results show that very significant part of the Arctic and West Asian acid pollution is produced by European countries. On the whole the Arctic pollution by SO_x, NO_x and NH_x comes from sources of Old World. The main source of sulphur pollution is located in Russia and of nitrogen compound - in Central and Northern Europe. About 50% SO_x, 70% NO_x and 40% NH_x deposition in Central Asia and Kazakhstan is imported from external sources. A similar situation is observed in European and Asian parts of Russia.

Key words: acidification, air pollution, air pollution modelling

1. Introduction

Results of the long-range air pollution transport studies carried out in Europe since 1978 revealed that a substantial part of acid pollutants can be transported within the whole Northern Hemisphere. The first model assessments of this phenomenon have been made by Tarrason and Iversen (1992). These estimates have been carried out for oxidised sulphur for 1 year. For several regions (e.g. Arctic) it is important to get the estimates for nitrogen compounds as well. In addition, transport conditions may vary from year to year which require multiannual estimates.

This paper is aimed at the evaluation of transcontinental and intercontinental transport of acid pollutants in the Northern Hemisphere using multiannual meteorological data and in particular at the evaluation of the Arctic basin pollution, the contribution of both North America to acid deposition in Europe and European sources to the pollution of Siberia and Central Asia.

Meteorological information used in model was prepared in Russian Hydrometeocentre on the basis of weather forecast modelling corrected by data of meteorological monitoring network.

2. General model description

The improved EMEP/MSC-E current model used for calculation is an Eulerian model with full splitting of all processes. The first version of the model was described by Pressman *et al* (1989). Common advection equation is solved

$$\frac{\partial Q}{\partial t} + \nabla \cdot (VQ) = E(t) - \lambda(t)Q, \tag{1}$$

where $Q(x,y,t)$ is the pollution mass in a cell, $V(x,y,t)$ is wind velocity vector, $E(t)$ is emission rate, $\lambda(t)$ is removal rate, x and y are horizontal co-ordinates and t is time.

In general the advection scheme used is similar to those of Egan and Mahoney (1972), Pepper and Long (1978), Pedersen and Prahm (1974):

$$Q_n(t+\Delta t) = [1-\lambda(t)\Delta t][E_n(t)\Delta t + \sum_{m=1}^{N} p_{mn}(t)Q_m(t)], \quad (2a)$$

$$Q_n(t+\Delta t)r_n(t+\Delta t) = [1-\lambda(t)\Delta t][E_n(t)r_{en}\Delta t + \sum_{m=1}^{N} p_{mn}(t)Q_m(t)r_m(t)] + Q_n(t+\Delta t)V_n\Delta t, \quad (2b)$$

where n and m are grid cell numbers, $r_n = x_n i + y_n j$ is radius-vector of mass centre position in the n-th gridcell, r_{en} is radius-vector of emission mass centre, $p_{mn} = \delta(X_m, X_n) \cdot \delta(Y_m, Y_n)$, δ is Kronecker's symbol, $X=[X]$ and $Y=[Y]$, Δt is time step. The equations (2a) and (2b) are conservation conditions of mass and the first moment. At each step calculations are made for each cell of the grid. Sums in the right parts are accumulated automatically with sequential addresses to cells at time step and their calculation is not required. Initial conditions of (2) are $p_{nn}(0)=0$, $Q_{nn}(0)=0$ for all grid cells.

The scheme errors are resulted from non-uniformity of mass distribution with grid cells (Gibbs effect). The suppression of oscillations can be also realized by the introduction of conservativity of the second moments and by means of smoothing filter, i.e. by distribution of mass with Eulerian grid.

Vertical distribution of concentration c is described by equation

$$\partial c/\partial t = \partial K_z \partial c/\partial z^2 \quad (3)$$

with boundary condition near surface

$$v_d c(0) = -K_z \partial c(z)/\partial z \big|_{z \to 0} \quad (4)$$

where z is a vertical co-ordinate, K_z is vertical mixing coefficient and v_d is dry deposition velocity. The approach is based on the moment conservation along the verticle while advection

$$Z_{0n}Q_n = \int_0^\infty zc(z)dz = h_{en}E_n + \sum_{m=1}^{N} p_{mn}Q_m Z_{0m} \quad (5)$$

where Q_m are the masses from equation (2a), h_{en} is source height.

A partial solution of eq.(3) in the form normalized qaussian distribution density is used:

$$C(Z) = Q\varphi(Z) \bigg/ \int_0^\infty \varphi(Z)dZ$$

where $\varphi(Z) = \exp[-(Z-\mu)^2/2\sigma^2]/\sigma\sqrt{2\pi}$, μ and σ are calculated at each time step and cell from equations (4) and (5):

$$\mu = V_d \sigma^2/K_Z; \quad \sigma = (Z_0 - \mu)\varphi(0) \bigg/ \int_0^\infty \varphi(Z)dZ$$

For vertical diffusion $\Delta\sigma \cong K_Z \Delta t/L_K$ and corresponding change of Z_0 are calculated (here L_K is scale factor, taken as equal to the boundary layer height). It is supposed that at the height of boundary layer value of K_z decreases by the factor of 10 in comparison with its maximum.

The gaussian form of vertical distribution makes it possible to consider the exchange with the free troposphere, to take into account effects of the surface (roughness, sea/land, orography extent of wetting, etc.), to distinguish cloud and subcloud washing out and to apply weighted mean advection velocity according to equation:

$$\overline{V} = \int_0^\infty c(z) V(z) dz \bigg/ \int_0^\infty c(z) dz$$

where $V(z)$ is distribution of the wind along the height. In the model 4-layer wind (at 1000, 925, 850, 700 mbar) is used for this procedure. Thus, in each cell at each time step horizontal transport is calculated at the height of main mass. This procedure is in force even if main pollution mass is placed above the boundary layer (in the free troposphere).

The model chemical scheme is shown in the fig.1. Ozone generation and decay imply the input of stratospheric ozone, natural sources and ozone decomposition due to solar radiation and surface contact. The symbol $(NH_4)_{1.5}SO_4$ indicates a mixture of equal NH_4HSO_4 and $(NH_4)_2SO_4$ fractions and NO_3^- and SO_4^{2-} designate the metal combination of nitrate and sulphate ions. More detailed description of oxidised nitrogen chemistry scheme is presented in (Pressman et al, 1989). For ozone creation potential assessment the results of Simpson (1995) are used. Chemical transformation parametrisation for ammonium is taken from (Finlayson-Pitts and Pitts, 1989).

Dry deposition is calculated in a common way: $\Delta Q = V_d c(0) \Delta t$.

It is taken into account that V_d depends on the surface type and state, in particular on surface moistering. Therefore V_d is calculated as $V_d = V_{d0} K(T,P,F)$, where T - temperature, P - precipitation amount and F - surface type, $V_{d0} = 0.3$ cm/s for SO_2 and NH_3, $V_{d0}=0.1$ cm/s for aerosols, NO_2 and PAN, $V_{d0}=1$ cm/s for HNO_3. For $T<2°C$ $K=1$ always and for wet land surface $K=3$ is taken. For sea surface it is assumed that at wind speed greater than 6 m s^{-1} V_{d0} is linearity growing from V_{d0} to 10 V_{d0} for SO_2, NH_3 and aerosols. These provisional estimates are close to results of Lindfors et al (1991).

While the calculations of wet deposition scavenging coefficient is taken as: $\Lambda = \beta S_0/V_p$, where S_0 - precipitation element surface, V_p - its volume, I - precipitation intensity, β - trapping efficiency factor (Engelmann, 1963; Galperin, 1989). It is assumed that above the boundary layer cloud scavenging takes place and $\beta=1$. In case of subcloud scavenging $\beta=0.25$. For SO_2 Λ depends on saturation of precipitation elements (Galperin, 1989). The dependence of S_0/V_p on rain intensity (Kelkar, 1959) is taken into account as well.

Model validation has been carried out on the basis of measurement data provided by monitoring network of EMEP (Co-operative Programme for Monitoring and Evaluation of the Long-Range Transmission of Air Pollutants in Europe) for 1987-92. Comparison results manifest good agreement between model and measurements (MSC-E, 1995).

3. Modelling results

The emission values are taken from official reports submitted by European and North American countries to UN Economic commission for Europe, from paper (Kato and Akimoto, 1992) and from (US EPA, 1993). The total oxidised nitrogen deposition averaged over 1991-1994 shown in the fig.2 as a sample.

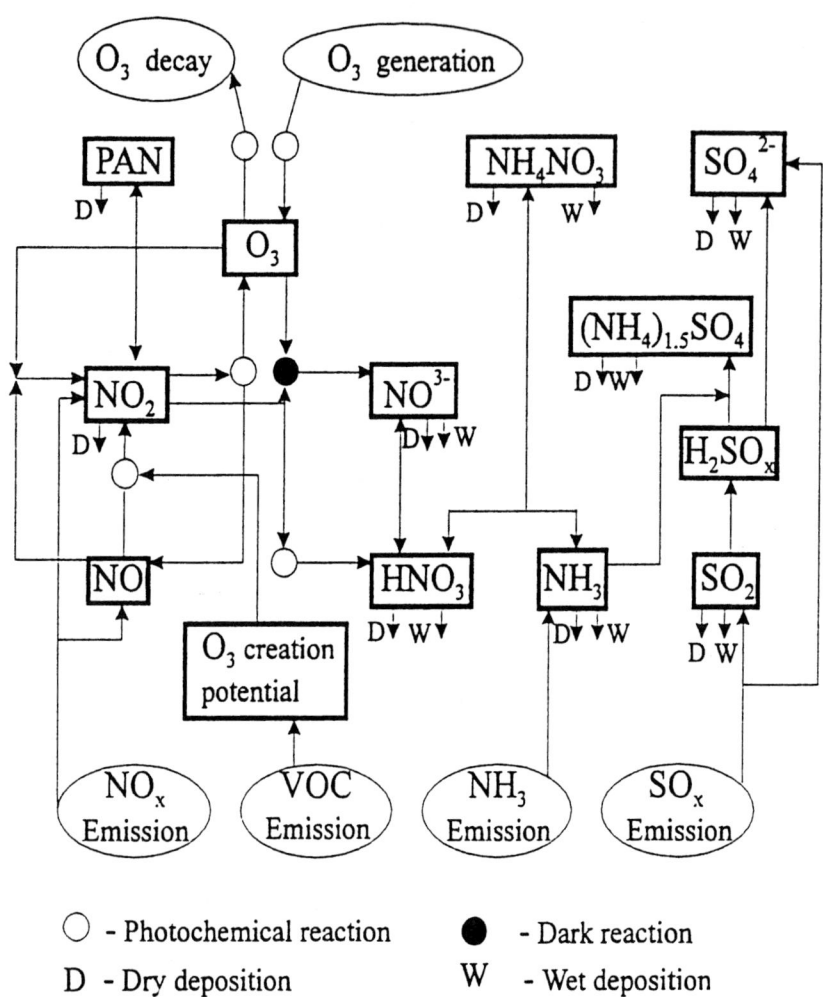

Fig. 1. Model chemical scheme

The calculation was made for 21 regions (countries)-emitters and 26 receivers. Summarized SO_x and NO_x deposition data for emitters and receivers are presented in Tables I and II. In the line "Total" the sum of deposition inside calculation area are given, "Export" is a pollution left the calculation area. The detailed information is in (MSC-E, 1995) and may be requested. The modelling has shown, that acid depositions in the USA, China, Western, Central and Southern Europe practically result from their own sources. In other regions the deposition is not only their "internal" problem but it depends to a great extent on the long-range transport.

For example, 70-85% of SO_x and NO_x deposition in Northern Europe are transported from external sources and only 20-30% of deposition is caused by internal sources. North American sources can impact on particularly sensitive regions on the coastal line of Norway, on Ireland and Great Britain and on North western part of France. About 50% SO_x, 70% NO_x and 40% NH_x deposition in Central Asia and Kazakhstan is imported from external sources. A similar situation is observed in European and Asian parts of Russia.

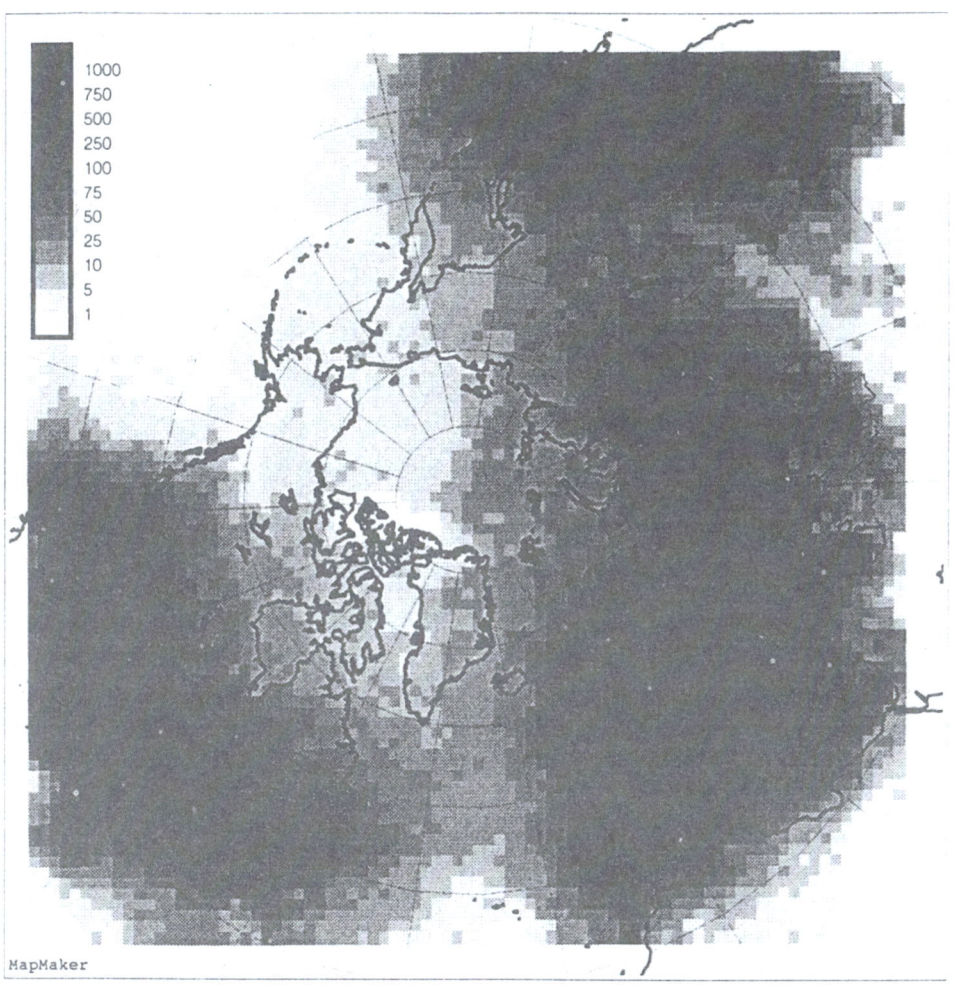

Fig. 2. The map of annual oxidised nitrogen deposition averaged for 1991-94.
Unit is 1 mg N m^{-2} · year^{-1}.

On the whole the Arctic pollution by SO_x, NO_x and NH_x comes from sources of Old World. The main source of sulphur pollution is located in Russia and of nitrogen compound - in Central and Northern Europe. 30% of the Arctic pollution by SO_x are from European sources located outside Russia, 40% of the pollution comes from sources located in Asian part of Russia, 22% - from European part of Russia and only 4% arrive from North American continent.

Quite another pattern is observed for NO_x Arctic pollution: sources in Central, Western and Southern Europe give 25% of deposition; Great Britain and Ireland - 22%; Scandinavian countries - 15%; the USA and Canada - about 7%.

50% of reduced nitrogen deposition in the Arctic are given by European sources, 25% - European part of Russia, the rest fraction comes from Central Asia and Kazakhstan. The USA and Canada fraction is negligible.

TABLE I
Oxidised sulphur deposition within North Hemisphere averaged for 1991-1994. Unit = 1000 t S per year.

Receivers	Emitters							Total deposition
	Europe[1]	North Europe[2]	Russia	Canada	USA	Far east[3]	Other	
Europe[1]	7262	48	93	0	0	0	97	7500
North Europe[2]	573	126	80	0	1	0	11	791
Russia	1283	46	2833	0	0	124	284	4570
Canada	12	0	32	899	710	0	1	1654
USA	2	0	8	239	7044	0	0	7293
Far East[3]	15	0	85	0	0	7730	41	7871
Arctic[4]	166	14	361	9	12	0	12	574
Oceans and seas	3790	33	215	362	2180	2042	302	8924
Other	444	4	365	0	0	141	888	1842
Total	13547	270	4072	1509	9955	10037	1636	41023
Export	290	0	17	6	364	773	247	1705
Emission	13837	270	4089	1515	10312	10810	1884	42731

TABLE II
Oxidised nitrogen deposition within North Hemisphere averaged for 1991-1994. Unit = 1000 t N per year.

Receivers	Emitters							Total deposition
	Europe[1]	North Europe[2]	Russia	Canada	USA	Far east[3]	Other	
Europe[1]	2307	65	50	2	3	0	53	2480
North Europe[2]	287	92	14	1	1	0	9	404
Russia	543	99	715	1	3	70	86	1517
Canada	6	1	4	253	475	0	12	751
USA	1	0	1	123	3174	1	13	3313
Far East[3]	5	1	33	0	0	1540	28	1607
Arctic[4]	120	39	50	7	12	4	23	255
Oceans and seas	1649	55	72	188	1309	822	156	4251
Other	245	19	225	5	14	50	235	793
Total	5161	370	1164	579	4990	2486	615	15371
Export	159	2	13	11	552	510	74	1321
Emission	5320	372	1177	590	5544	2998	690	16691

[1] Excluding Scandinavia and Russia
[2] Scandinavia and Baltic Sea
[3] China, Japan, South and North Korea
[4] Arctic basin to the North of 70° N and Greenland

References

Egan, A.B. and Mahoney, J.R.: 1972, *J.Appl.Met.*, **11**, 312-322.
Engelmann, R.J.: 1963, Rain scavenging of particulates, *USA EC Report HW-79382*, Hanford Atomic Products Operation.
Finlayson-Pitts, B.J. and Pitts J.N,:1988, Atmospheric chemistry fundamentals and experimental techniques, NY, John Wiley & Sons.
Galperin, M.V.: 1989, Adsorption-kinetic non-linear washout model of sulphur and nitrogen compounds from the atmosphere, in: *Air Pollution Modeling and Its Application VII*, N.Y. & London, Plenum Press, 475-484.
Kato, N. and Akimoto, H.: 1992, *Atmosph.Environ.*, **26A**, 2997-3017.
Kelkar, V.N.: 1959, *Indian J.Meteorol.Geophys.*, **10**, 2, 125-136.
Lindfors, V., Joffre, S.M., Damski, J.: 1991, Determination of the wet and dry deposition of sulphur and nitrogen compounds over Baltic sea using actual meteorological data, Helsinki, Finnish Meteorol. Inst., 14.
MSC-E: 1995, Atmospheric transport of acid compounds in the Northern Hemisphere for 1991-1994, Moscow, EMEP/MSC-E Report 8/95.
Pepper, D.W., Kern, C.D. and Long, P.E.: 1979, *Atmosph.Environ.*, **13**, 223-237
Pedersen, L.B. and Prahm L.P.: 1974, *Tellus*, **24**, 5, 594-602.
Pressman, A.Ya., Galperin, M.V., Popov, V.A. et al: 1991, *Atmosph.Environ.*, **25A**, 1851-1862.
Simpson, D.: 1995, *J. of Atmospheric chemistry*, **20**: 163-177
Tarrason, L. and Iversen, T.: 1992, *Tellus* **44B**, 114-132.
US EPA: 1994, National air pollutant emission trend 1900-1993, *EPA-454/R-94-027*.

VEGETATION ATMOSPHERE EXCHANGE OF AMMONIA: CANOPY CYCLING AND THE IMPACTS OF ELEVATED NITROGEN INPUTS.

M.A. Sutton, D. Fowler, J.K. Burkhardt and C. Milford

Institute of Terrestrial Ecology, Edinburgh Research Station, Bush Estate, Penicuik, Midlothian, EH26 0QB, UK.

Summary

Micrometeorological measurements of atmospheric ammonia (NH_3) exchange with semi-natural and agricultural plant communities were made using sensitive new instrumentation capable of determining NH_3 fluxes at <0.1 µg m^{-3}. The results are used to test hypotheses concerning the canopy cycling of reduced nitrogen (NH_x) and the existence of potential feedbacks between total N inputs (from agricultural sources or atmospheric deposition) and the net NH_3 flux. The measurements over cropland, together with a model calculating the 'canopy compensation point' for NH_3 indicate the importance of stomatal NH_3 emission and recapture of NH_3 by plant cuticles and water-layers. In contrast, measurements at an extremely clean upland moorland suggest that cuticular desorption of NH_3 is also possible at low concentrations. Interpretation of dew measurements suggests that leaf uptake of NH_4^+ may occur as a result of pH gradients between the leaf surface and apoplast. The combined conceptual model of NH_x exchange provides a useful basis for developing quantitative resistance models to predict NH_3 fluxes.

1. Introduction

Dry deposition of ammonia (NH_3) is an important component of the atmospheric nitrogen (N) load received by terrestrial ecosystems in Europe, contributing to both ecosystem acidification and eutrophication. To assess the impacts of N deposition, it is therefore essential to quantify the exchange of NH_3 between the atmosphere and the ground. This is complicated since both NH_3 emission and deposition may occur with time and with differences between ecosystem types (e.g. Duyzer *et al.* 1987; Sutton *et al.* 1993b, 1995b).

To quantify and predict NH_3 fluxes, it is necessary to develop a mechanistic understanding of the exchange process. A key factor is the relationship of NH_3 fluxes to N inputs and cycling. This may be demonstrated by the comparison between agricultural and semi-natural ecosystems, where the fertilizer N inputs to the soil of the former result in a shift from NH_3 deposition to bi-directional fluxes (Sutton *et al.* 1993b; Erisman & Wyers 1993; Schjørring *et al.* 1993a). This difference clearly has consequences for the impact of agricultural N inputs on the atmospheric transport of NH_3. However, there are also important potential feedbacks with total atmospheric N inputs (NO_y + NH_x). Ecosystems subjected to long-term elevated N deposition may respond with increased N concentrations in plant tissues, resulting in a reduced net input by dry deposition (Sutton *et al.* 1993b). Caution is therefore necessary in comparing total N deposition to critical loads, since altered NH_3 exchange may itself indicate exceedance of the critical load.

Over the past few years a number of micrometeorological measurements have been reported by the present authors and others indicating differences between ecosystems and also providing the basis for modelling the exchange process. In the present short paper, examples from recent measurements are used to test hypotheses of links between NH_3 canopy cycling processes and N inputs.

2. Canopy cycling of ammonia and the impact of nitrogen

Exchange of NH_3 between the atmosphere and plant canopies can take place either with leaf tissues (via stomata), or leaf surfaces (cuticles, surface water). In addition, soils, senescent leaves and possibly stems, may act as sources and sinks. Sutton *et al.* (1993b) provided a simple schematic overview of the main factors affecting the competing fluxes between each of these sites. A more detailed scheme focusing on the role of N is provided in Fig. 1. The magnitude of many of the fluxes noted here is uncertain, requiring further testing by field measurements, though this conceptual model is based on the interpretation of existing micrometeorological and throughfall chemistry measurements.

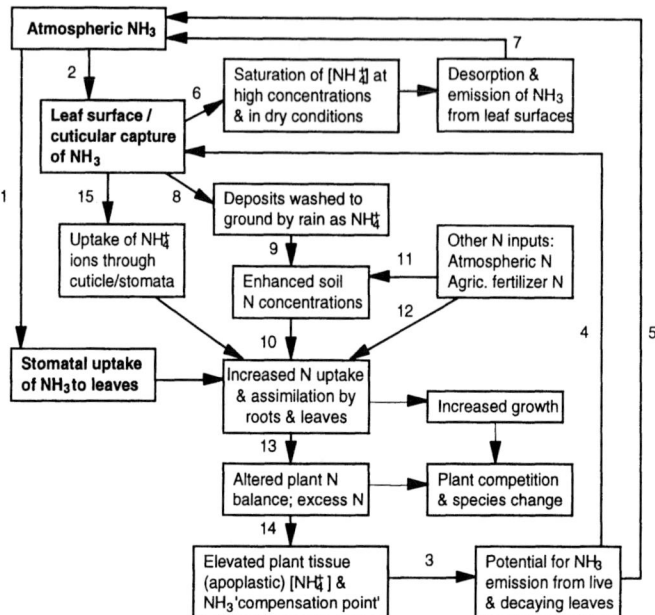

Figure 1. Hypothetical scheme of the effects of elevated NH_3 and other N inputs on the canopy cycling and exchange of atmospheric NH_3 with plant communities. Numbers refer to fluxes discussed in the text. Other wet and aerosol deposition, as well as dry deposition to bare soil surfaces is, for simplicity, summarized via fluxes 11 & 12.

Until recently, most of the N. European micrometeorological measurements of NH_3 fluxes over semi-natural ecosystems showed large rates of deposition, with small canopy resistances (R_c, see section 3) of the order 0-30 s m^{-1} (Duyzer et al. 1987; Sutton et al. 1993a). However, further measurements over a polluted Dutch heathland (NH_3 typically 1-40 µg m^{-3}) in transition from *Calluna* to *Molinia*, showed significant periods of NH_3 emission (Sutton et al. 1992; Erisman & Wyers 1993). Sutton et al. proposed that this could be due to a feedback due to long term N deposition (as noted above) and/or that these periods of emission could be due to evaporation of NH_3 deposited onto water layers. The work raised the question of the extent to which the N status of polluted sites was different to clean sites, as well as whether NH_3 emission from cuticular desorption could be observed from clean sites.

Where deposited NH_3 is either washed to the soil or absorbed via stomata, the extra N assimilation (from foliar and soil uptake) may result in increased growth or provide excess N, resulting in altered plant competition in semi-natural ecosystems. For healthy leaves a balance normally exists between NH_4^+ production and consumption regulating $[NH_4^+]$ in the apoplast, so that, according to the Henry equilibria, a 'stomatal compensation point' for $NH_{3(g)}$ (χ_s) may be defined. Where NH_4^+ is in excess, a larger χ_s provides increased potential for NH_3 emission. In senescing leaves of agricultural plants there is a limited sink for NH_4^+, so that larger NH_3 emissions are expected (Schjørring et al. 1993b). However, less is known concerning the response of native plants.

Although chamber studies have shown that the rate of NH_3 uptake through the cuticle is negligible (van Hove et al. 1989), Fig. 1 also notes the possibility of uptake of $NH_4^+_{(aq)}$ through the cuticle or stomata. Ample evidence from throughfall and laboratory measurements indicates that the canopy may take up NH_4^+ from rain, mists or NH_x dry deposited to waterlayers (e.g. Bowden et al. 1989; Draaijers 1993). Most of the measurements have been made for tree species, though available data for an agricultural canopy (maize, Römer et al. 1992) and heathland (Bobbink et al. 1992) also suggest efficient NH_4^+ uptake. It should be noted, however, that the opposite might be expected, particularly for agricultural canopies. Ammonia emission from crops is often suppressed in wet conditions (e.g. Sutton et al. 1993a), presumably because of the high solubility of NH_3 in water (dew, rain). Since stomata remain open it is to be expected that the potential for NH_3 emission from leaves remains and any NH_3 that is emitted by stomata is directly dry deposited back to water layers on leaves and other plant surfaces. Hence the NH_3 loss from stomata may be transferred to NH_4^+ on cuticles. If this is washed to the ground in rain, throughfall would be expected to contain substantial canopy leaching (enrichment) of NH_4^+. Clearly there is a need to resolve these conflicting results.

3. Micrometeorological theory and methods

The example micrometeorological measurements of NH_3 fluxes (F_t) reported here were made according to the classical aerodynamic gradient method, described in full elsewhere (Sutton et al. 1993a). The simplest resistance interpretation of the fluxes calculates the canopy resistance (R_c), where z is a reference height above the zero plane of the canopy:

$$-(\chi_{surface} - \chi\{z\})/F_t = R_t\{z\} = R_a\{z\} + R_b + R_c \qquad (1)$$

$R_t\{z\}$ is the total resistance, and $R_a\{z\}$ and R_b are the atmospheric and leaf boundary layer resistances respectively. To calculate R_c, $\chi_{surface}$ is assumed to be zero, so that this model is only really suitable to describe deposition fluxes, where R_c integrates the resistances of uptake by e.g. leaf cuticles (R_w) and stomata (R_s). An alternative model used here is therefore to calculate the 'canopy compensation point' (χ_c) which is the result of competition between bi-directional stomatal and cuticular fluxes (e.g. Sutton et al. 1995b):

$$F_t = (\chi_c - \chi\{z\})/(R_a\{z\} + R_b) \qquad (2)$$

$$\chi_c = \frac{[\chi\{z\}/(R_a\{z\} + R_b) + \chi_s/R_s]}{[(R_a\{z\} + R_b)^{-1} + R_s^{-1} + R_w^{-1}]} \qquad (3)$$

Ammonia concentrations were measured using sensitive continuous wet annular denuders as described by Wyers et al. (1993). The technique is labour intensive, but provides a time resolution of >2 minutes and is capable of determining concentrations > 0.02 µg NH_3 m^{-3}. Further methodological details are reported by Sutton et al. (1995a).

4. Results and interpretation

An example of the NH_3 fluxes measured over agricultural canopies is given in Fig. 2. This shows the typical pattern of bi-directional exchange, with emission during at least part of the day and deposition during evening and night. The fluxes were small at night due to restricted atmospheric turbulence, while the emissions occurred where χ_a was < 0.5 µg m^{-3}. The measurements are compared with an application of the model of Eqs. (2-3), which shows that the results can be well explained on the basis of emission from χ_s during the day when stomata are open, and parallel deposition/recapture to leaf cuticles as a function of relative humidity. This model, which has been tested for other periods and vegetation types (Sutton et al. 1995a,b), was based on an assumed apoplastic pH of 6.8. More recent unpublished data of Schjørring suggest smaller values typically around pH 6.3 or less. In this case the model may be fitted with similar results but providing larger apoplastic [NH_4^+]. These results support the picture of competing cuticular deposition and bi-directional stomatal exchange as outlined in Fig. 1 (fluxes 1-5) for an agricultural canopy well supplied with nitrogen. The question of solving the model with 2 unknowns (χ_s, R_w) has been addressed in the discussion following Sutton et al. (1995b).

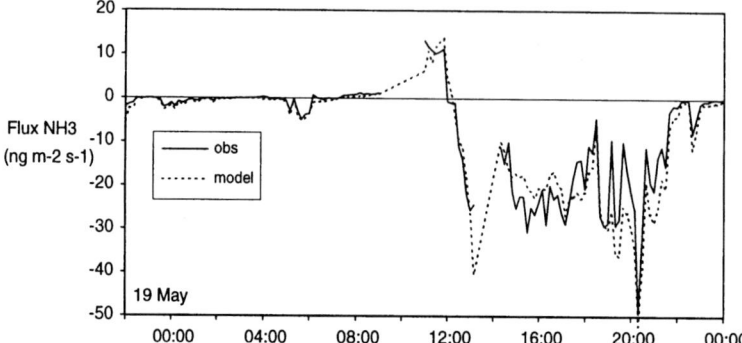

Figure 2. Example of bi-directional exchange of NH_3 over a wheat canopy at Sutton Bonington, C. England. The model (Eq. 3), calculates χ_s from leaf apoplast pH 6.8, 100 µM $[NH_4^+]$ and the Henry coefficient as a function of canopy temperature, and from R_w (s m^{-1}) = 2 exp([100-% relative humidity]/12). The relationship to χ_a and other parameters is discussed by Sutton and Fowler (1993) and Sutton et al. (1995a).

In order to test the influence of N inputs and concentrations on the exchange for semi-natural ecosystems flux measurements were made over an extremely clean site in N. Scotland (summer NH_3 typically 0.05-0.2 µg m^{-3}) (Fig. 3). These results are only possible because of the extremely low detection limit of the NH_3 analysis system, and are uncertain to ±0.2 ng m^{-2} s^{-1}. Nevertheless, they provide clear evidence of an event of NH_3 desorption associated with early morning drying (and warming) of the canopy (Fig. 1, flux 7). Warming of the canopy would also be expected to increase χ_s, favouring stomatal emission, however, this would not explain the short pulse of emission observed in the measurements. The magnitude of this emission is trivial compared with that observed at the Dutch sites (Sutton et al. 1992, 1995b; Erisman and Wyers 1993), but demonstrates that cuticular desorption may also occur at clean sites. A constraint may occur in environments with significant SO_2 concentrations, as oxidation to form SO_4^{2-} and reaction with NH_3 provides ammonium sulphates, which have negligible vapour pressure and are unlikely to re-volatilize.

A further chemical constraint may apply to the interpretation of these data, as it is possible that with significant warming and drying above the surface, evaporation of NH_4NO_3 aerosol may appear as a source of NH_3 in the air above the surface (cf. Zhang et al. 1995). This may occur where the air mass experiences a reduced relative humidity, although it should be noted that for the results in Fig 3, there was little effect on $\chi_a\{5\text{ m}\}$ above the surface, suggesting that this was not a significant factor. Similarly, this feature shows that storage errors, that might be expected around sunrise due to changing boundary layer depth, are expected to be trivial. Alternatively, above droughted vegetation in conditions of large sensible heat flux, a vertical gradient in relative humidity, with drier conditions near the surface, may result in aerosol dissociation and apparent NH_3 emission. Current models (e.g. Kramm and Dlugi 1994) have yet to test this effect, which deserves further investigation, though it is unlikely in the present measurements where the latent heat flux was large (Fig. 3) and relative humidity gradients small.

The data in Figure 3 also suggest that χ_s for this clean site is much smaller than more polluted areas. In the driest conditions a small rate of NH_3 deposition was observed. At these humidities cuticular uptake is expected to be small, suggesting that χ_s is unlikely to be >>0.1 µg m^{-3}. This contrasts with values of 2-15 µg m^{-3} for agricultural canopies and more polluted Dutch ecosystems (Sutton et al. 1995b).

5. Discussion and conclusions

The example results shown support a number of the fluxes outlined in the conceptual model of Fig 1 (fluxes 1-6) indicating the possibility for both substantial re-capture of NH_3 emitted by stomata (for the agricultural canopy) as well as the possibility of NH_3 desorption from the cuticle or associated water layers (the upland moorland canopy). Further evidence of desorption emission of NH_3 from water layers has also been discussed elsewhere for agricultural canopies and an initial model developed (Sutton et al. 1995a). The existence of periods of NH_3 desorption from a clean moorland site, begs the question of the extent to which emissions from N polluted semi-natural ecosystems sites are either desorption from surface deposited NH_3 (fluxes 6-7) or emission from leaf tissues and the soil as a consequence of long term N deposition (fluxes 5, 10-14). To answer this question more directly requires controlled experiments on the effect of N inputs on NH_3 fluxes. Nevertheless, the current hypothesis of a feedback between long term atmospheric N deposition and net NH_3 fluxes is consistent with data on the N status of moorland vegetation in relation to N deposition (Pitcairn et al. 1995), which show increased foliar %N at UK sites receiving larger amounts of total N deposition. In support of this picture, it may be noted that the plants at Dutch heathland site (Leende) showing significant NH_3 emission (Sutton et al. 1992), had a large %N content (1.65-2.1 % N) with values decreasing across the nature reserve with distance from nearby intensive agricultural land.

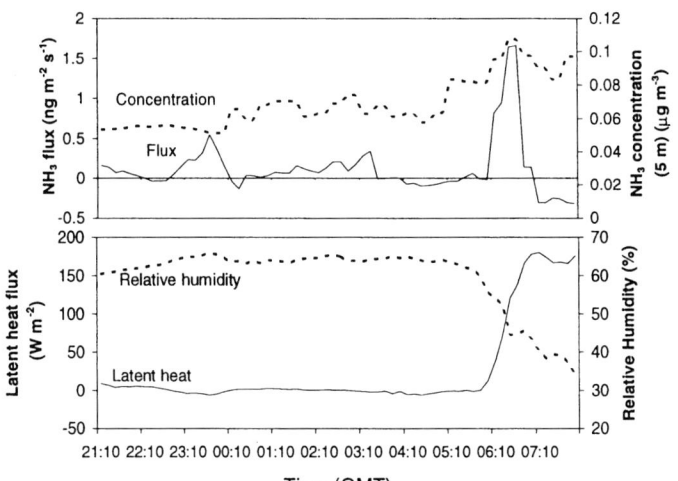

Figure 3. Example of bi-directional flux of NH_3 at a site in N. Scotland (Loch More, Sutherland, 1/6/1994). The extremely small concentration is typical for this remote upland region.

The bi-directional ammonia flux over the wheat crop supports the existence of an internal cycle within the canopy, where NH_3 emitted by stomata may be recaptured by plant cuticles and associated water layers (Fig. 1, flux 4). A result is that the NH_4^+ throughfall flux within an agricultural canopy or a semi-natural ecosystem subject to intense N deposition would be expected to include both wash-off from dry deposition to cuticles (fluxes 2, 8) and recaptured NH_3 emitted by the plants (fluxes 4, 8). Although, for the sake of clarity, Fig. 1 focuses mainly on the role of gaseous NH_3, it is evident that wet deposition and dry deposition of NH_4^+ aerosols also contribute to the fluxes shown, and provide an additional source of NH_4^+ to the leaf surface (flux 2). The quantitative contribution of a within-canopy cycle of NH_3 emission and recapture (c.f. Denmead et al. 1976) to throughfall in N polluted areas and for agricultural crop canopies warrants further investigation. Nevertheless, it is evident from the few available data, that despite periods of expected emission and recapture, NH_4^+ in throughfall may be substantially less than total deposition, and even in some cases less than wet deposition.

Such an internal cycle as discussed here necessarily results in a net flux above the canopy with only one direction. However it should be noted that different components of the canopy may act as source and sinks, even on single leaves. In addition, the distance of gaseous transfer may be minimal. Thus where a water drop lies over a stoma, NH_3 may pass directly from the substomatal cavity to leaf surface waterlayers.

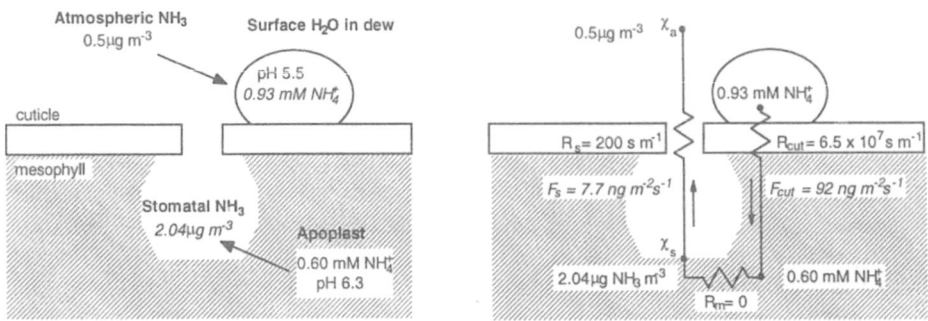

Figure 4. Example of possible fluxes of $NH_3(g)$ and $NH_4^+(aq)$ between a dew wetted agricultural plant and the atmosphere. Inferred values given in italics: χ_s is inferred from estimated apoplastic $[NH_4^+]$ (Sutton *et al.* 1995b) and pH (Schjørring, unpublished data); the dew $[NH_4^+]$ shown is that in equilibrium with an example χ_a and typical dew pH measured over a wheat canopy (all at 15 °C). The fluxes are calculated on a leaf area basis, with R_s a typical minimum for daytime conditions and R_{cut} for misted leaves derived from Bowden *et al.* (1989).

Two effects may explain the observation that throughfall NH_4^+ is less than atmospheric NH_x inputs, while at the same time canopy leaching of NH_x might have been expected: the first is that drying of rain wetted canopies may result in cuticular desorption of NH_3 (flux 7), although data are lacking to substantiate this as being the main fraction of the NH_x lost. The second possibility is that canopy uptake of aqueous NH_4^+ occurs through the cuticle or stomata, transferring the NH_x back to the leaf apoplast. Though often demonstrated for forest trees and other semi-natural vegetation, canopy uptake of NH_4^+ might seem counter-intuitive for agricultural plants where NH_3 is emitted by the leaves and recaptured by water layers (suggesting canopy leaching of NH_x rather than uptake). A possible key may lie in the existence of pH gradients between the apoplast and leaf surface water: the more acidic conditions on leaf surfaces would allow NH_4^+ to accumulate in surface water films, which is then able to diffuse back through the cuticle (or in water films around stomata) to the apoplast. This possibility is supported by measurements of dew over the wheat canopy of Fig. 2 (though made on a different day), which showed elevated $[NH_4^+]$ in the region of 1-2 mM, with typical dew pH 5.5. These data may be coupled with ^{15}N measurements of Bowden *et al.* (1989) which allow an $NH_4^+_{(aq)}$ leaf uptake resistance (R_{cut}) to be calculated. Their measurements were for red spruce needles which may have a smaller uptake rate and smaller χ_s than agricultural plant leaves, though the results are still useful to provide a first estimate of R_{cut}. A typical uptake rate for new needles was 0.45 µg g^{-1} leaf h^{-1}, which assuming a needle area of 0.004 m^2 g^{-1} leaf (L.J. Sheppard, pers. comm.), and given $[NH_4^+]$ in applied mists of 0.11 mM, provides an estimate of R_{cut} for $NH_4^+_{(aq)}$ of 6.5 x 10^7 s m^{-1}.

These values together with typical χ_a and daytime R_s are used in Fig. 4 to provide an example of the potential magnitude of the fluxes in and out of a dew wetted leaf. The value of χ_s is calculated to be 2 µg NH_3 m^{-3}, which provides a flux of NH_3 out of the leaf for these conditions of around 8 ng m^{-2} s^{-1}. In contrast, the more acidic dew allows increased $[NH_4^+]$, and despite the very large R_{cut}, the flux of NH_4^+ into the leaf is around 90 ng m^{-2} s^{-1}. In this calculation, the $[NH_4^+]$ of the dew is assumed to be in equilibrium with χ_a; if equilibrium with χ_s were assumed the flux of NH_4^+ back into the leaf would be much larger. Clearly, this provides only a very approximate indication of the magnitude of fluxes, and warrants more detailed investigation. An initial examination of Fig. 4. shows, for example, that a net flux into the leaf of 84 ng m^{-2} s^{-1} would not be maintained by atmospheric deposition alone, as a deposition velocity ($V_d = 1/R_t$) of 20 cm s^{-1} would be required. This suggests that R_{cut} may be underestimated or ΔpH overestimated. Nevertheless, Fig 4. serves to indicate that much of the stomatal emission of NH_3 that is deposited to wet surfaces within the canopy could be re-absorbed through the leaf surface as NH_4^+ (Fig 1. flux 15). A referee has questioned why a plant would emit NH_3 through stomata only to recapture it through the cuticle/surface waterfilms. However, this presupposes biological 'intention' of both these fluxes. In practice, it may be simply that either or both these external NH_x fluxes are the consequence of other biological and physico-chemical processes.

In conclusion, the present short paper has provided examples of micrometeorological measurements of NH_3 fluxes over semi-natural and agricultural ecosystems. These have been used to test hypotheses concerning canopy cycling of NH_x in relation to N inputs both from agricultural sources and atmospheric deposition. Application of a 'canopy compensation point' model supports the existence of emission of NH_3 by stomata and recapture by leaf surfaces at high humidity. In contrast, measurements at extremely low NH_3 concentrations show that desorption of NH_3 can also occur during drying conditions. The possibility of a feedback between long-term total N deposition and NH_3 exchange fluxes is, however, supported by the extremely small compensation point of the clean moorland site compared with forests in polluted areas, as well as by other evidence showing a link between N deposition and foliar % N content. Literature evidence of canopy NH_4^+ uptake by leaves may seem at variance with the observation of NH_3 emission and recapture onto wet leaf surfaces, though a possible explanation may be found in the existence of pH gradients between leaf surface water and the apoplast. Although measurements of throughfall NH_4^+ in agricultural crops are difficult to interpret in relation to N deposition, further measurements of this kind would be invaluable to understand canopy exchange of ammonia. In particular, coupling such studies with micrometeorological measurements would directly aid the parametrization of NH_3 deposition/emission fluxes.

Acknowledgements

The authors gratefully acknowledge financial support from the UK Department of Environment (Air Quality Division), as well as complementary EC Environment Programme funding via the 'EXAMINE' project. We thank R.J. Singles and D. Guerin for support in making the field measurements.

References

Bobbink R., Heil G.W. and Raessen M.B.A.G.: 1992. Environ. Pollut. **75**, 29-37.
Bowden, R.D., Geballe, G.T. & Bowden, W.B.: 1989. Can. J. For. Res. **19**, 382-386.
Denmead O.T., Freney J.R. and Simpson J.R.: 1976. Soil Sci. Biochem. **8**, 161-164.
Draaijers G. :1993. Ph.D. Thesis, University of Utrecht, Netherlands. 208 pp.
Duyzer, J.H., Bouman, A.M.H., Diederen, H.S.M.A. & Aalst, R.M. van: 1987. R 87/273. TNO, Delft, Netherlands.
Erisman, J.W. & Wyers, G.P.: 1993. Atmos. Environ. **27A**, 1937-1949.
Kramm G. and Dlugi R.: 1994. J. Atmos. Chem. **18**, 319-357.
Pitcairn, C.E.R., Fowler D. & Grace J.: 1995. Environ. Pollut. **88**, 193-205.
Römer, F., van Pul, A., Stolk, A.: 1992. In: Field measurements and interpretation of species related to acid deposition. (eds. G. Angeletti, S. Beilke & J. Slanina) 267- 272. Air Pollution Res. Rep. **39**, CEC, Brussels.
Schjørring, J.K., Kyllingsbaek A., Mortensen J.V. & Byskov-Nielsen S.: 1993a. Plant, Cell & Environ. **16**,161-167.
Schjørring, J.K., Kyllingsbaek A., Mortensen J.V. & Byskov-Nielsen S.: 1993b. Plant, Cell & Environ. **16**,169-178.
Sutton M.A. and Fowler D.: 1993. In: Proceedings of the WMO conference on the measurement and modelling of atmospheric composition changes including pollutant transport. World Meteor. Orgn., Geneva. 179-182.
Sutton, M.A., Fowler, D., Hargreaves, K.J. & Storeton-West, R.L.: 1992. In: Field measurements & interpretation of species related to acid deposition. (eds. G. Angeletti, S. Beilke & J. Slanina) 211-217. Air Pollution Res. Rep. **39**, CEC, Brussels.
Sutton, M.A., Fowler, D. & Moncrieff, J.B.: 1993a. Quart. J. Roy. Meteor. Soc. **119**, 1023-1045.
Sutton, M.A., Pitcairn, C.E.R. & Fowler, D.: 1993b. Adv. Ecol. Research. **24**, 301-393.
Sutton, M.A., Burkhardt, J.K., Guerin, D. & Fowler, D.: 1995a. In: Acid rain research: Do we have enough answers? (eds. Heij G.J. & Erisman J.W.) 71-80. Elsevier Science BV.
Sutton, M.A., Schjørring, J.K. & Wyers, G.P.: 1995b. Phil. Trans. Roy. Soc., London. Series A. **351**, 261-278.
Hove, L.W.A. van, Adema, E.H., Vredenberg, W.J. & Pieters, G.A.: 1989. Atmos. Environ. **23**, 1479-1486.
Wyers, G.P., Otjes, R.P. & Slanina, J.: 1993. Atmos. Environ. A **27**, 2085-2090.
Zhang Y. ten Brink H., Slanina S. & Wyers P.: 1995 In: *Acid Rain Research: Do we have enough answers?* (eds: Heij G.J. & Erisman J.W.) 103-112. Elsevier Science BV.

EXCHANGE OF OZONE AND NITROGEN OXIDES BETWEEN THE ATMOSPHERE AND CONIFEROUS FOREST

JAN DUYZER, HILBRAND WESTSTRATE and SAM WALTON

*TNO Institute of Environmental Sciences, P.O. Box 6011,
2600 JA Delft, The Netherlands*

Abstract. The deposition flux of O_3 to a Douglas fir forest in the Netherlands was monitored by eddy correlation during nine months. At the same time the concentration gradients of NO, NO_2 and O_3 were determined over the forest. The canopy resistance to O_3 uptake was calculated from the measurements and it compared well with model estimates. The sensitivity of the stomatal resistance to humidity calculated in the model was adapted to improve the comparison. A multi-layered model of canopy exchange which included the influence of chemical reactions between NO and O_3 and soil emissions was used to interpret the results for NO_2. The observed fluxes of NO_2 away from the surface into the atmosphere were probably caused by soil emissions of NO. The soil-emitted NO was converted to NO_2 in the trunk space and vented into the atmosphere. The model showed that the NO_2 flux above the canopy was either away or towards the canopy depending on the strength of the soil emission and the amount of NO_2 taken up in the canopy. A 'canopy compensation' point for NO_2 could be established above which deposition was the main process and below which emission was observed. The model calculations supported the observations which indicated a compensation point of approximately 10 ppb NO_2.

KEY WORDS : Dry deposition, forest, Ozone, Nitrogen oxides, Soil emission, Chemical reactions.

1. Introduction

Ozone and nitrogen oxides are key elements in tropospheric chemistry. Only very few data on the dry deposition velocity of these gases over forest are available. Hicks *et al.*, (1983) reported measurements of NOx (the sum of the concentration of nitric oxide, NO, and nitrogen dioxide, NO_2) fluxes over forests. The results of these eddy correlation measurements however were difficult to interpret. Fluxes were of variable magnitude and direction and no conclusions were drawn. Because of this limited knowledge the role nitrogen oxides play in European pollution climates is very uncertain (Duyzer and Fowler, 1994).
In 1992, a joint European study was started with the objective to study the exchange of O_3 and NOx between the atmosphere and forests.
In this paper an overview is given of three main areas from this study. More details are given in individual reports on:
- long term monitoring of fluxes of O_3 and NOx to a Douglas Fir forest at Speuld in the Netherlands (Duyzer and Weststrate, *et al.*, 1995a and b).

- a European joint campaign during which fluxes were measured above and below the canopy and at the forest floor at Speuld (Walton, *et al.*, 1995).
- the use of resistance and multilayer models to interpret and generalize the results of the measurements (Duyzer, *et al.*, 1995).

2. Methods

2.1 EXPERIMENTAL

Micro meteorological methods were used to measure fluxes above the canopy and below the canopy (Fowler and Duyzer, 1989).

The measurements were carried out in a 18 to 20 m high Douglas fir stand at Speuld, the Netherlands. In a 35 m tower a gradient and an eddy correlation system was mounted. At 30 m a sonic anemometer and a fast response ozone sonde (Güsten and Heinrich, 1992) were mounted. The concentration of NO, NO_2 and O_3 and air temperature were measured every 20 minutes at 24, 26.5, 30 and 35 m height above ground. The flux for ozone calculated from the gradient compared well with fluxes measured by the eddy correlation method provided local flux profile functions are used (Duyzer and Weststrate, 1995a).

In the absence of suitable procedures no corrections for chemical reactions between O_3 and NO or photolysis of NO_2 were applied to the data. Therefore the fluxes of NO and NO_2 calculated from the concentration gradients could still be biased by chemical reactions in air.

During the joint experiment the experiment was extended with measurements of turbulent fluxes of NO, NO_2 and O_3 at 35 and 25 m and 7 m. Exchange rates of NO and NO_2 at the forest floor were measured using dynamic chambers.

2.2 MODELS

Canopy resistances R_c calculated from the observed fluxes were compared with estimates of the canopy resistance from a resistance layer model described by Hicks *et al.* (1987) using parameters estimated from literature.

For NO_2 such a simple model appears not to be appropriate. Upward fluxes have been observed and these can not be modelled easily with a simple resistance layer approach. A more complex multi layer model, based upon a model originally described by Baldocchi *et al.*, (1988) for the estimation of SO_2 dry deposition to an oak forest, was adapted for Speuld. The model was extended with a description of chemical reactions between NO_x, O_3 and hydrocarbons above within and below the canopy. It also accounts for emission of NO and deposition of NO_2 at the forest floor.

3. Results and discussion

For a period of nine months the fluxes of O_3 and NO_x were monitored continuously. Figure 1 shows the observed canopy resistance on two days in 1993 compared with the estimates derived from calculations using the parameterisation scheme by Hicks et al. (1987). To improve the fit between model simulations and observations the sensitivity to VPD was increased considerably. This led to an improvement of the average difference between modelled and observed deposition velocities of a factor of three. The rise of the canopy resistance in the afternoon of July 10 however is still not simulated well by the improved scheme. This effect is probably a response of the trees to limit water losses on warm, sunny days. With large VPD's the canopy resistance is underestimated by the model. Similar dependencies were found for PAR and Temperature. However these parameters are highly correlated which complicates the analysis.

It also appeared difficult to simulate the nocturnal deposition rate well. At night the deposition rate was sometimes much larger than estimated using the model. Probably chemical reactions with NO or reactive hydrocarbons are important sinks for O_3. Chemical reactions with NO might be responsible for 50 % of the destruction of O_3 at night whereas the deposition at the cuticle and deposition at the soil seemed far less important (Duyzer et al., 1995).

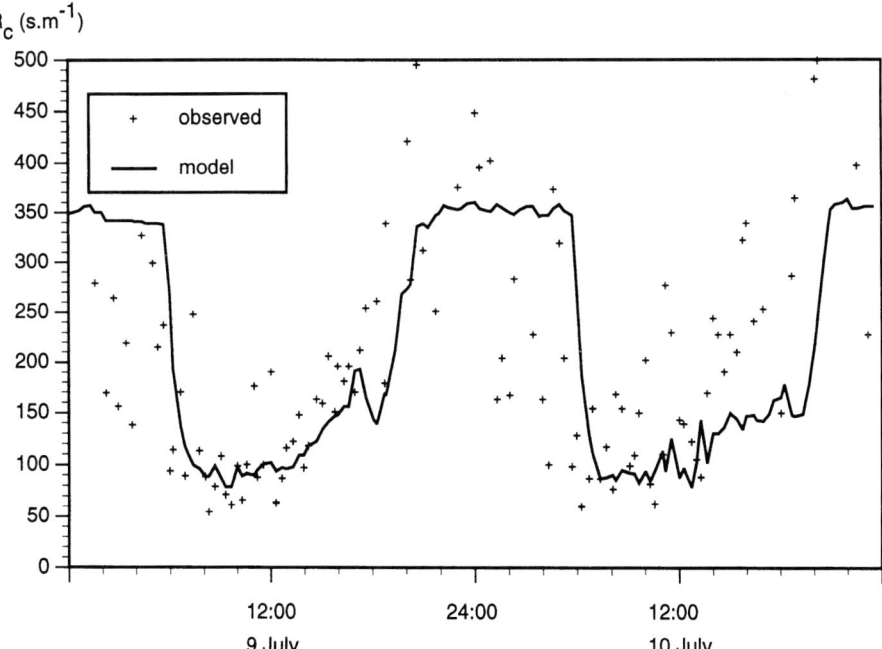

Figure 1 The canopy resistance to uptake of O_3 calculated directly from eddy correlation measurements at Speuld and the canopy resistance calculated using an improved parameterisation resistance model described in the text.

The results for NO and NO_2 were more difficult to interpret. The gradients of NO and NO_2 observed indicated an average downward flux of NO and an upward flux of NO_2. During the joint experiment NO_2 fluxes were measured by eddy correlation. Above as well as below the canopy these fluxes were often away from the surface and into the atmosphere. This result is contrary to the general understanding that NO_2 is taken up by forest canopies (Duyzer and Fowler, 1994; Meixner, 1994). Figure 2 shows the NO_2 flux observed during 9 months as a function of the NO_2 concentration. These results suggest a 'compensation' point above which NO_2 is deposited and below which emission is the main process (Johansson, 1987).

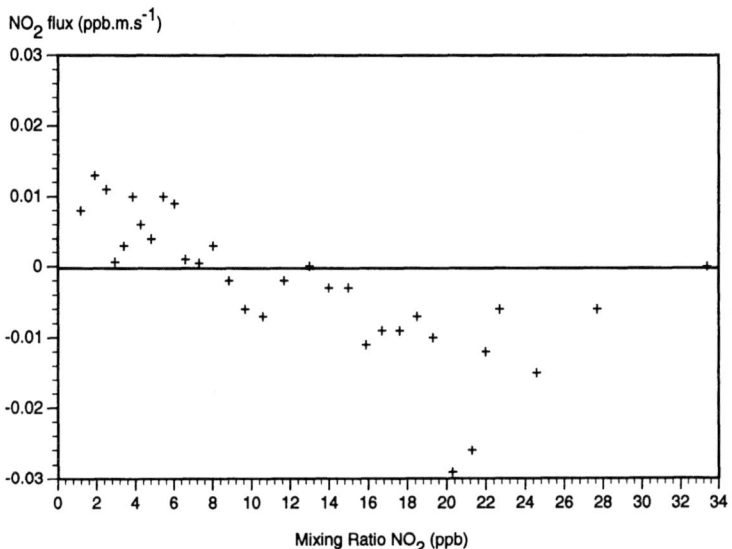

Figure 2 The flux of NO_2 above Speulderbos calculated from the concentration gradient as a function of the NO_2 concentration (the crosses indicate 20 minute values averaged over 9 months).

To interpret these results simulations with the multi layer model described above were compared with the detailed measurements carried out during the joint campaign. For ozone the agreement between calculations and measurements carried out above the canopy is good. The ozone flux observed and simulated below the canopy also matched reasonably well. For NO_2 and NO, the qualitative comparison was reasonable. The NO_2 flux was upward and the NO flux was downward in both the model calculations and the measurements. In the model the upward flux of NO_2 above the canopy was caused by emission of NO from the forest floor. In the trunk space, this NO was converted by O_3 to NO_2, which was partly taken up by the canopy and partly vented away in the atmosphere.

A sensitivity study with the multilayer model showed that the magnitude and direction of the flux of NO_2 observed above the canopy was dictated by two competing processes (Duyzer et al., (1995): 1) the uptake rate of NO_2 in the canopy and 2) the emission rate of NO from the forest floor. Process (1) is influenced by the stomatal resistance and the concentration of NO_2 in the air above the canopy. This indirectly leads to an influence of

VPD, PAR, LAI and temperature. Process (2) was probably influenced by soil water, nutrient status, and temperature. At Speuld high emission rates from the forest floor around 20 ng N m^{-2} s^{-1} (equivalent to 6 kg N ha^{-1} yr^{-1}) were observed. These high emission rates could be related to the large amount of nitrogen (more than 50 kg ha^{-1} yr) Speuld receives from the atmosphere (Hey and Schneider, 1991).

At Speulderbos the rates of the two competing processes (1) and (2) are almost alike. As a result, small variations in one of the driving parameters, such as R_c or the emission rate of NO from the forest floor might cause the NO_2 flux above the canopy to change direction. This explains many of the results obtained in Speulderbos and may also explain some of the variable results obtained in other studies (Hicks et al., 1983).

The influence of the NO_2 concentration on the deposition rate may lead to the occurrence of a 'canopy-compensation' point concentration above which downward fluxes and below which upward fluxes are observed. In Figure 3 simulated fluxes of NO and NO_2 are plotted as a function of height above ground for different atmospheric concentrations of NO_2. From the sensitivity analysis it appeared that values of this compensation point may lie between 17 and 35 ppb at night and 5 to 10 during the day. These values are supported by the long term experimental findings were a compensation point of 8 ppb NO_2 was observed (Figure 2).

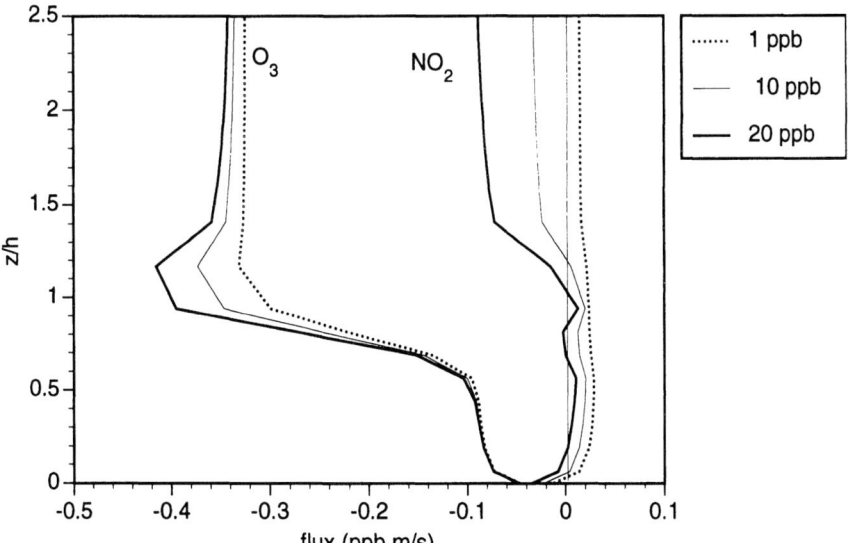

Figure 3 The flux of NO_2 and O_3 as a function of height calculated with the multilayer model for different atmospheric NO_2 concentrations (given between brackets).

4. Concluding remarks

It can be concluded that the dry deposition of ozone to forest can be modelled quite well using an existing resistance model framework. The agreement between model estimates and measurements could be improved considerably by changing the sensitivity of the stomatal conductance to vapor pressure deficit. To simulate nocturnal deposition proved more complex. Probably chemical reactions between O_3 and NO or hydrocarbons such as isoprene play a role in the nocturnal uptake.

The observed flux of NO_2 was often away from the surface using a multilayer model. This feature could be explained as a result of soil emission of NO, conversion of NO to NO_2 in the trunk space and uptake of NO_2 in the canopy. More research however is needed to quantitatively describe exchange of NOx between forest and the atmosphere. Especially more information is required on in-canopy turbulence, chemical reactions, and soil emissions. Studies in other forests receiving less nitrogen from the atmosphere are required to show how common the features observed at Speuld are.

Acknowledgements

The authors gratefully acknowledge the support received from the Commission of the European Communities (DGXII) and the British and Dutch Departments of the Environment. Sam Walton was on a six month visit at TNO and is now back at the University of Manchester (UMIST). The support he received from EERO is gratefully acknowledged.

References

Baldocchi, D. (1988), *Atmos. Environ.* **22**, 869-884

Duyzer, J., S. Walton, M. Gallagher, K. Pilegaard (1995), IMW TNO report R95/113, Delft, the Netherlands

Duyzer, J.H. and J.H. Weststrate (1995a), G.J. Heij and J.W. Erisman (eds.) Acid Rain Research. Do we have enough answers ? (Elsevier), 21-30

Duyzer, J.H, J.H. Weststrate (1995b). IMW TNO report R94/274, Delft, the Netherlands

Duyzer, J.H. and D. Fowler (1994), *Tellus*, **46B**, 353-372

Fowler, D., J.H. Duyzer (1989), M.O. Andreae, S.D. Schimel, editors. p. 189-207 John Wiley and Sons Ltd.

Gusten, H., G.Heinrich (1992), *J. of Atmospheric Chemistry* **14**, 73-84

Hey, G.J. and T. Schneider (1991). Acidification research in the Netherlands, Elsevier, Amsterdam.

Hicks, B.B., D.R. Matt, R.T. McMillen, J.D. Womack, R.E. Shetter (1983), In: The meteorology of acid deposition APCA, 189-201, 1983

Hicks, B.B., D.D. Baldocchi, T.P. Meyers, R.P. Hosker, JR., D.R. Matt (1987), *Water, Air and Soil Pollution*, **36**, 311-330

Johansson (1987), *Tellus*, **39B**, 426-438

Meixner, F.X. (1994). *Nova Acta Leopoldina*, **288**, 299-348

Walton, S., M.W. Gallagher, T.W. Choularton, J.H. Duyzer, J.H. Weststrate, K. Pilegaard, N.O. Jensen (1995), UMIST report to the EC

STUDIES OF PRECIPITATION CONTAMINATION LEVELS OVER THE NORTH-WESTERN FORESTS OF RUSSIA SUBJECT TO EMISSIONS FROM THE TWO NICKEL SMELTERS.

R.T. KARABAN and M.L. GYTARSKY
Institute of Global Climate and Ecology, 20-B Glebovskaya St., Moscow, 107258, Russia

Abstract The results of investigations carried out in the forests of Kola peninsula subject to long-term air pollution by the nickel industrial enterprises are presented. Samples of rainwater from the open sites, from under the coniferous (pine) trees crowns and of the stemflow were collected at various distances from the emission sources. The highest levels in pollution of rainwater are detected over the area adjacent to the smelters. Researches of contamination of the precipitation in the vicinity of the two nickel enterprises of Kola peninsula show that concentrations of pollutants vary significantly (up to an order of magnitude) depending upon the meteorological conditions. The area of impact on forests of Kola peninsula is restricted by the radius of 30-40 km from the emission sources.

Key words: air pollution, industrial emissions, sulphur dioxide, heavy metal aerosols, contamination of precipitation

1. Introduction

Chemical content and the levels of pH of rainwater were evaluated in the vicinities of two nickel enterprises of Kola peninsula "Severonickel" and "Pechenganickel" during the summer periods. The work was carried out according to the programme of survey of the forest damage caused by the industrial emissions (Gytarsky et al., 1995). Sulphur dioxide, sulphate, nickel and copper aerosols are the major products of these enterprises emissions. The annual emissions of sulphur dioxide from "Severonickel" and "Pechenganickel" smelters are about 212 000 tonnes and 211 000 tonnes respectively (Barrett and Protheroe, 1994).

The aim of the research was to study the industrial enterprises possible impact on the values of pH and the contaminants concentration in the rainwater collected at various distances from the emission sources.

2. Methods

The values of pH and the levels of pollution in rainwater were evaluated over the permanent sample plots set up in coniferous (pine) forests at various distances from the emission sources. While establishing the plots, the wind rose and the exposition of the plot with regard to the possible direction of the pollutants transfer were considered. Over the area of industrial impact of the "Severonickel" smelter 9 sample plots were established. The most remote plots to the south and to the north were respectively 47 and 24 km from the smelter. Sample plots were allocated in the forests with different forest damage degree. In the surroundings of "Pechenganickel" smelter the samples of rainwater were collected from the areas of severe, moderate and slight damage to forests at the distance from 4 to 42 km.

Over each plot the samples of rainwater were collected from the open sites (forest glades), under the crowns (throughfall) and from the stems (stemflow). The methods applied for the collection of the samples, their preliminary preparation and for the chemical analyses of the collected material are described elsewhere (Rovinsky and Wiersma, 1987; Karaban et al., 1992; Gytarsky et al., 1995). As a rule the samples were collected in 24 hours after the rain. In some cases the samples from several momentary heavy shower rains occurred during 2-3 days were generalised. So the obtained results present the average concentrations of the chemicals in rainwater during the definite period of collection of samples.

The systematic observations carried out over the local meteorological stations in Monchegorsk ("Severonickel" smelter) and Nikel ("Pechenganickel" smelter) were used for the analysis of the meteorological situation over the area of industrial impact.

3. Results and Discussion

The pollution of the rainwater with the industrial emission products may have the temporal and the spatial changes. The spatial variations result from the contribution of the local emission source to the pollution of the ground air layer. The temporal variations are stipulated with a number of factors such as the previous pollution of the cloud mass which has brought the rain to the area of research; the meteorological situation and the state of the atmosphere during the days prior to the rain; the input of the local emission sources to the cloud mass contamination occurred during the rain event, and finally the amount of precipitation.

Over the area of industrial impact of the "Severonickel" smelter the samples of rainwater were collected during the 7 rain events in 1986 and in 1989. In some cases the rains were after the periods of dry weather with the repeated temperature inversions during the night and in the morning. Figure 1 presents the temporal changes of the average values of hydrogen ions and sulphates concentrations in the samples of rainwater collected in the open sites, from under the crowns and in the stemflow. The mean concentration of sulphates in the samples of rainwater collected on the 6-th of August 1986 is approximately 10 times higher than that in the samples collected during the subsequent period (figure 1). That is due to a long period of dry weather and temperature inversions after which the rain occurred. The similar tendencies are observed in all the collected samples.

The total mineral content in the precipitation collected in the vicinities of "Severonickel" smelter varied from 15 mg/l in 1986 to 50-70 mg/l in 1989. The input of sulphates to the total mineral content was about 30-40 %. The alkaline elements (Na, K, Ca, Mg) gave 15 - 30 %, the microelements - 2 - 3 %, and the chloride ions - 6 - 30 %. The pH of the rainwater in the vicinity of "Severonickel" smelter in most cases was less than 5.0. Consequently, over the area of industrial impact of this enterprise the wet precipitation is slightly acidified.

The borders of the local impact of the "Severonickel" smelter were determined based on the spatial variations of nickel and copper depositions with rainwater. Figure 2 shows the values of nickel and copper depositions with the rainwater collected on the 6-th

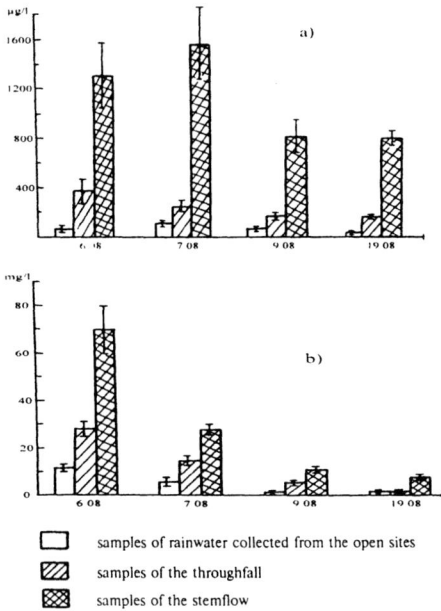

- ☐ samples of rainwater collected from the open sites
- ▨ samples of the throughfall
- ▩ samples of the stemflow

Fig. 1. Changes of the mean concentrations of hydrogen (a) and sulphates (b) ions in the samples of rainwater collected on August 6, 7, 9 and 19, 1986.

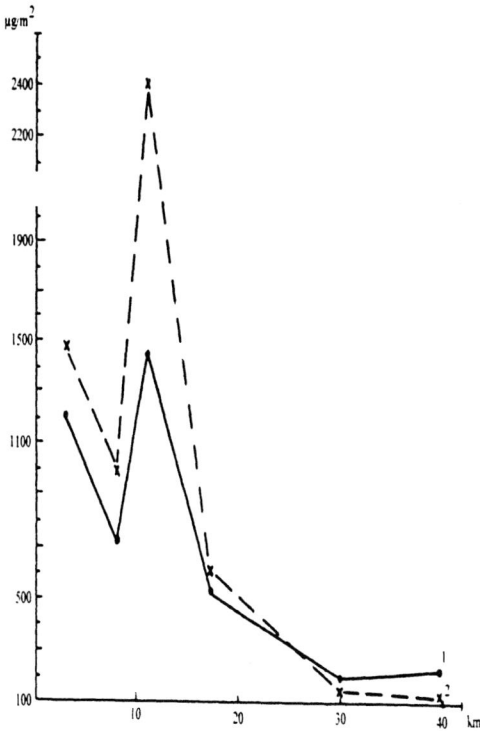

Fig. 2. Daily averaged depositions of nickel (1) and copper (2) with rainwater at various distance from the "Severonickel" smelter.

of August 1986. From figure 2 one can see that over the area adjacent to the emission sources (10-12 km from the smelter) the concentrations of nickel and copper in rainwater were respectively 3-8 and 8-20 times higher than those observed over the remote areas (40 km from the smelter).

Over the area of industrial impact of the "Pechenganickel" smelter samples of rainwater from the open sites, from under the crowns and of the stemflow were collected since July 26 to August 21, 1991. During this period of time there were 3 rain events. The periods of rain were short. Figure 3 shows the curves of the concentration of sulphates and the soluble nickel in the rainwater collected from the open site at various distances from the emission source. The highest values of sulphates and nickel concentrations in rainwater are observed at a distance of 6-8 km from the emission sources (figure 3). At 30 km from the emission sources the concentrations decrease significantly and become relatively constant (about 0.1 mg/l for nickel, 0.02 - 0.60 for copper and about 3.0 mg/l for sulphates).

Based on the obtained results it is possible to conclude that the local area with the markedly elevated levels of pollution in rainwater is limited by the distance of 14-16 km from the emission sources.

More complicated seems the explanation of the results of the analyses of the rainwater collected from under the crowns and of the stemflow. The samples of rainwater from the open sites concentrate the sediment particles from the emitter plus the material taken up by the rain drops, while the crowns and the stems accumulate contaminants during the dry periods since the last rain. Therefore in the vicinities of "Pechenganickel" smelter the concentration of sulphates in the throughfall was 5-6 times and in the stemflow - 15-20 times higher than the values collected over the open site. The concentration of nickel was 1.5-2.0 times higher. There was no strong reduction in the contaminants concentrations in the samples of the throughfall and in the stemflow with the distance increase. The even distribution of the concentrations in the rainwater sampled from under the crowns of trees probably may be explained by the decrease of the total canopy area of the coniferous trees as a result of the needles premature fall from the toxic effects of sulphur dioxide as well as the drying up of the trees shoots.

The highest values of pH in rainwater collected over the open sites in the vicinities of "Pechenganickel" smelter varied from 4.9 to 5.1, and the lowest values were from 4.1 to 4.2. No marked dependence was observed between the pH and the distance from the emission sources. Samples of rainwater collected from under the crowns and from the stems had the lower pH: 3.6 - 4.6 and 3.3 - 3.6 respectively. Probably, this is associated with the accumulation of sulphates and other acidforming components in the trees canopies and by the processes of alcalinization from the crowns.

The concentrations of the number of elements, that are not emitted by the "Pechenganickel" enterprise, or their emissions were small (Mg, Ca, K, etc.), did not depend on the distance from the emission sources. There was some increase of the nitrates and the ammonium content in the samples of rainwater collected from under the crowns and in the throughfall over the remote areas (30 - 40 km from the emission sources) which may be related to the physiological processes that occur in the plants.

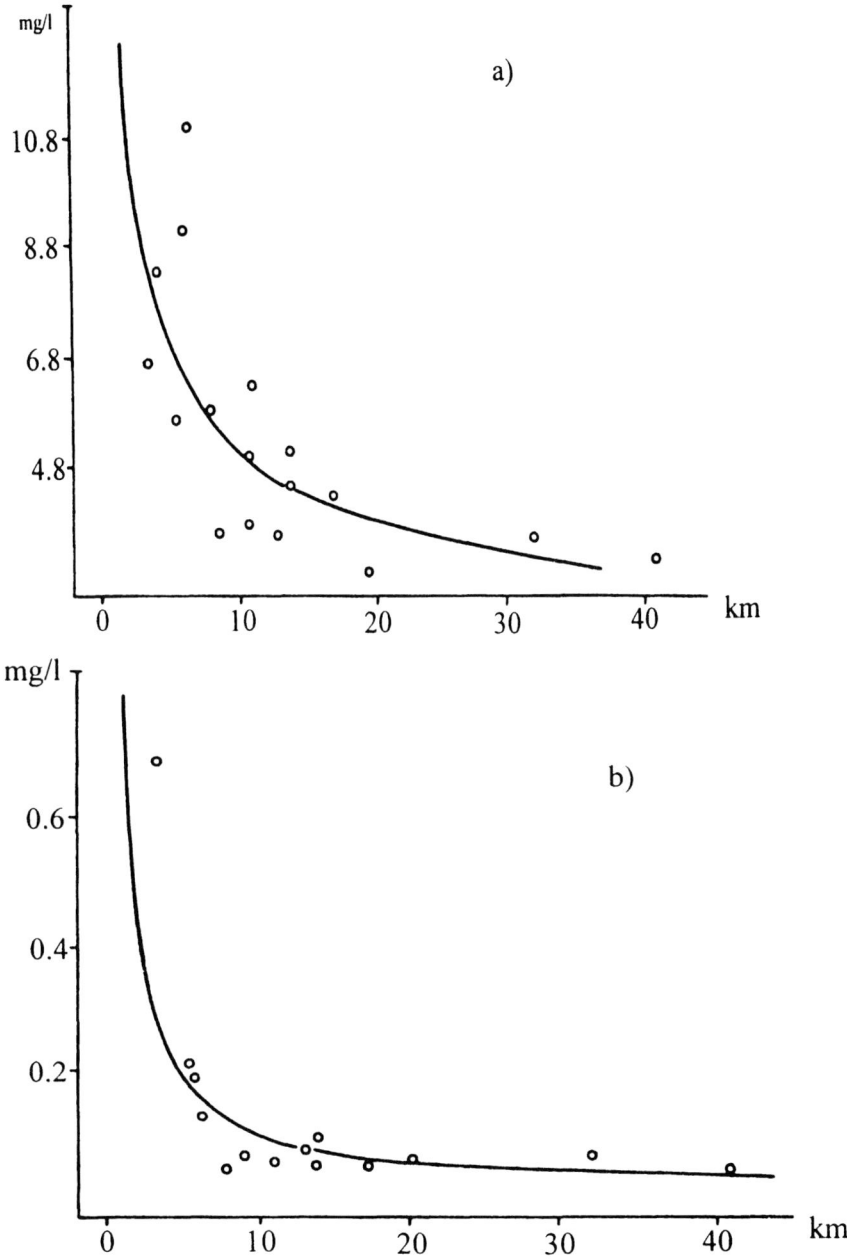

Fig. 3. Concentrations of sulphates (a) and nickel (b) in rainwater collected over the open sites in the forests growing in the vicinity of the "Pechenganickel" smelter.

4. Conclusions

- The field researches and the evaluations of pH and the chemical content in the samples of rainwater in the vicinities of Kola peninsula nickel enterprises were performed in 1986, 1989 and in 1991 according the programme of the studies of the coniferous forests damage by the industrial emissions.
- The rainwater pollution determined in the vicinities of two nickel enterprises showed the similar appropriateness of the temporal and spatial variations of the concentrations of the ingredients dominant for the emissions content.
- The significant increase of the concentrations of sulphates and nickel in the rainwater is observed over the local areas at a distance to 14-16 km from the smelter. The maximal values of the pollutant concentrations in the rainwater correspond to the distance of 6-8 km. At the distance of 30-40 km the levels of rainwater pollution notably decrease and become relatively constant.
- There is no clear dependence between the pH levels of the rainwater collected from the open sites and the distance from the smelter. The levels of pH were lower in the samples of the stemflow (min. pH=3.3, max. pH=3.6), and from under the crowns (min. pH=3.6, max. pH=4.6). That may be related to the accumulation of the acidforming components of the industrial emissions in the trees canopy during the dry periods as well as the alcalinization processes from the green mass of the tree crowns.

Acknowledgements

The authors thankfully acknowledge the financial assistance pro. ' ' by The John D. and Catherine T. MacArthur Foundation to support our participation in the International Conference "*Acid Reign' 95*". We are thankful to Vera Kuzmicheva and Nina Surkova for their valuable help during the implementation of the field research and preparing the obtained results for publication.

References

Barrett, M. and Protheroe, R.: 1994, "Sulphur Emission From Large Point Sources in Europe", 2-nd edition, Pollen Consultancy, Colchester.

Gytarsky, M.L., Karaban, R.T., Nazarov, I.M., Sysygina, T.I. and Chemeris, M.V.: 1995, "Monitoring of Forest Ecosystems in the Russian Subarctic: Effects of Industrial Pollution", The Science of the Total Environment, **164**, 57-68.

Karaban, R.T., Gytarsky, M.L. and Brouskina I.M.: 1992, "Forest Ecosystems Pollution Levels over Soviet-Norwegian Border Area", Effects of Air Pollutants on Terrestrial Ecosystems in the Border Area Between Russia and Norway, Proc. of the Intern. Symp.,TA-854/1992, 92:04, 66-76.

Rovinsky, F.Ya. and Wiersma, G.B.: 1987, "Procedures and Methods for Integrated Global Background Monitoring of Environmental Pollution", Environmental Pollution Monitoring and Research Progress Report Series, WMO/TD 178, **47**.

IONIC COMPOSITION OF RAINWATER AND ATMOSPHERIC AEROSOLS IN SARDINIA, SOUTHERN MEDITERRANEAN

GUERZONI S.[1], CRISTINI A.[2], CABOI R.[2], LE BOLLOCH O.[3], MARRAS I.[2] and RUNDEDDU L.[2]

[1] Ist. Geologia Marina/CNR, via Gobetti 101, 40129 Bologna, Italy, [2] Dip. Sc. della Terra/Università, via Trentino 51, 09126 Cagliari, Italy. [3] International Marine Centre, Torregrande (OR), Italy.

Abstract. The ionic composition of 55 aerosol samples and 31 precipitation events collected in a coastal site in southern Sardinia (Capo Carbonara, 39°06' N; 09°31' E) were compared. The samples were collected during one year period (Oct'90/Oct'91) and showed high variability in composition according to meteorological conditions. Rain and soluble part of aerosol showed a strikingly similar ionic composition: most significant anions were chlorine and sulphate, and sodium is the principal cations, followed by magnesium and calcium. The acid events are associated with N-NW trajectories (anthropogenic influxes from N. Europe) with avg. pH=4.65, non sea salt (nss) Ca=60 µeq/l and $NO_3/_{nss}SO_4$=0.6. Southern precipitations are influenced by Saharan dust alkaline effects, with avg. pH=6.75, $_{nss}Ca$=271 µeq/l and $NO_3/_{nss}SO_4$=0.4. Na/Cl ratio in rain is similar to sea water (0.87), whilst in aerosols there is a Cl loss (Na/Cl=1.10), probably due to reaction with nitric acid. Total fluxes of Ca, Mg, NO_3 and SO_4 were 104, 9, 64 and 113 µg/cm², and wet deposition exceeded (65-90%) dry deposition. Scavenging ratios (SR) as defined by the equation: SR=[(Ci)rain/(Ci)air]*d, (d= 1200g/m³) were calculated, using geometric means (Ci) of precipitation and aerosols collected concurrently during the period (a total of 23 samples). The SR values are Ca=3400, Cl=2400, Na, K, SO_4= 1700, Mg=1000 and NO_3 =750. These numbers could be useful to infer total fluxes by using simply rainwater ionic composition in Mediterranean semi-arid sites like Sardinia.

Key words; ionic fluxes, acidic deposition, Saharan influence, particulate fluxes, scavenging ratio.

1. Introduction

Several authors have studied the influence of Saharan dust transport to the Mediterranean and European regions (Chester et al., 1986, Loye-Pilot et al., 1985; Mamane, 1987). Among the numerous influences, that of Saharan dusts (i.e. dusts originating from North West Africa, namely from Algeria, Marocco, Tunisia and Libya), which act as neutralising agents, is of particular importance at this sampling location.

Previous studies (Correggiari et al., 1989) have shown direct transport of these dusts to Sardinia 3-4 times a year as well as indirect transportation which occurs more frequently at around 15-20 episodes a year. These events generally have pH values between 6 and 7, high concentrations of Ca and SO_4 and are high in particulate matter, with over 39% calcite (Guerzoni et al., 1993, Loye-Pilot et al., 1985). The mineralogical phases that characterise these events are: quartz, which is the dominant constituent of Saharan dusts; calcite and dolomite, which represent the major constituents of the soils of some regions North of the Sahara; feldspar, talc, illite, kaolinite and palygorskite (Molinaroli et al., 1993). Palygorskite and rounded quartz particles are regarded as the best indicators of Saharan origin.

The present work will compare the ionic composition of rainwater and soluble atmospheric aerosol, with emphasis on the role of the presence of particultes for specific transport events. The sampling site is located in the extreme south east of Sardinia, 118m above sea level at Capo Carbonara (39°06' N, 09°31' E), at an Italian Aeronautical Meteorological station. The dominant bedrock of the area is granitic, chosen for little or no transport of dusts from local sources.

2. Material and methods

Wet deposition was collected on a daily event basis in polythene containers, by means of a wet-dry sampler, between 4th October 1990 and 2nd December 1991. Due to the risk of contamination resulting from excavation work close to the sampling site, sampling was suspended between May and November 1991. Aerosols were sampled using a Sierra Anderson high volume sampler at the rate of 60 $m^3 h^{-1}$. Air was drawn through prewashed and weighed polyester fibre filters for a period lasting 48-72 hours per filter.

On rainwater samples, pH, conductivity, alkalinity and NH_4 were measured on unfiltered samples as soon as possible after sampling, using potentiometric methods. Samples were then filtered through 0.45 µm Nucleopore filter and anions (NO_3, SO_4 and Cl) were measured using ionic chromatography (Dionex 2000). Na and K were determined using flame emission spectrophotometry, whilst Ca and Mg were determined by AAS (Perkin Elmer 2100). Upon collection, aerosol samples were placed in a dessicator overnight and reweighed to obtain the total particulate weight. Analyses were subsequently carried out by first soaking the filters in 600 ml Milli-Q water and placing in an ultrasonic bath for 30 minutes. Blanks were treated in the same way for quality assurance purposes. The resulting solution was then analysed as for the rainwater samples. The ionic balance was checked for all samples and those with error greater than 5% were rerun or rejected. Concentrations were corrected for seasalt using Na as a reference species. The ratios of the various elements in seawater with respect to Na used were according to Wilson, in Keene *et al.*, 1986):

Ca/Na = 0.0439, Mg/Na = 0.2270, K/Na = 0.0218, Cl/Na = 1.160, SO_4/Na = 0.121

3. Results

3.1 RAIN AND AEROSOL

Table I shows the average composition of the rainwater and soluble fraction of the aerosol collected. The extreme pH values for rainwater samples were 3.96 and 7.23. Samples with pH values greater than or equal to 5.6 were considered 'alkaline', whilst those with pH less than this value were considered to be acidic. These results show that 58% are alkaline of which 39% are associated with winds prevailing from the west, while 42% are acidic. Among the alkaline group, 10 over 31 (30%) rain samples have pH values higher than 6.4 (Table II), and the ionic composition resulted enriched in $_{nss}$Ca and $_{nss}$SO$_4$.

TABLE I
Mean concentration of rainwater (μequ/l) and aerosol (μequ/m^3) components

Element	Rain A.M.	Rain G.M	Aerosol A.M.	Aerosol G.M.
NH4	20	15	4	2
NO3	26	21	37	34
HCO3	108	42	16	13
Ca	154	97	43	31
nssCa	136	80	35	22
Mg	115	44	49	45
nssMg	19	9	11	8
K	15	10	6	6
nssK	5	4	2	2
SO4	103	80	64	55
nssSO4	52	44	43	32
Cl	474	277	161	137
Na	424	241	171	158
pH	5.84	5.77		
H	9	2		
TDS*			24.5	23.6

a.m = arithmetic mean, g.m. = geometric mean
* total dissolved salts in μg/m^3

TABLE II
Average rainwater data divided into pH classes

pH	n° of samples	NH4 ueq/l	NO3 ueq/l	nssCa ueq/l	nssSO4 ueq/l	NO3/nssSO4	Na ueq/l	NO3/Na
<4,8	6	20	34	43	66	0.45	259	0.13
4,8-6,4	15	20	20	67	34	0.73	353	0.06
>6,4	10	21	33	294	73	0.47	630	0.05

From inspection of Table I, the most significant anions, both in the rain and aerosol samples were found to be Cl$^-$ and SO$_4^{2-}$, whilst the cation with the highest concentration was found to be Na$^+$, followed by Mg^{2+} and then Ca^{2+}. The most stark compositional contrasts between rainwater and aerosol are the enrichment of NO$_3$ in aerosol relative to rainwater and of NH$_4$ in rainwater relative to aerosols. The nitrate sampled in the aerosol appear to be in the form of sodium nitrate. A strong positive linear correlation of Cl and Na was found in both rainwater and soluble fraction samples, but the Na/Cl ratio was found to be 0.87 and 1.15 respectively, indicative of a loss in chloride from the soluble fraction as observed previously by Martens et al. (1973). The chlorine loss is around 30% and would appear to be correlated with nitrate which suggests a release of HCl.

TABLE III
Comparison of rainwater composition (μequ/l) at Capo Carbonara with other sites (Lim et al., 1991)

	Sardinia	Ireland	Bermuda	W. Atlantic
H	9	37	18	14
NH4	20	7	5	3
NO3	27	53	7	5
nssCa	155	11	15	10
nssMg	115	37	40	27
nssK	15	5	4	5
nssSO4	52	7	36	11
Cl	475	199	191	155
Na	424	163	148	13

TABLE IV
Comparison of aerosol concentrations (μg/m^3) at Capo Carbonara and other sites.

	Sardinia	Mallorca*	Cap Ferrat*
NH4	0.1	1.6	2.7
NO3	2	3.5	3.9
Ca	0.9	1.1	0.5
Mg	0.6	0.4	0.14
K	0.2		0.11
SO4	3.1	5	6.1
Cl	5.7	3.5	3.9
Na	3.9		0.8

* Mateu et al. (1993)

Table III compares the rain samples analysed in Sardinia with other marine and non-marine sites, and Table IV compares concentrations in air with other marine sites. The influence of the Saharan aerosol at Capo Carbonara is apparent with respect to other sites, due to the enrichment of Ca, Mg and SO_4, important components of these dusts and for the low content of the H^+ ion, largely neutralised by the Saharan carbonates.

3.2 CALCULATION OF FLUXES

To evaluate the total balance of substances deposited to the environment it is of fundamental importance to evaluate removal by dry deposition. While there is much data on the balances due to wet deposition (Caboi et al., 1992) data on dry deposition are scarce (Harrison & Pio, 1983). In addition, lack of dry deposition data may be due to difficulties in standardizing techniques (Eder & Dennis, 1990).

TABLE V
Deposition fluxes (µg/cm2) of the various elements at Capo Carbonara.

element	wet ug/cm2	dry ug/cm2	total ug/cm2	wet/total %
NH4	12	0.5	12.5	96
NO3	48	16	64	75
HCO3	180	14	194	93
nssCa	84	20	104	81
nssMg	5	4	9	56
nssK	6	3	9	67
nssSO4	84	29	113	74
Cl	457	121	578	79
Na	257	110	369	70

Fluxes were calculated for both rainfall and the soluble fraction of the aerosol for this work. For precipitation samples the total flux for the various elements ($\mu g/cm^2$) was calculated by multiplying the different concentrations (in µg/l) by the volume of precipitation (l) and divided by the area of the sampler (660 cm^2). For the soluble fraction of aerosols, the flux ($\mu g/cm^2$) was calculated according to the following equation: $F_x = C * VD * t$, where C is the concentration in $\mu g/m^3$, VD is the deposition velocity of the chemical species, and t is the sampling duration in seconds. The deposition velocities used according to the literature where 0.1, 0.5, 0.5 and 1 cm sec^{-1} for ammonia, sulfate, nitrate and carbonate respectively, and 2 cm sec^{-1} for all cations (Dulac et al., 1989, Slinn & Slinn, 1989). From Table V, it is apparent that the major part of the ions are transported by rain. However the dry fluxes are important, ranging from almost 50% for magnesium to less than 5% for ammonia. These results are in agreement with data for Spain and for France (Mateu et al., 1993). The only exception is the ammonia, whose flux is completely accounted for by precipitation in Sardinia, whilst it is almost 40% dry in France and Spain.

3.3 CALCULATION OF SCAVENGING RATIOS

Concentrations of elements in wet deposition have been shown to be extremely variable (Jaffrezo & Colin, 1988), which rends the quantification of these fluxes difficult. As many elements found in rainwater are derived from aerosols, a better understanding of the relationship between precipitation and aerosol could be obtained by calculation of "scavenging ratios" (SR). These ratios attempt to quantify the removal phenomena of trace constituents in the atmosphere by way of precipitation.

The precise determination of "scavenging ratios" requires the collection of rain and aerosol at the same location for a relatively long period, so that the influence of temporal variations are kept to a minimum (Savoie *et al.*, 1987). This work uses scavenging ratio defined as the ratio between the concentration of a given element in precipitation and the concentration of that same element in the aerosol, multiplied by the density of the air: **SR** = $([C]_{rain}/[C]_{air}) * d$, where $[C]_{rain}$ is expressed in µg/g; $[C]_{air}$ is expressed in µg/m^3 and d = 1200 g/m^3.

In the present work SR was calculated on the average calculated on the total rain (n=31) and aerosol (n=55) events sampled, as well as on the data from simultaneous rain and aerosol events only (n=23).

On the simultaneous data, as can be seen in Table VI, the washing effect is apparent. In fact the aerosols sampled simultaneously with precipitation (i.e. less than 48 hours before or after a rain event), have low concentrations of most elements, and therefore the SR increases. When in fact the calculation is carried out on the total of the events collected for the entire sampling period (i.e. using average data), the characteristics of the aerosol during dry periods is also taken into consideration. As these are normally more concentrated the SR is therefore lower.

TABLE VI
Scavenging ratios for 1) total events sampled and 2) simultaneous events only at Capo Carbonara, compared with SR values from other sites

Element	1 Sardinia total	2 Sardinia simultaneous	3 marine	4 rural non marine	5 urban
Ca	3459	4605	2100	1813	1579
Mg	1013	1524			816
Na	1708	1815	2900		744
K	1630	1845			1325
NO3	754	1052		3809	
SO4	1728	2173		690	
Cl	2419	2085	2800	735	7710

1 & 2 present work; 3 & 5 Jaffrezo & Colin (1988); 4 Pratt & Krupa (1985)

Scavenging ratio data from the literature are compared in Table VI with those obtained in this study. High scavenging ratios imply either efficient wet removal mechanisms or that the dry deposition results in low concentrations in the aerosol close to the surface with respect to the concentration where scavenging takes place (Pratt & Krupa, 1985). The phenomenon of removal appears to be more efficient for particles of crustal origin (e.g. Ca), and this result confirms the increasing scavenging efficiency with increasing particle size diameter (Pratt & Krupa, 1985; Jaffrezo & Colin, 1988), since crustal particles are in the range of 2-8 µm (Guerzoni *et al.*, 1993).

4. Conclusions

1. Rain and soluble part of the aerosol showed a strikingly similar ionic composition. Ionic balance studies show that the main ionic species in rainwater and

atmospheric aerosols are chlorine and sulphate for anions, and the principal cation is sodium, followed by magnesium and calcium.

2. The acid events are associated with N-NW trajectories (anthropogenic influxes from N. Europe) whilst southern trajectories transport alkaline Saharan dust. The buffering effect of this element on the rain was effectuated by way of a higher percentage of samples with pH>6 at Capo Carbonara (55%) with respect to other sites in Sardinia (25%). It seems that in this region the role of sodium is important for nitrate enrichment in aerosols.

3. Wet deposition largely prevails over the dry one, but still dry deposition is important for most of the crustal elements (20-45%), whilst it appears negligeable for ammonia.

4. The simultaneous sampling of rain and aerosol permitted the calculation of scavenging ratios for Ca, Mg, Na, K, NO_3, SO_4, and Cl. With careful consideration the SR could be used to infer dry deposition in Mediterranean sites where wet deposition is routinely measured.

Acknowledgements

The authors wish to thank the personnel of the meteorological station at Capo Carbonara for their help and G. Zini for the drawings. This work was partially supported by the Commission of the European Communities 'STEP' Program (STEP-CT 090-0080 DSCN). This work is contribution no. 998 of the IGM/CNR.

References

Caboi R., Cidu R., Cristini A., Fanfani L. and Zuddas P.: 1992, In: Proc. WRI-7, Y.K. Kharaka and A.S. Maest (eds.), Rotterdam: Balkema, 469-472. .

Correggiari A., Guerzoni S., Lenaz R., Quarantotto G. and Rampazzo G.: 1989, *Terra Nova*, **1**, 549-558.

Chester R., Murphy K.J.T., Towner J. and Thomas A.: 1986, *Chemical Geology*, **54**, 1-15.

Dulac F., Buat-Menard P., Ezat U., Melki S. and Bergametti G.: 1989, *Tellus*, **41B**, 362-378.

Eder B. and Dennis R.L.: 1990, *Water, Air, and Soil Pollution*, **52**, 197-216.

Guerzoni S., Landuzzi W., Lenaz R., Quarantotto G., Rampazzo G., Molinaroli E. and Cesari G.: 1993, *Water Pollution Research Reports*, **30**, 253 - 260.

Harrison R.M. and Pio C.: 1983, *Atmospheric Environment*, **17**, 2539-2543.

Jaffrezo J.L. and Colin J.L.: 1988, *Atmospheric Environment*, **22**, 929-935.

Keene W.C., Pszenny A.A., Galloway J.N. & Hawley M.E.: 1986, *Journal of Geophysical Research*, **91**, 6647-58.

Lim B., Jickells T.D. and Davies T.D.: 1991, *Atmospheric Environment*, **25A**,.745-762.

Loye-Pilot M.D., Martin J.M. and Morelli J.:1985, *Nature*, **321**,.427-428.

Mamane Y.: 1987, *The Science of the Total Environment*, **61**, 1-13.

Martens C. S., Wesolowsky J. J, Harris R. C. and Kaifer R.: 1973, *J. Geophysical Research*, **78**, 8778-8792.

Mateu J.,. Clom M, Forteza R. and Cerda V.: 1993, *Water Pollution Research Report*, **30**,.261-270.

Molinaroli E., Guerzoni S. & Rampazzo G.: 1993, *Geological Society of America, Special Paper*, **284**, 303-312.

Pratt G.C. and Krupa S.V.: 1985, *Atmospheric Environment*, **19**, 961-971.

Savoie D.L., Prospero J.M. and Nees R.T.: 1987, *Atmospheric Environment*, **21**, 103-112.

Slinn S.A. and Slinn W.G.N.: 1989, *Atmospheric Environment*, **14**, 1013-1016.

ATMOSPHERIC SULFUR AND NITROGEN IN WEST JAVA

G.P. AYERS[1], R.W. GILLETT[1], N. GINTING[2], M. HOOPER[3], P.W. SELLECK[1] and N. TAPPER[2]

1. Division of Atmospheric Research, CSIRO, PB1, Mordialloc 3195, Australia
2. Department of Geography, Monash University, Clayton, Vic. 3168, Australia
3. School of Applied Science, Monash University, Churchill, Vic. 3842, Australia

Abstract. Wet-only rainwater composition on a weekly basis was determined at four sites in West Java, Indonesia, from June 1991 to June 1992. Three sites were near the extreme western end of Java, surrounding a coal-fired power station at Suralaya. The fourth site was ~100 km to the east in the Indonesian capital, Jakarta. Over the 12 months study period wet deposition of sulfate at the three western sites varied between 32-46 meq m^{-2} while nitrate varied between 10-14 meq m^{-2}. Wet deposition at the Jakarta site was systematically higher, at 56 meq m^{-2} for sulfate and 20 meq m^{-2} for nitrate. Since sulfate and nitrate wet deposition fluxes in the nearby and relatively unpopulated regions of tropical Australia are both only ~5 meq m^{-2} anthropogenic emissions of S and N apparently cause significant atmospheric acidification in Java. It is possible that total acid deposition fluxes (of S and N) in parts of Java are comparable with those responsible for environmental degradation in acid-sensitive parts of Europe and North America.

1. Introduction

Acid deposition has only recently surfaced as an environmental issue in South-East (SE) Asia, where few atmospheric chemistry studies have been carried out in the past so few data are available with which to assess the possibility of significant environmental acidification via acid deposition (Ayers, 1991). However the anthropogenic sulfur and nitrogen emission inventories compiled independently by Foell and Green (1991) and Kato and Akimoto (1992) make clear the potential for such an issue to emerge since economic growth and population growth in the region combine to produce large secular trends in emission of both NO$_x$ and SO$_2$ (see also Galloway, 1989; Rodhe et al. 1992).

In the few locations where acid deposition has been studied, such as southwest China (Zhao and Xiong, 1988) and Hong Kong (Ayers and Yeung, 1995), the measured levels of atmospheric acidity and acid deposition are consistent with the predictions of Rodhe et al. (1992). Thus evidence at this stage, though limited, points to increasing perturbations to regional atmospheric acidity via growing atmospheric sulfur and nitrogen cycles.

Indonesia is one of the world's ten most populous countries, with considerable scope for economic development. Thus Indonesian emissions of anthropogenic N and S have the potential to rise dramatically in the coming decades, and significantly outstrip those of much smaller near neighbours, such as Malaysia and Singapore. Emissions estimates from Kato and Akimoto (1992) suggest a virtual doubling between 1975 and 1987 of both NO$_x$ (0.33 to 0.64 TgN y^{-1}) and SO$_2$ (0.20 to 0.49 TgS y^{-1}) emissions from Indonesia, while projections for 2010 from Foell and Green (1991) suggest this trend will continue.

Since there are essentially no data available at present to assess the extent of acid deposition in Indonesia the study described here was initiated with the aim of providing an initial perspective on the current situation, by generating data on rainwater composition and wet deposition of sulfate and nitrate at four sites in west Java over a 12 month period. The aim of this initial study was deliberately set at this relatively modest level, as logistical limitations, lack of appropriately trained staff and lack of chemically clean field laboratory facilities provided a limitation to the complexity of what could be attempted.

2. Scope of the Study

2.1 LOCATION AND DURATION

The study was carried out on the Island, Java, this island carrying the bulk of the Indonesian population and the majority of anthropogenic sources of NO_x and SO_2. The region studied was west Java, which includes the capital city, Jakarta (pop. >5 million), and a number of other major population and industrial centres. The four sites at which sampling took place were Jakarta, Serang, Cilegon, and Merak. The latter 3 sites were located 70-100 km to the west of Jakarta (Table I), at the western tip of Java, surrounding a major coal-fired power station located at Suralaya. This power station has an installed capacity of 1600 MW. No emissions figures for NO_x and SO_2 are available.

Table I. Relationship of sampling sites to Suralaya power station. Bearing is given from the Suralaya power station.

Site	Latitude, degrees	Longitude, degrees	Distance, km	Bearing, degrees
Jakarta	6.08	106.45	100	119
Serang	6.07	106.09	30	151
Cilegon	6.00	106.05	15	168
Merak	5.55	106.00	5	188

The Jakarta site was close to the city centre where atmospheric emissions are dominated by motor vehicles, some industry and local burning (rubbish; biomass). The Serang site was located on the outskirts (about 2 km from the urban boundary) of this small provincial city, close to rice fields but also near the major highway to Jakarta. The Cilegon site was close to the small township of Cilegon, in similar surroundings. The Merak site was in a location where population and motor vehicle densities are lower than at Serang or Cilegon, with the nearest human habitation a small local housing complex.

2.2 SAMPLING METHODS

Four sampling sites were employed, each having an Aerochem Metrics wet-only rainwater sampler plus a pair of 47 mm diameter filter samplers operating at about 15 litres per minute to sample aerosol and gas components at approximately 1.5 m above ground. One filter sampler contained a 1μm pore size Fluoropore filter for collection of aerosol for inorganic analysis. The other filter pack consisted of a pre-combusted quartz fibre filter to collect aerosol for subsequent total carbon (TC) analysis, followed by a pair of iodide/arsenite-impregnated Gelman 40 filters for collection of NO_2. Sample period was 1 week, with thymol biocide employed in the rainwater sampler to prevent biological degradation of the collected rain samples (Gillett and Ayers, 1991).

2.3 ANALYTICAL METHODS

All samples were shipped to Australia for analysis at CSIRO, with delays between sample collection and analysis varying from a week to a few months. However as a quality assurance procedure rainwater pH was measured on a small aliquot immediately upon sample retrieval in Indonesia, using a pH electrode and measurement system identical to that employed subsequently at CSIRO.

At CSIRO rainwater samples were analysed for pH (Orion Ross electrode and Orion low ionic strength buffers), and major ions by Ion Chromatography (Waters cation/ Dionex AS4 anion columns). Analytical precision, determined by "blind" duplicate analyses, was typically ±10% for anions but slightly higher for cations at the lowest levels.

Soluble aerosol components were determined by the same methods, after extraction of each aerosol filter in a 5 mL aliquot of Milli-Q water. The backup filters were also each extracted in 5 mL of Milli-Q water, the extracts subjected to a diazotization reaction followed by colorimetric determination to give an estimate of the atmospheric NO_2 level.

Total aerosol carbon was determined by combustion of quartz filter segments in an oxygen/helium stream at 800°C, conversion of the CO_2 produced into CH_4 using hydrogen over a palladium catalyst, and detection and quantification via a FID (Gillett et al., 1989).

3. Results

3.1 DATA QUALITY

Some shortcomings to the quality of the dataset must be acknowledged at the outset, arising from limitations to the local facilities available to the project.

First, rainfall amount was not determined independently at each site as planned, so had to be determined for each sample from the wet-only samplers. This may not be a serious deficiency, since our experience indicates that these devices sample heavy rainfall, typical for Indonesia, with high efficiency. However a second problem was that a number of samples were not shipped in their entirety to CSIRO, so complete rainfall amounts do not exist for 6 of 23 samples from Jakarta, 1 of 35 from Cilegon, 1 of 35 from Serang, and 7 of 35 from Merak. A third problem is that we cannot confirm the sampling of all rainfall events at all sites, although given the annual rainfall totals obtained (see below) this possible undersampling is unlikely to be > 20-30%. A fourth problem concerns the thymol added as biocide to the samplers to prevent biological consumption of organic acids. As shown by Gillett and Ayers (1991), thymol must be added to rainwater at ≥ 200 mg l^{-1} to be effective, but this threshold dose was not exceeded very often at the Indonesian sites, as indicated by a lack of significant organic acid levels in many samples, and a systematic lowering in H^+ concentration between sample collection in Indonesia and subsequent sample analysis in Australia. Thus organic acids probably make a larger contribution to overall rainwater composition at these sites than was revealed by the rainwater analyses.

In the case of the aerosol samplers a mistake was made during installation whereby the flowmeter was located between the filter and the pump, rather than after the pump, causing operation of the flowmeters at sub-ambient pressures for which they had not been calibrated. Although a post-study recalibration was performed at CSIRO, we estimate that the resultant correction applied to the sample volumes may be uncertain to ±25%.

Finally, although the aim of the study was to take weekly (*i.e.* 7 day) samples, continuously for 12 months, in practice this strategy proved impossible to adhere to at the local level. Thus instead of obtaining 52 aerosol/gas samples at each site, plus a number of rainwater samples corresponding to the number of weeks in which rain fell, only 23 aerosol samples were collected at Jakarta, with 35 collected at each of the other sites. However most of the year was actually covered, as a number of the samples were collected over periods of up to two weeks.

3.2 ATMOSPHERIC CONCENTRATIONS

With these caveats in mind, the mean rainwater, aerosol and gas data for sulfur and nitrogen species presented in Table II can be taken to provide a reasonable approximation to the annual average values encountered during the 12 months sampling period from June 1991 to June 1992. The additional values included for Jakarta are from a separate project carried out at the same site between September 1994 and January 1995, in which 14 weekly measurements of the gases shown were taken using the method of Ferm (1991).

Table II. Atmospheric concentrations of sulfur and nitrogen species. Rainwater components are volume-weighted means, in µmol l^{-1}; gas/aerosol components in nmole m^{-3}. TC is aerosol total carbon, in µg m^{-3}. Jakarta gas values highlighted by parentheses are from the period 1st September 1994 to 2nd January 1995 (see text for details).

	Jakarta	Cilegon	Serang	Merak
rainwater				
nss-SO_4^{2-}	37.3	30.8	21.2	21.5
NO_3^-	13.3	7.8	9.0	6.6
gas				
NO_2	770	470	74	200
(NO_2)	1150			
(SO_2)	290			
(HNO_3)	97			
aerosol				
nss-SO_4^{2-}	91	74	51	55
NO_3^-	23	11	8	13
TC	88	76	61	48

The data in the Table may be compared with similar data obtained in tropical Australia which, given the limited population in the region, can provide a "baseline" from an area of minimal anthropogenic emissions. Likens et al. (1987), Gillett et al. (1990) and Ayers et al. (1993) report annual average data on wet-only rainwater composition from Katherine, Jabiru and Darwin that yield seasonal mean nss-sulfate and nitrate concentrations averaging less than 5 µmol l^{-1} for both species, values clearly lower than those obtained from the Indonesian sites (Table II). Likewise, Ferm (1993) has reported SO_2 and NO_2 data from Jabiru for the period March 1991 to July 1992, with samples integrated over two months sampling periods. Mean values for SO_2 and NO_2 were, respectively 38 nmole m^{-3} (n=12) and 56 nmole m^{-3} (n=2). Although this comparison is limited by the small amount of Jabiru data involved, the systematically higher gas concentrations determined at the Indonesian sites are consistent with the hypothesis that there are significant anthropogenic perturbations to atmospheric S and N concentrations in west Java.

Finally, the Indonesian aerosol data also exhibit similar evidence of enhanced concentrations when compared with the small dataset available from northern Australia by Ayers and Gillett (1988).

3.3 DEPOSITION

Wet deposition fluxes of non-sea-salt sulfate and nitrate calculated from the data in Table II are given in Table III. The assumptions underlying this calculation are that total rainfall and volume weighted mean ion concentration were both accurately measured. As noted earlier the program of "blind" duplicate analyses confirmed the quality of the chemical analyses, however the annual rainfall totals were not well specified, and must be low since some samples at all sites were not returned in full to CSIRO. This is consistent with the fact that long-term rainfall records from the study region show average annual rainfall typically around 1700 mm, but total rainfall amounts deduced from the wet-only samples received at CSIRO totalled, respectively, 1275, 1312, 1305 and 1046 mm at Jakarta, Serang, Cilegon and Merak. In view of this, the wet deposition fluxes given in Table III must be taken to be "notional" only, rather than exact, as we assumed a rainfall total of 1500 mm at each site that is a compromise between the available data that is biased low, and the higher long-term average for the study region.

The Table also includes "notional" estimates of dry deposition fluxes at Jakarta, based on dry deposition velocities of 0.003 m s^{-1} for NO_2 and SO_2 and 0.01 m s^{-1} for HNO_3 and the gas concentrations listed in Table II (the lower NO_2 value is used). Justifications for selection of these values of deposition velocity for use with annual mean data may be found in, for example, the works of Meyers et al. (1991) and Hanson and Lindberg (1991).

Table III. Estimated annual depositions, meq m^{-2}.

	NO_3^- rain	SO_4^{2-} rain	NO_2 gas	SO_2 gas	HNO_3 gas
Jakarta	20	56	73	55	31
Cilegon	12	46			
Serang	14	32			
Merak	10	32			

4. Discussion

The deposition values in Table III must be acknowledged as being imprecise, for the reasons dealt with earlier. However the imprecision cannot mask the conclusion that the absolute values are greatly enhanced above "natural" fluxes, as represented by the tropical Australian rainwater data (Likens et al., 1987; Gillett et al., 1990; Ayers et al., 1993). The latter data all exhibit wet deposition fluxes of both NO_3^- and nss-SO_4^{2-} of ~5 meq m^{-2} y^{-1}. Thus the wet deposition data in Table III show apparent anthropogenic enhancements in NO_3^- flux of a factor of 2 - 4, and in nss-SO_4^{2-} by of a factor of 6 - 10. Galloway et al. (1984) identify such factors as clearly indicative of significant anthropogenic acidification.

The question of possible environmental consequences is unanswerable at present, since (a) these preliminary data cannot be considered to be adequately accurate and representative, (b) total deposition (wet plus dry) rather than wet deposition data alone are needed at a variety of locations, and (c) a knowledge of "critical loads" for Indonesian soil and surface water systems is required for comparison with accurately measured total deposition fluxes of mineral acid species

5. Conclusions

Datasets mainly on rainwater composition, but including some gas mixing ratios and aerosol composition information were gathered at four sites in Indonesia over 12 months from mid 1991. Despite some compromises in data quality, it is possible to conclude that : (1) organic acids may make a significant contribution to rainwater composition at these sites; (2) atmospheric composition at the study sites was significantly perturbed by anthropogenic emissions, leading to wet deposition fluxes enhanced by factors of 2 - 4 for nitrate and 6 - 10 for nss-sulfate over tropical Australian levels; (3) total acidic deposition (as S and N) at the study sites occurs at levels that has been responsible for adverse environmental consequences in the most sensitive regions elsewhere (*i.e.* in Europe and north America, see Hettelingh et al., 1991); and (4) a quantitative assessment of acid deposition is thus warranted in Indonesia, involving accurate determination of total deposition fluxes and a comparable determiantion of critical loads for regional ecosystems.

Acknowledgments

We acknowledge gratefully the contributions made by the Indonesian Coal Directorate and Indonesian Meteorological and Geophysical Agency. Funding was provided by Australian International Development Assistance Bureau, CSIRO and Monash University.

References

Ayers G.P., 1991, Atmospheric Acidification in the Asian Region, *Env. Monitoring Assessment*, **19**, 225-250.
Ayers G.P. and Gillett R.W., 1988, Acidification in Australia, in : *Acidification in Tropical Countries*, Rodhe H. and Hererra R. (eds) SCOPE Report 36, Wiley and Sons, Chichester, England, pp. 347-405.
Ayers G.P., Gillett R.W., Selleck P., Warne J.O., Huysing P., and Forgan B.W., 1993, A pilot study on rain-water composition at Darwin Airport. *Aust. Meteorol. Magazine*, **42**, 143-150.
Ayers G.P. and Yeung K.K., 1995, Atmospheric Acidity and Acid Deposition in Hong Kong, *Atmos. Environ.*, in press.
Ferm M., 1991, A sensitive diffusional sampler. Report L91-172, IVL, Box 47086, 402 58 Göteborg, Sweden.
Ferm M., 1993, Data from passive sampling of SO_2, NO_2 and NH_3, Proc. 2nd IGAC CAAP Workshop, Bhaba Atomic Research Centre, Bombay, 30th September - 2nd October 1992, Ed. G.P. Ayers.
Foell W.K. and Green C.W., 1991, Acid rain in Asia : an economic, energy and emissions overview, Proc. 2nd Workshop on Acid Rain in Asia, Asian Inst. Technol., Bangkok.
Galloway J.N., Likens G.E. and Hawley M.E., 1984, Acid precipitation : natural vs anthropogenic components, *Science*, **226**, 829-831.
Galloway J.N., 1989, Atmospheric acidification : projections for the future, *Ambio*, **18**, 161-166.
Gillett R.W., Ayers G.P. and Mainwaring S.J., 1989, Source contributions to atmospheric aerosols using organic tracers. In : *Man and his ecosystem : Proc. 8th World Clean Air Congress 1989*, Ed. L.J. Brasser and W.C. Mulder, Elsevier, Amsterdam, **Vol. 3**, pp. 551-556.
Gillett R.W., Ayers G.P. and Noller B.N., 1990, Rainwater acidity at Jabiru, Australia, in the wet season of 1983/84, *Sci. Tot. Environ.*, **92**, 129-144.
Gillett R.W. and Ayers G.P., 1991, The use of thymol as a biocide in rainwater samples. *Atmos. Environ.*, **25A**, 2677-2681.
Hanson P.J and Lindberg S.E., 1991, Dry deposition of respective nitrogen compounds : a review of leaf, canopy and non-foliar measurements, *Atmos. Environ.*, **25A**, 1615-1634.
Hettelingh J.-P., Downing R.R. and de Smet P.A.M., 1991, Mapping Critical Loads for Europe, CCE Technical Report #1, Coordination Center for Effects, National Inst. Public Health and Env. Protection, Bilthoven, Netherlands.
Kato, N. and Akimoto H., 1992, Anthropogenic emissions of SO_2 and NO_x in Asia: emission inventories, *Atmos. Environ.*, 26a, 2997-3017.
Likens G.E., Keene W.C., Miller J.M. and Galloway J.N., 1987, Chemistry of precipitation from a remote terrestrial site in Australia, *J. Geophys. Res.*, 92, 13299-13314.
Meyers T.P., Hicks B.B., Hosker R.P., Womack J.D. and Satterfield L.C., 1991, Dry deposition inferential techniques-II. Seasonal and annual deposition rates of sulfate and nitrate, *Atmos. Environ.*, **25A**, 2361-2370.
Rodhe H., Galloway J.N. and Zhao Dianwu, 1992, Acidification in Southeast Asia - prospects for the coming decades, *Ambio*, **21**, 148-150.
Zhao Dianwu and Xiong Jiling, 1988, Acidification in Southwestern China, in : *Acidification in Tropical Countries*, Rodhe H. and Hererra R. (eds) SCOPE Report 36, Wiley and Sons, Chichester, England, pp. 317-346.

DEPOSITION OF ACIDIC SPECIES AT A RURAL LOCATION IN NEW SOUTH WALES, AUSTRALIA

G.P. AYERS[1], H. MALFROY[2], R.W. GILLETT[1], D. HIGGINS[3], P.W. SELLECK[1] AND J.C. MARSHALL[1]

1. Division of Atmospheric Research, CSIRO, PB1, Mordialloc 3195, Australia
2. Environmental Services, Pacific Power, GPO Box 5257, Sydney 2001, Australia
3. Technical Services, Pacific Power, GPO Box 5257, Sydney 2001, Australia

Abstract. Wet-only rainwater composition on a daily basis, and atmospheric SO_2 and NO_2 concentrations on a monthly basis have been measured over a two year period at four sites ~100 km to the west of Sydney. Bulk aerosol composition on a monthly basis was also measured at one site. The study region is predominantly rural in character, but contains two coal-fired thermal power stations with a total installed capacity of 2320 MW, as well as several minor population centres, including a small city, with a total population of about 21,000. The measurement sites were located roughly on the perimeter of a circle of about 20 km radius having the power stations at its centre. Three of the sites were situated in rural settings, while the fourth was located on the outskirts of the small city of Lithgow. Atmospheric acid loadings at all sites were low by the standards usually associated with industrialised regions of Europe and North America, with about one third of rainwater total acidity provided by organic acids (formic, acetic and oxalic). At the three rural sites, total inorganic acid deposition, comprising measured wet deposition plus inferred dry deposition of acidic S and N species, averaged about 30 meq m^{-2} y^{-1}, a low figure by most standards. At the site located near the city of Lithgow total deposition of acidic S and N species averaged about 80 meq m^{-2} y^{-1}.

1. Introduction

Acid deposition has not received great attention in Australia, as the relatively small population (~17 million), large continent and geographically isolated centres of population mitigate against the build up of atmospheric acidity during long-range transport that occurs in Europe and north America. Nevertheless, atmospheric acidification processes do occur in Australia as elsewhere, albeit at lower levels (Ayers and Gillett, 1985; Ayers et al., 1986; Bridgman, 1989), so the study described here was commissioned near a pair of isolated, strong point sources of SO_2 and NO_2 so as to study acidic deposition under Australian conditions. An important aim of the study was to focus on total acid deposition, *i.e.* wet plus dry deposition of acidic sulfur and nitrogen species, an innovation for Australia since previous studies have dealt only with wet deposition of acidity. The timing was chosen by Pacific Power to give 12 months data prior to commissioning of one of the point sources, Mt Piper power station.

2. Scope of the Study

2.1 LOCATION AND DURATION

The study region straddles the continental divide 100 km WNW of the state capital city of Sydney (pop. ~3 million). The area is predominantly rural in character, with a few small towns and cities serving the regional population of about 21,000. In the centre of the study region are two coal-fired thermal power stations of 2320 MW total installed capacity, making these power stations (separated by about 6 km) the largest regional sources of SO_2 and NO_x. The study took place from May 1992 to May 1994, with one of the two power stations, Mt Piper, coming on line half way through the study, so one aim was to investigate any change in atmospheric acidity and deposition in the second year. Combined annual emissions of SO_2 and NO_x (as NO_2) from the power stations are approximately 33 and 16 ktonnes, respectively, emitted through stacks 177 metres (Wallerawang power station) and 250 metres (Mt Piper power station) in height.

2.2 SAMPLING STRATEGY

Four sampling sites were employed situated roughly 20 km SE, SW, NNW and E from the central point between the power stations. Each site contained a wet-only rainwater sampler, plus passive gas samplers for SO_2 and NO_2 (Ferm, 1991) located under a polyethylene rain shield 1.5 m above ground. Site 3 also contained a small bulk aerosol sampler operating at a few litres per minute flow, at a sampling height of ~ 1 m.

The rainwater samplers (prototypes of the Ecotech 200 sampler) employed a tipping bucket rain gauge (0.2 mm resolution) as rainfall sensor, and contained a data logger recording the time of each bucket tip as well as the time of each opening and closing of the wet-only sampler lid. The samplers employed a carousel containing eight polyethylene sample bottles, with the carousel rotating at 0900 each day, providing rainfall samples as 24-hour averaged wet-only samples. All surfaces in contact with rainwater were either high density polyethylene or Teflon. The samples were retrieved once per week, with sample preservation against biological degradation during the period prior to retrieval and analysis provided by addition of thymol to the sample bottles before these were installed in the sampler (Gillett and Ayers, 1991). The water volume in a separate 5 litre polyethylene bottle located under the tipping bucket rain gauge was also determined weekly, providing an in-situ check on tipping bucket rain gauge calibration.

The passive gas samplers at each site were exposed in duplicate, as recommended by Ferm (1991), and were exchanged at four-weekly intervals, yielding approximately monthly averaged data. The single bulk aerosol sample, collected on Fluoropore 1 µm substrate, was also retrieved at four weekly intervals.

2.3 ANALYTICAL METHODS

The rainwater samples were analysed for pH using Orion Ross electrode and Orion low ionic strength buffers, and sodium, potassium, magnesium, calcium, ammonium, chloride, nitrate, sulfate, phosphate, formate, acetate, oxalate and methanesulfonate, using a Dionex DX500 ion chromatograph and Dionex CS1 and AS11 columns. Precision of pH measurements was determined from routine checks against dilute acid standards to average ±0.046 pH units, while imprecision of the chemical analyses, determined from regular "blind" analysis of duplicate rain samples, was no worse than ±10% for any ion, and was routinely near ±5%. The SO_2 passive gas samples were analysed by ion chromatography, the NO_2 passive gas samples colorimetrically, with duplicate samples agreeing on average to within 10-20%, as found by Ferm (1991).

3. Results

3.2 ATMOSPHERIC CONCENTRATIONS

Annual and overall mean gas mixing ratios are presented in Table I, while Figure 1 contains representative time series plots. There is a very clear distinction shown in Table I between the NO_2, SO_2 and HNO_3 concentrations at Site 1, and those at Sites 2 - 4. The Site 1 values are clearly a factor of 2 - 4 higher than those at the other sites, and also show strong evidence in both NO_2 and SO_2 time series of a seasonal cycle with a winter maximum that is not evident at the other sites. Nevertheless, the mean values shown Table I are all low by the standards of polluted urban/industrial regions, even at Site 1, which is the site located in the city of Lithgow.

Table I. Overall mean Site values for acid gas concentrations, ppbv.

	SO_2			NO_2			HNO_3		
	Year 1	Year 2	mean	Year 1	Year 2	mean	Year 1	Year 2	mean
Site 1	2.5	2.0	2.2	5.8	6.7	6.2	1.8	1.2	1.5
Site 2	0.85	0.89	0.87	0.83	1.2	1.0	0.60	0.38	0.49
Site 3	0.63	0.52	0.58	1.2	1.4	1.3	0.45	0.38	0.42
Site 4	1.2	0.82	0.99	1.2	1.2	1.2	0.59	0.33	0.46

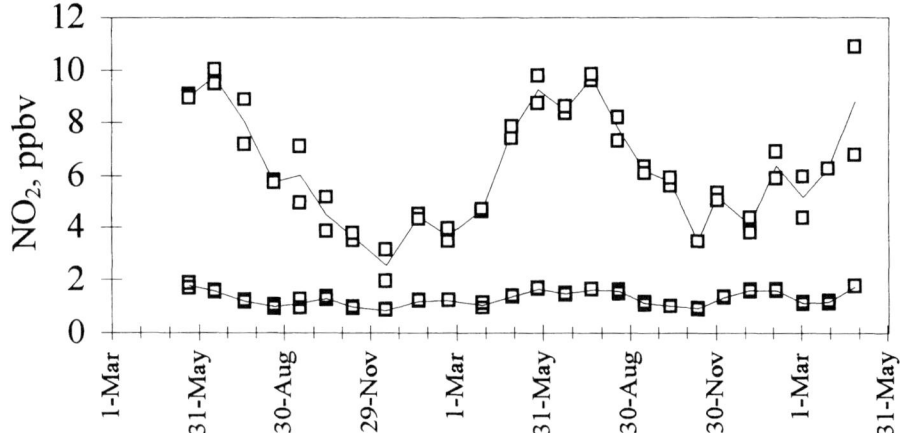

Figure I. Duplicate and mean NO_2 mixing ratios, ppbv; upper Site 1, lower Site 3.

No major interpretation of the aerosol data will be undertaken here, since the aerosol appears to be only a minor contributor to acid deposition flux. The mean aerosol concentrations of nitrate and sulfate averaged all samples, taken only at Site 3 were, 3.4 and 10.9 nmole m^{-3}.

Table II contains volume-weighted mean rainwater concentrations of all ions at each site, calculated using all available valid data. The data exhibit a number of notable features.

First, the mean pH values at all sites are significantly below 5, which as a raw statistic has often in the past been considered to provide prima facie evidence of significant acidification by anthropogenic activities. Second, however, organic acids (formate, acetate and oxalate) all occur in significant quantities, accounting for well over one third of total acid anions at all sites, indicating that pH cannot be taken alone to be a good indicator of acidification, since these acids generally are considered to be of natural origin in rural areas and are not considered to contribute to ecosystem acidification (Keene and Galloway, 1988). Third, the mean ion concentrations contained in Table IV are all low by the standards of acid-impacted regions of the northern hemisphere (NAPAP, 1991; Hettelingh et al., 1991). Fourth, the Table exhibits a general uniformity in the patterns of relative concentrations for many species across all four sites, with only a few notable exceptions. Among these is the fact that free acidity is clearly lowest at Site 2, apparently because both nitrate and nss-sulfate appear to be systematically low at this site. One other obvious difference between sites is the fact that non-sea-salt (nss) sulfate appears to be particularly elevated at Site 1 in comparison with the other three sites. As is apparent from the passive gas data, in which all acid gases were elevated at Site 1, this appears to be due to a consistent "Lithgow" (city) effect. Fifth, the availability of two full annual cycles of data provides some preliminary perspective on temporal (interannual) trends. The question of any possible effect of Mt Piper will be addressed later in terms of total annual acid deposition. Here it is sufficient to note from the annual concentration means that interannual variability of ±20-30% occurs for some species at some sites, indicating a need for caution in interpretation of short-term time trends.

Table II. Volume-weighted mean rainwater concentrations (μmol l^{-1}).

	Site 1			Site 2			Site 3			Site 4		
	Yr 1	Yr 2	**mean**	Yr 1	Yr 2	**mean**	Yr 1	Yr 2	**mean**	Yr 1	Yr 2	**mean**
pH	4.72	4.78	**4.75**	4.96	4.94	**4.95**	4.78	4.90	**4.84**	4.74	4.75	**4.75**
H$^+$	19.07	16.47	**17.82**	10.94	11.38	**11.19**	16.32	12.51	**14.18**	18.25	17.63	**17.92**
Na$^+$	4.66	5.91	**5.27**	3.84	3.07	**3.50**	4.60	5.00	**4.82**	5.68	9.71	**7.90**
NH$_4^+$	11.26	12.18	**11.71**	10.82	11.62	**11.34**	11.63	9.47	**10.42**	10.46	10.25	**10.35**
K$^+$	1.02	0.84	**0.93**	0.92	0.62	**0.79**	1.13	1.07	**1.09**	1.25	0.87	**1.05**
Mg^{2+}	1.05	1.32	**1.18**	0.82	0.52	**0.69**	0.95	1.24	**1.11**	1.20	1.41	**1.31**
Ca^{2+}	2.05	1.93	**1.99**	1.36	0.73	**1.08**	1.37	1.20	**1.28**	1.19	1.03	**1.10**
Cl$^-$	6.93	6.60	**6.77**	4.98	3.87	**4.36**	6.70	4.91	**5.64**	8.29	10.09	**9.27**
NO$_3^-$	8.34	7.40	**7.89**	5.99	6.02	**5.95**	8.01	7.11	**7.50**	7.64	8.09	**7.89**
PO$_4^{3-}$	0.26	0.33	**0.33**	0.45	0.41	**0.37**	2.16	0.49	**0.53**	1.79	0.41	**0.93**
SO$_4^{2-}$	7.99	7.68	**7.84**	3.54	3.07	**3.25**	4.14	3.55	**3.81**	5.33	6.08	**5.75**
nssSO$_4^{2-}$	7.72	7.32	**7.53**	3.26	2.90	**3.03**	3.87	3.25	**3.52**	4.98	5.49	**5.27**
C$_2$O$_4^{2-}$	0.55	0.42	**0.48**	0.59	0.36	**0.45**	0.62	0.42	**0.48**	0.55	0.42	**0.47**
Formic	7.91	5.89	**6.91**	7.02	7.89	**7.43**	9.38	7.49	**8.26**	7.79	7.20	**7.47**
Acetic	6.32	4.28	**5.31**	5.08	6.21	**5.61**	6.05	5.10	**5.49**	5.11	4.74	**4.91**
MSA	0.43	0.08	**0.18**	0.20	0.19	**0.19**	0.29	0.08	**0.11**	0.34	0.10	**0.20**

3.3 DEPOSITION

Given the availability of two twelve month periods of data it is possible to estimate two distinct sets of annual deposition fluxes for each of the rainwater, gas and aerosol species measured. It is important when doing this to be aware of the procedures used and the assumptions underlying these.

The wet deposition fluxes are calculated as the product of each twelve-monthly volume-weighted mean ion concentration and twelve-monthly rainfall total at the given site. The assumptions underlying this calculation are that total rainfall and volume weighted mean ion concentration were both accurately measured. We believe that the quality of the volume weighted mean ion data is indeed guaranteed by the fact that the chemical analyses were of demonstrably excellent quality, based on the program of "blind" duplicate analyses, and the rainfall totals are traceable via the tipping bucket tips count to the absolute, in-situ calibration provided by the tipping bucket water volumes determined each week throughout the study.

The validity of the dry deposition data for the gaseous species rests first on the adequacy of the passive samplers to provide data of acceptable quality, as discussed earlier on the basis of the original validation (Ferm, 1991) and the duplicate sampling strategy, and second on the use of an appropriate value for dry deposition velocity. For the latter parameters we adopt values of 0.003 m s^{-1} for NO$_2$, SO$_2$ and the aerosol components, and 0.01 m s^{-1} for HNO$_3$. Justifications for selection of values of this magnitude for use with annual mean data may be found in, e.g., the works of Meyers et al. (1991) and Hanson and Lindberg (1991). The deposition flux is then calculated as the product of species concentration and the assumed deposition velocity.

The validity of the aerosol deposition data likewise rests on both the deposition velocity, and the quality of the chemical analyses, which was discussed above was excellent. Again the deposition flux was calculated as the product of atmospheric concentrations and deposition velocity (Meyers et al., 1991), however the accuracy of the aerosol flux data is not of particular importance since as will be seen below it adds relatively little to the total S and N deposition fluxes. Estimated deposition fluxes for the key acidic species are tabulated in Table III.

Table III. Calculated annual depositions, meq m^{-2} y^{-1}.

	NO_3^- rain	SO_4^{2-} rain	NO_2 gas	SO_2 gas	HNO_3 gas	NO_3^- aerosol	SO_4^{2-} aerosol
Site 1	7.3	13.2	23.2	21.2	28.5		
Site 2	6.0	7.0	3.5	6.5	8.5		
Site 3	7.1	7.2	4.9	5.6	6.4	0.1	1.2
Site 4	5.6	7.8	5.3	9.8	8.8		

4. Discussion

Total calculated deposition of S plus N at Sites 1 - 4 was respectively : 95, 33, 33 and 39 meq m^{-2} y^{-1}, with the estimated dry deposition flux of gases exceeding the wet deposition flux at all sites. Aerosol dry deposition was a minor component (4%) of total deposition at the site where aerosol composition is measured.

The total deposition values at Sites 2 - 4 are low by European standards, and would constitute excessive deposition fluxes only in the unlikely event that regional ecosystems were as sensitive to acidification as the *most* sensitive regions identified in Europe and North America. It is unlikely that this is the case in the region where our study was centred, so the estimates in Table III point to an absence of significant acid deposition at Sites 2 - 4.

Site 1 is the only site at which total deposition fluxes are likely to be significant in terms of ecosystem acidification, and only then if the regional soil properties are such that at Site 1 the sensitivity to acidification is moderately high (*e.g.* level 3 in the European classifications; see Hettelingh et al., 1991). Since no soils sensitivity appraisal is available for the Lithgow region, this question must remain open. However the deposition fluxes at Site 1 are probably quite localised in extent; if as seems likely the high deposition fluxes are due to the influence of urban emissions from the small city of Lithgow, the high atmospheric concentrations of acid gases that yield most of the estimated total deposition are unlikely to extend far from the town perimeter.

The interpretation of the Site 1 data in terms of a localised "city" effect is consistent with the strong seasonal cycles in mixing ratio that occur only at this site and peak in winter when domestic use of coal and gas for heating also peaks. With this in mind it seems clear that the two power stations have little discernible influence on atmospheric acidity and acid deposition at the sites used in this study: the overall levels of wet deposition of sulfate and nitrate at Sites 2 - 4 are not discernibly different from levels measured elsewhere in southeastern Australia at Wagga Wagga, where to there is no proximity to coal fired power plant (Ayers and Manton, 1991).

Finally, it is clear from Tables II and III that there is no discernible change in atmospheric acidity in the second year of the study following the commissioning of the Mt Piper power station. This observation is consistent with that made above, that the power stations do not appear to significantly influence the atmospheric acidity at the study sites. However the study will be carried out for one additional year to verify this conclusion.

5. Conclusions

Representative datasets on rainwater composition, acid gas mixing ratios and aerosol composition have been gathered at four sites in SE Australia over a two year period from May 1992. The data suggest that :
- organic acids make a significant contribution to rainwater acidity at these sites, so rainwater pH is not a good indicator of acidification;
- dry deposition of S and N gases exceeds wet deposition of SO_4^{2-} and NO_3^- at all sites;
- dry deposition of aerosol components does not appear to be large in comparison with dry deposition of gaseous species and wet deposition of rainwater acidity;
- total acidic deposition at Sites 2 - 4 appear to be low, by world standards;
- total acidic deposition at Site 1 (Lithgow) is significant, though not large, by European and north American standards, but appears to be a product of localised, surface-level urban emissions rather than of emissions from two nearby coal-fired power stations.

7. Acknowledgments

We acknowledge with sincere thanks the contributions made to the very successful conduct of the study by all Pacific Power personnel involved.

8. References

Ayers G.P. and Gillett R.W., 1985, Some observations on the acidity and composition of rainwater in Sydney, Australia, during the summer of 1980-81, *J. Atmos. Chem.*, **2**, 25-46.
Ayers G.P., Gillett R.W., and Cernot U., 1986, Chemical composition of rainwater at New Plymouth, New Zealand in 1981-82, *Clean Air (Aust.)*, **20**, 89-93.
Ayers G.P. and Manton M.J., 1991, Rainwater composition at two BAPMoN regional stations in SE Australia, *Tellus*, **43B**, 379-389.
Bridgman H.A., 1989, Acid rain studies in Australia and New Zealand, *Arch. Environ. Contam. Tox.*, **18**, 137-146.
Ferm M., 1991, A sensitive diffusional sampler. Report L91-172, IVL, Box 47086, 402 58 Göteborg, Sweden.
Gillett R.W. and Ayers G.P., 1991, The use of thymol as a biocide in rainwater samples, *Atmos. Environ.*, **25A**, 2677-2681.
Hanson P.J and Lindberg S.E., 1991, Dry deposition of respective nitrogen compounds : a review of leaf, canopy and non-foliar measurements, *Atmos. Environ.*, **25A**, 1615-1634.
Hettelingh J.-P., Downing R.R. and de Smet P.A.M., 1991, Mapping Critical Loads for Europe, CCE Technical Report No. 1, Coordination Center for Effects, National Inst. Public Health and Env. Protection, Bilthoven, Netherlands.
Keene W.C. and Galloway J.N., 1988. The biogeochemical cycling of formic and acetic acids through the troposphere : an overview of current understanding, *Tellus*, **40B**, 322-334.
Meyers T.P., Hicks B.B., Hosker R.P., Womack J.D. and Satterfield L.C., 1991, Dry deposition inferential techniques-II. Seasonal and annual deposition rates of sulfate and nitrate, *Atmos. Environ.*, **25A**, 2361-2370.
NAPAP, 1991 Acid Deposition : State of Science and Technology, Published by: The NAPAP Office of the Director Washington, D.C., 2053.

SPATIAL AND TEMPORAL VARIABILITY IN PRECIPITATION CHEMISTRY IN THE URBAN AREA OF GREATER MANCHESTER.

D.E. CONLAN, S. J. LINDLEY and J.W.S. LONGHURST

Atmospheric Research and Information Centre, Department of Environmental and Geographical Sciences, Manchester Metropolitan University, Chester Street, Manchester M1 5GD. U.K.

Abstract. To investigate the spatial and temporal variability of acid deposition in the urban environment a small-scale intensive network of bulk collectors has been deployed around Greater Manchester, UK. This network has been in operation since 1986. The concentrations and deposition rates of non-marine (nm) sulphate, nitrate, ammonium, calcium and hydrogen are reported for 1994. Acidity was generally lower in the city centre of Manchester where calcium concentrations were highest. Calcium compounds in the urban atmosphere effectively buffer the precipitation acidity.

Key words: Acid deposition, air quality, urban

1. Introduction

Urban sampling sites have been specifically excluded from recent large networks for measuring chemical composition of precipitation, as there has been a wish to collect samples which reflect regional rather than local sources and deposition of air pollutants. In the UK, more people reside in the urban/suburban environment than in rural areas and it is these urban areas that have traditionally been subject to high concentrations of air pollutants. The effects of acid deposition and its precursors can often be seen in the urban environment.

The Greater Manchester Acid Deposition Survey (GMADS) has been in operation since 1986 and was established to determine urban background precipitation chemistry and establish the variation in some of the main species such as acidity, sulphur and nitrogen across a dense network of collectors in a highly industrialised conurbation. The county of Greater Manchester lies in the north west of England and is some 1300 km^2 in area and supports a population in excess of 2.5 million. A further 334,900 people live in an area of 855 km^2 outside of the County area. The whole survey area is intersected and circled by major communication routes and is one of the most important industrial regions in the UK. The area has a diverse industrial base contributing to a wide array of large and small point sources of pollution. In terms of altitude there is great variation ranging from 20 m above sea level in the south west of the survey area to over 600 m above sea level in the north east.

2. Methods

The GMADS network consists of 19 bulk precipitation collectors identical to those used in the UK national network. Sites were selected on the basis that they should characterise the regional precipitation chemistry as influenced by the urban area. The fundamental siting objective was therefore to site the collectors in the urban or near urban areas which were not unduly influenced by local sources of pollution. To achieve this, siting criteria were set requiring horizontal separations of 100 m between the collectors and small or mobile sources of pollution and 1 km between major roads or large point sources. Furthermore, the collectors were sited on undisturbed land to avoid local contamination by dust particles. Extremes of altitude such as valley bottoms or hill tops were avoided, be accessible under all weather conditions and site security was of utmost importance.

Samples were collected on a weekly basis and analysed for the following ions: sulphate, hydrogen, nitrate, ammonium, chloride, calcium, magnesium, zinc, sodium, potassium and hydrogen carbonate (Conlan, 1994). In conjunction with the monitoring of precipitation chemistry measurements of ambient nitrogen dioxide were undertaken. Simple, passive diffusion tubes were deployed at each site for an exposure period of one week. Such diffusion tube technology is used extensively in the UK for the monitoring of ambient nitrogen dioxide (Atkins, *et al.*, 1986; Campbell, *et al.*, 1993).

This paper describes the monitoring data for 1994 and investigates the spatial variability of the precipitation chemistry.

3. Results and Discussion

The spatial variability was significant for nm sulphate, nitrate, ammonium, calcium and hydrogen and in general maximum concentrations occurred in the north of the conurbation. nm sulphate concentrations were also high in the city centre of Manchester. The spatial pattern of nm sulphate concentrations was similar to that for calcium and the correlation between these variables was significant ($r=0.667$, $p<0.001$). The variability of sulphate is thought to be mainly due to sulphate aerosols in the atmosphere rather than gaseous sulphur dioxide. Analysis of the difference between nm sulphate concentrations in precipitation collected by a wet-only collector and that by a bulk collector indicated a significant correlation between dry deposition component of nm sulphate (Lee and Longhurst, 1992). Sulphur dioxide in this urban atmosphere may react with calcium bearing aerosols suspended in the atmosphere although it has been recognised that this gas-particle reaction may be slow and dependent on the gas diffusion into the particle matrix (Lee, 1993). It is thought that the main oxidation route for sulphur dioxide to sulphate is by aqueous phase reaction (Seinfeld, 1986) with ozone, hydrogen peroxide and also by transition metal catalysis (RGAR, 1990). Consequently, the most likely basis for the spatial variability of nm sulphate in this survey is the spatial variation in the dry deposition of nm sulphate to the bulk collectors.

Concentrations of hydrogen across the conurbation were variable with highest concentrations occurring in the fringes of the conurbation; acidity being lowest in the city centre. As acidity is a balancing term which closely relates to the concentrations of other ions, this spatial pattern highly reflects that of calcium. It is therefore likely that this spatial pattern of acidity is a function of the neutralisation of calcium in the city centre area rather than the elevation of acidity in the suburbs. The source of calcium in urban regions is probably the resuspension and deposition of urban dust which is mainly derived from construction and excavation work and also from fuel combustion especially from transportation sources (Irvine, et al., 1989).

The concentrations of nitrate were in general higher in the more urban parts of the study area where traffic flow rates were also highest. The major emissions of NOx in this urban area has been estimated to be derived from transportation sources (Longhurst, et al., 1994). However, the correlation between weekly measurements of nitrate concentrations in precipitation and weekly measurements of ambient nitrogen dioxide proved to be insignificant ($r = -0.109$, not significant). Consequently the spatial variability of nitrate is probably related to the dry deposition and scavenging of ammonium nitrate and nitric acid in the atmosphere.

The spatial pattern of ammonium concentrations in precipitations indicated highest levels in the north of the conurbation which rapidly decrease towards the most northerly and one of the more rural sites. Concentrations were also highest in the city centre of Manchester. Concentrations of the precursor gas ammonia have been found to be high in the city centre especially during the warmer summer months (Conlan, et al., 1993), with the majority of emissions coming from humans (Atkins and Lee, 1993). The region to the south west and to the north of the study area are predominantly agricultural and therefore a source of ammonia from livestock. The most likely atmospheric removal processes of ammonium are by irreversible gas phase reactions with hydrochloric acid and nitric acid and below cloud scavenging of ammonia (Allen, et al., 1988).

Table 1. Oneway analysis of variance tests for ion concentrations in precipitation with site location.

Variable	F ratio
nm sulphate	2.264*
calcium	4.808*
hydrogen	6.754*
nitrate	3.642*
ammonium	9.202*

* significance less than 0.01

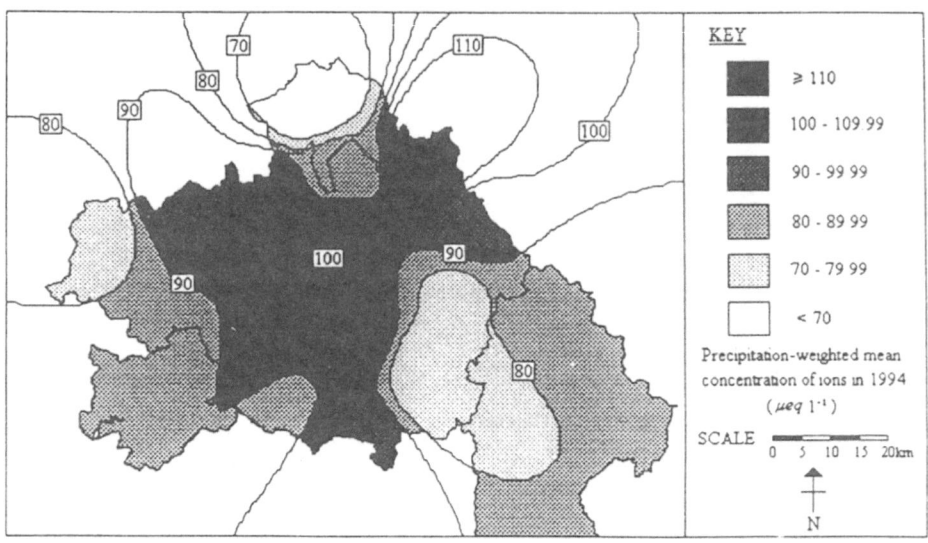

Fig. 1: Spatial variability of the mean concentration of non-marine sulphate in precipitation in 1994.

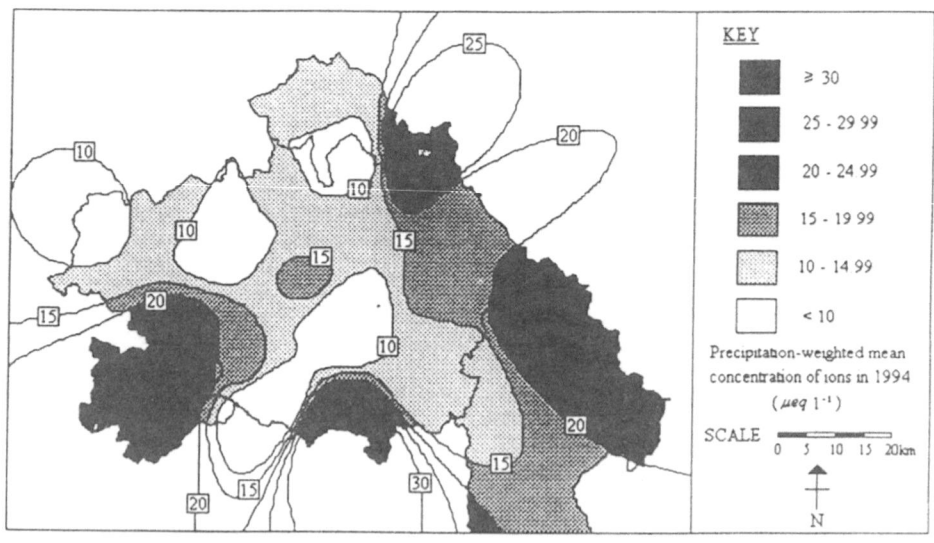

Fig. 2: Spatial variability of the mean concentration of hydrogen in precipitation in 1994.

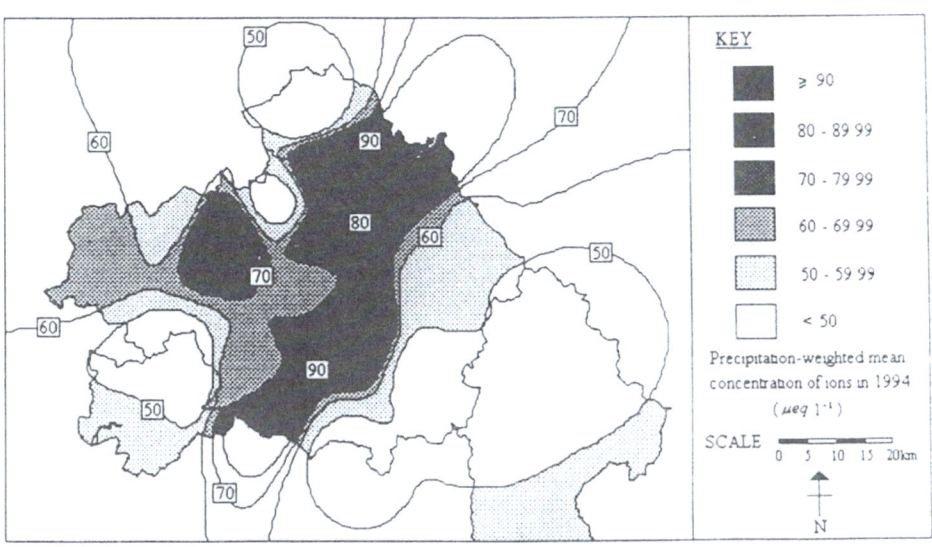

Fig. 3: Spatial variability of the mean concentration of calcium in precipitation in 1994.

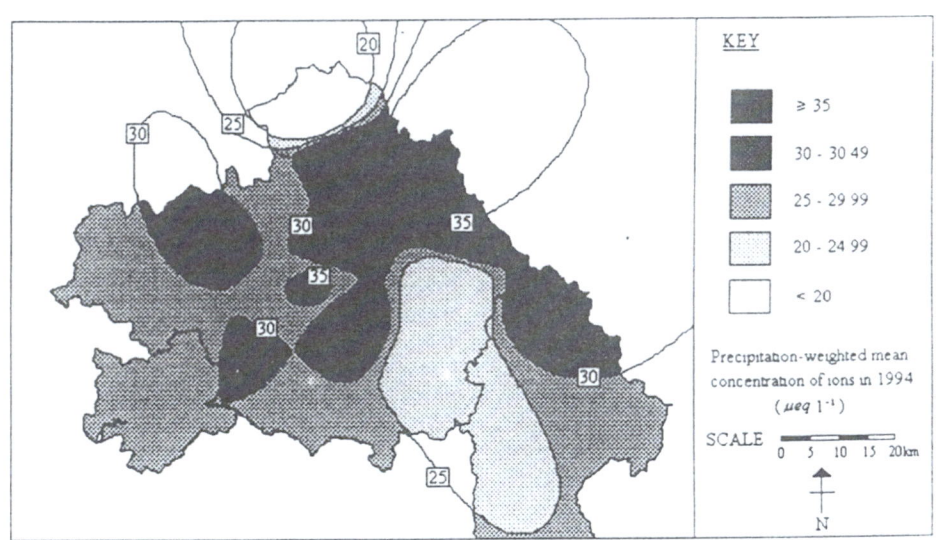

Fig. 4: Spatial variability of the mean concentration of nitrate in precipitation in 1994.

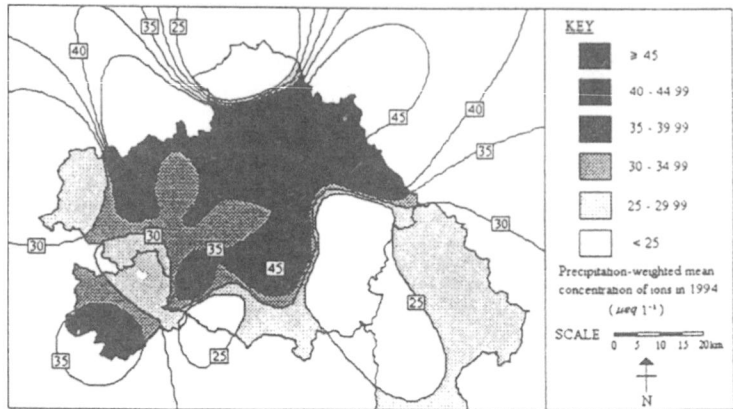

Fig. 5: Spatial variability of the mean concentration of ammonium in precipitation in 1994.

4. Conclusions

Concentrations of nm sulphate, calcium, hydrogen, nitrate and ammonium in precipitation were significantly variable across the conurbation of Greater Manchester in 1994. The most notable differences in concentration across the area were seen with acidity and calcium. The acidity of precipitation in the urban centre of Manchester was generally lower than the suburban and rural surrounds. This is principally due to the neutralising effect from the high concentrations of calcium in the city centre.

References

Allen, A.G.., Harrison, R.M. and Wake, M.T.: 1988, *Atmos. Environ.*, **22**, 1347-1353.
Atkins, D.H.F., Sandalls, J., Law, D.V., et al. : 1986, *The measurement of nitrogen dioxide in the outdoor environment using passive diffusion tube samplers*. Report AERE-R 12133, United Kingdom Atomic Energy Authority, Harwell, U.K.
Atkins, D.H.F. and Lee, D.S. :1993, *Atmos. Environ.* **27A**, 1-7.
Campbell, G.W., Steadman, J.R. and Stevenson, K.: 1993, *Atmos. Environ.* **28(3)**, 477-486. Conlan, D.E., Longhurst, J.W.S., Hare, S.E.. et al.: 1993, *Urban acid deposition results from the GMADS network, 1992*, Atmospheric Research and Information Centre, Manchester Metropolitan University, U.K.
Conlan, D.E., Longhurst, J.W.S., Bantock, J. et al.: 1994, *Urban acid deposition results from the GMADS network, 1993*. Atmospheric Research and Information Centre, Manchester Metropolitan University, U.K.
Irvine, K.N., Murray, S.D., Drake, J.J. et al.:1989, *Environ. Technol. Lett.*, **10**, 527-540.
Lee, D.S.: 1993, *Atmos. Environ.*, **27B**, 321-337.
Lee, D.S. and Longhurst, J.W.S.: 1992, *Water, Air and Soil Pollut.*, **64**, 635-648.
Longhurst, J.W.S., Rayfield, D. and Conlan, D.E.: 1994, *The impacts of road transport on urban air quality - a case study of the Greater Manchester region*. In Air Pollution II volume 1. Editors: Baldsano, J.M., Brebbia, C.A., Power, H. et al., 333-340.
Seinfeld, J.H.: 1986, *Atmospheric Chemistry and Physics of Air Pollution*.
Review Group on Acid Rain: 1990, *Acid deposition in the United Kingdom 1986-1988*. A third report of the United Kingdom Review Group on Acid Rain. 124 pages.

A GENERALISED DESCRIPTION OF THE DEPOSITION OF ACIDIFYING POLLUTANTS ON A SMALL SCALE IN EUROPE

J. W. ERISMAN, C. POTMA, G. DRAAIJERS, E. VAN LEEUWEN and A. VAN PUL

National Institute of Public Health and Environmental Protection (RIVM), P.O.Box 1, 3720 BA Bilthoven, the Netherlands

Abstract In describing the effects of acidification on the level of ecosystems, acid loads should be available at least on the size of ecosystems. No deposition maps on this resolution are available, hampering accurate estimation of exceedances of critical loads in Europe. Here, maps of small scale fluxes of Europe are presented. The maps are produced in close co-operation with EMEP. The acidifying components taken into account are oxidised sulphur and nitrogen and reduced nitrogen compounds. The method for estimating dry deposition is based on the combination of long-range transport model concentrations provided by EMEP and a detailed description of the dry deposition processes. Dry deposition velocities are calculated on a small scale using the inferential technique. Resistances are modelled using observations of meteorological parameters in Europe and parametrisation of surface exchange processes from deposition measurements. Wet deposition maps are derived using measured concentrations in precipitation in Europe together with precipitation amounts.

Key words: dry and wet deposition, Europe, local scale, modelling, measurements

1. Introduction

In Europe, sulphur compounds, and both reduced and oxidised nitrogen compounds, have been shown to acidify soils and surface waters. Furthermore, nitrogen deposition causes eutrophication (e.g. Heij and Schneider, 1991). The components considered in the acidification and eutrophication processes are SO_2 and aerosols of SO_4^{2-} (SO_x); NO, NO_2, HNO_2, HNO_3 and aerosols of NO_3^- (NO_y) and NH_3 and aerosols of NH_4^+ (NH_x). These pollutants are transferred to soil, vegetation and water surfaces by wet deposition, cloud and fog deposition and dry deposition (Erisman and Draaijers, 1995).

Investigations on abatement strategies based on the critical load concept require relevant deposition data on both local and regional scales (Nilsson and Grennfelt, 1988; Hettelingh *et al.*, 1991; Lövblad *et al.*, 1993). On the local scale, large variations in deposition over landscape features and their variations in sensitivity make it essential to compare the critical load value for a specific ecosystem with the actual deposition so as to determine the exceedance value. On the larger regional scale, the essential parameters are dispersion and deposition which must be estimated in order to assess the relevant abatement strategies. For pollution deposition over Europe and budget estimates the regional-scale approach is required (e.g. Tuovinen et al., 1994). The local-scale approach covers the calculation of the more site specific critical load exceedances (<1x1 km). The two approaches should be linked in order to evaluate the complete chain from emission to deposition and to develop relevant abatement strategies. This requires parametrisation of the deposition processes on ecosystem level (Erisman and Draaijers, 1995). On the national level there have been some studies showing the local variation in deposition: UK (UKGAR, 1990), Sweden (Lövblad et al., 1991) and the Netherlands (Erisman, 1993).

At this moment it is not feasible to estimate ecosystem input because of lack of data and of process descriptions on this scale. Here, a method is proposed to estimate local-scale (1/6x1/6°) fluxes in Europe by applying a combination of long-range transport modelling and local scale inferential deposition modelling (Erisman and Baldocchi, 1994, van Pul et al., 1995). In this paper the method is explained and results of local scale deposition fluxes are presented. Furthermore, the uncertainty in results is assessed.

2. Method description

The deposition model which was developed at RIVM is the EDACS model (Estimation of Deposition of Acidifying Components on a small scale in Europe, Van Pul et al., 1995). The basis for the estimates is formed by results of the EMEP long-range transport model. With this model dry, wet and total deposition is estimated on a 150x150 km grid over Europe using emission maps for SO_2, NO_x and NH_3 (e.g. Tuovinen et al., 1994). The local-scale approach used by RIVM depends strongly on LRT model results because modelled concentrations are used in EDACS to estimate the dry deposition. By using calculated concentration maps, the relationship between emissions and deposition is maintained and scenario studies, budget studies and assessments can be carried out on different scales. Wet deposition is added to the dry deposition to estimate total local scale deposition in Europe. The methods to determine the wet and dry deposition will be explained in the next sections.

2.1 WET DEPOSITION

Up to now, wet deposition maps on a European scale are based on long-range transport model results, whereas for most components wet deposition maps based on measurements are only available on national scales. Van Leeuwen et al. (this issue) used measurements to estimate the wet deposition in Europe. Acidifying components and basic cations were mapped on a 50x50 km scale over Europe for 1989, based on results of field measurements made at approximately 750 locations. Information on concentrations of ions in precipitation in 1989 was obtained from the EMEP database and from organisations responsible for wet deposition monitoring in their countries. Concentrations measured with bulk samplers were corrected for the contribution of dry deposition onto the funnels of these samplers. Sulphate concentrations were corrected for the contribution of sea salt. Point observations were interpolated to a field covering the whole of Europe using the kriging technique.

As an example, the wet deposition map of total potential acid (Fig. 1) is presented. Total deposition of potential acid is calculated as $NH_x+NO_y+2SO_x$ and is expressed in mol H^+ ha^{-1} a^{-1}. Potential acid can be used to determine the maximum acid input to ecosystems. The actual acid load depends on the input of alkaline components and the extend that NH_3 is nitrified in the soil (Heij and Schneider, 1991). Base cation deposition in Europe is described by Draaijers et al. (this issue). Fig. 1 clearly resemble European emissions of the three components and climate patterns. Highest input is found in areas with high rainfall (mountainous regions, coastal areas) and in areas with high sulphur emissions. Sulphur input yields the highest contribution to the potential acid input.

2.2 DRY DEPOSITION

The method for dry deposition has been developed by Van Pul et al. (1995). It is based on that used for the Netherlands (Erisman, 1993). Dry deposition in EDACS is inferred from the combination of long-range transport model concentrations provided by EMEP and parametrised dry deposition velocities. Concentrations at 50 m above the surface (blending height) are used. At this height it is assumed that concentrations and meteorological parameters are not influenced by surface properties to a large extent (Erisman, 1993). Dry deposition velocities of gases and particles at this height are calculated on a small scale using a land use map, routinely available meteorological data and a resistance model (Erisman et al., 1994). In the resistance model transport to and absorption or uptake of a component by the surface is described. using parametrisations of surface exchange processes. Meteorological parameters (wind speed, friction velocity, radiation, temperature, humidity and precipitation) are obtained from the ODS (Observational Data Set) for every 6 h and interpolated over Europe on a $1/6^0$ x $1/6^0$ grid.

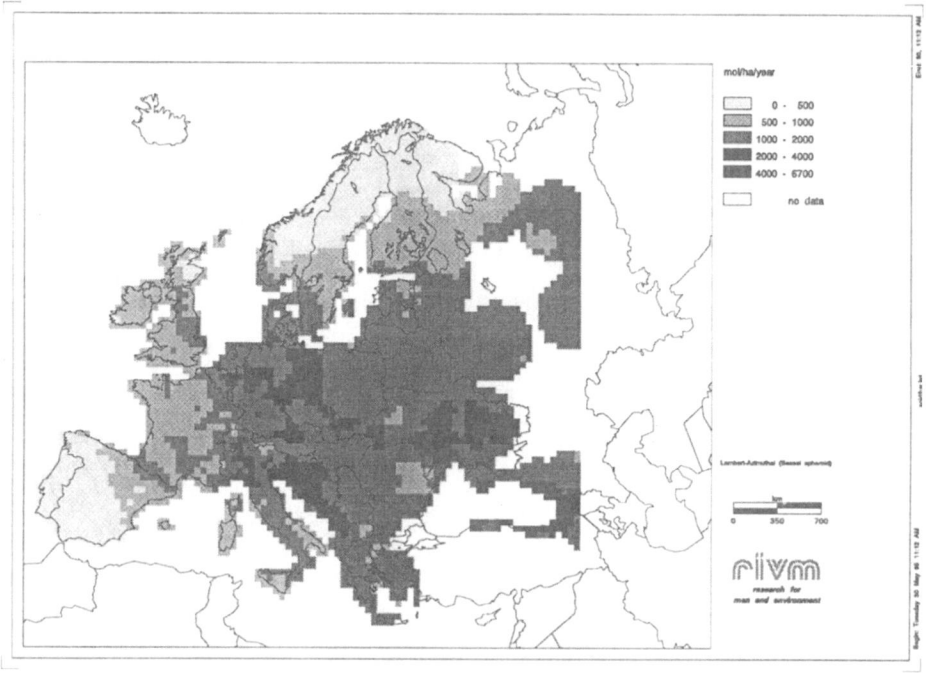

Fig. 1 Wet deposition of potential acid in Europe (50x50 km) in 1989 in mol ha^{-1} a^{-1} (Van Leeuwen *et al.*, 1995)

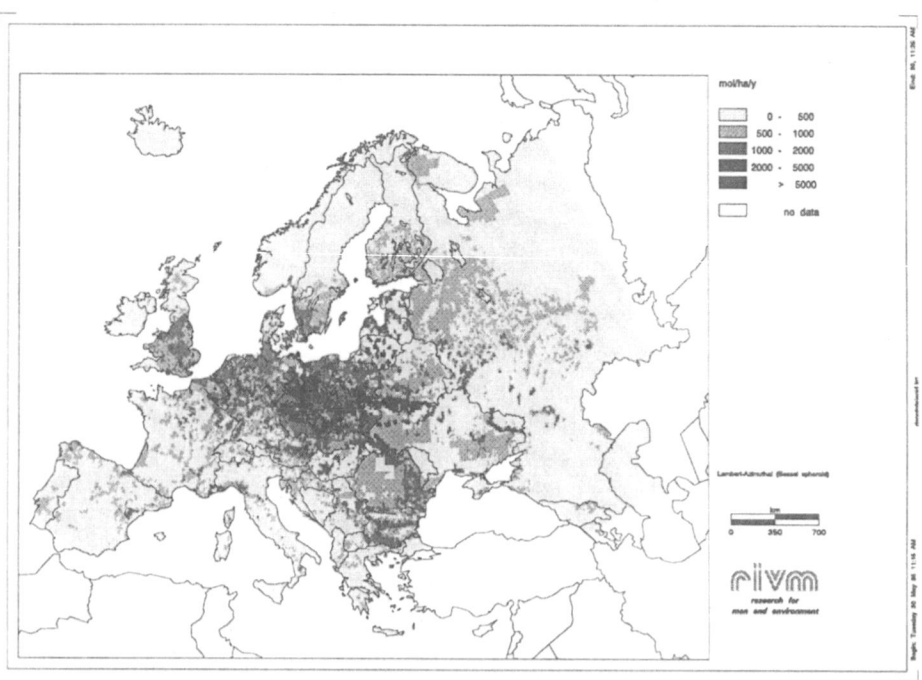

Fig. 2. Dry deposition of total potential acid in Europe on a 1/6°x1/6° scale in mol ha^{-1} a^{-1} (Van Pul et al., 1995)

The land use map of Europe, with a resolution of $1/6^0 \times 1/6^0$ (~10x20 km), is constructed by RIVM from ground-based and satellite observations (Van Velde et al., 1995). The resolution of this map determines the resolution of the deposition maps. Following land use classes are distinguished: coniferous and deciduous forests, grassland, arable land, permanent crops, inland waters and urban areas. Roughness length maps for the summer and winter season were derived from the land use map and z_0 classifications according to Erisman (1993). z_0 is used to estimate atmospheric transport to the surface. V_d is calculated for the land use class coverage within a grid cell. The average V_d for a grid cell is then calculated by weighting the land use specific deposition velocities with the surface area within that specific grid cell.

Fig. 2 shows a map of the annual average dry deposition of potential acid in Europe. The dry deposition of sulphur contributes most to the total potential input. The effect of land use (roughness) and the difference in V_d is clearly shown. In areas with forested terrain the dry deposition is higher than at low vegetation areas, and, for example, in dry areas, the dry deposition is decreased as a result of a difference in V_d estimates.

3 Uncertainty in the EDACS results

The wet deposition maps are subject to several sources of uncertainty, such as those associated with the measurements, assumptions and simplifications in the methods used, and the interpolation procedure. Van Leeuwen et al. (this issue) give an extensive description of these uncertainties. The main source of uncertainty is found to differ per region, varying from 50% in West, northwest and Central Europe where data quality is assumed to be good and sufficient (representative) data is present, to 60% in East, southeast and south-west Europe, where less data is available of which the representativeness and quality can be questionable and 70% in mountainous areas (e.g. the Alps) and upland areas, areas with complex terrain.

In most areas of Europe the highest contribution of uncertainty in dry deposition is the result of using the simple resistance formulation for a highly variable process, and, more specifically, the surface resistance parametrisations. It assumes a constant flux layer, i.e. there are no surface inhomogeneities, edge effects or chemical reactions. More and more accurate Rc parametrisations are needed for various vegetation species and surface types. Moreover, there is a lack of measurements which can be used to test these parametrisations, especially for southern and eastern European climates and surfaces.

Surface wetness is found to be one of the major factors influencing the deposition process of soluble gases. Processes leading to surface wetness should be better taken into account in greater detail. The overall uncertainty in the surface resistance due to these factors is different for each component and surface type. This uncertainty varies on an annual basis and is assumed to range between 20% to more than 100%.

In the current version of EDACS, the EMEP-LRT modelled concentrations on a 150x150 km grid are used. The uncertainty in the concentrations are estimated to amount 40 - 70% using a statistical analysis with EMEP measurements (Krüger, 1993). In source areas the uncertainty can be even higher. It is assumed that the concentration distribution within a grid cell is homogeneous. This is not the case in a grid cell which contains industrialised areas or many scattered sources such as with NH_3 and NO_x. For such conditions, sub-grid concentration variations are present. The uncertainty in the deposition in a grid cell due to these gradients is estimated at >25% (Van Pul et al., 1995).

The uncertainty in regional scale total deposition estimates strongly depends on the pollution climate and on landscape complexity of the area under study. The uncertainty is determined by the uncertainty in wet, dry or cloud and fog deposition. Fog and cloud water deposition is not taken into account. Furthermore, deposition estimates yield higher uncertainty in areas built up by complex terrain and with strong horizontal concentration gradients. (Erisman and Draaijers, 1995).

Acknowledgement

Anton Eliassen, Erik Berge and Helge Styve are thanked for providing the EMEP model concentration data.

References

Draaijers G., van Leeuwen E., Potma C., van Pul W., Erisman J.W., 1995) Water, Air and Soil Poll., this issue.
Erisman, J.W., 1993 Water Soil Air Pollut.,71:51-80.
Erisman, J.W., Pul, A. van and Wyers, P., 1994 Atmospheric Environment, 28:2595-2607.
Erisman, J.W. and Draaijers, G., 1995 Studies in Environmental Research, Elsevier, the Netherlands.
Heij, G.J. and Schneider, T., 1991 Report No. 200-09, RIVM, Bilthoven, The Netherlands.
Hicks, B.B., Hosker, R.P., Meyers, T.P. and Womack, J.D., 1991 Atmospheric Environment, 25A:2345-2359.
Iversen, T., Halvorsen, N., Mylona, S. and Sandnes, H., 1991 Norwegian Meteorological Institute, Oslo.
Krüger, O., 1993 In: proceedings CEC/BIATEX Workshop 4-7 May 1993, Aveiro, Portugal.
Leeuwen E. van, Potma C., Draaijers,G., Erisman,J.W., Pul W. van, 1995 Water, Air and Soil Pollut., this issue
Lövblad, G., Erisman, J.W. and Fowler, D., 1993 Report No. Nord 1993:573. Nordic Council of Ministers, Copenhagen, Denmark.
Pul A., Potma C., Leeuwen E. van, Draaijers G., Erisman J.W., 1995 Report 722401005 Blthoven, Netherlands
Tuovinen, J.P., Barrett, K. and Styve, H., 1994 EMEP/MSC-W, Report 1/94, Norwegian Met. Institute, Oslo.
UK Review group on acid rain, 1990 Warren Spring Laboratory, UK.

Velde, R.J. van, Faber, W.S., Katwijk, V.F. van, Kuylenstierna, J.C.I., Scholten, H.J., Thewessen, T.J.M., Verspuij, M., Zevenbergen, M., 1995 Report no. 712401001, RIVM, Bilthoven, the Netherlands.

UNITED KINGDOM: CONTINUOUS MONITORING

[1]FOWLER, D., [1]LEITH, I.D., [1]BINNIE, J., [1]CROSSLEY, A., [2]INGLIS, D.W.F.
[2]CHOULARTON, T.W., [2]GAY, M., [3]LONGHURST, J.W.S. AND [3]CONLAND, D.E.

[1]*Institute of Terrestrial Ecology, Edinburgh, Midlothian, U.K., EH26 0QB*
[2]*Department of Pure and Applied Physics, UMIST, PO Box 88, Manchester, U.K., M60 1QD*
[3]*Atmospheric Research and Information Centre, Manchester Metropolitan University, M1 5GD*

Abstract. Continuous monitoring of cloud and rain samples at three mountain sites in the UK has allowed consideration of the long term impact of the enhancement of the wet deposition of pollutants by orographic effects, specifically the scavenging of cap cloud droplets by rain falling from above (the seeder-feeder effects). The concentration of the major pollutant ions in the cloud water is related to the relative proximity of each site to marine and anthropogenic sources of aerosol. In general, the concentrations of major ions in precipitation at summit sites exceed those in precipitation to low ground nearby by 20% to 50%. Concentrations in orographic cloud exceed those in upwind rain by between a factor of five and ten. The results are consistent with seeder-feeder scavenging of hill cloud by falling precipitation in which the average concentration of ions in scavenged hill cloud exceed those in precipitation upwind by a factor of 1.7 to 2.3 for sulphate and nitrate respectively at Dunslair Heights and 1.5 to 1.8 for sulphate and nitrate at Holme Moss. The results suggest that the parameterisation of this relationship with scavenged feeder cloud water concentrations assumed to exceed those in seeder rain by a factor of two for the production of predictive maps of wet deposition in mountainous regions of the U.K. is satisfactory.

Keywords: wet deposition, orographic enhancement, seeder-feeder

1. Introduction

In upland areas of the UK the formation of mountain cap clouds during frontal rain events is common. This leads to the efficient scavenging of cap cloud droplets by the rain falling from aloft. This process is called the seeder-feeder process and was first identified by Bader and Roach (1977). The impact of this process on wet deposition was first discussed by Carruthers and Choularton (1984). The cap (feeder) cloud usually contains a considerably greater concentration of pollutant ions than the rain falling from above (seeder rain). There are two main reasons for this: Firstly the concentrations of ions in the atmospheric boundary layer tends to be greater than in the free troposphere where the seeder rain forms. This is due to strong surface sources of a range of pollutants associated with anthropogenic activity. When the boundary layer is forced to rise over a hill, cap clouds form on the atmospheric aerosol containing the soluble particulate material. Soluble trace gases dissolve in the cloud water contributing to the soluble mass to be scavenged and wet deposited. The second process concerns the rain production mechanism which usually occurs when the seeder-feeder process is operating. This process is the vapour growth of snowflakes in the frontal cloud which is inefficient at incorporating the dissolved particulate cloud droplets. These snow flakes melt to form rain as they descend to lower levels and then efficiently scavenge the cloud droplets in the feeder cloud by collision coalescence. Hence the seeder-feeder process is expected to enhance both the concentration of pollutants in rain and the volume of rain falling on upland terrain.

Much of the annual precipitation over upland areas in the UK occurs during the passage of frontal systems. Under these conditions the boundary layer is moist and the operation of the seeder-feeder process is favoured. Experiments at Great Dun Fell (Fowler *et al.*, 1988) and at Winter Hill (Inglis *et al.*, 1995) have confirmed the dominance of the seeder-feeder process in controlling the enhancement of pollutant deposition in rain over hills. However, the national network of rain collectors set up to monitor wet deposition (known as the secondary network, Devenish, 1986) does not include sites at high altitude because of the practical difficulties in operating sites in the uplands of the UK. Interpolating between low altitude sites to predict deposition at high altitude is inadequate since it does not take account of the seeder-feeder process. Dore et al. (1992) describe a method of revising the maps to include the seeder-feeder process. This involves the assumption (based on observations from Great Dun Fell) that the water scavenged from cap clouds has a pollutant concentration at least twice that of the rain which falls at low altitude. The revised maps are presented in RGAR (1990) for the period 1986-1988 and in CLAG (1994) for the period 1989 to 1991.

This paper describes work designed to test the method used to produce the revised maps against observations. Long term observations are used to investigate the relationship between feeder cloud and seeder rain at three sites of differing topography, geographical position and pollution climate.

2. Continuous Monitoring

The three mountain sites selected for continuous monitoring were Dunslair Heights in southern Scotland near Peebles, Great Dun Fell in Cumbria and Holme Moss in Yorkshire. Each site provided weekly samples of rain using bulk collectors and cloud water from passive collectors. These were analysed for the concentrations of the major ions. Continuous measurement of the major meteorological variables (rain, wind speed, wind direction air temperature) and the liquid water content of the cloud were also made.

2.1 HOLME MOSS

Holme Moss has an altitude of 550m above sea level (asl) and lies in the Peak District which comprises part of the Pennine Hills in northern England. To the west is the Mersey valley and to the east is west Yorkshire, both are heavily industrialised urban regions. Of the three sites Holme Moss is closest to major industrial and urban sources of pollutants. Also, for westerly wind directions it comprises part of the first upland terrain encountered by the boundary layer after crossing the coast.

Data shown here were collected between September 1993 and September 1994. Table 1 shows average concentrations of the major ions in rain and cloud water collected at the summit of Holme Moss. All averages for rain are precipitation weighted, those for cloud are not. For all ions the cloud is substantially more concentrated than the rain.

Estimates of the seeder rain concentration can be made from the network of rain collectors in Greater Manchester approximately 30 km west of Holme Moss (Longhurst et al., 1987). The concentrations of the major pollutant ions are taken from an average of five of the more rural network sites situated to the west of Manchester. These can be seen in Table 2. However, these data must be used with care. The strong spatial variability in the concentration of ions in rain in the network (Conlan and Longhurst, 1993) shows that the seeder rain composition is perturbed by proximity to strong sources of pollutants and other urban effects. The most locally pronounced of these appears to be connected to the production of aerosol containing calcium (Ca^{2+}) ions from building surfaces and quarrying work. The most important effects of this are to reduce the acidity of rain and to promote the rapid conversion of sulphur dioxide (SO_2) gas to sulphate (SO_4^{2-}) ions. Enhanced concentrations of Ca^{2+} and SO_4^{2-} and reduced acidity are evident in the Greater Manchester data when compared with the Holme Moss data and data from the secondary network RGAR (1990) which is comprised of rural sites. For this reason a seeder rain concentration of SO_4^{2-} ions is taken from the secondary network data as the sites in this network are unaffected by local pollution sources.

During the collection period there was a volume enhancement of 63% between the Greater Manchester network and Holme Moss. Knowing this and the increase in ion concentrations in rain, it is straightforward to calculate the concentration of ions in the

Table 1: Averaged concentrations of ions (µeq l^{-1}) in rain and cloud, Holme Moss Summit

	Cl$^-$	NO$_3^-$	SO$_4^{2-}$	Na$^+$	NH$_4^+$	K$^+$	Mg^{2+}	Ca^{2+}	H$^+$
Cloud	1762	349	528	1591	358	40	322	238	65
Rain	179	36	94	141	55	6	31	25	24

scavenged water. This, of course, assumes that the observed concentration enhancement is indeed due to the scavenging of cap cloud droplets. However this is justified by the findings of Inglis et al. (1995). During two month long experiments at another site in the southern Pennines deposition of anthropogenic ions was dominated by frontal rain events and the seeder-feeder process controlled deposition enhancement during these events. This work is also described briefly by Inglis et al. (1996).

The ratio of the concentration of ions in the scavenged water to the concentration of ions in the seeder rain is shown in Table 2. These results may be directly compared with the assumption incorporated in the revised maps which is that the concentration of ions in the scavenged water is twice that in the seeder rain. The calculation for SO$_4^{2-}$ must be regarded as approximate since the concentration of SO$_4^{2-}$ ions in seeder rain taken from the secondary network did not cover the same time period as the Holme Moss sampling. Although it was felt that the best option was to estimate the concentration of NH$_4^+$ and NO$_3^-$ from the Greater Manchester network data, urban enhancement effects still may result in the ratios shown in table 2 being somewhat under-estimated for these ions.

Table 2: Concentration of ions in seeder rain (µeq l^{-1}) and ratio of the concentration of ions in the scavenged water to that in the seeder rain, Holme Moss.

	Cl$^-$	NO$_3^-$	SO$_4^{2-}$	NH$_4^+$
Seeder rain	143	27	80	40
Scavenged water / seeder rain	1.7	1.8	1.5	2.1

2.2 DUNSLAIR HEIGHTS

Dunslair Heights has an altitude of 600m asl and lies in the Southern Uplands of Scotland. This is a rural region with no major sources of pollutants, the industrial region of central Scotland lies approximately 50km to the north and north west. Table 3 shows averaged concentrations of rain and cloud samples from the summit and rain samples from a nearby low altitude site (Ven Law). The concentration of all the major ions in cloud is lower at Dunslair Heights than at Holme Moss. The data shown here are from the period April 1992 to April 1993.

The volume enhancement for Dunslair Heights is 53%. This allows the calculation of the concentration of ions in the scavenged water. The ratio of the concentration of ions in the scavenged water to that in the seeder rain is shown in Table 4. The concentration of NH$_4^+$ ions in rain was observed to be lower at the summit than at the low altitude site resulting in the ratio of concentrations being less than one. This is probably due to contamination at the

Table 3: Average concentrations of ions (µeq l^{-1}) in cloud and rain from Dunslair Heights and rain from Ven Law.

	Cl$^-$	NO$_3^-$	SO$_4^{2-}$	Na$^+$	NH$_4^+$	K$^+$	Mg^{2+}	Ca^{2+}	H$^+$
Cloud	976	267	208	881	291	29	198	108	76
Summit Rain	144	28	61	131	30	21	22	22	21
Ven Law Rain	78	20	50	83	32	8	31	25	15

low altitude collector by NH$_3$ from a farm nearby. For the other ions the ratios are greater than for Holme Moss. Also shown is the ratio of the concentration of ions in the directly sampled cap cloud to that estimated to be in the scavenged water. For all ions, this ratio is greater than unity. This is because the cloud collector samples cloud close to cloud base where the liquid water content is relatively low (50-100 mgm^{-3}) and hence the concentration of ions in the cloud water is large. The seeder rain however, scavenges water through the whole depth of the cap cloud. The relationship between these concentrations will depend to a large degree on the physical characteristics of the hill.

Table 4: Ratios of concentrations of ions in scavenged water, seeder rain and cloud water, Dunslair Heights

Ratio	Cl$^-$	NO$_3^+$	SO$_4^{2-}$	NH$_4^+$
scavenged water / seeder rain	3.5	2.3	1.7	0.8
cloud water / scavenged water	3.6	6	2.5	11.4

2.3 GREAT DUN FELL

Great Dun Fell has an altitude of 850m asl and lies in the north Pennine ridge. To the west and south is the Eden Valley, an area of intensive agricultural activity. Further to the south west, the Cumbrian mountains of the Lake District lie between Great Dun Fell and the sea for the most common rain bearing wind direction. Rain was collected at the foot of the hill and cloud was collected at the summit. Table 5 shows the averaged concentrations. Concentrations of NO$_3^-$ and NH$_4^+$ in cloud are greater than at Holme Moss while concentrations of Cl$^-$ and SO$_4^{2-}$ are lower.

Rain was not collected at the summit because of considerable sampling difficulties in the very windy, exposed and cold conditions at this site. The composition of the scavenged water must therefore be estimated by an alternative method. The average liquid water content of the cap cloud sampled at the summit of Great Dun Fell is expected to be higher than that at the other sites due to its greater altitude. Indeed preliminary analysis of long term monitoring of the liquid water content suggests that the average value is in the region of 250 mg m^{-3} compared to 100 mg m^{-3} at Dunslair Heights. This is broadly consistent with the difference in the altitude of the hills.

On average at Dunslair Heights the concentration of ions in the directly sampled cloud water was approximately four times that in the scavenged water (excluding the figure for NH$_4^+$). Since we believe the liquid water content of the sampled cloud at Great Dun Fell to be greater than at Dunslair Heights by a factor of 2.5 it is reasonable to expect this ratio to be reduced from 4.0 to 1.6. Thus the relationship between the scavenged water and the seeder rain at Great Dun Fell can be calculated and is shown in Table 5. These values are somewhat higher than those for Holme Moss although the technique used to produce the Great Dun Fell values must be regarded as approximate especially in the case of NO_3^-.

Table 5: Concentrations of ions (μeq l^{-1}) in rain and cloud samples, and estimated ratio of the concentration of scavenged water to concentration of seeder rain at Great Dun Fell

	Cl^-	NO_3^-	SO_4^{2-}	Na^+	NH_4^+	K^+	Mg^{2+}	Ca^{2+}	H^+
Cloud	917	372	424	688	396	25	184	244	65
Rain	157	27	84	122	77	9	28	30	20
scavenged water / seeder rain	3.6	8.6	3.2	-	3.2	-	-	-	-

3. Discussion and Conclusions

Variations in the concentrations of ions in hill cloud between the three sites are clear and are in part linked to regional pollution climates. For example, the largest concentration of SO_4^{2-} and Cl^- ions in cloud were observed at Holme Moss the site closest to major sources of industrial pollutants and most directly exposed to marine air flows. Great Dun Fell cloud exhibited the largest concentration of NH_4^+ ions as a result of its proximity to regions of intensive agricultural activity. Dunslair Heights is the site most remote from sources of anthropogenic and marine ions and consequently the concentration of these ions is the lowest observed of the three sites.

The absolute values for concentrations of major ions in precipitation are consistent with the national concentration field. This shows a marked increase in the south Pennines, due to the proximity of major source areas and fewer directions from which uncontaminated precipitation may arrive. There are indications of effects of purely local sources (<20km) on the measurements, but these effects cannot be quantified from the current dataset.

If the ratio of the concentration of ions in scavenged water to that in seeder rain is averaged for the major pollutant ions at all three sites the mean figure is 3.0 with a standard deviation of 2.0. Although this average must be regarded as provisional it suggests that the value of 2.0 used for this relationship in the national mapping is a reasonable and in most cases conservative estimate.

Further work on meteorological data from the three sites is required. This will allow the long term monitoring to be linked to insights gained from other short term intensive experiments designed to examine the orographic enhancement of wet deposition in detail.

Acknowledgements

The work is supported by the UK Department of the Environment under contract EPG 1/3/30. The authors would also like to thank the staff at the Holme Moss BBC transmitter station for access to the site.

References

Bader, M.J. and Roach, W.T.: 1977, "Orographic rainfall in warm sectors of depressions"., Quarterly Journal of the Royal Meteorological Society, **103**, 269-280.

Carruthers, D.J. and Choularton, T.W.: 1983, "A model of the seeder-feeder mechanism of orographic rain including stratification and wind drift effects", Quarterly Journal of the Royal Meteorological Society, **109**, 575-588.

CLAG: 1994, "*Critical Loads of Acidity in the United Kingdom*", Critical Loads Advisory Group Summary Report, Ed. Pitcairn, C.E.R., ISBN 1 987393 20 1.

Conlan, D.E. and Longhurst, J.W.S.: 1993, "Spatial variability in urban acid deposition, 1990: Results from the Greater Manchester Acid Deposition Survey (GMADS) network in the UK", The Science of the Total Environment, **128**, 101-120.

Devenish, M.: 1986, "*The UK Precipitation Composition Monitoring Networks*", Report LR584(AP), Warren Spring Laboratory, Stevenage.

Dore, A.J., Choularton, T.W. and Fowler, D.: 1988, "An improved wet deposition map of the United Kingdom incorporating the seeder-feeder effect over mountainous terrain", Atmospheric Environment, **26A**, 1375-1381.

Fowler, D., Cape, J.N., Leith, I.D., Choularton, T.W., Gay, M.J. and Jones, A.: 1988, "The influence of altitude on rainfall composition at Great Dun Fell", Atmospheric Environment, **22**, 1355-1362.

Inglis, D.W.F., Choularton, T.W. and Wicks, A.J.: 1995, "The effect of orography on wet deposition in an industrial area", Quarterly Journal of the Royal Meteorological Society, in press.

Inglis, D.W.F., Choularton, T.W. and Wicks, A.J.: 1996, "Orographic enhancement of wet deposition in the UK: Case Studies and Modelling", A paper submitted to this volume.

Longhurst, J.W.S., Gee, D.R., Lee, G.S. and Green, S.E.: 1987, "The establishment of an Urban Acid Deposition Monitoring Network", The Environmentalist, **11**, 299-307.

RGAR: 1990. "*Acid Deposition in the United Kingdom 1986-1988*", Third Report on the United Kingdom Review Group on Acid Rain, ISBN 0 85624 650 6.

THE APPLICATION OF ^{210}Pb INVENTORIES IN SOIL TO MEASURE LONG-TERM AVERAGE WET DEPOSITION OF POLLUTANTS IN COMPLEX TERRAIN

[1]D. FOWLER, [2]R. MOURNE AND [2]D. BRANFORD

[1]Institute of Terrestrial Ecology, Bush Estate, Edinburgh
[2]Department of Physics, University of Edinburgh

Abstract. The radionuclide ^{210}Pb derived from gaseous ^{222}Rn is present in particle form in the atmosphere attached to the same aerosols which contain the bulk of the pollutant sulphur and nitrogen. When scavenged from the atmosphere by precipitation, the ^{210}Pb is readily attached to organic matter in the surface horizons of soil. The inventory of ^{210}Pb in soil can be used to measure the spatial variation in wet (or cloud) deposition within a region due to orography or land use, averaged over several decades (half life of ^{210}Pb is 22.3 years). Measurements of soil ^{210}Pb inventories along a transect through complex terrain in north Britain were made to quantify the orographic enhancement of wet deposition, at Great Dun Fell in Cumbria. At the hill summit (~800m asl) precipitation of approximately 2000 mm year^{-1} exceeds that on the low ground upwind by a factor of 2.0. The inventory of ^{210}Pb increases along the same transect by a factor of 3.3, due to seeder-feeder scavenging of orographic cloud. The measurements show that the average ratio of concentrations in scavenged orographic cloud to rain upwind of the hills is 2.2. These data are entirely consistent with the studies of the variation in major ion concentration with altitude at Great Dun Fell and elsewhere.

The modelling procedures developed to provide estimates of wet deposition throughout the UK uplands are shown to be consistent with these new field data.

Key words: ^{210}Pb, wet deposition, organic matter, soil complex terrain

1. Introduction

The deposition of sulphur and nitrogen throughout Europe and North America has been an important focus of scientific and political interest since the Stockholm conference on the Human Environment in 1972 (Sweden, 1972).

Central to the debate were the measurements of wet deposition of SO_4^{2-} and acidification of freshwaters and soil across the large areas of Canada and Southern Scandinavia in which effects were detected. The monitoring networks, both national and international rely on relatively simple methodology to quantify the input of major ions by wet deposition. In general, the measurement approach has been to quantify the spatial variability in the precipitation weighted annual mean concentration for the individual ions in precipitation and obtain wet deposition as the product of precipitation amount from large meteorological networks of collectors and the concentration fields of individual ions.

The methods have been adequate to demonstrate the regional magnitude of inputs and obtain trends with time. However, in complex terrain there has always been a difficulty in obtaining an adequate direct measure of wet deposition. The problem is particularly acute on mountains where exposure leads to severe problems in under-sampling precipitation, and in the operation and maintenance of equipment. These areas are also some of the most sensitive to acidic input in north and west Britain and in the uplands of Scandinavia.

The work described here provides methods of measuring the local variability in wet deposition of aerosol phase pollutants. The method averages over many thousands of precipitation events and may be used to validate wet deposition enhancement estimates in complex terrain

2. Methods

The method is based on the spatial variability of the inventory of ^{210}Pb in undisturbed soils. The source of ^{210}Pb is ^{222}Rn in the decay chain from ^{238}U to ^{206}Pb, the latter being the stable end product. The ^{210}Pb has a half life of 22.3 years and is quickly formed from short-lived Po daughters of ^{222}Rn the latter having a half life of 2.8 days. The gaseous ^{222}Rn emitted from soils to the atmosphere decays through Po to ^{210}Pb which is present as a particle and which rapidly becomes attached to sub-micron aerosol (Turekian et al., 1977). The mass median diameter of ^{210}Pb in the atmosphere has been estimated by Turekian to be in the range 0.3 to 0.4 μm. Measurements by Knuth et al. (1983) show that 85% of ^{210}Pb is associated with aerosols <1μm diameter at a site in New York State. The ^{210}Pb therefore 'tags' the aerosols which dominate the polluted atmosphere over Europe and North America. The aerosols are scavenged by cloud and precipitation process and deposited at the ground along with the ^{210}Pb. Dry deposition of ^{210}Pb can be shown to be a small contribution (<10%) to the total input, as a consequence of the small size of the ^{210}Pb containing aerosols and their small deposition velocities.

Within soils the Pb readily becomes attached to organic matter and is not readily leached. Lewis (1977) showed that organic matter in soils captures and retains almost 100% of the ^{210}Pb, and that erosion of the organic matter was the only significant removal process. He estimated the mean residence time of ^{210}Pb in organic soils to be of the order of 2000 y (2 orders of magnitude longer than its half life).

In principle therefore, the inventory of ^{210}Pb in soil provides a relative measure of total wet and dry deposition of aerosol which may be applied in the field to quantify local gradients in wet deposition due to land use or orographic effects. An important correction must be made to the inventory to compensate for any ^{210}Pb which is produced in situ. This fraction, termed the supported fraction may be calculated by measuring the parent ^{214}Pb activity in soil.

The ^{210}Pb inventory (I), corrected for the supported fraction then provides a measure of the atmospheric flux (at steady state) from the decay constant (λ) for ^{210}Pb according to

$$F_{Pb} = \lambda \times I$$
$$\text{Bq m}^{-2}\text{ y}^{-1} \qquad \text{y}^{-1} \qquad \text{Bq m}^{-2}$$

The method has been applied by Graustein and Turekian (1983, 1986, 1989), to estimate atmospheric inputs of atmospheric aerosols to a range of ecosystems.

The measurements reported here have been designed to quantify orographic enhancement of aerosols at a site with a large gradient in precipitation (from 1000mm to 2000mm), as a consequence of a ridge of hills with mean summit heights of about 800m downwind of a valley (the Eden Valley in Cumbria) with a mean altitude of 200m. The ridge of hills at Great Dun Fell experience very large wet deposition of SO_4^{2-}, NO_3^- and the summit of the hill is in cloud for some part of 230 days annually (Choularton et al., 1988). The orographic enhancement of the major ions in precipitation and wet deposition of pollutants by seeder-feeder scavenging has been shown to lead to inputs at the hill summit, a factor of 4 larger than in the upwind valley in short term studies (Fowler et al., 1988; Choularton et al., 1988). This site was therefore selected to validate the short-term precipitation chemistry measurements using the ^{210}Pb inventory methods. The advantage of this method is that the measurements average atmospheric inputs over typically 30 years (mean nuclear lifetime of ^{210}Pb is $\tau = 32.2$ y) which represent approximately 6000 precipitation events, and are entirely independent of the assumptions on which the precipitation chemistry studies were based. Furthermore, the collection problems of cloud water and precipitation on windy hill summits are eliminated as it is the inventory within the soil that is sampled.

2.1 SITES

The sampling sites at Great Dun Fell in Cumbria (north-west England) lie along the direction of the prevailing south-west winds and at right angles to the ridge. Samples were taken at 23 locations along the 14km transect of the ridge as indicated in Figure 1.

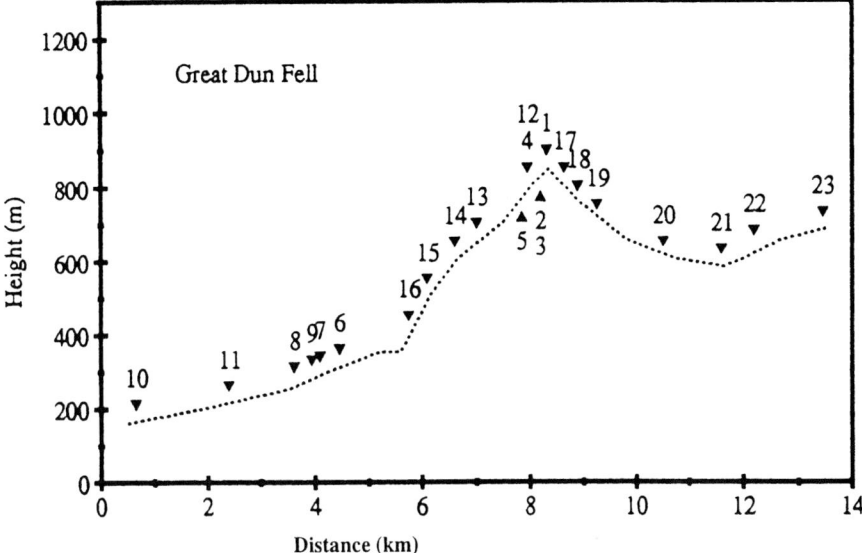

Figure 1: Hill profile and location of sampling sites for Great Dun Fell.

At each site, which is known to have been free of disturbance for at least 40 years, 5 soil cores each 10.3cm in diameter between the surface (including vegetation) and 20cm were taken at 3 locations. The cores were sub-divided into 0-1cm, 1-5cm, 5-10cm, 10-15 and 15-20cm sections to obtain the depth profile of radioactivity present within the soil profile.

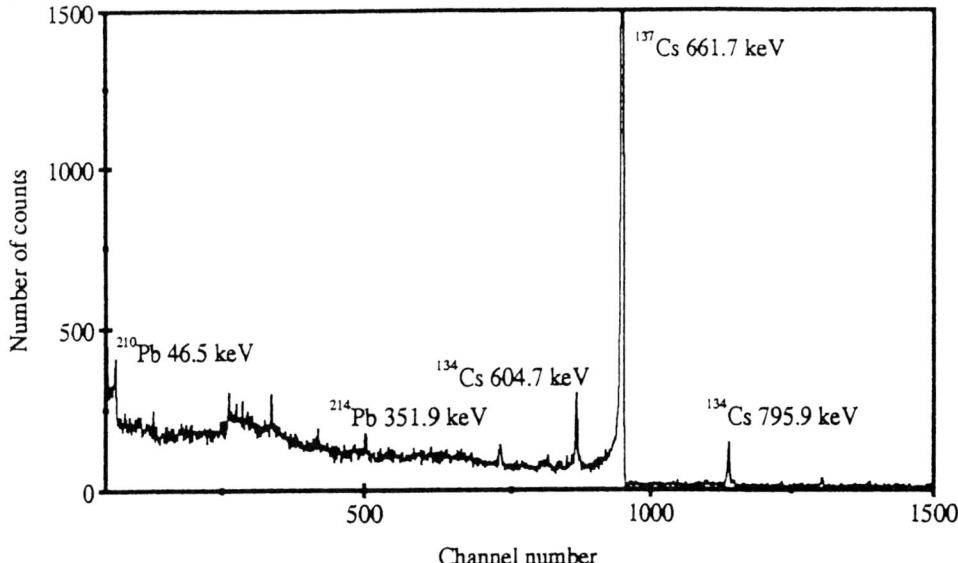

Figure 2: A γ ray spectrum obtained with a soil sample and the 19% Ge (Li) Detector.

Soil samples and vegetation were weighed, dried at 80°C, reweighed and ground after removing large stones, prior to packing into containers for the measurement of gamma-ray activity of the samples using Ge (Li) detectors. The γ ray spectrum of the individual soil samples showed the presence of a range of isotopes of interest including ^{210}Pb at 46.5k eV, ^{214}Pb at 351.9 k eV, ^{134}Cs (from the Chernobyl reactor accident) at 604.7k eV and ^{137}Cs (from both Chernobyl and atmospheric tests of atomic weapons during the 1960s). A typical spectrum is shown in Figure 2.

The activity of the radionuclide is given by

$$A = \frac{N_p}{T} \times \frac{1}{E_{full} \times b_\gamma} \times S$$

where N_p is the number of counts in the full energy peak, T is the counting period in seconds, E_{full} is the full energy peak efficiency of the detector for the γ ray of interest and bγ is the branching ratio if the radionuclide for the γ ray used, finally S is the correction factor for self adsorption within the sample.

Averages of the total ^{210}Pb inventories of the five cores taken at each site are shown in Figure 3, with uncertainty for each sampling site derived from the counting statistics of individual sample, the variations in the inventories at the site and the efficiency of detection

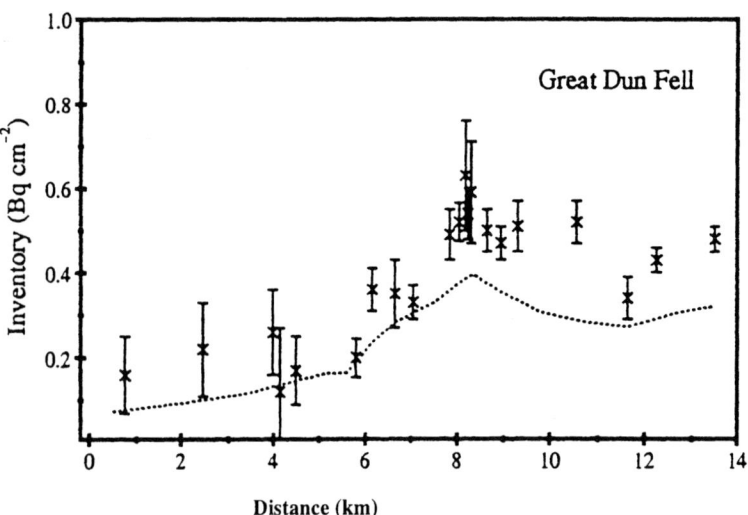

Figure 3:   Inventory of atmospheric ^{210}Pb (atmos) (Bq cm^{-2}) at Great Dun Fell.

The dotted line indicates the hill profile between the Eden Valley and Round hill (some 4km north east of Great Dun Fell summit).

The inventory of ^{210}Pb increases from the upwind valley site at 1.7 Bq cm^{-2} to 5.15 Bq cm^{-2} at Great Dun Fell summit, an increase of a factor of 3.3, whereas the rainfall increase over the same height range is only a factor of 2. There is clearly noise within the data set (Figure 3), especially at the hill summit, where some sampling sites indicate an inventory in excess of 6

Bq cm^{-2} almost a factor of 4 larger than in the Eden Valley. The results summarized in Table 1 may be used to compare the processes leading to increased wet deposition at the high elevations of Great Dun Fell with the measurements and modelling of wet deposition of SO_4^{2-}, NO_3^-, NH_4^+ by Fowler et al. (1988) and Choularton et al. (1988) respectively.

Table 1: Characteristics of the Great Dun Fell wet deposition data and ^{210}Pb inventory.

Altitude range (m)	160-840
Rainfall increase from valley to summit	2.0
^{210}Pb inventory increase between valley and summit	3.3
Concentration ratio of ^{210}Pb in precipitation summit/valley	1.7
Concentration ratios scavenged feeder cloud/seeder rain	2.2

The seeder-feeder scavenging process has been shown to modify the wet scavenging of pollutants on hills, and as a consequence has been used to modify wet deposition maps for S, N and acidity (Dore et al. 1992). The data presented here for ^{210}Pb, may be compared with the data used to modify the wet deposition maps. The main assumption made in the regional extrapolation was that the concentration of pollutants in the scavenged feeder cloud exceed the concentrations in the seeder rain (Figure 4) by a factor of 2. The increase in the ^{210}Pb inventory between the valley site and the land above 800m asl at Great Dun Fell may be used to estimate the concentration ratio (feeder cloud/seeder rain) for ^{210}Pb using the long term (30 yr) mean annual precipitation at the valley (1000mm) and hill summit (2000mm). The value

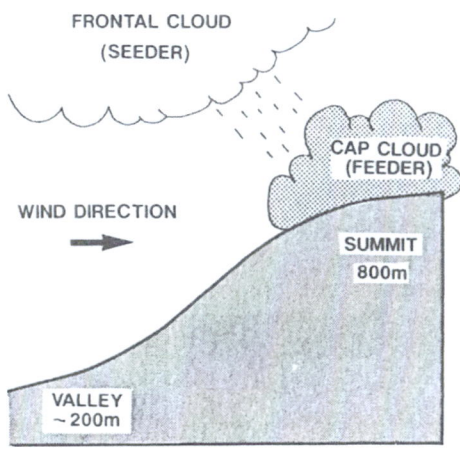

Figure 4: The main features of the seeder-feeder mechanism for orographic enhancement of precipitation and wet deposition.

obtained (2.2) is close to the value applied generally in the UK (2.0) and therefore strongly supports the procedure described by Dore *et al.* (1992). It is of course necessary to validate the procedure using ^{210}Pb for a range of other sites, and for more complex terrain than the simple ridge at Great Dun Fell, but the technique appears particularly valuable in quantifying aerosol and wet deposition processes at sites where conventional methods are not applicable.

Acknowledgements

This work was supported by a NERC CASE studentship held by R Mourne. The work forms a part of the Air Quality Research programme of the UK Department of the Environment under contract EPG 1/3/31

References

Choularton, T.W., Gay, M.J., Jones, A., Fowler, D., Cape, J.N. and Leith, I.D.: 1988, "The influence of altitude on wet deposition. Comparison between field measurements at Great Dun Fell and the predictions of a seeder-feeder model". *Atmos. Environ.* **22,** 1363-1371.

Dore, A.J., Choularton, T.W. and Fowler, D.: 1992, "An improved wet deposition map of the United Kingdom incorporating the seeder-feeder effect over mountainous terrain". *Atmos. Environ.* **26A(8),** 1375-1381.

Fowler, D., Cape, J.N., Leith, I.D., Choularton, T.W., Gay, M.J. and Jones, A.: 1988, "The influence of altitude on rainfall at Great Dun Fell". *Atmos. Environ.* **22,** 1355-1362.

Graustein, W.C. and Turekian, K.K.: 1983, "Pb^{210} as a tracer of the deposition of sub-micrometer aerosols". In: *Precipitation Scavbenging, Dry Deposition and Resuspension,* (Ed. Pruppacher, H.R., Semonin, R.G. & Slinn, W.G.N.) Elsevier, New York, 1315-1324.

Graustein, W.C. and Turekian, K.K.: 1986, "^{210}Pb and ^{137}Cs in air and soils measure the rate and vertical profile of aerosol scavenging". *J. Geophys. Res.* **91(D13),** 14355-14366.

Graustein, W.C. and Turekian, K.K.: 1989, "The effects of forests and topography on the deposition of sub-micrometer aerosols measured by lead^{210} and cesium^{137} in soils". *Agric. For. Meteor.* **47,** 199-220.

Knuth, R.H., Knutson, E.O., Feely, H.W. and Volchok, H.L.: 1983, "Size distribution of atmospheric Pb and Pb^{210} in rural New Jersey: Implications for wet and dry deposition". In: *Precipitation Scavenging, Dry Deposition and Resuspension,* (Ed. Pruppacher, H.R., Semonin, R.G. & Slinn, W.G.N.), Elseview, New York 1325-1334.

Sweden: 1992, *Air Pollution across national boundaries. The impact on the environment of sulfur in air and precipitation.* Sweden's case study for the United Nations Conference on the human environment.

Turekian, K.K., Nozaki, Y. and Benninger, L.K.: 1977, "Geochemistry of atmospheric radon and radon products". *Ann. Rev. Earth Planetary Sci.* **5,** 277-255.

Orographic Enhancement of Wet Deposition in the United Kingdom: Case Studies and Modelling

INGLIS D.W.F.[1], CHOULARTON T.W.[1], WICKS A.J.[1], FOWLER D.[2], LEITH I.D.[2], WERKMAN B[3] and BINNIE J.[2]

[1] Department of Physics, U.M.I.S.T., Manchester, P.O. Box 88, M60 1QD, U.K.
[2] Institute of Terrestrial Ecology, Bush Estate, Penicuik, Midlothian, EH26 0QB, U.K.

Abstract. Two field experiments to observe the detailed response of wet deposition to orography in a polluted environment are reported. Rain events were classed as frontal, convective or mixed on the basis of meteorological data. Analysis of the deposition enhancement and cap cloud composition confirmed that for the frontal events the seeder-feeder effect (scavenging of cap cloud by rain drops) dominates. The greater concentration of ions in the water scavenged from the cap cloud than in the rain means that deposition is enhanced for all ions. For marine ions the scavenged water was found to be between five and six times as concentrated as the rain and for anthropogenically produced ions it was about twice as concentrated.

A computational model of rainfall incorporating the seeder-feeder effect has been broadly successful in predicting enhancement although some details of the observed pattern remain to be explained.

Convective events were only important in the deposition of marine ions although this may not be the case in the summer months. Convective events were found not to be subject to the seeder-feeder effect.

Keywords. Wet Deposition, Orographic Rainfall, Pollution, Seeder-feeder Effect.

1. Introduction

The experiments described in this paper form part of a larger study intended to improve parameterisations of the wet deposition of pollutants to mountainous regions of the U.K. (Fowler *et al.* 1995). This study focuses on the importance of the seeder-feeder effect, first described by Bader and Roach (1977). Rain falling from aloft (seeder rain) washes out the small droplets from cap clouds (feeder clouds) which have formed as the moist boundary layer rises over mountainous terrain. The concentrations of pollutant ions in the cap cloud are often high leading to enhanced deposition of these ions.

Several studies have observed the increase of pollutant concentration in rain with altitude at mountain sites in the U.K. (e.g. Dore *et al.* 1992) and attributed it to the seeder-feeder effect. Questions remain, however, about the detailed response of the seeder-feeder effect to terrain and the local pollution climate. In this paper we describe results from detailed field experiments in a polluted environment along with modelling studies designed to address these issues.

2. The Winter Hill Experiments

Winter Hill summit (British National Grid reference SD 658 149) has an altitude of 456 m above sea level (a.s.l.) and lies on the northern edge of the industrial Mersey valley which typically has an altitude of 100 m a.s.l.. The Greater Manchester conurbation lies 30 km to the south east, Liverpool is 40 km to the south west and the Irish Sea coast is 40 km to the west (see Figure 1).

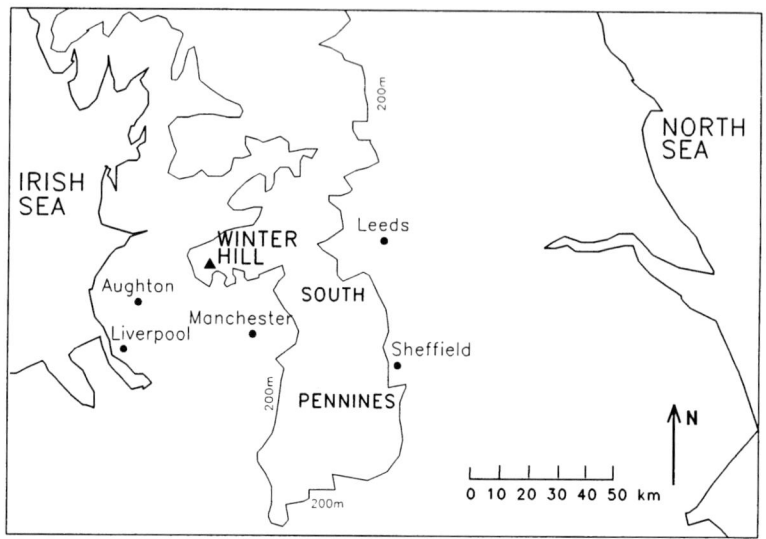

Fig. 1 Map of northern England showing the position of Winter Hill and major cities. The extent of the Pennine mountains is shown by the 200m contour. The meteorological station (Aughton) is also shown.

Although Winter Hill is not very high it is typical of the western Pennines. For the prevailing, rain-bearing wind direction it induces the first significant orographic uplift of the heavily polluted boundary-layer. The wet deposition profile of Winter Hill was expected to be typical of a medium size hill adjacent to an industrial region in the U.K..

The experiments ran during March and November 1992. Figure 2 shows the position of the instrumented sites. Rain collectors and passive cloud collectors were deployed at these sites along with automatic weather stations at sites 1 and 8. Rain and cloud samples were collected for chemical analysis on a daily basis. Further meteorological information was available from the Meteorological Office station at Aughton (Figure 1). Inglis et al.(1995) describe the experimental layout and methods in more detail.

3. Results

A total of nineteen rainfall events were analysed from the two experiments at Winter Hill. It is possible to separate these into two main categories in terms of synoptic meteorology. These are: rain events caused by frontal systems and those caused by convectively formed clouds.

The boundary layer through which frontal rain falls is usually very moist, favouring the formation of cap clouds over high terrain and hence operation of the seeder-feeder mechanism. The boundary layer in convective conditions, however, may have a low humidity making the formation of hill cap clouds unlikely. Thus, the altitude dependant deposition pattern may be expected to be characteristically different for the two types of event.

The distinction between the two types of rainfall event is made on the basis of automatic weather station data, soundings from Aughton (Figure 1) and surface charts. Rainfall events were classified as convective if two conditions were met. These are: the

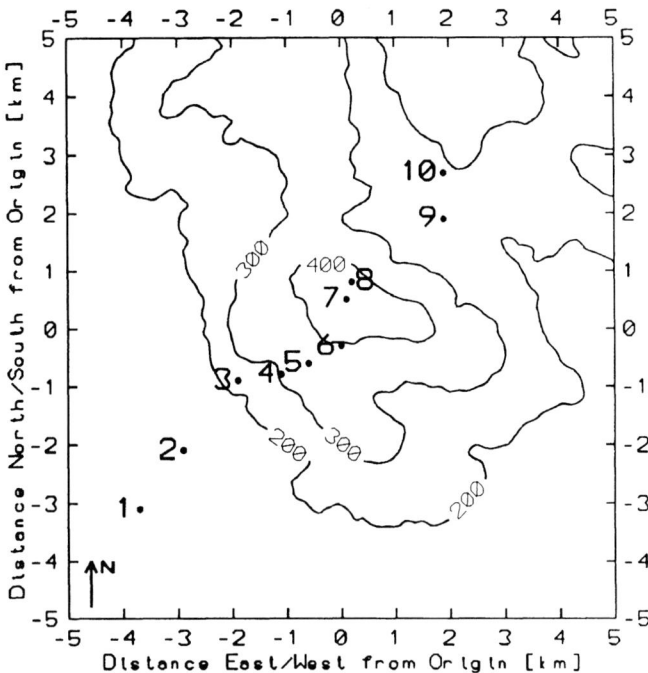

Fig. 2 Contour plot of Winter Hill terrain. The position and number of collector sites are shown in bold. The contours show altitude above sea level in metres.

absence of a weather front (warm, cold or occluded) near the site close to the time when rainfall was recorded, and the presence of a layer of potentially unstable air near the surface at least 1km deep. If one of these conditions is not met the event is classified as frontal. Events which appear to be comprised of significant amounts of convective and frontal rain are classified as mixed.

The chemical composition of the frontal rain is markedly different to that for the convective rain. Table I shows ion balances (the contribution of each ion to the balance is shown as a percentage) for the frontal and convective rain which falls in the valley and for sea water (Weast 1978). The balance for convective rain matches that for sea water very closely indicating that the convective rain events were formed almost exclusively on marine aerosol and as such are not important in the deposition of anthropogenic ions. However, the frontal rain contains substantial quantities of pollutant ions. The reasons for this difference and its implications are discussed in detail in Inglis et al.(1995).

3.1. Orographic Effects on Wet Deposition at Winter Hill

Figure 3 shows the modification of deposition of Nitrate (NO_3) ions. There is a marked tendency for total deposition in frontal events to increase with altitude although this is not monotonic (site 6). There is little altitude dependant change for the convective cases. The mixed cases show an intermediate deposition enhancement. We can express the change in rain volume as the ratio of rain volume at high altitude sites (sites 7 and 8) to the rain volume at low altitude sites (sites 1 and 2). This results in values of 1.23, 0.98 and 0.38 for the for frontal, mixed and convective rain event types respectively.

difference between the pollution climate in the boundary layer where the cap cloud forms and the free troposphere where the seeder rain forms. For instance, the boundary layer local to Winter Hill is abundant in marine aerosol which is incorporated into the cap cloud. This is scavenged by the seeder rain which having formed under different conditions contains far fewer marine ions. This relationship can vary ion to ion and here we see a different relationship for ions of anthropogenic origin, their relative abundance in the cap cloud compared to the seeder rain is less pronounced than for the marine ions.

Table II $\quad \dfrac{\text{Concentration of Scavenged Water}}{\text{Concentration of Seeder Rain}}$

Cl^-	NO_3^-	SO_4^{2-}	Na^+	NH_4^+	K^+	Mg^{2+}	Ca^{2+}	H^+
5.13	2.74	1.88	5.80	2.20	1.24	5.30	1.64	0.77

3.2 Modelling Orographic Enhancement

The rainfall model Rainstar, which is based on the model of Carruthers and Choularton (1983), treats the boundary layer as eight layers which flow over the terrain as dictated by a computational airflow model (Carruthers and Hunt 1990). Cap clouds condense and evaporate according to the vertical displacement of the layer. Raindrops which fall through the layers are subject to drift, evaporation, and growth by accretion with cap cloud droplets. The model is initialised with meteorological parameters and an aerosol loading. The rain volume and pollutant concentration can be predicted at any point in the model domain.

Figure 4 shows a typical comparison between the observed and modelled concentration of NO_3 in rain during a frontal rain event at Winter Hill. The increase in concentration between the low altitude sites and the high altitude sites is broadly predicted, in this case the concentration is underestimated by about 5% on average. However, much of the detail in the observed concentrations is not reproduced. This is especially true in the lee of high terrain where the model often over-predicts deposition. This issue is currently being addressed with further modelling and experimental work.

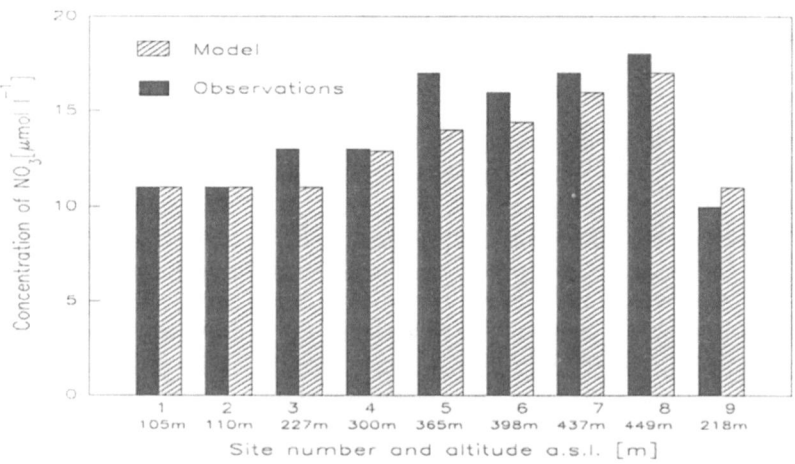

Fig. 4   Comparison between modelled rain concentrations of NO_3 and observed concentrations.

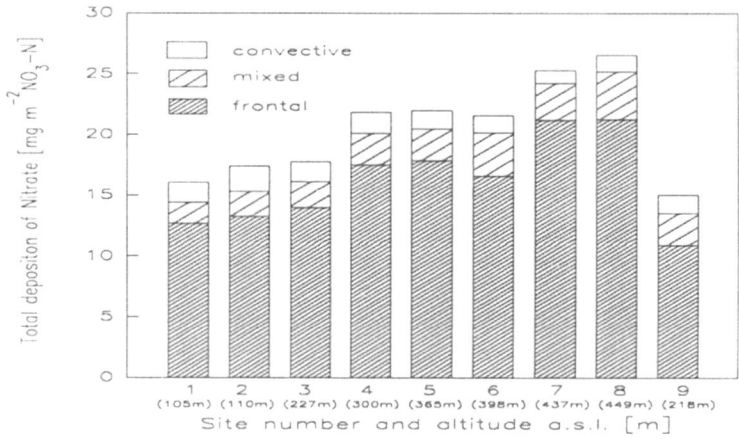

Fig. 3   Stacked bar chart showing total deposition of NO_3^- [mg N m^{-2}] in frontal, convective and mixed rain events.

If we assume that the deposition enhancement for the frontal events is caused by the seeder-feeder effect then it is straightforward to calculate the ionic composition of the "added" water from the observed enhancement to deposition and volume. This is shown in Table I. At sites 7 and 8 the cap cloud was sampled directly. The ion balance for this cloud water is also shown in Table I and can be seen to agree closely with the ion balance for the "added" water. This is very strong evidence that the "added" water was indeed scavenged from the cap cloud and that the seeder-feeder effect is responsible for the deposition enhancement in frontal rain at this site.

Table I Ion balances for sea water, convective and frontal rain, "added water" and cloud water at Winter Hill

	sea water	convective rain	frontal rain	"added" water	cloud water
Cl^-	45.2%	44.4%	23.3%	36.1%	38.7%
NO_3^-	-	0.4%	4.1%	3.4%	3.3%
SO_4^{2-}	4.8%	6.8%	22.8%	12.9%	12.7%
anion total	50.0%	51.6%	50.2%	52.4%	54.7%
Na^+	38.8%	33.8%	16.2%	28.4%	27.3%
NH_4^+	-	3.4%	13.7%	9.1%	7.9%
K^+	0.8%	1.1%	1.5%	0.6%	0.8%
Mg^{2+}	8.8%	6.9%	3.0%	4.8%	5.5%
Ca^{2+}	1.7%	2.7%	4.5%	2.2%	2.9%
H^+	-	0.5%	10.9%	2.5%	0.9%
cation total	50%	48.4%	49.8%	47.6%	45.3%

Table II shows the mean ratio of the concentration of the water scavenged during frontal rain events to the concentration of the frontal valley rain (which approximates to the seeder rain) for each ion. This relationship is important in terms of modelling the seeder-feeder effect (Fowler et al. 1995). Ions with a primarily marine source are between five and six times more concentrated in the scavenged water than in the seeder rain whereas primarily anthropogenic ions are roughly twice as concentrated. This is caused by the

4. Conclusion

Separation of the rain events at Winter Hill into frontal and convective types has been effective in explaining the deposition pattern. It has been shown that the seeder-feeder effect dominates the deposition enhancement in frontal rain events. The enhancement rate is substantially greater for marine ions than for pollutant ions although this relationship is expected to be strongly site dependant. A computational model of rainfall incorporating the seeder-feeder effect has been broadly successful in predicting enhancement although some details of the observed pattern remain to be explained.

During these experiments convective rain events are found to be important in the deposition of marine ions only. This may not be the case in the summer months when convection over land is stronger. Convective deposition is found not to be subject to the seeder-feeder effect.

Acknowledgements

The authors would like to thank the land owners in the Winter Hill area for their co-operation during the field experiments of 1992. We would also like to thank Mr. Martin Wells for helping with sample collection and Mrs. Anna Haley for the chemical analysis. This work was carried out under contract to the Department of the Environment.

References

Bader M.J. and Roach W.T.: 1977, "Orographic Rainfall in Warm Sectors of Depressions", Quarterly Journal of the Royal Meteorological Society, **103**, 269-280.

Carruthers D.J. and Choularton T.W.: 1983, "A model of the seeder-feeder mechanism of orographic rain including stratification and wind drift effects", Quarterly Journal of the Royal Meteorological Society, **109**, 575-588.

Carruthers D.J. and Hunt J.C.R.: 1990, "Fluid Mechanics of Air Over Hills: Turbulence, Fluxes, and Waves in the Boundary Layer", American Met. Soc., Meteorological Monographs, **23**, No.45.

Dore A.J., Choularton T.W., Brown R. and Blackall R.M.: 1992, "Orographic rainfall enhancement in the mountains of the Lake district and Snowdonia", Atmospheric Environment, **26**, No.3, 357-371.

Fowler D., Leith I., Binnie G., Crossley A., Inglis D.W.F. and Choularton T.W.: 1995, Submitted to this volume.

Inglis D.W.F., Choularton T.W. and Wicks A.J.: 1995, "The Effect of Orography on Wet Deposition in and Industrial Area", Quarterly Journal of the Royal Meteorological Society, **121**, to appear Oct 1995.

Weast R.C.: 1978, "CRC Handbook of Chemistry and Physics 59th Edition", CRC Press Inc., Boca Raton, USA.

DEPOSITION AROUND A COAL-FIRED POWER STATION DURING A WINTERTIME PRECIPITATION EVENT

K. JYLHÄ

Department of Meteorology, P.O.Box 4 (Hallituskatu 11), FIN-00014 University of Helsinki, Finland

Abstract. The contribution to local wet deposition of emissions from a coal-fired power station at Inkoo on the south coast of Finland has been investigated during a wintertime precipitation event. Making use of intensive radiosonde and weather radar observations of meteorological factors, concentrations of sulphur in deposition due to plume washout were predicted by a short-range deposition model. The model used the scavenging coefficient to parametrize the wet removal of pollutants, and it took into account the wind drift of falling precipitation particles within the plume. The model predictions were then compared with the chemical analysis results from snowfall samples collected within 10 km of the power station during the experiment.

The experiment was performed ahead of a deeply-occluded front during a period with strong advection of long-range transported pollutants. No reliable sign of the influence of the power station on the sulphate deposition could be identified. On the other hand, the deviations of acidity from the mean pH-value of 4.1 were concentrated in one sector near the expected area of deposited plume pollutants. If local emissions were responsible for these deviations, the explanation may lie in a slightly incorrectly estimated plume direction or the effects of alkaline fly ash. Nevertheless, definite conclusions cannot be drawn, because only a few collectors happened to be sited in the modelled sector of plume washout and none in its maximum area.

Key words: wet deposition, sulphate, pH, snowfall, Doppler weather radar, short-range deposition modelling.

1. Introduction

A considerable part of the man-made pollutants in the atmosphere results from fossil fuel power generation. Compounds emitted into the air are transported and mixed by airstreams, undergo chemical and physical transformation and are finally removed from the atmosphere by wet and dry deposition. Air pollutants may affect atmospheric properties and climate locally and possibly even globally, and are injurious to the health of human beings and animals. After being deposited onto the surface they may damage vegetation and cause acidification of lakes and soil.

When emission control strategies for any particular source are to be designed, it is essential to estimate, as accurately as possible, the influence of the source on ambient air quality and on the distribution of deposition. Besides using modelling, one may study the effects of the source by collecting deposition samples in its vicinity. However, deposition generally consists of substances originating from various sources. In many cases it is not easy to distinguish between short- and long-range transported pollutants in deposition because of the variation in time and space of the relevant chemical, physical and meteorological processes. Especially if the deposition collection period is a month or longer, it is hardly possible to make such a distinction, at least in the south of Finland (Nordlund and Tuovinen, 1988; Pohjonen, 1992).

Short-term wet deposition near coal- and oil-fired power stations has been empirically studied e.g. by Granat and Rodhe (1973), Dana *et al.* (1975), Li and Landsberg (1975), Patrinos *et al.* (1983), Stamm *et al.* (1984) and ten Brink *et al.* (1988). These field investigations were performed during single precipitation events, and according to most of them the local source had a discernible impact on wet deposition within the first 15 km downwind. However, in the case of a high background contamination, the contribution of locally-emitted sulphur compounds may be too small to be detected, as discussed by Dana *et al.* (1975) and ten Brink *et al.* (1988).

The purpose of the present paper is to analyse wet deposition by a single wintertime precipitation event in the vicinity of a 250 MW coal-fired power station unit at Inkoo on the south coast of Finland. In particular, any possible effects of plume washout on deposition values are sought after. In addition to chemical analyses of precipitation samples, meteorological measurements and a short-range deposition model have been utilized. Of special importance in this project are the frequent measurements of precipitation and wind by Doppler weather radar and radiosondes.

2. Measurements and modelling

The experiment for the collection and analysis of precipitation was carried out on 18-19 December 1991, during which a precipitation area affected Southern Finland. The deposition collection periods at 28 sampling sites lasted approximately 18 hours. Seventeen plastic buckets were placed in a circular array around the power station at distances of about 2 km. The rest of the collectors were sited in an arc northeast of the power station within a distance of about 9 km. Most of the deposition collection sites are shown in Figure 1. The precipitation samples were analysed for water amount, pH, electrical conductivity and concentrations of sulphate (SO_4^{2-}), total nitrogen (N), nitrate nitrogen (NO_3^--N) and chloride (Cl^-). In addition, three standard precipitation gauges were used as independent references for the water amount. Since the gauge-bucket ratio appeared to be as good as 1.0±0.2, no correction was made to the water amounts.

In order to analyse the meteorological factors influencing the wet deposition, frequent measurements were made by radiosondes and weather radar. The sounding system of the experiment consisted of two radiosounding stations, one at Inkoo and the other in Helsinki. The sounding interval was mainly as short as one hour. As a result, time-height sections of wind velocity, temperature and humidity up to an altitude of about 10 km were obtained. The time evolution of the precipitation area was in turn measured by Doppler weather radar. On the basis of the three-dimensional radar measurements at 15 min intervals, and of the reference measurements from the three standard precipitation gauges, hourly-averaged precipitation intensities R (mm h^{-1}) were computed at the sites of the deposition collectors.

Meteorological data from the radiosondes and the weather radar were utilized in a short-range deposition model to predict the horizontal distribution of the accumulated deposition of sulphur from the power station. The model used was based on a simple Gaussian plume equation and on the assumption that precipitation scavenging can be considered as an irreversible depletion process, the first-order decay rate of which is proportional to a factor known as the scavenging coefficient Λ (s^{-1}) (see e.g. Jylhä,

1991). The model also took into account the wind drift of precipitation particles within the plume. In this case, with the occurrence of snow and sleet, the wind may have transported precipitation particles considerably during their relatively long fall time. Their fall trajectory was therefore included in the model.

Sulphur (S) releases from the Inkoo power station consisted mainly of sulphur dioxide (SO_2), which was released at rates of about 0.2 kg s^{-1}. The emission rate of primary sulphates (SO_4^{2-}) was unknown, but it was assumed that roughly 5% of S was emitted as SO_4^{2-}. In the case under study wet deposition occurred during a winter night, and the distance between the power station and the deposition collectors within the range of 10 km corresponded to a travel time of less than 15 min. Because of the short travel time compared with the typical wintertime gas-phase oxidation rate of SO_2, the formation of secondary SO_4^{2-} in the gas phase was probably insignificant.

Part of the S content of the plume may have been collected by fly ash particles or by droplets of low-level stratus cloud below the main cloud base, but the deposition model ignored these possible removal mechanisms. Instead, only below-cloud scavenging of primary SO_4^{2-} by wet and dry snow and dissolution of SO_2 in the liquid fraction of the falling hydrometeors were taken into account using the approximate method of the scavenging coefficient Λ. Measured precipitation rates R downwind of the power station were utilized to evaluate Λ, applying formulae by Chang (1986) for SO_4^{2-} below and above the height of the 0°C isotherm. Below-cloud depletion of SO_2 was assumed to only have some significance at temperatures above zero: a linear dependence of Λ (s^{-1}) on R (mm h^{-1}) was assumed, with a roughly estimated ratio of 10^{-5} between them. Although Λ only poorly describes scavenging of SO_2, it is used in this paper to keep the algorithm for SO_2 wet removal consistent with that for sulphate.

3. Results and discussion

The deposition collection was performed ahead of a deeply-occluded front, where the wind direction varied only slightly with time at the estimated plume altitudes. Thus, the modelled sector of deposited plume pollutants remained rather narrow (Figure 1). Because of the slow fall speed of snow particles, the predicted area of maximum wet deposition was not located immediately downwind of the source but at a distance of 3-6 km, this distance depending on the estimated plume rise and on the wind speed compared with the fall speed. Unfortunately, only a few collectors happened to be sited in the modelled sector of plume washout and none in its maximum area.

Observations at Finnish air quality background stations (Leinonen and Juntto, 1992) indicate that during the case study period the contribution of long-range transported pollutants on deposition was larger than average. This diminishes the possibility of recognising any effects of the nearby source on sulphur deposition, even if plume-related scavenging of SO_2 or SO_4^{2-} did occur within the closest 10 km. Indeed, no additional deposition of S was found downwind of the power station, as can be seen in Figure 2a.

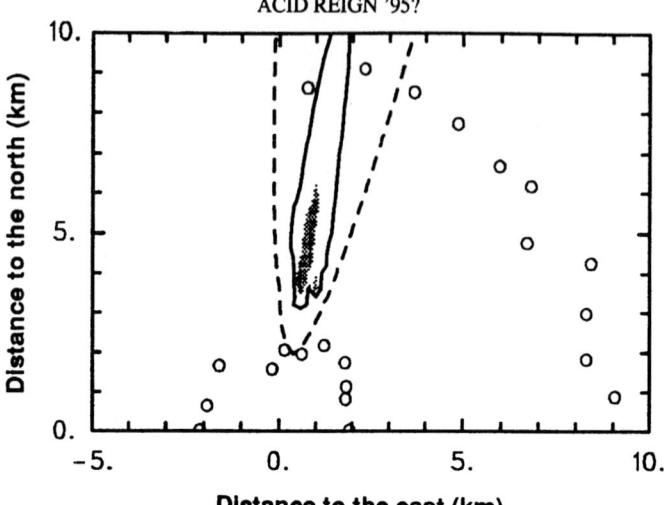

Fig. 1. The modelled horizontal concentration distribution of S in deposition due to emissions from the Inkoo power station between 1230 UTC on 18 Dec and 0630 UTC on 19 Dec 1991. Drifting of precipitation particles downwind is included assuming a terminal fall speed of 0.9 m s^{-1}. Isopleths are 0.1 mg l^{-1} (dashed line) and 1 mg l^{-1} (solid line), while values larger than 2 mg l^{-1} are indicated by shading. The point (0,0) denotes the site of the power station, while the deposition collectors are denoted by open dots.

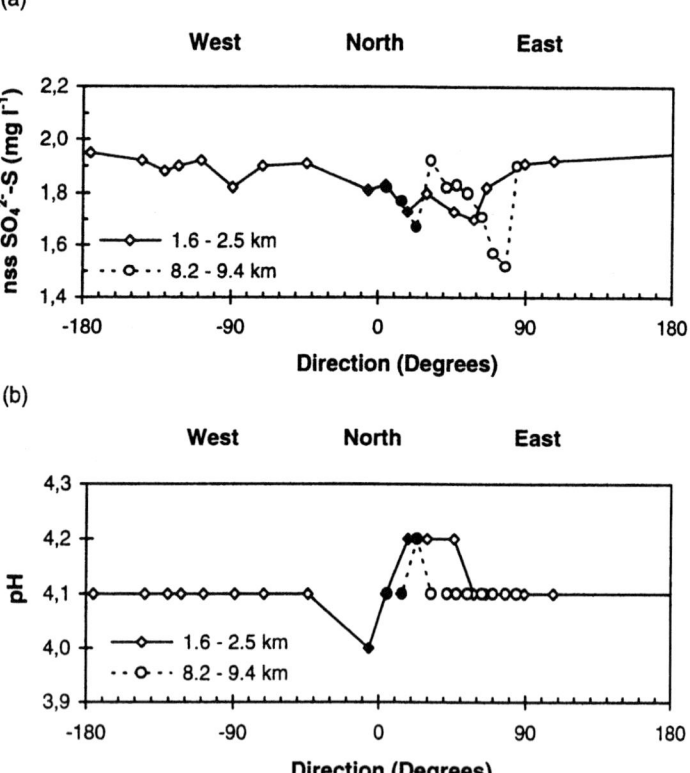

Fig. 2. Analysed (a) concentrations of non-sea-salt (nss) SO$_4^{2-}$-S and (b) pH in the snowfall samples collected at distances of 1.6–2.5 km and 8.2–9.4 km from the Inkoo power station. The horizontal axis indicates the direction of a collector with respect to the station. Values for the collectors near to the modelled local deposition area are denoted by solid marks.

Figure 2a shows concentrations of nonmarine SO_4^{2-}-S in the snowfall samples as a function of the direction with respect to the power station. The contribution of sea salt S has been estimated and eliminated with the aid of Cl^-, assuming that sea spray was the main source of Cl^-. The resulting values range between about 1.5 and 2.0 mg l^{-1}, with a mean value of 1.8 mg l^{-1}. Two curves for the observations are shown, one for the nearby collectors in the circle and the other for the more distant samplers in the northeastern arc. The values for those collectors which were sited in or near to the expected area of plume pollutants in deposition (Figure 1) are shown by solid marks. No reliable sign of the influence of the power station can be detected. The variation in Figure 2a is thus most probably caused by fluctuation of the background level and by slightly unequal sampling periods.

The snow samples were rather acid, their pH being 4.0 - 4.2 (Figure 2b). An interesting feature is that the deviations from the mean pH of 4.1, which exceeded the analysis accuracy of 0.1 pH units, were concentrated in one sector near the expected area of deposited plume pollutants. The only sample with a pH less than the average was collected on the western edge of the modelled plume sector, but somewhat unexpectedly, four samples of slightly less acidity were found inside and east of the sector. The question is whether these deviations were caused by emissions from the power station or by variations in concentrations of long-range transported pollutants.

If local emissions were responsible for the deviations in pH-values, the explanation may lie in a slightly incorrectly estimated plume direction or the effects of fly ash particles. Provided that scavenging of plume-related sulphur and nitrogen was efficient enough to make snowfall more acid, the deposition area was probably somewhat more to the west of that depicted in Figure 1. On the other hand, fly ash formed in a coal combustion process is clearly alkaline, with a pH of 10 - 11 (Keppo and Ylinen, 1980). Even if the alkalinity was slightly decreased in the atmosphere due to the possible formation of an acid sulphate layer on the surface of fly ash particles (Parungo et al., 1987), ash particles would tend to more or less neutralize the acidity of the collected precipitation.

4. Conclusions and discussion

This case study of wet deposition around the Inkoo coal-fired power station during a wintertime precipitation event combined chemical analysis of deposition samples taken within 10 km of the power station with meteorological data from radiosondes and a weather radar. No reliable sign of the influence of the power station could be found in the observed sulphate concentrations of the samples. Instead, the point source possibly slightly affected the acidity of the samples. However, definite conclusions cannot be drawn, because only a few collectors happened to be sited in the modelled sector of plume washout. Besides, the results may not be directly applicable to another snowfall event, since the experimental period was characterized by a strong advection of long-range transported pollutants.

The original goal of this experiment was to study wet deposition by rainfall. Since rain removes below-cloud SO_2 from the atmosphere more efficiently than snow does, there would probably have been a better chance of finding some signs of the emission from the power station in a rainfall situation. However, a considerable part of the yearly precipitation in Finland is attributable to snowfall. Especially in winter, when power generation is at a maximum, the freezing level is likely to be at or below a typical effective plume height. Hence in Finland it is also of interest to study scavenging by dry and wet snow.

In deposition models using the scavenging coefficient to parametrize the wet removal of pollutants, it is usually assumed that hydrometeors fall in a vertical direction onto the ground. However, when modelling wet removal due to ice crystals, their falling trajectory out of the vertical should be taken into account. The slower the terminal velocity compared with the wind speed, the more distant is the area of maximum deposition. The procedure developed in this study regards the fall speed as a constant. A more sophisticated method would embrace variations of the fall speed as function of time, height and hydrometeor size, together with assessed limits of error.

Acknowledgements

This research was carried out under contract 36DY92 from Imatran Voima Oy. Financial support was also received from the Jenny and Antti Wihuri Foundation, the Vilho, Yrjö and Kalle Väisälä Foundation at the Finnish Academy of Science and Letters, and the Kone Foundation. The kind help of numerous people at the Department of Meteorology of the University of Helsinki, at Imatran Voima Oy, in the Finnish Defence Forces and at the Finnish Meteorological Institute are acknowledged.

References

Chang, T.Y.: 1986, *J. Geophys. Res.*, **91**, 2805-2818.
Dana, M.T., Hales, J.M. and Wolf, M.A.: 1975, *J. Geophys. Res.*, **80**, 4119-4129.
Granat, L. and Rodhe, H.: 1973, *Atmospheric Environ.*, **7**, 781-792.
Jylhä, K.: 1991, *Atmospheric Environ.*, **25A**, 263-270.
Keppo, M. and Ylinen, P.: 1980, "Power plant ashes and their utilization. Part 1. The quantities of ashes produced in Finland and their utilization", Technical Research Centre of Finland, Concrete and Silicate Laboratory, Report 61, 63 pp + app. 78 pp.
Leinonen, L. and Juntto, S.: 1992, *Air quality measurements 1991*, Finnish Meteor. Institute, Helsinki, 220 pp.
Li T.-Y. and Landsberg, H.E.: 1975, *Atmospheric Environ.*, **9**, 81-88.
Nordlund, G. and Tuovinen, J.-P.: 1988, "Sadeveden sulfaattipitoisuus Inkoon voimalaitoksen lähiympäristössä: havaintojen ja mallilaskelmien vertailu", Finnish Meteorological Institute, Air Quality Dep., 13 pp.
Parungo, F., Nagamoto, C. and Madel, R.: 1987, *J. Atmos. Sci.*, **44**, 3162-3174.
Patrinos A.A.N., Dana, M.T. and Saylor, R.E.: 1983, *J. Geophys. Res.*, **88**, 8585-8612.
Pohjonen, M.: 1992, "Effects of acid deposition on the ecosystem of the drainage basin of lake Marsjön, Inkoo", Imatran Voima Oy, Research Reports, IVO-A-01/92, 114 pp. (Abstract in English)
Stamm, A.J., Caplan, P. and Hinrichs, R.A.: 1984, *Atmospheric Environ.*, **18**, 817-823.
ten Brink, H.M., Janssen, A.J. and Slanina, J.: 1988, *Atmospheric Environ.*, **22**, 177-187.

ACID POLLUTANTS IN AIR AND PRECIPITAION/DEPOSITION AT THE SUDETEN MOUNTAINS, POLAND

G.KMIEC[1], K.KACPERCZYK[1], A.ZWOZDZIAK[1] and J.ZWOZDZIAK[1]

[1] *Institute of Environment Protection Engineering, Technical University of Wroclaw, 50-370 Wroclaw, Wybrzeze Wyspianskiego 27, Poland*

Abstract. The results of several separately performed field studies on air quality in south-western Poland (Black Triangle) are presented. In the period 1988-1993 atmospheric aerosol measurement and precipitation/deposition samples were collected at three locations of mountain region (810-1490 m asl). Precipitation was monitored in forest ecosystem using a 24-gauge network was carried out during June-October 1992 and 1993. The occurence of cloud at ground level over hills is difficult to categorize and quantify in mountain terrain and has constituted a basic part of our studies. Precipitation/deposition samples (seven rain and three cloud collectors) in the Karkonosze Mountains were collected over the period 1994-1995 (four field campaigns were conducted) to examine aqueous chemical interactions between constituents in samples. Meteorological data measured at the site included temperature, humidity, wind speed and direction and the amount of precipitation.

Key words: aerosol, aerosol scavenging, rime, seeder-feeder, cloudwater chemistry, acid pollutants, "Black Triangle"

1. Introduction

Acidic aerosol in the atmosphere has become a problem of serious concern. It is a well-established fact that acidic aerosols have both direct and indirect effects on terrestrial and aquatic ecosystems. Atmospheric pollutions not only inhibit the growth of forest, but also account for their ever increasing damage at high elevation. The formation of "cap clouds" takes place due to the condensation of humid air that occurs when the air is forced to ascend over the mountain barrier. Analyses of water samples from cap clouds have shown that they are often heavily polluted. There is no rain from such clouds but their droplets can be deposited in the ground either directly through turbulent deposition or by being washed out by raindrops falling from the upper layer clouds. A process is known as the "seeder-feeder " [1,2].

Low clouds are observed along the Karkonosze ridge for approximately 240 days in the year, hence the process of turbulent cloud water deposition and the seeder-feeder effect is expected to be essential for determining the pollution deposition in the Karkonosze Mountains. The development of this phenomenon was confirmed by the preliminary results which aimed to investigate the rain, snow, rime pollutions and its effects on the forest ecosystem. Bulk precipitation was monitored monthly based on the samples taken at 24 sites in the Karkonosze National Park from 1992-1993 [3,4]. The investigations, however, did not take account of the so-called occult deposition from clouds/fogs. During four field campaigns: in February 1994, 1995 and July, October 1994 precipitation samples (rainwater, cloud/fogwater and rime were collected every day at 3 measuring points. Ambient aerosol samples were taken below at the summit of Szrenica. The aim of the investigations was to examine aqueous chemical interactions between constituents in precipitation water samples. Furthermore monthly rain and cloudwater samples were collected.

2. Materials and methods

2.1. SAMPLING SITES

Air chemistry and meteorology measurements were obtained at three sampling points differently located (810, 1362, 1490 m asl) in south-western Poland ("Black Triangle", the near-border region). The monthly rain samples were collected at 24 sites located between 1000-1500m asl on the north slopes of the Karkonosze Mts. (10.5x25 km) over two summer periods in 1992 and 1993. The rainwater data collected during summer months were used to define the relationships of chemical composition with altitudes in the Karkonosze National Park .Precipitation samples were collected at seven sampling sites located on the western slope of the Mount Szrenica at the altitudes: 1315, 1270, 1220, 1200, 1100, 1000 and 840 m asl from July to September 1994. There are also three cloudwater collectors placed at the altitudes: 1315, 1000 and 840 m asl. From October 1994 the cloudwater collectors were placed at the altitudes:1315, 1200 and 1000 m asl. During the field campaigns in February 1994 and 1995, July 1994 and October 1994 precipitation samples (rainwater, cloud/fogwater and rime) were collected every day at three measuring points. Each rain collector consisted of a polyethylene bottle with a funnel of 15.2 cm in diameter, placed 1.5 m above the ground. Cloudwater is collected using passive cloud collectors. Cloud samples, collecting on teflon, V-shaped strings (0.5 mm in diameter), are fed into a sample catchpot. A lid of 0.8 m in diameter prevents precipitation from entering.

2.2. ANALYTICAL PROCEDURES

Wet deposition samples (from rainwater and cloudwater collectors) were carried out directly in individual polyethylene bottles with funnels. Measurements of the pH of both cloudwater and precipitation were made directly after sample collection using pH-meter equipped with a combination electrode .Cloudwater and precipitation/deposition samples were analyzed for ions: sodium (Na^+), potassium (K^{++}), calcium(Ca^{++}), magnesium (Mg^{++}), and trace metals zinc (Zn^{++}), lead (Pb^{++}), iron (Fe^{+++}) using the ICP technique. Each sample was stored in a polyethylene sample bottle and refrigerated for the transport to the laboratory. Ion-selective method was used to determine the inorganic ions such as chloride (Cl^-), nirate (NO_3^-) and ammonium (NH_4^+). Sulphate ($SO_4^=$) ions were measured by the thorin method [7]. Concentrations of atmospheric aerosol compounds were determined by colorimetric methods as standard or reference methods based on absorption of the gas by chemical reaction [8].

3. Results and discussion

3.1. RIME - SNOW COMPOSITION COMPARISON

Measurements of an inorganic species in snow and rime were carried out simultaneously at three different sites along a west-east trunsect with elevations of 1315, 1200, 1000 m asl, respectively. Studies of winter precipitations sampling were conducted in February 1994, 1995 (two field campaigns). During the major sampling events, a total 27 fresh snowfall samples and 20 rime samples were carried out (12 snow and rime samples collected simultaneously). The variations in the concentration of the major ions in snow and rime are illustrated in Figures 1-4.

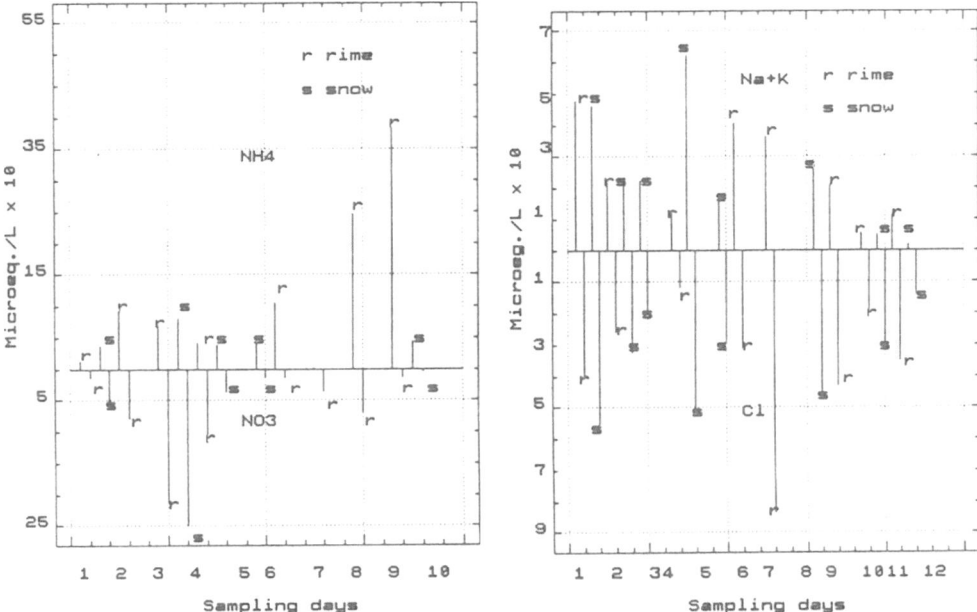

Fig. 1. Ammonium and nitrate concentrations in rime snow for the Mt. Szrenica (30.01.-10.02.1995)

Fig. 2. Sodium, potassium, chloride concentrations in rime and snow for the Mt. Szrenica (29.01-11.02.1994).

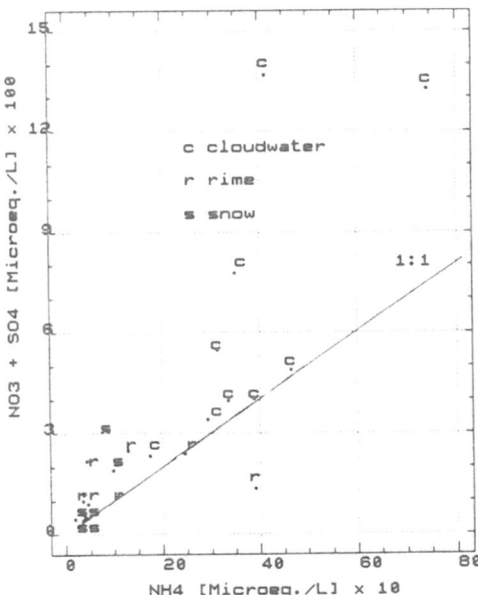

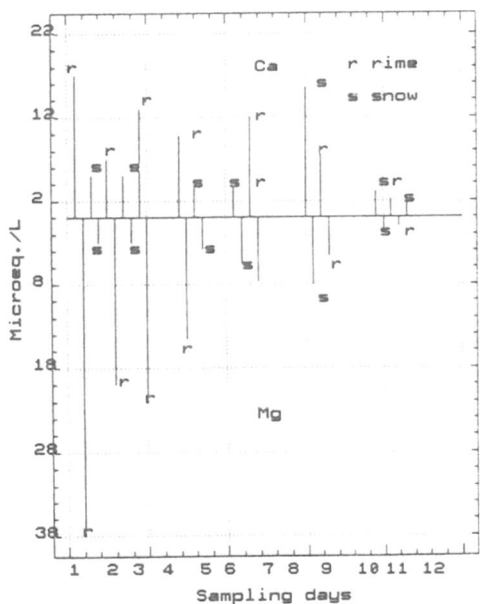

Fig. 3. Acid-base balance in rime, cloudwater and snow for the Mt. Szrenica. The line indicates the condition $NH_4^+ = SO_4 + NO_3^-$. Points below and to right of this line are alkaline, while points above and to left are acidic.

Fig. 4. Calcium and magnesium concentrations in rime and snow for the Mt. Szrenica (29.01 - 11.02.1994)

Table I provides a summary of statistics for the observations. The dual rime/snow sampling was designed to evaluate spatial variations in the chemical composition of rime and snow from the same event. The pH ranges between 3.3-5.9 for rime and 3.4-7.1 for snow samples reflect some spatial and temporal tendencies. Depending on the wind direction, contributions of maritime aerosols to the overall chemical budget in precipitation/deposition can be found. Sodium, potassium and chloride (Figure 2) clearly reflect this influence of sea salt species in samples collected in January/February (an obvious maritime trajectory).

TABLE I

Statistical data on snow and rime samples below the Szrenica Mt. (1315 m asl) in February 1994, 1995

Constituent [mg/L]	pH	N(NH$_4$)	N(NO$_3$)	Na	K	Ca	Mg	Cl	SO$_4$	Pb	Zn	Fe
RIME												
Avg. conc.	3.55	1.53	5.80	3.81	13.25	0.33	0.28	9.95	3.23	0.022	0.021	0.089
Min. conc.	3.31	0.10	bdl*	bdl	bdl	bdl	bdl	bdl	bdl	bdl	bdl	bdl
Mediane	3.42	1.22	5.22	3.50	1.02	0.17	0.13	9.04	6.11	0.013	0.015	0.072
Max. conc.	5.90	5.40	14.00	10.71	82.5	1.93	1.05	69.7	9.92	0.064	0.093	0.324
SNOW												
Avg. conc.	4.62	0.90	4.05	0.46	0.07	0.14	0.04	1.21	1.99	0.013	0.026	0.018
Min. conc.	3.36	bdl	bdl	bdl	bdl	0.03	bdl	0.50	bdl	bdl	bdl	bdl
Mediane	3.73	1.13	1.32	0.30	0.03	0.06	0.03	1.12	2.04	0.010	0.016	0.013
Max. conc.	7.10	2.42	19.00	1.32	0.20	0.42	0.09	3.50	3.80	0.032	0.093	0.053

* bdl - denotes below detection limit

The acid-base balance ($SO_4^= + NO_3^-$)/ NH_4^+ (calculated ratios) of collected snow, rime and cloud-water samples is illustrated in Figure 3. The samples with the highest acidity were collected at the Mt.Szrenica; this reflects the influence of anthropogenic emissions, within the Black Triangle Region (the accumulation of trapped pollutants which can be transported to higher elevated sites in the surrounding mountains). Observations of the compositon of simultaneously collected rime and snow samples indicate a disparity between ion concentrations with rime concentrations of most species generally several times higher. Ion concentrations in rime samples collected at two locations (1200m, 1000m asl), sometimes, were higher than those sampled at 1315m asl. The difference was attributed to lower values of cloud liquid water content.Cloud liquid water contents ranged between 0.3 and 2.34g/m^3 [5] at the Mt. Szrenica (30.01-06.02.95).

3.2. VERTICAL GRADIENTS IN CLOUD WATER COMPOSITION

Cloudwater samples were collected at three sites from 19 to 30 October 1994. A summary of statistics for the inorganic chemical composition of the cloudwater samples is given in Table II. A total of 15 cloud-water samples representing 11 events, were collected below the Mt. Szrenica (1315m asl). 4 samples representing two events, were collected at two different elevations (1200m and 1000m asl).

TABLE II

Statistical data on cloud water samples collected below the Szrenica Mt. (1315 m asl) in October 1994

Constituent [mg/L]	pH	$N(NH_4)$	$N(NO_3)$	Na	K	Ca	Mg	Cl	SO_4	Pb	Zn	Fe
CLOUDWATER												
Avg. conc.	4.80	5.00	5.90	2.30	1.30	2.40	1.40	2.50	8.30	0.22	0.22	0.52
Min conc	3.50	2.30	0.11	0.20	bdl	bdl	015	0.60	5.50	bdl	bdl	0.15
Mediane	4.40	4.90	4.20	1.00	0.50	0.71	0.65	2.10	8.90	0.15	0.12	0.59
Max conc.	7.20	10.20	17.00	10.00	6.50	9.50	8.60	5.00	10.10	1.09	0.64	0.90

Collection period was several hours. The events were frequently associated with the passage of cold fronts, although a variety of meteorological situations was encountered. Concentration variations were observed both between cloud interception events and within an individual event. Inputs of sodium, potassium (Na^+, K^+) and chloride (Cl^-)ions were also important to form neutral or partially neutralized important aerosol species that are scavenged by the cloud drops when strong of nitric acid (HNO_3) and/or sulphuric acid (H_2SO_4) blowed. Only twice cloudwater samples were collected simultaneously at three sampling sites, on October, 24 and 27, 1994. At that time concentrations of inorganic species found in cloudwater were all greater at the lower elevation sites. This behaviour is likely to be attributed to the combined effects of smaller volume of liquid water at lower elevations and to additional input of chemicals into the cloud layer at its base from the turbulently mixed air below the cloud. Since the lack of information on the latter process and a strong relationship between liquid water content (LWC) and the soluble content found [6], it is now considered that the decreasing ion concentrations in cloud water with the increasing altitude asl is the result of progressive dilution of ion concentrations by the increasing LWC with elevation.

3.3. CLOUDWATER VS PRECIPITATION (RAIN) COMPOSITION IN MONTHLY SAMPLES

During the measurement period in July - October 1994, rain (n=56) and cloudwater (n=10) samples were collected in a cycle of 14 days by means of rain and cloud collectors, respectively. The pH values of the rainwater at seven sites were ranged between 3.4-7.0 (mediane 4.5). The most frequently found pH ranges were: $3.5 < pH < 4.0$ (19%), $4.0 < pH < 4.5$ (24%), $4.5 < pH < 5.0$ (26%). The chemical composition of rainwater samples collected at seven different elevations was examined for pH ranges 4.0-4.5 and 4.5-5.0. Precipitation samples with pH values from 4.0 to 4.5 were observed in July, August and October at three elevations: 1000 m, 1200 m, 1220 m asl. Rainwater concentrations of $N(NH_4^+)$, $N(NO_3^-)$, Cl^-, Na^+, K^+, Ca^{++}, (Pb^{++}, Zn^{++}) were higher, however concentrations of $SO_4^=$ and Mg^{++} (Ca^{++}) were lower (mainly in July and October) than those sampled at at a lower location. For the pH range between 4,5-5,0 recorded in August and September at 1000 m, 1200 m, 1220 m and 1270 m asl., an increasing tendency of ion concentrations ($SO_4^=$, $N(NO_3^-)$, Na^+, Ca^{++}, Mg^{++} with decrease of site elevation was observed.

Concentrations of Cl^-, $N(NH_4^+)$, K^+ were higher at lower altitude. Ion concentrations in cloudwater samples, collected at elevations 1315 m and 1000 m asl., confirm mentioned above variability in concentrations of major ions with altitude for the pH ranges: 3,1-4,4 (1315 m) and 3,8-6,2 (1000 m): in the most cases major ions concentration levels were a factor 5-7 or more above those found in samples of rainwater.

4. Conclusions

The rainwater data collected during summer periods 1992 and 1993 were used to examine the regional variation in concentrations of major ions at different elevations in the Karkonosze National Park. The increase in concentrations of major ions with altitude was significant (by a factor 1.4) above 1300 m asl. The change of rainwater composition with altitude depends strongly on hill shape and atmospheric conditions. Measurements of chemical composition of monthly bulk precipitation/deposition along elevational gradient (840-1315 m asl) were assembled and used to defined the relationships with altitudes on Szrenica Mount. The results show that in most cases, ion concentrations in cloudwater ($\mu M/L$) were several times higher than those found in rain water samples.

Concurent measurements of cloudwater chemistry in event by event samples at below the Szrenica summit (1315) and at 1200 and 1000 m asl have revealed that ion concentrations were all greater at the lower elevation sites. A comparison of the acidity of cloudwater sampled at the Mt.Szrenica with that observed at other elevation sites show that the pH was not unusually low. Cloudwater pH values were less acidic at the higher altitudes. This difference occurs despite maxima in nitrates, sulphates and chlorides concentrations. The key reason for the pH difference between the sites is the amount of ammonium present in the cloudwater. More research is needed to improve understanding of variations in cloud water immersion frequency with altitude, LWC and chemistry in order to assess elevational gradient in cloud water chemistry in the region under study. These processes will be considered in the future within the EASE (Emission Abatement Strategies and the Environment) project.

References

Dore A.J., Choularton T.W. and Fowler D. 1992 : *Atmos. Environ.* **26A**, 1375-1381.
Fowler d.,Cape J.N., Leith I.D., Choularton T.W., Gay M.J. and Jones A. 1988: *Atmos. Environ.* **22**, 1355-1352.
Kmiec G., Kacperczyk K., Zwozdziak J. and Zwozdziak A.: *Karkonoskie badania ekologiczne* . Fischer Z.(ed.) Oficyna Wyd. Inst.Ekologii PAN (1993,1994), in Polish.
Kmiec G., and Kacperczyk K. 1994:*Procc.of Second Int.Conf. on Air Poll., Barcelona 27-29 Sep.1994*, Pollution Control and Monitoring ; Air Pollution II Vol.2, 245-252, Computational Mechanics Publications, Southampton Boston.
Kmiec G., Zwozdziak J., and Zwozdziak A. 1995 : *Environment Protection Engineering*, , Wroclaw Technical University Press, Wroclaw, Wybrzeze Wyspianskiego 27, Poland. In press.
Moeller D., Acker K. and Wieprecht W. 1995 :*Int.J.Sci.* In press.
Persson G.A.1966 : *Int.J.Air and Water Poll.* **10**, 645-652.
Warner P.O. 1976 : *Analysis of Air Pollutants*, John Wiley and Sons (ed.), New York 1976.

WET-ONLY AND BULK DEPOSITION STUDIES AT NEW DELHI (INDIA)

U.C. KULSHRESTHA[1], A.K. SARKAR[1], S.S. SRIVASTAVA[2] and
D.C. PARASHAR[1]

[1] *Chemistry Division, National Physical Laboratory*
Dr. K.S. Krishnan Road, New Delhi-110 012 (India)
[2] *Department of Chemistry, Dayalbagh Educational Institute, Dayalbagh, Agra-282 005*

Abstract. Rain water samples were collected at New Delhi during the monsoon of 1994 at a height of 30 m above the ground level using a wet-only collector. Simultaneously, bulk samples from two different heights at 30 m and 13 m were collected. Frequency distribution of pH in wet-only samples revealed that rain was mostly alkaline. Four out of 23 events were observed to be acidic where the ratio of $(Ca+Mg+NH_4)/(SO_4+NO_3)$ was very low. pH and ionic constituents were higher in bulk samples than in wet-only samples. On an average, the concentration in bulk samples at 30m height exceeded the wet-only samples by 13% while bulk samples collected at 13m height had 19% higher concentration than the bulk samples at 30m height and 32% higher than wet-only. The acidity of rain water was mainly contributed by sulphuric acid rather than nitric acid. At the height of 30 m, the acidity was primarily neutralized by NH_4 while at 13 m height, it was buffered by Ca and Mg indicating the influence of dust particles.

KEY WORDS: WET-ONLY, BULK SAMPLES, NEUTRALIZATION FACTORS, SOIL.

1. Introduction

In-cloud and below cloud scavenging processes are the prime mechanisms by which gaseous and particulate pollutants are removed from the atmosphere (Warneck, 1988). The importance of these two processes is different in various regions of the world. In western Europe and North America, the in-cloud scavenging process is important (Dean *et al.*, 1981; Strapp, 1988) while in the countries like India and China, below cloud scavenging process becomes more important as the surface layer of the atmosphere has higher concentration of pollutants viz. particulates and gases from local sources (Khemani *et al.*, 1985; Kulshrestha *et al.*, 1991; Meiyuan *et al*, 1993). Scavenging of pollutants of ambient air affects the chemical composition and pH of rain water. In India, several workers (Khemani *et al.*, 1989; Saxena *et al.*, 1991; Varma, 1989; Handa, 1982; Naik *et al.*, 1988) have studied the influence of particulate matter on the composition of rain water for different locations in the country.

This report presents the studies of chemical composition of rain water at Delhi where effort has been made to collect the samples at two different heights i.e. 30 m and 13 m above the ground to understand the influence of soil derived particles on the composition of rain water.

2. Experimental

2.1. SITE DESCRIPTION

Delhi is the capital of India and is among the highly polluted cities in the world. It lies in north-central India (76^0 50'E - 77^0 23'E, 28^0 12'N - 28^0 53'N) and is about 1400 km away from the sea. According to 1987-88 census, the number of various small, medium and large scale industries situated in the city is 73,000. These industries include ferrous and non-ferrous casting, engineering, chemicals, plastic, cloth dyeing, potteries, steel rolling, pharmaceutical and rubber pulverization. The vehicular strength of the city has grown from 0.93 million in 1985 to 1.57 million in 1990 (Aggarwal and Varshney, 1993). In the immediate vicinity of site, there are some experimental farms of the Indian Agricultural Research Institute.

2.2. SAMPLING AND ANALYSIS

The wet-only and bulk samples were collected on the top of the tower of the National Physical Laboratory at a height of 30 m from the ground using automatic wet-only and manual collectors, respectively. Simultaneously bulk samples were also collected using manual collector at the height of 13 m. The samples were collected on a 24 h basis (i.e. 10 am - 10 am). The ground near the site is covered with lawns and roads and the roofs on which the samples were collected are cemented. The diameters of the automatic and bulk collectors were 22 cm and 20 cm respectively. The duration of the showers varied between 15 minutes and 28 hrs. The collected samples were preserved with thymol (Gillet et al., 1992; Granat et al., 1992) and stored in prewashed polypropylene bottles. In this study, the bulk samples collected from 30 m and 13 m above the ground were termed $bulk_{up}$ and $bulk_{down}$ respectively and the 13 m and 30 m heights were termed as lower and high altitudes respectively..

Physical parameters viz. volume, pH and conductance were measured immediately after collection of samples. The anions, F, Cl, NO_3 and SO_4 were analyzed by ion chromatography (Kulshrestha et al., 1995). Na, K, Ca and Mg were analyzed by atomic absorption spectrophotometry (Varian Spectra AA-10) while NH_4 was estimated colorimetrically by the indo-phenol method (Weatherburn, 1967) using a UV-VIS spectrophotometer (Perkin Elmer Lambda 3b).

3. Results and discussion

3.1 WET-ONLY PRECIPITATION

Out of the 23 wet-only samples, only 15 samples had sufficient volume for the analysis of various ions These samples were checked for their ionic balance ($\Sigma cations/\Sigma anions$). None of these sample had ionic ratio greater than 1.3 indicating that there was no chance of contamination. Finally, linear regression of cation sum on anion sum gave a r value of .95 indicating that the quality of ion balance data is good. Table I gives the average rainfall, pH, conductance and volume-weighted average concentration of ions of the wet-only and bulk depositions. The average pH of wet-only was found to be 5.7. The pH of wet-only samples was more alkaline than the reference value 5.6 (Granat, 1972; Charlson and Rodhe, 1982; Khemani et al., 1985; Saxena et al., 1991; Singer et al., 1993) of rain water in more than 82% of the events. Only four events were acidic, the details of which are given in Table II. The first acidic event on 20-21 July, 1994 was the ninth event of the season and it had

TABLE I
Volume weighted concentration (μeq/L) of precipitation components

Variable	Wet-only (n=15)	Bulk$_{up}$ (n=15)	Bulk$_{down}$ (n=10)	Ratios Bulk$_{up}$/wet	Bulk$_{up}$/Bulk$_{down}$
F	1.57	2.63	2.47	1.67	1.06
Cl	2.42	2.82	2.60	1.16	1.46
NO$_3$	5.97	6.45	9.19	1.08	.70
SO$_4$	27.71	28.33	23.31	1.02	1.21
NH$_4$	30.55	28.33	34.33	0.93	0.82
Na	1.24	1.22	1.20	1.01	1.42
K	10.25	13.33	8.38	1.30	1.59
Ca	7.61	15.50	34.50	2.03	0.45
Mg	0.91	1.88	3.17	2.06	0.59
H$^+$	1.99	1.26	0.40	0.63	3.15
pH	5.7	5.9	6.41		
Rain (cm)	2.3	2.3	2.3		

TABLE II
Details of acidic events occurred during study

Date	Rain (mm)	pH	EC (μS/cm)	C$^+$ - A (μeq/L)	(Ca+Mg+NH$_4$)/(SO$_4$+NO$_3$) ratio
20-21 July 1994	7.8	5.20	13.51	-8.16	.58
28-29 July 1994	25.4	4.93	16.92	-11.87	.55
30-31 July 1994	34.3	5.22	7.24	-5.75	.50
25-26 August 1994	56.1	5.25	7.02	-13.33	.68*

* Concentration of Ca and Mg were below detection limits.

been raining continuously during two days prior to this event. A total of 131 mm rainfall was recorded before this event which comprised 28% of the total rainfall during monsoon. These results indicate that due to high rainfall, the lifting up of dust particles is suppressed and suspended particulate matter is washed out resulting in lower concentration of buffering species like Ca and Mg and dominance of SO_4 and NO_3. As a consequence the ratio of $(NH_4+Ca+Mg)/(SO_4+NO_3)$ is decreased leading to acidic pH of rain water. Similar situation prevailed for other three acidic events listed in Table II. In all four events the sum of anions (F, Cl, NO_3, SO_4) exceeds the sum of cations (H, Na, K, Ca, Mg, NH_4) which results in lowering the pH value.

TABLE III

Neutralization factors (NF)

	NH_4	Ca	Mg
Wet-only	0.50	0.19	0.02
Bulk$_{up}$	0.45	0.37	0.04
Bulk$_{down}$	0.61	0.83	0.08

The higher concentrations of SO_4 compared to NO_3 indicate that the acdity of rain water is mainly due to sulphuric acid. This acidity is buffered primarily by NH_4 followed by Ca resulting in the relatively alkaline pH of rain water. The role of NH_4, Ca and Mg has been validated by calculating neutralization factors (NF) using the formula-

$$NF_x = \frac{X}{NO_3+SO_4} \quad \ldots (1)$$

Where X may be Ca, Mg or NH_4.
The order of NF for NH_4, Ca and Mg suggests that the major neutralization in wet-only samples is NH_4 followed by Ca and Mg (Table III).

3.2 WET-ONLY VS BULK$_{UP}$ DEPOSITION

Figure 1 shows the pH of wet-only and bulk$_{up}$ samples. It is seen that pH of bulk$_{up}$ samples is always higher than the corresponding wet-only sample. The difference between wet-only and bulk components indicates the influence of dry deposition. The concentrations of F, Cl, NO_3, SO_4, K, Ca and Mg are higher by 67%, 16%, 8%, 2%, 30%, 103% and 106% respectively in bulk$_{up}$ samples than in wet-only samples. On an average, the bulk$_{up}$ samples have 13% higher concentration of components in comparison to wet-only. The remarkably higher concentrations of Mg and Ca in bulk seems to be due to the influence of dry deposition during long exposure of the collector for 24 hrs. This is further corroborated by higher NF for Ca and Mg in bulk samples than those of wet-only samples (Table III). However, the concentration of NH_4 is lower in bulk samples. Lower NH_4 concentration in bulk samples is probably due to possible retardation of its dry deposition because of comparatively higher pH of the bulk samples (Seinfeld, 1986).

3.3 BULK$_{UP}$ VS BULK$_{DOWN}$

The average pH of bulk$_{down}$ is higher than bulk$_{up}$ (Table I). The increased pH is due to the higher concentrations of Ca, Mg and NH$_4$. The concentrations of Ca, Mg and NH$_4$ are higher by 122%, 68% and 21% respectively in bulk$_{down}$ than in bulk$_{up}$. On an average the bulk$_{down}$ samples have 19% higher concentration of constituents than bulk$_{up}$.

The results of 10 individual bulk$_{up}$ and bulk$_{down}$ samples collected simultaneously revealed that Ca, NH$_4$ and NO$_3$ were higher in bulk$_{down}$ as compared to bulk$_{up}$ but this was not so for SO$_4$. The higher concentration of Ca, NH$_4$ and NO$_3$ at lower height seems to be due to their presence in coarse mode particles. NO$_3$ aerosols are reported in coarse mode as sodium nitrate (Yoshizumi and Asakuno, 1986) but the higher concentration of NH$_4$ at lower height is not very clearly understood.

Neutralization factors (Table III) reveal the dominance of NH$_4$ at higher altitude and that of Ca and Mg at lower altitude and further studies to understand this phenomenon are in progress using larger number of collectors.

4. Conclusion

This study reveals that the nature of rain water at Delhi is alkaline. The lower concentrations of Ca, NH$_4$ and Mg result in acidic rain. The free acidity of rain water is due to SO$_4$ rather than NO$_3$. This acidity is primarily buffered by NH$_4$ at higher altitudes and by Mg and Ca at lower heights. On an average the bulk deposition has 13% higher concentration of components than wet-only. In wet-only and bulk depositions at higher altitude, the main neutralization occurs by NH$_4$ while at lower height buffering of acidity is due to Mg and Ca. The bulk deposition at lower altitude has 19% higher concentration of components than at higher altitude.

Acknowledgements

Authors are grateful to Dr. E.S.R. Gopal, Director, NPL and Dr. K. Lal, Head, Material Characterization Division, NPL for their encouragements and Dr. A.P. Mitra, FRS, for his inspiration and advice. Benefit of useful discussions with Dr. L.T. Khemani, Assistant Director, IITM, Pune and Dr (Miss) Nandini Kumar, Department of Chemistry, Dayalbagh Educational Institute, Agra is gratefully acknowledged. Thanks to CSIR, India and CSC, London for their approval to carry out this work under project CREN and to Prof H Rodhe and Dr L Granat, IMI, Stockholm for supplying the collectors.

References

Aggarwal M., and Varshney C.K.: 1993, *Proc. of International Conference on Regional Environment and Climate Changes in East Asia*, Nov. 30- Dec. 3, 137-141.

Charlson R.J. and Rodhe H.: 1982, *Nature*, **295**, 683-685.
Dean A.H. and Hobbs P.V.: 1981, *Atmos. Environ.*, **15**, 1597-1604.
Gillet R.W. et al.: 1992, *Proc. of 2ndCAAP/IGAC/IGBP Workshop*, Bombay, Sep.29-Oct.2.
Granat L.: 1972, *Tellus*, **24**, 550-560.
Granat L. et al.: 1992, *Proc. of 2ndCAAP/IGAC/IGBP Workshop*, Bombay, Sep. 29-Oct.2.
Handa B.K et al.: 1982, *Mausam*, **33**, 485-488.
Khemani L.T. et al.: 1985, *Water, Air and Soil Pollut.* **24**, 365-376.
Khemani L.T. et al.: 1989, *Atmos. Environ.* **23**, 753-756.
Kulshrestha U.C. et al.: 1991, *J. Indian Assoc. for Environ. Management*, **18**, 161-164.
Kulshrestha U.C. et al.. 1995, *Environmental Monitoring and Assessment*, **34**, 1-11.
Meiyuan H. et al.: 1993, *Proc. of the International Conference on Regional Environ. and Climate Change in East Asia*, November 30- December 3, 334-338.
Naik M.S. et al.: 1988, *Acta Met. Sinica*, **2**, 91-99.
Satai P.D. et al.: 1993, *Indian. J. Radio & Space Physics*, **22**, 56-61.
Saxena A. et al.: 1991, *Environmental Pollution*, **74**, 129-138.
Seinfeld J.H.: 1986, *Atmospheric Chemistry & Physics of Air Pollution*. Wiley, NY.
Singer A. et al.: 1993), *Atmos Environ.*, **27**, 2287-2293.
Strapp J.W.: 1988, *J. Geophy. Res.*, **93**, 3760-3772.
Varma G.S.: 1989, *Atmos. Environ.*, **23**, 2773-2778.
Warneck P. (ed): 1988, *Chemistry of the Natural Atmosphere*, Academic Press, New York.
Weatherburn M.W.: 1967, *Anal. Chem.* **39**, 971-974.
Yoshizumi K. and Asakuno K.: 1986, *Atmos. Environ.*, **20**, 151-155.

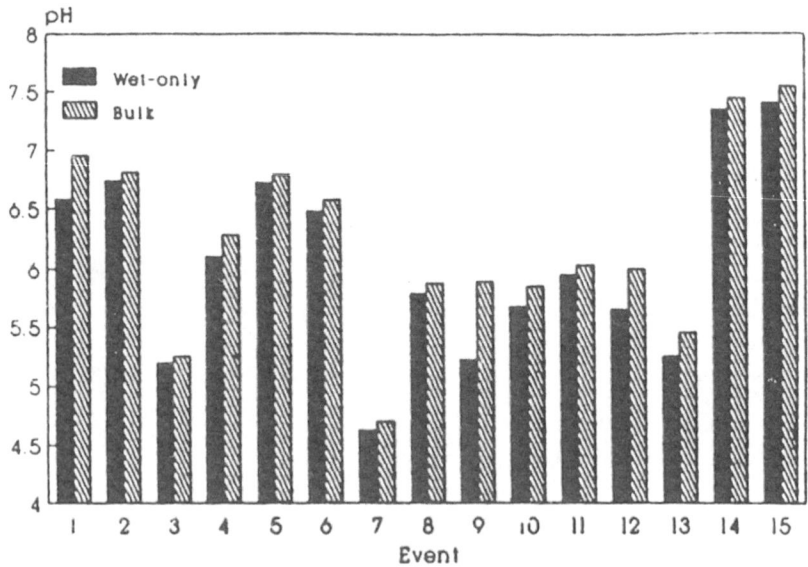

Fig. 1. Distribution of pH in wet-only and bulk samples.

A STUDY ON SHORT-TIME SAMPLING OF INDIVIDUAL RAIN EVENTS AT NEW DELHI DURING MONSOON, 1994

U.C. KULSHRESTHA[1], A.K. SARKAR[1], S.S. SRIVASTAVA[2] And
D.C. PARASHAR[1]

[1] *Chemistry Division, National Physical Laboratory*
Dr K S Krishnan Road, New Delhi-110 012 (India)
[2] *Department of Chemistry, Dayalbagh Educational Institute*
Dayalbagh, Agra-282 005 (India)

Abstract. Rain water samples of equal volume (50 mL) were collected from two convective showers at New Delhi on 28 July and 23 August during the monsoon, 1994. The variation of constituents of both the showers showed different trends which might have been due to different antecedent periods. The first shower occurred after an antecedent period of 2-3 hours while the second shower occurred after a 14 day interval. The first shower had acidic pH (<5.6) and relatively higher concentration of NH_4, SO_4, NO_3, Cl, F and K indicating insignificant below cloud scavenging. The second shower had alkaline pH (>5.6) and relatively higher concentrations of Ca and Mg. The higher concentration of Ca and Mg in the second shower were due to the loading with particulate matter during the preceding 14 days which made below cloud scavenging significant.

KEY WORDS: RAIN EVENT, pH, INTENSITY, CRUSTAL, ANTECEDENT PERIODS.

1. Introduction

Rain is the most effective scavenging factor in removing particulates from the atmosphere (Meszaros, 1991). It also dissolves gaseous pollutants during in-cloud and below cloud scavenging processes thus contributing to the acidity of rain water. The extent of acidity depends on the neutralization potential of rain water components and is controlled by particles scavenged during the two processes (Warneck, 1988). Higher concentrations of cations leads to alkaline rain water while dominance of anions generally results in acidic rain (i.e. pH< 5.6)(Charlson and Rodhe, 1982). In the Indian subcontinent the nature of rain water is alkaline (Khemani *et al*, 1989; Saxena *et al.*, 1991; Kulshrestha *et al.*, 1991) due to the influence of soil-derived particulate matter which is incorporated largely during below cloud scavenging process. The processes of scavenging below clouds may be insignificant when rain occurs daily or continuously for many hours during the monsoon. Hence, during such conditions the relative role of the two scavenging processes by which various constituents enter into rain water may be understood by collecting the samples at short-time intervals during a single rain event. Since studies on short time sampling of rain water within a single rain event are very limited (Gatz and Dingle, 1971; Raynor and Mc Neil, 1979; Kins, 1982; Naik *et al.*,

1994), we carried out the present study to understand the contribution of in-cloud and below cloud scavenging processes and variation of concentrations of various pollutants as a function of volume in two convective showers which occurred after different antecedent periods during the monsoon 1994.

2. Sampling and analysis

Samples of equal volume of 50 mL (1.5 mm rain) were collected from two convective showers on July 28, 1994 and August 23, 1994 on the terrace of the National Physical Laboratory, New Delhi at a height of 13 m above the ground. A total of 19 samples was collected (9 from the first shower and 10 from the second shower). The time interval between the first and last samples of first and second showers were 250 and 190 minutes respectively. The samples were collected using polypropylene bottle and funnel collectors (funnel diameter = 20 cm). After collection, the samples were transferred to storage bottles which were prewashed with double distilled deionized water and preserved with thymol (Granat et al., 1992; Koshy et al., 1993).

pH and conductance were measured immediately after the collection of sample using portable pH meter (Scandinovata, Model 102) and conductivity meter (Scandinovata, Model 122). The anions F, Cl, NO_3 and SO_4 were analyzed by Ion chromatography (Dionex 2000i/SP) using Na_2CO_3/$NaHCO_3$ eluent and H_2SO_4 as regenerant. Na, K, Ca and Mg were analyzed by atomic absorption spectrophotometry (Varian Spectra AA-10) while NH_4 was estimated colorimetrically using indo-phenol blue method (Weatherburn, 1967).

3. Results and discussion

The concentration of major ions, rainfall intensity, pH and conductance for two showers have been given in Table I. The rainfall intensity varied from 0.02 to 1.06 mm/min and reached a maximum during collection of third and fourth samples. Table I indicates that both the showers show a difference in the trend of variation of components. The concentration of crustal components in the first shower is lower whereas it is relatively higher in the second shower. This difference may be explained on the basis of antecedent periods. The first shower had an antecedent period of only 2-3 hours while the second shower occurred after 14 days from the previous event. Due to the shorter antecedent period for the first shower, the atmosphere was relatively free from particulates. The concentrations of crustal components (Ca and Mg) which have greater acid neutralization capacity, were lower resulting in a weak neutralization potential and acidic pH for all the samples in the first shower. The pH of second shower is alkaline except for the last two samples. It may be due to the loading with particulate matter during the 14 day antecedent period. The last two samples of the second shower which are acidic further validate that as the rain continues, the concentration of crustal components decreases enhancing the level of acidity of rain water. Table I indicates that the concentration of Ca in first shower is very low as compared to SO_4 and NO_3, and the concentration of Mg is below detection limits. In the second shower the concentrations of Ca and Mg are higher and the levels of the rest of the components are lower than in the first shower. This feature suggests that during the first shower, below cloud scavenging is insignificant but it contributes significantly to the second shower. It is corroborated by the findings of samples collected on 24 h basis for the same dates. Both these samples had similar values of pH (4.63 and 5.65 on July 28 and August 23, 1994 respectively) and ionic concentrations, to the individual events detailed in Table I.

TABLE I

Rainfall intensity*, pH, conductance** and major components***

S. No.	Inten sity	pH	EC	H	F	Cl	NO_3	SO_4	NH_4	Na	K	Ca	Mg
July 28, 1994													
E1	.10	4.42	48.2	38.01	6.10	20.26	46.00	127.35	65.23	3.82	37.83	47.12	10.70
E2	.05	4.53	24.5	29.51	4.56	10.82	23.20	30.70	44.29	.92	26.70	9.36	-
E3	.06	4.65	21.7	22.38	5.30	4.05	7.20	9.28	8.32	.87	28.80	10.80	-
E4	.31	5.27	9.6	5.37	1.43	5.36	6.80	34.00	14.08	1.77	30.68	55.76	-
E5	.09	5.51	8.2	3.09	3.28	9.08	5.20	14.50	9.18	.90	24.53	34.24	-
E6	.11	5.29	8.4	5.12	9.03	11.52	6.60	40.15	30.36	1.20	27.41	61.38	-
E7	.18	5.28	6.9	5.12	3.03	6.89	3.30	15.90	11.04	1.12	32.12	21.16	-
E8	.11	5.34	7.8	6.30	1.55	2.92	7.98	13.35	25.54	1.99	32.63	12.20	-
E9	.03	5.51	8.8	3.09	5.74	13.62	6.80	13.44	29.03	2.03	34.00	14.12	-
August 23, 1994													
E1	.10	5.76	16.8	1.73	2.06	3.86	7.27	27.15	38.05	1.06	1.25	34.74	3.59
E2	.56	5.87	8.1	1.34	2.60	1.73	2.12	7.33	12.56	.92	.20	32.65	.88
E3	1.06	5.95	6.1	1.12	3.15	7.60	2.42	6.11	8.83	1.08	1.02	29.00	1.00
E4	1.02	5.94	4.6	1.12	3.03	6.49	4.32	5.20	6.66	.87	.44	30.96	.64
E5	.99	5.93	3.6	1.17	1.98	.79	4.70	5.35	5.17	.98	.43	25.75	-
E6	.95	5.85	2.6	1.41	.94	2.53	5.27	4.00	3.35	.82	2.28	24.66	.60
E7	.62	5.89	3.3	1.51	1.43	1.93	6.42	2.65	1.25	1.02	1.06	28.42	-
E8	.38	5.92	2.4	1.51	1.92	2.32	4.29	2.60	3.26	.92	.67	26.30	.64
E9	.08	5.49	5.6	3.23	2.90	7.25	6.82	7.47	3.68	1.20	1.60	32.61	1.22
E10	.02	5.06	8.6	8.71	.61	4.08	4.30	11.35	6.05	.83	.07	25.23	-

*mm/min, **µS/cm, ***µeq/L.

Table 1 reflects that during the first shower, the intensity decreases as the pH increases while in the second shower, both pH and intensity decrease. The concentrations of Ca and SO_4 decrease in both the showers while the concentration of NH_4 increases in the first and decreases in the second shower. The concentration of NO_3 decreases in the first shower and increases in the second shower. This feature indicates that the increasing trend of pH in the first shower may be due to increase in NH_4 and decrease in SO_4 and NO_3 while the decrease in pH of the second shower may be due to a decrease in concentration of crustal components (mainly Ca) as the rain continues. In the case of the first shower the ratio of $(Ca+NH_4)/(SO_4+NO_3)$ increases as pH increases but this ratio decreases along with decrease in pH in the case of the second shower.

The correlation coefficients of various parameters for the first and second showers are given in Tables II and III respectively. The correlation coefficients of pH with intensity indicate that during the first shower pH did not vary as a function

TABLE II

Correlation matrix of various parameters for the first shower

	Intensity	pH	EC	H	F	Cl	NO_3	SO_4	NH_4	Na	K	Ca
Intensity	1.00											
pH	.23	1.00										
EC	-.24	-.87**	1.00									
H	-.31	-.97**	.95**	1.00								
F	-.22	-.58	.86**	.69	1.00							
Cl	-.33	-.34	.65	.49	.84*	1.00						
NO_3	-.20	-.77*	.95**	.87**	.87*	.74	1.00					
SO_4	.05	-.58	.84*	.68	.93**	.78*	.90**	1.00				
NH_4	-.30	.59	.77*	.71	.80*	.80*	.90**	.82*	1.00			
Na	.05	-.28	.64	.42	.78*	.63	.74	.82*	.72	1.00		
K	.08	-.19	.45	.28	.57	.41	.50	.56	.51	.86*	1.00	
Ca	.55	.11	.07	-.10	.36	.32	.15	.50	.15	.29	.01	1.00

1-tailed significance: * - .01, ** - .001

of intensity while during the second shower significant and positive correlation coefficients indicate as the intensity increases, the pH also increases due to more efficient scavenging of buffering species. Significant correlation coefficients of SO_4 and NO_3 with NH_4 suggest that the variation of both is highly dependant on NH_4 and vice versa. During rain out process H_2SO_4 and HNO_3 are converted into their corresponding salts by reaction with ammonia forming ammonium sulphate and ammonium nitrate.

TABLE III

Correlation matrix of various parameters for the second shower

	Intensity	pH	EC	H	F	Cl	NO_3	SO_4	NH_4	Na	K	Ca
Intensity	1.00											
pH	.72*	1.00										
EC	-.53	-.30	1.00									
H	-.62	-.97**	.22	1.00								
F	-.26	.47	.05	-.52	1.00							
Cl	-.03	-.20	.14	.12	.57*	1.00						
NO_3	-.45	-.19	.23	.05	-.26	-.01	1.00					
SO_4	-.53	-.29	.96**	.20	-.02	.11	.39	1.00				
NH_4	-.33	.01	.93**	-.09	.13	.04	.25	.94**	1.00			
Na	-.24	.06	.23	.22	.55	.42	.37	.22	.21	1.00		
K	.07	.16	-.09	-.30	-.08	.17	.52	.02	.02	.28	1.00	
Ca	-.30	.10	.64	-.26	.64	.37	.20	.56	.65	.56	.01	1.00

1-tailed significance: * - .01, ** - .001

4. Conclusion

This short-time sampling study reveals that if rain continues for a long period, below cloud scavenging becomes insignificant while in-cloud scavenging processes become significant. The pH of rain water is a function of particulate loadings in the atmosphere. Higher pH of rain water is observed when the rain occurs after a longer antecedent period, causing higher loading of atmosphere with components like Ca and Mg which have higher acid neutralizing capacity. However, in a cleaner atmosphere, after shorter antecedent period NH_4 neutralizes the acidity but does not raise the pH effectively. In brief, these results show that the acidity of rain water is highly dependent on the antecedental period before the sampling of rain water.

Acknowledgements

Authors are grateful to IMI, Stockholm for the supply of collectors and preliminary measurement devices to carry out the work under the auspices of Project on Chemical Research and Environmental Need (CREN) of CSC, London. Thanks are due to Prof E S R Gopal, Director, NPL, Dr K Lal, Head, Material Characterization Division for their encouragements and Dr A P Mitra, FRS for his inspiration. It is a pleasure to acknowledge the benefit of scientific discussion with Dr L T Khemani, Assistant Director, Indian Institute of Tropical Meteorology, Pune and Dr (Miss) Nandini Kumar, Department of Chemistry, Dayalbagh Educational Institute, Agra.

References

Charlson R.J. and Rodhe H.: 1982, *Nature*, **295**, 683-685.

Gatz D.F. and Dingle A.N.: 1971, *Tellus*, **23**, 14-27.

Granat L., Suksomsankh K., Simachaya S., Tabucanon M. and Rodhe H.: 1992, *Proceedings of 2nd CAAP/ IGAC/ IGBP Workshop,* Bombay, September 29- October 2.

Khemani L.T., Momin G.A., Prakash Rao P.S., Safai P.D., Singh G. and Kapoor R.K.: 1989, *Atmos Environ*, **23**, 757-762.

Kins L.: 1982, In *Deposition of atmospheric pollutants* (eds. H.W. Georgii and J. Pankrath), Reidel, Dordrecht, Holland.

Koshy K., Ayers G., Gillet R. and Selleck P.: 1993, *Proceedings of the International Conference on Regional Environment and Climate Change in East Asia.* 30 November- 3 December, 236-241.

Kulshrestha U.C., Saxena A. and Srivastava S.S.: 1990, *J. Indian Association for Environmental Management*, **18**, 161-164.

Meszaros E.: 1991, In *Atmospheric particles and nuclei* (eds. G. Gatz, E. Meszaros and G. Vali), Akademiai Kiado, Budapest, 68-74.

Naik M.S., Khemani L.T., Momin G.A., Rao P.S.P., Pillai A.G. and Safai P.D.: 1994, *Tellus*, **46B**, 68-75.

Raynor G.S. and Mc Niel P.: 1979, *Atmos. Environ.*, **13**, 149-155.

Saxena A., Sharma S., Kulshrestha U.C. and Srivastava S.S.: 1991, *Environmental Pollution*, **74**, 129-138.

Warneck P. (ed): 1988, *Chemistry of the natural atmosphere,* Academic Press, Inc, New York.

Weatherburn M W, *Anal Chem,* **39** (1967), 971-974.

NITROGEN COMPOUNDS IN ATMOSPHERIC PRECIPITATION

R.LAVRINENKO

A.I.Voeikov Main Geophysical Observatory, 7, Karbyshev Str.,
194018, St. Petersburg, Russia

Abstract. The paper presents results of a stoichiometric calculation of a nitrogen (N) compounds in precipitation of World Meteorological Organization's Global Atmosphere Watch (WMO GAW) stations. Long-term trends of ammonium sulphate (($NH_4)_2SO_4$) and ammonium nitrate (NH_4NO_3) contents in the North-West of Russia as well as in Byelorussia, Scandinavia, Western and Eastern Europe during the periods of 1958-1990 and 1972-1985 were investigated. A relatively, steady annual trend for the mean NH_4NO_3 concentrations was found typical of pure regions (5-15 μeq* l^{-1}). The concentrations in industrial regions are from 4 to 5 times higher than the background close to natural. The analysis of the trend for $(NH_4)_2SO_4$ content in precipitation shows a wide range of a variations of mean annual concentrations with an explicit tendency to their significant decrease in some European regions in the mid-eighties. Nitric acid (HNO_3) has not been discovered in precipitation from the European WMO GAW stations while calculations based on the US data revealed its remarkable content and tendency to its increase. Nitric acid and ammonium sulphate are not contained in precipitation over ocean, ammonium nitrate is present in insignificant amounts.

Key words: precipitation, ammonium sulphate, ammonium nitrate, nitric acid, concentration, trend

1. Introduction

Usually a change in content of nitrogen (N) and sulphur (S) in precipitation in time is considered as the trend of nitrate (NO_3^-), ammonium (NH_4^+) and sulfate (SO_4^{2-}) ions. This paper presents results of long-term trend investigations concerning the content of nitrogen in molecular forms ammonium nitrate (NH_4NO_3) and ammonium sulphate (($NH_4)_2SO_4$) on the bases of WMO GAW (World Meteorological Organization's Global Atmosphere Watch) data (WMO, EPA, NOAA, 1974-1989) and national network data (SCHM, MGO, 1970, 1985, 1986, 1989).

The calculation scheme of chemical compounds in precipitation consists of 5 blocks in which the contents of sea salts, nitrogen compounds, acids, bicarbonates and sulphur compounds are determined. These calculations are based on quantitative relations between ions in sea water and an equivalent form, expressing substances in those chemically equal units, proportionally to which they enter into reactions with each other and are bound in chemical compounds (Lavrinenko, 1979).

The disbalance of some ion in any block reveals the inaccuracy in analytical determining a particular ion. Sulphate percent ion concentration differences between measured values and values calculated using the ion pairing technique do not exceed 25% (Artz et al., 1993). But this is not the end in itself for such calculations. The main result is to estimate the anthropogenic components in precipitation and their sources, to find the natural background levels, using the international data bank.

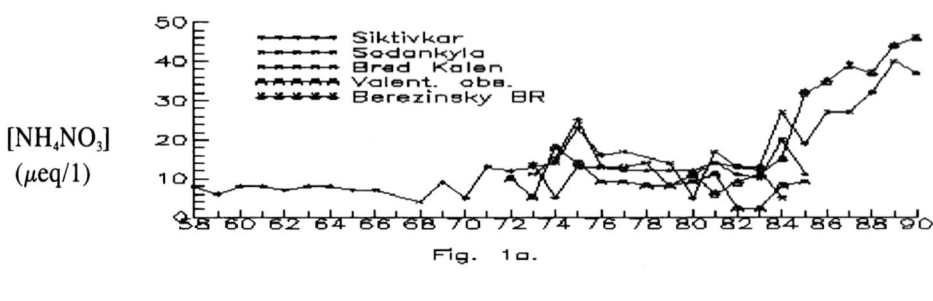

Fig. 1a,b NH4NO3 trend in precipitation WMO GAW sites

2. NH$_4$NO$_3$ precipitation trend

The content of NO$_3^-$ ion in precipitation is mostly due to the NH$_4$NO$_3$ compound. The connection between concentrations of NO$_3^-$ and NH$_4^+$ ions, which is observed in precipitation, supposes a common source for their genesis (Jones, 1971). Natural processes of vital functioning of microorganisms in air and soil, products of fuel combustion, as well as one of the forms of the nitric fertilizer may serve as such sources. The results of stoichiometric calculation of ammonium nitrate in precipitation are presented in Figure 1 a,b where two levels of observed concentration are reflected - in relatively clean (Figure 1a) and industrial regions of Europe (Figure 1b). A distinct trend to the increase of content of this compound in precipitation of the North-West region Russia may be observed as the example of the station Siktivkar, where the level of mean annual concentrations did not exceed 10-13 μeq* l^{-1} in 50-70-S. In the late 80-s the content NH$_4$NO$_3$ in this region increased from 4 to 5 times. During the short period of 1980-1990 NH$_4$NO$_3$ concentrations according to the data from Berezinsky Biospheric Reserve (BR), Byelorussia, increased by an order of magnitude. By 1990 their level had reached 40-50 μeq* l^{-1} (Figure 1a).

Such order of magnitude for the values was already characteristic of the Western and Eastern Europe in 70-s and the first half of 80-s (Figure 1b). However, in the clean regions of the European subcontinent the North of Scandinavia (Sonankyla, Bredkalen), South of Ireland (Valent. obs.), as well as in the North-West of Russia, the changes in the NH$_4$NO$_3$ content in precipitation in 70-s were not significant, and the order of magnitudes for the values made up 5-15 μeq* l^{-1} (Figure 1a).

[(NH$_4$)$_2$SO$_4$]
(μeq/l)

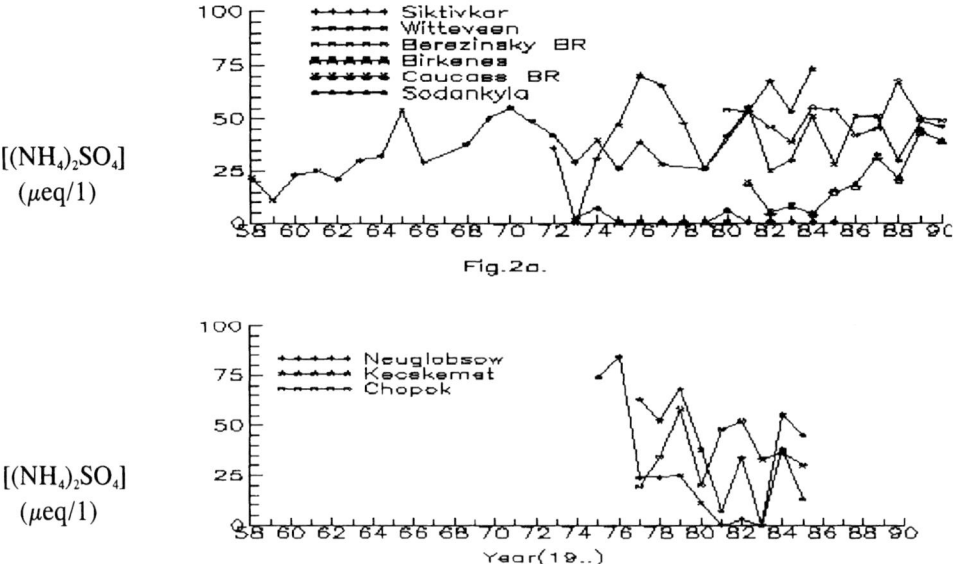

Fig.2a,b. (NH4)2SO4 trend in precipitation WMO GAW sites

These values correspond well enough with the calculated mean values of NH$_4$NO$_3$ concentrations in cloud water, which was investigated over the territory of the former USSR in 60-s (Petrenchuk, 1979). The long-term trend (1915-1945) of the NO$_3^-$ ion precipitation of the North-East of US (Ithaca) shows nearly the same range of 4-15 μeq* l^{-1}. This level increased 4-5 times by 1976 and made up about 45 μeq* l^{-1} (Galloway et. al, 1978).

The period since 1975 to 1985 was distinguished by a dynamic character of changes in mean annual NH$_4$NO$_3$ concentration with a large amplitude of oscillation in densely populated European regions (Fig. 1b). The maximum values there made up to 70-90 μeq* l^{-1}, which about 2 times exceeded the mean concentration level.

3. (NH$_4$)$_2$SO$_4$ and HNO$_3$ content in precipitation.

3.1 ((NH$_4$)$_2$SO$_4$) PRECIPITATION TREND

The known reactions in the atmosphere as well as particles of ash and ammonium fertilizers in the air may also be a source of ammonium sulphate in precipitation. At the same time, according to the data from American, Japanese and some European stations (Scandinavia, Ireland) this compound was not observed at all, or was only in insignificant accounts in separate months. As an example, results of calculations by the data from the stations Sodankyla and Birkenes are presented in Figure 2a.

In the North-West of Russia (Siktivkar) the level of mean annual (NH$_4$)$_2$SO$_4$ concentrations in the late 50-s and early 60-s made up 20-40 μeq* l^{-1}. Mean concentrations of this compounds in the water of frontal clouds over the territory of the former USSR in that period also corresponded to this order of magnitude and made up to 20-30 μeq l^{-1}. Starting from the mid-60-s and up to the late 80-s, the content of (NH$_4$)$_2$SO$_4$ in precipitation of the North-Western region of Russia changed within a stable interval from 30 to 50 µeq* l⁻¹. The data from Berezinsky BR confirm this stably high level of $(NH_4)_2SO_4$ content in precipitation (Figure 2a).

A high concentration level and a large amplitude of oscillations are characteristic also of the Western and Eastern Europe: data of Witteveen station in the Netherlands may serve as an example of this (Figure 2a). At the same time, according to the data from certain Eastern Europe stations, a trend to decrease of this compound content is revealed distinctly for the first half of the 80-s. This process was observed to be especially dynamic in the former East Germany (Neuglobsow) and Hungary (Kecskemet) (Figure 2b).

3.2 HNO_3 CONTENT IN PRECIPITATION

According to the WMO GAW data no nitric acid was discovered in precipitation of the European subcontinent. Nitric acid was calculated for precipitation of the American stations. A distinct trend to the growth of mean annual HNO_3 concentrations from 10 to 20 µeq* l⁻¹ was observed in the region of Washington during the period 1982-1987.

4. Nitrogen compounds in precipitation over ocean basins

The investigation of chemical composition of precipitation in open ocean allows to more reliably determine the content of some substance or other in the atmosphere in conditions close to natural, and to compare this with the results obtained on the continent. Data used for calculation of nitrogen compounds in precipitation over oceans were received by Soviet scientific-research expeditions to the Pacific and by American background stations located on some islands in that ocean. The results of precipitation analytical composition observations in Indian and Atlantic oceans were calculated using the published data (Keen et. al, 1986). The results of calculation of nitrogen compounds content in precipitation over oceans are presented in Table 1. The distribution of pH in precipitation samples gathered at Soviet expeditions ships in the Pacific Ocean since 1965 to 1972, closely corresponds in general to the normal law with frequency maximum near the mean value pH=5,86 (Selezneva, 1974). The major of samples, about 200 in number, gathered in different parts of the ocean in 1980-1989, possessed the values of pH also close to this mean value. However, in the Eastern part of the Pacific, near the American coast acid precipitation fall out had been observed with the minimum value of pH=3,7.

The nitrate-ion balance calculations does not reveal nitric acid in precipitation over the ocean basins. The acidity of the precipitation samples with pH<5,0 is provided for by a content of sulfuric acid. The sulphate-ion balance calculation shows that its content in precipitation is determined mainly by the sea salt-magnesium, calcium and potassium sulphates. An investigation of the sufficiently large amount of data, allow to conclude that either just traces of NH_4NO_3, or only insignificant amounts of it, about 1-4 µeq* l⁻¹ were seen in the precipitation. Ammonium sulphate also practically is not contained in the precipitation over oceans, the data of Soviet expeditions being an exclusion which can be explained by the effect from local pollution sources-combustion of fuel in tanks of the vessels.

Table 1.

Nitrogen compounds in precipitation over oceans

1,2,3 - Pacific Ocean: 1 - research ships, 2 - Samoa Islands, 3 - Hawaii, Mauna Loa
4 - Indian Ocean, Amsterdam Isl., 5 - Atlantic Ocean, Bermudas.

Region	Sampling period	Number of samples	pH	HNO_3	NH_4NO_3 $\mu eq/l$	$(NH_4)_2SO_4$
1	1965-1972	214	5.86	0	traces	15.0
	1980-1989	368	-	0	2.0	7.0
2	1975-1987	-	5.46	0	1.2	1.1
3	1980-1987	-	5.06	0	1.1	0
4	1980-1983	79	5.04	0	1.7	0
5	1982-1984	253	4.94	0	3.6	0

5. Conclusion

In the North of Scandinavia and North-West of Russia during a long period of time rather a stable range of NH_4NO_3 concentration, 5-15 $\mu eq * l^{-1}$, was kept. The order of magnitude of these values is in a good agreement with mean values of concentration of this substance in cloud water and may characterize the background level of its content in precipitation of continental stations. In industrially developed regions of Europe this level increased in the 70-s and in the former USSR in late 80-s from 4 to 5 times.

According to the data from a number WMO GAW stations, ammonium sulphate in precipitation is practically not observed. The presence of this compound in precipitation is an indicator of its anthropogenic pollution. On the territories of Western and Eastern Europe the trend of $(NH_4)_2SO_4$ concentrations is characterized by a wide range of their oscillations from zero values to the maximal, equal to 70-90 $\mu eq * l^{-1}$. Stable high level $(NH_4)_2SO_4$ content from 30 to 50 $\mu eq * l^{-1}$ in precipitation of the North-Western region of the former USSR is observed. There is a tendency to a decrease of the ammonium sulphate content in the former East Germany and Hungary.

Nitric acid is not observed in precipitation of the regions of European background stations. Its remarkable content in precipitation of American stations is likely to be due to the anthropogenic factor only. There is only very little nitrogen in precipitation over oceans, and it is mostly represented in the form of ammonium nitrate. An investigation of the sufficiently large amount of data obtained at open parts of the Pacific ocean allows to conclude that precipitation mean value pH is close to equilibrium (pH=5.65). In the East part of the Pacific and on the islands of Indian and Atlantic Oceans the acidity of the precipitation is provided for by a content of sulphuric acid.

Acknowledgment

The author wishes to thank Prof. E.Selezneva for many helpful discussions.

References

Artz, R.S. and Lavrinenko, R.F.: 1993, "Background Precipitation Chemistry Monitoring in the Soviet Union". Environmental Monitoring and Assessment, **26**, 1-25.

Galloway, J.N., Cowling, E.B.: 1978, "The Effects of Precipitation on Aquatic and Terrestrial Ecosystems: A Proposed Precipitation Chemistry Network". Air Pollut. Control, **28**, N 3, 229-235.

Jones, M.J.: 1971, "Ammonium and Nitrate Nitrogen in the Rainwater at Samaru, Nigeria". Tellus, **23**, 459-461.

Keene, W.C., Pszenny, A.P., Galloway, J.N., Hawley, M.E.: 1986, "Sea-Salt Corrections and Interpretations of Constituent Ratios in Marine Precipitation". Geoph. Research, **91**, N 6, pp. 6647-6658.

Lavrinenko, R.F.: 1979, "The Balance of Sulphates in Atmospheric Precipitation". Main Geophysical Observatory (MGO) Trudy, **418**, 34-42 (Russ.).

Petrenchuk, O.P.: 1979, "Experimental Investigations of Atmospheric Aerosol", Hydrometeoizdat, Leningrad, 263 pp. (Russ).

Selezneva, E.S.: 1974, "Some Physical-Chemical Parameters of Atmospheric Precipitation over the Pacific Ocean". MGO Trudy, **343**, 46-56 (Russ.).

USSR State Committee for Hidrometeorology and Control of Natural Environment, Main Geophysical Observatory (SCHM, MGO): 1970, 19985, 1986, 1989, "Monthly Data on the Chemical Composition of Atmospheric Precipitation", for 1962-1985.

World Meteorological Organization (WMO), Environmental Protection Agency (EPA), Natural Oceanic and Atmospheric Administration (NOAA): 1974-1989, "Global Atmospheric Background Monitoring for Selected Environmental Parameters BAPMoN Data", for 1972-1985.

ACID AND ALKALINE DEPOSITION IN PRECIPITATION ON THE WESTERN COAST OF SARDINIA, CENTRAL MEDITERRANEAN (40° N, 8° E)

O. LE BOLLOCH[1] and S. GUERZONI[2]

[1] *International Marine Centre, l.re E. d'Arborea 22, 09072 Torregrande, Oristano, Italy.* [2] *Istituto di Geologia Marina/CNR, via Gobetti 101, 40129 Bologna, Italy.*

Abstract: The ionic composition of 53 precipitation samples collected at a coastal site in western Sardinia (Torregrande) was measured in order to determine concentrations and fluxes of the major ions. Precipitation events were sampled over a two year period, Oct'92/Oct'93 and Oct'93/Oct'94. Despite the proximity to the African continent and the detection of Saharan inputs reflected in high pH values and high $_{nss}$Ca fluxes, there is evidence for acidic rain with over 54% of the events having pH values <5.6. This would suggest long-range transport of anthropogenic inputs, as there is no obvious local source. Within the study period, a marked decrease in median pH is recorded at Torregrande in '93/'94, decreasing from 5.94 to 5.18. This is coupled with an increase in total annual flux of NO_3 from 27 to 63 µg/cm², making the average flux of NO_3 in the two years comparable with that at Capo Carbonara (48 µg/cm²). Particulate fluxes at Torregrande show a marked inter- as well as intra-annual variability, from 70 to 200 µg/cm². Variations of pH seems to be partially due to the total particulate as well as $_{nss}$Ca fluxes. A statistic of Saharan enriched rain episodes of the two sites showed a peak transport season in April-May and October which is two times greater at Capo Carbonara. Rainfall on the west coast, therefore, appears to be less affected by Saharan dust than rainfall on the southern coast and therefore shows greater signs of acidification due to anthropogenic inputs.

Keywords: ionic fluxes, acidic deposition, Saharan influence, particulate fluxes, Mediterranean.

1. Introduction

Mediterranean sites have been recognized to be under the influence of alkaline deposition linked to desert dust outbreaks, that move from the Saharan desert towards Europe. Several authors have documented the role of calcium and carbonates in raising the pH of so called 'red rains'. In addition it has been documented that wet deposition accounts for almost 3/4 of the total (Glavas 1988; Guerzoni *et al.*, 1993; Rodà *et al.*, 1993). To date most acid deposition and modelling research efforts have focussed more on the influence of acid materials and less emphasis has been placed on the deposition of alkaline materials, such as Ca, Mg and K.

The southern Mediterranean basin should be highly affected by alkaline fluxes due to its proximity to North Africa, and in Sardinia prior work has demonstrated the role of Saharan dust in buffering the acidity of rain. Average pH values are typically above 5, and most of the rain episodes can easily reach pH values above 6.5 (Caboi *et al.*, 1992, 1995). The Central Mediterranean area is also affected by long-range transport of anthropogenic substances of 'European' origin, that can reach lower latitudes, resulting in acidic rains (Caboi *et al.*, 1992; Guerzoni *et al.*, 1995).

In this paper we present new data of samples taken at a remote coastal site on the western coast of Sardinia (40° N, 8° E) since October 1992. Data collected over a two year period, October 1992 to October 1994, is compared with data collected at a similar site in southern Sardinia (Capo Carbonara: 39° N, 9° E). We also provide new flux data for the Central Mediterranean.

2. Materials and Methods

Wet only precipitation samples were collected at the end of each rain event from a sampler with precipitation sensor. Samples were collected in polythene bottles and stored at 4°C until analysis. Analyses were carried out as soon as possible on filtered (0.45 μm) samples. Potentiometric methods using specific electrodes were used to determine conductivity, pH and ammonia. Ionic chromatography was used to determine Cl^-, SO_4^{2-} and NO_3^- and AAS (flame emission) for Na^+, K^+, Ca^{2+} and Mg^{2+}. Alkalinity was measured by titrating the sample with 0.01 M HCl using Gran's method. Ionic balances were then calculated and concentrations of the major ions were corrected for sea salt concentrations using Na as a reference species, in order to determine the fraction of salts in the sample which are attributable to inputs other than sea salt.

3. Results

The precipitation regime consists of a wet season from September to May and a dry period in June, July and August. The yearly precipitation for the period between October '92 and October '93 was 348 mm over 21 events and 378 mm over 33 events between October '93 and October '94. This is comparable with the other site in Southern Sardinia, where the total precipitation in 1991 was 357 mm. Table I lists the chemical data.

TABLE I
Mean concentration (μequ l^{-1}) of the major ions in rain samples collected in Western Sardinia (Torregrande) and Southern Sardinia (Capo Carbonara).

	W. Sardinia 92/93		W. Sardinia 93/94		S. Sardinia 1991	
no of events	21		32		31	
mm rain	348		378		357	
Cond (μS/cm)	56		55		88	
pH mean	5.94		5.18		5.84	
range	(5.17-7.21)		(3.82-7.89)		(3.96-7.23)	
	Total	nss	Total	nss	Total	nss
H	-	7	-	19	-	9
Ca	84	71	70	57	155	136
Mg	73	-	77	-	115	
Na	266	-	252	-	424	
K	12	9	17	14	15	5
NH_4	25	25	25	25	21	21
SO_4	75	60	90	76	104	52
NO_3	18	18	29	29	26	26
Cl	332	131	322	132	474	-
HCO_3	60	60	43	43	108	108

nss = non sea salt

The mean pH was 5.94 for 92/93 and 5.18 for 93/94. Six out of 21 samples may be classed as acidic, with a pH less than 5.6, in 92/93, whilst 25 out of 33 in 93/94. This is coupled with an increase in NO_3 and $_{nss}SO_4$ in 93/94. The marine contribution constitutes approximately 20% of the ionic composition for both years, except in the case of Mg where all is attributable to sea salt and Cl of which 40% is non sea salt. The seasalt fraction of the total dissolved salts is more than 60%. The ratio of Cl to Na is 1.22 and 1.96 respectively which is higher than the ratio found in seawater (1.16) suggesting an additional source of Cl. A medical waste incinerator nearby is a possible source of this excess Cl, however further study is currently underway to identify the source. The site in western Sardinia is similar in marine character to that in the southern part of the island, but less pronounced (Na conc. 266 compared to 424) as there are stronger winds at Capo Carbonara which affect seasalt content in rain (Lovett, 1978).

3.1 ACID DEPOSITION

The incidence of acidic events (pH < 5.6) is seen to increase dramatically from 30% to 70% of the total number of events in 93/94. Unlike in Northern Europe, where NH_4 is the important neutralising agent, Ca brought in on southerly winds carrying Saharan dusts, is believed to be more important at Torregrande. It is probable that anthropogenic ions (NO_3 and $nssSO_4$) are responsible for acidic rains and are transported from Northern Europe by the Mistral winds (N-NW). As sources largely depend on wind direction, the presence of these acidifying and neutralising agents is variable, which probably explains the wide range of pH values observed in 93/94. The ionic composition of the rainfall reflects the pH variation, with an increase in $nssSO_4$ and NO_3 concentrations in the latter year.

3.2 SAHARAN DUSTS

Samples with high pH values and enriched in calcium, supported by colour inspection of filters, were taken to be of 'Saharan' origin. Such episodes are seen to be more frequent at Capo Carbonara. From 1991 to 1994 in Southern Sardinia more than 30 episodes of red dust were detected, 13 in the western side and 20 in the southern one. This fact is reflected principally in the values of $nssCa$ and HCO_3, two times higher at Capo Carbonara than in Torregrande. By considering the 33 Saharan events which occurred at Torregrande and Capo Carbonara over a two year period, the seasonal frequency shows a peak in transport events in April-May and again in October (Figure 1), which is in accordance with previous observations of spring and autumn desert transport

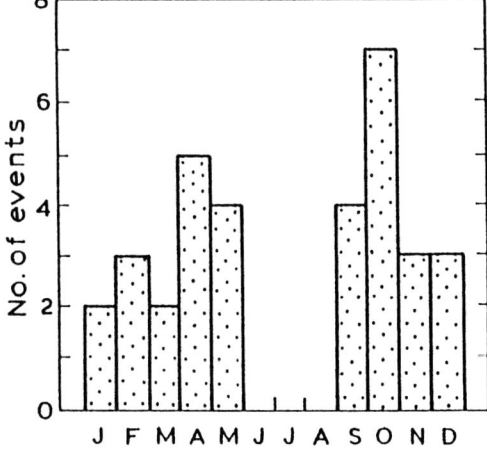

Fig. 1 Monthly distribution of the Saharan events in Sardinia in the period 1991-1994.

events (Bücher et al., 1983; Littman et al., 1990; Guerzoni et al., 1993).

3.3 FLUXES

Fluxes of the major ions were calculated (Table II). The effects of the Saharan episodes are clearly more visible in Southern Sardinia. Both Sardinian sites are marine sites, reflected in similar Na fluxes. Despite NO_3 fluxes also being similar at both sites, $_{nss}SO_4$ fluxes are greater on the western coast, although this is probably because neutralisation of $_{nss}SO_4$ by Ca is greater at Capo Carbonara due to the higher levels of this cation in the rainwater. The North American site in Florida is also a marine site with comparable fluxes. Despite the fact that Saharan events were noted to influence precipitation in Spain (Rodà et al., 1993), these inputs are not sufficient to buffer the anthropogenic fluxes, hence the high fluxes of these ions.

TABLE II
Fluxes of the major ions (μgcm^{-2}) in western Sardinia compared with a) Capo Carbonara (Guerzoni et al., 1995), b) NE Spain (Rodà et al., 1993) and c) Florida, (Savoie et al., 1987).

	W. Sardinia	S. Sardinia[a]	Spain[b]	America[c]
nsCa	39	84		
$nssSO_4$	115	84	360	85
NO_3	47	48	150	60
NH_4	16	12	183	
Na	226	258		223
HCO_3	87	108		

TABLE III
Average fluxes (μgcm^{-2}) between 1992 and 1994 of some of the major ions from 'Saharan' and non-Saharan episodes in Sardinia and the contribution of 'Saharan' episodes to these fluxes.

	Total	Saharan	Non-Sah.	Sah/total (%)
nssCa	40	26	14	65
$nssSO_4$	115	35	80	30
NO_3	47	11	36	28
NH_4	15	6	6	40
Na	225	30	195	13
HCO_3	87	60	21	76
Particles	135	96	39	71
pH		6.34	5.16	
mm rain		73	290	
no of events		13	40	

On examining the fluxes at Torregrande divided into Saharan and non-Saharan episodes (Table III) it is clear that the Saharan episodes have an important contribution to the input of ions to the rainfall composition. High $_{nss}Ca$ and HCO_3 fluxes in Saharan events are responsible for the higher average pH associated with these events, whilst $_{nss}SO_4$, NO_3 and NH_4 fluxes are mainly associated with non-Saharan events. However, 40% of the NH_4 flux is in Saharan events. Figures 2 and 3 show positive linear correlations of pH with $_{nss}$Ca content in precipitation and with particulate fluxes associated with each rain event.

Particulate fluxes, measured as total insoluble particles in rain, show marked inter- as well as intra-annual variability. The total annual flux for 92/93 was 69 μgcm^{-2}, with one episode accounting for more than 30% of this flux, whilst it was 200 μgcm^{-2} in 93/94

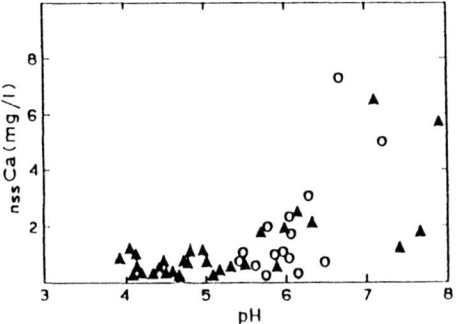 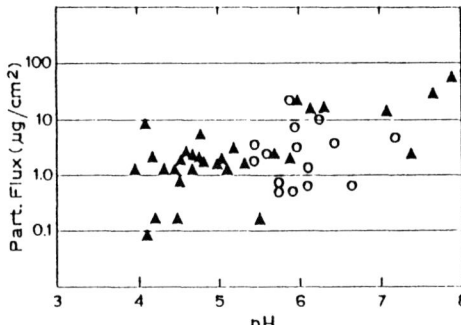

Fig. 2 - Plot of pH values and $_{nss}$Ca (mg l^{-1}) in rain samples (O = 1992/93; Δ = 1993/94).

Fig. 3 - Plot of pH values and particulate flux (mg cm^{-2}) in rain samples (O = 1992/93; Δ = 1993/94)

with 5 events accounting for 78% of the flux. Saharan events contribute 71% of the total particulate flux. This variability can also be seen at Capo Carbonara, although fluxes are much higher (920 μgcm^{-2} in 90/91 and 260 μgcm^{-2} in 92/93) due to the more frequent direct trajectories from the South and the closer proximity of the African continent.

4. Conclusions

A new station in operation for two years in western Sardinia permits the following conclusions:
- wet deposition between 39-40° N is under the neutralising effects of Saharan episodes: pH in rain resulted to be clearly correlated with particulate flux and the non marine Ca content.
- there is some variability between the two sampling sites, with the effects of crust-enriched deposition more visible on the southern coast: $_{nss}$Ca and HCO$_3$ fluxes are higher, NO$_3$ fluxes are similar. Non sea salt SO$_4$ fluxes appear to be lower at Capo Carbonara, although it is probable that $_{nss}$SO$_4$ has precipitated out of solution during neutralising reactions with Ca.
- A strong interannual as well as inter site variability in Saharan components was detected in Sardinia, confirming the "pulse" characteristics of the Saharan outbreaks.
- The western part of the island, appears to be more influenced by European anthropogenic influxes during 1993/94, precipitation showing signs of increasing acidification which deserves future attention.

Acknowledgements

Thank you to G. F. Obinu for starting collection and analysis of samples. Thanks are also due to Prof. A. Cristini, Mr. G. Contis and Mr. L. Muscas for help with the analyses. This work was partially supported by the EU STEP Programme (STEP-CT 090-0080 DSCN) and by the World Meteorological Organisation (Contract ACN/DS/0290). This work is contribution no. 997 of the IGM-CNR.

References

Bücher, A., Dubief, J. and Lucas, C.: 1983, *Revue de Geologie Dynamique et de Geographie Physique*, **24**, 153-165.
Caboi, R., Cristini, A., Fanfani, L., Frau, F., Pinna, R. and Zuddas, P.: 1992, In: *Water-Rock Interaction*, Kharaka & Maest (eds), Balkema, Rotterdam.
Caboi, R., Cristini, A. and Frau, F.: 1995, (in press)
Chester R., M. Nimmo, M. Alarcon, C. Saydam, K.J.T. Murphy, G.S. Sanders and P. Corcoran: 1993, *Oceanologica Acta*, **16** (3), 231-245.
Glavas, S.: 1988, *Atmospheric Environment*, **22**, 1505-1507.
Guerzoni S., Cristini A., Caboi R., Le Bolloch O., Marras I. and Rundeddu L.: 1995, *Water, Air and Soil Pollution* (this issue).
Guerzoni, S., Landuzzi, W., Lenaz, R., Quarantotto, G., Rampazzo, G., Molinaroli, E., Turetta, C., Visiv, F., Cesari, G. and Cristini, A.: 1993, *Water Pollution Research Report*, **30**, p. 253-260.
Littman, T., Steinrucke, J, and Gasse, F.: 1990, *Geookodynamic*, **11**, 163-189.
Lovett, R.F.: 1978, *Tellus*, **30**, 358-363.
Molinaroli E., Guerzoni S. and Rampazzo, G.: 1993, *Geological Society of America Special Paper* **284**, p. 303-312.
Rodà, F., Bellot, J., Avila, A., Escarré, A., Piñol, J. and Terradas, J.: 1993, *Water, Air and Soil Pollution*, **66**, 277-288.
Savoie, D.L., Prospero, J.M. and Nees, R.T.: 1987, *Atmospheric Environment*, **21**, 103-112.

Precipitation Chemistry at Sinhagad-a hill station in India

Medha S. Naik, G.A. Momin, A.G. Pillai, P.D. Safai, P.S.P. Rao
and L.T. Khemani

Indian Institute of Tropical Meteorology, Pashan, Pune 411 008

ABSTRACT The chemistry of precipitation in remote sites such as mountain tops is of interest in the study of atmospheric pollution and acid rain. The chemical composition measured at mountain site which is away from industrial and urban areas is useful as a reference level and it allows to determine the extent of anthropogenic contamination. Hence, rain water samples were collected at Sinhagad (18°21'N, 73°45'E, 1450 m asl) during the monsoon season (June-September) of 1992 and were analysed for major ions. The precipitation samples collected at Sinhagad were alkaline in nature and pH values ranged between 5.9 to 6.76. The ionic composition was dominated by soil dust. The concentration of Ca^{2+} was highest among all the ions. The concentrations of excess SO_4^{2-} and NO_3^- were small (23.8 and 15.2 μeq l^{-1} respectively) compared to the values of polluted regions in India. The correlation coefficient between the ions and pH values was calculated and it was found to be maximum in case of Ca^{2+}. Precipitation samples collected at Sinhagad were alkaline owing to higher concentration of Ca^{2+} and lower levels of acidic pollutants (SO_4^{2-} and NO_3^-).

Keywords : Precipitation chemistry, alkaline rain; sources

1. Introduction

The chemistry of precipitation in remote sites, such as high mountain emplacements, is of interest in the study of atmospheric pollution and acid rain for two reasons. One, chemical composition of precipitation measured in sampling points far from industrial and urban areas is useful as a reference level for the geographical area and second, it allows us to determine the extent of anthropogenic contamination.

Chemical data on precipitation at high altitude are available for some European mountains (Keller et al., 1987) but data for remote unpolluted area in Indian subcontinent mountains are scanty (Mahadevan et al., 1989). The present study is an attempt to characterize the chemistry of precipitation at Sinhagad, a remote unpolluted hill top. Sinhagad is a hill station situated in the Maharashtra state of India. It is one of the hill top of the Sahyadri mountains along the west coast of Indian Peninsula. Sinhagad, situated west of Deccan Plateau, is away from industrial and urban areas, hence low pollution levels should be expected. The calcareous lithology of Deccan Plateau influences the chemical composition of atmospheric aerosol (Khemani et al., 1982).

The aim of this study is to provide data on the chemical composition of precipitation and the pollution level at Sinhagad in order to assess the potential status of this place as a reference site.

2. Sampling and Analysis

Bulk samples were collected at Sinhagad (18°21'N, 73°45'E; 1450 m asl) which is about 90 km from the nearest coast. The sampling site is shown in Figure 1. The region is mostly free from industrial and vehicular activities, and sparsely

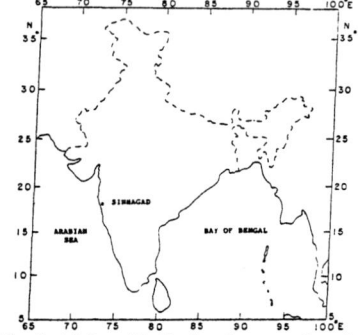

Fig. 1. A Map of India showing sampling site

populated. The methods of sampling and analysis of rain water were described (Naik et al., 1988).

3. Results and Discussion

3.1 CHEMICAL COMPOSITION OF SINHAGAD RAIN

The average concentration (μeq l^{-1}) of various chemical constituents together with pH and conductivity (μS cm^{-1}), their standard deviations and minimum and maximum values in rain water collected at Sinhagad during July to September 1992 is given in Table I. On an average, calcium (Ca^{2+}) contributed most (34%) of the total ionic concentration followed by sodium (Na$^+$) (17%) and chloride (Cl^{-1}) (18%). Sulphate (SO$_4^{2-}$) on an average, was low and accounted only for (11%) to the total ionic

Table I
Average concentration of major ionic components of rain water (μeq l^{-1}), pH values and specific conductivity (μS cm^{-1}).

	Cl$^-$	SO$_4^{2-}$	NO$_3^-$	NH$_4^+$	Na$^+$	K$^+$	Ca^{2+}	Mg^{2+}	pH	Cond.
Mean	47.2	29.3	15.2	7.2	46.1	6.4	91.0	22.5	6.39	26.2
S.D.	42.1	16.3	11.2	6.7	42.8	4.4	74.5	17.5	--	24.8
Min	8.7	2.5	1.0	0.6	6.9	2.6	9.5	4.2	5.9	5.0
Max	137.2	79.2	45.5	16.1	168.0	15.9	296.0	89.2	6.76	110.2

concentration. The nitrate (NO$_3^{-1}$) contribution was found to be less than that of SO$_4^{2-}$ which accounted for 6%. However, contribution of ammonium (NH$_4^+$), potassium (K$^+$) and magnesium (Mg^{2+}) were accounted by 14%. Chloride and Na$^+$ were present in the form of sea salt aerosols since their ratios (μeq l^{-1}) were close to that reported for sea water (1.16).

Sulphate and NO_3^- are mainly derived from anthropogenic sources and they decrease the pH of rain water. The concentrations of SO_4^{2-} and NO_3^- were lower in rain water at Sinhagad compared to those in rain water at the industrial region in Bombay. The values of both SO_4^{2-} (29.3 μeq l^{-1}) and NO_3^- (15.2 μeq l^{-1}) at Sinhagad were about 1/2 of the values reported by Khemani et al., (1994a) for industrial region in Bombay (70.4 SO_4^{2-} and 29.5 NO_3^-) which is about 100 km from Sinhagad. This indicates that there is no transport of pollutants from industrial region of Bombay to Sinhagad. The process of long-range pollutants (SO_4^{2-} and NO_3^-) does not seem to be effective in India during south-west monsoon period when meteorological factors are favourable for quick dispersal of pollutants whose sources are weak (Khemani et al., 1994b).

The constituents such as SO_4^{2-}, K^+, Ca^{2+} and Mg^{2+} in precipitation are derived either from marine or non-marine origins. The non-marine components of these elements were evaluated assuming that all the Na^+ ions originated from the sea. The marine and non-marine contribution and their percentages to the total ionic concentration in rain at Sinhagad was calculated. It showed that non-marine SO_4^{2-}, K^+, Ca^{2+} and Mg^{2+} were about 81, 85, 98, 49% respectively. This indicates that non-marine components such as K^+, Ca^{2+}, Mg^{2+}, SO_4^{2-} and NO_3^- constitute a major part of total ionic content.

The pH values of rain water were alkaline and varied between 5.90 and 6.76. This suggests the influence of non-marine components such as soil dust on pH of rain water.

3.2 GEOGRAPHIC COMPARISON

It is assumed that precipitation acidity originates primarily from sulphuric and nitric acids. The atmospheric alkaline species like NH_4^+ and Ca^{2+} mostly originate from soil and tend to neutralize the acidity in rain water samples. In assessing the impact of acid and alkaline species on rain water, it is important to know their relative concentrations in Sinhagad rain compared with those reported from other geographic regions. The average ionic composition of rain water (Table II) is compared with reported ionic composition of rain from the mountains of Snowdonia which are close to the west coast of north Wales; mountain stations in central and eastern Pyrenees in northeast Spain and remote inland station Mallikadevi hill in north India. Camarero and Catalan (1993) classified about 212 rain water samples into groups according to basic, neutral and acidic precipitation. Only two groups i.e. Group B (basic) and Group A (acidic) are compared in this study, excluding bicarbonate ion which was not analysed in the present study.

It is seen from Table II that the concentration of NO_3^- at Sinhagad is by and large, similar to that at Snowdonia and Pyrenees B but the concentration of SO_4^{2-} is 1.6 times more at Snowdonia and about double at Pyrenees (B) than that at Sinhagad. Hence, unlike in Sinhagad, acidic rain is expected at both these stations, namely Snowdonia and Pyrenees (B), but acidic rain is reported only at Snowdonia and not at Pyrenees (B). The pH of Snowdonia is acidic due to lower concentration of Ca^{2+} which is about 8 times less than that at Sinhagad and pH of Pyrenees (B) is alkaline due to higher concentration of Ca^{2+} which is about 2.6 times more than that at Sinhagad. The rain water of Pyrenees

(A) is acidic not only due to higher concentration of SO_4^{2-} and NO_3^- (about double than those at Sinhagad) but also due to lower concentrations of Ca^{2+} (about 1/2 of that at Sinhagad). Thus the above comparison shows that Ca^{2+} plays a major role in buffering

Table II
The ionic composition (μeq l^{-1}) of rain water collected at Sinhagad and from other geographic regions

Station	Cl^-	SO_4^{2-}	NO_3^-	NH_4^+	Na^+	K^+	Ca^{2+}	Mg^{2+}	pH	Cond.
Sinhagad* (India)	47.2	29.3	15.2	7.2	46.1	6.4	91.0	22.5	6.39	26.2
Snowdonia@ (U.K.)	106.5	49.3	14.6	19.6	100.2	2.5	11.5	24.9	4.61	--
Pyrenees B # (NE Spain)	30.3	54.0	16.8	13.3	31.8	13.7	240.9	15.3	6.31	20.7
Pyrenees A (NE Spain)	15.2	66.6	25.1	32.4	17.5	5.22	44.7	5.6	4.97	17.2
Mallikadevi $	31.0	32.7	--	19.4	41.3	23.1	55.0	25	6.39	--

* This study; @ Reynolds et al., 1990; # Camerero and catalon, 1993 ; $ Mahadevan et al., 1989

the acidity present in the atmosphere. The influence of alkaline components such as Ca^{2+}, Mg^{2+} and K^+ on the pH of rain, snow and cloud in India has been reported previously (Naik et al., 1988, 1994). The high concentration of alkaline components (K^+, Ca^{2+}, Mg^{2+}) are responsible for the high pH of precipitation in India and the alkaline pH of rain water at Sinhagad is consistent with these observations.

3.3 Ratio of ionic components

The ratio (μeq l^{-1}) of NO_3^- : nss SO_4^{2-} (non sea salt sulphate) indicates the relative importance of their contributions to the acidity of rain water samples (Calvert et al., 1985; Summers and Barrie, 1986). These acids in precipitation are originally formed in the atmosphere through pertinent gas or aqueous-phase chemical reactions (Finlayson and Pitts, 1986). The atmospheric alkaline species, namely Ca^{2+} and NH_4^+ neutralize a portion of acidity. The NO_3^- : nss SO_4^{2-} ratio (μeq l^{-1}) of rain water is 0.64 at Sinhagad. This suggests that contribution to the free acidity at Sinhagad rain water is more by sulphuric acid than nitric acid. This ratio for Snowdonia and Pyrenees B and A are 0.39, 0.34, 0.39 respectively which shows that sulphuric acid plays a major role in acidifying rain water.

Calcium and NH_4^+ play important roles in neutralizing the acidity and the extent of this can be seen from the ratio (NH_4^+ + nss Ca^{2+}):(NO_3^- :nss SO_4^{2-}) in μeq l^{-1}. If this ratio is less than unity, it shows that acidity caused by SO_4^{2-} and NO_3^- is not neutralized by the alkaline components present. Conversely, if the ratio is greater than unity, the acidity caused by SO_4^{2-} and NO_3^- is fully neutralized by the alkaline components. This ratio for Sinhagad is 2.47 which indicates that the acidity generated by SO_4^{2-} and NO_3^- is fully neutralized by the alkaline components such as NH_4^+ and Ca^{2+} resulting in high pH values. These ratios for Snowdonia, Pyrenees (B) and (A) are 0.51, 3.77, 0.85 respectively indicating rain water samples at Snowdonia and Pyrenees (A) are not completely neutralized by alkaline species and samples are acidic. But for the rain water in Pyrenees B group, the ratio is higher (3.77) indicating rain water is fully neturalized by the alkaline species present in high concentration. The ratio for Mallikadevi could not be calculated since NO_3^- is not reported.

4. Factor Analysis of the Data

In the present study, the identification of the sources of ions in the rain water at Sinhagad was carried out by principal component analysis (SPSS, 1983). This analysis indicated three principal components and the varimax rotated matrix of the chemical constituents is presented in Table III. In this table, only factor loadings above 0.5 are listed, as these are seemed to be statistically significant, the variance % associated with each factor is also included. Also, given in the Table III are the eigen values (>1) corresponding to the three factors. Factor 1 showed high loading for Cl^-, Na^+, Ca^{2+}, Mg^{2+} and NO_3^-. This result indicates that Factor 1 contains elements originating from soil and sea salt and accounts for the major % of variance (45.1%).

Table III
Varimax rotated factor loading and variance %

Components	Factor 1	Factor 2	Factor 3
Cl^-	0.78	--	--
SO_4^{2-}	--	--	0.70
NO_3^-	0.84	--	--
NH_4^+	--	0.93	--
Na^+	0.68	--	0.55
K^+	--	0.83	--
Ca^{2+}	0.90	--	--
Mg^{2+}	0.81	--	--
H^+	--	--	--
Var %	45.1	16.7	10.6
Eigen values	4.96	1.84	1.16

All of the components (Na^+, Cl^-, Ca^{2+} and Mg^{2+}) except NO_3^- in Factor 1 are normally associated with coarse particles. The NO_3^- particles are expected to be in the submicron size range, since these particles are formed by gas-to-particle conversion processes (Cox, 1974). Under normal atmospheric conditions, HNO_3 can initially form Aitken nuclei which will grow rapidly into submicron size and react with NH_3 to form NH_4NO_3. The formation of NH_4NO_3 in tropical countries is rare due to the persistence of higher temperatures. However, HNO_3 can react with soil/sea derived particles and become incorporated in rain water as the coarse aerosols mode (Wolff, 1984). The coarse NO_3^- observed in the rain water samples at Sinhagad appears to be a product of the reaction of airborne sea salt and soil-derived particles with gaseous HNO_3.

High loading for K^+ and NH_4^+ is observed in Factor 2. Generally sources of K^+ in atmospheric particles are the soil, sea and vegetation, But K^+ is not associated with other sea and soil oriented components in Factor 1. Hence K^+, in rain water may probably be from surrounding forest vegetation of the Sinhagad. The release of NH_3 by the forest vegetation is shown for various agricultural plants by Farquhar et al., (1980). Hence both K^+ and NH_4^+ in rain water may be due to emission from forest vegetation.

Factor 3 showed loading for Na^+ and SO_4^{2-}. The correlation of Na^+ with both first and third factor suggested one more source other than sea and soil. We cannot argue satisfactorily the interpretation of Na^+ in third factor. Excess of SO_4^{2-} (about 81%) showed very weak relationship with H ion. This indicates that SO_4^{2-} is in the form of salt of either sea or continental derived elements as a result of atmospheric reactions.

5. Conclusions

The study of the ionic composition of rain water collected at Sinhagad, a remote hill station, indicated higher concentration of alkaline elements (Ca^{2+}, Mg^{2+}) and lower concentration of acidic components (SO_4^{2-} and NO_3^-) as compared to other mountain regions of western countries. The rain water samples were alkaline and mostly influenced by Ca^{2+} and Mg^{2+}.

Rain water in India is by and large, alkaline in nature except in some industrial pockets. The acid rain is purely a local phenomenon. Long-rang transport of acidic pollutants is the main cause of acid rain in western countries. However, long-range transport of pollutants (SO_4^{2-} and NO_3^-) in the rain water was not observed at Sinhagad. Hence Sinhagad can be considered as one of the background stations from the Indian subcontinent.

References

Calvert, J.G., Lazrus, A., Kok, G.L., Heikes, B.G., Walega, J.G., Lond., J. and Cantrell, C.A., : 1985, *Nature*, **317**, 27-35.
Camarero, L and Catalan, J. : 1993, *Atmos. Environ.*, **27A, 1**, 83-94.
Cox, R.A. : 1974, *Tellus*, **26**, 235-240.
Farquhar, G.D., Firth, P.M., Wetscelar, R., and Weir, B. : 1980, *Plant Physiol.*, **66**, 710-714.
Finlayson-Pitts, B.J. and Pitts, J.N.Jr. (1986), *Atmospheric Chemistry*, Wiley-Interscience, New York.
Keller, H.M., Kloti, P. and Forster, F. : 1987, *In Proc. Int. Symp. Acidification and Water Pathways*, Bolkesjo, Norway May, pp. 237-248.
Khemani, L.T., Momin, G.A., Naik, M.S., Vijayakumar R. and Ramana Murty Bh.V. : 1982, *Tellus*, **34**, 151-159.
Khemani, L.T., Momin, G.A., Rao, P.S.P., Pillai, A.G., Safai, P.D., Mohan K. and Rao, M.G. : 1994a, *Atmos. Environ.*, **28**, 3145-3154.
Khemani, L.T., Tiwari, S., Singh, G., Momin, G.A., Naik, M.S., Rao, P.S.P., Safai, P.D. and Pillai, A.G. : 1994b, "Acid deposition in the vicinity of Super Thermal Power Plant in India", TAO Special Issue on Regional Environment and Climate Change in east Asia (in press).
Mahadevan, T.N., Negi B.S. and Meenakshy, V. : 1989, *Atmos. Environ.*, **23**, 869-874.
Naik, M.S., Khemani, L.T., Momin, G.A. and Rao, P.S.P. : 1988, *Acta. Meteorl. Sinica*, **2**, 91-99.
Naik, M.S. Khemani, L.T., Momin, G.A., Rao, P.S.P., Pillai, A.G. and Safai, P.D. : 1994, *Tellus*, **46B**, 68-75.
Reynolds, B., Williams, T.G. and Stevens, P.A. : 1990, *Sci. Total. Envir.*, **92**, 223-234.
SPSS (Statistical Package for the Social Sciences) : 1983, Version 3.0 Marketing Department SPSS Inc., 444 North Michigan Avenue, Chicago, IL 60611, 312/329-3300.
Summers, P.W. and Barrie, L.A. : 1986, *Water, Air, and Soil, Pollution*, **30**, 275-282.
Wolff, G.T. : 1984, *Atmos. Environ.*, **8**, 977-981.

RELATIONSHIP BETWEEN WET DEPOSITION OF SULFATE AND NITRATE AND RAINFALL AMOUNT IN JAPAN

S. SETO[1], M. KITAMURA, A. MORI, I. NOGUCHI, T. OHIZUMI, T. TAKEUCHI, T. DEGUCHI, and H. HARA

Data Analysis Group for the Acid Precipitation Survey over Japan

[1] *Hiroshima Prefectural Institute for Health and Environmental Sciences, 1-6-29 Minami-machi, Minami-ku, Hiroshima 734, Japan*

Abstract. A regression model of wet deposition on rainfall amount for non-seasalt sulfate (nss-SO_4^{2-}) and nitrate (NO_3^-) was applied to a data set obtained through a nationwide survey from April 1989 to March 1993. Wet-only samples on a biweekly basis were collected at 29 sites over Japan. Reparameterized bivariate lognormal distribution was employed to describe the joint distribution of concentration (C) and rainfall amount (R) for each site. Ranges of geometric mean (μ_D) of biweekly deposition (D = C R) for each site were 0.54-2.90 meq m^{-2} for nss-SO_4^{2-}, and 0.21-1.36 meq m^{-2} for NO_3^-; that of biweekly rainfall amount (μ_R) was 24.1-78.0 mm. Urban or industrialized areas had high values of μ_D for these ions. Ranges of estimates of the slope of the regression equation of $\log(D/\mu_D)$ on $\log(R/\mu_R)$, were 0.45-0.99 for nss-SO_4^{2-}, and 0.35-0.86 for NO_3^-; thus estimates of the slope for nss-SO_4^{2-} tend to be larger than those for NO_3^-. The present analysis, consequently, statistically clarified some differences between the two ions in deposition processes which is understood in the light of current knowledge of atmospheric chemistry.

Key words: Sulfate, nitrate, wet deposition, rainfall amount, precipitation scavenging, lognormal distribution, Japan.

1. Introduction

Precipitation chemistry in Japan has been gradually recognized from surveys with nationwide monitoring networks. Hara (1993) has assessed the acid deposition with respect to suspicions of its effects, precipitation chemistry, and research trends of related atmospheric chemistry. Central Research Institute of Electric Power Industry (CRIEPI) has also investigated the status of acid deposition and its effects. These studies indicated the spatial distributions and temporal variations for major chemical components in rainwater within Japan.

Because of the contribution to the acidity of rainwater through several scavenging processes, considerable efforts have been focused on sulfate (SO_4^{2-}) and nitrate (NO_3^-). Particularly, a number of studies conducted on the concentrations of SO_4^{2-} and NO_3^- in rainwater and rainfall amount in order to understand precipitation scavenging processes (e. g. Hicks and Shannon, 1979; Schnug and Holz, 1987; Seto et al., 1992).

The present paper discusses the relationships between wet deposition of non-seasalt sulfate (nss-SO_4^{2-}) and NO_3^-, and rainfall amount over Japan using a dataset obtained through Japan Environment Agency (JEA) Phase-II Survey. Assuming that the joint distribution of concentration and rainfall amount obeys bivariate lognormal one, we first describe spatial patterns of concentration for nss-SO_4^{2-} and NO_3^- in terms of mean and variance. Secondly, we then apply a regression model to wet deposition for these ions on rainfall amount for the purpose of understanding precipitation scavenging processes.

2. Materials and methods

2.1. EXPERIMENTAL

The JEA Phase-II Survey sites are shown in. Fig. 1. The monitoring sites of JEA Phase-II Survey are classified into two categories: National Air Pollution Surveillance Network (NASN) sites with four-year records, and remote island sites with one- or two-year records (Table I). Each site was provided with a wet/dry sampler. Biweekly wet-only precipitation samples were collected into a storage bottle in the sampler, and shipped to laboratories for chemical analysis. Samples were analysed for pH, electric conductivity and major ions; SO_4^{2-} and NO_3^- were determined by using ion chromatography.

The basic dataquality checks were performed by comparing of anion and cation equivalent sums, and of measured and calculated conductivities. All questionable results which can be assigned to reasonable causes were removed from the dataset. Non-seasalt sulfate was calculated by assuming sodium to be a conservative tracer for seasalt, and by estimating seasalt contributions from the known ionic ratios in seawater. The data during heavily snowy four months for NPR site were not included because a wet/dry sampler was replaced by a bulk sampler.

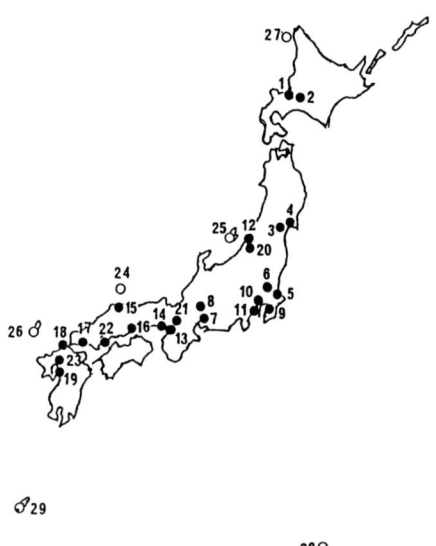

Fig. 1. Map showing the locations of JEA Phase-II Survey sites. (●): NASN site, (○): remote island site. Site numbers are the same as those shown in Table I .

2.2. STATISTICAL MODEL

A new reparameterized bivariate lognormal distribution derived from Iwase and Seto (1990), and Seto et al. (1992), is defined as

$$(1/\sigma_C \cdot \log(C/\mu_C), \; 1/\sigma_R \cdot \log(R/\mu_R)) \sim N(0, 0; 1, 1; \rho), \qquad (1)$$

where C is concentration (μeq L^{-1}), R is rainfall amount (mm), μ_C is the population median (or the population geometric mean) of C, μ_R is the population median (or the population geometric mean) of R, σ_C is the population standard deviation of $\log(C/\mu_C)$, σ_R is the population standard deviation of $\log(R/\mu_R)$, and ρ is the population correlation coefficient between $\log(C/\mu_C)$ and $\log(R/\mu_R)$.

The dimensions of μ_C and μ_R are the same as those of C and R, respectively; and σ_C, σ_R and ρ are dimensionless. Relation (1) can be rewritten as $(C, R) \sim LN(\mu_C, \mu_R; \sigma_C^2, \sigma_R^2; \rho)$. Wet deposition is defined as $D = C \cdot R / 1000$ (meq m^{-2}). The distribution of D also holds lognormality represented by $D \sim LN(\mu_D, \sigma_D^2)$ where

$$\mu_D = \mu_C \mu_R \text{ and } \sigma_D^2 = \sigma_C^2 + 2\rho\sigma_C\sigma_R + \sigma_R^2. \qquad (2)$$

The expectation and the standard deviation of $\log(D/\mu_D)$, given $R = r$, are calculated as

$$E[\log(D/\mu_D) \mid R = r] = \beta \cdot \log(r/\mu_R), \quad \beta = 1 + \rho \cdot \sigma_C / \sigma_R \qquad (3)$$

where β is the slope of the regression line, and

$$\sigma_{REG} = (\text{Var}[\log(D/\mu_D) \mid R = r])^{1/2} = (1-\rho^2)^{1/2} \cdot \sigma_C, \qquad (4)$$

respectively. The population correlation coefficient between $\log(D/\mu_D)$ and $\log(R/\mu_R)$ is also calculated as

$$\rho_{DR} = (\sigma_R + \rho \cdot \sigma_C) / (\sigma_C^2 + 2\rho\sigma_C\sigma_R + \sigma_R^2)^{1/2}. \qquad (5)$$

3. Results and discussion

3.1. CONCENTRATION

Many studies have been conducted on distributional models for concentration of major ions in rainwater (e.g. Lindberg, 1982; Warren et al., 1992). These studies reported that lognormal distribution was a candidate. This distribution was also fitted to rainfall amount (Hosking and Stow, 1987).

The lognormalities of concentrations for nss-SO$_4^{2-}$, NO$_3^-$, and rainfall amount were examined by plotting the data on the lognormal probability paper. It was indicated that both concentrations for these ions and rainfall amount were good-fitted to lognormal distributions excepting rainfall amount for a few sites. Hence, we assume that a joint distribution of concentration and rainfall amount obeys a bivariate lognormal.

The parameters in Relation (1) for each site were calculated using method of moments. The biweekly estimates of μ_C for nss-SO$_4^{2-}$ and NO$_3^-$ ranged from 7.5 to 75.0 μeq L^{-1} with a mean of 40.7 μeq L^{-1}, and from 4.1 to 32.9 μeq L^{-1} with a mean of 15.8 μeq L^{-1}, respectively. High estimates of μ_C exceeding 50 μeq L^{-1} for nss-SO$_4^{2-}$ were observed at SPR, KSM, TKY, UBE, KKS and OMT sites and those of exceeding 20 μeq L^{-1} for NO$_3^-$ at NGY, TKY, UBE and KKS sites. These sites locate in large urban or industrialized areas. Conversely, minimum estimates of μ_C for both nss-SO$_4^{2-}$ and NO$_3^-$

were observed at OGS site located at a remote island in the Pacific Ocean. The biweekly estimates of σ_C for nss-SO_4^{2-} and NO_3^- ranged from 0.190 to 0.394, and from 0.195 to 0.459, respectively. The estimates of σ_C for NO_3^- were greater than those for nss-SO_4^{2-} at 24 sites. Consequently, it was indicated that the mean values for nss-SO_4^{2-} are significantly higher than those for NO_3^-, but the variances for NO_3^- are greater than those for nss-SO_4^{2-} for the majority of the sites in Japan. The biweekly estimate of μ_R and σ_R ranged from 24.1 to 78.0 mm with a mean of 42.2 mm, and from 0.237 to 0.563, respectively. Thus, the estimates of σ_R are relatively greater than those of σ_C for these ions for most of the sites.

3.2. WET DEPOSITION

The deposition-rainfall amount relationships represented by eq. (3) for KKS site are shown in Fig. 2. It seems that log(D/estimate of μ_D) for both nss-SO_4^{2-} and NO_3^-, and log(R/estimate of μ_R) are linear and highly correlated each other, and that scatters from the regression lines for these ions do not depend on values of rainfall amount. These results suggest the validity of eqs. (3) and (4).

The biweekly estimates of μ_D, σ_D, β, ρ_{DR} and σ_{REG} in eqs. (2)-(5) are shown in Table I. The estimates of μ_D for nss-SO_4^{2-} and NO_3^- ranged from about 0.5 to about 2.9 meq m^{-2} and from about 0.2 to about 1.4 meq m^{-2}, respectively. High values of μ_D for these ions

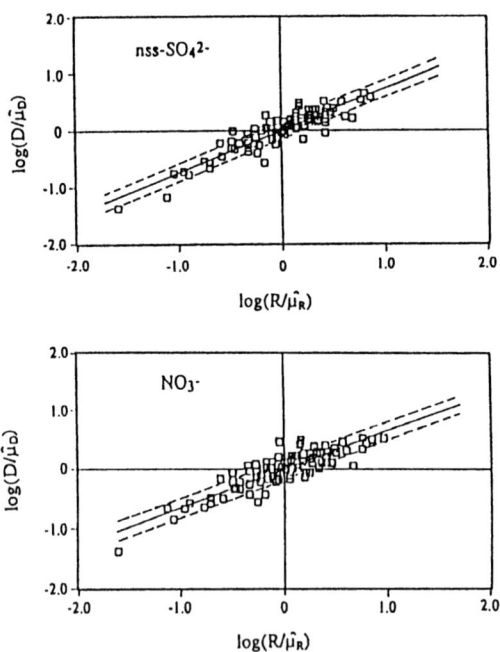

Fig. 2. The deposition-rainfall amount relationships for KKS site. A solid line and a broken line represent a regression equation of deposition (D) on rainfall amount (R), and a standard deviation from a regression equation, respectively.

were observed at the same areas as those of μ_C. Conversely, low values of μ_D for these ions were observed at RSR, OGS and AMM sites located at remote islands; but TSM, SAD and OKI sites had the same level of μ_D with NASN sites. Since these remote islands have no large anthropogenic emissions, this suggests the contribution of a long-range transport of acid pollutants from the outside of the islands.

TABLE I

The biweekly estimates of the parameters of deposition in eqs. (2) - (5).

No	Site	Code	Cat	nss-SO_4^{2-}						NO_3^-					
				n	μ_D (meq L-1)	σ_D (1)	β (1)	ρ_{DR} (1)	σ_{REG} (1)	n	μ_D (meq L-1)	σ_D (1)	β (1)	ρ_{DR} (1)	σ_{REG} (1)
1	Sapporo	SPR	N	89	1.26	0.363	0.729	0.832	0.201	89	0.36	0.381	0.706	0.768	0.244
2	Nopporo	NPR	N	60	0.79	0.313	0.549	0.570	0.257	61	0.34	0.291	0.455	0.513	0.250
3	Sendai	SDI	N	88	1.32	0.486	0.860	0.910	0.202	88	0.63	0.523	0.841	0.826	0.295
4	Nonodake	NND	N	83	0.54	0.463	0.741	0.830	0.258	83	0.38	0.439	0.649	0.768	0.281
5	Kashima	KSM	N	87	2.51	0.372	0.622	0.859	0.190	74	0.51	0.497	0.534	0.583	0.389
6	Tukuba	TKB	N	82	1.35	0.379	0.566	0.690	0.274	82	0.66	0.376	0.479	0.588	0.304
7	Nagoya	NGY	N	88	1.58	0.427	0.709	0.897	0.189	88	0.82	0.419	0.669	0.863	0.212
8	Inuyama	INY	N	95	1.35	0.416	0.676	0.740	0.280	94	0.85	0.396	0.648	0.750	0.262
9	Ichikawa	ICH	N	86	1.65	0.309	0.567	0.757	0.202	86	0.72	0.369	0.568	0.631	0.287
10	Tokyo	TKY	N	77	2.76	0.392	0.739	0.768	0.251	77	1.25	0.405	0.602	0.606	0.322
11	Kawasaki	KSK	N	82	2.30	0.365	0.773	0.825	0.206	81	0.82	0.389	0.504	0.503	0.336
12	Niigata	NGT	N	90	1.68	0.432	0.784	0.733	0.294	93	0.62	0.300	0.612	0.823	0.171
13	Oosaka	OSK	N	87	1.42	0.338	0.720	0.831	0.188	87	0.48	0.346	0.658	0.740	0.233
14	Amagasaki	AMG	N	77	1.66	0.328	0.593	0.750	0.217	77	0.68	0.304	0.488	0.665	0.227
15	Matsue	MTE	N	88	1.90	0.364	0.850	0.719	0.253	88	0.69	0.351	0.689	0.605	0.280
16	Kurashiki	KRS	N	81	1.37	0.364	0.743	0.788	0.224	81	0.59	0.352	0.705	0.773	0.223
17	Ube	UBE	N	95	2.83	0.339	0.617	0.791	0.207	95	0.88	0.314	0.595	0.825	0.178
18	Kitakyushu	KKS	N	93	2.90	0.356	0.732	0.909	0.149	96	1.36	0.327	0.631	0.871	0.160
19	Oomuta	OMT	N	85	2.82	0.426	0.661	0.874	0.207	85	0.61	0.422	0.609	0.812	0.246
20	Niitsu	NTS	N	85	1.85	0.318	0.890	0.796	0.192	86	0.82	0.287	0.777	0.768	0.184
21	Kyotobachiman	KHM	N	94	1.15	0.411	0.660	0.743	0.275	95	0.57	0.362	0.559	0.711	0.255
22	Kurahashijima	KHZ	N	88	2.20	0.324	0.616	0.745	0.216	88	0.72	0.329	0.578	0.688	0.239
23	Chikugoogori	COG	N	93	1.62	0.407	0.697	0.868	0.203	95	0.45	0.391	0.633	0.820	0.224
24	Oki	OKI	R	42	1.48	0.318	0.583	0.617	0.250	42	0.59	0.261	0.346	0.447	0.233
25	Sado	SAD	R	47	1.56	0.352	0.988	0.666	0.263	47	0.78	0.291	0.862	0.703	0.207
26	Tsushima	TSM	R	37	2.25	0.433	0.901	0.850	0.228	37	0.78	0.370	0.726	0.802	0.221
27	Rishiri	RSR	R	40	0.66	0.407	0.641	0.704	0.289	41	0.21	0.520	0.679	0.602	0.415
28	Ogasawara	OGS	R	11	0.58	0.370	0.448	0.510	0.319	9	0.27	0.293	0.535	0.742	0.197
29	Amami	AMM	R	17	0.83	0.389	0.678	0.733	0.265	19	0.46	0.394	0.641	0.672	0.292

Cat: Category of sites(N: National Air Pollution Surveilance Network(NASN), R: Remote island sites), μ_D: Geometric mean, (1): dimensionless.

The estimates of β for nss-SO_4^{2-} and NO_3^- ranged from 0.45 to 0.99 with the mean of 0.701, and from 0.35 to 0.86 with the mean of 0.620, respectively. It is shown that MTS, SAD, TSM and NTS sites located along the coast of or in the Japan Sea, and SDI site had high values (> 0.8) of β for nss-SO_4^{2-}.

Hicks and Shannon (1979) argued that a power law exponent (β) of about 0.6 might be a better estimate for wet deposition for SO_4^{2-}; and Seto et al. (1992) suggested a half-power relationship ($\beta = 0.5$) of wet deposition for SO_4^{2-} and NO_3^- on the basis of a single event. In general, scavenging processes will be strongly affected by the precipitation types such as convective or stratiform. When sampling duration become longer, collected rainwater would consist of several events having different precipitation types. This would affect on the values of β. Further efforts are also required to understand the effects of other meteorological parameters such as synoptic and local wind systems. It should be noted that the estimates of β for nss-SO_4^{2-} were rather greater than those of NO_3^- for the majority of the sites. This clarified statistically some differences in deposition processes between the two ions in Japan.

4. Conclusions

On the basis of bivariate lognormal distribution of concentration and rainfall amount, a regression model of wet deposition for nss-SO_4^{2-} and NO_3^- on rainfall amount, was applied to a dataset obtained through a nationwide survey in Japan. High values of biweekly geometric mean of concentrations and wet deposition for these ions were obtained at urban or industrialized areas. The slope of regression equation for nss-SO_4^{2-} tends to be greater than those for NO_3^-, which clarified statistically some differences in deposition processes between the two ions.

Acknowledgements

We are grateful to Japan Environment Agency (JEA) for permitting us to publish this paper. However, this publication has not been reviewed by the JEA; it does not reflect the views of the agency, and no official endorsement should be inferred.

References

CRIEPI Research Group of Acidic Deposition: 1992. *Acidic deposition in Japan*, Central Research Institute of Electric Power Industry, ET91005, 1-90.
Hara, H.: 1993, *Bull. Inst. Public Health*, **42**, 426-437.
Hicks, B. B. and Shannon, J. D.: 1979, *J. Appl. Meteor.*, **18**, 1415-1420.
Hosking, J. D. and Stow, C. D.: 1987, *J. Climate Appl. Meteor.*, **26**, 1530-1539.
Iwase, K. and Seto, S.: 1990, *Bull. Fac. Engng Hiroshima Univ.*, **38**, 153-161.
Lindberg, S. E.: 1982, *Atmospheric Environment* **16**, 1701-1709.
Schnug, E. and Holz, F.: 1987, *Atmospheric Environment* **21**, 1235-1241.
Seto, S., Oohara, M. and Iwase, K.: 1992, *Atmospheric Environment*, **26A**, 3029-3038.
Warren, W. G., Bohm, M. and Link, D.: 1992, *Atmospheric Environment* **26A**, 159-169.

THE COMPILATION OF MEASUREMENT BASED EUROPEAN WET DEPOSITION MAPS OF ACIDIFYING COMPONENTS AND BASE CATIONS

E.P. VAN LEEUWEN, G.P.J. DRAAIJERS, C.J.M. POTMA, W.A.J. VAN PUL and J.W. ERISMAN

National Institute of Public Health and Environmental Protection, RIVM, P.O. Box 1, 3720 BA Bilthoven, The Netherlands.

Abstract Precipitation concentrations in 1989 on a European scale were obtained from national organisations responsible for wet deposition monitoring in their countries and from the EMEP database. In total, results from about 750 monitoring locations scattered over Europe were gathered. Spatial analysis based on Regionalised Variable Theory revealed auto-correlation in all ion concentrations and reasonable bounded models were fitted to the experimental variograms. Maps of concentrations of acidifying components and base cations were compiled on a 50x50 km scale using the block-kriging interpolation technique. To obtain fluxes, concentrations were multiplied by long-term mean precipitation amounts from the EPA database. An extensive uncertainty analysis was performed to assess the quality of the maps.

Key words: wet deposition, Europe, acidifying components, base cations, measurements

1. Introduction

To date, wet deposition maps on a European scale have been based on long-range transport model results (e.g. EMEP model, Iversen *et al.*, 1991; TREND model, van Jaarsveld and Onderdelinden, 1990). Maps based on measurements have only been available on national scales, except for ammonium for which a map on a European scale was derived by Buijsman and Erisman (1988). Wet deposition maps of nutrients and potential acid based on measurements are needed for determining the actual input to ecosystems. Measurement based maps can also be used to validate the long-range transport models. In addition to the acidifying components, wet deposition of the base cations Ca^{2+}, Mg^{2+}, K^+ (and Na^+) also needs to be quantified (van Leeuwen *et al.*, 1995), because they too play an integral role in the chemical processes of acid deposition, since the acidity of any material is a function of both its acidic and basic content. Maps of base cations can be used for critical load mapping. When combined with dry deposition estimates on a small scale, local scale variations in total acidic input (van Pul *et al.*, 1995; Erisman *et al.*, this issue) and base cations (Draaijers *et al.*, this issue) can be quantified. In this paper, wet deposition of sulphate is mapped on a European scale for 1989 with a resolution of 50x50 km, based on results from field measurements made at about 750 locations. An uncertainty analysis is included. Maps of other components as well as a more detailed description of this study can be found in van Leeuwen *et al.* (1995).

2. Method description

Concentration data of ions in precipitation in 1989 was obtained from national organisations responsible for wet deposition monitoring in their countries and from the EMEP (European Monitoring and Evaluation Programme, Schaug et al., 1991) database. Figure 1 shows the locations of the measurement sites. Concentrations of SO_4^{2-}, NO_3^- and NH_4^+ were analysed at about 750 sites, concentrations of the other ions at about 600 sites. In east and south-west Europe spacing between sites is generally very large. Concentration data from 1989 were preferred, but in order to obtain the best spatial coverage, data from other years or the average of several years (mostly varying from 1985 to 1990) were used if 1989 data were not available. Information on approximately 300 sites did not originate from 1989.

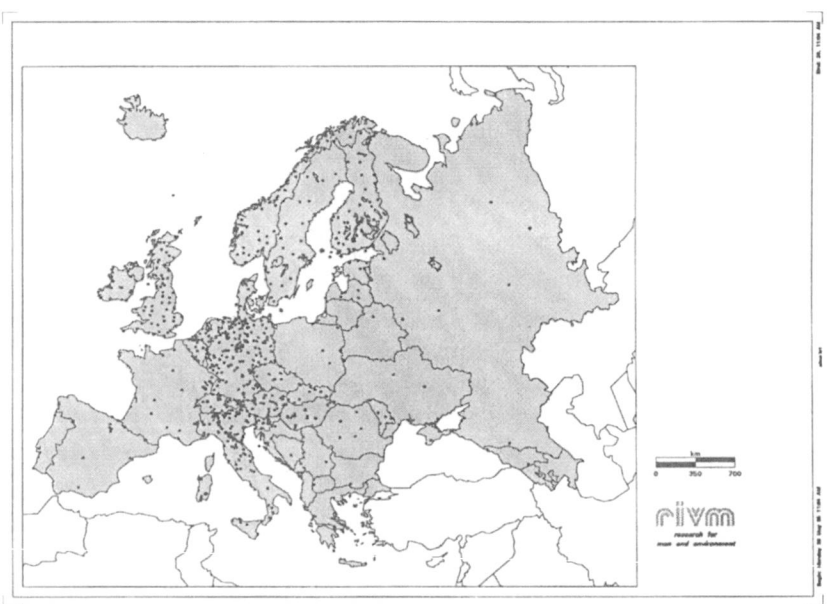

Figure 1 Locations of the wet deposition measurement sites.

Wet deposition was measured using bulk or wet-only samplers. The funnel surface of bulk samplers will also collect some dry deposition of gases and particles. Correction factors for the contribution of dry deposition onto the funnels were derived from several parallel measurements with bulk and wet-only samplers (van Leeuwen et al., 1995). Sulphate concentrations were corrected for the contribution of sea salt, by assuming that the ratio of sodium to sulphate in sea spray is the same as in bulk sea water, and that all sodium in a sample is of marine origin. At about 200 sites only information on the sulphate concentration was available while the sodium concentration was not measured. On these sites values from the interpolated sodium concentration map were used.

As the computed coefficient of skewness of the concentration data showed skew distributions, data were transformed to their common logarithms before interpolation. Interpolation was performed using the kriging interpolation technique based on the Regionalised Variable Theory (Matheron, 1965). Optimal variogram models were fit through the calculated experimental variograms. Estimates of concentrations were made on a regular grid of 50x50 km by ordinary block kriging. As annual mean concentrations show smaller spatial fluctuations over Europe than fluxes, interpolation was performed on concentration data. Back-transformation of the data (from log- to original values) was performed by taking the exponent. In this way the median value of the block is obtained (Journel and Huijbregts, 1978).

To obtain wet deposition fluxes, the interpolated annual mean precipitation concentrations were multiplied by precipitation amounts. As concentrations and precipitation amounts are physically linked, in principle only precipitation amounts from 1989 should be used. Precipitation amounts over Europe may show large variability over short distances, therefore several thousands of measurements are necessary to describe the variation to a reasonable extent. No useful database with validated data from 1989 were obtained. Therefore a database compiled by EPA (Environmental Protection Agency, USA) containing interpolated values of long-term mean precipitation amounts, based on several thousands of measurements (Legates and Willmott, 1990) was used. This EPA map is based on validated monthly mean precipitation amounts measured from 1920 to 1980. To remove systematic errors data were corrected by EPA (Legates and Willmott, 1990) and subsequently interpolated to a 0.5 degree of longitude by 0.5 degree of latitude grid (approx. 30x60 km). It was explored whether systematic differences between 1989 data from the ODS dataset (i.e. Observational Data Set which is a product of ECMWF (European Centre for Medium-range Weather Forecasts), containing unvalidated precipitation amounts measured at 1297 sites spread over Europe) and long-term mean data (EPA) could be revealed (van Leeuwen et al., 1995). As no systematic differences were found, EPA data were used to calculate wet deposition fluxes.

3. Results and discussion

For almost all elements the exponential model was used for variogram modelling in the kriging interpolation procedure, whereas for sodium and potassium the Gaussian model was applied. The effective range (r) of the ammonium variogram was smaller (about 350 km) than that of the other ions (about 600 km), indicating that for ammonium the concentration in precipitation varies more locally than the concentrations of the other ions. For sodium, potassium and chloride the 'nugget' variances in the variograms are large compared to the 'sill' variances, because these elements exhibit a spatial trend near the coast, hampering the calculation of one appropriate variogram that is valid for the whole of Europe. Large nugget variances may also be the result of measurement errors, as well as ion concentrations varying over shorter distances than those between measurements sites. To resolve the last two sources of spatial variability, a denser and more accurate network is required.

The method to map wet deposition presented in this paper is illustrated for non-marine sulphate. A concentration map is presented in Figure 2. Wet deposition of non-marine sulphate is presented in Figure 3. In some regions of Europe the distance between measurement sites is too large to obtain interpolated fields covering the whole of Europe. Interpolation proceeds until the maximum distance of spatial correlation is reached, which can be seen in the maps by some blank spots in eastern Europe. Concentrations and fluxes in Turkey should be interpreted with care because they are not based on measurements in Turkey itself, but solely the result of extrapolation from surrounding countries. The maps clearly resemble European emission and climate patterns. Large emission sources or source regions of sulphur can be recognised in the concentration map (Figure 2), whereas in the flux map (Figure 3) climate patterns and orographic effects are also observed. Large sulphate concentrations are mainly caused by SO_2 emissions from industry and power stations (burning of fossil fuels). Highest depositions of non-marine sulphate (500-2000 mol ha^{-1} yr^{-1}) are found in areas with high rainfall, i.e. mountainous regions (such as former Yugoslavia) and coastal areas, and in areas with high sulphur emissions (such as Ukraine and the Black Triangle, i.e. the border area between Germany, Poland and Czech Republic).

The wet deposition maps are subject to several sources of uncertainty. These are divided into three main categories: *(1)* uncertainty associated with the measurements, *(2)* uncertainty associated with assumptions in the methods used, and *(3)* uncertainty caused by the interpolation procedure. Van Leeuwen *et al.* (1995) give an extensive description of these uncertainties. The main source of uncertainty appears to be different for different regions of Europe. Therefore, a rough division into three areas is made, i.e. *(a)* west, north-west and Central Europe as 'good quality areas' where data quality is assumed to be high and sufficient (representative) data is available, *(b)* areas at the edges of the maps, i.e. east, south-east and south-west Europe as 'poor quality areas', where less data are available of which representativeness and quality can be questionable and *(c)* mountainous areas (e.g. the Alps) and upland areas (e.g. United Kingdom and Scandinavia), even though located in areas with many data, as 'complex terrain areas'. Uncertainties are presented as percentages reflecting the relative deviation from the estimated value for an average 50x50 km grid cell in terms of one standard deviation (σ), implying that the probability that the real value is the estimated value $\pm x\%$ is 67%. Uncertainties in concentrations and rainfall amounts have been estimated separately. Assuming uncertainties reflect random, i.e. non-systematic errors, the most dominating uncertainty in both variables was taken to calculate the uncertainty in the wet deposition fluxes. Using error propagation methods total uncertainty in the wet deposition fluxes is estimated to amount 50% in area *(a)*, 60% in area *(b)* and 70% in area *(c)*. These estimates are only valid if it is assumed that there is no correlation between precipitation concentration and rainfall amount. If these variables are correlated an extra term must be added: R times the product of individual errors, where R is the correlation coefficient between the variables. A propagation of errors with full positive correlation, i.e. the worst case, revealed uncertainties to amount 70% in area *(a)*, 80% in area *(b)* and 100% in area *(c)*. As the estimates are rather crude, no distinction between different components was made. It may be hypothesised that the uncertainty is highest for potassium and lowest for sulphate.

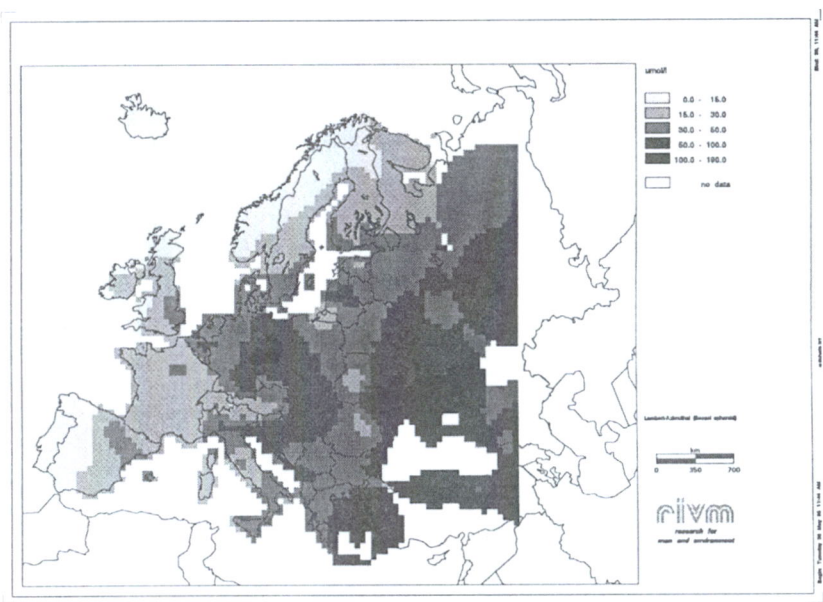

Figure 2 Non-marine sulphate concentrations on a 50x50 km basis in 1989 in µmol l^{-1}.

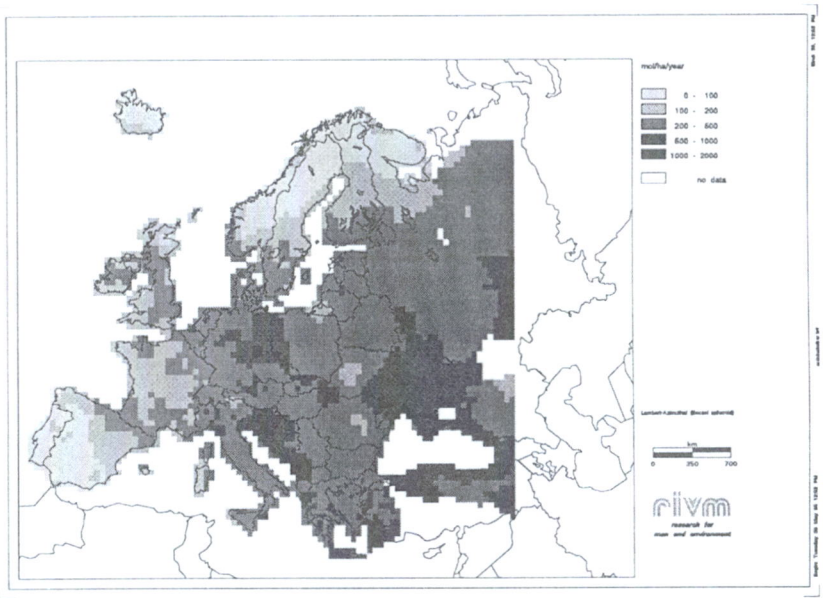

Figure 3 Wet deposition of non-marine sulphate on a 50x50 km basis in 1989 in mol ha^{-1} a^{-1}.

4. Conclusions

Information on concentrations in precipitation obtained from about 750 sites allowed to map regional-scale variations in wet deposition on a 50x50 km scale over Europe using the kriging interpolation technique. Effective ranges of the variogram models varied between 350 km for ammonium to 1200 for chloride, whereas the average effective range of all models equalled 600 km. To obtain fluxes, concentrations were multiplied by long-term mean precipitation amounts compiled by EPA. Patterns of distribution observed in the maps agree well with what would be expected from prior knowledge of European emissions and climate patterns, i.e. large depositions in the Black Triangle and mountainous and coastal regions. An extensive uncertainty analysis to assess the quality of the maps revealed total uncertainty for an average 50x50 km grid cell to range from (worst case between brackets) 50% (70%) in west, north-west and Central Europe to 60% (80%) in east, south-east and south-west Europe, and 70% (100%) in complex terrain (i.e. mountainous and upland) areas.

Acknowledgement

All organisations delivering data used to compile the European wet deposition maps are acknowledged for their contribution.

References

Buijsman E. and Erisman J.W.: 1986, Inst. for Met. and Oceanography, University of Utrecht, The Netherlands, Report R-86-5.

Draaijers G.P.J., Leeuwen E.P. van, Potma C., Pul W.A.J. van, Erisman J.W.: 1995 (this issue).

Erisman J.W., Potma C., Draaijers G.P.J., Leeuwen E.P. van, Pul W.A.J. van: 1995 (this issue).

Iversen T., Halvorsen N., Mylona S. and Sandnes H.: 1991, Norwegian Meteorological Institute, Oslo, Norway, EMEP/MSCW Report 2/92.

Jaarsveld H.J.A. van and Onderdelinden D.: 1990, National Institute for Public Health and Environmental Protection, Bilthoven, the Netherlands, RIVM Report 228603009.

Journel A.G. and Huijbregts Ch.J.: 1978 *Mining Geostatistics*, Academic Press, London.

Legates D.R. and Willmott C.J.: 1990, *International Journal of Climatology* **10**, 111-123.

Leeuwen E.P. van, Potma C., Draaijers G.P.J., Erisman J.W. and Pul W.A.J. van: 1995, National Institute for Public Health and Environmental Protection, Bilthoven, the Netherlands RIVM-Report 722108006.

Matheron G.: 1971, *The theory of regionalized variables and its applications,* les Cahiers du Centre de Morphologie, Mathematique de Fontainebleau, Ecole des Mines de Paris.

Pul W.A.J van, Potma C., Leeuwen E.P. van, Draaijers G.P.J., Erisman J.W.: 1995, National Institute for Public Health and Environmental Protection, Bilthoven, the Netherlands RIVM-Report 722401005

Schaug J., Pedersen U. and Skjelmoen J.E.: 1991, Norwegian Institute for Air Research, EMEP/CCC-Report 2/91.

DEPOSITION OF SULPHUR AND NITROGEN COMPOUNDS IN CROATIA

S. VIDIČ

Meteorological and Hydrological Service of Croatia, Grič 3, 10000 Zagreb, Croatia

Abstract. Since precipitation is an efficient scavenger of pollutants, concentrations of major ions in precipitation reflect changes in chemistry of the atmosphere and in the subsequent exposure of various ecosystems to deposition. The National Atmospheric Precipitation Chemistry programme was initiated in 1978 and operated by Meteorological and Hydrological Service of Croatia to provide needed information on geographical patterns and temporal trends in precipitation chemistry in Croatia. To accomplish this, a network of about 20 stations, settled in different geographical regions, operates on a daily basis for 15 years now. Some monitoring stations are site- and study- specific; others are included in long-term, regional, or European monitoring networks (EMEP, GAW, MEDPOL, GEMS). The purpose of this work was to summarise existing data from the whole network for the period 1981–1992 and to compare data from measurements with EMEP model calculations of acid deposition. Results presented here show that annual average concentration and deposition values at remote sites agree reasonably well compared to modelled ones.

Key words: monitoring network, chemical composition of precipitation, wet deposition, EMEP.

1. Introduction

The observed changes now taking place in the troposphere are caused by the gases, aerosols and particulate matter being released into the atmosphere from variety of sources and processes. The chemical composition of the natural atmosphere is determined by a dynamic balance between complicated sequences of inter-linked chemical, physical, biological and meteorological processes, keeping the concentrations of the large range of trace substances in the atmosphere at roughly constant levels. Despite international agreements and efforts made in various countries to reduce the emissions of major pollutants, the air quality in many parts of Europe is still poor.

A measurement programme of the chemical composition of precipitation in Croatia was initiated in late seventies with the main purpose to provide information on geographical patterns and temporal trends in precipitation chemistry, and to analyse the impacts of anthropogenic contribution to the chemical composition of the atmosphere.

In this work we analysed data from 12 stations and focused on the problem of sulphate and nitrate deposition in different geographical and climatic regions of Croatia.

2. Data and area of study

Monitoring network, established and held by Meteorological and Hydrological Service of Croatia (MHSC), consists of about 20 stations performing bulk daily sampling since 1981. Stations are located in suburban, urban and rural areas over the whole country. Precipitation quality measurements are performed mainly on meteorological stations and observatories with permanent staff. Measurement programme covers meteorological parameters, acidity, electrical conductivity and major ion concentrations. Sampling is organised on a daily basis, in accordance with precipitation measurement protocol (07–07 a.m. local time). According to the protocol polyethylene bottles should be changed and cleaned every day at a site. All data are analysed and stored at central analytical laboratory at MHSC in Zagreb. Data processing and storage is not fully computerised yet, which is one of the reasons why results of analysis of 12 stations are presented here. Second restriction was made on a data completeness criterion – preferably sites with more than 75 per cent coverage of analysed precipitation were selected, though it was not possible to follow this rule strictly at all sites (especially at coastal regions). Site location and measurement details are summerised at Table I and shown at Fig. 1 respectively.

3. Emissions of sulphur and nitrogen

Data on emissions of air pollutants are essential in transport processes understanding on regional and local scales and modelling. Up to now, the only available information on emission levels for Croatia, used for calculations in EMEP models, was the one assessed on the basis of the total emission levels for former Yugoslavia. Considerable efforts have been made to improve data on emissions and calculate their spatial distribution based on the EMEP 50 km x 50 km grid system in a last two years. Calculations were made by using CORINAIR and EMEP metodologies (Bouscaren,1992; McInnes et al, 1992; Pacyna and Joerss, 1991). Grided emissions of sulphur and nitrogen, calculated for the year 1990 (Jelavić, 1994), presumably overestimate actual emission levels in Croatia. They are presented in Fig. 2.

TABLE I

Characteristics of precipitation quality measurement sites ([1] EMEP, [2] MED POL sites)

Site name		Lat. (N)	Long. (E)	El. (m) a.m.s.l.	Meas. started	No of yr. anal.	Site type	Days with prec. per yr	Total prec. per yr (mm)
01–OS	Osijek	45° 32'	18° 44'	89	1981	6	Suburban	140	590
02–VA	Varaždin	46° 18'	16° 23'	167	1984	12	Suburban	167	849
03–PT	Puntijarka[1]	45° 55'	15° 58'	988	1978	13	Rural	147	1226
04–OG	Ogulin	45° 16'	15° 14'	328	1987	5	Suburban	166	978
05–PL	Plitvice	44° 53'	15° 37'	580	1980	10	Rural	138	1467
06–PU	Pula	44° 52'	13° 51'	30	1981	11	Urb./Ind.	105	833
07–RI	Rijeka[2]	45° 20'	14° 27'	120	1978	12	Urb./Ind.	129	1527
08–SE	Senj	44° 59'	14° 54'	26	1983	9	Urban	127	1267
09–ZZ	Zavižan[1,2]	44° 49'	14° 59'	1594	1978	14	Rural	168	1923
10–GO	Gospić	44° 33'	15° 22'	564	1981	12	Suburban	143	1376
11–ZA	Zadar[2]	44° 08'	15° 13'	5	1981	9	Urb./Mar.	106	791
12–DU	Dubrovnik	42° 39'	18° 05'	42	1981	9	Urb./Mar.	114	1054

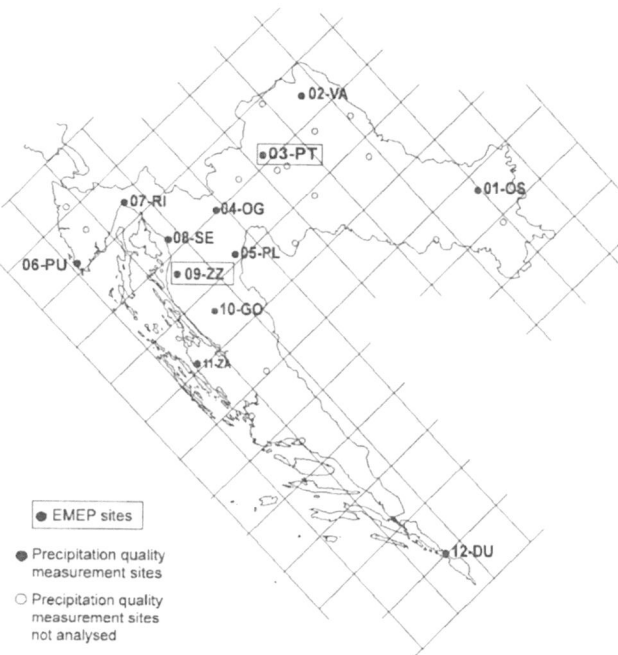

Fig. 1. Spatial distribution of precipitation quality measurement stations in Croatia

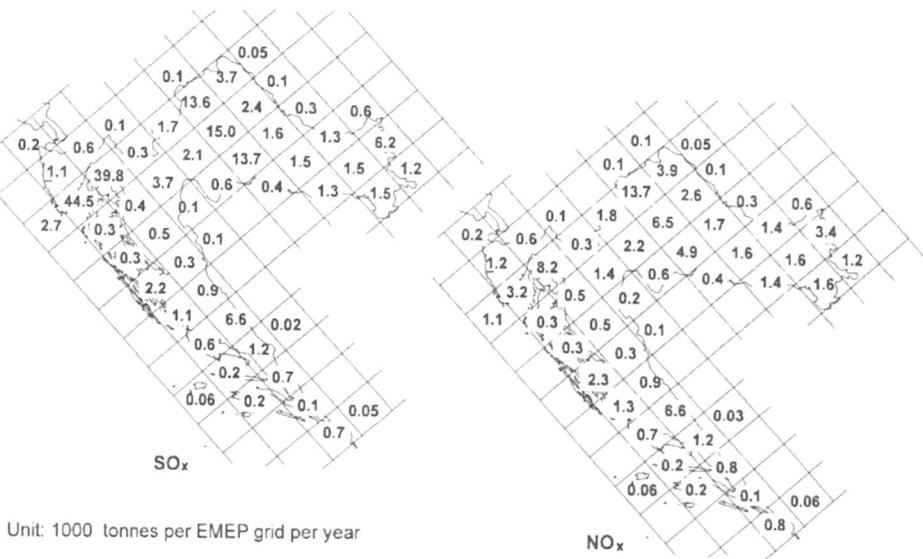

Fig. 2. 50 km x 50 km grided So_x and NO_x emissions in 1000 tonnes per year.

4. Results and discussion

Data sets taken for the extensive analysis were selected to cover different geographycal and climatic regions of Croatia: Northern continental part (01–OS, 02–VA and 03–PT), central mountainous region (04–OG, 05–PL and 10–GO) and the coastal area (06–PU, 07–RI, 08–SE, 11–ZA and 12–DU). Two sites, 03–PT and 09–ZZ are included in EMEP monitoring programme.

Compared to the grided emission data it can be noticed that major emission sources of sulphur and nitrogen are concentrated in Northern Adriatic and Northern continental regions which influence concentrations and deposition at sites in those regions directly (07–RI, 08–SE, 03–PT).

Elevated sites (03–PT, 05–PL, 09–ZZ and 10–GO) indicate presence of combined effects of large, regional and local scale transport of pollutants. Long-term average values of hydrogen, sulphur ($SO_4^=-S$) and nitrogen (NO_3^--N and NH_4^+-N) deposition and volume weighted concentrations are given in Table II. Concentration values $> \pm 3\ \sigma$ are excluded from annual statistics calculations.

The area, most affected by acidic deposition, is the central mountainous part. It is evident also that the EMEP site Puntijarka (03–PT) is under the influence of the urban and industrial area of Zagreb, especially with regard to nitrogen concentrations and deposition. Looking at concentrations for Zavižan (09–ZZ, EMEP site) we can see that they exhibit the lowest values in comparison to other sites. Considering the height of the station (1594 m a.m.s.l.) and prevailing meteorological conditions we might expect those values to be background ones for the Croatia.

TABLE II

Volume weighted average yearly concentrations and wet deposition from the period 1981-1992

Site	Concentrations of major ions in precipitation					Wet deposition of major ions			
	H^+ ($\mu g\ l^{-1}$)	pH (pH units)	$SO_4^{2-}-S$ ($mg\ l^{-1}$)	NO_3^--N ($mg\ l^{-1}$)	NH_4^+-N ($mg\ l^{-1}$)	H^+ ($mg\ m^{-2}$)	$SO_4^{2-}-S$ ($g\ m^{-2}$)	NO_3^--N ($g\ m^{-2}$)	NH_4^+-N ($g\ m^{-2}$)
01–OS	5.57	5.25	2.94	1.36	2.18	2.3	1.5	0.6	1.0
02–VA	3.69	5.43	3.94	1.05	1.37	3.0	3.3	0.9	1.1
03–PT	**9.83**	**5.01**	**1.67**	**1.52**	**0.67**	**12.1**	**2.1**	**1.9**	**0.8**
04–OG	8.65	5.06	2.03	0.94	0.89	11.5	2.3	1.1	1.1
05–PL	27.74	4.56	5.27	1.35	0.96	39.3	6.9	1.8	1.3
06–PU	2.38	5.62	3.51	1.39	0.95	2.1	2.5	1.0	0.7
07–RI	17.01	4.77	2.10	0.92	1.23	25.6	3.1	1.4	1.8
08–SE	5.85	5.23	2.09	1.39	0.55	6.2	2.1	1.4	0.5
09–ZZ	**5.14**	**5.29**	**1.32**	**0.42**	**0.45**	**10.1**	**2.5**	**0.8**	**0.8**
10–GO	2.42	5.62	2.97	0.83	1.37	3.2	3.6	1.1	1.6
11–ZA	5.26	5.28	3.14	1.15	0.86	4.4	2.4	0.9	0.6
12–DU	6.83	5.17	3.47	0.84	0.68	6.2	3.3	0.8	0.7

Deposition values show a slightly different distribution due to dependence on precipitation amount at each site. Nevertheless, wet annual deposition at site 09–ZZ is still 15–30% below the average value for the network (Fig. 3).

An attempt has been made to estimate trends for volume weighted annual concentrations for the period 1981–1992, though it was seeming that because of the high dispersion in concentrations at all sites, significant trends might not be expected. However, downward trends were obtained at three sites (at significance level $p < 0.05$): 03–PT (nitrates), 09–ZZ (sulphates) and 06–PU (sulphates and nitrates). Furthermore, except for ammonium, downward trends are indicated for most major ions in precipitation, but not at significant level (Fig. 4).

When compared to the EMEP MSC–W calculated annual total deposition budgets for years 1985–1992 (Sandnes and Styve, 1992, Sandnes, 1993, Tuovinen et al.,1994) it appears that calculated deposition of sulphur and nitrogen from nitrates slightly underestimate measured values at all sites but values for both EMEP sites are generally in a good agreement with modelled ones, except for the nitrogen at 03–PT. Modelled deposition of nitrogen from ammonium are underestimated by a factor of two.

5. Conclusion

Human activities have a profound influence on the atmospheric cycles of sulphur and nitrogen. Precipitation quality data at 12 sites were used in various ways to quantify the magnitude of anthropogenic impact at individual sites. The data are also analysed to assess regional patterns in these impacts. They suggest the exsistance of three regions of different pollution capacity (northern continental, central mountainous and coastal) indicating that different mitigation measures that should be taken to take into account major processes and effects.

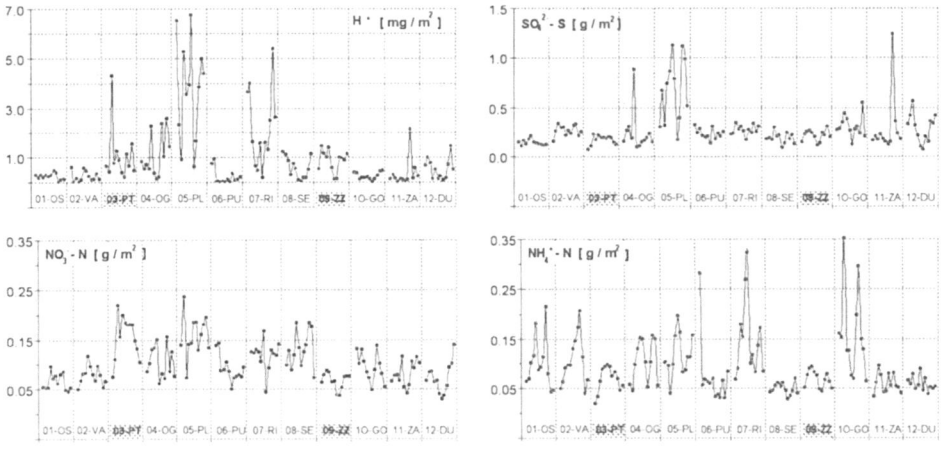

Fig. 3. Annual course of mean monthly deposition values of major ions in precipitation over the period 1981-1992 for 12 selected sites in Croatia.

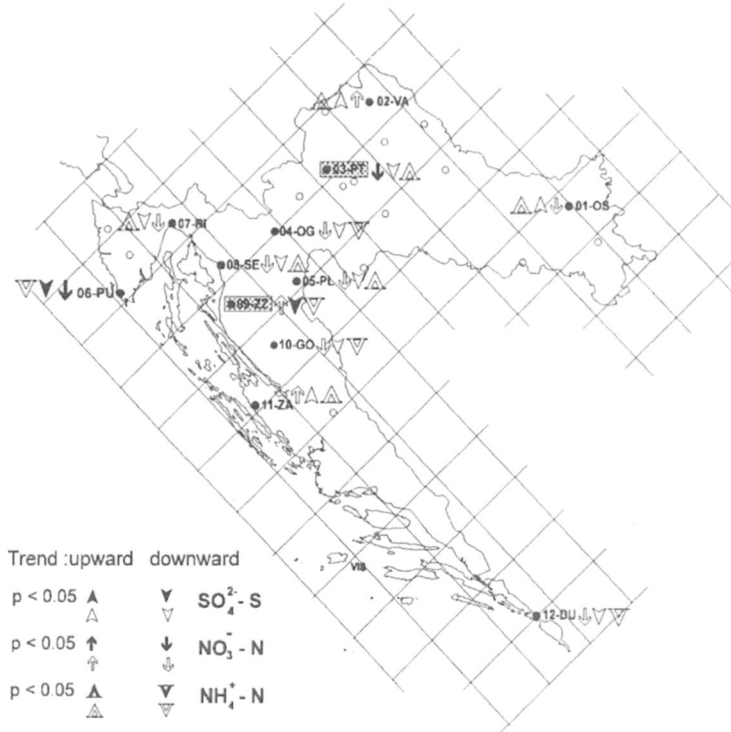

Fig. 4. Trends in sulphate, nitrate and ammonium ion concentrations in precipitation (1981-1992).

Concentrations of S and N species in precipitation at all sites vary interannually over a factor of about two. Significant downward linear trends were attained at few stations, though general impression is that concentrations decrease.

However, deposition of S and N compounds makes a slight inversion to that pattern. Highly affected areas are remote ones at central mountainous region, affected by pollution from both, local and regional sources.

References

Bouscaren, M.R.:1992, *Default emission factors handbook*, CITEPA, Paris.
Jelavić, V.: 1994, *Croatian air pollutant emission in the EMEP grid (1990), Report submitt.to the UNECE*.
McInnes, G., Pacyna J.M. and Dovland H.:1992, EMEP/CCC Report **4/92**.
Pacyna J.M. and Joerss K.E.:1991, EMEP/CCC Report **1/91**.
Sandnes, H. and Styve H.:1992, *Calculated budgets for airborne acidifying components in Europe, 1985, 1987, 1988, 1989, 1990 and 1991*, EMEP/MSC-W Report **1/92**.
Sandnes, H.:1993, *Calculated budgets for airborne acidifying components in Europe, 1985, 1987, 1988, 1989, 1990, 1991 and 1992.*, EMEP/MSC-W Report **1/93**.
Tuovinen, J.P., Barett, K and Styve H.:1994, *Transboundary acidifying pollution in Europe: Calculated fields and budgets 1985-93*, EMEP/MSC-W Report **9/94**.

SNOWPACK QUALITY AS AN INDICATOR OF AIR POLLUTION IN FINNISH LAPLAND AND THE KOLA PENINSULA, NW RUSSIA

A-J. LINDROOS[1], J. DEROME[2] and K. NISKA[2]

[1]*Finnish Forest Research Institute, Vantaa Research Centre, P.O.Box 18, FIN-01301 Vantaa, Finland,* [2]*Finnish Forest Research Institute, Rovaniemi Research Station, P.O.Box 16, FIN-96301 Rovaniemi, Finland*

Abstract. Bulk snow samples were collected from the snowpack in open areas along two sampling lines running to the west from the Cu-Ni smelters at Nikel and Monchegorsk, NW Russia, during 1991-1993. The aim of the study was to estimate the area affected by sulphur and heavy metal deposition from the smelters. Snowpack quality was used as an indicator of deposition during winter time. The total sulphur, copper and nickel concentrations in the snowpack decreased significantly ($p < 0.001$) with increasing distance from the smelters along the sampling line running directly to the west from Monchegorsk. The deposition pattern was similar each winter during 1991-1993. The pH values did not correlate with the corresponding sulphur concentrations, and there was no decreasing pH gradient in the snowpack on moving towards Monchegorsk. The effects of sulphur emissions from Monchegorsk on snowpack chemistry were not detectable on the Finnish side of the border. The 3-year mean of the total sulphur concentration was 0.27 mg/kg, and of the pH values 4.92, along the sampling line running to the west of Monchegorsk. The total sulphur concentrations near the smelters (< 20 km) varied between 0.37 and 0.95 mg/kg. The effect of the Cu-Ni smelters at Nikel on snowpack quality was not detectable in northern Finnish Lapland. The 3-year mean for total sulphur was 0.20 mg/kg and for pH 4.96 along the sampling line running to the west of Nikel.

Key words: Cu, Cu-Ni smelters, Kola Peninsula, Lapland, Ni, pH, S, snowpack

1. Introduction

The copper-nickel smelters in the Kola Peninsula, NW Russia, are significant emission sources of sulphur dioxide and heavy metals. The SO_2 emissions from Petshenganikel in 1990 were 190 000 tonnes, from the adjoining smelter at Zapoljarnyi 67 000 t, and from Monchegorsk 233 000 t. Annual copper and nickel emissions from Petshenganikel were 200 t and 300 t, respectively. The emissions for Monchegorsk were 2 200 t of copper and 3 100 t of nickel per year (Lapin lääninhallitus, 1992).

These industrial emissions have caused considerable changes in the forest ecosystems surrounding the smelter complexes (Kalabin *et al.*, 1990; Kauhanen and Varmola, 1992; Kozlov *et al.*, 1993; Kryuchkov, 1991; Tikkanen, 1995). The trees and ground vegetation were completely destroyed in the immediate vicinity of the smelters (Lukina *et al.*, 1993).

Emissions of sulphur dioxide and heavy metals from smelters have been reported to cause serious environmental damage (e.g. Hutchinson and Whitby, 1974; 1977). Sulphur deposition can also cause soil acidification (Berden *et al.*, 1987). According to Nordgren *et al.* (1985, 1986), the microbiological properties in the vicinity of smelters can also change.

The aim of this study was to investigate the extent of sulphur and heavy metal deposition in the area lying to the west of the smelter complexes at Nikel and Monchegorsk. The chemical composition of the snowpack was used as an indicator of air pollutants and their deposition.

2. Material and Methods

Snow samples were collected in March 1991, 1992 and 1993 from the snowpack along two lines running through Russia and Finnish Lapland to the west of the smelter complexes at Nikel and Monchegorsk (Figure 1). Because the sampling was carried out as early as in March, the effect of thawing episodes on snowpack concentrations was considered to be insignificant. The snow was sampled in open areas with a uniform snowpack as far away as possible from roads and settlements. The samples were taken at points not affected by the tree cover.

The samples were allowed to thaw in the laboratory and then filtered through filter paper. pH was measured on the samples, which were then conserved with nitric acid prior to analysis of total sulphur (S), copper (Cu) and nickel (Ni) by inductively coupled plasma atomic emission spectrophotometry (ICP/AES). The sampling procedure and analyses are described in detail in Derome *et al.* (1993).

The formation of the snowpack is naturally not completely the same over extensive areas owing to variations in snowfall, wind patterns and temperature. For this reason the study was restricted to the pH and concentrations of S, Cu and Ni, and no attempts were made to calculate actual deposition levels. Variations in the chemical composition of the snowpack were studied in relation to the distance to the emission sources using a linear regression model.

The snowpack in the study area represented snowfall for the period October to March, the annual precipitation in the area ranging from ca. 380 - 580 mm.

3. Results

3.1. EFFECT OF EMISSIONS FROM MONCHEGORSK ON THE ACIDITY AND SULPHUR, COPPER AND NICKEL CONCENTRATIONS IN THE SNOWPACK

No trends were found in the pH of the snowpack with increasing distance from the emission source along the line running to the west from Monchegorsk (Figure 2). The only clear exception to the rather constant pH level along the line were the relatively high pH values close to the open-cast mine at Kovdor, 100 km from Monchegorsk. The mean pH value for the three-year period along the line was 4.92.

Fig. 1. The sampling lines (A and B) extending to the west of Monchegorsk and Nikel.

Snowpack S concentrations fell sharply with increasing distance from Monchegorsk (Figure 2). The S concentration reached a constant level at a distance of about 40 km from the emission source. The mean S concentration for the three-year period along the line was 0.27 mg/kg, but at a distance of less than 20 km the S concentration varied from 0.37 to 0.95 mg/kg. The S concentration did not correlate with the snowpack pH along this line ($r = 0.06$, $n = 99$, $p > 0.05$).

The effect of distance from the emission source at Monchegorsk on snowpack S concentrations was investigated using regression analysis. The analysis was performed on the samples collected during the 3-year period at distances of 8 - 62 km from the emission source. Samples taken at distances greater than 62 km were omitted from the analysis because the emissions clearly no longer had any effect on snowpack composition. The concentrations were transformed logarithmically ($\log_{10}$) in order to give a better fit for the regression line. The distance from Monchegorsk explained 75% of the variation in the S concentration of the snowpack (Table I).

The Cu and Ni concentrations in the snowpack were determined along the line up to a distance of 120 km from the emission source during the 3-year period, but not on the Finnish side of the border (>120 km) in 1992 and 1993. The results for 1991 indicated that the Cu and Ni concentrations in the snowpack at a distance of more than 120 km from the emission source were below the detection limits of the analytical equipment. Both the Cu and Ni concentrations in the snowpack fell sharply with increasing distance from the smelter (Figure 3). The highest concentrations were recorded in 1991 close to the smelter (8 km), the Cu concentration being 421 µg/kg and Ni concentration 180 µg/kg. Regression analysis was performed on the samples collected over the distance 8 - 38 km from the smelter (Table I). According to the model, distance explained 70% of the variation in the Cu concentration and 80% of the Ni concentration. $\log_{10}$ transformations were used in the analysis.

TABLE I

Regression analysis between distance (km) from the Monchegorsk smelters and the snowpack S, Cu and Ni concentrations

Regression line	R^2	n	p*	Distance range from the smelters (km)
log S (mg/kg) = 0.27 + (-0.51 log km)	0.75	23	< 0.001	8 - 62
log Cu (µg/kg) = 4.41 + (-2.19 log km)	0.70	13	< 0.001	8 - 38
log Ni (µg/kg) = 3.29 + (-1.53 log km)	0.80	13	< 0.001	8 - 38

*F-test, H_0: ß=0

3.2. EFFECT OF EMISSIONS FROM NIKEL ON THE ACIDITY AND SULPHUR CONCENTRATION IN THE SNOWPACK

There was no clear gradient in the snowpack pH values with increasing distance from the Nikel smelter (Figure 4). The 3-year mean pH value was 4.96. The snowpack S concentration decreased gradually with increasing distance (Figure 4). The 3-year mean snowpack S concentration along the line was 0.20 mg/kg. Statistically significant correlation was found between the S concentration and the pH of the snowpack (r =-0.54, n = 31, p < 0.01). However this finding is of little meaning because the variations in pH and S concentration along this line were very small.

According to regression analysis performed on the untransformed values over the distance range 14 - 100 km, the distance to Nikel explained 42% of the variation in the snowpack S concentration (R^2 = 0.42, n=17, p < 0.01). In this case, too, the drop in the S concentration with increasing distance was relatively small.

4. Discussion

According to the results of this study, the effects of sulphur emissions from Monchegorsk on snowpack composition extend for a distance of 30 km from the emission point source. In this case sulphur deposition was not necessarily associated with acidic deposition because no gradient was found in the snowpack pH values. The copper and nickel emissions were also clearly reflected in the snowpack composition in the immediate vicinity of the smelter, although not as far as 30 km. The results rather well reflect the spread of emissions and their wet deposition from the Monchegorsk smelter towards Finland in the west, which in actual fact are relatively low owing to the prevailing wind directions (Anttila, 1995). Also, the high fell chain running in a north-south direction on the western side of Monchegorsk undoubtedly reduces the spread of emissions to the west.

The effects of emissions from Nikel on the chemical composition of the snowpack in northern Lapland were not as clear. The S concentrations increased with decreasing distance to the emission source, but nowhere near as sharply as close to Monchegorsk.

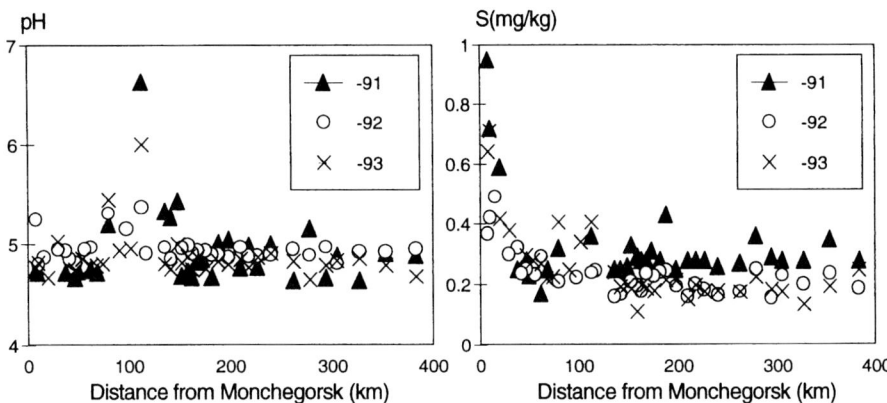

Fig. 2. Snowpack pH and S concentration along the sampling line running from Monchegorsk in 1991, 1992 and 1993.

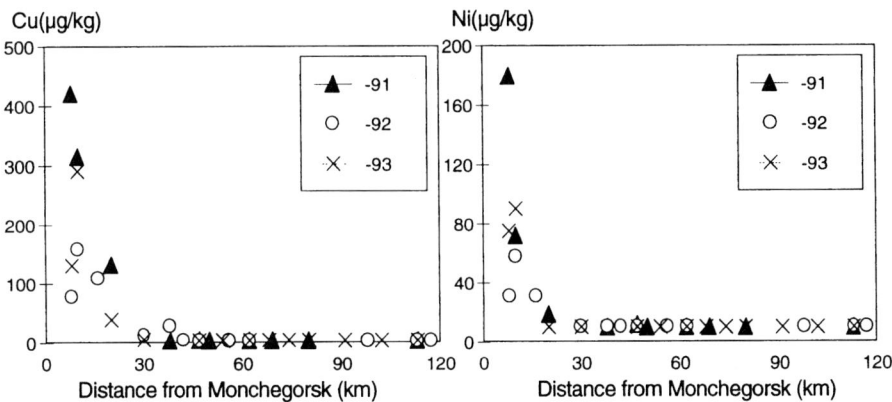

Fig. 3. Snowpack Cu and Ni concentrations along the sampling line running from Monchegorsk in 1991, 1992 and 1993.

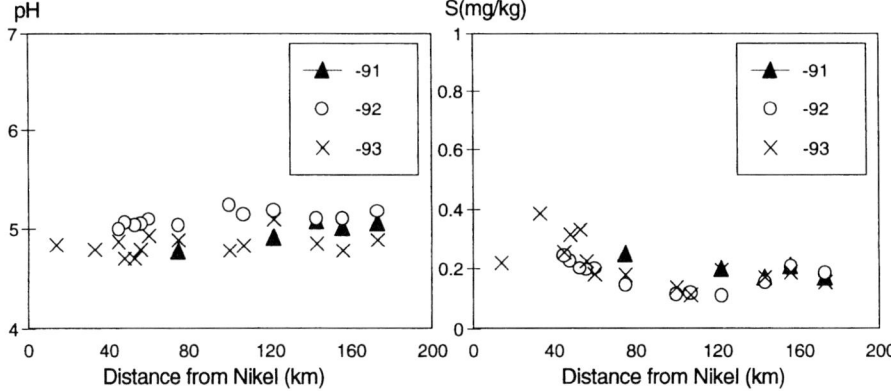

Fig. 4. Snowpack pH and S concentration along the sampling line running from Nikel in 1991, 1992 and 1993.

According to Tikkanen and Mikkola (1993), damage to the forest ecosystems is clearly visible in the surroundings of the smelters, and the severity of the damage increases sharply on moving towards the emission sources. The effects of emissions outside the major damage zones can only be identified using the most sensitive plant indicator techniques. The pattern in the chemical composition of the snowpack obtained in this study was in good agreement with the areal distribution of forest damage, especially in the Monchegorsk area. For instance, sulphur, copper and nickel concentrations in Scots pine needles in the area surrounding Monchegorsk decreased sharply with increasing distance from the point emission source (Raitio et al., 1993).

References

Anttila, P.: 1995, "Ilmanlaatu ja laskeuma", In Tikkanen, E. (ed.), Kuolan saastepäästöt Lapin metsien rasitteena, Metsäntutkimuslaitos, 43-75.

Berden, M., Nilsson, S. I., Rosen, K., and Tyler, G.: 1987, "Soil Acidification - Extent, Causes and Consequences", Report 3292, National Swedish Environmental Protection Board, Solna, 164 pp.

Derome, J., Nikonov, V., Lindroos, A.-J., Niska, K., and Välikangas, P.: 1993, "Snowpack Quality in Finnish Lapland and the Western Part of the Kola Peninsula in March 1991", In Derome, J. (ed), Russian-Finnish Cooperation Report, Finnish Forest Research Institute, 98-111.

Hutchinson, T.C., and Whitby, L.M.: 1974, Environmental Conservation 1, 123.

Hutchinson, T.C., and Whitby, L.M.: 1977, Water, Air, and Soil Pollut. 7, 421.

Kalabin, G., Kryuchkov, V. and Nikonov, V.: 1990, "Ecology Problems and Environmental Research Projects of the Kola North", In Kinnunen, K., and Varmola, M. (eds), Effects of Air Pollutants and Acidification in Combination with Climatic Factors on Forests, Soils and Waters in Northern Fennoscandia, Nord, Miljörapport 2, 26-30.

Kauhanen, H., and Varmola, M. (eds): 1992, Metsäntutkimuslaitoksen tiedonantoja 413, 269 pp.

Kozlov, M.V., Haukioja, E., and Yarmishko, V.T. (eds): 1993, "Aerial Pollution in Kola Peninsula", Proceedings of the International Workshop, April 14-16, 1992, St.Petersburg, Russia,Apatity, 417 pp.

Kryuchkov, V.: 1991, "Heavy Metal Accumulation in Spruce Needles and Changes of Northern Taiga Ecosystems", In Pulkkinen, E. (ed.), Environmental Geochemistry in Northern Europe, 177-184.

Lapin lääninhallitus: 1992, "Lapin läänin ja Murmanskin alueen sekä Pohjois-Norjan ja Pohjois-Ruotsin ilmansuojelu vuonna 1990", Ympäristönsuojelutoimisto, Rovaniemi, 25 pp.

Lukina, N., Liseenko, L., and Belova, E.: 1993, "Pollution-induced Changes in the Vegetation Cover of Spruce and Pine Ecosystems in the Kola North Region", In Derome, J. (ed), Russian-Finnish Cooperation Report, Finnish Forest Research Institute, 31-43.

Nordgren, A., Bååth, E., and Söderström, B.: 1985, Can. J. Bot., 63, 448.

Nordgren, A., Kauri, T., Bååth, E., and Söderström, B.: 1986, Environ. Poll. (Series A) 41, 89.

Raitio, H., Nikonov, V., and Lukina, N.: 1993, "Chemical Composition of Scots Pine Needles in the Industrial Region of Monchegorsk", In Derome, J. (ed.), Russian-Finnish Cooperation Report, Finnish Forest Research Institute, 23-30.

Tikkanen, E. (ed.): 1995, "Kuolan saastepäästöt Lapin metsien rasitteena", Metsäntutkimuslaitos, 232 pp.

Tikkanen, E., and Mikkola, K.: 1993, " The Lapland Forest Damage Project: A Gradient Study Approach in Investigating the Effects of Kola Pollutant Emissions on Forest Ecosystems", In Derome, J. (ed.), Russian-Finnish Cooperation Report, Finnish Forest Research Institute, 4-15.

EVALUATION OF THE CHANGES IN REGIONAL WINTERTIME DEPOSITION IN FINLAND DURING 1976 - 1993

J. SOVERI and K. PELTONEN

Finnish Environment Agency, Monitoring and Assessment Division, P.O.Box 140, FIN-00251 Helsinki, Finland

Abstract. Snowpack quality has been systematically monitored in Finland since winter 1975/1976 by the Finnish Environment Agency. Snow samples have been taken at the 53 groundwater observation stations in nearly natural state areas, where the impacts of local emission sources have been negligible. Thus the concentrations of impurities in snow can be considered as background levels. Samples were analyzed for conductivity, pH, major ions and trace metals (Cd, Cu, Pb, Ni, Zn and Hg).

The regional variation of the chemical composition of major ions in the snowpack was outlined by cluster analysis. Sampling sites were divided into six clusters by average linkage clustering method. The largest cluster - considered as the 'mean cluster' - was located in central and northern Finland. The lowest levels of the total amount of dissolved impurities occurred in the cluster formed by the most northern observation site exclusively. Highest rates of the acidifying components, sulphate (SO_4), nitrate (NO_3), ammonium (NH_4) and hydrogen ions (H^+), were encountered in southern Finland. Sampling locations of southern Finland were divided into three clusters. The acidity of snow varies among these clusters that in addition of deposition of SO_4 and NO_3 is considerably controlled by alkaline deposition.

The results of chemical analyses of snow were converted to mean monthly depositions using the water equivalent of snow and the time of deposition. Deposition of SO_4 during 1985-1993 was 32% lower compared with the period 1976-1984, which presumably results from 70% reduction of sulphur emissions in Finland since the beginning of the 1980s. The deposition level of NO_3-N seemed to stay relatively stable during the whole 18 year period while the deposition of NH_4-N has decreased 18%.

1. Introduction

It is well known that high concentrations of acidifying pollutants in snow causes the spring pH depression in surface waters. Areas of acid-intermediate rock association with shallow overlay have been found to be particularly sensitive to groundwater acidification (Thornton et al., 1981). Snow sampling offers a useful method for evaluation of atmospheric pollution and it has been applied to some extent in other studies in Scandinavia. According to Wright and Dovland (1977), high concentrations of H^+, SO_4, NO_3 and NH_4 in snow in southeastern and northernmost Norway are mainly due to long-range transport. Trace metal deposition in northern Sweden decrease from the Baltic coast towards the Swedish interior (Ross and Granat, 1986). Snowpack surveys have also been used in determination of environmental loads of specific point sources, such as concentrating plants, smelters and pulp mills (Ettala et al., 1986).

Sulphur emissions in Finland peaked in the mid-1970s and have gradually declined since (Kauppi et al., 1990). The level in 1992 was over 70% lower compared to 1980 (584 000 tons SO_2). The reduction is a consequence of changes in the structure of energy production, higher fuel oil quality and improvements in production engineering. NO_x emissions peaked in 1990 (290 000 tons NO_2) and have slightly reduced since (Statistics Finland, 1995).

Snow samples have been systematically taken at the sites of the groundwater observation network in Finland since winter 1975/1976. The purpose of this study is to delineate regional characteristics of snow chemistry and evaluate long term changes in amounts of wintertime deposition.

2. Materials and methods

Snow samples were collected from the sites of 53 groundwater observation stations located in areas representing background levels of atmospheric impurities. Sampling was carried out with Plexiglass cylinders. The sampler was pushed vertically into the snowpack to make sure that the samples would represent the entire period of deposition. This procedure was repeated within a 100 m^2 area until a sufficient amount of water had been obtained for the analysis.

A considerable amount of dissolved substances is released from the snowpack during the first stages of melting. According to Johannessen and Henriksen (1978) 44-76% of snowbound pollutants are released during the melting of the first 30% of the snow. Therefore it is essential to schedule the sampling before the ablation period. Long warm terms during the accumulation period of snow may cause midwinter melting, which consequently debilitates the representativity of samples.

To assure the representativity of the snow samples it is also important to select the sampling sites to match the average snow conditions of the area. Large open areas were avoided because of the their sensitivity to snow transport by wind (Kuusisto, 1986). The representativity of the snowpack was tested at the sampling site by measuring the depth or the water equivalent of snow.

Snowfall represents 30% of annual precipitation in southwestern Finland and more than 50% in northern Finland. The maximum water equivalent of snow is rougly 15% of the annual precipitation in southwestern Finland, in Lappland it usually exceeds 30%.

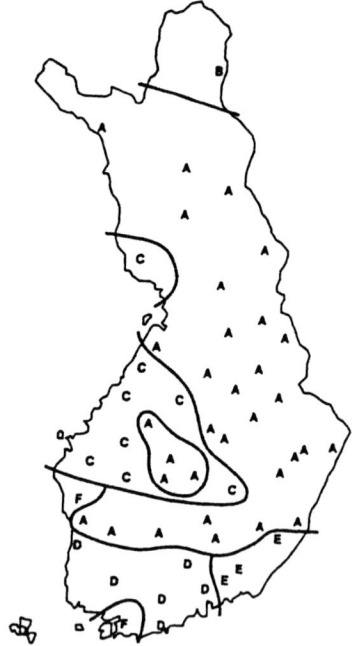

Fig 1. Cluster Analysis: The network of 53 observation sites and the regional distribution of clusters A to F.

The samples were stored in polythene bags and transported frozen to the laboratory. Prior to analysis the samples were melted at room temperature in polythene goblets, filtered with dressing gauze and homogenized by shaking. Unstable components were preserved for analysis. The field observations, sampling and sample handling were carried out in accordance with the instructions in the Finnish Environment Agency (Mäkelä et al., 1992).

In the laboratory the samples were analyzed for pH, conductivity (γ_{25}), total nitrogen (N_{tot}), nitrate nitrogen (NO_3-N), ammonium nitrogen (NH_4-N), total phosphorus (P_{tot}), phosphate phosphorus (PO_4-P) chloride (Cl), iron (Fe), manganese (Mn), sulphate (SO_4), sodium (Na), potassium (K), calcium (Ca), magnesium (Mg), silicate (SiO_2), fluoride (F), aluminium (Al), cadmium (Cd), copper (Cu), lead (Pb), nickel (Ni), zinc (Zn), mercury (Hg) and total organic carbon (TOC). The pH of the meltwater was determined with a Radiometer pH meter in 25 °C. N_{tot}, NO_3-N, NH_4-N, P_{tot}, PO_4-P, Fe and Mn were measured spectrophotometrically, Cl by a titration with mercury nitrate and SO_4 was measured by a turbidimetric method. SiO_2 was measured by automatic spectrophotometry with molybdate, F was measured potentiometrically and Na, K, Ca, Mg, Al, Cd, Cu, Pb, Ni, Zn and Hg were measured with AAS. TOC was measured by combustion-infrared method. The methods for analysis have been described by National Board of Waters (1981).

For statistical treatment of the data the values below detection limit (DL) were replaced by DL/2. Correlations were calculated for the entire data to reveal relationships between substances. The regional variation of the quality of snow was delineated by cluster analysis.

3. Results and discussion

3.1. REGIONAL CHARACTERISTICS OF SNOW CHEMISTRY

The snow samples were taken from 53 observation sites during 1976-1993, once a year in late winter when the water equivalent had reached its maximum. Correlations were calculated for the entire data. The most significant correlations were as follows: H^+-NO_3 (r = 0.710, p<0.001), NO_3-NH_4 (r = 0.698, p<0.001), SO_4-NO_3 (r = 0.691, p<0.001) and SO_4-NH_4 (r = 0.646, p<0.001). Acidity (H^+) is clearly better correlated to NO_3 than to SO_4 (r = 0.427, p<0.001). Also Ca and Pb correlates fairly good to NO_3, NH_4 and SO_4. These correlations may be interpreted as indications of common sources, energy production and traffic. On the other hand, the positive correlation between H^+ and NH_4 (r = 0.448, p<0.001) indicates that a great deal of SO_4 and NO_3 is incorporated in salts like ammonium nitrate (NH_4NO_3) and ammonium sulphate (($NH_4)_2SO_4$) (Joffre et al., 1990).

Cluster analysis, originated by Sokal and Michener (1958), was performed in order to condensate the data into more intelligible form. The aim was to categorize the snow data into relatively homogenous clusters in order to describe regional characteristics of snow chemistry. The analysis was carried out using the average linkage method included in the Statistical Analysis System (SAS) program package (SAS Institute Inc., 1985).

TABLE I

Cluster analysis: Mean concentrations, standard deviations, minimum and maximum values of clusters.

		Concentrations (μeq/l)							
		H	NH_4	Na	K	Ca	Mg	SO_4	NO_3	Cl
Cluster A	Mean	25.56	11.30	11.06	3.75	13.34	7.09	31.34	23.90	17.42
(n=32)	SD	5.50	2.83	2.22	0.87	3.74	1.05	6.31	3.34	3.79
	Min	16.62	4.46	6.80	2.42	7.65	5.55	19.79	15.40	11.67
	Max	41.44	16.28	17.40	6.49	24.72	9.87	43.23	29.57	26.48
Cluster B	Mean	12.36	2.27	18.03	2.46	8.20	8.23	14.71	13.03	18.51
(n=1)	SD	0.00	0.00	0.00	0.00	0.00	0.00	0.00	0.00	0.00
Cluster C	Mean	16.91	15.22	16.99	4.35	17.25	11.27	34.17	25.81	24.55
(n=8)	SD	4.10	2.32	5.58	0.83	4.40	2.58	6.55	3.42	5.91
	Min	12.20	12.20	10.66	2.94	12.75	6.91	26.27	21.58	14.39
	Max	24.03	18.10	26.10	5.48	26.42	14.26	41.94	31.56	29.94
Cluster D	Mean	16.21	22.86	17.95	6.72	25.83	10.88	45.78	27.11	30.75
(n=7)	SD	5.47	4.67	4.20	0.64	4.21	2.05	4.84	4.20	5.12
	Min	10.70	16.45	12.28	6.17	20.87	7.99	36.04	21.36	23.57
	Max	25.66	29.56	21.75	7.99	33.27	11.97	51.08	33.16	36.43
Cluster E	Mean	24.04	19.94	11.70	5.49	33.84	9.19	51.65	30.00	20.98
(n=3)	SD	1.75	3.65	0.07	0.46	7.72	1.89	2.41	2.32	2.01
	Min	22.02	15.89	11.67	5.11	27.45	7.48	49.01	27.46	19.14
	Max	25.18	22.96	11.78	6.00	42.42	11.22	53.74	32.00	23.13
Cluster F	Mean	30.75	24.70	17.66	5.41	17.56	10.90	49.46	38.37	29.62
(n=2)	SD	0.52	8.10	0.86	1.75	4.09	0.89	11.70	3.92	3.70
	Min	30.39	18.97	17.05	4.17	14.67	10.29	41.19	35.60	27.00
	Max	31.12	30.42	18.27	6.65	20.46	11.52	57.74	41.14	32.24
All	Mean	22.88	14.24	13.28	4.37	16.80	8.51	35.19	25.30	20.94
(n=53)	SD	6.66	5.84	4.31	1.35	7.23	2.37	9.54	4.82	6.51
	Min	10.70	2.27	6.80	2.42	7.65	5.55	14.71	13.03	11.67
	Max	41.44	30.42	26.10	7.99	42.42	14.26	57.74	41.14	36.43

The mean values of major components (H^+, NH_4, Na, K, Ca, Mg, SO_4, NO_3 and Cl) were calculated for cluster data. As a result six clusters were extracted from the data (Figure 1). Mean concentrations and standard deviations of these clusters are presented in Table I.

Cluster A (Figure 1) consists of 32 observation sites located in central and northern parts of Finland. Concentrations of major ions match fairly well with mean concentrations of entire data, thus cluster A can be considered as the 'mean cluster'.

The most northern site exclusively represents cluster B. Concentrations of all major constituents excluding Na, Mg and Cl, are lowest of all clusters. The relative importance of nitrate in acidification can be illustrated by the ratio (μeq/l) $NO_3/(SO_4+NO_3)$, which in this group is highest, 0.47, compared to other clusters (mean value 0.42).

The western parts of Finland are dominated by sites of cluster C. Concentrations of Mg, Na and Cl are relatively high, thus these constituents are apparently of marine origin.

TABLE II

Statistics of the wintertime deposition in Finland during 1976-1993

	Unit	Mean	Median	10th percentile	90th percentile	SD	Change 76-84 to 85-93
γ_{25}	mS/m	1.95	1.80	0.90	3.00	0.44	-14%
pH		4.78	4.70	4.40	5.30	1.49	+0.6%
N_{tot}	mg/m²/month	20.4	17.7	8.94	35.2	12.4	-9%
NO_3-N	mg/m²/month	10.0	9.16	3.96	16.7	5.85	-5%
NH_4-N	mg/m²/month	5.38	4.43	1.74	9.89	4.32	-18%
P_{tot}	µg/m²/month	542	383	147	1098	540	-5%
PO_4-P	µg/m²/month	259	181	54.5	532	345	-19%
Cl	mg/m²/month	21.9	18.5	8.73	37.5	15.5	-19%
Mn	µg/m²/month	722	529	160	1450	755	-16%
SO_4-S	mg/m²/month	16.9	14.0	4.33	32.0	12.8	-32%
Na	mg/m²/month	8.96	7.01	3.00	15.8	9.17	-2%
K	mg/m²/month	4.94	3.61	1.67	9.90	4.00	-32%
Ca	mg/m²/month	9.85	7.43	2.42	20.6	9.23	-30%
Mg	mg/m²/month	2.97	2.48	1.15	4.84	2.51	-39%
Cu	µg/m²/month	94.6	43.2	13.6	205	165	-57%
Pb	µg/m²/month	142	90.0	21.6	294	200	-65%
Zn	µg/m²/month	369	205	41.7	861	540	+8%
TOC	mg/m²/month	64.4	49.3	21.3	117	61.6	-27%

Cluster D in southern Finland consists of stations with high Cl concentrations, which is a typical feature in coastal regions. Concentrations of SO_4, NO_3, NH_4 and Ca are above the average. The ratio $NO_3/(SO_4 + NO_3)$ is 0.37, which is the lowest among clusters indicating a major effect of SO_4 to acidification. Because of the presence of a relatively large portion of base cations the concentration of H^+ is below the average.

Highest concentrations of SO_4 and Ca were found on observation sites of cluster E in southeastern Finland. SO_4 and NO_3 together represents 80% of major anions, which is an indication of vast anthropogenic emissions. As in cluster D, $NO_3/(SO_4+NO_3)$ ratio is 0.37. Concentration of Ca represents 56% of base cations (Na, K, Ca and Mg), therefore it has a significant role of the potential for neutralization.

Cluster F consists of two stations nearby the western coast of Finland. The proximity of the sea clearly contributes to high Cl and Na concentrations. H^+, NO_3 and NH_4 concentrations are highest of all groups.

In all, there is a clear geographical trend of acidifying constituents in Finland. Highest levels of anthropogenic pollutants (SO_4, NO_3 and NH_4) occur in southern Finland while terrigenous components are relatively more prevalent in northern parts of the country.

3.2. DEPOSITION AND TRENDS

The concentrations (mg l^{-1}) of snow samples were converted to deposition units (mg m^{-2} month^{-1}) using the areal water equivalent (Reuna et al., 1993) and the time of deposition. The method is described in detail by Soveri (1985). Statistics of the wintertime deposition for the period of 1976-1993 for selected constituents are presented in Table II. Substances with more than 30% of samples below detection limit have been neglected. To evaluate the changes in deposition during the 18 year period the mean values of deposition were calculated for periods of 1976-1984 and 1985-1993.

Deposition of SO_4 during 1985 - 1993 was 32% lower compared with the period 1976-1984, which presumably results from 70% reduction of sulphur emissions in Finland since the beginning of the 1980s. Deposition of NO_3-N has not been significantly reduced, instead NH_4-N deposition of the latter period is 18% lower. Trends of SO_4-S and N_{tot} depositions are presented in Figure 2. Trend curves (moving average of 4 years) indicate, that the ratio of SO_4-S and N_{tot} deposition was reversed at the beginning of the 1980s.

Acidity of snow has slightly reduced, though due to similar trend of base cation deposition to SO_4-S deposition, reduction (0.03 pH units) has not been that extensive. Since the detection limits of trace metals have improved during the observation period comparison of these results is somewhat difficult. Altogether, some notices of relative concentrations of trace metals can be accomplished. Concerning deposition of Pb, there is a steeper descending trend compared with the other trace metals, since the usage of unleaded gasoline has become common during the past decade.

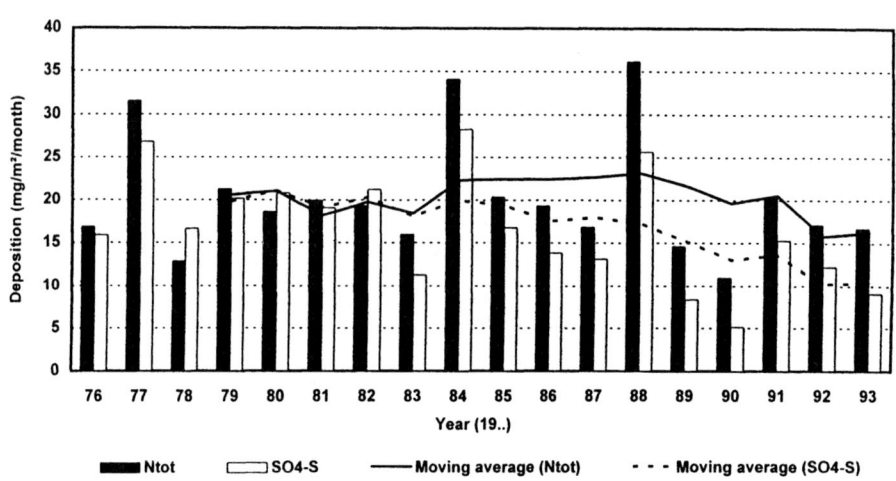

Fig. 2. Trend analysis of deposition of SO_4-S and N_{tot}.

4. Conclusions

The results of this study show that snow sampling provides useful information for evaluation of regional characteristics of deposition. The deposition of acidifying components in southern Finland is higher than in northern Finland. Although the amount of deposition determined by snow survey is somewhat lower compared to rainwater investigations (Järvinen and Vänni, 1990) snow sampling offers supplemental information for trend assessment of atmospheric pollutants. Deposition of SO_4 and NH_4-N showed descending trend while deposition of NO_3-N was rather stable during 1976-1993. Preconditions for applicability of snow method are a snow cover of sufficient duration, unsubstantial mid-winter melting and sampling before the spring thaw.

References

Babiaková, G. & Bodiš, D.: 1986, "Accumulation and evolution of sulphate and nitrate levels in snow", in Morris, E. M. (ed.) *"Modelling snowmelt-induced processes"*, International Association of Hydrological Sciences, Wallingford. Proceedings of a symposium held during the 2nd Scientific Assembly of the International Association of Hydrological Sciences at Budapest, Hungary, July 1986, 271.

Ettala, M., Kukkamäki, E. and Tamminen, A.: 1986, "The use of vertical snow sampling as an indicator of some emissions from point sources", *Aqua Fennica*, **16**, 91-108.

Joffre, S. M., Laurila, T., Hakola, H., Lindfors, V., Konttinen, S. and Taalas, P.: 1990, "On the effects of meteorological factors on air pollution concentrations and deposition in Finland", in Kauppi, P., Anttila, P. and Kenttämies, K. (eds.) *Acidification in Finland*, Springer-Verlag, Berlin, p. 43.

Johannessen, M. and Henriksen, A.: 1978, "Chemistry of snow meltwaters: Changes in concentration during melting", *Water Resources Research*, **14**, 615-619.

Järvinen, O. and Vänni, T.,1990.: "Bulk deposition chemistry in Finland", in Kauppi, P., Anttila, P. and Kenttämies, K. (eds.) *Acidification in Finland,* Springer-Verlag, Berlin, p. 151.

Kauppi, P., Anttila, P. and Kenttämies, K. (eds.) Acidification in Finland, Springer-Verlag, Berlin, p. XIII.

Kuusisto, E.: 1986, "The mass balance of snow cover in the accumulation and ablation periods", in Kane, D L. (ed.) *Symposium: Cold regions hydrology*, American Water Resources Association, Maryland, p. 397.

Mäkelä, A., Antikainen, S., Mäkinen I., Kivinen J. and Leppänen, T.: 1992, Vesitutkimusten näytteenottomenetelmät (Sampling methods of water research), *Publications of Water and Environment Administration - series B*, Helsinki, **10**, 86 pp. In Finnish.

National Board of Waters: 1981, Vesihallinnon analyysimenetelmät (Methods of chemical analyses of Finnish water administration). Report 213. In Finnish.

Reuna, M., Perälä, J. and Aitamurto S.: 1993, "Lumen aluevesiarvoja Suomessa vuosina 1946-1993" (Areal snow water equivalent values in Finland in the years 1946-1993), *Publications of Water and Environment Administration*, Helsinki, **165**, 284 pp. In Finnish.

Ross, H. B. and Granat, L.: 1986, "Deposition of atmospheric trace metals in northern Sweden as measured in the snowpack", *Tellus*, **38B**, 27-43.

SAS Institute Inc.: 1985, SAS/STAT^R User's Guide: Statistics, Version 6, SAS Institute Inc., Cary NC, p. 519.

Schöndorf, Th. and Herrman, R.: 1987, "Transport and chemodynamics of organic micropollutants and ions during snowmelt", *Nordic Hydrology*, **18**, 259-278.

Semkin, R. G. and Jeffries, D. S.: 1986, "Storage and release of major ionic contaminants from the snowpack in the Turkey Lakes Watershed", *Water, Air and Soil Pollution*, **31**, 215-221.

Skartveit, A. and Gjessing, Y. T.: 1979, "Chemical budgets and chemical quality of snow and runoff during spring snowmelt", *Nordic Hydrology*, **10**, 141-154.

Sokal, R. R. and Michener, C. D.: 1958, "A statistical method for evaluating systematic relationships", *University of Kansas Science Bulletin*, **38**, 1409-1438.

Soveri, J.: 1985, "Influence of meltwater on the amount and composition of groundwater in quaternary deposits in Finland", *Publications of the Water Research institute*, Helsinki, **63**.

Statistics Finland: 1995, Luonnonvarat ja ympäristö (Natural Resources and Environment). Statistics Finland, Helsinki, 29 pp. In Finnish.

Wright, R. F. and Dowland, H.: 1977, "Regional surveys of the chemistry of the snowpack in Norway late winter 1973, 1974, 1975 and 1976", *SNSF-project, Research report 12/77*, Oslo. 29 pp.

TRENDS IN SULPHUR DIOXIDE CONCENTRATIONS AND SULPHUR DEPOSITION IN THE URBAN ATMOSPHERE OF RIJEKA (CROATIA), 1984-1993

A. ALEBIC-JURETIC

Institute of Public Health, Kresimirova 52a, 51000 Rijeka, Croatia

Abstract. Trends in sulphur dioxide (SO_2) annual mean concentrations in the period 1984-1993 are given for two sites within the city of Rijeka. During this period a decline (in average 30%) is observed since mid-eighties up to now at both sites. Dry deposition of sulphur as sulphur dioxide (S-SO_2) follows the same trend. Deposition of total sulphur as sulphates (S-SO_4) and wet S-SO_4 exhibit similar pattern with a decline of 45%. Rain scavenging is found to be the main path of sulphate removal from the atmosphere. The decline of sulphur compounds in the urban atmosphere of Rijeka can be attributed to the use of fuel with lower sulphur content.

Key words: sulphur dioxide, sulphates, wet and dry deposition, trends

1. Introduction

Sulphur dioxide (SO_2) is considered the main atmospheric pollutant of anthropogenic origin. The principal sources of SO_2 are fossil fuel combustion (particularly coal and oil) as well as the industrial processes. After great London smog episode in 1952, particular attention was focused on sulphur dioxide and smoke emissions that resulted in significant reductions of these pollutants in mid-seventies. Acid rain problem risen at the same time led in 1979 to international convention on long-range transboundary pollution under which a protocol on reduction of SO_2 emissions by 30 % was signed in 1985. The second international protocol, signed in 1994., obliged participating countries to further reduction of SO_2 emission by 20-80 %, depending on existing sources.

In the atmosphere, SO_2 is converted to sulphate containing aerosol particles. Sulphates can be formed via:

(a) sulphuric acid vapour formation from homogeneous gas - phase reactions including photochemical and thermal steps,
(b) oxidation of SO_2 in fog and cloud droplets after its absorption
(c) oxidation on the surface of existing aerosol particles.

Sulphates are removed from the atmosphere by wet and dry deposition (Meszaros, 1981).

This work deals with trends in SO_2 annual mean concentrations and dry deposition of sulphur as sulphur dioxide (S-SO_2), as well as trends in total deposition (estimated from dust deposition) and wet deposition of sulphur as sulphates (S-SO_4, estimated from precipitation analyses) during the period of 1984 - 1993.

2. Material and Methods

2.1. LOCATIONS OF SAMPLING SITES

All sampling sites were located within the city center (Figure 1).

Sampling of SO_2 was performed at two sites: Site 1 is influenced by the emissions from an old petroleum refinery plant, while Site 2 is situated in a busy street. The rainwater samples were collected at Site 3, located at the Institute building in the street dividing the harbour from residential areas. The samples of dustfall were collected at Site 4 situated in a residential area, at the edge of the city center. The results are compared to those obtained at the remote site on the island of Cres, about 60 km south of Rijeka.

2.2. CHEMICAL ANALYSES

Average daily concentrations of SO_2 were determined by standard British (acidimetric) method (Standard Methods, 1976).

Total sulphates (wet and dry) were determined analysing a soluble fraction of dustfall by the turbidimetric method (Standard Methods, 1985). The sample collector was the one recommended in the literature (Lodge, 1989). Wet deposited sulphates (S-SO_4) were determined by the same turbidimetric method from the rainwater samples collected on a daily (event) basis in open polyethylene buckets. Due to the sampling protocol, a certain contribution of dry deposition (up to 20%) can be assumed (Dillon et al., 1988, Schuurkes et al., 1988)

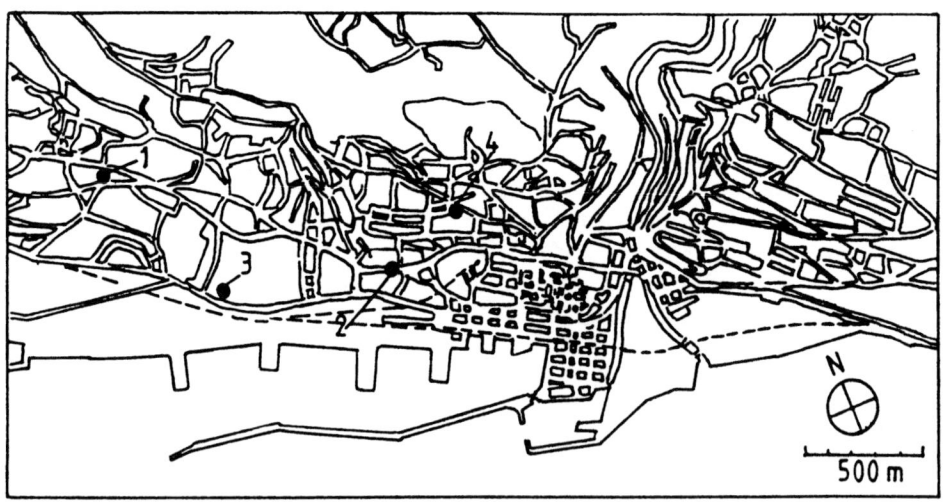

Fig.1: Location of sampling sites within the city center of Rijeka

3. Results and Discussion

Rijeka is an industrialized city of approx. 200.000 inhabitants. The first analyses of air pollutants (SO_2, smoke, dustfall) within the city started in 1973, before the extension of existing and new built industrial plants (coke plant, new petroleum refinery, oil fired power plant). Such a high industrialization resulted with a deterioration of air quality. According to an emission inventory (Matkovic and Alebic-Juretic, 1992), the principal sources of SO_2 within the city are industrial and power plants (Table I). The high emissions of SO_2 resulted in high ambient levels of SO_2 within the city.

TABLE I

Inventory of SO_2 emissions within the city area of Rijeka

No	Source	Emission (t/a)	%
1	Old Petroleum Refinery- Mlaka	3557	9.82
2	New Petroleum Refinery Plant - Urinj	16000	44.16
3	Power Plant Rijeka	8600	23.74
4	Coke Plant	4778	13.19
5	Paper Mill	1260	3.48
6	Municipal Heating Plants	690	1.90
7	Boilers	985	2.72
8	Domestic Heating	360	0.99
	Total	36230	100.00

In Autumn 1994 the coke plant was closed down, while the old petroleum refinery plant switched to a wider use of mixed gas for its energy supply.

3.1. SULPHUR DIOXIDE

Figure 2 shows annual means of SO_2 concentrations at two urban Sites 1 and 2.

At Site 1, under influence of petroleum refinery emissions, the annual averages are in the range of 66-106 $\mu g/m^3$ thus exceeding the WHO guideline for SO_2 of 50 $\mu g/m^3$ (Air Quality, 1987). Maximal annual averages (exceeding 100 $\mu g/m^3$) are registered in the period 1987-1989. After that a decline, approx. to 35% of the maximal values is observed at the beginning of nineties. The calculated 98 percentiles are within the range of 180-285 $\mu g/m^3$, while the maximal daily concentrations measured during this period are between 174 and 1084 $\mu g/m^3$.

At Site 2, a busy street in the very center of the town, the annual averages are in the range of 77-124 $\mu g/m^3$, and are also exceeding the WHO guideline for SO_2. The maximal annual means are obtained at mid-eighties (1987-1988) subsequent to a decline (approx. 25%) at the beginning of nineties. In the period 1984 -1993 annual averages above 100 $\mu g/m^3$ were registered six times at this Site, with maximum in 1987 (124 $\mu g/m^3$). The obtained 98 percentiles are in the range of 153-232 $\mu g/m^3$, while the highest daily concentrations measured are between 217 and 383 $\mu g/m^3$.

In the period of 1985 - 1986 samples of lichens were collected within the city area. The abundance of lichen species, as well as their cell membrane integrity indicated damages caused by high SO_2 concentrations (> 70 µg/m^3) during winter (Alebic-Juretic and Arko-Pijevac, 1989).

From 1990 on the annual means of SO_2 within the city of Rijeka follow the same trend as in other European cities, e.g. annual averages are below 100 µg/m^3 (though approaching this value at Site 2), with maximum daily concentrations between 200-500 µg/m^3 (mostly < 350 µg/m^3). In spite of that, the WHO guideline value for annual mean of SO_2 (50 µg/m^3) is reached and/or exceeded in the city area.

Average annual dry depositions of SO_2 for the period 1984-1993 are estimated (Katsoulis and Whelpdale, 1990) to be 0.97 g/m^{2}a S-SO_2 at Site 1 and 0.99 g/m^{2}a S-SO_2 at Site 2. Dry deposition of S-SO_2 within the city of Rijeka is estimated by its average values and is attributed to local sources. This value is five times higher compared to the result obtained for the remote site situated on the island of Cres in the period 1988 -1993 (0.20 g/m^{2}a).

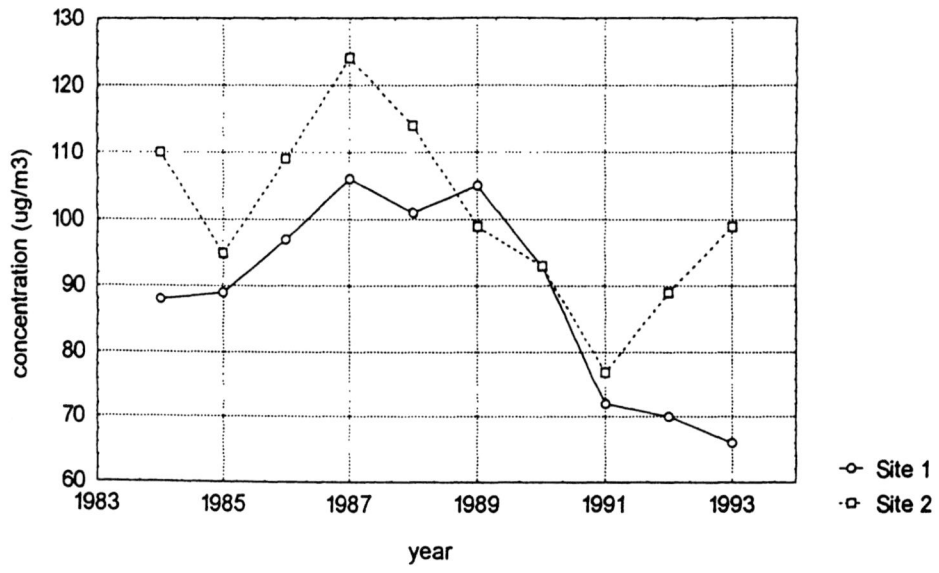

Fig. 2: Annual mean SO_2 concentrations at Sites 1 and 2

3.2. SULPHATES

Airborne sulphates are removed from the atmosphere by wet and/or dry deposition. Water soluble part of settable dust was analyzed to estimate deposition of total sulphur as sulphates (total S-SO_4). The obtained quantity of total S-SO_4 is the result of both, wet and dry deposition.

Analysis of precipitation chemistry within Kvarner Bay area indicated that high acidity and deposition of total S-SO_4 were registered within the city of Rijeka. This was explained by the local washout of the atmosphere (Alebic-Juretic, 1994). When comparing the wet deposited with the total deposited S-SO_4 a similar pattern is observed (Figure 3) showing a decrease since 1986. A good correlation between total and wet deposited S-SO_4 (r=0.924) indicate that the main path of the sulphate removal from the atmosphere is rain scavenging. For the period studied, wet deposition of S-SO$_4$ accounted in average for 75% of total deposited S-SO$_4$.

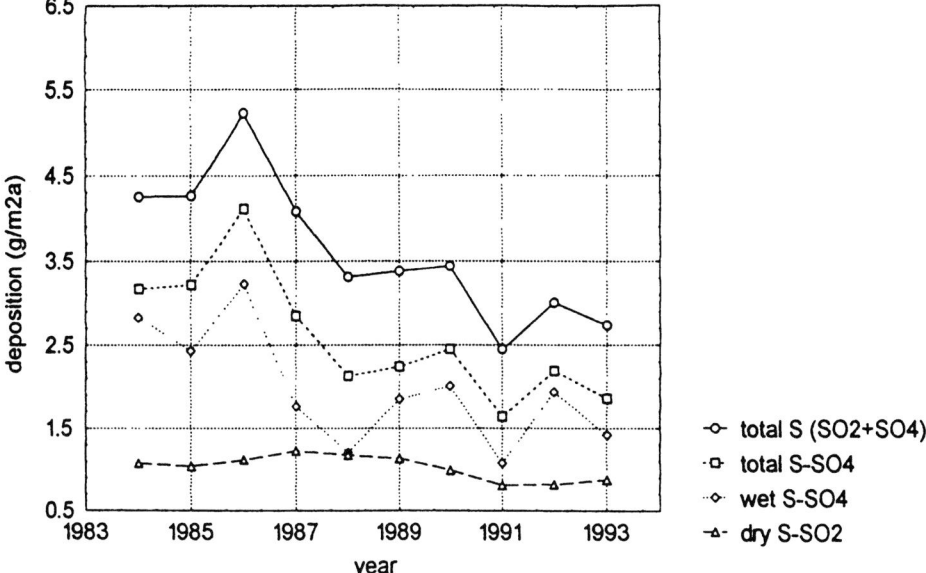

Fig.3.: Trends in sulphur deposition within the city of Rijeka

The quantity of total deposited S -SO$_4$ (wet and dry) varied from 1.64 -4.11 g/m^{2}a with an average of 2.59 g/m^{2}a. For the same period wet deposited S-SO$_4$ was in the range of 1.08 -3.23 g/m^{2}a, with an average of 1.98 g/m^{2}a. This value is close to the deposition rate of wet S-SO$_4$ equal to 2.17±1.99 g/m^{2}a, as estimated for Southeast Europe (Katsoulis and Whelpdale, 1990). Comparing the period from 1988-1993, the quantity of total deposited S-SO$_4$ at the remote site varied from 1.09 -1.70 g/m^{2}a, with an average of 1.44 g/m^{2}a. The deposited wet S-SO$_4$ for the same period varied from 0.45 - 1.16 g/m^{2}a with an average of 0.82 g/m^{2}a.

TABLE II

Total deposition of sulphur within the city of Rijeka (g/m^{2}a)

Location	Period	dry S-SO$_2$		wet S-SO$_4$		total S-SO$_4$		total S (SO$_2$+SO$_4$)
		x	range	x	range	x	range	
Rijeka								
	1984-1993	0.98	0.71-1.32	1.98	1.08-3.23	2.59	1.64-4.11	3.57
	1988-1993	0.97	0.71-1.23	1.59	1.08-2.01	2.09	1.64-2.45	3.06
Cres								
	1988-1993	0.20	0.15-0.27	0.82	0.45-1.16	1.44	1.09-1.70	1.64

Regarding the same period (1988-1993), the average deposition rates of total S-SO$_4$ and wet S-SO$_4$ in the city are considerably higher compared to those obtained at the remote site and are due to the local washout of the atmosphere (Table II).

It should be emphasized that in this study sea-salt sulphates were not taken into consideration, although precipitation chemistry data within Kvarner Bay area (Alebic-Juretic, 1994) show that sea-salt sulphates account to 30 % of wet S-SO_4 deposited within city of Rijeka, while on the island of Cres this percentage is even higher (55%).

4. Conclusion

Trends in sulphur dioxide concentrations in the period 1984-1993 show a decline at the two urban sites since mid-eighties. Trends in total (wet and dry) and wet S-SO_4 in last ten years show also a decline since mid-eighties. Rain scavenging is found to be the main pathway for sulphates removal from the atmosphere.

The decline in ambient SO_2 concentrations and deposition of sulphur (SO_2 and sulphates) observed in last years may be attributed to be the consequence of the use of fuel with lower sulphur content.

References

Air Quality Guidelines for Europe, WHO Regional Publications, European Series No 23, Copenhagen, 1987, p..357

Alebic-Juretic, A. :1994, "Precipitation Chemistry within Kvarner Bay Area, Northern Adriatic (Croatia), 1984 - 1991", Water, Air and Soil Pollut., **78**, 343-357

Alebic-Juretic, A. and Arko-Pijevac, M.:1989, "Air Pollution Damage to Cell Membranes in Lichens - Results of Simple Biological Test Applied in Rijeka, Yugoslavia", Water, Air and Soil .Pollut., **47**, 25 -33.

Dillon, P.J., Lusis, M., Reid, R. and Yap, D.: 1988, "Ten-Year Trends in Sulphate, Nitrate and Hydrogen Deposition in Central Ontario", Atmos. Environ., **22**, 901-905.

Katsoulis, B.D. and Whelpdale, D.M.: 1990, "Atmospheric Sulfur and Nitrogen Budgets for Southeast Europe", Atmos. Environ., **24**, 2959 - 2970.

Lodge, J.P.Jr., Ed. *Methods of Air Sampling and Analysis*, 3rd Ed., Intersociety Committee, Lewis Publ., Chelsea (USA),1989

Matkovic, N. and Alebic-Juretic, A.: *Study of Environmental Protection within Rijeka Municipality - Air*, Institute of Public Health, 1992 (in Croatian)

Meszaros, E.: *Atmospheric Chemistry*, Studies in Environmental Svience,Vol.11,1981, Elsevier, Amsterdam, p.72-76.

Schuurkes, J.A.A.R., Maenen, M.M.J. and Roelofs, J.G.M.: 1988, "Chemical Characteristcs of Precipitation in NH_3 Affected Areas", Atmos. Environ., **22** , 901-905

Standard Methods of Measuring Air Pollutants, WHO Offset Publication No 24, Geneva, 1976, p.43-47

Standard Methods for Examination of Water and Wastewater, 16th Ed., APHA, Washington DC, 1985; ibid, 15th Ed.,Baltimore,1980.

NINE YEARS OF MEASUREMENTS OF ATMOSPHERIC NITROGEN AND SULPHUR DEPOSITION TO DANISH FOREST

M.F. HOVMAND and H.V. ANDERSEN

National Environmental Research Institute
Post Box 358, DK 4000 Roskilde, Denmark

Abstract. Since 1985 measurements of gasses, aerosols, precipitation and throughfall have been carried out at three forest sites in Denmark with equal aged Norway Spruce plantations. The times series show a downward trend in the concentration of sulphur dioxide. Particulate sulphate, ammonia and particulate ammonium and the total nitrate seem to have a more constant concentration level. The wet deposition measurements show a decreasing trend in the content of acid (protons), sulphate, ammonium and nitrate, though for the nitrogen compounds it is only a slight fall. A decrease in concentrations of protons and sulphate is also seen in the throughfall measurements, in throughfall the nitrogen compounds hardly seem to decrease.

Key word index: Trend, gas, aerosol, bulk precipitation, throughfall, acid deposition.

1. Introduction

Since 1985 subsequent measurements of gasses, aerosols, bulk precipitation and throughfall have been carried out at three forest sites in Denmark with even aged Norway Spruce plantations. Estimation of sulphur- and nitrogen deposition to Danish forest and documentation of possible effects and changes in the forest ecosystem were the main aims of the study. Two of the sites are situated in the sandy soils of West Jutland : a north-western site situated 15 km from the North Sea and a south-western. One site is situated 250 km east of the Jutland sites, in the loamy soils of Sealand. Measurements make it possible to follow the trends in concentrations of atmospheric sulphur dioxide, particulate sulphate, nitrate and ammonium in air, in precipitation and in deposition to the forest canopy estimated by throughfall measurements.

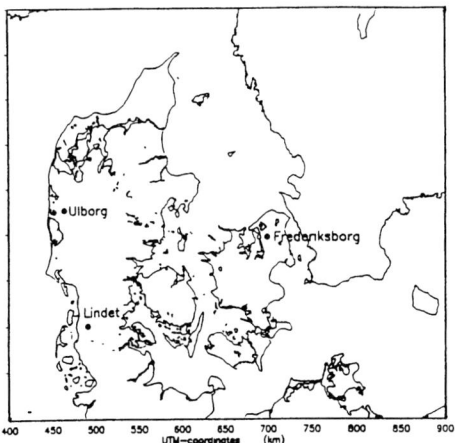

Fig. 1. Measuring sites in Denmark

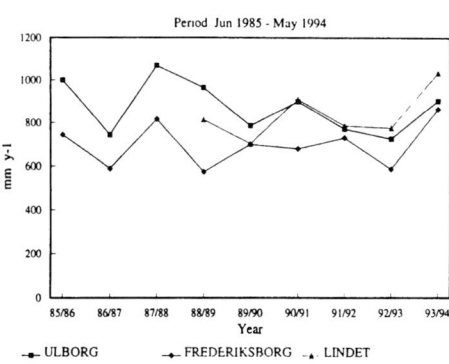

Fig. 2. Precipitation at the forest sites

Water, Air and Soil Pollution **85**: 2205–2210, 1995.
© 1995 *Kluwer Academic Publishers. Printed in the Netherlands.*

The location of the three experimental plots is shown in figure 1. The Ulborg site is located in an area of mixed heath land and conifer forest. The county is sparsely populated and no heavy industry or power plants are found within 50 km. The Frederiksborg site is located in mixed deciduous forest area with low agricultural activity and a low density of ammonia emissions. Copenhagen is 50 km south of the forest and major power plants are situated 50-100 km south and west of the site. The Lindet site is located in a forest surrounded by agricultural areas with a high animal production and high emission density of ammonia. The site is influenced by a close ammonia source southeast of the site. Major industries and power plants are situated more than 40 km away. The sites are part of tree trial experiment carried out at 13 locations in Denmark (Holmsgaard and Bang, 1977).

2. Experimental

2.1. GASES AND AEROSOLS SAMPLED BY FILTER PACK

Gases and aerosols have been sampled routinely as 24-h mean values by a filter pack, i.e. a holder mounted with a sandwich of four filters in series, the flow rate is 40 l min^{-1}. The first filter MF-Millipore, 1.2 µm, collects aerosols and three succeeding Whatman 41 (50 mm) filters are impregnated to collect gasses : NaF for nitric acid sampling, KOH for sulphurdioxide sampling and oxalic acid for ammonia sampling, respectively. Analysis of NH_4^+ is done by the indophenol method, while NO_3^- and SO_4^{2-} is determined by ion chromatography (Hovmand and Grundahl, 1991). The particle filter is also analyzed by PIXE (Proton Induced X-ray Emission) for multielement analysis. In Ulborg and Frederiksborg the sampling device for the filter packs is placed on a tower with the inlet at a height of 9 m above the ground in a clearing in the forests.

2.2. WET DEPOSITION AND THROUGHFALL MEASUREMENTS

Wet deposition and throughfall fluxes are sampled by a polyethylene funnel-flask system, the sampling area is 314 cm^2. A net mesh 700 um is placed in the funnel neck to prevent insects and leaf litter to enter the sampling bottle. At every site four samplers are mounted for open field sampling of bulk precipitation. Four samplers make it possible to identify and exclude contaminations by organic material. Representativity of the wet deposition samples is assessed by calculation of the relative standard variation for parallel collected samples, deviation for the whole period is 12%.

Wet deposition of sulphate and nitrogen compounds is estimated from bulk deposition measurements at the sites. Dry deposition to the funnel is also contributing to the bulk samples, by adsorption of gasses and impaction of aerosols probably playing a minor role in the Danish background areas with relatively low ambient gas and aerosol concentrations. Ammonia is in this relation an exception, and ammonium values in bulk samples are affected in areas with high ammonia concentrations as e.g. on the Lindet site.

Throughfall are sampled by minimum six randomly placed samplers at each site. The throughfall samplers are moved to a new position in the stand in order to minimize the variation in throughfall due to spacial deviation, e.g. distance to tree trunks, orientation of branches to prevailing winds and differences in activity between neighbouring trees. The sampling is described in detail by Hovmand and Kemp (1995).

3. Results and discussion

3.1. CONCENTRATION OF GASSES AND AEROSOLS

Sulphur-dioxide levels are shown in figure 3. A general decrease is seen at all station, though at Ulborg it is less pronounced. Ulborg is the most remote and north-westerly situated stations. The 50% reduction in Danish sulphur-emissions (Fenhann and Kilde, 1994) has only effected this station slightly. Concentration levels of particulate sulphate are similar at the different stations, and no significant trend is seen for this parameter during the 9 year period.

Gaseous ammonia and particulate ammonium are both sampled by the filter pack. This method only gives an estimate of the ammonia concentration (Andersen and Hovmand, 1994). However, as an yearly average the filter pack results on ammonia gives a reasonable estimate. The ammonia levels shown in figure 4 are much higher at Lindet compared to the other stations, the level is lowest at Frederiksborg, no significant trend is determined. Particulate ammonium are shown at figure 5, the concentration level is a little higher at Lindet. No significant trend is seen in the concentrations.

Total nitrate measured at the filter pack is the sum of nitric acid and particulate nitrate. Concentration levels are similar at Ulborg and Frederiksborg and no trend is seen in the concentration levels from these stations.

3.2. WET DEPOSITION

The yearly rain fall measured from 1985 to 1994 at the three forest sites Ulborg, Frederiksborg and Lindet, is shown in figure 2. The amount of precipitation sampled 2 meter or more above the ground is underestimated in average by 10-15 % due to wind generated turbulence around the funnel (Allerup & Madsen, 1979). The sampling stations in the forests were, however, well sheltered from the wind and an underestimation of precipitation of less than 12% is expected here. No obvious trend in the amount of precipitation is seen during the period. The western sites exhibit the highest precipitation rates in accordance with the general picture of the geographical distribution of precipitation in Denmark.

The wet deposition of acid (protons) and non marine sulphate are shown in figure 7 and 8, respectively. No significant difference are seen between the levels measured at Ulborg and Frederiksborg, but a downward trend are seen for both parameters at these stations.

Level of proton concentrations measured at Lindet is lower than at the other sites due to the elevated concentrations of ammonia in this area and . Sulphate deposition is a little higher at Lindet, but as the time series for this station only covers six year, possible trends are unresolvable. This applies for all parameters measured at Lindet.

Figure 9 and 10 show wet deposition of ammonium and nitrate. For Ulborg and Frederiksborg levels are close to each other for both parameters. A small downward trend could be suggested, but obviously to small to be significant, though the year to year variation is smaller than seen for the other parameters previously described. The ammonium level determined at Lindet is about 50 % higher compared to the other station. The elevated ammonium levels are probably mainly caused by dry deposition of ammonia to the funnel surface, and to a lesser degree caused by ambient ammonia wash-out by the rain drops. However, it is not possible from the present set up to reveal these phenomenons.

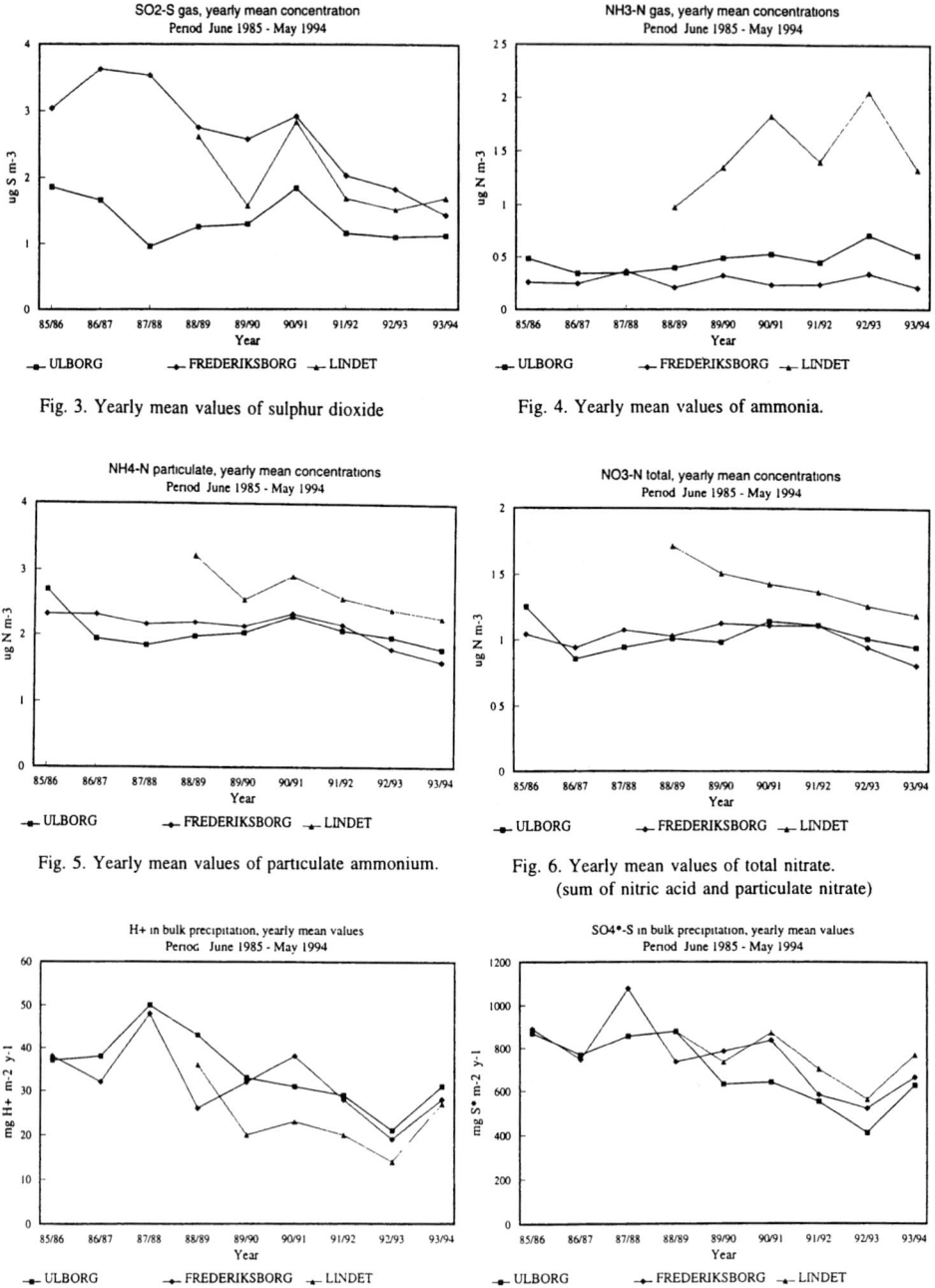

Fig. 3. Yearly mean values of sulphur dioxide

Fig. 4. Yearly mean values of ammonia.

Fig. 5. Yearly mean values of particulate ammonium.

Fig. 6. Yearly mean values of total nitrate. (sum of nitric acid and particulate nitrate)

Fig. 7. Acid deposition determined in bulk precipitation.

Fig. 8. Non marine sulphate in bulk precipitation.

3.2 THROUGHFALL

Yearly throughfall mean values of different compounds are shown in figure 11 to 14. Throughfall of protons, shown in figure 11, exhibit markedly lower flux rates than wet deposition of protons. This is obviously a result of a canopy interaction with a neutralizing effect on the rain water. Sulphate flux in throughfall, figure 12, is a factor of two higher than the corresponding wet deposition, implying a substantial dry deposition of especially sulphurdioxide. A decreasing trend on proton and sulphate throughfall is seen at all three stations, but with a substantial interannual variability. Throughfall of ammonium is at the same level at Ulborg and Frederiksborg, but Lindet has in average 50% higher values, figure 13. Throughfall of nitrate, figure 14, at Ulborg is about 20% lower compared to the other station, in accordance with the low NO_2 concentrations at this station.

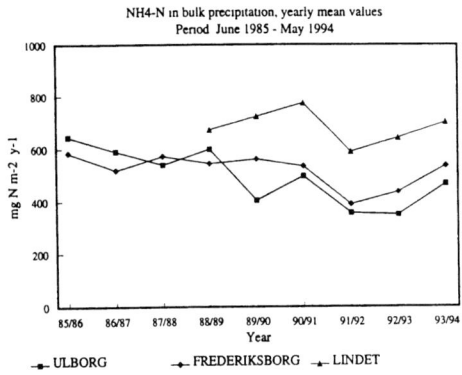

Fig. 9. Ammonium in bulk precipitation.

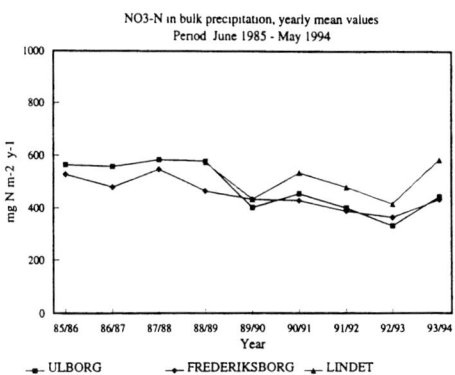

Fig. 10. Nitrate in bulk precipitation.

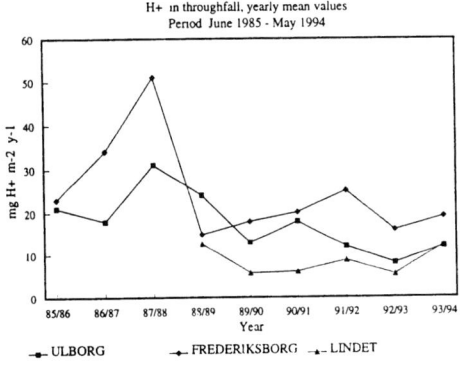

Fig. 11. Acid (proton) throughfall under Norway spruce.

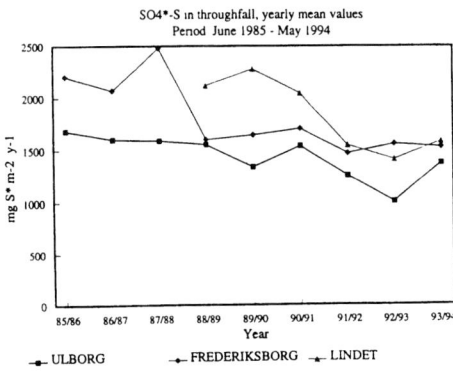

Fig. 12. Non marine sulphate throughfall under Norway spruce.

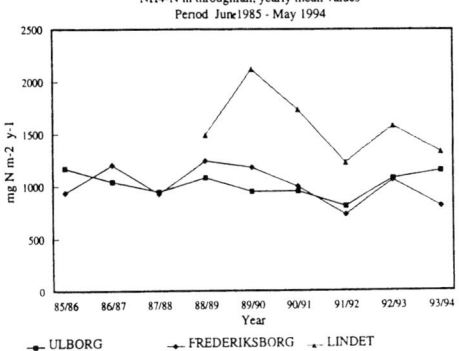

Fig. 13. Ammonium throughfall under Norway spruce.

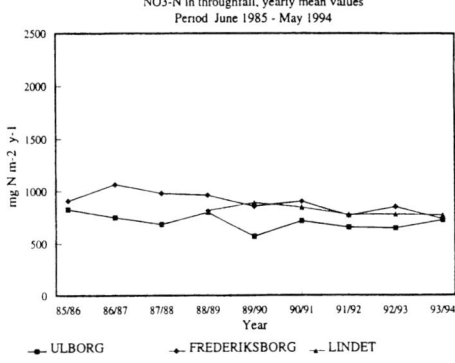

Fig. 14. Nitrate throughfall under Norway spruce.

The biological induced canopy interaction is believed to influence nitrogen throughfall fluxes. Especially ammonium fluxes was expected to increase from canopy leaching and organic litter fall. However, from these results the increase of ammonium flux seen, could be explained solely by the dry deposition of ammonia and ammonium. It is likely to assume an uptake by the trees of ammonia and nitric acid directly from the atmosphere, but the quantification of these fluxes is still poorly understood.

4. Summary and conclusions

In three Danish forests nine years of measurements of atmospheric gasses show a downward trend in the concentration of sulphur dioxide, while particulate sulphate, ammonia and particulate ammonium and the total nitrate seem to have a more constant concentration level, though with an interannual variability. The wet deposition measurements show a decreasing trend in the content of acid (protons), sulphate, ammonium and nitrate, for the nitrogen compounds it is only a slight fall. A decrease in concentrations of protons and sulphate is also seen in the throughfall measurements, though in throughfall the nitrogen compounds hardly seem to decrease.

Acknowledgments

This work was supported by The National Environmental Research Institute, The National Forest and Nature Agency, EEC (Forest and Forestry Division) and The Danish Environmental Research Programme.

References

Allerup P. and Madsen H.: 1979, "Accuracy of point precipitation measurements", Danish Meteorological Institute, Climatological Papers No. 5, ISBN 87 7478 168 5.

Andersen H.V. and Hovmand M.F.: 1994, "Measurements of the ammonia and ammonium by denuder and filter pack", Atmospheric Environment **28**, 3495-3512.

Andersen H.V. and Hovmand M.F.: 1995, "Ammonia and nitric acid dry deposition and throughfall flux", Submitted to Water. Air, & Soil Pollution.

Fenhann J. and Kilde, N.A..: 1994, Inventory of Emissions to the Air from Danish Sources 1972-1992. Risø National Laboratory, DK 4000, Roskilde.

Holmsgaard E. and Bang C.: 1977, "Et træartsforsøg med nåletræer, bøg og eg. De første 10 år", Forstlige Forsøgsvæsen Danmark, **35**, 159-196. Available at the Danish Forest and Landscape Research Institute, Skovbrynet 16, DK-2800 Lyngby, Denmark.

Hovmand M.F. and Grundahl L.: 1991, "Atmosfærisk nedfald af kvælstofforbindelser", NERI report no. 36, National Environmental Research Institute, Frederiksborgvej 399, DK-4000 Roskilde, Denmark.

Hovmand M.F., Grundahl L., Runge E.H., Kemp K. and Aistrup W.: 1993, "Atmosfærisk deposition af kvælstof og fosfor", NERI report no. 91, National Environmental Research Institute, Frederiksborgvej 399, DK-4000 Roskilde, Denmark.

Hovmand M.F. and Kemp K. (1995) Sulphur Deposition to Danish Spruce Forest. Submitted to Atmospheric Environment.

AMMONIA AND NITRIC ACID DRY DEPOSITION AND THROUGHFALL.

H.V. ANDERSEN and M.F. HOVMAND

National Environmental Research Institute
Post Box 358, DK 4000 Roskilde, Denmark

Abstract. Since 1991 measurements of fluxes of ammonia have been carried out periodically at a forest location in the western part of Denmark. The ammonia deposition velocities and fluxes are estimated from gradient measurements done by denuders and micrometeorology. The deposition velocities showed a large variation, ranging from deposition mainly governed by the atmospheric transport with fast adsorption at the surface to emission. Nitric acid deposition velocities and fluxes were measured in a period in May 1993 and the data indicate deposition mainly governed by the atmospheric transport and fast adsorption at the surface. The measured ammonia fluxes and an estimate for the particulate ammonium flux are compared to a nine year mean value of the net throughfall from Norway Spruce at the location.

Key words : ammonia, nitric acid, dry deposition, throughfall

1. Introduction

Atmospheric deposition of gaseous ammonia gives a substantial contribution to the total deposition of nitrogenous air pollutants, which is suspected to be a key factor in the disappearance of plant species in several ecosystems, forest decline and the transition of heathland into grassland (e.g. Nihlgård, 1985; Nilsson and Grennfelt, 1988; Grennfelt and Thørneløf, 1992). In order to obtain a better dry deposition estimate for ammonia at a monitoring site in a forest in western Jutland in Denmark (see Hovmand and Andersen, 1995), a programme with dry deposition measurements of ammonia was set up. Since 1991 measurements of the ammonia gradient have been carried out in five one-week periods at the forest site. The site is located in an area of mixed heathland and conifer forest. The forest is surrounded by agricultural areas to the west and north, the nearest having a distance of 1 km to the measuring site. To the south and east the forest proceeds more than 5 km. The forest is surrounded by areas with a relatively high emission rate of ammonia (Asman, 1990). The county is sparsely populated and no heavy industry or power plants are found within 50 km. Nitric acid gradients were measured simultaneously with the ammonia gradients in a period in May 1993. Throughfall has been measured at the site since 1985. Due to biological interaction the throughfall of nitrogen compounds are complicated to use as an estimate for the dry deposition.

2. Experimental

2.3. AMMONIA AND NITRIC ACID SAMPLED BY DENUDER

The vertical gradients of ammonia and nitric acid at the Ulborg site were measured by use of the denuder method. During the measuring periods, ammonia was sampled from a meteorological tower. In addition nitric acid was sampled in May 1993. Samples were taken during 3-h consecutive periods, except for the periods around sunrise and sunset. The denuders were placed at 18 and 36 m above ground (tree height are about 10-12 m). Samples at 24 m were also occasionally taken. Each sampling consisted of three parallel measurements at each level. The ammonia denuders have a flow of 3 l min^{-1} and the nitric acid denuders have a flow of 1 l min^{-1}. During exposure the denuders were heated a few degrees above ambient temperature to prevent condensation in the tubes. Sampling devices are described in detail by Andersen et al. (1993) and Bille-Hansen et al. (1994).

The ammonia sampling was done by oxalic acid coated denuders (Ferm, 1979). After exposure the tubes were extracted with 3 ml of deionized water and ammonia-N was determined by the indophenol blue method. The detection level for a 3-h exposure is about 0.05 µg NH_3-N m^{-3}, occasionally lower. The detection level is defined as three times the blank level.

The nitric acid sampling was also done by denuders. Soda glass tubes, 50-cm long and 4-mm i.d., were used. The nitric acid was sampled on the glass tube itself without use of a coating. Experiments have shown a sampling efficiency of 80 % using the soda glass tube itself as an absorbent. After exposure the tubes were extracted with 2 ml of deionized water and nitrate was determined by ion chromatography. The detection level was in the range 0.01-0.1 µg HNO_3-N m^{-3}, varying and occasionally relatively high due to contamination problems.

3. Results and discussion

3.4 DRY DEPOSITION VELOCITIES OF AMMONIA AND NITRIC ACID

In order to estimate the dry deposition velocity the gradient, i.e. the difference in concentration between the two heights 18 and 36 m, is used together with micrometeorological measurements. The procedure for estimation of the dry deposition velocity is described in Andersen et al. (1993). It is assumed that chemical reactions are not influencing the gradient and that the surface concentration is zero. Measurements with a coefficient of variation of more than 20% of the triple measurements in one or both heights are not used for estimation of dry deposition velocities and fluxes.

There is a large variation in the deposition velocities of ammonia (Table I), not only due to the turbulent conditions, but also due to other factors. 57 of 144 measured gradients show a negative profile, but only 10 of these profiles have a slope which differ significantly from zero at the 90% confidence level. 87 of the gradients show a positive profile and 49 of these are significantly different from zero at the 90% confidence level. The negative gradients might indicate emission, but can also be caused by chemical reaction in the air during sampling. However, the chemical reactions for ammonia is a rather unlikely interference during this relatively short time scale of sampling (Harrison et al., 1989). Figure 1 shows the flux of ammonia versus the ammonia concentrations. A downward flux is negative. In all measuring periods negative deposition velocities/positive fluxes were observed at higher ammonia concentration levels and conditions with a low-level turbulence. The observations during these conditions might relate to emission sources in the surrounding agricultural areas. In 1991 the negative deposition velocities/positive fluxes observed during conditions with well-mixed air were related to very low concentrations (0.02-0.09 µg NH_3-N m^{-3}). These observations might relate to some kind of compensation point of the plants, i.e. an air concentration below a certain point will cause an emission from the plants. In 1992 and 1993 negative deposition velocities/positive fluxes were also observed during conditions with well-mixed air and considerably higher ammonia levels. The measurements might indicate an emission from the forest or very close sources, though it has not be possible to identify any such close source. Duyzer et al. (1994) observed negative gradients above coniferous forest in The Netherlands. Their measurements of negative gradients were during day time and at relatively low ammonia concentrations. They found no clear explanation for these observations. From measurements of the ammonia concentration above a montane-subalpine forest in Colorado, Langford and Fehsenfeld (1992) observed, that the role of the forest as a source or sink depended on the atmospheric ammonia concentration. The canopy appeared to be a source of ammonia when exposed to low ammonia concentrations and a sink when exposed to air enriched by nearby agricultural sources. They found an average value of 0.8 ppb (1,1 µg NH_3 m^{-3}) at 20°C as the compensation point from the forest.

TABLE I
Ammonia and nitric acid concentration, dry deposition velocity and flux measured in different periods at the Ulborg site. * : negative flux is downwards

Period			conc.(18m) μg N m^{-3}	$v_{d(18 m)}$ m s^{-1}	flux* μg N m^{-2} h^{-1}
Mean of the NH_3 measuring periods		obs.	5	5	5
		mean	0.82	0.02	-21
24.05.-31.05.91 NH_3		obs.	42	25	25
		mean	0.38	0.03	-66
		min/max	< 0.02 / 1.82	-0.06 / 0.12	31 / -245
29.08.-05.09.91 NH_3		obs.	35	28	28
		mean	0.88	0.02	-67
		min/max	0.13 / 4.23	0.07 / 0.20	57 / -307
16.07.-23.07.92 NH_3		obs.	38	28	28
		mean	0.60	0.02	-8
		min/max	0.05 / 3.35	-0.09 / 0.13	671 / -868
04.02.-12.02.93 NH_3		obs.	33	18	18
		mean	0.36	0.05	-47
		min/max	0.04 / 1.76	-0.01 / 0.17	26 / -158
04.05.-14.05.93 NH_3		obs.	55	45	45
		mean	1.90	-0.01	82
		min/max	0.11 / 4.53	-0.07 / 0.03	624 / -217
04.05.-14.05.93 HNO_3		obs.	57	25	25
		mean	0.15	0.01	-15
		min/max	<0.02 / 0.42	-0.11 / 0.11	96 / -114

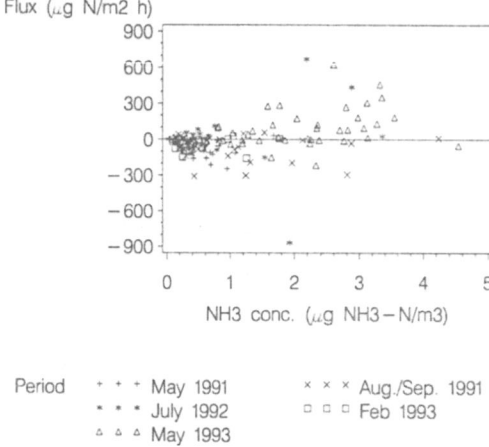

Figure 1. Flux of ammonia versus the ammonia concentration. The data are sorted out in relation to measuring period.

Figure 2a shows the flux of ammonia in the different 3-hour measuring periods in the measuring periods, though without the May 1993 data. The fluxes measured in May 1993 are shown in Figure 2b. The line is connecting the mean value of each period. In Figure 2a the overall flux is downwards and a slight maximum are seen in the late afternoon. In Figure 2b most of the mean values of the flux are positive, indicating a nett emission of ammonia in the period. Mean values of fluxes of each measuring period are given in Table I. The emission fluxes measured in May 1993 might be related to the high temperature, though no simple relation is found.

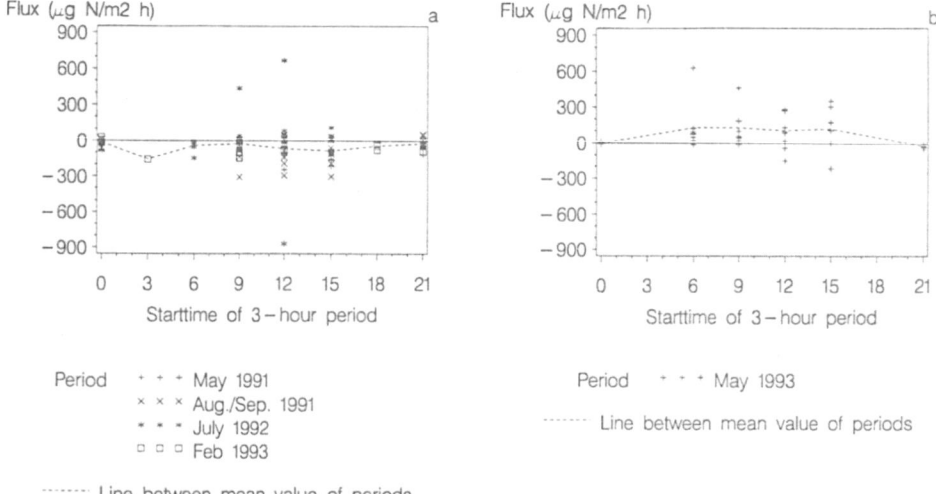

Figure 2. The ammonia flux at different times of the day, a) in the measuring periods in May and August/September 1991, July 1992 and February 1993. b) in the measuring period in May 1993.

The mean value of the concentration, deposition velocity and flux of nitric acid are given in Table I. The preliminary results of the dry deposition velocity of nitric acid show a relation to the friction velocity (Hovmand, 1994). The results indicate that the deposition of this gas is governed by the atmospheric transport and with a fast absorption at the surface. Of a total of 25 gradients observed, seven gradients are negative, though only three of these are significantly different from zero at the 90% confidence level). The negative deposition velocities relate to relative humidity/temperature, but no clear explanation is yet identified for these observations. The negative deposition velocities/positive fluxes are included in the values in Table I.

3.5 THROUGHFALL AND DRY DEPOSITION ESTIMATES

Net throughfall, defined as throughfall minus bulk deposition, can be taken as an estimate for dry deposition of elements like sulphur, which has little potential for canopy interactions. Bulk deposition and throughfall values are given in Hovmand and Andersen (1995). A mean dry deposition velocity of sulphur dioxide is estimated to 1.1 and 1.0 cm s^{-1}, respectively, on basis of the net throughfall data from Ulborg and Frederiksborg (forest location in North Sealand) (Hovmand and Kemp, 1995).

Net throughfall of nitrogen is obviously more difficult to interpret, due to the interactions of nitrogen in the canopy. The net throughfall of nitrate is differing substantially between the three Danish forest sites (Ulborg, Frederiksborg and Lindet). The average value from 1985 to 1994 is 229 mg N m^{-2} y^{-1} at Ulborg and 440 mg N m^{-2} y^{-1} at Fredriksborg. At Lindet the average value from 1989 to 1994 is 310 mg N m^{-2} y^{-1}. Our knowledge of ambient nitric acid concentrations are limited and it is not possible to give an estimate of the dry deposition of this compound with the view to document possible relation between net throughfall of nitrate and the dry deposition of nitric acid.

The nine years average values of net throughfall of ammonium at Ulborg and Frederiksborg are 528 and 492 mg N m^{-2} y^{-1}, respectively. At Lindet the mean value determined from 1989 to 1994 is 899 mg N m^{-2} y^{-1}. The average value of the ammonia flux to the canopy in Ulborg in the five periods with gradient measurements is estimated to 184 mg N m^{-2} y^{-1}, substantially less than stated before (Andersen et al., 1993). Assuming mean dry deposition velocities of 0.2-0.4 cm s^{-1} for particulate ammonium, the estimate for the dry deposition of particulate ammonium is 100-300 mg m^{-2} y^{-1}. These figures give a total amount of dry deposited ammonia and ammonium of about 300-500 mg N m^{-2} y^{-1} at Ulborg, which is comparable to the net throughfall of 528 mg N m^{-2} y^{-1}.

4. Summary and conclusions

Measurements of dry deposition velocities of ammonia and nitric acid have been carried out at a forest location in the western part of Jutland in Denmark. The ammonia deposition velocities showed a large variation, ranging from deposition mainly governed by the atmospheric transport with fast adsorption at the surface to emission. The nitric acid deposition velocities indicate deposition mainly governed by the atmospheric transport with fast adsorption at the surface. Bulk deposition and throughfall have been measured since 1985 at the location and the average value of the net throughfall of ammonium is comparable to an estimate of the dry deposited gaseous ammonia and particulate ammonium. The dry deposition estimate of ammonia is based on gradient measurements at the location.

Acknowledgments

This work was supported by The National Environmental Research Institute, The National Forest and Nature Agency, CEC (Forest and Forestry Division) and The Danish Environmental Research Programme. Thanks to P. Hummelshøj and N.O. Jensen, who delivered micrometeorology parameters for the estimation of dry deposition of ammonia and nitric acid.

References

Andersen H.V., Hovmand M.F., Hummelshøj P. and Jensen N.O.: 1993, *Atmospheric Environment* **27A**, 189-202.

Asman W.A.H.: 1990, "A detailed ammonia emission inventory for Denmark", DMU LUFT-A133. National Environmental Research Institute, Frederiksborgvej 399, DK-4000 Roskilde, Denmark.

Bille-Hansen J., Pedersen L.B., Hovmand, M.F., Andersen H.V., Jensen N.O., Hummelshøj P., Ro-Poulsen H. and Mikkelsen T.: 1994, "Background information for surveillance on the permanent observation plots in Denmark. Level II", EU-report available at the Danish Forest and Landscape Research Institute, Skovbrynet 16, DK-2800 Lyngby, Denmark.

Duyzer J.H., Verhagen H.L.M., Weststrate J.H., Bosveld F.C. and Vermetten A.W.M.: 1994, *Atmospheric Environment* **28**, 1241-1253.

Ferm M.: 1979, *Atmospheric Environment* **13**, 1385-1393.

Harrison R.M., Rapsomanikis S. and Turnbull A.: 1989, *Atmospheric Environment* **23**, 1795-1800.

Hovmand M.F.: 1994, "Atmospheric Nitrogen Deposition to Forest", In : Status Report 1992-1994. NECO, HEATH, International Projects. Center for Terrestrial Ecosystem Research. The Danish Environmental Research Programme. Available at the Center Secretariat at the Danish Forest and Landscape Research Institute, Skovbrynet 16, DK-2800 Lyngby, Denmark.

Hovmand M.F. and Andersen H.V.: 1995, Nine years of measurements of atmospheric nitrogen and sulphur deposition to Danish forest. This issue of *Water, Air, and Soil Pollution*.

Hovmand M.F. and Kemp K.: 1995, "Sulphur deposition to Danish Spruce forest", submitted to *Atmospheric Environment*.

Langford A.O. and Fehsenfeld F.C.: 1992, *Science* **255**, 581-583.

Nihlgård B.: 1985, *Ambio*, **14**, 2-8.

Nilsson J. and Grennfelt P., eds.: 1988, Critical loads for sulphur and nitrogen. Nordic Council of Ministers, Copenhagen. NORD 1988:97.

Grennfelt P. and Thørneløf E., eds.: 1992, Critical loads for nitrogen - a workshop report. Nordic Council of Ministers, Copenhagen. NORD 1992:41.

THE DRY DEPOSITION OF GASEOUS AND PARTICULATE NITROGEN COMPOUNDS TO A SPRUCE STAND

K. PETERS[1] and G. BRUCKNER-SCHATT[2]

[1] *Bayreuth Institute for Terrestrial Ecosystem Research, Dept. of Climatology, University of Bayreuth, D-95440 Bayreuth, Germany,* [2] *Plant Ecology, University of Bayreuth, D-95440 Bayreuth, Germany*

Abstract. The dry deposition of particle-bound NH_4^+ and NO_3^- and of gaseous NH_3 and HNO_3 to a 45 year old Norway Spruce forest was determined by the inferential method. Deposition velocities were calculated from meteorological data and plant morphology by a multi-layer model which combines the transfer of the trace substances within the canopy and their absorption probabilities due to a variety of deposition mechanisms. For gaseous deposition the same mathematical concept was adopted as for particles. That means that this approach is an alternative to the common concept of resistance analogy. The mean deposition velocities found were 0.47 cm s^{-1} for NH_4^+ and 1.85 cm s^{-1} for NO_3^-. While these values are in the range found in other studies, the mean deposition velocities for the gases were quite large. For NH_3 13 cm s^{-1} were found and for HNO_3 11 cm s^{-1}. It is suggested that the absorption of these gases on the plant surfaces is not as effective as assumed in the model.

Key words: dry deposition, coniferous forest, ammonia, ammonium, nitric acid, nitrate, mathematical model

1. Introduction

Many forest decline symptoms in Europe have been linked to elevated deposition of airborne nitrogen species (Schulze, 1989). The contribution of dry deposition to the total atmospheric input of nitrogen to forest ecosystems is still questionable due to the large difficulties in directly measuring these fluxes. In this study, we estimated the input of the trace gases ammonia and nitric acid and of particulate ammonium and nitrate to a Norway Spruce stand. For determining these fluxes F from measured airborne concentrations c we used the inferential method according to

$$F = v_d\, c.$$

Deposition velocities v_d were calculated by the model DEPOSITE which originally addressed the dry deposition of aerosol particles (Peters and Eiden, 1992). Modified versions of the model have now been developed which can be applied to gaseous compounds. Thus, with this approach is an alternative to the common concept of resistance analogy is presented.

After a brief introduction into the model, results from measurements made in 1992 are presented and critically discussed.

2. The Model

DEPOSITE is based on an analysis of the turbulent transfer of trace substances within the canopy and on the probability of absorption on plant surfaces. It is a multiple-layer approach. The conceptual framework was developed by Bache (1984).

For aerosol particles the surface absorption is due to sedimentation, inertial impaction and molecular diffusion through the boundary layers. For gases, besides the molecular diffusion, the uptake efficiency by the stomata and the cuticle is considered. It is generally assumed that for aerosol particles as well as for gaseous HNO_3 the plant surface represents a nearly perfect sink, *i.e.* the canopy resistances are negligible (Fowler et al., 1989, Erisman et al., 1994). Therefore, in the model HNO_3 deposition can be treated like the deposition of small particles, neglecting the uptake by the stomata. Nevertheless, for HNO_3 a mass accommodation coefficient α was included, which was set equal to 0.11 according to Ponche et al. (1993).

The canopy resistance for NH_3 was found to be a function of the relative humidity rh (van Hove et al., 1989). At larger humidities liquid water films will be established on the needle surfaces (Burkhardt and Eiden, 1994), and NH_3 deposition on the cuticle will be the most significant pathway. Therefore, a simplified approach was used in the model: For $rh < 70\%$ deposition of NH_3 on the cuticle was neglected, and the stomatal conductance was solely considered. On the other hand, for $rh \geq 70\%$ the stomatal uptake was neglected as compared to the deposition on the cuticle (like for particles and HNO_3). In this case $\alpha = 0.097$ was used (Ponche et al., 1993). Introducing accommodation coefficients α for the deposition of these gases on surfaces does not fully agree to the concept of zero canopy resistance, but numerous experiments revealed the existence of an uptake restriction at dry and wet surfaces.

The input parameters, which were continuously measured at a tower in the field, are the stomatal conductance and the vertical profiles of wind velocity, temperature and relative humidity. Plant morphology is characterised by the needle diameter, the microscale roughness of the needle surface and the vertical distribution of the leaf area.

3. Materials and methods

The investigation stand at the subalpine mountainous site Wülfersreuth in Germany is a spruce forest that is 45 years old and has a mean tree height of 20 m and a LAI of 11.4. The stomatal conductance for the whole canopy was estimated by Bauer (1993) using sapflow measurements according to the method of Cermak et al. (1973).

Airborne particle-bound concentrations c of the main inorganic ions as a function of the particle diameter d_p were measured by Berner cascade impactors mounted at the top of the canopy. The impaction plates were covered by polyethylene film, which after sampling was extracted with deionized water. The solution was analysed for NH_4^+ and NO_3^- by HPLC with conductivity detectors.

Two different types of impactors were used sequentially, one 5-stage impactor ranging from $d_p = 0.05$ μm to 10 μm and one 10-stage impactor ranging from 0.015 to 16 μm. For each sampling interval a mean value of v_d was calculated by the model based on 30-min averages of the meteorological input parameters. The measured concentrations of each stage i were directly multiplied by individual $v_d(d_{pi})$ where d_{pi} was set equal to the geometric mean of the d_p range of each stage:

$$F = \sum_{i=1}^{5(10)} c_i v_{di}.$$

HNO_3 and NH_3 were measured simultaneously by a wet annular denuder with 0.1 m formic acid as the absorption liquid (Keuken et al., 1990). Hourly measured concentrations were multiplied by appropriate values of v_d to get the deposition fluxes F.

From June to November 1992 a total of 23 aerosol samples were taken. The sampling times were on the average 1 day for the 5-stage impactor and 6 days for the 10-stage impactor, thus covering 23% of a whole year. The deposition of NH_3 was determined for a total of 21 days in July, August and October 1992, the HNO_3 deposition flux could be calculated for a total of 18 days within this period.

4. Results and discussion

In Figure 1 the input fluxes F of NH_4^+-N and NO_3^--N are shown for all sampling periods. The intensities of the fluxes cover a wide range. Maximum values occurred in periods where the ion concentrations in the larger particle size classes were relatively high. The results further reflect the wind velocity and the thermal stability of the air within the canopy. In scaling up the input fluxes F found for each sampling period to mean values for one year, the results listed in Table I were achieved. The total particulate N flux to the stand was calculated to be 5.4 kg ha^{-1} yr^{-1} which is in the range commonly found at forest sites (Hanson and Lindberg, 1991).

TABLE I

Yearly mean values of the dry deposition fluxes F of particle-bound ammonium and nitrate to the spruce forest.

	F (kg ha^{-1} yr^{-1})	v_d (cm s^{-1})
NH_4^+-N	2.6	0.47
NO_3^--N	2.8	1.85

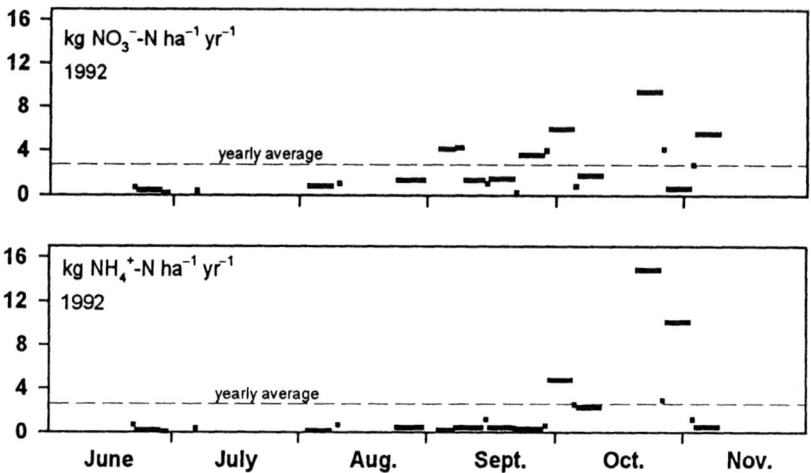

Fig. 1. Time course of the dry deposition fluxes F of particle-bound nitrate and ammonium to the spruce forest at Wülfersreuth, Germany.

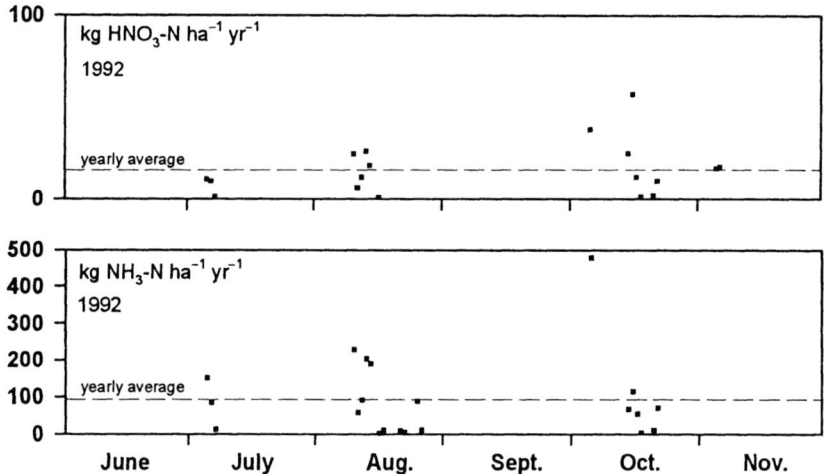

Fig. 2. Time course of the dry deposition fluxes F of nitric acid and ammonia to the spruce forest at Wülfersreuth, Germany.

In Figure 2 daily mean values of the gaseous deposition are shown. Yearly mean values are listed in Table II. In contrast to particle deposition the results are unreasonably high. For single days values of several hundreds of kg NH_3-N ha^{-1} yr^{-1} were determined (when scaled up to one year). The corresponding deposition velocities ranged between 30 and 40 cm s^{-1}. Though the maximum values were only found during days with extremely unstable thermal stratification, they are much greater than it could be expected from other studies. (See the review by Hanson and Lindberg, 1991).

TABLE II

Yearly mean values of the dry deposition fluxes F of gaseous ammonia and nitric acid to the spruce forest.

	F (kg ha^{-1} yr^{-1})	v_d (cm s^{-1})
NH$_3$-N	92	13
HNO$_3$-N	17	11

The fluxes of HNO$_3$ are much lower, which is mainly due to the lower airborne concentrations. The mean deposition velocity of HNO$_3$, 11 cm s^{-1}, is in the range that occasionally was found during other studies (Meixner et al., 1988, Fowler et al., 1989).

A comparison of the model used here with v_d values from other model approaches applied to our investigation stand would possibly help to clarify the discrepancies, but such an experiment is still missing.

5. Conclusion

While the deposition velocities of particle-bound nitrogen species that were determined in this study are in the range found in other investigations, the mean deposition velocities for the gases NH$_3$ and HNO$_3$ were quite large. As long as the high fluxes are not confirmed by further investigations, these values should be interpreted with caution. The reason for the discrepancy between our results and other reported values may be twofold:

- The accuracy of the measurements is not good enough. Especially during unstable conditions the modelling results very sensitively reflect small variations in the wind velocity and the temperature.
- The adsorption of the trace gases at the plant surfaces is not as effective as it is assumed by the model. From experiments with HNO$_3$ it was concluded that a surface saturation may be achieved limiting further deposition (Hanson and Garten, 1992). In experiments with NH$_3$ Bruckner et al. (1993) have shown that even at a relative humidity of 80% the uptake by the stomata was much larger than the adsorption at the cuticle.

Though there are accommodation coefficients α included in the model, it has to be concluded that the efficiency of the surface adsorption is still overestimated, especially for NH$_3$. This is equivalent to the fact that the canopy resistance R_c, as used in models based on the resistance analogy, should be significantly different from zero, even under wet conditions.

Acknowledgement

The project is funded by the BMBF (grant number PT BEO 51 - 0339476A).

References

Bache D.H.: 1984, "Prediction of the Bulk Deposition Velocity and Concentration Profiles within Plant Canopies", Atmos. Environ., **18**, 2517-2519.

Bauer G.: 1993, "Die Transpiration eines Fichtenbestandes (*Picea abies* (L.) Karst.) und ihre Abhängigkeit von bestimmenden Größen", Master thesis, Plant Ecology, University of Bayreuth.

Bruckner G, Schulze E.-D. and Gebauer G.: 1993, "^{15}N Labelled NH_3 Uptake Experiments and their Relation to Natural Conditions", in: Air Pollution Research Report 47 (ed. by J. Slanina, G. Angeletti and S. Beilke), E. Guyot, Brussels, pp. 261-270.

Burkhardt J. and Eiden R.: 1994, "Thin Water Films on Coniferous Needles", Atmos. Environ., **28**, 2001-2017.

Cermak J., Deml, M. and Penka M.: 1973, "A New Method of Sapflow Rate Determination in Trees", Biologica Plantarum (Prague), **23**, 171-178.

Erisman J.W., van Pul A. and Wyers P.: 1994, "Parametrization of Surface Resistances for the Quantification of Atmospheric Deposition of Acidifying Pollutants and Ozone", Atmos. Environ., **28**, 2595-2607.

Fowler D., Cape J.N. and Unsworth M.H.: 1989, "Deposition of Atmospheric Pollutants on Forests", Phil. Trans. R. Soc. Lond. B, **324**, 247-265.

Hanson P.J. and Garten C.T., Jr.: 1992, "Deposition of $H^{15}NO_3$ Vapour to White Oak, Red Maple and Loblolly Pine Foliage: Experimental Observations and a Generalized Model", New Phytol., **122**, 329-337.

Hanson P.J. and Lindberg S.E.: 1991, "Dry Deposition of Reactive Nitrogen Compounds: A Review of Leaf, Canopy and Non-Foliar Measurements", Atmos. Environ., **25A**, 1615-1634.

Keuken M.P., Otjes R.P. and Slanina J.: 1990, "Simultaneously Sampling of NH_3, HNO_3, HNO_2, HCl, SO_2 and H_2O_2 in Ambient Air by a Wet Annular Denuder", in: Physico-Chemical Behaviour of Atmospheric Pollutants (ed. by G. Restelli and G. Angeletti), Kluwer, Dordrecht, pp. 92-97.

Meixner F.X., Franken H.H., Duijzer J.H. and van Aalst R.M.: 1988, "Dry Deposition of Gaseous HNO_3 to a Pine Forest", in: Air Pollution Modeling and its Application VI (ed. by H. van Dop), Plenum, New York, pp. 23-35.

Peters K. and Eiden R.: 1992, "Modelling the Dry Deposition Velocity of Aerosol Particles to a Spruce Forest", Atmos. Environ., **26A**, 2555-2564

Ponche J.L., George C. and Mirabel P.: 1993, "Mass Transfer at the Air/Water Interface: Mass Accommodation Coefficients of SO_2, HNO_3, NO_2 and NH_3", J. Atmos. Chem., **16**, 1-21.

Schulze E.-D.: 1989, "Air Pollution and Forest Decline in a Spruce (*Picea abies*) Forest", Science, **244**, 776-783

van Hove L.W.A., Adema E.H., Vredenberg W.J. and Pieters G.A.: 1989, "A Study of the Adsorption of NH_3 and SO_2 on Leaf Surfaces", Atmos. Environ. **23**, 1479-1486.

SEASONAL AND DIURNAL VARIATION IN THE DEPOSITION VELOCITY OF OZONE OVER A SPRUCE FOREST IN DENMARK

K. PILEGAARD[1], N. O. JENSEN[2] and P. HUMMELSHØJ[2]

[1] *Environmental Science and Technology Department*, [2] *Meteorology and Wind Energy Department, Risø National Laboratory, P.O. Box 49, DK-4000 Roskilde, Denmark*

Abstract. The flux of O_3 was measured by the eddy-correlation method over Norway spruce in periods when the trees had a very low activity, periods with optimum growth, and periods with water stress. The aerodynamic resistance (r_a), viscous sub-layer resistance (r_b) and surface resistance (r_c) to O_3 were calculated from meteorological parameters and the deposition velocity. The canopy stomatal resistance to O_3 was calculated from measurements of the water vapour flux. The deposition velocities showed a diurnal pattern with night-time values of 3.5 mm s^{-1} and day-time values of 7 mm s^{-1}, when the trees had optimal growth conditions. The surface resistance was highly dominating in day-time and the influence of meteorology low. In night-time the surface resistance to O_3 was lower than the canopy stomatal resistance. A low surface resistance was also found in winter-time, when the activity of the trees was low. The surface resistance increased when the trees were subject to water stress. It is concluded that stomatal uptake is an important parameter for the deposition of O_3. However, other processes such as destruction of O_3 at surfaces, reaction with NO emitted from the soil, and reactions with radicals produced from VOC's emitted from the forest, should also be taken into consideration.

Keywords. Ozone, flux, deposition velocity, surface resistance, canopy stomatal resistance, water vapour flux, Norway spruce.

1. Introduction

Ozone (O_3) is mainly produced in the atmosphere by photochemical processes in polluted air. Some O_3 is also derived from the stratosphere. O_3 is toxic to plants and the wide-spread forest die-back has been linked to high concentrations of O_3 in areas normally regarded as only slightly polluted (Cape *et al.* 1994). O_3 is taken up by plants through the stomata, which is believed to be the major sink of boundary–layer O_3 (Galbally and Roy 1980). It is therefore important to study this process both as a means of removal of O_3 from the troposphere and as a potential source of damage to plants.

The aim of the present study is to quantify the flux of O_3 to coniferous forest, and to find the parameters governing the deposition velocity. The flux of O_3 to a Norway spruce plantation was therefore studied as a function of season and time of the day and related to simultaneously measured meteorological parameters.

2. Materials and Methods

Flux measurements were made in Ulborg (a forest in a remote rural area of western Jutland) during the periods 16-24 July, 1992, 4-12 February, 1993 and 7-17 June, 1994. The site is described in Andersen *et al.* (1993). The measurements were carried out from a 36 m tall mast placed in a Norway spruce (*Picea abies*) plantation with trees of a height of approximately 12 m and a good fetch in most directions except from a small sector towards SW.

The instrumentation for the eddy correlation measurements consisted of a 3D Solent Ultrasonic Anemometer (wind fluctuations), a GFAS OS-G-2 Ozone Sonde placed at 21 m height in the mast and a Scintrex LOZ-3 O_3 analyzer (chemiluminescence with Eosin Y), placed in a thermostated shed at the ground. The air was drawn down to this analyzer at 30 l min^{-1} through a Teflon tube (inner diameter 11.5 mm) with an inlet close to the sonic. The speed was sufficiently high to make the flow turbulent ($Re \geq 3000$), which is necessary to ensure that the attenuation in the tube was not a significant factor for introducing errors in the flux measurements (Lenschow and Raupach 1991). The total lag time was about 7 s. The LOZ-3 O_3 analyzer was calibrated immediately before and after the experiments and the concentrations adjusted accordingly. A large number of meteorological measurements were made simultaneously, the most important being wind speed (cup anemometers in different heights), wind direction (wind vane), temperature in different heights, water vapour flux (Ophir Infrared Hygrometer IR-2000) and surface temperature (Heimann KT 15.82 SMD).

Fluxes were calculated from the equation:

$$F_c = \overline{w' \cdot C'} \qquad (1)$$

where F_c is the flux of the compound in question, w the vertical wind velocity (after coordinate rotation of the sonic data) and C the concentration of O_3 at the measurement height (z). The prime indicates instantaneous deviation from the mean and the over-bar indicates the time average (0.5 h). The measured fluxes were corrected for errors due to changes in atmospheric density caused by heat and water vapour flux (Webb *et al.* 1980).

Deposition velocities (V_d) were calculated from the equation:

$$V_d(z) \equiv \frac{-F_c}{\overline{C}(z)} \qquad (2)$$

Resistances were calculated according to the model:

$$\frac{1}{V_d} = r_t = r_a + r_b + r_c \qquad (3)$$

where r_t is the total resistance, r_a the aerodynamic resistance, r_b the viscous sub-layer resistance and r_c the surface or canopy resistance. r_a and r_b can be calculated from meteorological observations; r_c is then calculated from equation 3 as the residual resistance. The aerodynamic resistance (r_a) was calculated by the expression analogous to equation 2:

$$r_a = \frac{\overline{u}}{u_*^2} \qquad (4)$$

where $\overline{u}$ is the mean wind speed (m s^{-1}) and u_* the friction velocity, calculated as $\overline{-u'w'}^{\frac{1}{2}}$.

The viscuos boundary layer resistance r_b was calculated from the parameterization of Jensen and Hummelshøj (1995):

$$r_b \propto \frac{\nu}{D} \left[\frac{c}{(LAI)^2} \left(\frac{lu_*}{\nu} \right) \right]^{\frac{1}{3}} \frac{1}{u_*} \qquad (5)$$

where ν is the kinematic viscosity of air ($15 \cdot 10^{-6}$ m^2 s^{-1}), D the diffusion coefficient of O_3 ($14 \cdot 10^{-6}$ m^2 s^{-1}), ℓ the leaf dimension, LAI the leaf area index, and c is a correction factor, determined empirically from the temperature profile and aerodynamic roughness. For the spruce forest the leaf dimension (ℓ) is 0.005 m, and the parameter c was found to be ≈ 100.

The surface resistance to water vapour was calculated according to equations 2 and 3 after subtraction of the saturated water vapour concentration at the surface temperature. The inferred canopy stomatal resistance to O_3 was calculated by multiplying the resistance to water vapour with the ratio of the molecular diffusivities of water vapour and O_3.

3. Results and Discussion

The June 1994 campaign gave typical results for O_3 and will therefore be presented as an example of the results. The concentrations of O_3 show a diurnal pattern with a minimum in the early morning and a maximum in the late afternoon (figure 1). The difference between the minimum and maximum is about 15 ppb, indicating that the air is only moderately polluted. The dominating wind-direction during the experiment was NW; this wind brings relatively clean air from the North Sea. The fluxes of ozone ranged from about -0.2 $\mu g \cdot m^{-2} \ s^{-1}$ to -0.6 $\mu g \ m^{-2} \ s^{-1}$ with a maximum around noon. The deposition velocities ranged from 3.5 mm s^{-1} to 7 mm s^{-1} with a sharp rise at dawn and a maximum already in the morning. This pattern indicates that stomatal processes could be important for the deposition of O_3. However, the rather high deposition velocities during the night indicate that other processes for removal of O_3 also play a significant role.

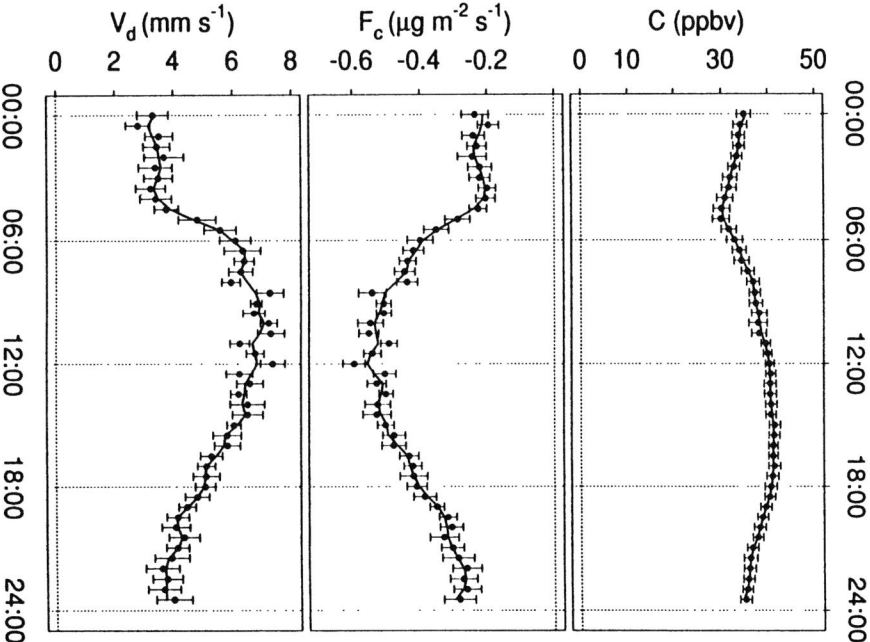

Figure 1: Average diurnal variation of ozone concentration (C), flux (F_c) and deposition velocity (V_d) over a Norway spruce stand during 7-17 June, 1994. The plots show $\bar{x} \pm s_{\bar{x}}$ and a smoothed line.

The results of the calculation of resistances are shown in table 1, where data from the other field campaigns in Ulborg are included for comparison. The resistances show a diurnal pattern with the lowest values during daytime. The surface resistance was highly dominating during daytime and the influence of meteorology low. Table 1 shows that low surface resistances to O_3 also can be found during winter time, when the activity of the trees is low. The surface resistance observed in July 1992 was somewhat higher than in in June 1994. This was due to the dry conditions during July 1992, where the water vapour deficit caused the stomata to close already at mid-morning resulting in a higher stomatal resistance (Jensen and Hummelshøj 1995).

	July 92		February 93		June 94	
	day	night	day	night	day	night
r_a	6	68	16	27	8	31
r_b	9	20	16	17	9	14
$r_c(O_3)$	368	800	208	179	151	288

Table 1: Average day (06:00 - 18:00) and night (18:00 - 06:00) values of resistances (s m^{-1}) over a Norway spruce plantation.

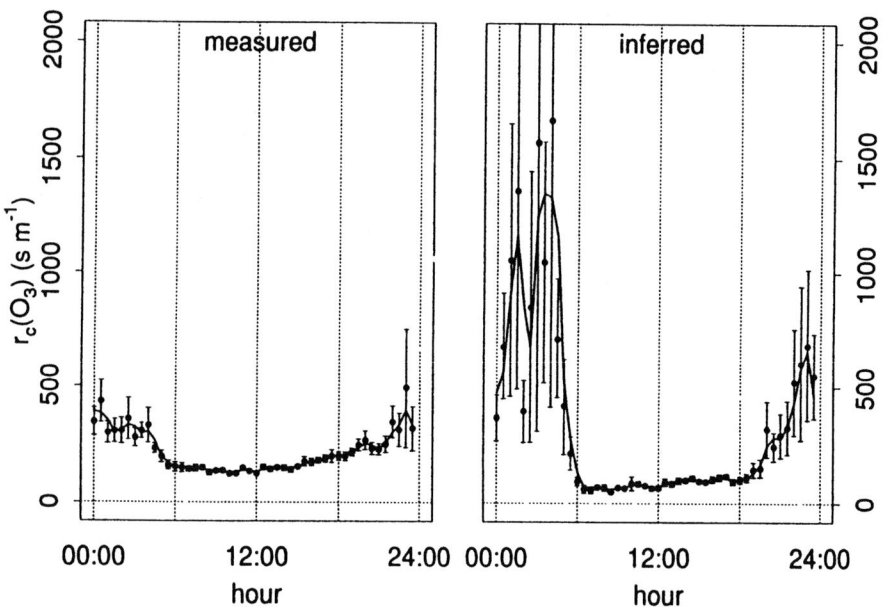

Figure 2: Average diurnal variation of measured surface resistance and inferred canopy stomatal resistance for O_3 to a spruce forest.

The measured surface resistance to O_3 was substantially lower than the inferred canopy stomatal resistance during the night. (figure 2). This is probably due to other sinks for O_3 than stomatal uptake. During the day the measured resistance was slightly higher than the inferred which may be a result of overestimation of the inferred conductance because above canopy water flux measurements in addition to transpiration via plant stomata may include evaporation from the soil (Gao and Wesely 1995).

As shown by Rondon *et al.* (1993) a clear diurnal variation in the surface resistance may not in itself be a proof of stomatal uptake since O_3 can be destroyed at photoactive sites on the surface of needles. The most important sinks for O_3 apart from stomatal uptake are reactions with nitric oxide (NO) (Kaplan *et al.* 1988), destruction at soil surface (Galbally and Roy 1980) and other surfaces and reactions with terpenes (Johansson and Janson 1993). Kaplan *et al.* (1988) found from profile measurements in a tropical forest that O_3 was removed rapidly from the atmosphere below the canopy. They attributed this to the reaction of O_3 with NO produced by bacterial activity in the soil. The emission of NO from this tropical forest soil was larger (10 ng N m^{-2} s^{-1}), that that of typical unfertilized coniferous forest soils in temperate areas (0.3 ng N m^{-2} s^{-1} (Johansson 1984)). The emission from fertilized forest soils could, however, be much larger, up to 60 ng N m^{-2} s^{-1}. Until now the NO emission from the forest floor in Ulborg has not been measured, but further research will show whether it can account for a significant amount of the total O_3 flux.

From a study of the O_3 fluxes to sparse grass and soil Massman (1993) concluded that the plant component received probably no more that 25% of the total O_3 depositional flux and that this percentage could be further decreased when available soil water decreased. In a study of air-surface exhange over a tallgrass prairie Gao *et al.* (1992) found that a considerable destruction of O_3 occurred at soil and leaf surfaces and that this non-stomatal removal of O_3 ranged from 25% of the total deposition for a healthy green canopy to close to 100% during winter time. Johansson and Janson (1993) found that under certain conditions reactions of O_3 with terpenes emitted from coniferous forests could be an important nighttime sink for O_3.

It is clear from the findings of this study and from other reported studies that more emphasis should be put into the study of the different sinks of O_3, especially in forested ecosystems.

4. Conclusion

The deposition velocity of O_3 showed a clear diurnal pattern which indicated stomatal uptake as an important removal process. However, the surface resistance to O_3 during the night was substantially lower than the resistance inferred from the water vapour flux. Furthermore low resistances were found during winter-time when the activity of the trees are supposed to be low. It is therefore concluded, that other sinks exist for O_3 and it is suggested that the most important of these are destruction at surfaces and chemical reactions with NO emitted from the forest floor and with terpenes emitted from the coniferous trees.

Acknowledgements

The field experiments were supported by the Danish Environmental Research Programme and the European Union (contract no. EV5V-CT92-0060). The National Forest and Nature Agency funded the permanent installations for meteorological observations in Ulborg for use in the "Ion balance project". The instruments for flux measurements were funded by the Danish Science Research Council and Risøs Integrated Environmental Project (RIMI).

References

Andersen, H. V., Hovmand, M. F., Hummelshøj, P. and Jensen, N. O.: 1993, Measurements of the NH_3 flux to a spruce stand in Denmark, *Atmospheric Environment* **27A**(2), 189–202.

Cape, J. N., Smith, R. I. and Fowler, D.: 1994, The influence of ozone chemistry and meteorology on plant exposure to photo-oxidants, *Proceedings of the Royal Society of Edinburgh* **102B**, 11–31.

Galbally, I. E. and Roy, C. R.: 1980, Destruction of O_3 at the earth's surface, *Quart. J. Roy. Met. Soc.* **106**, 599–620.

Gao, W. and Wesely, M. L.: 1995, Modeling gaseous dry deposition over regional scales with satellite observations - I. Model development, *Atmospheric Environment* **29**(6), 727–737.

Gao, W., Wesely, M. L., Cook, D. R. and Hart, R. L.: 1992, Air-surface exchange of H_2O, CO_2, and O_3 at a tallgrass prairie in relation to remotely sensed vegetation indices, *Journal of Geophysical Research* **97**(D17), 18,663–18,671.

Jensen, N. O. and Hummelshøj, P.: 1995, Derivation of canopy resistance for water vapour fluxes over a spruce forest, using a new technique for the viscous sublayer resistance, *Agricultural and Forest Meteorology* **73**, 339–352.

Johansson, C.: 1984, Field measurements of emission of nitric oxide from fertilized and unfertilized forest soils in Sweden, *Journal of Atmospheric Chemistry* **1**, 429–442.

Johansson, C. and Janson, R. W.: 1993, Diurnal cycle of O_3 and monoterpenes in a coniferous forest: Importance of atmospheric stability, surface exchange, and chemistry, *Journal of Geophysical Research* **98**(D3), 5121–5133.

Kaplan, W. A., Wofsy, S. C., Keller, M. and Da Costa, J. M.: 1988, Emission of NO and deposition of O_3 in a tropical forest system, *Journal of Geophysical Research* **93**(D2), 1389–1395.

Lenschow, D. H. and Raupach, M. R.: 1991, The attenuation of fluctuations in scalar concentrations through sampling tubes, *Journal of Geophysical Research* **96**(D8), 15,259–15,268.

Massman, W. J.: 1993, Partitioning ozone fluxes to sparse grass and soil and the inferred resistances to dry deposition, *Atmospheric Environment* **27A**(2), 167–174.

Rondon, A., Johannsson, C. and Granat, L.: 1993, Comparison of stomatal and cuticular uptake of ozone by Scots pine and Norway spruce, *in* P. Borrell (ed.), *The Proceedings of the EUROTRAC Symposium '92*, EUROTRAC, SPB Academic Publishers bv, The Hague, The Netherlands, pp. 723–726.

Webb, E. K., Pearman, G. I. and Leuning, R.: 1980, Correction of flux measurements for density effects due to heat and water vapour transfer, *Quarterly Journal of the Royal Meteorological Society* **106**, 85–100.

METHOD TO ESTIMATE ATMOSPHERIC DEPOSITION OF BASE CATIONS IN CONIFEROUS THROUGHFALL

M. FERM and H. HULTBERG

Swedish Environmental Research Institute
P.O. Box 47086 S-402 58 Gothenburg, Sweden

Abstract. A convenient non-electric method for estimating the dry deposition of base cations on a coniferous forest is presented. The dry deposition is estimated by multiplying the ratio of the base cation deposition to the sodium deposition on a surrogate surface with the dry deposition of sodium on the forest stand (throughfall technique). The surrogate surface is designed to resemble the needles in a coniferous forest with respect to particle deposition. Atmospheric non-marine dry deposition measured using the surrogate surface was compared to model calculated depositions. There was a good agreement for calcium but not for potassium.

Key words: base cations, surrogate surface, ion leakage, internal circulation, dry deposition.

1. Introduction

The base cations are defined as the most prevalent, exchangeable and weak acid cations in the soil. Base cations are essential for tree growth and development. Fluxes of base cations to and from the forest ecosystem are governed by atmospheric deposition and run-off, respectively. Within the ecosystem their transport is affected by weathering and ion exchange on soil particles, uptake by the roots and an ion leakage from the needles. The throughfall technique is widely used to estimate the atmospheric deposition of compounds that are neither retained by nor leached from the trees (such as sodium and sulphur). Throughfall measurements of base cations have been performed at the Swedish west-coast since 1980. The results in Figure 1 indicate that the wet deposition of Ca^{2+} has decreased in recent years, probably due to a decreased combustion of coal. The data show that the wet deposition of non-marine K^+ has decreased by 1% per year during the last 10 years, while the net throughfall has increased by 1%. The non-marine wet deposition of Ca^{2+} has decreased by 9% per year. The net change in Ca^{2+} deposition during 10 years is twice the absolute value of the standard deviation and the net change in K^+ is about equal to the absolute value of the S.D. The decrease in wet deposition of Ca^{2+} is supported by other observations (Granat, 1993; Lux, 1993; Matzner and Meiwes, 1993).

The net throughfall, NTF (throughfall- minus wet deposition) represent the dry deposition plus ion leakage, as the stem flow is negligible in comparison to the throughfall in a Norway spruce forest. Since the ion leakage from the trees is substantial especially for K^+, no conclusions can be made concerning the dry deposition trends of the base cations. For this reason and because it is important to compare the total atmospheric input with the run-off, a method to estimate the dry deposition of base cations is needed. The relationship between throughfall (TF), stem flow (SF), wet deposition (WD), dry deposition (DD), internal circulation (or ion leakage, IC) and uptake from the precipitation (UP) is given in equation 1.

$$TF + SF = WD + DD + IC - UP \qquad (1)$$

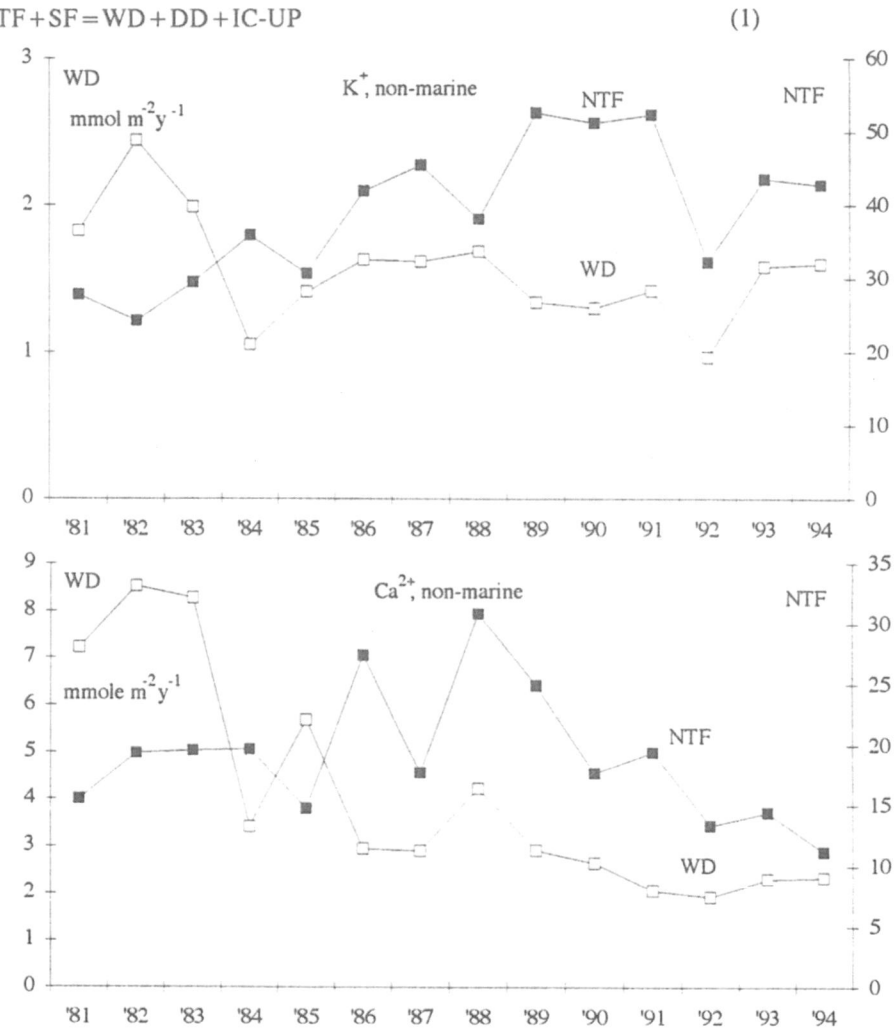

Fig. 1a and b Annual wet deposition and net throughfall of non-marine K^+ and Ca^{2+}. The throughfall data are taken from a Norway spruce forest in a reference catchment in the Gårdsjön research area and the precipitation data are taken from the Swedish deposition monitoring network.

2. Methods to estimate atmospheric deposition

Wet deposition can be measured by collecting rain in a funnel connected to a bottle. Some dry deposited material will, however, also be included if the funnel is not shielded from the air during dry periods, but this is not a severe problem in Sweden (Granat, 1988). Other errors like splash off from the funnel, disturbances of the air streams around the funnel, etc. can also affect the results (Lövblad and Westling 1989).

Dry depositions of base cations have previously been estimated by short time leaching of collected leaves (Lindberg and Lovett, 1985) and by using inert surfaces (Nihlgård 1970; Lindberg and Lovett, 1985). Graustein and Armstrong (1983) analysed the strontium concentrations and the $^{87}Sr/^{86}Sr$ isotopic ratio in throughfall, precipitation and tree bole to estimate the Sr deposition. By assuming that calcium behaves like strontium, the calcium budget can be estimated (Wickman and Jacks, 1992). The atmospheric dry deposition of base cations can also be estimated by a mathematical model using throughfall and precipitation concentrations (Ulrich, 1983; Westling et al., 1995).

3. Description of the method

3.1. Measurement principle

Sodium was considered as an inert ion with respect to the canopy, i. e. neither uptake nor leakage occurs. The net throughfall (NTF) of Na^+ is thus equal to the dry deposition. The method for estimating dry deposition was developed and applied at the Gårdsjön research area (Hultberg and Ferm 1995). The NTF of Na^+ was first plotted together with the deposition of Na^+ to a surrogate surface that mimics the surface of coniferous needles (see 3.2 materials) divided by 0.1 m^2 (empirically determined proportionality factor) versus time, see Figure 2. The material used in the surrogate sampler differs from the natural needles by lacking stomata and therefore misrepresents the gaseous uptake. They can therefore only be used to estimate the particulate fraction of the total deposition.

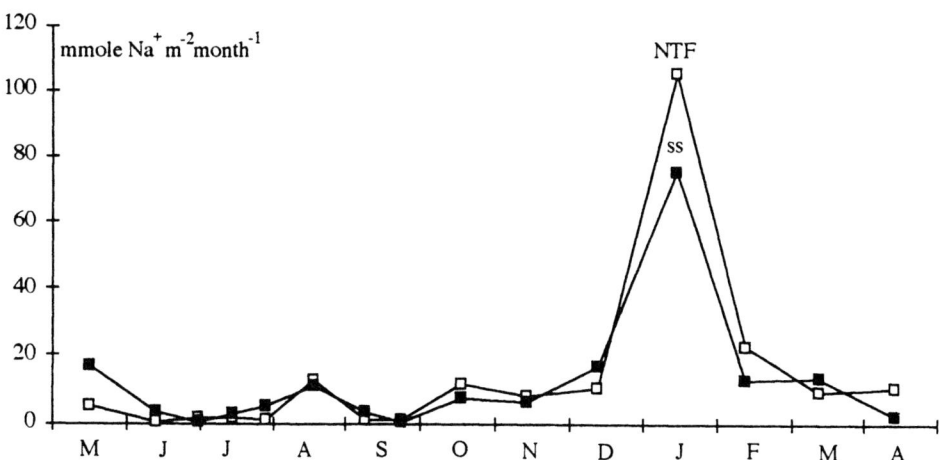

Fig 2. Net throughfall and dry deposition to the surrogate surface (divided by 0.1 m^2) of Na^+ during May 1992 to April 1993.

There is a good correlation between the NTF and the surrogate surface, with some exceptions. The particle depositions depend strongly on the wind speed. One explanation could be that the wind speed around the surrogate sampler was not proportional to the wind speed around the trees. Analogously the dry deposition of a specie (e.g. Na^+) during a certain period varies between individual trees.

The dry deposition of different species is therefore not estimated from the absolute deposition to the surrogate surface, but from the ratio between the specie in question and the sodium deposition to this surface. Effects from variation in meteorology are thus minimised. The following equation was used to estimate the dry deposition of a base cation X:

$$DD_X = \frac{[X]_{ss}}{[Na^+]_{ss}} \cdot TF\ Na^+ \tag{2}$$

ss indicates the leachate from the surrogate surface.

The internal circulation can then be calculated from equation 1, assuming that there is no uptake. The non-marine fractions are calculated using the base cation to Na^+ ratio in sea water.

3.2. Materials

The surrogate surface consists of a ca 70 m long Teflon string ($\varnothing=0.25$ mm) wounded 120 revolutions around two Teflon rings ($\varnothing=100$ mm). The device, which normally is used for collecting fog, is protected from rain by mounting it under a 1.25 x 1.55 m roof (1.8 m above ground). A 120 mm diameter funnel and a flask are mounted under the sampler to collect fog dripping down from the strings. At the end of the sampling period the strings are leached by rinsing them with ca 100 ml of de-ionised water.

3.3. Testing the sampler

It is very difficult to estimate the accuracy of the technique since at this site there is no compound that 1) is present only in coarse particles, 2) that doesn't interact with the canopy 3) that has a different origin than sea spray. Mg^{2+} and Cl^- fulfil the two first requirements, but they originate mainly from sea spray. NO_3^- is present in coarse particles as well as in the gas phase and fine particles. The canopy interaction is not so severe, but some losses can occur. Since HNO_3 has an extreme adsorption efficiency on most materials, it was felt worthwhile to compare the calculated DD with the NTF for NO_3^-.

Sampling was performed during one year from May 1992 to April 1993 at the Gårdsjön research area, ca 15 km from the open Swedish west-coast (58° 04'N, 12° 01'E). Throughfall, precipitation and deposition to the surrogate surface were measured on a monthly or bimonthly basis. The results are summarised in Table 1. As can be seen the data agree well for the whole period (one year), but the standard deviation for an individual measurement (about one month) can be large. This is not surprising since it is difficult to allocate a throughfall measurement to a certain small time period.

Table 1. Average NTF and DD (from equation 2) together with the standard deviation of the differences between the NTF and DD for three species (mmole m^{-2} month^{-1}).

	Cl$^-$	Mg^{2+}	NO$_3^-$
NTF	17.5	1.9	1.1
DD	17.3	1.7	1.3
S.D.	2.5	0.5	0.9

To further test the technique, the measured dry deposition of non-marine potassium using the surrogate surface was compared to a model calculated deposition (Fig. 3a and b). The model uses throughfall and precipitation data (Westling et al., 1995). The calculation method includes all major ions, and is based on ion balance of the calculated results and assumptions concerning the different behaviour of different ions

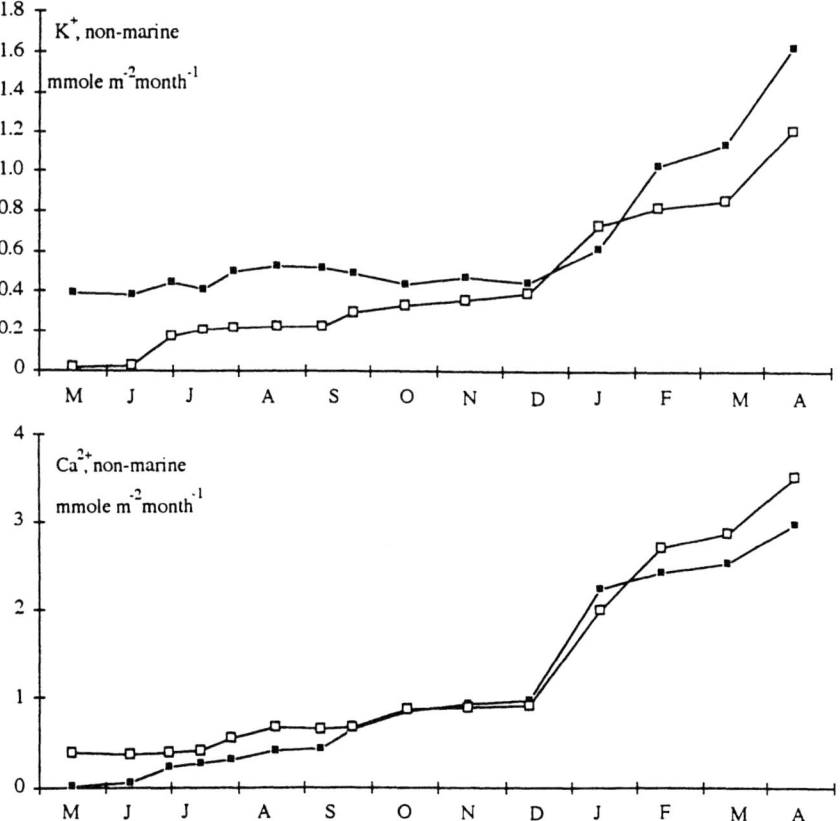

Fig. 3a and b. Accumulated dry deposition of non-marine K$^+$ and Ca^{2+}. The filled squares represent measurements and the open model calculated.

in the tree canopy. The agreement between measured and calculated DD is not as good for K^+ as it is for Ca^{2+}. Additional results will be given in subsequent article (Hultberg and Ferm, 1995). Reasonable results are also obtained for the dry deposition of SO_4^{2-} and NH_4^+.

4. Conclusion

The method is convenient to use and does not require electricity, which is an advantage at many forest stations. When calculating the dry deposition, the use of the deposition ratio of a reactive compound to an inert compound (Na^+) on the surrogate surface offers an improvement compared to the use of the deposition on the surrogate surface alone.

Acknowledgements

The precipitation data in Figure 1 was kindly supplied by Dr. Lennart Granat. Financial support has been obtained from the Swedish Environmental Protection Agency and from the EU project ENCORE.

References

Granat L.: 1988, Influence from sampling site surrounding on amount and composition of collected precipitation samples. Proc. of Expert meeting on sampling, chemical analysis and quality assurance. Arona (Italy) 11 to 14 October 1988. EMEP-CCC-Report 4/88, 237-241

Granat L.: 1993, Long term changes in precipitation chemistry. Workshop on integrated monitoring of air pollution effects on ecosystems. ed Jiri Cerny. Prague, September 17-20 1993.

Graustein W. C. and Armstrong R. L.: 1983, *Science* **219**, 289-292

Hedin L. O., Granat L., Likens G. E., Buishand T. A., Galloway J. N., Butler T. J. and Rodhe H.: 1994, *Nature* **367**, 351-354

Hultberg H. and Ferm M.: 1995, Water Air and Soil Pollut. this volume.

Lindberg S. E. and Lovett G. M.: 1985, *Environ. Sci. and Techn.* **19**, 238-244

Lux H.: 1993, Trends in air pollution, atmospheric deposition and effects on spruce trees in the eastern Erzgebirge. Workshop on integrated monitoring of air pollution effects on ecosystems. ed Jiri Cerny. Prague, September 17-20 1993.

Lövblad G. and Westling O.: 1989, Methods for determination of atmospheric deposition. In Methods for integrated monitoring in the Nordic countries. Nordic Council of Ministers, 1989:11 Copenhagen pp19-62.

Matzner E. and Meiwes K. J.: *J. Environ. Quality* **23**, 162-166

Nihlgård B.: 1970, *OIKOS* **21**, 208-217

Ulrich B.: 1983 In Ulrich B. and Pankrath J. (eds) Effects of accumulation of air pollutants in forest ecosystems, pp 33-45

Westling O., Hultberg H. and Malm G.: 1995, In L. O. Nilsson, R. S. Hüttl and U. T. Johansson (eds.) Nutrient uptake and cycling in forest ecosystems, pp 639-647

Wickman T. and Jacks G.: 1993, *Appl. Geochem.* **2**, 199-202

MEASUREMENTS OF ATMOSPHERIC DEPOSITION AND INTERNAL CIRCULATION OF BASE CATIONS TO A FORESTED CATCHMENT AREA

H. HULTBERG and M. FERM

Swedish Environmental Research Institute
P.O. Box 47086 S-402 58 Gothenburg, Sweden

Abstract. The dry deposition of base cations to a Norway spruce stand was estimated by multiplying the ratio of the ion deposition to the sodium deposition on a surrogate surface with the dry deposition of sodium on the forest stand. The method can in principle only be applied to species that are present only in particles, but the method gives reasonable results when tested on ions that are also dry deposited in other forms (SO_4^{2-}, NO_3^- and NH_4^+). The atmospheric input and especially the dry deposition of base cations is an important replacement for the loss of base cations from the soil by run-off. The calculated internal circulation of K^+ and Ca^{2+} showed maxima synchronously with rainfall maxima and constitute 71% and 53%, respectively, of the net throughfall deposition. The internal circulation of Ca^{2+} was almost equal to the SO_2 uptake.

Key words: base cations, surrogate surface, ion leakage, internal circulation, dry deposition.

1. Introduction

Decreased atmospheric deposition of base cations have been observed within the Swedish precipitation monitoring network as well as at Hubbard Brook in U.S.A. (Hedin *et al.* 1994). The decreasing input to the forest ecosystems may result in malnutrition among tree species and other vegetation as well as negative effects in aquatic ecosystems in acid sensitive areas.

Soil acidification, caused by atmospheric deposition, increases the loss of base cations from the topsoil by affecting the weathering and ion exchange processes in the soil. In Sweden the acidic deposition has caused a decreased content of base cations (Ca^{2+}, Mg^{2+} and K^+) in the soil (Warfinge *et al.*, 1993). The largest decreases are found in the originally least acid soils, indicating that the limit for further losses is being approached in the most acid soils (Falkengren-Grerup and Tyler, 1991). Calcium and magnesium deficiency have been observed at other countries (Raitio, 1993).

A net loss of base cations from the forest ecosystem, can lead to calcium and magnesium deficiency and to forest die-back (Tamm and Hallbäcken 1988; Warfinge *et al.*, 1993). The aim of this study is to quantify the atmospheric input of base cations and compare it to the loss by run-off in the Gårdsjön area (Hultberg and Grennfelt 1992), and to see if the calcium leakage from the Norway spruce is correlated to the SO_2 uptake.

2. Experimental

The experiments were carried out in the Lake Gårdsjön research area (Hultberg 1985), ca 15 km from the open Swedish west coast (58° 04'N, 12° 01'E). The deposition to a forested micro-catchment (G2, 0.5 ha) was investigated. The forest consisted mainly of Norway spruce and was part of the European Union project NITREX. 120 μmole NH_4NO_3 was experimentally added per square meter and year to the forest floor since April 1991.

Wet deposition (WD) was determined using bulk collectors for both precipitation amount and its chemistry. Throughfall (TF) i. e. the precipitation under the canopy was measured using 28 randomly distributed collectors. The collectors consisted of 200 mm diameter funnels (with a net in the bottom) threaded on a 5 l bottle. During the winter when the precipitation occasionally was deposited in the form of snow, buckets were used. Disposable plastic bags were used to collect the deposition in the bottles as well as in the buckets.

Cl^-, NO_3^- and SO_4^{2-} concentrations were determined using ion chromatography. pH and electrical conductivity were determined as soon after collection as possible. The base cations were determined using atomic absorption spectroscopy. Sampling was conducted on a monthly or bimonthly basis.

From May 1992 to August 1993 the deposition of base cations and other inorganic species were measured using a surrogate surface. The surface resembled the needles and consisted of a long Teflon string wounded around two Teflon rings. The method has been described earlier (Ferm and Hultberg, 1995). After a one month exposure period the strings were sprayed with de-ionised water and the leachate was analysed in the same way as WD and TF and is referred to as the surrogate surface (ss) in the text. The dry deposition is estimated by multiplying the ratio of the base cation deposition to the sodium deposition on the surrogate surface with the dry deposition of sodium on the forest stand (TF-WD). The internal circulation (ion leakage) can then be calculated from the throughfall and wet deposition of the base cation (Ferm and Hultberg, 1995). Run-off water was sampled using a flow proportional pump and analysed using the same techniques as described previously (see Table 1).

3. Results

The data from the surrogate surface and the wet- and throughfall deposition are summarised in Table 1. First the molar ratio between different ions and Na^+ is given for sea water, the same ratios for the material deposited on the surrogate surface and in the precipitation. The wet- and net throughfall depositions, the total atmospheric inputs (WD+DD) as well as outputs (run-off) are also included.

When the ratio between the deposition on the surrogate surface of an ion and Na^+ is higher than the same ratio in sea water, it indicates that there are other atmospheric sources than sea spray. This is the case for Ca^{2+} and K^+, but not for Mg^{2+}. For Na^+, Ca^{2+}, K^+, Mg^{2+} and Cl^- the reciprocal relations between Na^+ and the base cations are similar on the surrogate surface (ss) and in precipitation (WD), indicating that these ions are either present in the same particles or are present in different particles that behave in a similar manner with respect to dry and wet deposition. This fact is essential for justifying the principle behind this technique.

The difference between the net throughfall (NTF=TF-WD) and the estimated dry deposition (calc. DD) represents the ion leakage. It is assumed to be zero for Na^+. From Table 1 it can be seen that the ion leakage of Mg^{2+} and Cl^- are practically zero, while it is significant for Ca^{2+} and K^+.

Table 1. Ion ratios in the sea, on the material deposited on the surrogate surface and in precipitation (mole/mole). Wet-, net throughfall- and calculated dry depositions, atmospheric inputs (WD + calculated dry deposition) and outputs (run-off) are also given. Fluxes are given in mmole m^{-2} during one year starting May 1992.

	Na^+	Ca^{2+}	K^+	Mg^{2+}	Cl^-	SO_4^{2-}	NO_3^-	NH_4^+	H^+
X/Na^+, sea	(1)	0.022	0.021	0.122	1.17	0.061	0	0	0
X/Na^+, ss	(1)	0.061	0.067	0.124	1.00	0.109	0.303	0.064	0.013
X/Na^+, WD	(1)	0.063	0.050	0.117	1.04	0.579	0.834	0.632	0.769
WD	68	4	2.9	7.9	72	23	33	31	36
calc. DD	(201)	7	9.1	25	247	(17)	(20)	(7.5)	(1.9)
NTF	201	16	31	28	252	28	16	-1.7	8.4
input	269	11	12	33	319	51	>53	>38	44
output	213	5.2	4.8	20	248	58	3.9	0.9	54

In order to see if the surrogate surface gives reasonable results, the depositions of other ions were also measured, see Table 1. These compounds or their precursors lead to acidification and thereby an increased weathering rate and a loss of base cations. These compounds are present in both gas- and particle phases. The particles mainly consist of accumulation mode particles, which have smaller deposition velocities than the coarse particles containing the base cations. NO_3^- may, however, partly be present in coarse particles due to the reaction between gaseous HNO_3 and particulate NaCl (either in the atmosphere or on the Teflon strings). Since no measurements of gaseous or particulate concentrations in air were performed in this study, only comparisons with concentrations in precipitation, which is partly proportional to the total air concentrations, can be made.

The ratio between these ions and Na^+ is smaller on the surrogate surface (X/Na^+ ss) compared to the same ratio in precipitation (X/Na^+ WD) than it is for the base cations, because they are mainly present in small accumulation mode particles that have low deposition velocities. Another contributing factor is that the gaseous precursors are readily incorporated in precipitation, but not deposited on the Teflon string. NO_3^- has an intermediate behaviour between NH_4^+, SO_4^{2-} and the base cations. HNO_3 can be deposited on most materials, besides NO_3^- can be present in coarse sea spray particles. When HNO_3 reacts with NaCl, HCl is formed. If HCl evaporates the Cl^-/Na^+ ratio should be smaller on the surrogate surface than in the sea, as is the case in Table 1. An additional explanation may be that there are other Na^+ sources than sea spray. That is, however, unlikely because the Mg^{2+}/Na^+ ratio on the deposited particles is the same as in the sea. This is not only true for the average ratio but for all individual values (within ± 9 %).

The NTF for NH_4^+ is negative because it can be transformed to organic nitrogen compounds in the canopy and perhaps also be taken up by the needles (Ferm, 1993).

The surrogate surface technique cannot be applied to gaseous air pollutants because they behave differently on a Teflon surface than on needles. The calculated deposition will most likely be underestimated and the figures are therefore put within parenthesis. HNO_3 may be an exception because it efficiently adsorbed on all surfaces. The total inputs of base cations are estimated from the WD+calculated DD. For SO_4^{2-} the atmospheric input is calculated from the NTF instead of the calculated DD using the surrogate surface, since SO_2 is not deposited to Teflon.

In Table 2 the throughfall of base cations have been split up in marine and non-marine origins. As can be seen Mg^{2+} originates from the sea, the small value from other sources is most likely within the experimental error. 53% of the net throughfall (TF-WD) of Ca^{2+} comes from internal circulation and 71% of K^+. The non marine sources are important for the atmospheric input of both Ca^{2+} and K^+ at this site.

Table 2 Estimated origin of base cations and sulphate in throughfall. Chloride has been added to get an idea of the accuracy. The depositions are given in mmole m^{-2} y^{-1}. The recirculation figures for SO_4^{2-} represent the SO_2 uptake.

	Ca^{2+}	K^+	Mg^{2+}	SO_4^{2-}	Cl^-
WD, marine	1.5	1.4	8.2	4.1	79.4
WD, non-marine	2.6	1.4	(-0.4)	18.9	(-7.3)
DD, marine	4.4	4.3	24.5	12.2	236
DD, non-marine	2.9	4.8	(0.5)	5.0	(11.0)
recirculation	+8.3	+22.3	+(2.7)	+(10.7)	+(4.8)
throughfall	19.7	34.3	35.6	50.9	324

In Figure. 1 the calculated recirculation (NTF - calc. DD) of Ca^{2+} and K^+ is plotted as a function of time. As can be seen high ion leakage only occurs when there are high throughfall amounts of water. A high water deposition seems necessary to release all the different ions, even those which don't interact with the canopy (Na^+, Mg^{2+}).

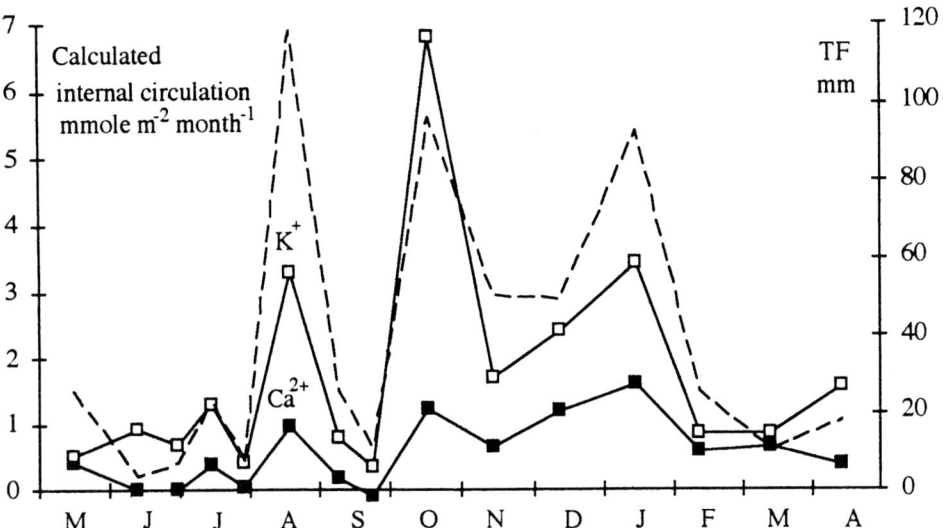

Fig 1. Calculated internal circulation of K^+ and Ca^{2+} together with throughfall amount of water (dashed line).

It has been noted that Ca^{2+} correlates with SO_4^{2-} in throughfall (Westling et al., 1995). A possible mechanism is that the SO_2 taken up through stomata brings Ca^{2+} as counter ion when it later on is released and removed by precipitation. The SO_2 deposition (stomatal uptake) was calculated from TF - WD - calc. DD SO_4^{2-}. Figure 2 shows that the two species not only co-variate, but are also similar in magnitude on a molar basis. The correlation between internal circulation of K^+ and uptake of SO_2 is not that good even though a certain co-variation exists due to the synchronous wash out of different ions when the throughfall amount is large.

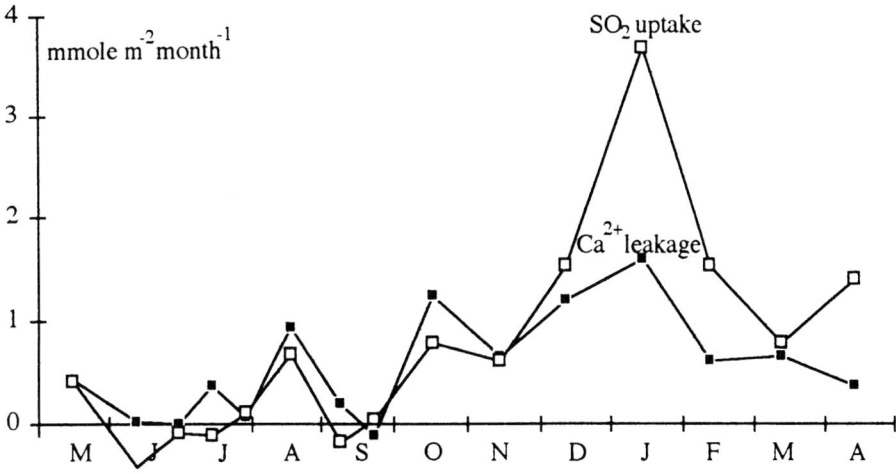

Fig. 2. Internal circulation of Ca^{2+} and uptake of SO_2 as a function of time.

4. Conclusion

The surrogate surface technique gives reasonable results for the atmospheric input of SO_4^{2-}, NO_3^-, and NH_4^+. When applied on the base cations it shows that most of net throughfall of K^+ and Ca^{2+} originates from internal circulation and that the internal circulation of Ca^{2+} is almost equal to the calculated SO_2 uptake. There is no measurable internal circulation of Mg^{2+}.

Acknowledgements

Financial support has been obtained from the Swedish Environmental Protection Agency and from the EU projects ENCORE and NITREX.

References

Falkengren-Grerup U. and Tyler G.: 1991, *Scandinavian J. of Forest Research* 6, 145-192

Ferm M.: 1993, *Intern.J. Anal. Chem.* 50, 29-43

Ferm M. and Hultberg H.: 1995, *Water Air and Soil Pollut*. this volume.

Hedin L. O., Granat L., Likens G. E., Buishand T. A., Galloway J. N., Butler T. J. and Rodhe H.: 1994, *Nature* 367, 351-354

Hultberg H.: 1985 *Ecol. Bull. (Stockholm)* 37, 133-157

Hultberg H. and Grennfelt P.: 1992, *Env. Pollut.* 75, 215-222

Raitio H.: 1993, Calcium and magnesium deficiency in young pines and the stand structure on the affected habitats. In Forest decline in the Atlantic and Pacific Region Eds. Hueller and Mueller-Dombois pp 133-143.

Tamm C. O. and Hallbäcken L.: 1988, *Ambio* 17, 56-61

Warfinge P., Falkengren-Grerup U., Sverdrup H. and Andersen B.: 1993, *Env. Pollut.* 80, 209-221

Westling O., Hultberg H. and Malm G.: 1995, In L. O. Nilsson, R. S. Hüttl and U. T. Johansson (eds.) Nutrient uptake and cycling in forest ecosystems, pp 639-647. Kluwer Academic Publishers, the Netherland.

THE USE OF MOSS, LICHEN AND PINE BARK IN THE NATIONWIDE MONITORING OF ATMOSPHERIC HEAVY METAL DEPOSITION IN FINLAND

H. LIPPO, J. POIKOLAINEN and E. KUBIN

The Forest Research Institute, Muhos Research Station, FIN-91500 Muhos, Finland

Abstract. Mosses, lichens and pine bark were compared as indicators of atmospheric heavy metal deposition in Finland. The samples were collected from the nationwide sampling network systematically covering the country as a whole. All three bioindicators showed a fairly similar result concerning heavy metal deposition. The major emission sources and the areas affected were reflected in the metal concentrations in the samples. However, there were differences between the accumulation of metals. The correlation between concentrations in mosses and lichens was generally higher than that between mosses and bark or lichens and bark. Concentrations in lichens were the highest and lichens reflected the regional differences in background areas as well as the local emission sources. The concentrations in the mosses were slightly lower than those in lichens and also the mosses pinpointed the emission sources and the extend of the areas polluted. Bark had the lowest concentrations and bark did not generally reveal regional differences as well as mosses and lichens. In spite of the differences, all three bioindicators proved to be suitable for monitoring atmospheric heavy metal deposition.

1. Introduction

The moss technique to survey atmospheric heavy metal deposition was developed in Sweden in the late 1960's (Rühling and Tyler 1970, Rühling 1971). Mosses, e.g. *Hylocomium splendens* and *Pleurozium schreberi*, obtain most of their nutrients from precipitation and from the impaction and sedimentation of airborne particles directly through their thin leaves. Besides nutrients, they also effectively absorb heavy metals and other chemical substances. Therefore they are suitable for the monitoring of airborne substances. The technique has gained wide acceptance and has expanded from a regional and national basis to the Nordic countries in 1985 and then to a large part of Europe in 1990 - 92 (Rühling et al. 1987, 1992, Rühling 1994).

The epiphytic lichen *Hypogymnia physodes* is a widely distributed lichenspecies. It obtains nutrients and e.g. heavy metals from deposition and it has also been used to survey regional heavy metal deposition (Pilegaard et al. 1979, Takala and Olkkonen 1985, Takala et al. 1985, Westman 1986, Kubin 1989, 1990, 1991). It is fairly resistant to impurities in the air (O'Hare and Williams 1975) and by that means it is a suitable monitor for both close to emission sources and for background areas (Guderian 1977).

The acidity of bark has an effect on the occurence of epiphytic lichens on the trunks and branches of trees (Barkman 1958). The bark has been used as an indicator for air pollution after it was found that there is a correlation between the acidity of bark and concentrations of atmospheric SO_2 (Staxäng 1969, Grodzinska 1971). The concentrations of SO_2 also has an effect on electrical conductivity and on the concentration of sulphur in the bark. Both of these have proven to be good indicators of sulphur emissions (Härtel and Grill 1972, Johnsen and Søchting 1973, Kienzl 1978, Grill et al. 1981). The bark absorbs also heavy metals and it has been used to survey heavy metal emissions from traffic (Laaksovirta et al. 1976, Lötschert and Köhm 1978) and industry (Swieboda and Kalemba 1979).

The Forest Research Institute established a network of 3009 permanent sites all over the country in 1985 and 1986 for forest condition monitoring (Reinikainen and Nousiainen 1985). One essential part of that has been to study the effects of air pollution, including heavy metal deposition on forests. Deposition has been monitored by collecting bioindicators and analysing the metal concentrations.

2. Material and methods

Samples of the moss species *Hylocomium splendens* and *Pleurozium schreberi* were collected and analysed according to the guidelines for the international survey (Nihlgård 1985, Rühling et al. 1987). Samples were dried, cleaned from extraneous plant material and the three youngest fully developed segments of each *Hylocomium* plant and corresponding green parts of *Pleurozium* were taken for analyses. Lichen samples of *Hypogymnia physodes* were collected from trunks and branches of trees. The majority of the samples were taken from pine trees. Once the dried samples had been cleaned, the whole thalli were taken for analyses (Kubin 1990). Bark (*Pinus silvestris L.*) samples were collected from at least three middle-aged trees from one plot in each cluster and the outermost cleaned three mm thick layers were taken for analyses (Poikolainen 1992). All samples were homogenised and then wet digested with a mixture of HNO_3 and $HClO_4$ and the concentrations of heavy metals were determined by ICP-AES-technique.

Sampling clusters from which the moss, lichen and bark samples were collected at the same time were included in the comparison. Mean concentrations for the clusters consisting of one to four sample plots for mosses and lichens were calculated for each of the metals. For bark there was only one sample plot representing each cluster. Detailed comparisons were done for Cr, Pb and Ni. The comparisons were based on maps where clusters were shown as different types of dots for the 50 highest and lowest concentrations of Cr and Ni and the 100 highest and lowest concentrations of Pb. The correlations between concentrations in different species based on the same sample clusters were calculated also for some other metals (Cd, Cu, Fe, V and Zn). Additionally, also the mean concentrations and coefficients of variation (cv) were calculated for each bioindicator. The differences between the mean concentrations were compared with Tukey's test ($p=0.05$).

3. Results and discussion

3.1 CHROMIUM

The highest concentrations of Cr in mosses and lichens were found around the Tornio steel mill at the northern coast of the Bothnian Bay (Fig. 1). The lowest concentrations in mosses and lichens were in northern Lapland. The highest and also the lowest consentrations in bark were distributed fairly even all over the country.

The highest mean concentration was in lichens and the lowest in bark (Table 1) and the differences between means were significant (Tukey's test, $p<0.05$). The differences in the mean concentrations may be caused by the variation in the lenght of the accumulation periods and the principle of accumulation. The content of metals in *Hypogymnia* reflects an average exposure of at least 10 years period of time and in *Hylocomium* and *Pleurozium* an average exposure of three years period of time. For pine bark it might be a steady state situation between the absorption and washing away of the pollutants during extended period of time. The cv's (Table 1) were at a fairly high level especially for lichens showing the existence of significant emission source.

The correlation between concentrations in mosses and lichens was 0.83 and those between both mosses and bark and between lichens and bark 0.39 and 0.45, respectively (Fig. 2). The high value between mosses and lichens indicates the similarity of the principle of accumulation for Cr.

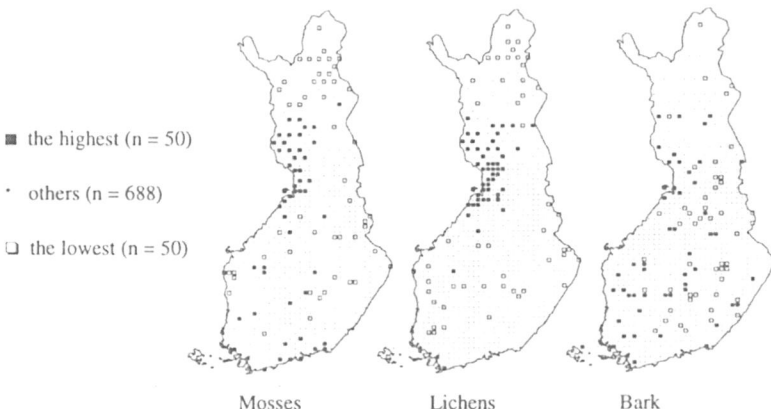

Fig. 1. Dots of the highest and the lowest chromium concentrations (µg/g) in mosses, lichens and bark.

TABLE I

Parameters calculated from the heavy metal concentrations in mosses, lichens and bark. $\overline{X}$ = average concentration (µg/g, n = 788) and cv = coefficient of variation (%).

Heavy metal	Mosses		Lichens		Bark	
	$\overline{X}$	cv	$\overline{X}$	cv	$\overline{X}$	cv
Cd	0.38	36	0.69	43	0.31	52
Cr	1.5	76	2.1	110	0.45	75
Cu	5.9	150	7.1	120	3.6	210
Fe	380	50	540	42	110	84
Ni	2.2	150	2.5	110	1.1	420
Pb	16	39	17	47	6.3	71
V	4.9	60	-	-	1.7	72
Zn	38	23	84	20	18	34

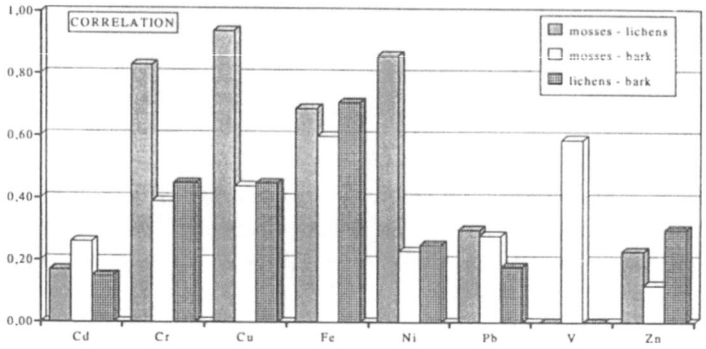

Fig. 2. Correlations between concentrations in different species.

3.2 LEAD

The highest concentrations of Pb in mosses were found in southern Finland (Fig. 3). Characteristic features were the discernible main roads between Helsinki, Tampere and Turku. The lowest concentrations in mosses were found in Ostrobothnia, Kainuu and Lapland. The highest concentrations in lichens and bark were found in southern Finland and also around Oulu, and the lowest were distributed fairly even in Lapland and southeastern Finland.

The mean concentration was the highest in lichens and the lowest in bark was (Table 1) and the differences between means were significant. The cv's were at low level especially for mosses and lichens indicating fairly even concentrations through the country as a whole. The correlation between concentrations in different species were at low levels (range 0.18 - 0.30) (Fig. 2).

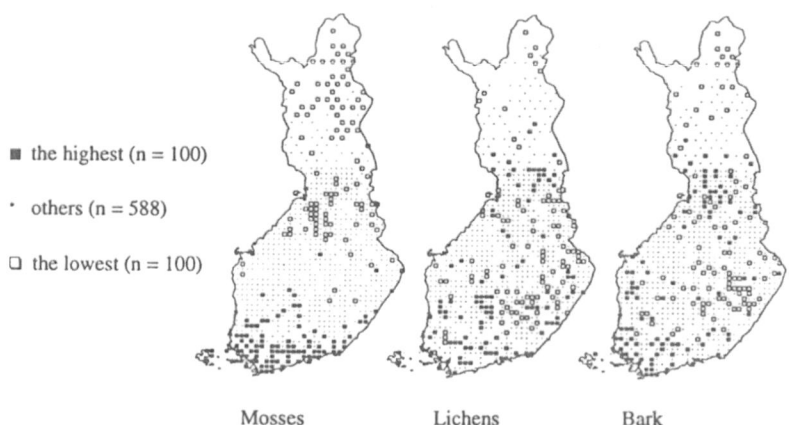

Fig. 3. Dots of the highest and the lowest lead concentrations in mosses, lichens and bark.

3.3 NICKEL

All three bioindicators showed a similar result concerning the highest levels (Fig. 3). The highest concentrations of Ni were found in samples around the Harjavalta smelters in southwestern Finland. Elevated values were also found in the north-eastern parts of Lapland due to the influence from Nikel and Monchegorsk smelters in Russia near the Finnish border. The lowest levels were found in western Lapland and Kainuu.

The mean concentration was the highest in lichens and the lowest in bark (Table 1). The differences between means were significant except means between mosses and lichens. The cv's were at a high level indicating the existence of significant emission sources. The correlation between concentrations in mosses and lichens was high (0,83) and those between both mosses and bark and lichens and bark were at the same fairly low level (0.23 - 0.26) (Fig. 2).

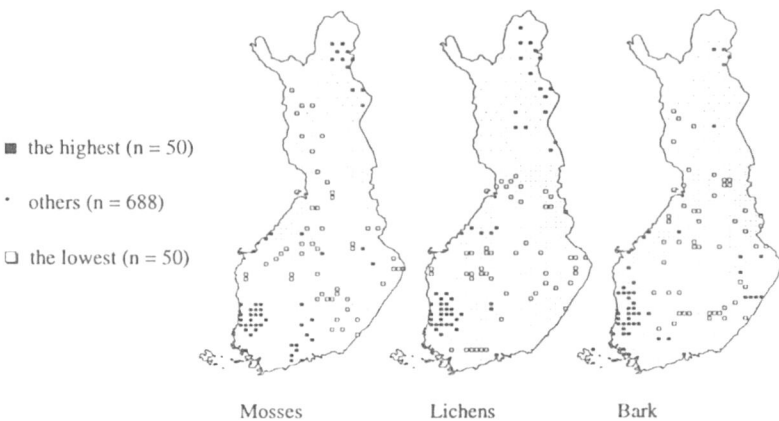

Fig. 4. Dots of the highest and the lowest nickel concentrations (µg/g) in mosses, lichens and bark.

3.4 CADMIUM, COPPER, IRON, VANADIUM AND ZINC

In addition to Cr, Pb and Ni, the concentrations of Cd, Cu, Fe, V and Zn were also taken under consideration. The mean concentrations were the highest in lichen and the lowest in bark for all those metals (Table 1) and the differences between means were significant.

The coefficients of variation (Table 1) were at low level especially for Cd, Fe and Zn indicating fairly even concentrations through the country as a whole. The cv's were high for Cu indicating the existence of significant emission sources.

The correlations between concentrations were at low level for Cd and Zn (range 0.12 - 0.30) and at fairly high level for Fe and V (range 0.59 - 0.72) (Fig. 1). Especially high value for Cu in mosses and lichens (0.94) indicates the similarity of the principle of accumulation for Cu.

References

Barkman, J.J.: 1958, "Phytososiology and ecology of cryptogamic epiphytes", Assen, 628 pp.
Grill, D., Härtel, O. and Krzyscin, F.: 1981, "Confining and mapping of airpolluted areas with coniferous barks", Archiwum Ochrony Srodowiska 2-4, 63-70.
Grodzinska, K.: 1971, "Acidification of tree bark as a measure of air pollution in southern Poland", Bulletin de L'Academie Polonnaise des Sciences, Serie des sciences biologiques, Cl II 19(3), 189-195.
Guderian, R.: 1977, "Air pollution. Phytotoxicity of acidic gases and its significance in air pollution control", Ecological studies 22, 127 pp.
Härtel, O and Grill, D.: 1972, "Die Leitfähigkeit von Fichten-borken- Extrakten als empfindlicher Indikator für Luftverun-reinigungen", European Journal of Forest Pathology, 2, 205-215.
Johnsen, I. and Søchting, U.: 1973, "Influence of air pollution on the epiphytic lichen vegetation and bark properties of deciduous trees in the Copenhagen area", Oikos 24, 344-351.
Kienzl, I.: 1978, "Baumborke als Indikator für SO_2-Immissionen", Thesis, Graz, 273 pp.
Kubin, E.: 1989, "Sulphur content of the epiphytic lichen *Hypogymnia physodes* as an indicator of atmospheric deposition in Finland", Medd. Nor. Inst. Skogforsk. 42(1), 109-120.
Kubin, E.: 1990, "A survey of element concentrations in the epiphytic lichen *Hypogymnia physodes* in Finland in 1985-86", In "Acidification in Finland", Kauppi, P., Anttila, P. & Kenttämies, K. (Eds.). Springer-Verlag, 421-446.
Kubin, E.: 1991, "A survey of element concentrations in the epiphytic lichen *Hypogymnia physodes* in northern Finland in 1986", Geological Survey of Finland, Special Paper 9, 185-194.
Laaksovirta, K., Olkkonen, H. and Alakuijala, P.: 1976, "Observations on the lead content of lichen and bark adjacent to a highway in southern Finland", Environmental Pollution 11, 247-255.
Lötschert, W. and Köhm, H.-J.: 1978, "Characteristics of tree bark as an indicator in high immission areas. II. Contents of heavy metals", Oecologia 37(1), 121-132.
Nihlgård, B.: 1985, "Survey of the heavy-metal deposition in the Nordic countries 1985 - Guidelines".
O'Hare, G.P. and Williams, P.: 1975, "Some effects of sulphur dioxide flow on lichens", Lichenologist 15(1), 89-93.
Pilegaard, K., Rasmussen, L. and Gydesen, H.: 1979, "Atmospheric background deposition of heavy metals in Denmark monitored by epiphytic cryptogams", Journal of Applied Ecology 16, 834-853.
Poikolainen, J: 1992, "Männyn kaarnan sähkönjohtokyky ja pH Itä-Lapin metsävaurioprojektin koealoilla", In Kauhanen, H. and Varmola, M. (Eds.) Itä-Lapin metsävaurioprojektin väliraportti, Metsäntutkimuslaitoksen tiedonantoja 413, 128-135.
Reinikainen, A. and Nousiainen, H.: 1985, "Biologien työohjeet VMI 8:n pysyviä koealoja varten", Metsäntutkimuslaitos, Moniste, 42 pp.
Rühling, Å. and Tyler, G.: 1970, "Sorption and retention of heavy metals in the woodland moss *Hylocomium splendens* (Hedw.) Br et Sch", Oikos 21, 92-97.
Rühling, Å.: 1971, "Regional difference in the deposition of heavy metals over Scandinavia", Journal of Applied Ecology 8, 497-507.
Rühling, Å., Rasmussen, L., Pilegaard, K., Mäkinen, A. and Steinnes, E.: 1987, "Survey of atmospheric heavy metal deposition in the Nordic countries in 1985", Nord 1987:21.
Rühling, Å. ,Brumelis, G., Goltsova, N., Kvietkus, K., Kubin, E., Liiv, S., Magnússon, S., Mäkinen, A., Pilegaard, K., Rasmussen, L., Sander, E. and Steinnes, E.: 1992, "Atmospheric heavy metal deposition in northern Europe 1990", Nord 1992:12.
Rühling, Å. (Ed.): 1994, "Atmospheric heavy metal deposition in Europe - estimation based on moss analysis", Nord 1994:9.
Staxäng, B.: 1969, "Acidification of bark in some deciduous trees", Oikos 20, 223-230.
Swieboda, M. and Kalemba, A.: 1979, "The bark of Scots pine (*Pinus silvestris L.*) as a biological indicator of atmospheric air pollution", Acta Societatis Botanicorum Poloniae 48, 539-549.
Takala, K., and Olkkonen, H.: 1985, "Titanium content of lichens in Finland", Annales Botanici Fennici 22, 299-305.
Takala, K., Olkkonen, H., Ikonen, J., Jääskeläinen, J. and Puumalainen, P.: 1985, "Total sulphur contents of epiphytic and terricolous lichens in Finland", Annales Botanici Fennici 22, 91-100.
Westman, L.: 1986, "Lavars indikatorvärde vid studier av luftföreningar och skogsskador", Statens Naturvärdsverk Rapp. 3187: 1-52.

THROUGHFALL DEPOSITION OF AMMONIUM AND SULPHATE DURING AMMONIA FUMIGATION OF A SCOTS PINE FOREST.

J. N. CAPE[1], L. J. SHEPPARD[1], J. BINNIE[1], P. ARKLE[1] and C. WOODS[2]

[1]*Institute of Terrestrial Ecology, Bush Estate, Penicuik, Midlothian EH26 0QB, UK*. [2]*Institute of Terrestrial Ecology, Merlewood Research Station, Grange-over-Sands, Cumbria, LA11 6JU, UK.*

Abstract. The estimation of the dry deposition of sulphur dioxide to forests is confounded by the possibility of co-deposition of SO_2 with NH_3 on leaf surfaces. A sector of Scots pine forest was selectively fumigated with NH_3 to give average concentrations up to 15 ppbV (nL L^{-1}) above ambient, in order to test the hypothesis that increased air concentrations of NH_3 would enhance the dry deposition of SO_2, and the consequent amounts of SO_4^{2-} measured in throughfall below the forest canopy. Ammonia gas, generated by evaporation of concentrated aqueous solution, was released above the canopy in proportion to wind speed when the wind direction was between south and west. Concentrations of NH_3 at canopy height were measured using passive diffusion tubes; throughfall was preserved with thymol and measured weekly. Meteorological data and SO_2 concentrations were recorded continuously, to permit the estimation of dry deposition input. Deposition of NH_4^+ in throughfall over 8 months was increased by up to 40 meq m^{-2} relative to 'control' sites upwind of the NH_3 release point, with largest values closest to the release point. Deposition of SO_4^{2-} in throughfall was also enhanced in the fumigated area, by up to 20 meq m^{-2}, even though average ambient SO_2 concentrations were 2.3 ppbV. The results are discussed in terms of the factors controlling SO_2 deposition on forest surfaces, the development of appropriate deposition models, and their relevance to using throughfall as an estimate of total S deposition.

Key words: throughfall, dry deposition, sulphur dioxide, ammonia, co-deposition.

1. Introduction

There has been some controversy over the origin of sulphur deposition to forest soils, and the use of throughfall measurements to quantify the deposition of sulphate (SO_4^{2-}) ions (Lindberg *et al.*, 1992). Several studies have been able to rationalise throughfall measurements with estimated inputs of sulphur in precipitation, as aerosol and by dry deposition of sulphur dioxide (SO_2), for example, the Integrated Forest Study (Lindberg and Lovett, 1992). However, there are several studies in Europe where the total input of sulphur to a forest floor, measured as the sum of inputs in throughfall and stemflow, greatly exceeds the estimates of total sulphur deposition by inferential techniques (e.g. Cape *et al.*, 1987; Ivens *et al.*, 1990). Either there is a significant component of internal cycling of sulphate in a forest, from soil to canopy and through foliar leaching to the soil, or the parameters used in the inferential method for estimating dry deposition of particles and gases to the canopy are wrong. Experiments using radioactive ^{35}S as a tracer have shown that internal cycling of S usually contributes a negligible fraction of the additional S measured below a forest canopy (known as *net throughfall*) (Garten, 1990; Cape *et al.*, 1992). Although dry deposition velocities for SO_4^{2-} containing particles have recently been shown to be an order of magnitude greater than expected on the basis of laboratory experiments (Wyers *et al.*, 1994), the use of the field-derived deposition velocities for particles in inferential models is still not sufficient to account for all of the S in net throughfall. The dry deposition of SO_2 to canopy surfaces, from where SO_4^{2-} ions may be readily washed by rain, would seem to be the only pathway by which the large deposition of S in net throughfall could occur.

Some of the earliest field measurements (Fowler & Cape, 1983) showed that the dry deposition of SO_2 to forest canopies was dominated by stomatal uptake, as had been observed for agricultural crops. The available field data also showed that deposition to rain-wetted surfaces was not much greater than to dry surfaces, which was interpreted in terms of phase equilibrium between SO_2 in rain drops and SO_2 in air, and slow oxidation (at night) on the canopy surface. The potential for enhanced rates of SO_2 deposition in the presence of NH_3, and *vice versa* ('co-deposition'), to dry or wetted plant surfaces is not easily incorporated into inferential models unless (i) ambient NH_3 concentrations are known; (ii) the degree of surface wetness is known; and (iii) the rate-determining concentration is known (i.e. whether SO_2 or NH_3 is in excess).

In the experiment described below, a sector of forest was fumigated with gaseous NH_3 at concentrations within the range of expected NH_3 concentrations in rural air. Measurements of the deposition of NH_4^+ and SO_4^{2-} ions in precipitation and in throughfall were made, to test the hypothesis that dry deposition of SO_2 would be enhanced in the sector of forest exposed to elevated NH_3 concentrations, and that such enhanced deposition would be apparent in the measured throughfall.

2. Methods

The study was established at Devilla Forest, Fife, central Scotland, in a 40-year old plantation forest of Scots pine (*Pinus sylvestris* L.). Gaseous NH_3 was generated by evaporating a saturated solution of NH_3 in water (density 0.88 g mL^{-1}) into a stream of air which was released from a pipe 5 m above the canopy surface. The rate of release was determined by the mean wind speed in the preceding 10 minutes; the flow rate of NH_3 solution was set at 1 mL min^{-1} for every 1 m s^{-1} windspeed, giving a release rate of 15 mg NH_3 s^{-1} / m s^{-1}. NH_3 was only released if (i) wind speed > 1 m s^{-1} (to avoid excessively large concentrations); (ii) the wind direction was between south and west (so that only a sector of forest was exposed to elevated concentrations); and (iii) a surface wetness detector was dry (to avoid the possibility of enhanced uptake of NH_3 and SO_2 by rain falling through the plume). The surface wetness detector was a curved aluminium plate (0.15 m x 0.15 m) with an interlaced conductive network etched on the upper surface. The electrical resistance of the detector, measured using a bridge circuit, went from > 800 kΩ when dry to < 400 Ω when wet. The detector was mounted at canopy height; it overestimated surface wetness during conditions when dew formed, having a thicker boundary layer than the canopy. Concentrations of NH_3 were measured at canopy height using passive diffusion tubes (4 replicates per location; Sutton, 1990) upwind, and within the fumigated sector. Concentrations of SO_2 were measured continuously at canopy height immediately upwind of the NH_3 release point, using a pulsed fluorescence monitor (ThermoEnvironmental model 43). Meteorological data were measured continuously at canopy height.

Wet deposition was collected upwind of the site using 3 bulk collectors, each mounted on a tower above the forest canopy. The towers were ca.100 m apart. Throughfall was collected (as 'control') at 4 locations upwind of the site, and at 10 locations within the fumigated sector. Each location had three 0.2 m diameter funnels mounted 0.5 m above ground and draining to a polypropylene bottle mounted in an opaque tube. Water samples from the 3 replicate collectors at each location were mixed to give a composite sample before analysis. Both rain and throughfall samples were preserved by the addition of sufficient solid thymol to provide a saturated solution (Hadi & Cape, 1995), to avoid degradation of NH_4^+ prior to collection and analysis. Water samples were collected weekly. The experiment started on 10 March 1993 and finished on 1 November 1993.

3. Results and Discussion

3.1 METEOROLOGICAL CONDITIONS DURING THE EXPERIMENT

The conditions required for the release of NH_3 were met for 23% of the time. The surface wetness detector was wet for 28% of the time. Total rainfall during the experiment, as recorded in the canopy-level precipitation gauges, was 533 mm, and the average throughfall was 312 mm, consistent with previous studies at this site. The total wet deposition of SO_4^{2-} in rain over the period (approx. 8 months) was 29 meq m^{-2}, in line with previous annual measurements in this forest of 25 to 35 meq S m^{-2} (Cape & Lightowlers, 1988). Deposition of SO_4^{2-} in throughfall in the 'control' area to the south-west of the emission point was 66 meq S m^{-2} over the 8 months, showing a comparable increase over wet deposition to that seen in previous studies (35 to 60 meq S m^{-2} yr^{-1} in throughfall).

3.2 CONCENTRATIONS OF NH_3 AND SO_2 ABOVE THE CANOPY.

Even with 4 replicate diffusion tubes at each location, there was considerable variability in the measured NH_3 concentrations. Given the possibility of systematic bias in measuring NH_3 concentrations by passive diffusion tubes (Sutton, 1990), the data should be regarded only as an indication of the relative concentrations above the forest. The measured concentrations were up to 15 ppbV above the ambient background concentration, which was less than or equal to the detection limit, based on analysis of blank diffusion tubes, of 3 ppbV. The average ambient SO_2 concentration was 2.3 ppbV.

3.3 DEPOSITION OF IONS IN THROUGHFALL

The spatial pattern of NH_4^+ deposition during the experiment is shown in Figure 1, where the 'excess' deposition (i.e. in excess of throughfall deposition in the 'control' area) is plotted as a function of the position of the throughfall collectors. The throughfall deposition of NH_4^+ in the 'control' area was equivalent to 7.4 meq N m^{-2}. There was a marked increase in NH_4^+ deposition in response to the fumigation treatment, reflecting dry deposition of NH_3 on the canopy, and subsequent wash-off by rain. There was net uptake of NH_4^+ from rain by the canopy in the 'control' area (Figure 3), which suggested that the fumigated trees might also use the gaseous NH_3 as a source of N, but chemical analysis of foliage at the end of the experiment showed no significant increase in foliar N concentrations compared to trees growing upwind of the fumigated plot.

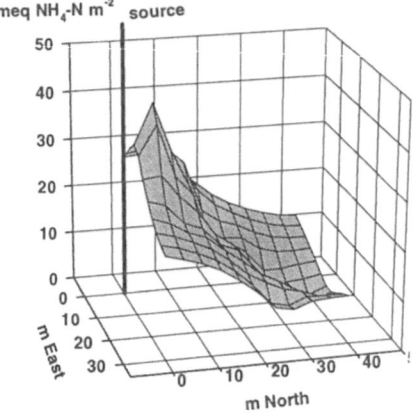 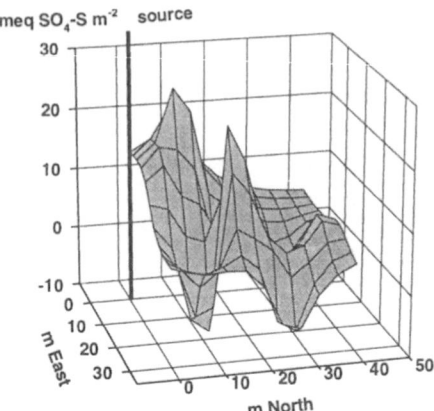

Figure 1. Excess deposition of NH_4^+-N in throughfall under NH_3 fumigated trees, compared to an adjacent non-fumigated area (March-October 1993).

Figure 2. Excess deposition of SO_4^{2-}-S in throughfall under NH_3 fumigated trees, compared to an adjacent non-fumigated area (March-October 1993).

The spatial pattern of 'excess' S deposition (i.e. in excess of the 'control' area) is shown in Figure 2. Apart from one set of collectors in the centre of the plot, there was an increase in excess S deposition in throughfall towards the fumigation source, and the pattern was broadly similar to that for 'excess' NH_4^+, supporting the original hypothesis that fumigation with NH_3 would stimulate dry deposition of SO_2, and increase the deposition of SO_4^{2-} in throughfall. The

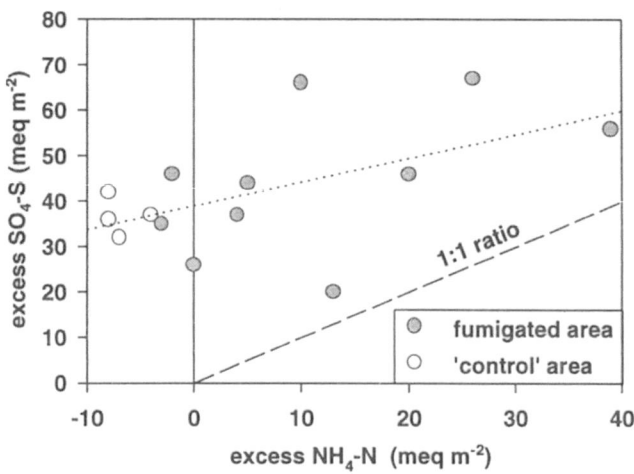

Figure 3. Comparison of NH_4^+ and SO_4^{2-} deposition in net throughfall (throughfall - rain): Devilla Forest, Fife, March-October 1993.

spatial variability in SO_4^{2-} deposition was greater than that for NH_4^+, not only for the whole period, but also for individual weekly collections. This finding points to spatial heterogeneity in the process of co-deposition which is not simply related to exposure to elevated concentrations of gases, but may involve differences in the duration of canopy wetness, the availability of reaction sites and/or the pathways by which water passes through the canopy.

If co-deposition of NH_3 and SO_2 were occurring, one might expect a 1:1 equivalent ratio of $SO_4^{2-}:NH_4^+$ in the 'excess' deposition. In practice, if one or other gas is in excess, there is scope for additional dry deposition to the canopy for that gas. At this site it would appear that NH_3 was in excess, so that the expected ratio would be less than 1. However, as noted above, there was evidence for NH_4^+ uptake by the canopy, which would bias the ratio of $SO_4^{2-}:NH_4^+$ towards a value greater than 1. If the net throughfall deposition of SO_4^{2-} is plotted against net throughfall NH_4^+ for each throughfall location (Figure 3), it can be seen that there was only a weak correlation (r=0.53; P=0.049), and the least-squares estimate of the overall $SO_4^{2-}:NH_4^+$ equivalent ratio was 0.5 ± 0.2 (s.e.).

4. Conclusions

The experiment has shown that it is possible to fumigate a defined area of forest with gaseous NH_3 at concentrations of the same order as those found in ambient air. As a result, the concentration and deposition of NH_4^+ in throughfall was enhanced downwind of the fumigation source, and the concentration and deposition of SO_4^{2-} was also enhanced. The degree of enhancement of SO_4^{2-} deposition over that found in the non-fumigated sectors of the forest was probably limited by the relatively small concentration of SO_2 compared with that of NH_3 in ambient air, i.e. deposition of SO_4^{2-} in throughfall below the non-fumigated forest may already have been enhanced by co-deposition of NH_3 and SO_2 to the canopy. It is possible that additional SO_4^{2-} deposition in the fumigated sector would only occur when SO_2 concentrations exceeded ambient NH_3 concentrations.

These results have important implications for modelling the dry deposition of SO_2 and NH_3 to vegetation; the deposition velocity is not simply related to the concentration of either gas, but depends on the relative concentrations. It may be possible to simplify the parametrisation of SO_2 deposition in inferential models in the limiting cases of very low or very high NH_3 concentrations. For example, in the IFS experiment, where gaseous NH_3 concentrations were assumed to be very small, co-deposition may be unimportant; conversely, in regions with very high NH_3 concentrations, as in the Netherlands, where NH_3 concentrations may be consistently greater than concentrations of SO_2, rates of dry deposition of SO_2 may be controlled more by the rate of transport to the canopy rather than by surface chemistry. In general, however, and especially over much of Europe where rural NH_3 and SO_2 concentrations are similar, the estimation of S and N deposition by inferential methods may not be possible without much better data on NH_3 concentrations. For forests, the best estimate of total S deposition at specific sites using currently available techniques may come from measurements of throughfall and stemflow, provided that sufficient care is taken in the experimental design to accommodate the great spatial variability in throughfall and stemflow deposition. The contribution of stemflow in this experimental study of processes was ignored, but stemflow may contribute a large proportion of the total amount of sulphate deposited to the forest floor (Cape et al.1987). However, the unpredictable amount of canopy uptake of NH_3 and NH_4^+ means that throughfall and stemflow measurements alone cannot be used to estimate total deposition of reduced nitrogen to forests.

Acknowledgements

This study was jointly funded by the UK Department of the Environment under contract PECD 7/12/49, and by the Natural Environment Research Council.

References

Cape, J.N., Fowler, D., Kinnaird, J.W., Nicholson, I.A. and Paterson, I.S.: 1987, "Modification of rainfall chemistry by a forest canopy." In: Pollutant Transport and Fate in Ecosystems (eds. P.J.Coughtrey, M.H.Martin & M.H.Unsworth) pp.155-169. Blackwell Scientific, Oxford.

Cape, J.N. and Lightowlers, P.J.: 1988, Review of Throughfall and Stemflow Data in the United Kingdom. Report to UK Department of Environment. I.T.E. Edinburgh.

Cape, J.N., Sheppard, L.J., Fowler, D., Harrison, A.F., Parkinson, J.A., Dao, P. and Paterson, I.S.: 1992, "Contribution of canopy leaching to sulphate deposition in a Scots pine forest." Environmental Pollution, **75**, 229-236.

Fowler, D. and Cape, J.N.: 1983, "Dry deposition of SO_2 onto a Scots pine forest." In: Precipitation Scavenging, Dry Deposition and Resuspension. (eds. H.R.Pruppacher, R.E.Semonin & W.G.N.Slinn) pp.763-774. Elsevier, New York.

Garten, C.T.: 1990, "Foliar leaching, translocation and biogenic emission of ^{35}S in radiolabelled loblolly pines." Ecology, **71**, 239-251.

Hadi, D.A. and Cape, J.N.: 1995, "Preservation of throughfall samples by chloroform and thymol." Int. J. Environmental Analytical Chemistry (vol. 62, in press)

Ivens, W.P.M.F., Kauppi, P., Alcama, J. and Posch, M.: 1990, "Empirical and model estimates of sulfur deposition onto European forests." Tellus, **42B**, 294-303.

Lindberg, S.E. and Lovett, G.M.: 1992, "Deposition and forest canopy interactions of airborne sulfur: results from the Integrated Forest Study." Atmospheric Environment, **26A**, 1477-1492.

Lindberg, S.E., Cape, J.N., Garten, C.T. Jr and Ivens, W.: 1992, "Can sulfate fluxes in forest canopy throughfall be used to estimate atmospheric sulfur deposition? - A summary of recent results." In: Precipitation Scavenging and Atmosphere-Surface Exchange (eds. S.E.Schwartz & W.G.N.Slinn) pp.1367-1377. Hemisphere, Washington.

Sutton, M.A.: 1990, "The surface/atmosphere exchange of ammonia." PhD Thesis, University of Edinburgh.

Wyers, G.P., Geusebroek, M., Wayers, A., Möls, J.J. and Veltkamp, A.C.: 1994, "Dry deposition of sub-micron aerosol on a coniferous forest." In: Transport and Transformation of Pollutants in the Troposphere. (eds. P.M.Borrell, P.Borrell, T.Cvitaš & W.Seiler) pp. 712-715. SPB Academic Publishing, The Hague.

A CANOPY BUDGET MODEL TO ASSESS ATMOSPHERIC DEPOSITION FROM THROUGHFALL MEASUREMENTS

G.P.J. DRAAIJERS and J.W. ERISMAN

National Institute of Public Health and Environmental Protection (RIVM), P.O. box 1, 3720 BA Bilthoven, The Netherlands

Abstract A canopy exchange model is presented which allows atmospheric deposition to be estimated from long-term throughfall and precipitation measurements. For a forest in the Netherlands, the combination of throughfall measurements and this model resulted in deposition estimates which were similar to deposition estimates derived from micrometeorological measurements and inferential modeling, deposition of NOy being the only exception. Unfortunately, several basic assumptions in the canopy exchange model are not properly evaluated, which up to now limits its application. Suggestions are made on how the model can be improved.

Key words: canopy budget modeling, throughfall, atmospheric deposition, canopy exchange

1. Introduction

A major problem when using throughfall and stemflow measurements for atmospheric deposition assessment is that a reliable method to distinguish between in-canopy and atmospheric sources of chemical compounds often is lacking. Lovett and Lindberg (1984) developed an empirical multiple regression model to estimate canopy exchange and atmospheric deposition on the basis of event throughfall and precipitation measurements. This model has proven very valuable for forests situated in areas with convective storms and extended dry weather periods (Lovett and Lindberg, 1984; Puckett, 1990) but was found less useful in areas characterised by frequent low-intensity rainfall and relatively short dry periods (Lindberg et al., 1990; Draaijers et al., 1994). An alternative for regression modeling is application of a canopy budget model which was developed by Ulrich (1983) and extended by Van der Maas & Pape (1991). This model allows a discrimination between canopy exchange and atmospheric deposition using long-term throughfall and precipitation fluxes. In this paper, deposition estimates for the Speulder forest (the Netherlands) derived from the canopy budget model in combination with throughfall measurements are compared to deposition estimates derived from micrometeorological measurements and inferential modeling. Throughfall and micrometeorological measurements were made over the same time period. Model assumptions and uncertainties are extensively discussed.

2. Model description

Model assumptions and a short overview of the calculation scheme are presented here. The following abbreviations are used: TF = throughfall flux, SF = stemflow flux, DD = dry deposition flux, BP = bulk precipitation flux, CL = canopy leaching, CU = canopy uptake, wa = weak acids, cat = total cations, an = total anions, bc = sum of base cations Ca^{2+}, Mg^{2+} and K^+. DDF = dry deposition factor and EF = excretion factor. An appropriate time step for running the model is 0.5-1 year but, in principle, monthly data can be used as well. In the model, Na^+ in throughfall is assumed not to be influenced by canopy exchange. Furthermore, particles containing Ca^{2+}, Mg^{2+}, K^+, Cl^- and PO_4^{3-} are assumed to have the same mass median diameter as Na^+ containing particles. Dry deposition of Ca^{2+}, Mg^{2+}, K^+, Cl^- and PO_4^{3-} can subsequently be calculated according to (Ulrich, 1983):

$$DD = DDF * BP$$

The dry deposition factor equals:

$$DDF = (TF_{Na}+SF_{Na}-BP_{Na})/BP_{Na}$$

Canopy leaching of these ions is calculated according to:

$$CL = TF+SF-BP-DD.$$

Canopy leaching computed for Cl^- is regarded as deposition of HCl(gas) as Cl^- leaching is generally assumed negligible (Draaijers, 1993). The total canopy uptake of H^+ and NH_4^+ is assumed to equal the total canopy leaching of Ca^{2+}, Mg^{2+} and K^+ minus canopy leaching of Ca^{2+}, Mg^{2+} and K^+ associated with foliar excretion of weak acids (canopy uptake should always balance canopy leaching). To calculate the latter, Van der Maas and Pape (1991) define an excretion factor equal to:

$$EF = CL_{wa}/(CL_{Mg}+CL_{Ca}+CL_K)$$

where CL_{wa} is computed according to:

$$CL_{wa} = TF_{wa}+SF_{wa}-BP_{wa}-DD_{wa}.$$

It is assumed that all organic acids are leached in a neutral salt form. For the calculation of the excretion factor it is very important that all ions significantly contributing to the cation-anion balance are measured, and also with the highest possible accuracy (Van der Maas and Pape, 1991; Draaijers, 1993). TF_{wa} is assumed equal to $TF_{cat}-TF_{an}$, SF_{wa} equal to $SF_{cat}-SF_{an}$ and BP_{wa} to $BP_{cat}-BP_{an}$ (e.g. Guiang et al., 1984). Dry deposition of weak acids is assumed equal to bulk precipitation of weak acids. The canopy leaching of base cations through exchange with H^+ and NH_4^+ is computed according to:

$$CL_{bc} = (CL_{Mg}+CL_{Ca}+CL_K) * (1-EF)$$

Canopy uptake of H^+ and NH_4^+ is subsequently calculated from the sum of exchanged ions of Ca^{2+}, Mg^{2+} and K^+ where it is assumed that, based on experiments in the laboratory (Van der Maas et al., 1991), H^+ has an exchange efficiency (= exchange activity) six times larger than NH_4^+:

$$CU_H = CL_{bc} / (1+ (1 / [6*(((TF_H/TF_{NH4} + (BP_H/BP_{NH4}))/2])))$$
$$CU_{NH4} = CL_{bc} - CU_H$$

Knowing their canopy uptake, the dry deposition flux of H^+ (from H_2SO_4, $(NH_4)HSO_4$, HNO_3 and HCl) and NH_4^+ (NH_3 and NH_4^+ aerosol) can be computed from TF+SF+CU-BP. Finally, it is assumed that canopy leaching of SO_4^{2-} and NO_3^- is zero allowing the calculation of dry deposition of SO_4^{2-} (SO_2 and SO_4^{2-} aerosol) and NO_3^- (NO, NO_2, HNO_2, HNO_3 and NO_3^- aerosol) according to TF+SF-BP (Van der Maas and Pape, 1991).

3. Throughfall and micrometeorological measurements at the Speulder forest

The Speulder forest research site is located at the national park 'de Hoge Veluwe' in the central part of the Netherlands (52°15'N, 5°41'E). De Hoge Veluwe is an approximately 100 m high ice-pushed morainic ridge with dry and sandy, nutrient poor podzolic soils. The measuring site consists of a homogeneous 2.5 ha monoculture of Douglas-fir, 35 years old with a stem density varying between 785 and 1250 tree ha^{-1}. Mean tree height equalled 21.6m. The canopy is well closed with a one-sided leaf area index varying from 9 in early spring to 12 at the end of the summer. Precipitation and throughfall fluxes were measured continuously on a weekly basis between 26 October 1992 and 21 July 1993 (Van Leeuwen et al., 1994). Bulk precipitation was sampled in a clearing approximately 300 m from the Speulder forest site by means of four continuously open funnels, each having a collecting area of 0.017 m^2. During several months wet-only precipitation was measured to estimate the contribution of dry deposition onto the funnels and to derive bulk to wet-only correction factors. Throughfall was sampled weekly by 25, 4m long gutters, each having a collecting area of 0.054 m^2. The stemflow ion flux was assumed to be on average 6% of the throughfall ion flux (Van Leeuwen et al., 1994). As throughfall and bulk precipitation samplers were stored in the field no longer than one week, biochemical transformation of N compounds were considered insignificant (Slanina et al., 1990). For this reason, no preservatives were used.

Between 23 November 1992 and 10 May 1993 the dry deposition flux of SO_2, NH_3 and NO_2 was estimated using the gradient technique. In the same period dry deposition of HNO_3 HNO_2 and HCl was inferred from measured air concentrations and parametrised dry deposition velocities (Erisman et al., 1994). A model describing stomatal conductance as a product of response functions for water vapour deficit, global radiation, temperature, soil moisture status and leaf area index (Bouten and Bosveld, 1992) was used to estimate gaseous uptake through stomata. Moreover, dry deposition of acidifying aerosols (SO_4^{2-}, NO_3^-, NH_4^+

and Cl⁻) and base cations (Na^+, K^+, Mg^{2+} and Ca^{2+}) was inferred using results of air concentration and particle size measurements (Römer and Te Winkel, 1994) and a parametrisation of the deposition velocity (Ruijgrok et al., 1994). During two campaigns in December 1992 and February 1993, respectively, fog deposition was estimated by measuring the turbulent water flux of fog droplets using the eddy correlation technique (Vermeulen et al., 1995). The fog water flux through sedimentation depends on the fog droplet radius and was estimated using Stoke's law. Fog deposition fluxes of SO_4^{2-}, NO_3^- and NH_4^+ were obtained by multiplying total fog water fluxes with the average chemical composition of the fog droplets measured by Römer and Te Winkel, (1994). Fog deposition estimates were extrapolated to the whole measurement period on the basis of fog duration measurements from a nearby meteorological station (Wyers et al., 1994; Vermeulen et al., 1995).

4. Results and discussion

Summed for the period 23 November 1992 - 10 May 1993, the dry and fog deposition estimate for NO_y is approximately twice as large as the net throughfall (= throughfall minus wet deposition) flux of NO_3^- (Table I). This difference is probably due to canopy uptake of oxidised nitrogen compounds. Dry and fog deposition estimates for SO_x, NH_x, Na^+, Cl⁻ and Mg^{2+} are not significantly different from corresponding net throughfall fluxes (paired t-test, one tailed, $\alpha = 0.05$). For K^+ and Ca^{2+}, dry and fog deposition estimates are significantly smaller (89% and 36%, respectively) than corresponding net throughfall fluxes. This is probably the result of the contribution of canopy leaching to the net throughfall flux of K^+ and Ca^{2+}. If net throughfall fluxes of K^+ and Ca^{2+} are corrected for canopy exchange using the canopy exchange model, very reasonable agreement is found with dry and fog deposition estimates (Table I). It must be stressed that observed differences between dry and fog deposition estimates from micrometeorological measurements and inferential modeling on the one hand and canopy exchange corrected net throughfall fluxes on the other hand can not be regarded exclusively due to canopy exchange but may also be the result of measuring artefacts. For the Speulder forest, the combination of throughfall measurements and the canopy budget model was found to result in dry and fog deposition estimates having an uncertainty of 30-40% (Draaijers et al., 1995). The uncertainty in deposition estimates from micrometeorological measurements and inferential modeling was estimated 30-50% (Erisman et al., 1994). As a result of these relatively large uncertainties, differences between the two deposition estimates can be regarded insignificant.

Up to now, several assumptions in the model are not properly evaluated. When a particular assumption is not valid, this propagates into successive calculations through which an accumulation of errors may arise. For instance, the dry deposition factor is computed using annual mean throughfall and bulk precipitation fluxes, thereby neglecting seasonal differences in pollution climate and canopy characteristics. Additional error will arise by using bulk precipitation data instead of wet-only deposition. The assumption that Mg^{2+}, Ca^{2+}, Cl⁻ and K^+ containing particles are deposited with equal efficiency as Na^+ containing particles may introduce an error as the particle size distribution of these constituents is not necessarily the same (Milford and Davidson, 1985). There will be a shift in the size distribution towards particles with smaller radii with increasing distance to source areas and/or lower relative humidity (Fitzgerald, 1975). Nevertheless, canopy leaching of Mg^{2+}, Ca^{2+} and K^+ is computed which, in turn, is used to calculate canopy uptake of H^+ and NH_4^+. The ratio of canopy uptake efficiency between H^+ and NH_4^+ used in the model is obtained from laboratory experiments with small Douglas-fir branches from the Speulder forest (Van der Maas et al., 1991). Whether these results may be extrapolated to field conditions, other tree species and other ecological circumstances still remains uncertain. The assumption that dry deposition of weak acids is equal to their wet deposition needs experimental evidence. The model assumption that for sodium and sulphur canopy exchange is negligible probably holds (Draaijers, 1993), but the assumption that canopy exchange is zero for oxidised nitrogen is not valid. Canopy foliage has been demonstrated experimentally to be capable of absorbing and incorporating gaseous NO_2 and HNO_3, as well as NO_3^- in solution (e.g. Reiners and Olson, 1984). Limiting condition for using the model is that the forest stand under consideration is not exposed to insect plagues or diseases. Such biotic stresses are found to enhance canopy leaching (Van Ek and Draaijers, 1994).

Table I

Dry and fog deposition estimates for the Speulder forest derived from micrometeorological measurements and inferential modeling for the period 23/11/92 - 10/05/93. Net throughfall estimates for this time period and net throughfall fluxes corrected for canopy exchange by the canopy budget model are presented as well. Results are presented in mol ha^{-1} a^{-1}. Scaling to one year was performed by multiplying fluxes in the measurement period with 365/169.

	SOx	NOy	NHx	Na	Cl	Ca	K	Mg
dry, gas	663	356	1443	0	0	0	0	0
dry, aerosol	216	414	645	599	885	101	34	118
fog water	34	23	96	2	4	1	1	0
total	913	793	2184	601	889	102	35	118

	SO4	NO3	NH4	Na	Cl	Ca	K	Mg
netTF	924	394	1728	692	802	159	305	138
netTF cor.	924	394	1983	692	802	86	35	98

5. Conclusions

For the Speulder forest, the combination of throughfall measurements and application of the canopy budget model resulted in deposition estimates which were more or less similar to deposition estimates derived from micrometeorological measurements and inferential modeling, deposition of NO_y being the only exception. Unfortunately, several assumptions in the model are up to now not properly evaluated, which limits its application. The model can be improved by *i)* incorporating canopy uptake of oxidised nitrogen compounds, *ii)* using wet deposition instead of bulk precipitation fluxes, *iii)* taking into account the different mass median diameters of Mg^{2+}-, Ca^{2+}- and K^+-containing particles compared to Na^+-containing particles, and *iv)* using shorter time periods (e.g. four weeks) when computing the annual mean dry deposition factor (Spranger, 1992). Additional research is recommended on the dry deposition of weak acids, and the ratio of canopy uptake efficiency between H^+ and NH_4^+ in relation to tree species, ecological setting and pollution climate.

References

Bouten, W and Bosveld, F.C.: 1992, *Stomatal control in a partially wet Douglas fir canopy*. Dutch Priority Programme on Acidification, report no. 791302-1

Draaijers, G.P.J.: 1993, *Ph.D. thesis*, University of Utrecht, The Netherlands.

Draaijers, G.P.J., Erisman, J.W., Spranger, T. and Wyers, G.P.: 1995, *Atmospheric Environment*, submitted.

Erisman, J.W., Draaijers, G.P.J., Duyzer, J. H., Hofschreuder, P., Van Leeuwen, N., Römer, F.G., Ruijgrok, W. and Wyers, G.P.: 1994, *Contribution of aerosol deposition to atmospheric deposition and soil loads onto forests.* RIVM report no. 722108005.

Fitzgerald, J.W.: 1975, *Journal of Applied Meteorology*, 14, 1044-1049.

Guiang, S.F., Krupa, S.V. and Pratt, G.C.: 1984, *Atmospheric Environment*, 18, 1677-1682.

Lindberg, S.E., Bredemeier, M, Schaefer, D.A. and Qi, L.: 1990, *Atmospheric Environment*, 24A, 2207-2220.

Lovett, G.M. and Lindberg, S.E.: 1984, *Journal of Applied Ecology*, 21, 1013-1027.

Milford, J.B. and Davidson, C.I.: 1985, *Journal for Air Pollution Control and Assessment*, 35, 1249-1260.

Puckett, L.J.: 1990, *Atmospheric Environment*, 24A, 545-555.

Reiners, W.A. and Olson, R.K.: 1984, *Oecologia*, 63, 320-330.

Römer, F.G. and Te Winkel, B.W.: 1994, *Droge depositie van aerosolen op vegetatie: verzurende componenten en basische kationen*. KEMA rapport 63591-KES/MLU 93-3243.

Ruijgrok, W., Tieben, H. and Eisinga, P.: 1994, *The dry deposition of acidifying and alkaline particles on Douglas fir.* KEMA report 20159-KES/MLU 94-3216

Slanina, J., Keuken, Arends, B., Veltkamp, A.C. and Wyers, G.P.: 1990, *Report on the contribution of ECN to the second phase of the Dutch Priority Programme on acidification*. ECN, Petten, the Netherlands.

Slinn, W.G.N.: 1982, *Atmospheric Environment*, 16, 1785-1794.

Spranger, T.: 1992, Ph.D. thesis, Christian Albrecht Universität, Kiel, Germany.

Ulrich, B.: 1983, *Interaction of forest canopies with atmospheric constituents: SO_2, alkali and earth alkali cations and chloride*. In: B. Ulrich and J. Pankrath (Eds.), Effects of accumulation of air pollutants in forest ecosystems, Reidel, Dordrecht, the Netherlands, 33-45.

Van der Maas, M.P. and Pape, Th.: 1991, *Hydrochemistry of two Douglas fir ecosystems and a heather ecosystem in the Veluwe, the Netherlands*. Dutch Priority Programme on Acidification, report no. 102.1.01.

Van der Maas, M.P., Van Breemen, N. and Van Langenvelde, I.: 1991, *Estimation of atmospheric deposition and canopy exchange in two Douglas fir stands in the Netherlands*. Internal publication, Department of soil science and geology, Agricultural University of Wageningen, The Netherlands.

Van Ek, R. and Draaijers, G.P.J.: 1994, *Water, Air and Soil Pollution*, 73, 61-82.

Van Leeuwen, N.P.M., Bleuten, W., Hansen, K.: 1994, *Deposition of acidifying and basic compounds measured at the Speulder forest by means of the throughfall method*. Department of Physical Geography, University of Utrecht, The Netherlands, report no. 941001.

Vermeulen, A.T., Wyers, G.P., Römer, F.G., Draaijers, G.P.J., Van Leeuwen, N.P.M. and Erisman, J.W.: 1995, *Atmospheric Environment*, in press.

Wyers, G.P., Veltkamp, A.C., Vermeulen, A.T., Geusebroek, M., Wayers, A. and Möls, J.J.: 1994, *Deposition of aerosol to coniferous forest*. ECN report no. ECN-C-94-051.

IN-CANOPY THROUGHFALL MEASUREMENTS IN NORWAY SPRUCE: WATER FLOW AND CONSEQUENCES FOR ION FLUXES

K. HANSEN

Danish Forest and Landscape Research Institute, Hørsholm Kongevej 11, 2970 Hørsholm, Denmark.

Abstract. Canopy throughfall was collected in funnels equipped with tipping buckets. The funnels were installed at 4 distances from the tree trunk and at 6 depths in the canopy of a Norway spruce forest at Klosterhede, Denmark. The throughfall water flux was registered during individual rain events. The smallest quantity of throughfall was sampled closest to the tree trunk, and the largest quantity of throughfall was sampled in the periphery of the canopy at all levels in the canopy. The quantity of throughfall was highest at the top of the tree and decreased down through the canopy. The intensity of the water flow decreased through the canopy which brought the throughfall water in contact with the foliage for longer periods in the lower canopy than in the upper canopy. The higher wettability caused a larger leaching of especially potassium in the lower canopy. Differences in rain intensity did not influence the distribution and the pattern of water flow in the canopy.

1. Introduction

Water flowing through a forest canopy plays a fundamental role in the transport of dissolved and suspended solids, and the water within the canopy influences the chemical, physical, and biological processes that occur on the vegetation surfaces (McCune & Boyce, 1992).

Knowledge of the hydrological characteristics of the canopy is required to understand the movement of water, nutrients, and atmospheric pollutants as well as the canopy interaction processes (leaching and uptake of nutrients and pollutants) that operate within the forest canopy.

In this study, the flow of throughfall water during individual rain events was examined on its way through different parts of the canopy. The effect of quantity and intensity of precipitation on the water flow and the consequences for leaching and uptake of mineral nutrients and pollutants in coniferous tree canopies were evaluated.

2. Materials and methods

The study was conducted in a 74 years old Norway spruce stand at Klosterhede Plantation in the western part of Jutland, Denmark. The forest site was located 15 km from the North Sea and dominated by strong westerly winds. The characteristics of the forest stand (Table I) are described in detail by Beier & Rasmussen (1993).

An 18 m high tower was raised in the forest. Horizontal girders were mounted at six levels in the tower (7, 9.5, 11, 13, 15, and 16 m above the ground) (Figure 1). The girders pointed towards the trunks of two trees (A and B).

On the horizontal girders, 31 funnels were installed in 4 distances from the trunks (0.1, 0.5, 0.9, and 1.3 m) (Figure 1). Filters were placed in the funnels to minimize contamination of the samples. Two additional funnels collected bulk precipitation.

TABLE I
Stand characteristics and climate (Beier & Rasmussen, 1993).

Tree species	Norway spruce (*Picea abies* (L.) Karst.)
Age (1993)	74 years
Trees per ha	860
Tree height	18 m
Basal area	29 m^2/ha
Leaf Area Index	4.8 m^2/m^2
Yearly precipitation	860 mm
Main wind direction	W, SW

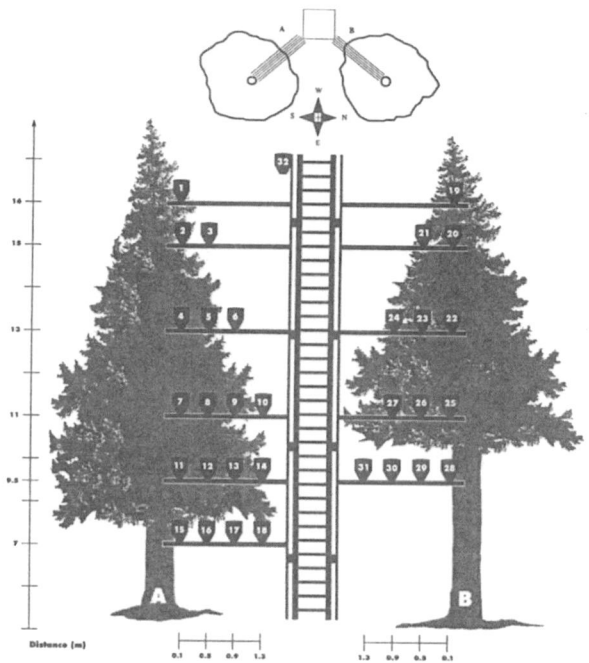

Fig. 1. The set-up of funnels at four distances from the trunks (0.1, 0.5, 0.9, and 1.3 m) and at six levels in the canopy (7, 9.5, 11, 13, 15, and 16 m above the ground) of tree A and B.

A tipping bucket (Rain-O-Matic, Pronamic A/S, Denmark) was installed below each funnel. The tipping buckets (polyethylene) had a resolution of 0.25 mm of rain per tip. The date and time for each tip of all funnels were recorded from January 1993 to January 1994. 60 individual rain events were sampled.

3. Results and discussion

3.1. POSITION IN THE CANOPY

The smallest quantity of throughfall was sampled closest to the tree trunk (0.1 m) and the largest quantity of throughfall was sampled in the periphery of the canopy (1.3 m) at all levels in the canopy (Figure 2). Such systematic, spatial variability in throughfall quantity was also observed by Beier et al. (1993) and Pedersen (1992), where throughfall was collected at the forest floor in a Norway spruce and a sitka spruce forest, respectively, corresponding to the lowest canopy levels in this study.

The quantity of throughfall was highest at the top of the tree and decreased down through the canopy (Figure 2), as water was intercepted in the canopy. In contrast, Fritsche et al. (1989) found that the quantity of water increased from the top to the bottom of a Norway spruce canopy with a maximum amount in the middle of the canopy. However, these results were based on only one sampling of throughfall (13 days).

The strong spatial variability of the throughfall quantity demonstrated an uneven distribution of the precipitation in the canopy. Reasons could be i) non-homogeneous distribution of rain drops to the canopy due to turbulent air flow just above and within the canopy, ii) horizontal translocation of water in the canopy where water is running from branch to branch, and/or iii) differences in interception related to the density of the canopy.

3.2. INTERCEPTION IN THE CANOPY

The quantity of throughfall beneath the canopy of each of the trees was approximately 50 % of the precipitation (Figure 2). The interception was therefore high compared to a study by Leyton et al. (1966), where interception was estimated in Norway spruce also by the use of throughfall measurements. The interception was probably overestimated since all funnels were placed directly beneath foliage. The forest was rather open (LAI = 4.8), and a correct estimate should include measurements beneath the open areas of the forest as well.

The results showed a large variability in the throughfall quantity according to the position of sampling in the forest. It is, therefore, important to account for the spatial variability of throughfall when interception is estimated using point measurements of throughfall. A large number of randomly placed throughfall collectors (Kostelnik et al., 1989) or specially designed integrating collectors or troughs (Beier & Rasmussen, 1989; Draaijers, 1993) are necessary for a correct representation of the throughfall.

3.3. FLOW OF WATER IN THE CANOPY

Sixty rain events, larger than four mm, were recorded in 1993. The events were analyzed to test whether the distribution and the flow of water through the canopy changed with different quantities and intensities of precipitation. Four random rain events with different quantities and intensities of precipitation were evaluated in detail: January 24th, 1993 (16 mm, 3 hr), October 9th, 1993 (17 mm, 10 hr), October 12th, 1993 (34 mm, 12 hr), and September 25-26th, 1993 (38 mm, 28 hr).

The quantity of accumulated throughfall water was generally higher in the upper than in the lower parts of the canopy at all distances from the tree trunk (Figure 3). The accumulated throughfall closely followed the pattern of accumulated rainfall despite a delay of water during the passage of the canopy (Figure 3). After the rain

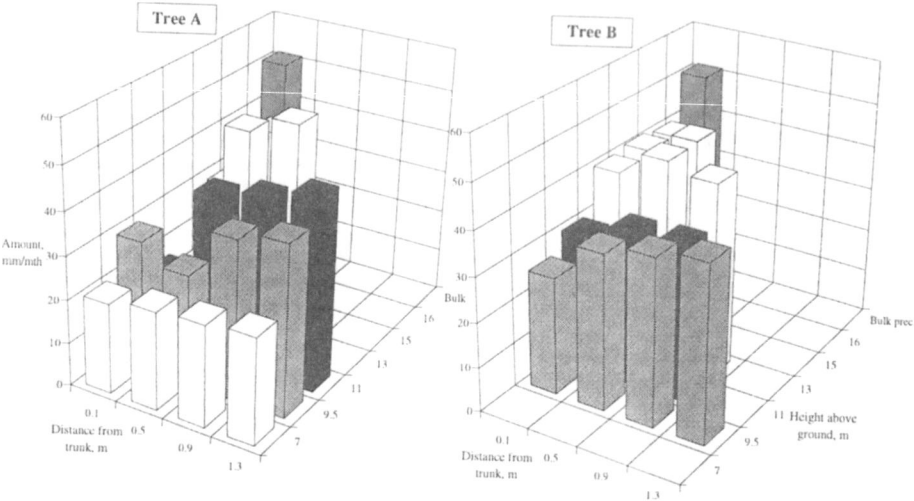

Fig. 2. Quantity of bulk precipitation and throughfall water (mm month^{-1}) as a function of the depth in the canopy (m) and the distance from the tree trunk (m) of the two trees A and B.

stopped, the last drips from the foliage were normally registered in the lower canopy.

The bulk precipitation funnel above the forest normally collected the highest quantity of water. Eventually, higher accumulated amounts of water were collected in other parts of the canopy (Figure 3). This was evident in the events of September 25-26th and October 12th where the quantity of precipitation was large. The amount of bulk precipitation was equal to or higher than the amount of throughfall until 15-20 mm of rain. Hereafter, the amount of accumulated throughfall rose and became higher than the accumulated quantity of bulk precipitation. Explanations for such unexpected high quantities of accumulated water in funnels situated lower in the canopy could be that: i) the funnels were placed in an open part of the canopy where they received both precipitation and throughfall, ii) after saturation of the canopy additional water was funnelled from other plant parts and into the collector (translocation), and/or iii) the tipping buckets tipped before actually being full which could happen in stormy weather. It is possible to view the passage of water through the canopy in three stages: an initial wetting stage, a continuing dripping stage, and a final stage when rain stops, dripping ends, and the canopy is drying. It seems possible that the canopy was totally saturated at 15-20 mm of precipitation. Hereafter, the storage of water within the canopy and the translocation of water in the canopy became as large as possible. Furthermore, throughfall to these specific funnels indicated a higher leaching of potassium, and suggested an extra dense canopy above the funnels and that throughfall water had longer contact with foliage because of translocation (Hansen, 1995). Therefore, explanation ii) seems most plausible.

In events with the same amount of precipitation, approximately the same amount of water reached the different layers in the canopy regardless of the intensity of the rain event (Figure 3). These results suggest that the distribution of water and the water flow in the canopy was independent on the rain intensity. The intensity of the water

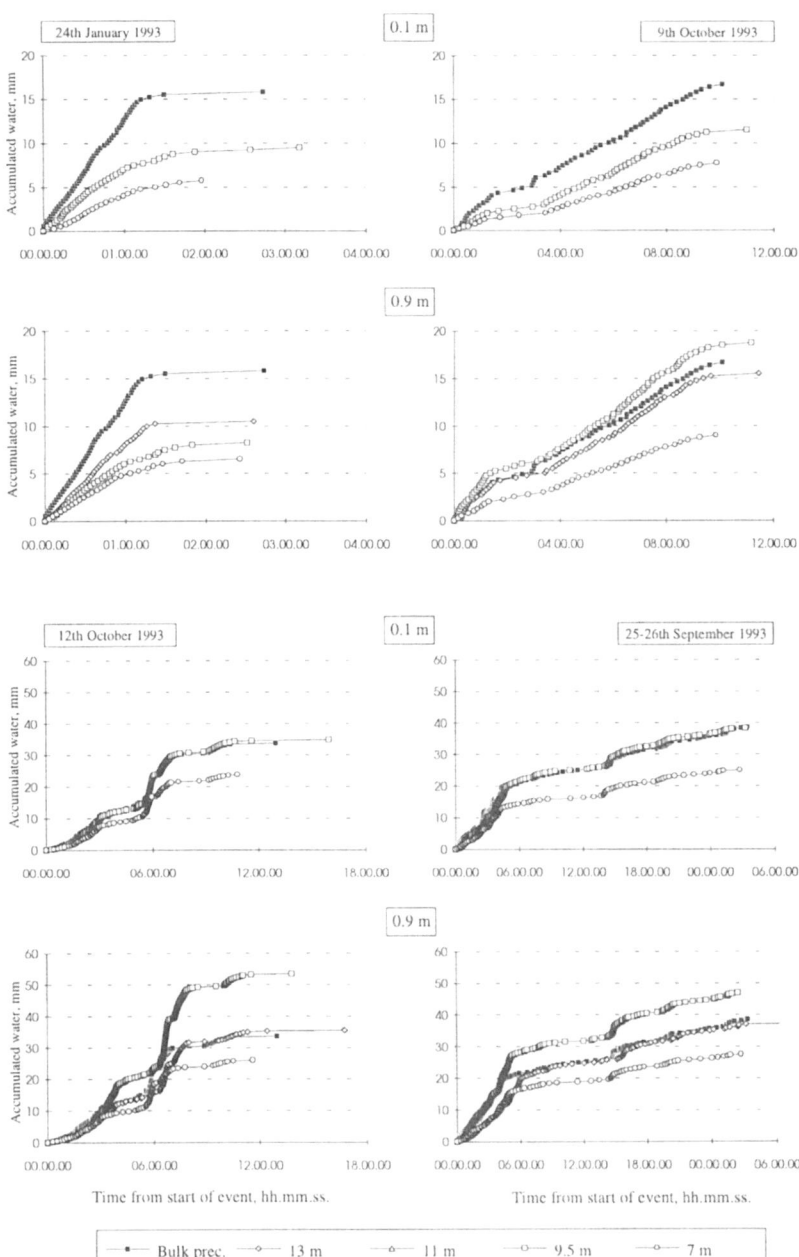

Fig. 3. Accumulated amount of water (mm) of four rain events sampled January 24th, 1993 (16 mm, 3 hr), October 9th, 1993 (17 mm, 10 hr), October 12th, 1993 (34 mm, 12 hr), and September 25-26th, 1993 (38 mm, 28 hr). The water is shown through the events for the distances 0.1 m and 0.9 m from the tree trunk and for each depth in the canopy (m).

flow generally decreased through the canopy, so that the duration of contact between water and foliage was longer in the lower parts of the canopy than in the upper parts.

3.4. CONSEQUENCES FOR CANOPY EXCHANGE PROCESSES

The smaller amount of water, the lower intensity of the water flow, and the translocation from the upper layers to the lower layers and outwards in the canopy generated a longer contact between water and foliage in the lower canopy which probably influenced the canopy exchange processes. Further, it was observed that the foliage in the bottom of a coniferous canopy was more wet (less hydrophobic) (Boyce et al., 1991). Since leaching is partially controlled by the wettability of the foliar surfaces (Tukey, 1970), a larger exchange in the lower canopy seems possible. In accordance to this, Hansen (1995) observed a higher leaching of K^+ in the lower canopy and Lovett et al. (1989) modelled a higher leaching in the lower canopy.

Acknowledgements

The author wish to thank all persons who assisted with the field work. A special thank to Preben Fiskbæk and Preben Jørgensen. Filip Moldan is acknowledged for his guidance in choosing a suitable datalogger. Thanks to Per Gundersen and Claus Beier for valuable suggestions on the manuscript. This research was financed by the Commission of the European Communities (grant ENV-892-DK), by the Danish Environmental Research Programme 1992-96, and by the Danish Forest and Landscape Research Institute.

References

Beier, C. and Rasmussen, L.: 1989, In: A.H.M.Bresser and P.Mathy (Eds.), *Monitoring Air Pollution and Forest Ecosystem Research*, Air Pollution Report Series **21**, 101-110.
Beier, C., Hansen, K. and Gundersen, P.: 1993, *Environmental Pollution*. **81**, 257-267.
Boyce, R.L., McCune, D.C. and Berlyn, G.P.: 1991, *New Phytologist*, **117**, 543-555.
Draaijers, G.P.J.: 1993, *Ph. D. Thesis*, University of Utrecht, the Netherlands, 199 pp.
Fritsche, U., Gernert, M. and Schindler, C.: 1989, *Atmospheric Environment*, **23**, 1807-1814.
Hansen, K.: 1995, *Atmospheric Environment*, (submitted).
Kostelnik, K.M., Lynch, J.A., Grimm, J.W. and Corbett, E.S.: 1989, *Journal of Environmental Quality*, **18**, 274-280.
Leyton, L., Reynolds, E.R.C. and Thomsen, F.P.: 1966, In: W.E.Sopper & H.W.Lull, *Forest Hydrology*, 163-177.
Lovett, G.M., Reiners, W.A. and Olson, R.K.: 1989, *Biogeochemistry*, **8**, 239-264.
McCune, D.C. and Boyce, R.L.: 1992, *TREE*, **7**, No. 1, 4-7.
Pedersen, L.B.: 1992, *Scandinavian Journal of Forest Research*, **7**, 433-444.
Tukey, Jr. H.B.: 1970, In: Machlis, L., Briggs, W.R. and Park, R.B. (Eds.), *Annual review of plant physiology*, **21**, 305-324.

ELEVATIONAL TRENDS IN THE FLUXES OF SULPHUR AND NITROGEN IN THROUGHFALL IN THE SOUTHERN APPALACHIAN MOUNTAINS: SOME SURPRISING RESULTS

J. SHUBZDA[1], S.E. LINDBERG[2], C.T. GARTEN[2], AND S.C. NODVIN[3]

[1]*Department of Ecology and Evolutionary Biology, University of Tennessee, Knoxville, Tennessee 37996-1610 USA.* [2]*Environmental Sciences Division, Oak Ridge National Laboratory, PO Box 2008, Oak Ridge, Tennessee 37763-6038 USA.* [3]*National Biological Service and Graduate Faculty in Ecology and Evolutionary Biology, The University of Tennessee, Knoxville, Tennessee 37996-1610 USA.*

Abstract. From 1986-1989, a team of scientists measured atmospheric concentrations and fluxes in precipitation and throughfall, and modeled dry and cloudwater deposition in a spruce-fir forest of the Great Smoky Mountains National Park which is located in the Southern Appalachian Region of the United States. The work was part of the Integrated Forest Study (IFS) conducted at 12 forests in N. America and Europe. The spruce-fir forest at 1740 m consistently received the highest total deposition rates (~2200, 1200, and 700 eq ha^{-1} yr^{-1} for SO_4^{2-}, NO_3^-, and NH_4^+). During the summers of 1989 and 1990 we used multiple samplers to measure hydrologic, SO_4^{2-}, and NO_3^- fluxes in rain and throughfall events beneath spruce forests above (1940 m) and below (1720 m) cloud base. Throughfall was used to estimate total deposition using relationships determined during the IFS. Although the SO_4^{2-} fluxes increased with elevation by a factor of ~2 due to higher cloudwater interception at 1940 m, the NO_3^- fluxes decreased with elevation by ~30%. To investigate further, we began year round measurements of fluxes of all major ions in throughfall below spruce-fir forests at 1740 m and at 1920 m in 1993-1994. The fluxes of most ions showed a 10-50% increase with elevation due to the ~70 cm yr^{-1} cloudwater input at 1920 m. However, total inorganic nitrogen exhibited a 40% lower flux in throughfall at 1920 m than at 1740 m suggesting either higher dry deposition to trees at 1740 m or much higher canopy uptake of nitrogen by trees at 1920 m. Differential canopy absorption of N by trees at different elevations would have significant consequences for the use of throughfall N fluxes to estimate deposition. We used artificial trees to understand the foliar interactions of N.

Key Words: atmospheric deposition, high elevation forests, foliar uptake, cloudwater, nitrogen, sulphur

1. Introduction

The Integrated Forest Study (IFS) developed uniform protocols for sampling and analyzing acid deposition and nutrient cycling in over a dozen sites located in the United States, Canada, and Norway. The study showed that high-elevation southern Appalachians forest stands received some of the highest of sulphate and nitrate loadings relative to all of the other IFS study sites(Johnson and Lindberg, 1992). The Smoky Mountain Tower site (ST) located at 1740 m in the Noland Divide Watershed (NDW) received an average of 2200, 1200, and 700 eq ha^{-1} yr^{-1} for SO_4^{2-}, NO_3^-, and NH_4^+, respectively, in total deposition (Johnson and Lindberg, 1992)(from 4/86 through 3/89).

Elevation plays a critical role in the amount of acid deposition measured in throughfall (TF) within the NDW (Lindberg and Owens, 1993). Lindberg and Owens (1993) found that an increase from ~1720 to ~1940m in elevation corresponded to a two-fold increase in water and SO_4^{2-} fluxes measured in TF. This increase was attributed to the enhancement of cloudwater interception at the upper site. During the IFS, cloud base was observed to be at or above ~1800m within the watershed (Johnson and Lindberg, 1992). Because cloudwater can exert a great influence on total deposition fluxes over small elevational gradients(Lovett *et al.*, 1982), sampling site location and elevation can greatly affect depositional flux estimates even within small catchments (Lindberg and Owens, 1993).

Here we report on the influence that elevation has on TF fluxes within the spruce-fir communities in the southern Appalachian Mountains. Using relationships formulated during the IFS, we estimate the annual cloud water input and compute annual net canopy exchange (NCE) of nitrogen (N) at two elevations and assess the importance of foliar uptake on nitrogen flux through the canopy.

2. Sites and Methods

Precipitation and TF sampling stations were located in the 17.4 ha. Noland Divide Watershed (NDW) (35°34'N, 83°28'W) in the Great Smoky Mountains National Park, North Carolina. The 1740-m elevation station (the lower site), was originally established in 1986 as part of the IFS to monitor ion fluxes in deposition. The upper site at 1920-m was established in the summer of 1993 to monitor the influence of cloud deposition to the upper portions of the catchment.

The overstory vegetation of the area is dominated by red spruce (*Picea rubens* Sarg) and interspersed patches of standing dead Fraser-fir (*Abies fraseri* (Pursh) Poir.) which has been devastated by the infestation of the balsam woolly adelgid (*Adelges piceae* Ratz.). A complete description of the sites can be found in Johnson and Lindberg (1992), Johnson *et al.* (1991), and Nodvin *et al.* (1995).

Precipitation and TF samples were collected from 3-Aug-93 to 3-Aug-94. Precipitation chemistry was sampled using an Aerochemetrics automatic wet-only collector located in natural gaps adjacent to canopy covered TF plots. During the freeze-free season (1-May until 31-Oct), TF was collected in 1 liter polyethylene bottles with polyethylene funnels of 3.5 cm diameter. Eight TF collectors were randomly placed beneath the canopy of mature red spruce and volume composited to account for the variability in TF. From November 1 to April 31, TF was collected at each site in 4 large-diameter plastic lined buckets located on platforms ~1 m from the ground and volume composited. Precipitation and TF volumes were measured using wedge-type rain gauges located adjacent to each TF collector in the freeze-free season or by weighing the collection buckets during the winter season. Sampling methodologies followed IFS protocols as given in Lindberg *et al.*, (1989).

Samples were collected twice weekly during the freeze-free period and were volume-composited for weekly totals. During winter, samples were collected weekly. The samples were analyzed for pH and conductivity immediately upon return from the field (within 24 hours), preserved with chloroform (10 μl per 30 ml sample), and stored at 4°C until analysis for major ions (using ion chromatography).

In addition, from June 23 to October 25, 1993, samples were collected beneath artificial trees located in adjacent forest gaps at 1700m and 1900m within the watershed. The use of inert artificial surfaces such as 'Christmas trees' can give a good estimate of total N and S deposition (Joslin *et al.*, 1990).

3. Results and Discussion

3.1 Water Flux

During the collection period, the amount of wet precipitation entering both sites was similar (Table I). The upper site received a slightly lower volume of precipitation which we attribute to a lower catch efficiency of the rain sampler during snow periods due to it's ridge-top location. Both sites received a significantly higher amount of rain compared to the three year average collected during the IFS (Table I). However, the average weighted mean concentrations at the lower site for all the major

Table I. Throughfall (TF) and precipitation deposition of SO_4^{2-}, NO_3^-, and NH_4^+, to the Noland Divide Watershed expressed as eq ha^{-1} yr^{-1}. The Lower Site and the IFS Site are the same.

	IFS Site (4/86-3/89)	Lower Site (8/93-7/94)	Upper Site (8/93-7/94)
Precipitation (cm)	203	298	281
SO_4^{2-}	596	740	770
Throughfall (cm)	215	288	342
SO_4^{2-}	2470	2480	3500
NO_3^-	866	1230	860
NH_4^+	220	410	310
Net Throughfall for SO_4^{2-}	1870	1740	2730

Net Throughfall = Throughfall - precipitation ; IFS Site and Lower Site are at the same location

ions were very similar for the two periods (data not shown). There was a much greater volume of TF at the upper site versus the lower sampling site, suggesting increased cloudwater input with elevation (Table I). Lindberg and Owens (1993) also reported similar hydrologic trends with increasing elevation at NDW during the summer of 1989.

Positive hydrologic fluxes in net-throughfall (NTF) indicate the presence of measurable amounts of cloudwater input (Lovett et al., 1982) (NTF =the flux in TF minus precipitation). By using all the positive NTF values, we estimate that the lower and upper sites received ~20 and 90 cm yr^{-1} of cloudwater deposition, respectively (Table II). Another way to calculate cloudwater input is to use "conservative" tracers of convenience such as Cl$^-$, SO_4^{2-}, and Na$^+$ in TF (Lovett et al., 1982). By combining the results of all methods, we estimate average cloudwater inputs to the lower and upper site of ~40 and ~70 cm respectively (Table II).

Table II. Cloud water volume estimates (cm) at the Noland Divide Watershed during 8/93-7/94

	Lower Site	Upper Site
NTF*	19	89
Cl^{-**}	43	62
SO_4^{2-**}	47	87
Na^{+**}	43	53
Average	**40 cm**	**70 cm**

* This estimate is based on positive net throughfall adjusted for evaporation and interception losses (Lindberg and Owens, 1993). **The estimates for Cl$^-$, SO_4^{2-}, and Na$^+$ are based on relationships formulated during the IFS.

3.2 Sulphate Deposition

For the year, there was no significant difference between the upper and lower sites in the amount of sulphate entering the watershed as wet deposition ($p<0.05$, $n=48$) (Table I). However, the annual sulphate flux measured in TF was ~40% greater at the upper site and was significantly greater across the entire sampling period ($p<0.05$, $n=48$) (Figure 1). Since TF represents a direct estimate of the total atmospheric deposition of sulphate entering a watershed and foliar leaching and/or uptake is minimal (Garten et al., 1988, Lindberg and Garten, 1988), the difference in fluxes between the two sites can be attributed to an increase with elevation of cloud and/or dry deposition. The IFS reported that cloud water represented 45-50% of the total sulphate loading

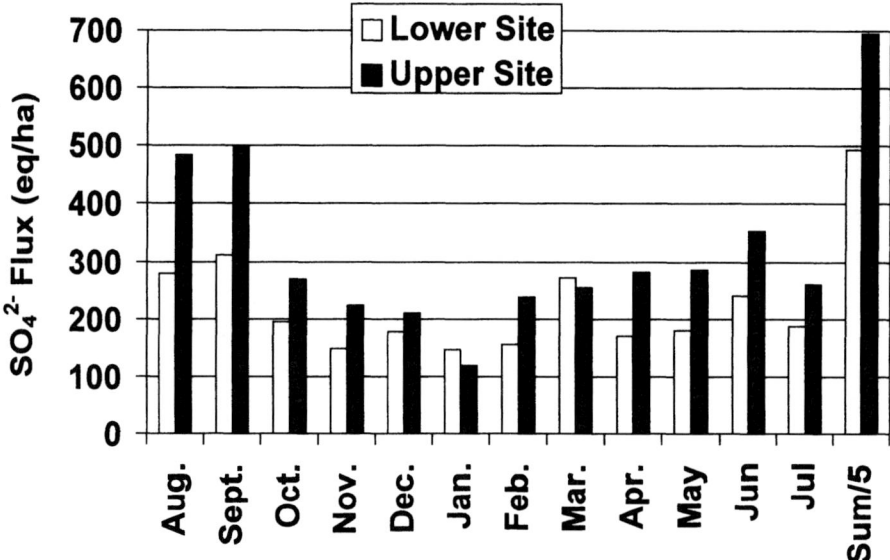

Fig. 1. Monthly total sulphate fluxes based on spatially and temporally composited weekly samples entering the NDW as throughfall. The collections are from 8/93-7/94.

(Lindberg and Lovett, 1992). We estimated cloudwater inputs of sulphate to the lower and upper sites based on the cloudwater inputs in Table II and cloudwater concentrations measured during the IFS (Johnson and Lindberg, 1992). Our cloudwater estimates are 950 and 1800 eq ha^{-1} yr^{-1} and represented ~40 and ~50% of the total deposition in TF to the lower and upper sites, respectively.

Nitrogen Deposition

Unlike sulphate, the fluxes of NO_3^- and NH_4^+ entering as TF were significantly lower ($p<0.05$, $n=48$) at the upper site (Figure 2). Lindberg and Owens (1993) noted a similar trend for NO_3^- in TF within this watershed. They found that NO_3^- fluxes at their lower site (at 1720-m located ~80-m from our lower site) were 30% higher than their upper site (at 1940-m located ~200-m from our upper site). This trend is unusual because one would expect the enhancement of cloudwater at the upper site to also increase the total deposition of N. Since cloudwater contains elevated concentrations of NO_3^- and NH_4^+ (Lovett et al., 1982) and we have determined that the upper site receives 75% more cloud deposition than the lower site, why do we not see higher N fluxes reflected in TF? There are two possible reasons for this "inverse" elevational difference: (1) dry deposition of N is actually much higher at the lower site, and/or (2) foliar N uptake is much greater at the upper site. Uptake may be estimated from net canopy exchange, defined as "NCE= TF -total deposition" (Johnson and Lindberg, 1992). Although we did not directly measure N dry deposition during the 1993-94 sampling period, we do not expect dry deposition to have been greater at the lower site based on the estimated total sulphate fluxes discussed above and the historical IFS dry deposition data. If anything, we would expect somewhat greater dry deposition to trees at the upper site due to higher wind speeds associated with the ridgetop location and the more open canopy at 1920 m (Lovett and Kinsman, 1990). The lower and upper site are estimated to have approximately a 70 and 50% canopy closure, respectively (Shubzda, unpublished data). The more canopy openings at the upper site would increase the "edge effect" which is known to enhance both cloud and dry deposition to individual trees (Lindberg and Owens, 1993).

An independent estimate of total inorganic N deposition to these two sites supports the expected trend of increasing N deposition with increasing elevation. Total inorganic N deposition collected under artificial 'Christmas trees' located in canopy gaps near ground level within the NDW was ~2 times greater at 1900m than 1700m.

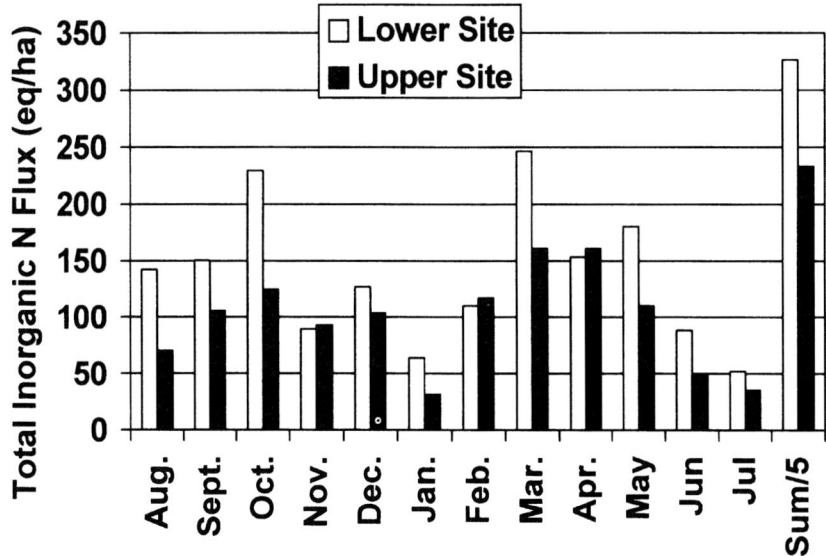

Fig. 2. Monthly total inorganic N fluxes in throughfall entering the NDW.

The NH_4^+ flux under the artificial trees showed a large increase (by an order of magnitude) with elevation, while the NO_3^- flux was actually slightly higher at the lower site. We believe that the greater in NH_4^+ flux with elevation is attributable to enhancement of cloud deposition at the upper site but that the trend in NO_3^- can only be explained by somewhat higher dry deposition at the lower site (cf. Lindberg & Owens, 1993). Nevertheless, the artificial tree results support our belief that total N deposition is greater at the upper versus the lower site.

For comparison with the TF flux, we estimated the total inorangic N deposition to the lower and upper sites to be 1950 and 2550 eq ha^{-1} yr^{-1}, respectively. These estimates are based upon: (1) measured wet deposition rates, (2) the assumption that the IFS 3-year mean dry deposition rates are reasonable approximations for both sites, and (3) calculated N flux in cloud deposition. Compared to the N flux in TF, these estimates of total N fluxes suggest a canopy uptake rate at the upper site that exceeds that at the lower site by a factor of four (i.e. NCE of -310 and -1380 eq ha^{-1} yr^{-1} for the lower and upper sites, respectively). Our estimates are based in part on the assumption that both dry deposition and cloudwater concentrations of N measured during the IFS from 1986-1989 can be applied to these same sites during 1993-1994. Our assumption is supported by the similar SO_4^{2-} fluxes measured in TF at the lower site measured during both periods (Table1).

Conclusions

In general, total deposition increases with increasing elevation (Lovett and Kinsman, 1990). This trend was expected because of the increase in cloudwater input entering the upper reaches of the NDW. The hydrologic and sulphate fluxes in TF support the enhancement of cloudwater input to the upper site. However, the total inorganic N flux in TF was lower at the upper site. Our analysis suggests that the observed difference in TF N flux was due to greater canopy uptake at 1920 m relative to 1740 m elevation in the Noland Divide Watershed. The reasons for the higher apparent uptake are unclear but site specific differences in canopy uptake of N in would confound the use of TF to estimate total N deposition in montane forests.

Acknowledgments

This is a contribution of the Great Smoky Mountains National Park Long-Term Inventory and Monitoring Program and was funded by the USDI National Park Service and National Biological Service. Special thanks to the staffs of the UT CPSU and the Great Smoky Mountains NP and especially Mark Mitch and Ellen Williams.

References

Garten, C.T., Bondietti, E.A., Lomax, R.D.: 1988, *Atmospheric Environment*, **22**,1425-1432.

Johnson, D.W., and Lindberg, S.E. (Eds): 1992, "Atmospheric deposition and nutrient cycling in forest ecosystems", Springer-Verlag, New York.

Johnson, D.W., Van Miegroet, H., Lindberg, S.E., Todd, D.E., and Harrison, R.B.:1991, *Canadian Journal of Forest Resreach*, **21**, 769-787.

Joslin, J.D., Mueller, S.F., and MH Wolfe.: 1990, *Atmospheric Environment*, **24A**, 3007-3019.

Lindberg, S.E., Johnson, D.W., Lovett, G.M., Taylor, G.E., Van Miegroet, H., and Owens, J.G.: 1989, *ORNL/TM*, **11214**, Oak Ridge National Laboratory, Oak Ridge, TN.

Lindberg, S.E., and Garten, C.T.: 1988, *Nature*, **336**, 148-151.

Lindberg, S.E., and Lovett, G.M.: 1992, *Atmospheric Environment*, **26A**, 1477-1492.

Lindberg, S.E., and Owens, J.G.: 1993, *Biogeochemistry*, **10**, 175-194.

Lovett, G.M., and Kinsman, J.D.: 1990, *Atmospheric Environment*, **24A**, 2767-2786.

Lovett, G.M., Reiners, W.A., and Olson, R.K.: 1982, *Science*, **218**, 1303-1304.

Nodvin, S.C., Van Miegroet, H., S.E. Lindberg, N.S. Nicholas, and D.W. Johnson.: 1995, *Water, Air and Soil Pollution*, this volume.

DEPOSITION OF ACIDIFYING COMPOUNDS IN SWEDEN

E. Hallgren Larsson, Knulst, J., Malm, G. and Westling, O

Swedish Environmental Research Institute, Aneboda Research Station, S-360 30 Lammhult, Sweden

Abstract. The Swedish Environmental Research Institute (IVL) has monitored deposition of acidifying compounds in Sweden. The monitoring programmes were initiated by various air quality protection associations, and regional forest and environmental authorities. The purpose is to quantify sulphur (S) and nitrogen (N) deposition to forests, and to illustrate possible acidification of the soil. Actual deposition of S and N is compared with critical loads.

Deposition is investigated by precipitation studies in open field areas and by throughfall studies in forest stands. Soilwater chemistry is examined in the forest stands and used as indication of soil conditions. For most of the study sites, data on needle loss, forest growth, and soil chemistry are available from the National Board of Forestry. All available data are combined in a computer database for evaluation.

Evaluation of data during 1985-94 shows that regional deposition monitoring illustrates the size and distributional pattern of S and seasalts. Monitoring data can identify certain regions receiving heavy loads of N, which can be found mainly in southern Sweden. Soilwater analyses indicates that large areas in Sweden have heavily acidified forest soils, low pH-values, low levels of calcium (Ca^{2+}), magnesium (Mg^{2+}), and potassium (K^+), and raised levels of inorganic aluminium (iAl). Forest sites in the coastal regions of southern and southwestern Sweden also showed raised inorganic N levels in soilwater. The relationship between deposition load and effects on soil chemistry is recognised by a correlation between S deposition and iAl levels in soilwater. Another correlation was found between N deposition (throughfall) and N levels in soilwater.

1. Introduction

The wet and dry deposition of sulphur (S) and nitrogen (N) to forest ecosystems in southern Sweden is presently greater than most ecosystems tolerate. Future reduction in forest growth by acidification of forest soils is a severe threat, especially in southern Sweden. Critical loads for S and N have been defined for different kinds of ecosystems to avoid harmful effects on the structure and functioning of the ecosystem. The critical load for S is mainly dependent on the weathering capacity of the soil, but also on the vegetation type, topography, and amount of base cations (BC) that is deposited (Nilsson & Grennfelt, 1988). Most common forest soils in southern Sweden originate from granites and gneiss, with low weathering capacity.

Large variations in deposition and in sensitivity of the ecosystems exist, even on a local scale. To make correct judgements of exceedances, the critical load for forest soils should not be compared with deposition integrated over a heterogeneous landscape but, instead, compared with the actual deposition to forests within an area. For S, throughfall monitoring has proved to be useful for determination of the total deposition, including fog deposition (e.g. Ivens *et al.*, 1990). The difference between throughfall measurements and bulk deposition from open field areas represent the amount of dry deposition, uptake in canopy and internal circulation within the tree. In areas with low S deposition internal circulation of S in vegetation might influence results from throughfall measurements significantly (Ivens *et al.*, 1990). This might be the case in remote parts of northern Sweden.

For N, canopy interaction is important. In large parts of Scandinavia, there is less N in throughfall than in precipitation, indicating canopy retention. Consequently, throughfall data will underestimate the atmospheric deposition of N compounds in this region.

Deposition of 2.5-3 kg S ha^{-1} and 3-5 kg N ha^{-1} has been stated by the Swedish Environmental Protection Agency (SNV) as annual target loads in Sweden.

In order to quantify deposition of S and N to forests in Sweden, and to illustrate possible acidification of soil and soilwater, a monitoring network was started by regional forest authorities in cooperation with regional environmental authorities and local industries. The network includes monitoring of bulk deposition, throughfall, and soilwater in 130 background locations (mainly spruce stands) spread throughout Sweden. Annual reports on the results (e.g. Hallgren Larsson *et al.*, 1995a, b, c) are available which can be used for abatement strategies. This article deals with some findings from monitoring 1985-94.

2. Materials and Methods

2.1 DEPOSITION

The open field collectors (bulk precipitation) consist of a funnel combined with a collector, and are placed on a pole in open areas in the vicinity of the forest collection sites. At the bottom of the funnel, there is a smaller funnel with a plastic netting (maze 2 mm) attached on top, in order to prevent contamination of the collected sample. During winter a snowsack is applied. A snowsack is a tubular plastic bag, mounted on PVC plastic rings, hanging in a holder on a pole (Lövblad & Westling, 1989).

Throughfall monitoring is performed by placing ten collectors with funnels on poles, at random along the diagonals of a square (30m·30m) within each forest stand. Between the funnel and the collector a piece of nylon netting is attached to prevent contamination of the collected sample by insects and forest litter. During winter the collectors and funnels are replaced by buckets for collection of snow samples. The ten sub-samples are combined and analysed as one sample.

All collectors are covered by aluminium foil in order to minimise effects of heat and sunlight on the chemical composition of the sample. Samples are collected once a month. The collected volume is registered and the samples are analysed for pH, conductivity, alkalinity, SO_4-S, Cl^-, NO_3-N, NH_4-N, and, in some cases, Na^+, K^+, Ca^{2+}, Mg^{2+}, and Mn^{2+}. Results are registered in a database. Calculated values in kg per hectare and month (kg·ha^{-1}·mo^{-1}) can be summarised for different periods. Data on stands are available, since most locations are permanent forest plots, established for forest observations.

2.2 SOILWATER

Soilwater from 0.5 m depth in mineral soil is sampled by 5 suction lysimeters with ceramic cups (P 80) three times per year (before, during and after vegetation season) at each forest stand (Beier et al., 1989). The five sub-samples are combined into one and analysed for pH, SO_4-S, Cl^-, NO_3-N, NH_4-N, Ca^{2+}, Mg^{2+}, Na^+, K^+, Mn^{2+}, Fe^{2+}, and total Al (TAl) which is fractionated into organic (oAl) and inorganic Al (iAl).

3. Results and Discussion

3.1 DEPOSITION OF S AND N

Results from some locations are shown in Figure 1. The regional pattern for S presents highest loads in southern Sweden where a gradient from Southwest towards Northeast has been found. In northern Sweden an opposite gradient, with higher loads in coastal (eastern) areas than further inland has been found. The S load of most investigated areas in southern Sweden exceeded target load several (4-7) times. Throughfall of S was higher in older (>50 yrs) spruce vegetation than in deciduous (birch and beech) and pine stands. This is probably due to varying canopy efficiency in collecting particles and gases from the atmosphere (van Ek & Draaijers, 1991). Deposition of S in open field areas was more uniform than throughfall. Bulk deposition of S was strongly related to precipitation amounts, but 2-3 times higher concentrations were found in precipitation from southern Sweden than in the northern parts.

N deposition contains about equal parts of ammonium and nitrate. Influence of local sources is more pronounced for ammonium than for nitrate, and somewhat higher values for ammonium were found in southern Sweden with more agricultural land use. In areas with low to medium loads of N (1-5 kg N·ha^{-1}·yr^{-1} as bulk deposition in open field areas) throughfall is often considerably lower than deposition in open field areas, due to canopy uptake. In areas with a heavy load of N relatively smaller portions are absorbed in the canopy and large quantities (>15 kg N·ha^{-1}·yr^{-1}) are deposited to the ground. Throughfall N in combination with bulk precipitation measurements, and studies of soilwater concentrations indicate the flux of N in the forest plot.

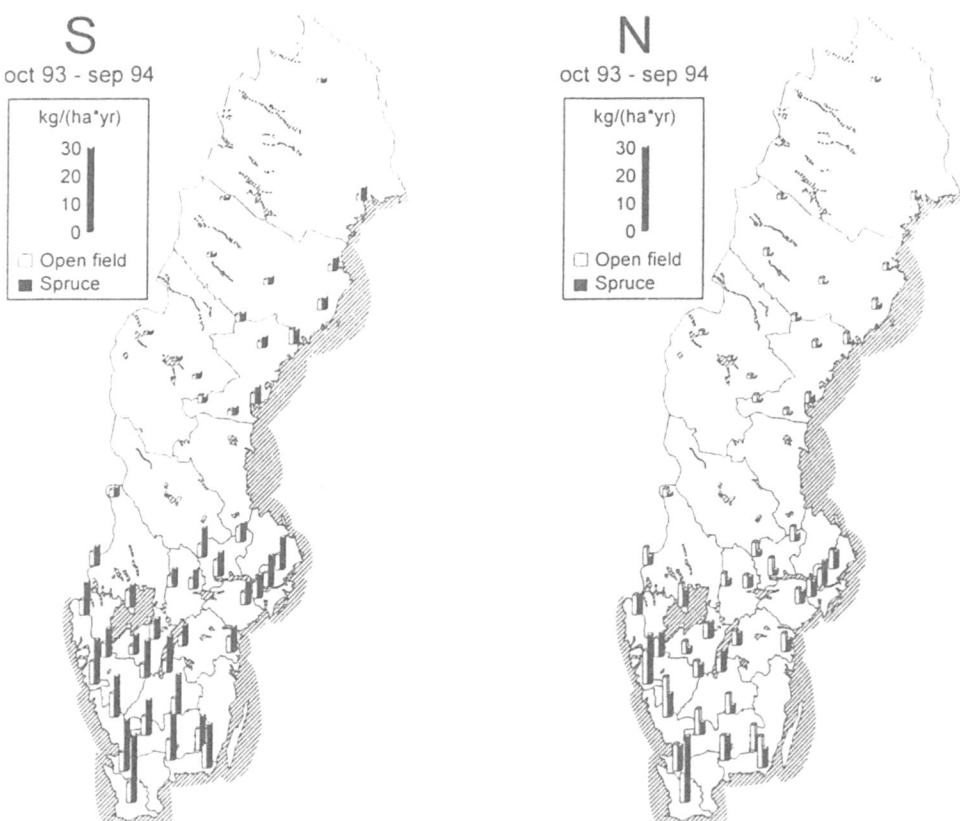

Fig. 1. Annual deposition of non-marine S and N (NH$_4$-N+NO$_3$-N) in open field and spruce forests illustrating the gradient with considerably higher loads, and greater dry deposition in southern Sweden than further north. Throughfall N varies more than S, due to N being a desirable component for vegetation.

Annually deposited amounts of S in open field and throughfall from spruce stands illustrate a decreasing trend in southern Sweden since monitoring started. This trend was broken in 1993/94 in southern and eastern parts of Sweden, as illustrated by data from Blekinge county, but not on the west coast, as illustrated by data from the county of Halland (Figure 2). Increased deposition of S in 1993/94 is probably mainly due to meteorological conditions.

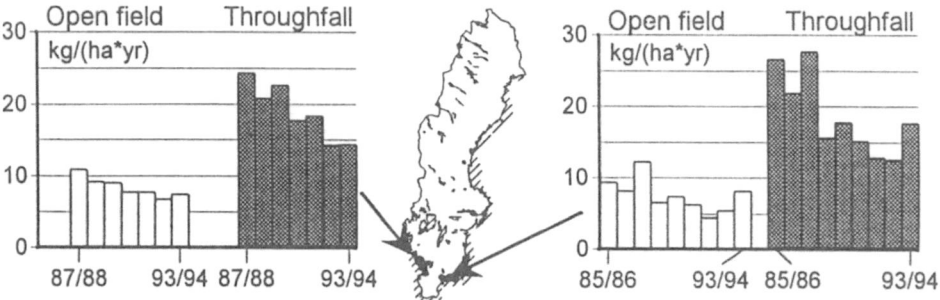

Fig. 2. Temporal changes of S deposition. Each bar represents an annual mean value for 6 open field areas and 4 spruce stands in each county.

3.2 EFFECTS OF S AND N DEPOSITION ON SOIL WATER

When acidifying compounds are deposited to the ground, exchangeable base cations (BC) are released from the soils, preventing acidification of the soilwater. This leads to soil acidification due to the loss of BC. For example, many forest areas on the west coast have acidified soils due to decades with high loads of S and N. A positive relationship between deposition of non-marine S and concentrations of inorganic aluminium (iAl) in soilwater has been found, which is most pronounced for the west coast. Another positive relationship has been found between throughfall N and amount of nitrate in soilwater. Figure 3 shows a threshold value of about 15 kg $N \cdot ha^{-1} \cdot yr^{-1}$ in throughfall, above which elevated concentrations of nitrate occurred in soilwater. This indicates leakage of nitrate from these forest areas to surrounding surface- and ground water.

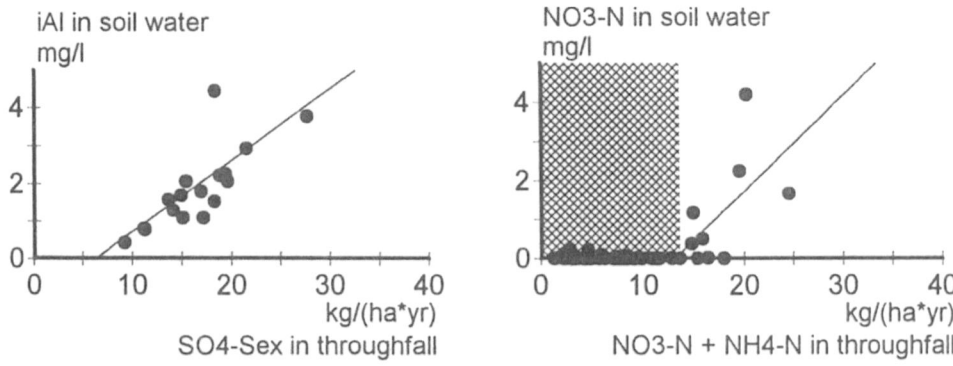

Fig. 3. Relationships between deposition and soilwater concentrations.

When low amounts of exchangeable cations in the soil is combined with high deposition of acidifying compounds, Al is released in order to prevent decrease of pH in the soil water. Al is known to have negative effects on plant growth. Inorganic forms are generally more toxic than organic compounds and the toxic effects are ameliorated by the presence of BC. The ratio between BC and iAl may therefore be used as indication of acidification status of the soil and risk for harmful effects on the ecosystem. Figure 4 shows many forest stands with low ratios (<1) between BC and iAl in south-western Sweden.

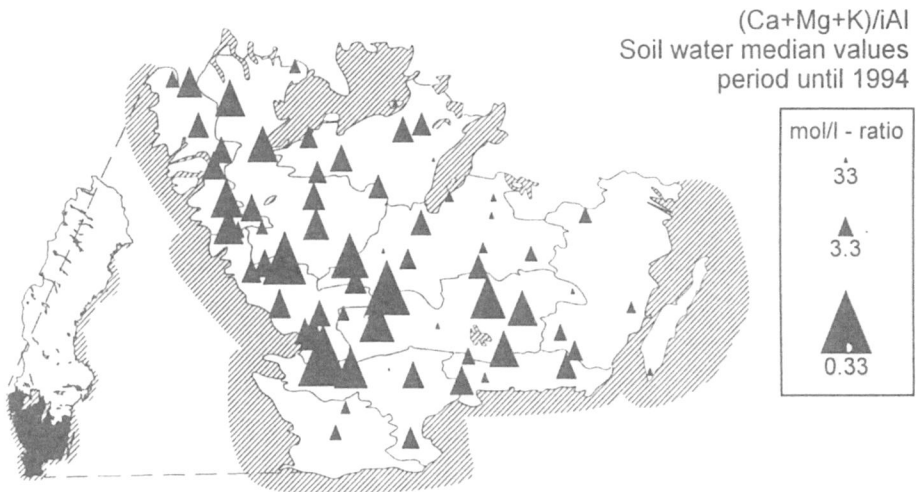

Fig. 4. Ratios between BC and iAl in soilwater, illustrating many forest stands in southwestern Sweden with low ratios (<1). A large triangle means a low ratio and great ecological risk.

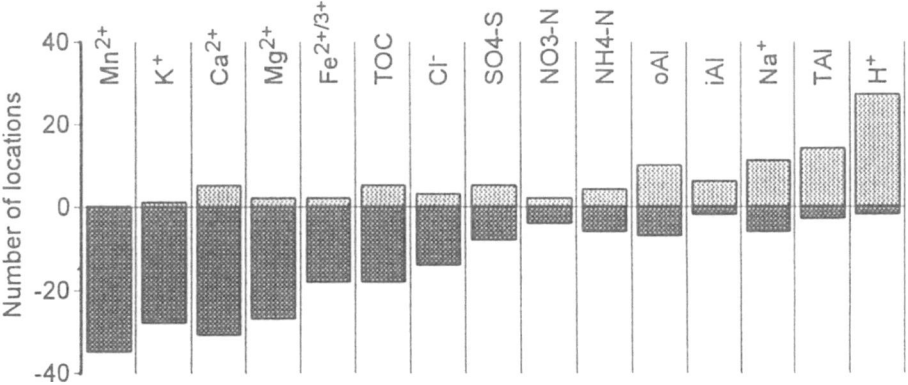

Fig. 5. Linear regression analyses of soilwater concentrations from 120 locations, indicating decreased concentrations of Mn^{2+} and BC at ~30 locations and increased concentrations of H^+.

Linear regression analyses illustrates the number of locations with increased or decreased concentrations in soilwater since the start of the survey with statistical significance at 95% confidence level. Most clear are decreased concentrations of Mn^{2+}, Ca^{2+}, Mg^{2+}, and K^+, and increased concentrations of H^+ (Figure 5).

4. Conclusions
- This monitoring illustrates the amounts and distribution of received S and sea salts
- Monitoring data can identify those regions receiving heavy loads of N (found especially in southern Sweden)
- Soil water analyses indicate that large areas in Sweden have heavily acidified forest soils, with low pH-values, low levels of Ca^{2+}, Mg^{2+}, K^+, and raised levels of iAl
- Some forest sites in coastal regions of southern and south-western Sweden have raised levels of nitrate in the soil water
- Positive relationships have been found between S deposition and iAl in soilwater and between throughfall N and nitrate in soilwater
- Soilwater concentrations of BC have decreased while H^+ have increased

Acknowledgements
These investigations were financed with grants from various regional air quality protection associations, county administrations, and regional forestry boards. Many field and laboratory personnel have in a devoted way collected and analysed the samples.

References
Beier, C., Butts, M., von Freiesleben, N. E., Høgh Jensen, K., Rasmussen, L., in Nihlgård B., Gyllin M., (Eds.): 1989. Methods for Integrated Monitoring in the Nordic Countries. The Working Group for Environmental Monitoring, Nordic Council of Ministers, Environmental Report 1989:11, Lund.

Ivens, W., Kauppi, P., Lövblad, G., Westling, O.: 1990. Throughfall Monitoring as a Mean of Deposition Monitoring: Evaluation of European Data. - Nordic Council of Ministers and UN-ECE 1990: 16. Copenhagen.

Hallgren Larsson, E., Sjöberg, K., Westling, O.: 1995a. Air Pollution in northern Sweden -Deposition, concentrations and effects. October 1993-September 1994. IVL Publication B 1185. (60 pages, In Swedish.)

Hallgren Larsson, E., Sjöberg, K., Westling, O.: 1995b. Air Pollution in central Sweden - Deposition, concentrations and effects. October 1993-September 1994. IVL Publication B 1186. (69 pages, In Swedish.)

Hallgren Larsson, E., Sjöberg, K., Westling, O.: 1995c. Air Pollution in southern Sweden -Deposition, concentrations and effects. October 1993-September 1994. IVL Publication B 1192. (74 pages, In Swedish.)

Nilsson, J., Grennfelt, P., (Eds.): 1988. Critical Loads for Sulphur and Nitrogen. Nordic Council of Ministers and UN-ECE 1988: 15, Stockholm.

Lövblad, G., Westling, O., in Nihlgård B., Gyllin M., (Eds.): 1989. Methods for Integrated Monitoring in the Nordic Countries. The Working Group for Environmental Monitoring, Nordic Council of Ministers, Environmental Report 1989:11, Lund.

van Ek, R., and Draaijers, G. P. J. 1991. Atmospheric Deposition in Relation to Forest stand Structure. Dep of Physical Geography. University of Utrecht.

ENERGY USE, EMISSIONS, AND AIR POLLUTION REDUCTION STRATEGIES IN ASIA

W. Foell[1], C. Green[1], M. Amann[2], S. Bhattacharya[3], G. Carmichael[4], M. Chadwick[5], S. Cinderby[5], T. Haugland[6], J.-P. Hettelingh[7], L. Hordijk[8], J. Kuylenstierna[5], J. Shah[9], R. Shrestha[3], D. Streets[10], Zhao D.[11]

[1] *Resource Management Associates,* [2] *IIASA,* [3] *AIT (Bangkok),* [4] *University of Iowa,* [5] *Stockholm Environment Institute,* [6] *ECON Energy (Norway),* [7] *RIVM (Neth.),* [8] *Wageningen Agricultural University (Neth.),* [9] *World Bank,* [10] *Argonne National Lab.,* [11] *Chinese Academy of Sciences.*

Abstract. In contrast to Europe and North America, air pollution in Asia is increasing rapidly, resulting in both local air quality problems and higher acidic depositions. In 1989, an east-west group of scientists initiated a multi-institutional research project on Acid Rain and Emissions Reduction in Asia, funded for the past two years by the World Bank and the Asian Development Bank. Phase I, covering 23 countries of Asia, focussed on the development of PC-based software called the Regional Air Pollution INformation and Simulation Model (RAINS-ASIA). A 94-region Regional Energy Scenario Generator was developed to create alternative energy/emission scenarios through the year 2020. A long-range atmospheric transport model was developed to calculate dispersion and deposition of sulfur, based upon emissions from area and large point sources, on a one-degree grid of Asia. The resulting impacts of acidic deposition on a variety of vegetation types were analyzed using the critical loads approach to test different emissions management strategies, including both energy conservation measures and sulfur abatement technologies.

Key words: Asia, Acid Rain, Energy, Emissions, Sulfur Deposition, Critical Loads, Sulfur Dioxide, Conservation

1. Introduction

In contrast to Europe and North America, emissions of air pollution species in Asia are increasing rapidly, resulting in both local air pollution problems and higher acidic depositions. In general, most Asian countries do not have a strong scientific nor public constituency for addressing potentially serious air pollution problems impacting important economic and cultural activities such as forestry, agriculture, and tourism. The political ramifications of trans-boundary air pollution in Asia have not yet been addressed.

In light of the above, in 1989 an east-west group of scientists initiated an integrated program of assessment and policy analysis to analyze long-term air pollution strategies at regional, national, and Asia-wide levels. This program led to the establishment of an international research project on Acid Rain and Emissions Reductions in Asia, funded for the past two years by the World Bank and the Asian Development Bank. A large number of Asian and western scientists and institutions participated in the project.

The first phase focussed on the development of PC-based software, the Regional Air Pollution INformation and Simulation Model (RAINS-ASIA). A model of this type has been used for many years in assessment of trans-boundary air pollution problems in Europe. While having some aspects in common with the European analysis, RAINS-ASIA was specifically designed to address the more dynamic and rapidly changing nature of the Asian energy system.

Phase I of the Project has recently been completed. In addition to its development of a powerful assessment tool, the analysis completed in this phase identifies the potential air pollution impacts associated with the anticipated growth of Asia's energy system. Initial results indicate that some regions of Asia could have excess acidic deposition similar or greater in magnitude to that experienced in the highly-polluted regions of central and eastern Europe.

2. The RAINS-ASIA Model

RAINS-ASIA is a policy-oriented model which provides a framework for integrated assessment of acid deposition. The region of study includes 23 countries of Asia, including all Asian countries east of Afghanistan. These countries are further divided into 94 sub-national regions and 355 Large Point Source (LPS) emission sources corresponding to large electricity generating plants. Of these 94 regions, 24 are large metropolitan areas (mega-cities). RAINS-ASIA consists of three sub-models: Energy and Emissions, Acid Deposition, and Ecosystems Impacts. A flow chart of RAINS-ASIA is shown in Figure 1.

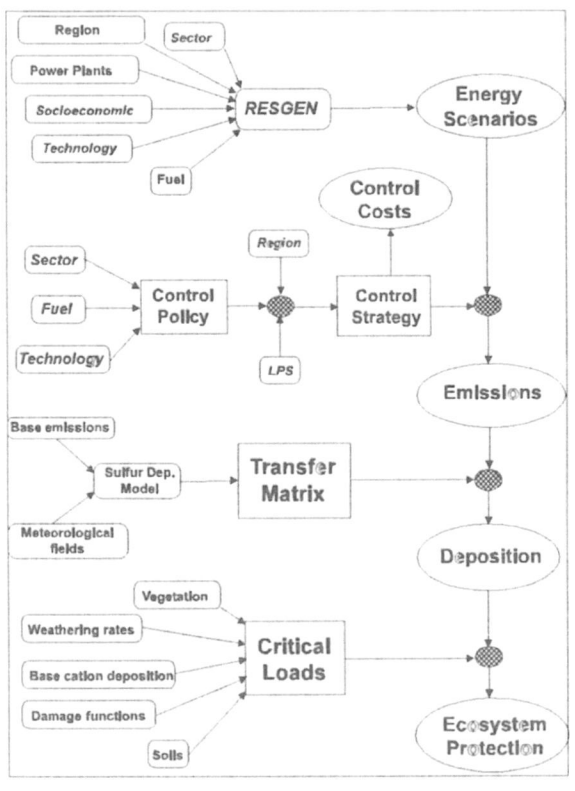

Fig. 1 - RAINS-ASIA Flow Chart

3. Energy Scenarios

To develop energy scenarios, a Regional Energy Scenario GENerator (RESGEN) was created to simulate energy use through the year 2020. RESGEN utilizes a simulation approach for generating regional energy scenarios. The goal of RESGEN is to estimate energy consumption contributing to SO_2 emissions on a sub-national scale. Output consists of regional and LPS energy consumption scenarios for each country. This information is available for input to the RAINS-ASIA model, where SO_2 emissions and acidic deposition are computed, and environmental impacts are assessed.

A Business-As-Usual (BAU) scenario was developed under the assumption that each country would follow existing energy policies without introducing additional measures to improve the

Fig. 2 - BAU Energy Scenario   Fig. 3 - LOW Energy Scenario

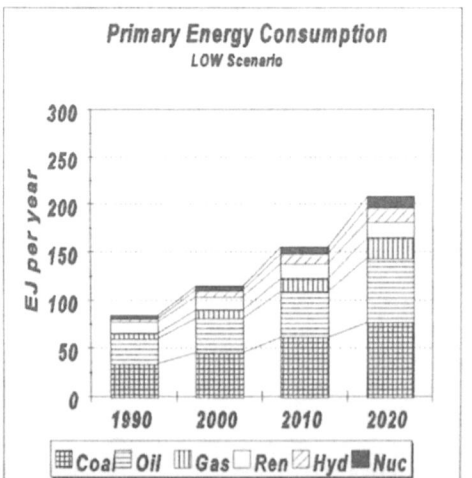
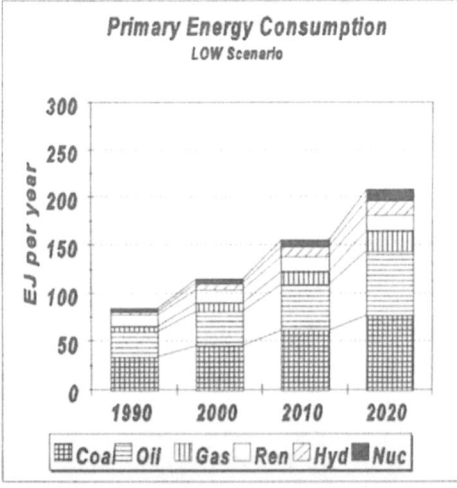

quality of the environment. Efforts were made to utilize the energy demand and supply forecasts made by the government planning organizations or research institutions in each country.

An alternative Energy Efficient Low Emissions (LOW) scenario was developed to explore the potential of emissions reductions through a strong effort to use energy more efficiently. The efficiency improvements possible through the year 2020 are based primarily on the experiences of the industrialized countries during the period 1973-1983, in response to the sharp rise in energy prices, as well as on general improvements in technology.

The Asia-wide results of the BAU and LOW scenarios are displayed in Figures 2 and 3. In the BAU scenario, energy consumption grows at an average rate of 4.0% per year. Total energy consumption in Asia would be 275 EJ by 2020. Although the LOW energy scenario incorporates assumptions regarding the implementation of significant energy efficiency improvements, energy growth remains at relatively high levels. However, the LOW scenario results in significant net energy savings of 11% below the BAU scenario in 2000, and 24% in 2020. The share of coal in primary energy was about 41% in 1990 and is projected to remain reasonably stable in the BAU scenario. Thus by 2020, total coal consumption would more than triple, reaching 110 EJ, or about 3.75 billion tons.

4. Emissions Scenarios

To provide a reference for this analysis, a baseline emissions scenario has been developed on the assumption that beyond the current emission controls in Japan and Taiwan, no further action will be taken for reducing SO_2 emissions. Due to strong economic growth and subsequently of energy consumption, the lack of measures to limit emissions leads to an increase of Asia total SO_2 emissions from 34 million tons in 1990 to more than 110 million tons in 2020.

Differences among countries in economic development and in energy supply structures generate a non-homogenous picture of future growth. For Japan, hypothetical uncontrolled SO_2 emissions would increase by about 34%, whereas for India and Indonesia the expansion exceeds a factor of four and five respectively. Such a differentiated picture not only emerges for national emissions, but also for the sectoral emissions. In 1990 about 16% of total SO_2 emissions in the region originated from large point sources, growing to 25% in 2020. Coal contributes about 75% of emissions in 1990 and about 73% in 2020.

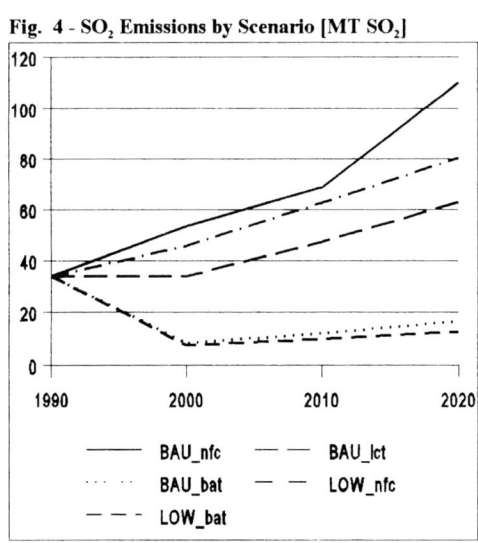

Fig. 4 - SO_2 Emissions by Scenario [MT SO_2]

BAU_nfc   BAU_lct
BAU_bat   LOW_nfc
LOW_bat

Four SO_2 emissions scenarios were developed for comparison with the above baseline scenario. Three scenarios were based on energy consumption from the BAU scenario, and two were derived from the LOW scenario. Each emissions scenarios used a different control strategy. The "best available control technology" (bat) strategy assumes the advanced control technologies, which represent the current technological standards in many industrialized countries, on all relevant emissions sources in Asia. Wet flue gas desulfurization (WFGD) is assumed for all industrial and power plant boilers burning coal and oil, including retrofits of existing plants. For other sectors, low sulfur coal and oil is assumed.

The "locally available control technology" (lct) is similar to the bat strategy, but assumes use of limestone injection for China, India and Pakistan. These technologies have an SO_2 removal efficiency of about 50% as opposed to the more than 90% removal efficiency of WFGD. The strategy, "no further controls" (nfc), assumes that beyond current emission controls in Japan and Taiwan, no further action will be taken to reduce SO_2 emissions.

The results of the emissions scenarios are compared in Figure 4. Not surprisingly, the BAU_bat scenario leads to a drastic decline of emissions. Between 1990 and 2020, emissions decline from 33.7 to 16.3 million tons, despite growth in energy consumption of 227%. The BAU_lct scenario results in a 43% reduction in emissions from the baseline scenario in 2020; however, emissions would still nearly double over the 30-year period. The LOW scenario with no further controls (LOW_nfc) results in emissions of 80 millions tons in 2020.

5. Acid Deposition Module

The second module of RAINS-ASIA, the ATMOS module, provides estimates of ambient levels of acid precursors and acid deposition loading throughout Asia. The projection of acid depositions is based on a transfer matrix for long range transport, calculated by using an atmospheric transport/deposition model described in R. Arndt *et. al.* 1995 (this volume).

Uncontrolled SO_2 emissions from the BAU energy scenario will cause a strong increase of sulfur deposition throughout the region. In 2020, virtually all eastern parts of China and some regions in India would experience deposition between two and five grams-S/m^2/yr. In many industrialized areas in Thailand, South Korea, and China, sulfur deposition will exceed five to ten grams-S/m^2/yr. Peak deposition of sulfur would escalate in some industrialized areas in China to about 27 grams-S/m^2/yr. For comparison, sulfur deposition observed in the industrial areas of Central and Eastern Europe peaked at about 15 grams-S/m^2/yr. In major areas in eastern and southern China, in eastern India, Indonesia and the Philippines, sulfur deposition will increase by a factor of four to five. In western India, Pakistan and in northern Thailand, calculations show deposition growth up to a factor of ten by the year 2020.

6. Impacts of Sulfur Deposition: Critical Loads and Exceedances

The IMPACT module of RAINS-ASIA assesses the risk of ecosystem impacts of sulfur deposition. Critical loads are compared with the estimates of sulfur deposition provided by the ATMOS module to determine which ecosystems may be at risk under various scenarios. Estimates of critical loads for Asia are shown in Hettelingh *et. al.* 1995 (this volume).

Estimates of critical loads are compared to estimates of sulfur deposition to obtain projections of the excess deposition (exceedances) above the critical load. The exceedances in 2020 for two emission scenarios are displayed in Figures 5 and 6. In the BAU-nfc case, critical loads will be exceeded in large parts of Asia. Pakistan, the western and central parts of India, western China, Myanmar and parts of Indonesia experience little or no excess deposition even under the highest emission scenarios considered in this paper. In other areas, e.g., in northern and eastern parts of India, in southern and eastern China, in Korea and in northern Thailand, widespread serious excess deposition could be expected. In addition, many 'hot spots occur on the local scale.

It must be emphasized that - under the assumptions of this scenario - excess deposition will reach unprecedented levels in some regions. The RAINS-ASIA model calculates that critical loads will be exceeded by five to ten grams-S/m^2/yr for large areas in central and eastern China, in northern Thailand, as well as in the surroundings of Bangkok, Manila and Singapore. For comparison, total sulfur deposition in many of these areas is currently in the range of two to three grams-S/m^2/yr. Although the current state of scientific knowledge does not permit

Fig. 5 - Excess Deposition in 2020, BAU_nfc Scenario

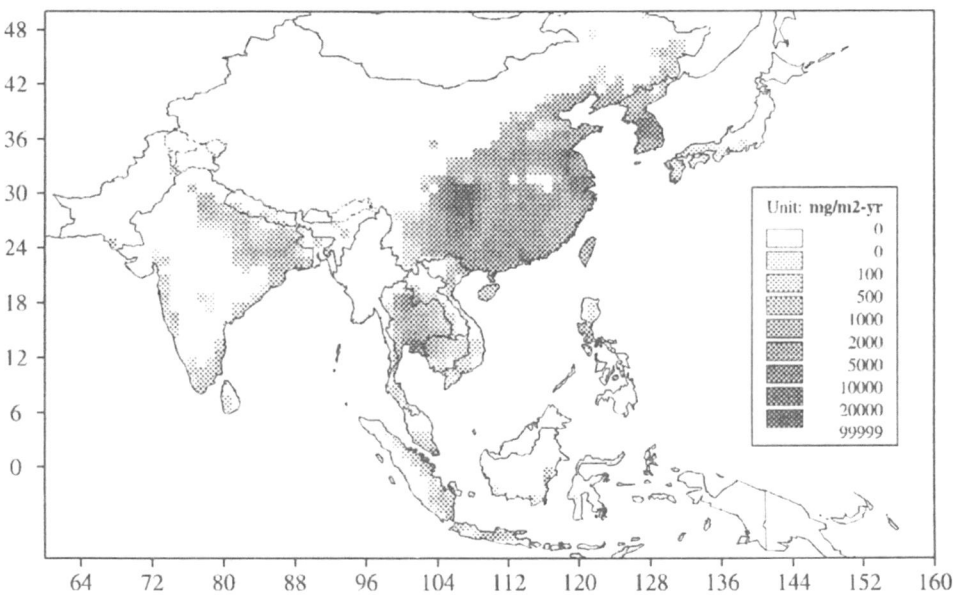

Fig. 6 - Excess Deposition in 2020, BAU_bat Scenario

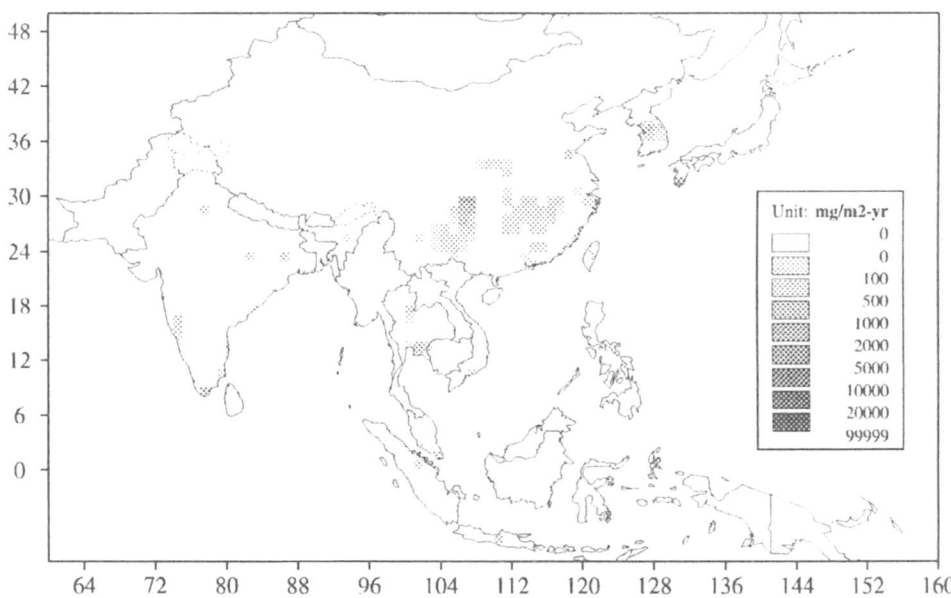

drawing conclusions about the environmental damage of such excess deposition, the fact that sulfur deposition will be more than ten times above the sustainable levels in large areas may give reason for serious concern.

7. Cost of Acid Rain Abatement

The RAINS-ASIA model provides the capability to search for cost-effective control strategies for achieving exogenously specified target emission and deposition levels. The emission control costs associated with the five emission scenarios described in this paper are shown in Figure 7.

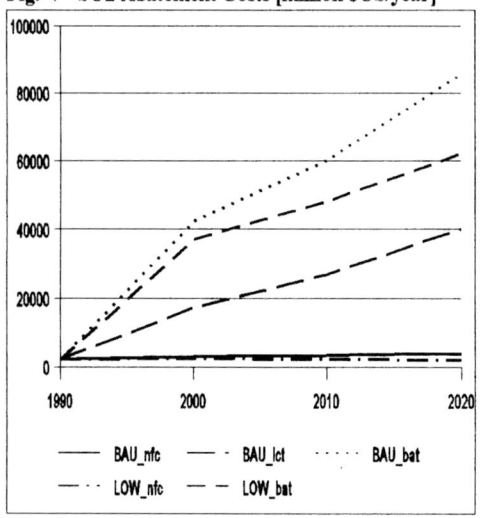

Fig. 7 - SO2 Abatement Costs [million $US/year]

The bat control strategy costs 42 billion US$ per year in 2000 for the BAU case; costs more than double by 2020. Adopting improvements in energy efficiency (LOW scenario) reduces the annual control costs by 12 and 27 percent respectively. Restricting control measures to locally available technologies (BAU_lct strategy) costs about 50 percent less than the BAU_bat strategy.

If no further action is taken beyond the existing regulations (nfc control strategy) the costs will reduce drastically to only 3 billion US$ per year in 2020 for the BAU scenario and 2 billion US$ per year for the efficiency case (LOW scenario), mostly contributed by control costs in Japan.

Acknowledgments

The project was funded by The World Bank and the Asian Development Bank. The views presented in this paper are those of the authors, and do not necessarily represent those of the funding institutions.

References

Amann, M.: 1990, Energy Use, Emissions and Abatement Costs in Europe , in Alcamo, J. et. al. (eds.), The RAINS Model of Acidification, Science and Strategies in Europe. Dordrecht: Kluwer Academic Publishers.

Arndt, R. and G.R. Carmichael: 1995, Long Range Transport and Deposition of SO_2 and Sulfate in Asia , in the Proceedings of Acid Reign '95, Kluwer Academic Publishers, this volume.

Asian Institute of Technology: 1995, Acid Rain and Emissions Reduction in Asia - Regional, AIT, Bangkok.

Bertok, I., et. al.: 1993, Structure of RAINS 7.0 Energy and Emissions Database, WP-93-67. IIASA, Laxenburg.

Bhatti, N., et. al.: 1992, "Acid Rain in Asia", Environmental Management, 16 (4), pp. 541-562.

Carmichael, G.R.: 1992, "Modeling of Acid Deposition in Asia", in the Proc. of the Third Annual Conference on Acid Rain and Emissions in Asia, W. K. Foell and C. W. Green (eds.), AIT, Bangkok.

ECON Energy: 1994, Energy Consumption in China, India, Indonesia, and South Korea (by T. Haugland and K. Roland), ECON Energy, Oslo, Norway. Report No. 303/94.

Foell, W. K.: 1992, "An International Research Program on Acid Rain and Emissions in Asia", in Acidification Research: Evaluation and Policy Applications, T. Schneider (ed.), Amsterdam: Elsevier Science Publishers.

Foell, W.K., and C.W. Green: 1991, "Acid Rain in Asia: An Economic, Energy and Emissions Overview", Proc. of the Second Annual Workshop on Acid Rain and Emissions in Asia, W.K. Foell and D. Sharma (eds.), AIT, Bangkok.

Hettelingh, J.-P. et. al.: 1995, "Deriving Critical Loads for Asia", in the Proceedings of Acid Reign '95, Kluwer Academic Publishers, this volume.

Hettelingh, J.-P.: 1992, "The Rains-Asia Impact Module: Computation, Mapping and Application of Critical Loads", in the Proc. of the Third Annual Conference on Acid Rain and Emissions in Asia, W. K. Foell and C. W. Green (eds.), AIT, Bangkok.

Hordijk, L.: 1991, "Use of the Rains Model in Acid Rain Negotiations in Europe", Environmental Science and Technology, 25 (4), pp. 596-603.

Kuylenstierna, J.C., et. al.: 1992, A Preliminary Mapping of Relative Sensitivity of Terrestrial Ecosystems to Acidic Deposition in Asia , in the Proc. of the Third Annual Conference on Acid Rain and Emissions in Asia, W. K. Foell and C. W. Green (eds.), AIT, Bangkok.

Resource Management Associates of Madison: 1993, Database Structure for the Energy and Emissions (ENEM) Module of the RAINS-ASIA Model, Working Paper.

LONG-RANGE TRANSPORT AND DEPOSITION OF SULFUR IN ASIA

R. L. ARNDT and G. R. CARMICHAEL

Department of Chemical & Biochemical Engineering
Center for Global & Regional Environmental Research
University of Iowa, Iowa City IA 52242 USA

Abstract. The long range transport of sulfur in Asia is analyzed through the use of a multi-dimensional acid deposition model. The air quality of this region is heavily influenced by the combination of Asia's growing population, its expanding economy, and the associated systems of energy consumption and production. These factors combined with a shift to using indigenous coal as the primary fuel source for the region, will result in increased emissions of pollutants into the environment. By the year 2020 sulfur emissions from Asia are projected to exceed the combined emissions from Europe and North America. We have estimated sulfur deposition in Asia on a one-by-one degree spatial resolution in the region from Pakistan to Japan and from Indonesia to Mongolia using a 3-layer Lagrangian model. Deposition in excess of 10 g S/m2 is predicted in south-central China. The relationship between emission source and receptor has been developed into a "deposition matrix" and examples of the source-receptor relationship are presented.

Key words: Asia, long-range transport, model, precipitation, source-receptor, sulfur deposition, transboundary

1. Introduction

The potential for large-scale pollution problems in Asia is great due to high emissions and the close geographical proximity of many of the major industrial and urban centers (e.g., Tokyo, Seoul, Taipei, Shanghai, Hong Kong, etc.). As a result of Asia's large and rapidly increasing population, its growing economy, and the associated systems of energy consumption and production, the future air quality of the region is at risk. Asia currently accounts for more than 55% of the world's population and is projected to reach four billion people by the year 2010. In addition, the economic growth rate for Asia is far higher than any other region of the world. These factors, combined with a shift to the use of indigenous coal as the region's primary fuel source, will result in increased emissions of pollutants into the environment (Foell et al., 1995).

In July 1992, researchers in Europe, the USA, and Asia began a collaborative project to study the effects of sulfur deposition in Asia. Based on the framework of an acid rain integrated assessment for Europe (RAINS) (Hordijk, 1991), data and models were brought together on energy use, emissions of sulfur dioxide, abatement of emissions, long-range transport and deposition, and environmental effects. This RAINS-ASIA project has produced the first estimates of the regional impacts of the expanding energy use in Asia. The project was funded by using multi-national trust funds through The World Bank and the Asian Development Bank.

In this paper the transport and deposition of sulfur oxides in Asia is modeled and the results are presented and discussed. Anthropogenic SO_2 emissions (Streets et al., 1995) and volcanic sources (Fujita, 1992; Spiro et al., 1992) are used as the emissions inventory. A 3-layer trajectory model used to study the sulfur cycle is presented. Country-to-country deposition relationships are presented for the region. These source-receptor relationships provide information on the character and importance of long range transport of sulfur compounds in Asia.

2. Model Methodology

The sulfur deposition patterns and source-receptor relationships were calculated using a regional-scale trajectory model which is a modified version of NOAA's Air Resources Laboratory Branching Atmospheric Trajectory (BAT) model (Hefter, 1984). The BAT model is a three-dimensional, multiple-layer Lagrangian model which calculates SO_2 and sulfate surface concentrations, and wet and dry deposition amounts. It was also modified to include the capability for modeling both elevated and surface emission sources.

The BAT model provides a 1° by 1° resolution of the concentrations and deposition of SO_2 and sulfate. The model's meteorological domain is 20° South to 60° North latitude and 39° East to 155° East longitude. The sulfur species modeling domain is 10° South to 55° North latitude and 60° East to 150° East longitude. Within the modeling domain SO_2 emission plumes are modeled as puffs released every three hours from the emission source location. Each puff is assigned a mass proportional to the source strength, and is assumed to mix uniformly in the vertical throughout an assigned layer (the model is separated into a boundary and upper layer in the daytime and into surface, boundary and upper layers at night), and to diffuse with a Gaussian distribution in the horizontal. Area emissions are modeled as surface sources (released at the center of the grid) while large point sources and volcanoes are treated as elevated sources. Individual emission puffs are followed throughout their transport and deposition "lifetimes". Each puff's transport is followed for up to five days (or until the mass falls below a cut-off value). Puffs which are transported beyond the modeling domain are no longer tracked. As the puffs are being transported, SO_2 is chemically converted to sulfate, and SO_2 and sulfate are deposited to the surface by wet and dry removal processes. Parameters used in the model and detailed discussion of the model's calculations are presented elsewhere (Carmichael and Arndt, 1995).

The meteorological data is used by the BAT model to calculate puff trajectories and to estimate inversion heights, which in turn are used in the determination of the heights of the three layers used in the model calculations. For this study 1990 meteorology was used. Specifically we used the data from the upper air sondings for 1990 provided by the National Meteorological Center (NMC) of the National Oceanic and Atmospheric Administration (NOAA). These data files contain rawinsonde and pibal vertical observations of wind speed and temperature from the surface to 500 hPa. The data is provided at six hour intervals. The meteorological data needed by the model at specific locations was obtained by interpolation between these observation points. If no observational data within a one degree radius is available, the information radius is increased at one degree intervals. Observational data up to a radius of five degrees of the puff may be used in the trajectory calculation.

Precipitation data used was NMC analyzed fields. This data was obtained from the National Center for Atmospheric Research (NCAR). The precipitation data consisted of accumulation values collected in six hour intervals throughout the region in 1990. This data was provided on a 1.4695° latitudinal by 1.4875° longitudinal grid spacing. Since these data are on a different grid than that used by the BAT model (i.e., 1° by 1°) it was necessary to interpolate the precipitation data. This was done inside the BAT model using a two-dimensional linear interpolation method (Carnahan et al., 1969). Furthermore, the BAT model requires hourly precipitation rates. At present the six hour accumulated values were assumed to be uniformly distributed within the six-hour interval. (The error in sulfur removal rates introduced by this assumption is uncertain but is addressed by Bullock, (1994).)

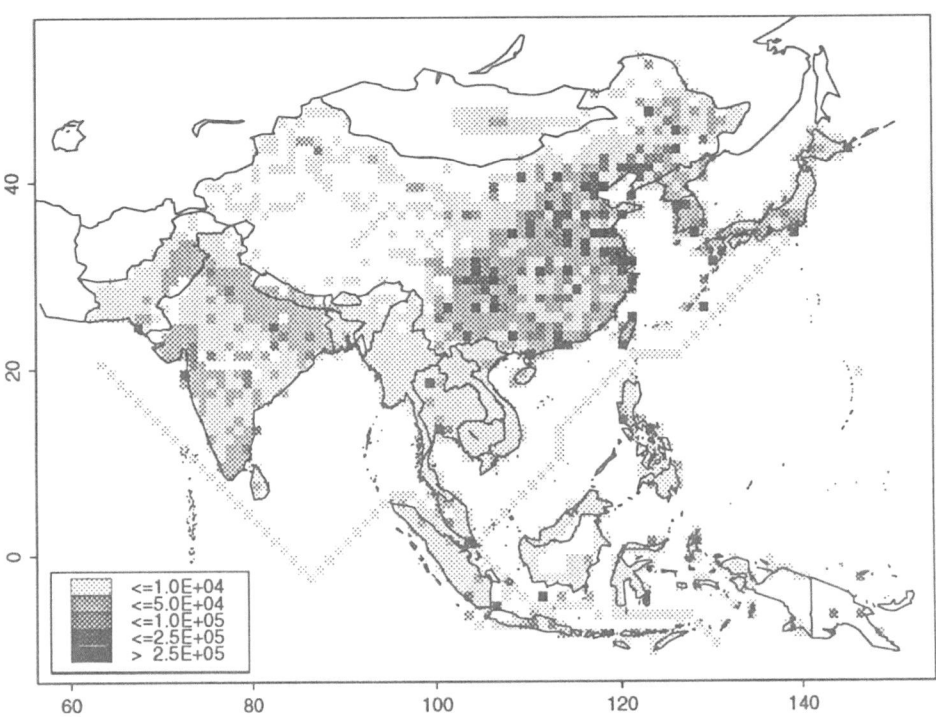

Fig. 1 Spatial distribution (1° x 1°) of the 1990 SO_2 emissions for Asia. Units in tonnes SO_2/year.

The emissions inventory was obtained from two data sources. Anthropogenic emissions for Southeast Asia, the Indian subcontinent, Mongolia, and shipping activity were provided by Streets et al., (1995) and for China, Hong Kong, Japan, Taiwan and North and South Korea data from Akimoto and Narita (1994) was used. Emissions from non-eruptive volcanic sources in the region were provided by Fujita (1992) and Spiro et al. (1992). The spatial distribution of these emissions is presented in Figure 1. High concentrations of emissions sources are indicated in Eastern and South-Central China, the Ganges River Basin, the Korean Peninsula, Northern Thailand, Bangkok, Singapore, and Taiwan. Volcanic sources can be seen throughout Indonesia, The Philippines, and Japan. These natural sources of sulfur contribute substantial emissions in South-East Asia and Japan. Emissions from sea lanes are shown traversing the seas and oceans of the region.

3. Results and Discussion

3.1 SULFUR DEPOSITION

The model calculated annual total deposition, in g-S/m^2-yr., is presented in Figure 2. Shown are the contributions from all anthropogenic sources included in the model and volcanoes. There are very few regions in Asia which are not impacted by sulfur deposition. The high sulfur deposition regions closely reflect the spatial distribution and the density of the emissions. For example, the dense emission regions in eastern and southern China, South Korea, Taiwan, northern Thailand, and eastern India all show

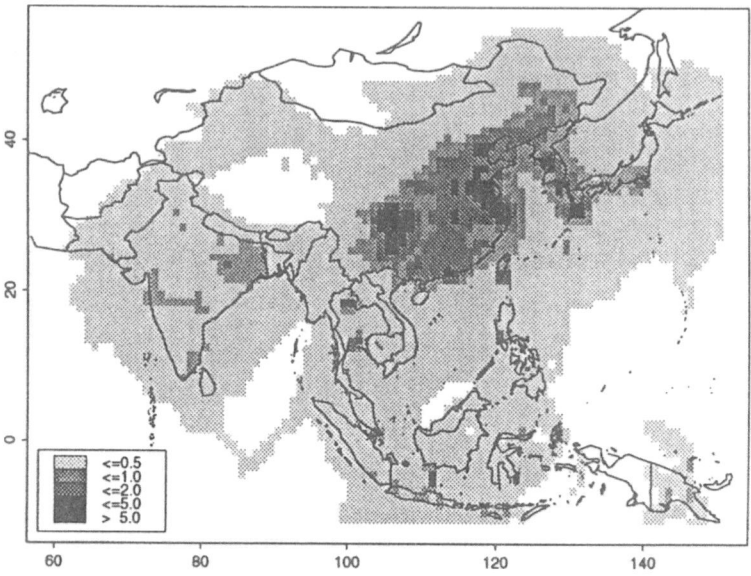

Fig. 2 Calculated annual sulfur deposition for 1990. Units in g-S/m^2/yr.

elevated sulfur deposition. The highest annual deposition (~10 g-S/m^2-yr.) occurs around the city of Chonqing in Sichuan Province, China. The strong continental outflow from east Asia is also clearly evident. Sulfur emissions in the latitude band 20° to 40° N result in high sulfur deposition virtually throughout the western Pacific Ocean at these latitudes. Transport and deposition off the eastern coast of India over the Bay of Bengal is shown as is the high sulfur deposition around Malaysia and the western parts of Indonesia. Elevated deposition levels are observed at several major urban sectors in the region, most notably: Bangkok, Hong Kong, Manila, Seoul, Singapore, and Tokyo. Deposition due to volcanic sources is evident throughout much of Southeast Asia and particularly from Mt. Sakurajima, on the island of Kyushu. The total deposition patterns are also heavily influenced by the annual precipitation patterns. High annual precipitation amounts are found to occur in northern India, Nepal, southeastern China, and Southeast Asia.

3.2 SOURCE-RECEPTOR RELATIONSHIPS

The BAT model calculates the deposition from the emissions for each grid cell directly and this information can be used to analyze a variety of policy-related questions. For example, the deposition from a specific source, region, or country can be viewed separately. This information can be aggregated to provide source-receptor information at a country-to-country or region-to-region level. Examples of the predicted country-to-country and region-to-region source-receptor relationships calculated by the BAT model are presented in Tables 1 and 2. The first two columns of Table 1 present where the sulfur deposited on South Korea originated from. For example, in Table 1, 82% of South Korea's deposition is due to the country's own emissions while 13% is from China, and the remaining 5% from other sources. The third and fourth column present where

TABLE I

Predicted deposition on S. Korea by each country and the resulting deposition on each country due to emissions from China.

DEPOSITION ON SOUTH KOREA	SULFUR (Tonnes)	DEPOSITION FROM CHINA	SULFUR (Tonnes)
South Korea	1.8E+05	China	6.0E+06
China	2.8E+04	Ocean	1.0E+05
Volcanoes	4.4E+03	All Others	9.0E+04
North Korea	2.6E+03	North Korea	6.0E+04
All Others	1.0E+03	Japan	3.9E+04
		South Korea	2.8E+04

emissions from China are deposited. These results make a significant point. While less than one percent of China's emissions are predicted to fall on South Korea, this accounts for 13% of the sulfur deposition in South Korea.

Transport and deposition can also be analyzed on a region-to-region basis. Table 2 presents the predicted deposition (within China) of two regions' emissions in southern China, specifically the Yunnan and Hunnan province. In the first two columns the regions where the sulfur emissions from Yunnan are deposited and deposition values in those provinces are presented. The same information for Hunnan is presented in columns three and four. It can be seen that deposition patterns for the two regions are quite different. Emissions from Yunnan are deposited over a much broader area with only 37% of its emissions being deposited within Yunnan. However, 63% of emissions from Hunnan are deposited within that region. These differences give insight into the factors that contribute to variations in transport and deposition. The differences between the two regions' deposition patterns can be accounted for, at least in part, by precipitation patterns and emission source types. For 1990, the year modeled, Hunnan received nearly twice the precipitation as Yunnan did. Hence, emissions from Hunnan would have experienced higher wet removal rates closer to their source locations and thus more sulfur was deposited within Hunnan. In Yunnan nearly 12% of the sulfur emissions are from elevated sources as compared to only 7% in Hunnan. Emissions from elevated sources can be expected to experience greater transport than their surface source counterparts. Thus Yunnan's higher percentage of emissions from elevated sources will result in more long range transport for its emissions.

TABLE II

Sulfur deposition on provinces in China resulting from emissions from the Yunnan and Hunnan provinces.

DEPOSITION FROM YUNNAN	SULFUR (Tonnes)	DEPOSITION FROM HUNAN	SULFUR (Tonnes)
Yunnan	9.9E+04	Hunan	9.0E+04
Sichuan	6.3E+04	Jiangxi	1.7E+04
Guizhou	5.0E+04	Hubei	1.3E+04
Guangxi	1.9E+04	Hebei-Henan-Anhui	5.5E+03
Hunan	1.7E+04	Guangxi	4.5E+03
Hubei	3.9E+04	Guizhou	3.2E+03
All Others	1.4E+04	All Others	1.1E+04

3.3 MODEL EVALUATION

There is limited observational data available for comparison with calculated deposition. However, Japan's EPA (Murano, 1994) has an extensive acid deposition data set for that country and comparison with model results demonstrates a similar spatial distribution of sulfur deposition (Carmichael and Arndt, 1995). Model comparison has also been done with a sulfur dioxide monitoring program that was developed as part of the RAINS-ASIA project. A comparison of these observed values and model predicted values shows the model captures the seasonal variation of SO_2 levels within the region (Carmichael et al., 1995).

4. Summary

To help quantify and anticipate environmental impacts associated with the expected growth in sulfur emissions in Asia it is imperative that we develop a greater understanding of the mechanisms of long range transport of pollutants in Asia. The source-receptor calculations presented in this paper help to demonstrate the role of long range transport in Asia. Transboundary pollutant transport issues appear to be of potential importance in east Asia (involving eastern China, Korea and Japan); southeast Asia, and among those countries bordering India. Acid deposition problems throughout Asia are anticipated as a result of the rapid expansion in energy consumption in the region. By developing an understanding of the relationship between sulfur emissions and their resulting deposition, we can better predict and assess the impact of Asia's growing energy demands on the environment.

Acknowledgments: This research was supported in part by funds from The World Bank and The Asian Development Bank as part of the RAINS-ASIA project. Special thanks to the collaborators on the RAINS-ASIA Phase-1 project.

5. References

Akimoto, H. and Narita, H.:Distribution of SO_2, NO_x, and CO_2 Emissions from Fuel Combustion and Industrial Activities in Asia with 1° x 1° Resolution, Submitted to *Atmos. Environ.*, 1993.
Carmichael, G.R., and Arndt, R.L. (1995) *Chapter 5. RAINS-ASIA: An Assessment Model for Acid Rain in Asia*, Phase-I Final Report.
Carmichael, G.R., Frem, M., Adikary, S., Ahmed, J., Mohan, M., Hong, M-S., Chen, L., Fook, L., Liu, C., Soedomo, M., Tran, G., Suksomsank, K., Zhao, D., Arndt, R., Chen, L..,Observations of the Regional Distribution of Sulfur Dioxide in Asia, Accepted to Water, Air, and Soil Pollution, special issue for Acid Reign '95.
Foell, W., Green, C., Amann, M., Bhattachrya, S., Carmichael, G., Chadwick, M., Hettelingh, J.-P., Hordijk, L., Shah, J., Shrestha, R., Streets, D., Zhao, D., Energy Use, Emissions, and Air Pollution Reduction Strategies in Asia,Accepted to Water, Air, and Soil Pollution, special issue for Acid Reign '95.
Carnahan, B., Luther, H., and Wilkes, J.: Applied Numerical Methods, John Wiley & Sons, New York, 1969.
Fujita, S. (1992) Acid Deposition in Japan, Technical Report ET91005, Central Research Institute of Electrical Power Industry, 1992.
Heffter, J.L.: (1983) Branching Atmospheric Trajectory (BAT) Model, NOAA Technical Memorandum, ERL ARL-121.
Hordijk, L. (1991) *Environ. Sci. and Tech.* 25, No. 4,596-603.
Murano, K. (1994) Activity of JEA for East Asian Acid Precipitation Monitoring Network. Presented at Workshop on Acid Rain etwork in South, East, and Southeast Asia, Malaysia, May 17-19, 1994.
Spiro, P., Jacob, D., and Logan, J. (1992) *J.of Geophys. Research* 97:6023-6036.
Streets, M., Amann, M., Bhatti, N., Cofala, J., Green, C.: (1995) *Chapter 4. RAINS-ASIA: An Assessment Model for Acid Rain in Asia*, Phase-I Final Report.

OBSERVED REGIONAL DISTRIBUTION OF SULFUR DIOXIDE IN ASIA

G. R. CARMICHAEL[1], M. FERM[2], S. ADIKARY[3], J. AHMAD[4], M. MOHAN[5], M-S. HONG[6], L. CHEN[7], L. FOOK[8], C.M. LIU[9], M. SOEDOMO[10], G. TRAN[11], K. SUKSOMSANK[12], D. ZHAO[13], R. ARNDT[1] AND L. L. CHEN[1]

[1]*Department of Chemical & Biochemical Engineering, Center for Global & Regional Environmental Research, University of Iowa, Iowa City, IA USA;* [2]*Swedish Environmental Research Institute, Gothenburg, Sweden;* [3]*Himalayan Climate Center, Nepal;* [4]*Jahangirnagar University, Bangladesh;* [5]*Center for Atmospheric Sciences, IIT, Delhi, India;* [6]*Department of Environmental Engineering, Ajou University, S. Korea;* [7]*Hong Kong Polytechnic University, Hong Kong;* [8]*Malaysia Meteorological Agency, Malaysia;* [9]*Department of Atmospheric Sciences, National Taiwan University, Taiwan;* [10]*Instituteof Technology, Bandung, Indonesia;* [13]*Eco-Environmental Center, Academia Scenica, China;* [11]*Institute for Chemistry, Hanoi, Vietnam;* [12]*Environmental Training Center, Bangkok, Thailand*

Abstract. SO2 concentrations have been measured for one year at forty-five locations throughout Asia using passive samplers. Duplicate samples were exposed at each site for one month intervals. The sites were selected to provide background information on the distribution of SO_2 over wide geographical regions, with emphasis on the regional characteristics around areas estimated to be sensitive to sulfur deposition. The annual mean values ranged from less than 0.3 $\mu g/m^3$ at Tana Rata, located at 1545 m on the Malaysia Peninsula, Lawa Mandau, (Borneo) Malaysia, and Dhankuta, Nepal, to values greater than 20 $\mu g/m^3$ at Luchongguan (Guiyang) China, Babar Mahal, Nepal, and Hanoi, Vietnam. In general high concentrations were measured throughout China, with the highest concentrations in the heavy industrial areas in Guiyang. The concentrations in east Asia around the Korea peninsula were ~5 $\mu g/m^3$. The concentrations in the southeast Asia tropics were low, with no station in Malaysia and Indonesia having average concentrations exceeding 1.7 $\mu g/m^3$. The observed SO_2 concentrations were found to display a distinct seasonal cycle which is strongly influenced by the seasonality of winds and precipitation patterns.

1. Introduction

The rapid expansion of energy use in Asia, combined with utilization of indigenous coal, will result in an increase in emissions of sulfur compounds. In addition, many countries are attempting to minimize their local pollution problems by installing taller stacks. Similar situations in Europe and North America have lead to significant regional problems related to acid deposition. The long-term and regional/local impact of these atmospheric emissions can affect not only the natural environment but also may have far-reaching implications for commercial and cultural activities such as forestry, agriculture, and tourism.

Monitoring of the sulfur dioxide and sulfate levels in Asia is critical to assessing the environmental impact of this rapidly expanding energy use. However, at present there is very limited information on the regional distribution of sulfur compounds in Asia. Most data in the region on sulfur dioxide and sulfate concentrations are confined largely to urban centers. In order to improve our understanding of the regional distribution of sulfur dioxide, a regional network using passive samplers was established. The new passive samplers offer a low-cost means of obtaining long-term ambient average sulfur dioxide concentrations throughout Asia. Forty-five stations were deployed in Bangladesh, China, Hong Kong, India, Indonesia, South Korea, Malaysia, Nepal, Taiwan, Thailand, and Vietnam. Sites were selected to give broad regional coverage, to provide information on the regional nature of the sulfur distribution, and were located away from the major urban areas and in areas identified to be sensitive to acid deposition.

In this paper results from one-year of monitoring are presented. The seasonal cycles as well as the annual-average values are presented and discussed. Particular attention is paid to the regional characteristics since this data set represents very diverse locations, with sites ranging from high altitude stations in Nepal, to tropical forests in Malaysia.

2. Network Design

As part of the *RAINS-ASIA* project (Foell et al., 1995)) funded by the World Bank and the Asian Development Bank a pilot monitoring program was conducted. (This aspect of the project was funded by the Asian Development Bank.) The objective of this monitoring activity was to obtain base-level data for future model evaluation. Sulfur dioxide was monitored at the surface at forty-five locations, for a 12-month period. The first sites began monitoring in December 1993, and others in February 1994. (Due to logistical reasons the monitoring in India began in November 1994.) Further details are presented below.

Sulfur dioxide was monitored using passive samplers developed by Martin Ferm at the Swedish Environment Research Institute (IVL) (Ferm, 1991). These new passive samplers offer a low-cost means of obtaining long-term ambient average sulfur dioxide concentrations. These simple, inexpensive (~$35 U.S./sampler including manufacturing, mailing and analysis) devices can be used to obtain weekly to monthly average sulfur dioxide levels. Extensive testing in Asia and Europe has proven these devices to be accurate and reliable (Ferm, 1991).

These short, sensitive, diffusive samplers consist of a polypropylene ring with two caps. A hole is made in the inlet cap where a membrane filter is placed. The membrane filter is made of Teflon and protected from mechanical destruction by a stainless steel net. Ambient SO_2 is trapped on a NaOH impregnated cellulose filter. On the filter sulphite is oxidised to sulfate during sampling and the sulfate amount is analysed using suppressed ion chromatography. In previous measurements in Asia an extra peak close to the sulfate peak often complicated the analysis. This problem has never been observed in samples from Europe. The cellulose filters also have a sulfate blank that can not be completely reduced. It was discovered that use of a very fine stainless steel net in parallel with the impregnated cellulose filters, eliminated the extra peak. Steel nets placed in the outer cap were used from the start of this project. The SO_2 concentration is calculated from Fick's first law of diffusion, using 11.5 mm as the diffusion distance; i. e., the length of the ring (10 mm) plus an average thickness of the laminar boundary air layer that the SO2 molecules have to pass by molecular diffusion.

Duplicate samples were exposed each month and then returned to IVL for analysis. Thus all samples were analyzed at the same laboratory. One field blank per country was sent out every month. The lab blank was about 0.17 µg sulfate per net. The blank increased during storage in the lab by about 40% during a few months. The sulfate measured on the field blanks nearly doubled before they were analysed. The field blanks were used in the calculation and gave a lower detection limit for one month sampling of about 0.2 µg /m^3 (0.09 ppb).

The measurement sites were selected to provide information on the regional aspects of sulfur in Asia. Thus sites were chosen to be away from major local sources and to be distributed in highly sensitive regions as determined from maps of ecosystem sensitivity developed as part of the *RAINS-ASIA* project. The location of the monitoring sites are shown in Figure 1. The distribution of sites was determined by a variety of considerations. For example, five sites were selected for China to complement monitoring programs already in existence. In China, four of the five sites were located in the south in highly sensitive areas. Practical considerations of security, access, etc., also played an important role in the final selection of sites. The final distribution of sites span a latitude range from 10°S to 40°N, a longitude range of 70°E to 130°E, and an altitude range from sea level to above 4000m.

For each country there was a principal contact person with responsibility to oversee all the measurements in that country (the country contact persons are co-authors on this paper). In many cases (e.g., Bangladesh, Nepal, Vietnam), this data represent some of the first monitoring activity outside of the major urban centers.

3. Results

The annual averaged observed SO_2 concentrations are summarized in Figure 2. The annual mean values range from less than 0.2 $\mu g/m^3$ at Tana Rata, located at 1545 m on the Malaysia Peninsula, Lawa Mandau, (Borneo) Malaysia, and Dhankuta, Nepal, to values greater than 20 $\mu g/m^3$ at Luchongguan (Guiyang) China, Babar Mahal, Nepal,

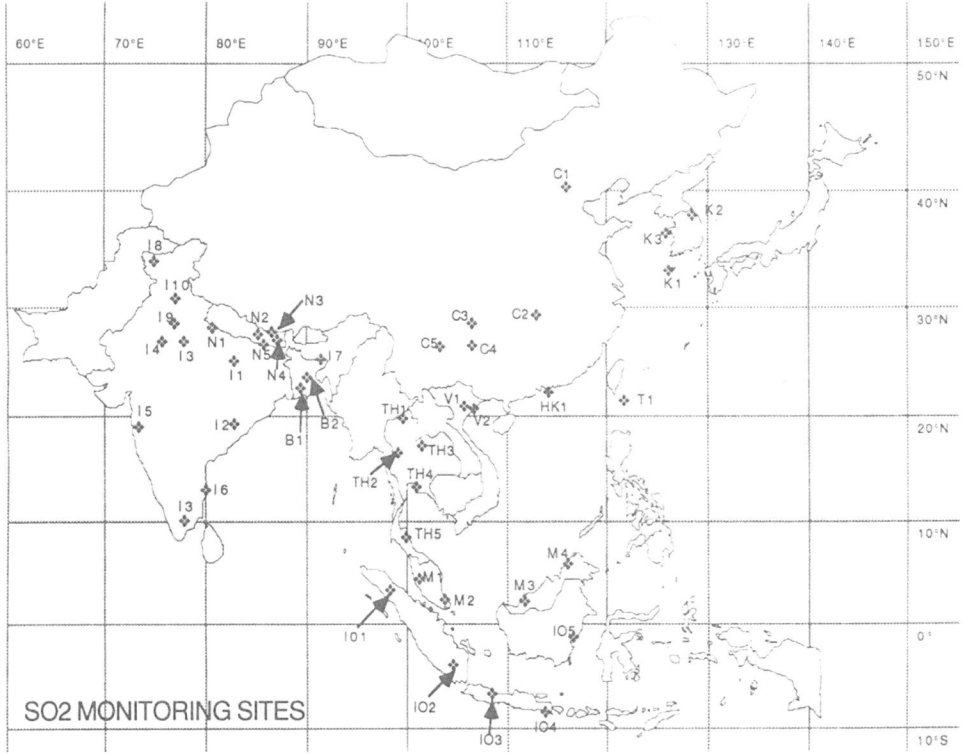

Figure 1. Location of SO2 passive sampler sites. The name of sites are presented in Figure 2.

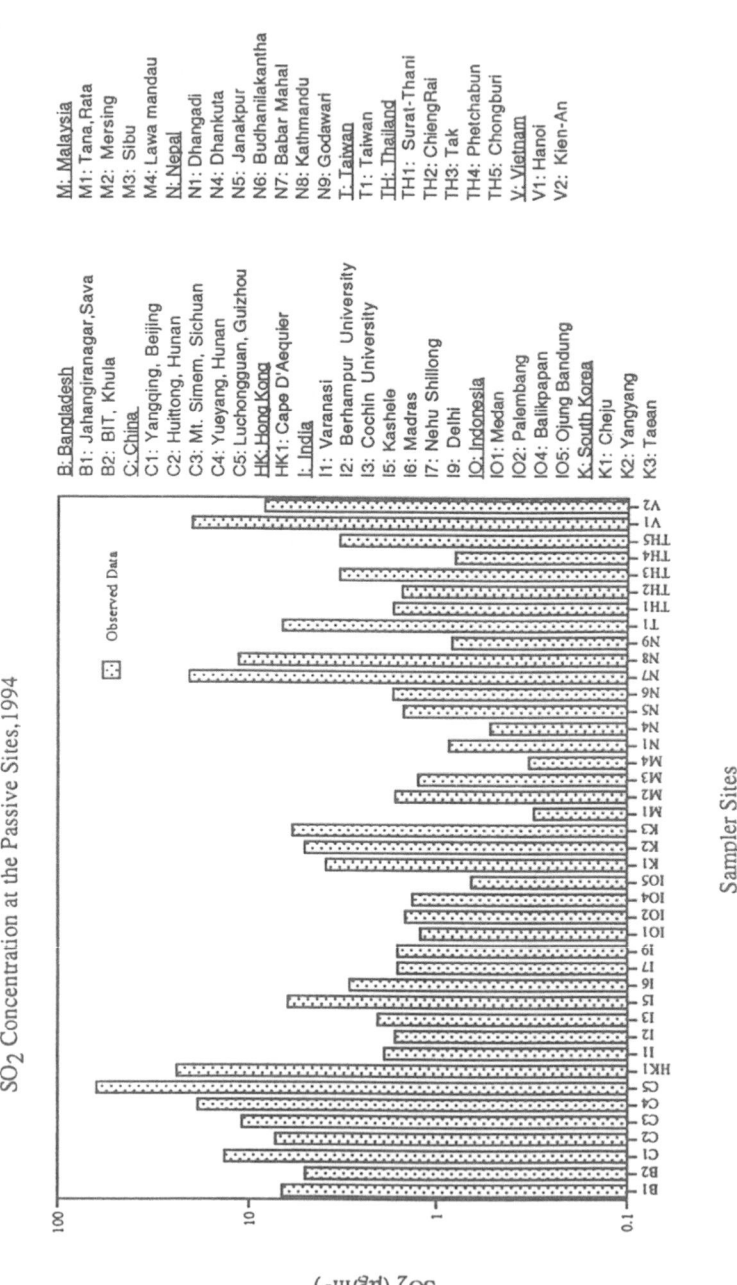

Figure 2. Annual averaged SO$_2$ surface concentrations (as SO$_2$) observed using the passive samplers. Each point is the average of duplicate samples exposed for month long periods.

and Hanoi, Vietnam. In general high concentrations were measured throughout China, with the highest concentrations in the heavy industrial areas in Guiyang. The concentrations in east Asia around the Korea peninsula were ~5 µg/m^3. The monitoring site at Cheju (station K1) is located on a volcanic island south of the peninsula. This site is excellent for characterizing air masses in east Asia, as it measures air from China, Korea and Japan at various times of the year. The average concentration at Cheju was 3.9 µg/m^3.

The concentrations in the southeast Asia tropics were low, with no station in Malaysia and Indonesia having average concentrations exceeding 1.7 µg/m^3. At the high altitude stations in Malaysia, i.e., Tana Rata (M1, 1500m) and Lawa Mandau (M4, 800m) the concentrations were low, ~0.3 µg/m^3, while the site at Mersing (M2, 40 m) located on the Malaysia peninsula observed values 5 to 10 times higher than those at Tana Rata. The concentrations in Nepal varied greatly from values less than 0.5 µg/m^3 in the Himalayas to greater than 20 µg/m^3 throughout the Katmandu valley.

Further insights can be found by examining the seasonal cycle of surface SO_2. The seasonal cycles of the observed SO_2 concentrations at four sites are presented in Figure 3. Shown are stations in Bangladesh, Korea, Malaysia and Nepal. All four stations show distinct seasonal cycles. For example, the seasonal cycle for station at Jahangimagar (station B1) in Bangladesh exhibits dry season concentrations exceeding 10 µg/m^3, falling to ~2 µg/m^3 during the monsoon season (May through October). This cycle is influenced both by the large wet removal of SO_2 during the monsoon season and also by the change in wind direction. During the monsoon season flows in this region are southerly from the Bay of Bengal area and are expected to have lower concentrations than those coming from the west and north during the dry periods. The station in Nepal shows a similar minimum value during the rainy season, followed by high concentrations during the winter inversion periods.

Yangyang (station K2), S. Korea, located on the northeast region of South Korea, shows a rainy season minimum in July and August and maxima in early winter and spring. In this region the flows in the winter are largely northerly, and those in the spring are west-northwesterly.

The stations in Malaysia show interesting seasonal cycles. The 2 stations on the Malaysia peninsula (M1 & M2) show similar seasonal cycles. For example, Mersing, Malaysia, located on the southern part of the Malaysia peninsula, experiences minimum values during the NE monsoons (November to March), and maximum values during the SW monsoon. During the NE monsoon air comes from the South China Sea, while the SW monsoon brings air from the major SO_2 emissions regions, e.g., Singapore, industrial regions around Kuala Lumpur, and the major shipping lane along the Straits of Malacca. The station at Lawa Mandau (M4) located on the island of Borneo shows an opposite seasonal cycle (not shown), with peak values in February to May (0.6 µg/m^3) and minimum values ~0.2 µg/m^3) during the SW monsoon.

Also shown in Figure 3 are model predictions. These results are obtained from a trajectory model developed and used as part of the *RAINS-ASIA* project. Details regarding the model are presented in (Arndt and Carmichael, 1995). At these four locations the model tends to overpredict the concentrations but does capture many of the important features in the observations. The model is being used to help explain seasonal variations in the transport and removal processes, and to estimate the contributions of various sources to selected receptor regions. For example, at Mersing we find that emissions from Kuala Lampur contribute ~15%, emissions from the rest of the Malaysia Peninsula contribute ~20%, and emissions from Singapore and the ship lanes accounting 50% and 5%, respectively. At Yangyang, South Korea, emissions from Seoul area contribute 20%, while emissions from the

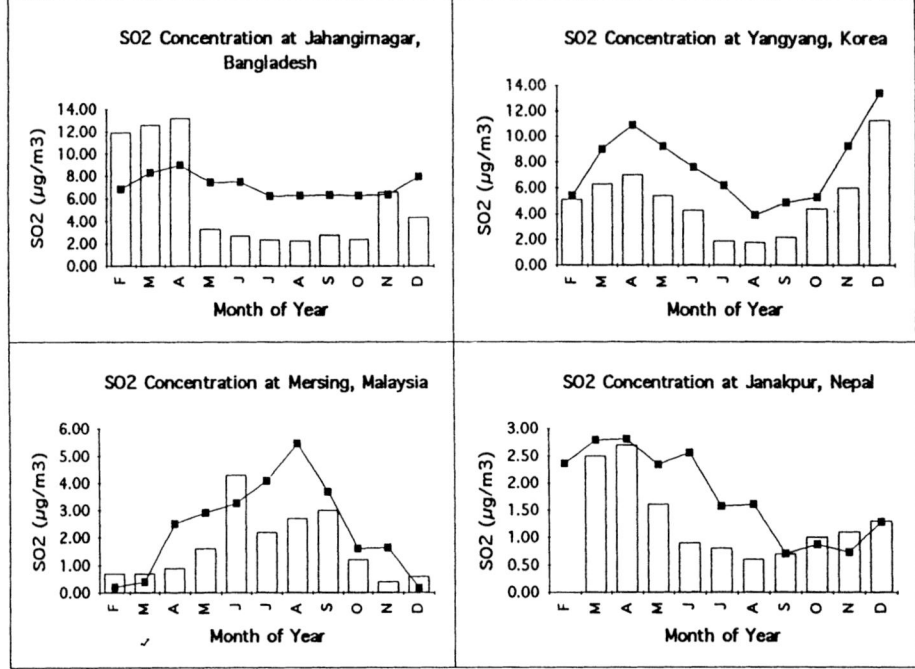

Figure 3. Seasonal cycle of observed surface-level SO_2 concentrations. Also shown are model predicted values. Details regarding the location of the site and the annual average are found in Figures 1 & 2.

remaining northern and southern regions of S. Korea contribute 25% and 5%, respectively. Emissions from N. Korea and China account for 30% and 20%, respectively.

4. Summary

Observed values of monthly surface level SO_2 concentrations at forty-five sites in Asia have been reported. This data is unique in the region in terms of the extent and regional focus of the network. Results show that SO_2 concentrations are elevated throughout Asia. The values are highest in the heavily populated and industrial regions in China, east Asia, and India. However, even at high altitude and fairly remote regions in Malaysia and Nepal appreciable amounts of SO_2 were detected. A distinct seasonal cycle was found at most sites.

Data from networks such as this one offer invaluable information on the distribution of SO_2 in Asia. Such information is essential in the assessment of the environmental impact of the rapid expansion of fossil fuel derived energy in Asia.

Acknowledgments

This work was performed as part of the *RAINS-ASIA* project. Funding was provided by the Asian Development Bank through a subcontract with the Asian Institute for Technology (AIT). Special thanks are extended to Jitendra Shaw at the World Bank and Drs. Bhatacharya and Shresta at AIT.

References

Arndt, R. and G. Carmichael: 1995, *Water, Air and Soil Pollut.* , this issue.
Ferm, M.: 1991, Report of the Swedish Environmental Research Institute, IVL B-1020.
Foell, W., C. Green, A. Amann, G. Carmichael:, 1995, *Water, Air and Soil Pollut.* , this issue.

ON THE ORIGIN AND THE TREND OF ACID PRECIPITATION IN CHINA

WENXING WANG

Chinese Research Academy of Environmental Sciences, Beiyuan, Beijing 100012, China

and

TAO WANG

Department of Civil and Structural Engineering, The Hong Kong Polytechnic University, Hung Hom, Hong Kong

Abstract. The acidity of the precipitation in China as well as the area affected by acidic rainfalls have increased in the years, along with the rapid economic growth in china. Field observations indicate that acid precipitation often occur are the southern par, of china ,although the emissiors of the precursors are stronger in the North.In this paper,we explain the geographical distribution of acid precipitation in china, based on the content alkaline ions in soil ,soil acidity, atmospheric particulate concentration, aerosol buffering capacity and atmospheric dispersion. It is further anticipated that continuous increasing of sulfur emissions in the North will eventually result in acid rain in the northern part of China. On the basis of the projection of SO_2 and No emissions ,it is argued that the problem of acid precipitation may get worse towars the year 2020,if expenditure on emission control maintains at current level.

1. Introduction

Accompanying the rapid economic development and population growth in east Asia, the world third largest acid rain area emerged in this region, following Europe and North America (Dovland, 1993; Winstanley, 1993; Cheng, 1993; Han, 1993). The main part of this acid rain region is located in the southern part of China, and the area affected is estimated to exceed one million km^2 (Wang et al., 1993). High acidity of rainfall and elevated levels of atmospheric SO_2 have caused severe damage to crops, forests and building materials. In the early 1980's, China initiated a national program to systematically study the acid rain problem in China. This program consists of monitoring, atmospheric processes, studies of effects on natural environments, materials damage, and control technology and strategy. Since then, numerous studies have been conducted. For examples, field observations including those obtained from aircraft found that acid rain in Chongqing and Guiyang areas in southwestern China was mainly caused by local emissions (Huang et al., 1988; Zhao et al. 1988); whereas results on the seasonal variations of rain water acidity and meteorological conditions in the FuJian province in southeastern China indicate that acid rain found in the Jiamen area was primarily caused by long range transport (Yu et al., 1992). Model calculations based on wet deposition data in the Spring in the Guangdong Province in South China also suggest that acid rain there seems to be caused by meso-scale and long range transport of pollutants from the northern provinces (Wang et al., 1992a). However, most of the previous work tends to focus on acid precipitation in relatively small spatial scales and are often limited to one or two factors. In the present study, we try to address the acid rain issue on a national scale, and consider several factors that could affect this complex environmental problem. In particular, our attempt focuses on explaining the fact that acidic rainfalls often are observed in South China although the North tends to have stronger sources of precursor emission.

2. Current State of Acid Rain in China

In 1982, the National Acid Rain Monitoring Network was established by the Chinese National Environmental Protection Agency and managed by the National Environmental Monitoring Center. One of the objectives of this network is to collect rain samples nationwide and to analyze them for chemical composition. Mean pH values of rain water obtained from this network in 1982 and 1992 are summarized in figure 1 (Cheng et al., 1993). It is obvious that the acid rain region is located in the South. In addition, it is estimated that the acid rain area has increased by about 600,000 - 700,000 km^2 since 1982 (Wang et al., 1993).

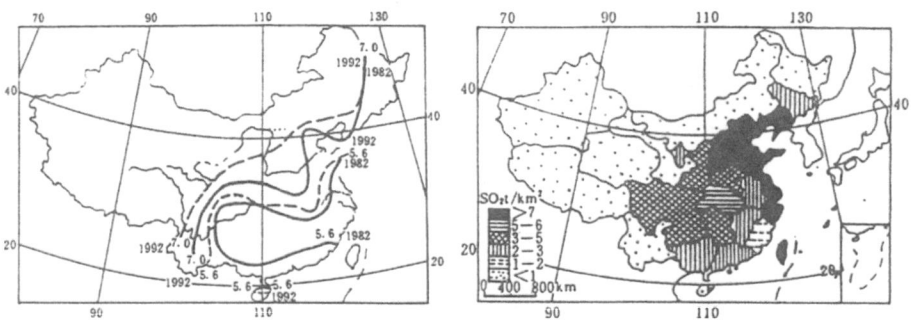

Fig. 1 Contours of annual volume weighted mean pH of rain water

3. Factors Affecting Acid Rain Formation

3.1 Emission intensity of precursors
As a part of the study of acid rain in China, emission intensities of SO_2 and NO_x were compiled for each economical sectors (Wang et al., 1993), based on the fuel consumption, sulfur content of the fuel and emission factors. The province-based SO_2 emission intensities are illustrated in Figure 2. Geographical distributions of NO_x emission intensities are similar to those of SO_2 and are not presented here. From figure 2, it can be seen that provinces with the strongest emission intensities

are located in the coastal regions facing Bo Hai and the Yellow Sea. These provinces include: Jiansu, Shandong, Liaoning, Hebei and Shanxi, all with emission intensity larger than 7.4 ton SO_2/km^2. Although these five provinces combined account for only 9% of the nation's land area, their SO_2 emissions in 1990 were 40 % of total emissions in China. However, acidic rainfalls have not yet been observed in these provinces. On the contrary, southern provinces including Jianxi, Fujian, Guangdong and Guangxi have relatively smaller SO_2 emission intensity (less than 2.6 ton/km^2), and yet they have experienced serious acid precipitation. As will be discussed in the following sections, factors other than precursor emission intensity appear to play more dominant roles in the formation of acid precipitation in the South.

3.2 Chemical composition of precipitation
The composition of rain water collected from 16 cites in the South and 11 cities in the North showed that the averaged concentrations of SO_4^{2-} and NO_3^- in the northern cities are 217.6 and 29.3 ueq/l, respectively. Their values in the South are 125.1 and 20.0 ueq/l, respectively. Total concentration of acidic ions in the North is about 1.7 times of that in the South. Meanwhile, the NH4- content in the rain samples from the South and North is comparable, but there is significant difference in the calcium and magnesium concentrations: total concentration of Ca^{2+} and Mg^{2+} in the North is approximately 3 times of that in the South. Therefore, acidity of rainfall in the North is significantly reduced due to abundant alkaline ions in the rain water. Thus, the acidity of rain water does not depend on absolute amount of acidic ions, but rather the amount of acidic ions relative to alkaline ions.

3.3 Atmospheric particulate matters and their buffering capacity
China has a large territory with distinct natural conditions across the nation. In the North, because of the relatively drier weather and less coverage of vegetation, concentration of atmospheric aerosols is higher than that in the South where it is typically moist and there is abundant plant coverage on soil.
In the non-acid rain region in the North, aerosols have high neutraling ability, thus have stronger buttering ability to acidic rain water. In the acid rain region in the South, however, aerosols tend to be of acidic nature, and thus they can hardly buffer the acidity of rain water.

3.4 The contents of alkaline matters in soil

Studies conducted in different regions in China indicate that 30-70% of urban airborne particulate matters originate from soils (Wang et al., 1992b). The distributions of alkaline matters in soils are closely correlated with the pH contour of rain samples shown in figure 3(Chen, 1995). Namely, regions with low calcium and sodium content are also those affected by acid rain. Therefore, chemical composition of aerosols and its subsequent impact on acidity of rain water may be largely determined by the nature of soils.

3.5 Meteorological Conditions

Meteorology can influence acidity of rain water in at least the following ways: it affects chemical reaction rates, dispersion, transport and deposition of air pollutants.

(1) Effects on photochemical reaction rates

The initial oxidation rate of SO_2 in China has been described by the following empirical formula (Wang et al., 1990)

$R = 0.175 \cdot RH + 2.03 \cdot \ln Io + 0.0704 \cdot [SO_2] - 2.35$

where R is the oxidation rate of SO_2 (ppbv.h-1), RH is relative humidity (%), Io is sunlight intensity (kw/m2), and $[SO_2]$ is mixing ratio of SO_2 (ppbv). This formula shows that stronger intensity of sunlight and higher relative humidity will increase the photochemical conversion rate of SO_2 to sulfate. In general, sunlight intensity and relative humidity decrease with increase of latitude, and thus the South tends to favor faster photochemical production of sulfate.

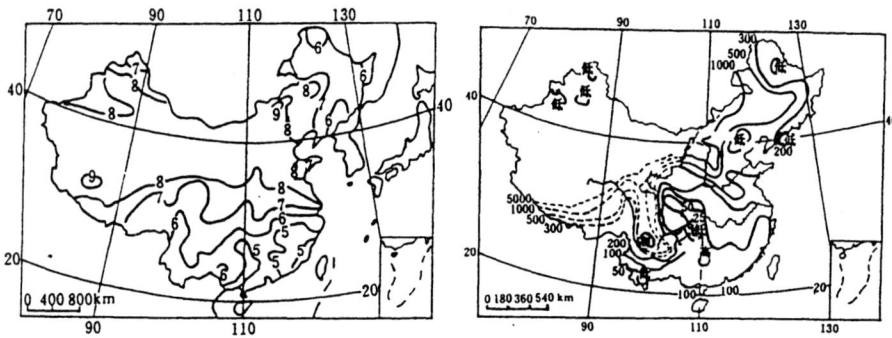

Fig 2 Annual emission intensity of sulfur dioxide, t / km^2 1990

(2) Effect on transport and deposition

Transport and deposition of acidic species are strongly affected by meteorological conditions. Xu et al. (1989) calculated ventilation and rainout capacity in China using a steady-state box model. Figure 4 shows the half value of rainout region length (HRRL) defined as the downwind extent of a box in which air pollutants are rained out as much as carried out of the box by winds, that is

$\Delta Xw = 53 V_E / R$

where ΔXw is half value of rainout region length (km), V_E is ventilation (m^2/s), and

R is precipitation (mm/year). It can be seen that ΔXw is proportional to ventilation and inversely proportional to precipitation and it is a measure of relative intensity of transport versus deposition of acidic species. From figure 4, it can be seen that areas with short HRRL overlap the acid rain regions. HRRLs are shorter than 100 km in the Sichuan Basin in South China, and are also short in the southern part of Yunan Province as well as in some parts of northeastern China. On the other hand, in northern- and large portions of northeastern China, HRRLs are general long and are between 300 to 500 km, suggesting that air pollutants may be easily transported to other regions as opposed to being washed out and removed from the atmosphere.

As discussed above, the distribution of acid precipitation in China can be explained according to precursor distribution, total suspended particulate matters and their buffering capability, alkaline matter contents in soils and meteorological conditions. Based on our current understanding, the future trend of acid rain in China can be speculated.

4. The Trend of Acid Rain in China

4.1 Projection of precursor emissions

From the increase of energy consumption and available desulfurization investment, SO_2 emissions in the year 2020 were projected by Han (1993) that SO_2 emissions in China would continue to increase through the year 2000, 2010 and 2020, reaching 20.98, 27.67 and 31.28 million tons, respectively. SO_2 emissions in the year 2020 would increase by 80% over that in 1990. At that time, China would probably be the largest SO_2 emitter among the three acid rain regions in the world. NO_x emissions are also projected to increase and probably at faster rates since NO_x emission control is more difficult and more expensive.

4.2 Historical trends of acid rain in typical areas in China

Historical trends of pH of rain water from five typical cities in China are shown that pH of rain water showed consistent decrease in all five places from 1982 to 1990. In the Wudu area of the Gansu Province in the Northwest, pH value decreased by 1.5 units in 8 years. The pH of rain water in all other cities have decreased about 0.4-1.0 pH unit in the same time.

4.3 Future trend of acid rain in China

Since the quantitative relationship between acidity of rain water and precursor emissions has not yet been fully established, it will be difficult for us to give a quantitative prediction about the acidity in rain water in China. However, on the basis of the acid rain formation mechanism, historical trends of rain water acidity and the projection of future precursor emissions, we argue that towards the year 2020 the acidity in rain water in China may continue to increase, the area affected by acid rain may continue to expand and may move toward the northern- and western parts of China.

5. Conclusion

5.1 In northern China, five provinces along the coast of Bo Hai and Yellow Sea account for only 9% of total national area, but 40% of total SO_2 emissions. Acid rain has not been observed in the region. It is believed that this due to alkaline nature of atmospheric particulate matters that exert strong buffer to acidity of rain water in the North.

5.2 Atmospheric particulate matters in northern China have very different characteristics than those in the South. Total suspended particulate concentration in the North is nearly twice of that in the South is about one order of magnitude higher than that in the South. Thus the aerosols in the North have stronger buffering ability to acidity of rain water. The difference in composition of airborne particulate matters is attributed to the difference in soil property: alkaline in the North, and acidic in the South.

5.3 Meteorological conditions are also different in the North than in the South: In the South, weaker atmospheric dispersion, higher humidity and more sunlight favor the formation of acidic species there.

5.4 It is projected that SO_2 emission in China will reach 31.78 million tons in the year 2020, or 80% increase from 1990. It is argued that acid rain pollution in China may get worse toward the year 2020. The acidity of rain water may continue to increase, the acid rain area may continue to expend, and new acid regions may emerge.

References

Chen C. (1995) Map of China soil environmental background (in Chinese), China Environmental Science Publishing House, Beijing, China.
Cheng Z. (1989) The state and trend of acid precipitation monitoring in China. in Proceedings of the Expert Meeting on Acid Precipitation Network in East Asia, Toyama, Japan, October 26-28. 1993, pp 209-237.
Dovland H. (1989) in Proceedings of EMEP-the European monitoring and evaluation programme. ibid, 84-47.
Galloway J. N., Likens G. E., Xiong J. L. and Zhao D. W. (1987) Acid rain: A comparison of China, United States and a remote area. Science, , vol 236, 1559-1562.
Han G. (1993) in Prediction of China environmental protection objectives (in Chinese), China Environmental Science Publishing House, Beijing, China.
Han J. (1989) in Proceedings of acid rain monitoring in Korea, ibid, 229-237.
Huang M. (1993) Measurements of acidity and chemical composition of rain water and cloud water. Atmospheric Science (in Chinese), 12(4) 389-395.
Wang W. (1990) Progress in atmospheric chemistry research in CRAES, in Proceedings of Environmental Sciences, pp 21-37, China Environmental Science Press, Beijing, China.
Wang W., Liang J. and Cheng Y. (1992a) Regional sources of acid deposition in South China. J. of China Environmental Sciences (in Chinese), 12 (1) 1-6.
Wang W., Jiang Z., Zhang M. and Wang W. (1992b) Characteristics of aerosol and effect on rain water acidity in south China, ibid 12(1) 7-15.
Wang W., Zhang W, Hong X. and Shi Q. (1993) Study on factors related to acidity of rain water in China. J. of China Environmental Sciences (in Chinese), 13 (1), 401-407.
Winstanley (1993) The United States National Acid Precipitation Program, in Proceedings of the Expert Meeting on Acid Precipitation Network in East Asia, Toyama, Japan, October 26-28. 1993, 46-60.
Xu D. and Zhu R. (1989) A study on the distribution of ventilation and rainout capacity in the Mainland. J. of China Environmental Sciences (in Chinese), 367-374.
Yu S. Cei X. and Cheng Z. (1989) The treads and control of acid rain in China, in Proceedings of formation mechanism of acid rain in Xiamen, China Science and Technology Publishing House (in Chinese), Beijing, China.
Zhao,D.Xiong,J.Xu, Chan, W.H.(1988) Acid rain in southeastern China.Atmos. Environ., 22,349-458

ACIDIC DEPOSITION AND ITS EFFECTS IN SOUTHWESTERN CHINA.

HANS M. SEIP[1], ZHAO DIANWU[2], XIONG JILING[3], ZHAO DAWEI[4], THORJØRN LARSSEN[1], LIAO BOHAN[1], ROLF D. VOGT[1]

1. Dept. of Chemistry, University of Oslo, P.O. Box 1033 Blindern, 0315 Oslo, Norway.
2. Research Center for Eco-Environmental Sciences, Chinese Academy of Sciences, P.O. Box 2871, Beijing, China.
3. Guizhou Inst. of Environmental Sciences, 148 Xinhua Road, Guiyang, China
4. Chongqing Inst. of Environmental Science and Monitoring, 37 Jia Ling VLG-1, Jiang Bei District, Chongqing, China.

Abstract. The emissions of SO_2 in China correspond at present to 8-10 TgSyr^{-1}. The rapid industrialization has caused a dramatic increase in the emissions in recent years and this increase is likely to continue. This paper describes studies of concentrations and effects of acidifying substances in parts of the Guizhou and the Sichuan provinces where the S-emissions are large.

A small catchment about 10 km from Guiyang centre was equipped with instruments for studies of soils, soil water and streamwater chemistry. The molar ratio $Al/(Ca + Mg)$ is > 0.8 in soil water in some places. Two small streams have median pH-values about 4.6 and 5.1. Laboratory studies with selected Chinese soils showed that the anion adsorption was low. These studies gave also important information on soil sensitivity.

The studies confirm that acid deposition may affect soils in parts of south-western China, but the sensitivity varies dramatically and there is a strong need for more information.

1. INTRODUCTION

The emissions of sulphur dioxide (SO_2) in China correspond at present to 8-10 TgSyr^{-1} (Zhao et al., 1994). The rapid industrialization has caused a dramatic increase in the emissions in recent years and this increase is likely to continue. Local effects on human health and on materials are severe in many areas. Emissions, transport, deposition and effects have been studied in several programs. In southern and south-western China soil acidification is indicated both by observations (Dai et al., 1995) and by model calculation (Zhao and Seip, 1991). In the present paper some recent studies in the Guizhou and the Sichuan provinces will be described briefly. The goal is to identify problems related to emissions of acidifying compounds and to provide basic information for formulating control strategies against acidification in China.

2. EMISSIONS AND CONCENTRATIONS

Emissions in the Guizhou Province (176 000km^2) are 0.36 TgSyr^{-1} (Zhao et al., 1995), and about the same (0.41 TgSyr^{-1}) in a 23 000 km^2 area including Chongqing city (Zhao et al., 1994). In both areas the main source is burning of coal with high sulphur content. These emissions result in high SO_2 concentrations in and around the cities Guiyang and Chongqing. Annual averages are 200 - 500 µg/m^3 in the cities, which is much higher than recommended guidelines for health effects: WHO recommends that in combination with other pollutants the 24h average should not exceed 125 µg/m^3 and in Norway a guideline of 40 µg/m^3 for the 6 months' average has been proposed (SFT, 1992). The precipitation is quite acid in large areas as illustrated in Fig. 1. Concentrations in precipitation near Guiyang and Chongqing are compared to values in Poland and southernmost Norway in Table 1.

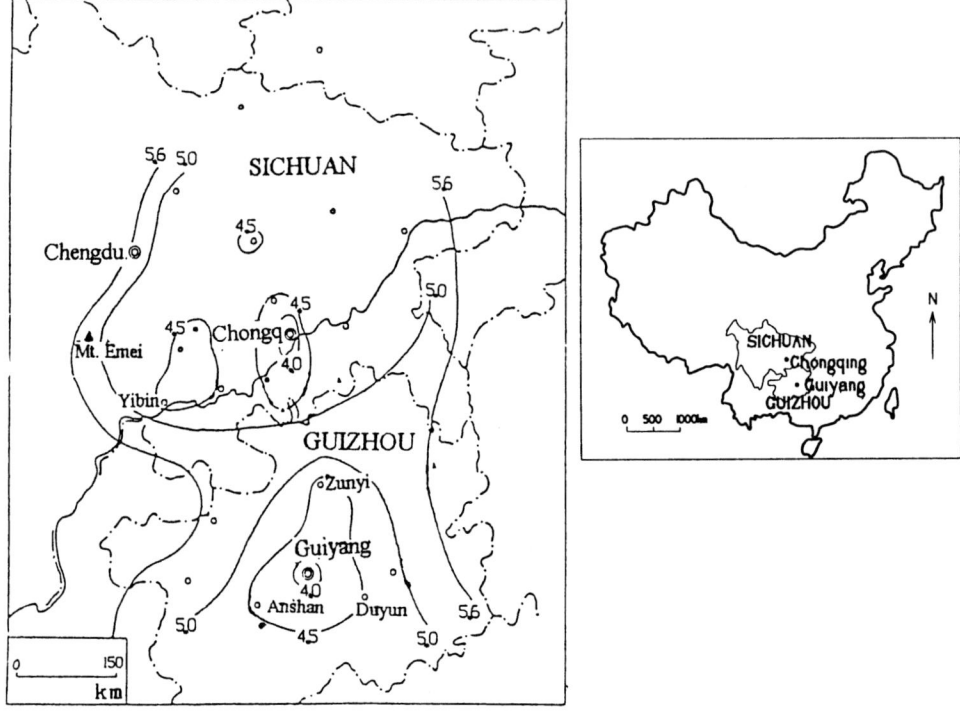

Fig. 1 pH in precipitation in the Sichuan and Guizhou provinces, 1988-89 (Zhao, 1991)

The sulphate concentrations are highest at the Chinese sites, while the opposite is true for nitrate. The acidity is very similar at the Chinese sites and the Norwegian site even if the sum of SO_4^{2-} and NO_3^- is much larger at the former. However, Ca^{2+} and NH_4^+ play a greater role at the Chinese sites resulting in similar pH values.

Table 1
Volume-weighted concentrations (µeq/L) in precipitation at two sites in China and two in Europe. Janow and Birkenes data from Vogt et al., 1994, Chongqing data from Zhao et al., 1994.

Parameter	Guiyang	Chongqing (rural)	Janow (Poland)	Birkenes (Norway)
H^+ (pH)	53 (4.28)	47 (4.33)	28 (4.55)	53 (4.28)
Ca^{2+}	132	74	65	9
Mg^{2+}	27	15	11	13
NH_4^+	25	116	(56)	41
SO_4^{2-}	213	200	110	61
NO_3^-	19	20	60	38

3. CATCHMENT STUDY

A small catchment (Liu Chong Guan) 10 km north-east of Guiyang centre has been instrumented for detailed studies of precipitation, soils, soil water and streamwater (Fig. 2). The area is about 7 ha, vegetation is mainly Chinese fur (*Cunninghemia Lanceolata*) and pine

(*Pinus Massoniana*) and wild rose bushes. Two first order streams join to form a second order stream about 100 m above a dam at the border of the catchment. Soils are generally dystric cambisol (yellow earth in the Chinese system). Seven plots were selected for sampling of soils and soil water. Soil characteristics are given in Table 2. pH (in water) is generally below 4 in the upper (O,A) horizons. The base saturation (BS) decreases from the O-horizons down to the C-horizons, while the aluminium saturation (AlS) shows the opposite trend. Generally AlS is very high; even in the A-horizon it may be about 90%.

Table 2
Effective cation exchange capacety (CEC_E), base saturation and aluminium saturation for soils in the Liu Chong Guan catchment.

Soil Horizon	CEC_E, meq/kg Range	Median	BS, % Range	Median	AlS,% Range	Median	n
O	137 - 223	180	43 - 59	51	38 - 39	39	2
A	58 - 188	99	4.2 - 28	21	67 - 93	77	7
B	58 - 142	84	6.2 - 22	15	76 - 92	82	7
B/C, C	25 - 94	86	3.7 - 40	6.1	58 - 94	92	6

Some results for stream and soil water are given in Table 3. The most acidic stream samples are from the western stream where pH down to about 4.3 has been measured. As to be expected, sulphate concentration values are high, while the nitrate concentrations in streamwater are very low. Concentrations in soil water are given for selected horizons of three plots. Sulphate concentrations vary considerably with the highest values in the C-horizon of plot A. Also the nitrate concentrations vary widely with median values ≈0 to more than 200 µeq/L.

At present we have less results for aluminium. However, for 21 soil water samples the median concentration of monomeric, inorganic aluminium (Ali) was 102 µM (2.76 mg/L) and three values exceeded 150 µM (from the A,E, and G plot). The median molar ratio Ali/(Ca+Mg) was 0.42, while 6 values were > 0.8 (4 from plot C, which has very high values for AlS, one from A and one from G). In spite of these high ratios, no apparent damage to the vegetation has been reported.

For more details about the study see Larssen (1994) and Xiong et al. (1995).

Table 3
Percentiles (10 and 90) and median for pH and concentrations of sulphate and nitrate in stream and soil water in the Liu Chong Guan catchment. Aluminium saturation for the included soils are also given. (The first letter denotes the plot, the second letter refers to the soil horizon).

	AlS, %	pH Low	High	Median	SO_4^{2-}, µeq/L Low	High	Median	NO_3^-, µeq/L Low	High	Median
Surface water:										
W. stream		4.3	6.8	4.6	418	807	535	0	8	2
E. stream		4.5	6.3	5.1	568	1228	766	0	12	4
Dam		4.7	6.7	6.1	436	826	549	0	10	2
Soil / soil-water:										
AA	67	3.7	4.3	3.9	1267	2924	2202	-	7	-
AC	85	3.9	4.4	4.0	1242	4625	3203	0	103	2
BA	76	4.0	5.1	4.2	378	1377	937	6	353	107
BB	76	4.8	6.0	5.2	478	921	721	1	37	8
CA	88	4.0	4.5	4.1	234	1137	481	37	425	131
CB	94	4.1	4.8	4.3	213	467	399	68	387	207

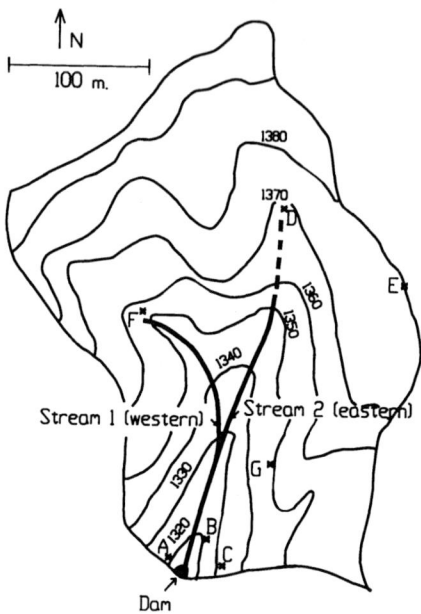

Fig. 2. The Liu Chong Guan catchment close to Guiyang (see Fig. 1).

4. LABORATORY EXPERIMENTS

Anion adsorption was determined for soils from the Guiyang catchment (A and E plots) as well as from a catchment in Nanchang (Liao et al., 1994). At concentrations corresponding to or somewhat higher than ambient levels in soil water in the Guiyang catchment, the sulphate adsorption was found to be low, typically 2 - 4 meq SO_4^{2-}/kg soil. Also some nitrate adsorption was observed; 2 - 3 meq NO_3^-/kg soil in some cases. Adsorption/desorption may be important in controlling short-term fluctuations of sulphate concentrations in soil water, but is not likely to prevent long-term acidification.

In these experiments we also studied cation release. Molar ratios Al/(Ca+Mg) in solutions obtained by mixing 5 g soil with 25 ml solution of $(NH_4)_2SO_4$ or NH_4NO_3 are shown in Fig. 3. Using ambient sulphate levels (about 400 - 3200 µeq/L) the ratio is similar or slightly higher than observed in the catchment indicating that the laboratory experiment is useful in studying soil sensitivity to acidification. (In this experiment fractionation of Al-species was not carried out. The nominator is therefore total Al.) The relationship between the ratio and the sulphate (or nitrate) concentration varies widely for different soils; for the A horizon in plot A the ratio is quite stable even for substantial increases in sulphate concentration.

Considering these relationships one must expect modelling of Al-levels in stream and soil water to be very difficult; a general experience from modelling exercises (see e.g. Stone and Seip, 1989).

5. CONCLUSIONS

Some tentative conclusions may be made by combining our results with those from other studies:

1. Concentrations of SO_2 in many areas of China are so high that harmful effects on health and materials must occur. Regional sulphur pollution seems to be particularly serious in the Sichuan and Guizhou provinces.

2. Compared to Europe and North America sulphate concentrations in precipitation are very high in large areas of China. pH in precipitation is low (< 4.5) in large parts of the Guizhou

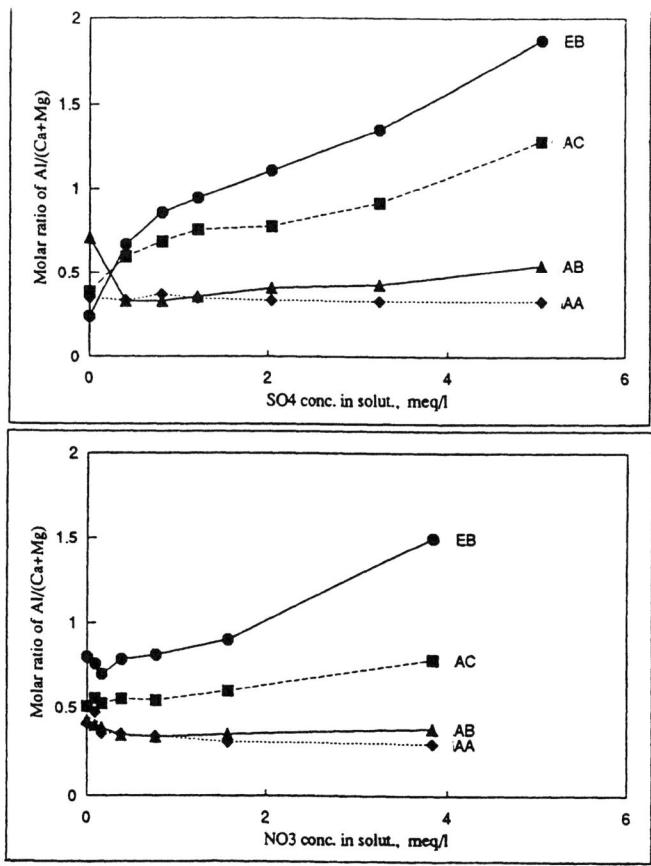

Fig. 3. Molar ratios $Al/(Ca^{2+}+Mg^{2+})$ in aqueous solutions obtained by mixing soil from the Liu Chong Guan catchment with solutions of $(NH_4)_2SO_4$ or NH_4NO_3. (First letter denotes the plot, the second letter refers to the soil horizon).

and Sichuan provinces even though high levels of Ca and/or NH_4^+ tend to decrease the acidity. On the other hand NH_4^+ may contribute to acidification of soils and waters, particularly if nitrification occurs.

3. Soils are often very heterogenous making general predictions difficult. Detailed studies of small catchments are important to increase the knowledge about Chinese soils with respect to acidification and test models to be used for larger areas.

4. Laboratory experiments measuring cation release for ambient and somewhat higher sulphate concentrations seem to be a convenient way to obtain information on soil sensitivity to increased deposition of acidifying compounds.

5. Although small streams may become acidified due to acid deposition, water acidification is not likely to become a major problem in China, because of neutralising processes in deeper soils and bedrock. However, soil acidification is probably going on in some sensitive areas with high deposition as indicated both by observations and model calculations. At present it is not known how serious this problem is likely to become. It will not only depend on future emission of acidifying S-and N-compounds, but also on emissions of basic metal compounds. Careful studies and monitoring are clearly necessary.

References:

Dai, Z., Liu, Y., Wang, X. and Zhao, D.: 1995, "Decrease in pH of soil in southern China within about 30 years", Ambio, accepted for publication.

Larssen, T.: 1994, "Acid deposition in southwestern China", Thesis, Dept. of Chemistry, University of Oslo.

Liao, B., Larssen, T., Seip, H.M. and Vogt, R.D.: 1994, "Anion adsorption and aluminium release from Chinese soils treated with different concentrations of $(NH_4)_2SO_4$ and NH_4NO_3", J. Ecol. Chem. **3**, 281-301.

SFT: 1992, "Effects of ambient air pollution on health and the environment. Air quality guidelines", Norwegian State Pollution Control Authority, Oslo, Norway, Report **93;18**, 179 pp.

Stone, A. and Seip, H.M.: 1989. "Mathematical models and their role in understanding water acidification: An evaluation using the Birkenes model as an example". Ambio, **18**, 192-199.

Vogt, R.D., Seip, H.M., Paw≈owski, L., Kotowski, M., Ødegård, S., Horvát, A., and Andersen, S., 1994. "Potential acidification of soil and soil water: a monitoring study in the Janow Forest, southeastern Poland". Ecol. Engin., **3**, 255-266.

Xiong, J., Larssen, T., Zhao, D., Seip, H.M. and Vogt, R.D.: 1995. "Studies of soils, soil water and stream water at a small catchment near Guiyang". (In prep.)

Zhao, D., (ed.), 1991. "Integrated report of acid rain research in southwest of China." Report from Research Center for Eco-Environmental Sciences, Beijing, China.

Zhao, D. and Seip, H.M.: 1991, "Assessing effects of acid deposition in southwestern China using the MAGIC model", Water, Air and Soil Pollution, **60**, 83-97.

Zhao, D., Seip, H.M., Zhao, D. and Zhang, D.: 1994, "Pattern and cause of acidic deposition in the Chongqing region, Sichuan Province, China", Water, Air and Soil Pollution, **77**, 27-48.

Zhao, D., Mao, J., Xiong, J., Zhuang, X. and Yang, J., 1995. "Critical load of sulfur deposition for ecosystem and its application in China". (To be published.)

PRECIPITATION CHEMISTRY IN JAPAN 1989-1993

HIROSHI HARA*, MORITSUGU KITAMURA, ATSUKO MORI,
IZUMI NOGUCHI, TSUYOSHI OHIZUMI, SINYA SETO,
TADASHI TAKEUCHI, AND TERUYUKI DEGUCHI

Data Analysis Group for the Acid Precipitation Survey over Japan

*Address for correspondence: Department of Community Environmental Sciences,
The Institute of Public Health, Shirokanedai 4-6-1, Minato-ku, Tokyo 108 Japan

Abstract. Precipitation chemistry in Japan was discussed on a wet-only sample database obtained in a nationwide survey from April 1989 to March 1993. Wet-only samples were collected at 29 stations over Japan on a biweekly basis. Commonly determined chemical parameters were measured in laboratories. The volume-weighted annual mean pH at each site ranged from 4.50 to 5.83 with a mean of 4.76. Concentration ranges and means (parenthesized) on an equivalent basis for major ions were as follows: nss-SO_4^{2-}; 5.2-58.9 (38.6), NO_3^-; 1.8-25.0 (14.1), NH_4^+; 0.55-29.8 (18.3), nss-Ca^{2+}; 2.0-34.5 (14.2), Na^+; 6.4-275.3 (49.1), Cl^-; 13.7-322.4 (63.5) μ eq L^{-1}. Acid-base relationships for Phase-II records were quantitatively discussed in terms of three measures: pH, fractional acidity, and our proposed pA_t.

Key words: Precipitation chemistry, sulfate, nitrate, ammonium, calcium, pH, Japan

1. Introduction

In 1992, Japan Environment Agency (JEA) started its Acid Deposition Survey Phase II with different monitoring techniques. Precipitation was sampled at 23 stations of JEA National Air Pollution Surveillance Network (NASN) where air pollutants including sulfur dioxide and nitrogen oxides were routinely measured, and also at newly built remote-island stations surrounding the Main Islands of Japan. In July, 1994, JEA issued the final report of this survey (JEA, 1994) whereas the interim report was published in 1992 (JEA, 1992) on which some scientific papers were presented elsewhere (Hara; 1993).

In this paper, a preliminary assessment is made on precipitation chemistry in terms of compositions, and concentrations of major ions, and further a semi-quantitative description of precipitation chemistry is presented.

2. Experimental

The monitoring sites were shown in Fig. 1 where Nos. 1-23 and Nos. 24-29 correspond to NASN and remote-island stations, respectively. These stations were provided with wet/dry samplers. Samples for chemical analysis were stored in a storage bottle housed in the sampler, and removed biweekly for shipping to analytical laboratories. Parameters determined in the precipitation samples were the major ion concentrations as shown in Table 1, conductivity, and precipitation amount. Details of the techniques can be found in the network document (Tamaki et al., 1993).

Data quality was assessed by calculating the ratio of cation sum to anion sum, R_1, and that of calculated to measured conductivity, R_2. Either of R_1 or R_2 different from unity by more than 20% led to re-analysis of the samples involved to resolve the discrepancy.

Non-seasalt (nss-) components were calculated in the usual way by assuming sodium to be a conservative tracer for seasalt, and estimating seasalt contributions from the known ionic ratios in seawater.

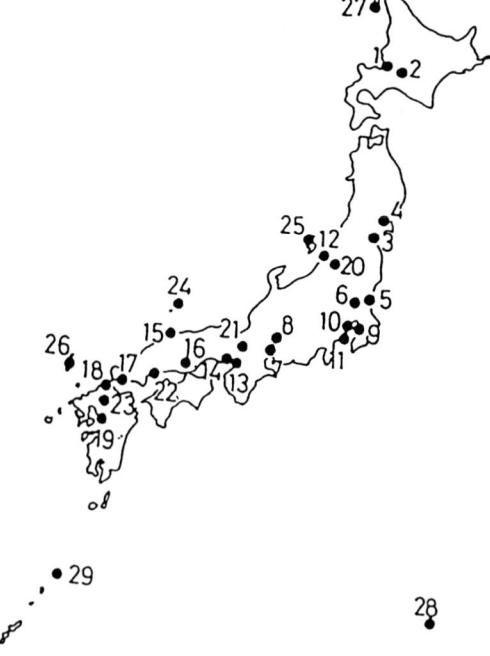

Fig. 1 Sampling Stations in Phase-II.

3. Results and Discussion

3.1. IONIC COMPOSITION AND CONCENTRATIONS

In order to assess the gross features of the precipitation chemistry, the volume-weighted annual mean ionic concentration was calculated as summarized in Table 1, together with their standard deviations and concentration ranges. The measured nine ions successfully explain electroneutrality where the mean ratio of cation sum (117.4 μ eq L^{-1}) to anion sum (122.0 μ eq L^{-1}) was 0.96 for the mean concentrations. Relative concentration of seasalt-derived components, Na$^+$, Cl$^-$, and Mg^{2+}, indicates that approximately half of the ions in the precipitation in Japan were originated from seasalt. Approximately 10% of SO$_4^{2-}$ and Ca^{2+} were of seasalt origin.

Table 1 Concentrations and Deposition of Major Ions

	Concentration/ μ eq L^{-1}				Deposition/meq m^{-2} y^{-1}			
	Mean	s. d.	Min	Max	Mean	s. d.	Min	Max
pH	4.8	0.7	4.5	5.8				
H$^+$	17.3	9.7	1.5	31.6	24.2	15.1	2.2	60.9
NH$_4^+$	18.3	6.7	0.6	29.8	25.9	12.3	1.1	55.4
Ca^{2+}	16.0	8.4	4.9	37.2	21.7	13.4	7.4	59.0
nss-Ca^{2+}	14.2	9.1	2.0	34.5	18.0	13.6	3.0	54.7
K$^+$	3.1	2.8	0.6	13.6	4.9	3.0	0.7	11.0
Mg^{2+}	13.6	15.9	1.8	60.9	22.9	21.2	2.7	80.7
Na$^+$	49.1	70.2	6.4	275.3	86.9	90.9	8.2	365.0
NO$_3^-$	14.1	4.1	1.8	25.0	19.4	7.4	3.1	40.8
SO$_4^{2-}$	44.4	12.5	20.5	63.6	62.5	22.8	22.5	105.0
nss-SO$_4^{2-}$	38.6	12.5	5.2	58.9	52.2	22.6	9.4	99.5
Cl$^-$	63.5	80.4	13.7	322.4	109.0	105.0	15.4	429.0
RF*					1403	319	590	2041

*Rainfall Amount/ mm y^{-1}

Mean pH values ranged from 4.5 to 5.8 with a volume-weighted mean of 4.8, and 70% of the sites had pH values less than 5.0. Although lower pH values were likely to occur in western Japan, the difference was not very significant. The mean pH values during the Phase-I survey period ranged from 4.4 to 5.5. The mean pHs in Phase-II were in the same level as in Phase-I. No distinct trend of pH lowering was inferred when the differences in monitoring techniques between these phases were taken into account.

Generally, pH of precipitation was determined by the acid-base balance relationship. Sulfuric and nitric acids originally formed will undergo neutralization by atmospheric basic species such as ammonia and calcium carbonate. Thus, nss-sulfate and nitrate concentrations are considered to represent the concentrations of these two acids before neutralization takes place in the atmosphere. Morgan (1982) designated the concentration sum of (nss-)sulfate and nitrate as input acidity. Daum et al. (1984) further discussed quantitatively neutralization in terms of their index, fractional acidity, $[H^+]/([nss-SO_4^{2-}]+[NO_3^-])$. In the case of precipitation chemistry in Japan, fractional acidity was 0.33, which indicates that most of the acids input into the atmosphere was neutralized.

Annual mean concentration ranges of nss-sulfate and nitrate were 5.2-58.9 and 1.8-25.0 μ eq L^{-1}, respectively (Table 1). Spatially, maximum concentrations of nss-sulfate and nitrate ions occurred in large urban or industrialized areas whereas considerably low concentrations were observed in remote islands. Non-seasalt sulfate and nitrate mean concentrations were 38.6 and 14.1 μ eq L^{-1} with concentration ranges of 5.2 to 38.6 and of 1.8 to 14.1 μ eq L^{-1}, respectively. Nss-sulfate peaked at more than 50 μ eq L^{-1} in three industrialized cities, (Sites 17, 18, and 19), and nitrate at more than 18 μ eq L^{-1} at Sites 8, 10, and 18.

The ratio of the standard deviation (s.d.) to the mean concentration would be a good quantity representing the concentration gradient over the country. The measured ions were grouped into three in terms of this ratio with their values in the parentheses: 1) the low ratio group; SO_4^{2-} (0.28), NO_3^- (0.29), nss-SO_4^{2-} (0.33), and NH_4^+ (0.37), 2) the medium ratio gourp; Ca^{2+} (0.53), H^+ (0.56), nss-Ca^{2+} (0.64), and K^+ (0.91), and the high ratio group; Na^+ (1.43), Cl^- (1.27), and Mg^{2+} (1.17). The species with high ratios are generally originated from seasalt, which means seasalt concentrations significantly varied from site to site although most of the stations were located on coastal areas.

Low ratio ions will be regarded as being evenly distributed over the country in comparison with seasalt species. This result suggests that these sulfur and nitrogen species would have been transported from long distant sources on a national scale and/or from a number of local sources densely distributed in areas surrounding stations, which happened to have resulted in rather homogeneous distributions of atmospheric sulfur and nitrogen species.

3.2. SEMI-QUANTITATIVE DESCRIPTION OF PRECIPITATION CHEMISTRY

In order to discuss precipitation chemistry at each site from the viewpoint of acid-base relationship, we present a quantitative index pA_i by taking the negative logarithm of input acidity which is the denominator of fractional acidity: $pA_i = -log\,([nss\text{-}SO_4^{2-}]+[NO_3^-])$. This measure, pA_i is physically the minimum pH under the condition that no neutralizing agents such as ammonia and alkaline calcium salts were incorporated into the water droplet in its history before hitting the earth's surface. A plot of pA_i with pH will be very informative.

In Fig. 2, annual mean pA_i values at each site were plotted as a function of corresponding pH. Annual pA_i values appeared in a highly limited range than pH

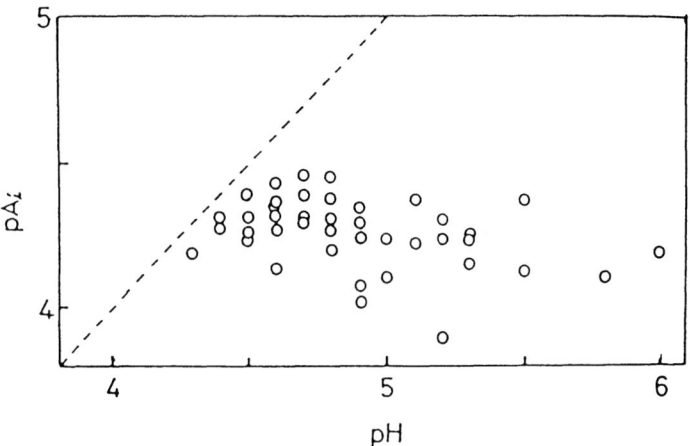

Fig. 2 pAi against pH for Phase-II Annual Means

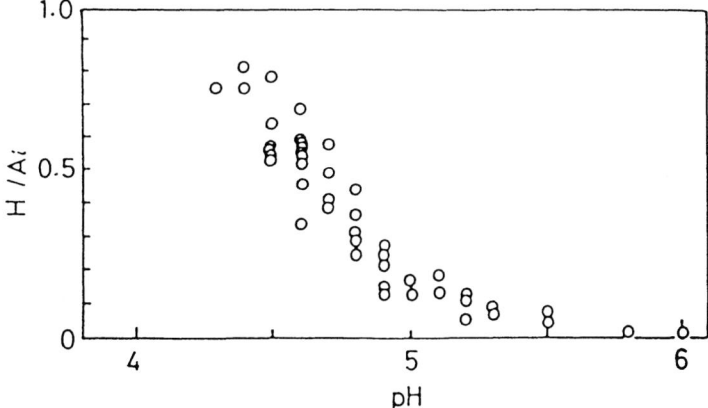

Fig. 3 Fractional Acidity, H^+/Ai against pH for Phase-II Annual Means

values: pA_i values ranged from 3.9 to 4.5 whereas pH from 4.4 to 6.0. This suggests that pH is rather controlled by basic species, which is further evidenced by plotting H/A_i against pH in Fig. 3 where pH values decreased with increasing fractional acidity, H/A_i.

A characteristic of precipitation chemistry in Japan was concluded as follows: concentration of input acidity was rather homogeneous in terms of pA_i within Japan, but pH varied considerably from site to site which resulted from different contributions of basic species.

References

Daum, P. H., Kelly, T. J., Schwartz, S. E., Newman, L. (1984) *Atmos. Environ.*, **18**, 2671-2684.

Hara, H. (1993) *Bull. Inst. Public Health*, **42**, 426- 437.

JEA (Japan Environment Agency) (1992) *Interim Report of Phase-II Survey of Acid Precipitation*, Japan Environment Agency, Tokyo, Japan.(in Japanese).

JEA (Japan Environment Agency) (1994) *Final Report of Phase-II Survey of Acid Precipitation*, Japan Environemt Agency, Tokyo, Japan (in Japanese).

Morgan, J. J. (1982) *Atmospheric Chemistry*, ed., Goldberg, E. D., pp. 17-40. Springer- Verlag, Berlin.

National Atmospheric Deposition Program (NADP) (1993) *NADP/NTN Annual Data Summary: Precipitation Chemistry in the United States, 1992*. NADP/NTN Coordinating Office, Colorado State University, Fort Collins.

Tmaki, M., Katou, T., Sekiguchi, K., Kitamura, M., Taguchi, K., Oohara, M., Mori, A., Wakamatsu, S., Murano, K., Okita, T., Yamanaka, Y., and Hara, H. (1991) *Nippon Kagaku Kaishi*, 1991, 667-674 (in Japanese).

Tamaki, M. (1993) In " *Sanseiu Chosa-ho(Precipitation Chemistry Measurement Techniques)*", Tamaki, M., ed., pp.183-184, Gyosei, Tokyo (in Japanese).

RAINWATER COMPOSITION AND ACID DEPOSITION IN THE VICINITY OF FOSSIL FUEL-FIRED POWER PLANTS IN SOUTHERN AUSTRALIA

G.P. AYERS, R.W. GILLETT, P.W. SELLECK AND S.T. BENTLEY

Division of Atmospheric Research, CSIRO, PB1, Mordialloc 3195, Australia

Abstract. Daily wet-only rainwater composition was determined at four sites in the Latrobe Valley, Victoria, between February 1990 and February 1992. The Latrobe Valley sits on vast resources of easily mined brown coal which at the time of the study fuelled five coal burning power plants of 5150 MW total installed capacity. A sixth gas-fired station of 400 MW capacity is also located in the Valley. Strong preference for down-valley westerly winds on raindays coupled with location of the four rainwater samplers along the central line of the Valley provided an ideal gradation in source-receptor relationships: from the westerly location of Site 1, upwind of all power stations, to Site 2 in the centre of the Valley, upwind of half the emissions sources, to Sites 3 and 4 at the eastern end of the Valley, downwind of all the power plants. Despite the ideal geographic layout the two years of wet deposition data exhibited no clear signal attributable to the power station emissions of sulfur and nitrogen, apparently because raindays in the Latrobe Valley are most often the product of frontal activity accompanied by high wind speeds, leading to high ventilation rates. The resultant wet deposition rates for sulfate and nitrate were in the ranges 7.0 to 14.5 and 3.1 to 4.7 meq m^{-2} y^{-1}, values that are low in comparison with values observed in populated mid-latitude regions of the northern hemisphere.

1. Introduction

The ten-year Latrobe Valley Airshed Study (LVASS), ending in mid-1988, had the aims of providing data on "Class 1 indicators" of air pollution in the Latrobe Valley, 100 km to the east of the city of Melbourne (pop. 3 million, lat. 38°S). Acid deposition was not a feature of the LVASS, because of the absence of any related Class 1 indicator and the paucity of prior work on acid deposition in Australia (Ayers and Gillett, 1984; Ayers et al., 1986; Bridgman, 1988). However with emissions of NO$_x$ and SO$_2$ in the Latrobe Valley in the range 10~100 ktonnes per annum, possibly quadrupling by the year 2005 (AEC, 1989; Marsiglio, 1988), the issue of acid deposition has since arisen. Thus the Latrobe Valley Rainwater Composition Study (LVRCS) was initiated by CSIRO and the State Electricity Commission of Victoria (SECV) in 1990, as a limited study based on rainwater composition within the Valley. Dry deposition and long-range transport were not included in this initial study.

2. Scope of the Study

2.1 AIMS

The study aims were: (1) to determine the acidic and alkaline species that regulate acid-base balance in rainwater in the Latrobe Valley, and (2) to use this information with air quality data available from the Latrobe Valley Air Monitoring Network (LVAMN) and other meteorological data to identify the source of rainwater acidity and acid deposition.

2.2 LOCATION AND DURATION

The Latrobe Valley is ~80 km long by ~30 km wide, with its long axis set east-west. Hills to the south rise to ~500 m, while the continental divide defines the northern boundary of the Valley, rising to over 1000 m. The area is mostly rural, with a few small towns and cities serving the regional population of ~100,000. Centrally located are four coal-fired thermal power stations and one gas-turbine power station of ~5550 MW total installed capacity. The largest power station, Loy Yang (2000 MW), is ~20 km to the east of the other power stations, clustered around the centre of the Valley. The study ran from March 1990 to March 1992.

2.2 SAMPLING STRATEGY

Historical meteorological data from a number of sites in the Valley indicated a very strong preference for westerly winds at 900 hPa on rain days (Ayers and Bentley, 1990). Accordingly four sampling sites were employed, Site 1 (Warragul) situated on the Valley centreline 40 km upwind (to the west) of the cluster of 4 power stations in the central Valley, Site 2 (LV Airport) situated between this cluster and Loy Yang (~10 km from each), with the remaining sampling sites situated downwind (to the east) of all power stations, each about 20 km from Loy Yang, Site 3 (Rosedale N.) to the north and Site 4 (Willung) to the south of the Valley centreline.

A tipping bucket rain gauge (0.2 mm resolution) was used as rainfall sensor, with the time of each bucket tip as well as the time of each opening and closing of the wet-only sampler lid logged electronically. The samplers contained eight polyethylene sample bottles, rotated automatically at 0900 each day, providing rainfall samples as 24-hour averaged wet-only samples. All surfaces in contact with rainwater were either high density polyethylene or Teflon.

The samples were retrieved weekly, with samples preserved against biological degradation by addition of thymol to the sample bottles prior to sample collection (Gillett and Ayers, 1991). Water volume in a separate 5 litre polyethylene bottle located under the tipping bucket rain gauge provided an in-situ, weekly check on tipping bucket rain gauge calibration.

2.3 ANALYTICAL METHODS

Samples were analysed for pH by electrode methods, metals by AA spectroscopy; NH_4^+ by the indophenol-blue method, and anions by suppressed IC. Detection limits were of order a few tenths of a µmole l^{-1} for all species. Precision for an individual analysis was 10% or better for concentrations above a few µmole l^{-1}. Due to insufficient funding for complete analyses only pH, Na^+, NH_4^+, Cl^-, NO_3^- and SO_4^{2-} were determined in samples having with pH>5.

3. Results

3.1 DATA QUALITY

The overall efficiency of the wet-only samplers for rainfall collection was readily determined by comparison of sampler rainfall with the tipping bucket rainfall totals, and was greater than 95% at all sites in both years. Rainfall data for the period of the study are displayed in Table I.

Table I. Numbers of raindays and tipping bucket rainfall totals.

	number of raindays				rainfall	(mm)		
	Site 1	Site 2	Site 3	Site 4	Site 1	Site 2	Site 3	Site 4
Year 1	183	189	198	175	1013	812	611	723
Year 2	216	195	186	190	982	765	588	626

The quality of the chemical analyses was assessed using the ion balance criteria employed by the US EPA (EPA, 1994). Overall 70% of samples yielded ion imbalance (calculated as difference between anion and cation sums expressed as a % of their mean) of less than ±20%, with only 2.7 % of samples yielding an ion imbalance of > 100%.

3.2 Rainwater Composition

Table II gives the volume weighted mean data for each site calculated over the 2 year period of study. The "non-sea-salt" (nss-) components were calculated assuming that sodium acts as a conservative tracer for sea-salt, with sea salt composition defined by Millero (1974).

Table II. Volume weighted mean ion concentrations (std error). Units : μmole l^{-1}.

	#1 Warragul	#2 LV Airport	#3 Rosedale N.	#4 Willung
H$^+$	6.8 (0.4)	8.4 (0.6)	9.2 (0.6)	9.4 (0.8)
Na$^+$	62.0 (7.0)	46.9 (6.2)	46.3 (7.0)	52.7 (7.6)
K$^+$	1.9 (0.3)	1.5 (0.2)	1.6 (0.3)	2.9 (0.4)
nss-K$^+$	1.1 (0.2)	0.9 (0.2)	0.9 (0.2)	2.0 (0.4)
Mg^{2+}	5.3 (0.9)	7.0 (1.0)	5.3 (0.9)	7.3 (1.0)
nss-Mg^{2+}	1.0 (0.2)	3.8 (0.6)	1.7 (0.3)	2.6 (0.5)
Ca^{2+}	2.8 (0.3)	4.7 (0.6)	3.2 (0.3)	4.1 (0.4)
nss-Ca^{2+}	1.9 (0.3)	4.0 (0.6)	2.5 (0.3)	3.2 (0.4)
NH$_4^+$	9.9 (0.5)	6.6 (0.6)	7.1 (0.6)	9.4 (0.9)
Cl$^-$	72.9 (8.1)	53.2 (6.3)	53.5 (7.7)	72.6 (9.2)
nss-Cl$^-$	0.1 (1.0)	2.5 (1.5)	1.0 (1.2)	7.0 (1.5)
NO$_3^-$	4.8 (0.4)	5.0 (0.4)	5.2 (0.5)	5.1 (0.5)
SO$_4^{2-}$	8.1 (0.7)	12.0 (1.0)	8.7 (0.6)	9.9 (0.8)
nss-SO$_4^{2-}$	4.7 (0.3)	9.2 (1.0)	5.9 (0.5)	6.7 (0.6)
CH$_3$SO$_3^-$	0.069 (0.012)	0.13 (0.02)	0.12 (0.02)	0.11 (0.02)
HCOO$^-$	7.5 (1.0)	8.3 (1.2)	7.5 (1.1)	7.9 (1.2)
CH$_3$COO$^-$	5.2 (0.6)	5.6 (0.7)	5.1 (0.7)	5.0 (0.8)

3.3 Wind Direction

900 hPa wind data obtained from the National Climate centre at 6 hourly intervals from Laverton (west of the Valley) and East Sale (east of the Valley) were each averaged to produce a vector mean wind direction for each 24 hour sample period in which rainfall was recorded at any LVRCS site (thus 4 or 5 data points were nominally available to make each 24 hour average).

Likewise a daily mean wind direction on LVRCS raindays was obtained from the data supplied by the SECV for surface level (10 m) winds at the Darnum North, Thoms Bridge, Minniedale Road and Rosedale South LVAMN sites. Frequency plots of resultant data for both 900 hPa and two of the four 10 m wind sites are presented in Figure I. As in the analysis of Ayers and Bentley (1990), raindays in the Latrobe Valley show 24 hour averaged wind direction at 900 hPa and 10 m predominantly from between southwest and northwest, with an additional minor preference for easterly winds at 10 m on some occasions of westerly 900 hPa winds.

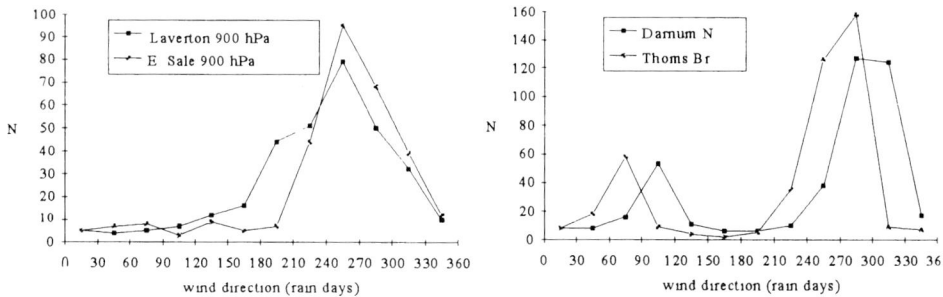

Figure I. Frequency plots for 900 hPa wind direction at Laverton and East Sale, left, and frequency plots for 10m wind direction at Darnum North and Thoms Bridge, raindays only.

4. Discussion

As a "worst case" scenario we assume that the acidifying cations H^+ and NH_4^+ in rain are associated entirely with nitric and sulfuric acids, which conceptually at least could arise mainly from anthropogenic sources. Thus in rain we would expect a simple acid-base balance between acid cation sum ($[H^+] + [NH_4^+]$) and acid anion sum ($[NO_3^-] + 2[nss-SO_4^{2-}]$). The illustrative plot for Site 1 data shown in Figure II and regression results in Table III suggest it is reasonable to assume that free acidity in the rainwater is contributed mostly by sulfuric and nitric acids. Thus we assess the levels of acid deposition in the LVRCS in terms of wet deposition fluxes of nss-sulfate and nitrate, and compare these fluxes with those from locations elsewhere in Australia that are removed from industrial sources of SO_2 and NO_x.

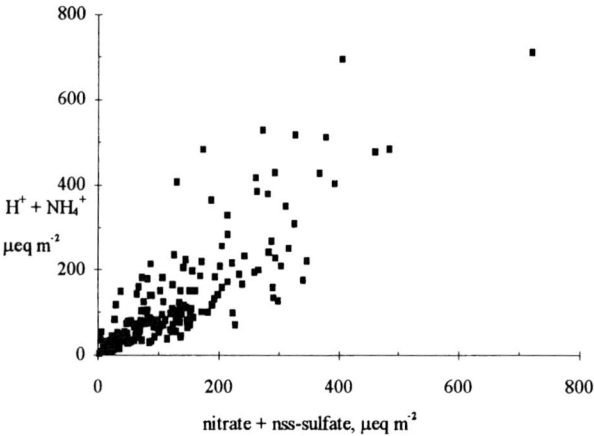

Figure II. Acid-base balance in LVRCS wet deposition data, Site 1.

Table III. Regression of "acid-base" cation sum, $[H^+] + [NH_4^+]$, against anion sum, $[NO_3^-] + 2[nss-SO_4^{2-}]$, in wet deposition ($\mu$mole m^{-2} day^{-1}).

	Site 1	Site 2	Site 3	Site 4
Slope	1.01	0.65	0.89	0.96
Stand. error	0.03	0.03	0.04	0.04
Number	202	171	142	137
r^2	0.85	0.76	0.77	0.79

The wet deposition fluxes of NH_4^+, nss-SO_4^{2-} and NO_3^- over the two years of the study are presented in Table IV. Despite the fact that has Site 1 is upwind of all the power stations, Site 2 is upwind of half the power stations, and Sites 3 and 4 are downwind of all power stations during the westerly winds on raindays, there is no clear trend in deposition. Moreover, the deposition fluxes are low in absolute terms by the standards of regions overseas impacted by acid deposition (Chadwick, 1990; Hettelingh et al., 1991), and are little elevated above background levels from the Australian sites of Katherine (Likens et al., 1987), Jabiru (Gillett et al., 1990) and Wagga Wagga (Ayers and Manton, 1991) that are remote from power stations.

Table IV. Annual average wet deposition of indicated species in meq m^{-2} during the LVRCS.

	Site 1	Site 2	Site 3	Site 4	Katherine	Jabiru	Wagga
NH_4^+	6.9	5.1	4.2	6.3	3.0	2.7	7.4
NO_3^-	4.7	4.0	3.1	3.4	4.3	5.0	5.6
nss-SO_4^{2-}	9.4	14.5	7.0	9.0	4.4	7.4	7.0

However assessment is more properly carried out in terms of total acidic deposition, *i.e.* wet plus dry deposition. Since measurement of dry deposition was outside the ambit of the LVRCS, only an indicative calculation of dry deposition fluxes in the Latrobe Valley is possible.

Calculation of gas deposition fluxes requires knowledge of deposition velocities, Vi, but these are unmeasured in the Latrobe Valley. The most recent compilation of well specified data from which to infer Vi is that of Hicks et al. (1991) and Meyers et al. (1991), who report seasonal mean and annual mean deposition data for S and N species determined continuously between 1984 and 1987 at nine US sites. Dry deposition of NO_2 was studied by Hanson et al. (1989) at the same sites (see also Hanson and Lindberg, 1991). From these studies we derive mean deposition velocities averaged over all sites, excluding the winter data as wintertime snow surfaces do not occur in the Latrobe Valley. Vi of 0.0036 m s^{-1} results for both SO_2 and NO_2.

To calculate dry deposition for the Latrobe Valley the annual average NO_2 and SO_2 mixing ratios for the 13 LVAMN sites from September 1986 to August 1987 were used (3.8 ppbv and 1 ppbv, respectively). Conversion of the mixing ratio units to molar units was carried out at 20°C and 1013.25 hPa ; combining these values with Vi given above yields the results presented in Table V, where the wet deposition values averaged over all LVRCS sites are also given.

Table V. Comparison of mean (annual average) wet and dry deposition of sulfur and nitrogen in the Latrobe Valley. Units are meq m^{-2} y^{-1}.

Dry deposition		Wet deposition	
SO_2	NO_2	nss-SO_4^{2-}	NO_3^-
9.4	18	10	3.8

The results given in Table V suggest that wet deposition may be outweighed appreciably by dry deposition in the Latrobe Valley, especially, since deposition of HNO_3 is an important component in the dry deposition data of Meyers et al. (1991), but is omitted here. Addition of the gas and rainwater values for each of the sulfur and nitrogen species in Table VI suggests total deposition values of order 50-60 millieq m^{-2} y^{-1} (assuming HNO_3 dry deposition equals NO_2 dry deposition, as suggested by the data of Hanson et al., 1989, and Meyers et al., 1991), of which three quarters is dry deposition. These values are comparable with the critical loads discussed by Chadwick (1990) for the most sensitive regions of the northern hemisphere, but are considerably less than those of less sensitive environments that are probably comparable with the Latrobe Valley (see *e.g.* Hettelingh et al., 1991).

5. Conclusions

The main conclusions to arise from the LVRCS are : (1) the acid-base balance in rainwater at all four LVRCS sites was determined by a mixture of acids including sulfuric, nitric, formic and acetic acids neutralised partially by ammonia and calcium carbonate; (2) no significant gradient in acidic deposition was observed from west to east across the LVRCS sites, which in view of the persistent westerly wind direction on raydays suggests that power station emissions do not have a dominant influence on wet deposition fluxes of acidity in the near field; (3) the absolute levels of acidic deposition of nitrate and non-sea-salt sulfate in LVRCS rainwater samples over the study period were low by the standards of industrialised regions in the northern hemisphere; (4) a qualitative estimate of dry deposition fluxes of NO_2 and SO_2 in the Latrobe Valley suggests that wet deposition is the less important pathway for acidic deposition in this region; and (5) a simple estimate of total acidic deposition (dry plus wet) is also low by the more polluted northern hemisphere standards, being comparable with levels assessed overseas as potentially significant only for the more sensitive of soil/surface-water environments.

Acknowledgments

We appreciate greatly the contributions made to the study by Bob Joynt and Bill Fitzgerald, of the SEC of Victoria. The financial support of the SECV is also gratefully acknowledged.

References

AEC, 1989, Acid Rain in Australia : a national assessment. Australian Environment Council Report No. 25, Australian Government Publishing Service, Canberra.
Ayers G.P. and Bentley S.T., 1990, Meteorological perspectives on site selection for a rainwater composition study in the Latrobe Valley, *Aust. Met. Mag.*, **39**, 95-103.
Ayers G.P. and R.W. Gillett, 1985, Some observations on the acidity and composition of rainwater in Sydney, Australia, during the summer of 1980-81, *J. Atmos. Chem.*, **2**, 25-46.
Ayers G.P., R.W. Gillett and U. Cernot, 1987, Rainwater acidity in Sydney, an addendum, *Clean Air (Aust.)*, **21**, 68-69.
Ayers G.P. and Manton M.J., 1991, Rainwater composition at two BAPMoN regional stations in SE Australia, *Tellus*, **43B**, 379-389.
Bridgman H.A., 1989, Acid rain studies in Australia and New Zealand, *Arch. Environ. Contam. Toxicol.*, **18**, 137-146.
Chadwick M.J., 1990, Relative sensitivity, critical loads and target loads and their inclusion in acidic deposition abatement strategy models. Proc. 2nd Workshop on Acid Rain in Asia, 19-22nd Nov. 1990 (Asian Inst. of Technology, Bangkok, Thailand).
EPA, 1994, Quality Assurance Handbook for Air Pollution Measurement Systems, Volume V: Precipitation Measurement Systems, United States Environmental Protection Agency, Office of Research and Development, Washington, EPA/600R-94/038e.
Gillett R.W. and Ayers G.P., 1991, The use of thymol as a biocide in rainwater samples. *Atmos. Environ.*, **25A**, 2677-2681.
Gillett R.W., Ayers G.P. and Noller B.N., 1990, Rainwater acidity at Jabiru, Australia, in the wet season of 1983/84, *Sci. Tot. Environ.*, **92**, 129-144.
Hanson P.J., Rott K., Taylor G.E. Jr, Gunderson C.A., Lindberg S.E. and Ross-Todd B.M., 1989, NO_2 deposition to elements representative of a forest landscape, *Atmos. Environ.*, **23A**, 1783-1794.
Hanson P.J. and Lindberg S.E., 1991, Dry deposition of respective nitrogen compounds : a review of leaf, canopy and non-foliar measurements, *Atmos. Environ.*, **25A**, 1615-1634.
Hettelingh J.-P., Downing R.R. and de Smet P.A.M., 1991, Mapping Critical Loads for Europe, CCE Technical Report No. 1, Coordination Center for Effects, Nat. Inst. Public Health and Env. Protection, Bilthoven, Netherlands.
Hicks B.B., Hosker R.P. Jr, Meyers T.P. and Womack J.D., 1991, Dry deposition inferential measurement techniques - I. Design and tests of a prototype meteorological and chemical system for determining dry deposition, *Atmos. Environ.*, **25A**, 2345-2359.
Likens G.E., Keene W.C., Miller J.M. and Galloway J.N., Chemistry of precipitation from a remote terrestrial site in Australia, *J. Geophys. Res.*, 92, 13299-13314, 1987.
Marsiglio J.A., 1988. Sources of pollutants important to the Latrobe Valley Airshed. *Clean Air*, **22**, 160-166.
Meyers T.P., Hicks B.B., Hosker R.P., Womack J.D. and Satterfield L.C., 1991, Dry deposition inferential techniques-II. Seasonal and annual deposition rates of sulfate and nitrate, *Atmos. Environ.*, **25A**, 2361-2370.
Millero F.J., 1974. The physical chemistry of sea-water. *Ann. Rev. Earth Plan. Sci.*, **2**, 101-150.
NAPAP, 1991 Acid Deposition : State of Science and Technology. The NAPAP Office of the Director Washington, D.C., 2053.
Tucker G.B., 1988. An overview of the Latrobe Valley Airshed Study. *Clean Air*, **22**, 129-132.

TERRESTRIAL ECOSYSTEM SENSITIVITY TO ACIDIC DEPOSITION IN DEVELOPING COUNTRIES

J.C.I. KUYLENSTIERNA[1], H. CAMBRIDGE[1], S. CINDERBY[1] and M.J. CHADWICK[2]

[1] *Stockholm Environment Institute at York, Biology Department, University of York, Box 373, York YO1 5YW, U.K.*
[2] *Stockholm Environment Institute, Lilla Nygatan, Box 2142, S-103 14 Stockholm, Sweden*

Abstract. Acidic deposition is considered a problem in Europe and North America but the potential for ecosystem damage from this pollution is also increasing rapidly in many developing countries. It is therefore important to assess current and future risks of ecosystem effects due to acidic deposition in these areas. It is possible to indicate risk areas by linking an assessment of sensitivity to net acidic input rates derived from deposition estimates for sulphur and nitrogen compounds and base cations. A method to assess and map a relative scale of terrestrial ecosystem sensitivity using international datasets is presented. The assessment relies on the determination of buffering mechanisms that prevent effects related to acidic deposition. Land-cover data, edaphic and climate datasets are combined using a GIS. Large areas are assessed as highly sensitive to acidic deposition in tropical regions of Asia, South and Central America and Africa, and also in the Boreal forests of northern Asia. Sensitive areas cover forest and non-forest ecosystems and some areas of agricultural production. Critical loads are not evaluated in this project but initial estimates will be applied to sensitivity classes at a further stage which will allow estimation of areas at risk by comparison with deposition.

key words: sensitivity; buffering; acidification; vegetation effects; soil; Al toxicity; GIS; mapping; nutrient

1. Introduction

Acidic deposition has affected Europe and North America for over a century and control measures are being implemented to reduce emissions. In contrast, there has been little focus on acidic deposition in developing countries as other problems have had higher priority. However, in recent years, the emissions of sulphur, nitrogen oxides and ammonia have been increasing rapidly as certain countries develop their economies and interest in air pollution effects has increased in these regions.

The impact of acidic deposition on the natural environment depends on the interaction between magnitude of deposition and sensitivity of the ecosystem. The sensitivity investigated here refers to the susceptibility of terrestrial ecosystem to changes in structure and function. This depends largely on buffering processes in the soil-plant system. Vegetation changes are assumed to occur as a result of alterations in soil water chemistry. Acidification of the soil involves the reduction in pH and leads to changes in nutrient and aluminium availability (potentially toxic to plant roots) (Ulrich & Sumner, 1991; Jordan, 1985). This can, under extreme conditions, lead to changes in plant community composition and decreases in growth rates. Sensitivity mapping gives a relative measure of the susceptibility of ecosystems to such changes.

Sensitivity mapping has predominantly been carried out in North America (e.g. Lucas & Cowell, 1984; McFee, 1980; Norton, 1980) and Europe (e.g. Edmunds & Kinniburgh, 1986; Chadwick & Kuylenstierna, 1991). It has recently formed part of the RAINS-ASIA project (Hettelingh *et al.*, 1995). These studies have described the sensitivity of soils, terrestrial ecosystems, ground- and surface waters and have used soil, geology, climate and land cover characteristics.

Various methods to calculate and map critical loads to inputs of acidity have been developed (e.g. Sverdrup, de Vries & Henriksen, 1989). These are defined as threshold deposition values above which deleterious changes to ecosystem structure and/or function arise (Nilsson & Grennfelt, 1988). Critical loads are largely determined by buffering rates and so are proportional to the sensitivity classification derived here allowing eventual application of critical loads to sensitivity classes.

This paper explains how relative sensitivity mapping has been carried out for Africa, Asia and South and Central America representing part of an initial investigation of the risk posed to terrestrial ecosystems of developing countries by acidic deposition. The sensitivity mapping is linked to projects determining atmospheric transfer and deposition of sulphur compounds in these continents (Robertson et al., 1995) which will allow eventual comparison to critical load estimates assigned to the sensitivity map.

3. Sensitivity evaluation for developing countries

The distribution of sensitivity is defined using three factors: soil type, land cover and a measure of soil moisture. Each data layer provides information concerning the buffering ability of the terrestrial ecosystem. The method reclassifies types of soil, vegetation and climate into a small number of categories (explained in the following paragraphs) which are combined using weights to produce a relative scale of sensitivity.

Soil types, based on the FAO soil map of the world (FAO-Unesco, 1988), were reclassified into four categories based upon the typical pH, base saturation and cation exchange capacity (CEC) in the rooting zone (Table 1). These values are derived largely from information supplied by FAO (FAO, 1992a). Base saturation and pH are intensity parameters used here to indicate the supply rate of base cations to the upper soil horizons and rate of neutralising weathering reactions. Weathering rates are related to soil mineralogy, and base cation enrichment from lower soil horizons increases base saturation and pH. The CEC is a capacity-limited parameter related to the rate at which soil base saturation will decrease due to acidic deposition rates above the weathering rate (McFee, 1980). Weights are assigned to the soil attributes (Table 1) which are summed and reclassified to give rise to four classes of buffering ability. Class 1 contains highly acidic tropical soils such as Acrisols and Ferralsols; Class 2 contains acidic soils such as Podzols and Histosols; Class 3 contains eutric, non-calcareous soil types and Class 4 contains the calcareous soil types and soils typically developed in arid regions.

The vegetation type indicates features relevant to site sensitivity such as rooting depth, occurrence of flooding, land management, mineralisation rates, leaf litter quality, nutrient cycling rates and tolerance of vegetation to acidic conditions. For example,

Table 1 Weights attached to soil attribute ranges used to assess the buffering ability of soil types

pH	weight	CEC (meq/100g)	weight	Base saturation (%)	weight
<4.0	1	<10	1	<25	1
4.0-5.5	2	10-20	2	25-50	2
5.5-7.2	3	20-50	3	50-75	3
>7.2	4	>50	4	>75	4

deeply rooted vegetation can enrich the surface soil layers with nutrients from lower horizons (Eyre, 1968). The nutrient demands and nutrient cycling rates indicate the potential for nutrient leaching from the soil and resulting buffering ability of the soil-plant system (Eyre, 1968). Land-use practices, such as liming, increase buffering, and seasonal flooding enriches the soil with high buffering rate minerals, both impacts leading to decreased sensitivity (McFee, 1980). The tolerance of vegetation to acidic conditions and high aluminium concentrations caused by acidic deposition will affect changes in the vegetation structure and function (Andersson, 1988). Vegetation types (land cover) have been assigned to one of four categories based on ecosystem buffering ability and tolerance of acidic conditions deduced from typical ecosystem characteristics. A digital land cover database (Rutgers University, 1990) has been used to infer vegetation type and land management practices.

The precipitation to potential evapotranspiration (P:PE) ratio is used to indicate site leaching characteristics which influence sensitivity. In arid and semi-arid regions little leaching occurs and the net annual movement of water is upwards leading to accumulation of alkaline weathering products (particularly $CaCO_3$) in the surface soil horizons resulting in pedocal formation (Eyre, 1968). The high buffering rate of minerals accumulating in these soils results in low sensitivity. The higher the soil moisture, the greater the potential for base cation leaching leading to pedalfer formation (Eyre, 1968) which represent soils susceptible to acidification. Prolonged, high leaching rates lead to acidic, highly weathered soils in the tropics and to highly organic acidic soils in temperate climates (Eyre, 1968), both with low buffering ability. The soil moisture has been characterised by three classes of P:PE ratio (Table 2). Precipitation and potential evapotranspiration were interpolated from point databases (FAO, 1992b).

4. The distribution of sensitivity

The distribution of buffering characteristics described by the three data layers were combined to form the final sensitivity map (within a GIS). The layers reinforce each other and give a more representative distribution of sensitivity than from one layer alone (Lucas & Cowell, 1984). For example, soil type gives no indication of rooting depths, land-use practices or tolerance and such information is supplied from the vegetation database. The soil moisture data will give an indication of current soil characteristics, whereas soil type shows the result of influences during soil development.

The combination method is a simple addition of the data layers with the exception of areas with calcareous or aridic soil types which are automatically assigned the lowest sensitivity class. The factors are combined with equal weights even though some factors may be of greater importance in defining sensitivity. A simple method of combining weights (shown in Table 2) has been used in the initial development of the methodology considering that these are large and relatively poorly researched areas. In the further development of the method higher weights may be assigned to some of the factors. The addition of the weights in Table 2 results in eight possible classes of sensitivity (3-10) which have been reduced to five as it is considered that a higher differentiation is not warranted by the uncertainties in data and methods.

Figure 1 shows the distribution of five classes of relative sensitivity in developing countries. The least sensitive ecosystems are located in desert areas, such as the Sahara,

Table 2 Weights assigned to categories of data layers for combination to form sensitivity classes

Data layer	Category	Description	Weight for combination
land cover/use	i	sensitive	1
	ii	sensitive-intermediate	2
	iii	intermediate-insensitive	3
	iv	insensitive	4
soil type	i	v. low buffering ability	1
	ii	low buffering ability	2
	iii	medium buffering ability	3
	iv	high buffering ability	insensitive
soil moisture	i	very humid/humid (P:PE <0.7)	1 (most sensitive)
	ii	moist sub-humid (P:PE 0.7-1.1)	2
	iii	semi-arid/arid (P:PE >1.1)	3 (least sensitive)

South-western Africa, the Middle East and North-western China. The most sensitive areas are in the old acidic tropical soil areas with high rainfall, such as in south-east Asia, Central Africa and the Amazon basin. In some of these tropical soils sulphate adsorption capacity will be very high and will mitigate against the effects of acidic deposition associated with sulphur deposition, for a time at least (McDowell, 1988). This aspect requires closer attention to determine the extent to which this mechanism will buffer. Sulphate adsorption would not, however, affect potential acidification by nitrate and ammonium deposition.

5. Discussion and conclusions

Broad classes of sensitivity have been distinguished for developing countries using methods that have previously been successfully applied in Europe. In southern Asia the sensitive areas overlap with areas of high population and high emission. The potential for effects is large and there are initial indications of acidification related effects in southern China (Zhao et al., 1994). In Africa the most sensitive areas are mainly in sparsely populated areas which currently have low emissions. The potential for effects may well be highest in South Africa where emissions are rising (Olbrich et al., 1995). In south America the sensitive regions are mainly in sparsely populated areas but the emissions in parts of the continent are likely to become high and may affect these sensitive ecosystems. The sensitive sites mainly consist of forest and other semi-natural ecosystems of the developing world. The potential for effects on agricultural lands in marginal soils of developing countries is greater than in developed countries where liming is more widely practised. The real risk is unknown as yet. In Figure 1 large areas of sensitive ecosystems in developing countries are identified. The lack of available data and uncertainties concerning effects on tropical and sub-tropical ecosystems necessitates a rather uncomplicated method for such regions providing an initial overview of potential problems. Distributions are considered preliminary and will change.

Critical loads will be assigned to the sensitivity classes by comparison with site-specific studies in various parts of the world in a further stage of this project. The critical loads will be set at the level of the weathering rate as it is assumed that this buffering will maintain the ecosystem structure and function at a reasonable level. Such critical

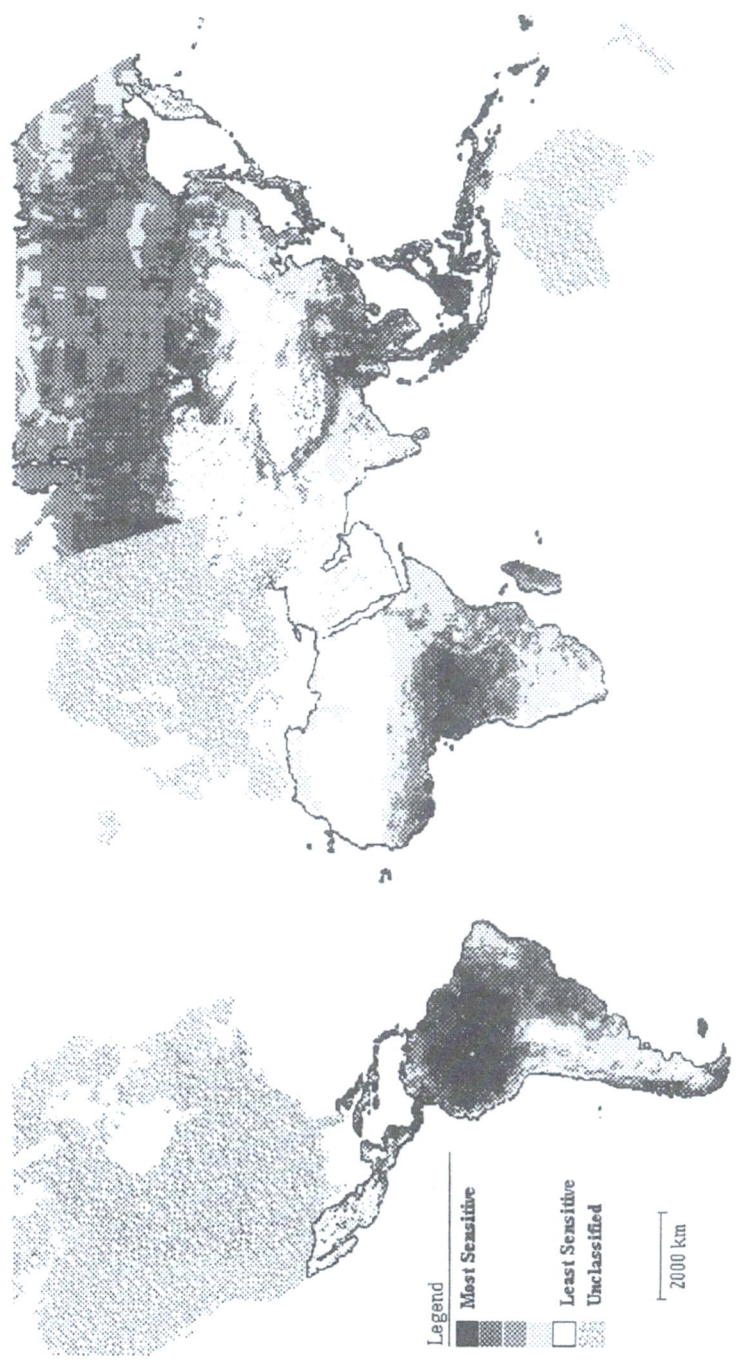

Figure 1 The sensitivity of ecosystems to acidic deposition in developing countries

load maps may be compared to net acidic deposition to indicate areas at risk. Neutralising base cation deposition estimates will be required for this exercise.

Figure 1 represents an initial attempt to map the sensitivity to acidic deposition at a global scale for use in raising awareness. This will allow policy development for avoidance of effects in the future rather than amelioration of effects once they have occurred. Future work will examine the validity of the map by comparing the results to national efforts. Efforts to map sensitivity have previously been made in China (Zhao et al., 1994), Japan (Yoshinaga et al., 1994) and South Africa (Olbrich et al., 1995). Comparable distributions are identifiable in national investigations and this global study. A verification procedure being undertaken with national experts will lead to modification of methods and data.

Acknowledgements

We would like to acknowledge the support of SIDA for the funding the developing country acidic deposition project and collaborators on this project, Henning Rodhe and Lennart Granat from MISU and Lennart Robertson from SMHI.

References

Andersson, M.: 1988, Water, Air and Soil Pollution **39**, 439-62.
Chadwick, M.J. & Kuylenstierna, J.C.I.: 1991, Perspectives in Energy **1**, 71-93.
Edmunds, W.M. & Kinniburgh, D.G.: 1986, Journal of the Geological Society **143**, 707-20.
Eyre, S.R. : 1968, "Vegetation and Soils, A World Picture". Edward Arnold, London.
FAO: 1992a, "Soil Properties and Qualities Estimation Based on Soil Groups of the Soil Map of the World". AGLS Soil Resources Group Working Paper. FAO, Rome.
FAO: 1992b, "FAO Agrometeorological Database". FAO, Rome.
FAO-Unesco. : 1988. "Soil Map of the World, Revised Legend". World Resources Report **60**. FAO, Rome.
Hettelingh, J-P, Chadwick, M.J., Sverdrup, H.U. and Zhao, D.(eds.): 1995, "Assessment of environmental effects of acidic deposition", Report from the "Acid Rain and Emission in Asia" project . RIVM, Bilthoven.
Jordan, C.F.: 1985, "Nutrient Cycling in Tropical Forest Ecosystems". John Wiley, Chichester.
Lucas, A.E. and Cowell, D.W.: 1984, In: "Geological Aspects of Acid Deposition" (ed. by O.P. Bricker), Acid Precipitation Series **7**, Ann Arbor. Butterworth, Boston, 113-129.
McDowell, W.H.: 1988, In "Acidification in Tropical Countries" (ed. by H. Rodhe and R.Herrera), SCOPE **36** John Wiley, Chichester, 117-139.
McFee, W.W.: 1980, In "Atmospheric Sulfur Deposition Environmental Impact and Health Effects" (ed. by D.S. Shriner, C.R. Richmond and S.E. Lindberg), Ann Arbor, Michigan, 495-506.
Nilsson, J. & Grennfelt, P. (eds.) :1988, "Critical Loads for Sulphur and Nitrogen". Miljørapport 1988:**15**. Nordic Council of Ministers, Copenhagen.
Norton, S.A. : 1980, In "Atmospheric Sulphur Deposition: Environmental Impact and Health Effects" (ed. by D.S. Shriner, C.R.Richmond and S.E. Lindberg), Ann Arbor, Michigan. 521-32.
Olbrich, K., van Tienhoven, M., Skoroszewski, R., Zunckel, M and Taljaard, J.: 1995, "A Prototype Atmospheric Deposition Risk Advisory System for the Eastern Transvaal Province". CSIR Report FOR-I 547, FORESTEK, CSIR, Pretoria.
Robertson, L., Rodhe, H. & Granat, L.: 1995, Water Air and Soil Pollution, this volume.
Rutgers University: 1990, "Global Grass I. Fifty global coverage maps". Rutgers University , New Jersey. USA.
Sverdrup, H., de Vries, W. and Henriksen, A.: 1989, "Mapping Critical Loads: A Guide to the Criteria, Calculations, Data Collection and Mapping of Critical Loads", UN-ECE and Nordic Council of Ministers.
Ulrich, B. & Sumner, M.E. (eds.): 1991, "Soil Acidity". Springer Verlag, Berlin.
Yoshinaga, S., Suzuki, Y., Matsukura, Y., Kobayashi, M. and Arai, T.: 1994, Journal of the Science of Soil and Manure, Japan **65**, 565-568.
Zhao, Dianwu., Xiong, Jiling and Mao, Jietai.: 1994, "Mapping Sensitivity, Critical Loads and Exceedance for China". RCEES, Beijing.

RAINWATER CHEMESTRY AT THE WESTERN SAVANNAH REGION OF THE LAKE MARACAIBO BASIN, VENEZUELA

J.A. MORALES[1], C. BIFANO[2] AND A. ESCALONA[2]

[1]Lab. Química Ambiental, Fac. Experimental de Ciencias, Universidad del Zulia, Maracaibo 4011, Venezuela. Fax: 58-61-420848. [2]Instituto de Geoquímica, Fac. de Ciencias, U.C.V., Caracas, Venezuela.

Abstract. The major part of Venezuela oil production is located in and around the Lake Maracaibo Basin. The samples were collected over a 1-year period at Catatumbo and La Esperanza sites. The rainwater was acidic, with a VWA-pH of 4.6 for Catatumbo and 4.2 for La Esperanza. This acidity is made up in 93% by inorganic acids (mainly H_2SO_4), and NH_4^+ is the major cation which buffer the acidity of precipitation. An excess of sulfate (SO_4^*) > 96% was obtained in both sites. Correlation analysis shows that H^+ is strongly correlated with SO_4^*. Anthropogenic air pollution from oil fields (H_2S) and the burning of sulphur-bearing fuels (SO_2) are probably the dominant sources; however, the lack of correlation between the H^+ and NO_3^- levels would appear to indicate that the SO_4^* is also of biogenic origin (H_2S-DMS from Sinamaica Lagoon-Lake Maracaibo-and the Caribbean). Statistical analysis of the pooled data indicated that the concentration differences between Catatumbo and La Esperanza sites are not significant at 99% confidence level.

Key words: Acid rain, Rainwater chemistry, Tropical atmosphere, Venezuelan savannah, Oil.

1. Introduction

Almost a third of the rains which falls on the Earth's continents is deposited within the narrow equatorial belt between 10° N and 10° S latitude. However, in spite of the resulting importance of tropical rainfall for global biogeochemical cycling, there are relatively few studies on precipitation chemistry in the wet tropics (e.g. Lewis, 1981; Galloway et al., 1982; Keene et al., 1983; Likens et al., 1987; Ayers and Gillet, 1988; Andreae et al., 1990; Sanhueza et al., 1994). Furthermore, some of the former studies have used techniques (e.g., long-term bulk collection) which do not produce reliable data on rain chemistry.

Over the last few years, the chemical composition of rainfall has been measured at everal sites in the central-eastern part of the Venezuelan savannah region. Volume-weighted average acids pHs (around or below 5.0), with a relatively high contribution (> 60%) of the organic formic and acetic acids have been reported (Sanhueza et al., 1994). On the other hand, in Venezuela most of the soils are acidic, with a high exchangeable aluminum content; terrestrial and aquatic ecosystems are sensitive to acid deposition (Sanhueza et al., 1988).

In Maracaibo Lake Basin, no other studies have been reported on the chemical composition of rainwater. In this paper, we report the chemical characterization of rain events colleted at two sites in the western part of the Venezuelan savannah region.

2. Experimental

2.1. SAMPLING AND CHEMICAL ANALYSIS

Collecting event rains was carried out at La Esperanza (10° N, 72° W at 99 m altitude) and Catatumbo (9° N, 72° W at 45 m altitude) rural sites in Lake Maracaibo Basin (Zulia State, Venezuela). These sites are located within a tropical savannah climatic area. Winds are predominantly northeasterly, east-northeast (northeast trades). Annual precipitation shows two peaks. Generally, the months with highest rainfall are April-May and September to November. The amount of rainfall (based on 30 year's data) is ~ 1000 mm (La Esperanza) and ~ 2500 mm (Catatumbo).

The sites are downwind of potentially important anthropogenic sources: Maracaibo, a city of about 1.8 millon inhabitants with a medium size industries such as a cement factory and a petroleum-burning electrical power plant; El Tablazo Petrochemical Complex with large amonnia, chlorine and olefin plants; and Cardon-Amuay and probably Aruba-Curacao refineries. Cattle ranching is the main activity of the region where the sites are located.

Rainfall samples were taken manually for each event using a plexiglass funnel, over a 1-year period between November 1988 and November 1989. For the analysis of anions and NH_4^+, a fraction of the rain samples was preserved inmediately after the rainfall with $CHCl_3$ and freezed at 4°C; another fraction was preserved with HNO_3 for the analysis of metallic cations.

Rainwater pH was measured never more than 3 days after the rainfall event with a Fisher mini pH-meter calibrated with standard buffer solutions (Fisher) of pH 4.01 and pH 7.41, before and after each measurement. The ions were analyzed using the following methods: NH_4^+ and Cl^- (ion selective electrodes); NO_3^- (colorimetric, after Cd reduction); $SO_4^=$ (turbidimetric method, following very carefully the standard addition indications for dilute samples. Some very dilute (< 0.6 mg/L) samples were analyzed by ion exchange chromatography (IEC) at the Atmospheric Chemistry Laboratory of the Instituto Venezolano de Investigaciones Cientificas (IVIC)); $H_2PO_4^-$ (Colorimetric, ascorbic acid); Na^+ and K^+ (F-AES); Ca^{++} and Mg^{++} (F-AAS). The organic anions formate and acetate were analyzed at IVIC by IEC using a Dionex model QIC ion chromatograph (details of the analysis are indicated in Morales et al., 1994).

3. Results and Discussion

3.1. CHEMICAL COMPOSITION

The results show (Table 1) that in both sites the ion levels distribution is over a broad range. High levels of H^+ and $SO_4^=$ with up 3 order of magnitud variability were registered. Statistical analysis of the pooled data collected at each site during the study period indicated that the concentrations differences between La Esperanza and Catatumbo sites are not significant at 99% confidence level.

An excess of sulfate (SO_4^*) > 96% was obtained in both sites. Correlation analysis shows that H^+ is strongly correlated with SO_4^* in La Esperanza (r = 0.85, n = 41) and

TABLE 1

Volume-weighted average concentrations. VWAC (µeq/L).

	La Esperanza (n=41)		Catatumbo (n=55)	
	Range	µeq/L	Range	µeq/L
pH	3.4-6.2	4.23[a]	3.0-6.3	4.64[a]
SO_4^*	2.9-396	63.6	0.1-1042	28.9
Cl^*	2.0-145.5	12.1	8.2-49.2	19.8
NO_3^-	3.5-59.0	12.0	3.2-147.6	12.0
PO_4^{3-}	0.1-8.6	0.7	0.1-11:1	0.7
$HCOO^-$	0.5-25.5	4.3	0.3-45.7	2.9
CH_3COO^-	0.1-19.8	3.2	0.2-13.3	2.8
NH_4^+	2.3-130	24.6	5.0-130	20.3
Na^+	0.7-116	9.6	1.3-95.7	8.4
K^*	0.5-9.4	1.3	0.4-15.2	1.7
Ca^*	0.4-32.4	2.5	0.4-38.3	5.3
Mg^*	0.01-5.0	0.2	0.01-0.3	0.4

a: pH values were previously converted to H^+, then a VWAC of H^+ calculated, and then this VWAC H^+ converted to pH units.

*: non-sea-salt concentration.

Catatumbo (r = 0.98, n = 55), and so the most acidic rain samples were those which also showed the highest concentrations of SO_4^*. Anthropogenic air polution from oil fields (H_2S) and the burning of sulfur-bearing fuels (SO_2) are probably the dominant sources. However, the lack of correlation (r=0.1-0.3) between the H^+ and NO_3^- levels would appear to indicate that the SO_4^* is problably also of biogenic origin (H_2S, DMS... from Sinamaica Lagoon-Lake Maracaibo-The Caribbean).

A comparison of our results with those obtained in another Venezuelan rural site unaffected by anthropogenic activities (La Paragua: Sanhueza et al., 1994) is presented in Figure 1. The concentrations of SO_4^*, NO_3^-, Cl^* and NH_4^+ are significantly larger (up to 15 (SO_4^*), 4 (NO_3^-), 10 (Cl^*) and 100 (NH_4^+) times higher) in western Maracaibo Lake Basin than in the eastern part of the Venezuelan savannah, suggesting that this region (western Maracaibo Lake Basin) is affected by significant air pollution. The relatively higher concentrations of NH_4^+ and Cl^* found in the La Esperanza and Catatumbo rains are likely related to the emissions of NH_3 and Cl_2 of the Ammonia and Chlorine Plants located upwind at the El Tablazo Petrochemical Complex. Also, the intense livestock activity in the region of study is related with the ammonium results.

The upward trend in nitrate concentrations in precipitation corresponds well with the trend in NO emissions. Tropical savannah soils seen to be an important source of NO which depends on soil gravimetric moisture (Cárdenas et al., 1993). In our monitoring sites the major predominance of optimum range humidity that favours NO biogenic emissions from the soil can be related to the higher levels of NO_3^- found in the rainwater.

The monthly concentrations are summarized in Table 2. The results show that Na^+ and Mg^{++} exhibited a similar trend (r ~ 0.95) to decrease in concentrations as rainfall amount increased. These ions come almost exclusively from large marine aerosol particles, which are effectively washed out by precipitation. The NH_4^+, Cl^* and NO_3^- concentrations show more variable values during the year. These ions can be controlled in rainwater by several processes occurring locally and/or at some distance from La Esperanza and

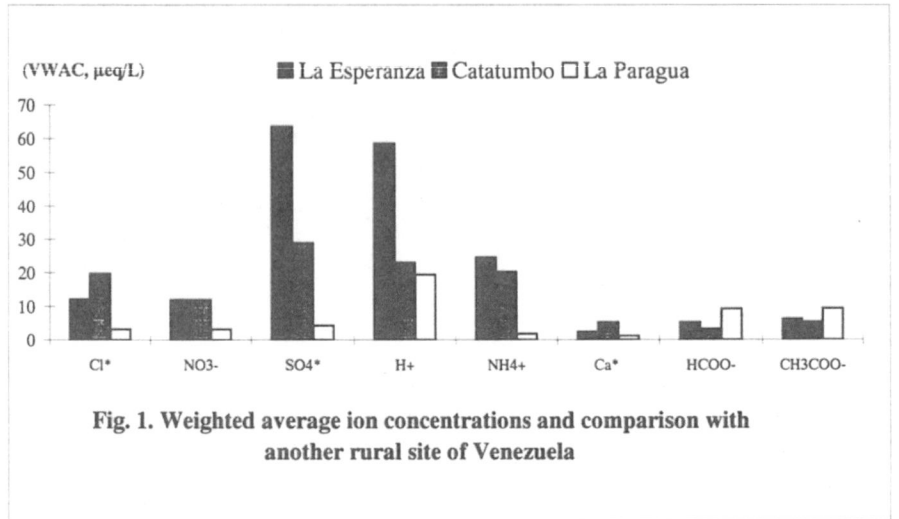

Fig. 1. Weighted average ion concentrations and comparison with another rural site of Venezuela

TABLE 2

Montly variation of precipitation chemistry at the two sites. VWAC (µeq/L)

Month	pH	Na^+	Mg^{++}	Ca^{++}	NH_4^+	SO_4^*	NO_3^-	Cl^*
La Esperanza								
J								
F								
M	4.5	40.7	11.6	10.0	31.1	49.9	19.6	35.6
A	4.4	22.7	6.3	7.2	50.0	42.1	13.3	58.1
M	4.8	7.8	1.5	3.3	25.9	19.3	12.0	24.3
J	4.8	36.4	5.1	3.2	29.0	16.8	11.8	0.1
J	5.2	4.5	1.2	3.1	18.4	8.2	10.0	5.9
A	5.1	16.1	3.9	4.7	32.7	32.8	16.8	2.4
S	4.8	21.4	2.8	2.4	17.2	26.1	10.1	3.3
O	5.4	24.5	3.0	2.3	36.2	19.3	11.1	17.2
N	3.8	8.3	0.9	1.5	17.6	146.2	10.9	12.1
D	4.0	11.9	1.7	1.8	20.4	100.7	10.8	11.0
Catatumbo								
J	4.8	21.0	2.6	2.7	10.0	15.3	6.6	19.8
F	4.7	32.0	7.2	10.3	36.1	50.3	25.4	22.1
M	4.0	27.8	3.7	6.0	35.8	120.0	13.6	23.1
A	5.3	64.0	14.5	22.4	45.8	73.6	21.8	17.5
M	4.5	21.2	3.3	5.8	29.1	24.3	12.1	33.0
J	4.3	18.5	2.7	5.4	32.9	45.8	38.7	31.7
J	4.5	14.5	2.4	4.2	27.6	38.0	21.4	18.3
A	4.8	11.8	2.2	3.4	24.3	33.3	11.0	10.1
S	5.1	7.5	0.9	3.6	24.7	20.2	8.7	11.5
O	5.1	10.8	3.0	5.4	19.6	23.8	9.3	3.6
N	4.4	2.5	0.8	20.0	7.1	11.4	5.7	48.0
D	4.8	9.4	2.1	1.9	7.4	11.7	4.7	19.6

Catatumbo sites, such as: air-sea exchange, air-land exchange, atmospheric emissions transport (e.g. El Tablazo) and atmospheric transformations.

In La Esperanza, the H^+ and SO_4^* concentrations peak in November. The lowest pH values (pH < 4.4 ; n=13 events) and highest SO_4^* concentrations (up to 410 µeq/L) were obtained in this month. It is important to note that in this month, the lowest pH (3.4) occurred in the largest rain event (78.2 mm). In La Esperanza monitoring site January and February were dry months without rainfall, and the highest storms were in November, when the Inter-Tropical Convergence Zone (ITCZ) is located over the central part of Maracaibo Lake Basin. Heavy rains are more representative of the chemical composition of cloud conditions, while light rains are strongly affected by below-cloud scavenging (Sanhueza et al., 1994). It is likely that a large proportion of H^+ and SO_4^* ions are incorporated to the rainwater by in-cloud scavenging processes during rainytime.

In La Esperanza and Catatumbo sites the first rains of the wet season (March-April), towards the end of the dry periods, when the atmosphere is heavily loaded with several compounds, show higher ion concentrations. In Catatumbo, the H^+ and SO_4^* distributions have a maximum in March (early rainy season). The highest acid rain (pH=3.0) was in one light event (4.5 mm in ≈ 45 min) occurred on March 05 with a SO_4^* maximum concentration of 1050 µeq/L, and the minimum occurred at the end of rainy season (November-December) in contrast with the results obtained in La Esperanza. This is probably due to variation of cloud types and of cloud amount over the different sites; and also to changes in emission rates of species and long-range transport over the region of the study.

3.2. CONTROL ON ACIDITY OF THE RAIN

Original acidity of the rain is defined as: H_2SO_4 ($SO_4^*/\Sigma Anions$), HCl ($Cl^*/\Sigma Anions$), HNO_3 ($NO_3^-/\Sigma Anions$), HCOOH ($HCOO^-/\Sigma Anions$) and CH_3COOH ($CH_3COO^-/\Sigma Anions$), where $\Sigma Anions= SO_4^* + Cl^* + NO_3^- + HCOO^- + CH_3COO^-$. The results show that at La Esperanza and Catatumbo sites, inorganic acids contribute with more than 90% to the original acidity of the rain (H_2SO_4 (44-67%), HCl (13-30%), HNO_3 (13-18%)). Measurements made in the central-eastern Venezuelan savannah region (Sanhueza et al., 1994) and other tropical sites (e.g. Keene and Galloway, 1986; Andreae et al., 1988; Ayers and Gillet, 1988; Andreae et al., 1990) show that precipitation acidity is dominated by HCOOH and CH_3COOH. In western Lake Maracaibo Basin, formic plus acetic acids only contribute less than 10% to the rainwater acidity.

The less than unity $H^+/(SO_4^* + Cl^* + NO_3^- + HCOO^- + CH_3COO^-)$ ratios (0.35-0.69) indicate that between 30 and 70% of the original acids have been neutralized by ammonia and calcium species. The monthly NH_4^+/Ca^{++} ratio (Table 2) is greater unity (3.1-15.7), suggesting that in the study region, NH_3 is the more important alkaline species that control the acidity of the rain.

4. Conclusion

Rainwater throughout the western Lake Maracaibo Basin region is acid and the average pH is between 4.23 and 4.64. The dominant anion is SO_4^* and the dominant cation is NH_4^+. Acid rain is mainly caused by sulfuric acid (44-67%); formic plus acetic acids only contribute less than 10%. Ammonia is the more important alkaline species that controls the acidity of the rain.

In Lake Maracaibo Basin, precipitation in large agricultural and natural vegetation areas is probably affected by up wind anthropogenic sources. This is evidenced by the higher level of H^+, SO_4^*, NO_3^-, Cl^* and NH_4^+ concentrations found in La Esperanza and Catatumbo sites in comparison with other eastern Venezuelan rural sites.

Acknowledgments

The Council for Human and Scientific Development (CONDES) of the University of Zulia is thanked for its financial support. IVIC Atmospheric Chemistry Laboratory for the analysis by ion chromatography.

References

Andreae, M.O., Talbot, R.W., Andreae, T.W., Harriss, R.C.: 1988, *J. Geophys. Res.* **93**, 1616-1624.
Andreae, M.O., Talbot, R.W., Berreshein, K.M., Beecher, K.M., Li, S.M.:1990, *J. Geophys Res.* **95D**, 16987-16999.
Ayers, G.P., Gillet, R.: 1988, In: Rhode, H., Herrera, R., Eds.: *Acidification in tropical countries*, SCOPE **36**, 347-402. John Wiley and Sons, Chichester, England.
Cárdenas, L., Rondon, A., Johansson, C., Sanhueza, E.: 1993, *J. Geophys. Res.* **98**, 14783-14790.
Galloway, J.N., Likens, G.E., Keene, W.C. and Miller, J.M.: 1982, *J. Geophys. Res.* **87**, 8771-8786.
Keene, W.C., Galloway, J.N., Holden Jr., J.D.: 1983, *J. Geophys. Res.* **88**, 5122-5130.
Keene, W.C., Galloway, J.N.: 1986, *J. Geophys. Res.* **91**, 14466- 14474.
Lewis Jr., W.M.: 1981, *Water Resources Res.* **17**, 169-181.
Likens, G.E., Keene, W.C., Miller, J.M., Galloway, J.N.: 1987, *J. Geophys. Res.* **92**, 13299-13314.
Morales, J.A., L. de Medina, H., G. de Nava, M., Velasquez, H., Santana, M.: 1994, *J. of Chromatography*, **A671**, 193-196.
Sanhueza, E., Cuenca, G., Gomez, J.M. et al.: 1988, In: Rhode, H., Herrera, R., Eds., *Acidification in tropical countries*, SCOPE, **36**, 197-248, John Wiley and Sons, Chichester, England.
Sanhueza, E., Arias, M.C., Donoso, L. et al.: 1994, *Tellus*, **44B**, 54-62.

SULPHATE DEPOSITION TO A SMALL UPLAND CATCHMENT AT SUIKERBOSRAND, SOUTH AFRICA

R.W.SKOROSZEWSKI

Division of Water Technology, CSIR, P.O.Box 395, Pretoria 0001, South Africa

Abstract. In 1992, a study was initiated by the Water Research Commission of South Africa, to investigate the relationship between atmospheric deposition and water quality in a small upland catchment. The selected catchment, which had a seasonal stream, was a pristine site at the Suikerbosrand Nature Reserve, which is 80 km south-east of Johannesburg. The catchment is 32.5 ha in extent and is characterised by having a quartz geology with sandy soils. Fifty-four percent of the catchment area is bare rock and the average soil depth is 15 cm. The climate is relatively arid when compared to other catchment studies in the northern hemisphere (Birkeness and Hubbard Brook) with long dry periods in the winter months and a low annual runoff (8.4 - 8.9% of mean annual precipitation). The measured inputs to the catchment included rainfall, rainwater chemistry, ambient SO_2 concentrations, rock runoff and bulk or particulate deposition. Outputs from the catchment included the measurement of runoff using a V-Notch weir and intensive sampling of a range of chemical water quality variables. During the wet summer months the dry deposition was estimated to be between 39 and 62% of the total atmospheric sulphate inputs into the catchment, whereas in the dry winter months this was estimated to be 90% of inputs. Over a complete annual cycle the net accumulation of sulphate on the catchment surface was estimated to be between 83 and 91% of inputs.

Key words: Water quality, Wet and dry sulphate (SO_4^{2-}) deposition, Gaseous deposition, Particulate deposition, Rock runoff, Stream flow, Suikerbosrand, South Africa.

1. Introduction

The Suikerbosrand study commenced in 1992 and forms part of a larger programme of the Water Research Commission of South Africa which aims to look at the salinisation of the Vaal Dam. The Vaal Dam is the most important source of water for the Gauteng Province, as it supplies more than nine million people with drinking water. It also supplies water for industry and agriculture.

One of the main purposes of this study was to examine the relationships between water quality and atmospheric deposition of sulphate (SO_4^{2-}) into a small, sensitive upland catchment. This included the measurement and estimation of both wet and dry deposition as well as water chemistry loads leaving the catchment.

The region is characterised by an annual mean precipitation of 700 mm per year, irregular rainfall patterns, dominated by summer thunder showers, long dry periods in the winter months when little or no rainfall occurs, and periodic drying out of the catchment between rainfall events.

There have been several well-documented calibrated catchment studies in the northern hemisphere such as Birkeness (Wright and Johannesen, 1980) and Hubbard Brook (Likens et al., 1977), but to the author's knowledge this is the first study of its kind in the southern hemisphere.

Considerable industrial development has taken place in the region around the Suikerbosrand catchment and over one million tons of SO_2 are emitted into the atmosphere annually (Tyson et al.,1988).

2. Materials and Methods

The criteria for selection of the study site included the following: the site should not be subjected to significant anthropogenic inputs other than atmospheric; it should be an upland catchment; and it should be less than 100 ha in extent. A suitable site was found at the Suikerbosrand Nature Reserve, located approximately 80 km south-east of Johannesburg in the Gauteng Province of South Africa. The catchment is 34.5 ha in extent, 1800 m above sea level, with a north-south axis of 700 m, an east-south axis of 500 m and a maximum elevation above the weir of 98 m.

The geology of the site is quartzite, which is an inert slow-weathering rock. Approximately 54% of the catchment was exposed rock with the remaining 46% consisting of grassland interspersed with areas of bare soil. The soils are characterised as loamy-sand with an average depth of 15 cm.

The rainfall over the two-year study period was measured using a tipping-bucket rain gauge (0.5 mm every 10 minutes). Wet deposition was collected for analysis using an automatic wet sampler and analyzed using ion-exchange chromatography.

The dry deposition component was estimated either directly by measuring ambient sulphur dioxide, whereby gaseous deposition was estimated using theoretical deposition velocities, or indirectly through the measurement of rock runoff and bulk sampling (particulate deposition).

Rock runoff was collected from five areas of exposed rock of between 0.54 and 0.74 m^2, which were demarcated using concrete strips. Each strip was sealed with an inert paint to prevent contamination. After each rainfall event, all wet and dry deposition was collected in polypropylene buckets. The net rock runoff was calculated by subtracting the wet deposition from the total rock runoff (Table 1).

Particulate deposition was measured using an inert bulk collector, which collected both wet and particulate deposition.

Ambient sulphur dioxide concentrations were measured using the peroxide method (Kemeny and Halliday, 1974). Measured volumes of air were drawn through a dilute solution of hydrogen peroxide where SO_2 was absorbed and oxidised to sulphuric acid. Test solutions were exposed for periods of either two or three days and the sulphate concentrations were determined using ion exchange chromatography. Theoretical deposition rates were calculated using deposition velocities for grassland of 0.3 and 1.3 $cm.sec^{-1}$ (Shepherd, 1974). The lower rate (0.3 $cm.sec^{-1}$) was used for the dry winter months of Period B, where the vegetation was dormant, and the higher rate (1.3 $cm.sec^{-1}$) was used in Periods A and C when the grasses were actively growing.

The total dry deposition (Table 2) was calculated by multiplying the net rock runoff by 0.54 (area of catchment covered by exposed rock) and then adding this to the sum of the gaseous and net particulate deposition multiplied by 0.46 (area of catchment covered by grassland).

A V-notch weir was constructed on site, where the level of water was recorded and used to calculate stream flow in $m^3.sec^{-1}$. Event-related sampling of the water chemistry was undertaken using automatic water samplers. Water samples collected were analyzed using ion exchange chromatography and atomic absorption spectrophotometry. Output of sulphate from the catchment was calculated by multiplying the measured sulphate concentrations by the flow.

3. Results and Discussion

The study was undertaken between October 1992 and March 1994. This comprised two "wet" summer periods and one "dry" winter period. These periods are referred to as follows: Period A (October 1992 - March 1993) - wet summer, Period B (April 1993 - September 1993) - dry winter, and Period C (October 1993 - March 1994) - wet summer.

The total rainfall recorded for the three periods was 491.5 mm (Period A), 62.0 mm (Period B) and 843.5 mm (Period C). Interestingly, 70% more rain was recorded in Period C than in Period A, both of which covered the same months of the year; this reflects the annual variability of rainfall in the region. The monthly data are graphically represented in Figure 1.

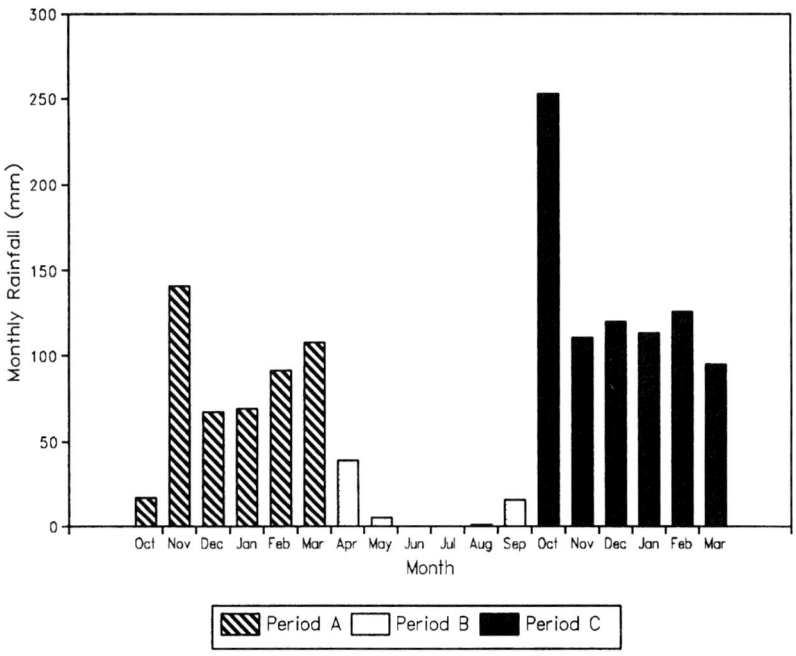

Figure 1: Monthly Rainfall at Suikerbosrand

The sulphate deposition measured from the wet sampler, the rock runoff plots, the particulate sampler and the calculated theoretical deposition using the atmospheric SO_2 data are shown in Table 1. The wet and total dry sulphate deposition rates are graphically represented in Figure 2.

The total wet deposition of sulphate (Table 1) was 11.8, 0.8 and 27.5 kg.ha^{-1} for periods A, B and C respectively. The highest monthly sulphate deposition occurred during the months of highest rainfall.

The highest ambient levels of SO_2 were found in Period B (11.4 μg.m^{-3}) when there was little or no rainfall. The SO_2 levels were significantly lower in Period C

Table 1: Calculated(C) and Measured(M) Sulphate (SO_4^{2-}) Deposition Rates

Period	Sulphate (SO_4^{2-}) Deposition in kg.ha^{-1}			
	Wet Deposition Rate (M)	Net Gaseous Deposition Rate (C)	Net Rock Runoff Deposition Rate (C)	Net Particulate Deposition Rate (C)
A (Oct 92 - Mar 93)	11.8	32.2	6.0	3.0
B (Apr 93 - Sep 93)	0.8	8.1	5.8	2.4
C (Oct 93 - Mar 94)	27.5	15.1	17.1	3.1

(4.9 $\mu g.m^{-3}$) which had 70% more rainfall than Period A (10.5 $\mu g.m^{-3}$.). Using deposition velocities of 1.3 (Periods A and C) and 0.3 m.sec^{-1} (Period B), the estimated sulphate deposition loads were 32.2, 8.1 and 15.1 kg.ha^{-1}, respectively.

The rock runoff plots gave sulphate depositions of 6.0, 5.8 and 17.1 kg.ha^{-1} for periods A, B and C respectively. The bulk samplers gave particulate depositions rates of 3.0, 2.4 and 3.1 kg.ha^{-1}.

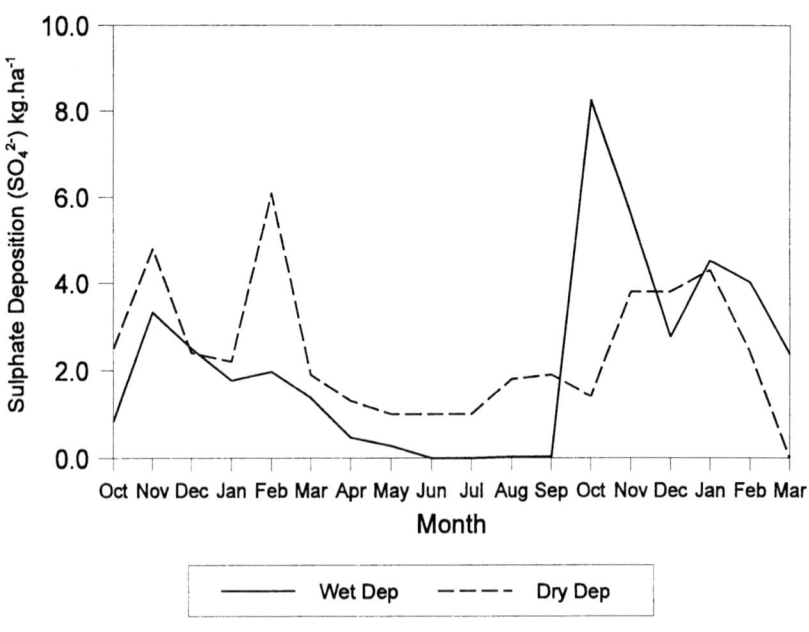

Figure 2: Wet and Total Dry Sulphate Deposition at Suikerbosrand

Table 2: Atmospheric Inputs and Outputs of Sulphate (SO_4^{2-}) for each sampling Period

Sampling Period	Total Wet Deposition (kg.ha^{-1})	Total Dry Deposition (kg.ha^{-1})	Total Catchment Output (kg.ha^{-1})
Period A	11.8	19.4	2.8
Period B	0.8	7.5	0.0
Period C	27.5	17.6	7.6

It must be noted that the rock runoff plots measured the total dry deposition (gaseous and particulate) whereas the bulk sampler measured primarily particulate deposition. Gaseous deposition is affected by, amongst other things, the nature of the receiving surface, wetness and temperature. Little or no gaseous adsorption would be expected using the bulk sampler, where the collecting funnel was constructed of inert polypropylene.

During Periods A and C, the proportion of dry deposition in the total deposition (wet + dry) amounted to 62% and 39% respectively. Periods A and B were both wet summer periods. In Period B, when relatively little rainfall was recorded, the proportion of dry deposition to total deposition was 90%.

Over an annual cycle (Periods A and B and Periods B and C), the proportion of dry to total sulphate deposition was 68% and 46%, respectively.

The sulphate deposition velocities used in the calculation of gaseous deposition (0.3 and 1.3 m.sec^{-1}) were theoretical values and have not been calibrated in the field. A deviation in these velocities of 50% would lead to a variation of 24%, 28% and 8% in the three estimates of total sulphate deposition during Periods A, B and C.

During the eighteen month study period, stream flow was recorded at the weir on fourteen occasions (seven in Period A and seven in Period C). The stream at Suikerbosrand was seasonal, only flowing after significant rainfall events of greater than 20 mm. No flows were recorded during Period B. The total flows in Periods A and B were 13 354 m^3 and 24 335 m^3, respectively. No stream flow was recorded between the recorded rainfall-related flow events. Eight flow events corresponded to individual storm rainfall events. The remainder of the flows responded to periods of general and widespread rainfall in the region. The combined runoff over the study period amounted to 8.7% of the total rainfall.

This relatively low runoff is characteristic of the region where the mean daily evaporation rate is 8mm during the summer months when most of the rainfall occurs (DWAF, 1986).

The levels of both sulphate and nitrate were often highest at the onset of flow and lowest at times of peak flow. Hydrogen ion concentrations were highest at times of peaks in flow, being inversely proportional to the acid neutralising capacity.

Mean levels of sulphate concentrations in the outflow were consistently between 9 and 10 mg.L^{-1}. These are significantly higher than those recorded from other calibrated catchments in Norway, Sweden and North America where average sulphate concentrations of between 1.1 and 5.7 mg.L^{-1} have been recorded (Hultberg, 1985). Similarly, mean levels of nitrate concentrations recorded at Suikerbosrand (2.5 mg.L^{-1}) were considerably higher than those measurements at other calibrated catchments (<0.5 mg.L^{-1}). The Suikerbosrand stream is however a seasonal stream with a relatively low mean annual runoff of between 8.4 and 8.9%. This contrasts with other northern hemisphere calibrated catchments which have mean annual runoff rates of 50% or higher.

Most of the sulphate deposited was not exported from the catchment. In period A, 9% of the total estimated inputs of sulphate was exported from the catchment, 0% in Period B (no flow) and 17% in Period C.

Acknowledgements

The author gratefully acknowledges the financial support of the Water Research Commission of South Africa.

References

DWAF: 1986, Management of the Water Resources of the Republic of South Africa. Department of Water Affairs and Forestry, Pretoria

Hultberg, H.: 1985, Ecological Bulletins 37: 133-157.

Kemeny, E., Halliday, E.C: 1974, CSIR.SMOG 5, pp1-15.

Likens, G.E.,Bormann, F.H.,Pierce, R.C., Eaton, J.S.,Johnson, W.M.:1977,Springer, New York, 146pp.

Shepherd, J.G.: 1974, Atmos Environ.,8,69-74.

Tyson, P.D., Kruger, F.J.,Louw,C.W.: 1988. South African National Scientific Programme Report No.150.

Wright, R.F., Johannesen, M.: 1980,In Drablos and Tollan, A, eds. SNSF-project, Oslo-As, pp 250-251.

MODELLING OF SULFUR DEPOSITION IN THE SOUTHERN ASIAN REGION

LENNART ROBERTSON
Swedish Meteorological and Hydrological Institute (SMHI)
S-601 76 Norrköping, Sweden

and

HENNING RODHE and LENNART GRANAT
Department of Meteorology, Stockholm University (MISU)
S-106 91 Stockholm, Sweden

Abstract. Acidification problems in developing countries are expected to become more prevalent in the coming decades. Assessments of means of abatement strategies are likely to become of vital interest. This paper presents some preliminary results of modelling of acidic deposition due to anthropogenic emissions of sulfur in the Southern Asian region. It is concluded that the study has some shortcomings, that has to be addressed in future work, such as lack of treatment of deep convection and that deposition and transformation rates used are not adapted to the tropics. Only very limited validation has been possible due to the lack of relevant measurements. Wet deposition data from rural Thailand are in fair agreement with calculated values. The study is one part of a larger project encompassing mapping ecosystem sensitivity to acid deposition, wet chemistry measurements and atmospheric transfer modelling.

Key words: acid rain, anthopogenic, sulfur deposition, modelling, Asia

1. Introduction

The countries in the South Asian region represent some of the most rapidly growing economies in the world today. Anthropogenic emissions are likely to increase substantially in the future and projections for the coming decades indicate that severe environmental problems are likely to occur (Rodhe *et al.*, 1992). A project funded by the Swedish International Development Authority (SIDA) carried out by the Stockholm Environmental Institute (SEI), the Department of Meteorology at Stockholm University and the Swedish Meteorological and Hydrological Institute, in cooperation with scientists from India and Thailand, examines acidification problems in South Asian countries. The project is divided into three different activities: (a) mapping of the ecosystem sensitivity to acidic deposition (critical loads), (b) wet chemistry measurements and (c) atmospheric transfer modelling. This paper addresses the release, transport, transformation and deposition of anthropogenic sulfur.

2. The MATCH Model

The MATCH (Mesoscale Atmospheric Transport and CHemistry model) model has been developed as a tool for air pollution assessment studies on different scales with support from the Swedish Environmental Protection Agency (Swedish EPA) (Persson *et al.*, 1995). It has primarily been used as a basis for decision making concerning environmental protection, such as assessment studies over Sweden (on a resolution of 20×20 km) and for

TABLE I

Sink parameters for SO_2 and SO_4^{2-}, respectively. The scavenging time-scale is related to a precipitation intensity of 1 mm/hour.

	SO_2	
Scavenging time-scale	Dry deposition velocities (cm/s)	
(hours)	Land	Sea
2.8	0.5 - 0.8	0.8

	SO_4^{2-}	
Scavenging time-scale	Dry deposition velocities (cm/s)	
(hours)	Land	Sea
1.0	0.1	0.05

high resolution applications within subregions within Sweden (5×5 km) on commission by local environmental authorities.

MATCH is a Eulerian 3-dimensional "off-line" model, which means that the physical atmospheric data are taken from some external source and fed into the model at regular time intervals (see section 4). The model includes horizontal and vertical transport, vertical diffusion, dry deposition, wet scavenging and chemical transformations. The design enables a flexible choice of horizontal and vertical grid, principally defined by the input meteorological data. For this application the model has 5 vertical layers extending up to 8.5 km on a latitude-longitude horizontal grid somewhat shifted towards the north, as shown in Figure 1. The transport is treated by a modified fourth order Bott scheme in the horizontal (Bott, 1989), and a zero order upstream scheme in the vertical. Vertical diffusion is based on K-theory where exchange coefficients follow suggestions by (Holtslag et al., 1991) for neutral and weakly unstable conditions. For convective conditions the vertical diffusion is assumed to be dependent on a convective turn-over time-scale defined by the boundary layer height and the convective velocity-scale. Vertical mixing by deep convection is not treated so far.

The chemistry scheme employed is a simplified linear transformation of SO_2 to SO_4^{2-}. The gase-phase transformation is explicitly treated by a sinusoidal bulk transformation rate with mean value $3 \cdot 10^{-6}$ s^{-1} and amplitude $2 \cdot 10^{-6}$ s^{-1}. The wet-phase transformation is parameterized by introducing a scavenging coefficient for SO_2 implicitly treating the the wet-phase chemistry at a bulk rate. Table I shows the deposition parameters applied. The dry deposition velocity for SO_2 over land areas has a diurnal variation within the range presented. The rates of this simplified scheme are similar to those used for European conditions. Further studies will be undertaken to derive more relevant rates for the tropics.

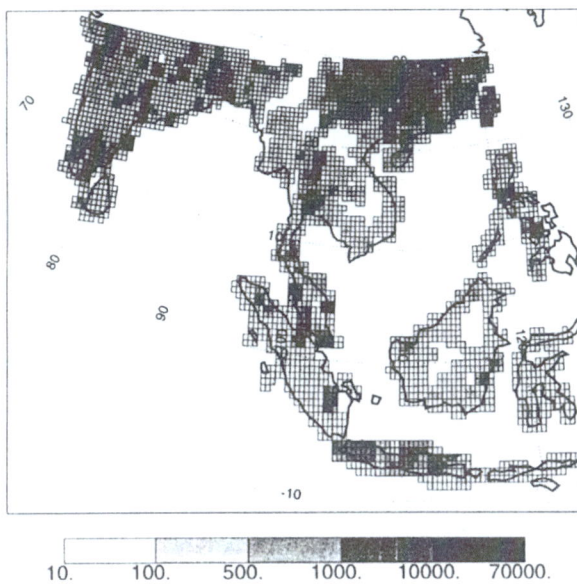

Fig. 1. Emission data used in the model projected on 0.5°× 0.5° resolution and a rotated latitude-longitude grid (tons S yr^{-1}). Based on Akimoto and Narita, 1994.

3. Emission data

Emissions of SO_x are taken from an inventory by Akimoto and Narita (1994). The inventory takes fuel combustion and industrial activities into account, and covers the geographical area of 60°E-150°E and 10°S-55°N with 1°× 1° resolution. Our work focuses on Southern Asia and the area (86°E-153°E and 19°S- 20°N) which defines the model domain with a 0.5°× 0.5° resolution into which the emission inventory has been transformed, see Figure 1. The total emissions of SO_x within the model area is 5.3 Tg S yr^{-1}. The emissions are assumed to be partitioned into 95% gaseous sulfur dioxide (SO_2) and 5% particulate sulfate (SO_4^{2-}).

4. Meteorological data

The forcing weather information is taken from ECMWF global analyses (available at 6 hour intervals). No precipitation analysis is available in the ECMWF data, so 6 hour ECMWF precipitation forecasts were utilized. For computational reasons we have limited the calculations to some short periods which roughly comprise a representative year. The selected periods are in January, April, June and September (1991-1992) with a total number of 67 days. These periods include winter and summer monsoons, and the intermediate dry periods. Provided that the selected periods make up a representative subset of a complete year, calculated depositions should arithmetically underestimate the annual deposition by a factor $365/67 = 5.45$.

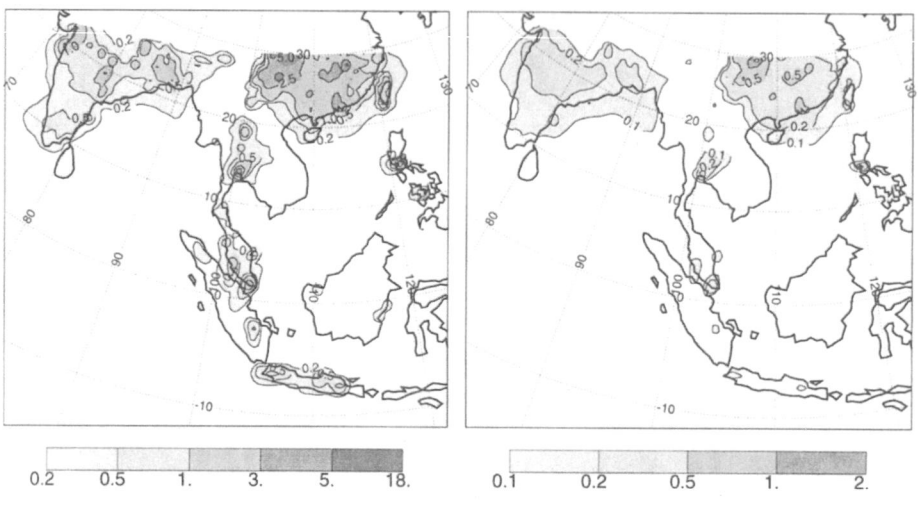

Fig. 2. Annual mean concentration of SO_2 (left) and SO_4^{2-} (right) in $\mu g\ S\ m^{-3}$.

5. Results and Discussion

5.1. Concentration and deposition of sulfur compounds

The results are presented in Figures 2 to 3 as annual SO_4^{2-} and SO_2 mean concentrations, annual wet deposition of SO_4^{2-} and annual total deposition of sulfur, respectively.

Table II shows the deposition load for each country within the model area. The load ranges from 30 to 1000 mg S $m^{-2} yr^{-1}$, with the largest load in southern China and the smallest in Indonesia.

The relative load with respect to each country's own emissions is shown in Table III. There is a prominent dominance by southern China and India which contributes to 82% of the total emissions. The less industrialized countries with low emissions import most of the sulfur deposited within these countries.

This study was restricted to anthropogenic emissions. Natural sources are expected to contribute of the order of $\sim$ 100 mg S $m^{-2} yr^{-1}$ over sea areas (Galloway and Gaudry, 1984).

5.2. Validation

We have had only a small number of data available (Granat et al., 1996) which overlap the period of calculation. Figure 4 shows some of these data and calculations in terms of SO_4^{2-} concentrations in precipitation. The simulations do not catch the measurements on a day to day basis, but the variability and the concentration level appear to be of the some order as in the data.

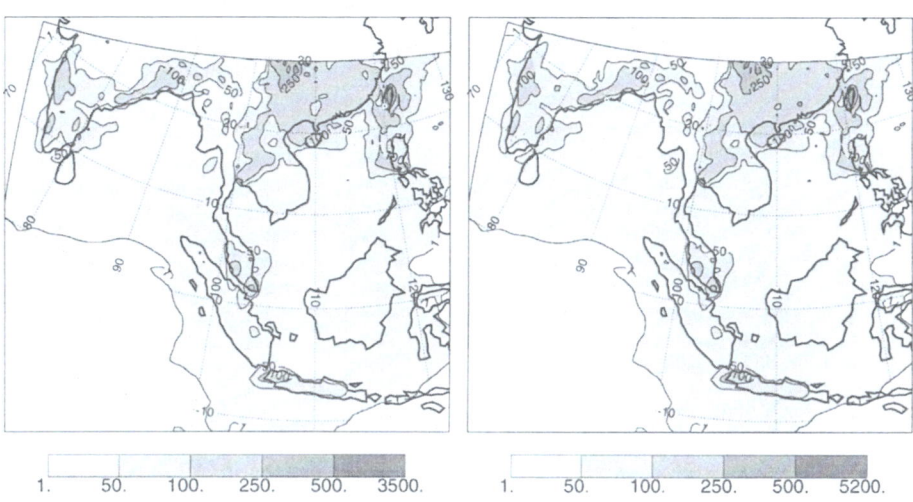

Fig. 3. Annual wet (left) and total (right) deposition of sulfur in mg S m^{-2}

TABLE II

Average deposition of sulfur for each country in mg S m^{-2}yr^{-1}.

Country	Dry deposition		Wet deposition	Total deposition
	SO_2	SO_4^{2-}		
Bangladesh	58	4.7	220	280
Cambodia	5.3	0.7	28	34
Southern China	260	12	680	950
India	100	6.0	180	280
Indonesia	19	0.5	95	120
Laos	16	1.5	190	210
Malaysia	72	1.4	210	280
Myanmar	26	3.1	140	170
Philippines	37	0.8	130	180
Thailand	92	3.5	360	460
Viet Nam	23	1.9	180	200

6. Conclusions

The results of this study provide a quantitative estimate of concentration and deposition of sulfur compounds due to anthropogenic emissions in southern Asia. Several shortcomings of this study need to be addressed in future work. In the first, place the model area has to

TABLE III
Deposition in relation to each country's own emissions (%).

Country	Dry deposition SO_2	SO_4^{2-}	Wet deposition	Total deposition
Bangladesh	36	2.8	130	170
Cambodia	37	4.9	200	240
Southern China	13	0.6	36	50
India	14	0.9	25	40
Indonesia	14	0.4	69	83
Laos	220	21	2700	3000
Malaysia	14	0.4	54	74
Myanmar	59	7.1	310	380
Philippines	10	0.2	36	48
Thailand	15	0.6	58	74
Viet Nam	37	2.9	270	310

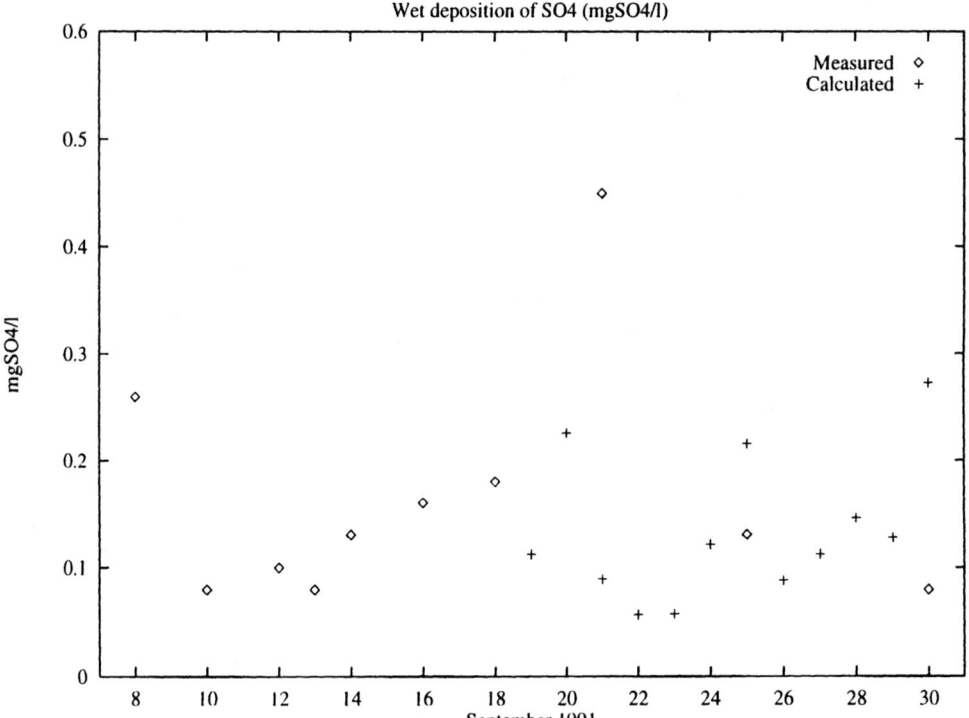

Fig. 4. A first attempt to validate the model results. The measurements are taken in Srigarind, 200 km northwest of Bangkok, (Granat, 1994). Measurements (crosses) and model (diamonds) are in mg $SO_4^{2-} l^{-1}$, for the period September 1991.

be extended to include the large emissions in northern China. Secondly, deep convection has to be included and transformation and deposition parameters valid for the tropics have to be derived. Natural sources should also be included.

Acknowledgements

Financial support was obtained from Swedish International Development Authority (SIDA) through Stockholm Environment Institute (SEI), under the project "Acidification studies in Asia" and from Stockholm University.

References

Akimoto, H. and H., Narita, 1994: *Atmos. Environ.* Vol **28**, No 213-225.
Bott, A.: 1989: *Monthly Weather Rev.*, **117**, pp. 1006-1015.
Galloway, J.N. and A. Gaudry: 1984, *Atmos. Environ.* **12**, 2649-2656.
Granat, L., Tabucanon, M., Suksomasanh, K., Simachay, S., and H. Rodhe, 1996: this volume.
Holtslag, A.A.M., Boville, B.A. and C.-H. Moeng: 1991, Proc. *Workshop on Fine-Scale Modelling and the Development of Parameterization Schemes*, ECMWF, Reading UK, 16-18 September.
Liljequist, G.: 1979, *Generalstabens Litografiska Anstalt*, Stockholm.
Persson, C., Langner, J. and L. Robertson: 1995, Proc. *Air Pollution Modelling and its Application*, Vol **X** (edited by E. Gryning), Plenum Press, New York, (in press)
Rodhe, H., Galloway J. and D. Zhao: 1992, *Ambio* **21**, pp 148-150.

The effects of organic carbon on acid rain in a temperate forest in Japan

M. INAGAKI[1], M. SAKAI[1] and Y. OHNUKI[1]

[1]*Forest Soil Laboratory, Kyushu Research Center, Forestry and Forest Products Research Institute, Kurokami 4-11-16, Kumamoto 860, Japan*

Abstract. From recent studies, we noticed that stemflow had an acidity that differed from that of precipitation or throughfall. Organic substances, supplied from the tree surface, would be one of the factors that modifies the acidity of rain. The objectives of this study were to determine the DOC concentration and to clarify the influence of dissolved organic carbon (DOC) on acidity in precipitation, throughfall and stemflow. Throughfall and stemflow were measured in sugi [*Cryptomeria japonica* D. Don], hinoki [*Chamaecyparis obutusa* Endl.] and kojii [*Castanopsis cuspidata* (Thumb.) Schottky.] stands. All samples were analyzed for their pH, electric conductivity (EC), major inorganic anions and cations and DOC concentration.

The annual average of DOC was highest in stemflow, and that of throughfall and precipitation were one-third and one-tenth of stemflow, respectively. The averages of DOC in stemflow in two coniferous, sugi and hinoki stands, were higher than that of broadleaved kojii stand. DOC concentration was low in summer and high in winter in all stands. In Stemflow, pH and DOC were negatively correlated, while EC and DOC in stemflow were positively correlated in all stands. However in throughfall, there was no evident relationship between pH, DOC and EC. This relationship was not explained by the cause of organic acid.

Key Words: precipitation, throughfall, stemflow, acidity, DOC, ion balances, Japan

1. Introduction

Recently, the monitoring survey for acid precipitation has been carried out in each part of Japan. From the data of past surveys, it was found that the properties of the throughfall and stemflow were different in each tree species (Sassa *et al.*, 1991). It was also found stemflow of sugi and hinoki, which are the major conifer species planted in Japanese artificial forest, had very low pH (Matsuura *et al.*, 1990).

Many studies had been carried out to investigate what regulates the acidity of stemflow. The major factor is thought to be acid deposition containing sulfuric and nitrogenous compounds, but not much attention was focused to organic matter. Torii & Kiyono (1992) thought the throughfall and stemflow properties are derived from substances leached from bark. According to them, one of the factors that determines the property of stemflow would be organic matter.

Dissolved organic carbon (DOC) in precipitation, throughfall and stemflow have been measured in several studies (Dalva and Moore, 1991; Edmonds *et al.*, 1991; Likens *et al.*, 1983; McDowell and Likens, 1988). Edmonds *et al.* (1988) studied the role of organic carbon in throughfall and stemflow acidity. They concluded organic acids that are supplied from the bark and the canopy, change the throughfall and stemflow pH.

The objectives of this study were to determine concentration values and seasonal changes of DOC in two conifer and one evergreen broadleaved stands, and to examine effects of organic substance to the acidity in precipitation, throughfall and stemflow.

2. Materials and Methods

The study was carried out at Kyushu Research Center and Tatsuda-yama experimental forest, which is a secondary forest located at the northern part of Kumamoto city (32°49'N, 130°44'E). The climate is moist-temperate. The average of annual precipitation during the last 30 years is 1968 mm. Throughfall and stemflow were measured in three stands; sugi [*Cryptomeria japonica* D. Don], hinoki [*Chamaecyparis obutusa* Endl.], kojii [*Castanopsis cuspidata* (Thumb.) Schottky.]. Sugi and hinoki are conifers and kojii is an evergreen broadleaf tree. The respective age of each stand in 1992 was 36, 37, 42.

Two types of precipitation were collected; wet-only deposition was collected in an open space in Kyushu Research Center by automatic rain collector (RT-5; Ikeda measurement instruments) since June 1992, and bulk deposition was measured by one polyethylene funnel led to 10 L polyethylene container which was installed adjacent to the automatic rain gauge since May 1993. The amount of precipitation was measured by tipping buckets rain gauge (US-300; Ogasawara measurement Instruments). Throughfall was collected in two 30 cm diameter polyethylene funnels at each stand, that is installed more than 1m far from the bark and placed 1.3 m above the soil surface, and it was led to a 10 L polyethylene container. Stemflow was collected by urethane form gutter enveloping the tree body at breast height drained to a 20 L polyethylene container. Stemflow was collected two points at each stands. Throughfall and stemflow measurement were started from June 1992 in kojii stand, August 1992 in sugi and hinoki stands. Water was usually collected after each rainfall. In this study, we report the data until November 1993.

pH was analyzed by glass electrode (HV-50V; TOA electronics ltd.). EC was analyzed by Electronic conductivity meter (CM-40S; TOA electronics ltd.). After a pH and EC measurement, rest samples were filtered through 1.0μm cellulose acetate filter and kept at 2 °C until other analyses were performed. Ca and Mg were measured by automatic absorption spectrophotometer (Z-6100; Hitachi ltd.) and Na, K, NH_4, Cl, NO_2, PO_4, NO_3, SO_4 were measured by ion chromatograph (IC-500; Yokogawa). DOC was measured by infrared carbon analyzer (TOC-5000; Shimazu).

3. Results and discussion

3.1. Averages and seasonal changes of pH, EC and DOC concentration during the sampling period

In the study period volume-weighted mean pH of precipitation was 4.6; both wet-only and bulk collector (Table 1). This result is almost comparable to averaged pH of precipitation in Japan. The pH in throughfall was higher than in precipitation in all stands. The pH in sugi stand was especially higher than in the other stands. The pH in stemflow in kojii

Table 1.

Volume-weighted means of pH, EC and DOC concentrations during studying term; EC(μS/cm), DOC (mg/L): The volume is weighted by precipitaion

	Precipitation		Throughfall			Stemflow		
	wet-only	bulk	sugi	hinoki	kojii	sugi	hinoki	kojii
pH	4.6	4.6	5.1	4.7	4.7	4.0	4.1	4.7
EC	14.3	12.0	24.1	28.4	27.3	63.3	65.9	39.3
DOC	1.4	1.0	4.3	2.9	3.1	11.5	12.3	7.1

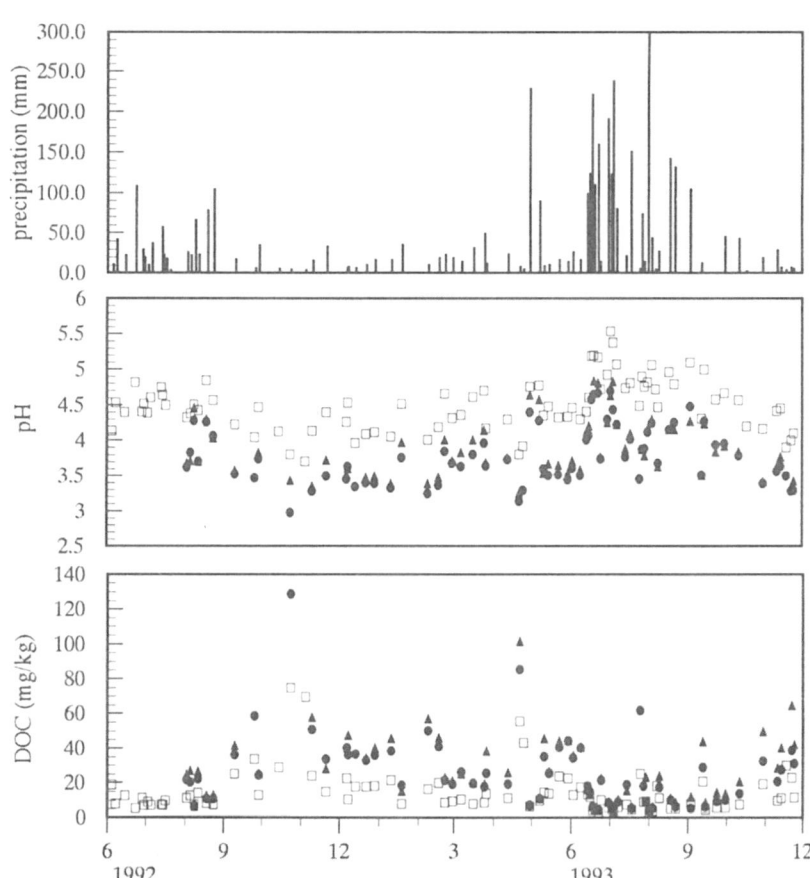

Fig. 1. Seasonal changes of precipitation (upper), pH in stemflow (middle) and DOC concentration in stemflow (lower) in sugi (●), hinoki (▲) and kojii (□) stands.

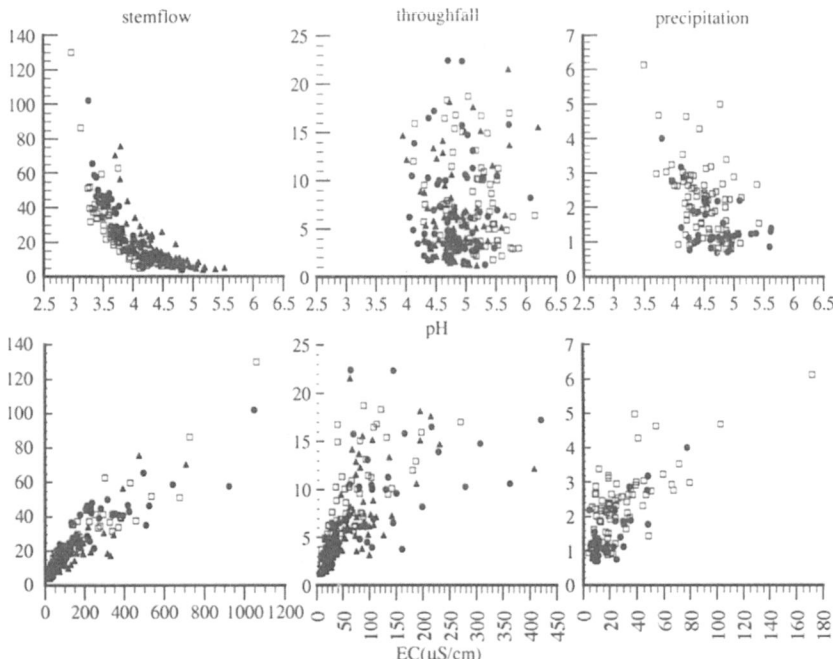

Fig. 2. Relationship between pH and DOC and between EC and DOC in each stands. Concentration of DOC values mg/L: stemflow and throughfall; in sugi(□), hinoki(●) and kojii(▲) stands: precipitation; wet(□) and bulk(●)

stand was higher than precipitation, but extremely low in sugi and hinoki stands.

DOC concentration was highest in stemflow. That in the two coniferous stands washigher than that of broadleaved kojii stand. DOC concentration in throughfall was higher than that in precipitation in all stands, and there was no significant difference between conifer and broadleaf stands. The value of this result was lower than in some other studies (Dalva & Moore, 1991). The volume-weighted mean of DOC was about one-second comparing to the arithmetic mean. During the sampling period from June 1992 to November 1993, the amount of precipitation was 4237 mm, for 130 % higher than the long-term mean. Above all, much rain fell during the summer of 1993. Low concentration of DOC is caused by the amount of precipitation.

pH of stemflow is higher in summer and lower in winter (Fig. 1). However, DOC concentration in stemflow had an opposite tendency; higher in winter and lower in summer. It is thought that pH becomes higher and DOC concentration becomes lower corresponding to magnitude of precipitation. The pH of stemflow was almost constant when precipitation exceeded 50 mm, while precipitation was less than that, pH of stemflow became lower. DOC concentration also fluctuated extremely along with precipitation. When precipitation exceeded 50 mm, DOC concentration was lower than 20 mg L^{-1}. Under 20 mm precipitation

Table 2.

Mean concentrations of chemical elements in precipitation, throughfall and stemflow during studying term.

		H	Na	NH_4	K	Ca	Mg	F	Cl	NO_2	PO_4	NO_3	SO_4	cations	anions	Cations-anions
							(μeqiv/l)									
Precipitation	wet-only	29.0	11.9	8.3	1.5	8.8	3.5	0.4	21.6	0.0	0.0	6.9	36.2	63.0	65.1	-2.1
	bulk	26.3	11.1	7.5	1.7	7.6	3.1	0.8	16.8	0.0	0.0	5.7	29.9	57.3	53.2	4.1
kojii	throughfall	19.8	34.6	41.4	27.5	35.5	25.2	3.2	75.7	0.0	0.0	22.7	75.5	184.0	177.1	6.9
	stemflow	21.2	40.0	48.6	70.7	48.2	30.3	10.8	114.2	0.0	0.0	18.8	102.4	259.1	246.2	12.9
sugi	throughfall	6.4	24.7	26.2	33.2	59.1	33.9	3.1	64.6	0.9	0.2	18.3	68.4	183.5	155.6	27.9
	stemflow	101.0	32.4	30.7	41.1	64.3	27.5	4.5	107.0	0.0	0.9	35.3	139.3	297.0	287.0	10.0
hinoki	throughfall	21.1	43.2	46.0	18.8	34.8	17.9	1.9	80.9	0.0	0.0	25.5	71.6	181.7	179.9	1.8
	stemflow	92.4	61.9	110.8	24.9	60.1	23.9	4.7	142.8	0.0	0.5	47.3	168.4	374.0	363.7	10.3

DOC concentration suddenly increased.

DOC concentration in stemflow would be regulated by retention time when precipitation is retained in bark. It seems morphological differences of barks affect on it. Kojii has smooth mono-layered bark. On the other side, sugi and hinoki have multi-layered rough fibrous bark. Therefore, it is assumed that stemflow of kojii flow down faster than sugi and hinoki, and that may determine DOC concentration. And that may also regulate the differences of averaged DOC concentration in stemflow in three stands.

3.2. Relationship among pH, EC and DOC

DOC concentration in stemflow had strong negative correlation with pH and positive correlation with EC in all species (Fig. 2). On the other hand, there were no significant relationship between pH and DOC concentration in throughfall. DOC in precipitation had correlation with pH comparing to the relationship between DOC in throughfall and pH.

Edmonds *et al.* (1991) concluded that anion deficit (cations - anions) in throughfall and stemflow means the amount of organic acid, and organic acids contribute their acidity. However, in this study anion deficit were very few except throughfall in sugi (Table 2). Last (1989) also reported ion balances of throughfall plus stemflow were in good agreement, and most of the acidity was associated with inorganic acids. Throughfall and stemflow acidity seemed not to be derived from organic acid.

Considering the relationship between pH, EC and DOC, organic substances might have a relationship with strong anions and influence throughfall and stemflow acidity indirectly.

Recently, composition of DOC in precipitaion and throughfall were measured (Likens *et al.*, 1983; McDowell and Likens, 1988). To clarify the relationship between organic substances and throughfall / stemflow acidity, the composition of stemflow needs to be analyzed. There are also some species which have high pH stemflow (Sassa *et al.*, 1991). To clarify the relationship acidity and organic substances, it should be measured stemflow DOC of such species.

4. Conclusions

DOC concentration decreased in the order stemflow > throughfall > precipitation in all stands. DOC concentration in stemflow is higher in winter and lower in summer. DOC concentration was also related with the magnitude of precipitation. DOC concentration in stemflow had a relationship with pH and EC, but it could not be explained by the cause of organic acid. Chemical analysis of composition of stemflow DOC should be performed.

Acknowledgments

We thank to S. Kaneko, M. Araki and A. Torii, in Kansai research center, F.F.P.R.I. for analysis of DOC, to T. Mizoguchi, F.F.P.R.I. for comments on this manuscript and to T. Kawazoe for his help to analyzing water samples.

References

Dalva, M. and Moore, T. R. : 1991, "Sources and Sinks of Dissolved Organic Carbon in a Forested Swamp Catchment", Biogeochemistry **15**, 1-19.

Edmonds, R. L., Thomas, T. B., and Rhodes, J. J. : 1991,"Canopy and Soil Modification of Precipitation Chemistry in a Temparate Rain Forest ", Soil Sci. Sol. Am. J. **55**: 1685-1693

Last, F. T. : 1989, "Acidic Deposition : Case Study Scotland", in Adriano D.C. and Havas M. , "Acidic Precipitation Volume I: Case Studies", Springer-Verlag, 237-274

Likens, G. E., Edgerton, E. S. and Galloway, J. N. : 1983, "The Composition and Deposition of Organic Carbon in Precipitation", Tellus **35B**, 16-24

Matsuura, Y., Hotta, I. and Araki, M. : 1991, "Surface soil pH Depression of Cryptomeria japonica Forests in Kanto District", Jpn. J. For. Environment **32(2)**, 65-69 (in Japanese)

McDowell,W. and Likens, G. E. : 1988, "Origin, Composition, and Flux of Dissolved Organic Carbon in the Hubbard Brook Valley", Ecological Monographs **53(3)**, 177-195

Sassa, T. , Gotou, K. , Hasegawa, K. and Ikeda, S. : 1991, "Acidity and Nutrient Elements of the Rainfall and Stem flow in the Typical Forests around Morioka City, Iwate Pref., Japan", Jpn. J. For. Environment **32(2)**, 43-58 (in Japanese)

Torii, A. and Kiyono, Y. : 1992, "Surface Soil pH Decrease under Sugi (*Cryptomeria japonica* D. Don) Trees on the Lowland in the Kinki District", J. Japan Soc. Air Pollut. **27(6)**, 325-328 (in Japanese)

ANALYSIS OF LONG-RANGE TRANSPORTED ACID AEROSOL IN RIME FOUND AT KYUSHU MOUNTAINOUS REGIONS, JAPAN

O. NAGAFUCHI[1], R. SUDA[1], H. MUKAI[2], M. KOGA[3] and Y. KODAMA[3]

[1] *Fukuoka Institute of Health and Environmental Sciences, 39 Mukaizano, Dazaifu, Fukuoka 818-01, Japan*
[2] *National Institute for Environmental Studies, Tukuba, Ibaraku 305, Japan*
[3] *University of Occupational and Environmental Health, Kitakyushu, Fukuoka 807, Japan*

Abstract. Rime-ice and snow samples were collected at mountain sites in Kyushu, Japan during the winter of 1994, and both soluble and insoluble substances in the melted rime-ice were analyzed by ion chromatography, inductively coupled plasma-mass spectrometry (ICP/MS) and analytical electron microscopy, in order to find the evidences of long-range transport of air pollutants from the cities of the East Asian region. The relationships between Al, which is most often used as an index of soil components, and other elements were examined. The positive correlation was found between Al and three elements (Ti, Mn, Ba). Therefore, behaviors of these elements in rime-ice may be similar to that of soil particles. Furthermore, it was possible to classify the elements into three groups with the relation between a lead isotope ratio and concentration ratio of lead and zinc (Pb/Zn). In addition, numbers of particles were found in the rime-ice. The particles around 1 μm in diameter were considered to be the combustion products of coal, whose Pb/Zn were similar to that of previous survey report in Korea. The existent forms of chemical species in rime-ice will become an important factor, when we consider the origin of air pollutants transported over long distances in the East Asian region.

Key words: rime and snow; acid aerozol; mountain site; Pb/Zn; lead isotope ratio; ICP/MS; electron microscopy; long-range transport; East Asian region.

1. Introduction

Rime-ice occurs on structures which are exposed to a cloud of supercooled droplets. The droplets impact on the structures and freeze causing the characteristic white rime-ice deposit. Rime-ice deposits are common on high elevation trees and mountain top structures. Therefore, it is conceivable that the components of rime-ice reflects atmospheric environment at the elevation rather than that of snowfall at the same site.

Rime-ice in the Kyushu mountainous regions repeatedly forms and then falls off. Namely, the rime-ice occurs under the influence of cold air masses and it separates from structures when the cold air masses leave. Consequently, it may be possible to identify the special air masses that affects to the component of rime-ice at the Kyushu mountainous regions. There are a number of papers regarding physics of rime-ice phenomenon (Langmuir, 1948; Macklin, 1962; Stallabrauss, 1978; Rogers et al., 1980; Auer and Veal, 1970; Hill and Woffinden, 1980; Cooper and Saunders, 1980). However, there are a few papers regarding chemical components of rime-ice (Utiyama, 1991; Duncan, 1992; Nagafuchi, 1993). Yet, only a few papers address long range transport of air pollutants using rime-ice components (Nagafuchi, 1993).

The objectives of this paper are to (1) analyze soluble and insoluble components in order to determine the origin of acidic composition in an atmosphere and (2) to prove long-range transport of acidic substance at the East Asian region.

2. Methods

2.1 Sampling

Rime and snow samples were collected in the Kyushu mountain region at the top of Mt. Ichifusa (elevation 1721m), Mt.Kuromi (elevation 1831m) and Mt. Hikosan (elevation 1200m), locations of which are shown in Fig. 1. The collection sites are isolated, are only accessible on foot. It was investigated at Mt. Hikosan to confirm urban effects from the city area of Fukuoka.

Rime collectors were set on the branches of the trees. They were fabricated using 2cmϕ polycarbonate tubing and Teflon net. Samples were physically scraped from rime collector and then scooped into prerinsed polyethylene bottles and were frozen until analysis. These surveys were carried out in Kyushu mountainous regions during the winter of 1994.

2.2 Analysis

Rime and snow samples were left at room temperature and melted. Samples were then separated into soluble component and insoluble component using membrane filter (Advantec, 25mmϕ, pore size 0.45µm). For the soluble components, pH, EC and major ion components were analyzed by a pH Meter (Horiba M-12), a Conductivity Meter (TOA CM-30ET) and an Ion Chromatograph (Yokogawa IC-7000), respectively. For the insoluble components, the filtrated matter was dried at room temperature and then carbon-gold evaporation was performed for the analysis by using an Electron Microscope (JEOL, JEM-1200EX, equipped with ASID 10, KEVEX-7000J). A part of the sample solution was filtrated by nylon membrane filter (Pall, 47mmϕ, pore size 0.2µm). The filters were placed into a 100 ml Teflon beaker, and 5 ml of concentrated HNO_3 (Tamapure A-1000)

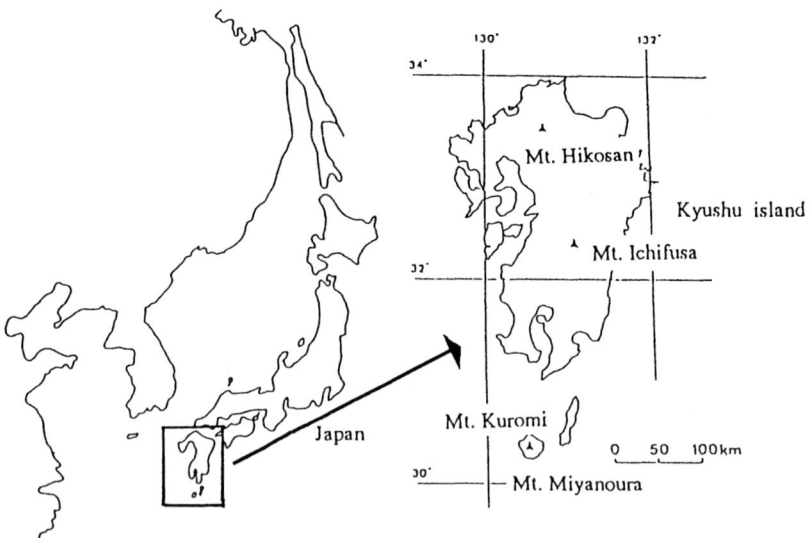

Fig. 1 Rime-ice and snow sampling sites in the Kyushu mountainous regions

was added. The mixture was covered with a Teflon lid and heated at 120 °C for 2-3 hr. $HClO_4$ (3-4 ml, Tamapure A-100) was then added, and the mixture was maintained at 180°C for 3-5 hr until carbon had been decomposed. To digest silicate rock, 2 ml of HF was added and then evaporated to nearly dryness at 200 °C. It was then dissolved into 0.4 N HNO_3 (5ml) and the sample solutions were stored in a 10 ml Teflon vial. Heavy metals and lead isotope ratio in the solution were measured by an inductively coupled plasma mass spectrometer (ICP-MS) (Perkin Elmer Elan-5000) (Mukai, 1993). SRM 981 (National Institute of Standards and Technology, Gaithersburg, Maryland) was used as a lead level isotope ratio standard. Correction factors for various lead concentration levels were derived from this standard and used in the calculation.

Isobaric back-trajectories were calculated for the 700-, 800-, 850-, 900-, 950- and 1000-hP pressure levels, based on the program by *Hayashida-Amano et al.* (1991). The trajectories had 3-day duration and ended at the each mountain top at 0000 and 1200 UT (0900 and 2100 LT) on the sampling day. Analyzed global wind fields data sets by the Japan Meteorological Agency were used as input data.

3. Results and Discussion

At the top of Mt. Ichifusa, the wind directions were measured from January to March in 1994. It was blowing from North or North West during the periods over 90%. Therefore, Kyushu mountain region appeared to be exposed mainly to north or northwest wind during the winter period.

3.1 Soluble components in rime-ice and snow

TABLE I
pH, EC and concentration of major ions in rime and snow

No.	Date	pH	EC S/cm	SO_4^{2-}	NO_3	Cl^-	NH_4^+ mg/l	Ca^{2+}	Mg^{2+}	K^+	Na^+
1	94.3.14	4.25	138	19.2	7.79	17.3	4.99	3.90	1.37	1.02	9.87
3	94.3.16	4.28	238	49.3	16.4	22.9	8.70	10.7	2.15	1.61	13.9
4	94.3.26	4.2	55.6	8.48	2.85	3.55	3.13	1.12	0.30	0.49	2.48
5	94.1.23	4.03	72.9	10.1	2.14	5.89	2.36	1.51	0.45	0.53	3.40
6	94.2.09	6.5	87.8	14.2	3.16	10.7	3.08	16.0	0.86	0.69	6.27
8	94.2.28	4.27	37.7	4.07	1.68	2.79	0.60	0.96	0.22	0.40	2.20
9	94.1.23	4.02	61.0	5.75	1.72	5.97	1.31	1.29	0.37	0.33	2.91
10	94.2.28	5.22	6.76	0.26	0.11	0.9	0.17	0.02	0.04	0.22	0.86
11	94.3.26	4.3	25.6	3.04	1.29	0.91	1.09	0.47	0.02	0.22	0.81

No.1-8: rime No.1,3,4,11: Mt.Hikosan(1200m) No.5,6,9: Mt.Ichifusa(1721m)
No.9-11: snow No.8: Mt.Kuromi(1850m) No.10: Mt. Miyanoura(1900m)

Soluble components in rime-ice and snow at Kyushu mountain sites were shown in Table I. Relationship between equivalent concentrations of anions and cations have a good ion balance except data of February 9 at Mt. Ichifusa at the time Kosa, that means "yellow sand" in Japanese, blows Japan from the deserts on the Asian continent. The component ratios of anions and cations in rime-ice at each mountain sites had a distinct characteristics. Furthermore, even component ratios in rime-ice and snow at the same site and the same day differed from one another. Relationship of equivalent concentrations between Na^+ and Cl^- was equal to a component ratio of sea water (1.17: 1), because the origin of Na^+ and Cl^- was conceivable as an ocean origin. Therefore, SO_4^{2-} and Ca^{2+} were classified as a sea salt origin or of a non sea salt (nss-) origin using Na^+ concentration. As a result, both nss-SO_4^{2-}/SO_4^{2-} and nss-Ca^{2+}/Ca^{2+} were over about 90%. Therefore, it appeared that both SO_4^{2-} and Ca^{2+} were not sea salt origin.

3.2 Insoluble components of rime and snow

Metal composition in the atmospheric aerosols has been utilized to evaluate the origin of aerosols. The Pb/Zn ratio, Te/Se ratio and Pb isotope ratio of air (Sekine, 1991; Mukai, 1993) have been used to trace long-range transported atmospheric pollutants. In the present studies, we propose a new index to evaluate the origin by analyzing metal components of particles in rime-ice. To examine an origin of metal in insoluble components, relationship between Al, that is most often used as an index of soil components, and other metals, Pb/Zn and Pb isotope ratio is shown in Fig. 2. Linear correlation is observed between Al and Ti, Mn and Ba. Relationship between Al and Pb was the same type of Ti, Mn and Ba except at the time of Kosa phenomena. This result indicates a possibility of the dilution effect by particles of a few content of Pb. Furthermore, Al concentration correlated negatively with Pb/Zn. It is reported that Pb/Zn ratio in aerosols of the East Asian Continent shows a high value (Sekine, 1991). However, when Kosa components in rime-ice were found, the ratio was a considerable small value. Therefore, it is not possible to specify the origin of aerosols with only high or low value of the ratio. Relationship Pb isotope ratio ($^{208}Pb/^{206}Pb$) and Pb/Zn ratio were classified into three groups. Therefore, if relationship between the correlation and air mass at the time of depositing rime appears, this index has possibility to be used for evaluation of an origin of the particles in rime-ice.

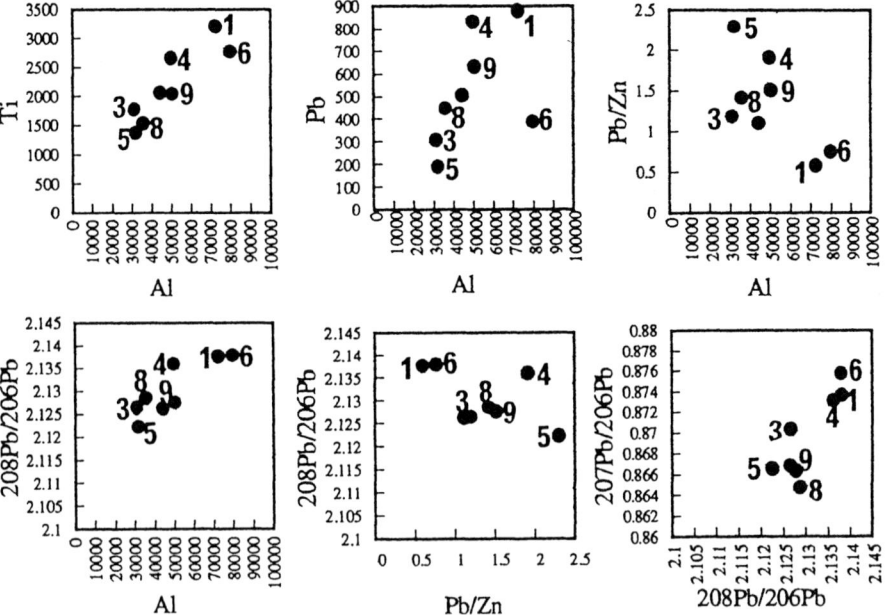

Fig.2 Relationships between insoluble Al (ng/mg) and the other elements (ng/mg), Pb/Zn and Pb ratio

3.3 Particles in rime and snow, and air trajectory analysis

The Electron Microscope was used to characterize the insoluble components of rime-ice and snow. Many sphere particles were observed in this components at each mountain sites (Fig. 3), and X ray analysis of each particle was carried out by EDS. Patterns of characteristic element components, that was detected by EDS, are shown in Fig. 3. Information regarding the intensity of X-ray spectrum with each element differs with each particle. Consequently, in this model, standardization of the values was carried out by making a full scale of maximum value of X-ray spectrum volume of each particle. Most types of sphere particles were Si-Al-Fe type. In other words, these particles are similar to coal combustion particles or soil particles. Furthermore, among these particles, the existence of Ni, that are specific elements of emission source particles of coal, was found. Therefore, it is strongly suggested that these particles are of coal combusition origin.

Fig. 4 shows air trajectories during the monitoring periods in 1994. Air trajectory of Mt. Kuromi on 28 February passed over the eastern part of the Korean peninsula. On the other

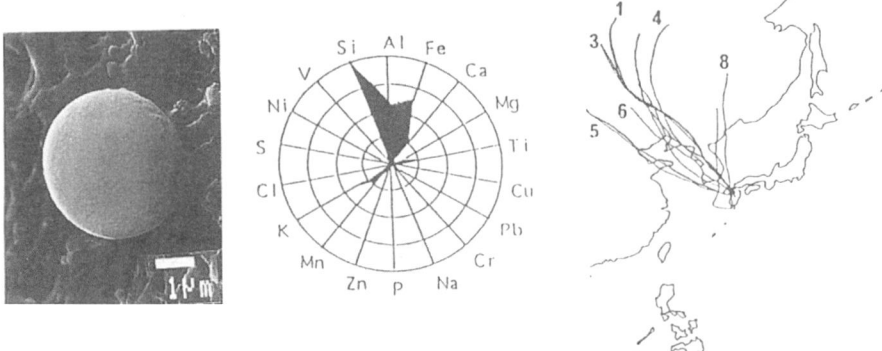

Fig. 3 SEM photograph and a model of particles in the rime at Mt. Ichifusa obtained from the result of X-ray spectra

Fig. 4 Air trajectory at each elevation of mountain. Each start at 0900 LT and 2100 LT (Japanese time) on the day before the collecting rime and lasts 3 days

hand, air trajectory of Mt. Ichifusa on 24 January passed over southern Korean peninsula and middle part of China. Soluble and insoluble components in rime that were collected at the Mt. Kuromi were very small. On the other hand, soluble and insoluble components in rime-ice that were collected at the Mt. Ichifusa were very large. In other words, small amounts of air pollutants were observed in case in which the air mass came from the eastern Korean peninsula, while larger amounts of air pollutants than the previous case were obtained from air masses which came from the middle part of China. This result may imply that there is a specific pollution source in this region.

4. Conclusion

Observations and analysis were made in rime-ice and snow at a mountain top site in the Kyushu mountainous regions during the winter of 1994. The pH values of soluble component in rime-ice were almost all around 4.0. However, we even observed the neutralization effects that was caused by Kosa components. Furthermore, a number of particles in rime were found out with the observation by electron microscopy. Also, Pb/Zn ratios and Pb isotope ratios of insoluble components in rime-ice were close to the values of east Asian continent rather than that of Japan.

From these limited data, it seems quite clear that acidic components in rime-ice deposits are not of local origin, but the acidic components were caused by pollutants that was carried by long-range transport, especially from the East Asian continent.

Acknowledgments

The authors are very grateful to S. Amano-Hayashida of National Institute for Environmental Studies and Y. Dokiya (Meteorological College) for their help in calculating air trajectories. The authors also wish to thank S. Tagami and T. Ishibashi (Fukuoka Institute of Health and Environmental Studies) for their assistance in sampling. This study was supported in part by Nippon Life Insurance Foundation.

Reference

Auer, A. H., J. and Veal, D. L., 1970, "An Investigation of Liquid Water-Ice Conten Budgets Within Orographic Cap Clouds", *J. Rech. Atmos.*, 4, 59-64.
Cooper, W. A. and Saunders, C. P. R., 1980, "Winter Storms over the San Juan Mountains. Part 2" : Microphysical Process, *J. Appl. Meteor.*, 19, 927-941.
Duncan, L. C., 1992, "Chemistry of Rime and Snow Collected at a Site in the Central Washington Cascades", *Environ. Sci. Technol.*, 61-66.
Hayashida-Amano, S., Sasano, Y. and Iikura, Y., 1991, "Volcanic disturbance in the stratosperic aerosol layer over Tukuba, Japan", *J. Geophys. Res.*, 15, 469-478.
Hill, G. E. and Woffinden, D. S., 1980, "A Balloon-Borne Instrument for the Measurement of Vertical Profiles of Supercooled Liquid water concentration,", *J. Appl. Meteor.*, 19, 1285-1292.
Langmuir, I., 1948, "Deposition of Rime on Cylinders, Spheres and Ribbons", *Occasional Report No. 1 Project Cirrus, G. E. Laboratory, Schenectady, NY*, 7-9.
Macklin, W. C., 1962,"The Density and Structure of Ice Formed by Accretion", *Quart. J. Royal. Metero. Soc.*, 88, 30-50.
Mukai, H., Furuta, N., Fujii, T., Ambe, Y., Sakamoto, K. and Hashimoto, Y., 1993, "Characterization of Sources of Lead in the Urban Air of Asia Using Ratios of Stable Lead Isotopes", *Environ. Sci. Technol.*, 27, 1347-1356.
Nagafuchi, O., Tagami, S., Ishibashi, T., Murakami, K. and Suda, R., 1993,"Evaluation of atmospheric environment by soluble components in rime",(in Japanese) *Chikyukagaku(Geochemistry)*, 27, 65-72.
Nagafuchi, O., Suda, R., Isibashi, T., Murakami, K. and Shimohara, T., 1993," Analisis of Long-range Transported Air Pollutants-Origin of Aerosol in Rime Found at Kyushu Mountainous Regions, Japan"(in Japanese), *NIPPON KAGAKU KAISHI*, (6), 788-791.
Rogers, D. C., Baumgardner, D. and Vali, D., 1983, "Determination of Supercooled Liquid Water Content by Measurement Rime Rate", *J. Appl. Meteor.*, 22, 153-162.
Sekine, Y. and Hashimoto, Y., 1991, "Long-range transport of airborne particulate pollutants in the eastern Asian area"(in Japanese), *J. Jpn. Soc. Air Pollut.*, 26, 216-226.
Stallabrass,. J. R., 1978, "An Appraisal of the Single Rotating Cylinder Method of Liquid Water Countent Measurement", *Report LTR-LT-92, Div. of Mech. Engr., National Reseach Council, Canada*, 37 pp.
Utiyama, M., Mizuochi, M., Yano, K. and Fukuyama, T., 1991, "Chemical Composition of Insoluble Substances in the Rime Sampled at Mt. Zaoh"(in Japanese), *NIPPON KAGAKU KAISHI*, (5), 517-519.

THE EFFECT OF ALKALINE DUST DECLINE ON THE PRECIPITATION CHEMISTRY IN NORTHERN JAPAN

I. NOGUCHI, T. KATO, M. AKIYAMA, H. OTSUKA and Y. MATSUMOTO

Hokkaido Institute of Environmental Sciences, Kita-19 Nishi-12, Kitaku, Sapporo, 060 Japan

Abstract. The precipitation chemistry in northern Japan, especially Hokkaido, has been investigated since 1982. This area has often been found to have high concentrations of alkaline road dust (asphalt dust) in the air, caused by the use of studded tires during the winter. It is well known that the composition of precipitation in these areas is often dominated by asphalt dust including calcium bicarbonate. However, recently the concentration of asphalt dust in the air has decreased owing to a ban on the use of studded tires. Simultaneously, in precipitation, the lowering of pH values and the increase of hydrogen ion depositions have been occurring owing to the decrease of non-sea-salt calcium ions (nss-Ca^{2+}) concentrations and depositions derived from asphalt dust. In addition, we found that a decrease of nss-Ca^{2+} firstly leads to a decrease of bicarbonate ions (HCO_3^-), the counter ion to nss-Ca^{2+} in asphalt dust. Therefore, the increase of H^+ concentrations and depositions was great in comparison with the decrease of nss-Ca^{2+} concentrations and depositions in areas where the HCO_3^- concentrations, varied by pH, and depositions had been low. Furthermore, this variation was mainly observed in the ionic composition of snow cover and snowfall at sites along the Japan Sea in northern Japan during winter. In this area, the Acid Shock effect may become a serious problem from the decline of pH values in melting snow. Moreover, we found that ammonium ions and non-sea-salt sulfate ions depositions have also been decreasing in response to a decrease of nss-Ca^{2+} depositions, derived from asphalt dust. It seems that this phenomenon is caused by the decrease of asphalt dust concentrations in the air.

Key words : Japan, precipitation, alkaline dust, studded tire, calcium, bicarbonate

1. Introduction

The precipitation chemistry of Asia is often dominated by alkaline dust, a large portion of which is soil dust containing calcium carbonate ($CaCO_3$) such as Kosa soil. As a result, the annual mean pH values of precipitation are often above 5.0 or 6.0 in spite of a high concentration of non-sea-salt sulfate ions (nss-SO_4^{2-}) in Asia (Zhao, D. et al., 1988; Quan, H., 1991; Ito, T., 1991; Naik, M. S. et al., 1994). Furthermore, in Asia in the future, it is considered that there will be a decrease of alkaline soil dust owing to mobilization and urbanization (Hedin, L. O. et al.,1994). For these reasons, a lowering of pH values in precipitation may occur from a decline of alkaline dust.

In Hokkaido and other northern areas of Japan, studded tires were used because most of these areas are covered with snow in winter. Thus a high concentration of alkaline road dust (asphalt dust) was found because studded tires scratch the surface of the asphalt roads. When asphalt pavement is made 11% lime is added over time, this lime is converted to $CaCO_3$ by carbon dioxides in the air. Since $CaCO_3$ in asphalt roads is dissolved little by little by precipitation water, the content of $CaCO_3$ in scratched asphalt dust depends on the condition of the pavement, weather conditions and other factors. As per our results, asphalt dust collected in the vicinity of roads was found to include 3% soluble $CaCO_3$. This $CaCO_3$ is 90- 96% of the soluble components of asphalt dust, other base cation is 1- 6%. During the winter season (Oct.- May) in Hokkaido, asphalt dust emmisions were calculated to be 96,100m^3 (220,000 t) (Kubo, H., 1983). In urban areas, the amount of asphalt dust in the total suspended particles were found to range from 30%

in autumn (before the use of studded tires) to over 90% in winter (Kato, T. et al., 1986). Precipitation composition is affected by asphalt dust in all seasons because this asphalt dust was observed not only in winter. The effect of asphalt dust on the precipitation chemistry has previously been reported on in Japan. For example, the Phase-I Survey report on acid precipitation by the Japan Environment Association showed that the annual mean pH values in northern Japan were high owing to the effect of asphalt dust (JEA, 1990). However, the concentration of asphalt dust in the air has decreased owing to the recent ban on the use of studded tires. A variation of precipitation composition has also been observed (Noguchi, I., 1994). This paper aims to show the change in composition of precipitation affected by alkaline dust using northern Japan as an example.

2. Experimental Section

We have carried out precipitation surveys in northern Japan since 1982 (Table 1, Fig. 1). In a fractional precipitation survey, wet only samples were collected for each 1 mm of rainfall in 6 fractions (1-5 mm and over 6 mm) per one rainfall event. In an other precipitation survey, bulk samples were collected in 1- or 2-week cycles. A Bulk sampler was equipped with a filtration system and a sample tank which were both shielded from the light. In winter, the snowfall bulk samples were melted at room temperature followed by filtration. In a snow cover survey, snow-core samples were collected using a steel pipe (12cm Φ) and they were treated by the same procedures as the snowfall samples.

Table 1 Situation of precipitation surveys in northern Japan

Survey	Term	Sampling method	Area	Sampling cycle	Site situation
Fractional survey	1982-1987 (Apr.-Nov.)	Wet only	9 sites in Hokkaido	1 rainfall	Urban site
Bulk survey in Sapporo	1984-	Bulk	Sapporo	1 - 2 week	Urban site
Bulk survey in Tohoku	1989-1993 (Dec.-Feb.)	Bulk	Sapporo and 8 sites in Tohoku area	2 week	Urban site
Snow covered survey	1988, 1992	Snow core	69(1988) sites, 78(1992) sites in Hokkaido	1 time before the snow thawed	Remote site

Samples were analyzed for precipitation, pH, conductivity and the concentration of major ions (SO_4^{2-}, NO_3^-, Cl^-, Na^+, K^+, Ca^{2+}, Mg^{2+} and NH_4^+). Measurements were made by a pH meter, a conductivity meter, ion chromatography, atomic absorption spectrometry, as well as an automated colorimetry or manual spectrometry. The analytical procedures and data quality control for reanalysis were carried out based on a manual for acid precipitation surveys prepared JEA. In addition, non-sea-salt (nss-) components were calculated from the Na+ composition of sea salt by the method generally used in Japan. The means were calculated based on volume-weighted average. The seasons were divided as follows; winter (Dec.-Feb.), spring (Mar.-May), summer (Jun.-Aug.) and autumn (Sep.-Nov.). The survey year started in December of each year. For example, data for winter 1988 included data from December 1987.

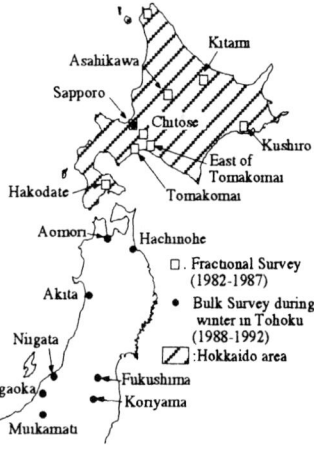

Fig.1 Sampling sites of precipitation surveys in northern Japan

3. Result and Discussion

3.1 Effect of asphalt dust
(1) Behavior of initial rainfall composition

In a fractional precipitation survey, the pH means ranged over 4.6 - 5.3. Sapporo (pH4.6) and Asahikawa (pH4.7), two large cities in Hokkaido tended to have lower pH means (Noguchi, I. et al, 1989). However, even at these low pH sites, the effect of asphalt dust on the precipitation chemistry showed. The pH means of the initial fractions were higher than those of later fractions, although the concentrations of the components were higher in the initial fractions (Table 2). This behavior of fractional pH means

Table 2 The concentration change of components in fractional precipitation (*: μ eq/l)

Site	Stage	pH	H⁺ *	nss-SO_4^{2-} *	NO_3^- *	NH_4^+ *	nss-Ca^{2+} *
Sapporo	1	4.79	16.2	101.8	35.1	40.2	111.0
	2	4.62	24.2	66.4	16.7	24.1	38.2
	3	4.59	25.5	59.9	13.7	21.7	27.4
	4	4.57	27.2	52.0	11.7	19.2	22.1
	5	4.56	27.7	49.8	10.4	18.7	13.7
	6	4.57	27.0	37.5	7.4	11.5	8.2
Asahikawa	1	4.78	16.6	147.1	50.6	35.2	215.8
	2	4.55	28.5	74.3	15.9	25.3	50.8
	3	4.62	23.8	79.7	15.8	25.8	38.9
	4	4.59	25.7	72.5	15.6	24.9	27.1
	5	4.70	20.0	52.8	9.1	15.9	23.1
	6	4.72	19.1	49.5	6.6	12.6	18.7

was caused by a rather high nss-Ca^{2+} concentration in the initial rainfall and by a more rapid decrease in the concentration of nss-Ca^{2+} derived from asphalt dust than the concentrations of acid ions (nss-SO_4^{2-}, NO_3^-) as rainfall progressed. This phenomenon was observed also at other sites in Hokkaido which also have high pH means.

(2) Unmeasured counter anion to nss-Ca^{2+} derived from asphalt dust

Samples in which the cation sum exceeded the anion sum, were observed in precipitation to have high concentrations of nss-Ca^{2+} derived from asphalt dust in a fractional survey (Fig. 2). These samples had been reanalyzed because these data were outside of the JEA reanalysis standard. These data were also often outside of the WMO reanalysis standard which use ion and conductivity balance. Therefore, we guessed that there were unmeasured anions. However, we could not detect a special peak from the ion chromatography chart of excess cation sum samples. Furthermore, we found a high concentration of inorganic carbon in the excess cation sum samples by the combustion-infrared absorption photometric method. The concentrations of CO_3^{2-} are also negligible in almost all samples which have pH values below 8. Thus, it was estimated that there was HCO_3^- as an unmeasured counter anion to nss-Ca^{2+} derived from asphalt dust.

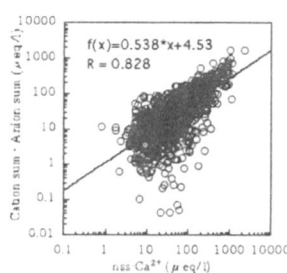

Fig.2 The concentrations of nss-Ca^{2+} and the difference between cation sum and anion sum

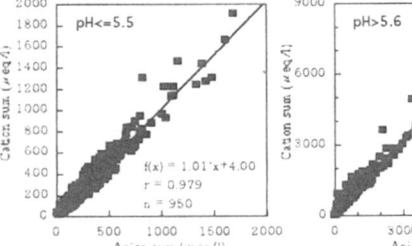

Fig.3 The ion balance by pH range

The concentration of HCO_3^- can be neglected in the lower pH ranges. Thus, it is observed that the cation sum balances with the anion sum in the lower pH ranges ($\leq$ pH5.5) and that the cation sum exceeds the anion sum in the higher pH ranges (>pH5.5) (Fig. 3). However, if the pH means of one rainfall event is below 5.5, we can have higher pH samples with high concentrations of nss-Ca^{2+} and HCO_3^- in each fraction. In this case, it is estimated that the decreased amount of nss-Ca^{2+} concentrations can not be linked to an increase in the amount of H^+ concentrations owing to a decrease in HCO_3^- concentrations.

3.2 The decline of asphalt dust in the air and the behaviour of precipitation components

The concentrations of asphalt dust in the air have decreased owing to the ban on the use of studded tires since the winter of 1990. At the Sapporo North monitoring station (Urban residential area, the distance from a nearest throughfare is 500m) the concentrations of asphalt dust before the ban on the use of studded tires (Dec. 1982) and after the ban (Dec. 1991) were 0.11mg/m³ (the contribution of asphalt dust total suspended particles: 80%) and 0.03mg/m³ (asphalt dust: 36%), respectively as shown in Fig. 4 (Kato, T. et al., 1986; Otsuka, H. et al., 1993). Furthermore, asphalt dust affects the concentration of PM-10 although most of the asphalt dust consists of large particles (>10 μm). The trends toward a decrease in PM-10 at the Sapporo North-station and the East-station (Urban residential area, in the vicinity of throughfare) since 1990 or 1991 are shown in Fig. 4 as examples. Small particles of asphalt dust which affect the PM-10 concentrations remain in the environment over a long period and their effects may be observed at cloud altitude.

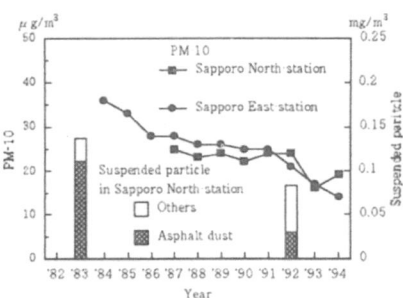

Fig.4 Concentrations of PM-10 and Suspended particles during winter in Sapporo.

In a previous bulk survey done in Sapporo (Noguchi, I., 1991; 1993; 1994), was found that the H^+ deposition had increased, especially in winter since 1990; in contrast, the nss-Ca^{2+} deposition has decreased, especially in spring and winter (Fig. 5). This variation was caused by a decrease of asphalt dust concentrations in the air. The reason why the H^+ depositions did not increase remarkably in the spring seasons, unlike the winter seasons, despite the remarkable decrease in the nss-Ca^{2+}, was that at first a decrease in nss-Ca^{2+} resulted in a decrease in HCO_3^- depositions, which had been rather higher in the spring seasons than in the winter seasons. This resulted in a small increase in the deposition of H^+, which was not as much as in the winters.

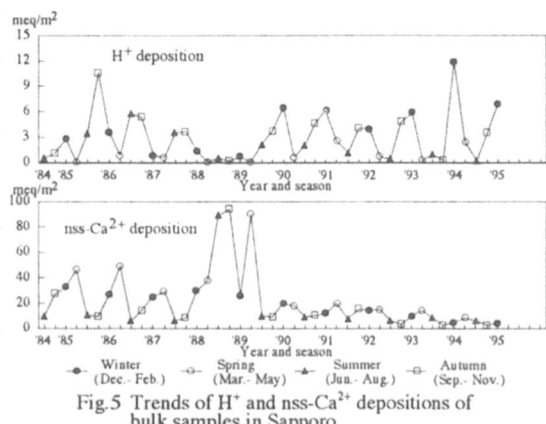

Fig.5 Trends of H^+ and nss-Ca^{2+} depositions of bulk samples in Sapporo.

In a snow cover survey (Araki, K. et al., 1989; Noguchi, I., 1993), the ratio of areas below pH 5.0 in total areas by the distributions of pH values using the Spline method, increased from 20% in 1988 to 75% in 1992. This variation was caused by a decrease of nss-Ca^{2+} concentrations from asphalt dust even in these remote areas. Moreover, the increase of H^+ concentrations based on the decrease of nss-Ca^{2+} concentrations along the Japan-Sea sites, which had very low HCO_3^- concentrations owing to low pH levels in 1988, was greater than in other areas, because the decrease of nss-Ca^{2+} directly affected the increase of H^+ in those sites.

In a bulk survey taken in the Tohoku area during winter (Hokkaido and Tohoku Environ. Labo. Associ., 1993), the H^+ and nss-Ca^{2+} depositions in this area showed respectively increasing and decreasing trends, because the ban on the use of studded tires in Tohoku started almost at the same time as in Hokkaido. In the same way as for the snow cover in Hokkaido, the increase of H^+ based on the decrease of nss-Ca^{2+} in almost all of the sites along the Japan-Sea were greater concerning the depositions and concentrations (Table 3). This variation means that the Acid Shock effect may become a serious problem from the decline of pH values in melting snow in these areas.

Table 3 The change rates of H+ and nss-Ca2+ depositions, pH and nss-Ca2+ concentrations during winter in northern Japan

Site	H^+ meq/m²/30days/yr	nss-Ca^{2+} meq/m²/30days/yr	Ratio of rates (H^+/nss-Ca^{2+})	pH (nssCa^{2+}) in 1987	pH (nss-Ca^{2+}) in 1992
☆Sapporo	0.68	-0.93	-0.73	5.91(99)	4.56(46)
☆Aomori	0.57	-0.52	-1.11	5.30(129)	4.58(69)
Hachinohe	0.47	-0.92	-0.51	6.38(193)	4.51(50)
☆Akita	0.62	-3.86	-0.16	6.50(198)	4.66(33)
Fukushima	0.42	-1.03	-0.40	6.23(238)	4.65(16)
Koriyama	0.11	-0.96	-0.11	7.21(342)	5.00(41)
☆Niigata	0.87	-0.89	-0.99	5.56(62)	4.77(26)
☆Nagaoka	1.71	-0.86	-1.98	5.24(26)	4.53(10)
☆Mukamachi	1.49	-0.73	-2.04	5.12(18)	4.49(7)

☆: site along the Japan Sea. ( ) μeq/l

On the other hand, we found that nss-SO_4^{2-} and NH_4^+ depositions of bulk samples taken in Sapporo had been decreasing since 1990 or 1991 when the decline of asphalt dust started (Fig. 6). This cause was discussed in winter (Dec. - Feb.), the effects of Kosa and other soil aerosols, gas and aerosol by microbiomass reaction. As a result, the relationship between nss-Ca^{2+} derived from asphalt dust, nss-SO_4^{2-} and NH_4^+ in concentration, deposition and composition, were statistically significant at P<0.05. Furthermore, the transport of atmospheric pollutants by Kosa aerosol has a similar composition to asphalt dust and the reaction between alkaline soil dust and atmospheric pollutants had previously been reported on (Altwicker, E. R. et al., 1984; Lee, M. H., 1990; Quan, H. et al., 1993; Nishikawa, M., 1993; Iwasaka, Y. et al., 1991). It was also reported that the riming ice in clouds in Sapporo during winter had a high level nss-SO_4^{2-} and NH_4^+ (Endo et al., 1994). Thus, it is necessary to study the relationship between riming ice in clouds and small particles of asphalt dust because it seems that small particles of asphalt dust have been observed at cloud altitude. Furthermore, the

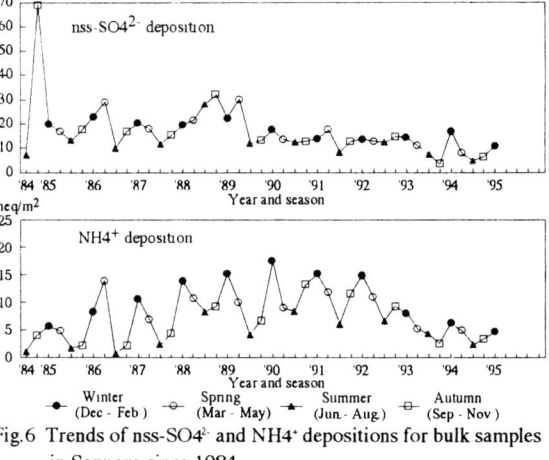

Fig.6 Trends of nss-SO_4^{2-} and NH_4^+ depositions for bulk samples in Sapporo since 1984.

effect of asphalt dust on the depositions of nss-SO_4^{2-} and NH_4^+ also seems to affect the model calculation of the scavenging process. Therefore, the behavior of the aerosol components since the 1980s is now under investigation.

4. Conclusion

In northern Japan, we found that asphalt dust, including $CaCO_3$, affects the precipitation ions even in wet samples, especially in initial rainfall by a fractional precipitation survey. Moreover, it is estimated that the precipitation chemistry was affected by not only nss-Ca^{2+} derived from asphalt dust but also HCO_3^-, the counter ion to nss-Ca^{2+}. Because this neutralization or antioxidation in precipitation chemistry has reduced after a ban on the use of studded tires, it was observed that the H+ depositions and concentration have increased owing to a decrease of nss-Ca^{2+} deposition and concentrations. This phenomenon was also observed in a snow covered survey in remote areas of Hokkaido. In addition, we found that the increase in the amount of H+ differed from the decrease in the amount of nss-Ca^{2+} owing to the effect of HCO_3^- decrease by season and area. It means that the increase in the amount of H^+ levels based on the decrease in the amount of nss-Ca^{2+} levels is large when HCO_3^- levels are low. As the result, this variation along the Japan-Sea during winter was great. Therefore, the Acid Shock effect may become a serious problem due to the decline of pH values in melting snow in these areas. Furthermore, although little known, it seems that asphalt dust affects the nss-SO_4^{2-} and NH_4^+ depositions which were included in riming ice in the clouds.

References

Altwicker, E. R. and Mahar, J. T. :1984, *Atomos. Environ.*, **18**, 1875-1883
Araki, K., Kato, T., Tabuchi, S., Noguchi, I., Takahashi, H., Sakata, K. and Aoi, T.: 1989, *Report of Hokkaido Research Inst. for Environ. Pollut.*, **15**, 73-80
Endo, T., Takahashi, T., Muramoto, K., Nakagawa, C. and Noguchi, I.: 1994, *American Metrological Society annual meeting 75th*, Dallas, 521-522
Quan, H.: 1991, *J. Japan Soc. Air Pollut.* **26**, 283-291
Quan, H. and Chen, Z.: 1993, *The 5th International Symposium on Acid Rain and Snow on the Japan-Sea Rim Abstracts*, 9-18
Hedin, L. O., Granat, L., Likens, G. E., Buishand, T. A., Galloway, J. N., Butler, T. J. and Rodhe H.: 1994, *Nature*, **367**, 351-354
Hokkaido and Tohoku Environmental Laboratories Association, 1993, *The Analysis report of Acid Deposition Survey in Hokkaido and Tohoku areas*
Ito, T.: 1991, *International Workshop on Acid Rain in East Asia Abstracts*, 19-25
Iwasaka, Y., Hayashi, M. and Yamato, M.: 1991, *Kosa*, Water Research Institute, Nagoya University, 250-255
Japan Environment Association, 1990, *Acid Precipitation in Japan, The report of Phase I Survey*
Kubo, H.: 1983, *Monthly Report of the Civil Engineering Research Institute*, **365**, 3-16
Kato, T., Noguchi, I. and Matsumoto, Y.: 1986, *Report of Hokkaido Research Inst. for Environ. Pollut.*, **12**, 13-21
Lee, M.H.: 1990, *The 2th International Symposium on Acid Rain and Snow in the Circum-Pam-Japan-Sea Area Abstracts*, 19-28
Naik, M. S., Momin, G. A., Rao, P. S. P., Safai, P. D., Pillai,A.G. and Khemani, L. T.: 1994, *Joint 8th CACGP Symposium / 2nd IGAC Conference Abstracts*, Fujiyoshida, Japan, 144
Nishikawa, M.: 1993, *Proceeding of the International Workshop on Development and Application of Biogeochemical Method in Acid Rain Research*, Japan, National Inst. for Environ. Studies, 71-77
Noguchi, I., Kato, K., Matsumoto, Y. and Araki, K.: 1989, *Report of Hokkaido Research Inst. for Environ. Pollut.*, **15**, 31-51
Noguchi, I.: 1991, *Journal of Environmental Laboratories Association*, **16**, 106-109
Noguchi, I.: 1993, *Hokkaido Journal of Public Health*, **16**, 113-125
Noguchi, I.: 1994, *Journal of Environmental Laboratories Association*, **19**, 35-39
Otsuka, H., Kato, T., Akiyama, M. and Matsumoto, Y.: 1993,*The winter meeting of the chemical research in Hokkaido Abstracts*, 51
Song, S. J.: 1991, *International Workshop on Acid Rain in East Asia Abstracts*, 27-31
Zhao, D., Xiong, J., Xu, Y. and Chan, W. H.: 1988, *Atomos. Environ.*, **22**, 349-358

SHORT-TERM VARIATION IN THE CONCENTRATION OF SELECTED IONS WITHIN INDIVIDUAL TROPICAL RAINSTORMS

M. RADOJEVIC and L.H. LIM

Department of Chemistry, University of Brunei Darussalam, Bandar Seri Begawan 2028, Brunei Darussalam

Abstract. Rainwater was collected at the campus of the University of Brunei Darussalam in Bandar Seri Begawan, Brunei Darussalam, using a funnel-in-bottle sampler. Polypropylene bottles were changed at intervals during rainstorm events. The pH and conductivity were determined immediately after collection on aliquots of the sample. Samples were refrigerated at 5 °C for subsequent chemical analysis. Analyses for Na, Mg, Ca, Zn and Fe were carried out by means of inductively coupled plasma atomic emission spectroscopy (ICP-AES); Cu and Mn were analysed by graphite furnace atomic absorption spectroscopy (GFAAS); K was analysed using flame atomic emission spectroscopy (FAES); and Cl^-, NO_3^- and SO_4^{2-} were analysed by ion chromatography (IC). Concentration versus time profiles are reported for three rainstorm events. All ions exhibited a decrease in concentration during the rainstorm. The first sample contained the highest concentration of ions, consistent with a "first-flush" effect. The contribution of the initial stages of the shower to the total quantity of ion deposited during the entire rainstorm is quite overwhelming; in many cases 20 to 30 % of the mass was deposited in less than 5% of rainstorm duration. On the other hand, the pH and conductivity variation during rainstorms did not exhibit a consistent pattern.

Key words: rainwater, deposition, acidity, sodium, potassium, magnesium, calcium, sulphate, chloride, nitrate.

1. Introduction

Rainwater is an important sink for water soluble atmospheric pollutants, and a source of both pollutants and nutrients in terrestrial and aquatic ecosystems. In recent years there has been a growing interest in the composition of rainwater in tropical regions and acid rain has been identified at several sites in Africa, Asia and Australia (Rodhe and Herrera, 1988). In past rainwater surveys samples have been collected over monthly (Keiding and Heidam, 1986; Low, 1988), weekly (Bruijnzeel, 1989; Lee and Longhurst, 1992) or daily (Camarero and Catalan, 1993; Lindberg, 1982; Galloway et al, 1982) sampling periods. Evaporation, dry deposition and chemical reactions can all alter the composition of samples left in the field for long periods. In order to minimise these effects, sampling on an event basis has become increasingly common (Singer et al, 1993; Sanhueza et al, 1987; Sanhueza et al, 1992; Andrea et al, 1990). Although the results of these studies can be used to evaluate deposition fluxes they provide no insight into the variation of composition during each rainfall event. Recently, near real-time rainfall pH measurements have been reported for several rainstorms using a continuous flow system (Jennings et al, 1992). The chemical composition of individual rain-drops has also been reported (Bachmann et al, 1993).

In this paper we report concentration/time profiles of several ionic species within three tropical rainstorms. This work was carried out as part of a wider survey into the composition of rainwater in Brunei Darussalam.

2. Experimental

Rainwater samples were collected on the roof of the Chemistry Department building at the campus of the University of Brunei Darussalam in Bandar Seri Begawan, Brunei Darussalam. There were no high buildings, or other tall objects, in the vicinity of the sampling site. Samples were collected using a polypropylene funnel (diameter = 20 cm) and 150 cm^3 polypropylene bottles. Sampling bottles were changed at intervals during the rainstorm events. The sampling periods tended to vary as the bottles were left out until there was sufficient sample for analysis. Prior to use, all bottles and funnels were rigorously cleaned by repeatedly rinsing with high purity, NANOpure, water obtained using a Type D4700 Deionisation System from the Barnstaed/Thermolyne Corporation. NANOpure water blanks which had been stored in the polypropylene bottles were analysed with all runs and no contamination was identified. Samples were neither filtered nor acidified prior to storage in order to avoid contamination.

The pH and conductivity were determined immediately after sample collection in the laboratory on aliquots of the sample using a Jenway 3045 Ion Analyser equipped with a combination glass electrode and a temperature compensation probe, and a Cole-Palmer Conductivity Meter, respectively. Samples were refrigerated at 5 °C for subsequent chemical analyses. Concentrations of Cl^-, NO_3^- and SO_4^{2-} ions were measured by ion chromatography (IC) using a Shimadzu Ion Chromatography system comprising a pump, oven, controller, conductivity meter and an automatic injector. Separation was achieved by means of a Shim-pack IC-A1 column. The mobile phase was 2.5 mM phthalic acid, 2.5 mM Tris(hydroxymethyl)aminoethane with a pH of 4. Concentrations of K were determined by means of flame atomic emission spectroscopy (FAES) using a Philips PU91000 Atomic Absorption Spectrophotometer. Analyses for Na, Ca, Mg, Zn, Fe and P were carried out by means of inductively coupled plasma-atomic emission spectroscopy (ICP-AES) using a Plasmascan 710 ICP instrument. Measurements of Cu, Mn and Pb were made using a Perkin Elmer 4100ZL atomic absorption spectrophotometer equipped with a graphite furnace (GFAAS). GFAAS and ICP analyses were carried out in the Microanalysis Laboratory of the Department of Chemistry, National University of Singapore. All the other analyses were performed in the Department of Chemistry, University of Brunei Darussalam. Working standards of low concentration, corresponding to levels of ions found in our rain-water samples, were prepared daily from 1000 ppm stock solutions. Analysis was completed within one to four months after sample collection.

3. Results and Discussion

Results of the measurements for the three rainstorms are given in Tables I, II and III. The tables show the concentrations of measured ions as a function of time from the beginning of the shower. Also shown is the precipitation amount in mm as calculated from sample volumes. Major anions were determined in all three events. Alkali metals and alkaline earth metals were determined comprehensively in two of the storms, and some measurements of selected transition metals were carried out in one storm. Levels of Pb were below the detection limit of 1×10^{-6} g dm^{-3} in all the samples. Concentrations of P were determined in only three samples and found to vary between 38×10^{-6} and 52×10^{-6} g dm^{-3}.

It is apparent that all the ions exhibited a similar variation with time during the three rain-storms, the concentration decreasing with time as the storm progressed. The highest concentrations were observed at the beginning of the storm, and this is consistent with a "first-flush" effect. Typical concentration/time profiles for some species during one of the rainstorms are illustrated in Figure 1.

The precipitation rate in mm min^{-1} can easily be calculated by dividing the rainfall amount by the time interval for each sample. The precipitation rate was very low towards the end of the storm when the rainfall was generally reduced to a slight drizzle. It is also possible to calculate the deposition rates of the ions. The deposition rates exhibited a similar pattern but the extent of decrease with time was even more pronounced then for concentrations. Some typical profiles of deposition rates are shown in Figure 2. The greater decrease is due to the combined effects of decreasing concentration and decreasing rainfall intensity during the storm. In some cases, more than 30% of the total mass of ion was deposited during the first five minutes of the storm. On the other hand, neither the pH nor the conductivity showed a consistent pattern; profiles varied considerably from one event to the next. The H^+ concentration depends on the difference between the sums of negative and positive cations. As complete analysis was not carried

Table I
Rainstorm of 23/5/1994. cond.= conductivity

time min.	rain mm	pH	cond. 10^{-6}mhos cm^{-1}	Cl$^-$	NO$_3^-$	SO$_4^{2-}$	K$^+$
				\multicolumn{3}{c}{concentration (10^{-6} g dm^{-3})}	
12	2.9	5.20	7.2	235	390	440	80
17	2.7	5.28	3.1	55	130	205	65
20	2.7	5.29	3.2	45	130	270	45
24	2.9	5.42	2.9	25	100	180	35
29	3.0	5.20	3.0	25	100	235	33
34	2.6	5.35	4.7	40	180	205	78
49	2.1	5.21	3.0	30	75	155	20
59	2.4	5.22	3.5	80	170	180	40
74	1.4	5.13	3.8	135	155	310	57

Table II
Rainstorm of 2/6/1994. ND= not determined. Conductivity units same as in Table I.

time min	rain mm	pH	cond.	Cl$^-$	NO$_3^-$	SO$_4^{2-}$	Na$^+$	K$^+$	Mg^{2+}	Ca^{2+}	Zn^{2+}
							\multicolumn{5}{c}{concentration (10^{-6} g dm^{-3})}
4	3.0	4.90	7.6	60	520	415	166	45	19	169	13
9	2.2	4.74	10.7	60	595	545	157	23	15	ND	ND
19	2.4	4.86	8.0	35	335	310	138	13	10	36	ND
29	2.4	5.10	3.4	9	105	130	99	6	3	15	ND
40	2.6	5.13	3.1	5	75	120	101	15	8	67	10
70	2.2	5.01	5.0	9	180	235	110	25	8	82	ND
90	2.6	4.88	7.6	10	340	130	108	7	5	41	10

Table III
Rainstorm of 2/8/1994. ND= not determined. cond = conductivity; same units as in Table I.

time min	rain mm	pH	cond	Cl$^-$	NO$_3^-$	SO$_4^{2-}$	Na$^+$	K$^+$	Mg^{2+}	Ca^{2+}	Mn^{2+}	Fe^{3+}	Cu^{2+}	Zn^{2+}
						concentration (10^{-6} g dm^{-3})								
5	2.7	5.10	17.8	470	1210	2420	340	350	117	1657	4.10	42	10	26
8	3.5	4.74	16.0	240	820	1580	236	202	66	1020	4.44	28	2	16
11.5	3.1	4.55	16.1	180	683	980	187	116	35	389	3.82	32	ND	21
15	2.4	4.56	15.5	180	570	830	165	97	29	124	ND	ND	ND	ND
19.5	2.6	4.50	14.5	70	440	520	150	74	18	183	0.69	ND	4	11
25	2.4	4.52	13.3	44	400	370	133	55	13	82	ND	ND	ND	8
35	2.3	4.66	11.1	44	350	430	130	45	13	148	0.62	14	1	9
52	3.0	4.66	9.4	10	210	220	90	40	8	64	0.55	7	2	6
237	2.9	4.56	11.4	44	370	280	115	30	10	99	0.55	4	8	6

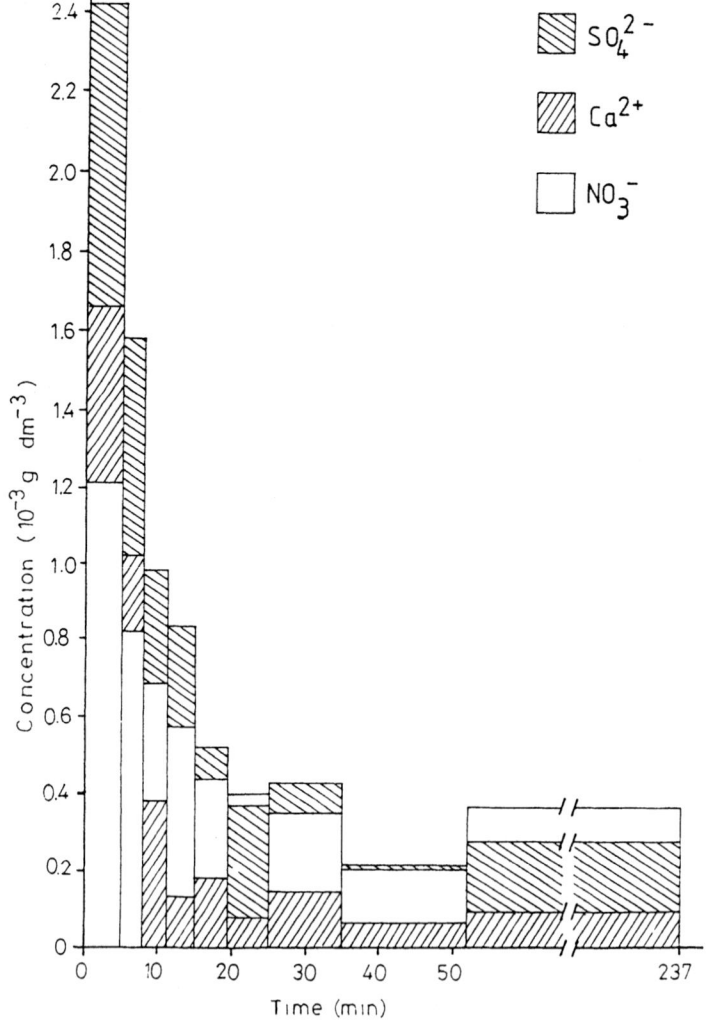

Fig. 1. Concentration versus time plot for rainstorm of 2/8/1994. The concentration of each ion is denoted by the top of its corresponding bar.

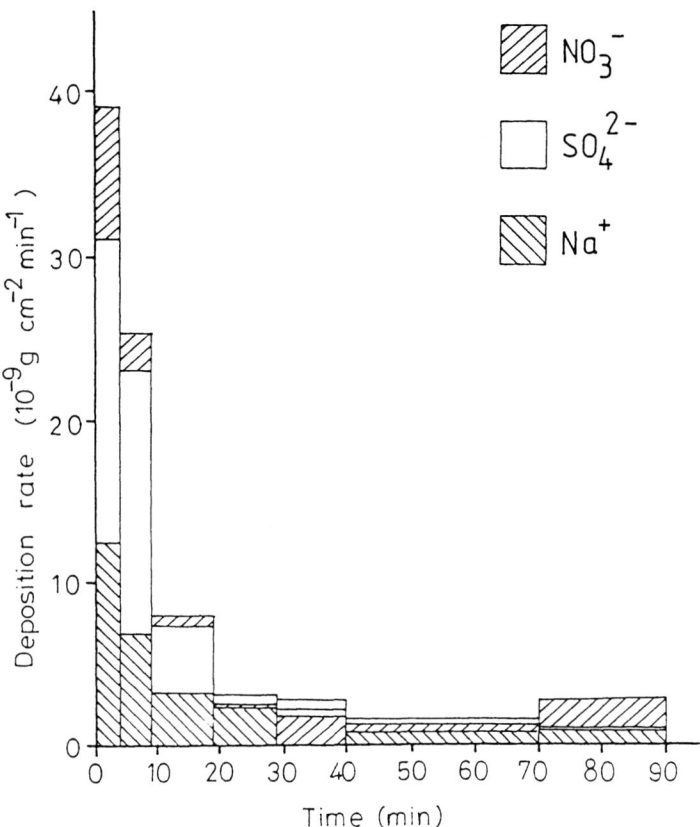

Fig.2. Deposition rate versus time plot for rainstorm of 2/6/1994. The deposition of each ion is denoted by the top of its corresponding bar.

out in this work it is not possible to correlate the pH, or the conductivity, with the chemical composition. It has been demonstrated in other studies that organic acids, such as formic and acetic acids, may contribute significantly to the acidity of rainwater at tropical and remote sites (Sanhueza et al, 1992; Andrea et al, 1990). These acids were not measured during the present study.

4. Conclusions

Concentration-time profiles for individual rainstorms exhibited a decrease in concentration during the events illustrating the cleansing action of rainfall on the atmosphere. Highest concentrations were observed at the start of the rainstorm, indicating a "first-flush" effect. Deposition rates of ions exhibited an even more pronounced decrease during the storms. The contribution of the early stages of the rain on the quantity of ions deposited was quite overwhelming; in many cases between 20 and 30% of the mass was deposited in less than 5% of rainfall duration. Conductivity and pH profiles exhibited no consistent pattern and varied considerably from one event to the next. Field measurements of the kind reported here could be used to validate the results of precipitation models.

Acknowledgements

The authors would like to thank the Microanalytical Laboratory of the Department of Chemistry, National University of Singapore, for allowing the use of the ICP and GFAAS instruments.

References

Andrea, M.O., Talbot, R.W., Berresheim, H. and Beecher, K.M.: 1990, *J. Geophys. Res.* **95**, 16987-16999.
Bachman, K.; Haag, I. and Roder, A.: 1993, *Atmos. Environ.* **27A**, 1951-1958.
Bruijnzeel, L.A.: 1989, *J. Tropical Ecology* **5**, 187-202.
Camarero, L. and Catalan, J.: 1993, *Atmos. Environ.* **27A**, 83-94.
Galloway, J.N., Likens, G.E., Keene, W.C. and Miller, J.M.: 1982, *J. Geophys. Res.* **87**, 8771-8786.
Jennings, M.M.; Perkins, T.D., Hemmerlein, M.T. and Klein, R.M.: 1992, *Water, Air, and Soil Poll.* **65**, 237-244.
Keiding, K. and Heidam, N.Z.: 1986, *Tellus* **38B**, 345-352.
Lee, D.S. and Longhurst, J.W.S.: 1992, *Water, Air, and Soil Poll.* **64**, 635-648.
Lindberg, S.E.: 1982, *Atmos. Environ.* **16**, 1701-1709.
Low, P.S.: 1988, *Sci. Total Environ.* **77**, 253-268.
Rodhe, H. and Herrera, R. (eds.): 1988, *Acidification in Tropical Countries, SCOPE 36*, John Wiley and Sons, Chichester, 405 pp.
Sanhueza, E., Graterol, N. and Rondon, A.: 1987, *Tellus* **39B**, 329-332.
Sanhueza, E., Arias, M.C., Donoso, L., Graterol, N., Hermoso, M., Marti, I., Romero, J., Rondon, A. and Santana, M.: 1992, *Tellus* **44B**, 54-62.
Singer, A., Shamay, Y., Fried, M. and Ganor, E.: 1993, *Atmos. Environ.* **27A**, 2287-2293.

A RAIN ACIDITY STUDY IN BRUNEI DARUSSALAM

M. RADOJEVIC and L.H. LIM

Department of Chemistry, University of Brunei Darussalam, Bandar Seri Begawan 2028, Brunei Darussalam

Abstract. Fifty-three bulk deposition samples were collected at the campus of the University of Brunei Darussalam between April and November 1994 using a funnel-in bottle sampler. The pH and conductivity were determined immediately after collection on aliquots of the sample. Samples were refrigerated at 5 °C for subsequent chemical analysis. Analyses for Cl^-, NO_3^- and SO_4^{2-} were carried out by means of ion chromatography, while Na^+, K^+, Ca^{2+} and Mg^{2+} were determined by atomic spectroscopy. The recorded pH values were in the range of 4.35 to 6.59. Seventy-seven percent of the samples had pH values below 5.6, demonstrating the occurrence of acidic deposition. The range, mean and standard deviation of measured concentrations are reported. There was very little difference between measurements in filtered and unfiltered samples. Correlation coefficients between pairs of parameters are reported and discussed.

Key words: rainwater, acidity, sodium, potassium, magnesium, calcium, sulphate, chloride, nitrate, conductivity.

1. Introduction

The occurrence of acid rain and its harmful effects in the industrialised countries of Europe and North America have been extensively studied and documented (Radojevic and Harrison, 1992). Over the last decade observations of acidic rainfall at tropical sites have been reported with increasing frequency and it is being recognised that acid rain is a global phenomenon which may affect remote areas far removed from industrial sources of acidifying compounds (Rodhe and Herrera, 1988). A cross-section of measurements of rainwater pH at tropical sites is shown in Table I. This acidity may be due to natural or anthropogenic causes. Some of the studies involved long-term sampling (over weekly periods) at remote sites with no on-site laboratory facilities. In some studies samples had to be stored and transported to distant laboratories for analysis (Galloway et al, 1982; Bruijnzeel, 1989; Andrea et al, 1990). There may, therefore, be some uncertainties in many of the reported measurements.

In order to expand further the data base on the pH and chemical composition of rainwater in tropical countries we analysed rainwater in Brunei Darussalam. No such study has been carried out previously in the country. Collectors were exposed for periods of between one and several days, and the pH was determined immediately upon sample collection in an on-site laboratory.

Brunei Darussalam is located on Borneo island. It has a land area of 5765 km^2 and a population of 261x10^3. The climate is tropical with high humidity and average temperature of 28 °C. Brunei lies outside the typhoon belt but it experiences high rainfall; typically around 2800 mm per annum. Although enjoying a high standard of living the country is devoid of major sources of industrial pollution. Sources of air pollution include motor vehicles, numbering about 100x10^3 in the whole country, a medical incinerator located 4 km from the sampling site, a small refinery located about 90 km from the sampling site, and forest fires. Other sources include transboundary pollution from industries, forest fires, and volcanoes in neighbouring countries. There is also considerable construction activity in the country, especially in and around the capital.

2. Experimental

Rainwater samples were collected on the roof of the Chemistry Department at the campus of the University of Brunei Darussalam located in the capital, Bandar Seri Begawan. There were no tall buildings or objects in the vicinity of the sampler. The site was close to a commercial area. Samples were collected using a funnel-in-bottle arrangement that collects bulk deposition (i.e. both wet and dry deposition during sampler exposure). Samples were collected on a daily basis and occasionally over several days. A polypropylene funnel 20 cm in diameter was used and all sampling bottles were made of polypropylene. Bottles were carefully cleaned by rinsing rigorously with high purity water several times and soaking in high purity water until required. The high purity, NANO pure, water was obtained using a Type D4700 Deionisation System from Barnstead/Thermolyne Corporation . We found this water to be purer than distilled, deionised or Milli-Q waters. NANOpure water blanks used for soaking the bottles were analysed with all runs and no contamination was noted. It was decided not to use any cleaning reagents in order to avoid any possible contamination. Bottles were weighed before and after sampling and the volume of rain was determined by difference.

Table I.
Measurements of rainwater pH at tropical sites. vwm = volume weighted mean. n/r = not reported

Reference	Site	Sampling frequency	pH	Comments
Galloway et al, 1982	Katherine, Australia	daily	4.2 - 5.4	vwm = 4.78
	San Carlos, Venezuela	daily	3.8 - 6.2	vwm = 4.79
Lacaux et al, 1987	Ivory Coast	event	4 - 6.5	mean = 5.0
Filho et al, 1987	Rio de Janeiro, Brazil	weekly	3.5 - 5.9	mean = 4.7 ± 0.4
Moreira-Nordemann et al, 1988	Brazil	n/r	3.6 - >9	range at 5 sites
Ayers and Gillet, 1988	Jabiru, Australia	event	4.27- 4.63	range of vwm for 3 yrs.
Sanhueza et al, 1987	Venezuela	event	4 - 5.8	range at 3 sites
Sanhueza et al, 1988	Venezuela	event	3.83 - 6.7	range at 10 sites
Sanhueza et al, 1989	La Paragua, Venezuela	event	4.03 - 5.6	vwm = 4.7
Sanhueza et al, 1991	Venezuela	event	4.8 - 5.8	range of vwm at 4 sites
Sanhueza et al, 1992	San Carlos, Venezuela	event	4.4 -5.4	range of vwm at 3 sites
Bruijnzeel, 1989	Java, Indonesia	weekly	5.75 - 7.1	vwm = 6.4
Andrea et al, 1990	Amazonia, Brazil	event	4.6 - 5.6	range of means at 2 sites

The pH and conductivity were determined immediately after sample collection in the laboratory on aliquots of the sample using a Jenway 3045 Ion Analyser equipped with a combination glass electrode and a temperature compensation probe, and a Cole-Palmer Conductivity Meter, respectively. Samples were refrigerated at 5 °C for subsequent chemical analyses. Concentrations of Cl^-, NO_3^- and SO_4^{2-} ions were measured by ion chromatography (IC) using a Shimadzu Ion Chromatography system comprising a pump, oven, controller, conductivity meter and an automatic injector. Separation was achieved by means of a Shim-pack IC-A1 column. The mobile phase was 2.5 mM phthalic acid, 2.5 mM

Tris(hydroxymethyl)aminoethane with a pH of 4. Concentrations of Na, Ca and Mg were determined by flame atomic absorption spectroscopy (FAAS) while concentrations of K were determined by flame atomic emission spectroscopy (FAES) using a Philips PU91000 Atomic Absorption Spectrophotometer. Analysis was completed between one and six months after sample collection. We observed that passage of samples and NANOpure water blanks through filter media, as well as acidification with Analar grade HNO_3, could lead to slightly elevated levels of some ions. Therefore, results reported here refer to samples that were neither filtered nor acidified.

3. Results and Discussion

Fifty-three samples were collected between 26/4 and 3/11/1994. The recorded pH values were in the range from 4.35 to 6.59. Seventy-seven percent of the samples had pH values below 5.6, the pH of rain-water in equilibrium with atmospheric CO_2, demonstrating the occurrence of acidic deposition in Brunei. This is within the range of pH values of rainwater reported at other tropical sites (see Table I).The frequency distribution of pH values of samples collected in the present survey is shown in Fig. 1.

The range, mean, and standard deviation of the pH, conductivity and major ions are given in Table II. These concentrations are generally higher than those measured in precipitation at remote tropical sites (Bruijnzeel, 1989; Ayers and Gillett, 1988; Galloway et al, 1982; Sanhueza et al, 1989; Sanhueza et al, 1992) but lower than those reported for rainwater in Europe and North America (Hendry and Brezonik, 1980; Martin and Barber, 1978; Granat, 1972; Clarke and Radojevic, 1987). Twenty-three of the samples were split upon collection with one fraction being filtered through Millipore HAWP filter papers. There was little difference between the pH, conductivity and ionic concentrations in filtered and unfiltered samples. For some of the ions, slightly elevated concentrations were observed in filtered samples indicating that filters may have acted as a source of contamination, as mentioned earlier. The pH values of filtered and unfiltered samples differed only at the second decimal place. The conductivities of filtered and unfiltered samples generally differed by less than 2%, and only occasionally by as much as 10 %. The results reported here refer to unfiltered samples.

Correlation coefficients between pairs of parameters are given in Table III. There are significant correlations between many of the ions. These may be due to a common source of these constituents or because of the inverse correlation between ionic concentrations and rainfall amount. High concentrations tend to be associated with low intensity rainfalls, whereas intense rainfalls tend to have low concentrations as shown by the negative correlation coefficients between concentration and rainfall amount in Table III. This dilution effect, a characteristic feature of rain-water also observed in other studies (Williams et al, 1988), could lead to a co-dependence between different chemical constituents. The correlartions were checked graphically and elimination of few data points with exceptional concentrations could improve the correlation coefficients significantly. For example, the value of r for the correlation between Cl^- and Na^+ was 0.692 for all the data points but elimination of a single sample with a very high Cl^- and low Na^+ concentration resulted in a value of r= 0.925. Ratios of Cl/Na were calculated for each sample. The mean of these individual Cl/Na ratios for all the samples was 2.6, however, elimination of the afore-mentioned sample resulted in a mean Cl/Na ratio of 1.3, only slightly higher than the seawater ratio of 1.16. High Cl/Na ratios may be due to a number of sources including anthropogenic or volcanic HCl, degradation of organochlorine compounds, soil dust, and washout of desorbed inorganic gaseous HCl from acidified sea-salt aerosols (Keene et al, 1986).

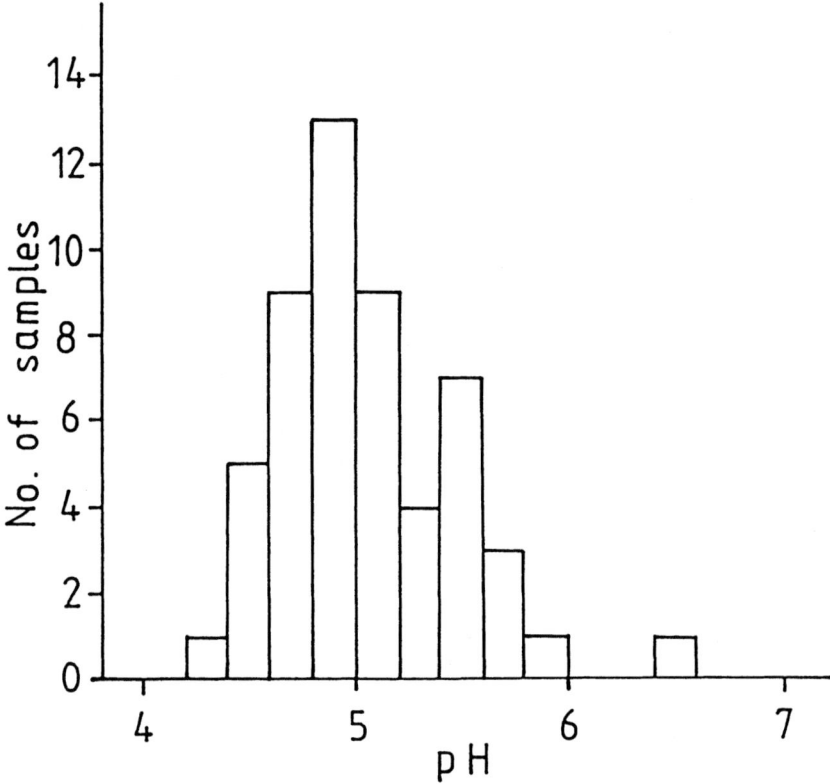

Figure 1. Frequency distribution of pH values for 53 bulk deposition samples collected in Brunei Darussalam between 26/4/1994 and 3/11/1994.

Table II.

Results of bulk deposition analysis in Brunei Darussalam (26/4 -3/11/1994). All concentrations are in 10^{-6} equivalents dm^{-3}. Conductivity is in 10^{-6} mhos cm^{-1}. (a) Calculated from volume-weighted H^+ concentration.

Parameter	n	min.	max.	average	s.d.	volume-weighted mean
pH	53	4.35	6.59	5.03	0.42	4.85 (a)
conductivity	53	2.2	29.5	10.9	6.5	-
Cl^-	35	0.96	125	22.1	30.5	15.5
NO_3^-	44	1.82	41.5	11.6	9.69	8.9
SO_4^{2-}	44	2.98	58.7	18.0	14.5	13.9
Na^+	48	0.74	68.3	14.7	13.3	10.9
K^+	41	0.26	13.6	3.58	3.40	2.15
Mg^{2+}	41	0.33	35.1	6.67	7.16	3.87
Ca^{2+}	45	1.50	90.0	22.1	19.6	11.9

Table III

Correlation matrix for the deposition chemistry data giving values of Pearson's correlation coefficient, r, for pairs of variables. Significant correlations (P<0.05) are underlined (mm-rainfall amount; cond.=conductivity).

	mm	cond.	H^+	Cl^-	NO_3^-	SO_4^{2-}	Na^+	K^+	Mg^{2+}	Ca^{2+}
Ca^{2+}	-0.585	0.671	-0.067	0.329	0.597	0.648	0.435	0.703	0.860	1
Mg^{2+}	-0.477	0.734	-0.006	0.506	0.362	0.555	0.698	0.732	1	
K^+	-0.508	0.696	0.030	0.359	0.413	0.486	0.458	1		
Na^+	-0.349	0.530	-0.022	0.692	0.075	0.406	1			
SO_4^{2-}	-0.337	0.619	0.369	0.582	0.831	1				
NO_3^-	-0.342	0.584	0.399	0.347	1					
Cl^-	-0.275	0.504	0.344	1						
H^+	0.129	0.247	1							
cond	-0.478	1								
mm	1									

Although there were significant correlations between H^+ and the measured anions (SO_4^{2-}, NO_3^- and Cl^-) it is unlikely that these are the only sources of acidity. It has been demonstrated that the contribution of organic acids, such as formic and acetic acids, to the acidity of precipitation at remote and rural sites may be equal to or greater than that of inorganic acids(Andrea et al, 1990; Sanhueza et al, 1992). However, weak acids such as formic and acetic acids can act to buffer the H^+ ions resulting from an increase in strong acids from fossil fuel combustion (Keene et al, 1983).These organic acids were not determined in the present work. The conductivity was highly correlated with all the ions except with H^+.

There was a close association between Ca^{2+}, Mg^{2+} and K^+, as indicated by the high values of the correlation coefficient. Ratios of these ions to Na^+ were calculated for each individual sample. The means of individual Mg/Na, Ca/Na and K/Na ratios for all rainwater samples were 0.58, 2.26 and 0.4 respectively, compared to 0.227, 0.044 and 0.021 for seawater. These high excesses, especially of Ca^{2+}, suggest a strong influence of wind-blown dust on our samples. These could result from local sources such as quarrying and construction activities which proliferate in the environs. Bulk samples, such as those collected in the present study, generally contain much higher levels of these dust-associated ions than wet-only samples (Jeffries, 1984; Lee and Longhurst, 1992).

Ion balance calculations were performed for each sample. The mean ratio of positive to negative ion concentrations was 1.49 (s.d.= 0.56) when the pH value determined in freshly collected samples was used to determine the H^+ concentration. However, pH of samples was found to increase during storage and it is the H^+ concentration at the time of ion analysis that should be used in ion balance calculations. The use of these higher pH values measured during ion analysis gave a mean ratio of positive to negative ions of 1.15 (s.d. = 0.34). One possible anionic species which was not measured is bicarbonate resulting from the dissolution of atmospheric CO_2. We calculated HCO_3^- concentrations assuming an atmospheric CO_2 concentration of 340 ppm. Inclusion of calculated HCO_3^- concentrations improved the ion balance slightly resulting in a ratio of positive to negative ions of 1.11 (s.d.= 0.32). However, the concentration of CO_2 in the laboratory may be higher then the ambient concentration assumed in the calculation and higher HCO_3^- concentrations may improve the ion balance. Nitrite was not observed in any of the samples. Traces of what may have been fluoride were observed in chromatograms of some samples, but these were not quantified. Phosphate was not determined in this study and neither was NH_4^+. Some samples were split into two fractions immediately upon collection, with one fraction being preserved with chloroform. The pH of the preserved samples remained unchanged even after long-term storage of one year. The increase in pH in the unpreserved samples may be due to the disappearance of organic acids such as formic and acetic acids from untreated samples (Keene et al, 1983). Formic and acetic acids will be addressed in future rainwater studies in Brunei Darussalam.

Conductivities were calculated for each sample on the basis of the determined ionic concentrations and known equivalent conductances, and these were compared with the measured conductivities. Given that NH_4^+ was not determined the agreement between measured and calculated conductivities was quite good; for most samples the discrepancy was <20% and only two samples had discrepancies of 50%.

References

Andrea, M.O., Talbot, R.W., Berresheim, H. and Beecher, K.M.: 1990, *J. Geophys. Res.* **95**, 16987-16999.
Ayers, G.P. and Gillett, R.W.: 1988, Acidification in Australia, in Rodhe, H. and Herrera, R (eds.), *Acidification in Tropical Countries, SCOPE 36*, John Wiley & Sons, Chichester, pp. 347 - 402.
Bruijnzeel, L.A.: 1989, *J. Tropical Ecology* **5**, 187 -202.
Clarke, A.G. and Radojevic, M.: 1987, *Atmos. Environ.* **21**, 1115-1123.
Filho, E.V.S., Ovalle, A.R.C. and Brown, F.: 1987, in Perry, R., Harrison, R.M., Bell, J.N.B. and Lester, J.N. (eds.) *Acid Rain: Scientific and Technical Advances*, Selper Ltd., London, pp. 331- 335.
Galloway, J.N. and Likens, G.E.: 1982, *J. Geophys. Res.* **87**, 8771-8786.
Granat, L.: 1972, *Tellus* **24**, 550-560.
Hendry, C.D. and Brezonik, P.L.: 1980, *Environ. Sci. Technol.* **14**, 843-849
Jeffries, D.S.: 1984, *Advances Environ. Sci. Technol.* **15**, 117-154.
Keene, W.C., Galloway, J.N. and Holden, J.D.Jr.: 1983, *J. Geophys. Res.* **88**, 5122-5130.
Keene, W.C., Pszenny, A.A.P., Galloway, J.N. and Hawley, M.E.: 1986, *J. Geophys. Res* .**91**, 6647-6658.
Lacaux, J.P., Servant, J. and Baudet, J.G.R.: 1987, in Perry, R., Harrison, R.M., Bell, J.N.B. and Lester, J.N. (eds.) *Acid Rain: Scientific and Technical Advances*, Selper Ltd., London, pp. 264-269.
Lee, D.S. and Longhurst, J.W.S.: 1992 *Water, Air and Soil Pollut.* **64**, 635-648.
Martin, A. and Barber, F.R.: 1978, *Atmos. Environ.* **12**, 1481-1487.
Moreira-Nordemann, L.M., Forti, M.C.,Di Lascio, V.L., Espirito Santo, C.M. and Danelon, O.M.: 1988, in Rodhe, H. and Herrera, R. (eds.), *Acidification in Tropical Countries, SCOPE 36*, John Wiley & Sons, Chichester, pp. 347-402.
Radojevic, M., and Harrison, R.M.: 1992, *Atmospheric Acidity: Sources, Consequences and Abatement*, Elsevier Applied Science, London.
Rodhe, H. and Herrera, R. (eds.) : 1988, *Acidification in Tropical Countries, SCOPE 36*, John Wiley & Sons, Chichester , 405 pp.
Sanhueza, E., Graterol, N. and Rondon, A.: 1987, *Tellus* **39B**, 329-332.
Sanhueza, E., Cuenca, G., Gomez, M.J., Herrera, R., Ishizaki, C., Marti, I. and Paolini, J.: 1988, in Rodhe, H. and Herrera, R. (eds.), *Acidification in Tropical Countries, SCOPE 36*, John Wiley & Sons, Chichester, pp. 197-255.
Sanhueza, E., Elbert, W., Rondon, A., Arias, M.C., Hermoso, M.: 1989, *Tellus* **41B**, 170-176.
Sanhueza, E., Ferrer, Z., Romero, J. and Santana, M.: 1991, *Ambio* **20**, 115-118.
Sanhueza, E., Arias, M.C., Donoso, L., Graterol, N., Hermoso, M., Marti, I., Romero, J., Rondon, A. and Santana, M.: 1992, *Tellus* **44B**, 54-62.
Williams, P.T., Radojevic, M., and Clarke, A.G.: 1988, *Atmos. Environ.* **22**, 1433-1442.

CRITICAL LOADS OF ACIDITY AND NITROGEN, BASED ON MULTIPLE CRITERIA FOR DIFFERENT SWEDISH ECOSYSTEMS

HARALD SVERDRUP[1], PER WARFVINGE[1] and KAJ ROSÉN[2]

[1] *Chemical Engineering II, Lund University, Box 124, 221 00 Lund, Sweden*
[2] *Forest Soils, Box 7007, Swedish Agricultural University, S-750 07 Uppsala, Sweden*

Abstract. The critical loads of acidity and nitrogen has been mapped for Swedish forest soils, using data from the Swedish Forest Inventory. The Swedish critical load map used in negotiations has been based on a number of ecological receptors. For terrestrial ecosystems criteria based on no adverse effect on growth, soil stability and groundwater quality was used. For surface waters, stream and lake biology was used as indicators for setting limits to acidification. A reduction of 75% of the acidity deposition in relation to 1988 is required in order to protect 95% of the forest resource in Sweden from effects of soil acidification. A reduction of 50% of the nitrogen deposition is required to avoid exceedance in more than 5% of the area. The mapping work was carried out by using the PROFILE model.

Keywords:Critical load, acidity, nitrogen, Norway spruce, Scots pine, ground vegetation,

1. Introduction

The critical loads maps are used as an input to optimizing the abatement strategies for S and N in Europe. The critical load was defined by the 1988 Skokloster critical loads workshop, as: **The highest deposition of acidifying compounds that will not cause chemical changes in soils leading to long term harmful effects on ecosystem structure and function.** Later "according to best current knowledge." was added. Take care to notice that if new results appear showing that the old values are inadequate, then a revision of the critical loads is permitted.

2. Methods

For each ecosystem, an indicator organism is chosen. A chemical limit is found for that indicator organism, and the chemical limit is entered into a chemical mass balance equation including all sinks and sources of acidity in the system. The chemical limit is applied to the solution concentration in the system. Thus ecosystem become connected through concentrations via mass balances to acid deposition. For critical loads of acidity, trees were taken as the indicator organism, for nitrogen trees and 4 different types of ground vegetation. Rearrangement of a mass balance for acidity yield the expression for the critical load (Sverdrup and Warfvinge, 1995a):

$$CL(A) = W - ANC_L \qquad (1)$$

ANC_L is the leaching of ANC from the system. It is determined by the limits set for Al concentration in the system. The BC/Al ratio come into the equation by setting the upper limit for Al. In Europe BC/Al=1 was used as the criterion for critical load. For other plants than conifers, other BC/Al limits can be applied. More information is found in Sverdrup and Warfvinge (1993, 1995a,b).

Table 1. Approximated limits of nitrogen for inducing vegetation changes, based on Swedish field surveys. The limits have been based on an evaluation of Swedish forest vegetation survey data. EX is excess soil nitrogen in the soil after removal by harvest has been accounted for.

Ecological change	[N] threshold mg N l^{-1}	EX kg N ha^{-1}yr^{-1}
Coniferous trees → Nutrient imbalances	0-0.2	0-0.8
Deciduous trees → Nutrient imbalances	0.2-0.4	0.8-1.6
Lichens vegetation type → Lingon-berry tupe	0.2-0.4	0.8-1.6
Lingon-berry vegetation type → Blueberry type	0.4-0.6	1.6-2.4
Blueberry vegetation type → Grass type	1-2	4-8
Grass vegetation type → Herbaeceous type	3-5	12-20

A stability criterion is applied; no Al leaching in excess of what is produced by weathering is permitted. If an excess of Al is leached, the soil will be depleted in Al-komplexes, and the structural stability may change.

The critical load of N is calculated from a mass balance. Inputs are balanced against sinks such as uptake, denitrification, immobilization and permitted leaching. Examples of different suggested values for permitted leaching are shown in Table 1.

When nitrogen appears in significant amounts in the soil, species composition of the ground vegetation may be changed. The long-term uptake of N is defined as N uptake modified with respect to what can be balanced by a long-term supply of base cations. Present growth may be enhanced by nutrients mobilized by acid deposition or management. The critical uptake is calculated from mass balances for N, Mg, K and Ca separately.

The critical load for Sweden is calculated by the Simple Mass Balance model (SMB) and the PROFILE model (Sverdrup and Warfvinge 1995a). The one-layer SMB model has been employed by all European countries. PROFILE also employs a mass balance approach to critical load, but contains a fuller description of the system and allow for chemical feedback on biological processes. In PROFILE, the soil profile is divided into 4 layers. Exceedance is defined as the excess of acid input over the critical load.

All input data is based on data and analysis of soil samples collected in the Swedish Forest Inventory, collected 1983-1985 at 1,804 stations. Deposition data for 1988 were prepared by the Institute of Water and Air Research (IVL) in Göteborg.

3. Results

Critical loads for forest ecosystems are shown in Fig. 1 and 2. The results from both models show good consistency. The weathering rate in the SMB calculations came from PROFILE calculations. The average critical load of acidity for Sweden is 0.58 keq/ha yr. Areas where critical loads have been exceeded are also shown in Fig. 1 and 2. The critical loads in Sweden are low, and for the 95% protection level for conifer ecosystems, nearly 85% of the Swedish foreted area has exceedance. The maps express the critical load in terms of acidity.

The role of forestry for the critical load can be seen in Fig. 1. The map was calculated by setting uptake by the forest to zero, equivalent to complete cycling

Figure 1: 1–The critical loads for Swedish soils if the objective is to protect 95% of the area. The number associated with the bars in the legend indicate the number of 50x50km squares in each category. 2–Areas where the critical loads have been exceeded for Swedish soils if the objective is to protect 95% of the area. 82% of Swedens forested area receives more acidic deposition than the critical load. 3–Areas where the critical loads have been exceeded for Swedish soils assuming complete return of all harvested N and base cations.

nutrients. The difference between the maps in Fig. 1 is significant. The role of forestry is less an acidifying effect, and more of an effect where net uptake of base cations make the soil more vulnerable to acidification. The role of forestry on soil acidification is therefore mainly an indirect effect. The difference corresponds to a decrease in exceedance of the critical load of approximately 40%. Forestry makes the soil more vulnerable to acid deposition, but it does not acidify the soil correspondingly to 40% of the deposition. Forestry without acid deposition does not acidify the soil significantly if base cation uptake is balanced by N uptake and base cation uptake remain less than weathering plus base cation deposition.

Without any acid deposition, soil acidity fluxes caused by forest growth will generally not exceed the neutralizing capacity of the soil. It can be expected that the forest soils will return to soil pH values significantly higher than today if the acid deposition is reduced below the critical load.

In order that 95% of the forested area has no exceedance, acid deposition will have to be reduced by at least 75% of the 1985 level. This implies an 80% reduction in S and 30% reduction in N, but the reductions may for the sake of economic optimization be distributed differently between S and N. It should be made quite clear that abatement of S alone will not be sufficient, S accounts only for approximately 65% of the deposited acidity in southern Sweden.

Fig. 3 shows critical loads for forest trees that have been calculated. It can be seen that the nitrogen deposition over Sweden is far to high from the perspective of long term sustainable forest yield. The present level will induce the forest to grow at a rate above what weathering and base cation deposition can sustain. This may lead the forest into a long term stress situation if this is permitted to continue. A long term permanent nutrient imbalance stress will cause the tolerance to additional stresses due to other partially natural factors to be less, and thus cause the system to have less stability.

Trees are not the only important species in a forest. Fig. 3 shows critical loads of nutrient nitrogen calculated for ground vegetation of the *vaccinium vitis-idaea* type. Similar critical loads for acidity using vaccinium vitis-idaea as indicator could in principle also be calculated. The pattern obtained shows remarkable good visual correlation with observed vegetation changes during the last 15 years.

4. Discussion

In the calculation of critical loads, trees for biomass production were used as the indicator plant for the ecosystem. This was done partly because trees are recognized as being among the more sensitive elements in the ecosystem with respect to acidification. But trees were also chosen because they have a monetary value that is recognized by policymakers.

The base saturation in a soil kept constant at $BC/Al=1$ will have a significantly lower base saturation than what it had historically. Coniferous trees are adequately protected against from growth effects at $BC/Al=1$, but many ground vegetation species will be subject to significant chemical stress even if critical loads for trees are applied as the maximum deposition level. Many endangered species require BC/Al ratios significantly higher than 1 (Sverdrup and Warfvinge 1993b). As a result, ground vegetation species composition may change over long time. The effect

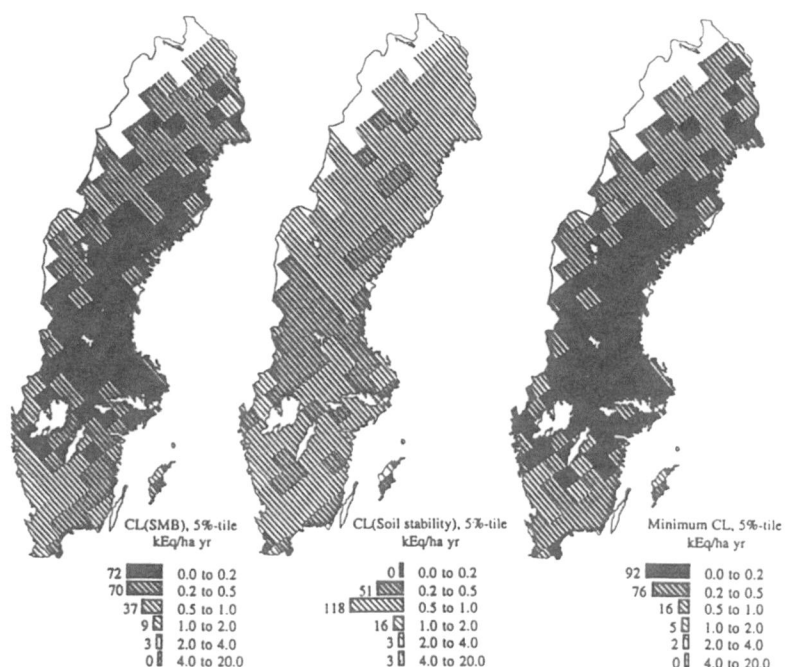

Figure 2: 1–The critical loads for Swedish soils if the objective is to protect 95% of the area as calculated with the SMB approach. 2–Critical loads calculated based on soil stability. 3–The minimum critical load, considering the results from PROFILE and SMB and the soil stability critical load.

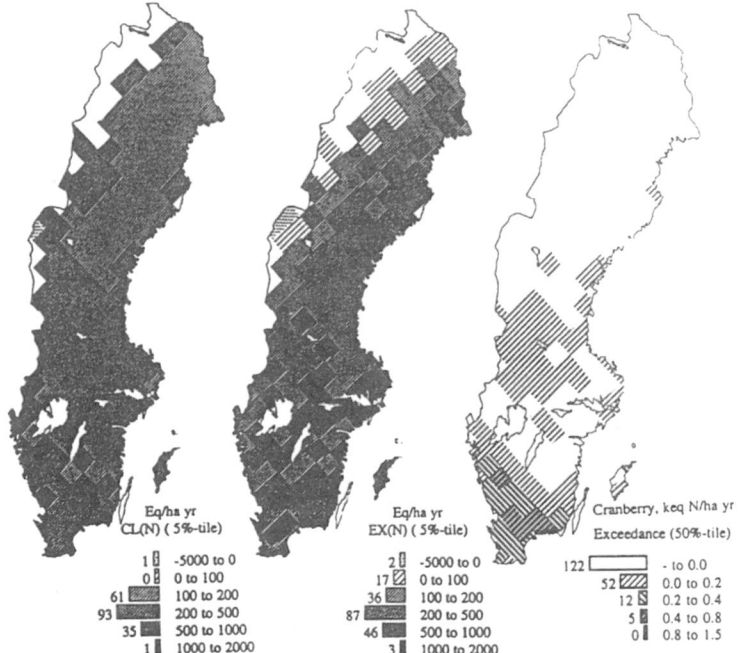

Figure 3: 1–Critical load of nutrient N for forest ecosystems, assuming no net N leaching, stem harvest only and 2-4 kg/ha yr immobilization. 2–Exceedance of critical load of nutrient N. 3–Exceedance of the critical load of nutrient N calculated with respect to lingonberry (*Vaccinium vitis-idaea*)

on such plants can be assessed by application of different BC/Al limits and response functions. Such response functions are available for a large number of plants (Sverdrup and Warfvinge 1993b).

The major effect of forestry on exceedance, is not through acidity input alone, but also by lowering of BC in the BC/Al response parameter. This makes the role of forestry for soil acidification difficult to quantify in a unique way. The effect can be seen by comparing maps in Fig. 1, where net uptake in the left map is set to zero (No harvesting at all). At present, very approximatively, exceedance is caused approximately 5-30% by forestry and landuse, 30-40% by N deposition and 50-60% by S deposition. The proportions vary between sites.

5. Conclusions

To obtain good environmental protection, multiple receptors must be introduced in the calculation. We can conclude that a minimum deposition reduction over Sweden of 80% of sulphur deposition and 30% of the nitrogen deposition is required in order to protect 75% of the groundwater resources, 95% of the forest resource and 90% of the surface water ecosystems. It must be concluded that the exceedance of the critical load of acidity arise mainly from deposition of S and N, and that the role of forestry is modest, but not insignificant. Sweden is receiving acid deposition in excess of critical load for more than 80% of its lake, stream and forest resources. Calculations indicate that the present level of acidic deposition is starting to cause significant economical damage. Calculations show that significantly more than 60% reductions in S depositions and 30% reduction in N deposition are required in order to reduce the forest area with excess from 210,000 km^2 to 11,000 km^2.

6. References

Hettelingh, J., Downing, R. and de Smet, P.: 1991, *Mapping Critical Loads for Europe*, Coordination Center for Effects, RIVM. CCE Technical Report No. 1, RIVM Report No. 259101001.

Sverdrup, H. U., de Vries, W. and Henriksen, A.: 1990, *Mapping critical loads*, Nordic Council of Ministers, Copenhage. Miljörapport 1990:15, Nord 1990:98.

Sverdrup, H. and Warfvinge, P.: 1993a, Calculating field weathering rates using a mechanistic geochemical model–PROFILE, *Journal of Applied Geochemistry* **8**, 273–283.

Sverdrup, H. and Warfvinge, P.: 1993b, Soil acidification effect on growth of trees, grasses and herbs, expressed by the (Ca+Mg+K)/Al ratio, *Reports in Environmental Engineering and Ecology* **1:93**, 1–184. Chemical Engineering II, Box 124, Lund University. 221 00 Lund, Sweden.

Sverdrup, H. and Warfvinge, P.: 1995a, Critical loads of acidity for Swedish forest ecosystems, in Staaf, H. and Tyler, G. (eds.), *Effects of acid deposition and tropospheric ozon on forest ecosystems in Sweden*, Ecological Bulletins 44, 75-89.

Sverdrup, H. and Warfvinge, P..: 1995b, Past and future changes in soil acidity and implications for growth under different deposition scenarios, in Staaf, H. and Tyler, G. (eds.), *Effects of acid deposition and tropospheric ozon on forest ecosystems in Sweden*, Ecological Bulletins 44, 335-362.

Warfvinge, P. and Sverdrup, H.: 1992, Calculating critical loads of acid deposition with PROFILE – a steady–state soil chemistry model, *Water, Air and Soil Pollution* **63**, 119–143.

THE USE OF CRITICAL LOADS IN EMISSION REDUCTION AGREEMENTS IN EUROPE

J.-P. HETTELINGH, M. POSCH, P.A.M. DE SMET, R.J. DOWNING

Coordination Center for Effects
National Institute of Public Health and the Environment (RIVM)
P.O.Box 1, NL-3720 BA Bilthoven, The Netherlands

Abstract: Critical loads have been used in the revision of the Sulphur Protocol of the Convention on Long Range Transboundary Air Pollution (LRTAP) of the United Nations Economic Commission for Europe (UN/ECE). Critical loads, i.e. maximum allowable depositions which do not increase the probability of damage to forest soils and surface waters, have been computed and mapped for Europe by means of the Steady-state Mass Balance Method, using national data and, if national data were unavailable, using a European database. Results show that areas with low critical loads are located mostly in northern and central Europe. The reduction of the excess of sulfur (S) deposition over critical loads was a starting point for negotiations leading to the Oslo Protocol on Further Reduction of Sulphur Emissions (the "Second Sulphur Protocol"). The new protocol protects about 81%, 86% and 90% of the ecosystems' area in 2000, 2005 and 2010, respectively. In addition, the total European area in which sulphur deposition exceeds critical loads by more than 500 eq ha^{-1}yr^{-1} will be reduced from about 19% in 1980 to practically zero in 2010. Besides these results, a methodology is presented which allows the combined assessment of the acidifying effects of S and N as well as the eutrophying effects of N deposition on ecosystems (so-called critical load functions and the protection isolines derived from them). This methodology is well suited to integrate ecosystem sensitivities into future negotiations on the reductions of nitrogen (N) compounds, taking into account present or anticipated S emissions.

Key words: critical loads, acid deposition, emission reductions, air pollution impacts, ecosystem sensitivity

1. Introduction

A critical load is defined as "a quantitative estimate of an exposure to one or more pollutants below which significant harmful effects on specified sensitive elements of the environment do not occur according to present knowledge" (Nilsson and Grennfelt, 1988). Critical loads for acidity and sulfur have been computed and mapped for forest soils and surface waters in Europe (Hettelingh *et al.*, 1991; Downing *et al.*, 1993). These critical loads are used in integrated models to assess emission reduction strategies with respect to abatement costs and the degree of environmental protection. Such integrated assessment models support negotiations to develop protocols on reductions of S and N emissions within the framework of the LRTAP Convention of the UN/ECE.

In June 1994, the Second Sulphur Protocol has been signed by 28 countries (UN/ECE, 1994). The starting point for negotiating national emission targets was to reduce the excess of the 1990 S deposition over critical loads for sulfur. A reduction of S emissions to obtain deposition levels not exceeding critical loads was not feasible for all ecosystems in Europe, and therefore it was decided to reduce the excess of S deposition over critical loads by 60% ("60% gap closure"). This was the first time that negotiations on emission reductions took into account environmental effects − by means of the critical load concept − in addition to technical and economic considerations of emission abatements.

Critical loads of N are required for the negotiations of a new nitrogen protocol.

However, N deposition leads not only to acidification, but also to eutrophication requiring the formulation of two kinds of critical loads. Furthermore, the contribution of both S and N deposition to acidification makes it advisable to consider both pollutants simultaneously. Elevated concentrations of nitrogen oxides also act as precursor in the formation of tropospheric ozone, but this aspect is not dealt with here.

This paper presents an overview of the methodology for computing critical loads, (i) as developed for the support of the Second Sulphur Protocol, and (ii) taking into account the acidifying aspects of S and N deposition (the so-called critical load function) as well as the effects of N as a nutrient. Furthermore, European maps of critical loads and their comparison with depositions ('exceedance' of critical loads) are presented and discussed.

2. Computation and Mapping of Critical Loads

In this section we summarize the methods for calculating critical loads for forest ecosystems. Detailed descriptions and justifications for the various simplifying assumptions can be found in Sverdrup *et al.* (1990), Downing *et al.* (1993) and Sverdrup and de Vries (1994), as well as in workshop reports (Nilsson and Grennfelt 1988, Grennfelt and Thörnelöf 1992).

2.1 THE CRITICAL LOAD OF SULFUR

The maximum allowable deposition of S, i.e. the highest deposition of S which does not lead to 'harmful effects' in the case of zero N deposition is given by

$$CL_{max}(S) = BC_{dep} - BC_u + CL(A), \qquad (1)$$

where BC_{dep} is the deposition and BC_u the net growth uptake of base cations and $CL(A)$ is the critical load of acidity, which is the sum of the base cation weathering and an acceptable (critical) leaching of acidity. This critical leaching of acidity (or alkalinity) links a soil-chemical criterium (e.g., a critical BC/Al ratio) with the potentially harmful effects.

Critical loads of acidity were not directly applicable for the evaluation of required S emission reductions, since current UN/ECE protocols concentrate on a single compound (S or N), rather than on acidity as a such. Therefore it was necessary to divide the critical load of acidity between S and N, making the following assumptions: (a) the fraction of the S deposition of the total acid deposition is used as proxy of the part of the critical load of acidity attributed to S, and (b) N deposition contributes to acidification only when it is not taken up or immobilized by the ecosystem. This lead to the definition of the so-called sulfur fraction, S_f, by

$$S_f = S_{dep}/(S_{dep} + N_{dep} - N_u - N_i) \quad \text{for} \quad N_{dep} > N_u + N_i \quad (S_f = 1 \text{ otherwise}), \qquad (2)$$

where S_{dep} is the S deposition and N_{dep} is the deposition of both oxidized and reduced N; N_u and N_i are net growth uptake and immobilization of N. The critical deposition of sulfur, $CD(S)$ - for a given N deposition - is then given by[1]

$$CD(S) = S_f \cdot CL_{max}(S), \qquad (3)$$

and it is this quantity which has been compared to S depositions in negotiating the Second Sulphur Protocol by computing the so-called exceedance of the critical deposition:

$$Ex(S) = S_{dep} - CD(S), \qquad (4)$$

and $Ex(S) \leq 0$ ensures that the ecosystem is protected.

2.2 THE CRITICAL LOAD FUNCTION FOR N AND S

The sulfur fraction was introduced to accommodate for the negotiations of the Second Sulphur Protocol which focussed on S rather than on acidity. The drawbacks of the sulfur fraction include its dependence on the deposition of S and N which change over time, making the critical load time-dependent. Future negotiations will focus on emission reductions for N, and in the following a methodology for estimating critical loads is presented (Posch et al., 1993) which allows the assessment not only of the eutrophying and acidifying effects of N, but also takes into account the effect of S, thus avoiding the introduction of ad-hoc factors.

A critical load of nutrient N, $CL_{nut}(N)$, preventing eutrophication is derived from a mass balance for N:

$$CL_{nut}(N) = N_u + N_i + N_{l,crit}/(1-f_{de}), \qquad (5)$$

where $N_{l,crit}$ is the allowable (critical) leaching of N from the rooting zone. The factor f_{de} ($0 \leq f_{de} \leq 1$) takes into account denitrification, N_{de}, which is not modeled as a fixed quantity, but depends on the net input of N:

$$N_{de} = f_{de} \cdot (N_{dep} - N_u - N_i) \quad \text{for} \quad N_{dep} > N_u + N_i \quad (N_{de} = 0 \text{ otherwise}). \qquad (6)$$

Inserting Eq.6 into the mass balance for N leads to Eq.5 for $CL_{nut}(N)$. The exceedance of the critical load of nutrient N is calculated as in Eq.4 by comparing it to the N deposition.

Besides S, the deposition of N also contributes to the acidification of ecosystems. Starting from the acidity balance, the excess deposition of S and N can be written as

$$Ex(N_{dep}, S_{dep}) = S_{dep} + (1-f_{de}) \cdot N_{dep} - (1-f_{de})(N_u + N_i) - CL_{max}(S), \qquad (7)$$

where $CL_{max}(S)$ is given by Eq.1. If the acidity balance in Eq.7 is negative or zero, i.e. $Ex(N_{dep}, S_{dep}) \leq 0$, for a given *pair* of deposition (N_{dep}, S_{dep}), we say critical loads are not exceeded. Note, that there are many values of N_{dep} and S_{dep} for which the exceedance

---

[1] For historical reasons (see Hettelingh et al., 1991) the term critical load of sulfur, $CL(S)$, is reserved for the expression $S_f CL(A)$

becomes zero, i.e. unique critical load values cannot be defined. Furthermore, for

$$N_{dep} \leq N_u + N_i = CL_{min}(N), \tag{8}$$

the maximum allowable deposition for S is given by $CL_{max}(S)$; and the maximum 'harmless' acidifying deposition of N is obtained by inserting $S_{dep}=0$ into Eq.7 and solving $Ex(N_{dep},0)=0$:

$$CL_{max}(N) = CL_{min}(N) + CL_{max}(S)/(1-f_{de}). \tag{9}$$

The relationship between N and S deposition and the exceedance of the critical load of acidity (Eq.7) is illustrated in Fig.1. The thick line separates the deposition values which cause no 'harmful effects' from those which require reduction. This function is termed the *critical load function* of the ecosystem.

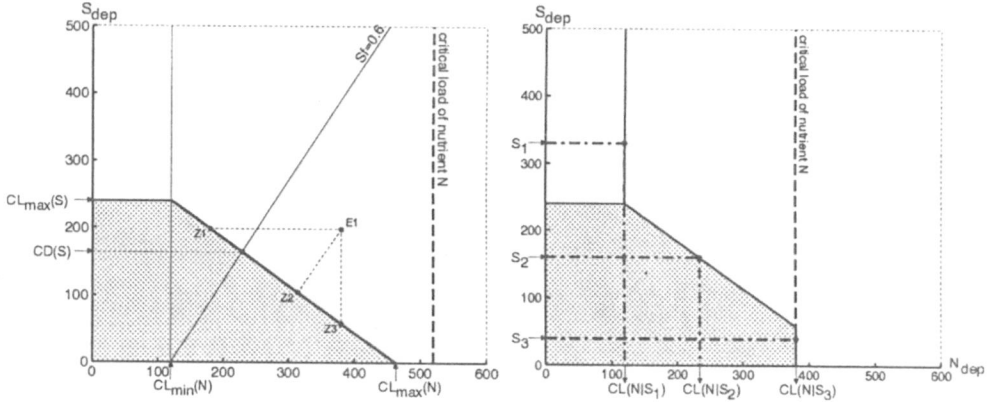

Fig.1. Left: Relationship between N_{dep} and S_{dep} and the exceedance of acidity (Eq.7) for a hypothetical ecosystem. Also illustrated is the way in which $CD(S)$ has been derived with the aid of the sulfur fraction S_f. Right: Examples of computing the conditional critical loads of N – $CL(N|S_1)$, $CL(N|S_2)$ and $CL(N|S_3)$ – for a hypothetical ecosystem for different S deposition values S_1, S_2 and S_3, respectively. The dashed vertical line in both figures shows the critical load of nutrient N for the two cases $CL_{nut}(N) \geq CL_{max}(N)$ and $CL_{nut}(N) < CL_{max}(N)$.

As mentioned above, unique critical loads of S and N cannot be defined, and consequently no unique exceedance of a critical load. This is best illustrated by an example (see Fig.1): Let the point E1 denote the current deposition of S and N; reducing N_{dep} substantially, one reaches the point Z1 and therefore non-exceedance without reducing S; on the other hand one can reach non-exceedance by only reducing S_{dep} till reaching Z2; finally, with a smaller reduction of both S_{dep} and N_{dep} one can reach non-exceedance as well (e.g. point Z3). In practice external factors, such as the costs of emission reductions, will determine which path will be followed to reach zero exceedance.

Considering also the critical load of nutrient N, two cases are possible: $CL_{nut}(N) \geq CL_{max}(N)$, in this case the nutrient critical load is not binding, or $CL_{nut}(N) < CL_{max}(N)$, in which case $CL_{nut}(N)$ determines the maximum allowable N deposition.

If future negotiations are not to involve S deposition reductions – and this is likely considering that a sulfur protocol has just been signed – the critical load functions can serve as a valuable tool for determining critical loads of N for a *fixed* S deposition (conditional critical load): By intersecting the critical load function for each ecosystem with the (present or any future) S_{dep} yields critical loads for N for the given S deposition, $CL(N|S_{dep})$ (see Fig.1, where the computation of conditional critical loads of N is illustrated for $S_{dep} = S_1$, S_2 and S_3). And these critical N loads can then be used in the 'traditional way', with the additional benefit that they are compatible with the present or any anticipated S deposition.

2.3 MAPPING CRITICAL LOADS

For presenting the spatial variation of critical load values, such as $CD(S)$, $CL_{nut}(N)$ or $CL(N|S_{dep})$, the cumulative distribution function of all values within a chosen grid cell is constructed and then a percentile, i.e. a protection percentage, is chosen as the representative critical load for that grid cell. In order to be useful for the analysis of emission reduction strategies, also for the numerous critical load functions an aggregation procedure is required for deriving a single function for every grid cell. This is done in the following way: for every pair of N and S deposition one checks, whether it lies below or above the given critical load functions, thus determining the percentage of protected ecosystems. Connecting those points in the (N_{dep}, S_{dep})-plane, which protect a prescribed percentage p of ecosystems, yields the *(100-p)-percentile function* (or *protection isoline*).

A single function, e.g. the 5th percentile function, replaces the individual critical load functions in the same way as a percentile is used as a surrogate for the set of critical load values within a grid cell. The selected percentile function for each grid cell can then be linked to models which (cost)optimize emission reductions for S and N. Both the costs of the individual abatement strategies and the shape of the percentile functions determine the optimal emission reduction strategy and thus the allocation of S and N deposition.

Ideally, critical loads are computed by each country and sent to the Coordination Center for Effects (CCE). The CCE then collates this data into European maps and computes critical loads for countries which do not provide national data, using a European database on forest soils (de Vries *et al.*, 1993, and references quoted therein).

The geographical representation of the critical loads has to be consistent with the resolution at which S and N deposition is modeled under the LRTAP Convention. The Co-operative Programme for the Monitoring and Evaluation of the Long Range Transmission of Air Pollutants in Europe (EMEP) uses a grid covering Europe with 150x150 km^2 grid cells. Consequently, critical loads have to be mapped on these EMEP-grid cells to be able to compute critical load exceedances. The value chosen to represent the cumulative distribution function of all critical load values within one EMEP-grid cell is the 5th percentile (thus protecting 95% of the ecosystems in the grid cell); and for the support of future protocols dealing with N (in conjunction with S), the 5th percentile protection isoline can be computed and mapped in each grid cell.

3. Results

Fig.2a shows the 5th percentile critical S deposition in each 150x150 km² EMEP-grid cell as used in the negotiations for the Second Sulphur Protocol. The map shows that the ecosystems most sensitive to S deposition cover large parts of Scandinavia, the British Isles and northern Germany. The most sensitive ecosystems ($CD(S) < 200$ eq ha^{-1}yr^{-1}) comprise approximately 14% of the European land area. Fig.2b displays the result of comparing the S deposition in 2010, due to emissions agreed in the Second Sulphur Protocol, with the 5th percentile $CD(S)$. It shows that large parts of Europe, about 90% of the mapped ecosystems, will be no longer subjected to risk of ecosystem damage (the numbers are 81% and 86% for the years 2000 and 2005, resp.). Areas where critical loads are still exceeded are scattered throughout Europe, with a very small fraction of the area exceeded by more than 500 eq ha^{-1}yr^{-1}.

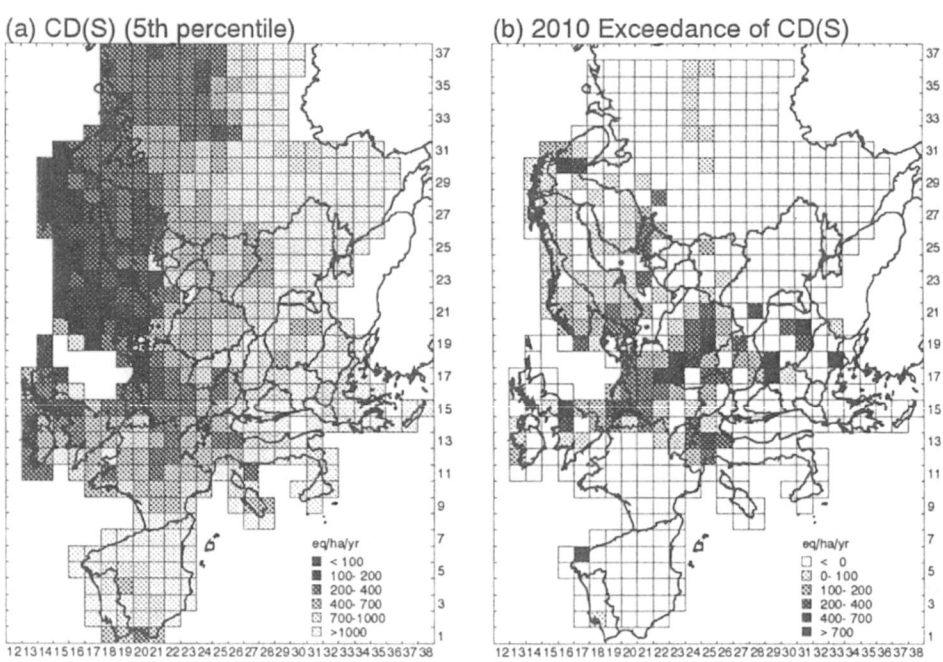

Fig.2. (a) The 5th percentile critical deposition of S, $CD(S)$, and (b) its exceedance in 2010 after implementing the emission reductions negotiated in the Second Sulphur Protocol.

However, these results are incomplete, because N is not taken into account and a fixed portion of the critical load of acidity is arbitrarily assigned to S. By contrast, the critical load function provides a flexible tool for evaluating the trade-offs between S and N deposition reductions and to meet the constraints set by the N critical loads of acidity and eutrophication simultaneously.

When comparing the N and S deposition in a grid cell with a protection isoline, five cases can arise: (0) the point (N_{dep}, S_{dep}) lies below the percentile function, i.e. no exceedance; (1) reductions in N_{dep} or S_{dep} are interchangeable, i.e. non-exceedance can be reached by either N_{dep} reductions or S_{dep} reductions alone; (2) some reductions in S depositions are mandatory; (3) some reductions in N_{dep} are mandatory; and (4) both reductions in N_{dep} and S_{dep} are required to achieve non-exceedance. Fig.3 shows which case occurs in each EMEP-grid cell for the 5th percentile protection isoline, comparing it with the 1990 S and N deposition, calculated from country emissions using transfer matrices provided by EMEP (Tuovinen et al., 1994). In producing Fig.3 the protection isoline for acidifying S and N has been intersected with the 5th percentile of $CL_{nut}(N)$ in every grid cell, in this way considering both effects of N. It can be seen that in the southern part of Scandinavia, in Poland, in the bordering area between Germany, Poland and the Czech Republic (the 'black triangle'), in the Netherlands and in central England both S and N deposition reductions are required to protect 95% of the ecosystems.

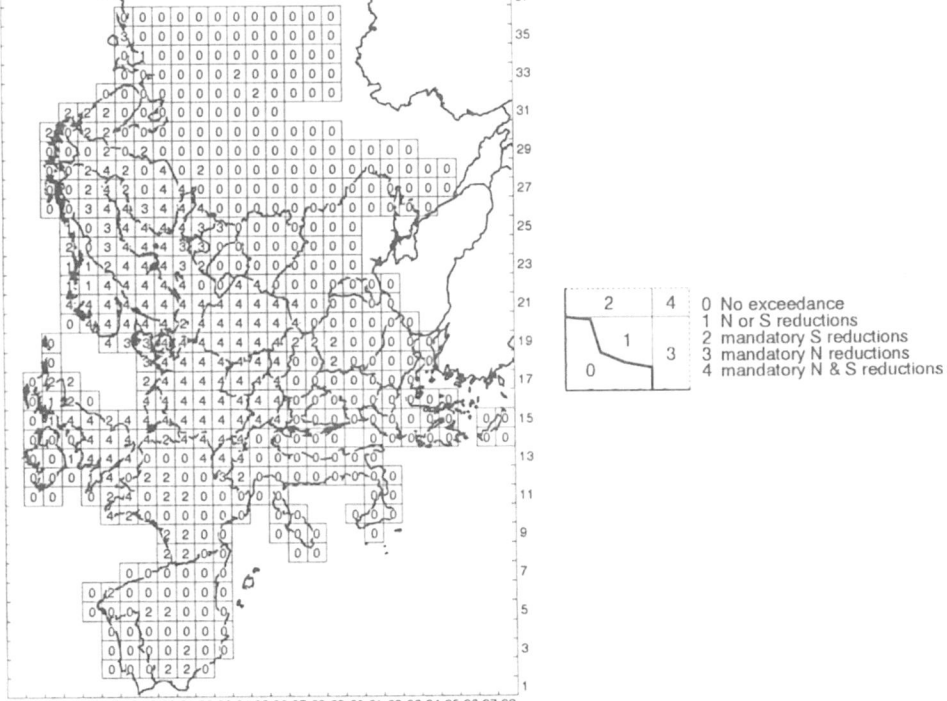

Fig.3. Type of N and/or S deposition reductions required in each EMEP-grid cell when comparing the 5th percentile function (95% protection isoline) with the 1990 S and N deposition (preliminary results).

4. Conclusions

Two approaches for the computation and use of critical loads are described in this paper. The earlier methodology is restricted to the computation of a critical deposition for S, which has been used in the negotiations on the revised UN/ECE Protocol on the further reductions of S emissions. The second, newly proposed methodology focusses on critical load exceedances rather than individual critical loads. Generalizing the notion of a percentile to critical load functions, ecosystem protection isolines can be computed for each grid cell in Europe reflecting combinations of S and N depositions at which no exceedance occurs of both the critical load of acidity and eutrophication. While the first methodology has paved the way for the incorporation of ecosystem protection targets into emission reduction agreements, the second methodology allows a more flexible allocation of emission reductions. The new method also indicates a way for the inclusion of other pollutants and/or effects.

Acknowledgements

This work is supported by the Directorate of Air and Energy of the Dutch Ministry of Environment. National Focal Centres and Working Groups under the LRTAP Convention of the UN/ECE are acknowledged for the collaboration enabling the European assessment presented in this paper.

References

Hettelingh, J.-P., Downing, R.J. and de Smet, P.A.M. (Eds): 1991, "Mapping Critical Loads for Europe", CCE Technical Report No.1, RIVM, Bilthoven, The Netherlands, 86 pp.

De Vries, W., Posch, M., Reinds, G.J. and Kämäri, J.: 1993, "Critical Loads and Their Exceedance on Forest Soils in Europe", Report 58 (revised version), Winand Staring Centre, Wageningen, The Netherlands, 116 pp.

Downing, R.J., Hettelingh, J.-P. and de Smet, P.A.M. (Eds): 1993, "Calculation and Mapping of Critical Loads in Europe", CCE Status Report, RIVM, Bilthoven, The Netherlands, 163 pp.

Grennfelt, P. and Thörnelöf, E.: 1992, "Critical Loads for Nitrogen", Nord 1992:41, Nordic Council of Ministers, Copenhagen, Denmark, 428 pp.

Nilsson, J. and Grennfelt, P. (Eds): 1988, "Critical Loads for Sulphur and Nitrogen", Nord 1988:97, Nordic Council of Ministers, Copenhagen, Denmark, 418 pp.

Posch, M., Hettelingh, J.-P., Sverdrup, H.U., Bull, K. and de Vries, W.: 1993, "Guidelines for the Computation and Mapping of Critical Loads and Exceedances of Sulphur and Nitrogen in Europe", In: R.J. Downing, J.-P. Hettelingh and P.A.M. de Smet, *op. cit.*, pp. 25-38.

Sverdrup, H., de Vries, W. and Henriksen, A.: 1990, "Mapping Critical Loads", Nord 1990:98, Nordic Council of Ministers, Copenhagen, Denmark, 124 pp.

Sverdrup, H. and de Vries, W.: 1994, "Calculating Critical Loads for Acidity with the Simple Mass Balance Method", Water, Air and Soil Pollution, 72, 143-162.

Tuovinen, J.-P., Barrett, K. and Styve, H.: 1994, "Transboundary Acidifying Pollution in Europe: Calculated fields and budgets 1985-93", EMEP/MSC-W Report 1/94, Norwegian Meteorological Institute, Oslo, Norway.

UN/ECE, 1994: "Protocol to the 1979 Convention on Long-Range Transboundary Air Pollution on Further Reduction of Sulphur Emissions", Document ECE/EB.AIR/40 (in English, French and Russian). New York and Geneva, 106 pp.

MAPPING BASE CATION DEPOSITION IN EUROPE ON A 10 x 20 KM GRID

G.P.J. DRAAIJERS, E.P. VAN LEEUWEN, C. POTMA, W.A.J. VAN PUL and J.W. ERISMAN

National Institute of Public Health and Environmental Protection (RIVM), P.O. box 1, 3720 BA Bilthoven, the Netherlands

Abstract Atmospheric deposition of base cations in Europe is mapped on a 10x20 km grid using the inferential modeling technique. Deposition fields are found to resemble the geographic variability of sources, climate and land use. In large parts of southern Europe, more than 50% of the potential acid deposition is found counteracted by deposition of base cations. In central and northwestern Europe, however, base cation deposition usually amounts less than 25% of the acid input. An uncertainty analysis to assess the quality of the base cation deposition maps revealed that for an average grid cell the deviation from the estimated value can be as large as 140%.

Key words: atmospheric deposition, base cations, Europe, acidification

1. Introduction

Acid deposition is widely recognized as a regional environmental problem, the chief causes of which are oxides of sulfur and nitrogen emitted during fossil fuel combustion and metal smelting, and ammonia emitted from animal manure (e.g. Heij & Schneider, 1991). Usually little emphasis is placed on the role of base cations such as Na^+, Mg^{2+}, Ca^{2+} and K^+ in relation to acid deposition. Base cations play an integral role in the chemical processes of acid deposition since the acidity of any material is a function of both its acidic and basic compounds. Besides their ability to neutralize acid input, base cations are important nutrient elements for ecosystems. The depletion of exchangeable base cations in sensitive soils is currently thought to represent a major detrimental effect of acid deposition on terrestrial ecosystems (De Vries, 1994). Atmospheric input may be a quantitatively important source of base cations to vegetation in nutrient-poor conditions.

Base cations in the atmosphere originate from both semi-natural and anthropogenic sources (Gorham, 1994). Semi-natural sources of base cations are associated with wind erosion of arid soils, forest fires and biological mobilization (pollen). Especially soils on calcareous bedrock may emit large amounts of alkaline particles to the atmosphere. In maritime areas, sea spray may be an important source of Na^+ and Mg^{2+}-containing particles. Anthropogenic sources of alkaline particles include agricultural tillage practices (ploughing, liming), traffic on unpaved roads, and oil/coal burning generating fly ash. Non-fossil fuel combustion (wood and peat) is found an important source for alkaline particles in e.g. Scandinavia (Antilla, 1990). Limestone quarries and cement factories are local sources of Ca^{2+}. Due to the large mass median diameters of alkaline particles and consequently high dry deposition velocities, most soil-derived particles are likely to deposit near the area of origin (Milford & Davidson, 1985), although long-range transport of, for example, desert dust has also been reported (e.g. Swap et al., 1992).

For accurate estimation of exceedances of critical acid loads which are used as a basis for emission control strategies, actual and future base cation deposition amounts in Europe have to be mapped. Up to now, no reliable information is available on the spatial pattern of base cation deposition over Europe. There is lack of knowledge on the spatial and temporal variation of emissions and concentrations, hampering accurate deposition mapping. In this paper, first base cation deposition maps of Europe are presented which are based on the so-called inferential modeling technique. An uncertainty analysis is included. Base cation deposition is compared to the deposition of potential acid ($SO_x+NO_y+NH_x$) as estimated from the EDACS model. The latter is described in detail in a companion paper (Erisman et al., 1995) and in Van Pul et al. (1995).

2. Methods

Total deposition is the sum of dry and wet deposition. The dry deposition flux of base cations is calculated as the product of the dry deposition velocity and air concentration at a reference height above the surface. In the inferential technique, the choice for a reference height (50 m) is a compromise between the height where the concentration is not severely affected by local deposition or emission and is still within the constant flux layer (Erisman, 1992). For this height, dry deposition velocity fields over Europe are constructed for every six hours from a land-use map and meteorological information, using a detailed parametrisation of the dry deposition process. Land-use information, based on topographical maps and satellite observations, was available on a 1/6°x1/6° lat./long. (approx. 10x20 km) grid (Van de Velde et al., 1994). Meteorological information was obtained from the Observational Data Set (ODS) which is a product of the European Center for Medium-range Weather Forecasts (ECMWF). The ODS dataset contains information of 1297 measurement sites spread over Europe (Potma, 1993). The parametrisation of the dry deposition velocity was based on the model of Slinn (1982) and tested with micrometeorological measurements recently performed at the Speulder forest in the Netherlands (Ruijgrok et al., 1994; Erisman et al., 1994). It includes both turbulent exchange and sedimentation of coarse particles. Six-hourly based dry deposition velocity fields were aggregated to annual means before they were combined with annual mean air concentration fields, yielding dry deposition estimates on a small scale over Europe.

Surface-level air concentrations were estimated from precipitation concentrations using scavenging ratios. These were derived from simultaneous measurements of base cation concentrations in precipitation and surface-level air performed at 23 sites in Canada (Eder and Dennis, 1990) and at 1 site in The Netherlands (Römer and Te Winkel, 1994). This approach to estimate air concentrations is based on the premise that cloud droplets and precipitation efficiently scavenge aerosols resulting in a strong correlation between concentrations within precipitation and the surface-level air. This assumption will only be valid for well-mixed conditions at sufficient distance from sources. Factors that will influence the magnitude and variability of scavenging ratios include particle size distribution and solubility, precipitation amount and rate, droplet accretion process and storm type (Galloway et al., 1993). Event scavenging ratios can range several orders of magnitude even for single species at a single location but scavenging ratios have been found reasonably consistent when averaged over one year or longer (Galloway et al., 1993). Therefore, annual mean precipitation concentrations were used to infer annual mean air concentrations of Na^+, Mg^{2+}, Ca^{2+} and K^+. Precipitation concentration data were taken from Van Leeuwen et al. (1995). They compiled measurement results for 1989 from about 600 sites scattered over Europe. Concentration maps were compiled using the block-kriging interpolation technique with blocks of 50x50 km (Van Leeuwen et al., 1995). It must be noted here that Hedin et al. (1994) found for several sites in Europe base cation concentrations in precipitation significantly reduced during the last 10 to 20 years.

To obtain wet deposition fluxes of base cations, above mentioned precipitation concentrations were multiplied by long-term mean precipitation amounts from the database of the USA Environmental Protection Agency (EPA). The EPA database contains validated monthly mean precipitation amounts measured from 1920 to 1980 at a several thousands of measurement sites scattered over Europe which were interpolated to a 0.5°x0.5° lat./long. (approx. 30x60 km) grid (Legates and Willmott, 1990).

Deposition of Na^+ and part of the deposition of Mg^{2+}, Ca^{2+} and K^+ will be the result of sea spray. In sea spray these ions are mainly associated with Cl^-, and thus do not contribute to neutralization of atmospheric acid. A correction procedure derived by Asman et al. (1981) has been used to estimate what fraction of Mg^{2+}, Ca^{2+} and K^+ is of non-sea salt origin.

3. Results and discussion

Besides land use and climate patterns, total deposition fields of base cations are found to reflect the geographic variability of sources over Europe. For Na^+, a clear pattern of increasing fluxes with decreasing distance to seas, in particular the Atlantic Ocean, is found. Large deposition of non-sea salt $Mg^{2+}+Ca^{2+}+K^+$ in south and southeast Europe (Figure 1) is mainly the result of wind erosion of calcareous soils, agricultural tillage practices and traffic on unpaved roads. In this part of Europe, the prevalence of warm and dry conditions in the summer period will enhance the suspension of alkaline particles in the atmosphere. Sahara dust might be an important additional source of alkaline particles. Antilla (1990) found 80% of the annual total deposition of particulate matter in Corsica caused by Sahara dust. The relatively high fluxes in the border area between Germany, Poland and the Czech Republic, as well as in e.g. Estonia, can be attributed to intensive industrial activity. In large parts of southern Europe, more than 50% of the potential acid deposition is counteracted by deposition of non-sea salt $Mg^{2+}+Ca^{2+}+K^+$. In central and northwestern Europe, however, base cation deposition usually amounts less than 25% of the acid input.

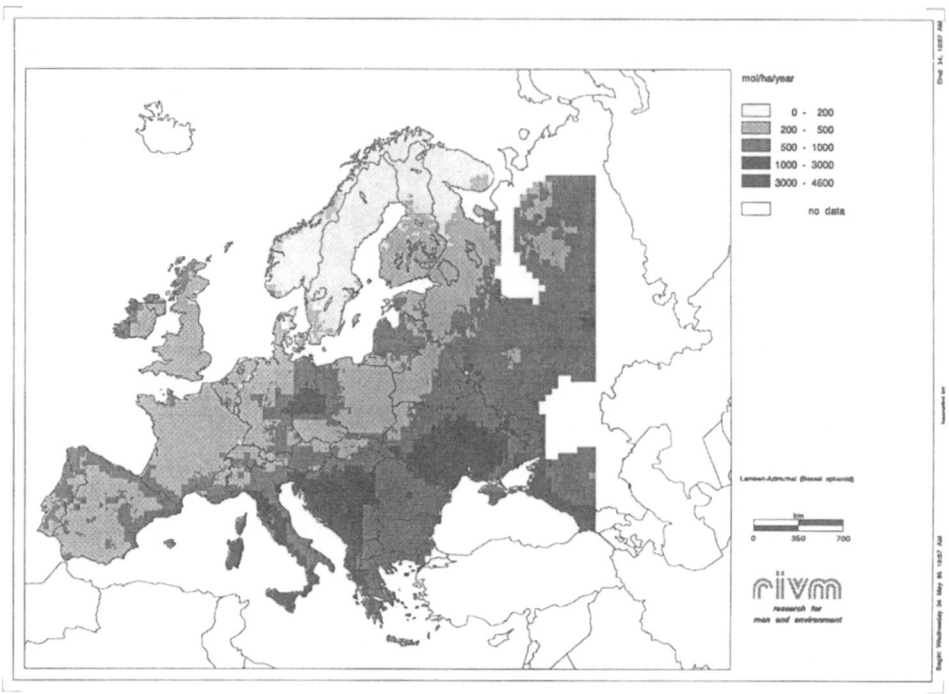

Figure 1 Spatial pattern of total deposition of non-sea salt $Mg^{2+}+Ca^{2+}+K^+$ over Europe in 1989.

Dry and wet deposition estimates are subject to relatively large potential error. For dry deposition this error is introduced by uncertainties in *i)* the parametrisation of the deposition velocity, and *ii)* the scavenging ratios and precipitation concentration maps used to estimate air concentrations. Ruijgrok et al. (1994) assessed the uncertainty of the model on which the parametrisation of the deposition velocity was based. The overall uncertainty in modeled deposition velocities integrated over the size distribution representative for alkaline particles at the Speulder forest was found to equal 60%. For other sites additional uncertainty will arise due to limited availability and accuracy of relevant land use information and meteorological parameters. The uncertainty in deposition velocity caused by variation in size distribution of alkaline particles amounts 30-50%, assuming an average mass median diameter (MMD) of 5 µm and taking a geometric standard deviation (σ_g) of 2-3 to represent the variation (Ruijgrok et al., 1994). The MMD of particles at a particular site will depend on the distance to sources and on e.g. ambient relative humidity (Fitzgerald, 1975).

Van Leeuwen et al. (1995) performed an extensive uncertainty analysis to assess the quality of the precipitation concentration maps used to estimate base cation air concentrations. The total uncertainty in average precipitation concentration per grid cell was estimated 30-50%. Largest uncertainty was found in areas with low measurement density leading to relatively large interpolation errors (i.e. southern and eastern Europe). Uncertainty was also introduced by e.g. using precipitation concentration data from different networks, all using their own sampling equipment and analytical procedures. Theoretical models (Slinn, 1983) and field measurements (Kane et al., 1994) suggest that the scavenging efficiency increases with particle diameter. Using the relationship between particle mass median diameter and scavenging efficiency presented by Kane et al. (1994), the uncertainty in estimated ambient air concentrations caused by variation in size distribution can be calculated to amount 50-100%, assuming a mean MMD of 5 µm and taking a σ_g of 2-3. Large errors in air concentrations will arise in areas very close or far from major sources and/or in areas with a strongly deviating precipitation climatology.

Using error propagation methods and assuming presented uncertainties in deposition velocities and air concentrations represent random errors, the total uncertainty in dry deposition for an average grid can be calculated to amount 80-120% (Draaijers et al., 1995). Systematic errors in dry deposition may arise from e.g. *i)* using scavenging ratios which are based on only a limited set of simultaneous ambient air and precipitation concentration measurements, *ii)* neglecting complex terrain effects in the parametrisation of the deposition velocity, *iii)* using annual mean air concentrations and deposition velocities for flux calculation, thereby neglecting temporal correlations. Van Leeuwen et al. (1995) estimated the uncertainty in wet deposition for an average grid at 50-70%. The uncertainty in total deposition of base cations can be calculated to amount 90-140% (Draaijers et al., 1995).

4. Conclusions

Atmospheric deposition of base cations in Europe is mapped on a 10x20 km grid using the inferential modeling technique. Deposition fields are found to resemble the geographic variability of sources, climate and land use. In southern Europe generally more than 50% of the acid deposition is counteracted by deposition of base cations. In central and northwest Europe, however, this is usually less than 25%. An uncertainty analysis to assess the quality of the base cation deposition maps revealed that for an average grid cell the deviation from the estimated value can be as large as 140%. Largest uncertainty is found in the dry deposition estimates. Calculated deposition velocities and air concentrations are very sensitive to variations in size distribution. The parametrisation of the deposition velocity which is based on results of micrometeorological measurements over forest in the Netherlands needs validation for other surfaces and climates to ensure applicability for the whole of Europe. Large uncertainty is introduced by using scavenging ratios to estimate air concentrations. Up to now, scavenging ratios are solely based on measurements performed in Canada and The Netherlands. They might be improved by incorporating results from other measurement sites and models describing the scavenging process. The uncertainty in deposition estimates will be reduced considerably if information on the temporal and spatial variability of base cation air concentration and/or emission becomes available through which air concentrations can be modeled more accurately and with a higher time resolution. In the near future, modeled deposition estimates will be validated on the basis of deposition measurements.

References

Antilla, P.: 1990, *Characteristics of alkaline emissions, atmospheric aerosols and deposition.* In: P. Kauppi, P. Antilla and K. Kenttämies (eds.), Acidification in Finland. Springer, Berlin, Germany.

Asman, W.A.H., Slanina, J. and Baard, J.H., : 1981, *Water, Air and Soil Pollution*, **16**, 159-175.

De Vries, W.: 1994, Ph.D. thesis, University of Wageningen, The Netherlands.

Draaijers, G.P.J., Van Leeuwen, E.P., Potma, C., Van Pul, W.A.J. and Erisman, J.W.: 1995, *Calculation and mapping of base cation deposition on a small scale over Europe.* RIVM report in preparation.

Eder, B.K. and Dennis, R.L.: 1990, *Water, Air and Soil Pollution*, **52**, 197-215.

Erisman, J.W.: 1992, Ph.D. thesis, University of Utrecht, The Netherlands.

Erisman, J.W., Draaijers, G.P.J., Duyzer, J.H., Hofschreuder, P, Van Leeuwen, N., Römer, F.G., Ruijgrok, W. and Wyers, G.P.: 1994, *Contribution of aerosol deposition to atmospheric deposition and soil loads onto forests*. RIVM report no. 722108005.

Erisman, J.W., Potma, C., Draaijers, G.P.J. , Van Leeuwen, E.P. and Van Pul, W.A.J.: 1995, *Water, Air and Soil Pollution*, this issue.

Fitzgerald, J.W.: 1975, *Journal of Applied Meteorology*, **14**, 1044-1049.

Galloway, J.N., Savoie, D.L., Keene, W.C. and Prospero, J.M.: 1993, *Atmospheric Environment*, **27A**, 235-250.

Gorham, E.: 1994, *Nature*, **367**, 321.

Hedin, L.O., Granat, L., Likens, G.E., Buishand, T.A., Galloway, J.N., Butler, T.J. and Rodhe, H.: 1994, *Nature*, **367**, 351-354.

Heij, G.J. and Schneider, T.: 1991, *Studies in Environmental Science*, **46**, Elsevier, Amsterdam, The Netherlands.

Kane, M.M., Rendell, A.R. and Jickells, T.D.: 1994, *Atmospheric Environment*, **28**, 2523-2530.

Legates, D.R. and Willmott, C.J.: 1990, *International Journal of Climatology*, **10**, 111-123.

Milford, J.B. and Davidson, C.I.: 1985, *Journal for Air Pollution Control and Assessment*, **35**, 1249-1260.

Potma, C.J.M.: 1993, *Description of the ECMWF/WMO Global Observational Data Set, and associated data extraction and interpolation procedures.* RIVM report no. 722401001.

Römer, F.G. and Te Winkel, B.W.: 1994, *Dry deposition of aerosols on vegetation: acidifying components and basic cations*, KEMA report no. 63591-KES/MLU 93-3243.

Ruijgrok, W., Tieben, H. and Eisinga, P.: 1994, *The dry deposition of acidifying and alkaline particles to Douglas fir: a comparison of measurements and model results.* KEMA report no. 20159-KES/MLU 94-3216.

Slinn, W.G.N.: 1982, *Atmospheric Environment*, **16**, 1785-1794.

Slinn, W.G.N.: 1983, *Air to sea transfer of particles.* In: P.S. Liss and W.G.N. Slinn (eds), Air sea exchange of gases and particles, Reidel, Dordrecht, The Netherlands.

Swap, R., Garstang, M. and Greco, S.: 1992, *Tellus*, **44B**, 133-149.

Van de Velde, R.J., Faber, W., Katwijk, V., Scholten, H.J., Thewessen, T.J.M., Verspuy, M. and Zevenbergen, M.: 1994, *The preparation of a European land use database.* RIVM report no. 712401001.

Van Leeuwen, E.P., Potma, C., Draaijers, G.P.J., Erisman, J.W. and Van Pul, W.A.J.: 1995, *European wet deposition maps based on measurements.* RIVM report no. 722108006.

Van Pul, W.A.J., Potma, C., Van Leeuwen, E.P., Draaijers, G.P.J. and Erisman, J.W.: 1995, *EDACS: European Deposition maps of Acidifying Components on a Small scale. Model description and results.* RIVM report no.722401005.

CALCULATION AND MAPPING OF CRITICAL LOADS OF S, N AND ACIDITY ON ECOSYSTEMS OF THE NORTHERN ASIA

V. N. BASHKIN[1], M. Ya. KOZLOV[1], I. V. PRIPUTINA[1],
A. Yu. ABRAMYCHEV[1] and I. S. DEDKOVA[2]

[1]*Institute of Soil Science and Photosynthesis RAS, Pushchino Moscow region 142292 Russia,*
[2]*MSC-E, Kedrova st, 8a, Moscow, Russia*

Abstract. On the basis of modified simple steady state mass-balance equations, the critical loads for nutrient and acidifying nitrogen as well as for sulphur and acidity have been calculated for various ecosystems of northern Asia using simplified expert-modelling GIS and grid cells 150x150 km. The minimal values of critical loads of nitrogen, CL(N), (<50eq/ha/yr) were shown for arctic and subarctic ecosystems and the maximal ones (>300eq/ha/yr) for ecosystems of chernozemic and chernozem-like soils in southern Siberia and the Far East. The minimal values of critical loads of sulphur, CL(S), as well as acidity were shown predominantly in the northern part of east Siberia and in the Kamchatka peninsula and the maximal ones for ecosystems having neutral and alcaline soils. The corresponding exceedances were indicated for many regions of the northern part of Asia with maximal values for regions of Ural mountains, frontery of Kazakhstan, Altai, lower Yenisei river flow, Far East, Sakhalin and South-Kurilean islands.

Key words: critical loads, nitrogen, sulphur, northern Asia

1. Introduction

Since the first Europen maps of critical loads were published in 1991 (Hettelingh *et al*, 1991), a number of revisions and improvements have been done with respect to data and methodologies. Experts from various countries have made efforts to enhance a scientific foundations and policy relevant to the critical loads program. The last report of RIVM Coordination Center for Effects and National Focal Centers for Mapping was published in 1993 (Downing *et al.*, 1993). It summarized the total work connected with calculation and mapping of critical loads in Europe as well as their exceedances at the modern state of pollutants emissons. However, there is agreement both nationally and internationally that long-range transboundary air pollution is not limited to the geographical limits of Europe and a transition of pollutants is spreading from Europe to North America and Asia as well as in the opposite directions. Consequently at present the calculation and mapping of critical loads outside of Europe are of great scientific and political interest. Until now only a few preliminary attempts have been made to calculate the critical loads for southern Asia (H. Swerdrup, pers. comm.). Therefore this study which has been carried out together with MSC-E EMEP (Head S.Dutchak) aims to estimate the critical loads of sulphur, CL(S), nitrogen, CL(N), and acidity, CL(A) as well as their exceedances for Northern Asia.

2. Materials and Methods

On the basis of modified simple steady state mass-balance equations, the critical loads for nutrient and acidifying nitrogen as well as for sulphur and acidity have been calculated for various ecosystems of northern part of Asia. Due to the large dimensions of the given area all calculation and mapping procedures have been carried out using geoinformation system with elements of simplified expert-modelling system (Bashkin et al., 1993). The initial information consisted of geobotanic, soil, hydrochemical, biogeochemical, hydrological etc regionalization. For every elemental taxon the main links of biogeochemical cycles of N, S and BC have been characterized quantitately on a basis of available case studies. The grid cells were 1 grad × 1 grad.

The algorithm for computer calculations of critical load of nutrient nitrogen (CL_{nutr} N) was based on the following equations:

$$CL_{nutr}(N) = {}^*N_u + {}^*N_i + {}^*N_{de} + {}^*N_{l(crit)}], \qquad (1)$$

where * means that each of the terms refers to the values at the actual total atmospheric precipitations at a side. N_u and N_i are permissible nitrogen uptake and soil immobilization, N_{de} is permissible denitrification and $N_{l(crit)}$ is permissible critical nitrogen leaching. Permissibe atmospheric nitrogen uptake (${}^*N_u$) was given as:

$$^*N_u = N_{upt} - N_u, \qquad (2)$$

where N_{upt}-annual accumulation N in biomass and N_u - annual uptake of N from soil.

N_{upt} was calculated accounting for the coefficients of annual biogeochemical turnover (Cb) the values of which varied from < 0,1 up to >25. Annual N_u from soil was calculated on a basis of nitrogen mineralizing capacity (NMC) of soils which was determined experimentally or calculated using regression equations (Bashkin et al, 1993). So,

$$N_u = (NMC - N_i - N_{de})Ct, \qquad (3)$$

where

$N_i = 0.15$ NMC,   if   C:N<10
$N_i = 0.25$ NMC,   if   10<C:N<14
$N_i = 0.30$ NMC,   if   14<C:N<20
$N_i = 0.35$ NMC,   if   C:N>20
$N_{de} = 0.145$ NMC+6.477,   if   NMC>60 kg/ha/yr
$N_{de} = 0.145$ NMC+0.900,   if   NMC<10 kg/ha/yr
$N_{de} = 0.145$ NMC+0.605,   if   10<NMC<60 kg/ha/yr

Permissible immobilization of atmospheric N deposition (${}^*N_i$) was found as:

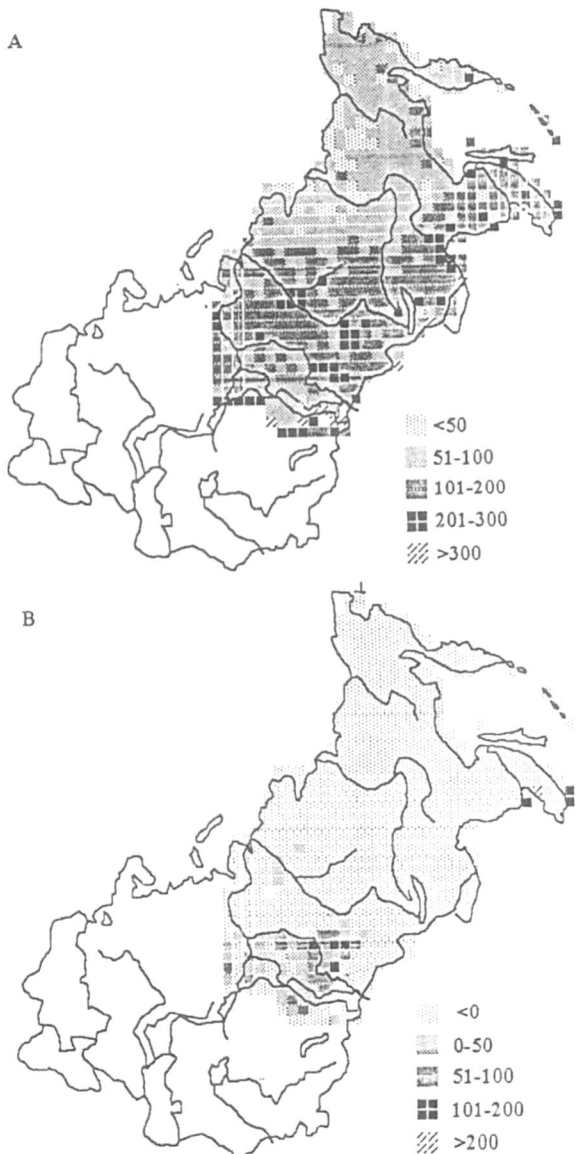

Fig. 1. Critical loads of nitrogen (A) and their exceedances (B) in the northern Asia (free space cell means zero exceedance)

$$^*N_i=[(0.2\ NH_4+0.1\ NO_3)/Cb]Ct, \quad \text{if}\ \ C:N<10 \qquad (4a)$$
$$^*N_i=[(0.3\ NH_4+0.2\ NO_3)/Cb]Ct, \quad \text{if}\ \ 10<C:N<14 \qquad (4b)$$
$$^*N_i=[(0.35\ NH_4+0.25\ NO_3)/Cb]Ct, \quad \text{if}\ \ 14<C:N<20 \qquad (4c)$$

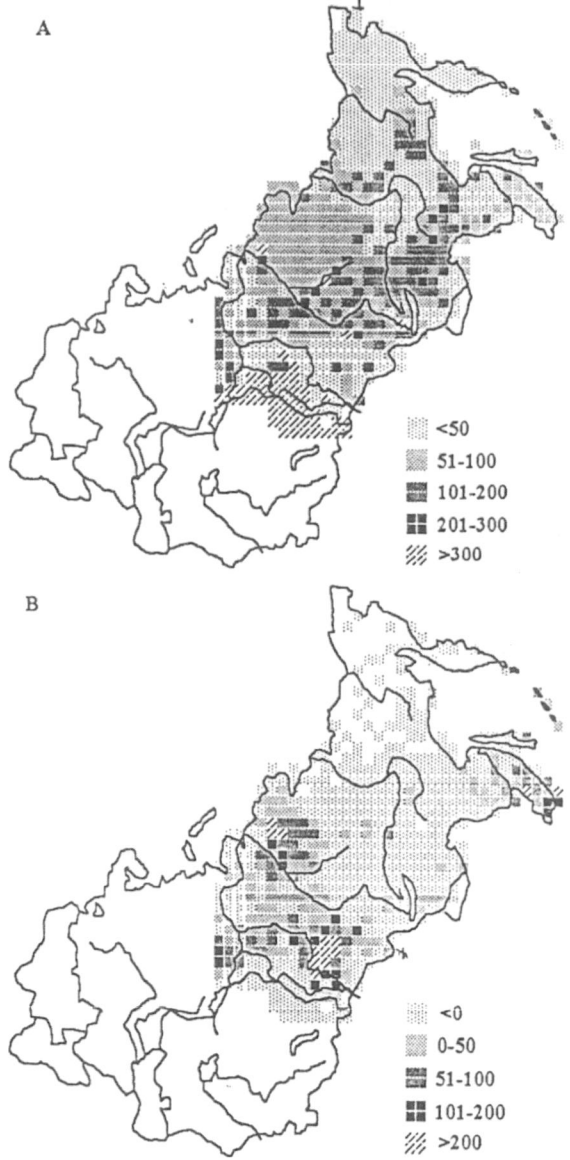

Fig. 2. Critical loads of sulphur (A) and their exceedances (B) in the northern Asia
(free space cell means zero exceedance)

$$^{\bullet}N_i = [(0.4\ NH_4 + 0.3\ NO_3)/Cb]Ct, \quad \text{if C:N} > 20, \quad (4d)$$

where Ct - hydrothemical coefficient given as a relative part of $\Sigma t > 5°C$ from the annual sum.

Permissible denitrification from atmospheric deposition ($^*N_{de}$) was found as:

$$^*N_{de}=(N_{de}/NMC)N_{td}\ Ct, \qquad (5)$$

where N_{de}/NMC - denitrification fraction, which depends on many features of soils and calculated on the basis of experimental data and N_{td} - total N deposition.

Finelly, permissible critical leaching of atmospheric nitrogen $^*N_{l(crit)}$ was given as:

$$^*N_{l(crit)}=Q\ C_{Ncrit}, \qquad (6)$$

where Q - annual surplus of precipitation (run off) and C_{Ncrit} - permissible nitrogen concentration in surface waters.

Critical loads for sulphur and acidity as well as exceedances for all studied parameters were calculated on a basis of Mapping Manual (Task Force on Mapping, 1993).

The calculations of sulphur and nitrogen depositions were made on the basis of meteodata and emissions for 1991 regarding the asian EMEP grid 150 × 150 km (Galperin et al., 1994). These model calculations do not differ principally from previous results (Rahn and Shaw, 1982).

3. Results and Discussion

Using above mentioned approaches it has been shown that the minimal values of CL(N) (<50 eq/ha/yr) were shown for arctic and subarctic ecosystems (Figure 1). The values of CL(N) in the limits of 50-100 eq/ha/yr are typical for the majority of ecosystems of permafrost area. So, these ecosystems are very sensitive to the excessive input of atmotechnogenic N. The maximal values of CL(N), >300 eq/ha/yr were noted for ecosystems of chernozemic and chernozem-like soils of the southern regions of Siberia and Far East. The exceedances of CL(N) are shown mainly in Ural mountains, in boundary regions with Kazakhstan steppes, in lower flow of Yenisei river (Norilsk industrial district) and in Far East. The minimal values of CL(S) as well as acidity were shown predominantly in the northern part of East Siberia and the Kamchatka peninsula (Figure 2). In the area between Yenisei and Ob rivers these values are shown to increase up to 50-100 eq/ha/yr and maximal values (>300 eq/ha/yr) were noted for ecosystems having neutral and alcaline soils. The corresponding exceedances were shown for many regions of the northern part of Asia with maximal values for Ural and Altai mountains, for boundary regions with Kazachstan, lower Yenisei river flow, Far East, Sakhalin and South-Kurilean islands.

4. Conclusions

The preliminary calculations of critical loads of S and N were shown that due to a number of peculiarities many ecosystems of Northern Asia are very sensitive to deposition of acid forming and eutrophying compounds. In total, the values of CL are significantly lower than those in Europe or South Asia. At the present level of atmotechnogenic S and N depositions accounting both local emission sources and transboundary pollution, the maximal exceedances of CL values were shown in such regions as Ural and Altai mountains, Far East and lower Yenisei river flow.

Acknowledgments

The authors wish to thank the Head of MSC-E S. Dutchak for the initiation and financial support of this research, Dr M. Hornung (ITE, UK) and T. Rusanova (ISSP RAS) for english text revision and technical assistance.

References

Bashkin, V. N., Snakin, V. V., Kozlov, M. Ya. *et al.*: 1993, In: *Calculation and Mapping of Critical Loads in Europe: States Report 1993*, (Downing, R. J., Hettelingh, J.-P. and de Smet, P. A. M.Eds) CCE, RIVM, 102-112.
Downing, R.J., Hettelingh, J.-P. and de Smet, P.A.M.(Eds): 1993, *Calculation and Mapping of Critical Loads in Europe: States Report 1993*, CCE, RIVM, 163 pp.
Galperin, M.V., Erdman, L.K., Subbotin S.P. *et al.*: 1994, *Modelling of Pollution of the Arctic by S and N Compounds and Heavy Metalls from the Sources in the North Hemisphere*, MSC-E Report, 33 pp.
Mapping Critical loads for Europe: 1991, *CCE Techn. report* No 1, RIVM, 86 pp.
Rahn, K. A. and Shaw, G. B.: 1982, *Sources and Transport of Arctic Pollution Aerosol: Chronicle of six Years of Research*, Natal Research reviens 34/3.
Sverdrup, H.V.: 1994, *Critical Loads for the Southern Asia*, Pers. Comm.
Task Force on Mapping: 1993, *Mapping Critical Levels/Loads*, Report 25/93. Umweltbundesamt, Bismarckplatz 1, Berlin.

ASSESSMENT OF CRITICAL LOADS IN LIUZHOU, CHINA USING STATIC AND DYNAMIC MODELS

SHAODONG XIE, JIMING HAO, ZHONGPING ZHOU, LING QI, HANHUI YIN

Department of Environmental Engineering, Tsinghua University, Beijing, 100084, P. R. China

Abstract. The project Comprehensive Control and Demonstration for Acid Deposition in Liuzhou area is a national key project in the 8th Five-year-plan, and the study on critical loads will provide scientific and quantitative accordance for formulating control strategy. In this paper, critical loads of acid deposition to soil in Liuzhou area, China, were calculated using the Steady State Mass Balance method (SMB and PROFILE) and dynamic modeling methods(MAGIC), based on data obtained from field investigations and physiochemical properties measured through experiments such as the organic content, cation exchange capacity, base saturation, sulfate adsorption capacity, gibbsite coefficient, biomass base cation uptake and selectivity coefficient for cations. Weathering rates necessary to calculate soil chemistry in applying SMB and MAGIC model were determined by computation with PROFILE using independent geophysical properties such as soil texture and mineralogy as the input data, or by the total soil base cation content correlation. The results have shown that the critical loads of acidity in this area are in the range of 0.7-6.0 keq ha^{-1} yr^{-1}, indicating sulfur deposition should be cut down by 50-90 percent of the present level. The upper soil layer is the most sensitive. The maximum allowable deposition loading of this area is also presented in the paper.

Key words: critical load, acid deposition, acidification model, sulfur deposition.

1. Introduction

Liuzhou, the industrial center of Guangxi Zhuang Autonomous Region, is one of the most serious acid rain districts in China, where frequency of acid rain ranges from 85 to 98% and the average precipitation acidity is pH 4.18. Therefore, the demonstration program of integrated acid deposition control in Liuzhou area has become one of the key projects in the national 8th-five-year plan. We have conducted research on the "Study of Critical Loads and Target Loads in Liuzhou", which is a sub project of the key project. The paper aims at predicting the maximum tolerable acid deposition of the ecosystem according to the physiochemical properties of soil and waterbodies in this area, thus to provide scientific and quantitative basis for control and decision-making.

In China, acid rain occurs mainly to the south of Yangzi River. The large districts are mainly covered by red earth and evergreen or deciduous angiospermous forests or coniferous and angiospermous mixed forests. The red earth amounts to 88.28 percent of the total soil area in Liuzhou. Parent geological materials are quaternary red earth, rattler, granite, sandstone, shale, siliceous shale, calcite and so on, among which red earth developed from rattler is distributed most widely. The principal clay mineral is kaolinite, followed by montmorillonite, quartz, hematite and illite while gibbsite is uncommon. *Pinus massoniana* forests account for 77 percent of the total forest area. *Pinus massoniana*, however, is quite sensitive to acid deposition and thus is suffering from its damage. Therefore, *pinus massoniana* was selected as the indicator organism to represent the ecology of the system studied. The critical chemical criteria proposed by European researchers were used because data showing their relations to pine decline were not completely available (Sverdrup and de Vries, 1994). Some researchers have studied and mapped the acidification sensibility for soil and waterbodies in some acid rain districts, but failed to present the quantitative critical loads (Zhou, et al., 1987). We studied the geological and physiochemical properties of red earth in Liuzhou and applied PROFILE (Warfvinge and Sverdrup, 1992), MAGIC (Cosby, 1985) and Simple Mass Balance Model (Sverdrup, et al., 1990; Hetteling, et al., 1991) to estimating the critical loads for acid deposition in Liuzhou.

2. Materials and Methods

2.1. METHODS FOR CRITICAL LOAD CALCULATION

The PROFILE model was applied to 10 localities distributed throughout Liuzhou City according to soil and parent geological material types. Critical loads were obtained when the output soil solution chemistry resulted from running PROFILE by altering acid deposition is in agreement with the set chemical limits. Also, Steady State Mass Balance (SMB) method was used for calculating the critical loads for acidity of soil in Liuzhou.

$$CL = BC_W + 1.5\left(\frac{Q^{2/3}}{K_{Gibb}} + 1\right)\left(\frac{0.7 \cdot BC_W + BC_d - BC_u - BC_{le,\min}}{((Ca + Mg + K)/Al)_{crit}}\right) \quad (1)$$

where BC_W is the weathering rate of base cation Ca, Mg, K and Na(eq ha^{-1} yr^{-1}), BC_d is the base cation deposition(eq ha^{-1} yr^{-1}), BC_u is the base cation uptake(eq ha^{-1} yr^{-1}), Q is the runoff rate(m^3 ha^{-1} yr^{-1}), $BC_{le,min}$ is the minimum leaching of base cation corresponding to the leaching caused by the residual concentration of base cation that the trees cannot take up (=0.015Q), K_{Gibb} is the gibbsite coefficient, $((Ca+Mg+K)/Al)_{crit}$ is the critical chemical value (=1.0).

To predict soil acidification and estimate critical loads, MAGIC model was applied to typical soil profile under different acid deposition strategies.

2.2. INPUT DATA (COLLECTED AND MEASURED)

Data about soil types and the corresponding land area, soil texture and total soil content analysis were obtained by collecting data from the Second National Soil General Survey in Liuzhou. We collected data on chemical components of precipitation in recent years and on meteorology and hydrology after the 1950's. Maps of soil mapping, geology and terrain were also derived. Field measurements were made to obtain those data unobtainable from the literature and research reports.

Weathering rates were estimated with PROFILE based on soil mineralogy and texture data, and subsequently used as input data for MAGIC model to testify the appropriateness of applying the models. Soil texture data were obtained from the Report of the Second Soil General Survey in Liuzhou. Mineralogy data were found in the above report, by special soil characterization studies, or/and by correcting and backcalculating the total soil analysis content data. In addition, soil mineralogy in typical soil profiles was measured with X-ray Diffraction (XRD). Experimental procedures can be found in the Lin, et al. (1992) and Sverdrup, et al. (1992).

Runoff rates were obtained from the Guangxi Hydrological Station. Deposition figures were received as wet deposition of sulfate, nitrate, ammonium and base cations. Total deposition was approximately derived from the literature and research reports related. For sulfate and nitrate, the following expression was used to calculate the total deposition:

$$tot_d = wet_d \cdot (1+1.5x_F) \qquad (2)$$

For ammonium and base cations:

$$tot_d = wet_d \cdot (1+2.0x_F) \qquad (3)$$

where x_F is forest cover fraction in the catchment.

The base cation uptake was calculated from the net growth rate and concentration of base cations in the tree. The calculation equation was:

$$BC_u = k_{gr} \cdot X_{BC} \qquad (4)$$

where, k_{gr} is the annual average net growth rate constant (kg ha^{-1} yr^{-1}), X_{BC} is the net uptake flux of base cation(keq kg^{-1}).

The uptake of total nitrogen was calculated in the same way as the base cation uptake.

An experiment was carried out to find the gibbsite coefficient. 40ml deionized water was added to 5g dried soil sample. By using H_2SO_4 (0.5%) the pH value was adjusted to the range of 3.5-5.0 at an interval of 0.3 unit. The pH was measured with a pH meter stirred by magnetic force. Adjustment was made every half hour until pH changed little (after about 4-5h). The solution was settled for 24 hours and filtered with mediated speed quantitative filter paper and then passed through 0.5 μ m micropore filter membrane. The aluminum content in the percolate was measured colorimetrically. A curve reflecting the correlation between pH and the solved aluminum was obtained from experiments. Thus the gibbsite coefficient could be also obtained. Experiments of soil from different forest sites in Liuzhou showed a value of pK_{Gibb}=6.5-7.0 in the upper organic layer (O-horizon), pK_{Gibb}=7.5-8.0 for the A-horizon, pK_{Gibb}=8.0-8.5 for the B-horizon, pK_{Gibb}=9.0-9.5 for the C-horizon,

The cation exchange capacity (CEC), base saturation (BS), organic content were measured with the standard methods established by the Agriculture Department of China.

3. Results and discussion

The physiochemical properties of 5 kinds of red earth in Liuzhou were measured. Weathering rates were calculated by employing PROFILE model and the total soil base cation analysis content correlation (Sverdrup and de Vries, 1994). The input data used by PROFILE model to calculate the sand and silt red earth chemistry were listed in Table I . Field measurements show there are great differences in mineralogy and texture among the 5 soils. Consequently, the weathering rates of the base cations are quite different, as shown in Table II. However, their CEC and BS are relatively low.

The acid deposition chemistry in Liuzhou is greatly different from that in Europe as deposition of sulfate, ammonium and base cations is higher while sea salts deposition is

TABLE I.

The input data for calculating the sand and silt red earth chemistry by PROFILE

Parameter	Unit	Soil Layer			
		1	2	3	4
Soil layer height	m	0.05	0.1	0.2	0.3
Moisture content	$m^3 m^{-3}$	0.29	0.3	0.3	0.29
Soil bulk density	$kg\,m^{-3}$	1020	1240	1220	1230
Surface area	$m^2 m^{-3}$	4.3×10^6	5.3×10^6	5.4×10^6	5.6×10^6
CO_2 pressure	time ambient	3	5	10	15
Inflow	% of precipitation	100	90	80	70
Percolation	% of precipitation	90	80	70	60
Ma+Ca+K uptake	% of total max	20	30	30	20
N uptake	% of total max	20	30	30	20
Dissolved organic carbon	$mg\,L^{-1}$	20	40	30	20
$\log K_{Gibb}$	$kmol^2 m^{-3}$	7.0	8.0	8.5	9.5
Mineral		% of total			
Plagioclase		1	1	1	1
Hornblende		0.1	0.1	0.1	0.1
Muscovite		12	13	19	25
Chlorite		2	3	4	3
Vermiculite		0.1	0.1	0.1	0.1
Kaolinite		15	19	15	16
Mixed layer Muscovite/Vermiculite		28	30	32	26
Anatase		0.8	0.8	0.8	0.8
Quartz		41	33	28	28

TABLE II.

The input data for claculating the critical loads of each forest red earth in Liuzhou

Soil type	Area (%)	Q	BC_w	BC_d	BC_u	N_u
		$m^3\,ha^{-1}\,yr^{-1}$	$keq\,ha^{-1}\,yr^{-1}$			
Sand and silt red earth	51.5	7000	2.9	0.5	2.0	1.3
Red silt earth	1.0	7000	5.9	0.5	2.6	1.5
Sand red earth	3.0	7000	1.2	0.5	1.3	0.93
Sand and silt yellow red earth	11.7	7000	0.52	0.5	0.72	0.55
Sand yellow red earth	21.1	7000	1.3	0.5	1.42	0.92

rather low. In 1989, the deposition was 4.06 keq $ha^{-1}\,yr^{-1}$ for sulfate, 0.66 keq $ha^{-1}\,yr^{-1}$ for nitrate, 0.324 keq $ha^{-1}\,yr^{-1}$ for chlorite, 1.93 keq $ha^{-1}\,yr^{-1}$ for ammonium and 2.3 keq $ha^{-1}\,yr^{-1}$ for Ca+Mg+K+Na. These figures indicate that the approximate deposition of acidity and potential acidity are 0.82 keq $ha^{-1}\,yr^{-1}$, 4.7 keq $ha^{-1}\,yr^{-1}$ respectively. Liuzhou is located in the subtropical area, with mild climate, plentiful rain and affluent trees. Its average temperature is 20.5 °C and precipitation is 1500mm. Accordingly, the biological organisms grow rapidly annually. At the same time, soil minerals are subject to weathering. The cation uptake is estimated at 0.7-3.0 keq $ha^{-1} yr^{-1}$

TABLE III.
The critical loads of total acidity for each forest red earth in Liuzhou

Soil type	Critical Loads of acidity		Critical loads of potential acidity		S critical loads	Required reduction of S deposition
	SMB	PROFILE	MAGIC	PROFILE	MAGIC	
	keq ha^{-1} yr^{-1}					
Sand and silt red earth	3.9	3.3	3.2	2.6	2.0	50%
Red silt earth	9.4	6.2	5.9	5.2	5.3	-30%
Sand red earth	1.3	1.5	1.4	1.2	0.8	80%
Sand and silt yellow red earth	0.72	0.87	0.82	0.7	0.4	90%
Sand yellow red earth	1.5	1.7	1.5	1.2	0.9	75%

As to the estimation of BC_d in Equation (1) to calculate the critical loads of acidity with SMB, we think it should be estimated under ambient air quality standards. The critical loads are an ecosystem's fundamental characteristics. When calculating the critical loads, we should take into account not only reducing acidic matter but also dust and particles in the atmosphere. Therefore, the total base cation deposition is estimated at 0.5keq ha^{-1}yr^{-1}, considering the environmental quality in Liuzhou, the current status of dust removal controlling and the environmental quality standards for cities in our country. Otherwise, the critical loads of acidity obtained from Equation (1) would be too great if we used the area wet deposition to calculate BC_d. That is to say, the ecosystem lies in a polluted air environment and is still tolerant to a greater acidity input.

After obtaining the above data, we used PROFILE, SMB and MAGIC to estimate the total acidity and potential acidity of the 5 soils. The results are given in Table III, showing that there is no great difference between the critical loads for forest soils in Liuzhou calculated from static method and those from dynamic method. Estimating critical loads in this area can be completed successfully by employing these methods. Compared with the northern Europe, the critical loads in Liuzhou are greater, mainly because of higher weathering rates of the cations resulted from higher temperature and more clay contained in the soil. According to Table III, the critical loads for potential acidity are lower than those for acidity. This is principally caused by the plants' lower ability of utilizing nitrogen. It is advisable to express the critical loads for acid deposition in the form of potential acidity in our country because it is lower and ammonium deposition is greater.

The pH values for each soil layer solution calculated from PROFILE are consistent with the values measured, showing that soils in the area are already in a steady state. Running MAGIC as a two-layer model under different acid deposition showed that (1) soil in this area is in the state of acidification and will be further acidificated; (2) the bottom soil layer is of greater buffering capacity, which has changed slowly in the past few decades except that in a few situations it has been deteriorating; (3) the upper soil layer is more sensitive to acid deposition.

4. Conclusions

The sand and silt red earth and red silt earth accounting for about 52 percent of the total area of Liuzhou show high weathering rates due to high clay content and high soil temperature. At present, sulfur deposition still exceeds the critical load. However, the critical loads are low for sand red earth, sand yellow red earth, sand and silt yellow red earth accounting for 35.8 percent of the total area. Critical loads are exceeded for about 80 percent of the forest area and the acid deposition should be cut by 50-90 percent. The maximum allowable sulfur deposition loading is 0.9keq ha^{-1} yr^{-1}(as shown in Table III) in order to protect 90% forest area in Liuzhou. The soil chemistry resulted from running dynamic MAGIC model in the area shows that the bottom soil layer has a greater buffering ability while the top soil is suffering from the damage of acid deposition. It is more appropriate to use the potential acidity as a measure of the critical loads for acid deposition in our country due to the high lever of ammonium deposition in the total deposition.

Acknowledgements

Professor Sverdrup and Professor Warfvinge from the Department of Chemical Engineering, Lund Institute of Technology deserve special thanks for their kind assistance in our work. They have supplied us with the running program of PROFILE model. We also wish to thank Professor Cosby from the Department of Environment Sciences, University of Virginia for their assistance with the MAGIC model.

References

Cosby, B.J., Homberger, G.H., Galloway, J.N. and Wright, R. F.: 1985, " Assessment of a Lumped-parameter Model of Soil Water and Streamwater Chemistry", Water Resources Research, 21, 51-53.

Hetteling, J. , Dowing, R. and de Smet, P.: 1991, Mapping Critical Loads for Europe, Coordination Center for Effects, RIVM, CCE Technical Report No. 1, TIVM Report No. 259101001.

Lin, G. Z., Liao, B. H. and Ding, R.: 1992, " A study on Buffering Capacity of the Forest Soil Against Acidic Precipitation in Several Areas of China", Chinese Geographical Science, 2(2), 126-132.

Sverdrup, H., de Vries, W. , Henriksen, A.: 1990, Mapping Critical Loads, Nordic Council of Ministers, Geneva, Miljorapport, 1990:15, Nord 1990:98.

Sverdrup, H., and de Vries, W.: 1994, "Calculating Critical Loads for Acidity with the Simple Mass Balance Method", Water, Air, and Soil Pollution, 72, 145-162.

Sverdrup. H., and Warfvinge, P. and Rabenhorst, M., et al.: 1992, " Critical Loads and Steady-State Chemistry for Streams in the State of Maryland", Environ. Pollut., 77, 195-203.

De Vries, W., Reinds, J., and Posch, M.: 1994, " Assessment of Critical Loads and Their Exceedance on European Forests Using a One-layer Steady-state Model", Water, Air, and Soil Pollution, 72, 357-394.

Warfvinge, P. and Sverdrup, H.: 1992, "Calculating Critical Loads of Acid Deposition with PROFILE- A Steady-State Soil Chemistry Model", Water, Air, and Soil Pollution, 63, 119-143.

Zhou, X., Jiang, J. and Qing, W.: 1987, "Discussion on the Sensibility of Soil to Acid Rain in Four Southern Provinces in China", Atmospheric Environment, 3, 42-48(in Chinese).

CRITICAL LOADS OF ACIDITY TO SURFACE WATERS IN THE VOSGES MASSIF (NORTH-EAST OF FRANCE)

JP. PARTY[1], A. PROBST[1], E. DAMBRINE[2], AL. THOMAS[2]

[1] Centre de Géochimie de la Surface (CNRS), 1 rue Blessig, F-67084 STRASBOURG,
[2] Centre de Recherches Forestières (INRA), Champenoux, F-54280, NANCY

Abstract.: Acid clearwater fishless streams have been identified in the Vosges mountains. In order to evaluate the relationships between acidifying factors (such as atmospheric deposition), buffering factors (such as bedrock and soil type), and surface water acidity, an exhaustive survey of streamwater acidity in the Vosges mountains (N-E France) was performed. A network of 11 measurement stations of atmospheric deposition was used to estimate and map deposition over the whole massif (total area ≈ 5000 km^2). Data on bedrock, soil, superficial deposits, and vegetation were collected from published studies. Sensitive areas as well as acidifying environment factors were derived from the corresponding maps. Over the whole massif, 19% of streams showed baseflow alkalinity below 30 µeq.l^1 and 7.5 % were identified as acid (pH < 5.4). Acid streams occur on the north-western side of the massif on quartz-rich sandstone and acid granites. In each of these areas, we could clearly point out on one hand, the negative influence of conifer vegetation and glacial soil abrasion or induration, and on the other hand the buffering effect of moraine deposits. A corresponding range of critical loads (< 0.2 to 2.0 Keq. ha^{-1}.yr^{-1}) for surface water was calculated using the Steady State Water Chemistry method (SSWC).

Keywords.: critical load, streamwater chemistry, acidity, sensitive areas, granite, sandstone, spruce, soil formation, North East of France

1. Introduction

The critical load is "the highest deposition of acidifying compounds that will not cause chemical changes in soil leading to long term harmful effects on ecosystem structure and function" (Nilsson and Grennfelt, 1988). In Europe, critical loads of acidity have been calculated and mapped over the past five years (Hettelingh et al., 1991). The purpose of critical load mapping was to give technical and scientific basis to reduce accurately the emission of atmospheric pollutants according to critical values for each country.
In France, surface water acidification has been detected in the late seventies particularly in the Vosges massif (Bourrié, 1978; Massabuau et al., 1987; Probst et al., 1990b). With the occurence of forest decline (Landmann and Bonneau, 1995), hydrochemical investigations began in small catchments of the Vosges mountains (Probst et al., 1987, 1990a, 1992). In this massif, a survey of surface water acidification was initiated in 1989 (Probst et al., 1990b) and exhaustively completed in 1992 (Party et al., 1993).
In this paper, we present the geographical distribution of streamwater acidification in relation to environmental factors. Then, for a set of streams, critical load of acidity are calculated by the SSWC method (Steady State Water Chemistry, Henriksen et al., 1990).

Values of critical loads of acidity are finally discussed in relation to the environmental factors and the present load of acidity.

2. Materials and methods

Data acquisition: For the purpose of this study, the data from two streamwater samplings were used low water flow and snowmelt. The sampling was mainly oriented toward fishless streams (Probst *et al.*, 1990b). Major elements were analysed. Then an exhaustive survey of pH and alkalinity (Gran titration) of 800 streamwaters in small catchments was performed (Party *et al.*, 1993). Deposition data were obtained from a network of 11 stations (open field and throughfall measurements) (Dambrine *et al.*, 1995). Bedrock and vegetation data as well as information on superficial deposits were obtained from regional maps (1/50,000) and litterature studies.

Critical load calculation: According to available data, the SSWC method was the most appropriate to calculate critical load values for surface water in the Vosges. However, it had to be adapted to the specific environmental conditions. A set of 38 sensitive streamwaters was used for the calculation.

CL acidity = $Q\,[BC]_0$,

where $[BC]_0$ is the "original" concentration of base cations (molc. ha^{-1}. yr^{-1})
$[BC]_0 = [BC]_t - F\,([SO_4^{2-}+NO_3^-]_t - [SO_4^{2-}+NO_3^-]_0)$,
where F is the ratio between change in base cation concentration and change in strong acid anion concentration (Henriksen *et al.*, 1990); $[SO_4^{2-}+NO_3^-]_t$ is the present-day (sulfate + nitrate) concentration and $[SO_4^{2-}+NO_3^-]_0$ is the "original" (sulfate + nitrate) concentration specially calculated for the Vosges Massif conditions; $F = \sin(90.\sum BC^*/S)$ where $S = 300$ µeq.litre^{-1}; $[SO_4^{2-}+NO_3^-]_0 = 0.15\,\sum BC^*$ (Probst *et al.*, 1995, adapted from Henriksen *et al.*, 1990). BC^* is the present-day sum of base cations corrected from natural atmospheric inputs.

3. Results and discussion

The Vosges Massif is composed of many sensitive silicate bedrocks like sandstone in the north-western part, granites, gneiss and schists in the north-eastern, center and south-eastern parts. In the center part of the massif, these bedrocks are essentially covered by glacial deposits characterized by accumulated material in the valley bottoms (moraine) and abrasion effect on the crests. Mixed stands of silver-fir and beech represent the natural forest but the massif is widely forested by Norway spruce since 1850.

Acid sensitivity was considered in relation to the main environmental factors of the Vosges mountains : bedrocks, superficial deposits, and vegetation. As a first step, we successively gathered maps of:

- bedrock geology and geochemistry (Fe_2O_3 and MgO+CaO contents),
- glacial deposits (glacial abrasion, soil induration and moraines), loess occurence, position in the landscape,
- vegetation types (conifers, deciduous, heathlands),
- surface water chemistry (pH and alkalinity).

The result of the superposition of these maps are summarized in table I. The diagonal represents the "usual mean" position of the streamwaters in relation to the chemistry of the underlying bedrock. However, some streams are located outside this usual relationship (see arrows). In these specific cases, we could particularly identify the crucial, positive or negative, past influence of glaciers in the control of the streamwater buffering capacity. Among the other environmental factors, the relation with deciduous trees and conifers (particularly scots pine, Norway spruce and associated peat-bog with sphagnum moss) had been clearly detected (Probst *et al.*, 1995).

Over the whole massif, among the 800 catchments, 19% of streams presented baseflow alkalinity below 30 μeq/l. Acid streams (7.5 %) occur in the north-western side of the massif on quartz-rich sandstone (5%) and acid granites (2.5%).

TABLE I

Relationships between geochemical characteristics of bedrocks and surface water chemistry. The potential shift from the usual relationship is given by the arrows.

BEDROCK		WATER pH / Alk	< 5,4 / < 0	5,4-5,7 / 0-10	5,7-6,2 / 10-30	6,2-6,6 / 30-100	> 6,6 / > 100
Fe_2O_3	MgO + CaO						
< 1 %	< 2 %		Sandstone →			→	
1-2 %	< 2 %			Acid binary granite →		→	
2-4 %	2-5 %			←	Argillaceous sandstone →	→	
≥ 4 %	≥ 5 %				←	Syenitic granite	
>> 4 %	>> 5 %						Granite with amphibole

Positive influence of glacial accumulation (moraine and loess) ⟶ +

← Negative influence of glacial abrasion, crest position, spruce and peat-bogs

▨ Usual correspondance between surface water and bedrock

Alk : alkalinity, μeq.litre^{-1}

On the basis of this exhaustive survey, the chemical data from 38 catchments, among which about 70 % had an alkalinity lower than 20 µeq.litre^{-1} (Probst et al., 1990a, 1995) have been used to calculate critical loads of acidity to surface waters. According to these results, a first map has been tentatively drawn up (Fig. 1): critical load values range between 0.2 and more than 2.0 Keq.ha^{-1}.yr^{-1}. For 29 streamwaters, critical load is below 1.2 Keq.ha^{-1}.yr^{-1} and for 17 streamwaters below 0.5 Keq.ha^{-1}.yr^{-1} whereas in 8 catchments, values are around 0.2 Keq.ha^{-1}.yr^{-1} or less. Since streamwater samplings mainly concern acidified streams (Probst et al., 1990b), these critical load values are mainly representative of the most sensitive areas and therefore should not be extended to the whole massif. In this small study area, even if the values fall in the same range as those found at the european scale (Hettelingh et al., 1991), the occurence of the extreme values seems to be greater.

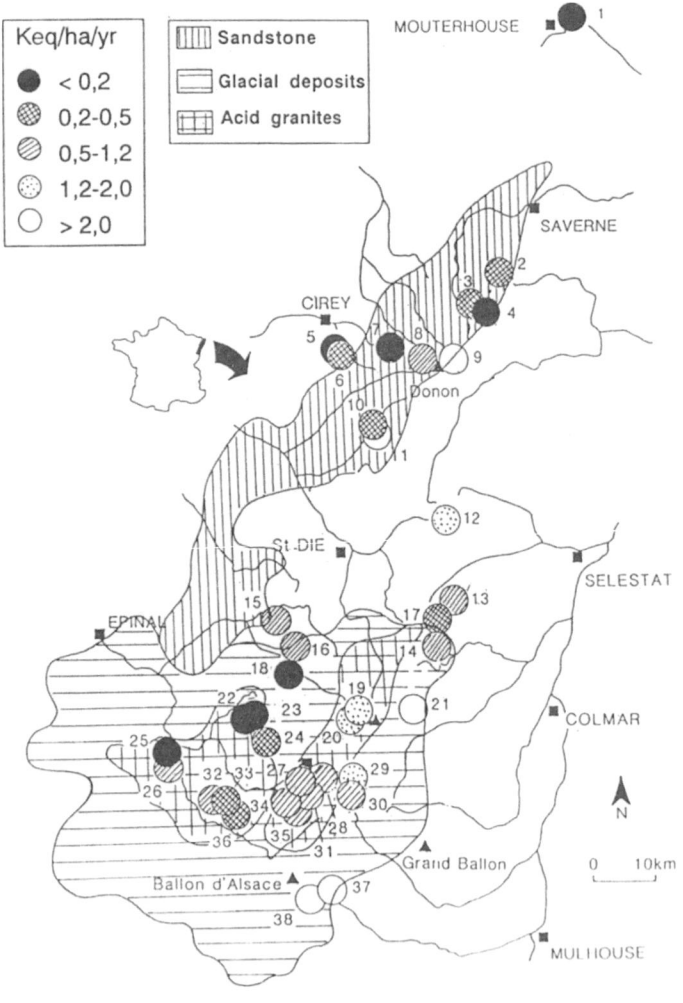

Fig. 1. Map of critical load of acidity to streamwater and corresponding sensitive areas.

In the southern-part of the Vosges, the lowest critical loads for streams generally correspond to the combination of acid granites, local periglacial soil induration (favourizing lateral flow) and continuous Norway spruce cover. In the north-western part, the most sensitive area is related to the presence of sandstone covered by scots pine forest and secondary by peat bogs (Probst *et al.*, 1995). Within these units, local variability in streamwater acidity could be at least partly attributed to soil and bedrock heterogeneity, to their distribution within the drainage catchments, and possibly to historical land use. Particularly, the influence of textural and structural differenciation of horizons within the soils on surface water acidity, as for example the presence of compact basal soil layer has also been mentioned in Germany (Feger, 1994).

Measured proton inputs to forests (taking into account atmospheric inputs and production by forest growth) vary from 1 Keq.ha^{-1}.yr^{-1} for the north-west and east sides of the Vosges, 1.5 Keq.ha^{-1}.yr^{-1} in the south-western part and 2 Keq.ha^{-1}.yr^{-1} on the central part edge. Exceedance values show that catchments on sandstone covered by scots pine in the north-western part of the Vosges have certainly lossed their buffering capacity for a long time.

4. Conclusion and perspectives

Bedrock characteristics and surface water chemistry show the great sensitivity to acidity of the western side of the Vosges. Moreover, a crucial effect of superficial deposits was observed as already mentioned in the nearby Black Forest (Feger, 1994). Integration of superficial formations into european data bases in order to produce maps of critical load at the European level should then be recommended. Following investigations will be focussed on evaluation of groundwater and forest soil critical loads in the Vosges followed by an extension to the whole France.

Acknowledgments

This work has been supported by the ADEME (Agence de l'Environnement et de la Maitrise de l'Energie) and the Ministère de l'Environnement. Special thanks to MMr. Ch. Elichegaray, D. Savanne (ADEME) and to Mrs P. Ebner (Ministère de l'Environnement).

References

Bourrié, G.: 1978, *Sci. Géol. Mém.*, Strasbourg, **52**, 1-174

Dambrine, E., Probst, A., Party, J. P.: 1993, *Pollution atmosphérique*, N° **spécial**, 21-28

Dambrine, E., Ulrich, E., Cénac, N., Durand, P., Gauquelin, T., Mirabel, P., Nys, C., Probst, A., Ranger, J., Zéphoris, M.:1995, in: Landmann, G., Bonneau, M. eds: *Forest decline and atmospheric deposition effects in the French mountains.* Springer. Berlin. Heidelberg New York, 177-200

Feger, K. H.: 1994, in: Steinberg C. E. W., Wright R. F. eds: *Acidification of freshwater ecosystems: implications for the future*. J. Wiley & sons. Berlin, 67-82

Henriksen, A., Lien, L., Traaen, T.: 1990, *Acid Rain Research Report*. NIVA, Oslo, **22**, 1-45

Hettelingh, J. P., Downing, R. J., de Smet, P. A. M. (Eds): 1991, *CEC Tech. Rep. 1*. National Institute of Public Health and Environmental Protection, Bilthoven, 1-86

Landmann, G., Bonneau, M. (Eds): 1995, *Forest decline and atmospheric deposition effects in the French mountains*. Springer. Berlin. Heidelberg. New York, 1-461

Massabuau, J. C., Fritz, B., Burtin, B.: 1987, *C. R. Acad. Sci. Paris*, **305**, série III, 121-124

Nilsson, J., Grennfelt, P. (eds): 1988, *Critical loads for Sulphur and Nitrogen*. Miljøreport, Nordic council of ministers, Copenhagen, **15**, 1-418

Party, J. P., Probst, A., Dambrine, E.: 1993, *Détermination et cartographie des charges critiques en polluants atmosphériques dans les Vosges*. Rapport scientifique ADEME, année 1992, 1-68

Probst, A., Dambrine, E., Viville, D., Fritz, B.: 1990a, *J. Hydrol.*, **116**, 101-124

Probst, A., Fritz, B., Ambroise, B., Viville, D.: 1987, *Proc. Vancouver Symp. Ass. Int. Hydrol. Sc.*, **167**, 109-120

Probst, A., Massabuau, J. C., Probst, J. L., Fritz, B.: 1990b, *C. R. Acad. Sci. Paris*, **311**, série II, 405-411

Probst, A., Probst, J.L., Massabuau, J.C., Fritz, B.: 1995, in: Landmann, G., Bonneau, M. eds: *Forest decline and atmospheric deposition effects in the French mountains*. Springer. Berlin. Heidelberg. New York, 371-386

Probst, A., Viville, D., Fritz, B., Ambroise, B., Dambrine, E.: 1992, *Water Air Soil Pollut.*, **62**, 337-347

NITROGEN CRITICAL LOADS FOR NATURAL AND SEMI-NATURAL ECOSYSTEMS: THE EMPIRICAL APPROACH

ROLAND BOBBINK and JAN G.M. ROELOFS

Department of Ecology, Researchgroup Environmental Biology
University of Nijmegen, Toernooiveld 1, 6525 ED Nijmegen The Netherlands

Abstract. One of the major threats to the structure and the functioning of natural and semi-natural ecosystems is the recent increase in air-borne nitrogen pollution (NH_y and NO_x). Ecological effects of increased N supply are reviewed with respect to changes in vegetation and fauna in terrestrial and aquatic natural and semi-natural ecosystems. Observed and validated changes using data of field surveys, experimental studies or, of dynamic ecosystem models (the 'empirical approach'), are used as an indication for the impacts of N deposition. Based upon these data N critical loads are set with an indication of the reliability. Critical loads are given within a range per ecosystem, because of spatial differences in ecosystems. The following groups of ecosystems have been treated: softwater lakes, wetlands & bogs, species-rich grasslands, heathlands and forests. In this paper the effects of N deposition on softwater lakes have been discussed in detail and a summary of the N critical loads for all groups of ecosystems is presented. The nitrogen critical load for the most sensitive ecosystems (softwater lakes, ombrotrophic bogs) is between 5-10 kg N ha^{-1} yr^{-1}, whereas a more average value for the range of studied ecosystems is 15-20 kg N ha^{-1} yr^{-1}. Finally, major gaps in knowledge with respect to N critical loads are identified.

1. Introduction

Mans activities pose a number of threats to the structure and the function of natural and semi-natural ecosystems, and thus to the natural variety of plant and animal species. One of the major threats is the increase in air-borne N pollution (NH_y and NO_x) in recent decades. Nitrogen is the limiting nutrient for plant growth in many of the natural and semi-natural ecosystems and most of the plant species from these habitats are adapted to nutrient-poor conditions, and can only compete successfully on soils with low N levels (e.g. Chapin, 1980; Tamm, 1991). Nitrogen is, furthermore, the only nutrient of which the cycling through the ecosystem is almost exclusively regulated by biological processes. To establish reliable critical loads for N, it is essential to understand the effects of N upon these ecosystem processes. The most important impacts of increased atmospheric N deposition upon biological systems are: (i) short-term effects of N gases and aerosols to individual species (critical levels); (ii) soil-mediated effects of acidification; (iii) soil-mediated effects of N enrichment; (iiii) increased susceptibility to secondary stress factors and, (iiiii) changes in (competitive) relationships between species, resulting in loss of diversity.

Empirical critical loads for N deposition to natural and semi-natural ecosystems have been discussed and established at the 1992 UN/ECE workshop at Lökeberg (Sweden; Grennfelt and Thörnelöf, 1992) and updated in 1994 at Grange-over-Sands (UK; (Hornung *et al.*, 1995) using the working definition adopted at Lökeberg. These N critical loads will be incorporated in near future in the WHO Air Quality Guidelines for

Europe (WHO, 1996) and used in the negotiations for a new NO_x protocol.

The critical N loads have been set on the basis of observed and published changes in the structure or function of ecosystems using experimental data, field observations and/or dynamic ecosystem models (the 'empirical approach') (Bobbink et al., 1992; Bobbink and Roelofs, 1995). Changes in plant development and in species composition or dominance have been used as an 'detectable change' for the impacts of excess N deposition, but in some cases a change in ecosystem function, such as N leaching or N accumulation has been used (e.g. ombrotrophic bogs; coniferous forests). The N critical loads have been established within a range per ecosystem, because of (i) real intra-ecosystem variation, (ii) intervals between experimental treatments, and (iii) uncertainties in deposition values. Furthermore, the reliability of the presented figures is indicated:
- reliable: when a number of published papers of various studies show comparable results;
- quite reliable: when the results of some studies are comparable;
- best guess: when no data are available for this type of ecosystem. The N critical load is then based upon knowledge of ecosystems which are likely to be more or less comparable with this ecosystem.

The N critical loads clearly depend on (i) the abiotic conditions, especially those which influence the nitrification potential and immobilization rate in the soil, and (ii) the land use and management in the past and present. Therefore, the empirical N critical loads have been evaluated for specific groups of related ecosystems. The following groups of semi-natural and natural ecosystems have been treated: softwater lakes, wetlands & bogs, species-rich grasslands, heathlands and forests. In this paper the N critical loads for European soft-water lakes are discussed in detail, whereas the N critical loads for the other ecosystems are summarized. Finally major gaps in knowledge and recommendations with respect to empirical nitrogen critical loads are identified.

2. Effects of N deposition on softwater lakes

In the lowlands of Western Europe many soft waters are found on sandy soils which are poor in calcium carbonate or almost devoid of it. The waters are poorly buffered and the concentrations of calcium in the water layer are very low; they are shallow and fully mixed water bodies, with periodically fluctuating water levels, and are mainly fed by rain water, and thus oligotrophic. These softwater ecosystems are characterized by plant communities from the phytosociological alliance LITTORELLION (Schoof-van Pelt, 1973; Wittig, 1982; Roelofs, 1986; Vöge, 1988; Arts, 1990). The stands of these communities are characterized by the presence of rare and endangered isoetids, such as *Littorella uniflora*, *Lobelia dortmanna*, *Isoetes lacustris*, *I. echinospora*, *Echinodorus* species, *Luronium natans* and many other softwater macrophytes. For more details on these communities see Schoof-van Pelt (1973) and Schaminée et al. (1992). These soft waters are nowadays almost all within nature reserves and have become very rare in Western Europe. This decline may be illustrated by the fact that *Littorella uniflora* was known from more than 230 sites in the Netherlands in the early 1950s, of which only about 40 still exist. Furthermore, a strong decline in amphibians has been observed in these soft waters (Leuven et al., 1986).

The effects of nitrogen pollutants on these soft waters have been intensively studied in the Netherlands both in field surveys and experimental studies. Field observations in about 70 soft waters (with well-developed isoetid vegetation in the 1950s) showed that the waters in which these macrophytes were still abundant in the early 1980s, were poorly buffered (alkalinity is 50 to 500 µeq l^{-1}), circumneutral (pH=5-6) and very low in dissolved nitrogen (Roelofs, 1983; Arts et al., 1990). The softwater sites where these plant species had disappeared, could be divided into two groups. In 12 of the 53 softwater sites eutrophication, resulting from inflow of enriched water, seemed to be the cause of the decline. In this group of non-acidified waters plant species such as *Lemna minor* had become dominant, with high concentrations of phosphate and ammonium in the sediments. In some water bodies a dense plankton bloom was observed.
In the second group of lakes and pools (41 out of 53) another development had taken place: the isoetid species were replaced by dense stands of *Juncus bulbosus* or aquatic mosses such as *Sphagnum cuspidatum* or *Drepanocladus fluitans*. This clearly indicates acidification of these soft waters in recent decades, probably caused by enhanced atmospheric deposition. In the same field study, it has been shown that the nitrogen levels of the water layer were higher in ecosystems where the natural vegetation had disappeared, compared with ecosystems where the LITTORELLION stands were still present (Roelofs, 1983). This strongly suggests the detrimental effects of atmospheric nitrogen deposition in these softwater lakes.

A number of ecophysiological studies has revealed the importance of (i) inorganic carbon status of the water as a result of intermediate levels of alkalinity, and, (ii) low nitrogen concentrations, for the growth of the endangered isoetid macrophytes. Furthermore, almost all of the typical softwater plants had a relatively low potential growth rate. Increased acidity and higher concentrations of ammonium in the water layer clearly stimulated the development of *Juncus bulbosus* and submerged mosses such as *Sphagnum* and *Drepanocladus* species (Roelofs et al., 1984; Den Hartog, 1986). It has also been shown in cultivation experiments that the nitrogen species involved (ammonium or nitrate) differentially influenced the growth of the studied species of water plants. Almost all of the characteristic softwater isoetids developed better with nitrate instead of ammonium addition, whereas *Juncus bulbosus* and aquatic mosses (*Sphagnum* & *Drepanocladus*), were clearly stimulated by ammonium nutrition (Schuurkes et al., 1986). The importance of ammonium for the growth of these aquatic mosses is also reported by Glime (1992).
The effects of atmospheric deposition have been studied in small-scale softwater systems during a 2-year treatment with different artificial rainwaters. Acidification, without airborne nitrogen input (sulphuric acid), has not resulted in a mass growth of *Juncus bulbosus* and a diverse isoetid vegetation remains present. However, after increasing the nitrogen concentration in the precipitation (as ammonium sulphate), similar changes in floristic composition as under field conditions have been observed: a dramatic increase in dominance of *Juncus bulbosus*, of submerged aquatic mosses and of *Agrostis canina* (Schuurkes et al., 1987). It became obvious that the observed changes occurred because

of the effects of ammonium sulphate deposition, leading to both eutrophication and acidification. The increased levels of ammonium in the system stimulated directly the growth of plants such as *Juncus bulbosus*, whereas the surplus of the extra ammonium will be nitrified in these waters (pH≥4.0). During this nitrification process H$^+$-ions are produced, which increase the acidity of the system. The results of this study demonstrate that the changes in composition of the vegetation have occurred after 2-year treatment with ≥19 kg N ha^{-1} yr^{-1}. A reliable critical load for nitrogen deposition in these shallow softwater lakes is thus most likely below 19 kg N ha^{-1} yr^{-1}, and most probably between 5 and 10 kg N ha^{-1} yr^{-1} (Table 1). This value is supported by the observation that the strongest decline in the species composition of the Dutch LITTORELLION communities has coincided with nitrogen loads of 10 to 13 kg N ha^{-1} yr^{-1} (Arts, 1990).

3. Conclusions and Gaps in Knowledge

In this paper the effects of N deposition on softwater ecosystems have been evaluated in detail with published evidence and a N critical load for these sensitive systems are presented. Based upon observed changes in structure and function of natural and semi-natural ecosystems and, sometimes with the use of dynamic ecosystem models, N critical loads have been formulated and set for all treated systems (Table 1). Most of earths biodiversity is present in semi-natural and natural ecosystems and nowadays many anthropogenous stress factors threaten this natural heritage. One of these threats is the air-borne N pollution (NH$_x$ and NO$_y$). It is crucial to control the N load, in order to prevent negative effects on these ecosystems. In this document critical loads for N are given as reliable as possible. As most of the research efforts have focused on acidification in forest systems, serious gaps in knowledge exist on the effects of enhanced N deposition on natural and semi-natural terrestrial and aquatic ecosystems. The following gaps in knowledge are important:
- quantified effects of enhanced N deposition on fauna in all reviewed ecosystems are extremely scarce;
- the critical load for N deposition to arctic/(sub)alpine systems is largely speculative;
- more research is needed on the N effects on the forest floor, because most of the research had focused on the trees only.
- a serious gap in knowledge is the effect of N on forests on calcareous bedrock, which are not sensitive to acidification;
- more long-term research is needed, especially in montane/subalpine meadows, species-rich grasslands, ombrotrophic bogs and (sensitive) freshwater ecosystems;
- the long-term effects of enhanced atmospheric N in grassland and heathland of high nature conservation importance under different management regimes are insufficiently known.
- long-term effects of nitrogen eutrophication in (sensitive) aquatic ecosystems (freshwater and marine) need further research;
- the possible differential effects of the deposited N species (NO$_y$ or NH$_x$) are insufficiently known to make a differentiation between these N species for critical load establishment.

To establish reliable critical loads, it is crucial to understand the long-term effects of increased N deposition on ecosystem processes in a representative range of communities. It is thus very important to quantify the effects of N loads on natural and semi-natural terrestrial and freshwater ecosystems by manipulation of N inputs in long-term ecosystem studies in unaffected and affected areas. These data are essential to validate the presented critical loads and to develop robust dynamic ecosystem models, which are reliable enough to calculate critical loads for N deposition in semi-natural and natural ecosystems.

TABEL I

Summary of the empirical N critical loads (kg N ha^{-1} yr^{-1}) to (semi-)natural freshwater and terrestrial ecosystems. ## **reliable;** # **quite reliable and** (#) **best guess** (after Bobbink et al,. 1992; Bobbink and Roelofs, 1995).

	Critical load	Indication
Shallow softwater lakes	5-10 ##	Decline isoetid species
Mesotrophic fens	20-35 #	Increase tall graminoids, decl. diversity
Ombrotrophic (raised) bogs	5-10 #	Decrease Sphagnum and subordinate species, increase tall graminoids; N accumulation
Calcareous species-rich grassland	14-25 ##	Increase tall grass, decline diversity
Neutral-acid species-rich grassland	20-30 #	Increase tall grass, decline diversity
Montane-subalpine grassland	10-15 (#)	Increase tall graminoids, decl. diversity
Lowland dry-heathland	15-20 ##	Transition heather to grass
Lowland wet-heathland	17-22 ##	Transition heather to grass
Species-rich heaths/acid grassl.	7-15 #	Decline sensitive species
Upland Calluna moorland	10-20 (#)	Decrease heather
Arctic and alpine heaths	5-15 (#)	Decline lichens, mosses and evergreen dwarf-shrubs, increase in grasses and herbs
Coniferous forest (acidic, managed)	10-20 ##	Changes ground flora; N leaching
Deciduous forest (acidic managed)	15-20 #	Changes ground flora
Deciduous forests (calcareous)	15-20 (#)	Changes ground flora
Coniferous tree health	10-15 #	Nutrient imbalance, low nitrif. (Boxman et al., 1988)
Coniferous tree health	20->50 #	Nutrient imbalance, intermed/high nitrif. (Boxman et al., 1988)
Deciduous tree health	15-20 #	Nutrient imbalance; shoot-root ratio

References

Arts, G.H.P.: 1990. *Deterioration of atlantic soft-water systems and their flora, a historical account*. PhD thesis, University of Nijmegen.

Arts, G.H.P., Van Der Velde, G., Roelofs, J.G.M. and Van Swaay, C.A.M.: 1990. *Freshwater Biol.* 24, 287-294.
Bobbink, R., Boxman, D., Fremstad, E., Heil, G., Houdijk, A. and Roelofs, J.: 1992. In: *Critical loads for Nitrogen.* Nord (Miljörapport) 41, 111-159.
Bobbink, R. and Roelofs, J.G.M.: 1995. In: *Mappjng and modelling of critical loads for nitrogen - a workshop report.* Institute of Terrestrial Eoclogy, Peniciuik , 9-19.
Boxman, D., Van Dijk, H., Houdijk, A. and Roelofs, J.: 1988. In: *Critical loads for sulphur and nitrogen.* Miljörapport 15, 295-323.
Chapin, F.S.: 1980. *Ann. Rev. Ecol. Syst.* 11: 233-260.
Den Hartog, C.: 1986. In: *Proceedings 1st. Internat. Symposium on water milfoil (Myriophyllum spicatum) and related Haloragaceae species.* Vancouver, Canada, pp. 51-58.
Glime, J.M.: 1992. In: *Bryophytes and lichens in a changing environment.* Clarendon Press, Oxford, pp. 333-361.
Grennfelt, P. and Thörnelöf, E. (eds.): 1992. *Critical loads for nitrogen.* Nord (Miljörapport) 41, 428 pp.
Hornung, M., Sutton, M.A. and Wilson, R.B.: 1995. *Mapping and modelling of critical loads for nitrogen - a workshop report.* Institute of Terrestrial Eoclogy, Peniciuik , 207 pp.
Leuven, R.S.E.W., Den Hartog, C, Christiaans and Heyligers, W.H.C.: 1986. *Experientia* 42: 495-503.
Roelofs, J.G.M.: 1983. *Aquat. Bot.* 17: 139-155.
Roelofs, J.G.M.: 1986. *Experientia* 42: 372-377.
Roelofs, J.G.M., Schuurkens, J.A.A.R. and Smits, A.J.M.: 1984. *Aquat. Bot.* 18: 389-411.
Schaminée, J.H.J., Westhoff, V. and Arts, G.H.P.: 1992. *Phytocoenologia* 20: 529-558.
Schoof-van Pelt, M.M.: 1973. *Littorelletea, a study of the vegetation of some amphiphytic communities of western Europe.* Ph.D. Thesis, University of Nijmegen.
Schuurkens, J.A.A.R., Kok, C.J. and Den Hartog, C.: 1986. *Aquat. Bot.* 24: 131-146.
Schuurkes, J.A.A.R. Elbers, M.A., Gudden, J.J.F. and Roelofs, J.G.M.: 1987. *Aquat. Bot.* 28: 199-225.
Tamm, C.O.: 1991. *Nitrogen in terrestrial ecosystems. Questions of productivity, vegetational changes, and ecosystem stability.* Springer Verlag, Berlin.
Vöge, M.: 1988. *Limnologica* (Berlin) 19: 89-107.
World Health Organisation (WHO): 1996. *Air quality guidelines for Europe.* WHO, Geneva, pp xxx.
Wittig, R.: 1982. *Decheniana* (Bonn) 135: 14-21

CRITICAL LOADS OF ACIDITY FOR SURFACE WATERS -
Can the ANC$_{limit}$ be considered variable?

A. HENRIKSEN[1], M. POSCH[2], H. HULTBERG[3] and L. LIEN[1]

[1] *Norwegian Institute for Water Research (NIVA), P.O.Box 173 Kjelsås, N- 0411 Oslo, Norway.* [2] *National Institute of Public Health and the Environment (RIVM), P.O.Box 1, NL-3720 BA Bilthoven, The Netherlands.* [3] *Swedish Environmental Research Institute (IVL), P.O.Box 47086, S-402 58 Gothenburg, Sweden*

Abstract. The critical load of acidity for surface waters is based on the concept that the inputs of acids to a catchment do not exceed the weathering less a given amount of ANC. The Steady State Water Chemistry (SSWC) Method is used to calculate critical loads, using present water chemistry. To ensure no damage to biological indicators such as fish species a value for ANC$_{limit}$ of 20 µeq/l has been used to date for calculatiing critical loads. The SSWC-method is sensitive to the choice of the ANC$_{limit}$. In areas with little acid deposition the probability of acid episodes leading to fish kills is small even if the ANC$_{limit}$ is set to zero, while in areas with high acidic deposition fish kills may occur at this value. Thus, the ANC$_{limit}$ can be a function of the acidifying deposition to the lake, nearing zero at low deposition and increasing to higher values at higher deposition. A formulation for such an ANC$_{limit}$ has been worked out, and we have tested the effect of the ANC$_{limit}$ as a linear function of the deposition, assuming ANC$_{limit}$ = 0 at zero deposition with a linear increase to 50 µeq/l at a deposition of 200 meq.m^{-2}.yr^{-1}. For areas with high deposition the effect of a variable ANC$_{limit}$ is small, while in areas with low deposition the effect is significant. For Norway the exceeded area decreases from 36 to 30% using a variable ANC$_{limit}$ instead of a fixed value of 20 µeq/l.

Key words: critical load, acidity, water acidification, soil acidification, fish, empirical models, sulphur deposition, acid episodes,

1. Introduction

In recent years large areas of Europe and eastern regions of North America have suffered from acid precipitation resulting in the acidification of surface waters, increased fish mortality and other ecological changes. The concept of critical load has come into wide use in connection with international negotiations on reducing emissions of nitrogen and sulphur compounds. It was first put into practical use in Canada in the last part of the 70's in relation to the problem of lake acidification. It was further developed by working groups established by the the Nordic Council of Ministers in 1985 and used in the Scandinavian countries as a method for quantifying the extent and spatial dimension of the acidification problem. Since then it has been developed on a European basis by a number of international cooperative programmes and activities under the United Nations Economic Commission for Europe's Convention on Long Range Transboundary Air Pollution, signed in 1979.

2. The Steady State Water Chemistry Method

The critical load of acidity for surface waters can be estimated on the basis of present water chemistry by means of a simple steady-state method. The critical load of acidity is calculated as:

$$CL(Ac) = Q*([BC^*]_0 - ANC_{limit}) \quad (1)$$

where Q (m·yr^{-1}) is the runoff, $[BC^*]_0$ the pre-industrial seasalt corrected base cation concentration and ANC_{limit} is the selected critical ANC (Acid Neutralization Capacity) threshold (Henriksen et al. 1993).

The Acid Neutralization Capacity is used as the chemical criterion for sensitive indicator organisms (usually fish) in surface waters. ANC is defined as the difference between base cations and strong acid anions. Including present-day nitrate leaching the present exceedance of the critical load of acidity is defined as:

$$Ex(Ac) = S_{dep} + N_{le} - BC^*_{dep} - CL(Ac) \quad (2)$$

where S_{dep} is the total S deposition and BC^*_{dep} is the non-marine base cation deposition (Henriksen et al. 1993). N_{le} is present nitrate leaching from the lake and its catchment, which is estimated from yearly runoff and present lake nitrate concentration minus an estimated background concentration of nitrate in lakes not affected by acid deposition (which for Norway has been estimated as 4 µeq/l, (Kämäri et al. 1992).

In order to calculate the critical load of acidity to surface waters a value for the ANC_{limit} is needed. This value has been derived from the information on water chemistry and fish status obtained from the 1000 lake survey carried out in Norway in 1986 (Henriksen et al. 1988, Lien et al. 1992). The Scandinavian countries have so far used ANC_{limit} = 20 µeq/l as the critical chemical value for fish in surface waters (Henriksen et al. 1990). The natural ANC in lakes can, however, be equal or less than 20 µeq/l in areas with granitic and gneissic bedrock with thin soil cover. For such lakes the ANC_{limit} has so far been set to the original ANC-value of the lake. The critical load for acidity will thus be zero for such lakes. For Norway 163 (16%) of the lakes included in the 1000-lake survey in 1986 have "negative" critical loads using ANC_{limit} = 20 µeq/l. The corresponding figures for Sweden and Finland are 18 (0.4%) out of 4015 and 4 (0.3%) out of 1450, respectively (Henriksen et al. 1993). The much higher percentage of lakes with negative critical loads in Norway is due to the larger number of lakes with low base cation concentrations. These Norwegian lakes are all located in areas where the bedrock consist of granites and gneisses and the soil covers are thin, as is also the case for the Swedish and Finnish lakes.

The critical load is thus sensitive to the choice of the ANC_{limit}. For Norway, a value of ANC_{limit} = 0 µeq/l will give a critical load exceedance in 25% of the area of Norway, whereas an ANC_{limit} = 20 µeq/l will result in an exceeded area of 36%. In this paper we suggest a variable ANC_{limit} that adresses some of the shortcomings of a fixed value.

3. Fish and ANC

The value of 20 µeq/l for ANC_{limit} was chosen to ensure no toxic episodes during the year. In areas with little acid deposition, however, the probability of acid episodes leading to fish kills is small even if ANC is close to zero, while in areas with high acidic deposition fish kills may occur at this value. Thus, the ANC_{limit} could be considered variable, e.g. a function of the deposition to the lake, nearing zero at low critical loads and increasing to an upper limit at higher critical loads. There are good biological arguments for considering ANC_{limit} as a variable: The pH-range 5.5 to 6.0 is regarded as safe under natural conditions for most fish species. One of the first attempts on setting a critical level for effects of acidification on aquatic systems was made in a Swedish case study (1971). The study used a critical limit of pH = 5.5 for salmon fish species. The toxicity of inorganic Al was unknown at this time (Schofield 1977). The toxicity of Al in the pH range 5.5 to 6.0 depends on the concentration and chemical form of Al, the concentration of Ca as well as temperature (Brown 1982; Rosseland and Hindar 1991). In areas with higher critical loads, deposition of sulphur to acidified forest soils causes leaching of Al. This will result in the presence of inorganic Al-forms in lakes and streams also at deposition levels of sulphur at the critical load as long as the soils remain dominated by Al and H^+ (Hultberg 1988). Inorganic Al causes toxic effects on young life stages and to adults of many fish species and other aquatic animals also at low concentrations (salmonid fishes ≥ 30 to 50 µg/l; lake plankton 100 µg/l) in the pH-range 5.5 to 6.0 (Fivelstad and Leivestad 1984; Henriksen et al. 1984; Hultberg 1988).

In lakes roach (*Leuciscus rutilus*) and Arctic char (*Salvelinus alpinus*) are two of the most sensitive fish species. Crayfish species in lakes (*Astacus astacus and Pacifastacus leniusculus*), glacial relicts of crustaceans as well as molluscs (*Margaritana margaritifera*) and mayfly species (*Ephemera sp.*) among the insects are very sensitive to acid episodes and inorganic Al (Hultberg 1977; Nyberg et al. 1986). In streams with sea-running brown trout (*Salmo trutta*) and rivers with Atlantic salmon (*Salmo salar*) draining of upstream acid lakes or soils, later running into areas with non-acid soils with resulting high pH (5.5 to > 6.0) in the water are still affected by low survival of overwintering young year classes as well as fish kills of adult fish (Rosseland et al. 1986; Degerman et al. 1986). This is caused by inflow of water with inorganic Al into water with higher pH and ANC of 20 to ≥ 50 µeq/l at periods with high waterflow and low temperatures. Liming of such upstream acid areas results in decreased transport of inorganic Al which in turn has caused dramatic increases in survival of overwintering young year classes of sensitive fish populations as well as invertebrates in the non-acid downstream parts of the water system (Nyberg et al. 1986). Decreased concentrations of inorganic Al was caused by precipitation of Al in lakes, wetlands and headwater streams in upstream limed parts of the catchment. Inflow of inorganic Al from acid catchments into limed lakes and streams may result in toxicity to fish at high pH in the mixing zone (Dickson 1978, 1983; Skogheim et al. 1984; Rosseland et al. 1992; Poleo et al. 1994).

The acidification recovery process at critical load will be slower for soils than for surface waters. Fish populations and other aquatic biota in lakes and rivers may therefore be exposed to toxic Al-forms also at very low sulphur load. In areas with higher critical loads ANC-values up to 50 µeq/l are necessary to avoid negative effects on fish and other aquatic fauna. The ANC_{limit} of 20 µeq/l was set to the reaction of brown trout, the most abundant fish species in Norway. In other countries other species are more abundant and thus a variable ANC_{limit} seems to be required to protect most aquatic organisms.

4. Method

In order to develop an equation for a variable ANC_{limit} we can rewrite the critical load equation (equation 1) in the following way:

$$[BC^*]_0 - CL(Ac)/Q = ANC_{limit} \qquad (3)$$

which can be visualized by intersecting the line $y = [BC^*]_0 - x/Q$ with the horizontal line $y = ANC_{limit}$, where y is the concentration-axis and the x-axis is the deposition axis. The x-coordinate of the point of intersection of these two straight lines is the critical load.

This way of deriving the critical load can be generalized for an ANC_{limit} which depends on the deposition, i.e. x. Then $y = ANC_{limit}$ is no longer a horizontal line, but a function of the deposition. The critical load is then computed by intersecting the line $y = [BC^*]_0 - x/Q$ with the curve $y = ANC_{limit}(x)$ (see Figure 1). As argued above, the ANC_{limit} can be low in areas with low deposition (backround areas) and should be higher in areas with high depostion. the simplest way to express this is to assume a constant ANC_{limit} below and above given deposition values and a linear dependence between those limits:

$$ANC_{limit} = \begin{cases} A_1 & \text{for } Ac_{dep} \leq Ac_1 \\ k*Ac_{dep} + d & \text{for } Ac_1 < Ac_{dep} < Ac_2 \\ A_2 & \text{for } Ac_{dep} \geq Ac_2 \end{cases} \qquad (4)$$

where $k = (A_2 - A_1)/(Ac_2 - Ac_1)$ and $d = A_1 - k*Ac_1$. If $A_1 > 0$, it must be ensured that $[BC^*]_0 > A_1$. Depending on $[BC^*]_0$ and Q we have three cases, and for each of them we can derive explicit formulae for the critical load and the ANC_{limit} (see Figure 1):

(a) $Q*([BC^*]_0 - A_1) \leq Ac_1$:
Then we simply have: $CL(Ac) = Q*([BC^*]_0 - A_1)$ and $ANC_{limit} = A_1$ (5a)

(b) $Q*([BC^*]_0 - A_1) > Ac_1$ and $Q*([BC^*]_0 - A_2) < Ac_2$:
In this case we obtain after a few calculations:
$CL(Ac) = Q*([BC^*]_0 - d)/(1 + k*Q)$
and $ANC_{limit} = (k*Q*[BC^*]_0 + d)/(1 + k*Q)$ (5b)

(c) $Q*([BC^*]_0 - A_2) \geq Ac_2$:
Again, one simply gets: $CL(Ac) = Q*([BC^*]_0 - A_2)$ and $ANC_{limit} = A_2$ (5c)

The critical load, i.e. the deposition obtained by solving equation 3 after inserting the ANC_{limit} given by equation 4, depends on the parameters of the ANC-function (Ac_1, A_1, Ac_2 and A_2, see equations 5a-c), which have to be derived from biological criteria, i.e. the (fish) species to be protected. The expression for the ANC_{limit} is not restricted to the broken linear function as defined in equation 4, but any monotonously increasing function, e.g., a S-shaped curve, could be chosen. However, in such a case equation 3 would become non-linear, and the critical load would have to be computed by an iterative method.

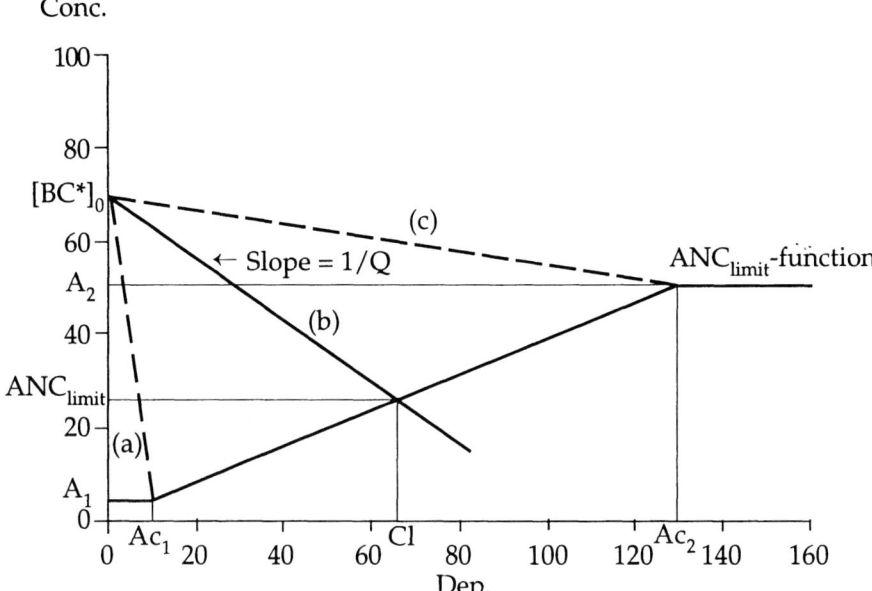

Figure 1. Determination of the critical load for a deposition dependent ANC_{limit} (equation 4). Three cases are possible, depending on the parameters of the function (Ac_1, A_1 and Ac_2, A_2), $[BC^*]_0$ and the runoff Q (see text).

It should be noted that, although now ANC_{limit} is not a fixed value for all lakes, each lake will have its own value for all depositions, given by the characteristics of the lake ($[BC^*]_0$ and Q).

5. Results

To test the concequences of a variable ANC_{limit} for the critical loads in Norway, Sweden and Finland we used $A_1 = 0$ for $Ac_1 \leq 0$, and $A_2 = 50$ for $Ac_2 = 200$. Using the critical load databases for Finland, Norway and Sweden (Henriksen et al 1990), we have calculated the critical loads and their exceedances (Table 1).

Table 1. Percent of area (Norway) and percentage of lakes (Finland and Sweden) for which critical loads of acidity is exceeded for a variable ANC_{limit} and a fixed ANC_{limit} (20 µeq/l).

Country	Fixed ANC_{limit}	Variable ANC_{limit}	Difference
Sweden	30,2	28,6	1,6
Finland	17,2	14,3	2,9
Norway	36,0	30,0	6,0

The effect of a deposition dependent ANC_{limit} is larger in Norway than in Sweden and Finland, because the number of sensitive lakes are highest here, as pointed out above.

Using a fixed $ANC_{limit} = 20$ µeq/l, the critical load of present acidity for surface waters is exceeded in 36% of the area of Norway. The variable ANC_{limit} (0-50µeq/l) reduces the exceeded area to 30%, and most of these areas receive little acididic deposition (central and northern Norway). The distribution of areas where the critical load of acidity is presently exceeded using the variable ANC_{limit} corresponds better to those where fish populations are damaged than using the fixed ANC_{limit} (Henriksen and Hesthagen, in prep.).

Acknowledgement

This work has been partly supported by the Nordic Council of Ministers (NMR).

References

Brown, D. J. A.: 1982, Water Air and Soil Pollut. **18**, 343.
Degerman, E., Fogelgren, J. E., Tengelin, B. and Tîrnelîv, E.: 1986, Water Air and Soil Pollut. **30**, 665
Dickson, W.: 1978, Verh. Int. Verein. Limnol. **20**, 851.
Dickson, W.: 1983, Vatten. **30**, 400.
Fivelstad, S. and Leivestad, H.: 1984, Institute of Freshwater Research, Drottningholm, Report **61**, 69.
Grahn, O.: 1980, 'Fish-kills in Two Moderately Acid Lakes Due to High Aluminum Concentration', in D. Drabløs and. A. Tollan (eds.), Ecological impacts of acid precipitation, SNSF-project, Proc. Int. Conf. Sandefjord, Norway. p 310.
Henriksen, A. Forsius, M., Kämäri, J., Posch, M. and Wilander, A. 1993. *Exceedance of critical loads for lakes in Finland, Norway and Sweden: Reduction requirements for Nitrogen and Sulphur deposition.* Acid Rain Research Report 32/1993. Norwegian Institute for Water Research (NIVA) Oslo, Norway. 46 pp.
Henriksen, A., Kämäri, J., Posch, M., Lövblad, G., Forsius, M. and Wilander, A. 1990. *Critical loads to surface waters in Fennoscandia*. Nordic Council of Ministers. Miljørapport 1990:124.
Henriksen, A., Lien, L., Traaen, T.S., Sevaldrud, I.S. and Brakke, D. 1988. *Ambio* **17**, 259-266.
Henriksen, A., Skogheim, O. K. and Rosseland, B. O.: 1984, Vatten, **40**, 225.
Hultberg, H.: 1977, Water Air and Soil Pollut. **7**, 279.
Hultberg, H.: 1988, , In Nilsson, J. and Grennfeldt, P.(eds.), 'Critical Loads for Sulphur and Nitrogen', NORD 1988: 15. Nordic Council of Ministers,.
Kämäri, J., D.S. Jeffries, D.O. Hessen, A. Henriksen, M. Posch and M. Forsius 1992. Nitrogen Critical Loads and their Exceedance for Surface Waters. In: Grennfelt, P and E. Thörnelöf (eds.) 1992. *Critical loads for nitrogen*, NORD 1992:41, Nordic Council of Ministers, Copenhagen, Denmark.
Lien, L., Raddum, G.G. and Fjellheim, A. 1992. *Critical Loads of Acidity to Freshwater - Fish and Invertebrates*. Report from the project "Naturens Tålegrenser" no.23. Norwegian Institute for Water Research, Oslo, Norway.
Poleo, A. B. S., Lydersen, E., Rosseland, B. O. Kroglund, F., Salbu, B., Vogt, R. D., and Kvellestad, A.: 1994, Water Air and Soil Pollut. **75**, 339.
Rosseland, B. O. and Hindar, A.: 1991, 'Mixing Zones - A Fishery Management Problem', In: Olem, H., Schreiber, R. K., Brocksen, R. W. and Porcella, D.(eds.), Int. Lake and Watershed Liming Practices, Lewis Publishers, Chelsea, Michigan, ISBN I-8080786-00-7, p 161.
Rosseland, B. O., Skogheim and Sevaldrud, I, 1986, Water Air and Soil Pollut. **30**, 65.
Rosseland, B. O.. Blakar, I. A., Bulger, A., Kroglund, F. Kvellstad, A., Lydersen, E., Oughton, D. H. Salbu, B., Staurnes, M. and Vogt, R.: 1992, Environ Pollut. **78**, 3.
Schofield, C. L.: 1977, Research Technical Completion Report A-072-NY, Office of Water Researched Technology, Dept. of the Interior, Washington D. C. , p 27.
Skogheim, O.K., Rosseland, B.O. and Sevaldrud, I. 1984. Rep. Inst. Freshw. Res. Drottningholm, **61**, 195.
Sweden's Case Study ,: 1971, UN Conference on the Human Environment in Stockholm 1972.

A PALAEOLIMNOLOGICAL ASSESSMENT OF THE IMPACT OF ACID DEPOSITION ON SURFACE WATERS IN NORTH-WEST SCOTLAND, A REGION OF HIGH SEA-SALT INPUTS

T.E.H.ALLOTT[1], P.N.E.GOLDING[1] and R.HARRIMAN[2]

[1]Environmental Change Research Centre, University College London, 26 Bedford Way, London WC1H 0AP, UK. [2]Freshwater Fisheries Laboratory, Pitlochry, Perthshire PH16 5LB, UK.

Abstract. Recent critical loads assessments suggest that sensitive surface waters in the north-west of Scotland have acidified, whereas earlier surveys indicate little chemical or biological evidence of acidification. It has been suggested that regionally high sea-salt inputs are affecting either critical loads calculations or the susceptibility of surface waters to acidification. We use palaeolimnological techniques to test the hypothesis that the critical load exceedances in north-west Scotland are real. Pre-industrial and present day loch-water pH are inferred from diatom assemblages in sediment cores from 21 lochs in order to estimate recent pH change. The results indicate consistent post-1800 declines in loch-water pH, although the magnitude of this decline is small (<0.4 pH unit) and in most cases within the error of the technique. It is concluded that although slight acidification might have taken place, this has not been of sufficient magnitude to significantly effect most biological communities (e.g. higher plants, invertebrates and fish).

Key words: Surface waters, north-west Scotland, acidification, critical loads, sea-salts, palaeolimnology, diatoms.

1. Introduction

The north-west of Scotland is a region sensitive to acidification, but remote from emissions sources and therefore experiencing relatively low levels of acid deposition. Previous studies in the area found little evidence of surface water acidification or associated biological problems (e.g. Battarbee 1989, Harriman 1989, Flower *et al.* 1993). However, critical loads maps prepared for the UK indicate significant numbers of exceedance squares in the region, implying that acidification has taken place (Allott *et al.* 1995). Significantly, these exceedances are also apparent in scenarios of future emissions reductions, as a high proportion of deposition in the region cannot be attributed to specific sources. It is therefore important to assess whether the exceedances are real or reflect an artefact of the critical loads models and their assumptions.

Several hypotheses have been put forward to explain the exceedances (Allott *et al.* 1995, Harriman *et al.* 1995). Suggestions that the exceedances are false because of over-estimates in deposition values are unlikely, as recent studies of acid deposition in the region indicate that current input values are under-estimates rather than over-estimates (Prof. David Fowler, *pers. comm.*). Other hypotheses focus on sea-salt inputs. The region is characterised by significant inputs of sea-salts, and the marine contribution to total cation levels is correspondingly high. Harriman *et al.* (1995) reject the hypothesis that the exceedances are false due to errors in sea-salt corrections within critical loads calculations. They suggest that the exceedances are false because catchments with historically high inputs of neutral sea-salt sulphate may have an inherent capacity to assimilate small additional loadings of sulphate without changing the acidification status of surface waters.

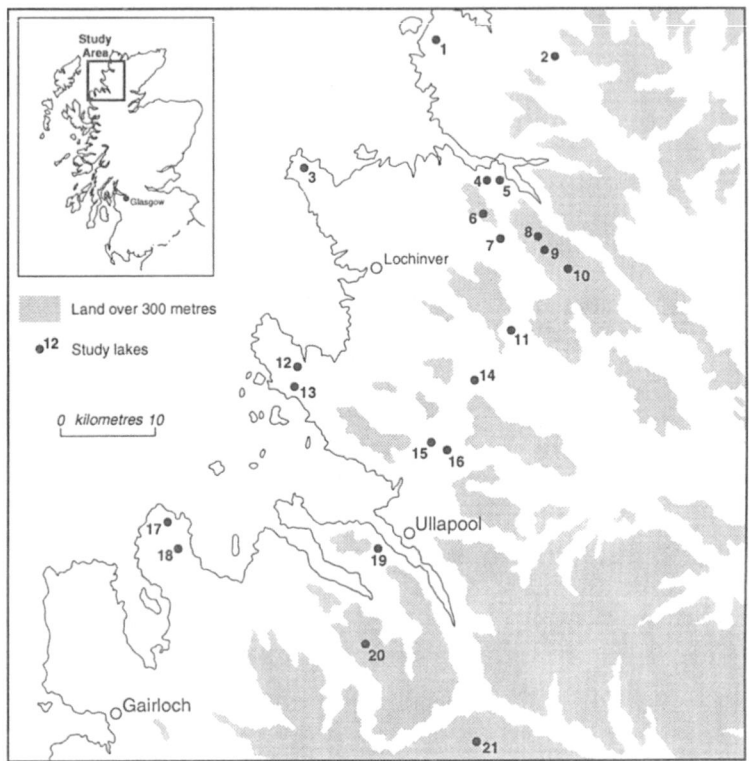

Fig. 1. Map of the study area showing the locations of the study lochs. 1, Un-named loch (UN02); 2, Loch a' Cham Alltain (CHAM); 3, Loch Cul Fraioch (CULF); 4, Loch nan Eun (NEUN); 5, Loch Coire a' Bhaic (BHAI); 6, Loch Bealach Cornaidh (CORN); 7, Lochan Feoir (FEOI); 8, Loch Bealach na h-Uidhe (NAHU); 9, Loch Fleodach Coire (FLEO); 10, Loch nan Cuaran (CUAR); 11, Lochan Fhionnlaidh (FHIO); 12, Loch na Beiste (NABE); 13, Loch na Creige Duibhe (CREI); 14, Loch na Gruagaich (GRUA); 15, Lochanan Dubha (DUBH); 16, Loch na Criche (CRIC); 17, Loch Dubh Camas an Lochain (DCAL); 18, Loch na Beiste (BEIS); 19, Loch nam Badan Boga (BOGA); 20, Loch Toll an Lochain (LOCH); 21, Loch a' Mhadaidh (LAMH).

This paper focuses on the hypothesis that the exceedances are real, and that the acidity of sensitive surface waters in the region has increased as a result of acid deposition. Critical loads and critical load exceedances are calculated for 21 generally low-alkalinity lochs in north-west Scotland, and palaeolimnological techniques are used to evaluate post-1800 changes in loch-water acidity.

2. Methods

WATER CHEMISTRY AND CRITICAL LOADS

Water chemistry was sampled in May 1992 from 90 small (>20 ha) lochs on sensitive geologies in north-west Scotland. From this dataset 21 lochs with low critical loads were selected for more detailed study across a chloride and elevation gradient (Figure 1). Water chemistry was sampled from each loch on three or four occasions during the period May 1992 - July 1993 and analysed at the Freshwater Fisheries Laboratory, Pitlochry using standard methodologies (Harriman et al. 1990). Acid neutralising capacity (ANC) was determined using the method of Cantrell et al. (1990). Critical loads for sulphur were calculated for each sample using both the steady state water chemistry (SSWC) model (Henriksen et al. 1992) and the diatom critical loads (DCL) model (Battarbee et al. in press). Critical load exceedances were calculated using national 20 x 20 km grid square based mean sulphur deposition data for the period 1989-92.

PALAEOLIMNOLOGICAL EVALUATION OF POST-1800 pH CHANGE

Short (c.30 cm) sediment cores were obtained from each loch in July 1993 using a Glew gravity corer (Glew 1989) and extruded in the field. In the laboratory the cores were analysed for spheroidal carbonaceous particles (SCPs) according to the methods of Rose (1990). SCPs are produced by high temperature fossil fuel combustion and the SCP profiles of lake sediments can be used as an indirect dating technique (Renberg & Wik 1985). Rose (1991) demonstrated that SCPs are continuously present in Scottish lake sediments after c.1850 and that there is a marked increase in SCP concentrations at c.1950. These features have been used to provide an approximate chronology for each sediment core. Standard techniques (Battarbee 1986) were used to analyse the diatom assemblages from two levels in each sediment core; the core top (0-0.5 cm interval; present day sediment) and a sample from pre-1800 sediments (typically >20 cm). Diatom-inferred pH (DpH) was calculated for these levels using the weighted averaging regression and calibration techniques described by Birks et al. (1990). Weighted averaging was implemented using WACALIB version 3.3 (Line et al. 1994) and species optima from the SWAP dataset (Stevenson et al. 1991). Previous palaeolimnological studies of acidified lakes in the UK have shown that in pre-industrial times (pre-1800) diatom assemblages and inferences of water chemistry are relatively stable (e.g. Flower et al. 1987). Consequently, comparison of the present day and pre-industrial DpH values was used to provide a measure of recent pH change (ΔpH) (cf. Cumming et al. 1992).

3. Results and Discussion

A summary of the chemical characteristics and critical loads of the study lochs is shown in Table I. The lochs have low alkalinities (<40 µeq l^{-1}) with the exception of UN02, which has a much higher alkalinity and acts as a non-sensitive control site. Chloride values range from <100 µeq l^{-1} to >2000 µeq l^{-1}, reflecting varying sea-salt inputs related to distance from the sea and altitude (Figure 1). The lochs have very low levels of labile monomeric aluminium, with the exception of BHAI where mean concentrations exceed 50 µg l^{-1}. Critical loads are generally low. The DCL model results indicate that 18 of the lochs have a critical load of <0.5 keq S ha^{-1} yr^{-1}, emphasising the sensitivity of these surface waters to acidification. Comparison with deposition data shows that at these 18 lochs the DCL critical load has been exceeded, implying that acidification has taken place.

TABLE I

Summary of chemical characteristics and critical loads for the study lakes. Lakes are arranged in order of current measured lake-water pH. Values for chemical variables represent means of samples collected between 1992 and 1993; Alk = alkalinity, Al-L = labile monomeric aluminium, TOC = total organic carbon, ANC = acid neutralising capacity. Critical load values represent means of critical loads calculated from individual water chemistry samples; SSWC = steady-state water chemistry model, DCL = diatom critical loads model. ΔpH represents the diatom-inferred estimate of post-1800 pH change (see text). Measurements are in μeq l^{-1} except pH, ΔpH, aluminium (μg l^{-1}), TOC (mg l^{-1}), critical loads (keq S ha^{-1} yr^{-1}) and critical loads exceedances (keq S ha^{-1} yr^{-1}).

Code	NGR	Loch Name	pH	Alk	Cl	Al-L	TOC	ANC	Critical Load SSWC	Critical Load DCL	Critical Load Exceedance SSWC	Critical Load Exceedance DCL	ΔpH
BOGA	NH 100930	Loch nam Badan Boga	4.82	-17	503	4	4.1	0	.05	.09	.48	.44	-0.26
DCAL	NG 871972	Loch Dubh Camas an Lochain	4.90	-11	1005	4	6.8	16	.33	.25	.13	.21	-0.35
FHIO	NC 191103	Lochan Fhionnlaidh	4.97	-9	429	6	3.5	5	.17	.15	.28	.30	-0.15
GRUA	NC 243158	Loch na Gruagaich	5.17	-4	282	4	2.9	8	.25	.09	.38	.54	-0.37
NABE	NC 004125	Loch na Beiste	5.22	-3	1312	3	3.5	12	.37	.31	.08	.14	-0.18
CRIC	NC 166037	Loch na Criche	5.27	0	460	2	3.6	15	.23	.23	.22	.22	-0.42
CULF	NC 025330	Loch Cul Fraioch	5.50	5	2094	0	4.2	23	3.85	.61	-	-	-0.41
CREI	NC 005118	Loch na Creige Duibhe	5.55	7	1068	15	3.4	21	.42	.35	.03	.10	-0.32
BHAI	NC 247295	Loch Coire a' Bhaic	5.59	10	282	51	4.0	26	.42	.17	.30	.55	-0.19
CHAM	NC 283446	Loch a' Cham Alltain	5.60	6	530	9	3.0	17	.49	.33	-	-	-0.22
DUBH	NC 147055	Lochanan Dubha	5.63	9	741	2	3.3	23	.44	.36	.01	.09	-0.23
NEUN	NC 232298	Loch nan Eun	5.68	8	383	1	3.1	21	.39	.19	.33	.53	-0.15
LOCH	NH 074833	Loch Toll an Lochain	5.99	16	260	0	1.1	20	.22	.14	.31	.39	-0.12
BEIS	NG 885943	Loch na Beiste	6.05	33	982	2	4.1	43	1.06	.74	-	-	0.00
NAHU	NC 264256	Loch Bealach na h-Uidhe	6.08	17	283	2	1.2	22	.42	.23	.30	.49	+0.04
FLEO	NC 275248	Loch Fleodach Coire	6.09	21	287	0	1.5	27	.61	.31	.11	.41	-0.06
FEOI	NC 229252	Lochan Feoir	6.13	36	498	1	4.6	55	.95	.46	-	-	-0.15
CORN	NC 208282	Loch Bealach Cornaidh	6.14	18	426	3	1.3	24	.55	.29	.17	.43	-0.09
CUAR	NC 292238	Loch nan Cuaran	6.20	19	294	1	.8	22	.48	.29	.24	.43	-0.30
LAMH	NH 199732	Loch a'Mhadaidh	6.22	27	192	1	1.2	25	.95	.36	-	.35	-0.40
UN02	NC 168478	Un-Named	7.00	160	1741	2	4.2	177	2.75	2.37	-	-	-0.08

Diatom-inferred estimates of post-1800 pH change are shown in Figure 2. Although there are clear between-site differences, the majority of lochs show a decline in DpH. The exceptions to this pattern are NAHU, which shows an increase in DpH, and BEIS, whichshows no change in DpH. The general consistency in ΔpH suggests that recent acidification has taken place, and that the critical load exceedances are therefore real. However, there are several reasons to view these results with caution.

Firstly, the magnitude of the decline in DpH is small (typically <0.4 pH units). At only six of the lochs is ΔpH greater than the error of pH prediction (Figure 2). Secondly, there are several mismatches between the ΔpH values and predictions of critical load exceedance (Table I). Critical load exceedances in the DCL model effectively predict response of diatom assemblages to lake-water acidification (Battarbee *et al.* in press), and this should be reflected in declines in DpH. This is not the case at NAHU where the critical load has been exceeded (Table I) but there is a slight increase in DpH. Conversely the lochs where the DCL is not exceeded (Table I) should show no change in DpH. This is true of BEIS, but CULF shows a significant decline in DpH. Such apparent mismatches between critical load and palaeolimnological assessments of acidification status possibly result from the scale problems inherent when using 20 x 20 km scale deposition data to make calculations of critical load exceedance at a catchment scale. Of additional significance to interpretation

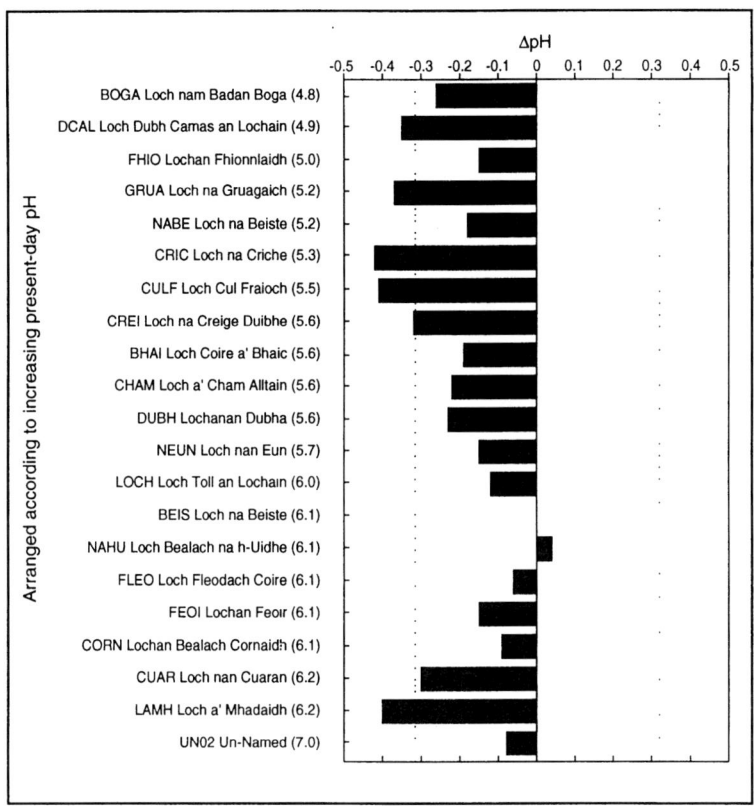

Fig. 2. Diatom-based estimates of historical change (pre-industrial to the present) in pH from the 21 study lakes. The estimates represent the differences in diatom-inferred pH between the top and pre-1800 sediment levels. The lakes are arranged in order of current measured lake-water pH. The dotted lines represent +/- one root mean squared error after bootstrapping ($RMSE_{boot}$) of pH prediction.

of the palaeolimnological data is the slight decline in DpH indicated at UN02, the non-sensitive control site. The data from this loch highlight the dangers in over-interpreting ΔpH values within the error of pH prediction. More detailed studies of post-1800 diatom biostratigraphy are required before the palaeolimnological data can be interpreted with greater confidence, particularly for those lochs where ΔpH is below the error of pH prediction.

The critical load exceedances, even if real, equate to relatively slight changes in loch-water pH (<0.4 pH unit). This contrasts with typical post-1800 acidification of >1 pH unit in sensitive surface waters in more heavily impacted regions of Britain such as Wales, Cumbria, Galloway and the Trossachs (e.g. Battarbee et al. 1988). Mean ANC values at all the study lochs are above zero, and levels of labile aluminium are generally very low (Table I). From these considerations it can be concluded that even if slight surface water acidification has occurred in the study region, it is unlikely to have resulted in significant impacts to most biological communities (e.g. higher plants, invertebrates and fish).

Acknowledgements

We acknowledge the help of the various colleagues at the Freshwater Fisheries Laboratory and Environmental Change Research Centre who helped with field sampling. We are also grateful to Prof. David Fowler and Jane Hall of the Institute of Terrestrial Ecology for providing sulphur deposition data. Tim Aspden drew the figures. This research was funded by the UK Department of the Environment. However the views expressed are entirely those of the authors.

References

Allott, T.E.H., Battarbee, R.W., Curtis, C., Harriman, R., Hall, J., Bull, K & Metcalfe, S.E.: 1995, in *Critical Loads of Acid Deposition for UK Freshwaters*, Critical Loads Advisory Group, Institute of Terrestrial Ecology, Penicuik, 25-33.
Battarbee, R.W.: 1986, in *Handbook of Holocene Palaeoecology and Palaeohydrology*, Berglund, B.E., ed., John Wiley and Sons, 527.
Battarbee, R.W.: 1989, *Geographical Journal* **155**, 353.
Battarbee, R.W., Anderson, N.J., Appleby, P.G., Flower, R.J. et al.: 1988, *Lake Acidification in the United Kingdom 1800-1986*, Ensis Publishing, London, pp 68.
Battarbee, R.W., Allott, T.E.H., Juggins, S., Kreiser, A.M., Curtis, C. & Harriman, R.; in press, *Ambio*.
Birks, H.J.B., Line, J.M., Juggins, S., Stevenson, A.C. & ter Braak, C.J.F.: 1990, *Phil. Trans. Roy. Soc. Lond.* B **327**, 263.
Cantrell, K.J., Serkiz, S.M. & Perdue, E.M.: 1990, *Geochem. Cosmo. Acta* **54**, 1247.
Cumming, B.F., Smol, J.P., Kingston, J.C., Charles, D.F., Birks, H.J.B., Camburn, K.E., Dixit, S.S., Uutala, A.J. & Selle, A.R.: 1992, *Can. J. Fish. Aquat. Sci.* **49**, 128.
Flower, R.J., Battarbee, R.W. & Appleby, P.G.: 1987, *J. Ecol.* **75**, 797.
Flower, R.J., Jones, V.J., Battarbee, R.W., Appleby, P.G., Rippey, B., Rose, N.L. & Stevenson, A.C.: 1993, Research Paper No. 8, Environmental Change Research Centre, University College London, London.
Glew, J.R.: 1989, *J. Paleolimnol.* **2**, 241.
Harriman, R.: 1989, in *Acidification in Scotland, Proceedings of a Symposium Organised by the Scottish Development Department*, Scottish Development Department, Edinburgh, 72.
Harriman, R., Gillespie, E., King, D., Watt, A.W., Christie, A.E.G., Cowan, A.A. & Edwards, T.: 1990, *J. Hydrol.* 116, 267.
Harriman, R. Christie, A.E.G. & Watt, A.W.: 1995, in *Acid Rain and its Impact: the Critical Loads Debate*, Battarbee, R.W., ed., Ensis Publishing, London, 108.
Henriksen, A., Kämäri, J., Posch, M. & Wilander, A.: 1992, *Ambio* **21**, 356.
Line, J.M., ter Braak, C.J.F. & Birks, H.J.B.: 1994, *J. Paleolimnol.* **10**, 520.
Stevenson, A.C., Juggins, S., Birks, H.J.B., Anderson, D.S., Anderson, N.J., Battarbee, R.W., Berge, F., Davis, R.B., Flower, R.J., Haworth, E.Y., Jones, V.J., Kingston, J.C., Kreiser, A.M., Line, J.M., Munro, M.A.R. & Renberg, I.: 1991, *The Surface Waters Acidification Project Palaeolimnology Programme: Modern Diatom/Lake-Water Chemistry Data Set*, Ensis Publishing, London, pp 86.
Renberg, I. & Wik, M.: 1985, *Ecological Bulletins* **37**, 53.
Rose, N.L: 1990, *J. Palaeolimnol.* **3**, 45.
Rose, N.L.: 1991, *Fly-ash particles in lake sediments: extraction, characterisation and distribution*. Unpublished Ph.D. thesis, University of London.

THE USE OF CRITICAL LOAD EXCEEDANCES IN ABATEMENT STRATEGY PLANNING

GUN LÖVBLAD[1], PERINGE GRENNFELT[1], OLLE WESTLING[1], HARALD SVERDRUP[2] and PER WARFVINGE[2]

[1]*Swedish Environmental Research Institute (IVL), P.O. Box 47086, S-402 58 Göteborg, Sweden* [2]*Lund Institute of Technology, Department of Chemical Techn., P.O. Box 124, S-221 00 Lund, Sweden*

Abstract. Critical loads for sulphur and nitrogen are defined to produce effective control strategies over Europe, such as those of the new sulphur protocol. To determine the critical load exceedances on the European scale it is necessary to simplify and generalize. The spatial variation on a scale smaller than the 150 x 150 km EMEP grid squares is considered for critical loads, via a cumulative frequency distribution and the 95 percentile for the grid square is determined. The deposition is assumed to be uniform over the area and the exceedance over the 95 percentile critical load is determined. In reality, the spatial variation is considerable for critical loads as well as for deposition. Calculations based on the frequency of local critical load exceedances have been made for two grid squares in southern Sweden. Local critical loads for acidity are compared to local deposition. Deposition variations due to pollution gradients within the square and to ecosystem structure have been considered. The results are similar for the two squares. The calculations based on local exceedances on 50 x 50 km grid squares and consideration to landuse variability, indicate that in order to protect 95% of the ecosystems in the square, emission reductions 25% greater than the large-scale European approach are needed. The effect of enhanced deposition at forest edges is of relatively small importance for the total exceedance.

Keywords: Critical loads, acidity, sulphur, variability, abatement strategy.

1. Introduction

1.1. USE OF CRITICAL LOADS FOR ABATEMENT STRATEGIES

Critical loads are used as a base for abatement strategies over Europe in the agreements signed as protocols to the Convention of Long-range Transboundary Air Pollution. In the new sulphur protocol, critical loads for acidity are calculated in all parts of Europe and compared to the actual load of acid pollution (Downing et al., 1993). The abatement strategies of the different countries are calculated with the aim of reducing the exceedance of critical loads in critical areas.

However, in the mapping of critical loads and deposition over this large area, simplifications and generalisations are necessary and for practical reasons, the spatial resolution can not be too detailed. Even if it is not possible to go into a much finer scale over the whole area, it is essential to know if calculations of exceedance made on a finer scale will influence the resulting abatement strategies in any way. If necessary, calculations can be made on a finer scale in those areas where a better resolution is of importance.

1.2. FACTORS OF IMPORTANCE FOR THE UNCERTAINTY OF THE EXCEEDANCE OF CRITICAL LOADS

To determine the critical load exceedance, accurate data are necessary for critical loads as well as deposition of pollution. At present there is a lack of data for both of these.

Deposition to the ecosystems is not sufficiently well kown, as some of the deposition processes, dry deposition and fog deposition are not easy to estimate. In principle, generally applicable measurement methods are lacking and model estimates do not agree with measured data. The deposition of some essential compounds are more difficult to estimate than others, for example the deposition of base cations is an important factor. There is, at present, only European data on wet deposition of base cations. More about the gaps of knowledge as regards deposition can be found in the report from the workshop on deposition (Lövblad et al. eds, 1993).

Critical loads are calculated from soil characteristics, landuse data and vegetation parameters. For Sweden, critical loads are calculated based on soil data at approximately 1800 sites. Inaccuracies are due to the uncertainty of the representativeness of these sites for the whole of Sweden and to the fact that data on some essential parameters in the critical loads equation are not available or not available for all types of ecosystems. Nitrogen immobilisation is an example of a parameter for which data are very uncertain. In Sweden most critical load data are available for forested land. In other types of sensitive natural ecosystems, sufficient base data are lacking.

Finally, there are large local variations, temporally and spatially, in deposition as well as in the factors determining the critical loads. The temporal variations are less important since by definition critical loads are used to protect long-term negative effects on the ecosystems. An evaluation of the local variability indicates that the variability in deposition is not less than for critical loads (Lövblad, 1995). The spatial variability has to be considered in some way in the calculations.

2. Methods

2.1. CALCULATION OF LARGE-SCALE CRITICAL LOAD EXCEEDANCE

The calculation of critical load exceedance is described in detail in Downing et al. (1993). The exceedance is calculated for a net with grid cells 150 x 150 km and the data are presented as critical sulphur deposition in Tuovinen et al., (1994). They are recalculated to critical load, using the EMEP data on nitrogen deposition (Tuovinen et al., 1994) and estimated base cation deposition (Lövblad et al. 1992).

The variability of the critical loads is in the large-scale approach considered using a cumulative frequency distribution of critical loads within the grid cell and the critical load value for the grid cell should protect 95% of the ecosystems. The most sensitive ecosystems i.e. the remaining 5%, can be considered as not possible to protect at present.

The variability of deposition is not considered in the present way of calculating exceedance on the Europe scale. In Sweden, since the sulphur emissions are low and the sources are few and far between, the variability between nearby areas is larger due to variability in landuse rather than due to influence of local emission sources. Each grid cell is ascribed the deposition calculated by the EMEP model, as a mean deposition over the square, independent of the variability in landuse.

2.2. CALCULATION OF CRITICAL LOAD EXCEEDANCE WITH A FINER RESOLUTION

An alternative calculation of critical load exceedance has been made for two Swedish EMEP grid squares, 19;21 and 20;21. The squares are chosen since they contain to a large degree Swedish land area and since the dry deposition processes contribute considerably to the total deposition. The calculations have been made for sulphur in order to compare the results from the fine scale method with the European calculation results (see for example figure in Tuovinen et al. 1994). However, deposition data for nitrogen deposition as well as base cation deposition are necessary for this calculation.

Each square was first subdivided into nine 50 times 50 km subgrid squares. Landuse data were collected for each subgrid square. Calculated critical loads data for 105 and 158 forest sites within the two squares were ascribed to the size of area and forest type for which it was assumed to be representative. For other types of natural ecosystems, the frequency distribution was assumed to be equal to that of forested land within the subgrid square. Urban and agricultural areas were defined as insensitive areas. No critical load values were defined, but the exceedance of critical load value was defined to be the lowest within the grid square. This resulted in a compilation of areas having a certain size and a certain critical load.

Deposition of sulphur to forest land was estimated from throughfall measurement results within the grid square and for the type of forest considered (Table 1, 2). Deposition to unforested natural ecosystems was estimated to be equal to measured bulk deposition in open fields in the actual grid square. Deposition data were then estimated, one value per subgrid square and landuse category.

TABLE I Mean deposition of non-marine sulphur (kg $S \cdot ha^{-1} \cdot year^{-1}$) to forests and to open fields (bulk deposition) (Hallgren-Larsson and Westling, 1993 a, b, c). In some regions, wet deposition is measured also at sites where there are no throughfall measurements.

County	Wet (bulk) deposition	No. of sites	Spruce forest	No. of sites	Pine	No. of sites	Broad-leaves	No. of sites
Skåne	5.7	9	15.6	6	-	0	9.1	1
Blekinge	5.4	6	12.7	4	7.0	1	8.7	1
Kronoberg	5.4	9	12.2	6	7.6	2	5.1	1
Halland	6.7	6	14.2	4	8.5	2	-	0
Jönköping	6.0	9	10.6	5	6.1	3	10.3	1
Älvsborg	6.3	12	10.7	8	-	0	-	0
Gbg o Bohus län	6.5	12	12.5	10	10.4	2	-	0

TABLE II Ratios between total and wet (bulk) deposition of sulphur used in the calculations.

EMEP grid square	Area	Ratio between total and wet deposition of sulphur	
		Spruce	Pine and broadleaves
19;20	Kronoberg,	2.2	1.3
	Jönköping	1.9	1.1
20;21	N Halland,	2.1	1.5
	Gbg o Bohus län,	1.8	1.4*
	Älvsborg	1.8	1.4*

* No data, assumed value from earlier measurement results.

Deposition to forest edges were generalized from measurements in southern Sweden as 20% enhanced deposition in relation to forests, although larger values have been observed at wind exposed and agricultural areas. Studies of maps over the area of interest indicated that the part of the forest area which could be regarded as forest edges (0 - 50 m from the edge) were 5 - 15%, somewhat different in different areas (Table 3).

TABLE III Forest edges and slopes as a part of the forest covered area. Results from map studies.

Mapped area		Forested part of the total land area	Forest edge part of forested area (%)		Slopes as part of the forested area (%)	
	km²		Edges towards S	Edges towards N	Slopes towards S	Slopes towards N
Uppsala	275	0.625	8.1	8.3	<0.1	0.19
Åmål SV	288	0.711	7.3	7.5	1.2	1.1
Örebro NV	287	0.639	6.4	6.3	<0.1	<0.1
Norrköping SO	288	0.666	8.0	8.3	0.34	0.26
Göteborg NO	138	0.666	7.6	7.5	1.5	0.63
Kungsbacka SO	550	0.567	4.8	4.8	1.7	1.9
Tingsryd SV	120	0.615	12.2	11.6	<0.1	<0.1
Lessebo SV	576	0.782	8.1	7.7	<0.1	<0.1
Trelleborg NO Malmö SO	581	0.085	13.5	11.1	<0.1	<0.1

The deposition over the EMEP grid square was finally normalized to be equal to that calculated by the EMEP model. In reality, there is no general agreement between model calculated deposition over 150 times 150 km squares. There are large variations between nearby squares as regards landuse. The forest cover in the square is the dominant factor influencing the magnitude of deposition to the square.

Deposition data for nitrogen and base cations were estimated, one value for each grid and ecosystem category (Lövblad et al. 1992) from the national mapping.

The subgrid squares were finally added together to 150 times 150 km squares. Within each large square, the exceedance of critical loads for acidity was calculated for areas with a certain size, critical load and deposition. The exceedance values calculated for different magnitudes of landarea were arranged in increasing order. The 95% percentile of the exceedance values was determined and defined as the need for abatement for that grid square.

3. Results and Discussion

The results from the fine scale calculations of critical loads exceedance and thus, "need for abatement" for the two EMEP squares are presented in Figure 1. The need for reduction in deposition is in this way 45 and 46 $meq \cdot m^{-2} \cdot year^{-1}$ in comparison with the results from large scale measurements 36 for both squares, i e a 25% higher need of abatement in order to protect the ecosystems from acidification than the large-scale calculation over Europe (Figure 2). The calculations indicate further that the enhancement of acid deposition to forest edges is of minor importance for the exceedance of critical loads. The major reason is that there is also an enhancement of base cation deposition. The enhancement of deposition at forest edges also give a small contribution to the total deposition over a 150 x 150 km grid square, less than 5% in the two squares studied.

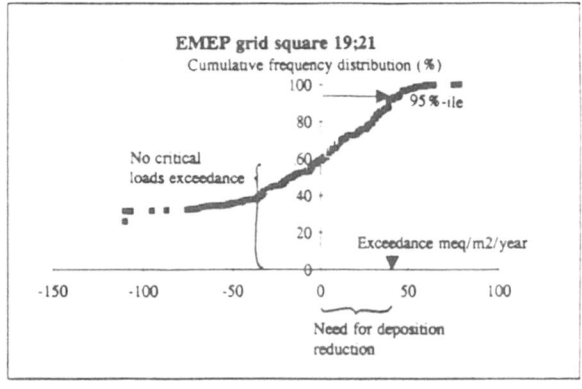

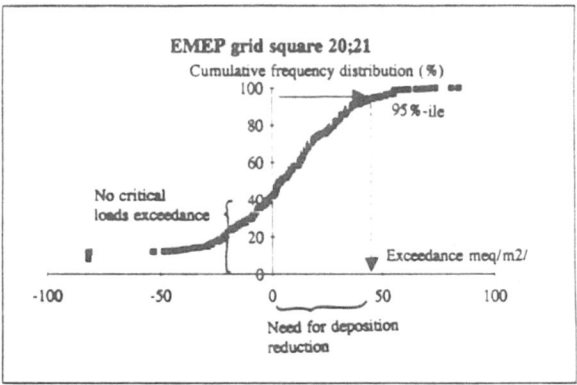

Fig. 1. Cumulative frequency distribution (%) of critical load exceedances (meq·m⁻²·year⁻¹) within EMEP squares 19;21 and 20;21 in relation to % of the total land area.

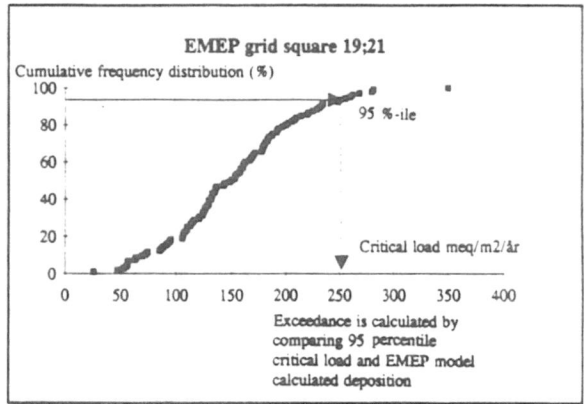

Fig. 2. Cumulative frequency distribution of critical loads in forest soil within EMEP grid square 19;21 (meq·m⁻²·year⁻¹).

The deposition is estimated for a nine year period 1985 - 1993. If only the deposition in 1992 is studied, the exceedance has decreased by 26%. The deposition has been reduced by approximately 40% in the area of interest due mainly to control of air pollution. Further control measures in accordance with the new sulphur protocol will further reduce the exceedance, even if it cannot be expected that all ecosystems will be protected. It is of no importance for the exceedance calculations of 95% percentiles of exceedance whether agricultural land and urban areas are included or not.

The results obtained for the two Swedish EMEP grid square are not generally applicable to all Europe. However, they indicate that the critical load exceedance calculated on 50 x 50 km scale, and with special consideration to variations in land use, enhances the need for deposition reduction in areas with a large forested part of the land area and a relatively high load in relation to critical loads. The effect of variability is expected to be lower in areas where the variation in land use is less pronounced and in areas where the wet deposition dominates. In many areas the local emission variations are quite large and as important for the deposition variation as the landuse. Fine scale calculations are not necessary on the same scale for all parts of Europe. In some areas calculations the large scale calculations may be sufficient. In other areas, calculations on even finer scale 20 x 20, 10 x 10 or even 1 x 1 km should be carried out. It is of interest for the European abatement strategy planning, that the influence of the local variations in both critical loads and deposition is estimated.

References

Downing, R.J., Hettelingh, J-P. & de Smet, P.A.M. (Eds.) (1993) Calculation and Mapping of Critical Loads in Europe: Status Report 1993. RIVM Report, Bilthoven, the Netherlands.

Hallgren-Larsson, E. & Westling, O. (1993a, b, c) Luftföroreningar i Sverige. Nedfall och effekter okt. 1992 till sept. 1993. IVL Reports B-1133, 1139 and 1150. Swedish Environmental Res. Inst., IVL-Aneboda, Sweden. In Swedish, English summary.

Lövblad, G., Pedersen, U., Andersen, B., Hovmand, M., Joffre, S. & Reissell, A. (1992) Mapping deposition of sulphur, nitrogen and base cations in the Nordic countries. IVL Report B-1055, Swedish Environmental Res. Inst., Göteborg, Sweden.

Lövblad, G., Erisman, J.W. & Fowler, D. (Eds.) (1993) Models and Methods for the Quantification of Atmospheric Input to Ecosystems. Nordiske Seminar- og Arbejdsrapporter 1993:573. The Nordic Council of Ministers, Copenhagen.

Lövblad, G., Grennfelt, P., Westling, O., Sverdrup, H. & Warfvinge, P. (1995) Hur påverkas bedömningen av ett områdes kritiska belastning av variabilitet i tid och rum? IVL-Report for the National Swedish Environmental Protection Agency, Swedish Environmental Res. Inst., Göteborg 1995-01-30. In Swedish, English summary.

Tuovinen, J-P., Barrett, K. & Styve, H. (1994) Transboundary Acidifying Pollution in Europe: Calculated fields and budgets 1985-1993. EMEP/MSC-W Report 1/94. The Norwegian Meteorological Inst., Blindern, Norway.

CRITICAL LOADS OF ACIDITY TO SURFACE WATERS -
How important is the F-factor in the SSWC-model?

A. HENRIKSEN

Norwegian Institute for Water Research, P.O.Box 173 Kjelsås, N-0411 Oslo, Norway

Abstract. The critical load of acidity to surface water is based on the condition that the inputs of acids to a catchment do not exceed the weathering rate less a given amount of ANC (Acid Neutralizing Capacity). The Steady State Water Chemistry (SSWC) Method is used to calculate critical loads of acidity, using present water chemistry. To calculate the weathering, the so-called F-factor is used to estimate the part of the base cation flux that is due to soil acidification. The F-factor has been estimated empirically from historical data comparisons from Norway, Sweden, U.S.A. and Canada and is considered to be a function of the base cation concentration by the formula: $F = \sin(BC^*/S)$, where BC^* is the present base cation concentration and S the base cation concentration at which $F = 1$. At higher values for BC^* F is set to 1. For Norway, Sweden and Finland S has been set to 400 µeq/l (ca. 8 mg Ca/l), giving F-values in the range 0.05-0.2. The importance of the F-factor in the calculations of the critical loads of acidity for Nordic surface waters was tested by calculating the magnitude of the area where the critical load of acidity is exceeded in Norway for different values of S. Similar calculations were carried out for the Finnish and Swedish lake data. Varying S from 100 µeq/l to 1200 µeq/l, the exceeded area in Norway decreases from 31,9 to 28,3%. For $F = 0$ ($S = \infty$, i.e. assuming no soil acidification), the exceeded area is reduced to 27,2%. For Finland and Sweden the percent of lakes exceeded are reduced from 16,6 to 12,9% and 30 to 23,6%, respectively. For $F = 0$ the percent of lakes exceeded are reduced to 11,4 and 16,4, repectively. These results indicate that the F-factor is not of great importance for calculating critical load and critical load exceedances in Norway, Finland and Sweden.

Key words: Critical load, acidity, water acidification, soil acidification, empirical models, base cation fluxes

1. Introduction

The critical load of acidity to surface water is based on the condition that the inputs of acids to a catchment do not exceed the weathering rate less a given amount of ANC. The Steady State Water Chemistry Method (SSWC) (Henriksen *et al.* 1992) is used to calculate critical loads of acidity to a lake, using its present water chemistry. To calculate the weathering rate, the so-called F-factor is used to estimate the part of present base cation flux that is due to soil acidification (Henriksen 1984, Brakke 1990). The F-factor expresses the fraction of base cations in runoff that has been ion exchanged in the soil, and thus the effect of soil acidification on the runoff water. The F-factor has been critisized in connection with use for calculations of critical load of acidity for surface waters, and considered to be the most uncertain element in the SSWC-method (i.e. Sullivan *et al.* 1990). We have tested the importance of F in the calculations of the critical loads of acidity to Norwegian, Swedish and Finnish surface waters by varying the variables in its equation.

2. Method

The F-factor is defined as the ratio of change non-marine base cation concentration (denoted by asterisk) due to changes in strong acid anion concentrations, and is calculated as follows (Brakke et al. 1990):

$$F = \Delta[BC^*]/(\Delta[SO_4^*] - \Delta[NO_3$$

$$= ([BC^*]_t - [BC^*]_0)/(([SO_4^*]_t - [SO_4^*]_0) + ([NO_3]_t - [NO_3]_0)) \qquad (1)$$

where the subscripts t and 0 refer to present and background concentrations of non-marine base cations($[BC^*]$), non-marine sulphate ($[SO_4^*]$) and nitrate ($[NO_3]$), respectively (Henriksen et al. 1992).

From equation 1 the non-marine background, or preacidification, base cation concentration can be derived:

$$[BC^*]_0 = [BC^*]_t - F(([SO_4^*]_t - [SO_4^*]_0) + ([NO_3]_t - [NO_3]_0)) \qquad (2)$$

The critical load of acidity to a lake is then calculated by the formula:

$$CL(Ac) = Q \cdot ([BC^*]_0 - [ANC]_{limit}) \qquad (3)$$

where Q is the runoff (m·yr^{-1}), $[BC^*]_0$ the original seasalt corrected base cation concentration and $[ANC]_{limit}$ is the selected critical ANC threshold (Henriksen et al. 1992, Henriksen et al. 1995). The unit is meq/m^2/yr.

If F=1, all incoming H+ is neutralized in the catchment (only soil acidification), at F=0 none of the incoming H+ is neutralized in the catchment (only water acidification). Since extensive water acidification has been documented in many areas of the world F must be <1 in those areas. The F-factor was estimated empirically to be in the range 0.2-0.4 based on historical data comparisons from Norway, Sweden, U.S.A. and Canada (Henriksen 1984). Brakke et. al. (1990) suggested later that the F-factor should be considered to be a function of the base cation concentration by the formula:

$$F = \sin(\Pi/2 \cdot [BC^*]/S) \qquad (4)$$

where $[BC^*]$ is the present base cation concentration (µeq/l) and S the base cation concentration (µeq/l) at which F = 1. For $[BC^*] > S$ the ratio $[BC^*]/S$ is set to 1.

For Norway S has been set to 400 µeq/l (ca. 8 mg Ca/l) (Brakke et al. 1990). For acidified lakes in Norway calculated F-values lie in the range 0.05-0.2.

In equation 4 present base cation concentration is used for practical reasons. Posch et al (1993) suggested a non-linear relationship between F and the original base cation concentration $[BC^*]_0$:

$$F = 1 - \exp(-[BC^*]_0/B) \tag{5}$$

where B is a scaling factor estimated to be 131 µeq/l fra paleolimnological data from Finland (see Posch *et al.* 1993). Inserting this expression into Equation 3 gives a non-linear equation for $[BC^*]_0$ which can be solved by an iterative procedure.

The two expressions for F, equations 4 and 5, in fact give similar results when applied to calculation of critical loads for surface waters in Norway. Figure 2 show the cumulative distribution functions of $[BC^*]_0$ one computed with equation 4 and one computed with equation 5, indicating that either of the equations can be used for calculation of critical loads. For simplicity, all critical load calculation using the Norwegian database has been carried out using equation 4.

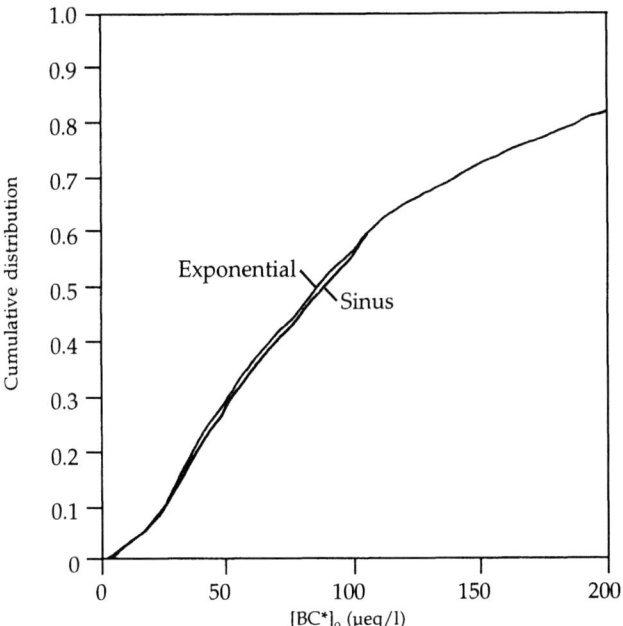

Figure 1. Cumulative distribution functions of $[BC^*]_0$, once computed with equation 4 (lower line) and once computed with equation 5 (upper line) in equation 2 for 2305 lakes in Norway.

3. Results

To test the importance of the F-factor in the calculations of the critical loads of acidity to Nordic surface waters, we have for Norway calculated the magnitude of the area where the critical load of acidity is exceeded for different values of S (Table 1). Similar calculations have been carried out for the Finnish and Swedish lake data bases used for the critical load calculations (Henriksen *et al.* 1990) (Table 1). Varying S from 100 µeq/l to 1200 µeq/l (Table 1), the exceeded area in Norway decreases from 31.9 to 28.3%. For

$F = 0$ ($S = \infty$, i.e. assuming no soil acidification), the exceeded area is reduced to 27.2%. For Finland and Sweden the the percent of lakes exceeded are reduced from 16.6 to 12.9% and 30 to 23.6%, respectively. For $F = 0$ the percent of lakes exceeded are reduced to 11,4 and 16.4, repectively.

TABLE 1.
Percent of area (Norway) and no. of lakes (Finland and Sweden) for which critical loads of acidity is exceeded for some values of S in equation 4. S=400 µeq/l has been used to calculate the critical load databases for Finland, Norway and Sweden.

S-value	Exceeded area Norway		Percent of lakes exceeded	
µeq/l	km^2	Percent	Finland	Sweden
100	102158	31.9	16.6	30.0
200	99715	31.1	16.6	30.0
300	97333	30.4	16.2	29.9
400	*95148*	*29,7*	*15.4*	*29.6*
600	92623	28.9	14.4	28.3
800	91248	28.5	13.7	26.5
1200	90644	28.3	12.9	23.6
∞ (F=0)	87308	27.2	11.4	16.4

These results, considering the large range of S-values tested, indicate that the F-factor is not of great importance for calculating critical load and critical load exceedances in Norway, Finland and Sweden.

As pointed out above, S=400 µeq/l has empirically been chosen for the calculations of critical loads in the three Nordic countries. Data from the Risdalsheia site of the RAIN-project (Reversing Acidification in Norway), where the acid rain was removed, showed that the F-factor for the 8-year period the experiment lasted was 0.18, not very different from the calculated value of 0.10 (Wright *et al.* 1993). The measured value should be expected to exceed the calculated one, because Risdalsheia has not yet reached steady state. At the Sogndal site of the RAIN-project, wher acid rain was added over an 8-year period the experimental F-value was found to be 0.35. Here also, the calculated value was 0.10. The Sogndal site, also, has not reached equilibrium. With further soil acidification, the measured F is expected to decrease (Wright *et al.* 1993). The results from these large scale, long term experiments indicate strongly that the measured and calculated values for F are within an acceptable range, confirming the empirical values used to calculate F according to equations 4 and 5.

References

Brakke, D.F., Henriksen, A. and Norton, S.A.:1990, *Verh. Internat. Verein. Limnol.* **24**, 146-149.
Henriksen, A. Posch, M., Hultberg, H. and Lien, L. 1995. *Water Air Soil pollut* (submitted*)*.
Henriksen, A., Kämäri, J., Posch, M. and Wilander, A. 1992. *Ambio*, **21**, 356-363.

Henriksen, A., Kämäri, J., Posch, M., Lövblad, G., Forsius, M. and Wilander, A. 1990. Critical loads to surface waters in Fennoscandia. Nordic Council of Ministers. Miljørapport 1990:124.
Henriksen, A.: 1984, *Verh. Internat. Verein. Limnol.*, **22**, 692-698.
Henriksen, A: 1984. *Verh. Internat. Verein. Limnol.* **22**, 692-698.
Posch, M., Forsius, M. and Kämäri, J.: 1993, *Water Air Soil pollut.* **66**, 173-192.
Sullivan, T.J. Charles, D.F., Smol, J.P., Cummings, B.F.; Selle, A.R., Thomas, D.R., Bernert, J.A. and Dixit, S.S. 1990. Nature, **345**, 54-58.
Wright, R.F., Lotse, E. and Semb, A. 1993. Can. J. Fish. Aquat. Sci. **50**, 258-268.

PREDICTING FRESHWATER CRITICAL LOADS FROM NATIONAL DATA ON GEOLOGY, SOILS AND LAND USE

J.R. HALL[1], S.M. WRIGHT[1], T.H. SPARKS[1], J. ULLYETT[1], T.E.H. ALLOTT[2] and M. HORNUNG[3]

[1] Institute of Terrestrial Ecology, Monks Wood, Abbots Ripton, Huntingdon, PE17 2LS, UK, [2] Environmental Change Research Centre, University College London, 26 Bedford Way, London, WC1H 0AP, UK, [3] Institute of Terrestrial Ecology, Merlewood, Grange-over-Sands, Cumbria, LA11 6JU, UK.

Abstract. Using information on geology, soils and land use, a map has been generated for Great Britain which indicates five classes of sensitivity of surface waters to acidification. This map has been used for designing sampling strategies for mapping critical loads of acidity for freshwaters. This paper evaluates the freshwater sensitivity map using a data set of water chemistry collected as part of the UK critical loads programme. Discriminant analysis was used to predict five critical load classes from information on geology and soil sensitivity for freshwater sites. This showed geology and soil information can correctly predict approximately 50% of all critical loads classes. In addition, 77% of sites fall within one critical loads class of that predicted. Predictions may be improved by including other variables eg altitude and geographical location. Differences between lake, stream and reservoir sites are also examined. Ranges of critical loads values were determined for each of the five classes of surface water sensitivity. While a trend in critical load values was evident between classes, there was significant overlap. A simplified sensitivity map with only three classes related more closely to critical loads values. The paper demonstrates the usefulness of the surface water sensitivity map for assessing acidification at a national scale, but highlights the difficulties of predicting critical loads for individual sensitive catchments using national data.

Keywords: critical loads, freshwaters, geology, soil, land use, sensitivity, acidification, catchments.

1. Introduction

Soils, geology and land use are all important factors influencing the effects of acid deposition on surface waters (Hornung et al., 1995). Information on these factors may be used to map the sensitivity of surface waters to acidification. The development of one such map is described by Hornung et al. (1995). Two additional versions of this freshwater sensitivity map have been generated (Ullyett et al., 1995), one excluding the effects of land use and the other using a new land use database derived from satellite imagery. The results described here are based on the latter map.

In the UK, water samples have been collected and critical loads calculated from single sites (a lake or headwater stream) in approximately 1500 10km squares (Battarbee et al., 1995). Sites were selected in the most sensitive area, in terms of acidification, of each square. The results from this survey were used in this study. The main aim was to examine the extent to which the critical loads class for individual freshwaters can be correctly predicted from the input variables on geology, soil and land use used for creating the freshwater sensitivity map. A second aim was to explore the possibility of assigning critical loads values to classes of the freshwater sensitivity map.

2. Methods

The map showing freshwater sensitivity to acidification was generated by overlaying maps of geology and soils within a geographic information system (GIS). A map of sensitivity of groundwaters to acidification (Kinniburgh & Edmunds, 1986) provided the basis for the geological input. This map gives four "acid susceptibility" classes based on the mineralogy and geochemistry of the dominant rock types in each map unit of the 1:625000 geology map of Great Britain. For soils, the dominant soil series in each 1km square of Great Britain was assigned to one of three sensitivity classes on the basis of soil mineralogy, texture and base saturation. These sensitivity classes reflect the ability of geology and soils to buffer inputs in the short term (a rate process) as well as in the longer term (the capacity). A land use modifier was applied to the soil sensitivity map to take account of the impact of agricultural liming on soil chemistry. The agricultural areas in which lime is likely to be applied at regular intervals were identified from the ITE Land Cover Map v1.0 (Fuller *et al.*, 1994) which is derived from satellite imagery. For these, the soil sensitivity was re-assigned to the lowest class. The modified soil map was overlaid with the geology map to produce the freshwater sensitivity map used for this study.

Combining the four classes of "acid susceptibility" for geology with the three soil sensitivity classes described above, gives 12 classes of freshwater sensitivity. Hornung *et al.* (1995) aggregated these to five categories of sensitivity (Table I).

Table I

Table showing freshwater sensitivity classes (italics) resulting from combinations of geology and soil classes

Soil sensitivity classes	Geology "acid susceptibility" classes			
	High	Medium	Low	Non-sensitive
High	*high*	*medium-high*	*medium-low*	*low*
Medium	*medium-low*	*low*	*low*	*low*
Low	*non-sensitive*	*non-sensitive*	*non-sensitive*	*non-sensitive*

The freshwater critical loads data are the results of a UK survey of acid-sensitive lakes and headwater streams (Battarbee *et al.*, 1995), in which approximately 1500 freshwaters throughout the UK were sampled. The freshwater sensitivity map described by Hornung *et al.* (1995) incorporating land use effects was originally used as an aid to site selection in some parts of the country, ensuring that waters in the most sensitive area of each 10km square were chosen for sampling. Critical loads for samples were calculated using two models. The Diatom model is an empirical model based on a dose-response relationship between sulphur deposition and changes in diatom composition, taking into account variations in site sensitivity as represented by water calcium values (Battarbee *et al.*, 1995). This model provides the base critical load for a site. The Henriksen steady-state water chemistry model (Henriksen & Brakke, 1988) is used to derive critical loads for individual species or species assemblages; the presence/absence of the species is related to the acid neutralizing capacity values of the waters. Both models give broadly similar results; however, the critical loads calculated using the Diatom model tend to be lower.

Using GIS, the freshwater sensitivity classes, geology and soil classes were identified for each freshwater sampling point. These data together with critical loads values, site altitude and geographical location were imported into MINITAB for statistical analysis.

Discriminant analysis was used to determine the best variables (eg geology, soils, altitude etc) for predicting critical load "classes". Traditionally, 5 ranges of critical load values, termed "classes", have been used for mapping critical loads of acidity in the UK and Europe (eg Battarbee et al, 1995). These classes, namely 0-0.2, 0.2-0.5, 0.5-1.0, 1.0-2.0 and >2.0 keq H^+ ha^{-1} $year^{-1}$, have been used for the discriminant analysis in this study. The analysis was repeated for lakes, streams and reservoirs separately.

To examine the distribution of critical loads across the freshwater sensitivity map classes, the mean values of the critical loads from the Diatom and Henriksen models were determined for each combination of geology and soil classes shown in Table I.

3. Results

The discriminant analysis for the Diatom model critical loads showed that information on geology and soils alone can predict approximately 50% of all critical loads classes correctly for all site types (Table II). The results were generally the same for individual site types, with streams giving the poorest predictions (46%) probably due to the use of a single water sample and fluctuations in stream chemistry and critical loads. For each type, including information on site altitude and geographical location had little effect on the ability to predict all critical loads classes (less than +5%). However, for all site types, 77.4% fell within one critical loads class of the predicted class. Broadly similar results were seen for individual site types.

Prediction of individual critical load classes for all sites, using geology and soil information alone, varies from one class to another (Table II). Predictions are the most accurate for the less sensitive freshwaters ie those in critical loads class 1, in which 52.4% of streams are correctly predicted, increasing to 77.8% for lakes and reservoirs combined. Inclusion of altitude and location slightly improves the results for streams (56.5%) in this class. A relatively high success rate was also achieved for lakes (73.5%) and streams (60.7%) in critical loads class 3. The predictive power was poorest for the more sensitive waters with lower critical loads ie classes 4 and 5. For both of these classes, inclusion of the additional site information increased the percentage of all sites correctly assigned to their critical loads class eg +7% for class 5. Ordination of sites using principal components analysis (PCA) suggests that there is a potential to discriminate between the critical loads classes although a wide overlap between them was evident.

The mean critical loads values for the 12 combinations of soil and geology classes shown in Table I were summarised to ranges of mean critical loads for the five classes of the freshwater sensitivity map (Table III). Although a clear sequence of critical load ranges across the sensitivity classes was evident, there is substantial overlap between these ranges.

TABLE II

Discriminant analysis based on geology and soils and Diatom model freshwater critical loads for a) all sites, b) lakes, c) streams, d) reservoirs, e) lakes + reservoirs, indicating the number and proportion of sites (percentages in parentheses) predicted in each critical loads class

		Critical loads class*					
		1	2	3	4	5	Total
a) all sites	actual	538 (71.2)	22 (13.3)	133 (69.4)	0	21 (25.3)	714 (50.7)
	+/- one class	597 (79.0)	140 (84.3)	147 (77.0)	185 (87.3)	21 (25.3)	1090 (77.4)
	total number	756	166	191	212	83	1408
b) lakes	actual	261 (76.1)	7 (6.8)	108 (73.5)	0	15 (25.4)	391 (47.3)
	+/- one class	272 (79.3)	87 (84.5)	113 (76.9)	162 (93.1)	15 (25.4)	649 (78.6)
	total number	343	103	147	174	59	826
c) streams	actual	108 (52.4)	8 (22.2)	17 (60.7)	0	4 (57.1)	137 (46.0)
	+/- one class	145 (70.4)	28 (77.8)	21 (75.0)	12 (57.1)	4 (57.1)	210 (70.5)
	total number	206	36	28	21	7	298
d) reservoirs	actual	139 (67.1)	13 (48.1)	0	7 (41.2)	10 (58.8)	169 (59.5)
	+/- one class	178 (86.0)	17 (63.0)	5 (31.5)	11 (64.7)	12 (70.6)	223 (78.5)
	total number	207	27	16	17	17	284
e) lakes + reservoirs	actual	428 (77.8)	14 (10.8)	116 (71.2)	0	17 (22.4)	575 (51.8)
	+/- one class	450 (81.8)	112 (86.2)	126 (77.3)	173 (90.6)	17 (22.4)	878 (79.1)
	total number	550	130	163	191	76	1110

* critical load classes:
ranges of critical load values (keq H^+ ha^{-1} $year^{-1}$) in classes: 1: > 2.0, 2: 1.0 - 2.0, 3: 0.5 - 1.0, 4: 0.2 - 0.5, 5: <=0.2

Simplifying the freshwater sensitivity map into three classes by combining the medium-high and medium-low classes and the low and non-sensitive classes (Table III), results in a better separation of critical load values for both Diatom and Henriksen models but, reduces the resolution of the map.

TABLE III

Mean critical load values for each soil-geology combination summarized for the five freshwater sensitivity classes given in Table I and for a simplified three-class system derived from the five sensitivity classes

Freshwater Sensitivity class	Mean critical load values (keq H^+ ha^{-1} $year^{-1}$)	
	Diatom model	Henriksen model
a) using 5 classes derived from the 12 combinations of soil and geology as in Table I		
High	1.361	1.613
Medium-high	3.685	2.913
Medium-low (2 values)	2.888 - 4.366	2.803 - 3.162
Low (4 values)	8.679 - 45.940	3.547 - 9.997
Non-sensitive (4 values)	9.012 - 42.395	6.201 - 10.105
b) using 3 classes		
High	1.361	1.613
Medium* (3 values)	2.888 - 4.366	2.803 - 3.162
Low** (8 values)	8.679 - 45.940	3.547 - 10.105

* combining medium-high and medium-low classes in a)
** combining low and non-sensitive classes in a)

4. Conclusions

The paper illustrates the potential of using national data sets, together with simple site information to predict freshwater critical loads, and the usefulness of the sensitivity map in making a general assessment of freshwater acidification at a national scale.

The results show that national data used for defining the sensitivity of freshwaters to acidification may be effective in predicting a critical load class in about 50% of cases. However, 77.4% of all sites fell within ± one critical loads class of that predicted. The failure to predict a higher proportion of critical loads classes more accurately, highlights the difficulties in attempting to relate national data to individual sensitive catchments. Although additional data on site altitude and geographical location make little difference overall, they enhance the percentage of individual critical load classes that can be correctly predicted, in particular those classes with the lowest critical loads. The ability to predict classes with higher critical loads in about 70% of cases highlights the usefulness of the sensitivity map in identifying non-sensitive areas. Critical loads for freshwaters reflect the weathering rate of the geology and soils of the surrounding catchment. Prediction of these critical loads could be enhanced by including information on soil depth and soil moisture, but these are not available at a national scale.

This study compares point data from national maps with critical loads that are catchment related. Since many of these catchments are small areas, the geology and soil sensitivity may not be expected to differ across the catchment from that at the sampling point. However, as local variations in soil and/or geology can change freshwater chemistry from that predicted by the small scale maps used in this study, a more accurate prediction of critical loads for individual waterbodies may be obtained by using higher resolution, catchment-specific data. The freshwater sensitivity map used in this study was not intended to be site specific; it was designed to identify the main areas of the country with a high probability of waters susceptible to acidification.

Relating the mean freshwater critical loads to the five classes of the freshwater sensitivity map resulted in substantial overlap between the ranges for each class. This can be improved by reducing the number of classes on the map from five to three, although this also reduces the detail of the map. In this way it may be possible to assign broad ranges of critical loads values to classes of the freshwater sensitivity map. However, it is only the most sensitive class which typifies freshwaters most likely to become acidified, where the low critical loads for acidity may be exceeded by sulphur deposition.

Acknowledgements

The authors acknowlege the Department of the Environment for funding this work (PECD7/10/90). However, the views expressed are entirely those of the authors.

References

Battarbee, R.W., Allott, T.E.H., Bull, K.R., Christie, A.E.G., Curtis, C., Flower, R.J., Hall, J.R., Harriman, R., Jenkins, A., Juggins, S., Kreiser, A., Metcalfe, S., Ormerod, S.J. & Patrick, S.T. 1995, Critical Loads of Acid Deposition for UK Freshwaters, Department of the Environment. London, UK, 139 pp.
Henriksen, A. & Brakke, D.F. 1988, Water, Air & Soil Pollution, **42**, 183.
Fuller, R.M., Groom, G.B. & Jones, A.R. 1994, Photogrammetric Engineering & Remote Sensing, **60**, 553-562.
Hornung, M., Bull, K.R., Cresser, M., Ullyett, J., Hall, J.R., Langan, S. & Loveland, P.J. 1995, Environmental Pollution, **87**, 207-214.
Kinniburgh, D.G. & Edmunds, W.M. 1986, Hydrogeological Report 86/3, British Geological Survey, Wallingford, UK.
Ullyett, J., Hall, J.R & Bull, K.R. 1995,In: Acid Rain and its Impact: the Critical Loads Debate. University College London,UK, 103-106.

SULPHUR DEPOSITION AND CHANGES IN SWEDISH LAKE CHEMISTRY 1988-1993

L. RAPP and K. BISHOP

Department of Forest Ecology, Swedish University of Agricultural Sciences
S-901 83 Umeå, Sweden

Abstract. Acidification of surface waters in northern Europe due to anthropogenic sulphur (S) deposition has led to new emission restrictions based on Critical Loads (CL). There is likely to be considerable interest in documenting the effect resulting S deposition changes have on surface water quality. This paper will focus on how the chemistry of 134 reference lakes in Sweden has changed between 1988 and 1993 in response to a decline in S deposition. Only 10% of the reference lakes had significant declines in sulphate during the 5 year study period. A similar number of lakes had an increase in the acid neutralizing capacity (ANC), but few of those with an increase in ANC were also lakes with significant sulphate decreases. Since there is good evidence that S deposition decreases will eventually result in ANC increases, a five year period is probably too short for evaluating the S protocol in terms of changes in lake chemistry. It takes a number of years to equilibrate to new deposition levels, and weather patterns may also obscure longer term trends.

Key words: sulphur deposition, lake chemistry, response, Sweden, critical loads

1. Introduction

Anthropogenic sulphur deposition has already acidified many surface waters in Northern Europe and continued deposition threatens to aggravate the situation. The concept of Critical Loads (CL) has been widely accepted for bringing acid deposition to levels that do not damage natural ecosystems. The second international protocol on sulphur (S), signed in Norway during 1994, represents a major commitment to reducing S deposition to levels approaching the CL for S deposition in Europe.

A time scale for recovery from acidification is not specified in current CL methodology. Instead a steady state concept was used in which the final result of a certain S deposition level on the acid neutralizing capacity (ANC) of surface waters is considered (Henriksen et al. ;1992). In the steady state approach, the long term average sources of acidity and alkalinity in the system are assessed to determine the maximum acid input that the ecosystem can tolerate without damaging the most sensitive components of the aquatic and terrestial ecosystem.

Despite the long-term steady state basis of the S protocol, there is likely to be considerable interest in documenting what effects deposition changes brought about by the S protocol will have on surface water quality. In fact the S protocol includes a provision for review after 5 years.

Between 1985 and 1989 the S deposition to Sweden was relatively stable (Lövblad, 1990). Then a large decline in S deposition was observed between 1989 and 1992. Trend studies from five Swedish EMEP (European Monitoring Evaluation Program) stations (Kindbom et al.; 1994) show downward trends for sulphur deposition after 1989 at four of the stations (the fifth station lacks data after 1989). The decline is about 25-45% where the higher decline is observed in northern Sweden. A large decline (30%) in atmospheric deposition is also observed in the lake Gårdsjön catchment in southern Sweden between 1989 and 1992 (Hultberg et al).

Although it is not clear to what extent the decline in S deposition to Sweden can be related to emission reductions and to what extent it was a feature of weather patterns, it does provide an opportunity to examine the short-term effects which deposition reductions have had on lake chemistry. This paper will focus on how the chemistry of a set of reference lakes in Sweden has changed between 1988 and 1993. The change in two parameters, the sulphate concentration and ANC, will be investigated. Sulphate is the anion in acid deposition which has done most to acidify European surface waters, although there is now concern that increasing nitrogen emissions will contribute more to acidification in the future. ANC is the criteria of surface water quality which CL are based upon.

2. Materials and Methods

2.1 STUDY LAKES

Water chemistry data for 134 lakes were taken from the Swedish system of reference lakes. These are used for following the development of acidification in lakes which have not been subjected to liming. All reference lakes that had measurements every year between 1988 and 1993 were included in the study. The October-March measurements were used in calculating average annual values for lake chemistry, as this is a period when lake chemistry is more stable. The lakes were sampled on average 20 times during the study period.

2.2 SULPHUR DEPOSITION

The atmospheric deposition of sulphur was estimated using a data set from the National Critical Load Mapping for Sweden (Lövblad et al.1992). This data set represents the deposition during 1985-1989 when no obvious trend in measured wet deposition was observed so that the total deposition was relatively stable (Lövblad, 1990). The data includes S deposition to open field, S_{op}, spruce forest, S_{sp}, and the combination of pine and deciduous forest, S_{pd}.

In order to take into account the different types of vegetation in the catchment surrounding a reference lake, a data base from the National Forest Inventory was used.

The resolution of this data set is 25 x 25 km, and within each cell there is an estimate of the fractions of land area covered by spruce (F_{sp}), pine and deciduous forest (F_{pd}), and other land categories. It is assumed that Sop is connected to everything but spruce, pine and deciduous forests (F_{op}). From this the total land-use weighed S deposition, S_{tot}, (1988-1989) for the individual lakes could be calculated.

$$S_{tot} = S_{sp} \cdot F_{sp} + S_{pd} \cdot F_{pd} + S_{op} \cdot F_{op} \quad meq/m2/year$$

2.3 STATISTICAL APPROACH

In the statistical analyses the concentration of non marine sulphate, SO_4^*, and ANC, defined as $BC^* - SO_4^* - NO_3$, were investigated. BC^* refers to non marine base cations, $Na+Mg+Ca+K$. In order to evaluate whether changes had occurred during the period from 1988 to 1993, linear regression was used. For each of the 134 lakes a yearly average from all measurements between October and March was used in the statistical analyses. The criteria for significance was $p < 0.05$.

3. Results and Discussion

A similar number of increases and decreases in lake SO_4^* was observed (Table 1, Figure 1). Significant decreases in lake SO_4^* were observed in 14 lakes mostly located in northern Sweden. On the other hand, there were 17 lakes which had a significant increase in lake SO_4^* in southern Sweden. Decreases in SO_4^* occurred, in general, in areas exposed to low S load.

Significant increases in ANC were observed in 13 lakes and no decreases were observed. One third of the lakes with significant S decreases had also significant ANC increases, so it seems to be a mix of recovery.

TABLE 1
Significant Changes in SO4* and ANC of 134 reference lakes between 1988 and 1993

	Change	Number of Lakes	S Deposition meq/m²/year	
			Average	Standard Deviation
SO_4^*	Increase	17	70	15
	Decrease	14	45	20
ANC	Increase	13	55	20
	Decrease	0	-	-

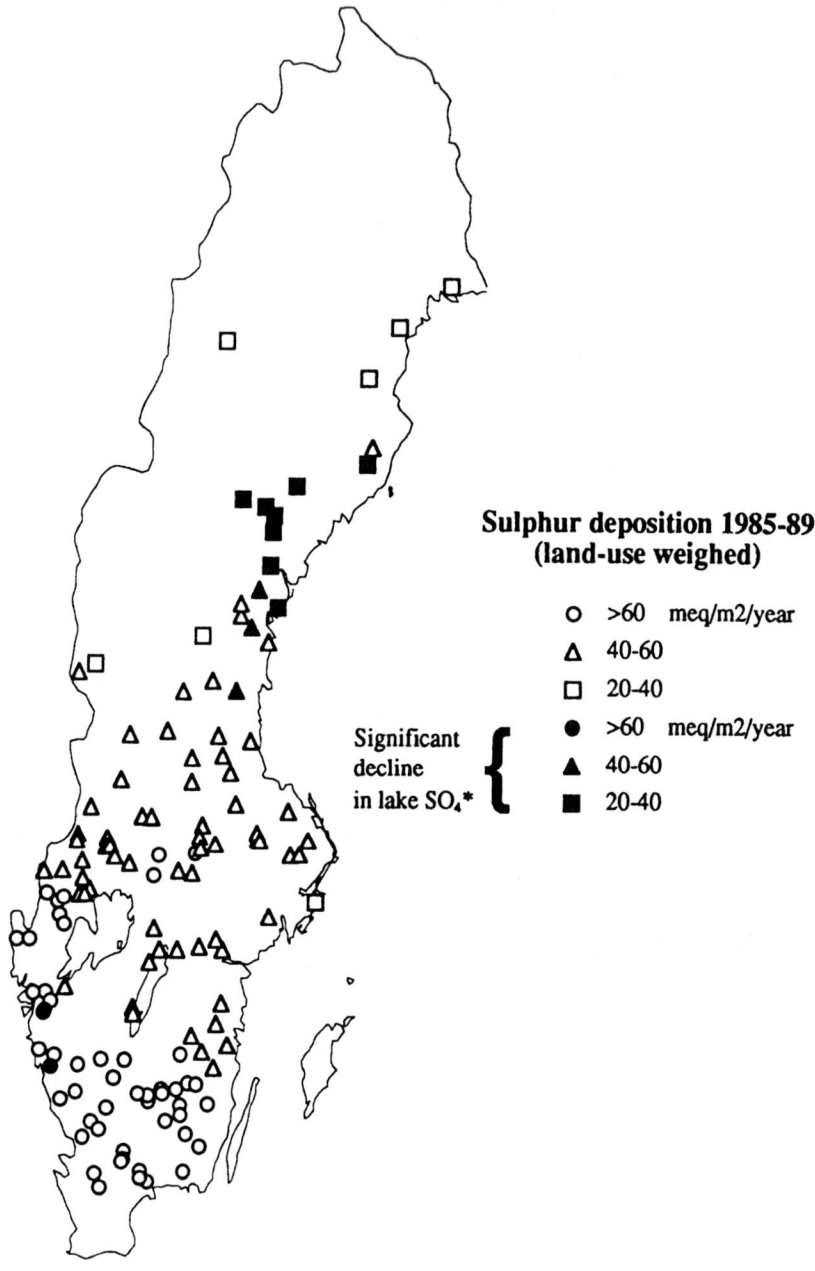

Fig. 1. Study lakes with S deposition indicated by symbol shape. For lakes where the decline in lake SO_4^* was significant the symbols are filled.

Therefore, even if trend studies indicate a decrease in S deposition for the whole of Sweden no major changes in sulphate and ANC have taken place. In fact, decreases as well as increases in lake SO_4^* have occurred. There is, however, considerable evidence that declines in S deposition will lead to declines in runoff S (Wright and Hauhs, 1991).

In catchment manipulations in southern Norway and Sweden where air pollution was almost completely removed, the full response to S deposition decreases takes a number of years, during which time some S release from the catchment occurs (Bishop and Hultberg, 1995). However, differences in the residence time of lakes will also affect the rate of response as will weather pattern periods. Therefore it will be difficult to evaluate the effects of deposition declines on lake chemistry in a short term perspective, i e, during the space of 5 years.

4. Conclusions

This study revealed that in spite of the fact that there has been a large decrease in S deposition for the whole of Sweden between 1988 and 1992, no major changes in sulphate and ANC can be observed in lakes. This is consistent with catchments manipulations which reveal that recovery takes several years as some of the S stored in the catchment is released. Therefore, as current CL calculations deal with long-term steady states, a five year period is too short for evaluating the S protocol in terms of changes in lake chemistry since it takes a number of years to equilibrate to a new deposition level.

Acknowledgements

This study was supported by the Swedish Environmental Protection Agency. The authors would like to acknowledge their colleagues at the department of Forest Ecology for good advice regarding the statistical approach.

References

Bishop, K. and Hultberg, H. Reversing Acidification in a Forest Ecosystem: The Gårdsjön Covered Catchment, March 1995, *Ambio* Vol. 24 No2,
Henriksen, A., Kämäri, J., Posch, M. and Wilander, A. Critical Loads of Acidity: Nordic Surface Waters, August 1992, *Ambio* Vol. 21 No 5, 356-363
Hultberg, H., et al., to be published
Kindbom, K., Lövblad, G. and Sjöberg, K. 1994, Sulphur and Nitrogen Compounds in Air and Precipitation in Sweden 1980 - 1992, IVL Report B1144, Swedish Environmental Research Institute, Göteborg Sweden
Lövblad, G et.al, MAPPING DEPOSITION OF SULPHUR,NITROGEN AND BASE CATIONS IN THE NORDIC COUNTRIES, Swedish Environmental Research Institute, Report B 1055, 1992

Lövblad, G., 1990, LUFTFÖRORENINGSHALTER OCH DEPOSITION I BAKGRUNDSLUFT, Swedish Environmental Protection Agency, Report 3812 (in Swedish), 1990:14

Wright, R.F and Hauhs, M. 1991, Reversibility of acidification: soils and surface waters. *Proc. Roy. Soc.* Edinbugh 97B, 169-191

THE RELATIONSHIP BETWEEN SALMONID FISH DENSITIES AND CRITICAL ANC AT EXCEEDED AND NON-EXCEEDED STREAM SITES IN SCOTLAND

R. HARRIMAN[1], E.E. BRIDCUT[1], AND H. ANDERSON[2]

[1]*Freshwater Fisheries Laboratory, Faskally, Pitlochry, Perthshire, Scotland PH16 5LB.*
[2]*Macaulay Land Use Research Institute, Craigiebuckler, Aberdeen, Scotland AB9 2QJ*

Abstract. The critical load concept is now accepted throughout Europe as a means of estimating the sensitivity of key components of aquatic and terrestrial ecosystems to atmospheric inputs of sulphur (S) and nitrogen (N). Current UK freshwater maps, based on steady-state water chemistry, are derived using a critical acid neutralising capacity (ANC_{LIM}) value of zero μeql^{-1}, which is based on the probability of occurrence of salmonid fish in lakes. In practice most acidification damage to salmonid fish occurs in nursery streams at the emergence and first feeding stages. In general a clear relationship exists between salmon (*Salmo salar* L.) and trout (*S. trutta* L.) densities in Scottish streams and ANC values. However, differences between sites depend on which ANC value is used (eg maximum, minimum or mean). By contrast, when the exceedance of critical loads is compared with salmonid densities the relationship is less clear because many exceeded sites have good salmonid densities. Many of these latter sites are found in north-west Scotland where sea-salt inputs are high and ANC is usually greater than zero μeql^{-1}, although diatom-based studies indicated slight acidification of these waters, with a point of change in diatom flora close to ANC = 20 μeql^{-1}. These false exceedances are probably due to preferential adsorption of acidic SO_4 deposition which results in an overestimate of exceedance values. All sites with a mean ANC ≤ 0 are fishless but some sites with negative minimum ANC values had normal salmonid densities. Consequently a mean ANC_{LIM} value of zero in the critical load equations for UK freshwaters appears to be too low to protect salmonid stocks. Values between 20-50 μeql^{-1} represent a more realistic range if prevention of long term damage to salmonid stocks is to be achieved.

Key words: Acid neutralising capacity, Scottish streams, critical loads, salmonids, exceedances.

1. Introduction

With the general acceptance by scientists and governments of the cause-effect link between acidic deposition and freshwater acidification at the end of the last decade, recent research efforts have been directed towards cost-effective emission reduction strategies which provide the maximum environmental benefit. The output from these studies, which is now accepted by European governments, is the "Critical Load" concept which has been used to produce European sensitivity maps for key ecological targets (ie vegetation, soils and fresh waters) (Nilsson and Grennfelt, 1988). The methodology for freshwater critical load calculations was developed by Henriksen (1988) using empirical relationships to derive pre-acidification weathering rates. Fish are chosen as a general biological indicator for fresh waters and an ANC (acidic neutralising capacity) value is selected to represent the threshold below which fishery problems would be expected. In Norway an ANC limit of 20 μeql^{-1} was selected, based on a 50% probability of adequate fish stock in Norwegian lakes. In Sweden an ANC value of 50 μeql^{-1} was used but, because most of the salmonid stocks in the UK reside in upland streams and rivers, an ANC value of zero μeql^{-1} was considered more appropriate. In this study we test the efficacy of using an ANC_{LIM} of zero μeql^{-1} in a range of Scottish streams and attempt to establish which ANC value (eg maximum, minimum, mean) should be used in the critical load equation.

2. Materials and Methods

The study was carried out in four major catchments along the sulphur deposition gradient in Scotland: Halladale; Spey; Loch Maree and Water of Fleet (Figure 1). Within each catchment

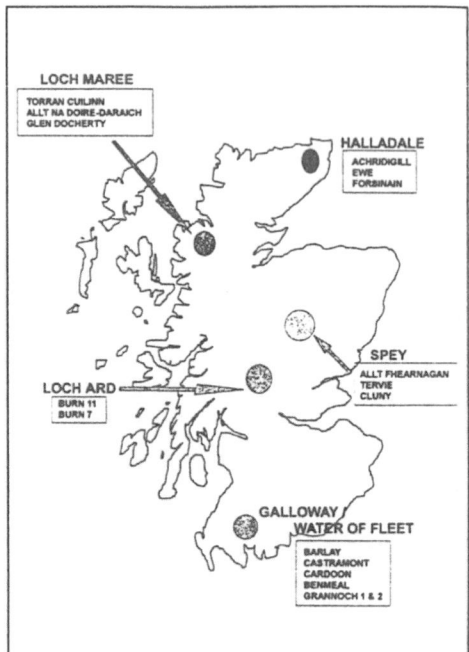

Figure 1. Study areas and sampling sites in Scotland

three streams of differing sensitivity (based on calcium levels) and similar levels of sulphur depositions were selected. At each site, variation of habitat parameters known to be influential in salmonid distribution (eg water velocity, water depth, substrate and overhanging vegetation) was kept to a minimum. An additional five sites (Benmeal (BM), Grannoch 1 (GR1) and Grannoch 2 (GR2)) in Galloway, south west Scotland and (Burn 7 (B7) and Burn 11 (B11)) in the Loch Ard catchment, central Scotland (Figure 1) were selected. These sites, which had mean ANC values less than zero μeq^{-1} and were devoid of salmonids, were chosen for comparative purposes only. The Halladale, Water of Fleet and Spey catchments were sampled during summer and autumn 1991 and spring, summer and autumn 1992. The Loch Maree catchment was sampled only in autumn 1991 and spring and summer 1992, due to weather conditions (with the exception of the Torran cuilinn (TC) stream which was sampled in the autumn of 1992).

On each sampling occasion stop nets were used to enclose 100 m^2 of the selected stream. Within this area fish were captured using 400 V pulsed DC electrofishing apparatus. Three electrofishing runs were carried out in the selected area, fish captured from each fishing were kept separate. The Zippin (1956) method of analysis was used to determine fish population densities for all salmon (*Salmo salar* L.) and trout (*S. trutta* L.) age classes. The biomass of each salmonid species age class was calculated as a product of the mean weight and fish density (m^{-2}) (Shackley and Donaghy, 1992). The proportions of 0+, 1+ and 2+ trout and salmon captured at each site and season were compared using a two-way ANOVA. Season and site were used as the variables. Significance was based at the 0.05 level. Average densities of 0+, 1+ and 2+ trout and salmon were based on two winter and autumn values and one spring value.

TABLE I

Biomass (g/m) of brown trout, *Salmo trutta* and salmon, *S. salar* at the 12 study sites during 1991 and 1992

Site	Summer 1991		Autumn 1991		Spring 1992		Summer 1992		Autumn 1992	
	Trout	Salmon	Trout	Salmon	Trout	Salmon	Trout	Salmon	Trout	Salmon
Water of Fleet										
Barlay	8.29	1.6	3.32	1.07	0.59	2.25	8.7	3.06	2.88	1.81
Castramont	3.63	0.36	1.43	0.23	1.63	0.69	4.58	0.37	1.43	0.27
Cardoon	2.42	0	2.3	0	1.78	0	1.5	0	2.35	0
Spey										
Allt Fhearnagan	1.85	4.04	1.28	1.56	0.11	0.91	1.4	3.26	0.34	1.16
Tervie	5.52	0.65	10.66	0.74	2.24	0.07	9.79	1.62	9.24	1.13
Cluny	2.61	1.35	3.31	1.05	1.67	0.31	2.84	1.98	1.78	1.97
Halladale										
Achridigill	2.84	0.21	2.29	0.55	2.01	0.25	3	1.15	1.67	1.56
Ewe	1.39	1.41	2.56	1.32	1.9	0.94	0.78	1.68	1.85	2.92
Forsinain	1.96	5.58	1.18	2.89	1.29	3.54	0.86	12.99	0.46	6.03
Loch Maree										
Torran Cuilinn			0.29	1.47	0.1	0.9	0.22	0.68	0.32	1.58
Allt na Doire-Daraich			0.33	1.46	0.48	0.87	0.11	0.93		
Glen Docherty			0.54	0.55	0.53	0.1	0.52	0.5		

Dip water samples were collected at each site on each fish sampling occasion and during a wide range of flow conditions throughout the study period. Standard analytical methods were used (Patrick *et al.*, 1991). ANC was calculated according to:

$$\text{ANC } (\mu eql^{-1}) = Alk_2 + (TOC \times 4.5)$$

where Alk_2 is the equivalence alkalinity and TOC is the total organic carbon content of the stream sample (Cantrell *et al.*, 1990). The critical load value ($Keqha^{-1}yr^{-1}$) was calculated using the steady state water chemistry method described by Henriksen *et al.* (1992) and Harriman and Christie (1993). Exceedance of critical load was calculated by subtracting the annual deposition of sulphur from the critical load.

3. Results and Discussion

Mean density of both 0+ and 1+ trout exhibit a significant, positive relationship with minimum ANC (P<0.04). Mean 0+ trout density also show a significant, positive regression with mean ANC and exceedance of critical load at corresponding minimum ANC values (P<0.0008). For maximum ANC the only significant relationship was with 2+ salmon (P=0.02).

At a number of sites it was apparent that a large salmon biomass coincided with a low trout biomass (eg the FO and FH streams) (Table I). Conversely at the TE site the high trout biomass coincided with a low salmon biomass. Salmon are generally more abundant than trout in shallow, faster flowing waters (Jones, 1975; Kennedy and Strange, 1982). Therefore, even though every effort was made to select similar physical and habitat characteristics at each site, differences in the proportions of salmon and trout biomass may have been influenced by flow and depth variations. This may have affected the relationship between mean ANC and total salmonid biomass (c.f. BA<FO).

The lower salmon biomass at AC, CA and GD is unlikely to be a result of salmon inaccessibility due to physical barriers as salmon were recorded in all sites except the CD stream. Furthermore there is no evidence of high trout densities causing a density dependence effect on salmon densities in these streams as no significant linear relationships were found between salmon and trout densities (particularly GD where trout biomass was similarly low) (Bridcut and Harriman, 1994). Streams in the Loch Maree area are known to have problems with low trout egg deposition due to a significant decline in sea trout stocks throughout the western highlands (Walker, 1993).

Overall trout distribution and age class structure at the different study sites appear to correlate better with the mean or minimum ANC and exceedance of the critical load (associated with the minimum and maximum ANC) values than the salmon in the same study streams. However anomalies in the overall relationship between mean ANC, exceedance of critical load and salmonid densities were found at the Halladale sites and FH, where a negative minimum exceedance value was obtained (ie exceeded), while a positive ANC and good salmonid densities were found. The TC and DD streams similarly had occasional negative minimum exceedance values which did not reflect the relatively high salmon biomass at these Loch Maree streams. Also large ANC variability in the Halladale catchments was not evident in the Loch Maree catchments. This therefore suggests that the anomaly is not related to ANC but to the exceedance values. The Loch Maree and Halladale catchments, receive relatively low non-marine sulphur inputs and high sea-salt sulphur depositions and the false exceedance values may have arisen because: (a) of an error in the sea-salt correction factor to determine the non-marine base cation concentrations in the critical load equation or; (b) these catchments do not exhibit a steady-state water chemistry. The latter is more likely as the first hypotheses has been tested and rejected (Harriman *et al.*, 1995). The steady-state critical load estimates assume that sulphur inputs are in balance by their outputs. Studies in the Halladale catchments have shown that acid inputs of sulphur are preferentially adsorbed by catchment soils producing much lower SO_4 outputs than inputs. Details of the experimental studies of the relative adsorption of neutral marine SO_4 compared to acidic SO_4, and their effects on SO_4 run-off, are described in detail by Anderson *et al.*, 1996 and Harriman *et al.*, 1996). It appears that false exceedances are mainly caused by the large imbalance between sulphate inputs and sulphate run-off which affects the exceedance value rather than the critical load value. It should be emphasised that the false exceedances are based on an ANC_{LIM} of zero. Allot *et al.* (1996) showed slight acidification at similar sites in north-west Scotland but diatom changes were occurring at an ANC level around 20 µeql^{-1}.

To extend the relationship between total salmonid density and mean or minimum ANC_{LIM} we included five additional sites. These sites were fishless with mean ANC<0. It is apparent from Figure 2 that some sites with a negative minimum ANC value support salmonid populations eg, CD and DD. However, no salmonids are present in streams when the mean ANC value is less than or equal to zero µeql^{-1}. The mean ANC value as opposed to the minimum ANC value produces a more accurate and realistic biological indicator of salmonid survival as minimum ANC gives no indication of the frequency of ANC values less than zero µeql^{-1}. The data indicate that a mean ANC_{LIM} value of zero correspnds to a complete loss of salmonid fish. If ANC_{LIM} is used in the context of the true definition of critical loads ie a "no damage" scenario, then an ANC_{LIM} value of 20 µeql^{-1} would seem to be more appropriate for low TOC streams and closer to 50 µeql^{-1} for high TOC streams.

4. Conclusions

Although proportions of salmon and trout varied from stream to stream there was a positive relationship between ANC and total salmonid densities. In Scottish streams, a mean ANC of zero µeql^{-1} appears to be an inappropriate ANC_{LIM} value in the critical load equation. To protect salmonids from damage a value between 20-50 µeql^{-1} should be used, depending on TOC levels. At sites with high background sea-salt levels salmonid densities and ANC values do not correlate with calculated exceedance values due to significant retention of "acidic" sulphate in these catchments.

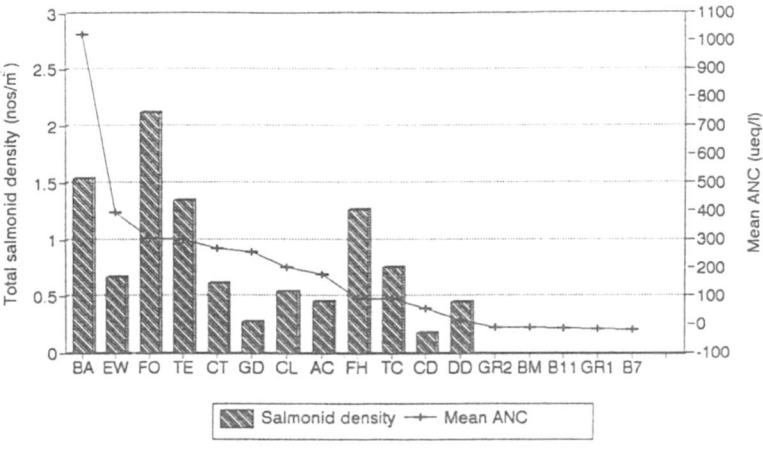

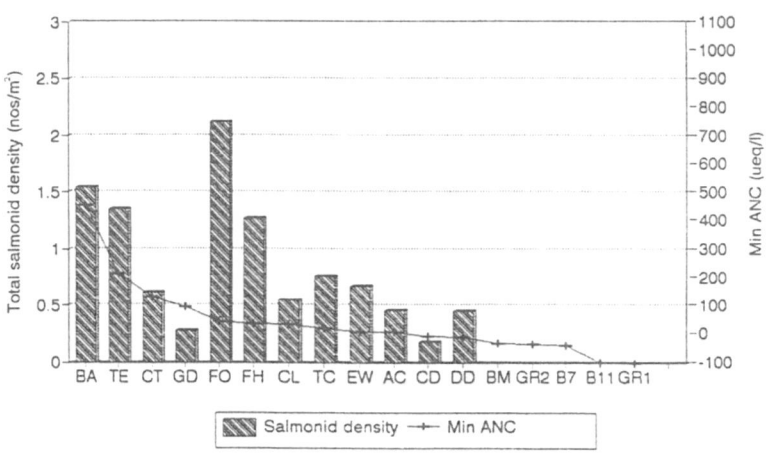

Figure 2. Relationship between total salmonid density and a) Mean ANC, b) Minimum ANC at study sites, BA - Barlay; CT- Castramont; CD - Cardoon; FH - Fhearnagan; TE - Tervie; CL - Cluny; AC - Achridigill; EW - Ewe; FO - Forsinain; TC - Torran Cuilinn; DD - Allt na Doire-daraich; GD - Glen Docherty; BM - Benmeal; GR1 - Grannoch 1; GR2 - Grannoch 2; B7 - Burn 7; B11 - Burn 11.

Acknowledgments

Sincere thanks are expressed to staff at the Freshwater Fisheries Laboratory, Pitlochry and collaborators at selected field sites for assistance during this project.

References

Allott, T.E.H., Golding, P.N.E. and Harriman, R.: 1996, this volume.
Anderson, H.A., Peacock, S., Berg, A. and Ferrier, R.C.: 1996, this volume.
Bridcut, E.E. and Harriman, R.: 1994, *The relationship between critical loads of fresh water and juvenile salmonid communities in Scotland*. Report to the Department of the Environment. ENSIS London.
Cantrell, K.J., Serkiz, S.M. and Perdue, E.M.: 1990, *Geochim. Cosmochim. Acta* 54, 1247-1254.
Haines, T.A.: 1981, *Trans. Am. Fish. Soc.* 110, 669-707.
Hansen, L.P., Naesje, T.F. and Naedhuis, I.: 1986, *ICES North Atlantic Salmon Working Group Report.*
Harriman, R. and Christie, A.E.G.:1993, *Critical loads: concept and applications.* 103-108. Institute of Terrestrial Ecology.
Harriman, R., Christie, A.E.G and Watt, A.W.: 1995, *Acid Rain and its Impacts: The critical loads debate.* 108-114. ENSIS London.
Harriman, R., Anderson, H. and Miller J.D.: 1996, this volume.
Henriksen, A.: 1988, *Critical loads for nitrogen and sulphur*. Copenhagen: Nordic Council of Ministers.
Henriksen, A., Kamari, J., Posch, M and Wilander, A.: 1992, *Ambio* 21, 356-363.
Jones, A.N.: 1975, *J. Fish Biol.* 7, 95-104.
Kennedy, G.J.A. and Strange, C.D.: 1982, *J. Fish Biol.* 20, 579-591.
Morris, R. and Reader, J.P.: 1990, *The Surface Water Acidification Programme. Freshwater Ecosystems. Effects of Acid Rain* 357-368.
Nilsson, J. and Grennfelt, P. (eds): 1988, *Critical loads for sulphur and nitrogen* Copenhagen: Nordic Council of Ministers.
Patrick, S., Waters, D., Juggins, S. and Jenkins, A.:1991, *The United Kingdom acid waters monitoring network: site descriptions and methodology report.* Report to the Department of the Environment and Department of the Environment (Northern Ireland). ENSIS London.
Shackley, P.D. and Donaghy, M.J.: 1992, *Scottish Fisheries Research Report* 51.
Walker, A.F.: 1993, In: *Prolbems with sea trout and salmonin the western highlands.* Atlantic Salmon Turst, Pitlochry, Perthshire.
Zippin, C.: 1956, Biometrics 12, 163-189.

MACROINVERTEBRATE STATUS IN RELATION TO CRITICAL LOADS FOR FRESHWATERS: A CASE STUDY FROM N.E. SCOTLAND.

D. TURNBULL[1], C. SOULSBY[1], S. LANGAN[2], R. OWEN[3] AND D. HIRST[4].

[1] *Department of Geography, University of Aberdeen, Aberdeen, Scotland AB9 2UF.* [2] *Macaulay Land Use Research Institute, Cragiebuckler, Aberdeen. Scotland* [3] *North East River Purification Board, Aberdeen. Scotland* [4] *Scottish Agricultural Statistical Service, Rowett Research Institute, Aberdeen, Scotland.*

Abstract. The Critical Load concept provides a method for the assessment of an ecosystem's sensitivity to acidification. This paper examines how variations in critical loads for freshwaters are reflected by the diversity and abundance of macroinvertebrates. The results indicate that acidified sites, those with the lowest critical loads, have significantly fewer species than less sensitive sites. The data are discussed in terms of ordination analysis relating catchment attributes to critical loads and macroinvertebrate status. It is concluded that although critical loads provides a good predictor for biotic status it is not as sensitive as parameters such as pH or alkalinity.
Key words: macroinvertebrates, critical loads, acid exceedence, acidification index, ordination, Scotland.

1. Introduction

Large areas of the Scottish uplands are characterised by base-deficient bedrock, such as granite, and thin acidic soils. These regions are particularly susceptible to surface water acidification. Critical load can be defined as "the highest deposition loading which will not cause chemical changes leading to long-term harmful effects on the most sensitive ecological systems" (Nilsson and Grennfelt, 1988). It has therefore been suggested as a useful tool for classifying a catchment's susceptibility to acidification.

Macroinvertebrate status should partly reflect the physico-chemical conditions of a site. It has been demonstrated that acid waters generally result in low macroinvertebrate species richness (for example Rosemond *et al.*, 1992), hence they have been proposed as indicators (Wade *et al.*, 1989). This paper examines the relationship between stream acidity and macroinvertebrate composition. It also investigates the value of using freshwater critical loads for predicting biotic status.

2. Study area and methods

The study sites (Figure 1) are all headwater subcatchments of the River Dee in N.E. Scotland whose areas range from between 5 to 61 km^2. The streams drain granite dominated catchments including the Cairngorm and Lochnagar massifs, though there are also some small calcareous intrusions. There are four main soil types: alpine soils, gleys, hill peats and podzols. The land-use of the area is primarily open heather moorland. A relatively diverse habitat for vegetation is provided by patches of montane blanket bogs, flushed mire, wet heath and remnant scrub.

In 1983 the North East River Purification Board (NERPB) sampled 82 headwater catchments in the Grampian and Cairngorm Mountains (Pugh, 1985). Of these 20 were selected on the basis of their alkalinity levels to represent a range of acid sensitivities; the 10 which are in the Dee catchment are the focus of this study.

Streamwater chemistry was sampled for 11 years on a monthly basis. Water samples were analysed for 20 determinands including pH, alkalinity, anions, cations and metals according to the methods specified by Doughty (1989). Critical loads were calculated for each sample using the Henriksen equation (Henriksen and Brakke, 1988). Discharge was obtained from correlations between gauging board stage and discharge at nearby flow stations. Other catchment characteristics were obtained from maps (Table I).

Benthic macroinvertebrate 'kick' samples were taken during spring (April), summer (July) and autumn (September) between 1984 and 1994, for methodology see Davidson et al. (1985). The total number of species recorded at each site during this period is given in Table I. As a result of monitoring by the Norwegian Invertebrate Acidification Monitoring Program Raddum, Fjellheim and Hesthagen (1988) derived an acidification index based on acid-sensitive species. Many of the species are common between Norway and highland areas of Scotland so it was possible to apply the index to the Dee sites.

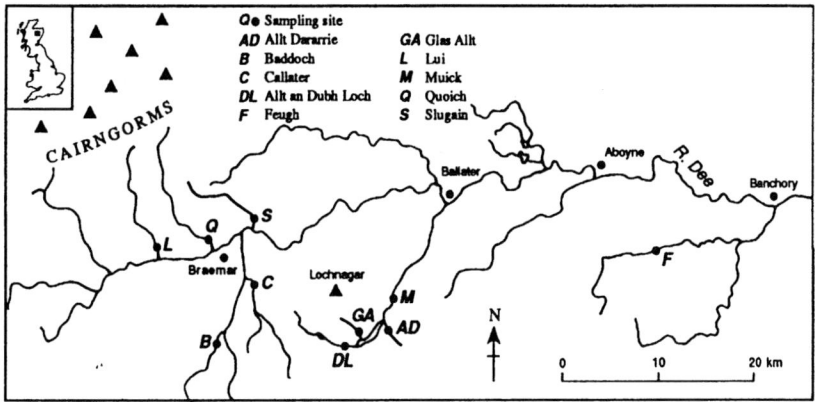

Fig. 1. Dee catchment and study sites.

TABLE I
Catchment characteristics of River Dee sites

Site	NERPB code	National Grid Reference	Area (km^2)	Highest altitude (m)	Mean water temperature (°C)	Acid deposition loading (keq H$^+$ ha^{-1} yr^{-1})	Mean daily discharge (m^3 s^{-1})	Mean daily unit discharge (l s^{-1} km^{-2})	Total no. of species
Allt an Dubh Loch	104	NO267820	14	1110	5.92	2.20	0.65	48.07	51
Glas Allt	92	NO274825	5	1150	5.28	2.20	0.24	48.07	47
River Muick	89	NO299841	38	1150	7.02	2.20	1.74	45.79	65
Quoich	79	NO117911	60	1196	7.09	2.20	2.67	44.55	77
Lui	101	NO069898	61	1240	6.92	2.20	2.55	41.74	64
Feugh	97	NO524903	21	723	7.41	1.20	0.53	25.30	90
Slugain	103	NO159932	15	860	7.43	2.20	-	-	82
Allt Darrarie	91	NO309852	13	832	5.91	2.20	0.29	21.95	81
Callater	102	NO155883	34	980	7.13	2.20	1.10	32.37	69
Baddoch	80	NO135832	23	975	6.99	2.20	0.79	34.24	71

Acid deposition, from precipitation data (1986-88) for 20 km grid squares (UKRGAR, 1990). - missing data.

The index ranges from 0 and 1; if only tolerant species are present a score of 0 results and the site is considered highly acidified, whilst if a score of 1 is obtained the site is less acidified due to the presence of acid sensitive species (Fjellheim and Raddum, 1990).

3. Results and Discussion

3.1. CRITICAL LOADS FOR FRESHWATERS

On the basis of critical load the 10 Dee sites could be classified into three groups (Table II): (1) the Allt an Dubh Loch, Glas Allt and River Muick sites had mean critical loads of less than 0.6 keq H^+ ha^{-1} yr^{-1} which were all exceeded by mean deposition levels, hence these sites can be considered as acidified; (2) The Quoich, Lui, Feugh, and Slugain had critical loads of between 0.9 and 1.7 keq H^+ ha^{-1} yr^{-1} and may be considered acid sensitive and thirdly (3) the Allt Darrarie, Callater and Baddoch sites had mean critical loads of between 2.4 and 4.0 keq H^+ ha^{-1} yr^{-1} which were not exceeded by acid deposition levels and are thus well-buffered catchments. Summary statistics for the 11 years sampling programme are given in Table II where it can be seen that the three low critical load sites also had the lowest concentrations of calcium (Ca) and alkalinity and pH level.

TABLE II
Summary physico-chemical statistics

Water parameter	Acidified sites			Acid sensitive sites				Well-buffered sites		
	Allt an Dubh Loch	Glas Allt	River Muick	Quoich	Lui	Feugh	Slugain	Allt Darrarie	Callater	Baddoch
CL mean	0.47	0.51	0.57	0.92	0.98	1.30	1.63	2.40	2.48	3.93
max	2.13	2.67	1.84	1.95	3.20	5.42	8.00	5.21	6.26	9.69
min	0.15	0.13	0.28	0.27	0.38	0.42	0.35	0.32	0.83	0.85
CL* mean	1.73	1.69	1.63	1.28	1.22	-0.10	0.57	-0.20	-0.28	-1.73
max	0.07	-0.47	0 36	0 25	-1.00	-4.22	-5.80	-3.01	-4.06	-7.49
min	2.05	2.07	1.92	1.93	1.82	0.78	1.85	1.88	1.37	1.35
Ca mean	37.50	34.16	57.08	72.90	84.66	121.25	122.61	158.05	217.64	303.23
max	115.77	192.61	207.58	146.21	329.34	280.94	318.86	387.22	563.87	833.33
min	14.47	4.79	8.98	35.93	35.43	17.96	36.43	21.96	80.84	89.32
Alkalinity mean	0.38	0.52	0.96	2.67	3.21	5.44	6.64	12.57	10.85	17.28
max	5.50	13.30	6.40	6 40	13.00	21.58	18.50	33.83	26.00	33.72
min	0.00	0.00	0.00	0.00	0.00	0.00	0.00	0.00	2.50	0.00
pH mean	5.33	5.60	5.91	6.45	6.51	6.60	6.76	6.88	7.00	7.15
max	6.27	7.10	6.61	7.01	7.14	7 40	7 46	7.74	7.68	7.82
min	3.80	4.20	4.00	5.00	5.29	4.80	4.27	4.92	6.10	4 30
Q/A mean	48.07	48.07	45.79	44.55	41.74	25.30	-	21.95	32.37	34.24
max	492.72	492.72	692.03	673.35	629.61	469.78	-	379.23	434.91	500.50
min	5.99	5.99	3.67	3.04	2.86	3.17	-	1.50	2.55	2.44

CL = Critical Load, CL* = CL exceedence, Q = discharge, A = catchment area, - no data. Units are: critical loads (keq H^+ ha^{-1} yr^{-1}), Ca (μeq l^{-1}), alkalinity (mg l^{-1}), pH (pH units) and discharge per unit area (Q/A) (l s^{-1} km^{-2}).

3.2. CRITICAL LOADS AND TAXON RICHNESS

A total of 138 taxa were found and the mean number of taxa (taxon richness) varied between 11 and 27 taxa per stream. As in other studies (e.g. Sutcliffe and Hildrew, 1989) those streams with low mean pH and Ca concentrations were the most impoverished in taxa. Figure 2 illustrates that those sites whose mean critical loads were exceeded by acid deposition levels generally had low taxon richness. Given their relatively low mean critical loads the Feugh and Slugain have high taxon richness, however these two sites have relatively low altitudes and high water temperatures (Table I), hence high productivity and diversity was expected.

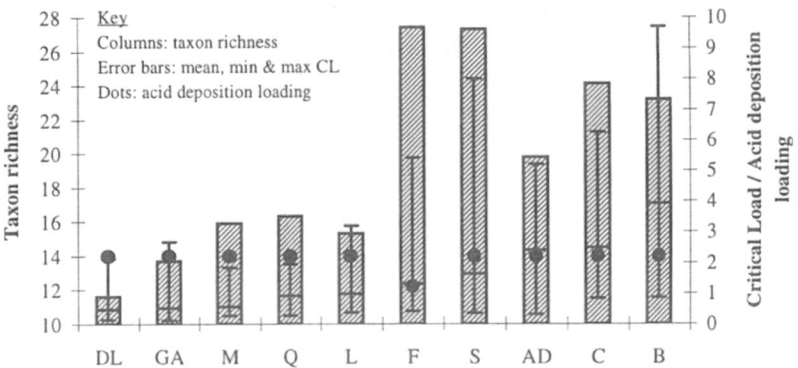

Fig. 2 Taxon richness in relation to Critical Load and acid deposition (units keq H^+ ha^{-1} yr^{-1}, sites as in Fig. 1)

Taxon richness is a measure for assessing macroinvertebrate status in relation to acidity. However, a more sensitive approach would be to consider individual species responses. Applying the invertebrate acidification index to the Dee data resulted in those sites with low critical loads and low taxon richness also having low mean acidification index scores (Figure 3). Though useful for distinguishing sites of different acidity this index was insensitive at detecting temporal trends for macroinvertebrate recovery as site acidity fell, Ephemeroptera abundance proved a better indicator for this (Soulsby et al., 1995).

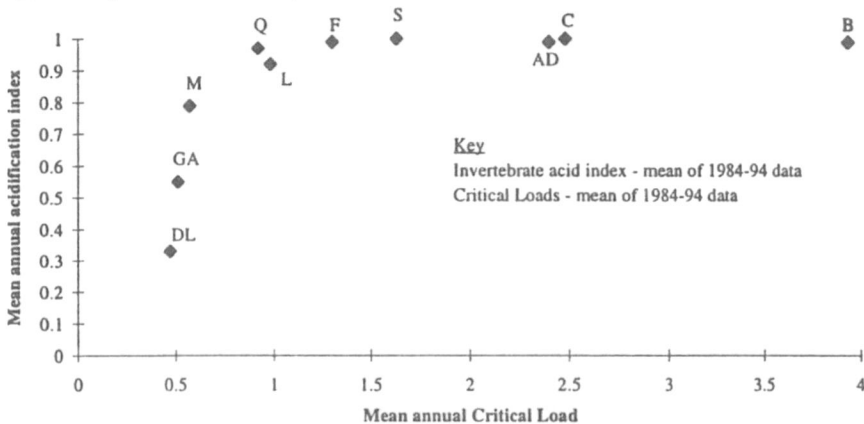

Fig. 3. Invertebrate acid index in relation to Critical Load (units keq H^+ ha^{-1} yr^{-1}, sites as in Fig. 1 key).

3.3. CRITICAL LOADS AND COMMUNITY COMPOSITION

A de-trended correspondence analysis (DECORANA) of all the invertebrate data resulted in a strong seasonal clustering which masked any between site differences. To remove the seasonality the data were reanalysed according to seasons. The 138 taxa belonged to four taxonomic groups, the Plecoptera (31 species), the Ephemeroptera (19 species), the Trichoptera (38 taxa) and a group (50 taxa) which included Diptera, Coleoptera, Mollusca and Annelida. The abundances and taxon richness of these were correlated with seasonal averages of the physico-chemical data (Table III). The Ephemeroptera taxa richness and abundance most strongly correlated with those parameters relating to the acidity of the sites, particularly pH and alkalinity. Though also significant critical load and Ca were less strong predictors. Natural factors such as the river's regime will also influence the biotic communities. The negative relationships between Q/A and taxon richness and abundance suggests the importance of river 'spatiness'.

TABLE III
Product-moment correlations between macroinvertebrates and environmental variables
(spring, summer and autumn 1984-94)

Variable	Taxon richness				Abundance	
	Plecoptera	Ephemeroptera	Trichoptera	Total	Ephemeroptera	Total
pH	0.22, 0.08, 0.41	0.82, 0.81, 0.81	0.22, 0.40, 0.38	0.50, 0.64, 0.61	0.60, 0.55, 0.48	0.50, 0.64, 0.61
Alkalinity	0.19, 0.02, 0.28	0.61, 0.63, 0.67	0.09, 0.24, 0.29	0.32, 0.46, 0.49	0.62, 0.43, 0.44	0.27, 0.29, 0.31
Critical Load	0.15, -0.06, 0.32	0.53, 0.65, 0.67	0.05, 0.28, 0.67	0.25, 0.44, 0.52	0.55, 0.37, 0.39	0.21, 0.22, 0.28
Ca	0.15, -0.10, 0.30	0.59, 0.64, 0.67	0.15, 0.29, 0.36	0 34, 0.43, 0.52	0.58, 0.36, 0.34	0.23, 0.22, 0.25
Q/A	-0.17, -0.05, -0.29	-0.52, -0.17, -0.43	-0.20, -0.17, -0.23	-0.31, -0.25, -0.22	-0.44, -0.21, -0.33	-0.26, -0.15, -0.27
$r > 0.20$	$p < 0.05$,		$r > 0.25$	$p < 0.01$,	$r > 0.32$	$p < 0.001$.

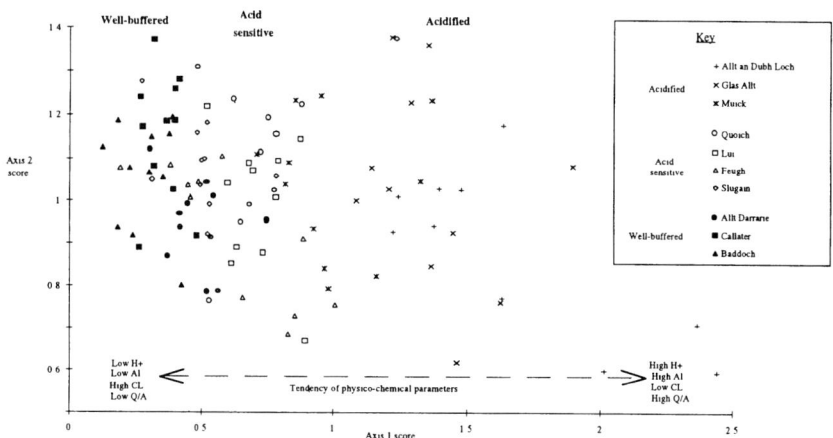

Fig. 4. Canonical ordination of Ephemeroptera and Plecoptera from spring samples.

Figure 4 illustrates a canonical ordination analysis (CANOCO) (Ter Braak, 1992) of log transformed spring Ephemeroptera and Plecoptera data in which hydrogen ion concentration (H^+), Critical Load, aluminium (Al) and Q/A were the best parameters in explaining the variation between site biology. These variables were all strongly correlated ($p<0.001$) with the first ordination axis (critical load, r=-0.66; Q/A, r=0.66; H^+, r=0.91 and Al, r=0.78). Only Al was significantly correlated with axis 2 (r=-0.79). Though there is considerable overlap between samples three clusters can be identified which broadly correspond to the critical load classes introduced in 3.1. The 'acidified' sites group according to high values on the H^+ gradient and low critical loads; samples from the three 'well-buffered' streams cluster according to low H^+ and high critical loads and the 'acid sensitive' sites form an intermediate group. Similar ordinations were also found for the summer and autumn samples. The Trichoptera and the 'other invertebrate' group clusters were significantly correlated with physico-chemical indicators of acidity, however these relationships were weaker than for the Ephemeroptera and Plecoptera.

4. Conclusions

Macroinvertebrate status was strongly related to site acidity, particularly to pH and alkalinity. Though not studied in this paper factors such as biological interactions, disturbances from intermittent floods and habitat differences are also important in determining community structure. The relationships between critical load and biotic status may provide a useful tool for relating acid exceedence limits to 'ecosystem health'. However, the high variability in the critical loads estimates suggests the importance of hydrological conditions at the time of sampling and needs further investigation. On the basis of both the chemical and biological methods used it was possible to classify the study sites into three categories based on their 'degree of acidification'. Future work will test this classification system on the neighbouring River Spey catchment.

Acknowledgements

Funding from the Leverhulme Trust is gratefully acknowledged. Thanks is also given to the NERPB staff involved in this study, particularly Mike Davidson and Derek Fraser.

References

Davidson, M.B., Owen, R.P. and Young, M.R.: 1985, In: *The Biology and Management of the River Dee* (Ed. D. Jenkins), NERC, Monks Wood, 64-82.
Doughty, C. .: 1989, *Baseline Study of Acidified Waters in Scotland*, Report to the Department of the Environment - Research Contract No. PECD7/10/104, 72 pages.
Fjellheim, A. and Raddum, G.G.: 1990, *Sci. Total Environ.* **96**, 57-66.
Henriksen, A. and Brakke, D.F.: 1988, *Env. Sci. and Technol.* **22**, 8-14.
Nilsson, J. and Grennfelt, P. (Eds.): 1988, *Critical Loads for Sulphur and Nitrogen*. Nordic Council of Ministers, Copenhagen.
Pugh, K.B.: 1985, In: *Biology and Management of the River Dee* (Ed. D. Jenkins), NERC, Monks Wood, 34-41.
Raddum, G. G., Fjellheim, A. and Hesthagen, T.: 1988, *Verh. Internat. Verein. Limnol.* **23**, 2291-2297.
Rosemond, A.D., Reice, S.R., Elwood, W. and Mulholland, P.J.: 1992, *Freshwater Biology* **27**, 193-209.
Soulsby, C., Turnbull, D., Langan, S.J., Owen, R. and Hirst, D.: 1995, *This volume*.
Sutcliffe, D. W. and Hildrew, A. G.: 1989, In: *Acid Toxicity and Aquatic Animals* (Eds. Morris, R., Taylor, E.W., Brown, D.J.A. and Brown, J.A.), Cambridge University Press, Cambridge.
Ter Braak, C. J. F.: 1992, *CANOCO - a FORTRAN program for Canonical Community Ordination*. Microcomputer Power, Ithaca, NY.
UK Review Group on Acid Rain (UKRGAR).: 1990, *Acid deposition in the United Kingdom 1986-1988 (Third report)*, HMSO, London.
Wade, K. R., Ormerod, S. J. and Gee, A. S.: 1989, *Hydrobiologia* **171**, 59-78.

VALIDATION OF THE UK CRITICAL LOADS FOR FRESHWATERS: SITE SELECTION AND SENSITIVITY

C.J. CURTIS[1], T.E.H. ALLOTT[1], R.W. BATTARBEE[1] and R. HARRIMAN[2]

[1] *Environmental Change Research Centre, University College London, 26 Bedford Way, London WC1H 0AP, UK*
[2] *S.O.A.F.D. Freshwater Fisheries Laboratory, Faskally, Pitlochry, Perthshire PH16 5LB, UK*

Abstract. Critical loads maps for UK freshwaters have been produced on a 10 x 10 km grid square basis, and used to map critical load exceedances under various deposition scenarios. A single lake or stream site was selected to represent the most sensitive water body in each grid square using predefined criteria. In the UK a major programme of data screening and validation has been undertaken in order to address issues of accuracy and validity. A major part of this validation exercise, the within-square variability study, is designed to test the extent to which the site chosen for mapping represents the most sensitive water body within each grid square or mapping unit. Sampling of all lake sites in thirty-two randomly chosen 10 x 10 km grid squares has shown that in two thirds of cases, the selection exercise has identified a site in the lowest critical load class within a square. However, up to a third of all sites selected to represent grid squares could be replaced by more sensitive sites with a critical load smaller by at least one Skokloster class. The mean overestimate of "diatom model" critical loads for sulphur in the within-square variability study is 0.188 keq ha^{-1} yr^{-1}. This means that current critical load maps show overestimates for some grid squares. In order to determine where the most sensitive site has not been identified, further work on catchment scale classification of freshwater sensitivity is being carried out.

1. Introduction

Critical loads maps for UK freshwaters have been produced on a 10 x 10 km grid square basis, with a single lake or stream site selected to represent each grid square using predefined criteria to obtain the most sensitive site (Kreiser *et al.* 1993). The critical load maps therefore present, in theory, the lowest critical load of any of the water bodies within each grid square, and do not provide information on the remaining population of lakes within the grid squares. This approach differs from the mapping of percentile critical loads for soils in the UK; it is not currently possible to calculate percentile critical loads for freshwaters because of the lack of a national lakes inventory or ecosystem area data.

In the UK a major programme of data screening and validation has been undertaken in order to address issues of accuracy and validity. This paper addresses an important part of the validation work; the assessment of the reliability of the site selection methods used, i.e. how often the most sensitive site has been chosen to represent a 10 x 10 km grid square. A simple way of estimating this factor is to sample all the water bodies in a grid square and determine whether the site selected for mapping has the lowest critical load.

2. Methods: random square selection and sampling strategy

The study dataset represents a ten percent subsample of grid squares which fall into the three highest exceedance classes of the diatom model for sulphur, stratified by exceedance class (Table I, e.g. see Allott *et al.*, 1995). These sites are of the greatest interest and importance because they are the most heavily impacted group in the UK national dataset, and critical loads calculated in the diatom model (DCL: see Battarbee et al., in press) are generally lower than values derived from the steady-state water chemistry model, or SSWC model (Henriksen and Brakke, 1988; Harriman and Christie, 1995).

In a single sample survey, it is most appropriate to determine the within-square variability of lake sites; streams are temporally much more variable, and critical loads can reflect flow conditions. Grid squares containing only streams, or with only one lake, were therefore excluded; in the latter cases the most sensitive lake site must have been selected by default.

The final selection of 32 squares contains 311 lakes in total with a surface area greater than 0.5 hectares (the minimum size specified for the original national survey). All these lakes, regardless of origin or location, were dip sampled (at the outflow, if present) during autumn or spring between September 1993 and November 1994. For any one 10 x 10 km square, all sites were sampled within 48 hours of each other, and usually on the same day, to ensure comparable depositional and hydrological conditions at related sites. The 0.5 litre samples were sent within a week to the Scottish Office Freshwater Fisheries Laboratory, Pitlochry, for chemical analysis according to the methodology described previously in Harriman et al. (1990). Critical loads were then calculated with Paradox database software for all sites, using both the diatom and Henriksen models for sulphur and total acidity.

TABLE I

Number of sites in the three highest exceedance classes of the diatom model critical load for sulphur (September 1994 data). Note that in 1995 these figures have changed with the use of updated deposition and runoff data.

Exceedance Class	Number of sites	Number of streams	Number of lakes	10% subsample
Black (>1.0 keq ha^{-1} yr^{-1})	90	12	78	8
Red (0.5 - 1.0 keq ha^{-1} yr^{-1})	125	16	109	11
Yellow (0.2 - 0.5 keq ha^{-1} yr^{-1})	141	20	121	13
TOTAL:	356	48	308	32

3. Results and Discussion

The results of the within-square variability study are presented in Fig.1 and summarised in Table II. The predefined site selection criteria were successful in identifying the most sensitive lake site in twelve of the thirty-two study squares according to the DCL, and eleven squares according to the SSWC. In two thirds of cases, the selected site falls within the lowest class of the DCL for sulphur. In seven out of every eight cases, the site selection technique has identified a site with a critical load within one class of the lowest value recorded in the square.

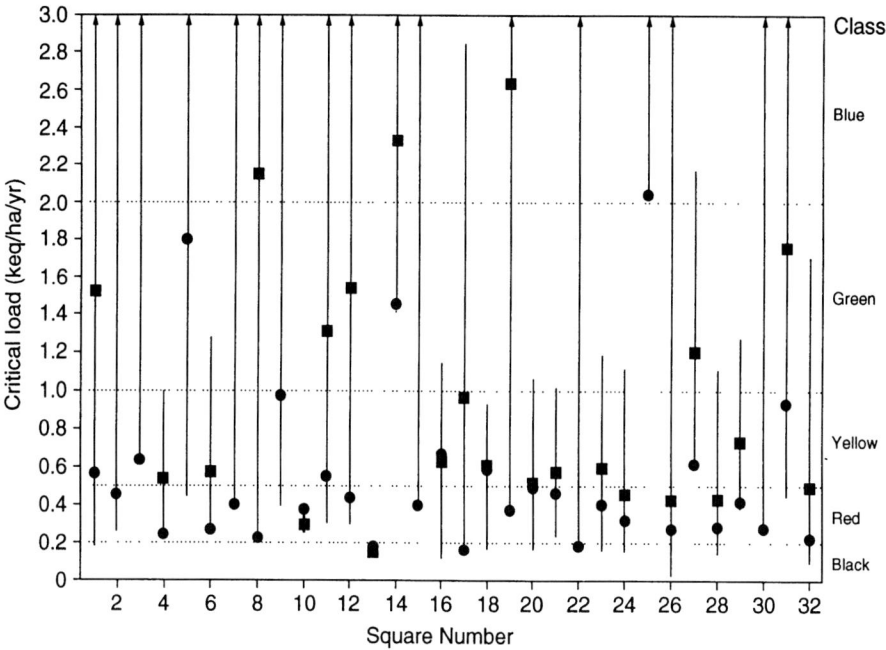

Fig. 1: The within-square variability of diatom model critical loads for sulphur. Vertical lines show the range of critical loads; ● marks the critical load at the selected site; ■ marks the mean critical load within the square.

Critical load class differences are useful in describing changes to national datasets, but are not an effective measure of absolute differences in critical load, because the boundaries of the "Skokloster" classes used are somewhat arbitrarily defined by the UNECE for mapping purposes, and the classes are not of equal size (Harriman et al., 1995; Nilsson and Grennfelt, 1988). For example, the black critical load class represents a value of <= 0.2 keq ha^{-1} yr^{-1}, while the yellow critical load class covers a range of values from 0.5 to 1.0 keq ha^{-1} yr^{-1}. The class change represented by a difference in critical load also depends on the proximity of a critical load value to a class boundary. A difference in critical load of 0.1 keq ha^{-1} yr^{-1} could represent the difference between 0.45 keq ha^{-1} yr^{-1} (red) and 0.55 keq ha^{-1} yr^{-1} (yellow), i.e. a class change. It could also represent the difference between 0.35 keq ha^{-1} yr^{-1} and 0.45 keq ha^{-1} yr^{-1}, which would not result in a change of class.

One way to evaluate the impact of the failure to select the most sensitive site is to determine the absolute difference in critical loads between the selected and most sensitive sites in a grid square, i.e quantify the extent to which critical load maps would change if the most sensitive sites had always been selected in the national datasets. The mean difference between diatom model critical loads for sulphur at selected and most sensitive sites within the thirty-two study squares is 0.188 keq ha^{-1} yr^{-1} (Table II). If this figure is taken as the mean overestimate of DCL for sulphur for the national dataset, then subtraction of this figure from currently mapped values will create an adjusted dataset which shows a much greater number of sites in the most sensitive class (Table III), but fewer sites within all the other classes. The pattern of increased exceedances is more evenly spread across the range of classes, with a seventeen percent increase in the number of exceeded sites. A breakdown of the mean overestimate of critical load according to the original class applied to each square shows that, for the diatom model, the overestimate appears to be greatest for the most impacted sites (Table II). These sites are not necessarily the most sensitive sites, since critical load exceedance is also a function of deposition.

TABLE II
Success of site selection strategy according to comparison of critical loads at site selected for mapping against most sensitive site in grid square

ATTRIBUTES OF SELECTED SITE:	DIATOM MODEL CRITICAL LOAD		HENRIKSEN MODEL CRITICAL LOAD	
	SULPHUR	TOTAL ACIDITY	SULPHUR	TOTAL ACIDITY
Best sites chosen /32 (%):	12 (37.5)	12 (37.5)	11 (34.4)	11 (34.4)
In same class as lowest /32 (%):	21 (65.6)	19 (59.4)	16 (50)	17 (53.1)
Within 1 class of lowest /32 (%):	28 (87.5)	28 (87.5)	27 (84.4)	27 (84.4)
Within 2 classes of lowest /32 (%):	32 (100)	32 (100)	31 (96.9)	31 (96.9)
Mean difference from lowest value (keq ha^{-1} yr^{-1})				
All 32 squares:	0.188	0.182	0.444	0.421
"Black" exceedance squares only:	0.245	0.223	0.464	0.416
"Red" exceedance squares only:	0.195	0.190	0.474	0.448
"Yellow" exceedance squares only:	0.146	0.150	0.410	0.402

The use of the mean overestimate of critical load in the thirty-two study squares to "correct" the national dataset is an oversimplification, because the mean differences are heavily weighted by a few of the selected sites which have much higher critical loads than the most sensitive site in their related grid square. This point is illustrated by consideration of the class differences in Table II. Two thirds of selected sites fall within the same class as the most sensitive site, but one third of sites should be replaced by alternatives which are more sensitive by at least one class. In over a third of squares, the most sensitive site has been selected and no "correction" is required. In the grid squares where the difference between critical loads at the selected and the most sensitive site is greatest, the subtraction of the mean difference will not give a small enough figure for the lowest critical load. Also, the subtraction of a mean overestimate nationally has an inherently greater effect on the most sensitive sites where the class boundaries are closer together. The result would be a map on which some squares showed an underestimate of minimum critical load, and other squares showed an overestimate. Until it becomes possible to predict where the greatest overestimates of critical load occur, it would be misleading to use the mean overestimate to make changes to the national map.

It is difficult at this stage to evaluate the strength of the relationship between freshwater sensitivity predicted by the available maps (Hornung et al., 1995) and observed variations in sensitivity within grid squares. The study squares were selected at random from only the most heavily impacted areas of critical load exceedance, and no direct account was taken of sensitivity. There is no evidence to suggest that success in identifying the most sensitive site is related either to regional sensitivity or to the number of lakes within a square. There is no correlation between the range of critical loads and the number of sites, although in areas where there is a great density of sites, there are likely to be small, local variations in chemistry and critical load which cannot be predicted from the available data.

TABLE III
Class changes in UK national map of diatom model critical load for sulphur inferred from mean difference in within-square variability study

Critical Load Class ($keq\ ha^{-1}\ yr^{-1}$)	Number of sites (mapped)	Number of sites (adjusted)	Exceedance Class ($keq\ ha^{-1}\ yr^{-1}$)	Number of sites (mapped)	Number of sites (adjusted)
Black (<= 0.2)	93	252	Black (>1.0)	75	101
Red (0.2 - 0.5)	236	166	Red (0.5 - 1.0)	122	183
Yellow (0.5 - 1.0)	210	163	Yellow (0.2 - 0.5)	154	156
Green (1.0 - 2.0)	180	159	Green (0.0 - 0.2)	94	79
Blue (>2.0)	854	833	Blue (<0.0)	1128	1054
TOTAL:	1573	1573	TOTAL:	1573	1573

Closer inspection of the data suggests that many local factors may be contributing to the relative sensitivity of sites in the same area. Topographical factors such as lake : catchment ratio, residence time, relative relief within the catchment and drainage density will affect the deposition and flux of sulphur. Other physical factors which will affect sensitivity are the type, depth and extent of soils, local geology, vegetation and land use. The movement within the UNECE towards the use of critical loads for total acidity, rather than for sulphur alone, will affect the criteria which need to be applied in identifying the most sensitive sites. The introduction of nitrogen into the calculation of critical loads has meant that catchment specific biological processes take on a greater significance. The relevant parameters have only been partially accounted for during previous site selection exercises by the use of freshwater sensitivity maps derived from soil, geology and land-use data (Hornung et al., 1995).

4. Conclusion

The within-square variability study shows that there may be a significant overestimate of the minimum critical load in around one third of the 10 x 10 km grid squares of the UK. The ten percent subsample of exceedance squares indicates a mean overestimate of DCL for sulphur by 0.188 keq ha^{-1} yr^{-1}. The overestimates appear to be even greater for the SSWC critical loads, especially for total acidity. However, at this stage it is not possible to say where the overestimates of critical load are likely to occur. The within-square variability of critical loads indicated here shows the need for a better resolution of catchment and deposition data if we are to be more confident of adhering to the principle of mapping the most sensitive sites in the UK.

Acknowledgements

This work was funded by the UK Department of the Environment. We gratefully acknowledge colleagues within the DoE Critical Loads Advisory Group (freshwaters sub-group) for helpful discussions, and members of the Environmental Change Research Centre, especially Martin Kernan, for help with fieldwork. The views expressed here are entirely those of the authors.

References

Allott, T.E.H., Battarbee, R.W., Curtis, C.J., Harriman, R., Hall, J., Bull, K. and Metcalfe, S.E.: 1995, in Critical Loads Advisory Group (Sub-group on Freshwaters), *Critical Loads of Acid Deposition for UK Freshwaters*, ITE, Penicuik, pp. 25-33.

Battarbee, R.W., Allott, T.E.H., Juggins, S., Kreiser, A.M., Curtis, C. and Harriman, R.: in press, *Ambio*.

Harriman, R., Gillespie, E., King, D., Watt, A.W., Christie, A.E.G., Cowan, A.A. and Edwards, T.: 1990, *Journal of Hydrology* **116**, 267-285.

Harriman, R., Allott, T.E.H., Battarbee, R.W., Curtis, C.J., Hall, J. and Bull, K.: 1995, in Critical Loads Advisory Group (Sub-group on Freshwaters), *Critical Loads of Acid Deposition for UK Freshwaters*, ITE, Penicuik, pp. 19-24.

Harriman, R. and Christie, A.E.G.: 1995, in Critical Loads Advisory Group (Sub-group on Freshwaters), *Critical Loads of Acid Deposition for UK Freshwaters*, ITE, Penicuik, pp. 7-8.

Henriksen, A. and Brakke, D.F.: 1988, *Environ. Sci. Technol.* **22**, 8-14.

Hornung, M., Bull, K.R., Cresser, M., Ullyet, J., Hall, J.R., Langan, S., Loveland, P.J. and Wilson, M.J.: 1995, *Environmental Pollution* **87**, 207-214.

Kreiser, A.M., Patrick, S.T. and Battarbee, R.W.: 1993, in M. Hornung and R.A. Skeffington (eds.), *Critical loads: concept and applications*, ITE Symposium No.28, HMSO, London, pp. 94-98.

Nilsson, J. and Grennfelt, P. (eds.): 1988, *Critical Loads for Sulphur and Nitrogen*, Miljørapport 1988:15, Nordic Council of Ministers, Copenhagen.

CRITICAL LOADS FOR SOILS AND WATERS IN A SELECTED SCOTTISH CATCHMENT

[1]K.P.MACPHEE, [1]S.J.LANGAN and [2]M.F.BILLET

[1] *Macaulay Land Use Research Institute, Craigiebuckler, Aberdeen, AB9 2QJ.*
[2] *Department of Plant & Soil Science, University of Aberdeen, AB9 2UE.*

Abstract: In the UK, critical loads have been mapped for both soils and freshwaters and the maps indicate that discrepancies may occur between these two receptors over sensitive areas of the UK. Freshwater critical load maps were prepared by calculating the Henriksen critical load for the most sensitive water body in each 10 km grid square. Critical loads for soils were calculated according to the mineralogy and associated soil properties of the dominant soil at a 1 km resolution. To examine the differences between the soil and freshwater data sets it is necessary to calculate critical loads at a smaller scale using the catchment as the focus for study. This was done by selecting a catchment on granitic parent material in the North of Scotland. Data on water chemistry, collected on a weekly basis, was used to calculate temporal variations in critical loads for freshwaters using the Henriksen method. Soil sampling across the catchment was conducted on a grid based system to provide estimates of spatial variability in sensitivity. Profile characteristics and soil chemical data obtained from detailed soil sampling programmes were used in the PROFILE model to determine the spatial variation in critical loads for soils. In general, the results show that the critical loads for soils tend to be lower than those for freshwater. The spatial variation in the soil critical load tends to be small whilst the temporal variation in critical load for freshwaters is large. In order to account for these differences it is important to identify the key processes within the catchment which play a major role in controlling streamwater chemistry. This procedure improves the relationship between critical loads for soils and waters.

Key words : critical loads, Henriksen, PROFILE, freshwaters, soils, catchment, temporal variability, spatial variability.

1.Introduction

Nilsson and Grennfelt (1988) define the critical load as : 'a quantitative assessment of one or more pollutants below which significant harmful effects on specified sensitive elements of the environment do not occur according to present knowledge'. This concept has been developed as a method of evaluating the effectiveness of abatement policies on limiting the impact of acidic atmospheric deposition, mainly in the form of sulphur dioxide and oxides of nitrogen. In order to reduce the impact of these acidifying pollutants member states of the United National Economic Commission for Europe (UN-ECE) devised a sulphur reduction protocol in 1985 which required a blanket reduction of emissions by 30% of the 1980 levels by all members. The 1993 negotiations on revision of this protocol were based on application of the critical loads concepts to provide a targeted reduction policy.

Acidification within an ecosystem is determined by the demand and supply of base cations. Therefore where supply of base cations is less than the demand, acidification occurs. The critical load therefore is set at the point at which base cation supply and demand balance or at an appropriate chemical limit specific to a chosen indicator organism for each ecosystem type. European critical load calculations use the soil solution ratio between Ca and Al as the critical parameter (Sverdrup and Warfvinge, 1992). This assumes that maintaining an appropriate Ca:Al ratio will protect vegetation communities (Ulrich et al, 1984, Ulrich,1985). In the UK, national critical loads maps were produced for both soils and freshwaters (Critical Loads Advisory Group, 1994). The methodology used to map the freshwaters data involved application of the Henriksen method for calculation of critical loads. This takes the form of a steady state equation and is based on the principle that base cation production should be greater than or equal to the acidic ion input. Thus the critical load of the chosen receptor is calculated against an acceptable alkalinity (Henriksen, 1984, Henriksen et al, 1988 and Sverdrup et al, 1990). The measured water chemistry data was obtained from a spot sample of the most sensitive standing water body in each 10 km critical load map square. Sensitivity was determined based on geology using the geological sensitivity map of Edmunds and Kinniburgh (1986), national maps and soil and land use maps. Other criteria included selection of the highest altitude site within the area of greatest sensitivity with a minimum size of 0.5 ha. The critical load obtained was applied to the entire 10 km grid square.

The soils critical load maps were produced by allocating a Skokloster class to the dominant soil at 1 km resolution. The Skokloster classification system identifies five classes of critical load for soils defined by their parent material. Those soils derived from highly resistant parent material are awarded a low critical load whereas those of a high critical load are derived from highly weatherable parent rock (Langan and Wilson, 1994).

The aim of this work is to examine the differences observed between the soils and freshwaters maps of Scotland and to attempt to clarify at a catchment level the nature of these differences by identifying important processes and properties within the catchment which play a major role in controlling streamwater chemistry. In doing this the catchment is adopted as the focus of study as opposed to the map grid squares. An extensive study of the waters and soils of the catchment provides a large dataset with which to apply the Henriksen method and the PROFILE model to calculate critical loads for freshwaters and soils across the catchment.

2. Catchment Characteristics

The Bhealach catchment is located 58 20 N 3 52 W (NGR NC 908405) on the northerly flank of Cnoc nam Bo Rhiabhach in Western Caithness, Scotland. An upland catchment with an area of 1.5 km^2, it drains northwards to the River Halladale falling from an altitude of 360 m at the catchment summit boundary to 170 m at the stream sampling point. It is situated on the most northerly granitic intrusion in Scotland and the soils are classified as of the Charr and Spinneag series within the Countesswells Association. A detailed soil survey indicates that the soils of the upper catchment are dominated by deep peats, peaty rankers and peaty gleyed podzols.

Vegetation is dominated by *Narthecio-Ericum tetralicis* and is characterised by northern and common bog heather moorland. Precipitation data from Warren Spring Laboratory (UKRGAR, 1990) indicates annual mean precipitation estimated in the order of 1130 mm for the period 1986-1988.

3. Sampling procedure and analysis

Water chemistry was monitored by taking spot samples weekly from January 1994 to January 1995. Each water sample was analysed for Ca, Na, K and Mg on an ARL 3580B ICP Analyser and on a Dionex Ion Chromatograph for Cl and SO_4_S. Flow data were obtained from the HRPB gauging site on the River Halladale and results were weighted accordingly. Freshwater critical loads were calculated from the water chemistry for each sample, using the Henriksen model. The model is discussed further in Henriksen (1984) and Henriksen et al (1988).

Soil was sampled on a grid based system across the catchment and six profiles representative of each soil complex were selected for application of PROFILE. Soil profiles were fully described and a sample was taken from each horizon. Soil characteristics such as horizon depth, texture and structure were measured. The soils were air dried at 30°C and sieved to 2 mm. Mineralogy was determined on each soil sample, after ballmilling, by XRD analysis on a Philips PW1130/90 2kW x-ray generator using Co $K\alpha$ radiation. Partical size analysis, moisture contents, loss on ignition, pH values and exchangeable cations in 1.0 M ammonium acetate at pH 7 were determined as described in Soil Conservation Service (1972) and Sheldrick (1984). The critical loads for soils were calculated using the steady state soil chemistry model PROFILE which evaluates the soils' final chemical status for a given set of conditions using either equilibrium relationships or kinetic equations. The input requirement functions for the model include information on soil stratification, atmospheric deposition, climatic conditions, water balance and nutrient uptake. A full description of the model's input requirements, functions and outputs can be found in Sverdrup and Warfvinge (1988,1992).

4. Results and Discussion

From the UK Freshwaters Maps (Critical Loads Advisory Group, 1994) and Skokloster soil classification (Langan and Wilson, 1994), the critical load values applied to the mapping unit within which the study catchment lies are 2.0 keq H^+ ha^{-1} yr^{-1} for freshwaters and 0.2 to 0.5 keq H^+ ha^{-1} yr^{-1} for soils. The critical load value of 2.0 keq H^+ ha^{-1} yr^{-1} assigned to the 10 km square of the freshwater critical load map was not calculated for the same catchment as in this study however the results are comparable as the value allocated by the national mapping procedure is applied to the entire grid square including this catchment. This highlights the potential differences between the soils and waters critical loads when trying to interperet or use the national map at a more regional and local scale.

The temporal variation in freshwater critical loads from the Bhealach catchment resulted in an annual flow weighted mean value of 0.6 keq H^+ ha^{-1} yr^{-1} with values ranging between 0 and 2.0 keq H^+ ha^{-1} yr^{-1}. The critical loads calculated using PROFILE based on a Ca:Al ratio of 1 in the B/C horizon for the Bealach soil profiles ranged from 1.6 to 2.2 keq H^+ ha^{-1} yr^{-1} with a mean value of 1.9 keq H^+ ha^{-1} yr^{-1}. In examining the relationship between the soils and the waters at a catchment level, the assumption is made that the water chemistry reflects the chemistry of soils from which it has drained between deposition and entering the stream (Billet and Cresser 1992). The data shows a much greater variability in freshwater critical loads than in the soils, which is to be expected due to predominant soil flow pathways varying between high and low discharge. The high flow waters show higher sensitivity than is mapped and the range of values is large, ranging across four of the five critical load classes. This range of values suggests that the critical load of 2.0 keq ha^{-1} yr^{-1} allocated by the UK mapping system does not reflect the extent of the sensitivity of these waters which illustrate temporal variation between 0 - 2 keq ha^{-1} yr^{-1}. It is significant however that the mapping programme used standing waters in the majority of sites including this 10 km square whereas the study examines flowing waters. The results suggest a higher sensitivity in running waters and this is a subject of ongoing work in this study catchment. In considering the soils, again there is variation with the Skokloster class suggesting a substantially higher sensitivity than is calculated using PROFILE. This may be explained by the high content of plagiocalse feldspar throughout the catchment soils. Mineralogical analysis data show values of plagoicalse feldspar ranging from 24 to 58% of the total mineral content.

Figure 1. illustrates the range of critical loads for the freshwaters and the critical load for each soil horizon sampled from the catchment profiles There is little variation between the critical loads for soil horizons, suggesting the hydrological routing during

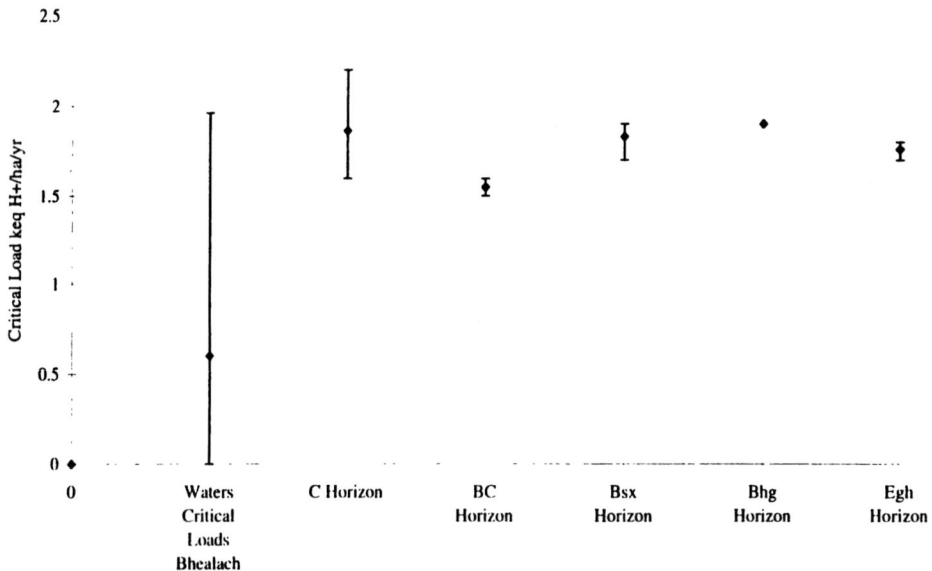

Fig.1. Critical load for each soil horizon from the Bealach profiles and the freshwaters critical load range.

base flow is not dominated necessarily by flow through a single horizon but may have contributions from some or all of the horizons below the organic horizon. As the PROFILE model is mineralogically driven, the organic horizons have the model default values applied and are therefore disregarded in terms of the critical load values produced by the model which is significant as they are of critical importance at high discharge. The results also indicate very little spatial variability within the soil critical loads across the catchment soils. The points above 2 keq H^+ ha^{-1} yr^{-1} for the C horizon probably reflect the presence of the indurated horizon present over much of the catchment. Within this indurated horizon there is may be a greater buffering capacity present but due to the high compaction properties of this horizon the water does not infiltrate much of this material. The higher range of sensitivities shown in the range of freshwater critical loads measured throughout the year suggests significantly variable hydrological routing regimes in the catchment between high and low flow. The results suggest that streamwater is derived from more than one horizon during periods of low flow and is dominated by the contribution from the organic horizons and possibly surface runoff during periods of high flow.

In order to incorporate the element of varying hydrological regimes into the critical loads work, it is suggested that the HOST (Hydrology of Soil Types) classification system may be used to evaluate the catchment soils susceptibility to flooding using estimates of percentage runoff. The HOST system predicts the hydrological response of a soil based on conceptual models of the processes taking place within the profile using soil physical properties and hydrological variables at a catchment scale (Boorman et al, 1995). Using this method the HOST classes may be used to weight the soil critical loads according to their dominant hydrological regime and percentage coverage of the catchment thus altering the sensitivity of the soil unit and improving the relationship between critical loads for soils and waters.

5. Conclusions

This work suggests that the apparent differences in sensitivity between soils and waters may be attributed to varying hydrological regimes within the catchment and it may indicate that the range of lower freshwater critical load values are reflecting periods of organic horizon throughflow and surface runoff. It is possible that by incorporating the HOST classification system the sensitivity of the soil units may be altered and the differences between the soils and waters critical loads reduced.

In continuing this research, two further catchments are being studied to incorporate a pollution gradient over Scotland. This will also provide a greater data set with which to propose a universally applicable method with which to improve the soil/waters critical load relationship.

Acknowledgements

This work is funded by the U.K. Department of the Environment. Thanks are also due to Dr. Mark Hodson for his advice in the writing of this paper and Prof. Malcolm Cresser for his comments.

References

Boorman, D.B., Hollis, J.M. and Lilly,A. : 1995, I.H. Report No. 126 published by the Institute of Hydrology.
Billet, M.F., Cresser, M.S. : Environmental Pollution, Vol. 77, Nos 2 & 3, 1992.

Critical Loads Advisory Group. : 1994, Critical Loads of Acidity in the United Kingdom, Department of the Environment, London.
Henriksen, A. : 1984, Verhandluger der Verein Limnology, **22**, 692-698.
Henriksen, A., Lien, L., Traeen, T.S., Sevaldrud, I.S. and Brakke, D.F. : 1988, Ambio, **17**, 259-266.
Langan, S.J. and Wilson, M.J. : 1994, Water, Air and Soil Pollution, **75**, 177-191.
Nilsson, J. amd Grennfelt, P. (Eds) : 1988, Published by the Nordic Council of Ministers, Copenhagen.
Sheldrick, B.H. (Ed) : 1984, Land Resource Research Institute, Ottawa, Ontario.
Soil Conservation Service : 1972, US Dept. of Agriculture, Washington DC.
Sverdrup, H. and Warfvinge, P. : 1992, Lund University Press.
Sverdrup, H. De Vries, W. and Henriksen, A. : 1990, Task Force on Mapping - UN-ECE, Umweltbundesamt, Berlin, Federal Republic of Germany.
Sverdrup, H. and Warfvinge, P. : 1988, Water, Air and Soil Pollution, **38**, 387-408.
Ulrich,B. : 1985, In Cooley, J. and Golley, F. (Eds) : pages 217-237. Nato conference Series 1985.
Ulrich, B., Meines, K.J., Konig, N. and Khanna, P.K.: 1984, In Andersen, F. and Kelly,J.M. (Eds) : pages 69-70. Documentation of an International workshop in Uppsala, May 14-17, 1984. Section of Systems Ecology, Swedish University of Agricultural Sciences, Uppsala.
United Kingdom Review Group on Acid Rain (UKRGAR) : 1990, Third report of UKRGAR, published by H.M.S.O.

THE USE OF CATCHMENT ATTRIBUTES TO PREDICT SURFACE WATER CRITICAL LOADS: A PRELIMINARY ANALYSIS

M.KERNAN

Environmental Change Research Centre, University College London,
26 Bedford Way, London WC1H 0AP, UK

Abstract. Current applications of the critical loads concept are geared primarily towards targeting emission control strategies at a regional and international level. Freshwater critical maps in the UK have been produced at a resolution of 10 km grid squares and do not take into account variations of water chemistry within the mapping unit. They are therefore of limited use at the catchment scale. This paper assesses the potential for the development ofn empirical statistical model to predict catchment critical loads using readily available secondary data. Multivariate statistical analysis of existing critical loads chemistry data together with data obtained from the Institute of Terrestrial Ecology (ITE) identifies strong relationships between surface water chemistry composition and a variety of site-specific catchment attributes, particularly rainfall, altitude and site sensitivity. Although there were problems with the data used, particulary in terms of noise, collinearity and spatial resolution, the strength of the relationships indicates that accurate prediction of catchment scale critical loads should be possible using a higher resolution, catchment specific dataset.

1. Introduction

Freshwater critical loads maps for the UK have been produced at a national scale by the DoE Critical Loads Advisory Group (CLAG) (Harriman *et al*, 1995). These are based on a single water chemistry sample from the most sensitive area within each 10 km grid square from the national grid (Kreiser *et al.*, 1993). This limits these maps in that they do not relate to the majority of freshwater bodies and no conclusions can be drawn about lakes outside the sample population. At this resolution the use of the critical loads approach for the management of individual catchments (for example by forestry or conservation organisations) is precluded.

Relationships between surface water chemistry and the characteristics of contributing catchment areas have been well documented (Ruess and Johnson, 1986, Hornung *et al*, 1990). Attempts have been made to use various environmental parameters to predict where acid waters (and soils) may occur at a regional level (Kinniburgh and Edmunds, 1986, Langan and Wilson, 1992). Other studies have related surface water chemistry to soil (Rees *et al.*, 1989) geology (Duarte and Kalff, 1989) and land use (Hornbeck, 1992) individually, or using an integrated approach (Hornung *et al.*, 1995). This work has been undertaken at a variety of spatial scales with data at varying resolutions. Hitherto prediction of surface water chemistry at the catchment scale has required the paramaterisation of complex dynamic modelling (Cosby *et al.*, 1990) or mapping at very high resolutions (Billet and Cresser, 1992).

The use of regional models is likely to lead to inaccuracies at the catchment scale while those requiring high resolution data will only apply to catchments for which such data is available. An intermediate approach is required whereby water chemistry can be predicted at a local scale without the need for very high resolution data or expensive water sampling programmes. This paper describes the preliminary analysis which precedes the development of an empirical statistical model for this purpose. The results are used to gauge the efficacy of using such an approach with more detailed catchment specific data.

2. Data and Methodology

2.1 SECONDARY DATA

The data used in this analysis were provided by the Critical Loads Advisory Group (CLAG) Freshwaters subgroup and the Institute of Terrestrial Ecology (ITE). A number of chemical determinands (and derived critical loads) were obtained from the CLAG database, together with data relating to altitude, sulphur (S) and nitrogen (N) deposition and rainfall. Digital data relating to soil critical load, land cover and site sensitivity were provided by ITE. These related to the 1km square in which the water sample was taken rather than the entire catchment area. The soil critical load was extracted from that used by Hornung (1993) to produce a provisional map of critical loads for acidity of soils for Great Britain. The classification system is shown in Table I. The ITE land cover data are derived from satellite imagery together with field surveying and comprise 18 classes (aggregated by ITE from 25). The data are based on the dominant land cover class per 1km^2 (Fuller et al, 1993). To facilitate the application of multivariate statistical techniques, the data were further aggregated into the classes shown in Table II. The site sensitivity classification combines soil (Hornung et al, 1995), geology (Kinniburgh and Edmunds, 1986), and land class (Bunce and Heal, 1984). This was used to produce a sensitivity map which attempted to predict surface water acidification at a regional level (Hornung et al, 1995). Five sensitivity classes were identified which predicted the likelihood of acid waters ranging from occurrence to occurring at all flows.

2.2 STATISTICAL ANALYSIS

Prior to multivariate analysis the data were log transformed, firstly, to fulfil the assumption of normality required for parametric techniques and second, because ionic concentrartions are often characterised by log-normal distributions (Ott, 1990). Additionally the categorical data from ITE were converted to dummy variables to facilitate analysis. Ordination techniques were employed to examine the relationships between these multivariate datasets. These arrange sites along axes based on the relationship between variables within those sites (Ter Braak, 1987a). Principle components analysis (PCA) and redundancy analysis (RDA) (Van den Wollenberg, 1977) were undertaken using CANOCO Version 3.10, a FORTRAN program for canonical community ordination (Ter Braak, 1990) to assess the relationships between critical load and the various catchment attributes. Analysis of variance (ANOVA) and linear regression techniques were applied to selected variables using SAS, a statistical application suitable for handling large datasets. This shows the relationship between individual catchment characteristics and critical load indicating those variables which may drive a more sophisticated statistical model.

TABLE I

Mineralogical and petrological classification of soil material and critical loads of soils (after Nillson & Grennfelt, 1988)

Class	Minerals controlling weathering	Parent material	Critical load (keq H^+ $ha^{-1} yr^{-1}$)
5	Quartz K-feldspar	Granite Quartzite	.1
4	Muscovite Plagioclase Biotite (<5%)	Granite Gneiss	.5
3	Biotite Amphibole (<5%)	Granodiorite Greywacke Schist Gabbro	1.0
2	Pyroxene Epidote Olivine (<5%0)	Gabbro Basalt	2.0
1	Carbonates	Limestone Marl	4.0

TABLE II

Aggregated land cover classes (modified from Fuller *et al*, 1994)

Class	Code	Description
1	LV1	Water & Built/bare ground (inc beach)
2	LV2	Agricultural grass
3	LV3	Arable
4	LV4	Deciduous woodland
5	LV5	Coniferous woodland
6	LV6	Lowland semi-natural grass/moor a)grass
7	LV7	Lowland semi-natural grass/moor b)dwarf shrub
8	LV8	Upland semi-natural grass/moor a)grass
9	LV9	Upland semi-natural grass/moor b)dwarf shrub

3. Results and discussion

PCA of the chemistry determinands indicated that the sites varied along a chemical gradient strongly dominated by calcium (Ca^{2+}) concentration and alkalinity. This is evidenced by the high correlation between these determinands and the first ordination axis (r values of .96 and .91 respectively). The eigenvalue of the first PCA axis (.6299) shows how this gradient dominates the structure of the data. The diatom critical load (DCL) (Battarbee *et al*, 1995) is also highly correlated with the first axis (r = .95)

An RDA biplot (figure 1) illustrates the relationship between the chemistry determinands (solid vectors) and both the continuous (dotted vectors) and categorical (filled circles) predictor variables. In an RDA correlation biplot the angle between vectors (in this instance representing the continuous data) determines the direction of correlation. From this it can be seen that DCL is inversely correlated with rainfall and altitude ($P<0.01$). The position of each class of the categorical variables is at the centroid of all those sites characterised by that class. To interpret the relationships between the categorical variables and the chemistry vectors correlations can be gauged by projecting the centroids orthogonally onto the vectors (Ter Braak, 1994). Thus the correlation of SCL with DCL decreases from positive and increases negatively in a sequence SCL1, SCL2, SCL3, SCL4, SCL5. Similarly, sites in SS class 1 are likely to be characterised by high DCL as this is at the positive extreme of the DCL gradient with the converse applying to SS5.

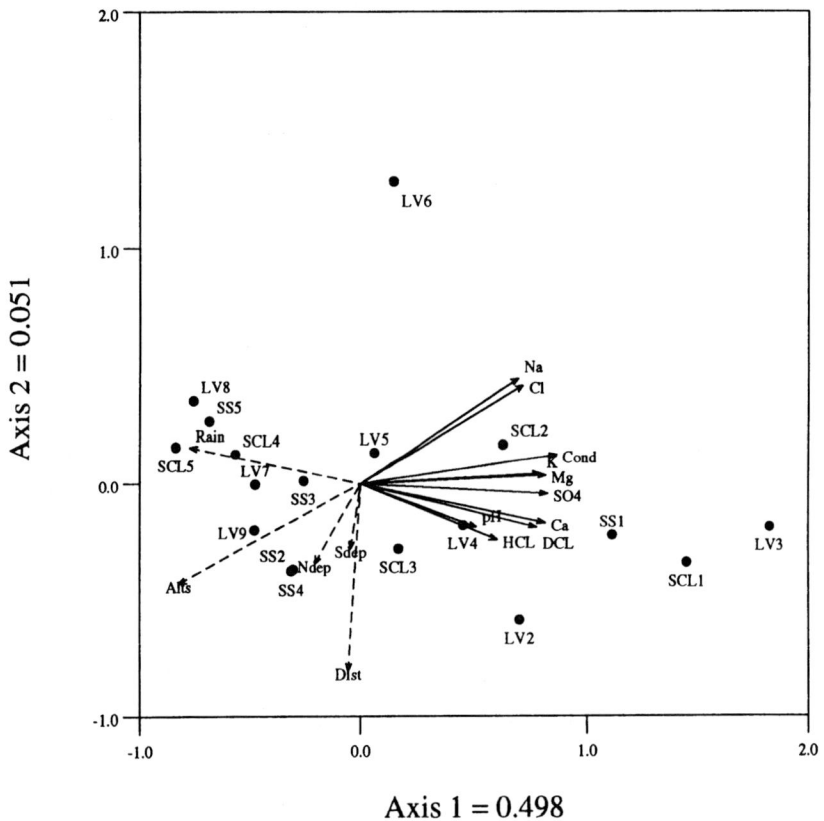

SS : Site sensitivity
SCL : Soil Critical Load
LV : Land cover class
Rain: Rainfall

Dist: Distance from sea
Ndep: N deposition
Sdep: S deposition
Alts: Site altitude

HCL: Steady state critical load
DCL: Diatom critical load

Fig. 1. RDA correlation biplot of chemistry and environmental variables
(plotted using CALIBRATE - Juggins and ter Braak, 1993)

Sites in SCL class 1 and LV class 3 (arable) sites will also have high DCL values. Both the SS and the SCL classifications reflect the increasing sensitivity of those sites with low critical loads. The biplot also shows that these sites tend to be in upland moor areas (LV8). As with the PCA the first axis is much stronger than the second. Both the chemistry and catchment paramaters vary predominantly along this gradient indicating that the latter may be used to explain differences in the former. An examination of the eigenvalues from the RDA shows that the catchment variables account for 56% ($P<0.01$) of the variation in the water chemistry. The most influential of these, as determined by forward selection (Ter Braak, 1988) are rainfall, altitude and site sensitivity.

ANOVA and regression analysis were applied to DCL to assess the strength of the relationships with some of the more influential environmental variables (Table III). This illustrates the importance of altitude and rainfall as potential predictors.

TABLE III

ANOVA and regression analysis with DCL as the response variable

Analysis	Variable	r^2	F Stat	Prob > F
ANOVA	Site sensitivity	.43	188.5	0.0001
ANOVA	Soil critical load	.44	154.4	0.0001
ANOVA	Land cover	.45	102.4	0.0001
Regression	Rainfall	.41	660.7	0.0001
Regression	Site altitude	.31	436.2	0.0001

4. Conclusion

It is clear that critical load is related to the surrogate catchment variables used in this analysis and that rainfall received and altitude are important parameters. It is also clear that the data used to represent catchment attributes have inherent problems, particularly in terms of collinearity, noise, resolution and spatial applicability. Given these problems, the amount of variation in the chemistry data explained by the environmental variables suggests catchment attributes may be used to predict catchment scale critical loads. The use of explicit soil and geology data rather than the surrogates employed here and the use of high resolution land cover data, all included as catchment parameters rather than relating to 1km squares should improve the strengths of these relationships and attain the accuracy required for management purposes. Additionally, the water chemistry used in this analysis varied along a lengthy gradient (for example, calcium concentrations range from 12 to 8810µeq/l). The relationships between water chemistry and catchment parameters might not be the same for more sensitive sites along a reduced calcium gradient. Surface waters in areas with low natural buffering capacities are of greatest concern in management terms. The development of a model to predict catchment scale critical loads will therefore need to incorporate potential differences in the water chemistry/catchment realtionships among these more sensitive sites.

Acknowledgements

This paper forms part of the research undertaken as part of a NERC funded PhD studentship (Ref GT4/92/17/R). I would like to thank both the Critical Load Advisory Group for the use of the chemistry database held at the Environmental Change Research Centre and Jane Hall at The Institute of Terrestrial Ecology for providing the digital data for each of the sampling sites. I would also like to Rick Battarbee, Tim Allott and Steve Juggins for useful discussion.

References

Battarbee R.W., Allott, T.E.H., Juggins, S., Kreiser, A.M.:1995 In: R.W. Battarbee (Ed.), *Critical loads of acid deposition for UK freshwaters.* HMSO, London.
Billett, M.F., Cresser, M.S.:1992 *Environ. Pollut.,* **77,** 263-268
Bunce, R.G.H., Heal, O.W.:1984 In R.D. Roberts and T.M. Roberts (Eds.) *Planning and Ecology.* Chapman and Hall, London, pp164-168
Cosby, B.J., Jenkins. A., Miller, J.D., Ferrier., R.C., Walker, T.A.B.: 1990, *J. Hydrology*, **120,** 143-162.
Duarte, C.M., Kelffe, J.:1991, *Arc. Hydrobiol.* **115,** 27-40.
Fuller, R.M., Groom, G.B., Jones, A.R.:1994, *Photogrammetric Engineering and Remote Sensing* **60,** 553-562.
Harriman, R., Allott, T.E.H., Battarbee, R.W., Curtis, C., Hall, J., Bull, K.:(1995), In: R.W. Battarbee (Ed.), *Critical loads of acid deposition for UK freshwaters.* HMSO, London.
Hornbeck, J.W.:1992, *Environ. Pollut.* **77** 151-155.
Hornung, M., Le-Grice, S., Brown, N., Norris, D.:1990 In: R.W. Edwards, A.S. Gee, J.H. Stoner (Eds.) *Acid Waters in Wales.* Kluwer, Dordrecht, pp55-66.
Hornung, M.:1993 In: M. Hornung & R.A Skeffington (Eds.) *Critical loads: concepts and applications.* ITE Symposium no. 28, HMSO, London, pp31-33
Hornung, M., Bull, K.R., Cresser, M., Ullyet, J., Hall, J.R., Langan, S., Loveland, P.J.: 1995, *Environ. Pollut.* **87** 207-214.
Juggins, S., ter Braak, C.J.F.:1993 *CALIBRATE v 0.52 (Beta Test) - a computer program for species-environment calibration by [weighted-averaging] partial least squares regression,* ECRC, University College London.
Kinniburgh, D.G., Edmunds, W.M.: 1986, *The susceptibility of UK groundwaters to acid deposition.* Hydrogeological Report, 86/3. British Geological Survey, Wallingford, UK pp208.
Kreiser, A.M., Patrick, S.T., Battarbee, R.W.:1993. In: M. Hornung & R.A Skeffington (Eds.) *Critical loads: concepts and applications.* ITE symposium no. 28, HMSO, London, pp 94-98
Langan, S., Wilson, M.J.:1992, *J. Hydrology* **138,** 515-528.
Nillson, J., Grennfelt, P.:1988, *Critical loads for sulphur and nitrogen.* Nordic council of Ministers, Copenhagen.
Ott, W.R.:1990, *J.. Air Waste Management Assoc.* **40,** 1378-1383.
Rees, R.M., Parker-Jervis, F., Cresser, M.:1989, *Water Resources* **23,** 511-517.
Reuss, J.O., Johnson, D.W.:1986, *Acid Deposition and the Acidification of Soils and Waters.* Springer-Verlag, New York, 119pp
Ter Braak, C.J.F,:1994, *Ecoscience* **1,** 127-140
Ter Braak, C.J.F.:1987, In R.H.G. Jongman, C.J.F. ter Braak, O.F.R. van Tongeren (Eds.), *Data analysis in community and landscape ecology.* Pudoc, Wageningen, pp 91-173.
Ter Braak, C.J.F.:1990, *CANOCO - a FORTRAN Program for CANOnical Community Ordination by [Partial] [Detrended] [Canonical] Correspondence Analysis, Principle Components Analysis and Redundancy Analysis (Version 3.1),* Agriculture Mathematics Group, Wageningen.
Ter Braak, C.J.F., Prentice, I.C.:1987, *Advances in Ecological Research,* **18,** 213-238.
Van den Wollenburg, A.L.:1977, *Psychometrika,* **42,** 207-219.

CRITICAL LOADS - A VALUABLE CATCHMENT MANAGEMENT TOOL?

D J TERVET[1], D A RENDALL[1] and A B STEPHEN[2]

1 Solway River Purification Board, Rivers House, Irongray Road, Dumfries DG2 0JE, United Kingdom
2 West Galloway Fisheries Trust, Bladnoch Distillery, Newton Stewart DG8 9AB, United Kingdom

Abstract. Planning advice for forest planting in acid sensitive areas suggests that, where calculated critical loads for acidity are exceeded at a catchment level, new conifer planting may not be appropriate. In south west Scotland, acid waters are currently found in areas where critical loads are not exceeded.

The rivers Cree and Bladnoch show a decline in pH of about one unit since 1970, when major afforestation of the headwaters began. No equivalent decline in pH was observed in the adjacent Water of Luce, although it receives similar inputs and has similar geology and soils. Little of the Luce catchment is afforested.

Recent surveys of water quality, invertebrate fauna and salmonid fish reveal a picture of widespread acid conditions, impoverished benthos and absence of young salmon. 25 streams (total catchment $>150km^2$) recorded pH <4.5 in high flow conditions. Critical loads for acidity were $>1.5 keqha^{-1}yr^{-1}$ for 12 and $>2 keqha^{-1}yr^{-1}$ for 6 of the 25 streams. Published deposition data suggested that one stream with pH <4.5 and 7 streams with pH <5 were in areas where critical load was not exceeded. In 22 catchments, forestry was a major land use.

To be effective as planning and management tools, systems must be robust and easy to operate. Critical load exceedance calculations remain research tools at the catchment level where deposition data is generally inadequate. The uncertainties inherent in critical load exceedances render them sources of argument and not beacons of enlightenment.

Key Words: Critical loads, water, catchment assessments, forestry, acidification.

1 Introduction

In Galloway, in south-west Scotland, there are many rivers and streams which have become acidified over the past 40 years. Waters with pH <6.0 are found in an area exceeding 1400 km^2 and with pH <5.0, in >700 km^2. In 1984, in view of their long history as successful fisheries, most of these streams were designated under the EC Freshwater Fisheries Directive as suitable for salmonids. The River Cree (catchment 368 km^2) has failed to comply with the pH standards of the Directive in most years since and significant parts of the rivers Luce, Dee, Fleet and Bladnoch have failed regularly (Solway RPB,1994).

The increase in acidification has been paralleled by an increase in the extent of coniferous afforestation, which now covers about 30% of Dumfries & Galloway Region. There has also been a significant reduction in the success of local salmon fisheries. It is now generally accepted that, until the mid-1980s, the forestry practices of ploughing and road building were incompatible with the maintenance of healthy fisheries. Although there is now much greater awareness of the physical needs of fishery streams, reliable methods to plan new forests or plantations in acid sensitive areas have not been fully developed.

The Forestry Authority in the UK has recommended that catchment-based assessments of critical loads for acidity based on the Henriksen steady-state water chemistry model should be used as a guide to the selection of suitable sites for new coniferous forests (The Forestry Authority, 1993). The concept of critical loads was developed originally for determining a European-wide strategy for acid gas emission reductions, in order to have the most beneficial effect on water quality; the application proposed by The Forestry Authority is a catchment-based methodology. This paper assesses the suitability of this methodology for streams in south west Scotland.

2. Data sources and methodology

Areas subject to acidification, principally the Luce, Bladnoch, Cree, Fleet and Dee catchments, are sparsely populated and records are limited. However, from 1956 onwards, pH data, from the Solway River Purification Board (Solway RPB), are available for one site on the Cree, at Bargrennan, and, for other catchments, from the mid-1960s. From 1984, major ion chemistry was determined at all acid sites.

Critical loads for acidity were calculated using the procedures recommended by the Critical Loads Advisory Group (CLAG) (CLAG Freshwaters sub group, 1992) and the methodology was confirmed by Curtis (personal communication). No data for dissolved organic carbon were available and the modification, suggested by CLAG, was not made. The value of excess base cations in precipitation was taken as 8ueql^{-1}, the standard value used by CLAG for Scottish freshwater sites, although values as high as 26ueql^{-1} have been measured at Loch Dee, in Galloway. The limit value for acid neutralising capacity (ANC) was taken as zero.

Fishery data were taken from the West Galloway Fisheries Trust records. Minimum salmonid population densities were calculated from single electrofishing passes. Benthic invertebrates were sampled by Solway RPB using the standard three minute kick sample.

3. Results and discussion

The widespread planting of conifer forests by the Forestry Commission began in about 1948 in the Cree catchment. Upstream of Bargrennan, minor development took place between 1948 and 1954, covering a total of about 7% of the catchment, which was followed by a regular planting of about 4% of the catchment per year until 1965 (Forestry Commission, personal communication). No further development occurred until 1977 and, between that date and 1987, planting extended to about 65% of the catchment, or about 74 km^2.

From 1956 to 1994, pH data at Bargrennan was subjected to cusum analysis, which identifies periods when mean values were steady (Table I).

TABLE I
Mean pH values in the R Cree at Bargrennan

Period	Mean pH	No of values
1956-1967	6.19	61
1967-1973	6.89	22
1973-1984	6.18	48
1984-1994	5.64	87

Each change of mean was highly significant with $p < 0.2\%$

These mean pH values are shown in relation to the extent of forest planting and UK sulphur dioxide emissions (UKRGAR, 1990)(Figure 1). The decline in pH coincides with increasing maturity of the forest, suggesting the increased scavenging efficiency of mature conifers during a period when there was a general reduction in acid gas emissions. Peat

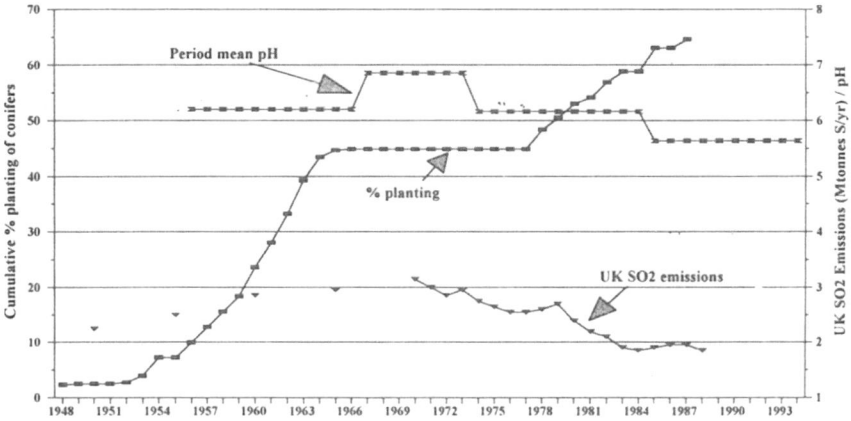

Fig. 1. UK SO$_2$ emissions and catchment planting and pH at Bargrennan (R. Cree)

is the predominant soil in the Cree catchment and the increase in mean pH from 1967 - 1973 may relate to the absence of ploughing. In the nearby Water of Luce catchment, which has little afforestation there was a constant period mean pH of 6.6 in the river at Glenluce from when records began in 1964.

One of the essential requirements of critical load exceedance calculations is data on deposition quality. Solway RPB has been monitoring rainfall quality at Loch Dee for many years and it has been possible to test some of the assumptions made by CLAG in their calculation of critical load values. Two "constants" in the calculations are annual rainfall and an estimate of excess base cations in precipitation. In the absence of measured values, CLAG recommend a value of 8ueql^{-1} for excess base cations. Table II details the annual variation in rainfall and excess base cations measured at the site.

TABLE II
Annual variation in rainfall and excess base cations at Loch Dee

Year	Rainfall (mm)	Excess base cations (ueql^{-1})
1984	2130	11.5
1985	2663	13.9
1986	2740	N/A
1987	2574	14.9
1988	2819	14.9
1989	2248	16.5
1990	2671	26.5
1991	2374	21.8
1992	2856	19.3

For the Green Burn, a Loch Dee tributary, the difference between critical loads calculated using (i) long-term average rainfall and 8 ueql^{-1} excess base cations and (ii) measured annual rainfall and excess base cations, have been plotted as a time series (Figure 2), with positive values indicating that "standard" calculations result in an enhancement of critical load, a situation which implies greater security from acidification. Over 10% of the critical load values calculated from weekly Green Burn water data are below $1 ueql^{-1}$ (using standard "constants"), which emphasises the significance of differences of up to about $0.3 ueql^{-1}$ for extended periods.

Fig. 2. Difference in Critical Loads using standard and actual rainfall

Critical loads were calculated for about 80 sites, sampled at high flow, using the standard "constant" values proposed by CLAG. Half the streams had an ANC < 0 and the calculated average reduction from pre-acidification ANC was about 60 ueq/l. Rainfall data, from the UK Secondary Precipitation Network, adjusted to account for forest cover, was provided by the Institute of Terrestrial Ecology (Hall, personal communication). This was used to determine whether the critical load was exceeded and the results compared with recorded pHs and assessed against available data for benthic invertebrates and salmonid fish densities from the streams (Figures 3-5).

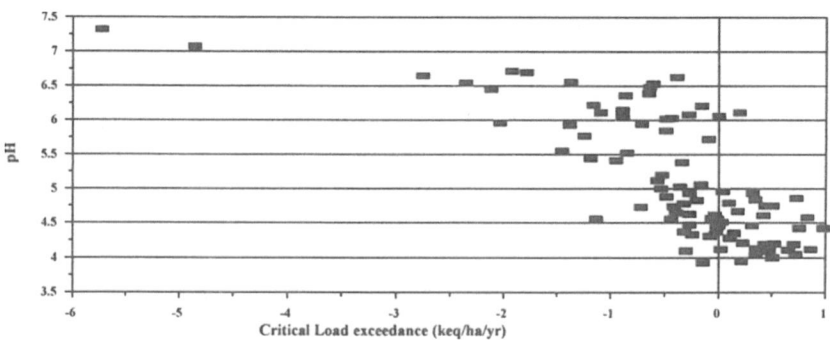

Fig. 3. High flow pH in relation to Critical Load exceedance

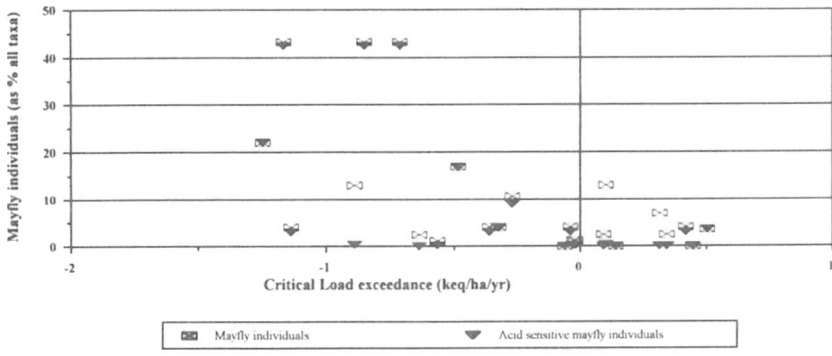

Fig. 4. Number of mayfly individuals in relation to Critical Load exceedance

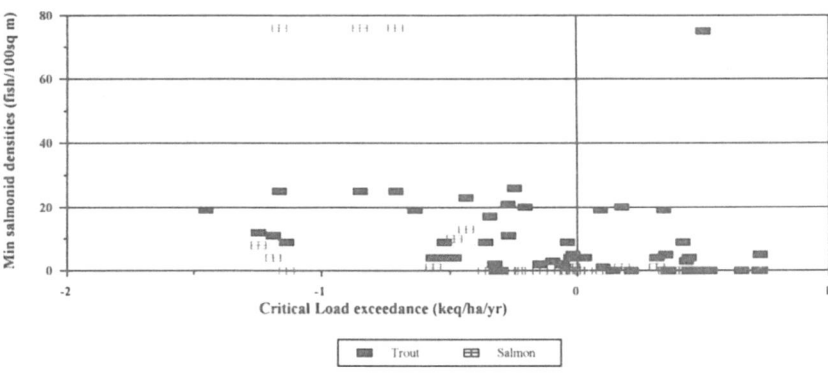

Fig. 5. Minimum density of salmonids in relation to Critical Load exceedance

In many streams with pH < 4.5, and several with pH < 5, the critical load is exceeded. However, the number of streams where the critical load is not exceeded, and pHs are < 5, indicates an uncertainty with the calculation procedures which should be allowed for in any planning application for forest planting. The importance of this allowance is demonstrated from the relationships shown between critical load exceedance and salmonid fish densities or mayfly presence. Where critical loads were exceeded, salmon are generally absent and trout present in only small numbers. Only when critical loads are above deposition estimates by about $0.5 keqha^{-1}yr^{-1}$ do salmon begin to be found. Mayflies disappear as critical loads are exceeded, which suggests that they may not be a simple surrogate for the assessment of fish stocks.

The critical load calculations have been carried out assuming that an acceptable ANC is zero. This is presumed to coincide with a steady-state pH of 5.5 and a minimum pH of about 4.5, and is regarded as the value at which there is a 50% chance of finding trout

(CLAG, 1993). CLAG have also suggested that ANC values substantially greater than zero may be required to protect key species of conservation interest (CLAG, 1992). Data presented here indicates that even when critical loads are not exceeded, by up to about 1keqha^{-1}yr^{-1}, there is no certainty of the presence of trout and that salmon may well be absent. Since the calculations reflect a far from steady state condition, using annual average deposition data and high flow instantaneous water quality, it is not surprising that the link is imprecise. Uncertainties are also introduced by the use of deposition data modelled from values obtained at sites which may be remote from the location of interest. At Loch Dee in 1983, modelled wet sulphur deposition was 1.0 g Sm^{-2}yr^{-1} when the recorded value was 1.9 g Sm^{-2}yr^{-1}(UKRGAR,1987). When 1987 data derived from emission based models was compared, the uncertainty became greater, with estimates from 0.5-1.0 g Sm^{-2}yr^{-1} compared with a measured value of 1.2 g Sm^{-2}yr^{-1} (UKRGAR,1990) depending on the model selected. Since it is these models which will be required to determine whether critical loads are likely to be exceeded up to 15 years ahead of any forestry development, the need for caution is apparent.

These calculations have been based only on estimates of sulphur emissions and deposition. Galloway also suffers from critical load exceedances due to nitrogen deposition, an aspect of the acidification process which is still being developed (UKRGIAN, 1994). For critical load exceedances to be used at face-value at the catchment level to make decisions on developments that may damage high-value salmon fisheries, requires greater confidence in the methodology and confirms the need for the selection of a higher limit for ANC or a margin of safety in the estimate of critical load exceedance.

4 Conclusions

Critical loads may have a place in the general assessment of water quality but the use of the steady state model as a catchment management tool requires considerable caution. The combination of uncertainties in the calculation of critical loads and in the estimates of deposition mean that streams should be regarded as potentially at risk unless the critical load exceeds deposition by at least 1 keqha^{-1}yr^{-1}. A higher safety factor might be appropriate in areas of high nitrogen deposition. Only with these caveats should the procedure be used as a method for forestry planning in acid sensitive areas.

References

CLAG, Freshwaters sub-group: 1992, *Critical Loads and Acid Deposition for UK Freshwaters*, Research Paper No 5, ECRC, University College, London, 115pp.
Solway River Purification Board: 1994, *Annual Report 1993-94*, Solway RPB, Dumfries, 66pp.
The Forestry Authority: 1993, *Forests and Water Guidelines,* 3rd Edition, HMSO, London, 32pp.
UK Review Group on Acid Rain: 1987, *Acid Deposition in the United Kingdom 1981-1985*, DTi, Stevenage, 104pp.
UK Review Group on Acid Rain: 1990, *Acid Deposition in the United Kingdom 1986-1988,* DoE, 124pp.
UK Review Group on Impacts of Atmospheric Nitrogen: 1994, *Impacts of Nitrogen Deposition in Terrestrial Ecosystems,* DoE, London, 110pp.
CLAG, Freshwaters sub-group: 1993, *Critical Loads and Acid Deposition for UK Freshwaters (Summary),* Research Paper No 10, ECRC, University College, London, 28pp.

TOWARDS A NEW METHOD OF SETTING A CRITICAL LOAD OF ACIDITY FOR OMBROTROPHIC PEAT

E. J. WILSON[1], R. A. SKEFFINGTON[1], E. MALTBY[2], P. IMMIRZI[2], C. SWANSON[2] & M. PROCTOR[2]

[1]*National Power Research & Engineering, Windmill Hill Business Park, Whitehill Way, Swindon, Wilts SN5 6PB, UK.*

[2]*Wetland Ecosystems Research Group, Department of Geography, University of Exeter, Exeter EX4 4RJ, UK.*

Abstract. Critical loads of acidity for mineral soils can be set according to the capacity of the underlying bedrock to replenish the base cations leached by acid deposition. Unfortunately, this relatively simple approach cannot be applied to peat, one of the most widely occurring soil types in the wetter, western areas of Europe. These organic soils depend on atmospheric deposition for their supply of base cations rather than mineral weathering. We aim to develop a critical load methodology for ombrotrophic peat, using a combination of field observations and laboratory experiments. Simulated rain has been applied to intact cores of peat to determine the key chemical processes governing the response of these soils to both increases and decreases in acid deposition. It is evident that peat does not behave as a simple ion exchanger; the complex reactions of decomposition, sulphate reduction, nitrate uptake and organic acid production also control the response to acid inputs. This paper looks at some of the results from these experiments and considers the implications for setting critical loads.

1. Introduction

The critical loads approach is now widely used as a tool for developing pollution control strategies (Bull, 1991). It determines the amount of pollutant - the "critical load" - that an ecosystem can tolerate without damage in the long-term (Nilsson & Grennfelt, 1988). The aim is to maximise environmental benefits by targeting emission reductions to reduce deposition in ecologically sensitive areas, rather than uniformly.

Critical loads of acidity have been set for mineral soils based on the capacity of the bedrock to supply base cations. The aim is to protect soils from acidifying any further, rather than restore them to some pre-industrial condition. A shortcoming of this approach is that it cannot be applied to the highly organic, ombrotrophic peat soils which are widespread in the wetter, western areas of Europe. In these soils, the supply of base cations in the surface layers depends on atmospheric inputs rather than weathering of deep bedrock. Most countries have arbitrarily allocated these soils to the lowest (most sensitive) critical load class, although in the UK, a provisional method developed by the University of Aberdeen has been used (Smith *et al.*, 1992). This approach involves the repeated centrifugation and re-suspension of small pellets of peat in order to determine the reaction to acid loading (Skiba & Cresser, 1989). Since the structure of peat is likely to be crucial to its response, we have developed the method using larger intact cores of the hydrologically active portion of the peat (acrotelm). We aim to determine whether these naturally acidic soils are vulnerable to acid deposition and how they might respond to increases and decreases in loading. Data from a combination of field and experimental investigations are being used to define a critical load for ombrotrophic peat.

2. Methods

Eight upland sites with ombrogenous blanket peat were selected to encompass the wide range of pollutant deposition in the UK. The less polluted sites were in Exmoor (SW England) and Beinn Eighe (NW Scotland), the most polluted sites in the South Pennines and North York Moors (N England), and intermediate sites in Berwyns (N Wales) and the North Pennines (N England). Acid inputs range from 34 meq H^+ m^{-2} yr^{-1} at Beinn Eighe, to 269 meq H^+ m^{-2} yr^{-1} in the South Pennines. At each site, 30 intact cores of acrotelm peat (including the vegetation) were taken using a 10 cm diameter drainpipe. In the laboratory, the cores were treated with solutions which mimicked the chemical composition of rainfall where the peat was sampled. Six treatments simulated current H^+ deposition, reductions and increases in acid loading (either 0.2x, 0.5x, 1x, 2x, 4x, and 6x; or 0.2x, 0.5x, 0.8x, 1x, 2x and 4x H^+ deposition) by varying the amount of sulphuric acid in the rain solution. Two years rainfall was applied to 4 replicate cores per treatment over 12 weeks (except Exmoor where it was 9 weeks). The volume and pH of leachate from the peat core was measured daily and a volume-weighted sub-sample bulked on a weekly basis for cation and anion analysis. At the end of the experiment, pH and exchangeable cations were determined for the peat itself. Details of the sampling procedure and experimental methods can be found in Skeffington *et al.* (1995). In addition to the cores, ten replicate samples of the acrotelm were taken at each site to examine the relationship between current atmospheric deposition and peat chemistry in the UK. This part of the project will be discussed in a forthcoming paper.

3. Results

EFFECTS OF ACID TREATMENT OF PEAT CORES

Applying different amounts of acidity to the peat cores had a significant effect ($p<0.05$) on the *net* release or retention of H^+, Ca^{2+} and Mg^{2+} by the peat, based on the flux of ions in core leachate. Figures 1-3 show the net flux (leachate minus rain treatment) of these ions for selected sites. Positive values denote a net release or loss of the ion from the peat and negative values a net gain or retention by the peat core. Generally, Ca^{2+} and Mg^{2+} are retained by the peat in the less acid treatments, with a concomitant release of H^+. As acid inputs are increased, H^+ starts to be retained by the peat and Ca^{2+} and Mg^{2+} are released. It can be seen From Figure 1 that peat taken from Beinn Eighe in Scotland is in approximate equilibrium (ie H^+ flux in equals H^+ flux out) with rain between 4x and 2x current H^+ deposition. Other treatments showed no movement towards equilibrium during the experiment (Fig. 1). A similar equilibration pattern between rain and peat is seen for Ca^{2+} and Mg^{2+}, following an initial 'flush-out' peak (Figs 2 & 3). The rain treatment closest to equilibrium (for H^+) varies from site to site; for South Pennines and Exmoor it is close to current H^+ deposition (1x treatment), whereas for North York Moors, North Pennines and Berwyns it ranged from 2x to 4x H^+ deposition. The flux of cations other than Ca^{2+} and Mg^{2+} was not related to acid treatment. Na^+ behaved conservatively as might be expected, with the net flux stabilising around zero. Principal Component Analysis indicated that the fluxes of K^+ and NH_4^+ were associated; both ions were leached from the peat at all sites except Exmoor which retained NH_4^+.

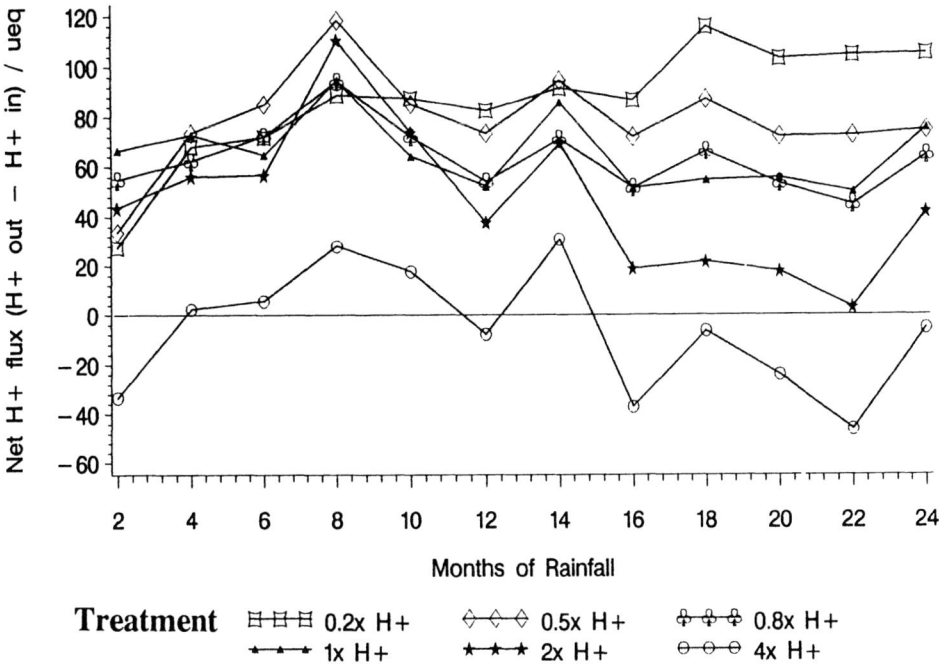

Fig. 1. The net H^+ flux (H^+ in leachate leaving core minus H^+ in rain treatment applied) for *Calluna* peat cores collected from Beinn Eighe, Scotland. Points are means of 4 replicate cores.

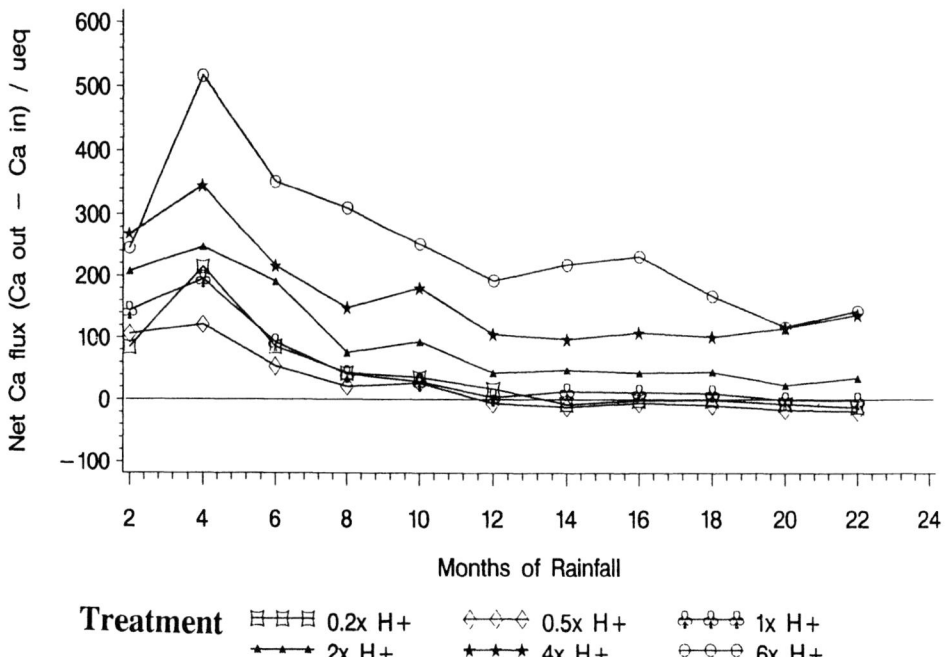

Fig. 2. The net Ca^{2+} flux (Ca^{2+} in leachate leaving core minus Ca^{2+} in rain treatment applied) for *Calluna* peat cores collected from the South Pennines (Redmires). Points are means of 4 replicate cores.

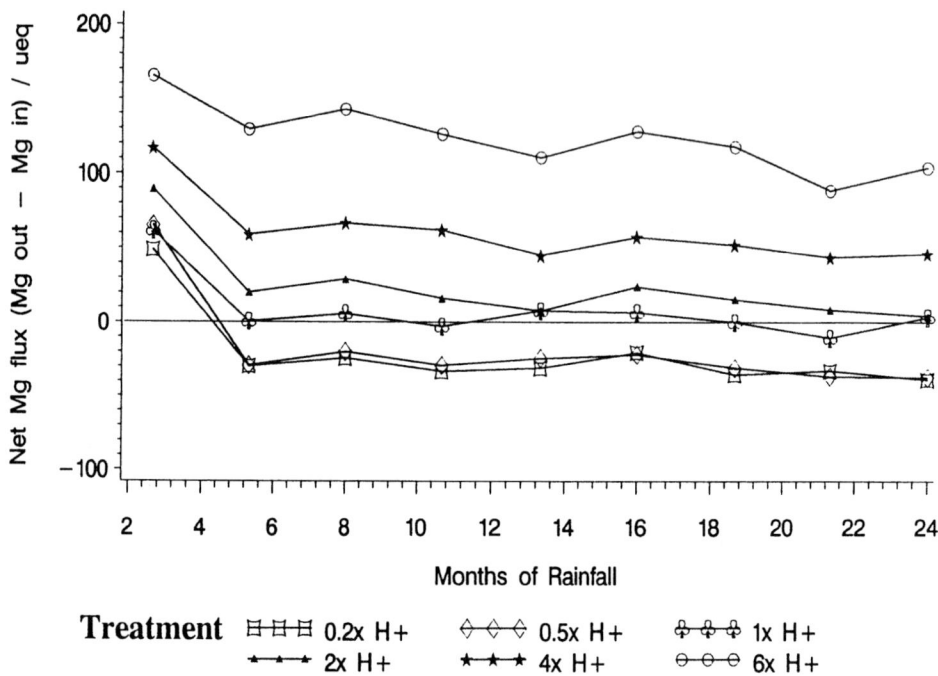

Fig. 3. The net Mg^{2+} flux (Mg^{2+} in leachate leaving core minus Mg^{2+} in rain treatment applied) for *Eriophorum* peat cores collected from Exmoor. Points are means of 4 replicate cores.

Acid treatment had a significant effect on the amount of SO_4^{2-} released or retained by the cores for most sites. In general, sulphate was released from the peat in the lower treatments but differences between treatments were not significant. In the higher treatments (2x-6x ambient SO_4^{2-} deposition), sulphate was generally *retained* by the peat, and retention increased significantly with input. Only in the Beinn Eighe and Berwyns peat was SO_4^{2-} released in all treatments. The effect of acid loading on SO_4^{2-} flux is shown for South Pennines (Holme Moss) in Figure 4. Treatment had no effect on either NO_3^- or Cl^- flux. NO_3^- was taken up by the peat in most sites except North Pennines and Beinn Eighe where it was approximately in equilibrium with rain inputs. Cl^- generally behaved conservatively with average net fluxes of zero, although there was a small net release (Berwyns, North York Moors) and retention (Beinn Eighe) in some sites.

Ion budget analysis and multiple regression have been used to identify the key processes controlling the retention or release of H^+ in response to changes in acid loading. In Exmoor peat for example, ion exchange with Ca^{2+} and Mg^{2+} is quantitatively an important H^+-consuming process, with SO_4^{2-} retention providing an equal or larger sink for H^+ at high acid (and SO_4^{2-}) loadings (Figure 5). It is base cation exchange however, that best explains the treatment-related variation in H^+ flux for Exmoor. At other sites (eg South Pennines), the change in SO_4^{2-} flux explains a greater % of the variation in H^+ retention/production with acid loading than does $Ca^{2+} + Mg^{2+}$ exchange. Other processes such as NO_3^- retention and K^+ and NH_4^+ leaching are quantitatively important sinks for H^+ in all sites, but do not explain the increase in retention of H^+ by the peat at higher acid loading.

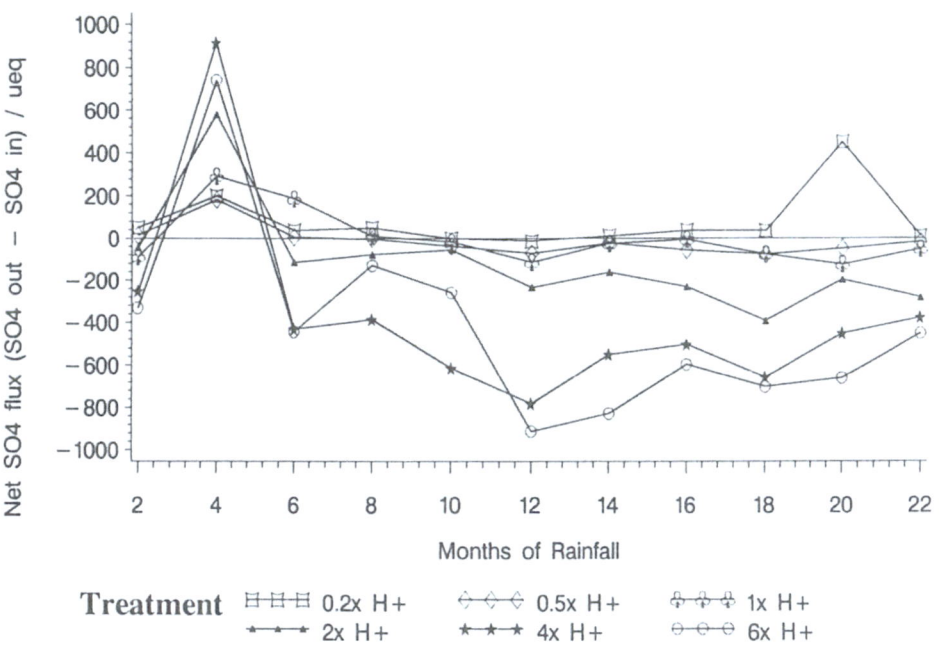

Fig. 4. The net SO_4^{2-} flux (SO_4^{2-} in leachate leaving core minus SO_4^{2-} in rain treatment applied) for *Eriophorum* peat cores collected from the South Pennines (Holme Moss). Points are means of 4 replicate cores.

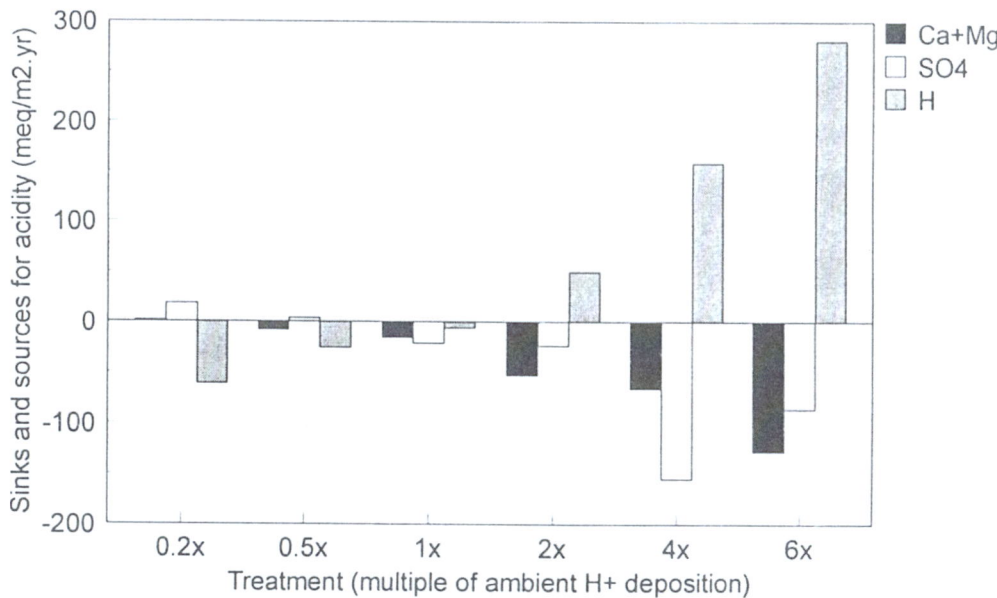

Fig. 5. The effect of acid treatment on the net production/retention of H^+ by peat and the contribution from two major controlling processes ($Ca^{2+}+Mg^{2+}$ exchange and S transformations). Positive values for Ca+Mg and H indicate retention of these ions in the peat, while negative values denote their release. In the case of SO4, negative values indicate reduction to S^{2-} and positive values oxidation to SO_4^{2-}. *Sphagnum* peat collected from Exmoor.

4. Discussion

The peat cores in this experiment did not attain chemical equilibrium with the applied rain solutions, suggesting that soil quality is unlikely to respond rapidly (within a few years) to changes in deposition chemistry. Even if peat were to behave as a simple ion exchanger, changes in the composition would be slow since the exchangeable pool of H^+ is so much larger than the atmospheric flux of H^+. In practice, the response is complicated by a number of internal processes which act as net sinks or sources of acidity, modifying the equilibrium position. The effect of treatment on SO_4^{2-} retention in the peat cores suggests that the reduction of SO_4^{2-} to H_2S, FeS/FeS_2 and organic-S is likely to be an important sink for H^+ deposition, especially in more polluted sites. The release of K^+ and NH_4^+ does not appear to be controlled by ion exchange and most likely originates from decomposition, of either the peat itself or moribund vegetation on the surface of the core. Most peat cores retained all the nitrate applied, irrespective of their original pollution climate or type of vegetation cover. This has also been reported for catchment scale experiments (eg Hemond, 1980) and is probably a result of assimilation by biomass or denitrification. Both decomposition and NO_3^- uptake/denitrification will consume a certain amount of incoming H^+ deposition although results indicate that the magnitude of this sink is independent of acid loading.

One criterion for setting a critical load for ombrotrophic peat might be to prevent any further deterioration in base saturation, analogous with the approach used for mineral soils. The loss of Ca^{2+} and Mg^{2+} by exchange with incoming H^+ will depend partly on the deposition of base cations and the base saturation % of the peat itself. The response will be modified by other processes which generate or consume acidity, and all of these need to be taken into account when setting a critical load. The insight we have gained from this experiment into the key processes involved in acid metabolism is currently being used to explore different approaches for calculating a critical load for peat. These will be reported in a later paper.

Acknowledgements

This contribution forms part of the National Power-PowerGen Joint Environmental Programme and is published by permission of the Programme Committee.

References

Bull, K. R.: 1991, *Environ. Pollut.* **69**, 105-123.
Hemond, H. F.: 1980, *Ecol. Mon.* **50** (4), 507-526.
Nilsson, J., Grennfelt, P. (Eds): 1988, *Critical Loads for Sulphur and Nitrogen.* Report 1988:15 Nordic Council of Ministers.
Skeffington, R. A., Wilson, E. J., Immirzi, P., Maltby, E.: 1995, *Hydrology and Hydrochemistry of British Wetlands.* (Eds. J. M. R. Hughes & A. L. Heathwaite). Chap. 5, 183-198. John Wiley & Sons Ltd.
Skiba, U., Cresser, M.: 1989, *Water Res.* **23**, 1477-1482.
Smith, C. M. S., Cresser, M. S., Mitchell, R. D. J.: 1992, *Ambio* **22**, 22-26.

THE CALCULATION OF BASE CATION RELEASE FROM THE CHEMICAL WEATHERING OF SCOTTISH SOILS USING THE PROFILE MODEL.

SIMON J. LANGAN[1], HARALD U. SVERDRUP[2] and MALCOLM COULL[1]

1 Macaulay Land Use Research Institute, Craigiebuckler, Aberdeen, AB9 2QJ, UK
2 Department of Chemical Engineering, University of Lund, S-22100, Sweden

Abstract: To investigate the weathering rates of different soil parent materials which occur in Scotland, a study has been undertaken in which detailed soil mineralogy has been used to calculate base cation release. To calculate base cation release, this data, and supplementary soil chemistry and physical attribute data, have provided the input to the PROFILE model. The model is a multi-layer, steady state, deterministic model in which the soil is represented by a series of mixed tank reactors, each of which has the mineralogical, physical and chemical attributes measured for individual soil horizons.

The major parent materials from which Scottish soils have developed are glacial till, derived from acid to basic igneous rocks, schist and other metamorphic types, Lower Palaeozoic greywackes and shales, Old Red Sandstone sediments, Carboniferous sediments and Permo-Trias sediments. For each of the parent materials, three soil profiles were analysed and used with the PROFILE model. The base cation release rates, calculated for these parent materials in the top 50cm of the soil profile, varies between 0.2 and 3.2 keq/ha/yr, although, for a given parent material, the range was usually quite small. In general, these results compare very favourably with those suggested for the calculation of critical loads using an empirical approach proposed at Skokloster. In comparison with current rates of deposition, this suggests many of these soils are being acidified and that for many soil-plant combinations, the critical load may be exceeded.

1. Introduction

The rate at which base cations are released from the chemical weathering of soil parent materials is vital to our understanding of soil fertility, plant nutrient supply, soil and surface water acidification. In the study of acidification the balance between base cation release through weathering and loss by biological uptake or leaching will determine the sensitivity of the ecosystem to enhanced acidification resulting from acid deposition.

Regional and national assessments of sensitivity to acidification have frequently incorporated a measure of the weathering rate. For example, Norton (1980), Kinniburgh and Edmunds (1986), and Hornung et al (1995) based ecosystem assessments on the classification of geology and soils according to weathering rates and soil chemistry. These approaches simply ranked geology and soils according to different estimates of buffering capacities. However with progression towards a new European sulphur protocol and the proposals for a targeted ecosystem approach based on critical loads (Bull 1992) there was a need to assign values to these different sensitivities.

Much of the work to date on critical loads has largely been based on empirical data and evidence, for example, Henriksen *et al* (1986), Chadwick and Kuylenstierna (1991) and Langan and Wilson (1994). In order to develop and advance the use of weathering rates as an indicator of environmental sensitivity (for acidification), it is necessary to consider independent and non empirical methods for their calculation. One such approach is the use of mathematical models.

The aim of this paper is to show the calculation of soil weathering rates by the application of the PROFILE model to a range of soils and parent materials common to Scotland. The results of these calculations are reviewed and briefly discussed in terms of differences between the soils weathering rate and critical load suggested by Langan and Wilson (1994) using the empirically based Skokloster, Nilsson and Grennfelt (1988), approach.

The major parent materials from which most Scottish soils have developed are ultimately derived from acid to basic igneous rocks and till, schist and other metamorphic types, Lower Palaeozoic greywackes and shales, Old Red Sandstone sediments, Carboniferous sediments and Permo-Triasic sediments (after Wilson et al, 1984).

2. Methods and materials

The PROFILE model is a steady state deterministic model in which the soil is represented by a series of mixed tank reactors, each of which has the mineralogical, physical and chemical attributes measured for individual soil horizons. The release of base cations from silicate minerals to soil solution is described in the model as a result of the reactions with a) hydrogen, aluminium and the cation content of the parent material, b) water and aluminium, c) carbon dioxide and, d) organic acids. Constraints on the rate of release in the model are included as a function of complexing on active dissolution sites (Sverdrup and Warfvinge, 1993) and the availability of moisture as a function of soil water content. A more comprehensive description and discussion of the model structure can be found in Warfvinge and Sverdrup (1992) and Sverdrup and Warfvinge (1993). The model has been suggested as one of the 'level 1' methods for the calculation of critical loads (Sverdrup et al, 1990) and increasingly the model is being used in Europe as part of the UN-ECE critical load programme.

For the present study, 17 Soil Associations have been selected which represent the spectrum of parent materials and of geological units occurring in Scotland (Table I). For each association, three soil profiles were selected from the soil archive at MLURI. In recognition of the well documented problems of acidification of soils and waters associated with granites and greywackes of the Countesswells and Ettrick associations, respectively data from six soil profiles were used.

The criteria for soil selection were that the soil should be representative or typical of the most extensive soil series within the association, wherever possible the soil should be from under a natural/semi-natural vegetation type and that there should be supplementary physical and chemical information available which describe the soil profile from which the samples were taken.

For each soil profile, three horizons were selected for X-ray diffractometry (XRD) determination of the soil mineralogy. The methods and techniques of Wilson (1987) and Chung (1974) were used to give a semi-quantitative estimate of the percentage concentration of each of the minerals present. This data was then used as an input to the model. The model additionally requires data which give site and soil profile details in terms of vegetation, temperature and number of soil horizons. These were available from soil surveyors field records and soil monographs. Precipitation and deposition data required as model input were from the UK Acid Deposition national monitoring database (UKRGAR, 1990). From these data, the model calculates steady state base cation release rates as well as other soil chemistry (such as base saturation, base cation/aluminium ratios). This information is provided on a horizon by horizon basis or as a total for the soil.

3. Results

Table I shows the Skokloster class for each of the associations considered along with the range and mean base cation release calculated by PROFILE. In order to compare the results, the weathering rate calculated is the cumulative weathering down to a depth of approximately 50 cm.

Of the seventeen associations analysed and modelled, PROFILE calculated thirteen associations to have a weathering rate within the range allocated according to the Skokloster scheme. For three of the associations where there was a discrepancy between the two methods of calculation, the PROFILE results underestimated the weathering rate in comparison with Skokloster. For these associations, the soil mineralogy could be expected to have a small dark mineral content, which may not have been detected by the methods used in the study. Some of these dark minerals have high rates of dissolution and, if present, even in small amounts, will give rise to elevated weathering rates. It is also possible that the soil mineraoogy is not identical to the inerred parent material mineralogy. The fourth association in which there was a discrepancy between the two methods was the Balrownie association. The parent material of this association is described as sandstones, which have frequently been water modified. As a consequence of this modification, the resultant soils exhibit a large range in mineralogy and texture, which is reflected in the input data and the results of the model. The results for the Balrownie association show the largest range of weathering rates of those associations studied. It would be difficult for any approach to classify such a parent material's weathering rate within a limited range.

The model results indicate that weathering rates, for those parent materials considered, vary from <0.2 keq ha^{-1} yr^{-1} for the most sensitive soils developed from quartzites, to in excess of 1.5 keq ha^{-1} yr^{-1} for soils on water modified sandstones. The majority of soils, for those areas which are uncultivated and support both natural and semi-natural ecosystems of high conservation value, are sensitive to acidification with critical loads in the range 0.2 to 0.5 keq ha^{-1} yr^{1}. These are the soils derived from granites, greywackes and shales and metamorphic schists and gneisses. A general comparison of both current and predicted future deposition (modelled on a regional basis (UKRGAR, 1990) to the areas of study suggest that acid inputs are within the range of base cation release from soil weathering. This implies that the ecosystems developed on these parent materials may still be at risk from increased soil acidity.

For the majority of associations, the Skokloster class defines a range of weathering rates within which the results from PROFILE are the same. It is possible to suggest, at this stage

Association	Parent material geology	% cover	Skokloster keqH⁺ha yr⁻¹	PROFILE 50cm keqH⁺ha yr⁻¹	Soil type	Association	Parent material geology	% cover	Skokloster keqH⁺ha yr⁻¹	PROFILE 50cm keqH⁺ha yr⁻¹	Soil type
Arkaig	Schists, gneisses, granulites and quartzites principally from the Moine Series	16.22	0.2-0.5	0.3		Hobkirk	Sandstones and marls of Upper Old Red Sandstone age	0.75	0.5-1.0	0.4	BE
				0.2	PG					0.6	BE
				0.3	PG					0.1	BE
				0.3	PG					0.5	BE
Balrownie	mainly Sandstones of Lower Old Red Sandstone age, often water modified	1.83	0.5-1.0	1.9		Insch	Gabbros and allied igneous rocks	0.67	1.0-2.0	0.5	BE
				0.9	GBE					0.6	BE
				1.5	GBE					0.2	BE
				3.2	GBE					0.6	BE
Corby	Fluvioglacial and raised beach sands and gravels derived from acid igneous rocks	3.08	0.2-0.5	0.3		Lochinver	Lewisian gneiss	4.67	0.2-0.5	0.2	PG
				0.3	HIP					0.1	PG
				0.3	HIP					0.1	PG
				0.2	HIP					0.4	PG
Countesswells	Granites and granitic rocks	5.75	0.2-0.5	0.3		Rowanhill	Carboniferous sandstone, shales and limestone's	3.05	0.5-1.0	0.5	BE
				0.3	HIP					0.2	BE
				0.5	HIP					0.6	BE
				0.3	HIP					0.6	BE
				0.2	PPG		Sourhope	Old Red Sandstone intermediate lavas	1.71	0.5-1.0	0.6	BE
				0.1	PGP					0.3	BE
				0.3	AP					0.5	BE
Darleith	Basaltic rocks	3.53	1.0-2.0	0.7						1.1	
				0.9	BE	Strichen	Arenaceous schists and strongly metamorphosed argillaceous schists of Dalradian age	7.98	0.2-0.5	0.4	HIP
				0.4	BE					0.3	HIP
				0.8	BE					0.3	HIP
Durnhill	Quartzites and quartzose grits	1.6	<0.2	0.1						0.7	HIP
				0.1	PG	Tarves	From intermediate rocks or mixed acid and basic rocks, both metamorphic and igneous	2.07	0.5-1.0	0.8	BE
				0.1	PG					0.9	BE
				0.0	HIP					0.7	BE
Ettrick	Lower Palaeozoic greywackes and shales	9.26	0.5-1.0	0.8						0.7	BE
				0.7	BE	Thurso	Middle Old Red Sandstone flagstones and sandstone's	1.35	0.5-1.0	0.7	NCG
				0.8	BE					1.0	NCG
				0.8	BE					0.7	NCG
				0.4	PP					0.5	
				0.6	PP	Torridon	Torridonian Sandstone's and grits	2.25	0.2-0.5	0.2	PG
				0.7	PG					0.2	PG
				0.7	HIP					0.1	PG
Foudland	Slates, phyllites and other weakly metamorphosed argillaceous rocks	3.25	0.2-0.5	0.3	HIP					0.2	
				0.2	HIP						
				0.4	HIP						
				0.4	HIP						
				0.8	HIP						

Table 1: Details of soils, parent materials and their base cation release as predicted by PROFILE and Skokloster approaches
(Soil types: AP Alpine Podzol, GBC Gleyed Brown Earth, PGP Peaty Gleyed Podzol, BE Brown Earth, PG Peaty Gley, NCG Non Calcareous Gley, HIP Humus Iron Podzol, PP Peaty Podzol)

of the analysis, that the weathering rate results from the model provide (with the noted exceptions) an independent verification of the use of a limited range weathering rate to describe the supply of base cations from specific generic parent materials. However, because of the variation in the weathering rates shown, it is more difficult to sustain the argument for setting a single weathering rate and critical load for different parent materials.

One outcome of this work is, therefore, to suggest that the empirical soil based critical load maps for the UK and Europe be reconsidered, with the critical load for each Skokloster class redefined in terms of a range of weathering rates. If this approach is adopted, it will convey a level of uncertainty in the results which, would answer the earlier criticism of the UK and European maps from some non governmental organisations.

A consequence would be to create a revised exceedance map(s) in which it would also be possible to show some uncertainty. A suggested reconfiguration of the exceedance map would be similar to that suggested by Langan and Hornung (1992) in which soils and ecosystems could be classed as: not exceeded, within the range of exceedance or exceeded by, and specify a range.

Further validation and comparison of the model performance, in calculating weathering rates, with other available methods is essential. As outlined above, many upland UK soils have weathering rates which are comparable to atmospheric inputs of acidity. Without accurate estimates of soil weathering rates it will be difficult to predict the ecosystem response which results from emission abatement. To achieve better estimates of weathering rates, work is in hand to expand both the number and scale (soil profile, catchment and regional applications) of applications on which the model has been run together with other methods of weathering rate determination.

4. Conclusion

For the majority of parent materials considered, there seems to be a general agreement between the weathering rates used in the UK empirical critical load map based on the Skokloster method and those calculated by the PROFILE model. These two methods seem to represent a reasonable approximation of soil weathering rates, given our current understanding, by which we can provide a synthesis of regional weathering rates. Further more detailed analysis and comparison of the model with other calculations of weathering rates, at a variety of geographical scales, is necessary. The results of this analysis should be considered and incorporated into the empirical national soils critical load and exceedance maps. This could be achieved by substituting the current single values for critical load with a range of critical loads. On this basis revised exceedance maps should be drawn, and further work undertaken, to identify how this range of critical loads and exceedances relates to different ecosystems viability. This would help to portray a level of uncertainty associated with critical loads which is currently not shown on critical load and exceedance maps.

Acknowledgement

This work was carried out under funding from the Department of the Environment (Air Quality Division) critical load programme and The Scottish Office Environment Department, to whom the authors are grateful. The views expressed are however those of the authors.

References

Bull, K.R. 1992 1991 *Environmental Pollution* **69**, 109
Chadwick, M.J. and Kuylenstierna, J.C.I., 1991 in *Acid Deposition: Origins, Impacts and Abatement Strategies.* Ed. J.E. Longhurst. Springer-Verlag.353pp.
Chung, K.H., 1974 *Journal of Applied Crystallography* **7**, 519
Henriksen, A., Dickson, W., and Brakke, D.F. 1986 *in Critical Loads for Nitrogen and sulphur.* Ed. J. Nilsson Nordic Council of Ministers Report 86:11.
Hornung, M., Bull, K.R, Cresser, M.C, Ullyett, J., Hall, J.R., Langan, S.J., Loveland, P.J., and Wilson, M.J. 1995 *Environmental Pollution* **87**, 207
Langan S.J., and Hornung M., 1992 *Environmental Pollution*, **77**, 205
Kinniburgh, D.G., and Edmunds, W.M., 1986 The susceptibility of UK groundwaters to acid deposition. *Hydrogeological Report 86/3* British Geological Survey, Wallingford, UK.
Langan, S.J, and Wilson, M.J, *1994 Water Air and Soil Pollution* **74**, 177
Nilsson, J., and Grennfelt, P., (Eds.) 1988 Critical loads for sulphur and nitrogen. *Report from a workshop held at Skokloster, Sweden* March 1988, Published by Nordic Council of Ministers, Copenhagen.
Norton, S.A. in *Atmospheric Sulphur Deposition: Environment Impact and Health Effects.* pp.539-53. Ann Arbor, Michigan.
Sverdrup, H.U., De Vries, W., and Henriksen, A. *1990 Mapping critical loads: A guidance to the criteria, calculations data collection and mapping of critical loads.* Nordic Council of Ministers report. 1990:14.
Sverdrup, H. U. and Warfvinge, P. *Journal of Applied Geochemistry* **8**, 273
United Kingdom Review Group on Acid Rain (UKRGAR) 1990. *Acid deposition in the United Kingdom 1986-1988.* Third report of UKRGAR, published by H.M.S.O, London.
Warfvinge, P., and Sverdrup, H., 1992 *Water, Air and Soil Pollution* **63**, 119
Wilson, M.J., Bain, D.C., and Duthie, D.M.L., 1984 *Clay Minerals* **19**, 709
Wilson, M.J., (ed.) 1987 *A handbook of determinative methods in clay mineralogy.* Blackie, Glasgow.

ESTIMATING UNCERTAINTY IN THE CURRENT CRITICAL LOADS EXCEEDANCE MODELS

R. I. SMITH[1], J. R. HALL[2] AND D. C. HOWARD[3]

[1]*Institute of Terrestrial Ecology, Edinburgh Research Station, Bush Estate, Penicuik, Midlothian EH26 0QB, Scotland.* [2]*Institute of Terrestrial Ecology, Monks Wood, Abbots Ripton, Huntingdon PE17 2LS, England.*
[3]*Institute of Terrestrial Ecology, Merlewood Research Station, Grange-over-Sands, Cumbria LA11 6JU, England.*

Abstract. The critical loads approach to quantifying areas at risk of damage requires deposition and critical loads data at the same spatial scale to calculate exceedance. While maps of critical loads for soil acidification are available at a 1 km scale no monitoring networks in Europe measure wet and dry inputs at this scale and, further, the models currently used to estimate deposition incorporate a number of assumptions which are not valid at the 1 km scale. Simulations of 1 km deposition from 20 km data show that the uncertainty introduced by using 20 km scale estimates of deposition is small, except in mountain areas where it can give misleading results, but a major problem is the uncertainty in estimates of deposition at the 20 km scale produced by the current models.

Key words: critical load, deposition model, spatial scale, uncertainty, probability distribution

1. Introduction

In Europe, assessments of the deposition of various chemical species and their effects on vegetation, soil and freshwaters are used to develop policies for control of pollutant emissions. A critical loads approach has been adopted to identify the geographical distribution of sensitive receptors and to assess the size of the pollutant effects. The critical load exceedance, the excess deposition above the critical load, is the statistic used to identify areas at risk of damage.

Deposition of pollutants throughout Europe is estimated at a range of spatial scales varying from 5 km to 150 km but the spatial variability of land surface sensitivity to pollutants is dominated by sub 1 km scale differences over much of the area. The differences in spatial scales may therefore be an important area of uncertainty in the estimate of critical load exceedances. For the UK, deposition is estimated at a 20 km scale using models for wet, dry and cloud droplet deposition (RGAR, 1990). Constraints on the availability of monitoring data, various simplifying assumptions within the models and an incomplete understanding of the processes preclude the estimation of deposition for the UK at a 1 km scale, since these estimates would reflect regional generalisations rather than true local heterogeneity.

The critical load for acidity of soils can be determined by a number of methods and, for the UK, is linked to soil maps at a 1 km scale. A simple approach allocates empirical critical load classes based on the soil sensitivity to acidic deposition. More complex methods using mass balance equations or dynamic process-based models are being developed. Only the use of soil sensitivity classes, known as the empirical soil critical load and used in current international negotiations, is independent of the deposition of base cations, principally calcium and magnesium, or of acidifying components, like sulphur and nitrogen, and the uncertainty introduced from these deposition estimates. Apart from difficulties in mapping, the current protocols produce unique empirical soil critical load values for each 1 km square, but the underlying uncertainty is hidden and generally unknown.

In this study the empirical soil critical load at the 1 km scale is used and will be assumed to be exact. Modified models of the deposition processes and data at a 20 km scale will be used to estimate the likely distribution of 1 km deposition values within a 20 km square. This permits estimation of the area of critical load exceeded along with a range of uncertainty at the 1 km scale, although not an exact locational match of deposition and critical load. The results will be compared with the standard method where the same 20 km deposition value is used for all 400 of the 1 km squares within the 20 km square.

2. Estimation of deposition

2.1. THE 20 KM DEPOSITION MODELS

The deposition models considered in this paper are those used to provide the deposition of non-marine sulphur to 20 km squares in the UK with output expressed as deposition of acidity assuming $2H^+$ per non-marine SO_4^{2-}. Total deposition is the sum of deposition through 3 pathways, wet, cloud droplet and dry, and the importance of the pathways varies considerably across the UK, depending on region and landscape.

Wet deposition is modelled as a product of rainfall and rain ion concentration, both enhanced at higher elevations by the presence of polluted orographic cloud. These enhancements are modelled in a simple form suitable for average rather than detailed topography.

Both cloud droplet deposition and dry deposition are modelled using a resistance analogy 'big-leaf' model. Cloud droplet deposition is modelled to high elevation land uses at deposition rates close to those for momentum with only an aerodynamic resistance used to calculate the deposition velocity. For dry deposition 3 resistances, an aerodynamic, a quasi-laminar boundary layer and a canopy resistance, are required to calculate the gas deposition velocity. Deposition is obtained as the product of deposition velocity and concentration.

The models are described in more detail in Smith *et al* (1995) where, following discussion of the accuracy of the inputs and the sensitivity of the models, it is suggested that the estimates of total sulphur input to a 20 km square could have an uncertainty of ±40% in central England increasing to ±80% in the west of Scotland and in Wales. The basis on which these estimates are made is quite fragile as there are many uncertainties in the system which cannot be quantified directly without an extensive investigation. Various catchment studies suggested that predictions within ±30% were possible with more detailed modelling, even in areas where high uncertainty was expected. For the purposes of this study, the estimated total sulphur deposition to any 20 km square in the UK is assumed to have an uncertainty of ±40%.

2.2. SIMULATED 1 KM DEPOSITION

A set of possible 1 km deposition values were generated from data held at the 20 km scale by modifying the 20 km scale models. Deposition was also estimated using these modified models at both the 20 km and the 100 km scale for comparison purposes.

All 20 km squares in England, Scotland and Wales with at least 25% of the square as land were selected. Using a detailed altitude map all 400 1 km squares within each of the 652 selected 20 km squares were categorised into 8 altitude bands. Gas and rain ion concentrations were assumed constant over the 20 km square but wind speed and rainfall were adjusted for altitude. The orographic enhancement of wet deposition was modelled by assuming all rainfall in excess of 800 mm had twice the normal rain ion concentration. The distribution of land uses within each 20 km square was used to infer likely distributions of land uses within each altitude band for each 1 km square by assuming that moorland occurred at the highest altitudes, then forest, grass, arable and urban in descending altitude order. Distributions of deposition values at the 1 km scale were then generated for each altitude band in each 20 km square.

The modified models gave results at the 20 km scale with a similar spatial pattern to those from the original 20 km models and with 5% less accumulated deposition over the country.

Over the whole country the distribution of differences between each of the simulated 1 km deposition estimates and the corresponding 20 km estimate was positively skewed. The mean percentage difference was 2% and the range was -70% to +330%. 10% of the 1 km estimates were more than 32% greater than the 20 km estimate and 5% more than 50% greater. In absolute terms about 21000 km^2 have simulated 1 km deposition about 0.3 keq H$^+$ ha^{-1} y^{-1} above the 20 km estimate and 7000 km^2 about 0.5 keq H$^+$ ha^{-1} y^{-1} above.

3. Critical loads and critical load exceedances

The soil critical loads for acidity were determined by dividing soil materials into five classes defined on the basis of their dominant weatherable minerals (Nilsson and Grennfelt, 1988). Critical loads were then assigned to these classes according to the amount of acidity which could be neutralised by base cations released from weathering of the relevant minerals. The class values used in this paper were set at 0.2, 0.5, 1.0, 2.0 and 4.0 keq H$^+$ ha^{-1} y^{-1}.

The critical loads at the 1 km scale provided a distribution of critical load values for each altitude band in each of the 652 selected 20 km squares. The critical load exceedances were then calculated using a series of deposition estimates: the 100 km deposition estimate for current deposition and both 20 km and simulated 1 km estimates for current deposition, ±20% of current deposition and ±40% of current deposition. To match the simulated 1 km deposition estimates against 1 km critical loads, maximum and minimum areas of exceedance were calculated for each 20 km square. In many cases the distributions of critical load values and of deposition values within an altitude band in a 20 km square did not overlap, so the area of critical load exceedance was uniquely defined (i.e. all squares exceeded or no squares exceeded). In cases where the two distributions overlapped 100 random matchings of deposition to critical load in each altitude band within each square were investigated.

4. Areas of critical load exceedance

4.1. NATIONAL ESTIMATES

The areas of critical load exceedance for the 652 selected 20 km squares calculated using the different deposition estimates are shown in Figure 1. The current 100 km estimate gave 74413 km^2 compared to the current 20 km estimate of 82472 km^2. The range of areas of critical load exceedance from current 1 km deposition estimates, that is the 'best case' scenario to the 'worst case' scenario, was 78248 to 85034 km² or about 8% of the current 20 km estimate. Using 100 random matches where there were possible changes in the area of exceedance gave a shorter likely range of 81348 to 81525 km², indicating that the likely estimate from the current 1 km simulations was about 1000 km² less than the current 20 km estimate.

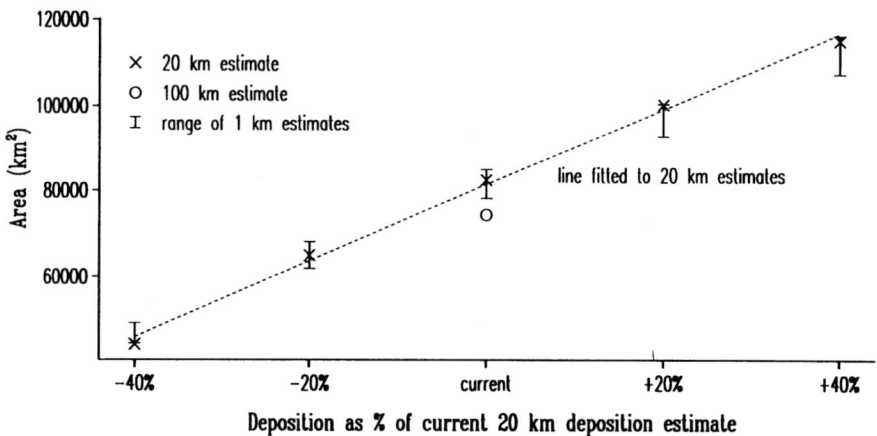

Fig. 1. Area of critical load exceedance for all 20 km squares (total area 230000 km²).

The most striking feature of figure 1 is that the potential uncertainty caused by spatial scale mismatch is much smaller than the potential uncertainty in the output of the regional scale models as reflected by the assumed ±40% uncertainty in the 20 km scale estimates used in this study. In the range of -40% to +40% of the current 20 km deposition estimate there is an almost linear relationship between deposition and area of critical load exceedance, with each 1% change in deposition causing a change in area of critical load exceedance of 900 km², 1.1% of the current 20 km deposition. The relationship is non-linear beyond these specified ranges.

The areas of critical load exceedance from the 20 km deposition estimates were within the ranges of areas from the simulated 1 km deposition values except at -40% of current deposition. Using the simulated 1 km deposition estimates and including the ±40% uncertainty at the 20 km scale gave the estimated national area of critical load exceedance as 80000 km² with a range of uncertainty of 50000 km² to 110000 km², or ±38%.

4.2. REGIONAL ESTIMATES

To investigate whether type of landscape affects the results, 5 areas of ten 20 km squares covering an area of 40 km E-W x 100 km N-S were analysed separately. The comparisons of areas of critical load exceedance from 20 km and simulated 1 km deposition estimates are shown in figure 2.

In East Anglia, a fairly flat landscape, the areas of exceedance estimated from the 20 km deposition and the 1 km deposition agreed well. The model adjustments used to simulate the 1 km deposition values would not give large variability in this landscape and the uncertainty in 20 km deposition estimates was a more important problem. The area of exceedance was approximately 25% of the land area with a range of 18% to 32%.

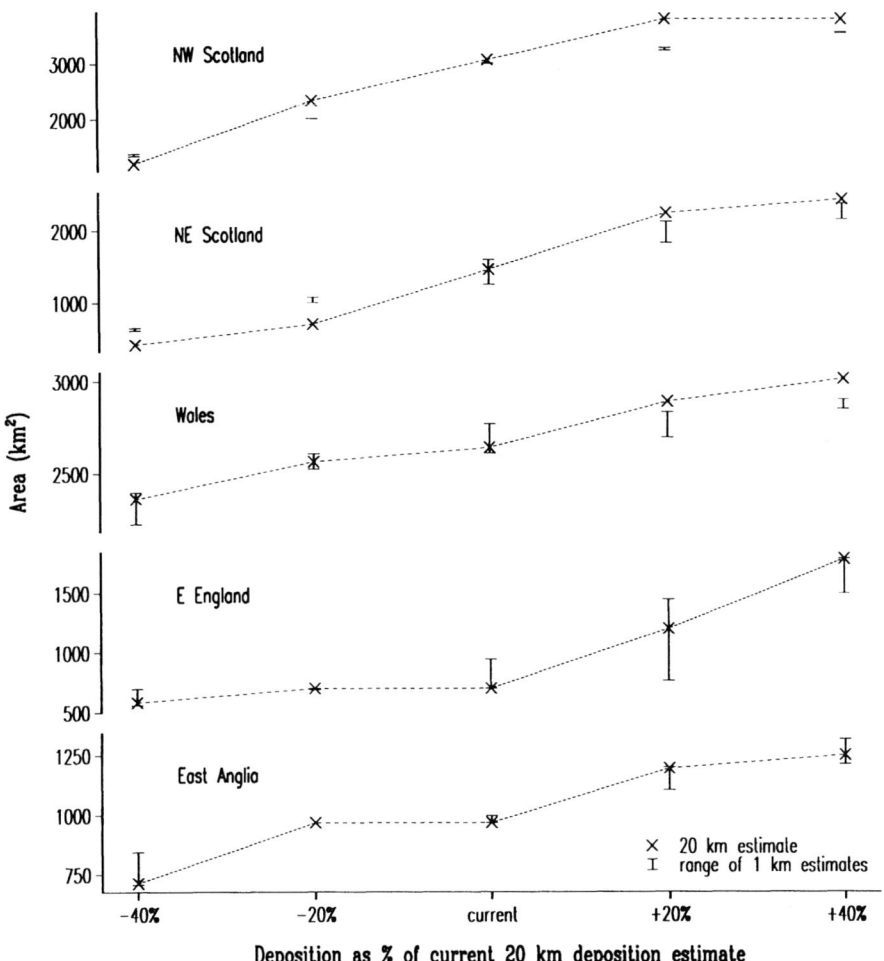

Fig. 2. Area of critical load exceedance for five 40 km x 100 km regions.

In east England, dominated by dry deposition and close to a number of major sources of sulphur, the -40% and -20% 20 km deposition estimate gave little reduction in area of exceedance. The 1 km and 20 km estimates agreed and the range of 1 km estimates was similar to the range of uncertainty in the 20 km deposition estimate. The area of exceedance was approximately 18% of the land area with a range of 14% to 40%.

In a mountainous region of Wales the +20% and +40% 20 km deposition estimates both appeared to overestimate the area of critical load exceedance. Both uncertainty in the 20 km estimate and the spatial scale matching problem were important. The area of exceedance was approximately 68% of the land area with a range of 57% to 73%.

In the north of Scotland the two areas, one near the wetter west coast and one in the east, were both mountainous and in both the 20 km deposition estimates usually lay outside the range of the 1 km deposition estimates. At higher levels of deposition the 20 km estimate seemed to overestimate the area of exceedance and at lower levels it seemed to underestimate exceedance. In the west the area of exceedance was approximately 75% of the land area with a range of 32% to 90% and in the east approximately 38% with a range of 15% to 58%.

The use of 20 km deposition estimates appeared satisfactory for the two flatter landscapes but did not agree with results from the simulated 1 km deposition estimates in mountain regions. Greater uncertainty was observed in the mountain regions but it was reduced by using 1 km rather than 20 km deposition estimates.

5. Conclusions

For national estimates of the area of critical load exceedance in the UK the use of 20 km deposition estimates rather than 1 km deposition estimates introduces an uncertainty with a range of about ±4%. However, the uncertainty in the output of the deposition models may be ±40% which is the major component of the combined uncertainty in the area of exceedance of ±38%. For regional critical load exceedance estimates the use of a 20 km deposition gives acceptable results except in mountain areas where it will give an extended range of uncertainty. In particular, the reduction in areas of critical load exceedance following reduced sulphur emissions appears to be overestimated in mountain areas by using 20 km deposition estimates.

In this study the use of critical load classes may hide problems, especially since the minimum class interval at 0.3 keq H^+ ha^{-1} y^{-1} is of the same order as potential changes in deposition estimate from a 20 km to a 1 km scale. There will be a potentially large, but undefined, uncertainty in critical load values which is ignored by the current protocols and has been ignored in this study. For the more complex critical load procedures the deposition of base cations is a required input and it has a larger uncertainty than that for sulphur deposition.

This study shows that the approximate range of uncertainty for an estimated area of critical load exceedance for a region is 20% of the total land area of the region in England and Wales rising to 50% in north Scotland while at a national level the range of uncertainty is 60000 km^2 or about 25% of the land area in the study. Clearly the major problem is the uncertainty in estimating average deposition to areas of the order of 20 km rather than the problem of matching deposition and critical load at different spatial scales.

Acknowledgments

We gratefully acknowledge the support of the Air Quality Research Programme of UK Department of the Environment, contract number EPG 1/3/31.

References

Nilsson, J. and Grennfelt, P. (editors): 1988, *Critical loads for sulphur and nitrogen*, Nordic Council of Ministers (Report 1988:11).

RGAR: 1990, *Acidic deposition in the United Kingdom: The Third Report of the Review Group on Acid Rain*, UK Department of the Environment.

Smith, R.I., Fowler, D. and Bull, K.R.: 1995, In: *Acid Rain Research: Do we have all the answers?* Elsevier Science, 175-186.

REGIONAL ASSESSMENT OF THE TEMPORAL TRENDS IN SOIL ACIDIFICATION IN SOUTHERN SWEDEN, USING THE SAFE MODEL

MATTIAS ALVETEG, HARALD SVERDRUP and PER WARFVINGE
Chemical Engineering II, Lund University, P.O. Box 124, S-221 00 Lund, Sweden

Abstract. The dynamic soil acidification model SAFE was applied to 44 forested sites in Skåne, southern Sweden, using available Swedish databases on present soil status, vegetation and deposition. Time series of deposition were derived for each site from present deposition in a generalized fashion by dividing deposition into different classes and scaling with deposition trends from the literature. This study connects the current status of the soil and the soil development with critical load maps calculated with the steady-state model PROFILE.
The model was calibrated against measurements of present base saturation from the Swedish Forest Inventory. Model output was compared with available measurements of soil water chemistry.
Model output was used to assess the time delay between changes in acidic input and system response in terms of exchangeable base cations and pH. The model was also used for scenario analysis, applying the reductions agreed in the Oslo Protocol to assess the environmental benefits of the agreement.

Key words: Acidification, Dynamic modeling, regional

1 Introduction

When deciding about abatement strategies to decrease the adverse effects of acid rain, it is important to assess not only the effect on single sites, but also the effect at a regional scale. A common approach is to estimate the emission reduction needed in order to protect 95 % of the studied ecosystems. The regional pattern of acidification has previously been assessed with PROFILE and with other steady-state models such as the simple mass-balance method. This approach can not, however, shed any light on questions regarding the time-frame of acidification. The purpose of this paper is to show how temporal trends of soil acidification in Skåne, southern Sweden, can be studied with the dynamic soil chemistry model SAFE, using the same data and assumptions that previously has been used for nation-wide PROFILE calculations (Sverdrup et al., 1992a).

2 Method

The SAFE model is a dynamic, multi layer, soil chemistry model that calculates the weathering rate and the soil water chemistry (Warfvinge et al. 1993). The SAFE model needs physical and chemical input regarding the soil profile as well as time series of deposition, uptake and precipitation. The veg-

TABLE I

Default values for soil density, DOC, apparent gibbsite constant, carbon dioxide pressure and relative factors for deriving mineralogy data for all layers. Mineral group one consists of K-feldspar, oligoclase and albite, group two consists of hornblende, pyroxene, epidote, calcite, muscovite, chlorite and apatite.

variable	layer 1	layer 2	layer 3	layer 4	Unit
density	600	1200	1300	1400	kg m^{-3}
DOC	10	5	2	0	mg l^{-1}
log K$_{gibbsite}$	6.5	7.5	8.5	9.2	concentrations in mol l^{-1}
P$_{CO_2}$	3	5	10	20	relative to ambient pressure
surface area	0.4	0.8	0.9	1.0	relative to B-layer data
Mineral group 1	1.0	1.0	1.0	1.0	”
Mineral group 2	0.0	0.33	1.0	1.0	”
Vermiculite	0.0	0.33	1.0	0.33	”

etation at the sites in this study range from pure confierous to mixed forests and pure decidous forests.

2.1 CONSTANT INPUT

As in the critical load exercise, input data were taken from the Swedish Forest Inventory (SFI), except for data on present deposition that were taken from The Cooperative Program for Monitoring and Evaluation of the Long Range Transmission of Air Pollutants in Europe (EMEP) and the Swedish Environmental Research Institute (IVL). Available data included total chemical analysis, current uptake of nutrients, layer thickness, texture, moisture classification, current deposition and precipitation. The UPPSALA model (Sverdrup et al., 1992b) was used to derive mineralogy from total chemical analysis. It was assumed (see table 1) that all sites had the same depth-dependent concentrations of DOC, P$_{CO_2}$, the same apparent gibbsite solubility constants and that the upper 50 cm of the soil determines the effect on vegetation.

Cation exchange is a vital process in the dynamic approach and measurements of cation exchange capacity (CEC) and present base saturation are therefore needed in order to run the SAFE model. Missing data regarding exchangeable cations for some of the SFI database sites limited the number of studied sites to 44. Measured base saturation in layer 3 ranged from 2 to 72 %, with 80 % of the sites having a base saturation between 4 and 45%, while CEC ranged from 12 to 210 mEq kg^{-1}. Like in most other areas, there is unfortunately no measurements on historic base saturation. The initial base saturation was therefore calibrated, i. e. dynamic calculations were made with different initial base saturations until SAFE produced a base saturation for 1988 that was in agreement with the measurements. The CEC was assumed to be constant over time. Soil water chemistry was available from SFI for 34 revived soil samples, i. e. dried soil equilibrated with water. In the available measurements 25 mg of dried soil, sampled 1988, had been equilibrated with 50 ml of water. The measured soil water chemistry was used for evaluation of model output only and was not used for any kind of calibration.

2.2 TIME SERIES INPUT

A number of assumptions had to be made to derive appropriate time series to be used as input to the dynamic SAFE model. Nutrient circulation, i. e. litter fall and canopy exchange, was assumed to be negligible, an assumption that theoretically leads to underestimated nutrient concentrations in the upper horizons. Since nitrogen (N) dynamics, although not well known, has a great impact on soil water chemistry, two extreme sets of time series of N net uptake was used. In the first set, N net uptake was assumed to be the minimum of N deposition and N net uptake 1988. This implies that when N deposition is increased above the current N net uptake, all excess N leaches out. In the second set of times series, all N entering the soil is either immobilised or taken up by the vegetation. This implies that leaching of N never occurs. The difference in model output using these two sets of uptake input may be viewed as an estimate of the uncertainty in the model approach suggested here.

Net uptake of base cation was assumed to be constant over time. At times when the derived (see above) N net uptake was lower than during 1988, however, net uptake of base cations was assumed to be 70 % of the N net uptake expressed in mol_c.

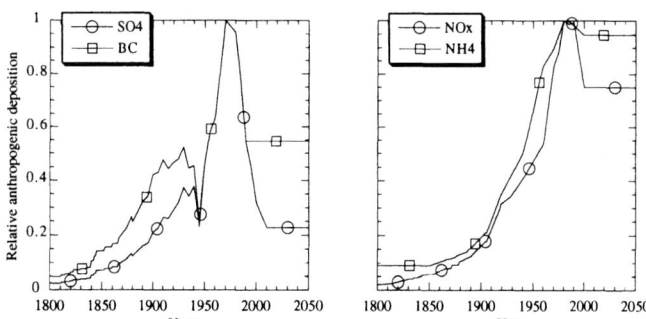

Fig. 1. Standard curves used for scaling anthropogenic deposition at all sites.

In order to create time series of deposition, the present deposition was divided into anthropogenic and sea salt deposition assuming that all sodium (Na) comes from sea salts. Deposition of sea salts was assumed to be constant over the years, while anthropogenic deposition was scaled according to a set of standard curves (see figure 1). The standard curve for anthropogenic Sulphur deposition was based on data from Rothamsted UK (Sverdrup et al. 1995), work by Mylona (1993), and RAINS calculations. Future anthropogenic deposition of S was taken from RAINS calculations, based on the Oslo Sulphur Protocol (United Nations 1994), while future deposition of other elements was held constant. A description of the RAINS model can be found in Alcamo et al. 1990.

3 Results and Discussion

When evaluating model output at the regional scale, it should be kept in mind that the assumptions for model inputs are made at the regional scale, not at the catchment scale. Model output from regional applications are therefore not likely to show as good a fit as single site applications.

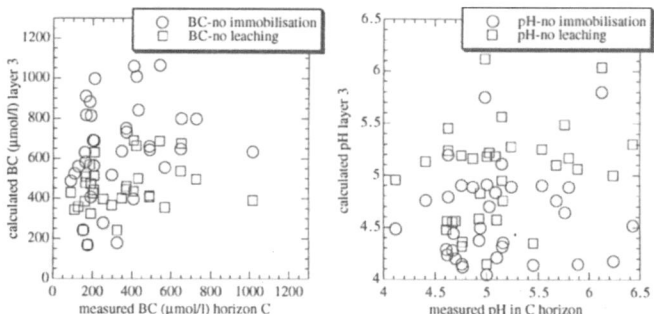

Fig. 2. Calculated pH and base cation concentrations as compared to measurements 1988.

As can be seen in figure 2 calculated pH values in layer 3 is in the same range as measured pH for horizon C, although there is a slight tendency to underestimate pH. Calculated base cation (Ca^{2+}, Mg^{2+} and K^+) concentrations are in better agreement with measurements although there is a tendency to overestimate the concentrations. The set of input where no N is allowed to leach out gives a slightly better overall fit for both pH and base cations. Further studies are needed to determine whether this is due to the fact that revived soil samples were used or due to the modeling approach.

As can be seen in figure 3 and 4, the different sites have different dynamics. The dynamics is basically determined by deposition, weathering rate and CEC. The greater the CEC, the longer time it takes for the soil to respond to a change in deposition. The dynamics also varies with layers, with more abrupt changes in base saturation in the lower layers. The depletion of the upper layers may temporarily increase the flux of base cations to the lower layers, thus temporarily sustaining a relatively high base cation concentration and base saturation in the lower layers. In this simulation, the upper layers started to get depleted at the same time as there was a peak in the acidity load. The global effect was a sudden change in base saturation in the lower layers (Figure 4).

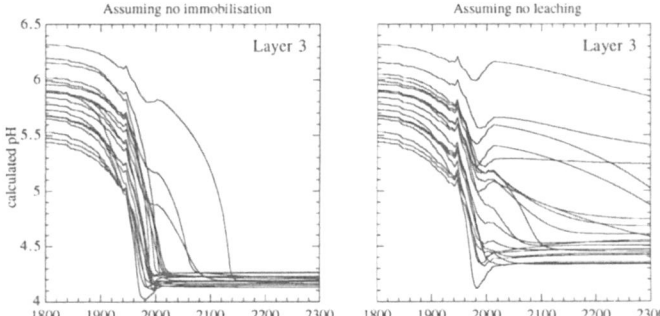

Fig. 3. Calculated pH in layer 3 for 20 different sites using the Oslo Sulphur Protocol as a future S deposition scenario and holding future deposition of other elements constant. The left graph is without immobilisation while in the right graph a 100% immobilisation of excess N is assumed. It can be noted that some sites show a dip in calculated pH around 1990, thus suggseting that the Oslo Sulphur Protocol will result in slightly increased pH at these sites.

As can be seen in Figure 3, the reduction of S deposition as agreed in the Oslo Sulphur Protocol results in stabilized pH values in the lower layers. At some sites there is even a slight increase in pH after 1990 (Figure 3). At sites having a high CEC, the lower layers are not as acidified as in the low CEC sites. A change in acid deposition changes the steady-state of a site but does therefore not necessarily change the trend in pH and base saturation at the studied sites. The two sets of N net uptake inputs produce similar output for the most sensitive site, although the no N leaching input gives slightly higher steady-state pH and a greater difference between sites. Since the two sets of input on N net uptake produces different pH for 1988, and since SAFE models cation exchange using a gapon adsorption isotherm, the calibrated historic base saturations are different in the two runs (Figure 4).

According to model output, the Oslo Sulphur Protocol will give a slight improvement in soil status in Skåne as compared to the *present* situation and a break in the trend of decreasing pH at most sites. It is, however, not enough to restore the base saturation to historic values (figure 4). To do so, the only reasonable mitigation strategy is to substantially reduce the deposition of N compounds. It should be noted, however, that all results are preliminary at this stage of model development.

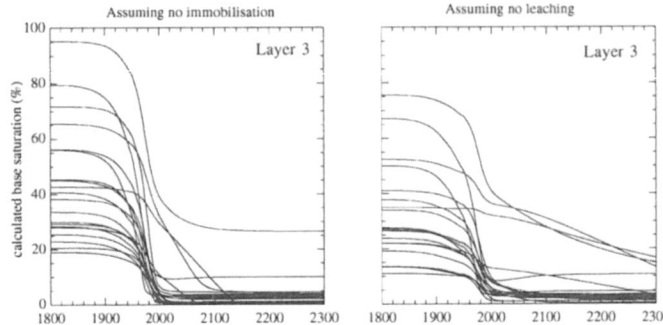

Fig. 4. Calculated development of base saturation in layer 3 for 20 different sites using the Oslo Sulphur Protocol as future S deposition scenario and holding future deposition of other elements constant. The left graph is without immobilisation while in the right graph a 100% immobilisation of excess N is assumed.

Acknowledgements

This work was done with support from the Swedish Environmental Protection Board (SNV) and, in the initial phase, from Wolfgang Schöpp in the Transboundary Air Pollution (TAP) group at the International Institute of Applied Systems Analysis (IIASA), Austria.

References

Alcamo, J., Shaw, R. and Hordijk, L.: 1990, *The RAINS model of Acidification Science and strategies in Europe*, Kluwer Academic Publishers.

Mylona, S.: 1993, Trends of sulphur dioxide emissions, air concentrations and depositions of sulphur in Europe since 1880, *Technical report*, EMEP/MSC-W.

Sverdrup, H., Warfvinge, P., Blake, L. and Goulding, K.: 1995, Modelling recent and historic soil data from the Rothamsted experimental station, UK, using SAFE, *Agriculture Ecosystems and Environment* **53**, 161–177.

Sverdrup, H., Warfvinge, P., Frogner, T., Håøya, A., Johansson, M. and Andersen, B.: 1992a, Critical loads for forest soils in the nordic countries, *AMBIO* **XXI**, 348–355.

Sverdrup, H., Warfvinge, P., Rabenhorts, M., Janicky, A., Morgan, R. and Bowman, M.: 1992b, Critical loads and steady-state chemistry for streams in the state of Maryland, *Environmental Pollution* **77**, 195–203.

United Nations: 1994, Protocol to the 1979 convention on long-range transboundary air pollution on further reductions of sulphur emissions.

Warfvinge, P., Falkengren-Grerup, U., Sverdrup, H. and Andersen, B.: 1993, Modelling long-term cation supply in acidified forests stands, *Environmental Pollution* **80**, 209–221.

REGIONALIZATION OF CRITICAL LOADS UNDER UNCERTAINTY

ANDREAS BARKMAN, PER WARFVINGE and HARALD SVERDRUP

Chemical Engineering II, Lund University, P.O. Box 124, S-221 00 Lund, Sweden

Abstract. The steady-state model PROFILE was used to perform Monte Carlo simulations of critical loads of acidity and exceedances of forest soils for 128 sites in the province of Scania, southern Sweden. Statistical tests showed that 100 sites had normal distributed critical loads and exceedances and that the variance of these parameters was statistically equal for all sites. Pooled estimates of the standard deviation was 0.19 and 0.31 $kmol_c$ ha^{-1} yr^{-1} for the critical loads and exceedances, respectively. Introduction of uncertainties, expressed as confidence intervals, in the cumulative distribution function for critical loads showed that overlaps between percentiles were substantial. The 5%-ile was systematically equal to the 57%-ile using a 67% confidence interval and equal to the 87%-ile when a 95% confidence level was chosen. The overlaps of percentiles cause a reduction of acidic deposition according to the mean value of the 5%-ile to protect only 68% of the ecosystem area with an 84% probability and not a guaranteed protection of 95% as if uncertainties did not exist. Thus, uncertainties make it possible to advocate reductions to levels of deposition below the 5%-tile of critical loads.

Keywords: Regionalization, critical loads, forest, uncertainty, risk assessment

1. Introduction

In the southern most province of Sweden, Scania, the observed forest damages, measured as needle loss, almost doubled during the period 1986-1991 (Schlyter 1993). The explicit cause-effect relationships are complex and not fully understood, but the high deposition of acidic compounds is likely to constitute one major pre-disposing factor. A more detailed investigation of the critical load of acidity and its exceedance in the region has therefore been regarded as vital in order to provide information for the assessment of regional mitigation options such as liming or/and N-free fertilization.

Application of the critical load concept on a small regional scale poses high demands on the quantity and quality of input data to the mathematical models estimating the critical load.

One factor, often neglected when applying the concept of critical loads on the national and international scale, is the level of uncertainty in the calculations due to uncertainties in the input data. A high level of uncertainty may impair the desired effect of mitigation actions such as reduction of acidic deposition, but uncertainty may also obliterate any difference between sites within a region. This reduces the possibility to identify and classify forest areas subject to different probabilities of adverse effect.

The objective of this study is to develop a statistical methodology which can assess the uncertainty of the calculated critical loads and their exceedances and the implications of uncertainty on the use of the 5%-ile as a basis of reduction of acidic deposition. The aim is also to investigate the constraints uncertainty poses on the number of identifiable classes of critical loads within Scania.

2. Methods

2.1 MODEL AND INPUT DATA

The critical load and their exceedances of forest soils were calculated on the basis of a BC/Al-ratio=1. The regionalized PROFILE model (Warfvinge and Sverdrup 1992; Sverdrup and Warfvinge 1992) was applied according to the procedure outlined in Sverdrup et al. (1992b). In total 128 sites have been available for simulation. 89 sites originate from the National Forest Inventory (Ståndortskarteringen) performed 1983-87, the remaining 39 sites are the permanent regional monitoring sites run by the Regional Board of Forestry (Skogsvårdsstyrelsen). The sites simulated are assumed to be representative for the forest soils in Scania.

PROFILE requires input data on atmospheric input of major anions and cations, precipitation and runoff, mineralogy, uptake of nitrogen and base cations and physical properties of the soil such as texture and density.

Atmospheric deposition (1988 data) was calculated by the Swedish Environmental Research Institute on a 50 by 50 km grid according to Lövblad et al. (1992). The precipitation, precipitation surplus and annual average temperature were taken from national maps. The Department of Forest Survey at the Swedish University of Agricultural Sciences in Umeå calculated the long-term stem growth/uptake for all sites using the HUGIN model (Hägglund 1981). The mineralogy was determined by back calculations from a total analysis using the UPPSALA model (Sverdrup et al. 1992a), Other soil parameters were determined as described by Jönsson et al. (1995).

2.2 UNCERTAINTIES

The uncertainties in the calculated critical loads and their exceedances were estimated by Monte Carlo simulations. Each of the 128 sites in the region were simulated 500 times, randomly varying the input data according to the procedure described by Jönsson et al. (1995).

For each site the generated output distributions of critical loads and exceedances were analyzed with a Kolmogorov-Smirnov test, $p = 0.05$, and a Wald-Wolfowitz runs test, $p = 0.05$, to identify which of the sites were normal distributed (Siegel 1956). 100 sites passed both tests for both the critical loads and the exceedance.

In order to control whether the 100 normal distributed sites were representative for the total population of 128 sites a Kolmogorov-Smirnov test was performed on the distributions of mean values of the sites in the two populations. The population of 128 sites was found to be well represented by the 100 normal distributed sites with respect to the range and variability of critical loads and their exceedances.

A Bartlett's test of sphericity, $p = 0.05$, indicated that there is no significant difference between the variances of the critical load distributions and the distributions of exceedances. Thus, there are adequate conditions for a pooling of the variances according to:

$$s_p^2 = \frac{1}{k}(s_1^2 + \cdots + s_k^2)$$

where $k = 100$. The pooled estimate, s_p, was 0.19 and 0.31 kmol$_c$ ha^{-1} yr^{-1} for the critical load and exceedance, respectively.

In order to check whether the number of calculations for each site was sufficient, 95% confidence intervals, I, for the expectation values, μ, of critical load and their exceedances were calculated according to:

$$I_\mu = \bar{X} \pm \frac{s_p}{\sqrt{N}} \cdot t_{\frac{\alpha}{2}} \sum_{i=1}^{k}(n_i - 1)$$

where $k = 100$, $N = 500$ and $n_i = 500$. Due to the high number of degrees of freedom (represented by $\sum_{i=1}^{k}(n_i-1)$) and the number of simulations per site, the 95% confidence intervals were not wider than ± 0.017 and ± 0.027 kmol$_c$ ha^{-1} yr^{-1} for the critical loads and exceedances, respectively. Thus, the uncertainty regarding the mean values of critical loads and exeedance is more than 2 magnitudes smaller than the uncertainty of the actual variable. A cumulative distribution of mean values, based on 500 simulations per site, is therefore an appropriate way of characterizing the distribution of critical loads and exceedances in Scania.

3. Results

Figure 1 and figure 2 show the cumulative distribution functions for the critical load of acidity (CL) and exceedance (EX) of forest soils in Scania based on mean values from the 100 sites which were normal distributed. The figures also show the individual standard deviation for each site.

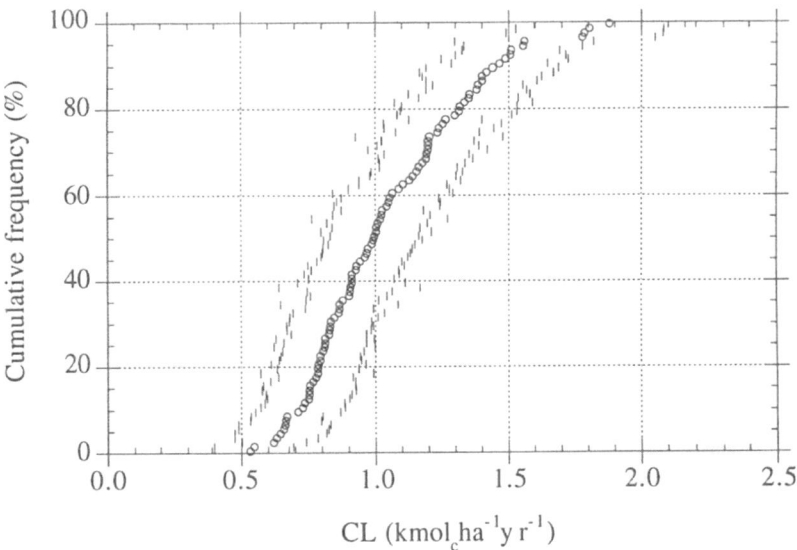

Figure 1: Cumulative distribution of the calculated critical loads for the 100 sites with $\pm$ one standard deviation indicated.

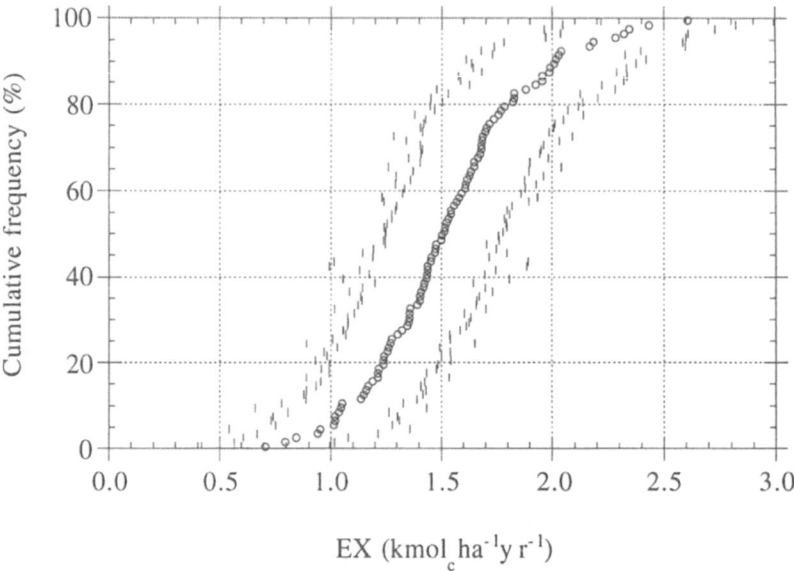

Figure 2: Cumulative distribution of the calculated exceedances for the 100 sites with ± one standard deviation indicated.

Figure 1 and figure 2 also show that the standard deviations is rather constant throughout the range of critical loads and exceedances respectively. There seems to be a tendency of increasing standard deviation with increasing mean value. However, the tendency is too weak to be statistically significant according to the Bartlett test in the previous chapter. The figures also display that the uncertainties in the calculations cause a substantial overlap between different percentiles - a fact which questions the area protected by a reduction to a certain percentile but also inflicts constraints on the number of separable classes of critical loads.

4. Discussion

Figure 3 shows the cumulative distribution function of mean values for the critical loads in conjunction with the pooled standard deviation, i.e. a 67% confidence interval, and a 95% confidence interval. It also illustrates how a higher level of confidence impairs the resolution between percentiles. At a 67% confidence interval there is an overlap between the 5 and the 57%-ile (indicated by letter A in figure 3). At 95% confidence level the overlap has increased to the 87%-ile (B). Thus, a requirement of high probability of sites being systematically different reduces the possibilities to identify distinct classes of critical loads. It is therefore hazardous to design mitigative treatment programs based on site classifications, as well as to assess whether mitigation actions applied on a regional scale are excessive or not.

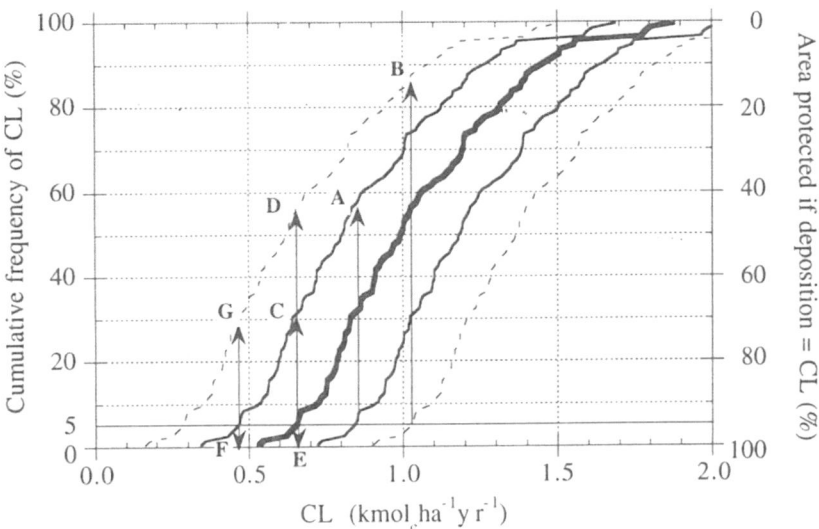

Figure 3: Cumulative distributions of the calculated critical loads with ± one standard deviation and a 95% confidence interval for each site.

Reducing the deposition to the level of the 5%-ile CL (i.e. 0.67 kmol$_c$ ha^{-1} yr^{-1}) would, if there were no uncertainties, ensure protection of 95% of the ecosystem area. This reduction would protect 68% of the ecosystem area with an $68 \pm \frac{32}{2} = 84\%$ probability (C), and 45% of the area with 97.5% probability (D). On the other hand the probability that a reduction of acid deposition to the 5%-ile critical load will yield 100% protection is substantially greater than 16% (E).

The uncertainty analysis also makes it possible to advocate reductions to levels of deposition below the 5%-tile CL. Reduction according to the lower part of the 67% confidence interval for the 5%-ile CL (i.e. 0.45 kmol$_c$ ha^{-1} yr^{-1}, indicated by letter F) will lead to 100% area protection with approximately 65% probability, but only protect 70% of the area with a 97.5% probability (G). This illustrates how uncertainties apply to mitigation measures and reduction of acidic deposition at an arbitrary level of risk.

The current approach accounts for uncertainties related to the number of sites calculated, the number of runs on each site and uncertainties in input parameters. It does not, however, account for uncertainties in the model structure and the critical BC/Al-ratio applied. The latter issue has been addressed by Cronan and Grigal (1995). An investigation of overlaps between confidence intervals for critical loads calculated for different BC/Al-ratios would add an important dimension to the process of assessing critical loads at different levels of risk.

5. Conclusions

The statistical framework in this study seems to be sufficiently stringent and encompassing for the investigation of uncertainties in critical loads and exceedances calculated by the PROFILE model.

It can be concluded from this study that an increased number of sites increases the spatial resolution but does not offset effects induced by uncertainty. Hence, it is more important to reduce the uncertainties in the input parameters than to increase the total number of sites or the number of runs for each site.

With our assumptions on the range of variability/uncertainty of the input parameters applied the uncertainty in the estimations of critical load are relatively large compared to the best estimated value.

In Scania the standard deviation for the 5%-ile CL is much lower than the typical exceedance. This implies that the uncertainties do not question whether a reduction of acidic deposition is necessary or not, but rather pose constraints on the size of the region protected by reductions according to a certain %-ile of critical loads.

Acknowledgment

This study was performed with support from the Swedish National Board of Forestry (contract 9306), the Swedish Natural Science Research Council, and the Swedish Environmental Protection Agency (contract 802–509–93–Ff).

References

Cronan, C. and Grigal, D.: 1995, Use of calcium/aluminum ratios as indicators of stress in forest ecosystems, *Journal of Environmental Quality* **24**, 209–226.

Hägglund, B.: 1981, *Forecasting growth and yield in established forests*, Report 31, Department of Forest survey, Swedish University of Agricultural Sciences, Umeå.

Jönsson, C., Warfvinge, P. and Sverdrup, H.: 1995, Uncertainty in predicting weathering rates and environmental stress factors with the PROFILE model, *Water Air and Soil Pollution* **81**, 1–23.

Lövblad, G., Amann, M., Andersen, B., Hovmand, M., Joffre, S. and Pedersen, U.: 1992, Present and future deposition of sulphur and nitrogen in the Nordic countries., *Ambio* **21**, 339–347.

Schlyter, P.: 1993, *Operational areal forest damage survey*, Department of Environmental and Energy System Studies, Lund University, Lund.

Siegel, S.: 1956, *Nonparametric statistics for the behavioural sciences*, McGraw-Hill, Tokyo.

Sverdrup, H. and Warfvinge, P.: 1992, Critical loads, in P. Sandén and P. Warfvinge (eds), *Modeling acidification of groundwater*, Swedish Meteorological and Hydrological Institute (SMHI), Norrköping, pp. 171–186.

Sverdrup, H., Warfvinge, P., Rabenhorst, M., Janicki, A., Morgan, R. and Bowman, M.: 1992a, Critical loads and steady-state chemistry for streams in the state of Maryland, *Environmental Pollution* **77**, 195–203.

Sverdrup, H., Warfvinge, P., Johansson, M., Frogner, T., Håöja, A.-O. and Andersen, B.: 1992b, Critical loads to forest soils in the Nordic countries, *Ambio* **5**, 348–355.

Warfvinge, P. and Sverdrup, H.: 1992, Calculating critical loads of acid deposition with PROFILE – a steady-state soil chemistry model, *Water Air and Soil Pollution* **63**, 119–143.

THE EFFECTS OF SCALE AND RESOLUTION IN DEVELOPING PERCENTILE MAPS OF CRITICAL LOADS FOR THE UK

J. HALL[1], K. BULL[1], M. BROWN[1], H. DYKE[1], J. ULLYETT[1] and M. HORNUNG[2]

[1] *Institute of Terrestrial Ecology, Monks Wood, Abbots Ripton, Huntingdon, PE17 2LS, UK,* [2] *Institute of Terrestrial Ecology, Merlewood, Grange-over-Sands, Cumbria, LA11 6JU, UK.*

Abstract. Critical loads are estimated in the UK by the Department of Environment's Critical Loads Advisory Group and sub-groups. The Mapping and Data Centre at ITE Monks Wood acts as the National Focal Centre for the UNECE programme for mapping critical loads. The centre is responsible for the generation of UK data sets and their application for national and European purposes. To make effective use of these data, it is necessary to draw upon other environmental data and examine the issues of scale, uncertainty and the way that data are presented. This paper outlines the methodologies which have been employed to derive national maps. Early critical load maps were not vegetation specific, but now critical loads for acidity and for nutrient nitrogen for soils, critical levels maps for ozone and sulphur dioxide, and sulphur deposition maps, have been generated on a vegetation or ecosystem specific basis. These have been used to derive a number of different types of critical load and exceedance maps. The results show the importance of the method selected and the data used for the interpretation. The visualisation of critical loads and the corresponding exceedance data is an important aspect in producing information for pollution abatement strategies.

Keywords: Critical loads, percentiles, scale, resolution, uncertainty, UK, UNECE.

1. Introduction

Critical loads are estimated by members of the UK Department of Environment's Critical Loads Advisory Group (CLAG) (CLAG, 1994). The Mapping and Data Centre at ITE Monks Wood acts as the National Focal Centre for the UNECE programme for mapping critical loads (CCE, 1993). In the UK critical loads data are calculated and mapped at different scales for different receptors. The national soil critical loads are derived from a 1km database of the dominant soil types (Hornung *et al.*, in press). Freshwater critical loads have been calculated for a single catchment within 10km squares and mapped using this grid resolution (Battarbee *et al.*, 1995). Data that are being developed for vegetation critical loads use a combination of data sets at 10km, 1km and 25m resolution. In addition, national measured and modelled deposition are available on a 20km grid (CLAG, 1994). This paper examines the issues of scale in the generation of national maps of critical loads and exceedances at 1km and 20km resolution and also the effects of mapping UK data using the UNECE European Monitoring and Evaluation Programme (EMEP) grid. A simple approach to visualising uncertainty is also presented.

2. Methods

Early UK soil critical loads and deposition data were not vegetation specific, but now both have been generated on a vegetation or ecosystem specific basis (Hornung *et al.*, 1995). This study made use of a recently calculated 1km data set of critical loads of acidity for soil-woodland ecosystems, calculated using the Simple Mass Balance (SMB) model

(Sverdrup et al., 1995), to derive percentile maps at different scales. National exceedances are calculated using 20km deposition data (Fowler et al., 1995). For this study similar deposition data (non-marine (wet + dry) sulphur) estimated for woodland ecosystems were used. For exceedances at the EMEP scale, EMEP calculated oxidised sulphur deposition (Tuovinen et al., 1994) were applied.

The 1km woodland-soil critical loads data were calculated for the whole of the UK assuming that woodland exists across the whole country. The ITE Land Cover Map (LCM) (Fuller et al. 1994), a land use map derived from satellite imagery, and the CORINE land use map of Northern Ireland were used to provide values for woodland ecosystem areas within different size grid squares. The data used were held in a geographic information system (GIS) as grid-based (ie raster) maps at different resolutions, in which values of critical loads etc were assigned to each square.

"Percentile" maps were calculated to show the critical loads which offer protection to a given percentage area (CCE, 1991). An "n-percentile" map protects (100-n)% of the area eg the "5-percentile" critical load protects 95% of the area.

To create a 5-percentile 1km map of critical loads for soils, critical load values were mapped only where more than 5% of a 1km square was woodland. Squares with less than 5% woodland were assigned a high critical load value (10 keq H^+ ha^{-1} $year^{-1}$).

5-percentile critical load maps for the 20km and EMEP grids were generated from the 1km data. In each case, "grid-area" and "ecosystem-area" critical loads were derived. In the former, the 5-percentile critical load was calculated to protect 95% of the total grid square area (ie where > 5% of the grid square was woodland). Again, grids with less than 5% woodland were assigned a value of 10 keq H^+ ha^{-1} $year^{-1}$. The 5-percentile ecosystem-area critical load was calculated to protect 95% of the total woodland ecosystem area within each grid square.

Exceedance maps were created by subtracting the critical loads maps from the deposition maps in the GIS. National 20km sulphur deposition was used for calculating exceedances of the 1km and 20km critical loads. For the 1km exceedance map, all 400 1km squares within each 20km square of the deposition map are assumed to have the same value. Exceedances were also generated using EMEP deposition with 20km and EMEP 5-percentile critical loads.

To examine, in a simple way, the issue of uncertainty in exceedances, maps were produced showing the ratio of deposition to critical loads. When the uncertainty in deposition or critical load is proportionally high (eg > 100%) then a high ratio (eg > 2) would indicate probable exceedance. Conversely a low ratio (eg < 0.5) indicates a low probability of exceedance. For intermediate ratio values exceedance is uncertain.

3. Results

3.1. CRITICAL LOADS

The 1km grid-area and ecosystem-area maps of critical loads are identical as areas <5% grid-area are not mapped. The maps show the lowest critical loads in the New Forest, the Weald, the Brecklands, south Wales, North York moors, southern and north-eastern Scotland and small areas throughout Northern Ireland.

At other resolutions the grid-area maps are less sensitive than the ecosystem-area maps (Figure 1) as usually less woodland area is being protected. The grid area-map at the 20km scale still highlights the major sensitive areas of the country seen on the 1km map. The ecosystem-area map shows lower critical loads (<1.0 keq H$^+$ ha^{-1} year^{-1}) over much of the country drawing attention to small woodland areas of high sensitivity.

These effects are further exaggerated at the EMEP scale, where small sensitive areas become "lost", as a critical load is assigned to a larger area. The grid-area map results in high critical loads (>1.0 keq H$^+$ ha^{-1} year^{-1}) for all grid squares and the ecosystem-area map lower critical loads (<1.0 keq H$^+$ ha^{-1} year^{-1}) for most squares. Ecosystem-area critical load maps at the EMEP scale are used in the European protocol discussions. It is therefore important that comparisons are made between national maps and the European maps. Table I summarises the areas in each of five critical loads classes for the 5-percentile grid-area and ecosystem-area maps at the 1km, 20km and EMEP scales.

Table I

Table showing the percentage area in each critical loads class for the different 5-percentile maps

Critical load map	Percentage area in each critical loads class*				
	1	2	3	4	5
1km	62.2	14.2	7.9	1.2	1.8
20km grid-area	64.6	20.1	11.1	1.8	2.4
20km ecosystem-area	6.4	25.0	34.7	9.8	24.1
EMEP grid-area	77.2	22.8	0	0	0
EMEP ecosystem area	0.59	0.19	17.7	32.7	48.9

* Critical load ranges (keq H$^+$ ha^{-1} year^{-1}) for classes: 1: >2.0, 2: 1.0-2.0, 3: 0.5-1.0, 4: 0.2-0.5, 5: <=0.2

3.2. EXCEEDANCES

Exceedance maps of grid-area critical loads at the 1km and 20km scale give similar patterns, when comparing exceedances using the same deposition field. Exceedances calculated using the ecosystem area critical loads result in larger areas of the country being exceeded and higher exceedances due to the lower critical load values.

The areas exceeded at the 20km scale by 20km national deposition and by EMEP deposition give broadly similar patterns but with differences in detail at a local scale (Figure 2). Using the EMEP deposition results in very high exceedances in central Britain and fewer areas of exceedance in Cumbria (north-west England) and Scotland than found with the national 20km deposition data. However, the total percentage area exceeded using the two deposition fields is very similar (Table II). Exceedances calculated at the EMEP scale only result in exceedance with the ecosystem area critical loads, where 90% of the country is estimated to be exceeded.

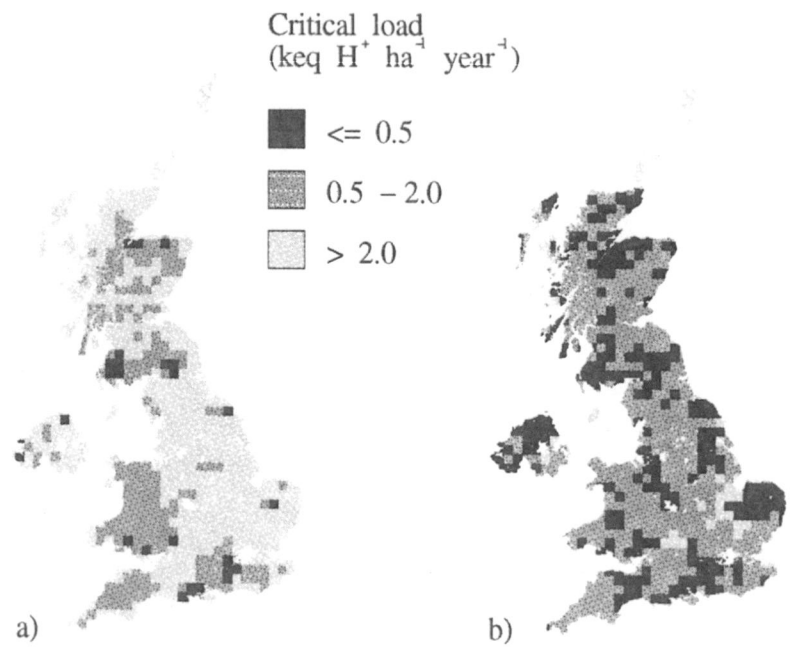

Fig. 1. 5-percentile critical load of acidity for soil-woodland ecosystems a) grid-area, b) ecosystem-area.

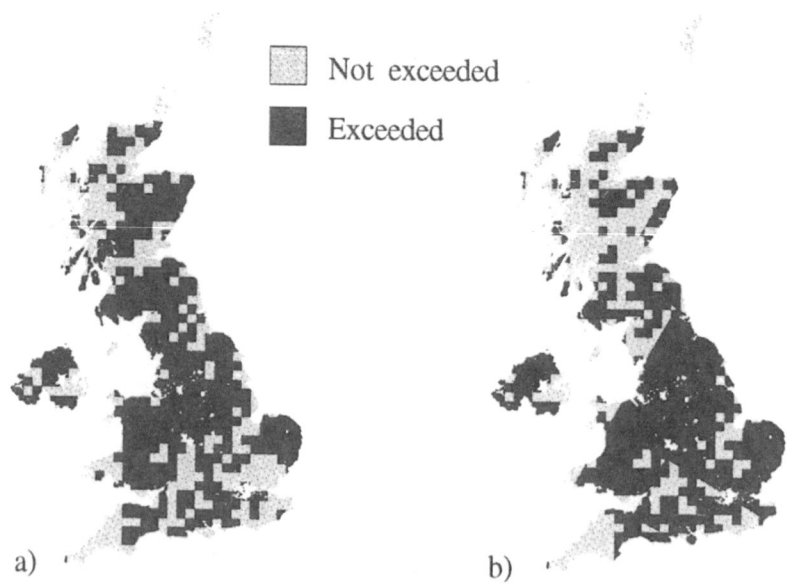

Fig. 2. Exceedance of 20km 5-percentile ecosystem-area critical load of acidity for soil-woodland ecosystems by a) 20km non-marine sulphur deposition 1989-92, b) EMEP oxidised sulphur deposition 1985-93.

Table II

Table showing the percentage area in each exceedance class

Exceedance map	Percentage area in each exceedance class*					Total exceeded
	1	2	3	4	5	
20km grid-area exceeded by 20km deposition	87.0	5.6	2.7	3.1	1.6	13.0
20km grid-area exceeded by EMEP deposition	86.8	2.8	2.9	2.8	4.7	13.2
20km eco-area exceeded by 20km deposition	36.5	13.1	13.8	23.1	13.5	63.5
20km eco-area exceeded by EMEP deposition	36.8	3.0	12.9	16.4	31.0	63.2
EMEP eco-area exceeded by EMEP deposition	10.0	2.8	14.8	27.9	44.5	90.0

* Exceedance ranges for classes (keq H^+ ha^{-1} $year^{-1}$): 1: not exceeded, 2: 0-0.2, 3: 0.2-0.5, 4: 0.5-1.0, 5: > 1.0

3.3. UNCERTAINTY

The map in Figure 3 was calculated using the 20km non-marine sulphur deposition and the 20km 5-percentile ecosystem area critical load for soils. It shows few areas where exceedance is unlikely to occur and a larger area where exceedance is probable, mainly in the regions where critical loads are lowest. For a significant area of the country the likelihood of exceedance is uncertain.

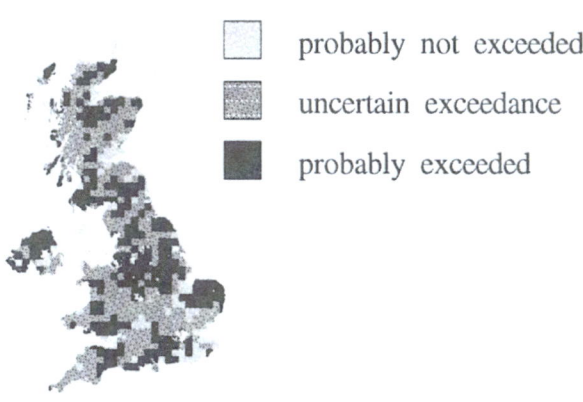

Fig. 3. Uncertainty expressed as 20km non-marine sulphur deposition divided by the 5-percentile ecosystem-area critical load of acidity for soil-woodland ecosystems.

4. Conclusions

Maps of grid-area critical loads at the 1km and 20km scales give similar results, although some small sensitive areas may disappear using the larger grid squares. The 20km grid-area map still highlights the more sensitive areas of the country, which are not so evident on the ecosystem-area map which has low critical loads for much of the country. The ecosystem-area maps provide a more sensitive pattern of critical loads. The EMEP scale grid-area and ecosystem-area maps result in a very generalised picture of the national critical loads. The implications of scale on national maps of critical loads have been discussed elsewhere (CLAG, 1994).

Comparisons of exceedances calculated using national and EMEP scale deposition show broadly similar patterns of exceedance, although there are differences in detail and exceedance patterns across the country. This reflects the difference in national and EMEP scale deposition estimates. However, the percentage area exceeded using the two deposition fields are very similar.

It is recognised that there are uncertainties in critical load and deposition estimates and that there is a need to improve the accuracy of both. However, the simple approach used here helps to identify areas where exceedance is most likely. It also demonstrates that exceedance is uncertain over wide areas of the country.

Acknowledgements

The authors acknowledge the Department of the Environment for funding this work (PECD7/10/90). However, the views expressed are entirely those of the authors.

References

Battarbee, R.W., Allott, T.E.H., Bull, K.R., Christie, A.E.G., Curtis, C., Flower, R.J., Hall, J.R., Harriman, R., Jenkins, A., Juggins, S., Kreiser, A., Metcalfe,S., Ormerod, S.J. & Patrick, S.T. 1995, Critical Loads of Acid Deposition for UK Freshwaters, Department of the Environment. London, UK, 139pp.
CCE. 1991. Mapping Critical Loads for Europe, Bilthoven, Eds. J-P. Hettelingh, R.J. Downing, P.A.M. de Smet. Netherlands.
CCE. 1993. Calculation and Mapping of Critical Loads in Europe: Status Report 1993, Eds. J-P. Hettelingh, R.J. Downing, P.A.M. de Smet. Bilthoven, Netherlands.
CLAG. 1994, Critical Loads of Acidity in the United Kingdom. Department of Environment, London.
Fowler, D., Leith, I.D., Smith, R.I., Choularton, T.W., Inglis, D. & Campbell, G. 1995, In: Acid Rain and its Impact: the Critical Loads Debate, Battarbee, R.W., ed. University College London, UK,17-25.
Fuller, R.M., Groom, G.B. & Jones, A.R. 1994, Photogrammetric Engineering & Remote Sensing, **60**, 553-562.
Hornung, M., Bull, K.R., Cresser, M., Hall, J., Loveland, P.J., Langan, S.J., Reynolds, B. & Robertson, W.H. 1995, In: Acid Rain and its Impact: the Critical Loads Debate,Battarbee, R.W., ed. University College London, UK, 43-51.
Hornung, M., Bull, K.R., Cresser, M., Hall, J., Langan, S.J., Loveland, P. & Smith, C. Environmental Pollution, in press.
Sverdrup, H., De Vries, W., Hornung, M., Cresser, M., Langan, S.J., Reynolds, B., Skeffington, R. & Robertson, W. 1995, In: Mapping and Modelling of Critical Loads for Nitrogen. Institute of Terrestrial Ecology, Edinburgh, UK.
Tuovinen, J-P., Barrett, K. & Styve, H. 1994, Transboundary Acidifying Pollution in Europe, Meteorological Synthesizing Centre-West, Norway.

CRITICAL LOADS FOR NITROGEN DEPOSITION FOR GREAT BRITAIN

K R BULL[1], M J BROWN[1], H DYKE[1], B C EVERSHAM[1],
R M FULLER[1], M HORNUNG[2], D C HOWARD[2], J RODWELL[3], D B ROY[1]

[1]*Institute of Terrestrial Ecology Monks Wood, Abbots Ripton, Huntingdon, PE17 2LS, UK.* [2]*Institute of Terrestrial Ecology Merlewood, Grange-over-Sands, Cumbria, LA11 6JU.* [3]*Unit of Vegetation Science, Lancaster University, Lancaster, LA1 4YQ*

Abstract. There is currently much interest in mapping critical loads for nitrogen deposition as part of a strategy for controlling nitrogen emissions. While nitrogen deposition may cause acidification and excess nutrient effects, the former were considered previously in studies of sulphur deposition. In the UK, work on developing nutrient nitrogen critical loads maps has used several methods and databases. Two approaches are described here, one a steady state calculation using a nitrogen saturation limit for soil systems, the other an empirical estimate of critical loads set to prevent changes to vegetation communities. The empirical method uses national species records and land cover data derived from satellite imagery. Maps drawn from the available data are dependent upon a number of factors which reflect the approach used. To apply the nutrient critical loads to a strategy for future abatement measures, the nutrient nitrogen values for soils have been incorporated within a "critical loads function" which takes into account both acidity and nutrient effects as related to deposition loads for sulphur and nitrogen. This function may be used with deposition data to identify the need for sulphur and nitrogen emission reductions.

Key words: Nitrogen deposition, critical loads, abatement strategies, soils, vegetation.

1. Introduction

The recent negotiation of a new UNECE sulphur protocol (UNECE, 1994) has shifted the emphasis of air pollution control to address emissions of nitrogen oxides and ammonia. The UK is addressing this issue both nationally and internationally using a similar critical loads approach to that used for sulphur (S). However, for nitrogen (N), not only acidity, but also excess nutrient effects on biological systems, have to be considered, and these have recently been reviewed in relation to effects in the UK (INDITE, 1994). While acidification due to N can be dealt with in the same way as that resulting from S deposition (CLAG, 1994), a new methodology is required to estimate critical loads for nutrient N.

Two approaches for estimating and mapping critical loads for nutrient N are presented here and these form the basis of national maps which are being proposed for policy purposes. Both approaches are based on methods discussed and agreed upon at two international workshops held in Lokeberg, Sweden (Grennfelt & Thornelov, 1993) and Grange-over-Sands, UK (Hornung *et al.*, 1995). In the first approach, empirical estimates of N critical loads are made for a number of vegetation types based upon known changes identified from field observations, experimental evidence, etc (Bobbink & Roelofs, 1995). Appropriate values are mapped for areas where the most sensitive vegetation types exist. The second approach uses a mass balance calculation to estimate the critical load at which N saturation of the soil system will occur (Hornung *et al.*, 1995); the calculation is made for particular soil-vegetation systems taking into account relevant N processes such as uptake by vegetation and denitrification and immobilisation in the soil.

This paper aims only to outline the various methods used in the generation of the current national maps. The detailed methodology for each of the approaches will be the subject of later papers. However, example maps will be given to illustrate the approaches used. Finally, a means for identifying areas with excess N and/or S over the critical loads for acidity and nutrient N (the critical loads function) will be presented for the present data sets. This method is being used to explore the significance of future emission reduction scenarios (Metcalfe *et al.*, 1995).

2. Data and Methods

To map critical loads, two sets of information are required. First, an estimation of the critical loads value or values for a particular receptor or receptors. Second, information on the geographic distribution of the receptors to which those critical loads values are to be applied. For the UK, it has been accepted that the critical loads values for nutrient N shall be defined as agreed at the Lokeberg and Grange-over-Sands workshops (see above). National data for vegetation distributions have been used to identify the areas for mapping the critical loads values.

For empirically based estimates of critical loads for nutrient N, values for mapping have been suggested for several vegetation types, while for others ranges have been identified (Hornung *et al.*, 1995). For each vegetation type, the recommended or most appropriate value from the workshop range was used for all UK maps.

To define the geographic distribution of receptors there are three major data sets available in the UK. A Land Cover Map of Great Britain (LCMGB) showing the distribution of 25 major landcover types across Britain has been derived from Landsat imagery (Fuller, 1994). The National Vegetation Classification (NVC) identifies the distribution of vegetation communities and their species compositions in Britain (Rodwell, 1991a, 1991b, 1992, 1994). The Biological Records Centre (BRC) database holds information on the distribution of individual species for the UK; the data are usually mapped on a 10km grid. These data sets provide sufficient information to (a) identify suitable species associated with sensitive vegetation communities identified at Grange over Sands (using the NVC), (b) map a representative selection of these species using geographically referenced records (using BRC records) (Eversham & Roy, in press) and (c) map the main areas of the country where these species occur by relating them to major areas vegetation as shown by land cover (using the LCMGB).

For the mass balance approach, critical loads are calculated using an equation which takes into account the various inputs, long term sinks and the fluxes out of the plant-soil system. The critical load is set to avoid excess N in soil solution which may lead to N leaching (Hornung *et al.*, 1995). For the UK, three vegetation types, woodland, heathland and acid grassland, were identified as important sensitive receptors for N deposition effects. For each of these, available soil and vegetation data were used to estimate such parameters as immobilisation, denitrification, N removal by harvesting. Suitable default values have been listed for UNECE calculations (CCE, 1993) and these have been recently reviewed by Hornung (*pers. comm.*).

To map data for this paper, information was summarized at 20km. For this, cumulative frequency distributions of the critical loads values of the ecosystems within each 20km area have been derived and the 5-percentile area statistic estimated for mapping. Maps therefore show critical loads giving protection to 95% of the ecosystems within a grid. With deposition at the 5-percentile critical load, 5% of the area of ecosystems, with lower critical loads, remain unprotected. This procedure has been used extensively for mapping critical loads in Europe (CCE, 1991), and was used as the basis for the maps used in the discussions for the Oslo Protocol (UNECE, 1994).

3. Critical loads maps for nutrient N

Empirical maps of critical loads for nutrient N are based upon distribution maps of species which are representative of the communities for which critical loads values have been listed (Hornung et al., 1995). Figure 1 shows a typical multi-species distribution map; it is evident that, while some species are ubiquitous, some species indicating key vegetation types have a much more limited distribution. A method for selecting representative species has been devised which avoids those species which are widespread (Eversham & Roy, in press). Using the LCMGB and the species distributions for indicative heathland species a nutrient N critical loads map of Great Britain can be drawn by applying heathland critical loads. The resulting 5-percentile map (Figure 2) indicates that there are few areas of Britain where heathland vegetation indicator species are absent (unshaded areas) although the map does not differentiate between dense and sparse areas of heathland. While there are no very low critical loads there is a clear difference shown between the lowland and upland areas where different critical loads are applied.

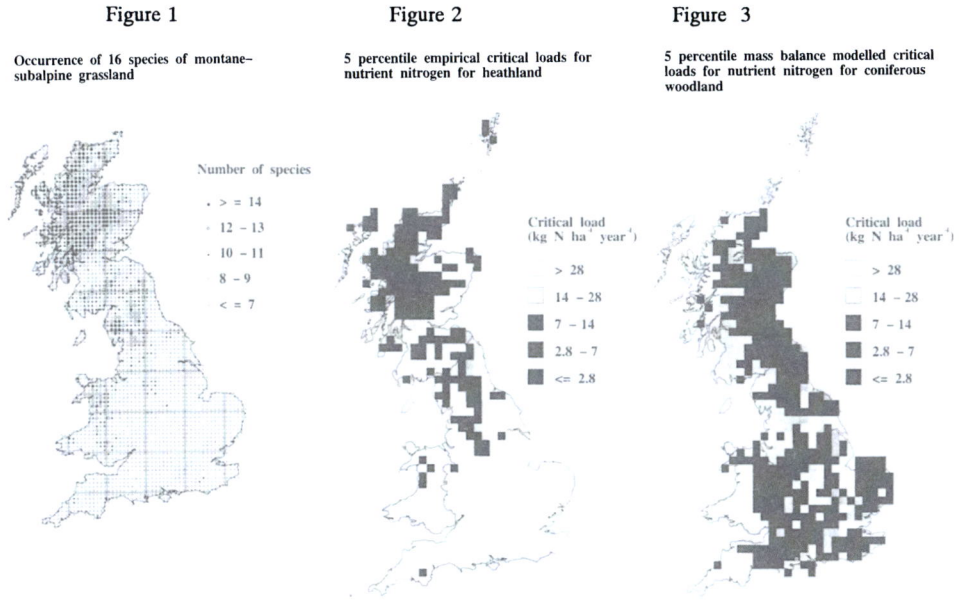

Figure 1
Occurrence of 16 species of montane–subalpine grassland

Figure 2
5 percentile empirical critical loads for nutrient nitrogen for heathland

Figure 3
5 percentile mass balance modelled critical loads for nutrient nitrogen for coniferous woodland

For mass balance calculations, it has been assumed that there is total cover of the vegetation types of interest. Calculations are made for vegetation-soil combinations for every 1km grid square using the dominant soil type within a grid area. These critical loads values are then applied to appropriate areas as determined by the ITE LCMGB. The resulting 5-percentile map for coniferous woodland (Figure 3) demonstrates how critical loads values are influenced by input parameters. Some of these influences are indirect, such as rainfall (the wetter western part of Britain is clearly different from the drier east); others such as soils may have a more direct and detailed influence.

Both of the above approaches have been used for mapping critical loads for nutrient N for Britain and are providing data for use within the UNECE relating to a future N protocol. For grassland and heathland the empirical approach has been applied, while for woodland, the lowest critical loads value of either the empirical estimate or the mass balance calculation is applied to each 1km cell.

4. Using critical loads maps to show areas of excess N and S

In order to be able to take account of both acidity and excess nutrient N effects it is necessary to incorporate the effects of S and N in a single "critical loads function". The method has been described in some detail in Posch *et al.* (1993) where the procedures for calculating the necessary parameters are explained. The function enables S and N deposition to be assessed in relation to critical loads of acidity and nutrient N (Figure 4). It describes an "envelope of protection" for an ecosystem which defines those

Table I

Deposition reduction requirements for different regions of the critical loads function

Deposition reductions required

1 Only S
2 S required before option of S or N
3 Both S and N must be reduced before options available
4 N required before option of S or N
5 Only N
6 Either S or N
7 None - ecosystem protection

Figure 4

The critical loads function for sulphur and nitrogen deposition, incorporating acidity and nutrient loads, showing the different regions for control options

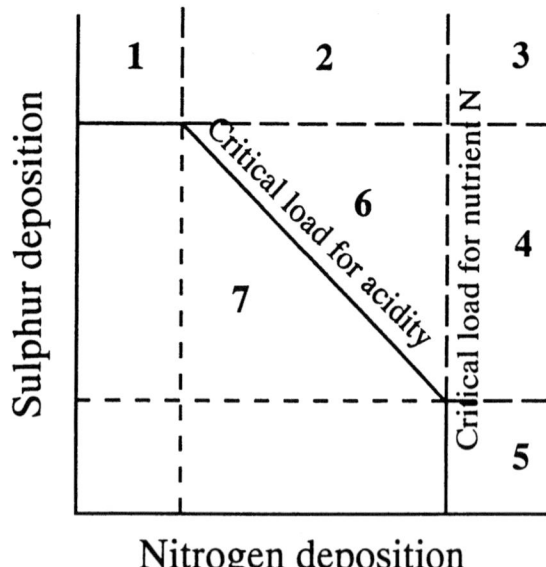

combinations of S and N deposition which will not exceed the critical load. The function consists of two parts:-

(i) the acidity function, this is both N and S dependent (since both can acidify), but the simple linear function is shifted along the N axis to allow for N transformations (eg immobilisation) which reduce the acidifying effects of N;

(ii) the nutrient N function, this is independent of S and represented by a line parallel to the S deposition axis, this function is ignored if the critical load of nutrient N is greater than the critical load of acidity resulting from N deposition only (ie acidity determines the critical load).

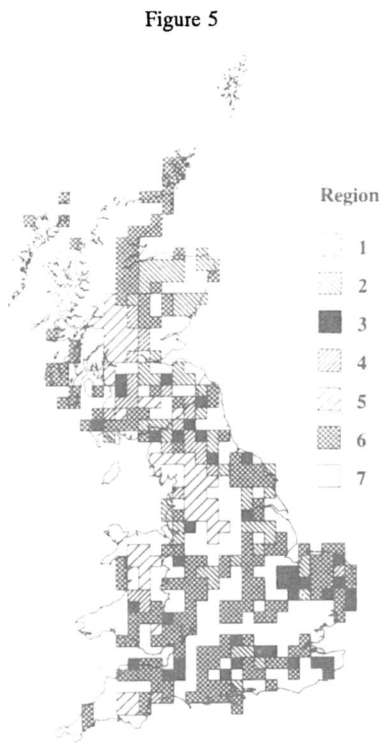

Figure 5

Region
1
2
3
4
5
6
7

Using the critical loads function it is possible to define those areas on Figure 4 where N and/or S deposition reductions are required to achieve ecosystem protection, ie reduce deposition below the critical loads of acidity and nutrient N. For some combinations of N and S there is an option to reduce either pollutant to achieve protection; for others one or other (or sometimes both) must be reduced before protection is possible. The different areas on the graph are described in Table I.

Using critical loads data for acidity and nutrient N for Great Britain, together with current deposition estimates (Metcalfe *et al.*, 1995) it is possible to draw maps showing the various deposition reduction requirements as listed in Table I. To simplify presentation in this paper results are again presented for 20km grids. For this, a simplified approach has been adopted; a 5-percentile critical loads function was estimated by calculating the 5-percentile of each part of the function separately. While the resultant function may not relate directly to any particular ecosystem, it gives a good indication of protection to the grid square.

A map for acid grassland (Figure 5), showing the 7 deposition classes, indicates that there are different strategies required for achieving critical loads in different parts of the country. In some areas there are options for reducing N or S. In others the emphasis is on N or on further reductions of S before options of either become possible.

This approach has great potential for exploring options for emission controls and identifying their effectiveness in affording ecosystem protection. It is being used in relation to future scenarios (Metcalfe *et al.*, 1995), while a similar approach is providing the means for discussion of abatement options in Europe.

5. Conclusions

The methods which are described here are those which are being used for generating national critical loads for nutrient N. The results demonstrate the need for taking nutrient effects of N into account when considering control strategies for N emissions since critical loads for acidity are often of the same order of magnitude. By using the critical loads function approach it is possible to identify the controls of S and N which may be required thus providing a powerful tool to the policy maker.

Acknowledgements

The authors acknowledge the support of the UK Department of the Environment for funding this work (contract PECD7/10/90)

References

Bobbink, R. & Roelofs, J.G.M.: 1995, Empirical nitrogen critical loads: update since Lokeberg 1992. In:Hornung, M., Sutton, M. & Wilson, R. B. (Eds): 1995, Mapping and modelling of critical loads for nitrogen -a workshop report. 9-19, Institute of Terrestrail Ecology: Edinburgh.

CCE: 1993, Calculation and mapping of critical loads in Europe. Coordination Centre for Effects Status Report 1993 (Eds. R. J. Downing, J.-P. Hettelingh & P. A. M. de Smet). RIVM report 259101003. RIVM: Bilthoven.

CLAG: 1994, Critical loads of acidity in the United Kingdom. Report by the Department of Environments Critical Loads Advisory Group (Edited by Pitcairn, CER).

Eversham, B. C. & Roy, D. B.: in press, Biodiversity research using national data. In: Data for Action on Biodiversity. Proceedings of the National Federation for Biological Recording Seminar.

Fuller, R. M., Groom, G. B. & Jones A. R.: 1994, The Land Cover Map of Great Britain: an automated classification of Landsat Thematic Mapper data. Photogrammetric Engineering and Remote Sensing, 60, 553-562.

Grennfelt, P. & Thernelov, E. (Eds.): 1993, Critical loads for nitrogen - a workshop report. Nordic Council of Ministers: Copenhagen.

UKRGIAN: 1994, Impacts of N Deposition in Terrestrial Ecosystems. Report of the United Kingdom Review Group on Impacts of Atmospheric N (Edited by Pitcairn, CER).

Hornung, M., Sutton, M. & Wilson, R. B. (Eds): 1995, Mapping and modelling of critical loads for nitrogen - a workshop report. Institute of Terrestrail Ecology: Edinburgh.

Metcalfe, S. E., Whyatt, D., Bull, K. R.: 1995, Spatial variability in emissions reduction strategies for sulphur and nitrogen in the UK. Water, Air and Soil Pollution. In press.

Posch, M., Hettelingh, J.-P., Sverdrup, H. U., Bull, K. R. & de Vries, W.: 1993. Guidelines for the computation and mapping of critical loads and exceedances of S and N in Europe. In: Calculation and mapping of critical loads in Europe (edited by Downing, R. J., Hettelingh, J.-P. & de Smet, P. A. M.), RIVM report 259101003. RIVM: Bilthoven.

Rodwell, J.S. 1991a, 1991b, 1992, 1994. British Plant Communities. 1. Woodland and scrub. 2. Mires and heaths. 3. Grasslands and montane communities. 4. Aquatic communities, swamps, and tall-herb fens. Cambridge: Cambridge University Press.

UNECE 1994. Protocol to the 1979 Convention on Long-Range Transboundary Air Pollution. United Nations: Geneva.

EVALUATING CRITICAL LOADS OF ACIDITY FOR SWISS FOREST SOILS: COMPARISON OF TWO CALCULATION METHODS

D. KURZ[1], U. EGGENBERGER[2] and B. RIHM[3]

[1] *EKG Geo-Science, Aebistr. 2A, CH-3012 Bern, Switzerland,* [2] *Min.-petr. Institute, Univ. of Bern, Baltzerstr. 1, CH-3012 Bern, Switzerland,* [3] *METEOTEST, Fabrikstr. 29, CH-3012 Bern, Switzerland*

Abstract. A variant of the European model *Simple Mass Balance* (SMB) and the regionalized *PROFILE* model have been used to calculate critical loads of acidity for Swiss forest soils. The single layer SMB has been applied to 11,800 receptor points and the multi-layer PROFILE to 720 forest sites. Weathering rates used in SMB calculations were assessed by means of a modified de Vries soil classification, and calculated from physical properties of the soil system with PROFILE.
Cumulative frequency distributions of the results at the national resolution show that PROFILE predicts lower critical loads. Up to the 65-percentile critical load PROFILE percentile values are average 65% of the SMB percentile values. The upper percentiles of the PROFILE critical loads are merely 25% of the respective SMB predictions.
The analysis of the model predictions on a regional scale implies that the multi-layer model should be used for the assessment of critical loads for areas with potentially calcareous forest soils. The inherent inability of the SMB method to properly account for processes in the carbonate system and to estimate adequate weathering rates results in significantly higher critical loads compared to PROFILE predictions. Both models estimate critical loads in better agreement for low weathering forest soils in high-precipitation areas where hydrogen and aluminum leaching govern the model result.

1. Introduction

Within the framework of the Convention on Long Range Transboundary Air Pollution (LRTAP) of the United Nations Economic Commission for Europe (UN ECE) emission reductions of sulfur and nitrogen compounds are being negotiated. Reductions are being established on the basis of the critical load concept. A critical load has been defined as the highest deposition of a compound, that will not cause chemical changes leading to long-term harmful effects on ecosystem structure and function (Nilsson, 1986). The sensitivity of a variety of ecosystems with regard to acid deposition has been mapped in terms of critical loads of acidity across Europe (Hettelingh *et al.*, 1991; Downing *et al.*, 1993). A formal procedure has been established for calculating critical loads of acidity (Sverdrup *et al.*, 1990). This involves selecting a receptor type having an intrinsic capacity to neutralize acid inputs, identifying a biological indicator and determining a critical chemical value above which the indicator organism will show an adverse response (Sverdrup and Warfvinge, 1993). Relating environmental data to a single quantitative value entails the use of models. Regional estimates of critical loads of acidity are currently assessed by a static modeling approach. In this approach, which is usually referred to as steady state mass balance method, a steady state is assumed for ecosystem relevant chemical processes, such that all sources of acidity are balanced by sources of alkalinity. In practice the objective is to avoid the selected critical chemical value being exceeded at critical load.

At present critical loads are calculated by various implementations of the steady state mass balance approach for soil and water chemistry, e.g. the Simple Mass Balance model (SMB) (Downing *et al.*, 1993) and PROFILE (Warfvinge and Sverdrup, 1992). SMB model variants have been employed by all European countries. Since 1993 Switzerland has additionally been applying a four layer steady state mass balance model called regionalized PROFILE. The SMB method has been used to calculate critical loads for Swiss forest soils with a 1x1km grid resolution resulting in 11,800 receptor points. The regionalized PROFILE model was applied to 720 forest sites representing a spatial resolution of a 4x4km grid. Both models have been run with identical model independent input data in particular climate, deposition and vegetation data for the 720 sites. A base cation to Al (BC/Al) molar ratio in the soil solution equal to 1 has been used as critical chemical value.

The objective of the present contribution is to compare model predictions and to evaluate differences with respect to the use of the two model variants in determining critical loads.

2. Materials and Methods

2.1. MODEL CONCEPTS COMPARED

The critical load of acidity for forest soils is derived from a mass balance in terms of acid neutralization capacity (ANC) demanding all inputs of acidity to the system to be balanced by all sources of alkalinity (Table I) and permitting no net depletion of base saturation. Omitting the exchange alkalinity term (ANC_X) and terms dependent upon landuse activities (ANC_{BC}, ANC_N), the critical load of acidity is calculated as the weathering rate (ANC_W) minus the permitted alkalinity leaching (ANC_L). The alkalinity concentration in the leaching soil solution at critical load corresponds to a soil solution having a BC/Al ratio at the limit specified.

In the SMB critical ANC_L is determined from a simplified expression for ANC as the maximum permitted leaching of aluminum and protons. In accordance with the BC/Al criterion, the critical aluminum leaching is derived from a mass balance of the relevant base cations accounting for weathering, deposition and uptake processes and certain physiological limitations. The critical proton leaching is connected to the aluminum concentration via the gibbsite equilibrium.

In PROFILE the individual terms of the balance are no longer independent and processes are linked via the soil solution. PROFILE takes into account chemical weathering of soil minerals, uptake of base cation, ammonia and nitrate by the vegetation, nitrification of ammonia to nitrate as well as solution reactions involving the carbonate system, Al species and organic acids. Deposition input of sulfate ammonia, nitrate and base cations is also considered. Soil processes are represented by mass balances for ANC, base cation, nitrate and ammonia and kinetic equations for weathering and nitrification inside the PROFILE model (Table I). Weathering of base cations

TABLE I

Characteristics of the SMB and PROFILE models (ANC sinks in italic)

Process		SMB-Model	PROFILE
Atmospheric deposition	ANC_D	input	input
Hydrology		input; precipitation excess	input; variable flow with depth
Soil stratification		input; one layer	input; multi-layer
Geochemical processes:			
Production of base cations by weathering of carbonate and silicate minerals	ANC_W	input; classification	kinetic
Production of Al-species by weathering of Al-hydroxide		forcing function	equilibrium
Solution reactions	ANC_L	simplified charge balance	equilibria; carbonate and Al species, organic acids
Cation exchange reactions	ANC_X	-	steady state
Biological processes:			
Nitrification	ANC_{NNI}	-	kinetic
Denitrification	ANC_{NDE}	-	input
Mineralization/immobilization of N-compounds	ANC_{NIM}	-	input
Uptake of nitrogen compounds	ANC_{NU}	-	input; forcing function
Uptake of nutrient cations	ANC_{BCU}	input	input; forcing function
Mineralization/immobilization of nutrient cations	ANC_{BCIM}	-	input

is calculated from independent geophysical properties of the soil system.

Both models consider the soil compartment chemically isotropic and the soil solution perfectly mixed. In the SMB the soil compartment is identical with the system boundary, the rooting depth, whereas in PROFILE it is one out of four soil horizons within the system boundary.

2.2. INPUT DATA

Input data on the national scale were collected according to the UN ECE Task Force on Mapping guidelines (Sverdrup et al., 1990). The data acquisition strategy consisted of deriving input data from either national or regional survey information or single spot measurements, using transformation functions or computerized submodels. From the steady state approach it follows that seasonal variations cannot be addressed, therefore long-term average values of the input variables are used. Hydrology data were derived from the Hydrological Atlas of Switzerland. Deposition of acidifying compounds was extrapolated from survey (Nationales Beobachtungsnetz für Luftfremdstoffe) and field measurements, using individual deposition models for wet and dry deposition (Rihm, 1994). Bulk deposition loads of base cations were supplied by the Coordination Center for Effects (CCE) with a 1° longitude x 1/2° latitude resolution. Data for tree species distribution and basic information to assess biomass base cation and nitrogen uptake are derived from the National Forest Inventory. Soil properties including bulk density, surface area and moisture content were extrapolated from a series of reference soil profiles by means of a soil classification. The soil mineral composition was derived from total elemental soil analyses by a normalization procedure (Eggenberger *et al.*, in prep). Remaining parameters were drawn from Sverdrup *et al.* (1990).

3. Results and Discussion

Critical loads of acidity calculated for Swiss forest soils range between 0.1 and 10.3 (PROFILE) and 48.5 keq/ha/yr (Standard SMB, SSMB), respectively (Plate I/B). 65% of the sites and receptor points considered yield values below 2 (PROFILE) and 3 keq/ha/yr (SSMB). PROFILE generally predicts substantially lower critical loads than the SMB variant employed. Almost 90% of the PROFILE spot values are below the respective SMB estimates.

Cumulative frequency distributions of the two critical load populations (Plate I/B) confirm this tendency. Up to the 65-percentile critical load PROFILE percentile values are average 65% of the respective SSMB values. The upper percentiles of the PROFILE critical loads are merely 25% of the SSMB predictions. The latter discrepancy is explained by the different derivation of the ANC_L in the two models. The simplified expression for aqueous ANC used with the SMB neglects the presence of other species than aluminum and protons, whereas in PROFILE additional solution components participating in acid-base reactions have been included in the solution ANC definition. ANC_L in the SMB becomes by definition negative and is not applicable if there are any doubts that aluminum species and protons predominate over carbonate species in the soil solution. Therefore large discrepancies between the two model results are observed in northern and northeastern areas of Switzerland where calcareous soils are present (Plate I/D).

The discrepancy in the lower percentiles of the critical load cumulative frequency distribution mainly results from differences in the weathering rates estimated by a modified de Vries classification (SMB) and those calculated by PROFILE (Plate I/A). Using the much lower PROFILE weathering rates with the SMB (Modified SMB, MSMB) improves the agreement of the critical loads estimated by the two methods (Plate I/C). This is particularly true for weathering rates ranging below 1 keq/ha/yr, which do not completely outweigh the net effects of base cation deposition and uptake considered in the ANC_L term of the SMB model. Remaining differences in the model predictions may be related to the soil stratification and processes such as Al speciation and complexing of Al with organic acids, additionally considered in the PROFILE model.

The principles of the modeling procedure may affect the 5-percentile critical load of acidity at 150x150km EMEP grid cell resolution, commonly used under the convention (LRTAP) for policy decisions on emission reductions. PROFILE 5-percentile critical loads are 47, 46, 36, 92 and 85% of the SMB values for the cross sections of the EMEP grid cells 23/12-14 and 24/13-14 with the area of Switzerland (Plate I/D-G).

PLATE I
Discrepancies in the weathering rates and critical loads calculated with SMB and PROFILE.

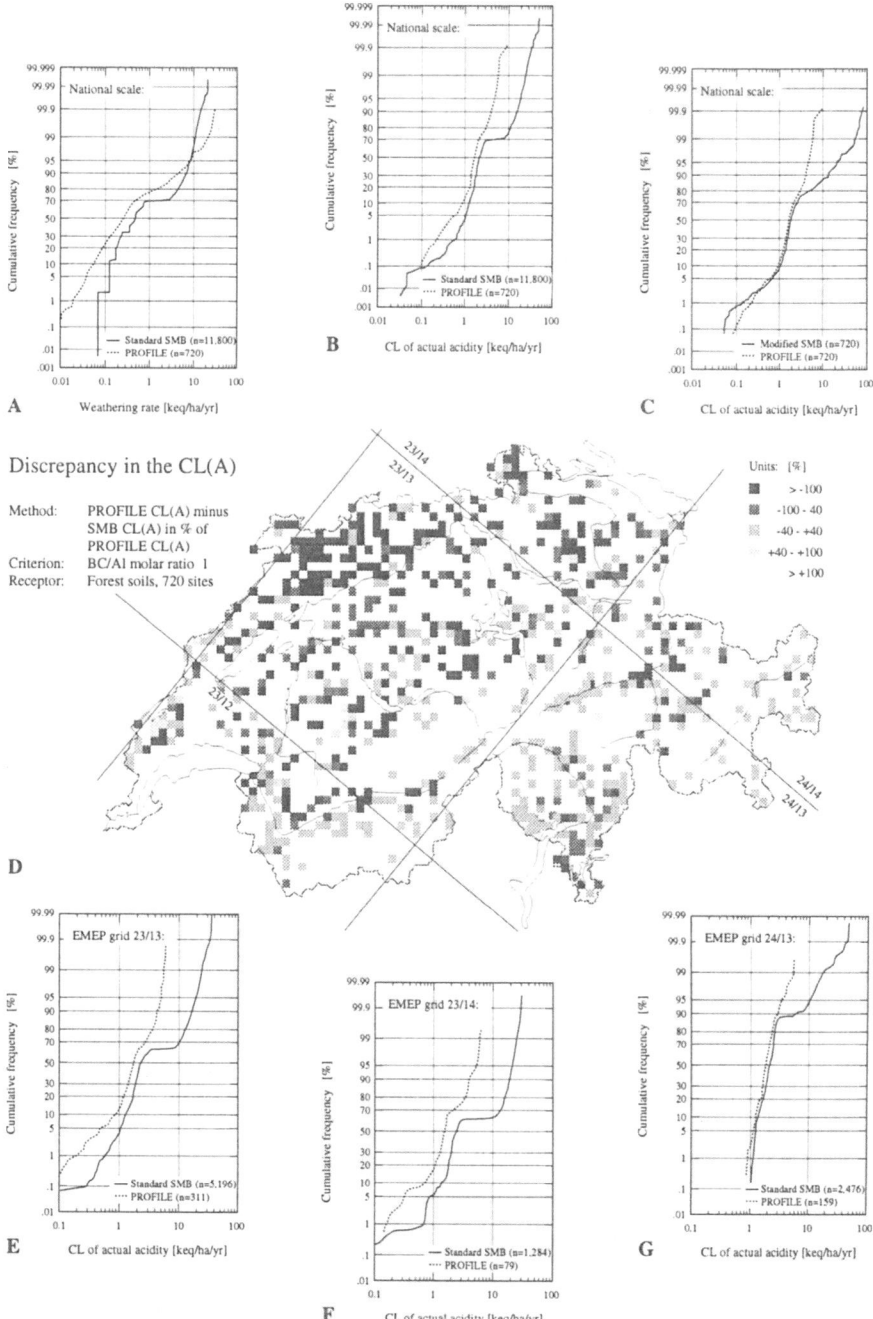

4. Conclusions

An analysis of critical load predictions of the two model approaches indicates that the SMB model should only be applied to areas, forest soils of which fulfill the confining conditions introduced by the various assumptions and simplifications necessary for its derivation. For other areas, particularly areas comprising calcareous soils, a more complex model such as PROFILE has to be used to get realistic critical loads of acidity. Additionally weathering rates estimated by means of a soil classification, frequently used with the SMB, may introduce large uncertainties into the critical load calculations. 5-percentile critical loads calculated with the discussed SMB variant have to be considered as conservative estimates with respect to their use for policy decisions on emission reductions. However critical acid loads for Swiss forest soils are, despite uncertainties in their derivation, frequently (~60% of the sites) substantially exceeded by present loads and emission reductions are needed to attain a long-term sustainable forest ecosystem.

Acknowledgments

The study is based on the Swiss critical loads mapping activities receiving appreciative support from P. Blaser, S. Zimmermann (WSL/FNP), S. Braun (IAP), H. Sverdrup and P. Warfvinge (Lund University). The program is guided and financed by the Swiss Federal Office of Environment, Forest and Landscape. The contents of this paper do not necessarily reflect the views of the FOEFL and no official endorsement should be inferred. The manuscript profited from elaborate comments of two anonymous reviewers.

References

Downing, R. J., Hettelingh, J.-P., de Smet, P. A. M. (eds): 1993, RIVM Report 259101003, Coordination Center for Effects, Natl. Inst. Publ. Health and Environ. Prot. Bilthoven, Netherlands.

Eggenberger, U., Kurz, D., Achermann, B., Blaser, P., Rihm, B., Sverdrup, H., Warfvinge, P., Zimmermann S.: in prep, Environmental Series Air, Federal Office of Environment, Forests and Landscape (FOEFL) Berne, Switzerland.

Hettelingh, J.-P., Downing, R. J., de Smet, P. A. M. (eds): 1991, RIVM Report 259101001, Coordination Center for Effects, Natl. Inst. Publ. Health and Environ. Prot. Bilthoven, Netherlands.

Nilsson, J.: 1986, Miljørapport 1986:11, Nordic Council of Ministers Copenhagen, Denmark.

Rihm, B. 1994. Environmental Series Air 238, Federal Office of Environment, Forests and Landscape (FOEFL) Berne, Switzerland.

Sverdrup, H.; de Vries, W., Henriksen, A.: 1990, Miljørapport 1990:14, Nordic Council of Ministers Copenhagen, Denmark.

Sverdrup, H., Warfvinge, P.: 1993, Reports in ecology and environmental engineering 1993:2, Lund University Lund, Sweden.

Warfvinge, P., Sverdrup, H.: 1992, Water, Air, and Soil Pollut. 63, 281-291.

SULPHATE SULPHUR CONCENTRATION IN VEGETABLE CROPS, SOIL AND GROUND WATER IN THE REGION AFFECTED BY THE SULPHUR DIOXIDE EMISSION FROM PŁOCK OIL REFINERY (CENTRAL POLAND)

W. MIKUŁA

Warsaw Agricultural University, Faculty of Horticulture, Department of Environment Protection, Nowoursynowska 166, 02-787 Warsaw, Poland

Abstract. Research was carried out in 1984-1990 in the region affected by the sulphur dioxide emission from one of the greatest oil refineries in Europe (Płock, central Poland). The sulphate sulphur concentration in the vegetable crops (red beet, carrot, parsley, bean, cabbage and dill), the soil and in ground water was defined in selected allotment gardens of Płock city and in a household garden located in the rural area about 25 km from the town. The highest amount of sulphate sulphur was found in the vegetable crops cultivated in the garden situated in the closest vicinity of the refinery. Sulphate sulphur contents harmful for plants (above 0.50 per cent d.m.) were noted in cabbage and carrot leaves in almost all the gardens (except one). The soil in all examined gardens was characterised by high sulphate sulphur concentration, which considerably exceeds the maximum amount admissible for light soils in Poland, i.e. 0.004 per cent d.m. The sulphate sulphur concentration in ground water in all the gardens exceeded the highest permissible content in drinking water in Poland (200 mg*dm^{-3} of sulphate or about 67 mg*dm^{-3} of sulphate sulphur). The sulphate sulphur content in the soil and ground water was not significantly dependent on the garden's distance from the refinery. Generally, the abovenormal sulphate sulphur concentrations occurred quite universally in the examined region and they concerned all the considered environmental components (vegetable crops, soil, ground water) and all the gardens.

Key words: Sulphate sulphur, abovenormal concentrations, vegetable crops, soil, ground water, allotment gardens, household garden, oil refinery.

1. Introduction

Płock and its immediate surroundings (central Poland) has been recognised as one of 27 environmental hotspots in Poland. Undoubtedly, the main cause of the environmental degradation in this area are the Mazovian Refinery and Petrochemical Works in Płock (MR&PW), which has been in operation for over 30 years. This enterprise, the second largest of its kind in Europe, manufactures as many as about 120 types of raw materials, semi-finished and final products, using a wide variety of technological processes (Karaczun, 1991). It emits about 400 noxious substances into the air (Niedomagała, 1985/1986), of which, gaseous sulphur dioxide, emitted in very large amounts, is one of the most hazardous compounds affecting the environment. E.g., in 1988 the Płock refinery emitted as much as 67 261 t of SO_2 into the atmosphere, accounting for 59 per cent of the total of gaseous pollutants (excluding carbon dioxide) entering the air from these works (Ochrona Środowiska 1989, 1989).

Studies which have been conducted so far have shown many unfavourable changes in the composition and function of, e.g., crop plants, soil and ground water (Biernacka, 1984 and 1989; Nowicki, 1985 a and b; Lenart, 1990 and 1991; Indeka, 1991). There have been very alarming signals regarding the excessive accumulation of sulphur pollutants, including sulphate sulphur, in vegetable crops, soil and ground water (Nowakowski, 1982; Zimny *et al.*, 1988; Mikuła and Budzikowski, 1991; Mikuła, 1991). In 1984, a 7 year long study project got under way covering several allotment gardens in Płock and a household garden in a village nearby. The aim of the project was to determine whether the sulphate sulphur concentrations in six vegetable species, soil and ground water may be recognised as abovernormal, exceeding the relevant standards. In addition, the correlation between the sulphate sulphur concentration in the particular organs of the vegetable crops, and in soil and ground water was also studied.

2. Materials and methods

The research was conducted in four allotment gardens in Płock, situated about 0.5 to about 6 km south and south-east from the Refinery and Petrochemical Works, and in a household garden in the village of Sanniki, situated about 25 km south-east from Płock. The dominating wind direction in Płock region is W and the share of winds from NW and N directions is low (Mikuła, 1993). The gardens under consideration were located therefore in an area with low direct inflow of air pollutants from the works. In all the gardens the soils are light. They are loamy sand or sandy loam with slightly acidic or close to neutral reaction (Zakrzewska, 1992).

The following six vegetable species were cultivated in the research gardens: red beet (*Beta* L.) var. Czerwona Kula; carrot (*Daucus* L.) var. Perfekcja; parsley (*Petroselinum* L.) var. Berlińska; bean (*Phaseolus* L.) var. Wiejska; white cabbage (*Brassica* L.) var. Kamienna Głowa; and garden dill (*Anethum* L.). The former three species were grown in 1984-1990, bean in 1989-1990, and cabbage and dill in 1986-1987. The vegetables were grown at special experimental plots, in succession on three units of 4 m^2 each. In all the gardens, seeds were sown into natural soil at the same time. The same agrotechnical measures were applied on a parallel basis everywhere. The vegetables were grown on plots which had earlier been fertilised by green factors (lupin and phacelia). No pesticides were used. The vegetables were harvested in September.

The leaves and roots of red beet, carrot and parsley as well as the shoots and pods of bean, cabbage leaves and dill shoots were first thoroughly cleaned with a small brush to remove the dust settled on their surface. Then, they were dried: first at 60°C, to prevent sap leakage from the highly hydrated tissues, and then at 100°C. Three subsamples of cleaned vegetable organs from the same year and garden were combined into one sample.

Soil and ground water were sampled at the same time as were the vegetables. Soil was collected from an arable layer 0-30 cm at nine spots on each plot, and, after meticulous mixing, one representative soil sample was obtained. Two-litre ground water samples came from piezometers installed at all the experimental plots.

Soil samples were dried at room temperature and passed through a sieve (mesh 1.02 mm). After removing insoluble organic pollutants, ground water samples were stored in polypropylene canisters at 5°C.

The sulphate sulphur was determined in prepared vegetable, soil and ground water samples using the nephelometric method as modified by Nowosielski (1974) for the wave length of 490 nm. This method consists in the reaction of barium chloride with sulphate ions contained in the sample resulting in the formation of barium sulphate suspension. In the case of plant and soil samples, the sulphate sulphur content was defined in 2 per cent of acetic acid extracts, and in that of ground water, directly in the samples collected.

Altogether, each set of 35 samples of leaves and roots of red beet, carrot, parsley, soil and ground water was analysed. And so was each set of 10 samples of bean shoots and pods, cabbage leaves and dill shoots.

A two-ways variance analysis was carried out for the values representing the vegetables, treating the vegetable organs and gardens as factors. And so was a one-way variance for the values of soil and ground water, with gardens being recognised as a factor. Using Duncan multiple tests, the differences between the mean sulphate sulphur concentrations in the particular organs of vegetables and gardens were found to be significant. The two analyses were carried out for the confidence level of 95 per cent. To determine the correlation between the sulphate sulphur content in vegetables and its concentration in soil and ground water, linear regression was performed. The statistical analyses were carried out using the statistical package Statgraph 2.0.

In addition, the standard deviation was calculated, as a measure of the spread of the results round the mean values, for the sulphate sulphur concentrations in the vegetables (in terms of the particular organs and gardens), and in soil and ground water.

3. Results and discussion

3.1. SULPHATE SULPHUR CONCENTRATION IN VEGETABLE CROPS

The sulphate sulphur concentration in the particular vegetable organs and gardens was found to be variable; these two factors essentially affected its content in the vegetable crops. A significant impact was also exerted by the interaction between the vegetable organs and gardens (Table I).

A distinctly higher concentration of sulphate sulphur (0.24 per cent of dry mass on average) was found in the garden closest to the refinery (about 0.5 km away). In the other gardens, the mean sulphate sulphur content in vegetable crops was very similar. The spread of results was largest in the garden situated closest to the Refinery and Petrochemical Works, and smallest in that located about 2 km from the plant (Table II).

Regarding the vegetable organs, the sulphate sulphur content was highest in cabbage and carrot leaves (0.63 per cent and 0.43 per cent of d.m. on average). A high concentration of sulphate sulphur was also found in dill shoots and red beet leaves. A distinctly lower content was identified in bean pods. The widest spread of results was noted for carrot leaves (Table III). Thus, the data obtained confirmed the information from other authors, indicating that the assimilation organs of plants are more susceptible to sulphur accumulation (Czarnowski, 1987; Godzik, 1991).

A content of sulphate sulphur harmful for plants, i.e., above 0.5 per cent of d.m. (Mikuła, 1991) was found in cabbage and carrot leaves in almost all the gardens, except the one situated about 2 km from the refinery. The maximum concentration was noted in carrot leaves in a garden situated closest to the Refinery and Petrochemical Works, amounting to as much as 1.27 per cent of dry mass.

The comparison between sulphate sulphur content in vegetable crops grown in the gardens in the Płock region and in vegetable crops cultivated at allotment gardens close to other industrial plants in Poland shows that this content in Płock was lower than

TABLE I

Variability of sulphate sulphur content in vegetable crops

Source of variability	$F_{emp}/F_{0.05}$
Vegetable organs	39.928
Gardens	2.400
Vegetable organs x gardens	1.057

TABLE II

Sulphate sulphur concentration in vegetable crops for particular gardens

Approximate distance from Works (km)	Mean content (per cent of d.m.)	Standard deviation
2	0.17 a	0.11
25	0.18 a	0.17
6	0.18 a	0.17
3	0.20 a	0.20
0.5	0.24 b	0.26

* Means contents followed by same letters are not significantly different at 95 per cent confidence interval according to Duncan multiple test for the differences between means.

TABLE III

Sulphate sulphur concentration in particular vegetable organs

Vegetable organ	Mean content (per cent of d.m.)	Standard deviation
Bean pods	0.01 a	0.01
Red beet roots	0.08 b	0.02
Carrot roots	0.08 bc	0.03
Bean shoots	0.09 bc	0.04
Parsley roots	0.13 cd	0.03
Parsley leaves	0.17 de	0.07
Red beet leaves	0.18 e	0.07
Dill shoots	0.21 e	0.10
Carrot leaves	0.43 f	0.25
Cabbage leaves	0.63 g	0.11

* Mean contents followed by same letters are not significantly different at 95 per cent confidence interval according to Duncan multiple test for the differences between means.

in the vicinity of the artificial textile factory in Torun and the cellulose and paper works in Ostrołęka (Zimny et al., 1988).

3.2. SULPHATE SULPHUR CONCENTRATION IN SOIL AND GROUND WATER

The sulphate sulphur concentration in soil and ground water did not depend significantly on the distance to the refinery (Table IV).

The soils in the gardens under study were characterised by a high concentration of sulphate sulphur. Excessive sulphate sulphur contents, i.e., more than 0.004 per cent of d.m. (Drożdż-Hara, 1978), were found in all the gardens. Even a mean sulphate sulphur content in soil (0.009 per cent of d.m.) exceeded above twice the highest permissible concentration (Table IV), and the maximum value identified at a garden situated about 3 km from the Mazovian Refinery and Petrochemical Works was as much as 0.030 per cent of dry mass.

The ground water pollution level was determined by comparing the sulphate sulphur concentration with its maximum permissible content in drinking water. It was a result of the lack of relevant standards for drinking water and the considerable use of ground water for this purpose in Poland. It turned out that in the gardens under study the sulphate sulphur concentration level was very high. The relevant Polish standard of 200 mg*dm^{-3} of sulphates, i.e., about 67 mg*dm^{-3} of sulphate sulphur (Dziennik Ustaw RP, No. 35, 1990), was exceeded in all the gardens. The highest sulphate sulphur concentration (370 mg*dm^{-3}) was found in a garden about 2 km distant from the refinery. Even the mean sulphate sulphur concentration in ground water (105 mg*dm^{-3}) was already higher than the maximum permissible content in drinking water (Table IV).

Compared with the other industrial regions in Poland, the sulphate sulphur concentration in ground water in the gardens of the Płock region was high, e.g., much higher than in the ground water in the north-western Province of Szczecin (Gałamon et al., 1988).

3.3. CORRELATION BETWEEN THE SULPHATE SULPHUR CONCENTRATION IN VEGETABLE CROPS AND THAT IN SOIL AND GROUND WATER

Sulphate sulphur concentration in vegetable crops did not increase according to the increased concentration of sulphate sulphur in soil and ground water. Positive values of correlation coefficients were noted only for the relations between bean shoots and soil, between carrot leaves and soil, between red beet leaves and ground water and between carrot roots and ground water, but all these relations were not significant (F_{emp}/F values lower than 1). All the relevant coefficients of correlation and determination were very low, without exceeding, respectively, 0.260 and 7 per cent (Tables V and VI). Therefore, it may be presumed that the decisive factors accounting for the excessive sulphate sulphur concentration in vegetable crops in the area under study are air pollutants, absorbed by plants through the surface of the aboveground organs, mainly leaves. It requires, however, experimental confirmation.

TABLE IV

Sulphate sulphur concentration in soil and ground water and variability of contents depending on garden

	Mean content	Standard deviation	$F_{emp}/F_{0.05}$
Soil	0.009 (per cent of d.m.)	0.006	0.975
Ground water	105 (mg*dm^{-3})	96	0.913

TABLE V

Correlation between sulphate sulphur content in vegetable crops and its concentration in soil

Organ	Correlation coefficient	Determination coefficient (per cent)	F_{emp}/F
Red beet leaves	- 0.445	19.77	2.464
Red beet roots	- 0.032	0.10	0.010
Carrot leaves	0.042	0.18	0.018
Carrot roots	- 0.069	0.47	0.028
Parsley leaves	- 0.230	5.28	0.334
Parsley roots	- 0.225	5.08	0.107
Bean shoots	0.236	5.55	0.118
Bean pods	- 0.957	91.59	21.778
Dill shoots	- 0.174	3.03	0.063
Cabbage leaves	- 0.337	11.33	0.128

TABLE VI

Correlation between sulphate sulphur content in vegetable crops and its concentration in ground water

Organ	Correlation coefficient	Determination coefficient (per cent)	F_{emp}/F
Red beet leaves	0.258	6.64	0.712
Red beet roots	- 0.005	0.00	0.000
Carrot leaves	- 0.292	8.51	0.931
Carrot roots	0.203	4.10	0.257
Parsley leaves	- 0.439	19.30	1.435
Parsley roots	- 0.948	89.85	17.711
Bean shoots	- 0.478	22.89	0.594
Bean pods	- 0.255	6.52	0.139
Dill shoots	- 0.056	0.31	0.006
Cabbage leaves	- 0.719	51.74	1.072

4. Conclusion

In the area affected by sulphur dioxide emissions from the Płock refinery (central Poland), abovenormal sulphate sulphur concentrations were quite common This was true of all the environmental components, i.e., vegetable crops (more specifically, cabbage and carrot leaves), soil and ground water as well as all the gardens under study, situated up to about 25 km from the works in the direction with low direct inflow of air pollutants. It must be noted that the sulphate sulphur content in soil and ground water did not depend significantly on the distance to the refinery. The highest sulphate sulphur concentration in vegetable crops was found in a garden closest to the works, but abovenormal concentrations were also identified in other gardens.

References

Biernacka, E.: 1984, ,,Microelements in soils and plants and environment contamination", Aura, (3), 9-11 (in Polish).
Biernacka, E.: 1989, ,,Influence of refinery-petrochemical works emissions onto trace elements content in the meadow vegetation", Zesz. Probl. Post. Nauk Roln., **325**, 251-257 (in Polish).
Czarnowski, M.: 1987, ,,Photosynthesis of deciduous trees in industrial regions", Studia Naturae - Ser. A. (Zakład Ochr. Przyr. i Zasob. Natur. PAN), **31**, 11-28.
Drożdż-Hara, M.: 1978, ,,Studies on sulphur pollution influence on cultivated soils transformation in the neighbourhood of sulphur mine. Part II. Changes of chemical and physicochemical properties of cultivated soils polluted with sulphur", Roczn. Glebozn., **29** (2), 135-150 (in Polish).
Dziennik Ustaw RP ,No 35, 1990, ,,Decree of Health and Social Care Minister, changing a decree on quality standards of drinking and industrial water", 479-480 (in Polish).
Gałamon, T., Paszkiewicz, Z. and Stańkowska-Walczak, D.: 1988, ,,Determination of the selected pesticides, nitrates, nitrites, ammonium ions, sulphates and carbamide in surface and underground water and in selected agricultural products. Part V.", Roczn. PZH, **39**, 145-150 (in Polish).
Godzik, S.: 1991, ,,Air pollutants and their influence on the plants", In: Proc. of Conf. ,,Polluted environment and plant physiology", PTB, Sekcja Fizjol. i Biochem. Rośl. Oddz. Warsz., Warsaw, 25-30 (in Polish).
Indeka, L.: 1991, ,,Contamination of cereals with heavy metals in the region affected by the emissions from Mazovian Refinery and Petrochemical Works", GEA, **1**, 18-27 (in Polish).
Karaczun, Z.: 1991, ,,Effect of refinery and petrochemical industry on level of heavy metal and sulphur accumulation and biological activity of some enzymes in cultivated soils of Płock area", Doctor's thesis, Department of Environment Protection, Warsaw Agricultural University-SGGW, Warsaw (in Polish).
Lenart, W.: 1990, ,,Płock sozological syndrome", In: Proc. from conf. ,,Natural environment protection and treatment in University of Warsaw research", UW, Warsaw, 51-62 (in Polish).
Lenart, W.: 1991, ,,Environment protection in Płock voivodeship", GEA, **1**, 28-42 (in Polish).
Mikuła, W. and Budzikowski, H.: 1991, ,,Influence of the immissions from Mazovian Refinery and Petrochemical Works on the chemical composition and selected physical properties of soil, ground water and dustfall in Płock allotment gardens", GEA, **1**, 76-96 (in Polish).
Mikuła, W.: 1991, ,,Accumulation of sulphur and other macroelements in plants cultivated in Płock allotment gardens", GEA, **1**, 97-113 (in Polish).
Mikuła, W.: 1993, ,,Analysis of allotment gardens productive functions in the zone affected by the refinero-petrochemical industry", Treatises and Monographs, Wydaw. SGGW, Warsaw (in Polish).
Niedomagała, J.: 1985/1986, ,,Selected problems of environment degradation on the area of Central Poland", Stud. Region., **9/10**, 45-64 (in Polish).
Nowakowski, W.: 1982, ,,Influence of emission from refinery-petrochemical industry on growth of vegetable crops and their chemical composition", Treatises and Monographs, Wydaw. SGGW-AR, Warsaw (in Polish).
Nowicki, W.: 1985a, ,,Research on dynamics of the natural environment transformation in the zone affected by Mazovian Refinery and Petrochemical Works". In: ,,Mazovian Refinery and Petrochemical Works in Płock and environment", PAN, Warsaw, 17-27 (in Polish).
Nowicki, W.: 1985b, ,,Estimation of the natural environment research state in the zone affected by Mazovian Refinery and Petrochemical Works in Płock", In: ,,Mazovian Refinery and Petrochemical Works in Płock and environment", PAN, Warsaw, 29-40 (in Polish).
Nowosielski, O.: 1974, ,,Nephelometrical determination of $S-SO_4$", In: ,,Methods of fertilisation requirements determination", PWRiL, Warsaw, 349-350 (in Polish).
,,Ochrona Środowiska 1989": 1989, GUS (High Statistical Office), Statistics of Poland, Warsaw, 82-83 (in Polish).

Zakrzewska, M.: 1992, „Physicochemical properties of soils located near Płock urban-industrial complex", Doctor's thesis, Department of Environment Protection, Warsaw Agricultural University-SGGW, Warsaw (in Polish).

Zimny, H., Mikuła, W. and Nowakowski, W.: 1988, „Influence of selected chemical, timber, and paper plant emissions in Poland on the total sulfur and sulfate content in plants, soil and ground water", Environ. Protect. Eng., **14**, 117-126.

CRITICAL LOADS MAPPING IN POLAND: LESSONS LEARNED

W. MILL

Institute for Ecology of Industrial Areas, ul. Kossutha 6, 40-833 Katowice, Poland

Abstract. Since 1990 the Institute for Ecology of Industrial Areas, acting as National Focal Center, is actively involved in an international research programme aimed at the calculation and mapping of critical loads of acidifying compounds. Following the methodological guidelines elaborated under the leadership of UN/ECE Task Force on Mapping and Coordination Center for Effects, national maps of critical loads and their exceedances for acidity, sulphur and nitrogen have been produced. These maps have already been utilized in derivation of European maps of critical loads of acidity and sulphur submitted to the UN/ECE LRTAP Convention as scientific input to the negotiations on the Second Sulphur Protocol.
The lessons learned from the critical loads mapping exercise can be summarized as follow:
- the majority of Polish territory is covered with forest soils sensitive to acidification at an average Central European level;
- the exceedances of critical loads, estimated on the basis of national deposition data reveal the time changes of ecological risks on the territory of Poland as a reflection of economic transition. The significant difference in the scale of those risks (measured by the percentage of the country territory with the maximum exceedances of critical loads) that appear in the period between 1987, representing the period of central planned economy and 1990, representing the early transition phase to a market economy, is particularly notable.

Key words: critical loads, acid deposition, forest soils, mapping, mathematical modelling

1. Introduction

For the last five years, the Institute for Ecology of Industrial Areas (IETU) in Katowice has been acting as the National Focal Center (NFC) in the activities conducted under the UN/ECE Working Group for Effects (WGE), Task Force on Mapping (TFM) and the RIVM's Coordination Center for Effects (CCE). The research program which started in IETU in 1991 resulted in maps of critical loads and exceedances of acidity, sulphur and nitrogen for Polish terrestrial ecosystems (Mill et al., 1992,1993,1994). These maps, aggregated by CCE into European maps of critical loads and their exceedances, have been successfully applied in the negotiations of the Second Sulphur Protocol.
  This paper describes the progress and results of the studies carried out by the Polish National Focal Center.

2. Methodology of calculations and mapping

The development of methodologies to calculate critical loads dates from the late eighties and was guided by TFM and scientifically supervised by CCE. A number of revisions and improvements have been made with respect to data and methodologies during workshops organized by the CCE. These workshops provided forums to review work conducted to date and to discus methods and other issues concerning future work. The resulting methodology of calculating critical loads of acidity, sulphur and nitrogen as well as their exceedances have been described by Sverdrup et al.(1990) and Downing et al.(1993). The guidelines for mapping of critical loads of acidity in Europe are described in an updated Mapping Vademecum (Hettelingh and de Vries, 1992).

2.1. RECEPTORS MAPPED

Basically, forest soils were the subject for calculating and mapping critical loads because of the nearly 30% of the Polish territory is covered in forests. Surface waters were considered as a receptor of acid deposition only for Tatra Mountains lakes, which sensitivity is the highest in all Poland. Because of the small area and very specific geographical character, calculations of critical loads for the Tatra lakes were treated as a separate case study.

2.2. CALCULATION METHOD

The steady-state mass balance equations recommended by CCE for calculating critical loads for forest soils were found to be satisfactory for the majority of the Polish territory, which is low in elevation. For the mountainous part of Poland a special procedure described in Sverdrup (1992) has been used.
Note that all quantities provided in this chapter are expressed in [eq ha^{-1}yr^{-1}], unless otherwise noted.

Critical loads of acidity

The basic equation used to calculate critical loads of acidity $CL(A)$ is:

$$CL(A) = ANC_w - [ANC_{l(crit)}] \cdot Q \tag{1}$$

where:
ANC_w = acid neutralizing capacity produced by weathering
Q = runoff [m^3 ha^{-1}yr^{-1}]

Critical alkalinity leaching $ANC_{l(crit)}$ is defined as the ANC consumed by the maximum acceptable alkalinity leaching at critical load.

Substitution of empirical parameters and rearrangement of equation (1) leads to the following two final equations:

$$CL(A) = \begin{cases} ANC_w + 0.09 \cdot Q + 0.2 \cdot Q & \text{(the AL criterion) or} \quad (2) \\ ANC_w + 0.09 \cdot Q + 1.5 \cdot (BC_d + ANC_w - BC_u) & \text{(the Al:Ca criterion)} \quad (3) \end{cases}$$

The lower of the two values respectively calculated by equations (2) and (3) has been used.

Critical loads of sulphur and nitrogen

$$CL(S) = S_f \cdot CL(A) \qquad (4)$$

where S_f is the so-called "sulphur fraction".

$$CL(N) = N_u + N_i + (1-S_f) \cdot CL(A) \qquad (6)$$

The exceedance of the critical loads of acidity, sulphur and nitrogen is obtained by subtracting the critical loads from the deposited acidity, sulphur and nitrogen as well as the acidity produced by soil processes.

2.3. DATA SOURCES AND MAPPING METHOD

Most of the geographical data used to determine the critical loads values were taken from the national data sources apart from branch to stem ratio and relative soil moisture saturation data.
Soil data. The dominating types of soil in particular grids were adopted on the basis of the data from "Map of Polish Soils 1:300,000" (1961). 40 types of predominant soils in Poland were applied to the calculations.
Meteorological data. The data concerning precipitation, runoff and average annual temperature were obtained from "Hydrological Atlas of Poland" published by the Institute of Meteorology and Water Management for the years 1951-1975.
Forest data. The data concerning the spatial location of forests were based on "Forest Map of Poland", edited by Forest Management and Geodesy Office. The data concerning resources, forest growth and age of trees were obtained from the data bank of the Forest Management Office. For calculations of critical loads those forestry areas were chosen where the percentage of forests in grid surface was larger than 20 per cent.
Uptake data. The uptake data: BC_u, N_u, N_i, N_l were determined on the basis of forest growth in particular grids and the contents of particular elements in stem and branch.
Deposition data. The data concerning sulphur and nitrogen deposition were taken from the national data sources provided by the Institute of Environmental Engineering of the Warsaw University of Technology (Abert *et al.*, 1992, 1993). Deposition data for 1987 and 1990 as well as for 5 year period (1987 - 1991) were used. The base cation deposition data were adopted from EMEP data provided by CCE-RIVM.

The spatial distribution of critical loads of acidity, sulphur and nitrogen as well as their depositions and exceedences are presented in form of maps. These maps have been produced in accordance with CCE recommendations. Mapping resolution is compatible with NILU grid. It has been set at 0.2° longitude and 0.1° latitude which corresponds to grid size about 10 × 10 km for Poland. There are 2170 receptor points, nearly 930 of which covered by forests.

A sample of the produced maps, concerning sulphur critical loads and exceedances is presented in Figure 1.

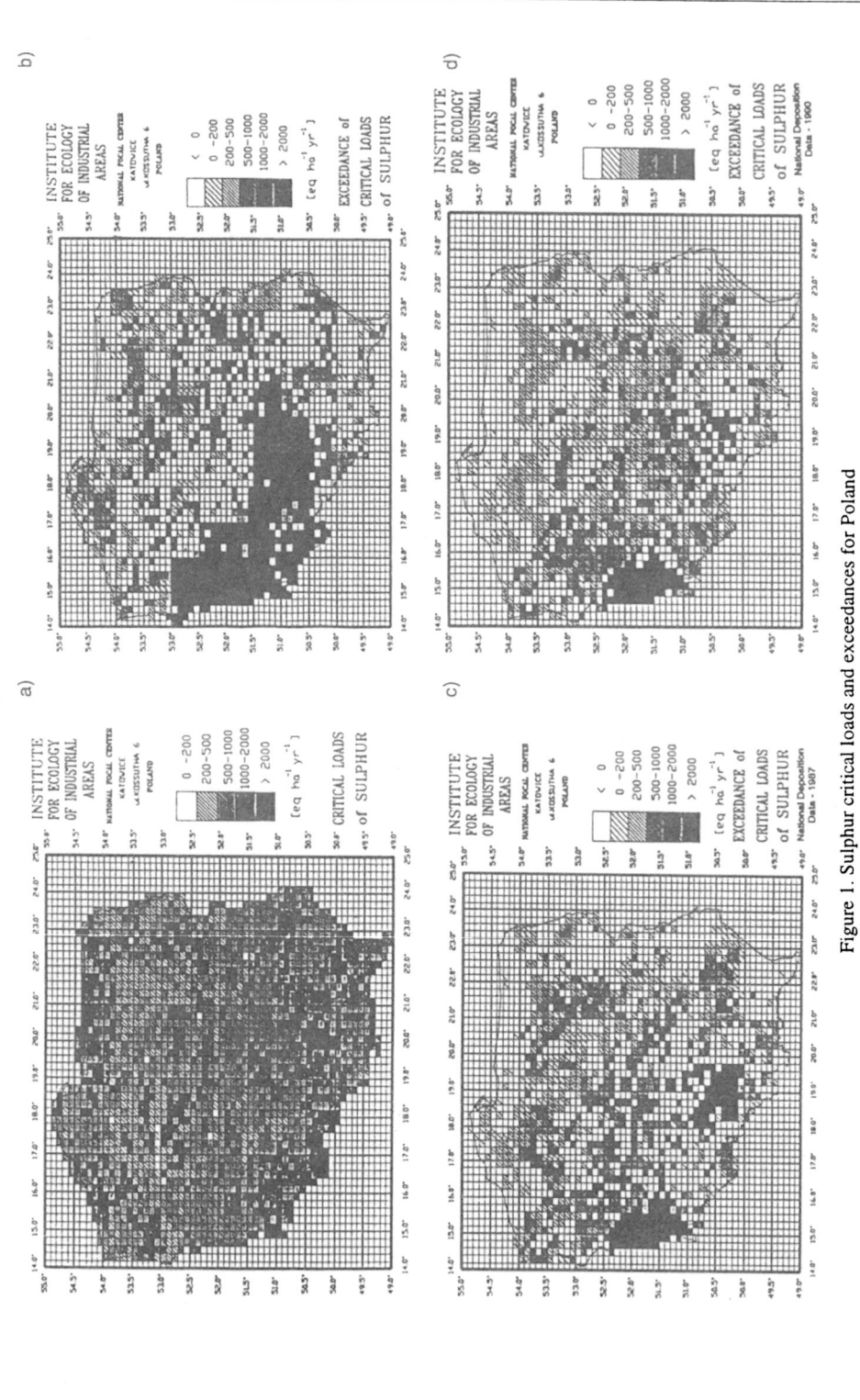

Figure 1. Sulphur critical loads and exceedances for Poland

3. Results and discussion

The values obtained for critical loads and their exceedances were stored in the form of data bases and presented as maps generated using GIS technology. For each set of output data a frequency analysis has been performed to derive the specificity of them.

Critical loads. Considering the distribution of the number of grids attributed to particular classes of critical loads values, one can assume, that the majority of Polish territory is covered with soils, for which the critical loads of acidity, sulphur and nitrogen are within the range of 500-1000 eq ha^{-1} yr^{-1} (Figure 1.a.). It is a feature characteristic for the region of Central Europe. However it is only a "photographic", not processed picture of the forest ecosystems sensitivity to acidity deposition. When comparing the estimation of this sensitivity with the spatial resolution applied by CCE on the territory of the whole Europe, i.e. grids sized 150km x 150km, and attribute only one critical load to each of them, represented by the 5 percentile value taken from the whole collection of load values in a grid, the picture will change significantly. One of the basic rules of the critical loads conception is protection of the most sensitive ecosystems of the considered area, in this case EMEP grid. This is the reason for attributing the 5 percentile value to the whole grid as the representative value of the critical load. The consequence of such interpretation, in case of Poland, is the change of class of critical loads values for the majority of Polish territory from the range of 500-1000 eq ha^{-1} yr^{-1} to the range of 200-500 eq ha^{-1} yr^{-1} (Downing et.al.,1993).

Exceedances. Within the described project an interesting experiment was carried out, consisting of comparing the ecological results measured by the value of critical loads exceedances and depositions of acidifying compounds on the territory of Poland assessed by EMEP with the deposition assessed on the basis of national inventory of emissions and national model of air pollution transport. This comparison leads to the following conclusions:

- the area of the highest exceedances (>2000 eq ha^{-1} yr^{-1}) of critical loads of sulphur, resulting from the mean deposition estimated by EMEP for the territory of Poland for 1987-1990 (25% of coverage) is significantly larger than the area of those exceedances resulting from the deposition estimated on the basis of national data and using the national model (5% of coverage) as indicated in Figures 1.b.,c.,and d. The reason for this distinguishable discrepancy can be explained by the difference in precision of the acidogenic compounds emission inventory for Poland. Another source of the observable difference is the precision of air pollution transport mapping in both models. The EMEP model applies to the territory of whole Europe, functioning within the spatial resolution with grid sized 150km x 150km, whereas the national model makes simulations on the territory divided into grids sized 30km x 30km;

- the EMEP model roughness results in blurring of the areas of the highest impact of the acidogenic compounds deposition caused by transboundary transport and regions under the local emission impact (see Figures 1.b. and 1.c.). The picture of both those regions is much more clear on the maps of critical loads exceedances elaborated on the basis of national deposition data. The area influenced by the transboundary acidogenic

substances transport from Germany and Czech Republic is easily distinguishable, and covers the territory of adjoined to border, south-western part of Poland as well as the Upper Silesia region remaining under the pressure of direct, local impact of highly developed industry and dense urbanization;
- the maps of critical loads exceedances, elaborated on the basis of national deposition data present also the time changes of ecological risks on the territory of Poland. The attention should be paid to the significant difference in the scale of those risks (measured by the percentage of the country territory with the maximum exceedances of critical loads), that appear in the period between 1987 (Figure 1.c.), representing the period of central planned economy, and 1990 (Figure 1.d.), representing the early transition phase to market economy. The coverage of areas of the highest exceedances of critical loads of sulphur were as follows: 8% in 1987, and 4% respectively for 1990. All the exceedances in 1987 are the result of the economy of People's Republic of Poland, directed first of all on heavy industry. The year 1990 was the first year of rapid change into market economy, what resulted in sudden decrease of industrial production and advancing economic recession;
- the comparative experiment, carried out within this project, confirms the importance of national data in estimating ecological effects of the acidification of the atmosphere as well as in the process of elaborating the strategies for acidifying emission reduction on the European scale within the LRTAP Convention. The results of this project, and especially the data bases and maps of critical loads and their exceedances for Poland have been submitted to CCE in Bilthoven in order to enclose them into the all-European project of critical loads mapping.

References

Abert,K., Budziński K., Juda-Rezler,K.: 1992, *Testing of Regional Eulerian Grid Model For Concentration and Deposition of Sulphur and Nitrogen Compounds for The Improvement of The EMEP model*. Warsaw University of Technology, Warsaw, Poland. Polish Contribution-in-kind to EMEP for 1992, 47 pp.

Abert K., Budziński K., Juda-Rezler,K.: 1993, *Regional scale air pollution models for Poland*. (Paper accepted for publication in Ecological Engineering, The Journal of Ecotechnology).

Downing R.J., Hettelingh J-P., de Smet P.A.M.: 1993. *Calculation and mapping of critical loads in Europe: Status Report 93'*, RIVM Report No.259101003, Bilthoven, The Netherlands.

Hettelingh J-P., de Vries W.: 1992. *Mapping Vademecum*. National Institute of Public Health and Environmental Protection, Coordination Center for Effects, Bilthoven, The Netherlands.

Hettelingh J-P., Downing R.J., de Smet P.A.M.: 1991. *Mapping critical loads for Europe:* CCE Technical Report No.1, RIVM Report No.259101001, Bilthoven, The Netherlands.

Map of Polish Soils: 1961. Instytut Uprawy, Nawożenia i Gleboznawstwa, Wydawnictwa Geologiczne, Warszawa.

Mill W., Rzychoń D., Wójcik A.: 1992. *Mapping critical loads for Poland:* National Focal Center Report No.1, Institute for Ecology of Industrial Areas, Katowice, Poland

Mill W., Wójcik A., Rzychoń D.: 1993. *Mapping critical loads for Poland:* National Focal Center Report No.2, Institute for Ecology of Industrial Areas, Katowice, Poland

Mill W., Wójcik A., Rzychoń D.: 1994. *Mapping critical loads for Poland:* National Focal Center Report No.3, Institute for Ecology of Industrial Areas, Katowice, Poland

Sverdrup,H., de Vries W., Henriksen A.: 1990. *Mapping Critical Loads: A guidance manual to criteria, calculations, data collection and mapping*. In: UN ECE Mapping Manual.

CRITICAL LOADS OF ACID DEPOSITION FOR FOREST ECOSYSTEMS IN THE KOLA PENINSULA

G. KOPTSIK[1] and S. KOPTSIK[2]

[1] *Faculty of Soil Science,* [2] *Faculty of Physics, Moscow State University, Moscow 119899, Russia*

Abstract. We assessed critical loads of acid deposition and their exceedance for soils in the Kola Peninsula using a simple balance method and mapped them within 1.0° x 0.5° longitude/latitude grid cells. Critical loads of acidity vary from 200 to 800 mol$_c$/ha/y with the type of soil, parent rock, vegetation and climatic conditions. The critical deposition values are dominated by S contribution. Present sulphur depositions are higher than critical values in the large part of the Kola Peninsula (about 40% of total area). The greatest excess (800-1200 mol$_c$/ha/y) occur in north-western and western parts, especially in surroundings of nickel smelter in Nickel. Terrestrial ecosystems in the north-western Kola Peninsula are particularly susceptible to acid deposition damage due to relatively high soil sensitivity and heavy sulphur deposition.

1. Introduction

Acid deposition is a major environmental hazard over large areas in Europe, contributing to observed damage of forests and surface water. Eastern Finnmark and the Kola Peninsula receive high loads of sulphur, a primary component of acid deposition (Fig. 1a). The boreal forests of the northern Finnmark-Kola are among the northernmost coniferous forests of the world. Under the prevailing extreme growth conditions, when trees and vegetation are under strong natural stresses, it can be presumed that even minor loads of air pollutants may have severe effects upon forest vitality. The areas surrounding Pechenganickel smelter, the north-western part of the Kola Peninsula, are especially polluted from emissions of sulphur dioxide. The capacity of ecosystems to withstand and buffer the effects of acid deposition varies through a wide range according to their physical, chemical and biological properties. The critical load concept is now developed and widely used to assess the sustainability of ecosystems, to compare critical loads with the present pollutant deposition and to connect them with emission reduction strategy in the frame of the Convention on Long-Range Transmission of Air Pollution. A critical load is the highest deposition of acidifying compounds that will not cause chemical changes leading to long-term harmful effects on ecosystem structure and function (Nilsson and Greenfelt, 1988). The purpose of this paper is to assess and map critical loads of acid deposition for soils in the Kola Peninsula and to identify the areas where critical loads are exceeded.

The preliminary critical load assessment and mapping for forest ecosystems in the Kola Peninsula have been done by Koptsik et al. (1991, 1992). The presented revised maps are the next step in critical load assessment for this region on the basis of new experimental results and improved methodology.

2. Materials and methods

The studied area is located in the north-western part of Russia, in Murmansk region. The critical loads for acid deposition were calculated with the simple mass balance model (SMBM) according to the procedure outlined in the CEC guidelines (Sverdrup et al., 1990; Hettelingh and de Vries, 1990). The present calculations are made for acidification of forest soils (0-50 cm) as receptor. According to SMBM

$$CL(A) = ANC_w - ANC_l , \qquad (1)$$

where $CL(A)$ is critical load of acidity; ANC_w is ANC produced by chemical weathering of primary and secondary soil minerals; ANC_l is ANC consumed by the maximum acceptable alkalinity leaching at critical load.

Weathering is the main process used to quantify critical acid loads. In our calculations we have used the equation of Olsson and Melkerud (1990) to estimate the weathering rate based on easily measurable soil variables like the total base cation content of the parent material and the sum of temperatures above 5°C.

The critical acidity leaching flux was calculated as the sum of aluminium leaching and hydrogen leaching. Critical aluminium leaching was obtained from the molar BC/Al ratio of 1.0 for forests and 2.0 for tundra ecosystems (Sverdrup and Warfvinge, 1993), and the hydrogen leaching was calculated from a gibbsite equilibrium.

Critical deposition of sulphur, $CD(S)$, was obtained from the critical loads of acidity (Downing et al., 1993):

$$CD(S) = S_f (CL(A) + BC_{dep} - BC_u) , \qquad (2)$$

where S_f is the sulphur fraction; BC_{dep} is the total non-marine base cation deposition; BC_u is the net uptake of base cation in the tree biomass.

The exceedance of the critical sulphur deposition, $Ex(S)$, was obtained by subtracting the critical sulphur deposition from the deposited sulphur, S_{dep}:

$$Ex(S) = S_{dep} - CD(S) , \qquad (3)$$

Critical loads and their exceedances are estimated and mapped as 5 and 50 percentile values with a 1.0° x 0.5° grid resolution and it's 4 x 4 subgrid for soils of the Kola Peninsula and for soils of surroundings of Nikel, respectively. In connection with the absence of the regular monitoring net it is impossible to present adequite supporting information for the large spatial variability of soil and climatic conditions in the Kola Peninsula. Present investigation is based on the soil cover subdivision on soil areas with similar bioclimatic conditions, relief development and parent rocks (Belov and Baranovskaya, 1969). Available information was collected for different soils in each soil area to give the quantitative estimation of various parameters and constants for critical load assessment. The soil and vegetation data were taken both from experimental measurements and literature (Belov and Baranovskaya, 1969; Manakov, 1972; Manakov and Nikonov, 1981; Nikonov, 1987 and other). Obtained results were interpreted on the whole territory of each soil area according to the soil and vegetation distribution. The areal distributions of soils and vegetation were taken from Soil Map of the World (1978) and USSR Forest Atlas (1973). The average deposition data was taken from CCE data, as adapted from EMEP data, and from Sivertsen et al. (1991).

3. Results and Discussion

Soils of the Kola Peninsula are characterized by low values of weathering rate. They range from 0 to 360 mol_c/ha/y for 0.5 m soil layer depending on soil type, parent rock, texture and climatic conditions. The maximum values of weathering rate are typical for soils located in mountain areas and soils developed on eluvo-deluvium of gabbro, diabase and nepheline syenites as the most chemical rich rocks. Used method yields an underestimated results for eluvo-deluvium of nepheline syenites due to the high content of alkali metals oxides which are not considered in the calculations. The weathering rate of widely distributed moraine-derived soils varies through a wide range with the type of underlying rocks. Soils derived from kyanite schist, alkaline granite and sandstone are notable for minimum weathering rate thanks to the low content of alkali-alkaline earth metals. More than a half (65%) of the Kola Peninsula area is occupied by soils with low weathering rate (< 200 mol_c/ha/y). Among them there are histosols with weathering rate close to zero.

Critical loads of acidity vary mainly between 200 and 800 mol_c/ha/y with soil and vegetation types and with hydrology. In sandy soils the critical acidity leaching is the most important proton sink. So critical loads depend greatly on runoff and the chosen acidity leaching criteria. The use of critical aluminium concentration, critical BC/Al ratio and Al depletion criterion for calculation of the critical acidity leaching yield rather different results. We use the minimum values calculated according to critical BC/Al ratio. In forest ecosystems both increased leaching due to acid deposition and uptake of base cations by forest contribute to acidification.

The critical deposition values are dominated by S contribution since NO_x is not a problem in most of the Kola Peninsula area (Fig. 1b, 2a). The lowest critical deposition values are found in areas with histosols in the south-eastern part of the Kola Peninsula. For these soils, critical deposition values are underestimated. The critical loads of acidity and critical sulphur deposition increase in the direction histosols < coarse textured podzols < medium textured podzols. Relatively sensitive soils occupy about 44% of the studied area.

According to SMBM critical sulphur deposition is exceeded in 40% of the total area of the Kola Peninsula (Fig. 1c), especially in north-western part. In industrial areas exceedances reach up to 1200-1900 mol_c/ha/y (Fig. 2b).

The uncertainty in the estimated critical load values can be rather large due to uncertainty in critical chemical values applied, assessment method and data (de Vries, 1994). For example, the critical aluminium concentration values directly influence the critical load. We used critical aluminium concentrations calculated from critical molar BC/Al ratio of 1.0 for forest soils. Used values seem to be very low comparing with natural aluminium concentration values observed in background area in northern Fennoscandia by Motova et al. (1993) for similar soils. Toxic effects of Al are mainly connected with acidification pushes due to mineralization and nitrification (Ulrich, 1983) whereas average annual values are used in SMBM. Uncertainties in the assessment method are associated with model assumptions. For instance, uncertainty of estimated weathering rates is dominated by the use of correlation equation (Olsson and Melkerud, 1990) for assessment. The linear relation of total content of base cations (Ca and Mg) and base cation release was found for podzols on sandy loamy till composed of Precambrian granitic basement

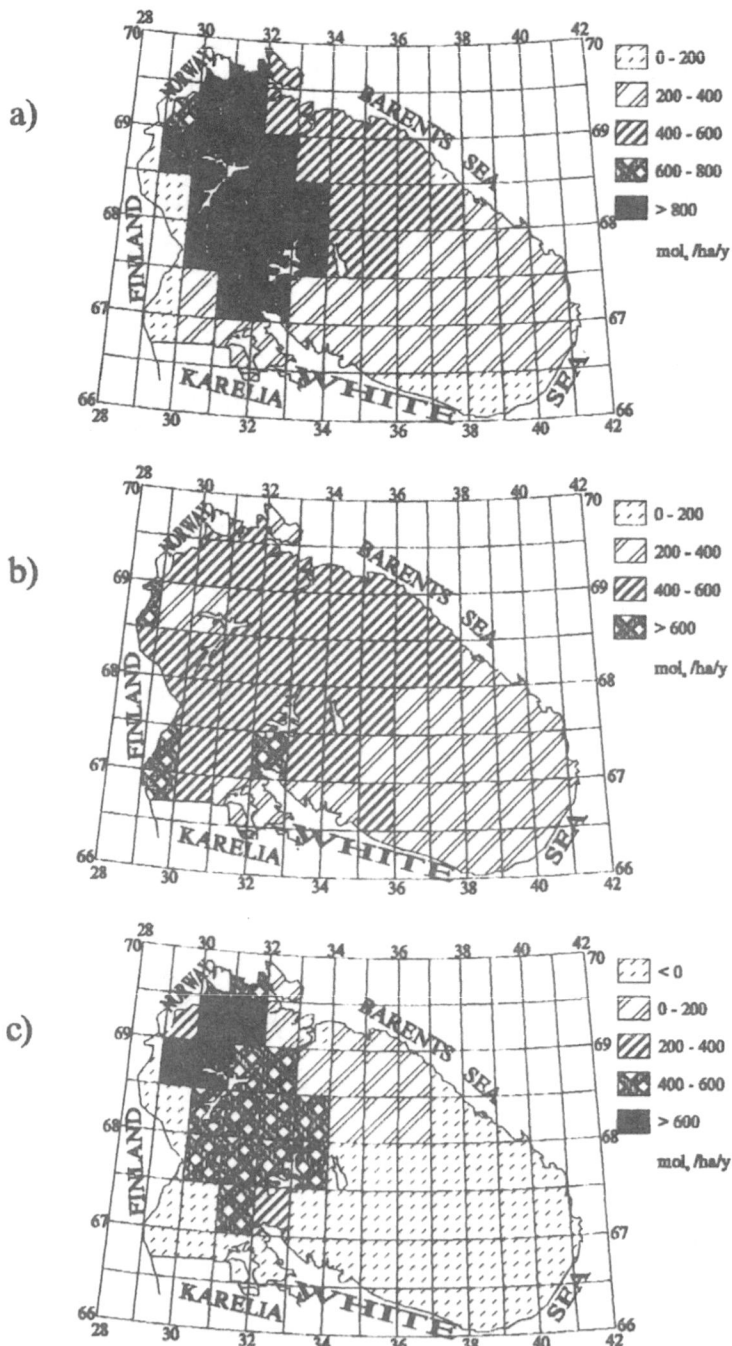

Fig. 1. Deposition in 1990 (a, EMEP data), critical deposition (b), and exceedance in 1990 (c) for sulphur for soils in the Kola Peninsula (5 percentile).

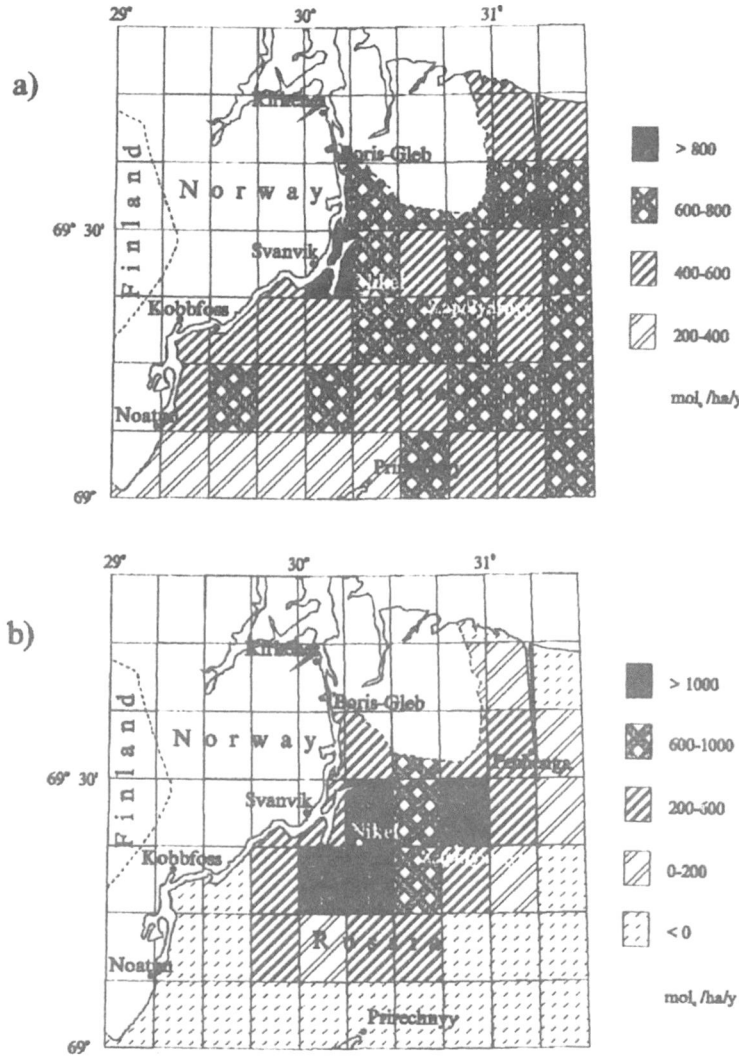

Fig. 2. Critical deposition (a) and exceedance (b) for sulphur for soils in Nikel area (5 percentile).

rocks in the case study in Sweden. The correlation is based on the fact that Mg and Ca are associated mainly with easily weatherable minerals in these soils. In other soil types it seems useful to take into account Na and K, which can often be associated to more weatherable minerals. The assumption that weathering does not take place at the daily mean temperature less than $5°$ C is also rather discussible. Both limited number of experimental data and their spatial variability determine uncertainties in data.

Obtained results are certainly preliminary and represent the current state of critical load assessment problem in the Kola Peninsula. These maps can be revised as new experimental input data become available and as assessment method will be improved. Complex process-oriented multi-layer models seem to be reasonable to apply for acidification assessment and prediction as it is done in Nordic countries, The Netherlands, USA (Cosby et al, 1985; Sverdrup et al., 1990; de Vries, 1994; Kämäri et al., 1995). For these purposes permanent complex monitoring should be conducted.

Acknowledgments

This study was partially supported by EERO grant and Programme "Ecological Safety of Russia".

References

Belov, N.P. and Baranovskaya, A.V.: 1969, *Soils of Murmansk Region*, Nauka, Leningrad, 147 pp. (in Russian).
Cosby, B.J., Hornberger, G.M., Galloway, J.N. and Wright, R.F.: 1985, *Water Resour. Res.* **21**(1), 51-63.
Downing, R. J., Hettelingh, J.-P., and de Smet, P.A.M. (eds.): 1993, *Calculation and Mapping of Critical Loads in Europe: Status Report 1993*, CCE, RIVM, Bilthoven, The Netherlands, 163 pp.
Nilsson, J. and Greenfelt, R. (eds.): 1988, *Critical Loads for Sulphur and Nitrogen*, UN-ECE, NCM, 8-57.
De Vries, W.: 1994, *Soil Response to Acid Deposition at Different Regional Scales: Field and Laboratory Data, Critical Loads and Model Predictions*, Wageningen: DLO Winand Staring Centre, The Netherlands, 487 pp.
Hettelingh, J.-P. and de Vries, W.: 1992, *Mapping Vademecum*, RIVM, Bilthoven, The Netherlands, 58 p.
Kämäri, J., Posch, M., Kähkönen, A.-M., and Johansson, M.: 1995, *Sci. Total Environ.* **160/161**, 687-701.
Koptsik, G.N., Makarov, M.I., Sokolova, T.A. and Morgun, L.V.: 1991, *Assesment and Mapping of Weathering Rate of forest Soils in the European Part of Soviet Union*, Technical Report, MSU, Moscow, 73 pp. (in Russian).
Koptsik, G.N., Sokolova, T.A. and Terekhin, V.G.: 1992, in: V.Kismul, J.Jerre and E.Løbersli (eds.), *Effects of Air Pollutants on Terrestrial Ecosystems in the Border Area between Russia and Norway*, 177-184.
Manakov, K.N.: 1972, *Productivity and Biological Turnover in Tundra Biogeocenoses*, Leningrad, Nauka, 147 pp. (in Russian).
Manakov, K.N. and Nikonov, V.V.: 1981, *Biological Cycle of Mineral Elements and Soil Formation in Spruce Forests of the North*, Nauka, Leningrad, 195 pp. (in Russian).
Motova, A.D., Nikonov, V.V. and Derome, D.: 1993, *Pochvovedenie* **12**, 52-56 (in Russian).
Nikonov, V.V.: 1987, *Soil Formation in the Areas of Northernmost Pine Ecosystems*, Nauka, Leningrad, 141 pp. (in Russian).
Olsson, M. and Melkerud, P.A.: 1990, in: Pulkkinen, E. (ed.), *Proceedings of the Conference on Environmental Geochemistry in Northern Europe*, **34**, Finland, 45-61.
Sivertsen, B., Makarova, T., Hagen, L.O. and Baklanov, A.A.: 1992, *Air Pollution in the Border Areas of Norway and Russia. Summary Report 1990-1991*, Lillestrnm (NILU OR 8/92), 14 pp.
Soil Map of the World: 1978, FAO-UNESCO.
Sverdrup, H., de Vries, W. and Henriksen, A.: 1990, *Mapping Critical Loads*, UN-ECE and NMR, 122 pp.
Sverdrup, H. and Warfvinge, P.: 1993, *Reports in Ecology and Environmental Engineering*, 1993:2. 93 p.
Ulrich, B.: 1983, in: B.Ulrich and J.Pankrath (eds), *Effects of Accumulation of Air Pollutants in Forest Ecosystems*, The Netherlands, pp. 1-29.
USSR Forest Atlas: 1973.

UNCERTAINTY ANALYSIS OF CRITICAL LOADS FOR TERRESTRIAL ECOSYSTEMS IN RUSSIA

M.Ya.KOZLOV, V.N.BASHKIN and O.M.GOLINETS

Institute of Soil Science and Photosynthesis RAS, Pushchino, Moscow region 142292 Russia

Abstract. The goal of this study is to give a comprehensive and quantitative estimation of the uncertainty of computed in different scale nitrogen (N) and sulphur (S) critical loads (CL) values for terrestrial ecosystems of the Northern Asia, European part and the North-Western regions of Russia. The CL values are used to set goals for future deposition rates of acidifying compounds so that the environment is protected. In this research CL values for terrestrial ecosystems are determined using the expert-modelling geoinformation system (EM GIS) approach. UNCSAM software package is used as the tool for uncertainty analysis. The analysis presented here focuses on the estimation and effect of the input source uncertainties and sensitivities on the CL values in various regions under study. In spite of the region, nitrogen uptake by vegetation, nitrogen leaching from terrestrial ecosystems and the difference between deposition and uptake by plants of base cations (BC) are the most influential factors for all terrestrial ecosystems of Russia.

Keywords: uncertainty analysis, computed critical loads, nitrogen, sulpher, acidity.

I. Introduction

Until now the majority of researches devoted to uncertainty analysis has been concentrated on the assessment of critical loads of sulphur and nitrogen compounds as acid forming and eutrophying agents. Many of these studies of uncertainty analysis of acidification models dealt with specific sites (Beck, 1987). Analysis of regional variability and uncertainty especially in different scale are quite rare (Hettelingh, 1989; De Vries, 1991; Hettelingh and Janssen, 1993) and the majority of these investigations is devoted to water quality data (Hettelingh *et al.*, 1992; Posch *et al.*, 1993; Kamari *et al.*, 1993). It is considered that on a regional level it is often hard to estimate intuitively the most influential parameters (Kozlov et al.,1995). However, regional uncertainty analysis during the calculations of critical loads of acid forming compounds at various ecosystems is of special interest for such countries as Russia, Ukraine, China, India etc where very high spatial variability of natural conditions coincides with high degree of incorrectness, unsufficiency and contradictions of initial information.

So, the purpose of this paper is to describe the influence of the uncertainty in input variables on the critical loads of S and N acid forming compounds for the terrestrial ecosystems in different regions of Russia.

2. Materials and Methods

The main methodological and conceptual ideas in calculation of critical load values of nitrogen and sulphur for terrestrial ecosystems of Russia as well as the uncertainty analysis have been described earlier (Bashkin *et al.*, 1993; Bashkin *et al.*, 1995; Kozlov *et al*, 1995). Three different scale databases have been used for the goals of this study: DB for the calculation of critical loads for the European part of Russia in LoLa grid scale (1grad. longitude and 0.5grad. latitude); DB for the Northern Asia in LoLa grid scale (1grad. longitude and 1grad. latitude) and DB for the North-Western part of Russia (St-Petersburg, Karelia and Murmansk regions) in LoLa grid scale (15min. longitude and 30min latitude).

Quantitative assessment of CL using EM GIS is related to the subdivision of the whole area under study into "elemental" taxones with relatively homogeneous characteristics and to the collection of all available information for use in steady-state mass balance (SSMB) equations. Accordingly, an area comprising the whole European part of Russia, characterized by great complexity of natural and anthropogenic conditions was chosen. This area was subdivided into 23 soil types and subtypes. Regarding the Northern part of Asia, this territory was divided into 48 soil types and subtypes and 45 vegetation types but for the aims of the given study the ecosystems of Regosols, Humic Cambisols, Cryic Gleysols, Gelic Podzols, Andosols, Dystric Cambisols, Eutric Cambisols, Luvic Phaerozems, Chernozems and Kastanozems soils were characterized. In the North-Western part of Russia 39 soil types and subtypes with corresponding ecosystems of tundra and forest types were divided. All subdivided united soil and vegetaton types and subtypes are characterized by relatevely homogenous background conditions (climatological, hydrological etc.) that permits us to carry out an uncertainty analysis.

Consequently, the complexity of soil cover, for ex., the European part of Russia is associated not only with a great number of soil types and subtypes but also with their different spatial distribution. One can see that the most widespread soils are Podzols(S10), Luvisols(S11), complexes of Podzols and Histosols(S12) and Chernozems(S17) and we have used only these predominant soils as examples during subsequent speculations and discussions. The most significant spatial distribution in the Northern Asia is connected with Cambisols and Podzols as well as with corresponding ecosystems. In the third region under study various Histosols were predominant.

As a based equations, the modified SSMB model has been used for calculation of different critical loads (Bashkin et al., 1993):

actual acidity: $CL(A) = ANC_w - ANC_{l(crit)}$, (1)

where: ANC_w - acid neutralizing capacity produced by weathering,

$ANC_{l(crit)}$ - acid neutralizing capacity consumed by the maximum acceptable alcalinity leaching at critical load;

maximum CL(S): $CL_{max}(S) = CL(A) + (BC_{dep} - BC_u)$, (2)

where: BC_{dep} - base cation deposition

and BC_u - base cation uptake by vegetation;

minimum CL(S): $CL_{min}(S) = CL_{max}(S) - N_l$, (3)

where: N_l - critical N leaching;

maximum CL(N): $CL_{max}(N) = N_u + N_i + CL_{max}(S)$, (4)

where: N_u - critical N uptake by vegetation,

N_i - critical N immobilazation;

minimum CL(N): $CL_{min}(N) = N_u + N_i$ (5)

and nutrient CL(N): $CL_{nut}(N) = (N_u + N_{de} + N_i + N_l) * C_t,$ (6)

where: N_{de} - nitrogen denitrification,

C_t - hydrothermic coefficient.

The parameters of the given equations characterize the main links of biogeochemical cycle of S, N and BC in the natural ecosystems related to their sustainability to the atmotechnogenic input of acid forming and eutrophying compounds.

The total assessment of an uncertainty (sensitivity) of model parameters can not be achieved easily by analytical manipulations and by the determination of elemental statistics, even for such simple models such as the SSMB model for CL calculations. Therefore, in recent years the approaches such as Monte Carlo methods and Latin Hypercube Sampling technique, which increase an effectiveness of sampling, have been widely applied. For these purposes, the software package UNCSAM (UNCertainty analysis by Monte Carlo SAMpling techniques) has been developed by Janssen et al.(1992). This was used in our studies.

A contribution of input variables of the uncertainty of the model's results was estimated by using regression and correlation analysis of model outputs from corresponding inputs. For ex., in the case of the $CL_{nut}(N)$ model output is CL(N) and input variables are N_u, N_l, N_{de}, N_i. For different CLs, the analyses were carried out separately for various soil types. The uncertainty analyses were based on the assumption of a linear relationship between model outputs and input parameters. As a rule, the existing criteria are connected to an assessment of parts of the variance, influenced by any source in relation to the total variance value. Moreover, the measure of interrelation of input sources has an important meaning. Firstly the assessment of the uncertainty of input parameters based on the methods of regression analysis has been done. Since it is impossible to prove that input sources are not correlated, absolutely, use of the criterium RTU (RooT of the Uncertainty contribution) is mostly correct (Kros et al., 1993). This critirium expresses the relative changes in the uncertainty of model outputs in relation to the relative changes in the uncertainty of input sources. The relative measure of sensitivity was assessed by criterium SRC (Standard Regression Coefficient) which characterized an alteration of computed outputs as a standard deviations from a corresponding alterations of input parameters.

3.Results and Disscusion

Table I shows the uncertainty (RTU) and sensitivity (SRC) contribution of all input variables which were included into SSMB equations of the critical loads of sulphur, nitrogen and acididy for various ecosystems of the Northern Asia, the European part and the North-Western regions of Russia.

Table I

Uncertainty contribution of input variables to the critical loads of S, N and acidity for various ecosystems in different regions of Russia, percentage.

Uncertainty and Sensitivity			Variables						
			Nu	Nl	Nde	Ni	ANCw	ANCl	(BCd-BCu)
Northern Asia	RTU	min	22	6	1	8	-	-	-
		mean	47	23	17	14	-	-	-
		max	64	39	21	22	-	-	-
(1° * 1°)	SRC	min	40	0	0	0	-	-	-
		mean	67	21	5	6	-	-	-
		max	97	55	30	29	-	-	-
European part of Russia	RTU	min	25	19	8	10	-	-	-
		mean	38	32	14	15	-	-	-
		max	49	46	18	22	-	-	-
(1° * 0.5°)	SRC	min	21	25	1	3	-	-	-
		mean	48	43	2	5	-	-	-
		max	61	71	3	10	-	-	-
North-Western part of Russia	RTU	min	51	14	10	17	17	14	20
		mean	53	18	10	25	40	17	44
		max	57	21	10	42	84	21	60
(15' * 30')	SRC	min	56	8	6	12	22	5	0
		mean	69	11	6	23	43	10	42
		max	84	13	6	43	85	14	65

Regarding the Northern Asia, one can see the following. The strongest influential factor for the values of CLnut(N) is the parameter Nu. For the biggest part of Siberia and Far East territory this parameter has the first rank and only in the case of Humic Cambisols and Cryic Gleysols ecosystems is its rank second. Among others, N_i and N_{de} parameters, which were closely intercorrelated, are the weakest ones. The given parameters do not practically have any influence on the values of CLnut(N), except Dyctric Cambisol ecosystems. The influence of Nl values is decreasing in the following row of ecosystems: Regosols > Humic Cambisols > Cryic Gleysols = Gelic Podzols > Andosols > Dystric Cambisols > Eutric Cambisols > Luvic Phaerozems = Chernozems > Kashtanozems. These results reflect, in significant degree, the geographical change of ecosystems from north to south and correspondingly an alteration of relationship of temperature and moisture constituents in hydrothermic coefficient Ct. So, the main impact to the assessment of the influence of inputting parameters (N_u, N_i, N_{de}, N_l) on both uncertainty and sensitivity of outputting values of $CL_{nut}(N)$ belongs to N_u. It is connected firstly, with a deficit of N as the main nutrient in all studied ecosystems as well as an existing spatial and temporal variability of this parameter that relates to a significance and correctness of experimental and computed values of N_u. In accordance with relatively better knowledge of hydrological picture and relatively homogenous values of critical concentration of nitrogen in surface waters, $c_{l(crit)}$, included in the calculation of N_l values, the input of the given parameter into the uncertainty of $CL_{nut}(N)$ is expressed in a lesser degree. Furthermore, the runoff processes are practically not significant for ecosystems of Luvic Phaerozems, Chernozems and Kashtanozems due to an exceedance of evapotranspiration above precipitations. During the calculations of $CL_{nut}(N)$ for ecosystems of the Northern Asia, the values of critical immobilazation and denitrification of N depositions both in relative and absolute meanings played a subordinate role that obviously reflects their minor contribution into uncertainty and sensitivity analysis of the computed output values.

In regards to the European part of Russia, one can see that the most important contribution in the uncertainty, based on this criterion for the majority of soils, is connected with N_u and the second most important rank varies for different conditions (Table I and II). The analogous conclusion can be made on a basis of SRC criterion which shows that the model is most sensitive to parameters N_u and N_l for all soil types and to parameters N_l and N_{de} for Luvisols.

Table II

The relative ranking of some input variables for various soils in the European part of Russia according to RTU

variables	S10	S11	S12	S17
Nu	1	4	1	1
Nl	2	1	3	2
Nde	3	2	4	4
Nim	4	3	2	3

Regarding the parameters used for calculation of Indifferent Exceedance Curve (Downing *et al*, 1993), the analysis of uncertainty and sensitivity shows that during the assessment of acidifying influence of S and N depositions [CL(A), $CL_{max}(S)$, $CL_{min}(S)$], the neutralyzing capacity of base cations (BCd - BCu), related to a difference between the deposition of BC and their uptake by plants has the maximal influence on outputting model values. Consequently, this value has the first rank for model outputs in equations (1), (2), and (3). During the assessment of eutrophying N values [$CL_{min}(N)$, $CL_{nut}(N)$, $CL_{max}(N)$], the predominant influence is connected with the parameter of critical nitrogen uptake, N_u.

It should be noted, however, that the application of different assessment of uncertainty/sensitivity must not be absolute, because a number of possibilities, whose correctness is not obvious, are suggested by this analysis. Nevertheless, the demonstration of the "importance" of the uncertainty of the source, N_u to the uncertainty of model outputs for various soils is scarcely likely to be accidental.

4. Conclusions

Use of various statistical approaches in an assessment of values of critical loads of S, N and acidity in terrestrial ecosystems of different regions of Russia allows for improvement in the interpretation of obtained results and the determination of sources of discrepancies related with N_u, N_l and (BC_d-BC_u). To a large extent, the interpretation of data can be done through the application of uncertainty and sensitivity analyses which permit the estimation of both the uncertainty of input variables and the uncertainty of model outputs.

Acknowledgements

The authors wish to thank Drs J.-P.Hettelingh for helpful discussion of scientific results, P.Hueberger and P.Janssen (RIVM, Bilthoven, The Netherland) for giving the opportunity to use the UNCSAM model for uncertainty analysis and Russian Fund of Basic Researches for financial support of this study.

REFERENCES

Bashkin V N., Kozlov M.Ya., Priputina I.V.,*at al.*,1993.Status Report 1993,CCE/RIVM, **102-112 p**.

Bashkin V.N., Kozlov M.Ya., Priputina I.V., and Abramychev A.Yu., 1995, *Geo-and Biosphere Modeling*, in press.

Beck ,M. 1987. *Water Resources Research*, 23 (8), **1393-1442**.

De Vries, W. , 1991. Report 46 , DLO The Winand Staring Centre, Wageningen, The Netherlands.

Downing, R.J., J.-P. Hettelingh, P.A.M. de Smet. 1993. Status Report, CCE/RIVM, Bilthoven, The Netherlands.

Hettelingh, J.-P., 1989. Ph.D. dissertation, RR-90-3, IIASA, Laxenburg, Austria.

Hettelingh, J.-P., R.Gardner, and L.Hordijk ,1992 *Environmental Pollution*,77, **177-183**.

Janssen, P.H.M, P.S.C. Heuberger, and R. Sanders , 1992. RIVM Report, Natl. Inst. Environ. Prot. (RIVM), Bilthoven,The Netherlands.

Kamari, J., M.Forsius, and M.Posch , 1993. *Water Air and Soil Pollution*, 66, **173-192**

Kozlov M.Ya., Bashkin V.N., and Golinets O.M.,1995. *Environmetric*, in press

Kros J., W. De Vries, P. Janssen, and C. Bak , 1993. *Water Air and Soil Poll.* 66: **29-58**.

Posch, M., M.Forsius, and J Kamari, 1993. *Water Air and Soil Pollution*, 66, **173-192**

DERIVING CRITICAL LOADS FOR ASIA

JEAN-PAUL HETTELINGH[1], HARALD SVERDRUP[2], DIANWU ZHAO[3]

[1]*Nat. Inst. for Public Health and the Environment (RIVM), P.O.Box 1, NL-3720 BA Bilthoven, The Netherlands*
[2]*Department of Chemical Engineering II, Lund University, P.O.Box 124, S-22100 Lund, Sweden*
[3]*Research Center for Eco-Environmental Sciences, P.O.Box 2871, 100085 Beijing, China*

Abstract. Critical loads have been computed and mapped in Southeast Asia, comprising China, Korea, Japan, The Philippines, Indo-China, Indonesia and the Indian subcontinent. The methodology involved the Steady-State Mass Balance (SSMB) method, originally developed for Europe. In contrast to Europe, where critical loads were computed for forest soils and surface waters, in Asia critical loads for 31 different vegetation types have been computed. Critical chemical limits as well as soil stability criteria were derived for each of these vegetation types, which include both natural and managed ecosystems. Results show that low critical loads in Asia occur in Bangla-Desh, Indo-China, Indonesia and the southern part of China. Uncertainties of the results are mainly due to uncertainties in base cation deposition. The critical loads are part of the impact module of the Asian version of the Regional Air pollution INformation and Simulation model (RAINS-Asia), a model used to assess abatement strategies for sulfur emissions which are rapidly increasing in this part of the world. The difference in the level of detail between European and Asian critical load maps enables different applications. In Europe, critical loads for sulphur were used in comparison to actual sulphur deposition with the aim of decreasing the excess of sulphur deposition over critical loads through optimal emission abatement. In Asia in general and China in particular the geographical distribution of critical loads of sensitive ecosystems, with some emphasis on crops, is likely to be used as a basis for future emission (re-)allocation.

Key words: acid deposition, air pollution impacts, critical loads, integrated modeling, ecosystem sensitivity.

1. Introduction

Due to high economic growth, emissions of SO_2 in Asia are rapidly increasing from about 34 Mt in 1990 to about 110 Mt by 2020 when no control measures are taken (Foell *et al.*, 1995). Historical data in Europe provide evidence of the relationship between soil and surface water acidification and increasing emissions, transport and deposition of sulfate. Therefore, it is appropriate to investigate the risk of damage due to acidification in Asia (see also Rodhe and Herrera, 1988; Zhao and Xiong, 1988; Zongwei and Kyunfeng, 1991; Xue and Schnoor, 1994). A Regional Air pollution, INformation and Simulation model for Asia (RAINS-Asia) has been developed to assess the relationship between the energy system, sulfur emissions, long-range dispersion and environmental impacts. The impact module of RAINS-Asia (Hettelingh *et al.*, 1995a) is used to provide information on the probability of damage from acid deposition exceeding critical loads for emission reduction alternatives (Foell *et al.*, 1995). The critical load concept has been used in negotiations of the reduction of sulfur emissions in Europe (Hettelingh *et al.*, 1995b) with the aim of decreasing the excess of sulfur deposition over critical loads through cost-optimal emission abatement. The derivation of critical loads in Asia, which is described in this paper, was improved in comparison to the European application, e.g. by including significantly more vegetation types. This improved resolution may also enable the use of critical loads as a basis for future emission (re-)allocation.

2. Computation and Mapping of Critical Loads in Asia

Critical loads were computed for Asian ecosystems by means of the Steady-State Mass Balance (SSMB) method. This method was applied in conjunction with the semi-quantitative method of relative sensitivity (Kuylenstierna and Chadwick, 1989) in order to compare the areas with stock-at-risk. The application of the method of relative sensitivity to Asia is described elsewhere (Cinderby et al., 1995). This paper focusses on the quantitative assessment of critical loads in Asia using the SSMB method.

2.1 THE STEADY-STATE MASS BALANCE METHOD

The SSMB method is extensively described elsewhere (Downing et al., 1993; Sverdrup and de Vries, 1994). In Europe critical loads were computed with the SSMB method mainly for forest soils and surface waters. For Asia an improvement to the European exercise was made by including 31 ecosystems in the critical load assessment. The computation of critical loads is based on plant response criteria and soil stability criteria. A critical molar ratio of the concentrations of base cations to aluminum in soil solution, $(BC/Al)_{crit}$, for each plant species is used as indicator of plant response. Using these critical BC/Al ratios in the SSMB method it is possible to compute maximum allowable acidifying deposition, i.e. the critical load:

$$CL = ANC_w + \left(1.5 \frac{x_{Ca+Mg+K} ANC_w + BC_d - BC_u}{(BC/Al)_{crit} K_{gibb}}\right)^{1/3} Q^{2/3} + 1.5 \frac{x_{Ca+Mg+K} ANC_w + BC_d - BC_u}{(BC/Al)_{crit}} \quad (1)$$

where ANC_w is the alkalinity produced by weathering, $ANC_{l,crit}$ is the critical alkalinity leaching, K_{gibb} the gibbsite solubility constant, Q is the runoff, BC_d the base cation deposition, BC_u the base cation uptake and $x_{Ca+Mg+K}$ the fraction of weathering as Ca, Mg and K. The soil stability criterion is introduced to avoid that acid deposition leads to Al leaching in excess of Al produced by weathering and other processes, e.g. in high precipitation areas. The critical load for acidity based on the soil stability criterion is computed by:

$$CL = 3ANC_w + (2\frac{ANC_w}{K_{gibb}})^{1/3} Q^{2/3} \quad (2)$$

Critical loads for each ecosystem are computed by taking the minimum of equations (1) and (2).

2.2 DATA AND MAPPING

Several different types of response expressions, using the BC/Al ratio as explanatory variable, can be derived from assuming an antagonism between base cations and Al through different types of ion exchange at the roots. The response equation can be made specific for (a) particular species, and (b) a particular ion exchange mechanism, i.e. Vanselow, Gapon or Gaines-Thomas (Sverdrup and Warfvinge, 1993). An example of relationships between root growth reduction and BC/Al ratios in soil solution is given in Fig.1 for five tree species. Critical values for BC/Al as thresholds for root growth reduction have been established for many plant species by Sverdrup and Warfvinge (1993). An overview of the ecosystems for which critical loads have been computed is provided in Table 1.

Other data, i.e. climate, soil, geology and vegetation data, and their geographical distribution, required to compute and map critical loads have been derived from a variety of sources which are summarized in Cinderby *et al.* (1995). The uncertainty of the critical load estimates is largely due to the uncertainties of base cation weathering and base cation deposition estimates which require further verification.

In order to be able to compare sulfur depositions, which are computed by the RAINS-Asia model on a raster of $1°\times1°$ grid cells, with critical loads the original polygon map of critical load has to be converted to the same raster. Each grid cell contains several critical loads reflecting the variety of ecosystems and their areas. A cumulative distribution of these critical loads is derived for each grid cell allowing the computation of a pth percentile (protecting $100-p$ percent of the ecosystem area) for comparison with the deposition in that grid cell (see Hettelingh *et al.*, 1995a).

TABLE 1
Values used for the calculation of critical loads with the SSMB method in Asia (see Hettelingh *et al.*, 1995a; some vegetation types of a total of 31 have been lumped for convenience).

Vegetation type	$(BC/Al)_{crit}$	Root depth (m)	$\log_{10}K_{gibb}$	BC_w (eq ha^{-1}yr^{-1})
Polar or rock desert	6	0.1	8.1	0
Tundra	2	0.2	8.1	0
Cool semi-desert/ scrub	2	0.3	8.1	0
Montane cool scrub/grass	2	0.3	8.1	0
Cool scrub/grassland	2	0.3	8.1	0
Main + southern taiga	1	0.5	8.5	250
Coniferous forest	1.5	0.5	8.5	250
Mixed forest	1	0.8	8.5	250
Temperate broadleaf forest	0.6	0.8	8.5	250
Interrupted temperate woods	1	0.5	8.5	100
Dry/highland woods	2	0.5	8.5	100
Mediterranean woodland	1	0.5	8.5	100
Interrupted tropical woods	2	0.5	8.1	0
Subtropical dry forest	2	0.5	8.5	100
Subtropical wet forest	1	0.2	8.1	0
Tropical dry forest	1	0.5	8.5	100
Tropical wet forest	0.6	0.2	8.1	0
Tropical savanna	10	0.5	8.5	100
General farmland	10	0.3	8.3	0
Coastal wetland	10	0.5	8.5	0
Hot scrub/grassland	10	0.3	8.3	50
Succulents/thorn dry woods	10	0.3	8.3	50
Semi-arid forest	10	0.5	8.3	50
Non-polar rocky vegetation	10	0.2	8.5	0
Sand and semi desert	10	0.5	8.5	0

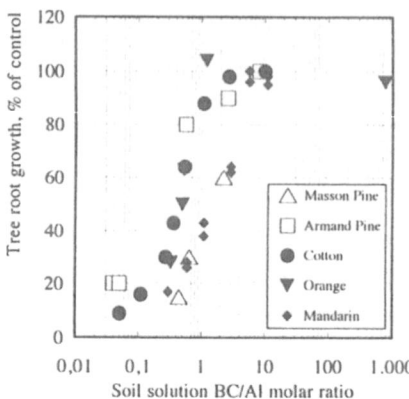

Fig.1.The relationship between root growth reduction and BC/Al ratios for five tree species growing in China. For example, a root growth reduction of less than 20% occurs for *BC/Al > 1* (Sverdrup and Warfvinge, 1993).

3. Results

Fig.2 shows the map of critical loads for acidity in Asia. Low critical loads (i.e. below 500 eq ha^{-1}yr^{-1}) are found in south-eastern Asia, parts of the Himalayan range and the Tibetan plateau and in parts of the boreal forest in northern China. Other areas with low critical loads (high sensitivity) include the rain forest strip in south-western India.

These results do not imply that soils are acidified yet. Fig.2 illustrates that some areas with low critical loads are subject to a higher risk of acidification than other areas. The map provides indications of where continued excess of critical loads may, in some areas earlier than in others, ultimately exhaust the buffering capacity of the soils. Areas with an increased risk of growth reduction, yield loss or dieback can thus be located. The dry regions in most of India and north-western China show relatively high critical loads (exceeding 2000 eq ha^{-1}yr^{-1}). The 25th percentile critical load, i.e. protecting 75% of the ecosystems in each grid cell against sulfur based acidification has been chosen as basis for comparison with sulfur deposition. The reason of choosing the 25th percentile is that the resolution of the vegetation data, used as basic map in the SSMB method, does not allow reliable estimates of ecosystem protection when lower percentiles are chosen. Fig.3 shows the 25th percentile critical load map of acidity for every 1°×1° grid cell covering Asia.

4. Concluding Remarks

A first critical load map for Asia has been derived by applying the Steady-State Mass Balance method to 31 vegetation types, taking into account information on soil types and climatic and meteorological conditions which are specific to the Asian region. The map of critical loads presented in this paper (a) enables large scale assessments of areas at risk due to long-range dispersion of acidifying pollutants, and (b) may assist in decisions on emission (re-)allocations. However, the calculation of critical loads for much of Asia with its highly varying environmental conditions and ecosystems is a considerable effort.

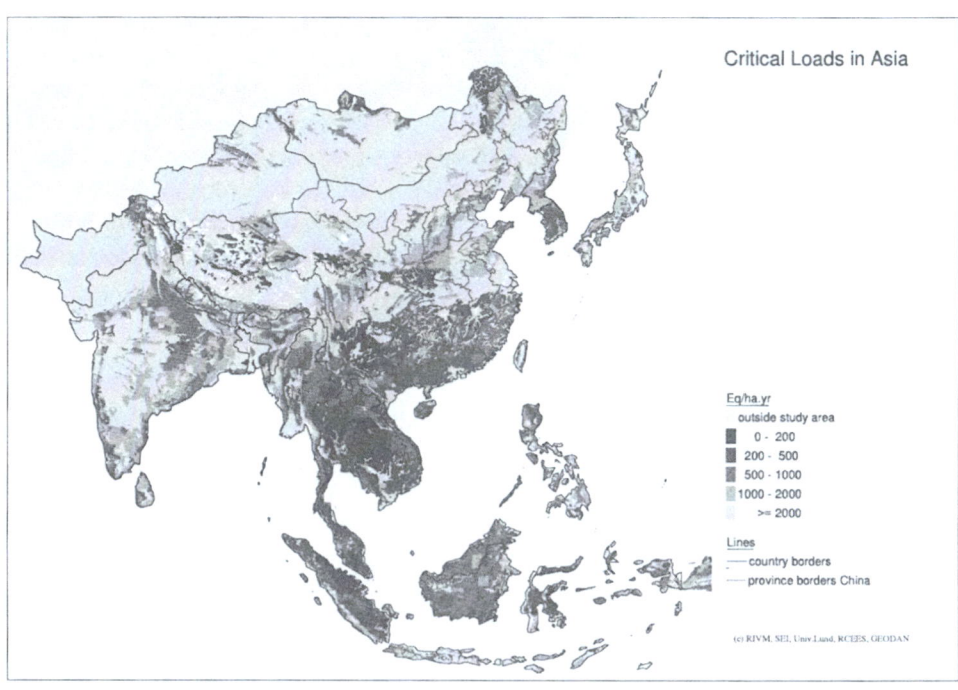

Fig.2. Critical loads in Asia computed with the Steady-State Mass Balance Method.

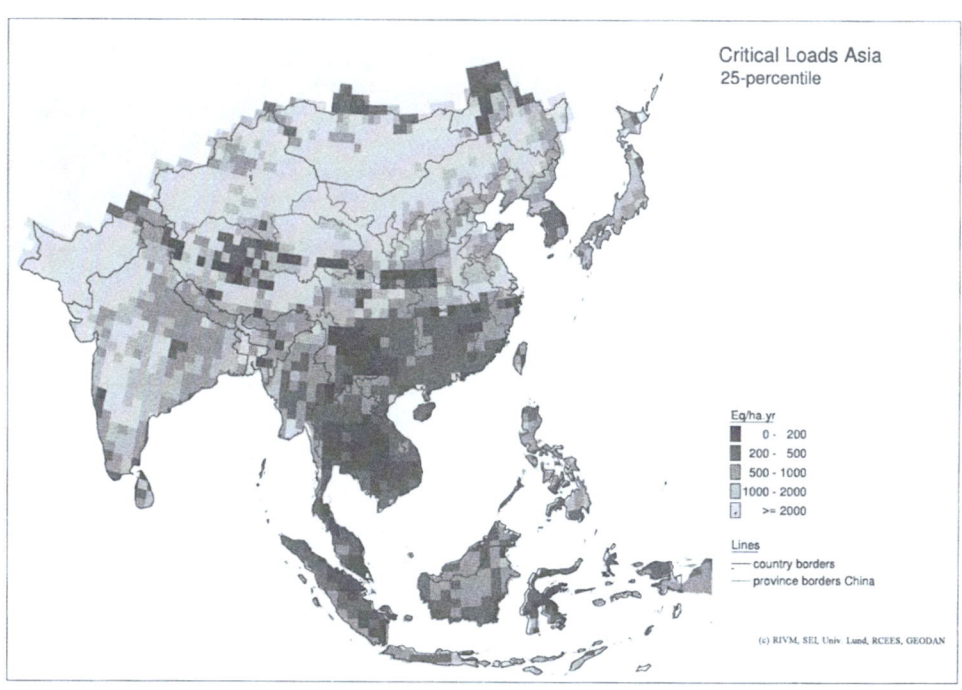

Fig.3. The 25-percentile critical load, protecting 75% of the ecosystems in each 1°×1° grid cell in Asia.

They encompasses tropical rain forest to near desert conditions and mangroves to areas of taiga: there is an associated range of soil types and a wide range of cropping systems and land use. Therefore, extensive validation procedures are required to improve the assessment of risk of damage on local scales. Further work is also required to include other causes of risk of damage in addition to deposition. Atmospheric concentrations of SO_2, NO_x, and O_3 have been shown to cause direct damage to natural ecosystems and crops, as well as having health effects in large urban areas. In fact in many parts of Asia the concern for urban air quality is increasing and air concentrations of SO_2 may already reach hazardous levels.

Acknowledgements

The World Bank (J.Shah) and the Asian Development Bank (A.Azimi) are acknowledged for supporting the project "Acid rain and emission reductions in Asia". Our colleagues in the scientific team for the development of RAINS-Asia are M.Amann, R.L.Arndt, G.R. Carmichael, J.Cofala, W.K.Foell, C.Green, L.Hordijk, W.Schöpp and D.Streets. The design of the impact module in particular has benefited from the collaboration with M.Chadwick, J.Kuylenstierna and S.Cinderby (SEI-York), J.U.Ahmad (Bangla-Desh), T.X.Gian (Vietnam), S.Lee (Korea), C.M.Liu (Taiwan) and J.Shindo (Japan).

References

Cinderby, S., Kuylenstierna, J.C.I. and Chadwick, M.J.: 1995, "Background Data and Mapping the Sensitivity in Asia", In: Hettelingh *et al., op. cit.* pp. 27-42.

Downing, R.J., Hettelingh, J.-P. and de Smet, P.A.M.: 1993, "Calculation and Mapping of Critical Loads in Europe", CCE Status Report 1993, RIVM, Bilthoven, The Netherlands, 163 pp.

Foell, W.K., Green, C., Amann, M., Bhattacharya, S., Carmichael, G., Chadwick, M., Cinderby, S., Haugland, T., Hettelingh, J.-P., Hordijk, L., Kuylenstierna, J., Shah, J., Shresta, R., Streets, D. and Zhao, D.: 1995, "Energy Use, Emissions, and Air Pollution Reduction Strategies in Asia", *this volume*.

Hettelingh, J.-P., Chadwick, M., Sverdrup, H., and Zhao, D.: 1995a, "Assessment of Environmental Effects of Acidic Deposition", RIVM, Bilthoven, Netherlands, 145 pp.

Hettelingh, J.-P., Posch, M., de Smet, P.A.M., Downing, R.J.: 1995b, "The Use of Critical Loads in Emission Reduction Agreements in Europe", *this volume*.

Kuylenstierna, J.C.I. and Chadwick, M.J.: 1989, "The Relative Sensitivity of Ecosystems in Europe to the Indirect Effects of Acidic Depositions", In: Kämäri *et al.* (Eds), Regional Acidification Models, Springer, Heidelberg, 306 pp.

Rodhe, H. and Herrera, R. (Eds): 1988, "Acidification in Tropical Countries", SCOPE 36, John Wiley, Chichester, 405 pp.

Sverdrup, H. and de Vries, W.: 1994, "Calculating Critivcal Loads for acidity with the Simple Mass Balance Method", *Water, Air, and Soil Pollution*, **72**, 143-162.

Sverdrup, H. and Warfvinge, P.: 1993, *"Effect of Soil Acidification on Growth of Trees and Plants as Expressed by the (Ca+Mg+K)/Al Ratio"*, Report 2:1993, Department of Chem. Eng. II, Lund University, Lund, Sweden.

Xue, H.B. and Schnoor, J.L.: 1994, "Acid Deposition and Lake Chemistry in Southwest China", *Water, Air, and Soil Pollution*, **75**, 61-78.

Zhao, D. and Xiong, J.: 1988, "Acidification in Southwestern China", In: H. Rodhe and R. Herrera (Eds), *op. cit.*, pp. 317-347.

Zongwei, F. and Kyunfeng, S.: 1991, "Relative Sensitivities of Woody Plants to Acid Deposition in South Areas of China", *Journal of Environmental Sciences (China)*, **3(2)**, 61-68.

EVALUATION OF ESTIMATION METHODS AND BASE DATA UNCERTAINTIES FOR CRITICAL LOADS OF ACID DEPOSITION IN JAPAN

J. SHINDO[1], A. K. BREGT[2] and T. HAKAMATA[1]

[1] *National Institute of Agro-Environmental Sciences, Kannondai 3-1-1, Tsukuba Ibaraki 305 Japan,*
[2] *DLO Winand Staring Centre for Integrated Land, Soil and Water Research(SC-DLO), P.O.Box 125, Wageningen, The Netherlands*

Abstract. Simplified steady state mass balance model for critical load (CL) estimation was applied to a test area in Japan to evaluate applicability of the model. Three different criteria for acidification limits (1: $[Al^{3+}]<0.2$ eq. m^{-3}, 2: Al/BC<1.0 mol mol^{-1}, 3:Al depletion criterion) were used. Mean values and spatial distribution patterns of CL values calculated by these criteria were extremely different from each other. The first criterion produced much higher CL than the second criterion in the Japanese condition with high annual precipitation. Improvements including definition of the criterion were considered necessary. As quantitative data of the base cation weathering rate (BC_w) was lacking, the rate was specified based on surface geology and the soil type of each site. To evaluate uncertainty of BC_w used for CL calculation, ion content and particle-size distribution were measured for the soils collected from the test area, and BC_w was estimated with PROFILE model based on these measurements. It appeared that BC_w estimates by surface geology were adequate as a mean value, but they had uncertainty of about 50% of the average values due to variability of ion contents within the same surface geology group.

1. Introduction

The concept of the critical load (CL) introduced by Nilsson(1986) is now widespread among the scientists and decision makers as a mean of assessment of acid deposition. CL is defined as "the highest deposition of acidifying compounds that will not cause chemical changes leading to long term harmful effects on ecosystem structure and function" (Nilsson and Grennfelt, 1988). Several models were proposed to estimate CL and large scale maps of CL estimates were already made for Europe and for Asia (Hettelingh et al., 1991; Hettelingh et al., 1993). In Japan, CL has got great attention recently, because yearly averages of rain pH is low, that ranges 4.2 to 5.0 throughout the country and considerable amounts of sulfate and nitrate are observed in rain water. Forest decline has also been observed in several areas in the latest decade. Acidic deposition and soil acidification are considered to cause the forest decline along with the air pollution, pathogen effect, drought etc. In such situations, quantitative estimation of ecosystem sensitivity to acid deposition is required and the critical load is expected to be useful for this objective.

We applied the steady state mass balance model to a test area in Japan. The first objective of this study is to prove the applicability of the existent model to the Japanese condition where meteorology, vegetation, soil etc. are different from those in Europe. The second objective is to evaluate the accuracy and uncertainty of base data needed for the CL estimation. We concentrate to the base cation weathering rate of minerals in soils, because quantitative data on it or mineralogy data in the national or regional scale are lacking in Japan, in spite that base cation weathering is one of the most important processes to control the acidity of soils.

2. Materials and methods

We applied the following steady state mass balance model for calculation of CL(eq.ha^{-1}yr^{-1}):

$$CL = BC_w + H_{l(cnt)} + Al_{l(cnt)} = BC_w + [\{Al_{l(cnt)}/Q\}/K_{gibb}]^{1/3} Q + Al_{l(cnt)} \quad (1)$$

where BC_w : base cation weathering (eq.ha^{-1}yr^{-1}), $H_{l(cnt)}$ and $Al_{l(cnt)}$: critical hydrogen and aluminum ion leaching (eq.ha^{-1}yr^{-1}), K_{gibb} : gibbsite equilibrium constant (m^{6}eq^{-2}) and Q : runoff (m^{3}ha^{-1}yr^{-1}). The maximum allowable deposition (CD: critical deposition) of acidic substances (H$^+$+2NH$_4^+$) is estimated as follows:

$$CD = CL - acid_{biol} \quad (2)$$

where $acid_{biol}$ denotes the net production of acidity in the ecosystem mainly by biological processes, which includes acid production by base cation uptake (BC_u), acid neutralization by nitrogen uptake (N_u), immobilization (N_{im}), denitrification (N_{de}) etc. We took only BC_u and N_u into consideration. A criterion on an acidification limit must be defined to estimate CL and CD. Three criteria have been proposed and $Al_{l(cnt)}$ is formulated based on each criterion respectively (Table I). The formula of $Al_{l(cnt)}$ based on the criterion 2 is slightly modified by us. While the second criterion has been used most frequently, features of CL estimates based on different criteria and their adequacy have not been discussed yet. We applied these three formulas to estimate CL and CD and the results were compared with each other.

The test area is Hiroshima and Shimane Prefectures located in the west of the Main-land of Japan (Fig.1). The data used for CL and CD estimation were derived from several national data bases whose spatial resolutions are about 1 square kilometer. BC_w was estimated based on the acidity of parent material and soil texture according to the method described by De Vries (1991). Surface geology data was used to classify the parent material into acid, intermediate and basic and soil type data was used to derive texture classes. Q was calculated as the difference between annual precipitation and annual evapotranspiration estimated by the Thornthwaite method. BC_u and N_u were calculated from net growth rates evaluated from annual temperature and BC and N concentration in trees derived from literatures for each forest type (deciduous and evergreen, needle and broad leaf etc.). For a natural forest net uptake was assumed to be zero, as the uptake by trees is in equilibrium with the release of BC and N due to litter decomposition. As spatial distribution of BC_d was not available, 500 eq.ha^{-1}yr^{-1} that was rough

TABLE I

Criteria of acidification limit and formulas to express $Al_{l(cnt)}$ used for estimation of critical loads.

Criteria and critical values	Formula to express $Al_{l(cnt)}$	Reference
Criterion 1 [Al^{3+}]$_{cnt}$ = 0.2 eq.m^{-3}	[Al^{3+}]$_{cnt}$ Q	Hettelingh et al.(1991)
Criterion 2 (BC/Al)$_{cnt}$ = 1 mol mol^{-1}	1.5 max{$f_{BC}BC_w+BC_d$-BC_u , [BC]$_{l(min)}$Q}	Sverdrup et al.(1994),
	(BC/Al)$_{cnt}$	Hettelingh et al.(1993)
Criterion 3 (Al$_l$/Al$_w$)$_{cnt}$ = 1 mol mol^{-1}	$f_{Al/BC}$ (Al$_l$/Al$_w$)$_{cnt}$ BC_w	Sverdrup et al.(1994)

BC_d: Base cation deposition(eq.ha^{-1}yr^{-1}),   [BC]$_{l(min)}$:Minimum concentration of BC leaching(eq.m^{-3}) =0.002
f_{BC}= (Ca+Mg+(K))$_w$/BC$_w$=0.7 eq.eq.$^{-1}$, $f_{Al/BC}$=Al$_w$/BC$_w$=2.0 eq.eq.$^{-1}$

estimation of average BC_d for whole Japan was used for every grid cell of the area. Other parameters in the formulas and critical values were set to the same values as in European and Asian applications.

In order to evaluate the uncertainty of BC_w estimation, a soil survey was carried out at 42 sites in the area. For each site topsoil just below a litter layer and subsoil (at 5 to 15 cm depth) were collected at three sampling points that were 3 to 5 m apart each other. Particle-size distributions were measured and total ion contents in <2mm and >2mm fractions were analyzed with ICP spectrometry. BC_w was calculated for each site with the PROFILE model version 3.2 (Sverdrup et al., 1993) by assuming two soil layers of 10 cm and 40 cm. The mineral contents and surface area of soil particles needed by PROFILE model were derived from results of the above measurements according to the method of Sverdrup et al. (1992). Other input data were the same for all sites that were rough averages of the area. Base cation weathering estimated by PROFILE is indicated as $BC_w(P)$. The sites were divided into six surface geology groups (granite, rhyolite, andesite, gabbro, volcanic ash and sediment rock), and into six soil type groups (Brown forest soils(dry), Brown forest soils, Andosols, Regosols, Red soils and Yellow soils). For the contents of some elements and $BC_w(P)$, total sum of squares (TSS) was divided into the following sum of squares (SS): between-groups indicating the magnitude of difference between group means, within-group due to variability of the values within a group and within-site that is SS of sampling errors due to difference between values at three sampling points. We regarded the within-group SS as an index of the uncertainty, as it denotes the magnitude of the difference between values at the sites belonging to the same group. The GLM (General Linear Model) procedure of SAS was used for the statistical analysis.

3. Results and Discussion

3.1 ESTIMATION OF CL AND CD

Spatial average CLs estimated by criteria 1, 2 and 3 in the test area were extremely different from each other that were 4993, 2117 and 5869 eq.ha^{-1}yr^{-1}, respectively. CDs were generally lower

Fig. 1 The test area for CL and CD estimation.

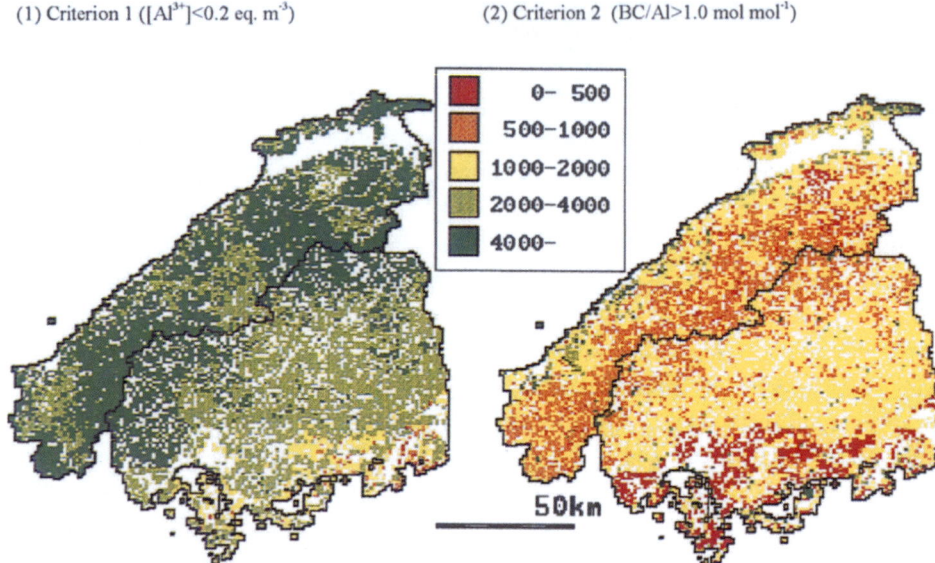

Fig. 2 Spatial distribution of CD estimates (eq.ha^{-1}yr^{-1}).

than CLs because of the higher BC_u than N_u. Fig. 2 shows the spatial distribution of CD estimates by criteria 1 and 2. Spatial distribution patterns were also different each other. Criterion 1 produced CL and CD showing clear correlation with runoff distribution and having little similarity to the maps made by the other two criteria. In the map by criterion 2, the major factor to determine CL and CD distribution was BC_w, and sensitive areas were located in the southern part covered by Regosols with low clay content and mountainous area around the border between two prefectures, whose surface geology is mainly granite and rhyolite. Spatial distributions of CL and CD by criterion 3 were similar to those by criterion 2 though absolute values were extremely different.

Annual average of acidic deposition (H$^+$+2NH$_4^+$) ranged from 480 to 1050 eq.ha^{-1}yr^{-1} at two locations in Hiroshima Prefecture in 1986 and 1987. These deposition rates are considered to effect some part of the test area adversely from the CD map based on criterion 2, while no effects will be expected from the CD values by criteria 1 and 3. Therefore, it is crucially important to determine which criterion is most realistic for the area where the model is applied and also to set the suitable critical values to the criterion that may depend on vegetation type. In spite that the critical value for criterion 1 ([Al^{3+}]<0.2 keq.m^{-3}) is the almost minimum value of Al toxicity thresholds for various tree species (de Vries, 1991) reported so far, it produced much larger CL values than those by criterion 2. It suggests that criterion 1 has a possibility to yield too high CL values in a high precipitation area like Japan.

3.2 EVALUATION OF UNCERTAINTY OF BC_w ESTIMATE

Fig. 3 shows average BC_w(P) for each surface geology group. Average rates were low for granite and rhyolite and they were high for other surface geology type. In our calculation of CL,

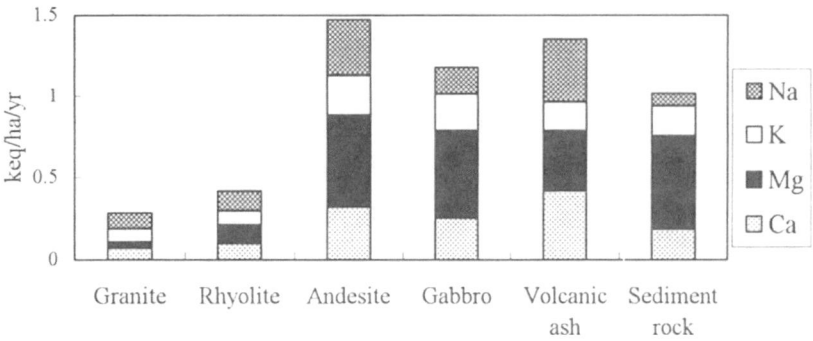

Fig.3 Base cation weathering estimated with the PROFILE model

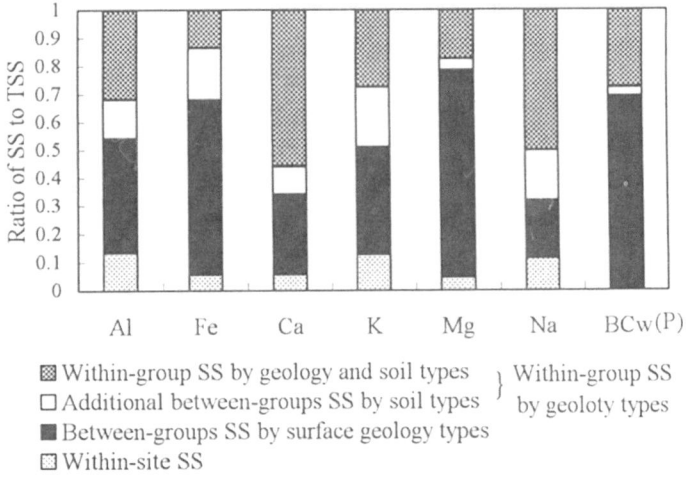

Fig. 4 Ratio of SS due to surface geology and soil types

granite and rhyolite were classified as the acidic group, andesite and volcanic ash were intermediate and gabbro was put into the basic group. Sediment rock was assigned to acidic and intermediate groups according to the texture. Roughly speaking, the classification of acidity seems to be realistic from Fig.3 except for gabbro.

Fig.4 shows ratios of SS due to several factors to TSS regarding contents of some elements and $BC_w(P)$. Within-site SSs were less than 15% of TSS for all elements. Ratios of between-groups by surface geology types were about 60% to 70% for magnesium and iron, while they were low for calcium and sodium. For $BC_w(P)$, where within-site SS is not shown because $BC_w(P)$ was calculated based on the three points means of ion contents, the ratio of between-groups SS was more than 70%. Within-group SS by surface geology types, which indicates the uncertainty of BC_w estimation caused by inhomogeneity of the values within a surface geology group, varied from 20% to 60% by the elements and it was reduced to 13% to 50% by taking the soil type into consideration. For $BC_w(P)$, within-group SS by surface geology types was about 30% and pooled standard deviation of $BC_w(P)$ corresponding to this within-group SS was 321 eq.ha^{-1}yr^{-1}, coefficient of variance of which was about 50% of the mean value (621 eq.ha^{-1}yr^{-1}). Variability of $BC_w(P)$ within a group was different by surface geology types as shown in Table II. BC_w estimation at locations assigned to gabbro and sediment rock had large uncertainty. Uncertainty of estimation of $BC_w(P)$ with PROFILE model should also be considered to evaluate the entire uncertainty of BC_w.

TABLE II

Mean $BC_w(P)$ for surface geology group and standard deviation of within each group (eq.ha^{-1}yr^{-1})

Surface geology type	Granite	Rhyolite	Andesite	Gabbro	Volcanic ash	Sediment rock	Pooled
Group mean	283	418	1472	1180	1357	1015	621
Standard deviation	170	309	124	696	363	683	321

Conclusions

In the application of steady state mass balance model to estimate the critical load of acidity, definition of an acidification criterion and its value are crucially important. Different criteria produce completely different results. Estimating BC_w based on surface geology and soil types appeared to be adequate as a mean value. Values at individual points, however, have large uncertainty caused by variability within the same surface geology group. Evaluations of accuracy and its improvement of other base data such as runoff, growth uptake rate and base cation deposition rate are also necessary.

References

De Vries, W.:1991, "Methodologies for the assessment and mapping of critical loads and of the impact of abatement strategies on forest soils", Report 40, DLO the Winand Staring Centre, the Netherlands, 109pp.

Hettelingh, J.P., Downing, R.J., de Smet, P.A.M.:1991, "Mapping critical loads for Europe", CCE Technical report No.1, RIVM, Bilthoven, the Netherlands, 86pp.

Hettelingh, J.P., Sverdrup, H., Meijer, E.: 1993, "RAINS Asia impact module", In papers for the RAINS Asia effects task force meeting, Beijing, China.

Nilsson. J. (ed.) :1986, "Critical loads for nitrogen and sulfur", Nordic Council of Ministers. Report 1986:11. Copenhagen, Denmark.

Nilsson, J. and Grennfelt, P.(ed.):1988, "Critical loads for sulfur and nitrogen", Nordic Council of Ministers. Report 1988:15. Copenhagen, Denmark.

Sverdrup, H., Warfvinge, P. and Jonsson, C.:1992, "Critical loads of acidity for forest soils, groundwater and first order streams in Sweden", HMSO Proceedings from the British Critical Loads conference, Institute of Terrestrial Ecology, Merlewood Research Station, Cumbria, Great Britain.

Sverdrup, H. and Warfvinge, P.:1993, "Calculating field weathering rates using a mechanistic geochemical model PROFILE", Applied Geochemistry **8**, 273-283.

Sverdrup, H and de Vries, W.:1994, "Calculating critical loads for acidity with the simple mass balance method", Water, Air and Soil Pollution **72**, 143-162.

APPLICATION OF THE CRITICAL LOADS APPROACH IN SOUTH AFRICA

A.M. VAN TIENHOVEN[1], K.A. OLBRICH[1], R. SKOROSZEWSKI[2], J. TALJAARD[3] and M. ZUNCKEL[3]

[1]Forestek, CSIR, Private Bag, X11227, 1200, Nelspruit, South Africa.
[2]Watertek and [3]Ematek, CSIR, PO Box 395, 0001, Pretoria, South Africa.

Abstract. South Africa is the most industrialised country in southern Africa and stands at some risk from negative pollution impacts. To the authors' knowledge, this paper presents the first attempt to apply the critical loads approach on the African continent; although sensitivity mapping has been performed for Africa and the rest of the world (Kuylerstierna *et al*, this conference). Actual sulphate and base cation deposition loads in Mpumalanga (formerly the Eastern Transvaal province of South Africa) were mapped from 16 monitoring sites. The region is characterised by long, dry periods with little rain, high evaporation (up to 8 mm per day) and low run-off (15% of MAP). Provisional critical load and exceedance maps were produced for the surface waters using the Steady-State Water Chemistry Model and the Diatom model. Maps of soil sensitivity to acid deposition, based on bedrock lithology, soil chemical characteristics and land cover, were produced. A weathering rate of 0.39-0.86 keq/ha/year was calculated for the most sensitive sites and taken as the critical load, based on the assumption that the weathering rate represents the buffering ability of the system. The critical loads were contrasted with measures of actual deposition to examine potential scenario's for critical load exceedances. A key factor in refining the sensitivity maps, and allowing estimation of the critical loads, is the accurate calculation of weathering rates under the warmer and more arid environmental conditions prevalent in South Africa. In a developing country such as South Africa, where research resources are limited, the critical loads approach is a valuable means of assessing the risk of potential impacts of atmospheric deposition.

Key words: South Africa, Eastern Transvaal, Mpumalanga, critical loads, acidification.

1. Introduction

South Africa is recognised as having comparatively high levels of air pollution emissions, particularly in the Eastern Transvaal Highveld. The total SO_2 emission rate for the whole of South Africa is estimated to be 2.9 million tons per annum, of which 1.2 million tons are produced by coal-burning power stations. 75 % of these power stations are located in Mpumalanga (formerly the Eastern Transvaal) (Annegarn *et al*, 1994; Tyson *et al*, 1988, Turner, pers. comm.). If the needs of South Africa as a developing nation are considered, the requirements for economic growth and associated employment, education, housing, and primary health care, greatly overshadow issues relating to air pollution and its impacts. Since the critical loads approach can provide a basis on which cost-effective decisions can be made by regional and national authorities with respect to air quality, its potential use in South Africa was investigated.

This paper presents the results of a pilot study to investigate the application of the critical loads approach in South Africa.

2. Materials and methods

The area chosen for the pilot study is the region presently defined as the Eastern Transvaal Province, but now renamed Mpumalanga (Figure 1.). This region encompasses large areas of commercial forest plantations, key natural environments such as the Kruger National Park, and large industrialised areas. Topographically, the area is divided into a plateau in the west (1 400 - 2 000 mamsl) which is relatively dry and cool with frost in the winter months; and a broad gently undulating plain to the east (800 - 150 mamsl) which is generally hot and dry. The two zones are separated by a steep escarpment zone varying between 20 to 40 km wide. The mean annual temperature ranges from 12 °C to 23°C - depending on local topography. Most of the rain falls in the summer months as thundershowers. Rainfall ranges between 500 mm to 2 000 mm per year, but the average rainfall for the region is 880 mm - approximately 70% higher than the mean annual rainfall of 500 mm for South Africa.

In order to estimate actual deposition loads, rain chemistry data was collected from a coordinated monitoring network made up of 16 stations. The actual wet sulphate deposition map was created on a GIS, by finding the product of the mean annual rainfall coverage and the mean volume-weighted sulphate concentration coverage. Pollutant dry deposition was not monitored.

To calculate the critical loads of surface waters, Mpumalanga was divided into grid squares corresponding to the area covered by 1:50 000 Ordinance Survey maps. The most sensitive water body in each grid was selected - based on low base cation concentration and as few sources of anthropogenic pollution, other than atmospheric, as possible. One sample from each grid was analysed for base cations, acidic anions, Cl^{-1} (chlorine), alkalinity and pH.

Both the diatom model (Battarbee, 1993) and the steady state water chemistry model (Henriksen *et al.*, 1986) were utilised in the assessment of the critical loads for surface waters. Two versions of the steady state water chemistry model were tested. The first method (Battarbee, 1994) considers the levels of sulphate in the surface waters, whereas the second method (Sverdrup *et al*, 1990) does not include this sulphate in the calculation.

The estimates for soil critical loads were derived by producing a sensitivity map of the region according to the guidelines of the Stockholm Environment Institute (Chadwick *et al.* 1991), using factors such as land cover, geology (or lithology) and soil features such as pH, cation exchange capacity and base saturation. Critical loads were then assigned to the final sensitivity classes.

Although climate is an important determinant of sensitivity to acidification, we elected to ignore rainfall, as its effects are implicit in both the land cover coverages and the soil coverages.

Soil critical loads were assigned to the sensitivity classes by assuming that if the mineral weathering rate provides sufficient cations to balance incoming acid deposition, then the ecosystem will not acidify. All other inputs and outputs of acidity and alkalinity were ignored.

Most of the soils of Mpumalanga contain inert or very slow weathering minerals, so much of the region is sensitive to acidification (Task Force on Mapping, 1993, Sverdrup, 1990).

Weathering rates were derived by scaling European weathering rates to local weathering conditions. Data from the Lake Gårdsjön catchment served as the standard against which local environmental parameters were adjusted (Sverdrup *et al*, 1990). The Sabie River catchment was chosen to represent conditions in Mpumalanga since there is readily accessible environmental data for the catchment (Schutz, 1990). The catchment also falls within a high rainfall, forestry area with sensitive soils and geology. In terms of acidification potential the Sabie River catchment can be described as a sensitive ecosystem.

The weathering rate for the Sabie River catchment was calculated to range from 0.7 to 1.54 keq/ha/yr. This estimate was improved by considering the exposed mineral surface (Sverdrup *et al*, 1990) and adjusting for significant differences in soil moisture saturation, texture, acidity and temperature between Gårdsjön and Sabie to give a final rate of 0.39-0.86 keq/ha/yr.

3. Results and Discussion

The generalised approaches used and the results obtained are summarised in Table I.

Table I. Synthesis of approaches and calculations adopted in applying the critical loads approach in South Africa

	Approach	Results	
		Classes	Critical load
Actual deposition of sulphate	Product of mean annual rainfall and mean volume-weighted sulphate concentration	Actual deposition ranged from 0.1-0.7 keq SO_4^{2-}/ha/yr	
Soil sensitivity	Derived from: Land cover (forestry, woodland, grassland, cultivated lands) Soil features (CEC, pH, base saturation) Lithology (eg granite, quartz, shale, etc)	1: sensitive 2 3 4 5: resistant	Estimated weathering rate of a sensitive site (class 1) to be the critical load for class 1 sites ie. 0.39-0.86 keq/ha/yr is the critical load for sensitive sites
Surface water critical loads	Critical load = (pre-acidification flux of base cations) - (base cation concentration in rainfall) - (alkalinity leaching)	<0.2 keq H^+/ha/yr - Sensitive 0.2-0.5 0.5-1.0 1.0-2.0 >2.0 keq H^+/ha/yr - Resistant	

Actual wet deposition loads of sulphate range from 0.1 to 0.7 keq sulphate/ha/yr with the greatest deposition occurring in the western and southern portions of the province - areas where most of South Africa's power-generating plants are found. Actual deposition is lowest in the eastern parts, where large natural areas, such as the Kruger National Park, are situated.

A recent South African study estimated dry deposition of sulphate (SO_4^{2-}) over a forested area to be 0.109 keq/ha/yr compared with 0.025 keq/ha/yr over grassland (Piketh *et al*, 1994). Closer to an industrialised area dry deposition estimates ranged between 1.02 - 1.69 keq/ha/yr.

The diatom and steady state models for determining critical load exceedances for surface waters presented quite different pictures. The diatom model showed that critical loads for surface waters of only a small part of Mpumalanga were exceeded, and these were restricted to the south-eastern part of the region. Few surface waters were found to be acidified - pH values were mostly greater than 6.0. In contrast, the steady state method showed that significantly more of the region's surface waters were exceeding their critical loads. The calculation which included sulphate concentration in the surface waters resulted in many negative values - suggesting much variability, which could be attributable to analytical errors, or because sulphate concentrations in the surface waters are not in balance with the atmospheric inputs.

Surface water critical loads (Figure 1.) less than or equal to 0.5 keqH$^+$/ha/yr were largely restricted to the upland areas of the province, or areas where the soil type was sandy with low levels of base cations. The highest critical loads and hence the lowest sensitivities (> 2.0 keqH$^+$/ha/yr) were found primarily in the south-western parts of the province.

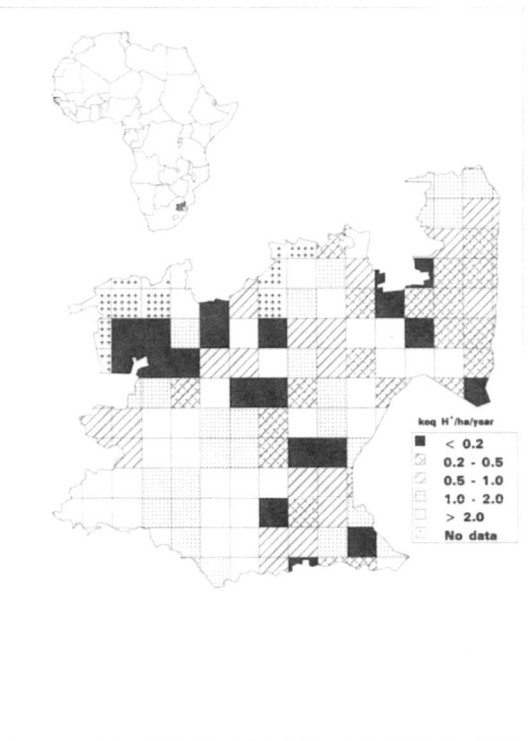

Fig.1 Provisional critical loads for the surface waters of Mpumalanga province, South Africa, calculated with the steady state water chemistry model.

The critical load exceedances were calculated by subtracting critical loads from the measured deposition levels of sulphate. The areas where these exceedances occurred corresponded significantly with sites where the critical loads were less than or equal to 0.5 keqH$^+$/ha/yr, being largely confined to upland areas which have a steep topography. The surface waters where the critical loads were exceeded were often not acid i.e. had a pH level of greater than 7.0. Even though exceeded areas may show no evidence of damage at present, the time lag of responses needs to be considered - damage effects may only become apparent after some time.

Sulphate retention in the studied catchments varied between 5 and 95% (mean of 81%). It is not known how long this sulphate remains in the soil, what the lag time is before the sulphate appears in the surface waters, or whether the sulphate concentrations in the surface waters are in a steady state with the atmospheric inputs.

The region is also characterised by a high evaporation rate. Consequently, there is a low runoff percentage ranging between 3% and 74% (mean 15%) which is far less than the runoff rates experienced in Europe and North America where runoff often exceeds 50%.

The soil sensitivity map showed the most sensitive soils to largely correspond with the escarpment area, where there is a combination of high rainfall and acid soils. The south western parts of the province are more resistant to acid deposition- this area is predominantly farmland.

The weathering rate for the Sabie River catchment was roughly calculated to range from 0.39 to 0.86 keq/ha/yr. This value was then considered to be the amount of buffering capacity available to neutralise the atmospheric inputs of acidity and so it represents an estimate of the potential critical load for the **most sensitive** site classification.

Actual deposition over Mpumalanga was shown to range from 0.1 to 0.7 keq/ha/yr but was in the range 0.4 to 0.5 on the most sensitive sites. Thus the weathering rates given for the Sabie catchment, i.e. 0.39 to 0.86 keq/ha/yr should be adequate to balance acid inputs to these sensitive sites on an annual basis.

We conclude there is no immediate cause for concern over potential negative impacts of acid deposition on soils in Mpumalanga. However the ranges calculated are broad and exceedance can occur if the larger deposition rates are set against the slowest weathering rates.

In addition, the available estimates of dry deposition in Mpumalanga vary between approximately 1.5 and 1.9 times that of wet deposition (Skoroszewski, 1995; and Olbrich and Du Toit, 1993). If all wet sulphate deposition load classes are scaled up by 1.9 times, then there would be critical load exceedances even at the upper limit of the weathering rate. It is therefore important that estimates of dry deposition for the whole region be considered as this has a substantial impact on the estimated exceedances.

4. Conclusions

Estimates of critical loads and their potential exceedances have been calculated for both the surface waters and the soils of Mpumalanga. Consideration of sensitivity of both surface waters and soils indicates that there is a large degree of parity between the areas where both environmental receptors are identified as sensitive to acidification.

At present the indications are that there is little threat of acidification through **wet** deposition. However, dry deposition may prove to be a more significant contributor to acidic deposition in the area. Future research efforts should focus on accurately quantifying this component.

The critical loads approach, incorporated into a wider system to define air pollution policies, should be adopted at a national level. It is the authors' firm belief that this system can provide the most cost effective solution to the country's needs in terms of air quality and air pollution impacts.

Acknowledgements

This work was funded through the CSIR's STEP (Scientific and TEchnological Positioning) scheme. Bob Scholes' assistance in quantifying soil weathering rates is gratefully acknowledged.

References

Annegarn, H.J., Kneen, M.A., Piketh, S.J., Crouch, A., Maenhaut, W. and T.A. Cahill.: 1994. "Remote aerosols over southern Africa and global warming." Proceedings of the 1994 Clean Air Conference, Cape Town, South Africa.

Battarbee, R.W.: 1993. Critical loads and acid deposition for UK freshwaters (summary). A report to the Department of the Environment for the Critical Loads Advisory Group for Freshwaters.

Battarbee, R.W.: 1994. " Critical Loads of Acidity for Freshwaters." In: Critical Loads of Acidity in the United Kingdom. Critical Loads Advisory Group. Report to the Department of the Environment, U.K. pp 19-27.

Chadwick, M.J., Kuylenstierna, J.C.I. and Gough, C.A.: 1991. "The Stockholm Environment Institute map of relative sensitivity to acidic depositions in Europe". In: Mapping Critical Loads for Europe Eds Hettelingh, J., Downing, R.J. and de Smet, P.A.M. Coordination Center for Effects Technical Report No. 1. RIVM Report No. 259101001, National Institute of Public Health and Environmental Protection, the Netherlands. pp 49-57.

Henriksen, A., Dickson, W. and Brakke, D.F.: 1986. "Estimates of critical loads to surface waters." In Critical loads for sulphur and nitrogen. Ed Nilsson, J. Nordic Council of Ministers, Copenhagen. pp 87-120.

Kuylenstierna, J.C.I., Cambridge, H., Cinderby, S. and M.J. Chadwick.. 1995. "Assessing the sensitivity of ecosystems to acidic deposition in developing countries." Proceedings of the 5th International Conference on Acidic Deposition, 26-30 June, Göteborg, Sweden.

Olbrich, K.A. and B. Du Toit.: 1993. "Assessing the risks posed by air pollution to forestry in the Eastern Transvaal, South Africa". Confidential report to Eskom, Division of Forest Science and Technology Report FOR-C 214, CSIR, Pretoria.

Piketh, S.J. and H.J. Annegarn.: 1994. "Dry deposition of sulphate aerosols and acid rain potential in the Eastern Transvaal and lowveld regions. Proceedings of the 25th Clean Air Conference, 24-25 November, Cape Town..

Schutz, C.J.: 1990. "Site relationships in Pinus patula in the Eastern Transvaal escarpment area". PhD thesis, University of Natal, South Africa.

Skoroszewski, R.W.: 1995. "Sensitivity of surface waters in the Eastern Transvaal province." CSIR report WM 769.

Sverdrup, H.U.: 1990. The kinetics of base cation release due to chemical weathering. Lund University Press, Sweden. 246pp.

Sverdrup, H., de Vries, W. and A. Henriksen. 1990. Mapping critical loads - a guidance to the criteria, calculations, data collection and mapping of critical loads. Environmental Report 1990:14 prepared for the Workshop and Task Force on mapping critical loads and levels. 124pp.

Task Force on Mapping.: 1993. Draft Manual on Methodologies and Criteria for mapping Critical Levels/Loads and Geographic area where they are exceeded. Convention on Long-Range Transboundary Air Pollution, UN-ECE. Geneva. 109pp.

Tyson, P.D., Kruger, F.J. and C.W. Louw.: 1988. Atmospheric pollution and its implications in the Eastern Transvaal Highveld. South African National Scientific Programmes Report 150, CSIR, Pretoria. 114pp.

ASSESSMENT OF THE ECONOMIC COSTS OF DAMAGE CAUSED BY AIR-POLLUTION

M. R. HOLLAND

ETSU, Harwell, Oxfordshire OX11 0RA, UK.

Abstract. Cost-benefit analysis is one of the fundamental tools for the development of economic instruments for pollution control. The costs of various abatement measures are reasonably well characterised. However, assessment of the economic costs of pollutant impacts is less well developed. This paper reports on two studies carried out for DGXII of the European Commission, the ExternE-Project and the Green Accounting Research Project. Both studies have been performed by international, multi-disciplinary research teams.

Analysis of the effects of emissions of PM_{10}, SO_2, NO_x and VOCs (as ozone precursors) has included assessment of human health, materials, crops and other terrestrial ecosystems, and freshwater fisheries. The analysis follows the 'impact pathway' approach, linking dose-response functions, valuation data and other models. It differs significantly to earlier 'top-down' approaches that made only very limited use of the wealth of scientific data available. Most success has been achieved in analysis of impacts on human health, building materials and crops. Significant uncertainties exist for these receptors, though these have been identified and are now being addressed. Assessment of impacts on other receptors, perhaps most notably forests, is more limited. The methodology is particularly applicable for analysis of impacts on receptors for which the critical loads approach is not appropriate.

Key Words: Air pollution, external costs, economic instruments, health, building materials, acidification

1. Introduction

The trend towards economic valuation of environmental and social damages, particularly those effects associated with the energy sector, is apparent in the European Union, the USA, Canada, Australia and elsewhere (European Commission, 1995a). It is being driven by several factors, the most important of which include: the need to integrate environmental concerns when selecting different energy technologies and fuels; a need for evaluation of the costs and benefits of stricter environmental standards; and increased attention to the use of economic instruments for environmental policy.

In recent years the critical loads approach (Downing *et al*, 1993) has been adopted as the basis for discussion of plans for reducing transboundary air pollution within Europe (UN ECE, 1993), allowing consideration to be given to the spatial variability of sensitive receptors. This approach is more economically efficient than the adoption of uniform emission reductions regardless of the likely impact from a given source. However, the critical loads approach suffers several problems. In particular, it is only truly applicable for cases where there is a threshold for damage. Also, it does not seek to quantify damage but more simply identifies areas where impacts are likely.

Economic valuation of 'goods' as fundamental as human life remains controversial. However, the valuation process seeks only to quantify society's preference for allocation of resources, not to quantify the 'intrinsic worth' of a life, an ecosystem, or whatever. It is thus 'willingness to pay' for environmental improvement, or 'willingness to accept' increased degradation or risk that is assessed, rather than absolute 'worth'. Decision makers have to deal with problems of this kind everyday, when they decide on budgets for health care, education, environmental protection, defence, etc. Cost-benefit analysis is a means of increasing the transparency of the reasoning underlying such processes through the provision of a numeric framework for decision making.

Since 1991, the ExternE Project (European Commission, 1995a) has been developing the 'impact pathway' methodology for detailed assessment of the environmental and social impacts and costs of technologies for electricity generation. Amongst many other burdens imposed by the energy sector, the damages caused by emissions of PM_{10} (particulate matter less than 10 μm in diameter, including nitrate and sulphate aerosols), sulphur dioxide (SO_2), oxides of nitrogen (NO_x) and volatile organic compounds (VOCs, through their role in ozone, O_3, formation) have been assessed. In recognition of the influence of the technologies used and their location, analysis has been conducted for individual power stations of specified site and design. A companion study, the Green Accounting Research Project has used the same methodology to estimate the damages associated with these emissions at the national level in four countries (Germany, Italy, the Netherlands and the UK). The receptors that have been considered include human health, materials, crops, forests and other semi-natural terrestrial ecosystems, and freshwater fisheries.

This paper reports on the state of development of the impact pathway methodology for the assessment of the environmental and social impacts of air pollution and associated costs. The merits of using this approach for analysis in support of the development of future policy are compared with those of using the critical loads approach.

2. Methodology

2.1 The Impact Pathway Approach

The 'impact pathway' methodology follows a logical progression from emission to impact and subsequent valuation (Figure 1). The number of stages in the analysis reflects the complexity of the individual impact being assessed. Groups of experts from across Europe were brought together to identify the most appropriate exposure-response models available for each of the receptors considered. These models are being updated as new information becomes available. A more detailed discussion of the methodology used for both impact assessment and valuation in the ExternE Project is presented elsewhere (European Commission, 1995b).

2.2 Status of Assessment for Different Receptors

The methodology is most complete for assessment of damages to human health, materials and crops, and least complete for effects on forests and natural ecosystems (Table 1). Indeed, for natural ecosystems the critical loads approach seems likely to be the only option for general application for the foreseeable future (though models are available for certain geographic areas). Assessment of damage to freshwater fisheries is possible for the UK as far as quantifying losses of fish over time for different deposition scenarios. Research on valuation of impacts to recreational fisheries is on-going.

Models are available to describe the atmospheric chemistry and transport of PM_{10} and the acidifying pollutants. Experience shows that for a single power station, or an

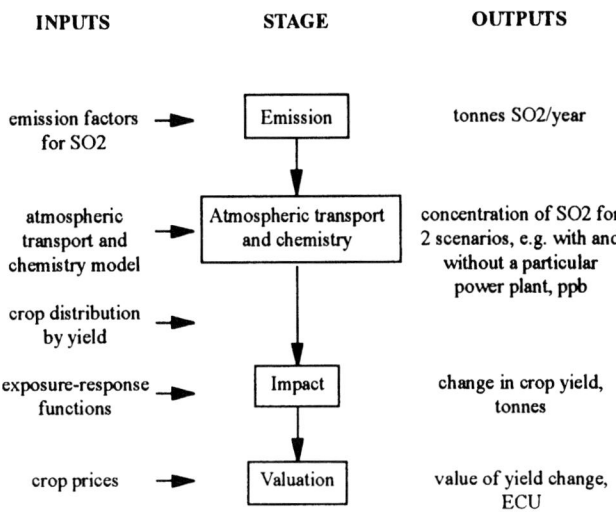

Figure 1. An example of the application of the impact pathway approach; assessment of the direct effects of SO_2 from a single power station on crops.

Receptor	Extent of analysis	Key uncertainties and research needs
Human health	Pan-European, through to impact assessment and monetary valuation for SO_2, PM_{10} and O_3	1. Existence of thresholds 2. Speciation of particulates 3. Chronic effects 4. Valuation of life
Materials	Pan-European for damage from acidic deposition through to impact assessment and monetary valuation for 'utilitarian' buildings	1. Human behaviour with respect to maintenance of property 2. Stock at risk in Southern Europe 3. Exposure-response data for ozone 4. Stock at risk for cultural buildings
Crops	Pan-European through to impact assessment and valuation for direct effects of SO_2 and O_3 on some crops	1. Lack of data for some crop species 2. Limited data at realistic SO_2 levels 3. Interactions between pollutants and other stresses
Forests	Experimental application of some exposure-response functions relating acidity to forest damage, and assessment of mitigation costs	1. Spatial variation in response to pollution stresses 2. Long term effects 3. Valuation of non-timber benefits
Other terrestrial ecosystems	Application of the critical loads approach to identify areas and ecosystem types at risk in the UK	1. Lack of dose-response data 2. Lack of valuation data
Freshwater fisheries	Impact assessment limited to the UK at present, but data allowing extrapolation to Scandinavia is believed to be available	1. Valuation of impacts

Table 1. Extent of economic analysis of damages of air pollution for different receptor types.

individual country, it is necessary to run the analysis to 1000 km or more to account for 80% of the total impact. Modelling of O_3 is more complex, because of its episodic nature and the complexity of the conditions which govern its formation. The models that are available for O_3 tend to require much computing time and also tend only to provide results for single events rather than seasonal or annual data.

Uncertainty arises from several sources, including: experimental error; transference of data from one situation to another (e.g. the use of dose-response functions developed in the USA for describing ozone effects on crops in Northern Europe); our inability to calculate damages for all types of impact; biases inherent in valuation procedures; and the need to incorporate ethical and political judgements (e.g. the selection of discount rate). Whilst some of these aspects can be described using standard techniques, others cannot, and alternative methods are necessary. Sophisticated techniques based on decision analysis are being assessed within the ExternE Project. These involve the derivation of probability distributions around each input, based on available data wherever possible, but also expert judgement where necessary.

3. Results

Analysis of the air pollutants of interest to this paper is typically dominated by effects on public health. This is shown in Figure 2, which considers some effects of a coal fired power station in the English midlands. For consistency only damages within the UK are shown in the figure, though damages caused by the power station were assessed over the whole of Europe for some impacts. Inclusion of impacts outside of the UK would significantly increase the figures shown, as would consideration of a plant more typical of UK coal stations, which are not fitted with the same level of abatement technology. The next most important category is typically effects on building materials. Damage to crops is trivial in comparison. Damage to forests is underestimated in the Figure, as the methodology used does not take proper account of long term effects.

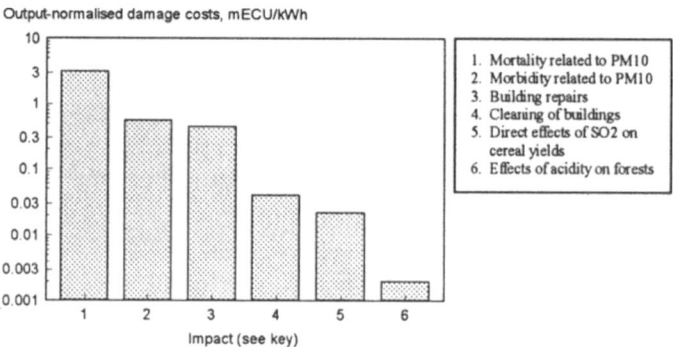

Figure 2. Selected impacts within the UK for a coal fired power station in the English midlands. The plant has 90% flue gas desulphurisation, low NOx burners and 99.7% efficient electrostatic precipitators.

Given the importance of the damage associated with health effects it is necessary to consider the major uncertainties present. Our most important assumptions are that there is no threshold for effects; that estimates of the value of statistical life (derived for an average member of the population on a willingness to pay/accept basis) are applicable to mortality associated with air pollution, which will mainly affect people with a very short life expectancy; and that selected response functions are generally applicable. If the first two of these are incorrect, results will be too high. However, we have not here included impacts such as chronic effects of PM_{10} on mortality, which if present could inflate damages significantly (Pope et al, 1995).

Figure 3 demonstrates the relative importance of mortality in determining estimated health damages across the UK, and of the estimated contribution for different pollutants. A proportion of the damage allocated to each pollutant are attributable to natural (background) levels of these substances, rather than human activity.

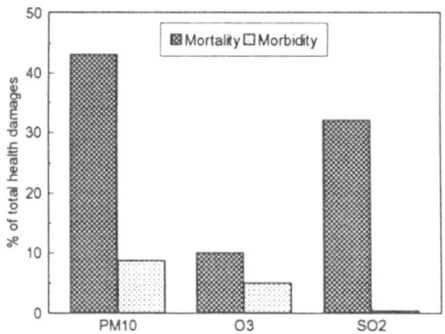

Figure 3. Contribution of PM_{10}, SO_2 and O_3 to total acute effects on health associated with these pollutants in the UK. Effects of SO_2 and PM_{10} may not be additive, in which case PM_{10} estimates are to be preferred. No account is taken here of possible chronic effects of long term exposure to these pollutants.

4. Discussion

Detailed assessment of the impacts of the major air pollutants, through to an economic assessment of damage, is now possible for a number of important receptors. An important feature of the methodology (if properly applied) is its transparency, deriving from the manner in which it proceeds logically, step by step, through the analysis, highlighting the level of confidence in each of the parameters required. Further work is clearly needed in order to combine uncertainties in a thorough and consistent manner.

One obvious advantage of economic valuation is the fact that it introduces a common unit to cover a wide range of variables. By doing so all stages of the analysis can be expressed quantitatively, including the comparison between technology or policy options. This allows debate to focus more on the underlying parameters used and assumptions made rather than on the derivation of weightings for different effects, which is seldom clear if the weighting process is purely qualitative.

It is evident that neither the critical loads approach nor our impact pathway approach are individually capable of analysis of all potentially sensitive receptors for the major air pollutants. The critical loads approach is most applicable to natural and semi-natural ecosystems including forests and freshwater fisheries. Application of critical loads to materials is slightly misleading as there is typically no critical load as such - materials degrade even in pristine environments, and hence have no physical or chemical threshold that is relevant to the policy debate. In the context of the pollutants of interest to this paper the impact pathway approach is currently most applicable to analysis of effects on human health, materials, and crops. Application to more natural ecosystems requires further work, particularly on valuation. Overall it is thus apparent that the two approaches complement each other well, together providing coverage of all of the major receptors of concern.

Given the concentration of people (and therefore buildings) in urban areas, there is a spatial separation of those receptors for which the valuation methodology is most advanced and tends to provide the largest numbers, and the natural and semi-natural ecosystems for which critical loads is more applicable. Optimisation of abatement strategies to deal with the full range of reported effects is thus likely to be more complex than previously recognised. It also seems likely that analysis will require data at a higher level of spatial disaggregation than has hitherto been used to inform negotiations on pollution abatement within Europe.

Acknowledgements

This work has involved researchers from more than 30 institutes through the ExternE Project and the Green Accounting Research Project. Particular thanks are due to Pierre Valette of the European Commission, Nick Eyre, Fintan Hurley of the Institute of Occupational Medicine, Anil Markandya of Metroeconomica, Mike Hornung at the Institute of Terrestrial Ecology, Wolfram Krewitt and Petra Mayerhofer at IER, and my colleagues at ETSU. Both studies have been generously funded by DGXII of the EC under JOULE II. ETSU's contribution to ExternE has been co-funded by the UK Department of Trade and Industry. The views expressed here are those of the author and do not necessarily reflect those of either the funding bodies or fellow workers.

References

Downing, R.J., Hettelingh, J.-P. and de Smet, P.A.M.: 1993, "Calculation and Mapping of Critical Loads in Europe". RIVM Report No. 259101003. Coordination Center for Effects, Bilthoven, the Netherlands.
European Commission, DGXII, Science, Research and Development, JOULE Programme: 1995a, "Externalities of Fuel Cycles, 'ExternE' Project, Report Number 1, Summary".
European Commission, DGXII, Science, Research and Development, JOULE Programme: 1995b, "Externalities of Fuel Cycles, 'ExternE' Project, Report Number 2, Methodology".
Pope, C.A. III, Thun, M.J., Namboodiri, M.M., Dockery, D.W., Evans, J.S., Speizer, F.E., and Heath Jr., C.W.: 1995, "Particulate Air Pollution as a Predictor of Mortality in a Prospective Study of US Adults. American Journal of Respiratory Critical Care Medicine, **151**, 669-674.
PORG: 1993, "Ozone in the United Kingdom 1993", Third report of the United Kingdom Photochemical oxidants review group, for the Department of the Environment, HMSO, London.
TERG: 1988, "The Effects of Acid Deposition on the Terrestrial Environment of the United Kingdom", First Report of the UK Terrestrial Effects Review Group, Department of the Environment, HMSO, London.
UNECE: 1993, "The State of Transboundary Air Pollution, 1992 Update", United Nations Economic Commission for Europe, Geneva.

ESTIMATING THE IMPACT OF AIR POLLUTION ON ENVIRONMENTALLY VALUABLE SITES

M. BROWN[1], H. DYKE[1], S.M. WRIGHT[1], R.A. WADSWORTH[1], K.R. BULL[1], A. FARMER[2], S. BAREHAM[3], S.E. METCALFE[4], D. WHYATT[4] and C. POWLESLAND[5]

[1]*Institute of Terrestrial Ecology, Monks Wood, Abbots Ripton, Huntingdon, Cambridgeshire. PE17 2LS. (Contact address),* [2]*English Nature, Northminster House, Peterborough. PE1 1UA,* [3]*Countryside Council for Wales, Plas Penrhos, Fford Penrhos, Bangor, Gwynedd. LL57 2LQ,* [4]*School of Geography & Earth Resources, University of Hull, Hull. HU6 7RX,* [5]*Her Majesty's Inspectorate of Pollution, 2 Marsham Street, London, SW1P 3EB*

Abstract. Concern about the environmental effect of air pollution on areas of high conservation value in the UK has prompted the statutory agencies to initiate an investigation on these areas. For this, critical loads maps have been used together with predicted air pollution data, monitored air pollution data and remotely sensed land cover information within a geographic information system (GIS). Additional information on designated Sites of Special Scientific Interest (SSSI) for England and Wales have also been incorporated. This provides the framework for examining potential impacts to these sites under various current and future scenarios. The approach allows for the investigation of the impacts of individual point sources as well as complete national scenarios. Preliminary results are provided from analysis of a single pollutant (sulphur). These indicate that nationally up to 52% of the area of SSSI's (5000 km^2) are at risk from soil acidification. Using this approach it has been possible to apportion the load on any SSSI, thereby enabling the ecological impacts of each point source to be identified.. This information can then be used to assess priorities for regulatory controls.

Key words: Conservation areas, geographical information systems (GIS), sulphur deposition, critical load

1. Introduction

Acidification of the natural environment by anthropogenic emissions of sulphur and nitrogen is well documented in many parts of Europe (Last & Watling, 1991). At present in Britain, point source emissions of sulphur account for approximately 70% of the total load (RGAR, 1990). Although the UK is committed to reduce these emissions in line with the Oslo protocol (UNECE, 1994), nitrogen oxides may still be increasing and could continue to do so in the future.

In Britain, Sites of Special Scientific Interest (SSSI) are statutorily protected under the Wildlife and Countryside Act 1981 for the purpose of conserving the best examples of wildlife habitats and native species. The potential impacts of acidic deposition on nature conservation in Britain has been extensively documented (Woodin & Farmer, 1991; Woodin & Farmer, 1993). However there has been little attempt to quantify the degree of damage to these important habitats and relate it to emission sources.

In this investigation the potential relative impacts of sulphur emissions from large point sources is considered in relation to SSSI. For this, the precautionary principle has been adopted, in which it has been assumed that "excessive deposition" onto a SSSI will result in subsequent acidification and changes that will adversely affect the biological status of the site. The critical loads approach (CLAG, 1994; Bull, 1990) has been used to define the threshold for effects and "excessive depostion" in line with UK government's policy of pollution control (HMSO, 1990).

2. Methods

The study made use of geographical information systems (GIS). These are computer based systems for collecting, storing, retrieving, transforming and displaying spatial data (Burrough, 1986). They enable large geographically referenced datasets to be efficiently maintained and manipulated. More importantly, GIS provide the facilities for integrating disparate datasets for the analysis of spatial phenomena. The analyses for this study have been carried out using the GRID module in ARC/INFO GIS. This module contains a set of tools specifically designed for the handling and analysis of raster (grid-based) data. Numerical analyses using map algebra were performed and automated in AML (ARC Macro Language).

The data for this study included: modelled sulphur deposition fields from major point sources in Great Britain, SSSI boundary information for England and Wales, a soil critical load database for GB and land cover information. The GIS has been used to attribute the source of sulphur deposition on SSSI where the critical load for acidity of soils is exceeded for a number of current and future emission scenarios. Land cover data were used to quantify stock at risk.

Deposition footprints at 20km resolution for every major point source of sulphur in Great Britain were generated by the Hull Acid Rain Model (HARM 7.2) and incorporated into the GIS. HARM 7.2 is a linear source-receptor, statistically based, model which provides average deposition estimates for each 20km grid square in the UK (Metcalfe et al.,1989). This linearity allows each of the individual deposition fields to be scaled and then summed to produce national deposition fields for a range of emission scenarios. Other components of the deposition field (diffuse sources), such as natural background, low - medium level (sources other than power stations and oil refineries) and European sources (from 1990 EMEP data) have been included at the same resolution.

Exceedance maps were generated by overlaying the deposition field onto a soil critical loads for acidity map in the GIS. The critical loads map (Hornung et al., 1995 in press) was generated by allocating each 1km square of Great Britain to one of five critical load classes (Nilsson and Grennfelt, 1988), according to the amount of acidity that would be neutralised by base cation weathering of minerals from each dominant soil unit

The analysis was performed in two stages. Source attribution from emission sources to exceeded SSSI areas was estimated and an exceedance map generated. This exceedance map was used identify the land cover types within exceeded SSSI. For this the Institute of Terrestrial Ecology's (ITE) land cover map of Great Britain (Fuller et al., 1993) was used. This has been derived from satellite imagery from Landsat Thematic mapper data. The classification combines summer and winter data, to distinguish 17 cover types on a 25m grid.

A GIS macro was used to estimate source attribution of deposition onto SSSI for each deposition scenario. This consisted of i) generating a deposition field by scaling each of the point sources to an estimated emission for a particular scenario, ii) summing them, and iii) adding of the diffuse fields to produce a 20km total sulphur deposition field. A 1km exceedance map was then generated by subtracting the 1km critical loads values from the total deposition field.

The exceedance map was used as a mask to exclude those areas not exceeded. The contribution from each component of the deposition field (point source, European source etc) to each 1km grid square in GB was then estimated using:

$C_i = l_i / l_a$ Where: C_i - contribution of the ith source to a grid square
l_i - load deposited by the ith source
l_a - total load for all sources to the grid square

By summing the values of C for all 1km cells the source contribution to deposition was determined.

SSSI areas were generated as a raster dataset suitable for combining with the pollution and critical loads data. The final stage of the work was to calculate the percentage contribution of the individual sources to exceedance within the SSSI, by summing C_i over exceeded SSSI grid squares using the exceedance mask.

A number of emission scenarios were investigated to determine the effect of operating conditions on deposition to SSSI. These included: 1993 emission rates, unit emission rate and 2001 modelled emission rate. The components of the deposition field under investigation consisted of: individual power station emissions, low - medium level emissions, European emissions and natural (biogenic) emissions. Emission rates for 1993 were based on the generators reported emissions. The "unit" emission rate (set at 100000 tonnes of sulphur per year to reduce scaling errors in the GIS) indicated the importance of the geographic location of the point sources. This is particularly important when considering source attribution since smaller power stations situated close to sensitive habitats may do proportionately more damage per tonne SO_2 emitted than large stations some distance away. The future emission scenario for the year 2001 was modelled using HARM 7.2. The diffuse source fields for this scenario were scaled in accordance with expected 2001 values. This scenario was derived assuming the new plant standards needed to achieve the Large Combustion Plant Directive (LCPD).

3. Results and Discussion

The areas of exceedance for each scenario within SSSI for England and Wales are presented in Table I. It can be seen that in England, 38.3% of the area of SSSI are at risk of increasing acidification, in Wales the story is even bleaker with 60.7% of SSSI areas at risk. Although the situation is improving by 2001, over a quarter of the area of English SSSI will still be at risk, whilst in Wales the figure is nearer a half.

TABLE I

Areas within SSSI exceeded for the three scenarios

Exceeded areas within SSSI (km^2)
percentage of total SSSI area shown
in parentheses

Scenario	England	Wales
1993 emissions	3070 (38.3%)	1323 (60.7%)
Unit Emissions	3499 (43.6%)	1441 (66.2%)
Future emissions (2001)	2017 (25.2%)	1013 (46.5%)

Summary results of the source attribution values for the current and 2001 scenarios are shown in Table II, where twenty four point sources whose individual deposition is less than one percent of the total have been amalgamated into a single group. It is known that European and low-medium level sources account for over half of the total sulphur load onto Britain and the results show that these sources make a similar contribution to exceeded SSSI areas. The reductions for individual point sources identified for 2001 result in a proportionate increase of the European and, more significantly, low-medium level emissions. By 2001, few power stations are predicted to contribute individually more than 1% of the deposition to exceeded SSSI. Unit emissions show very different patterns of contributions reflecting the geographic position of receptors with respect to each source.

TABLE II

Percentage of deposition onto exceeded SSSI areas

Source	1993 emissions England	Wales	unit emissions England	Wales	2001 emissions England	Wales
European deposition	39.3	43.1	28.8	31.5	43.7	46.6
Low - med deposition	16.3	16.4	12.0	12.0	26.4	24.4
Natural background	5.4	5.9	3.9	4.3	10.5	11.4
Other UK point sources	8.9	7.5	35.5	28.4	7.6	7.7
Drax	3.8	2.0	1.3	0.7	1.8	0.8
Ratcliffe on Soar	3.3	2.7	1.3	1.1	0.9	0.7
Fiddlers Ferry	3.0	2.9	1.9	1.9	1.4	1.2
Ferrybridge	2.7	2.5	1.3	1.1	0.9	0.8
Cottam	2.8	1.9	1.2	0.8	0.9	0.6
West Burton	2.7	2.6	1.2	1.2	0.9	0.7
Rugeley	2.3	2.1	1.9	1.6	0.6	0.5
Didcot	2.2	1.6	1.7	1.2	1.0	0.7
Eggborough	2.1	1.8	1.3	1.1	0.9	0.8
Aberthaw	2.1	2.4	2.0	2.4	0.9	1.1
Ironbridge	1.4	1.7	1.7	2.0	0.6	0.6
Ince	1.4	1.8	1.8	2.2	0.4	0.4
Ballylumford	0.8	1.4	1.2	2.0	0.6	1.1

Summary land cover statistics within the exceeded SSSI for the two countries are presented in Table III. For England and Wales the results show a 36% reduction of land cover types within exceeded SSSI between 1993 and 2001. It should be noted that SSSI contain a disproportional amount of habitats important for conservation as well as rare or threatened species. The distribution of exceeded areas within SSSI for Wales are shown in Figure 1.

TABLE III

Land cover types within exceeded areas for England and Wales

ITE Land Cover Type	Area of land cover (km^2)					
	1993 emissions		unit emissions		2001 emissions	
	England	Wales	England	Wales	England	Wales
Coastal bare ground	31	5	36	9	7	2
Saltmarsh	20	2	26	3	5	1
Bog	74	36	78	65	68	44
Marsh/Rough grass	93	52	93	67	49	47
Managed grass	504	156	471	163	202	102
Lowland grass - heath	158	41	134	41	62	29
Montane/Hill grass	760	341	735	342	484	273
Open shrub heath/moor	445	276	472	247	361	198
Dense shrub heath/moor	258	200	274	206	202	145
Bracken	219	69	205	71	131	49
Deciduous/mixed	461	91	535	95	241	54
Coniferous	65	20	111	23	52	14
Arable/tilled	130	10	128	11	69	6
Total area	3217	1298	3299	1343	1933	964

Fig. 1. Sites of special scientific interest in Wales showing areas of exceedance for 1993 sulphur deposition as modelled by HARM 7.2

4. Conclusion

The results suggest current sulphur deposition is resulting in extensive exceedance of SSSI in England and Wales. To quantify any effects of excess sulphur onto these areas, however, would require information on the species present, together with knowledge of dose-response relationships. Whilst some information is available for the former, the latter is more problematic. A more useful approach might be to use the current predictions as the basis for future detailed field investigations. In the meantime, the current infomation may be used in accordance with the precautionary principle as the basis for implementing controls on point sources. Thus the above results serve as a useful first estimate and provide data for targeting further detailed studies. However, the results do not take into account uncertainties in: sulphur deposition, critical loads estimates, contribution of nitrogen deposition, the moderating effects of base cation deposition and dynamics of acidification and recovery. These too merit further study in the future

Acknowledgements

The authors would like to thank English Nature for funding this work, Her Majesty's Inspectorate of Pollution for providing emissions data, and the Department of the Environment's Critical Loads Advisory Group for critical loads data.

References

Bull, K.R. 1990. The critical loads approach to gaseous pollutant emission control: In Environmental Pollution. **69**, 105 - 123

Burrough, P.A.:1986, Principles of Geographical Information Systems for Land Resources Assessment. Clarendon Press, Oxford.

Critical Loads Advisory Group. 1994. Critical Loads of Acidity in the United Kingdom, Summary Report. Prepared at the request of the Department of the Environment.

Department of the Environment. 1990. *This common inheritance*: Britain's environmental strategy. London HMSO.

Hornung, M., Bull, K.R., Cresser, M., Hall, J., Langan, S.J., Loveland, P. and Smith, C: 1995, *An empirical map of critical loads of acidity of soils in Great Britain*. Environmental Pollution, in press.

Fuller, R.M., Groom, G.B.,Jones, A.R. and Thompson, A.G. 1993. *Countryside Survey 1990. Mapping the land cover of Great Britain using Landsat imagery: a demonstrator project in remote sensing. Final report*. August 1993. 71pp. British National Space Centre.

Last, F.T. & Watling, R.: 1991. Acidic deposition: Its nature and impacts. Proceedings of the Royal Society of Edinburgh, 97.

Metcalfe, S.E., Atkins, D.H.F., and Derwent, R.D.: 1989. *Acid deposition modelling and the interpretation of the United Kingdom secondary precipitation network data*. In: Environmental Pollution. **23**, 2033 - 2052.

Metcalfe, S. E. & Whyatt, J. D.: 1995, Modelling future acid deposition with HARM. In:Acid Rain and its impacts: the critical laods deabate (Ed. R. W. Battarbee). Ensis: London.

Nilsson, J. & Grennfelt, P. (eds.) 1988. *Critical Loads for Sulphur and Nitrogen*. Report of workshop held in Skokloster, Sweden 19-24 Marsh 1988. (1988:11), Copengagen: Nordic Concil of Ministers.

RGAR 1990 Review Group on Acid Rain. Third report. Acid Deposition in the United Kingdom 1986 - 1988. Dept. of the Environment & Dept. of Transport.

United Nations. 1994. Protocol to the 1979 convention on long range transboundary air pollution on further reduction of sulphur emissions. ECE/EB.AIR/40

Woodin, S.J. & Farmer, A.M. 1991. The effects of Acid Deposition on Nature Conservation in Great Britain in Great Britain, Focus on Nature Conservation No 26. Peterborough, Nature Conservancy Council.

Woodin, S.J. & Farmer, A.M. 1993. *Impacts of sulphur and nitrogen deposition on sites and species of nature conservation importance in Great Britain*. Biological Conservation, **63**: 23 - 30.

INTEGRATED ASSESSMENT OF EMISSION CONTROL SCENARIOS, INCLUDING THE IMPACT OF TROPOSPHERIC OZONE

M. AMANN, M. BALDI, C. HEYES, Z. KLIMONT AND W. SCHÖPP

International Institute for Applied Systems Analysis, A-2361 Laxenburg, Austria

Abstract. The RAINS (Regional Air Pollution INformation and Simulation) model was developed at IIASA as an integrated assessment tool to assist policy advisors in evaluating options for reducing acid rain. In recent years, the European implementation of this model has been used to support the negotiations on an updated, effect-based Sulphur Protocol under the Convention on Long-range Transboundary Air Pollution. The development of future strategies for reducing the environmental damage caused by air pollutants requires a multi-pollutant, multi-effect approach. In this context, the RAINS model is being further developed to include ozone. This paper outlines the development of an integrated assessment model for tropospheric ozone, which combines information on the emissions of ozone precursors (NO_x and VOCs), the available control technologies and abatement costs, the formation and transport of ozone and its environmental effects in Europe.

Key words: integrated assessment model, RAINS, tropospheric ozone, NO_x, VOCs, emission control scenarios.

1. Introduction

The RAINS (**R**egional **A**ir Pollution **IN**formation and Simulation) model (Alcamo *et al.*, 1991) was developed as an integrated assessment tool to assist policy advisors in evaluating options for reducing acid rain. Such models help to build consistent frameworks for the analysis of abatement strategies. They combine scientific findings in the various fields relevant to strategy development (economy, technology, atmospheric and ecological sciences) with regional databases. The environmental impacts of alternative scenarios for emission reductions can then be assessed in a consistent manner ('scenario analysis'). Integrated assessment models also enable the identification of those strategies that minimise the costs required to achieve a set of environmental targets.

In recent years, the European implementation of the RAINS model has been used to support the negotiations on an updated Sulphur Protocol under the Convention on Long-range Transboundary Air Pollution (LRTAP). RAINS and other integrated assessment models indicated that flat-rate, source-oriented approaches, as used in earlier protocols, do not necessarily produce cost-effective solutions (UN/ECE, 1990). For the first time, the Second Sulphur Protocol made use of an alternative, effect-oriented approach, in which the extent of emission reductions is guided by the impacts that emissions from a given source have on sensitive ecosystems. The cost-minimal allocation of abatement measures was identified with the optimisation features of the RAINS model.

Although several pollutants contribute to acid deposition - RAINS includes sulphur dioxide (SO_2), nitrogen oxides (NO_x) and ammonia (NH_3) - the Second Sulphur Protocol limited its scope to emissions of SO_2 only. For the future, the highest priority is being given to the development of a strategy for the second step of the NO_x Protocol.

Reducing nitrogen emissions based on environmental effects will be a rather complex process. The interrelation of several environmental effects (acidification, eutrophication, tropospheric ozone, human health, etc.) constitutes a multi-effect, multi-pollutant problem. The RAINS model is being developed further to maintain its applicability within this framework. A major avenue of development is the inclusion of ground-level ozone (O_3).

This paper outlines the development of an integrated assessment model for tropospheric ozone, which combines information on the emissions of ozone precursors, the available control technologies and abatement costs, the formation and transport of ozone, and its effects in Europe. The aim is to describe the source-receptor relationships in such a way that the costs and effectiveness of emission reduction strategies can be quantified.

2. Integrated Assessment of Ozone

2.1. EFFECTS AND CRITICAL LEVELS

Studies of the impacts of ozone indicate that critical levels to protect agricultural crops and forests can best be established with long-term exposure measures, in particular, by the 'accumulated excess ozone' concept. Currently, a threshold concentration of 40 ppb is proposed for both crops and trees. This exposure index is referred to as AOT40, the accumulated exposure over a threshold of 40 ppb (Fuhrer and Achermann, 1994). For agricultural crops the accumulated exposure should be calculated for daylight hours (mean global radiation ≥ 50 Wm^{-2}), and for three months (generally May-July). For trees the cumulative exposure is calculated for 24 hours a day over six months. Most current air quality standards to protect human health are defined as short-term concentrations. An integrated assessment model needs to take these different measures into account.

2.2. ATMOSPHERIC TRANSPORT AND CHEMISTRY

Ozone formation involves chemical reactions between NO_x and volatile organic compounds (VOCs) driven by solar radiation. It occurs on a regional scale in many parts of the world.

An integrated assessment model for ozone needs to relate ozone exposure to changes in the emissions of ozone precursors. These source-receptor relationships must be valid for long-term ozone exposure, in line with the proposed critical levels, and applicable on the scale of Europe. One photochemical model designed for long-term, European ozone calculations, which can, therefore, provide the basis of the necessary source-receptor relationships, is the EMEP ozone model (Simpson, 1992a, 1993; Labancz, 1993).

The EMEP model has been used to investigate the relationships in different areas of Europe between mean O_3 concentrations and changes in precursor emissions, using 1989 emissions and meteorological data. Two broad patterns may be distinguished, as shown by the isopleth diagrams in Figure 1. The six-month summer mean ozone concentration is shown as a function of uniform reductions in emissions of NO_x and VOCs. The top right-hand corner of each diagram represents the base case without any emission reductions. The grid squares investigated are also identified in Figure 1.

In northern, southern and eastern areas of Europe, the typical pattern of ozone formation behaviour corresponds to isopleth (a). In these regions the NO_x / VOC ratio is relatively

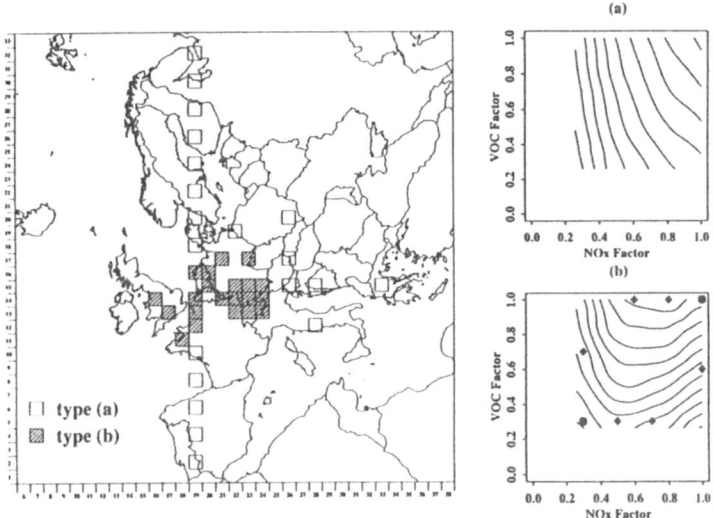

Fig. 1. Typical patterns of ozone formation behaviour in Europe.

low and there are sufficient peroxy radicals to convert nitric oxide (NO) to nitrogen dioxide (NO$_2$), leading to O$_3$ production. Decreasing the available NO$_x$ leads directly to a decrease in ozone. Ozone formation is limited by the availability of NO$_x$. In such regions, reductions in emissions of NO$_x$ are likely to be effective in reducing ozone concentrations, but ozone is less sensitive to reductions of VOC at constant NO$_x$.

For NW Europe, however, isopleth diagram (b) shows a different picture. The isopleths form a ridge dividing the diagram into two areas. On the left of the ridge, corresponding to the greatest reductions in NO$_x$ emissions, the system tends towards the NO$_x$-limited case. On the right of the ridge, the NO$_x$ / VOC ratio is relatively high and the NO$_2$ concentrations are sufficiently great that NO$_2$ competes with VOCs for reaction with the OH radical. In this region, reducing VOC emissions results in lower ozone concentrations; to a large extent, ozone shows a linear dependence on VOC emission changes (Simpson, 1992b). However, ozone concentrations may be increased, at least initially, by NO$_x$ reductions in the absence of concurrent reductions in VOC emissions.

2.3. SIMPLIFIED DESCRIPTION OF OZONE FORMATION

For practical reasons, the calculation of O$_3$ concentrations in an integrated assessment model needs to be computationally efficient. Our simplified approach uses a statistical model to summarise many EMEP model calculations. Two model versions are described, one designed for estimating daily O$_3$ concentrations, the other for seasonal mean values.

Both versions can reproduce the observed range of O_3 formation behaviour; their relative advantages and disadvantages for use in an integrated model are being assessed further.

These models use the concept of "effective" emissions, suggested by recent studies with the EMEP model (Simpson, 1994) which showed that processes by which boundary-layer air is mixed with the free troposphere often determine the extent to which emissions during the earlier stages of a trajectory affect the final O_3 concentration. To allow for this, emissions along the trajectory are weighted by the amount of subsequent dilution to give the dilution-weighted or "effective" emissions used in the statistical models.

In subsequent sections, the following abbreviations are used for model variables:

n, v	-	emissions of NO_x and VOCs, respectively
en, ev	-	"effective" emissions of NO_x and VOCs, including natural sources
J_{NO2}	-	mean NO_2 photolysis rate during 4-day trajectory
$\bar{x}$	-	seasonal mean value of variable x

2.3.1. Daily Model

The daily model is a regression model relating the early afternoon ozone concentrations calculated by the EMEP model at a particular site to predictor variables related to the appropriate air mass trajectory arriving at that site. The model formulation is:

$$[O_3] = f(en_{trj}, ev_{trj}, J_{NO2})$$

where trj denotes a trajectory-related quantity and function f() is either a non-parametric, local regression or a linear regression with quadratic variables (Heyes and Schöpp, 1995).

The daily model has been tested most extensively at Schauinsland, a site in SW Germany within the non-linear behaviour region. For this site, the residual variance of the daily model is 7.3 ppb. In the base case, without emission reductions, the daily model predictions have a similar distribution to the EMEP model results, although the mean O_3 concentration exceeds the EMEP mean (60.5 ppb) by 1.2 ppb.

Further refinements of the daily model will improve its treatment of the different effects of NO_x emissions, take account of the time available for VOC species to react, and include the influence of changes in free tropospheric concentrations of O_3 and its precursors.

2.3.2. Seasonal Model

The seasonal model is a regression model that relates the mean ozone concentration over a three- or six-month period to national annual emissions of NO_x and VOCs. It provides a summary of results obtained from many runs of the EMEP model; the points at which the model is fitted are shown on the isopleth diagram in Figure 1(b).

The mean ozone concentration at receptor j, $[O_3]_j$, is assumed to be a function of the VOC and NO_x emissions from each emitter country i and the mean "effective" emissions experienced at the receptor over the period in question. The model is formulated as:

$$\overline{[O_3]}_j = k_j + \sum_{i=1}^{M}(a_{ij}v_i + b_{ij}n_i + c_{ij}n_i^2) + \alpha_j \overline{en}_j^2 + \overline{en}_j \sum_{i=1}^{M} d_{ij}v_i$$

where M is the number of emitter countries.

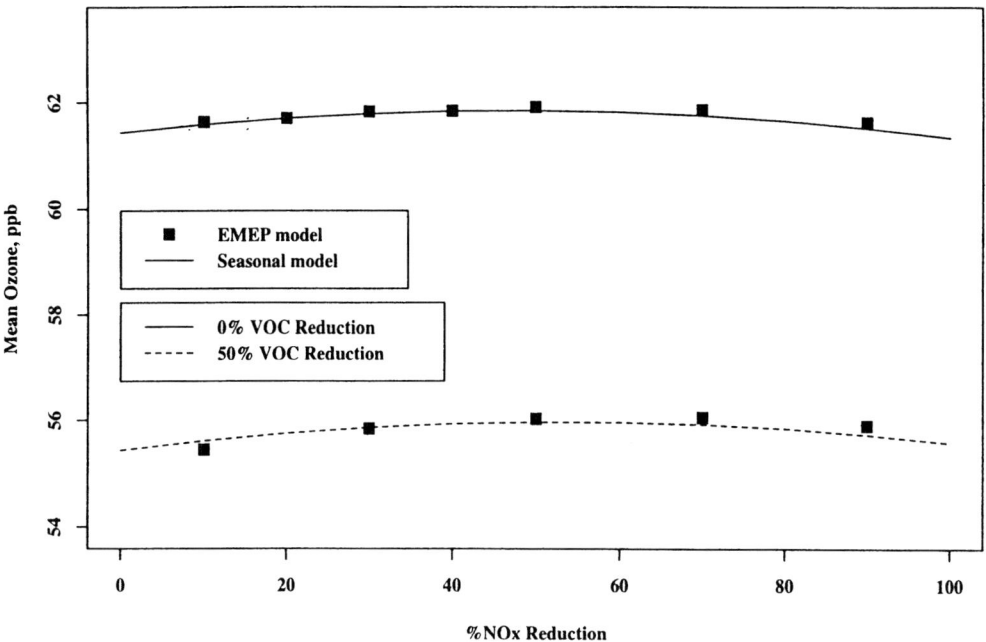

Fig. 2. Model predictions of mean ozone at Schauinsland as a function of NO_x reductions in the UK.

The seasonal model's predictions of mean ozone at Schauinsland as a function of NO_x reductions in the UK are compared in Figure 2 with corresponding EMEP model results. The agreement between the two models is satisfactory, both for base case VOC emissions and when all VOC emissions are reduced by 50%.

2.3.3. Local Factors

The simplified models predict mean O_3 concentrations in the atmospheric boundary layer. Local factors will also influence ozone exposure (eg AOT40) at ground-level. European ozone measurement data have been analysed (Kettunen *et al*, 1994; Baldi *et al*, 1995) to identify the important factors, such as local NO_x emissions, altitude, and distance from the coast, and investigate ways of incorporating them into the integrated model.

2.4. EMISSIONS AND COSTS

The current RAINS model already provides data for the estimation of NO_x emissions from energy use scenarios. A design study (Olsthoorn, 1994) of VOC emission inventories, scenarios and emission control options proposed using the CORINAIR90 inventory as the basis of the VOC part of the integrated model. The availability and costs of technologies to reduce VOC emissions have been reviewed (Caliandro, 1994).

Within Europe, transport is the largest single source sector for the emissions of NO_x and VOCs, contributing 50-60% of European NO_x emissions and about 40-50% of man-made, non-methane VOCs. Because of this sector's importance, a new transport module is being developed for RAINS. The design of the new module allows the forecast of fuel demand on the basis of socio-economic factors, and should improve the estimation of future emissions from this source sector. This approach will also facilitate the assessment of different policy, as well as technical, measures when developing abatement scenarios.

3. Conclusions

The RAINS integrated assessment model has been used with success in the negotiations leading to the signing of the Second Sulphur Protocol to the LRTAP Convention. Further development of RAINS for application to multi-effect, multi-pollutant scenario analysis is underway. This development includes the construction of a model for ozone to assist in the evaluation of effect-based strategies for the reduction of NO_x and VOC emissions.

Acknowledgements

The development of the simplified ozone formation model has been carried out in collaboration with EMEP, MSC-W, Oslo. We are very grateful to David Simpson for making the EMEP ozone model available to us, and for many helpful discussions. We are also indebted to Steffen Unger (GMD, Berlin) for converting the EMEP model to run on a parallel processing system and efficiently providing us with the model results.

References

Alcamo, J., Shaw. R. and Hordijk, L. (Eds): 1991, *The RAINS Model of Acidification*. Kluwer, Amsterdam.
Baldi, M. and Calori, G.: 1995, *Tropospheric Ozone: A data base for Europe and statistical analysis*. WP-95-xxx. International Institute for Applied Systems Analysis (IIASA), Laxenburg, Austria.
Caliandro, B.: 1994, *Control Technology Options and Costs for Reducing Volatile Organic Compounds*. WP-94-80. International Institute for Applied Systems Analysis (IIASA), Laxenburg, Austria.
Fuhrer, J. and Achermann, B. (Eds): 1994, *Critical Levels for Ozone, a UN-ECE workshop report*. Swiss Federal Research Station for Agricultural Chemistry and Environmental Hygiene, Liebefeld-Bern, Switzerland.
Heyes, C. and Schöpp, W.: 1995, *Towards a simplified model to describe ozone formation in Europe*. WP-95-34. International Institute for Applied Systems Analysis (IIASA), Laxenburg, Austria.
Kettunen, A., Schöpp, W. and Klimont, Z.: 1994, *Statistical Analysis of Tropospheric Ozone Concentrations*. WP-94-88. International Institute for Applied Systems Analysis (IIASA), Laxenburg, Austria.
Labancz, K.: 1993, *Evaluation of the EMEP MSC-W oxidant model: comparison with measurements*. In Anttila, P. (Ed.), *EMEP workshop on the control of photochemical oxidants in Europe*. Finnish Meteorological Institute, Helsinki, Finland.
Olsthoorn, X.: 1994, *Towards an Integrated Assessment Model for Tropospheric Ozone. Emission Inventories, Scenarios and Emission Control Options*. WP-94-27. International Institute for Applied Systems Analysis (IIASA), Laxenburg, Austria.
Simpson, D.: 1992a, "Long period modelling of photochemical oxidants in Europe. Calculations for July 1985", *Atmos. Environ.*, **26A**, 1609-1634.
Simpson, D.: 1992b, *Long Period Modelling of Photochemical Oxidants in Europe: A) Hydrocarbon reactivity and ozone formation in Europe; B) On the linearity of country-to-country ozone calculations in Europe*. EMEP MSC-W Note 1/92. MSC-W, Norwegian Meteorological Institute, Oslo, Norway.
Simpson, D.: 1993, "Photochemical model calculations over Europe for two extended summer periods: 1985 and 1989. Model results and comparisons with observations", *Atmos. Environ.*, **27A**, 921-943.
Simpson, D.: 1994, "Biogenic VOC in Europe. Part II: Implications for Ozone Control Strategies", *J. Geophys. Res.*, to be published.
UN/ECE: 1990, *Integrated Assessment Modelling*. EB.AIR/WG5/R.7, United Nations Economic Commission for Europe, Geneva, Switzerland.

DEVELOPING OPTIMAL ABATEMENT STRATEGIES FOR THE EFFECTS OF SULPHUR AND NITROGEN DEPOSITION AT EUROPEAN SCALE

C. A. GOUGH, M. J. CHADWICK, B. BIEWALD, J C. I. KUYLENSTIERNA, P. D. BAILEY & S. CINDERBY

Stockholm Environment Institute at York, PO BOX 373, University of York, YORK YO1 5YW, UK.

Abstract. Integrated Assessment Models were successfully used to provide input to the negotiations for the Oslo Protocol on Further Reductions of Sulphur Emissions, finalized within the United Nations Economic Commission for Europe Convention on Long-range Transboundary Air Pollution in Oslo in June 1994. The techniques developed within this framework will be extended now to the simultaneous analysis of sulphur and nitrogen deposition. In addition to acidification, atmospheric deposition of nitrogen contributes to eutrophication of certain ecosystems, through a nutrient effect, and originates from the long-range transport of emissions of both oxidised and reduced nitrogen (NO_x and NH_3). Modelling reductions in nitrogen deposition thus introduces a need to establish multi-pollutant multi-effect modelling techniques. This paper investigates the development of a model set up to examine reductions of these pollutants in an economically and environmentally efficient manner. The control of nitrogen deposition encompasses action across several economic sectors, particularly the power, transport and agricultural sectors. Combining sulphur and nitrogen deposition limits on a European scale will require a flexible modelling approach and the issues governing possible approaches are presented.

Keywords: Integrated Assessment Modelling, eutrophication, acidification.

1. Introduction

The 1988 Sofia Protocol on the Control of Emissions of Nitrogen Oxides or their Transboundary Fluxes was aimed at stabilizing emissions at their 1987 levels by 1994 and this, largely, has been achieved. The next stage in the negotiations for reductions of nitrogen deposition will address the possibility of developing a combined Protocol for reductions in sulphur and nitrogen emissions together, or a new NO_x Protocol, according to the critical loads approach and using the results of integrated assessment models (IAM). One of these models, CASM (Co-ordinated Abatement Strategies Model) (Gough *et al.*, 1994) is used in this paper to illustrate the challenges facing modellers in developing optimisation procedures for targeting sulphur and nitrogen emission reductions.

The purpose of this paper is to present some of the issues associated with IAM of sulphur and nitrogen together, for the protection of natural and semi-natural ecosystems against acidification and eutrophication. Examples of the types of inputs required are presented to illustrate the nature and challenges of this IAM activity.

2. Sources and Their Control

The CASM model requires certain input data in order to generate and evaluate economically and environmentally efficient strategies for the abatement of transboundary pollutants. Emission estimates for each source region are projected for future years, and from these estimates, the costs of applying available abatement technologies are calculated. These costs are brought together to construct a cost curve from which total annual costs of different emission reduction levels may be estimated. Emissions and cost curves are used in the model to estimate European deposition levels and the potential for their reduction using atmospheric transfer coefficients derived from EMEP (European Monitoring and Evaluation Programme) data (Eliassen *et al.*, 1988). Cost curves include technological measures and exclude structural or efficiency measures that would also lead to reduction in emissions, the modelling of such measures is treated using a scenario approach.

The transport sector is the most significant source of NO_x emissions in many European countries and transport activity and energy use is growing at a faster rate than many other sectors in the majority of European economies. A detailed analysis of present and forecast road transport activity and consideration of other mobile sources of NO_x has been carried out (Bailey, 1996). These data and a database of power stations (SEI, 1994) are used in conjunction with energy use data, derived from official energy forecasts (IEA, 1993), and emission factors (a function of the typical combustion technologies in each sector) to estimate future emissions, as a basis for the cost curve development. An example of a marginal abatement cost curve is illustrated below showing estimated costs for reductions in Sweden applied across all sectors.

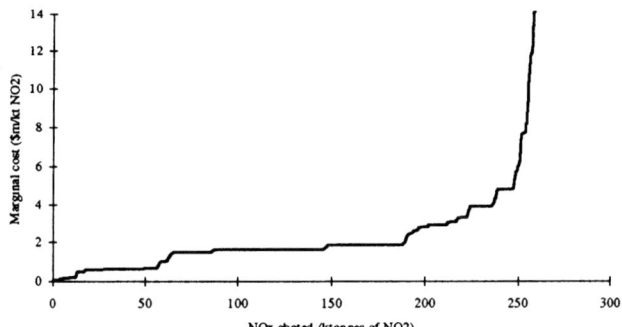

Fig.1. Cost Curve for NO_x reduction for Sweden

The above example shows the marginal costs of technologies that reduce emissions of NO_x, similar cost curves may be prepared for SO_2 and NH_3. There are certain technologies, e.g. fluidised bed combustion, coal gasification, that reduce both SO_2 and NO_x emissions from a combustion process. Representing these technologies on a single curve for each pollutant would imply that some arbitrary proportional allocation of the cost is applied for each pollutant curve, this would also lead to a loss of consistency within an IAM, where cross-checking would not be possible. Consequently, methods are being developed with CASM that allow the multi-dimensional cost curves to be represented in the model; these describe costs and reduction levels for sulphur and nitrogen in a single function, thus avoiding problems associated with those technologies that abate both pollutants.

3. The Sensitivity Of Ecosystems To The Effects Of Nitrogen Deposition

Nitrogen deposition contributes to both acidification and eutrophication of terrestrial and aquatic ecosystems. Nitrogen deposited as ammonium or nitrate will cause acidification if nitrate ions leach from the soil (Reuss and Johnscn, 1986). As nitrogen is deposited on ecosystems it enters the nitrogen cycle and may be taken up by plants or micro-organisms, be immobilized in the soil organic matter, denitrified, or leached from the system as nitrate, in some cases after mineralization and nitrification. The acidifying effect of nitrogen deposition may be quantified from estimates of nitrate leaching. Nitrogen is also an important plant nutrient, often limiting in temperate terrestrial and marine ecosystems, hence addition of nitrogen may give rise to eutrophication and associated changes in structure and function. Non-marine surface waters are mostly insensitive to nutrient effects due to increased nitrogen inputs.

There are two main approaches for the calculation of nitrogen critical loads: the mass-balance approach (Posch *et al.*, 1993) and the empirical approach; results from the latter are used here. Empirical critical loads are based upon observations of vegetation changes at known nitrogen loads (Bobbink *et al.*, 1992). These are documented for a limited number of vegetation types and thus could not be used, as yet, to provide complete cover for a European critical load map. The assessment and mapping of the relative sensitivity of vegetation communities to nitrogen are underway at the Stockholm Environment Institute (SEI), in order to apply empirical critical loads. This method is based upon the hypothesis that sensitivity is proportional to the range of nitrogen availability for a given vegetation type. The most sensitive vegetation types are those with a narrow range of availability such that small additions of nitrogen give rise to large changes in the character of the vegetation. In less sensitive vegetation types the range of nitrogen availability is larger and additions of nitrogen, whilst giving rise to changes in relative abundance of plant species, do not alter the character of the community type. A vegetation map being compiled at SEI, currently covering all European countries excluding the former Soviet Union, has been reclassified to form a preliminary map with five classes of relative sensitivity to which critical load values have been assigned, according to the classification shown in Table 1. The critical load ranges are based on compilations of empirical observations (Bobbink *et al.*, 1992). The values for exceedance calculations are set at the low end of the ranges because of uncertainties in the long term critical load value and the potential for underestimates in calculated deposition, which are partly due to differences in resolution between the receptors and EMEP grids.

TABLE I

Vegetation types assigned to sensitivity classes (in kg N ha^{-1} yr^{-1})

Sensitivity class	Vegetation types	Critical load range	Value used
1	Arable	>50	50
2	Improved grazing/ market gardening	30-80	30
3	Neutral/ acid species rich grassland/ dry unimproved grassland/ Mediterranean scrub/ deciduous forest	15-30	15
4	Heath/ acidic unimproved wet grasslands, species-rich heath/ swamp/ marshes/ conifer and mixed forest	10-20	10
5	Bogs/ Arctic heaths/ tundra/ alpine meadows/ heaths/ peatland	5-15	5

4. Nitrogen deposition and its modelling

Figure 2 shows average annual exceedance of the nitrogen critical load, for part of Europe, resulting from total nitrogen deposition (NO_x and NH_3) estimated using deposition data from EMEP (Tuovinen *et al.*, 1994) for the period 1985-93. Critical load exceedance provides an indication of where damage from the effects of excess nitrogen deposition is likely to occur and is estimated by subtracting critical loads from the deposition data. This data will be used in CASM to direct abatement strategies aimed at reducing exceedance in an efficient manner using Linear Programming optimisation techniques (Gough *et al.*, 1994).

Fig.2. Exceedance of preliminary nitrogen critical loads

These methods have been applied successfully in the analysis of SO_2 pollution - a reasonably straightforward exercise since the acidifying impact of sulphur deposition is comparatively simple to characterise. Requirements by non-crop plants for sulphur are small and sulphate adsorption capacity is generally low relative to inputs and, given the long deposition history in Europe, it may be assumed that all sulphur deposited will acidify. Therefore, sulphur deposition can be directly transformed to equivalents of acidity and simply used as input to an integrated assessment model. Modelling nitrogen inputs is considerably more challenging; the nitrogen cycle is complex and there are two effects associated with deposition of nitrogen to ecosystems: acidification and eutrophication. In order to determine the proportion of the nitrogen deposition that acidifies (N_{acid}) the proportion of nitrate that leaches from the ecosystem has to be estimated. This requires a spatially representative accounting framework that calculates the long term removal (N_r), denitrification (N_{dN}), immobilisation (N_i) in the soil, nitrogen fixation (N_{fix}) and "natural" nitrate leaching rate N_l (Kuylenstierna and Chadwick, 1992). From these values the proportion leaching (N_{acid}) may be calculated and converted to equivalents of acidity input and compared to the critical loads presented earlier, according to equation 1.

(1) $$N_{acid} = N_d + N_{fix} - N_r - N_i - N_{dN} - N_l$$

The map of exceedance of nitrogen critical loads (nutrient effects) illustrates the significance of this effect as an important consequence of nitrogen deposition as an additional effect to that of acidification. The "60% Gap Closure" targets adopted for the 1994 Oslo Protocol on Further Reductions of Sulphur Emissions imply that critical loads for acidity will not be met across Europe; the deposition targets implied a 60% reduction in the gap between 1990 deposition and the 5 percentile critical load in every receptor. Thus to achieve cost-effective solutions against the effects of nitrogen deposition, the optimisation of sulphur and nitrogen abatement together becomes important.

5. Optimisation and future directions

IAMs have been used to model optimal solutions for the reduction of exceedance of critical loads for acidity. If emission reductions to eliminate exceedance across Europe can be implemented according to the abatement cost curves, a simple solution of the most cost-effective strategy to reach critical loads can be derived. However, given the levels of exceedance, the models may be used to develop intermediate strategies that lead to a maximum reduction in exceedance for the least cost, or some similar objective. There are several ways in which the models may achieve this; for example, by adopting higher deposition targets, e.g. critical loads that protect a lower area of a receptor (higher percentiles), or by developing an approach that optimises the gap between deposition and critical loads using a technique known as goal programming. The latter approach led to efficient abatement strategies for sulphur reductions which provided minimal overall levels of exceedance for the least cost; the development of this technique for combined sulphur and nitrogen reduction strategies could be particularly valuable, exploiting increased opportunities for optimisation arising from the inclusion of an additional pollutant. A short term solution could be to consider the reduction of nitrogen deposition with sulphur emissions fixed at those levels agreed under the Oslo Protocol, in which case approaches similar to those used for sulphur could be applied. However, this should be seen as a preliminary approach and will not avoid the long term issues associated with modelling the two pollutants and effects together.

Whatever approach to optimisation is adopted a decision must be taken concerning the relative importance of avoiding each effect, i.e. a priority weighting system must be developed that reflects a particular view. A useful initial approach could be to carry out a series of model runs adopting different weights in order to investigate the characteristics and sensitivity of system and the implications of adopting particular combinations of weights.

6. Summary

In this paper some of the input parameters for an IAM of sulphur and nitrogen deposition were presented. These include marginal abatement cost curves incorporating measures fitted to mobile sources, critical loads for the effect of eutrophication and additional mapped data necessary for quantifying exceedance of critical loads for acidity due to nitrogen deposition. Further analysis is required in order to establish a model to optimise strategies against the effects of acidification and eutrophication in the near term.

References

Bailey, P. D.: 1996, this volume.
Bobbink, R., Boxman, D., Fremsted, E., Heil, G., Houdijk, A. and Roelofs, J.:1992, "Critical loads for nitrogen eutrophication of terrestrial and wetland ecosystems based upon changes in vegetation and flora", In *Critical Loads for Nitrogen* (ed. by P. Grennfelt and E. Thörnelöf), Nordic Council of Ministers, Copenhagen.
Eliassen, A., Hov, Ø., Iversen, T., Saltbones, J. and Simpson, D.: 1988 "Estimates of Airborne Transboundary Transport of Sulphur and Nitrogen over Europe", The Norwegian Meteorological Institute, MSC-W of EMEP.EMEP/MSC-W Report 1/88. Oslo.
Gough, C. A, Bailey, P. D., Biewald,B., Kuylenstierna, J. C. I. and Chadwick, M. J.: 1994, "Environmentally Targeted Objectives for Reducing Acidification in Europe", Energy Policy, **22**(12), 1055-1066.
IEA: 1993, "Coal Information 1992", International Energy Agency, OECD, Paris.
Kuylenstierna, J.C.I. and Chadwick, M.J.:1992, "The acidifying Input of Nitrogen Depositions to Ecosystems: A Preliminary Method to Allow Comparison to Critical Loads for Acidity", In *Critical Loads for Nitrogen - a Workshop Report* (ed. by P. Grennfelt and E Thörnelöf). Nordic Council of Ministers Report, Nord 1992:41.
Posch, M., Hettelingh, J.-P., Sverdrup, H.U., Bull, K and de Vries, W. :1993, "Guidelines for the computation and mapping of critical loads and exceedances of sulphur and nitrogen in Europe", in *Calculation and Mapping of Critical Loads in Europe: Status Report 1993* (ed. by Downing, R.J., Hettelingh, J.-P. and de Smet, P.A.M.), RIVM, Bilthoven, 25-38.
Reuss, J.O. and Johnson, D.W.:1986, "Acid Deposition and the Acidification of Soils and Waters", Ecological Studies **59**. Springer-Verlag. New York.
SEI: 1994, "The European Fossil-fuelled Power Station Database Used in the SEI CASM Model", Stockholm Environment Institute, Stockholm.
Touvinen, J-P., Barret, K. and Styve, H.: 1994, "Transboundary Acidifying Pollution in Europe: Calculated fields and budgets 1985-93", EMEP/MSC-W Report 1/94. Oslo

TESTING, IMPROVEMENT, AND CONFIRMATION OF A WATERSHED MODEL OF ACID-BASE CHEMISTRY

T.J. SULLIVAN[1] AND B.J. COSBY[2]

[1] E&S Environmental Chemistry, Inc., P.O. Box 609, Corvallis, OR 97339, USA
[2] Department of Environ. Sciences, Univ. of Virginia, Charlottesville, VA 22903, USA

Abstract. Strategies to control the emission of atmospheric pollutants such as sulfur and nitrogen, are generally based in large part on projections using models that simulate the influence of sulfur and/or nitrogen deposition on the acid-base chemistry of surface waters. One of the principal models used throughout Europe and North America for such assessment is the Model of Acidification of Groundwater in Catchments (MAGIC). All watershed models are simplified representations of reality, and as such require careful testing to establish their veracity prior to use for making policy projections. This is particularly true where the use of these model projections has the potential for serious environmental or economic consequences. During the past five years, we have tested the MAGIC model in a large variety of settings and under quite varying environmental conditions. This work has included comparing model hindcast simulations with diatom-inferences of historical acidification, sensitivity analyses to examine the response of the model to alternative assumptions and formulations, and detailed testing of model forecasts by comparing simulated chemistry with the results of catchment-scale and plot-scale experimental acidification and deacidification. Our analyses have elucidated a number of potentially-important deficiencies in model structure and method of application. These have resulted in changes to the model and its calibration procedures. Our work has included in-depth evaluation of issues related to regional aggregation of soils data, background sulfur deposition, natural organic acidity, and aluminum mobilization. The result has been an improved and more thoroughly-tested version of MAGIC. The process we have followed to improve and confirm the MAGIC model has been iterative and time consuming. It required the availability of large volumes of data from experimental manipulation and paleolimnological studies. We believe that such model testing and confirmation efforts should be a critical prerequisite for regional or national assessment activities that are based largely on the results of environmental models.

Key words: acidification, model testing, modeling, acid deposition

1. Introduction

Simulation models have found increased usage for projecting environmental effects of ecosystem perturbations. Such projections are frequently used as the basis, or justification, for public policy and legislation. A need has arisen to test the veracity of model projections, especially in cases where policy and/or economic interests are at stake. As Oreskes et al. (1994) pointed out, however, verification and validation of mathematical models of natural systems are impossible, because natural systems are never closed and model results are nonunique. Model confirmation is possible, and entails demonstration of agreement between prediction and observation. Such confirmation is inherently partial. It is critical that policy-relevant models be verified under a variety of conditions.

The Model of Acidification of Groundwater in Catchments (MAGIC, Cosby et al. 1985) has been widely used throughout North America and Europe to project changes in the chemistry of drainage waters impacted by atmospheric sulfur deposition. MAGIC projections of the effects on surface water chemistry of various sulfur emissions scenarios formed the technical foundation for a large part of the National Acid Precipitation Assessment Program's Integrated Assessment (IA, NAPAP 1991). We report here on the results of a five-year research effort to improve the performance of MAGIC and to provide testing and confirmation of the model. Our model evaluations have included hindcast comparisons with diatom reconstructions of pre-industrial lakewater chemistry in the Adirondack Mountains of New York, and tests of the veracity of model forecasts using the results of whole-catchment acidification experiments in

Maine (Norton et al. 1992) and Norway (Gjessing 1992) and whole catchment acid-exclusion experiments in Norway (Wright et al. 1993). The results of specific components of this research have been described in recent manuscripts (Driscoll et al. 1994, Sullivan et al. in press a, Sullivan and Cosby in review, Cosby et al. in press a,b). The purpose of this paper is to summarize the results of this integrated model testing effort to date and to describe the cumulative influence of the consequent model revisions.

2. Methods

MAGIC is a lumped-parameter model of intermediate complexity (Cosby et al. 1985) that is calibrated to the watershed of an individual lake or stream and then used to simulate the response of that system to changes in atmospheric deposition. MAGIC includes a section in which the concentration of major ions is governed by simultaneous reactions involving sulfur adsorption, cation weathering and exchange, aluminum dissolution/precipitation/speciation, and dissolution/speciation of inorganic carbon. A mass balance section of MAGIC calculates the flux of major ions to and from the soil in response to atmospheric inputs, chemical weathering inputs, net uptake in biomass and losses to runoff. Detailed testing of model performance, on which to base modifications to model structure or method of application, requires availability of substantial data bases that document the response of watersheds to changing levels of acidic deposition. Data sets that were critical to the integrated MAGIC model testing exercise described here included several studies of lakes in the Adirondack Mountains (Charles and Smol 1990) and also the results of watershed acidification and de-acidification experiments in Maine and Norway (Gjessing 1992, Norton et al. 1993, Wright et al. 1993).

The MAGIC model represents the horizontal dimension of the watershed as a homogeneous unit and the vertical dimension as one or two soil layers. Watershed and soils data required as model inputs are aggregated to provide weighted-average values for each soil layer. Within the U.S. EPA's Direct Delayed Response Project (DDRP, Church et al. 1989), which formed the technical foundation for NAPAP modeling efforts in the Northeast, soil characteristics were aggregated on the basis of attributes of soil sampling classes across the entire northeastern United States. Subsequent to the DDRP, there was concern that Adirondack soils might differ in their chemical properties from similar soils in other areas of the Northeast, and that MAGIC projections for Adirondack watersheds might be biased because they were based on soil attributes that actually reflected conditions elsewhere than the Adirondacks. The DDRP soils data were therefore reaggregated to characterize Adirondack watershed attributes using only soil data collected from pedons in the Adirondacks (Sullivan et al. 1991). Modeling for the DDRP and Integrated Assessment also assumed that the deposition of sulfur in pre-industrial times was limited to sea salt contributions. Based on analyses presented by Husar et al. (1991), this assumption was modified such that pre-industrial deposition of sulfate was assumed equal to 13% of current values (Sullivan et al. 1991).

Recalibration of MAGIC to the Adirondack lakes database using the regionally corrected soils and background SO_4^{2-} data resulted in approximately 10 μeq L^{-1} lower estimates of current ANC. A substantial shift was also observed in predicted pre-industrial and current lakewater pH (~0.25 pH units) for lakes having pH greater than about 5.5. This shift was attributed to the higher pCO_2 values estimated for Adirondack lakes, compared with the Northeast as a whole (Sullivan et al. 1991).

Concern was raised subsequent to the IA regarding potential bias from the failure to include organic acids in the MAGIC model formulations used by NAPAP. MAGIC hindcasts of pre-industrial lakewater pH showed poor agreement with diatom-inferences of pre-industrial pH. Sullivan et al. (in press a) performed revised MAGIC simulations for these Adirondack lakes that included the triprotic organic acid analog model of Driscoll et al. (1994). The fitted pK_a values were 2.62, 5.66, and 5.94, and the calibrated site density was 0.055 mol sites per mol C. The revised MAGIC hindcasts of pre-industrial lakewater pH showed considerably closer agreement with diatom inferences. The mean difference was reduced from 0.6 pH units to 0.2 pH units, and individual lakes improved by a full pH unit. The importance of organic acids in achieving reliable model results increased with increasing concentration of dissolved organic carbon (DOC). All study lakes for which estimates of ΔpH (diatom-inferred current pH minus pre-industrial pH) decreased by more than 0.5 pH units, upon inclusion of organic acids in the model, had DOC > 400 μM. At the Lake Skjervatjern and Risdalsheia sites, both of which have DOC well above 400 μM, the inclusion of organic acids in the modeling efforts had a large effect (0.2 to 0.5 pH units) on model predictions of pH (Sullivan in press).

MAGIC simulates Al solubility based on an assumed gibbsite equilibrium:

$$Al(OH)_3 (s) + 3H^+ \rightleftharpoons Al^{3+} + 3 H_2O \quad (1)$$

whereby the relationship between Al^{3+} and H^+ activities is cubic and is determined by the assumed solubility product (K_{SO}) for a form of gibbsite:

$$[Al^{3+}]/[H^+]^3 = K_{SO} \quad (2)$$

The model calculates the concentration of acidic cations on the basis of simulated concentrations of base cations and mineral acid anions using mass balance and electroneutrality constraints. The acidic cations are partitioned between H^+ and Al^{n+} using the gibbsite equation, thermodynamic equations describing the speciation of dissolved aluminum, the partial pressure of CO_2, and the triprotic organic acid analog formulation. This partitioning is important because inorganic Al in solution can be toxic to aquatic biota, even at low concentrations (Baker and Schofield 1982). Surface waters and many soil waters are generally undersaturated with respect to gibbsite, however (Seip et al. 1989), and model simulations often overpredict the change in Al^{3+} concentration (Sullivan et al. in press b). Sullivan and Cosby (in review) proposed modifying the Al formulation in the MAGIC model to better reflect empirical relationships between Al^{3+} and H^+. A power term of 2, rather than 3, was found to represent more satisfactorily the Al^{3+}/H^+ relationship in Equation 2 for low-pH lakes and streams in a variety of databases. The revised formulation (expressed as $[Al^{3+}]/[H^+]^2 = K_{SO}$) was used to predict Al^{3+} concentrations in runoff at Risdalsheia and Bear Brook, and in both cases yielded closer agreement with measured values than had the original MAGIC predictions (Sullivan and Cosby in review).

3. Results and Discussion

In order to evaluate the incremental and cumulative impact of the model modifications, a suite of model simulations was conducted for the Adirondack DDRP lakes. The changes are considered cumulatively. The baseline structure (structure 0) was that used in the DDRP and NAPAP IA studies. The change from structure 0 to structure 1 was accomplished by modifying the assumption regarding background sulfur deposition (Husar et al. 1991) and reaggregating the soils data and recalibrating the model specifically for the Adirondack subregion, as opposed to using values for the northeastern United States as a whole (e.g., Sullivan et al. 1991). Structure 2 was derived from Structure 1 by adding the triprotic organic acid analog model to the surface water compartment (using parameters calibrated for the Adirondacks; e.g., Driscoll et al. 1994). Structure 3 was derived from Structure 2, with the aluminum/ hydrogen ion relationship changed from cubic to quadratic and re-calibrated for the Adirondack lakes (Sullivan and Cosby, in review).

The suite of simulations was based on the application of an assumed deposition scenario to derive a 50 year forecast using each model structure. The deposition scenario assumed constant sulfur deposition from 1984 (the calibration year) to 1994, followed by a 30% decrease in sulfur deposition from 1995 to 2009, with constant deposition thereafter until 2034. The modeled responses of thirty three Adirondack lakes to this deposition scenario were considered. The impacts of the changes are illustrated by tabulating the percentage of lakes predicted to have pH, ANC or Al values in excess of commonly accepted thresholds of potential biological effects (Table 1).

The overall effect of the various changes to the model structure and application procedures was an increase in the percentage of lakes exceeding various biological thresholds with respect to pH, Al, and ANC subsequent to an hypothesized 30% decrease in sulfur deposition. The largest changes were observed for pH and Al; ANC projections were less affected (Table 1). The modifications to the model that caused the greatest changes in projected output were the recalibration of the model to the Adirondack subregion and modification of the assumption regarding background SO_4^{2-} (Structure 0 to Structure 1) and the incorporation of the triprotic organic acid model into MAGIC (Structure 1 to Structure 2). The modification of the Al algorithm (Structure 2 to Structure 3) caused fewer lakes to be projected to exceed Al threshold values in response to the reduced deposition scenario; this change was quantitatively less important than the previous changes.

The magnitude of effect of the cumulative modifications to the model was considerable. Thirty-two percent of the lakes had measured pH less than 5.5 in 1984, whereas only 8% were projected to still have pH less than 5.5 after the reduction in sulfur deposition, using the original MAGIC application (Structure 0). In contrast, Structure 3 (the improved version of MAGIC) projected that 32% of lakes would still have pH less than 5.5 in the year 2034. Similarly, of the 30% with measured $Al_i > 50$ μg L^{-1} in 1986, the original model structure projected only 4% would still have $Al_i > 50$ μg L^{-1} in 2034 compared to 30% projected to continue to have high Al_i by Structure 3. Based on model projections using the improved version of MAGIC (Table 1), little recovery of Adirondack lakes would be expected subsequent to a 30% reduction in sulfur deposition. The number of lakes having pH < 6 is actually projected to increase, and the number of lakes projected to have ANC < 0 only decreases slightly in response to lower deposition. These estimates are independent of any possible increases in NO_3^- leaching that might occur. The lack of recovery suggested by these model projections is attributable partly to a decrease in the modeled base saturation of watershed soils. These results may affect expectations of recovery in response to sulfur emission controls mandated by the Clean Air Act Amendments of 1990.

MAGIC and other models of acid-base chemistry have generally not considered, or only considered in a cursory fashion, the effects of changes in land use on the chemistry of drainage waters. It has become clear in recent years that landscape processes can affect acid-base chemistry in a variety of ways. Some processes increase the pH and ANC of surface waters, whereas others contribute to acidification or reduce the base saturation of the soils, thereby

Table 1. MAGIC predictions of the percentage of Adirondack DDRP lakes (Ñ = 610) having pH, ANC, and Al above or below threshold values for the unmodified MAGIC performed by NAPAP (1991) and the revised model procedures described here.

	Percentage of Lakes Having pH Below Value			Percentage of Lakes Having ANC Below Value ($\mu eq\ L^{-1}$)			Percentage of Lakes Having Al Above Value ($\mu g\ L^{-1}$)		
Data Type	5	5.5	6	0	25	50	50	100	200
Model Application[a]									
Structure 0	0	8	20	6	34	44	4	0	0
Structure 1	4	18	40	14	40	44	10	2	0
Structure 2	4	30	42	14	40	44	34	18	10
Structure 3	8	32	44	14	40	44	30	10	4
Measured 1984 Values	12	32	38	18	48	59	30	18	10

[a] Four sets of model scenarios were conducted as follows: Structure 0 - unaltered from DDRP and NAPAP; Structure 1 - revised assumed background SO_4^{2-} in lakewater and soils aggregation to Adirondack subregion; Structure 2 - added organic acid representation to Structure 1; Structure 3 - added revised Al algorithm

increasing their sensitivity to acidic deposition (Sullivan et al. in review). One important feature of landscape change that can alter the acid-base chemistry of drainage waters is forest growth. Removal or planting of a forest can affect hydrology, the deposition of sulfur, nitrogen, and marine salts, nitrogen cycling, and the uptake of base cations to support tree growth. On-going research is investigating the role of forest growth processes in the acidification of Adirondack lakes. It is hoped that sufficient data will become available with which to incorporate into MAGIC base cation uptake by the forest in a more quantitative fashion. In addition, information on the dynamics of forest growth will be needed to parameterize models to simulate nitrogen release in Adirondack watersheds.

MAGIC contains an extremely simplified representation of nitrogen dynamics within catchment soils. The model simulates net nitrogen retention as a linear process. That is, retention of either NO_3^- or NH_4^+ is assumed to be linearly proportional to the input fluxes of these ions. There are no processes controlling the details of N cycling in the model. In light of the increasing concern about N saturation in forested ecosystems, this is a serious shortcoming in the model, and is currently being addressed by considering a number of modifications for the model.

4. Conclusions

The research described above resulted in a number of changes to the structure of the MAGIC model and the manner in which the model is recommended to be applied. Each modification resulted in varying degrees of change to model predictions of the response of key variables to acidic deposition inputs (Table 1). The largest effects resulted from reaggregation and recalibration of the model to sub-regional datasets and the inclusion of organic acids; other modifications have been less important to model output. The various model modifications have affected key variables in different ways, but the cumulative result has been that the revised model predicts smaller acidification responses and smaller recovery responses than did earlier applications of MAGIC, at least as applied to Adirondack lakes.

References

Baker, J.P. and Schofield, C.L.: 1982. *Water Air Soil Pollut.* **18**, 289-309.
Charles, D.F. and Smol, J.P.: 1990. *Verh. Internat. Verein. Limnol.* **24**, 474-480.
Church, M.R., Thorton, K.W., Shaffer, P.W., Stevens, D.L., Rochelle, B.P., Holdren, R.G. Johnson, M.G., Lee, J.J., Turner, R.S., Cassell, D.L., Lammers, D.A., Campbell, W.G., Liff, C.I., Brandt, C.C., Liegel, L.H., Bishop, G.D., Mortenson, D.C. and Pierson, S.M.: 1989. *Future Effects of Long-Term Sulfur Deposition on Surface Water Chemistry in the Northeast and Southern Blue Ridge Province* (Results of the Direct/Delayed Response Project). U.S. EPA, Environmental Research Lab, Corvallis, OR.
Cosby, B.J., Norton, S.A., and Kahl, J.S.: In press a. *Using a paired-watershed manipulation experiment to evaluate a catchment-scale biogeochemical model.*
Cosby, B.J., Wright, R.F., and Gjessing, E.: In press b. *An acidification model (MAGIC) with organic acids evaluated using whole-catchment manipulations in Norway. J. Hydrol.*
Cosby, B.J., Wright, R.F., Hornberger, G.M., and Galloway, J.N.: 1985. *Water Resour. Res.* **21**, 51-63.
Driscoll, C.T., Lehtinen, M.D. and Sullivan, T.J.: 1994. *Water Resour. Res.* **30**, 297-306.
Gjessing, E. 1992. *Environ. Internat.* **18**, 535-543.
Husar, R.B., Sullivan, T.J., and Charles, D.F.: 1991. In: Charles, D.F. (ed.). *Acidic Deposition and Aquatic Ecosystems. Regional Case Studies.* Springer-Verlag, New York. pp. 65-82.
NAPAP.: 1991. *National Acid Precipitation Assessment Program 1990 Integrated Assessment Report.* National Acid Precipitation Assessment Program, Washington, DC. 520 pp.
Norton, S.A., Kahl, J.S., Fernandez, I.J., Schofield, J.P. Rustad, L.E., Haines, T.A., and Lee, J.: 1993. In: Rasmussen, L., Brydges, T., and Mathy, P. (Eds.). *Experimental Manipulations of Biota and Biogeochemical Cycling in Ecosystems.* ECSC-EEC-EAEC, Brussels. pp. 55-63.
Norton, S.A., Wright, R.F., Kahl, J.S., and Scofield, J.P.: 1992. *Environ. Pollut.* **77**, 279-286.
Oreskes, N., Shrader-Frechette, K., and Belitz, K.: 1994. *Science* **263**, 641-646.
Seip, H.M., Anderson, D.O., Christophersen, N., Sullivan, T.J., and Vogt, R.D.: 1989. *J. Hydrol.* **108**, 387-405.
Sullivan, T.J.: In press. *Evaluation of biogeochemical models using data from experimental ecosystem manipulations.* Proc. Third Intern. Symp. Ecosys. Manipulation.
Sullivan, T.J. and Cosby, B.J.: In review. *Modeling the concentration of aluminum in surface waters.*
Sullivan, T.J., McMartin, B., and Charles, D.F.: In review. *Re-examination of the Role of Landscape Change in the Acidification of Lakes in the Adirondack Mountains, New York. Sci. Tot. Environ.*
Sullivan, T.J., Cosby, B.J., Driscoll, C.T., Charles, D.F. and Hemond, H.F.: In press a. *Influence of organic acids on model projections of lake acidification. Water Air and Soil Pollution.*
Sullivan, T.J., Cosby, B.J., Norton, S.A., Charles, D.F., Wright, R.F., and Gjessing, E.: In press b. *Multisite testing and evaluation of a geochemical model of acid-base chemistry: Confirmation of the MAGIC model using catchment manipulation experiments and historical diatom inferences.* Proc. Third International Symposium on Ecosystem Manipulation.
Sullivan, T.J., Cosby, B.J., Driscoll, C.T., Hemond, H.F., Charles, D.F., Norton, S.A., Seip, H.M., and Taugból, G.: 1994. *Confirmation of the MAGIC Model using independent data: influence of organic acids on model estimates of lakewater acidification.* Report DOE/ER/30196-4. U.S. Department of Energy.
Sullivan, T.J., Bernert, J.A., Jenne, E.A., M. Eilers, J.M., Cosby, B.J., Charles, D.F., and Selle, A.R.: 1991. *Comparison of MAGIC and diatom paleolimnological model hindcasts of lakewater acidification in the Adirondack region of New York.* Prepared for the U.S. Department of Energy under Contract DE-AC06-76RLO 1830. Pacific Northwest Laboratory, Richland, WA.
Wright, R.F., Lotse, E., and Semb, A.: 1993. *Can. J. Fish. Aquat. Sci.* **50**, 258-268.

A COMPARISON OF DIFFERENT RANKING SCHEMES FOR ASSESSING THE EFFECT OF LARGE POINT SOURCES OF POLLUTION IN THE UK

R.A. WADSWORTH[1], M.J. BROWN[1], K.R. BULL[1], S.E. METCALFE[2], D. WHYATT[3] and C. POWLESLAND[4]

Institute of Terrestrial Ecology, Monks Wood, Abbots Ripton, Huntingdon, Cambridgeshire, PE17 2LS (Contact address), [2] Department of Geography, University Edinburgh, [3]School of Geography & Earth Resources, University of Hull, Hull, HU6 7RX, [4]Her Majesty's Inspectorate of Pollution, 2 Marsham Street, London, SW1P 3EB

Abstract. The availability of national maps of critical loads for soils, vegetation and freshwaters helps enable the assessment of the effects of large point sources of pollution in the UK. The deposition "footprint" of most major sources has been modelled and combined in a GIS with a national critical loads database. As part of an integrated pollution control strategy (IPC) it may be helpful to rank point sources in order of their effects on the environment. A comparison of the discriminating power and effectiveness of several ranking schemes has been carried out. A variety of ranking schemes were investigated, such as; total area where sulphur deposition exceeds the critical load or average mass deposited on areas where the critical load is exceeded. Their relative merits were compared for several "current" and future scenarios, such as, actual 1993 emissions or predicted emissions for 2001. Rankings for the unit emissions provided a measure of the pollution potential of each source and were a complex function of the location of sensitive areas and meteorological conditions. Rankings under other scenarios tended to be dominated by the relative magnitude of the emissions. Comparison between the ranking schemes was made using non-parametric statistics. The comparisons reveal complex interactions between different schemes. The approach is providing practical solutions to a pollution control strategy based on maximising environmental benefits.

Key words: Ranking, integrated pollution control, critical loads, sulphur, point source emissions

1. Introduction

Since the introduction of the Second Sulphur Protocol (Oslo 1994) there has been increasing pressure to reduce the emissions of sulphur oxides into the atmosphere. Coal and oil fired power stations are major sources of sulphur emissions. As part of the integrated pollution control (IPC) strategy carried out by Her Majesty's Inspectorate of Pollution (HMIP) an assessment of every major source was needed. National maps of critical loads for soil, vegetation and freshwaters can be used to help assess the significance of pollution from different sources. There are numerous criteria which could be used to assess the comparative effect of each source, but what was not known was how sensitive ranking sources might be to the choice of criteria or operating scenario. Similarities and differences in rankings may have a significant effect on the perceived importance of individual sources. Information on the rank of different sources is supplied to HMIP as part of the environmental information they use in their assessments. In this paper a number of different ranking procedures are considered for several different emission scenarios. The similarity in ranks is considered using long and short range transport models and for the short range model long and short time periods.

2. Method

The fate of emissions from each power station was calculated using two models: a long range, long duration, deposition model and a short range, variable duration, atmospheric concentration model. The long range model, the Hull Acid Rain Model (HARM) (Metcalfe et al 1995) provides average deposition in each 20km by 20km square across the UK. The short range model, PLUMES was written specifically for HMIP to provide an interface to a three dimensional Gaussian plume model based on the "R91" model (NRPB 1979). PLUMES provided atmospheric concentrations at a resolution of one kilometre for an area 80 by 80km an annual concentration field is calculated using the actual emission rate and the hourly concentration field is calculated using the maximum emission rate. Topographic or land cover influences are not modelled, the differences between sources which can be modelled are: height of chimney, temperature, volume and speed of emissions and the local weather pattern.

Footprints and concentration fields for the sources from both models were integrated with the unmodified critical loads for acidity of soils of the UK (Hornung et al 1995) within a GIS. The analysis reported in this paper concentrates on the use of the soils critical loads as the most well established critical loads map. The similarity between the different critical loads maps means this choice is not likely to be significant. Soils critical loads, derived from 1km soils data, have been used for policy discussions at national and international levels. Subtracting the critical load from the acid deposition indicates whether the critical load is exceeded and damage may occur. The excess over the critical load is termed the exceedance. The GIS enables statistics of areas exceeded, amount of sulphur deposited and degree of exceedance to be calculated for each source with and without the contribution from non-power station sources. Within each grid cell the deposition or concentration is assumed to be uniformly distributed.

Several criteria to assess the effect of individual sources were devised, together with a range of operating scenarios. For each combination of criteria and scenario the power stations were put into descending rank order. The similarity of the ranks was assessed using Spearmans Rank Correlation Coefficient.

2.1 THE CRITERIA USED FOR RANKING

There are many criteria which could be used to rank the sources. For HARM the criteria selected use the soils critical loads map. It is assumed that the significance of the deposition is dependant on the sensitivity of the receptor; deposition to insensitive areas is relatively unimportant. Criteria used to assess the significance of the deposition fall into three main types: the area where the critical load is exceeded, the amount deposited on exceeded areas and the degree of exceedance. The criteria used for each source with the HARM were:
1. area where the critical load is exceeded with only that station;
2. area where the critical load is exceeded with that station plus the deposition contributed from non-power station sources;

3. average mass of sulphur per unit area deposited on areas where the critical load is exceeded by current sulphur deposition levels;
4. average mass of sulphur per unit area deposited on areas where the critical load is exceeded under the "70% reduction" scenario (the scenario assumes total emissions of sulphur will be reduced by 70% of their 1980 emissions);
5. the ratio of the deposition divided by the critical load is summed for all squares in the UK;
6. the ratio of the deposition divided by the critical load is summed for areas where the critical load is currently exceeded;
7. the ratio of the deposition divided by the critical load is summed for areas where the critical load will be exceeded under the "70% reduction" scenario.

Criteria 1 and 2 identify those sources large enough to cause deposition which exceeds the critical load. Criteria 3 and 4 provide measures of the amount of pollutant deposited on sensitive areas. Criteria 5, 6 and 7 provide measures of the significance of the deposition, high values indicate high deposition on areas with low critical loads (the most sensitive areas). The 70% reduction scenario is used to emphasis those areas at risk in the future.

PLUMES does not generate a deposition footprint so the criteria used for HARM cannot be applied. For PLUMES only the peak concentration in air is used to rank the sources. Over a small area there is no equivalent concentration field to the low level and European deposition fields used with the long range model.

2.2 SCENARIOS INVESTIGATED

The five scenarios investigated for HARM were: current emission rates, emission rates predicted for the year 2001, authorised emission rates, emission rates under the new plant standard and a unit emission rate. The current emission rates are those calculated from the returns of the generators for 1993. Predicted emission rates for 2001 are currently subject to discussion between the generators and HMIP. In cases where the 2001 rate would exceed the current authorised load, the authorised load is used instead. The authorised load is imposed by HMIP and provides a legal rather than a physical upper limit to emissions. The new plant standard is the emission rate which is equivalent to emission standards for new plant as set out in the Large Combustion Plant Directive (88/609/EEC). The "unit" of emission was set at 10000 tonnes per year to reduce the degree of scaling error within the GIS. The use of a unit emission allows the importance of geographic location to be examined. The problems of attributing the source of deposition for sensitive areas are discussed in Brown et al (1995).

2.3 STATISTICS USED

By ranking sources it is possible to set priorities and optimise benefits, providing of course that the rankings represent real differences and are not an artifact of the technique used. By comparing ranks produced by different schemes it is possible to identify consistencies and anomalies between them and gain confidence in their use.

One of the most widely used statistics for comparing ranks is the Spearman Rank Correlation Coefficient. This measure has the advantage that it is non-parametric and is independent of the assumptions about the underlying distribution of the data used in parametric comparisons such as Pearsons Correlation Coefficient (Canavos 1984). It has the additional advantage that the coefficient is not restricted to revealing only linear relationships.

Spearmans rank correlation coefficient is bounded between plus and minus one. Values sufficiently close to one (or minus one) indicate that the series are monotonically increasing (or decreasing) together. The significance of the value of r_s can be obtained from standard statistical texts.

3. Results

Table I compares the rankings produced by HARM (for the 1993 emission scenario) with those produced by PLUMES for long and short duration simulations. With 24 sources the absolute value of r_s is significant if greater than 0.343 at the 0.05 confidence level and 0.485 at the 0.01 confidence level. Of the thirty six unique combinations twenty one are positively correlated at the 0.01 level of significance, seven are positively correlated at the 0.05 level and the remaining eight are positive but not significant. Seven of the non-significant values involve the annual rankings from PLUMES.

The long range model with seven criteria and five scenarios produces 35 possible rankings, space does not permit the publishing of all the ranks. Spearmans correlation was calculated for all 595 unique combinations, 415 (69.7%) yielded a significant (at 0.05 or better) positive correlation, 68 yielded no result because there were too many tied ranks and 112 produced a value which was not statistically significant. The majority of "problem" results are associated with the fifth scenario (unit emissions). Ninety eight of the non-significant comparisons and all of the not calculable comparisons involve the fifth scenario. Of the 378 unique comparisons involving the first four scenarios, 357 (94%) were significantly positively correlated at the 0.01 level of confidence or better, six were significantly positively correlated at the 0.05 level, twelve were positive but not significantly, two comparisons yielded a negative correlation one of which was statistically significant.

For HARM the ranking schemes are remarkably insensitive to the criteria used to determine their ranks. One reason for the stability of the rankings must be the considerable range in the size of the sources, for the new plant standard scenario the largest source was nearly nine times larger than the smallest, in the actual emission scenario the largest source was 43 times larger than the smallest. There are two exceptions to the consistency of these results. The first exception comes from the use of a "unit" emission rate. As might be expected for this scenario the geographic location of the source has a much more significant role than with any of the other scenarios. Rankings within the unit emission scenario are consistent for all criteria but are inconsistent with the other scenarios. This inconsistency is encouraging as it demonstrates that generally the largest and most active sources are located so that deposition occurs away from the most sensitive areas. The second exception to consistency is for the first and second criteria where the large number of sources with a zero or very low exceeded area prevent proper analysis.

For PLUMES the two rankings are consistent and statistically significant. Comparisons between the long and short range models exhibit a certain degree of correspondence. Models of the long range dispersion of pollutants are remarkably difficult to validate, while the comparison with a short range model is not offered as validation inconsistent results between the two models would be a cause for further investigation.

TABLE I

Comparison of ranks using Spearmans r_s; ranks produced of data from long range transport model (HARM) and short range transport model (PLUMES).

	HARM						PLUMES	
	criterion 2	criterion 3	criterion 4	criterion 5	criterion 6	criterion 7	annual	hourly
criterion 1	0.694	0.703	0.694	0.711	0.697	0.698	0.092	0.201
criterion 2		0.961	0.958	0.959	0.958	0.960	0.329	**0.351**
criterion 3			0.986	0.997	0.998	0.990	0.300	**0.343**
criterion 4				0.991	0.992	0.995	0.278	**0.373**
criterion 5					0.997	0.993	0.315	**0.367**
criterion 6						0.994	0.299	**0.361**
criterion 7							0.299	**0.377**
annual								**0.355**

Notes
see text for explanation of criterion
HARM simulations used actual emission rates
PLUMES simulations used actual emissions for ranking annual maxima and maximum emission rate for ranking maximum hourly concentration
Numbers in bold significant at 0.05

4. Conclusions

The results of this investigation are encouraging, if different criterion led to wildly different rankings it would be difficult to justify any of them. Information on the relative importance of different sources, based on their rank, has been supplied to HMIP for use in their assessments and decision making. Within the assessment process other environmental information and information on technology and economics are also considered; the ranks provide one way of determining what is environmentally desirable. The work described in this paper has demonstrated the robustness of the ranking process. With any scenario based on actual emissions or proposed emissions the relative importance of the different sources varies very little. With the unit emission scenario the importance of the geographic location of the site is revealed. If the siting of new source were an issue then ranking locations on the basis of a unit of emission might be of some importance. It is possible that the effect of the location of a unit source could eventually be extended to produce a map which shows the sensitivity of source locations. Such a map would provide a companion to the existing maps showing the sensitivity of receiving locations.

References

Brown M., Dyke H., Wright S.M., Wadsworth R.A., Bull K.R., Farmer A., Barham S., Metcalfe S.E., Whyatt D. & Powlesland C.: 1995., Estimating the impact of air pollution on environmentally valuable sites. paper submitted to Acid Reign '95

Canavos G.C.: 1984 Applied probability and statistical methods, Little, Brown & Co. Boston, Toronto

Hornung M., Bull K.R., Hall J.R., Langan S.J., Loveland P. & Smith C.: 1995, An empirical map of critical loads of acidity for soils of Great Britain, Accepted for Environmental Pollution

NRPB: (National Radiological Protection Board) 1979 The first report of a working group on atmospheric dispersion: a model for short and medium range dispersion of radionuclides released to the atmosphere.

Metcalfe S.E., Whyatt J.D.. & Derwent R.G.: 1995, A comparison of model and observed network estimates of sulphur deposition across the United Kingdom for 1990 and its likely source attribution. Quarterly Jo of the Royal Meteorological Society. in press.

SPATIAL VARIABILITY IN EMISSIONS REDUCTION STRATEGIES FOR SULPHUR AND NITROGEN IN THE UK

S.E.METCALFE[1], J.D.WHYATT[2], R.G.DERWENT[3], K.BULL[4] AND H.DYKE[4]

[1] *Department of Geography, University of Edinburgh EH8 9XR, UK,* [2] *School of Geography and Earth Resources, University of Hull, Hull HU6 7RX, UK,* [3] *Meteorological Office, London Road, Bracknell RG12 2SZ, UK,* [4] *I.T.E. Monks Wood, Abbots Ripton, Huntingdon PE17 2LS, UK*

Abstract. The roles of sulphur (S) and nitrogen (N) in causing critical loads exceedance across the UK show considerable spatial variability at the present time. Over much of lowland Britain it appears that the environment can only be protected by reducing N deposition, whilst in upland areas (e.g. most of Scotland and Wales) reductions in S deposition are the primary requirement. Using the Hull Acid Rain Model (HARM) the effects of current and possible future emissions control legislation on critical loads exceedance can be explored. Based on HARM output, the implementation of the UNECE Sulphur Protocol (1994) will bring about a substantial reduction in the amount of S being deposited in the UK, especially in central and southern parts of the country. Some areas will remain where additional reductions in S are required. Over most of the country, however, the need to reduce N deposition will become paramount. The changing contributions and significance of non-UK sources can be estimated.

Key words. sulphur, nitrogen, critical loads function, emissions reductions

1. Introduction

The implementation of the UNECE Second Sulphur Protocol, signed in Oslo in 1994, is expected to bring about a significant decrease in the amount of atmospheric sulphur (S) being emitted and deposited across Europe. However, in order to minimise the environmental damage resulting from total acid deposition, the effects of total nitrogen (N) deposition must also be considered. As sulphur dioxide (SO_2) emissions decrease, so the relative importance of N will increase (UKRGIAN, 1994).

The critical loads (CL) approach forms the basis for emissions abatement policies within the UK (CLAG, 1994) and the UNECE (UNECE, 1994) and critical loads functions may be defined which take into account both nutrient N and the critical load of acidity (S + N) that are believed to provide reasonable estimates of damage thresholds for particular ecosystems (Posch *et al.*, 1993). The success of current and future emissions control legislation will be gauged by its ability to reduce critical loads exceedance against both these measures. In this paper we use the Hull Acid Rain Model (HARM) to explore the effects of the UNECE S protocol and possible reductions in emissions of oxides of nitrogen (NO_x) on critical loads exceedance across the UK.

2. The Hull Acid Rain Model

HARM is a receptor orientated Lagrangian statistical model which describes the behaviour of air parcels which arrive at locations across the UK. The model is driven by emissions of SO_2, NO_x, NH_3 and HCl from both anthropogenic and natural sources. Emissions are held for the EMEP area in grids of 150 x 150 km and for the UK in 20 km grids; they are disaggregated by source type. The disaggregation of emissions allows the relative importance of different sources to be assessed (source attribution).

Details of the S chemistry in the model and of its validation against observations from the UK's monitoring networks are given in Metcalfe *et al.* (in press). The nitrogen chemistry is still largely as described in Metcalfe *et al.* (1989) although the version used in this paper (HARM10.4) incorporates some changes to improve the representation of both dry and wet removal. The chemistry of the different species in HARM is usually coupled. In this paper, however, we have removed the usual coupling between S and NH to simplify the description of the change in the CL exceedance maps (see below).

Deposition fields from three model runs have been used (Figure 1). The three runs are: a) full emissions based on the latest emissions inventories available to us; b) a S protocol run which assumes that the emissions reductions agreed in Oslo are fulfilled and c) a multi-pollutant run which reduces both SO_2 and NO_x emissions. Under the S protocol, UK emissions of SO_2 are reduced by 71.4% overall and EMEP emissions by 40.5% relative to full emissions. The same S reductions apply under the multi-pollutant scenario, but in addition NO_x emissions are reduced by 49% for the UK and 30% for the EMEP area. Under the scenarios, total S deposition to the UK is reduced by 60% and total N deposition by 20%. The contribution of EMEP area anthropogenic sources is predicted to rise.

Figure 1
Total S and N deposition modelled by HARM 10.4 for a) "current" emissions,
b) S protocol emissions and c) S protocol and NO_x emission reductions

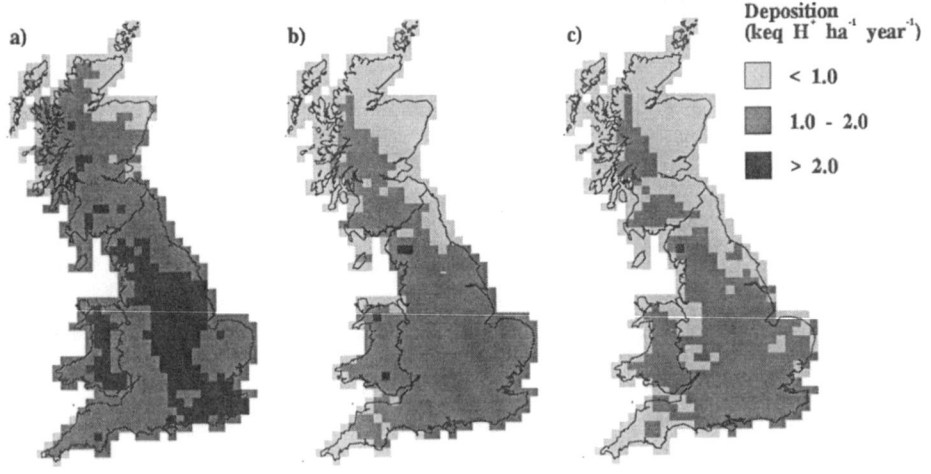

3. The Critical Loads Function

The critical loads function provides a means of estimating in graphical terms the critical loads for acidity and nutrient nitrogen in relation to deposition of sulphur and nitrogen. The use of the function is described in detail in Bull *et al.* (1995) in relation to the inclusion of nutrient nitrogen critical loads in a strategy for future abatement. Calculation of the parameters necessary for using the critical loads function are detailed in Posch *et al.* (1993). The critical loads function defines those combinations of sulphur and nitrogen which will or will not cause exceedance of the critical load. It distinguishes between areas of the graph (Figure 2) where the acidity and/or nutrient nitrogen critical loads are exceeded and where neither is exceeded - the "envelope of protection" (Bull *et al.* 1995). Table 1 summarises the reductions of nitrogen and/or sulphur necessary for ecosystem protection to be achieved.

Table I
Deposition reduction requirements for different regions as estimated by the critical loads function

Deposition reductions required

1 Only S
2 S reduction required before option of S or N
3 Both S and N must be reduced before options available
4 N reduction required before option of S or N
5 Only N
6 Either S or N
7 None - ecosystem protection

Figure 2
The critical loads function for sulphur and nitrogen deposition, incorporating acidity and nutrient loads, showing the different regions for control options

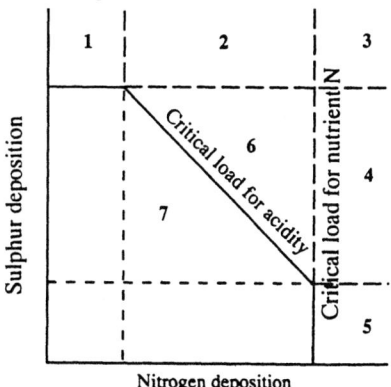

In this paper the critical loads function is applied at a 20 km scale corresponding to the resolution of the HARM modelled deposition data. Five percentile (20 km) critical loads for each part of the function are calculated separately for GB on an ecosystem-area basis (see Bull *et al.* 1995 for further details). These are combined to estimate the function for a particular soil-vegetation type, e.g. heathland. The effect of future changes in deposition of S and N can easily be investigated by recalculating the functions for different scenarios.

4. Critical Loads Exceedance

Exceedance maps are shown for two receptor ecosystems: heathland, typified by the occurrence of species such as *Calluna vulgaris*, and woodland (Norway spruce). Under full emissions, the exceedance map for heathland (Figure 3) indicates that 399 out of 740 20 km squares are not exceeded (category 7). Of the remainder a majority appear to require reductions in S deposition (primarily category 2) for protection. There are also large areas of northern England, southern Scotland, upland Wales and southern England in category 3 (reductions in S and N). Only 41 squares fall into categories 4 and 5 where a decrease in N deposition is the primary requirement.

Figure 3

Exceedances for a) current, b) S protocol and c) additional N reduction scenarios as defined by the critical loads function for heathland

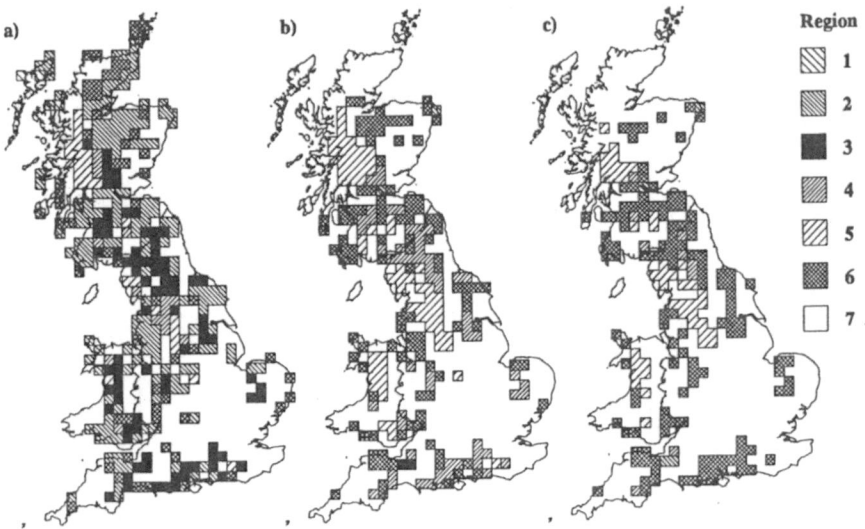

Exceedances for a) current, b) S protocol and c) additional N reduction scenarios as defined by the critical loads function for woodland

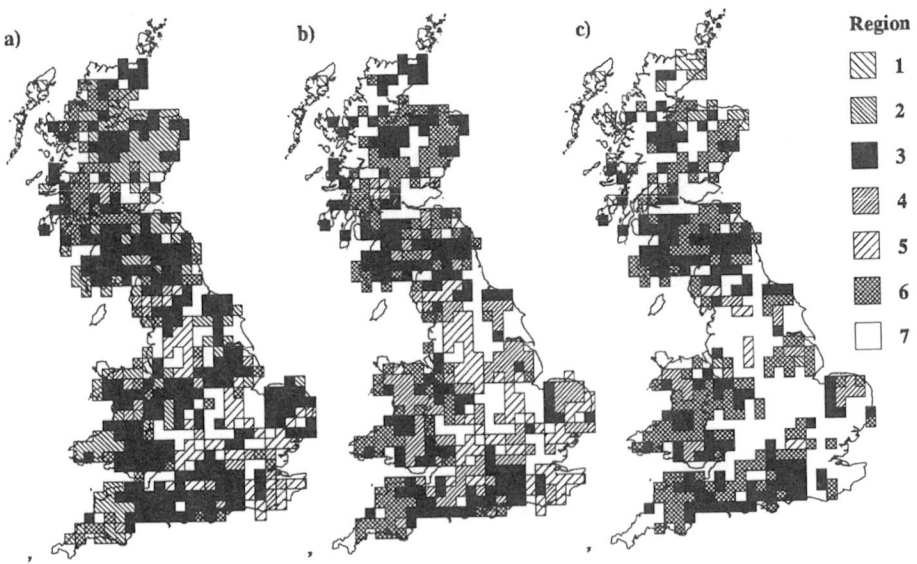

With the implementation of the S protocol, the number of protected squares rises to 507. Only 2 squares remain in category 2, with the majority requiring a reduction in N (122 squares in categories 4 and 5). There is also an increase in the number of squares in category 6. The implementation of NO_x reductions again increases the protected area (558 squares), but exceedance remains widely distributed across Great Britain. Areas of upland Wales, the Pennines, Cumbria and western Scotland fall into category 5 (protected only by reducing N), but the majority of exceedance is assigned to category 6. Although no indication of the magnitude of exceedance is given by these maps, it seems likely that category 6 areas are close to being protected.

The picture for woodland (Figure 4) is, as expected, more dramatic with larger areas of exceedance persisting under all scenarios. Under full emissions, only 224 out of 740 grid squares are protected and this only increases to 391 under the combined S + N reduction scenario. The full emissions map shows a majority of squares (259) in category 3. Interestingly the number of squares in category 2 is lower for woodland than for Calluna heath. In an area extending from Cumbria to Kent the need to reduce N deposition (category 5) is paramount.

Following the implementation of the S protocol, all category 2 squares disappear. One square remains in category 1, but given its location there may be considerable uncertainties associated with this estimate. A major feature is the increase in areas assigned to category 4, often changing from category 3. These areas occur extensively throughout England and Wales. The implementation of NO_x reductions leaves few squares in categories 4 and 5. As might be expected, category 6, the area of options, expands. An interesting feature of this map, however, is the reappearance of squares requiring less S deposition (categories 1 and 2), particularly in the far north east of Scotland.

5. Conclusions

HARM output can be used to assess the possible success of emission reduction strategies in meeting CL for acidity and nutrient N. There are clearly significant differences between receptors. For Calluna heath, once the S protocol is implemented, the main emphasis will be on reducing N deposition. The apparent sensitivity of this ecosystem to N deposition in the UK has been described elsewhere (UKRGIAN, 1994). In the case of woodland, however, further reductions in S remain an important objective over much of the UK and in some places it is clear that the emissions reductions in the current S protocol will not be sufficient.

Acknowledgements

This work was supported by Air Quality Division, DoE under contract number PECD 7-12-64 (SEM & JDW), PECD 7-10-90 (KRB & HD) and EPG1/3/17 (RGD). R. Warren, Imperial College supplied EMEP area emissions estimates for the scenarios.

References

Bull, K.B., Brown, M.J., Dyke, H., Eversham, B.C., Fuller, R.M., Hornung, M., Howard, D.C., Rodwell, J. & Roy, D.B.: 1995, Critical loads for nitrogen deposition for Great Britain, *Water, Air and Soil Pollution*, in press.

CLAG: 1994, Critical loads of acidity in the United Kingdom. Report by the Department of the Environment's Critical Loads Advisory Group (Edited by Pitcairn, C.E.R.)

Posch, M., Hettelingh, J-P., Svedrup, H.U., Bull, K.R. and de Vries, W.: 1993, Guidelines for the computation and mapping of critical loads exceedances of S and N in Europe. In: Calculation and mapping of critical loads in Europe (edited by Downing, R.J., Hettelingh, J-P. and de Smet, P.A.M.), RIVM report 259101003. RIVM: Bilthoven. l

Metcalfe, S.E., Whyatt, J.D. and Derwent, R.G.: in press, *Q.J.R. Meteorol. Soc*, **121**

Metcalfe, S.E., Atkins, D.H. and Derwent, R.G.: 1989, *Atmos. Environ.*, **23**, 2033-2052.

UKRGIAN: 1994, Impacts of Nitrogen deposition in Terrestrial Ecosystems. Report of the United Kingdom Review Group on Impacts of Atmospheric N (Edited by Pitcairn, C.E.R.).

UNECE: 1994, Protocol to the 1979 Convention on Long-Range Transboundary Air Pollution. United Nations: Geneva.

EMISSIONS OF ACIDIFYING AIR POLLUTANTS IN THE NORTH WEST REGION OF ENGLAND

J.W.S. LONGHURST, S. J. LINDLEY and D.E. CONLAN

Atmospheric Research and Information Centre, Department of Environmental and Geographical Sciences, Manchester Metropolitan University, Chester Street, Manchester M1 5GD, UK

Abstract. Most estimates of emission are concerned with the nation state level. This paper will discuss methods utilised in the estimates of emissions to the atmosphere of sulphur dioxide, volatile organic compounds and oxides of nitrogen from a densely populated and heavily industrialised region of the United Kingdom. Data on power generation, industrial plant, fuel usage, air, sea and road transportation, and human population statistics have been integrated into a method to provide regional emission estimates. The resulting emission patterns are described in terms of sources and emission density. Spatial and temporal patterns are identified and major sources of emissions discussed in terms of national control programmes. Transportation is the dominant source of oxides of nitrogen emissions whilst power generation is the dominant source of sulphur dioxide. The relative importance of the North West as an emission source within the UK is assessed. The change in the strengths of acidifying emissions between 1987 and 1992 is discussed and the rate of change in emission magnitudes between the North West region and the UK as a whole compared.

Key words: Emission estimation, north west England.

1. Introduction

The North West of England is one of the most important industrial and commercial regions of the UK. It has a diverse range of heavy and light engineering as well as increasingly important service, commercial and financial enterprises. The region comprises of the urban counties of Greater Manchester and Merseyside and the shire counties of Lancashire and Cheshire. The region covers 7342 square kilometres and has some 6.4 million people in 37 administrative districts including the cities of Manchester and Liverpool and the Boroughs of Wigan, Macclesfield and Burnley. Average population density is 872 persons per square kilometre with 415 in Cheshire and 2001 in Greater Manchester (Church, 1994). In 1991 the gross domestic product (GDP) of the North West was £48 880 million of which Greater Manchester contributed £20 155 million. This is the second largest regional contribution to the GDP after the South East (Church, 1994). The urban core of the region straddles the River Mersey. This area is home to traditional heavy industries such as glass making and is also one of the UK's most important locations for the chemical industry. The communications network is well developed containing major north-south and east-west road and rail communication routes. The urban and inter-urban road network experiences heavy traffic flows. In particular high traffic flows are experienced on the M6, M62, M63 and M56. To the south of the Greater Manchester conurbation is Manchester Airport, the UK's third busiest. Notable pollutant sources along the Mersey estuary include Fiddler's Ferry power station (coal fired, base load, 1800 MW), Stanlow oil refinery and a variety of petro-chemical complexes. The area thus provide a multiplicity of point, mobile and area sources of acidifying emissions.

The physical geography of the region includes the low land along the west coast and central Cheshire and the low lying land of the Mersey flood plane which penetrates as far as the Greater Manchester conurbation. Upland areas of the Pennines form the eastern boundary to the region. The region has a long history of air pollution and was the first area in which acid deposition was identified (Longhurst *et al*, 1993).

Estimates of emission of acidifying pollutants have been made by taking a 'top-down' approach. This involved dissaggregating national emission estimations to a local level through the use of indicators of the proportion of a particular polluting activity occurring in the specified region (Lee and Longhurst 1993, Longhurst *et al*, 1994).

2. Emissions of Acidifying Pollutants in the North West

Estimates of emissions have been made for the North West for a number of key pollutants (Lee and Longhurst, 1993, Longhurst *et al*, 1994). Lee and Longhurst (1993) estimated emissions of nitrogen oxides (NO_x), sulphur dioxide (SO_2), hydrogen chloride (HCl) and ammonia (NH_3) for 1987. These were made on a *per capita* basis *pro-rata* with nationally derived emissions estimates for 1987 (DoE, 1988) except for the source area of power stations. National emissions estimates from power station sources were spatially disaggregated according to the contribution of North West fossil fuelled plants to the total UK fossil fuelled electricity generation. Emissions from other sources were summed and apportioned according to the percentage of the UK's population resident in the region. Emissions of HCl were estimated for power station and incinerators through the use of primary data and other sources by *pro rata* calculations. Ammonia emissions were estimated for activities based on agricultural and human population data.

Longhurst *et al* (1994b) made emission estimates for carbon monoxide (CO), volatile organic compounds (VOCs), black smoke and lead from transportation, sulphur dioxide (SO_2) and oxides of nitrogen (NO_x) for 1987 and also for 1992 using national emissions data (DoE, 1992). A refined method was designed in an attempt to reduce some of the uncertainties associated with the *per capita pro-rata* approach of Lee and Longhurst (1993). Whereas the power station contribution could be estimated with some certainty, other estimates based on population as a surrogate were less certain. Therefore, in addition to isolating power stations as a source, use was also made of the other source breakdowns to provide a more sensitive representation of activities within the North West region. This was possible for air traffic, shipping, refineries and road transport which are all significant activities within the region (DoT, 1993, DTI, 1993). Again, the remaining source categories for which individual calculations could not be made were summed and apportioned according to population. Statistics and sources used in the estimation of emissions are shown in Table I.

3. Emission Sources and Strengths in the North West

Emissions estimates for SO_2, NOx and VOCs for 1987 and 1991 and NH_3 and HCl for 1987 are given in Table II. Trends in emissions over the period 1987 to 1991 for SO_2, NOx and VOCs are given in Table III in which they are contrasted to the UK national situation. The data for completing the NH_3 and HCl estimations were not available.

3.1. SULPHUR DIOXIDE

SO_2 is emitted during fossil fuel combustion and its largest national source in 1991 was power generation at 71% (DoE, 1992). The 1987 emission estimate of SO_2 in the North West is 287 kt compared to 280 kt in 1991 which indicates that emissions of SO_2 in the North West have reduced by 7 kt (2.4%). This is less pronounced than the national reduction of 8.5% which may be due to the relative increase in the proportion of power generating activity occurring in the North West from 5.8% to 6.31% between 1987 and 1991. Spatially, power generation is concentrated in a central band across the region and due to the closure of 2 plants, production has become concentrated in a fewer number of point sources. The location of these point sources to the west of Greater Manchester may have an important influence over air quality in this densely populated area due to prevailing south-westerly winds.

3.2. NITROGEN OXIDES

The principal sources of NO_x are fossil fuel combustion, the two largest sources being transportation and power generation at 51% and 26% of UK emission totals respectively for 1991 (DoE, 1992). Estimates for 1987 suggest that the North West region emitted 237 kt of NO_x and 262 kt in 1991. There has been an increase in emissions of NO_x in the North West of 25 kt over the period 1987 to 1991, representing a 10.9% increase compared to a national rise of only 5.5%. Spatially, the sources of NO_x emissions are more disperse than those of SO_2 despite 20% being from power station point sources. Transportation emissions have increased in importance from 56% to 62% of North West emissions between 1987 and 1991 and these are likely to be largely associated with the regions main high-speed transportation routes, such as the M56, M63, M6 and M62 and the two large metropolitan areas of Merseyside and Greater Manchester.

3.3. VOLATILE ORGANIC COMPOUNDS

Industrial processes and solvents account for almost half of national and regional emissions with another third from road transport. VOC estimates for 1987 attribute 260 kt to the North West region and 269 kt in 1991. Regional VOC emissions determined using the more recent national data indicates a 3.9% rise over the 4 years compared to a slight decrease at the national scale. The distribution of emissions is strongly related to the urban areas of the region.

Table I

Statistics and sources used in the estimation of emissions

National Emission Source Category	North West Statistical Indicator	North West Share of UK Source Total (%) 1987	1991
Power Stations	GWh energy from fossil fuelled plants. 1987- 12105 1991- 14467	5.8	6.3
Refineries	Stanlow Oil Refinery. 1987- 13 Mt oil. 1991- 12 Mt oil.	14.3	13.8
Aircraft	Manchester Airport passenger traffic	11	11
Shipping	Liverpool Docks cargo 1987- 25.3 Mt, 1991- 37.4 Mt.	5.5	7.6
Road Transport	Vehicle registrations 1991 - 2,348,000	10.6	10.6
Others	Population	11.5	11.7

Source: Longhurst et al (1994)

Table II

Estimates of Emissions for 1987 and 1991 in kt

Species	1987 North West Estimate	1991 North West Estimate
SO_2*	287	280
NO_x*	237	262
VOCs*	259	269
HCl†	28	n/a
NH_3†	19	n/a

Sources: Longhurst et al (1994)* and Lee and Longhurst (1993)†

Table III

Emissions trends in the North West region between 1987 and 1991

Species	National Change (%)	Regional Change (%)
SO_2	- 8.5	- 2.4
NO_x	5.5	10.9
VOC	-0.3	3.9

Source: Longhurst et al (1994)

4. Emissions in Greater Manchester

There have been a number of emissions estimates that have been produced for the metropolitan county of Greater Manchester itself (Longhurst et al, in press). Table IV shows a summary of emissions estimated from the road network of Greater Manchester for NO_x and VOCs. This shows 24 hour average annual weekday traffic (AAWT) values for different road types in the county, for all vehicles and for cars only. Motorways are highlighted as responsible for the largest emission of NO_x with 'A' roads more important for the emission of VOCs. The importance of cars in the overall emission total for all species is also very clear.

Table IV

AAWT Emissions of VOC and NO_x from roads in Greater Manchester

Vehicle Type	Road Type	Emissions (tonnes)	
		NO_x	VOC
All Vehicles	'M'	69	14
	'B'	11	11
	'A'	48	43
	Total	129	68
Cars Only	'M'	30	6
	'B'	7	8
	'A'	26	29
	Total	63	42

Source Longhurst et al (in press)

5. Conclusion

Emissions of SO_2 from the North West region have reduced over time and are still reducing. The emissions of other pollutants are either showing no substantial changes or are rising over the period 1987-1991. In the majority of cases, the regional trend mirrors the national trend but the rate of change is often different. This has important implications for the spatial expression of the benefits of acidifying emission control programmes.

References

Church, J. (Ed) 1994 *Regional Trends 29. 1994.* Central Statistics Office, HMSO, London.
Department of the Environment 1988 *Digest of Environmental Protection and Water Statistics* No. 11 HMSO, London.
Department of the Environment 1992 *Digest of Environmental Protection and Water Statistics* No. 15 HMSO, London.
Department of Trade and Industry 1993 *Digest of Energy Statistics* HMSO, London.
Department of Transport 1993 *Digest of Transport Statistics* HMSO, London.
Lee, D.S. and Longhurst, J.W.S. 1993 *Environmental Pollution* **79** pp. 37-44.
Longhurst, J.W.S., Raper, D., Lee, D Heath, B. Conlan, D.E. & King H.: 1993 *Fuel* **72** 1261 -1280.
Longhurst, J.W.S., Lindley, S.J., Conlan, D.E. & Watson, A.F.R. 1994 Emissions of air pollutants in the north west region of England. In Baldasano, J.M., Brebbia, C.A., Power, H. & Zanetti, P. (Eds) *Computer Simulation. Air Pollution 11, Volume 1.* Computational Mechanics Publications, Southampton. pp 99 - 106.
Longhurst, J.W.S, Lindley, S.J., Conlan, D.E. Rayfield, D.J., and Hewison, T. In press Air Quality in Historical Perspective: a case study of the Greater Manchester conurbation. In Power, H. (Ed) *Urban Air Pollution Volume 2,* Computational Mechanics Publications, Southampton.

CHANGING PUBLIC INTEREST IN, AND AWARENESS OF, ACID DEPOSITION: SOME EVIDENCE FROM THE UK

J.W.S. Longhurst, J.Bantock, S.E. Hare & D.E. Conlan
Atmospheric Research and Information Centre, Department of Environmental and Geographical Sciences, Manchester Metropolitan University, Chester Street, Manchester M1 5GD, UK.

Abstract. It is fundamental that the general public have access to usable environmental information on which they can base their decisions. Since 1984 the Atmospheric Research and Information Centre (ARIC) has operated a public information programme for the UK on the subject of acid deposition. The objective of the programme is to disseminate information on acid deposition without advocacy. ARIC provides enquirers with a broad range of authoritative and accurate facts and opinions from a wide range of parties from all sides of the debate. These sources include pressure groups, governmental bodies and industrialists from the UK and overseas. By deconstructing complex technical material and reassembling it for dissemination in a user friendly form, ARIC assists those receiving information to obtain a balanced perspective. This enables personal decision making within the context of the fullest information resource ARIC is able to provide.

Key Words: Public awareness; public information; acid deposition.

1. Introduction

Public environmental awareness is informed by, and responds to, media interest and there are a number of examples of rises and falls of public awareness in specific subject areas. The environmental agenda of the media is, in part, related to the absence of more pressing economic or political concerns, although this has become less so in the more recent past, but certainly was the case in the time period during which acid deposition moved from the scientific domain into a publicly debated issue. Acid deposition or "acid rain" came to prominence during the 1970s and throughout the 1980s was subjected to extensive media coverage which effectively introduced the concept of acid deposition to the general public. Today, however, acid deposition is no longer an overt headline issue. This position has been replaced by "new" environmental concerns such as global climate change and urban air quality. However, the extensive, overt publicity received during the 1970s and 1980s created a situation whereby the subject of acid deposition moved through a series of publicity peaks, up the agenda. Public recognition of the control of acid deposition has led to the issue no longer receiving media attention, this in turn reinforces the public's perception.

The international dimensions of acid deposition led to the adoption in 1979 of the UNECE Convention on Long-Range Transboundary Air Pollution. This placed pressure on the UK to make a major public commitment to reduce acidic emissions, however despite mounting evidence concerning the effects of acid deposition, the UK did not initially support a number of the international initiatives (Longhurst, 1990, Longhurst *et al* 1993a, 1993b). At this time the UK's record in international negotiations related to acid deposition, which was acknowledged to have been defensive, was widely publicised. This adverse publicity prompted radical action from pressure groups and individuals within the UK and Europe and put further pressure on the government to take action. In 1985 the Helsinki Protocol, to reduce SO_2 emission by 30% of 1980 levels by 1993, was not supported by the UK (Longhurst, 1990, Longhurst *et al* 1993b). By the end of 1986 policy changes were taking place within the UK. During 1986 the British Prime Minister, Mrs Thatcher, after a visit to Norway, announced a new initiative to cut acidic emissions by introducing an FGD retrofit programme for UK fossil fuel power stations, and agreed to sign the Sofia protocol (1988) which was a commitment to reduce NO_X emissions to 1987 levels by 1994. A number of countries, concerned at the limitations of the Sofia protocol agreed an Informal Declaration (1988) to cut NO_X emissions by 30% by 1998, from a base year of between 1980 and 1986, this again was not supported by the UK (Longhurst, 1990, Longhurst *et al* 1993b). By this time negotiations on the European Community Directive on large combustion plant were well advanced and the directive was adopted in 1988 (Longhurst, 1990). Each of these initiatives was the subject of considerable media attention resulting in enhanced public awareness of acid deposition.

This paper will assess the level of public awareness of acid deposition and discuss in outline the effectiveness of information programmes in enabling informed decision making by the general public.

2. Acid Deposition and Public Awareness

In support of international negotiations on controlling acid deposition a number of public information programmes were established in the 1980s. In Europe these included the Stop Acid Rain Campaign by Swedish and Norwegian NGOs, the National Environmental Protection Board in Sweden, pressure groups in, amongst others, West Germany, The Netherlands and the UK (see issues of Acid News, 1982 - 1995), in North America and Canada governmental and non governmental public information and advisory activities, and in 1984 the Acid Rain Information Centre was established to provide public information.

The transient media attention of the 1980s created a situation whereby public interest and perception of environmental issues increased, and the constantly increasing negative effects of acidification on the environment attracted the attention of increasing numbers of concerned individuals and institutions. Amongst the first to react were education providers, this resulted in the topic of environmental protection in general, and acidification in particular, finding its way increasingly onto teaching curricula, both nationally and internationally. This has effectively led to issues associated with acid deposition retaining a high profile and being institutionalised within educational establishments. The information programmes and advisory services established during the 1980s, however, are dynamic and have evolved over recent years, and today acid deposition is only one of a number of atmospheric issues which compete on the information agenda. The Stop Acid Rain campaign operated by the former National Environmental Protection Board, Sweden, for example, during 1990 began to incorporate a broad range of environmental issues in their publications, which was reflected in the name change of their prominent publication, from Acid Magazine to Enviro. Whilst, in 1992, the Acid Rain Information Programme of ARIC was renamed the Air Quality Information Programme. There is, however, evidence to show that issues associated with acid deposition have retained a relatively high profile amongst the general public in the UK. Public attitudes surveys in the UK have been carried out by the Department of the

Environment at regular intervals over the past decade (DoE 1986, 1989 & 1994). The surveys have been based on a random sample of the population within England and Wales. It was acknowledged that individual responses to questionnaire surveys can be biased by high profile media coverage of events occurring at the time in question, therefore, the DoE surveys were carried out, as far as possible, to avoid high profile, awareness raising events such as the Earth Summit in Rio. The 1993 survey was carried out by NOP Market Research. A sample population of some 3000 individuals was identified from the electoral register in 160 English and Welsh parliamentary constituencies. The sample was stratified by region, urban/rural mix, educational attainment, car ownership and social class. The sample population was interviewed face to face using a structured interviewing technique avoiding leading questions. Hence, the sample population results can be considered representative of the population of England and Wales. The major qualification to this statement is that the sample population excluded those less than 18 years of age. Hence, the views and experiences of age group recognised as most aware of environmental issues are not represented.

In the 1993 survey respondents were asked to express their level of concern about a number of individual environmental issues, and by assessing atmospheric / air quality responses it can be seen how public attitudes have changed over time, Table I.

TABLE I
Percentage of respondents "very worried" about each Atmospheric / Air Quality issue : 1986, 1989, 1993

Environment Issue	1986	1989	1993
Ozone Layer Depletion	..	56	41
Fumes & Smoke from Factories	26	34	35
Traffic Exhaust Fumes & Urban Smog	23	33	40
Global Warming	..	44	35
Acid Deposition	35	40	31

.. Not included in the earlier surveys   (Source : DoE, 1986, 1989, 1994)

Levels of concern related to acid deposition were highest during the late 1980s which coincided with extensive media coverage. Throughout the latest survey acid deposition was no longer a prominent headline issue, however, it was still identified as an issue which caused public concern and was identified as being more important to the general public than issues such as, for example, inner city decay (DoE, 1994). The 1986 survey attempted to assess public knowledge of acid deposition, and the same questions were repeated in 1989 to measure changing attitudes. Respondents were asked what they thought caused acid deposition. They were encouraged to give as many causes as they could, but in both cases about 30% were unable to suggest any cause. These data are shown in Table II. The findings of the survey indicate that there has been little change in public perception of the causes of acid deposition between 1986 and 1989 with the welcome exception of a better public understanding of the role of nuclear power. Unfortunately this question was not asked during the latest survey in 1993. The demise of this sector of the survey is, in part, indicative of the decision makers' perception that the issues has been solved and it no longer requires political attention.

The 1993 survey asked respondents to identify environmental issues or trends they felt would cause them the most concern in the future, 20 years hence. Respondents were not prompted with any suggestions. Five major issues associated with atmospheric pollution / air quality were identified, Table III.

TABLE II
Perceived causes of acid deposition : 1986 and 1989 (Source : DoE, 1986, 1989)

Cause	1986	1989
Factories	37	37
Chemicals/chemical plants	..	25
Coal and oil power stations	23	24
Car exhausts	10	15
Nuclear power stations	19	8
Sprays/fertiliser	..	4
Others	20	11
Don't know	32	29

TABLE III
Future Atmospheric / Air Quality Concerns (Source : DoE, 1994)

Environment Issue	1993
Traffic fumes (including congestion and noise)	36
Level of air pollution	29
Global Warming	27
Depletion of ozone layer	20
Acid Deposition	5

These results show that acid deposition remains a publicly identified environmental issue, although less important than issues that have received media attention during the early 1990s. Most significantly, as part of the 1993 survey respondents were asked how they would obtain information on air pollution in their area. The results of the analysis showed that overall 40% of respondents were not aware of any information supplier on air pollution. Local government was identified by 45% as their most likely source of information. Some 85% of the sample population believed that information on pollution was inadequate and that it is a duty of both government and industry to provide public information. The overall results of the 1993 survey indicated that acid deposition remains an environmental issue in terms of public opinion, however the level of knowledge and the ability of the general public to obtain information on the subject gives cause for concern, thus reflecting the importance of increasing the availability and effectiveness of publicly available environmental information sources.

3. The Acid Rain and Air Quality Information Programme

The Acid Rain Information Centre, established in 1984 with the support of the DoE, soon became recognised as a major national source of information on all aspects of acidification. At the beginning of 1992 the information programme changed its focus and was renamed the Air Quality Information Programme to reflect the changing public demand for information. The primary objective, on establishment in 1984 and today, is the dissemination of information on acid deposition and more recently atmospheric pollution, without advocacy, to enable individuals to make their own informed decisions on the nature of the problems. In turn they are informed and empowered to take responsibility for the environmental consequences of their life style. It is the view of ARIC that only with a thorough appreciation of the conflicting perspectives can an individual begin to understand the complex issues associated with air quality (ARIC, 1994). To achieve the objectives of the programme, ARIC has established contacts with numerous organisations, both national and international, who supply resources for dissemination. ARIC also produce information resources of a factual nature on subjects not well documented. The major users of the programme fall into six categories: the secondary education sector; the higher education sector; the media; the general public; libraries and environmental organisations and consultancies (ARIC,1994). Demand for information has been recorded since 1984 and demand continues to increase rising from 2164 in 1987 to 5148 in 1994. Originally the majority of these information requests were for information on acid deposition but over the years there have been a greater variety of requests covering a wider range of air quality issues. Table IV compares acid deposition requests for 1991-1994. The data indicate a small year on year decline in demand for information on acid deposition. The majority of requests related to acid deposition emanate from the secondary education sector, reflecting the prominence of the subject within national curricula. It should be noted that this is the group excluded from the DoE's public attitude surveys. The limited number of requests from the general public reflect, in part, the lack of recent vigorous political and media attention due to the subject now being perceived as an environmental problem that has been resolved. It can be assumed that the relatively high level of requests for information on acid deposition will be maintained as long as the subject remains part of the national curriculum (ARIC,

1994).

TABLE IV
Percentage of Information Requests for Acid Deposition Received by ARIC

Year	Percentage	Total Information requests
1991	61	4152
1992	61	4601
1993	50	4871
1994	44	5148

4. Conclusion

The rise and fall of public interest in acid deposition reflects the changing media attention given to the issue. Currently the issue is perceived as a problem which is being solved by effective international action. Consequently, the public's environmental attention is focused upon more urgent or newsworthy issues. However, acid deposition has become incorporated into the curriculum of many subjects in secondary schools and demand for information is generated by this institutionalising process. Hence, there remains a significant demand for information on acid deposition which ARIC will endeavour to satisfy. ARIC will continue to expand the Public Information Programme to cover a broader range of air quality issues, with emphasis on acidification. Due to the high demand for information on acid deposition, ARIC would welcome contacts from the *Acid Reign '95* community who have information resources which could be disseminated by ARIC to a wide ranging audience.

5. References

ARIC.: 1994, *Annual Report*. Atmospheric Research and Information Centre. Manchester Metropolitan University.
Department of the Environment.: 1986, *Digest of Environmental and Water Statistics* 9. HMSO, London.
Department of the Environment. :1989, *Digest of Environmental and Water Statistics* 12. HMSO, London.
Department of the Environment. :1994, *Digest of Environmental and Water Statistics* 16. HMSO, London.
Longhurst J.W.S.: 1990, 'Acid deposition', in P. Buckley (ed) *Longmans world guide to environmental issues and organisations*. Longmans, London. pp 3 - 23.
Longhurst, J.W.S., Raper, D., Lee, D Heath, B. Conlan, D.E. & King H.: 1993a *Fuel* **72** 1261 -1280.
Longhurst, J.W.S., Raper, D., Lee, D Heath, B. Conlan, D.E. & King H.: 1993b *Fuel* **72** 1363 -1380.
Swedish NGO Secretariat on Acid Rain. : 1982 - 1995, *Acid News*.Gothenburg.

POLICY OF AIR PROTECTION IN POLAND

Z.M. KARACZUN

Warsaw Agricultural University, Department of Environment Protection, Nowoursynowska 166,
02-766 Warsaw, Poland

Abstract. The changed political situation and recognition of the acute destruction of the natural environment in Poland have caused a series of actions aimed at preventing further deterioration of Polish environment. One of the most important events which took place in the last few years was the enactment by the Polish Parliament in May 1991 of the Act on the National Ecological Policy. The basic assumption of the new environmental policy is a declaration that sustainable development will in future direct economic development in Poland.
The aim of the presented paper is to introduce existing policy of air protection and instruments which have been implemented to protect the air. Special attention will be paid to legislation instruments, introduction and enforcement of proper economic mechanisms strengthening air protection and foreign policy aiming at increasing foreign assistance for this objective.

Key words: Air protection, emission, environmental policy, sulfur dioxide, nitrogen dioxide, carbon dioxide

1. Introduction

Poland is one of the countries in Europe which emit the greatest amounts of air pollutants. To a large extent, this fact is related to the political and economic system which existed in Poland after World War Two. The change in the political situation in Poland caused a number of measures to be taken, aimed at preventing the further deterioration of the state of Polish nature.

One of the most important events which have taken place in recent years was the passing of a resolution on the environmental policy of the state by the Parliament of the Republic of Poland in May 1991 (Parliament of the Republic of Poland, 1991). The basis of the environmental policy is the adoption of sustainable development as the road towards further economic dvelopment of Poland. The implementation of the objectives of this policy should contribute to improved air quality, in particular to (Ministry of Environmental Protection, Natural Resources and Forestry, 1991): reduction of dust emissions into the air by about 50% with respect to the state in 1980; reduction of SO_2 and NO_x emissions by 80% with respect to 1980; reduction of emissions of volatile organic compounds, hydrocarbons, heavy metals and other air pollutants; reduction of carbon dioxide emissions to the level agreed upon at the international level.

The purpose of this paper is to evaluate the legal and economic tools for the implementation of the environmental policy in the field of air quality in Poland.

2. Premises and method

The basic premises of the work were as follows:
- the environmental policy is implemented by creating an appropriate legal framework which can be enforced, introducing and applying economic instruments in environmental protection;
- air pollution causes a number of damages - both economic (e.g., losses in crop size, or in timber growth etc.) and social (poorer public health), and this should force the government to conduct efficient air protection policies.

To achieve the objective of the study, documents and authorized statements which may affect the environmental policy in Poland were analyzed, and the existing legal regulations and economic instruments for environmental protection were reviewed. Foreign assistance programmes in Poland were also analyzed in terms of their efficiency as instruments for the implementation of the national environmental policy.

3. Results of the study and discussion

In Poland, there are several legal acts which are valid for air protection; many of them, however, emerged in completely different political and economic conditions, therefore, they need to be urgently amended. One of the regulations is the 1980 Law on Environmental Protection and Shaping of the Environment, a fundamental act for the environmental law in Poland. Amendments to this law have been discussed since the early 1990s. The neccesity of creating a new law on environmental protection results above all from the new political situation in Poland after 1989, ownership transformations in the econom and a different approach to the private property. The new law should also make it possible to introduce in Poland the economic instruments for environmental protection which are applied in the capitalist economy market creation, emission trading, or the possibility of insuring a technological process from the liability for the impact of emergencies degrading the state of the natural environment etc. At present, it is impossible to apply these types of solutions pursuant to the Polish law. A new approach to air protection should also make it possible for the public to become widely involved in environment management.

The necessity of changing the Polish environmental law also results from Poland's willingness to tie the Polish law with the directives of the European Union. To date, the Polish law on preventing pollutants has addressed only the particular environmental media (the air, water, or vegetation), introducing specific threshold (permissible) pollution standards which must not be exceeded and creating an administrative permitting system for given pollutant emissions into the environment. This system should, however, be considered inefficient, for in many regions of Poland the allowable concentrations still continue to be exceeded even up to the level which threatens the residents health but, despite this, the plants which are responsible for hazards of this type are seldom closed down. Moreover, the legal system in Poland prevents a negotiated determination of the emitted pollutant level (making it difficult for plants to phase in environmental controls); it also makes it difficult, or even impossible, for the public to be notified that the permissible contamination standards have been exceeded and who is responsible for it (Karaczun, 1993). Therefore, for some time now, the Ministry of Environmental Protection,

Natural Resources and Forestry has conducted studies on the possibility of establishing in Poland a permitting system based on the use of the Best Available Technology (BAT) and other recommendations related to the implementation of the Directive of the Council of the European Communities 93/C 311/06 of September 30, 1993, on the prevention and reduction of pollution (Council of Europe, 1992). It should be expected that the work on the unification of the Polish environmental law with the legislative acts of the European Union should be finalized by the year 2000.

For several years in Poland there has been in place a fairly efficient system of economic instruments for environmental protection (Ministry of Environmental Protection, Natural Resources and Forestry, 1994a). It includes such instruments as fees (e.g., for pollutant emissions into the air or waters), enforcement incentives (e.g., fines for exceeding the permissible emission levels), subsidies (endowments and preferential credits), and others. The payments from fees for the use of the environment and fines for contravention of the conditions of this use are collected by special funds, mainly by the National Fund for Environmental Protection and Water Management. The funds thus collected are then allocated for cofinancing environmentally friendly projects, in the form of subsidies or preferential credits and loans.

A phenomenon which may hamper work in air protection is the absence of commercial and investment banks in funding projects in this field. Despite the existence of the National Fund for Environmental Protection and Water Management and the Bank for Environmental Protection, the latter was founded in 1992, with the National Fund for Environmental Protection and Water Management as the majority shareholder. Its statutory objectives include, funding projects in environmental protection. In Poland there are no sources of funds for financing substantial environmentally friendly projects (Stodulski, 1994), for the maximum size of credits granted by these institutions amounts to about 20 billion ZL (about 1 million USD. In the middle of 1994 1 USD = 21.000 ZL). The maximum value of credits offered by commercial banks is much higher than those granted by the National Fund for Environmental Protection and Water Management and the Bank for Environmental Protection, and should be sufficient for cofinancing medium-sized air protection projects (the maximum credit size offered by commercial banks in Poland reaches about 250 billion ZL - i.e., about 12 million USD). However, even the amounts offered by commercial banks are too low compared with the costs of some substantial air protection projects.

The system of economic instruments in environmental protection in Poland is threatened by the poor exaction of due fees and fines by the National Fund for Environmental Protection and Water Management (in 1992, it managed to exact only 72% of fees and 22% of fines due).It limits the funds at its disposal; moreover, it creates an atmosphere of impunity among the executives of the plants which refuse to pay the fees calculated and the fines levied (Karaczun 1994).

The change in the political situation in Poland in 1989 made this country more active on the international scene, increasing the interest for Poland to shape a new environmental deal in the world. At the end of 1991, foreign assistance to Poland for environmental protection (Ministry of Environmental Protection, Natural Resources and Forestry, 1994b) encompassed 146 projects with the total value of 215 million USD (with about 111 million USD allocated for air protection programmes). Assistance projects which consist only in the planning phase are, however, more and more often criticised. After the study costs have been covered and the project designed there is usually no money left for its actual implementation. Moreover, although often valuable and necessary, foreign assistance will never replace domestic funds allocated by Polish institutions and organizations, and its proportion in the total expenditures for air protection does not exceed 3 - 5% (Ministry of Environmental Protection, Natural Resources and Forestry, 1994b).

After the political changes in 1989 Poland took an active part in international negotiations on environmental conventions. An effect of this was Poland's signing and ratification of such international agreements as the Montreal Protocol and the Vienna Convention (since 1990), the Framework Convention on Climate Change (since 1994). Now work is under way on the possiblity of Poland's implementation of its obligations imposed by the IInd Sulfur Protocol (under auspices of the UN ECE Convention on Long Range Transboundary Air Pollution) and the Copenhagen Amendments to the Montreal Protocol. Poland's adherence to these agreements imposes on it a number of obligations the satisfaction of which, although not a priority from the point of view of environmental protection in Poland, is the implementation of the principle of joint responsibility for solving global environmental problems (Karaczun 1994).

It should be borne in mind, however, that one of the basic objectives of the environmental policy of the State is to reduce the amounts of pollutants emitted into the air. Therefore, the results of this policy in the field of air protection should be evaluated in terms of change in the size of emissions.

A drop in the industrial output and a deep economic recession, which Poland has gone through over several recent years has brought about a decrease in the use of primary energy factors and basic fuels. Therefore, there emerged the essential question whether the economic growth which was noted for the first time in 1992 would cause increased energy consumption, and thereby higher air emissions, since the energy generating industry is the main source of emissions in Poland. Practically, these misgivings have not turned out to be true. The increase in the industrial output (and also in the domestic product) is faster than the mean growth rate of the consumption of energy and fuels, meaning that it is achieved by drawing on simple reserves (in terms of the effective use of energy) rather than generating more energy. A positive aspect of changes in recent years has also been a new structure of using basic energy factors, for there has been an increase in the consumption of fuels which are considered to be less environmentally harmful (natural gas, oil), and there has been a drop in the consumption of those that stress the environment (hard and brown coals).

In 1993 the total gas emissions into the air, calculated from the fuel consumption and technological indicators, were (Main Statistical Office, 1994):

- sulfur dioxide - 2 725 000 t, lower by 1 455 000 t than in 1988 (a decrease by about 35%). The power industry is the main source of these gas emissions in Poland (accounting for 62.1% of the total SO_2 emissions).

- nitrogen dioxide - 1 130 000 t, lower by 270 000 t than in 1988 (a decrease by about 17%). The power industry was the main source of these gas emissions (accounting for 40.1% of the total emissions) and moving sources (vehicles) were responsible for 37.5%.

- carbon dioxide - 397.1 million t, lower by 112.3 million t than in 1988 (a decrease by 22%). The fuels and power industry is responsible for most carbon dioxide emissions: accounting for about 65% of the total CO_2 emissions.

In commenting on these data, it should be said that the picture of pollutant emissions into the air seems to indicate a success of the air protection policy being implemented - the size of these emissions has considerably dropped over the last three to five years. It seems, however, that on the other hand, it is too early to formulate conclusions of this type, for a good deal indicates that, to the greatest exetent, lower emissions were caused by a drop in the industrial output than came as an effect of the implementation of environmental programmes. Moreover, there is a number of grounds which indicate that the favourable tendency to reduce emissions may be reversed. Particularly, they are: a rapid increase in emissions caused by the municipal sector and transport, a local elevation of sulfur dioxide concentrations caused by sales of cheaper coal with greater sulfur content to individuals and a higher number of vehicles etc. Therefore, what is very important is how the situation will change in the next two to three years. If the favourable tendency for emissions to fall remains, this will indicate that the environmental policy being conceived is an efficient one and that Poland has a real chance of sustainable development of its economy.

4. Conclusions and recommendations

At present, air protection is not considered a priority, therefore the amounts of funds allocated for projects in this field are not satisfactory. Although over the last five years in Poland air emissions of most pollutants have distinctly dropped, it cannot be considered a success of the air protection policy being implemented, for it has been above all a result of a deep recession in the economy. It is uncertain, however, whether it is a permanent process, and the fact that the Poland is recovering from the recession poses a danger that in a few years emissions will start to grow again.

To improve the efficiency of the air protection policy in Poland and to guarantee that in the next years it will lead to the reduction of air emissions of pollutants, it is necessary to undertake the following actions:

- to accelerate the legislative work in the field of air protection from pollution. It is necessary to regulate these problems in a complex way and to pass a number of detailed executive orders on the protection of the ozone layer, the mitigation of greenhouse gas emissions and those that generate an acid stream. Work on the unification of the Polish law with the legal acts of the European Community is also essential. Here, there are, large discrepancies in the possibility of public participation, actions to be performed when the permissible ambient quality standards are exceeded and the size of emissions from moving sources;
- improve the efficiency of exacting fees for the commercial use of the environment and penalties for noncompliance with the conditions of this use. The low exaction rate not only decreases the amounts of funds collected by the National Fund for Environmental Protection and Water Management, but also provides for a feeling of impunity among the executives of industrial plants which have an adverse impact on the environment;
- to incorporate commercial and investment banks in joint funding, along with the National Fund for Environmental Protection and Water Management, of projects in air protection, and to create a system which would make it possible to finance large projects in air protection. The main difficulty in undertaking actions for this purpose is the high cost of such projects, which it is impossible for single industrial plants to cover, even with support from the National Fund for Environmental Protection and the Bank for Environmental Protection;
- to play an active part on the international scene in order to create joint programmes for air

protection, to acquire funds for environmentally friendly restructuring of the industry in the process of ownership transformation and also to gain access to state-of-the-art technologies for air protection.

Acknowledgements

The present author would like to thank Mr Reinhold Pape of the Swedish NGO Secretariat on Acid Rain and the Institute for Sustainable Development for making it possible to conduct studies on the air protection policy in Poland.

References

Council of European Communities, 1993: Directive of the Commission on Integrated Prevention and Control of Pollution. 93/C 311/06. Final text, pp. 32
Karaczun Z., 1993: Policy of Air Protection in Poland. Parts I and II - in Polish. Report of the Institute for Sustainable Development. No. 4. Warsaw, pp. 63
Karaczun Z., 1994: Policy of Air Protection in Poland. Part III. Report of the Institute for Sustainable Development, Warsaw, pp. 78
Main Statistical Office:Environmental Protection - in Polish 1994. Main Statistical Office, Warsaw, pp. 518
Ministry of Environmental Protection, Natural Resources and Forestry, 1991: Premises of a New Environmental Policy of the State - in Polish. Typescript, Warsaw, pp. 28
Ministry of Environmental Protection, Natural Resources and Forestry, 1994a: Economic Instruments in Environmental Protection - in Polish. Typescript at the Economic Department, Warsaw, pp. 6
Ministry of Environmental Protection, Natural Resources and Forestry, 1994b: Foregin projects in environment protection in Poland - in Polish. Typescript at the Project Implementation Unit, Warsaw, pp. 8
Parliament of the Republic of Poland, 1991: Resolution of the Parliament of the Republic of Poland of May 10, 1991, on the environmental policy of the State- in Polish. Parliament handout no. 659, p. 1
Stodulski W., 1994: Banks and Sustainable Development in Poland - in Polish. Paper no. 4. Institute for Sustainable Development. Warsaw. pp 84

POLLUTION PREVENTING EFFORTS AND STRATEGIES FOR THE KATHMANDU VALLEY

PRAMOD K. JHA

Central Department of Botany, Tribhuvan University, Kirtipur, Kathmandu, Nepal

Abstract. The Kathmandu valley, nestled in the midst of the mighty Himalaya, is one of the worth visiting places in the world. Unplanned urbanization, increasing population, polluting vehicles and industries have started degrading the environment in the valley. Centralized development activities in the last two decades resulted in undesirable environmental change. If the current trend continues, valley will loose its importance. Realizing the importance of healthy environment, His Majesty's Government of Nepal has formed the Environment Protection Council under the chairmanship of Prime Minister to formulate environmental planning and policies. Over two dozen active NGOs are involved in environmental awareness and management in the Kathmandu valley. Some major steps have been undertaken by GOs and NGOs. Present paper deals with the state of pollution, efforts made to minimize it, and major actions/strategies for preventing pollution in the Kathmandu valley.

Introduction

The Kathmandu valley is the political, cultural, touristic, educational, administrative, commercial, and financial capital of Nepal. It is situated at an average altitude of 1330 m ASL and covers an area of about 351 km^2, and is surrounded on all sides by mountains up to 2785m ASL. The valley known for its exquisite environment and cultural heritage, has been engulfed in environmental problems (Jha, 1993). Unplanned urbanization, increasing population pressure, polluting vehicles and industries, centralized development activities, etc. have started degrading the environment in the valley, and resulted in bad international publicity. The environmental scenario is alarming, however, timely planned action can help in regaining the environmental reputation of the Kathmandu valley.

This paper is aimed to communicate the state of pollution and efforts made to minimize pollution in the Kathmandu valley. Some suggestions are also made for concerned organizations to prevent pollution in the valley.

Policy Background

The environmental protection work was accelerated through a discussion forum for pollution preventing strategies for the Kathmandu valley, jointly organized on June 6,1993 by the Royal Nepal Academy of Science and Technology (RONAST) and the Central Department of Botany, Tribhuvan University. Thereafter a few important seminars such as "traffic management and pollution control "(organized by the valley Traffic Police)," urban air quality management in Kathmandu valley" (organized by Urban Environment Management Committee, Kathmandu), and projects have been conducted to evaluate the environmental quality of the valley. In the present paper, resolutions of the seminars, findings of the projects, essence of the literature and personal experiences have been summarized.

About a dozen major projects (Greater Kathmandu Municipality Environment Improvement Project (MEIP), Bagmati Basin Environmental Study, Kathmandu

Greenary Project, Vehicular Emission Project, Alternative Energy Study, Urban Air Quality, Pollution Control Project, Water Pollution Baseline Study Project, Shivpuri Integrated Watershed Development Project, Bagmati Watershed Project, Environmental Research Project for Pollution in the Kathmandu valley, etc.) were undertaken to improve the deteriorating urban environment in the Kathmandu valley. Besides these, government and non-governmental organizations have made efforts to minimize environmental problems in the valley (Table 1). Some of these efforts were successful, partially successful but some could not achieve success. The short and long term plans for the Kathmandu valley were decided and the land use plan for Kathmandu valley was formulated. However, these have never come in light with strong political commitment.

Table 1. Major efforts done to minimize pollution in the Kathmandu valley

S.N.	Efforts	Organization	Remarks
1.	Formation of Environment Protection Council (EPC) (October 1992)	HMG	needs institutionalization
2.	Publication of Nepal Environmental Policy and Action Plan (NEPAP) (1993)	EPC	complete
3.	Formation of Urban Environment Management Committee for the Kathmandu Valley (April 1993)	EPC	partially successful
4.	Creation of Environment Protection Trust (1994	EPC	requires Fund
5.	Vehicle emission control project funded by UNDP in 1993 and by HMG in 1994-95	EPC/NPC TP	complete in progress
6.	Bagmati river (partial) rehabilitation project in 1994-95	EPC/MOHPP	in progress
7.	Shivpuri Watershed (source of Bagmati river) Development Project (1992-97)	HMG/FAO HMG/MOF	in progress in progress
8.	Bagmati Watershed Project (1985-95)	MOFSC	partially successful
9.	Development of Bishnumati river corridor	MUNICIP	Partially successful
10.	Installation of pollution control device in Himal cement factory (1994-95)	MOI	in progress
11.	Solid Waste collection, recycling, dumping, etc.	SWMRMC	unsatisfactory
12.	Survey of diverting Melamchi and Roshi Khola to increase water supply for Kathmandu	MOWR	survey complete, implement. expensive
13.	Exploitation of ground water for Kathmandu	MOWR/JICA	partially successful
14.	Environment Research Project (1989-94)	RONAST	complete
15.	Land Resource Mapping project for the Kathmandu	HMG, Dept of Survey	complete
16.	Environment indicators and indices (1995)	KMTNC	in progress
17.	Industrial pollution control management project (1994-96)	MOI	in progress
18.	Development of electrified vehicles	EPC	started on experimental basis
19.	Waste-water treatment plant at Balaju Industrial District (proposed)	EPC	awaited

His Majesty's Government of Nepal created an Environment Protection Council (EPC) in October 1992 under the chairmanship of Prime Minister. The EPC functions as an apex body responsible for formulating policies, preparing guidelines, establishing procedures, and coordinating environmental conservation efforts. In the last two years, the EPC concentrated on environmental issues of the Kathmandu valley, and initiated to minimize pollution in the valley. Some of the steps were : formation of Urban

Environment Management Committee under the chairmanship of Minister of Housing and Physical Planning, installation of pollution abating device in Himal cement factory, categorization of polluting vehicles (75 Haritage Smoke Unit (HSU) for diesel vehicles and 3% carbon monoxide emission for petrol vehicles), Bagmati river rehabilitation, etc. The EPC has also prepared Nepal Environmental Policy and Action Plan (EPC, 1993), and has plan to prepare environmental policy and action plan for the Kathmandu valley.

Observations

As the country's capital, Kathmandu is suffering from the effects of an over concentration of public and private sectors. The basic services such as water, electricity, road network, etc. can support 600,000 and one million population in the urban area and the Kathmandu valley, respectively (TRN, 1994), that covers only 60 percent of the human population. There is an energy crisis in the valley, and the demand for electricity is increasing at 10% annually. Drinking water scarcity is another problem, and only one third of required quantity has been served. The total demand for fuel wood in the valley was 500,670 mt in 1991 while the production from the existing forest in the valley was about one fifth (TRN, 1994).

Major air pollutants are emitted out through old vehicles. The UNDP project report (1994) presented to EPC indicates that 91.2 percent diesel vehicles emit more than 75 HSU whereas 49,23 and 38 percent 4 wheelers, 3 wheelers and 2 wheelers respectively emit over 3 percent carbon monoxide. Around 90,000 vehicles run on the unmaintained roads (54% roads are still gravel and earth top) of the Kathmandu valley. Pollutant contribution of road vehicles into the atmosphere was about 40,000 mt per year (after NPC/IUCN, 1991). The team led by D.H. Stedman concluded that the main vehicular emission problem is because of lack of vehicular maintenance (NCS/IUCN, Nepal 1993). There are few reports(Bhattarai and Shrestha, 1981; Manandhar *et al.*, 1987; CEDA, 1989; Devkota, 1992; Sharma *et al.*, 1992) on the atmospheric pollution level in the Kathmandu city. Recent detail report by Karmacharya and Shrestha (1993) reveal that the total solid particles (TSP) are too high in the atmosphere of the Kathmandu city. It was 182-555 μ/m^3 (in 9 hr collection) and 789 to 2250 μ/m^3 (in 24 hr collection), while WHO fixed standard for TSP as 120 μ/m^3. They found lead content in atmosphere at warning level (0.18 to 0.53 μ/m^3 in 9 hr collection and 0.2 to 1.2 μ/m^3 in 24 hr collection while WHO standard is 0.5 - 1.0 μ/m^3). Other vehicular emission gases (SOx, NOx CO) were found below the WHO standard. Poor quality of gasoline and diesel used in Nepal is another factor for increased pollutants. Various industries mainly cement, brick kiln and diesel plants, burning of fossil fuels at home and dust particles from solid wastes thrown on roads, etc were identified as polluting factors (Shrestha, 1993).

The river system in the valley has turned to drainage and self-purification capacity has declined (Jha, 1993). Aquatic biodiversity, particularly in rivers, has declined. About a decade back, upstream of Bagmati river in the Kathmandu valley had more than 36 species of fish, however now about a dozen species exit in the river. Important fish species viz *Puntius* sp, *Garra* sp., *Danio* sp, *Esomus* sp. etc are non-existant in the upper pat of the river (Sharma and Pant, 1992). Discharge of the domestic liquid waste directly or indirectly into the river without any treatment is one of the major factors responsible for the water pollution.

The major industries discharging effluents into rivers in Kathmandu are: Bansbari tannery, Balaju industrial district, carpet factories, Jawalakhel distillery, Patan industrial district, Himal cement factory, etc. The concentration of water pollution became high in summer due to low water flow. Drying of rivers has been occurring mainly due to deforestations, secondly due to the decrease of water table level as there is no recharging system present in the Kathmandu valley (Bista, 1993). The water quality in the Kathmandu valley is also very unsatisfactory and even chlorinated water of Kathmandu is found heavily contaminated (Jha, 1992). Over eighty percent of drinking water samples contains very high concentration of bacteria. Bottino *et. al.* (1991) reported that most of the chemical parameters of the tap water sample were less than the permissible values except $N-NH_3$ and $N-NO_2$ ion concentrations whereas total coliform counts in most of the samples were high, and still this situation exists.

Analysis of available meteorological data reveals that average Kathmandu city temperature has increased about 1°C in the last 25 years. The process of valley warming has been accelerated and fluctuations in rainfall have been noticed (Jha, 1993). Noise level at working places, industries and main roads crosses the permissible level, *i.e.* above 75 db. and reaches unpto 120 db. Agricultural land in the valley is under fast encroachment for construction. Pesticide use has increased largely in the valley. The problem of solid waste disposal is mounting with each passing day. Every day 550 m^3 of garbage is collected from the two cities (Kathmandu and Patan) of the valley, i.e. over 300 mt per day, out of that three fourth of the solid waste is biodegradable. One of the problems with solid waste management is the lack of permanent dumping sites.

Suggestions

The Kathmandu valley needs environmental policy and action plan. Here some short term (with in 2 years) and long term (with in 5 years) actions are suggested to concerned organizations to prevent or minimize pollution (Table 2). Most of the short term proposed activities are possible with the national resources, however, some of the long term proposed activities require technical and financial support from the developed countries. Pollution control act was prepared two years back by the Environmental Protection Council, but could not get priority by the government. There was a solid waste recycling plant in the centre of the city, but due to smell pollution, local people opposed its existence in the city. It was closed but not transferred to other site in periphery of the valley. Likewise there is always problem with the dumping sites, and till date local administration has not identified suitable long-time dumping site. These issues can easily be solved with efficient administrative system.

Development of an international airport outside the valley, decentralization of the administrative, educational institutions, industrial activities, use of clean fuel and maintained clean roads can reduce the pollution level by more than fifty percent. Political commitment, sincere efforts and a sense of responsibility can minimize and prevent the pollution in the Kathmandu valley.

Table 2. Strategies / actions suggested to concerned organizations to prevent pollution in the Kathmandu valley.

S.N.	Strategies/ Actions	Concerned Organization
	Short term	
1.	Develop and implement urban management plan	MOHPP/ MUNICIP
2.	Formulate pollution control act	HMG/EPC
3.	Establish pollution control agency	EPC
4.	Restrict high emission vehicle with in ring-road of the valley	TP
5.	Reduce auto traffic volume in congested urban area	TP
6.	Maintain roads and plantation on the roadsides	DOR/MUNICIP
7.	Improve water supply (quality and quantity)	NWSC
8.	Treat waste water (industrial as well as domestic) before their discharge in water bodies	MOI, Industries MUNICIP
9.	Recycle solid wastes and improve solid waste collecting and dumping system	SWMRMC MUNICIP
10.	Identify dumping sites for regular solid waste disposal	MOHPP
11.	Enhance environmental awareness, education and research programmes, including creation of department of environmental studies at TU	EPC, MOECSW TU, RONAST
	Long term	
12.	Decentralize administrations, educational institution, industrial activities, etc	HMG
13.	Develop an international airport outside the valley (at Biratnagar, Pokhara, Janakpur, or Bhairhawa)	HMG
14.	Encourage efficient and clean vehicles by reducing tax	MOWT/MOF
15.	Improve alternative of private driving by developing network to roads and public transportation	MOWT/MOR
16.	Introduce lead-free gasoline in the valley	NOC
17.	Provide electricity, improved stoves, solar heaters, etc at low price to reduce consumption of biomass fuel	NEC Industries, NGOs
18.	Improve the Bagmati and Bishnumati rivers by checking sewage disposal and diverting other streams in it.	MOWR MUNICIP
19.	Encourage use of biofertilizer and biocontrol measures, as soil quality and agro-products are getting polluted	MOA

DOR : Department of Roads, HMG, EPC : Environment Protection Council, HMG : His Majesty's Government, MOA : Ministry of Agriculture, MOECSW : Ministry of Education, Culture and Social Welfare, MOF : Ministry of Finance, MOFSC : Ministry of Forest and Soil Conservation, MOHPP : Ministry of Housing and Physical Planning, MOI : Ministry of Industries, MOWR : Ministry of Water Resources, MUNICIP : Municipalities (Kathmandu, Lalitpur, Bhaktapur), NOC : Nepal Oil Corporation NWSC : Nepal Water Supply Corporation, NGOs : Non-Governmental Organizations RONAST : Royal Nepal Academy of Science & Technology, SWMRMC: Solid Waste Management and Resource Management Centre, TP : Traffic Police, TU : Tribhuvan University.

Conclusion

Developing countries have limited resources and that should not be wasted in self-created problems. Technologies are available to prevent and control pollution. Prevention is better than control. Government alone cannot solve environmental problems, and therefore people's participation is equally important. Pollution prevention is a challenging job, but not beyond reach.

The Kathmandu valley can regain its old reputation of Shangri-la or natures paradise. Government is giving emphasis on road maintenance, environmental awareness, vehicles testing, traffic management, water quality, solid waste management, waste water treatment, etc. If suggested strategies are followed, the 21st century Kathmandu may become a clean, green and healthy valley.

Acknowledgements

Author is thankful to the organizers of the Acid Reign conference for generous support provided to attend the conference, and authorities of the Swedish Environmental Research Institute Gothenburg , Sweden for facilitating my visit. I record my sincere thanks to Prof. Dr. D. Bajracharya, Head, Central Department of Botany, Tribhuvan University for encouragement, and to Environment Protection Council Secretariat for information.

References

Bhattarai, P.R. and Shrestha, P.R.: 1981, "Lead contents in the dust of Kathmandu city roads", Nep. Chem. Soc. Proc. 1: 47-50.
Bista, P.R. : 1993, "Water pollution." Paper presented in the discussion forum on pollution preventing strategies for the Kathmandu valley, Organized by the Royal Nepal Academy of Science & Technology and the Central Department of Botany, Tribhuvan University, Kathmandu, June 6, 1993.
Bottino, A., Thapa, A., Scatolini, A., Ferino, B., Sharma, S. and Pradhananga, T.M.: 1991, 'Pollution in the water supply system of Kathmandu city." Jour. Nepal Chemical Society 10: 33-44.
CEDA . : 1989, " A study on the environmental problem due to urbanization in some selected municipalities of Nepal," Report submitted to UNDP, Kathmandu by the Center for Economic Development and Administration, Tribhuvan University, Kathmandu.
Devkota, S.R. : 1992, "Energy utilization and air pollution in Kathmandu valley, Nepal," M.S. thesis submitted at Asian Institute of Technology, Bangkok.
EPC. : 1993, "Nepal Environmental Policy and Action Plan : Integrating Environment and Development", Environment Protection Council, HMG, Kathmandu.
Jha, P.K. : 1992, "Environment and Man in Nepal", Craftsman Press Ltd. , Bangkok.
Jha, P.K. :-1993, "Environmental hazards in Kathmandu Valley," Science Universal. 3:22-24.
Karmacharya, A.P. and Shrestha, R.K.: 1993, "Air quality assessment in Kathmandu valley", Environment and Public Health Organization, Kathmandu.
Manandhar, M.S., Ranjitkar, N.G., Pradhan, P.K. and Khanal, N.R.: 1987, Study on health hazard in Kathmandu valley. Report submitted to Nepal National Committee for the Man and Biosphere, Kathmandu.
NCS/IUCN.: 1993, " Fuel Efficiency Automobile Test," The newsletter of the National Conservation Strategy Implementation Project, IUCN, 4: 11.
NPC/IUCN. : 1991, "Environmental Pollution in Nepal", National Planning Commission and the World Conservation Union, National Conservation Strategy Implementation Programme, Kathmandu.
Sharma, A.P. and Pant, M.B.: 1992, "New field and issues in fisheries development in Nepal," Proc. Workshop on Human Resources Development in Fisheries Research in Nepal, Fisheries Development Division, Department of Agriculture Development, HMG, Nepal, pp. 195-216.
Sharma, U., Shahi, R.R., Shrestha, S., Thapa, A., Sijapati, J., Rana, P. and Pradhananga, T.M.: 1992, "Atmospheric pollution in Kathmandu city: I. Particulate matter in the Kathmandu city and study of mycoflora in it," Jour. Nepal Chemical Society, 11:1-8.
Shrestha, M.B.: 1993, "Air pollution", Paper presented in the discussion forum on pollution preventing strategies for the Kathmandu valley, Organized by the Royal Nepal Academy of Science & Technology and the Central Department of Botany, Tribhuvan University, Kathmandu, June 6,1993.
TRN.: 1994, "Study on Kathmandu's long term policies", The Rising Nepal, June 19,1994.

A SPATIAL DECISION SUPPORT SYSTEM TO ALLOW THE INVESTIGATION OF THE IMPACT OF EMISSIONS FROM MAJOR POINT SOURCES UNDER DIFFERENT OPERATING POLICIES

R.A. WADSWORTH & M.J. BROWN

Institute of Terrestrial Ecology, Monks Wood, Abbots Ripton, Huntingdon, Cambridgeshire. PE17 2LS.

Abstract. A Spatial Decision Support System (SDSS) is being developed for Her Majesty's Inspectorate of Pollution (HMIP). It enables the investigation of the spatial implications of different operating procedures from large point sources of pollution. The environmental effect of emissions is assessed using the critical loads methodology developed at ITE and modelled deposition "footprints". This approach allows an "effect per unit emission" or "pollution potential" to be determined for each source. Individual sources are modelled and included within the SDSS if their current emissions are above a given threshold. The SDSS provides a graphical user interface (GUI) which facilitates a fast, efficient and effective means to specify and to examine the effect of different operating policies. Mapping, statistical and optimization facilities are provided to help describe the effect of any specified strategy. Maps may be produced as deposition rates or exceedance values. Statistics may be visualised as histograms and scatter plots. The optimization facility uses linear programming to minimise the total environmental impact (estimated from emissions and critical loads) or maximise power produced within environmental limits. The SDSS has been written in the Arc/Info Macro Language (AML) and provides an invisible interface with standard GIS facilities and programmes written in "C".

Key words: Decision support system, optimization, air pollution, sulphur, critical loads

1. Introduction

Despite steadily increasing use of electricity there is currently considerable over capacity in the electricity generating industry in the UK (Newbery 1994). Operators of coal and oil fired power stations are being squeezed between the base load providers (nuclear power and international transfers) and modern, flexible gas powered generators. The UK has agreed to reduce the total amount of sulphur emitted by signing the Second Sulphur Protocol (Oslo 1994). Coal and oil fired stations currently account for approximately 70% of the sulphur dioxide emitted in the UK (RGAR 1994).

Her Majesty's Inspectorate of Pollution (HMIP) is responsible for authorising emissions of sulphur and nitrogen oxides from power stations. Authorised limits provide an upper bound on the operation of each station. In addition, individual generating companies allocated a mass emission quota under the UK National Plan. Quotas may be switched between stations provided the company "bubble" is not exceeded. In practice few power stations approach their annual authorised limits. The power industry was privatized in 1992 since then the daily operation of each station depends on an internal market which may be characterised as a "multiunit, single-price auction". The market is designed to be economically efficient although Green & Newbery (1992) discuss why it may not be optimal. Each day the National Grid Company plc (NGC) estimates the power needed nationally for each half hour period of the following day. Managers of each power station submit a price schedule for each half hour slot. The NGC selects bids from the price schedule in order to minimise the total cost. Selected stations are paid at the rate of the highest accepted bid for that half hour period. Additional complications arise due to limitations in the ability of the grid to transmit power so a reasonable spatial distribution of supply and demand must be maintained. The existence of an economically efficient market does not imply environmental efficiency, stations which invest in cleaning up emissions by fitting flue gas de-sulphurisation may be priced out of the market.

Periodically the power generating companies submit to HMIP their estimates of the loads which they wish each power station to operate at in future. It is HMIP's task to assess the coherence and reasonableness of the proposals. Several approaches may be adopted to foster the use of spatial data and models in decision making. An approach exemplified by NELUP (O'Callaghan 1995) or RAINS (Alcamo et al 1990) is to build a tailor made decision support system (DSS) from scratch. The advantages of this approach is that it is possible to construct a transparent, fast, efficient and potentially robust system which can fit directly into the decision making context. The disadvantages of the approach is that development time is often measured in tens of man-years and modifications may take days or weeks to implement. An alternative approach is to use a geographic information system (GIS). This approach allows access to hundreds of analysis routines and an active user community, but means investing many months in becoming familiar with a GIS, being satisfied with a relatively slow system, having little knowledge of how algorithms are implemented and risking inconsistent analysis. Watson and Wadsworth (in press) discuss these two conflicting approaches in greater detail. For the HMIP DSS, a compromise approach between these two extremes has been adopted. The compromise approach is to tailor a specific front end to a GIS using a macro-language. This combines the advantages of ease of use, consistent and coherent analysis with rapid development and reasonable control over the functionality. The advantages are thought to outweigh the disadvantages of slowness and dependence on proprietary software.

2. Description of the DSS

In recent years, computerised decision support systems have been created to bridge the gap between policy makers and complex computerised models (Fedra et al 1990, Abel et al 1992). A DSS may be defined as "a computer based system that helps decision makers confront poorly structured problems through direct interactions with data and analysis models" (Sprague et al 1986). A DSS should be used by policy makers so results should be available quickly without the need to consult "experts" and it must be easy to use and unambiguous. A DSS differs from other computer systems such as expert systems and information systems in that it is not designed to provide a single definitive answer but to assist in the decision making process. Simon (1960) provides a robust and convenient model for a cyclical decision process as the use of intelligence (identifying the problem), design (possible processes) and choice (the selection of a satisfactory process). The design of the DSS for HMIP is based on a number of design criteria principally: friendly, flexible and useful interface, coherent and consistent data, robust and appropriate models and clear, unambiguous displays of information.

The system is designed to run on a UNIX platform. The three basic components of the system: data, models and interface will be described separately.

2.1 DATA

The main sources of data used in the system are: operating rates, thermal efficiency and installed capacity at each power station, predicted deposition footprints and critical loads for acidity of soils in the UK.

The sulphur deposition fields from each station are predicted using the Hull Acid Rain Model (HARM) which provides figures at a 20km resolution (Metcalfe et al 1989). Changes in emissions produce different footprints by assuming a linear response between deposition rate and emission rate, for large deviations from actual emissions this assumption may not be reliable. The critical loads for acidity of soils are provided by the Critical Loads Advisory Group (CLAG) at 1km resolution (Hornung et al 1995). The soils critical loads map has been empirically derived from 1km soils data. Where deposition is greater than the critical load there is a risk of damage occurring, the excess of deposition over critical load is termed the exceedance. Critical loads are used in national and international policy discussions. Details of power station operations were provided by HMIP from data submitted by the generating companies. Emissions, predicted depositions and power generated were used to estimate a "pollution potential" for each source which can used in the objective function in the optimization routine.

2.2 MODELS

A linear programming (LP) module written in C is integrated into the system. Two "flavours" of optimization are offered, the minimisation of environmental damage subject to economic constraints and maximisation of energy production subject to environmental constraints.

The pollution potential minimization problem may be constrained by: existing authorised loads, thermal efficiency, total sulphur emitted and UK power requirements. As fuel accounts for around 60% of generating costs the thermal efficiency can be interpreted as an approximate measure of financial cost (exact costings for power stations are not available).

The use of the optimization routines is illustrated by two figures. Figure 1 shows the deposition pattern resulting from maximising power production subject to the total sulphur emitted being less than half a million tonnes per year. Using the power generated in the first optimization as the only constraint in an environmental damage minimisation optimization results in the deposition pattern shown in Figure 2. Although the main deposition still occurs around London there is a decrease in deposition on sensitive areas in Wales, the Pennines and the Lake District. Note, however, that the total amount of sulphur emitted has increased as this was not used as a constraint. Increased realism in the optimization requires proper costings of individual power stations and a representation of transmission constraints.

2.3 FUNCTIONALITY AND GRAPHICAL USER INTERFACE

Seven main facilities are available to the user: case sensitive help, retrieval of existing scenarios, inclusion of non-power station sources of pollution, specification of scenarios by altering individual power stations, specifying a region of interest, plotting maps,

Fig. 1. Screen dump of results of economic optimization

Fig. 2. Screen dump of results of environmental optimization

histograms and graphs and running optimization models. The facilities may accessed in any order although only one may be active at any time. The interface is primarily controlled with a pointing device (a mouse) and makes use of standard XView facilities for entering values and controlling the system. Across the top of the screen is the main control panel containing a number of icons giving access to the main options. Below and to the left of the control panel is the mapping area and to the right are areas for graphs and other statistics.

Some scenarios are provided with the system others are produced by the user. Non-power stations sources such as European or domestic emissions can be added to the power station footprints. A scenario can be generated by specifying a load factor for each individual source or by using the linear programming model (which is provided with its own GUI).

3. Error and Uncertainty

A problem with all DSS is that they allow a user freedom to combine measured and modelled data in potentially unexpected ways to answer questions that were not anticipated by the designer of the system. Errors and uncertainties which exist in the underlying data can be compounded in a manner which is difficult to estimate even when the precision and accuracy of the underlying data can be quantified. The DSS makes use of the best currently available data to provide support for decisions which cannot be delayed. Work is being carried out in a parallel project on ways of estimating and communicating the inherent uncertainty to potential users.

The output from the HARM model has been accepted by the Department of Environment for use in national assessments. HARM results are in good agreement with other atmospheric dispersion models operating at a similar scale and with existing measured data. Detailed estimates of the expected spatial uncertainty are not available. Critical loads used in the system are based on empirical estimates. Critical loads for the UK are being revised using a simple mass balance approach it is unclear whether any estimates of reliability will be distributed with the data. The linear programming model contains a series of assumptions about how power stations can operate. Assessing the relative importance of the different assumptions is impossible without access to commercially confidential data.

4. Conclusions

The DSS provides a useable and useful means to assess the national and regional impacts of different generating scenarios. Maps and statistics can be generated easily and quickly without the user having to master the use of a GIS. Invisible links to an interactive optimization model and access to the results of a complex air quality model are provide. The modular design adopted means that an interactive development process which is so necessary to the successful design of DSS can be carried out and the system can easily be remodelled to meet the current and projected needs of HMIP. Work is currently under way on extending the choice of optimizations offered, and the presentation of error estimates.

A reactive or proactive stance to decision making can be adopted using the DSS. Proactive analysis may be made by the user exploring different patterns of generation or by attempting to optimise operations using a linear programming approach. Reactive analysis responding to specific proposals put forward by the generators and other interested parties may be carried out quickly.

The system has demonstrated an attractive compromise to fostering the inclusion of spatial considerations within decision making between: large, specialist, expensive, tailor made DSS and the arcane and potentially inconsistent GIS world. It is hoped that this work will be carried forward to strengthen and facilitate efficient decision making about the management of power production and atmospheric pollution.

References

Abel D.J., Yap S.K., Ackland R., Cameron M.A., Smith D.F. & Walker G.: 1992, Environmental decision support system project: an exploration of alternative architectures for geographic information systems. *Int. Jo. of Geographic Information Systems*, **6** 193-204

Alcamo J., Shaw R. & Hordijk L.: (eds) 1990 The RAINS model of acidification. Science and Strategies in Europe Dordrecht, Netherlands Kluwer Academic Pub.

Bull K.R.: 1991 The critical loads/levels approach to gaseous pollutant emission control. Environmental Pollution **69** 105-123

Fedra K. & Reitsma F.R. 1990 Decision support and GIS. Reprinted from Scholten, H.J. & Stillwell J.C.H. *GIS for Urban and Regional Planning*, KLUWER Acc. Pub. Netherlands

Green R.J. and Newbery D.M.: 1992 Competition in the British electricity spot market, *Jo. of Political Economy* **100** 929-953

Hornung M., Bull K.R., Hall J.R., Langan S.J., Lovaland P. &Smith C.: 1995, An empirical map of critical loads of acidity for soils of great Britain. Accepted for Environmental Pollution

Metcalfe S.E., Atkins D.H.F. & Derwent R.G.: 1989, Acid deposition modelling and the interpretation of the United Kingdom secondary precipitation network data. Atmospheric Environment **23** No.9 2033-2052

Newbery D.M.: 1994 The impact of EC environmental policy on British Coal. Oxford Review of Economic Policy **9** No.4 66-95

O'Callaghan J.R.: 1995 NELUP: An introduction. Jo. of Environmental Planning and Management **38** No.1 5-20

RGAR: 1990 Review Group on Acid Rain. Third Report. Acid deposition in the United Kingdom 1986 - 1988. Dept. of the Environment & Dept. of Transport

Simon H.: 1960, The new science of management decisions. New York, Harper & Row

Sprague R & Watson H.J.: 1986, A framework for the development of DSS. In: *Decision Support Systems*, edited by R. Sprague & H.J.Watson Prentice Hall, Englewood Cliffs New Jersey

Watson P. & Wadsworth R.A.: (in press), A computerised decision support system for rural policy formulation, *Int. Jo. of Geographic Information Systems* (in press)

THE UNITED KINGDOM NATIONAL MATERIALS EXPOSURE PROGRAMME

R.N.BUTLIN, T.J.S.YATES, M.MURRAY AND G.ASHALL

The Building Research Establishment, Garston, Watford, UK, WD2 7JR

Abstract. In 1986, the National Materials Exposure Programme was set up within the United Kingdom to investigate the effects of acid deposition on buildings and building materials. Thirty sites were chosen, which represented a range of geographical and pollution climates. Each site met a minimum meteorological and pollution monitoring regime (including SO_2, NO_2). After four years, other sites were included (with less frequent data collection) and some sites removed. At each site, samples of 3 types of stone, mild steel, painted steel, Cu, Al and galvanised steel were exposed, with some of the stone sheltered from direct precipitation. Samples were removed periodically for analysis and dose-response relations derived for different materials. The empirical relationships derived are in the form of:

$$\text{decay rate} = a\,[SO_2] + b\,[H^+] + c\,[\text{rainfall}] + d$$

These dose response relations have been used to develop critical load maps for materials for the United Kingdom. Eight years of data have been collected, some for the UNECE task force programme. Laboratory tests using an Atmospheric Flow Chamber were also undertaken. Since the beginning of the programme addition materials have been exposed on some sites including mortars. A further set of eight sites has been used to assess the effects of ozone on a range of organic materials (for example polyvinyl chloride, polycarbonate, sealants). The paper presents up-to-date findings for the programme and confirms the dominance of dry deposition of sulphur dioxide as the main decay process for sensitive materials in areas of significant pollution.

Key Words: materials, buildings, stone, metal, pollution, decay, acid deposition

1. Introduction

In 1987 the National Materials Exposure Programme (NMEP) was set up in the United Kingdom to investigate the effects of pollution on building materials. It comprised two four-year exposure periods and was commensurate with an International programme (starting September 1987) for which the Building Research Establishment (BRE) was the lead laboratory for stone tests. One of the original aims of the NMEP was to produce a data set that could be used to develop predictive equations that relate the atmospheric concentrations of SO_2 and acidic rain to the observed degradation of a range of building materials. The first phase of the NMEP took place between April 1987 and 1991. It consisted of 29 sites. The sites were selected according to a set of requirements agreed by the Building Effects Review Group (BERG 1989) to provide a range of pollution and meteorological climates and were located on buildings (modern and historical) and on open sites. Four of the sites are also part of the International Materials Exposure Programme co-ordinated by the United National Economic Commission for Europe (UN/ECE). A second phase took place between 1991 and 1995 with some samples being exposed for the whole of the eight year period. The sites have been managed by BRE, the Warren Spring Laboratory (WSL), British Coal Corporation, Coal Research Establishment (CRE), National Power, Technology and Environmental Centre (CERL) and by PowerGen.

The materials exposed in the NMEP included metal (mild steel, galvanised steel, painted steel, aluminium and copper) and stone (Portland Limestone, Monks Park limestone and White Mansfield Dolomitic Sandstone). The metal samples were mounted on sloping racks facing south, and exposed to direct rainfall. Stone samples were mounted on freely rotating carousels in rain exposed and rain sheltered positions. In addition, samples of mortar and glass were exposed at some sites. Further details on the preparation of the stone and metal specimens can be found in the report of the Building Effects Review Group (BERG) (1989). The analysis of stone tablets is discussed in more detail in Butlin 1992, 1992a, and metal samples in Butlin 1992b. Samples have now been collected at 2, 4 and 8 years exposure. Results from the first phase of NMEP (i.e. after 4 years exposure) show that rates of degradation of new materials have declined in line with the reduction of SO_2 concentrations of ground level-sulphur. The results of have been represented as dose response functions that correlate to atmospheric pollution.

2. Development of Dose-response Relationships

2.1 CALCAREOUS STONE

Analysis of the data from the first phase of the NMEP has shown that it is possible to use regression analysis to fit a range of dose response functions to both the data sets for the sheltered and unsheltered tablets. Linear equations containing annual average atmospheric SO_2 (mg m^{-3}), total rainfall (mm) and rainfall acidity (gm^{-2} H$^+$) for the exposure period can account for >70% of the observed variation in calculated surface recession rates of unsheltered tablets (based on the weight change). It is also possible to predict the sulphate and nitrate contents of the sheltered tablets with similar accuracy. The following equations were obtained for Portland limestone. r= correlation coefficient.

4 year surface recession (mm) = 8.4 + 1.36(SO_2) + 0.0048 (Rainfall) + 29 (H$^+$) $r^2 = 0.73$
2 year concentration of SO_4^{2-} (mg g^{-1} sample) = 2079 + 191(SO_2) $r^2 = 0.75$
2 year concentration of NO_3^- (mg g^{-1} sample) = 55 + 3 (NO_2) $r^2 = 0.61$

2.2 METALS

Simple linear regression equations involving annual average SO_2 (mg m^{-3}), and total rainfall for the exposure period (mm) can account for up to 84% of the total observed variation. This make it possible to predict rates of corrosion (expressed as surface recession calculated from the weight change) to within $\pm 20\%$ at the majority of sites without including terms for deposition of chloride or particulate matter. R= rainfall

Mild steel 4 year surface recession (mm) = 41 + 1.9 (SO_2) + 0.012R $r^2 = 0.76$
Copper, 4 year surface recession in (mm) = 2.2 + 0.025 (SO_2) + 0.0037R $r^2 = 0.41$
Aluminium 4 year surface recession in (mm) = 0.72 + 0.012 (SO_2) - 0.00018R $r^2 = 0.43$
Galvanised mild steel 2 year surface recession in (mm) = 1.7 + 0.077 (SO_2) + 0.0011R
$r^2 = 0.85$

3. Discussion of the results from the first phase

In general, dose-response functions are only valid for the data set on which they are based. However, the wide range of pollution climates included in the NMEP make it likely that these dose-response functions can be applied to the UK. This assumption is being tested in the second phase of the NMEP (1991-95). It is possible to compare the surface recession predicted by the dose-response functions to the measured value. There is a significant correlation ($r^2 = 0.77$) between predicted and calculated degradation for Portland limestone. The correlation between the predicted and calculated rates indicates that it is reasonable to use these dose-response functions to predict rates of degradation and surface recession. The NMEP dose-response function for Portland limestone has been used to predict the ratio of the total degradation (pollution induced + natural) to the natural degradation for 1987 and for the modelled data assuming reductions in emissions of SO_2 of 60%, 70% and 90%. This data can be used to estimate the spatial area of the UK where the pollution induced damage is above the acceptable degradation rate. Details of the predictions and mapping procedures are given in a separate paper for this conference.

4. Preliminary results for the 8-year exposure trials

Limited preliminary analysis has been undertaken of the stone samples. Figures 1 and 2 indicate the profile of weight changes with time of the stone samples in rain exposed and rain sheltered positions at the BRE York site. For exposed samples, the linear reduction in weight of all three stone types (Mansfield, Portland, Monks Park) continued over 8 years. This reflects stone loss by acid deposition and rain water dissolution. In the case of the sheltered samples there are obvious anomalies in behaviour. For the Portland samples, the more or less linear increase in weight has continued. This is due to a build up of calcium sulphate on the surface which is not removed by leaching with rainwater. However for White Mansfield and Monks Park stone there is an increase in rate of weight gain after two years followed by a decrease, especially for White Mansfield. This could be due to movement of the samples after 3 years to a more exposed position which appears to give greater exposure to wind driven rain onto the 'sheltered' samples. The sharper drop in weight for White Mansfield can be explained by the existence of Magnesium Sulphate as product of reaction with sulphur dioxide. This is more soluble in rainwater than Calcium Sulphate, and so is washed out faster. It could also be possible that the anomalous behaviour of two out of the three stone types is due to the position of the carousels with respect to wind driven rainwater. Analysis of other samples from other sites will indicate which interpretation is correct.

5. Associated Studies

5.1 MORTARS

A series of four cement mortar types, both carbonated and uncarbonated were exposed at Birmingham, York and Wells in decreasing order of atmospheric sulphur dioxide (annual average) concentrations. At each site, samples were placed in rain exposed and rain sheltered positions. The mortars were selected as typical masonry mortars that would be susceptible to sulphur dioxide/sulphate attack. Some of the mortars were carbonated, others were not. They were cut into tablets and exposed in a situation simulating the effect of brick masonry using a non-absorbing insulant. Control samples were also used in the trials exposed to 30 ppb by dry deposition in a closed chamber. Analysis was via a ranking visual assessment system using chemical analysis of soluble salts and determination of total sulphate via acid extraction (uncarbonated samples only). The results for a 3 year exposure trial indicated that sulphate concentrations in the mortars correlated positively with local pollution levels. However after 3 years exposure, sulphate concentrations were not high enough to cause crystallisation damage.

5.2 CHAMBER STUDIES

The use of a controlled atmospheric investigation into the effects of sulphur dioxide on building material has been part of the overall programme to supplement data on dose-response functions obtained empirically from site investigations. The details of the equipment used and the results from a wide range of investigations using different stones, sulphur dioxide levels and humidities have been published elsewhere. (Grossi et al 1994, Lewry et al 1994). Perhaps the most significant findings from chamber studies relating to the field studies are the form and magnitude of the dose-response functions. The deposition of sulphur dioxide onto a fresh dry surface is an order of magnitude lower than into a wet one and is dependent on humidity. The form of the curve relating SO_2 to decay rate also needs further investigation. It is likely that the field trials in the NMEP give linear dose response functions on the early part of a SO_2 versus decay curve, which then becomes non linear at higher SO_2 concentrations.

6. The Effect of Ozone on Building Materials

A programme was set up to assess the effect of ozone on building materials. Initially a desk study drew the following conclusions:
- O_3 has no significant effect on surface coatings and sealants, the exception to this being polysulphide sealants and bituminous materials.
- Paints containing carbonate filters are not affected by O_3
- The degradation of rubber by O_3 is well known and the cost of preventive measures to the manufacturer and subsequently to the consumer is significant

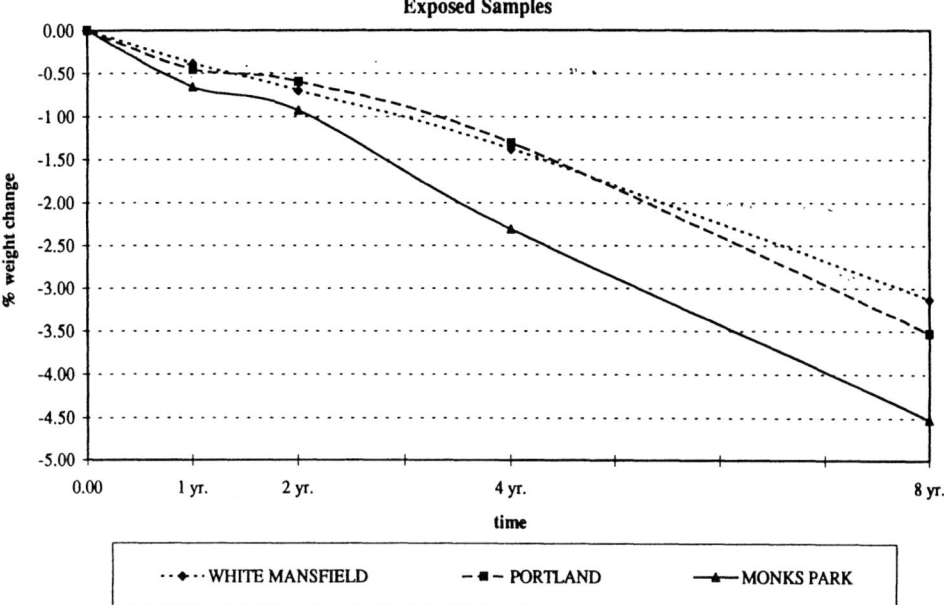

Figure 1. The weight losses of exposed stone samples at York (primary NMEP site).

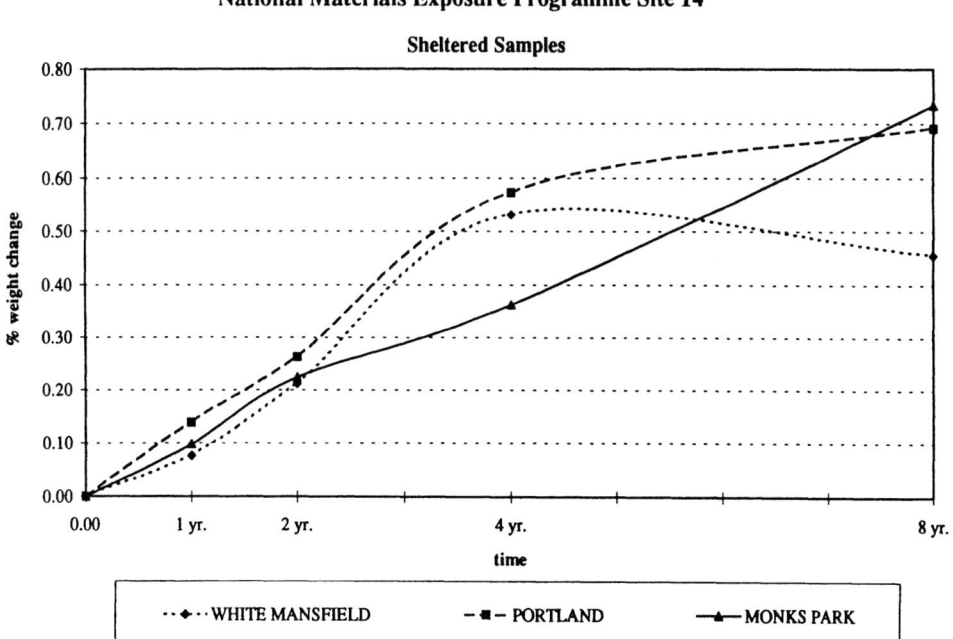

Figure 2 The weight increases of sheltered stone samples at York (primary NMEP site).

- O_3 can also degrade other polymers and bituminous materials
- the presence/absence of Ultra-Violet (UV) light may contribute to ozone effects

To determine the relationships between the rate of degradation, the ozone levels and other meteorological variables, an ozone exposure programme was set up. A range of organic building materials including plastics (structural and non-structural), sealants and roofing materials were exposed at nine sites around the UK. O_3 levels ranging from 11-33 ppm. The first years sample analysis is still underway.

7. Conclusions

The National Materials Exposure Programme in the UK has produced dose-response functions after four years exposure. These can be used to predict future areas in the UK still likely to be susceptible to further materials damage after current international abatement scenarios have been introduced. Preliminary results from the second phase of the NMEP indicate that the current dose-response functions are unlikely to change significantly. Investigations of mortar samples after 3 years exposure show pollution product analysis in the mortars correlate with SO_2 levels, although no sulphate damage was noticed.

Acknowledgements

The work has been funded by the AQ and CD Divisions of the UK Department of the Environment. The paper is Crown Copyright and published with the permission of the Chief Executive of BRE.

References

Building Effects Review Group (BERG). 1989. The Effects of Acid Deposition on Buildings and Building Materials. London HMSO.
Butlin et al. 1992. A Four-year Study of Stone Decay in Different Pollution Climates in the United Kingdom. Proceedings of the VIIth Congress on Stone Deterioration and Conservation, Lisbon, 1992, 345-353.
Butlin et al. 1992(a). Preliminary results from the Analysis of Stone tablets from the National Materials Exposure Programme (NMEP). Atmospheric Environment Vol 26B, No 2, pp 189-198. 1992.
Butlin et al. 1992(b). Preliminary results from the Analysis of Metal samples from the National Materials Exposure Programme (NMEP). Atmospheric Environment Vol 26B, No 2, pp 199-206. 1992.
Grossi, C.M., Lewry, A.J., Butlin, R.N. and Esbert, R.M. 1994. Laboratory studies on the Interaction between SO_2 Polluted Atmospheres and Dolomitic Building Stone. International Symposium on the Conservation of Monuments in the Mediterranean Basin, Venice, 1994, pp 227-232.
Grossi, C.M., Esbery, R.M., Lewry, A.J. and Butlin, R.N. 1994(a). Weathering of Building Carbonate Rocks under SO_2 Polluted Atmospheres. 7th International IAEG Congress, Lisbon, 1994, pp 3573-3582.
Lewry, A.J., Asiedu-Dompreh, A., Bigland, D.J. and Butlin, R.N. The Effect of Humidity on the Dry Deposition of sulphur dioxide onto Calcareous Stones. Construction and Building Materials 1994, Vol 8, No 2 pp 97-100.
Yates, T.J.S., Butlin, R.N. and Coote, A.T. Predicting the Degradation of Building Materials in the UK. Durability of Building Materials and Components 6. 1993. Pp 579-589.

EFFECT OF ACIDIFICATION ON ATMOSPHERIC CORROSION OF STRUCTURAL METALS IN EUROPE

D. KNOTKOVA[1], P. BOSCHEK[2] and K. KREISLOVA[1]

[1] *SVÚOM Praha a.s., U měšťanského pivovaru 4, 170 04 Prague, Czech republic,*
[2] *Charles University, Ovocný trh 5, 110 00 Prague, Czech republic*

Abstract. Results of atmospheric corrosion tests for structural metals in the open air and under shelter gained within multilateral European programs are analyzed from the viewpoint of pollution effects. New results of statistical analysis of corrosion and environmental data for the UN ECE ICP and ISOCORRAG programs are compared, trends in Europe for the period 1987 - 1992 are shown. In the new pollution situation it will be necessary to consider combined effects with factors such as O_3 and NO_2 with specific aspects according to the type of exposure (open air, shelter, indoor).

Key words: atmospheric corrosion, structural metals, effect of acidification, trends in corrosivity

1. Introduction

This publication deals with atmospheric corrosion of basic structural metals in free exposure atmospheric conditions and under simple shelters in climatic conditions in Europe proceeding from the results of multilateral European programs and their statistical treatment in the view point to evaluation of importance of acidifying pollution for the resulting corrosion effect. There are trends in corrosivity and industrial pollution during the last decade mentioned.

2. Materials and methods

Exposed materials (flat specimens of unalloyed carbon steel, weathering steel, copper 99,5 %, zinc 98,5 %, aluminium 99,85 %) and methods for environmental factors measurement and corrosion losses evaluation are presented in detail in the Manuals for the UN ECE ICP (ICP, SCI, 1988) and ISOCORRAG program (ISO SVÚOM, 1986). Methods are in agreement with ISO standards prescriptions for atmospheric field tests (ISO 8565:1992), corrosion loss evaluation (ISO 9226:1992) and removal of corrosion products (ISO 8407:1991).

3. Results and discussion

3.1. EFFECTS OF ACIDIFYING POLLUTION ON CORROSION OF METALS

Results of three multilateral world wide corrosion test programs are being summarized for deeper evaluation including deriving of damage functions related to environmental variables, providing data for different climatic and pollution situations for the exposure interval 1 - 4 years (ISOCORRAG, MICAT, UN ECE ICP), (ICP UN ECE, 1995a; Knotková, 1994; and Morcillo, 1994).

This publication is not aimed at further derivation of damage functions.

Contribution of single environmental factors to the corrosion attack can be considered according to the values of standardized regression coefficients listed in Table I, based on the analysis of 1-year results. A simple linear additive calculation model was applied for this treatment (Corrloss = a + B_1SO_2 + B_2TOW + B_3Cl + ...).

The results, however, show some distinct effect differences. This is affected by diversity of the environmental information in both exposure programs. A certain extent of statistical dependence of variables influenced obtained results too.

The comparison of the results of both exposure programs is easier for reduced environmental information (criterions given in ISO 9223 only). Results are evident from Table II. The similarity of results is better. The main difference is in the higher importance of TOW for UN ECE ICP results.

TABLE I

Contribution of environmental factors to corrosion losses of exposed metals - 1 year exposure (standardized regression coefficients derived from linear regresse model)

Metal	standardized regression coefficients for environmental factors							
	RH	TOW[1]	SO_2	NO_2	NaCl	Cl^{-}[2]	pH[2]	cond[2]
ISOCORRAG program								
carbon steel [3]	-	0,19	0,41	-	0,41	-	-	-
zinc [3]	-	0,15	0,38	-	0,39	-	-	-
copper [3]	-	0,07	0,25	-	0,55	-	-	-
aluminium [3]	-	0,19	0,55	-	0,17	-	-	-
ICP UN ECE								
weathering steel [3]	0,02	0,46	0,29	0,36	-	- 0,19	- 0,08	0,51
weathering steel [4]	0,31	- 0,01	0,51	0,02	-	- 0,21	0,01	0,45
zinc [3]	0,19	- 0,21	- 0,21	0,09	-	0,20	- 0,43	0,23
zinc [4]	0,46	- 0,49	- 0,24	0,16	-	0,07	- 0,36	0,16

[1] Time of wetness - period with relatively humidity above 80 % and temperature above 0°
[2] measured in precipitation
[3] unshetered exposure
[4] sheltered exposure

TABLE II

Partial correlation for corrosion losses and reduced environmental information (4 years of exposure)

program	UN ECE IPC (N = 37)		ISOCORRAG (N = 29)
metal	steel - unsheltered	steel - sheltered	steel - unsheltered
TOW	0,568 **	0,290	0,089
SO_2	0,545 **	0,697 ***	0,417 *
Cl	0,092	- 0,068	0,279
metal	zinc - unsheltered	zinc - sheltered	zinc - unsheltered
TOW	0,157	0,112	0,111
SO_2	0,641 ***	0,605 ***	0,782 ***
Cl	0,182	0,042	0,080

* 5 % significance level; ** 1% significance level; *** 0,1% significance level

Expressions of the partial correlation coefficients for more extensive set of environmental data for UN ECE ICP are presented in Table III. Inclusion of larger pollution characterization (NO_x, O_3) will be important for description of the contemporary types of atmospheres. More experimental results will be needed for derivation of single and combined effects (ICP, SCI, 1995a).

TABLE III

Partial correlation for corrosion losses and extended environmental information (UN ECE IPC - 4 years exposure)

	steel		zinc	
	unsheltered	sheltered	unsheltered	sheltered
TOW	0,550 ***	0,343 *	0,028	0,054
SO_2	0,385 *	0,725 ***	0,565 ***	0,540 ***
NO_2	- 0,249	- 0,388 *	- 0,064	- 0,080
pH	- 0,391 *	- 0,166	- 0,339 *	- 0,159
Cl^-	0,110	- 0,013	0,354 *	0,140
O_3	- 0,367 *	- 0,179	0,091	0,059

* 5 % significance level; ** 1% significance level; *** 0,1% significance level

3.2. TRENDS IN CORROSIVITY AS STUDIED IN EUROPE

The first multilateral corrosion testing program in Central- and East-Europe was that of the COMECON countries started in 1968 under the coordination of SVÚOM. Corrosion tests were organized in open exposure conditions and in sheltered louvre boxes.

Results give information about atmospheric corrosivity in Europe (Knotková, 1995) in the past decades.

Long term changes in SO_2 pollution in the atmosphere of the main corrosion test sites of SVÚOM (Letnany and Kopisty) and the Hungarian test site Orgovanyi are evident from Table IV.

TABLE IV

Mean annual concentration of SO_2 in open atmosphere

test site	SO_2 (mg.m^{-2}.d^{-1})										
	period										
	1976	1977	1978	1979	1980	1981	1982	1983	1984	1985	1986
Letnany	80	82	76	76	79	90	85	83	79	87	83
Kopisty	126	129	123	95	116	120	108	80	89	96	90
Orgovanyi	45	53	66	78	64	43	35	33	31	25	-

In connection with new environmental acts and with restructuralization of industry in the Central Europe the pollution level decreased significantly during the last period. SVÚOM as subcentre for structural metals within the ICP UN ECE for materials expressed the relations between the pollution and corrosivity changes in the period 1987 - 1992 (ICP UN ECE, SVÚOM, 1995b).

Statistic analysis focused firstly on the analysis of trends of selected corrosion factors whose importance had been proved in preceding stages of solution. These factors were measured systematically in the course of six years of observation, not just for 2 one-year exposure periods as it was in the case of corrosion losses. That is why the trends can be statistically assessed and the result, even for other factors than SO_2, will help at the analysis of highly complex problem.

Change in average value of SO_2 concentration on the 35 test sites of the ICP UN ECE is evident from Figure 1.

The importance of differences of variable averages for individual years was tested and a test of the linear trend was carried out in connection with the extent and the character of the data.

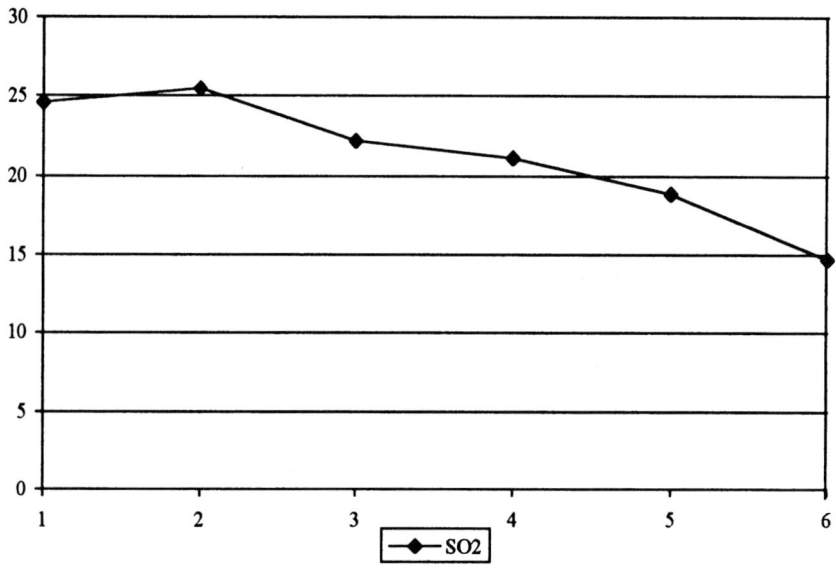

Figure 1. UN ECE ICP program - Mean values of SO_2 concentration ($\mu g.m^{-3}$) for 6 years period

Analysis of environmental factors trends

Data from 1988-1993 period are available for the statistic analysis of the importance of changes, respectively trends, of environmental factors values. In this case, a single-factor model of analysis of variance with repeated measures is suitable.

Corresponding F-test values and importance levels derived from them are mentioned in Table V, from which a significant trend of SO_2 concentration is evident as well as in the case of precipitation pH, which is however a dependent characteristic. The decreasing trend of NO_x concentration is outlined too.

TABLE V

Testing of trends in environmental characteristics

year	SO$_2$	NO$_2$	TOW	Cl	O$_3$	pH
1	24,53	35,36	0,41	1,76	44,27	4,58
2	25,46	34,09	0,38	2,19	45,47	4,69
3	22,18	32,90	0,39	2,30	47,47	4,70
4	21,10	32,27	0,39	1,85	41,93	4,76
5	18,87	30,14	0,43	1,77	46,00	4,94
6	14,67	29,04	0,42	2,34	41,67	4,85
N	32	30	20	19	15	21
F	10,51	3,81	1,44	1,34	3,27	3,09
sign.	$p<0,01$	$p<0,01$	n.s.	n.s.	$p<0,05$	$p<0,05$
F (lin)	17,46	7,33	1,10	0,28	1,62	8,44
sign	$p<0,01$	$p<0,05$	n.s.	n.s.	n.s.	$p<0,01$
F(6-1)	22,19	5,89	0,15	1,44	2,01	7,21
sign	$p<0,01$	$p<0,05$	n.s.	n.s.	n.s.	$p<0,05$

Analysis of corrosion losses trend (change)

We used matched samples t-test for testing the difference importance between means of repeated observations.

The Table VI presents means of corrosion losses and t-test values with their significance levels.

TABLE VI

Testing of trends in corrosion losses

year	St/un	St/sh	Zn/un	Zn/sh
1	252,3	92,7	13,7	9,7
6	190,2	53,4	7,9	4,3
N	31	31	31	31
t (6-1)	5,60	7,13	6,10	5,62
sign.	$p<0,01$.	$p<0,01$	$p<0,01$	$p<0,01$

Test of the difference of repeated evaluation of corrosion losses significance (Table VI) has rather an orientation importance. Certain difference in fall of means of repeated 1-year corrosion losses for carbon steel and zinc can be attibuted to the differences in corrosion mechanism to a certain extent. The difference in corrosion losses has been tested as significant for both metals and both exposure conditions.

4. Conclusion

4.1. EFFECTS OF ACIDIFYING POLLUTION ON CORROSION OF METALS

Statistical analysis of 1-year exposure results of the ISOCORRAG program represents a good documentation of decisive effects of SO$_2$ pollution and of chlorides in marine regions on the resulting corrosion attack of structural metals in temperate climatic conditions. The significance of time of wetness, expressed according to ISO 9223, is rather low.

An analogous analysis of the UN ECE ICP data with an larger environmental characterization does not lead to explicitly comparable conclusions. The significance of TOW was in some cases comparable with that of SO_2 pollution. The significance of NO_2 is much below that of SO_2.

4.2. TRENDS IN CORROSIVITY

Extent of data for the comparative analysis is relatively small, but some trends in atmospheric corrosivity and values of important environmental characteristics are evident:
- corrosivity categories (ISO 9223) derived from Scandinavian test sites are in majority of cases lower (C3 to C2), lower corrosivities for Scandinavian test sites correspond with the decline of SO_2 concentration and lower conductivity of precipitations,
- analysis of environmental factors trends proved both the importance of average values differences and a linear trend of these values, particularly for SO_2, then for pH and also for NO_x to a certain extent,
- average values of TOW, Cl^- and O_3 has not proved nor the significance of average values differences, neither the linearity of the trend,
- analysis of corrosion losses trends proved the significance of corrosion losses differences between mean values for two exposure periods tested.

Comparison of one-year corrosion losses for two exposure periods can serve only for the first informative estimation of trends on corrosivity.
It will be necessary to focus interest on evaluation of significance of combinations and levels of pollutions species typical for the contemporary types of atmospheres.

Aknowledgements

The multilateral exposure programs are the result of cooperation between organizations and namely national contact persons responsible for gathering meteorological and pollution data, for providing sites for exposure of materials and for cooperation in evaluation of results.

References

ICP UN ECE Effects on materials, including historic and cultural monuments - Main Research Centre (SCI): 1988, *Report No 1 - Technical manual*
ICP UN ECE Effects on materilas, including historic and cultural monuments - Main Research Centre (SCI): 1995a, *Report No 18 - Statistical analysis of 4 year materials exposure and acceptable detorioration and pollution levels*,
ICP UN ECE Effects on materilas, including historic and cultural monumets - SVÚOM - subcentre for structural metals: 1995b, *Report No 19 - Trends in corrosivity based on corrosion rates and pollution data*,
ISOCORRAG - collaboration testing program for validization of corrosivity classification systems, main research centre SVÚOM, Praha
Knotková D., Vrobel L.: 1988, *Degradation of metals in atmosphere*, **ASTM STP 963**, 248 - 263
Knotková D., Barton K.: 1992, *Atmospheric Environment Vol. 26A*, **17**, 3169 - 3977
Knotková D., Boschek P., Kreislová K.: 1994, *Atmospheric Corrosion*, **ASTM STP 1239**, 38 - 55
Knotková D., Vlčková J., Kreislová K.: 1995, *Corrosion 95, Paper 234*
Lipfert F.W., Bennari M., Daum M.L.: 1985, *Environmental Systems Analyses Group. Report*
Morcillo M.: 1994, *Atmospheric Corrosion*, **ASTM STP 1239**,

MAPPING OF URBAN MATERIAL DEGRADATION FROM AVAILABLE DATA

S.E. HAAGENRUD[1], J.F. HENRIKSEN[1], T. SKANCKE[2]

[1]Norwegian Institute for Air Research, P.O. Box 100, N-2007 Kjeller, Norway
[2]NORGIT-senteret A.S, P.O. Box 229, N-1601 Fredrikstad, Norway

Abstract. An assessment study of air quality and metallic material degradation was carried out in the Sarpsborg/Fredrikstad region in 1982-83. Based on air quality modelling, the developed dose-response function for steel was also modelled and mapped for the region (Haagenrud et al., 1985). This type of modelling can now be improved.

In the UN ECE International Cooperating Programme for Materials dose-response functions for a range of materials and mapping procedures based on the" critical/acceptable load" concept have been developed. By using these functions and procedures together with the available air quality data and dispersion models the corrosion is modelled and mapped for Oslo for 1985 and 1993. The corrosion modelling in the ArcView geographical information system is presented in the ENSIS integrated environmental surveillance and information system as a new ENSIS CORROSION module. Integrated in the ENSIS CORROSION module is also information from the recently completed digitalized building register GAB, containing all buildings in Norway. Application of such mapping for cost benefit assessments and maintenance planning is discussed.

1. D/R functions and mapping procedures from UN ECE ICP on materials

To evaluate the effect of airborne acidifying pollutants on corrosion of materials, the Executive Body for Convention on Long-Range Transboundary Air Pollutants decided to launch an international cooperative programme (ICP) within the United Nations Economic Commission for Europe (UN ECE). The programme started in September 1987 and involved exposure at 39 test sites in 12 European countries and in the United States and Canada (Kucera and Fitz,1995).

Samples have been withdrawn after 1, 2 and 4 years exposure and dose-response functions (D/R) are in the process of being elaborated (Kucera and Fitz, 1995). For unsheltered exposure most of the D/R-functions have the same form as for zinc:

$$ML = a + b\ TOW\ [SO_2][O_3] + c\ Rain\ [H^+]$$

where:
ML = mass loss after 4-year exposure, g/m^2
TOW = time of wetness (RH>80%, T>0°), as time fraction of a year (8760 hours)
$[SO_2]$ = concentration, µg/m^3
$[O_3]$ = concentration, µg/m^3
Rain = amount of precipitation, m/year
$[H^+]$ = concentration, mg/l

For zinc, which is used as an example in the present paper, the function is

$$ML = 14.5 + 0.043\ TOW\ [SO_2][O_3] + 80\ Rain\ [H^+]$$

The equations should at present be seen as provisional and may be subject to further elaboration when the results from the 8-year exposure will be available in 1996.

As for the *mapping procedure* the concept of "Critical levels" is not as easily defined and applied as for some natural ecosystems (Downing et al, 1993). This is due to the continuously occurring natural corrosion for materials. Within the same context the UN ECE ICP Working Group has therefore defined an "acceptable rate of corrosion" linked to pollution values, but which is considered acceptable based on technical and economic considerations. This provides the basis for *mapping* "acceptable areas" and "areas of exceedance" where the pollution level/load is unacceptable, in an analogous way to the maps produced for "critical levels" in other ecosystems. The following concepts and maps are defined at *level 1:*

Corrosion rate and acceptable corrosion rate. Corrosion rates are mapped from dose-response functions. It is recommended that *acceptable corrosion rate* should be clearly related to corrosion rates in areas with background pollution. As such it is recommended that background corrosion (K_{10}) is taken as the lower 10%-il of the observed corrosion rates in the ongoing ECE ICP (Kucera and Fitz, 1995). Acceptable corrosion rates are then defined as being a percentage increase in the background corrosion rate.

Acceptable exposure load/level is the concentration or load which does not lead to an unacceptable increase in the rate of corrosion. From this definition it is possible to calculate the acceptable pollution level from the acceptable corrosion rate and dose-response functions.

Environmental degradation parameters. In addition to these maps it is also recommended to *map* the geographical distribution of some of the most *important environmental degradation parameters* and their combinations.

Mapping at level 2 is similar to level 1, but includes data on the stock of material in each mapping unit and the economic costs associated with corrosion of these materials (e.g. replacement or repair costs).

2. Available data and air pollution models for Oslo city

For the corrosion mapping of Oslo the following data and models have been used:

For 1985, modelling has been performed for grid size 1x1 km^2. Calculation of SO_2 and NO_2 concentrations in the grid are based on information of emission sources (industry and heating, traffic in main and local roads) and modelling of the concentration by use of the NILU model system "KILDER" (Gram and Bøhler, 1993). For the corrosion mapping the existing grid winter concentrations were reduced to yearly average values by multiplying the grid values with 0.5 for SO_2 and 0.3 for NO_2, respectively.

The O_3 grid values were formed by using the UN/ECE-ICP established relation between O_3 and NO_2:

$$[O_3] = 60.5 \exp. -0.014[NO_2].$$

For TOW, Rain and the [H$^+$] concentration the yearly values measured at the Oslo UN/ECE-ICP exposure site were used.

The 1993 corrosion mapping was conducted using the 1985 values for TOW, Rain and [H$^+$] and a new inventory for the SO$_2$ and NO$_x$ emissions. A new model-system "EPISODES" with a grid square of 500m×500 m was used for calculation of the SO$_2$ and NO$_x$ concentrations (Grønskei et al, 1993). The values were transformed to yearly average values in the same way as for the 1985 data.

ENSIS "CORROSION". The integrated environmental surveillance and information system ENSIS was developed and demonstrated for the Winter Olympic Games in Lillehammer (Sivertsen and Haagenrud, 1994). Since then ENSIS is under further development and implementation in major Norwegian cities.

ENSIS contains a number of different applications and tools for air and water quality. Based on the needs and requirements of the user, the ENSIS-concept with the ArcView geographical information system, can be used to establish a tailor made application for any user need of environmental information. In that context the new ENSIS CORROSION module has been developed as a pilot demonstrator for the Oslo area.

GAB Building register. For cost assessments information on the materials and buildings exposed (stock at risk) must be included. The Norwegian building register, GAB, with parcels, properties, buildings and addresses, were established in 1978 and completed with digitalized information on almost all 4 mill buildings in Norway. Information from GAB is integrated into the ENSIS CORROSION module for Oslo.

3. Results

The *environmental degradation factors* NO$_2$, SO$_2$, O$_3$ and the combined factor SO$_2$ x O$_3$, have been modelled and mapped for 1985 and 1993, showing the considerable decrease in the SO$_2$ exposure levels from 1985 to 1993, which is due to regulations.

Level -1 corrosion maps have been produced for weathering steel, zinc, aluminium, copper and bronze. Figure 1a shows the map for zinc corrosion year 1985, where the highest 4-year corrosion of zinc in the centre is shown to be >32 g/m² (= 4.5 µm). Figure 1b shows the zinc corrosion mapping for year 1993, exhibited in the ArcView ENSIS CORROSION module, showing also results from the UN/ECE ICP test site.

According to the dose-response relationship the corrosion adds up from contributions from natural-, wet pollution- and dry pollution corrosion. The relation between these contributions can be calculated and exhibited for each grid in the GIS-system. As shown in Figure 1a the corrosion 23 g/m² adds up from 14.5- background (63%), 1.4- wet pollution (6%) and 7.1 g/m² dry pollution corrosion (21%), while for corrosion 32 g/m² the figures are: background 14.5- (45%), wet 1,4- (5%) and dry pollution 16.1 g/m² (50%).

For zinc and the other materials maps are produced showing areas with *exceedance of acceptable corrosion rates*.

For each of the materials *acceptable SO$_2$-levels* for present level of ozone and H$^+$ concentration and n = 1.2, 1.5 and 2 will be produced, as well as maps for *acceptable O$_3$ levels* and *acceptable H$^+$ levels*.

Assessment of corrosion costs-level 2. Maps have been produced showing exposure environment and the exposed built environment from the building register GAB This is the first phase of a planned cost assessment study for Norway based on this new information data and tools. By use of the tools available in the ArcView system counting of types of buildings and building materials can be done in each grid.

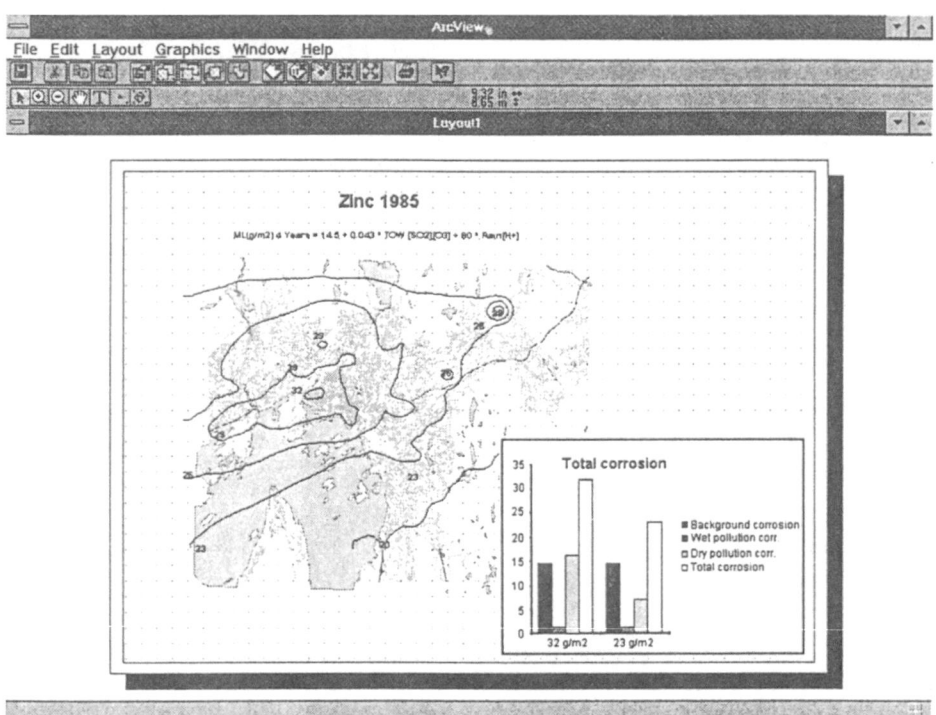

Figure 1a: *Mapping of zinc corrosion in the Oslo area from ECE ICP dose-response function and available data and models for Oslo.*
Mapping year 1985, where window shows contribution from various terms in the dose-response function at corrosion rates 23 g/m^2 and 32 g/m^2 respectively.

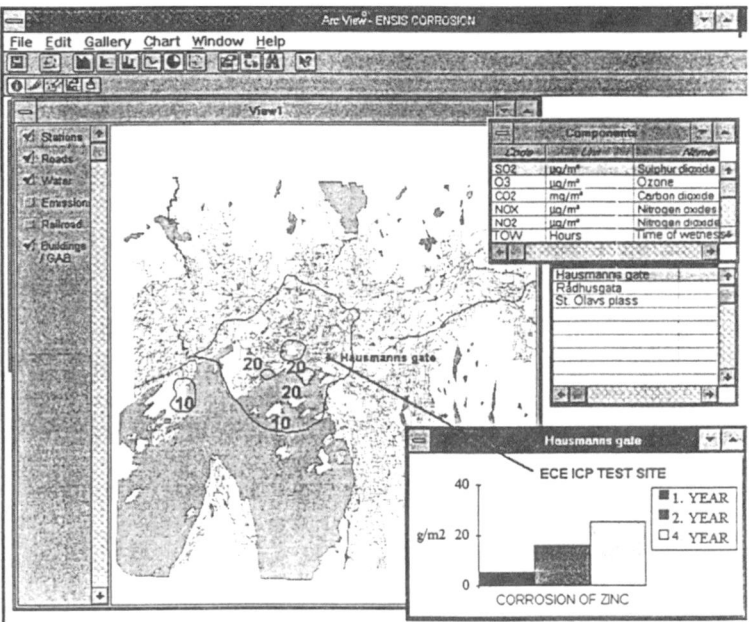

Figure 1b: *Mapping of zinc corrosion in the Oslo area from ECE ICP dose-response function and available data and models for Oslo.*
Mapping year 1993 exhibitied in the ArcView ENSIS CORROSION. Windows show components, measuring sites and zinc corrosion results from the ECE ICP test site in Oslo.

4. Discussion and conclusion

Corrosion of important materials have been modelled for the urban Oslo area according to the established UN ECE procedure and based on an adaptation of best available data and models. The quality of the modelling of course relies on the quality of the data. This quality is discussed elsewhere, but just as important is the development of user-friendly tools for modelling the degradation rate of the built environment in cities. This is a major step forward as regards more accurate cost assessment of materials corrosion as input to policy decisions on abatement strategies. By use of these models the effect of various pollutant emission reduction scenarios in the mapping area can be assessed.

Mapping of the corrosion rates can easily be transformed into maps for service life and maintenance intervals if the performance requirement is defined for the material in question. It could then serve as a tool for maintenance planning for individual users etc. In this respect the question of transformation and validation of these D/R functions to the micro-environment on the building surface has to be addressed. Such mapping would be of extreme value for standardization work going on within CEN and ISO/TC59 WG 9 "Design life of buildings" on durability of building materials and components.

Acknowledgement

The work providing data and tools for Oslo has been heavily supported by the Norwegian Pollution Control Authority (SFT) and the Norwegian Mapping Authority. Geir Bakke Nielsen, NORGIT, Sverre Solberg and Sam Walker, NILU have done much of the modelling and development of the ENSIS CORROSION demonstrator.

References

Downing, R., Hettelingh, J.P. and de Smet, A.M.: 1993, *Calculation and Mapping of Critical Loads in Europe. Status Report 1993.* RIVM Report No. 259101003.

Gram, F. and Bøhler, T.: 1993, *User's Guide for the "KILDER" Dispersion Modelling System.* Lillestrøm (NILU TR 5/92).

Grønskei, K.E., Walker, S.-E. and Gram, F.: 1993, *Atmos. Environ.*, **27B**, 105-120.

Haagenrud, S.E., Henriksen, J.F. and Gram, F.: 1985, Electrochemical society. Symposium on corrosion effects of acid deposition. Las Vegas 14-15 October 1985 (NILU F 53/85).

Kucera, V. and Fitz, S.: 1995, *Direct and indirect air pollution effects on materials including cultural monuments.* Acid Reign '95, 5th International Conference on Acidic Deposition, Gøteborg, 26-30 June 1995.

Sivertsen, B. and Haagenrud, S.E.: 1994, *EU 833 ENSIS '94. An environmental surveillance system for the 1994 winter Olympic Games.* Presented at: Vision Eureka ITEM Conference, Lillehammer, 14-15 June 1994 (NILU F 10/94).

ENVIRONMENTAL ASPECTS OF ATMOSPHERIC CORROSION

MIKHAILOV A A, SULOEVA M N, VASILIEVA E G

*Institute of Physical Chemistry, the Russian Academy of Sciences
Leninski Prospect 31, Moscow 117915, Russian Federation*

Abstract. Issues pertaining to the environmental potential in decreasing the damage due to atmospheric corrosion and the evaluation of atmospheric corrosivity for materials on the territory of the former USSR and Russia are considered. The most significant results obtained under the UN ECE Programme on Effects on Materials, including Historic and Cultural Monuments are discussed. The evaluation procedures and the first results of the mapping of the European part of Russia according to the relative air humidity and yearly values of time of wetness are reported. It is concluded that the multiplicative approach to the creation of damage functions for materials and the developed conception of acceptable levels are good basis for mapping of the European part of Russia. The problems of mapping of the Asian part of Russia are not solved for extremely cold regions with low values of the time of wetness.

Key words: Atmospheric Corrosion, Russia, Corrosivity, Mapping, Time of Wetness.

1. Introduction

The atmospheric corrosion of metals is an important industrial and environmental problem. The harmful effects of air pollution, particularly that of sulphur dioxide (SO_2), resulted in the acidification of soils and fresh water, alterations in flora and fauna, and an abrupt increase in losses due to atmospheric corrosion. Following the Convention on Long-Range Transboundary Air Pollutions, the UN/ECE International Co-Operative Programme on Effects on Materials, Including Historic and Cultural Monuments was elaborated and begun in 1987. The results of this Programme will be used for the assessment of the economic damage and the development of the measures aimed at decreasing atmospheric pollution. This communication considers the issues of fighting atmospheric corrosivity and summarizes the most important results of the four-year tests under the UN/ECE Programme.

2. Environmental potential in decreasing the damage due to atmospheric corrosion

The material damage caused by corrosion is enormous. Thus, in the former USSR, about 10 - 12% (or 20 - 25 million tons) of the annual production of metals was lost as a results of corrosion; i.e., almost each eighth blast furnace worked to compensate for corrosion losses (Tomashov, et al, 1986). Note that in addition to direct metal losses, the damage is also associated with the replacement or repairs of products and objects, which increases the economic losses manyfold. The damage due to the premature atmospheric destruction of historic and cultural monuments can hardly be estimated at all.

According to estimates made in developed industrial countries, the economic damage due to corrosion can be decreased using modern means and methods of protection only by 25% (Cabrillac et al., 1987). This means that any further decrease in these losses can

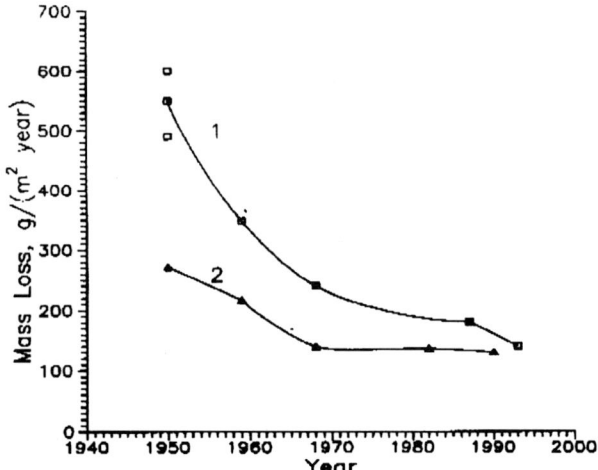

Fig. 1. Trend-effect of corrosion of Carbon Steel in Moscow (1) and near the Moscow at the Zvenigorod corrosion station (2).

be achieved mainly by decreasing the extent of air pollution. This may result in a significant economic effect.

In 1950 in Moscow, the corrosion rate for carbon steel, zinc, and copper was as high as 600, 19, and 12 g/(m^2 year), respectively (Berukshtis et al., 1971). A gradual (since 1953) transition to gas heating (instead of coal heating) in Moscow decreased the corrosion rate in the open atmosphere more than 4-fold for carbon steel, Figure 1, and about 2.5 - 3-fold for zinc and copper.

Over the past few years, the atmospheric corrosivity in Russia (and former USSR) has significantly decreased (Table 1). This decrease is due to a decrease in the release of atmospheric pollutants as a result of an abrupt reduction of the industrial production. Because Russia has enormous material resources, the decrease in corrosion losses can hardly be overestimated.

These examples show the importance of a well-substantiated approach to the problem of decreasing the effects of atmospheric pollution on materials as part of the work aimed at the development of a proper environmental policy.

3. UN/ECE International Co-operative Programme on Effects on Materials, Including Historic and Cultural Monuments

The major objective of this programme is the development of the dose-response functions for the calculation of the rates of the material deterioration, including the estimate of the dry and wet deposition (UN ECE ICP, 1988). Models allowing for the physicochemical peculiarities of atmospheric corrosion and taking into account the results of previous modeling (Stockle et. al., Knotkova et. al., Coote et. al., Henriksen et al., Tidblad et al., 1993) were developed on the basis of the results of four-year tests.

TABLE I

Statistical data on the distribution of the air pollution control stations in urban areas and industrial centers of the FSU (until 1989) and Russian Federation (1991) according atmospheric corrosivity with respect to Carbon Steel, Zinc and Copper. Corrosivity is given in categories of the ISO 9223 standard (Mikhailov et al., 1994).

Period	Material	Number of sites with specified corrosivity						Total number of test sites
		2-3	3	3-4	4	4-5	5	
1983-1987	C. Steel	230	228	246	457	27	205	1424
	Zinc-Copper		918	264	10	232		1424
1989	C.Steel	447	401	78	180		25	1131
	Zinc-Copper		846	239	21	21	4	1131
1991	C.Steel	303	238	17	13		1	572
	Zinc-Copper		563	7	1	1		572

The most significant results of this work are:
1. The development of dose-response functions for materials on the basis of a multiplicative approach involving strong synergistic effects of sulphur dioxide and ozone for metals (V. Kucera et. al., 1995), found under laboratory conditions (J. Tidblad, 1994, J.-E. Svensson, 1995). For an unsheltered four-year exposure, as an example, the equations for zinc and sandstone can be written as:

Zinc: $$ML = 14.5 + 0.043\ TOW[SO_2][O_3] + 80\ Rain[H^+] \qquad (1)$$

Sandstone: $$ML = 29.2 + 6.24\ TOW[SO_2] + 480\ Rain[H^+] \qquad (2)$$

where ML — mass loss after 4-year exposure, g/m^2;
TOW — time of wetness (RH > 80%, T > 0 °C), time fraction;
$[SO_2]$ and $[O_3]$ — average annual concentrations, $\mu g/m^3$;
Rain — amount of precipitation, m/year;
$[H^+]$ — annual concentration, mg/l;

Note that the TOW is an integral component of the equations for all metals, except for copper (which is quite unexpected because SO_2 is only active when surface of metal is wet). One of the reason for this can be the fact that the generally accepted TOW (at relative air humidity (RH) > 80% and temperature (T) > 0 °C) is for copper inconsistent with the real value. This circumstance may affect the constant term of the equation, value of which for some materials is comparatively high. The use of TOW for the description of damage to stone is also a new approach.

The equations universal for practically all territories of Europe were obtained over a wide range of environmental parameters, including TOW from mainly 2000 to 6000 h/year. Therefore the lower limit for the use of these equations can be the TOW value of 2000 h /year.

A significant territory of the Asian part of Russia is situated in the zone of cold and very cold climate and is characterized by low TOW (less than 2000 h/year; in a number of places, 1000 h/year) and consequently, by corrosion rate lower than those obtained under the UN ECE Programme. The obtained equations are hardly applicable to these regions.

2. The effect of the wet deposition (the total amount of deposited [H$^+$] in rain) on the zinc, copper, sandstone, and limestone corrosion. The results of eight-year tests will allow us to reveal the presence of the effect of the wet deposition parameters on the corrosion of other materials (steel, bronze, aluminium).

3. The development of regression equations for the assessment of the ozone levels from the data on nitrogen dioxide (NO$_2$) levels:

$$[O_3] = 60.5 \exp(-0.014[NO_2]) \tag{3}$$

and the assessment of the TOW from average annual data on T and RH:

$$TOW = -10700 + 176\, RH + 120\, T \tag{4}$$

where [NO$_2$] and [O$_3$] - average annual concentrations, ug/m^3;
TOW - time of wetness, h/year;
RH - relative humidity (annual mean), %;
T - temperature (annual mean), °C;

Equation (4) is valid for the temperature range $0 < T < 16$ °C (Mikhailov A.A. et al., 1994). These equations will be used for the assessment of the ozone levels and time of wetness in places, where these values are not measured. Thus, we used equation (4) for the assessment of the TOW in the European part of Russia. From the data of the national programme (Panchenko Yu.M. et al., 1984), we obtained an analogous equation for the assessment of TOW for the range of negative annual temperatures ($-17 < T < 0$ °C):

$$TOW = -1016 + 45\, RH - 92\, |T| \tag{5}$$

The average annual temperature in the European part of Russia changes from -8°C in the north eastern part to +15 °C in the south. The range of the changes in the relative air humidity is from 65 to 89%. Accordingly, the TOW in this part of Russia changes over a wide range from 1000 to 5000 h/year. Figures 2 shows the results of the mapping of the European part of Russia in terms of RH and TOW (mapping scale 150 x 150 km).

In the large cities the RH is normally lower than in rural regions. Therefore, for the mapping, each square of the grid was given the RH and TOW of a city with the largest population or (in the absence of such a city) the maximum values.

In the north eastern part, relatively low TOW values are mainly due to the temperature factor, whatever high the relative air humidity is (up to 80 - 89%). In the south eastern part (Saratov-Orenburg), low TOW values are associated with low air humilities. In the central, western, and southern part of the European part of Russia, increased TOW values contribute to corrosion processes.

In mountain regions (e.g., the Caucasus), the TOW values within one square may vary over a wide range from 1000 (high mountains) to 4000 h/year (piedmont).

The above-presented data are part of the preparatory work aimed at the mapping of the territory in terms of material damages and acceptable levels of pollutants for materials

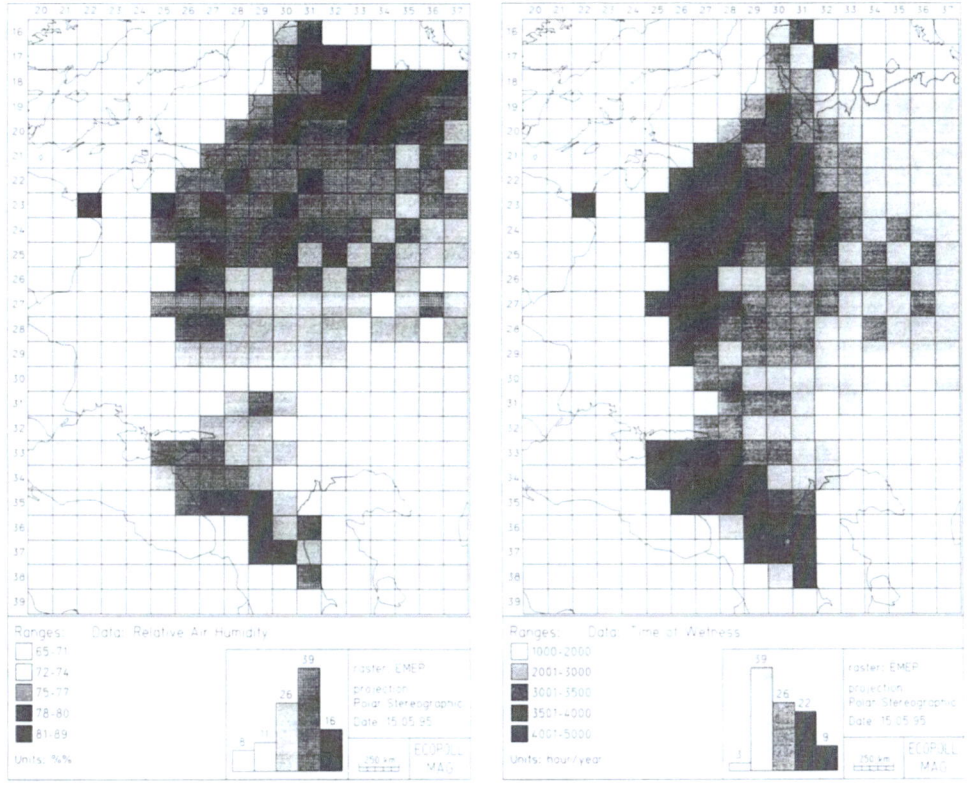

Fig. 2 Maps of the relative air humidity and time of wetness for the European part of Russian Federation

and buildings in accordance with the Manual on Methodologies and Criteria for Mapping Critical Levels/Loads (Manual, 1993).

4. Conclusion

On the whole, the results obtained under UN/ECE Programme on Effects on Materials, including Historic and Cultural Monuments, serve as a good basis for the assessment of material losses and acceptable levels of aggressive factors in Europe (and European part of Russia) and for the related mapping. At present, the problems of mapping of Asian part of Russia remain unsolved for extremely cold regions with low values of the time of wetness.

Acknowledgement

The UN/ECE Programme is a result of co-operation between the organizations listed below. Each of the organizations has been responsible for gathering meteorological and pollution data and for providing sites for the exposure of materials: SVUOM, Society for corrosion protection and surface treatment, Czech Republic; Technical Research Centre of Finland-VTT; BayerischesLandesamt fur Denkmalpflege, Germany; Agency for Energy Sources (ENEA), Italy; TNO Division of Technology for Society, Dept. of Environmental Chemistry, The Netherlands; Norwegian Institute for Air Research, Norway; Swedish Corrosion Institute, Sweden; Building Research Establishment, Dept. of Environment, United Kingdom; Ministerio de Obras Publical Y Urbanismo (MOPU), Spain; Institute of Physical Chemistry, Russian Academy of Sciences, Russian Federation; Ministry of the Environment, Estonia; Institute of Technology, Laboratory of Mineralogy and Petrology, Portugal; National Research Council of Canada and the Ministries of the Environmental of Canada and of Ontario; United States Environmental Protection Agency USA.

References

Berukshtis G.K., Klark G.B.: 1971, Corrosion Resistance of Metals and Metal Coatings in Atmospheric Conditions (in Russian), Nauka, Moscow, 159 p.
Cabrillac C., Leach J.S.L., Marcus P., Pourbaix A.: 1987, Metals and Mater.,3, p. 533.
Coote A., Yates T.J.S., Chakrabarti S., Murray M.J., Butlin R.N.: UN/ECE on Effects on Materials. 1993, Buiding Research Establishment, Watford, United Kingdom. Report N 13, Part 2.
Henriksen J.F., Arnesen K., Anda O., Rode A.: UN/ECE ICP on Effects on Materials. Evaluation of decay to paint systems for wood, steel and galvanized steel after 4 years of exposure. 1993, Norwegian Institute for Air Research, Lillestrom, Norway.
Knotkova D., Kreislova K., Holler P., Vlckova J.: 1993, UN/ECE ICP on Effects on Materials. 1993, SVUOM, Prague, Czech Repablic. Report N 12.
Kucera V., Tidblad J., Henriksen J., Bartonova A., Mikhailov A.A.:UN/ECE ICP on Effects on Materials, 1995, Swedish Corrosion Institute, Report N 18.
Manual on Methodologies and Criteria for Mapping Critical Levels/Loads and Geographical Areas where they are exceeded: 1993, Task Force on Mapping with the Coordination Center for Effects and Secretariat of UN ECE, Federal Environmental Agency, Berlin, Germany., p. 51.
Mikhailov A.A., Suloeva M.N., Vasilieva E.G.: 1994, Protection of Metals, 30, p. 329.
Panchenko Yu.M., Shuvakhina L.A., Mikhailovski Yu.N. 1984, Zaschita Metallov, 20, p.851 (in Russian).
Stockle B., Reisener A., Snetlage R.: UN/ECE ICP on Effects on Materials. 1993, Bavarian State Conservation Office, Munich, Germany. Report N 11.
Svensson J.-E.: 1995, The Influence of Different Air Pollutants on the Atmospheric Corrosion of Zinc. Doctoral Thesises, Department of Inorganic Chemistry, Goteborg, 62 p.
Tidblad. J.: 1994, Atmospheric Corrosion of Ni, Cu, Ag and Sn by Acidifying Pollutants in Sheltered Environments. Doctoral Thesises, Department of Materials Science and Engineering. Division of Corrosion Science. Royal Institute of Technology. Stockholm, 31 p.
Tidblad J., Leygraf C., Kucera V.: UN/ECE ICP on Effects on Materials. Corrosion attack on electric contact materials. Evaluation after 4 years of exposure. 1993, Swedish Corrosion Institute, Stockholm, Sweden.
Tomashov N.D., Chernova G.P.: 1986, Theory of Corrosion and Corrosion-proof Structural Alloys (in Russian), Metallurgia, Moscow, 358 p.
UN/ECE ICP on Effects on Materials. 1988, Swedish Corrosion Institute, Stockholm, Sweden. Report N 1.

LABORATORY STUDY OF SO_2 DRY DEPOSITION ON LIMESTONE AND MARBLE: EFFECTS OF HUMIDITY AND SURFACE VARIABLES

E.C. SPIKER[1], R.P. HOSKER Jr.[2], V.C. WEINTRAUB[1], and S.I. SHERWOOD[3]

[1] *U.S. Geological Survey, 956 National Center, Reston, VA 22092 USA*
[2] *U.S. National Ocean and Atmospheric Administration, Oak Ridge, TN 37831 USA*
[3] *U.S. National Park Service, Washington, D.C. 20013 USA*

Abstract. The dry deposition of gaseous air pollutants on stone and other materials is influenced by atmospheric processes and the chemical characteristics of the deposited gas species and of the specific receptor material. Previous studies have shown that relative humidity, surface moisture, and acid buffering capability of the receptor surface are very important factors. To better quantify this behavior, a special recirculating wind tunnel/environmental chamber was constructed, in which wind speed, turbulence, air temperature, relative humidity, and concentrations of several pollutants (SO_2, O_3, nitrogen oxides) can be held constant. An airfoil sample holder holds up to eight stone samples (3.8 cm in diameter and 1 cm thick) in nearly identical exposure conditions. SO_2 deposition on limestone was found to increase exponentially with increasing relative humidity (RH). Marble behaves similarly, but with a much lower deposition rate. Trends indicate there is little deposition below 20% RH on clean limestone and below 60% RH on clean marble. This large difference is due to the limestone's greater porosity, surface roughness, and effective surface area. These results indicate surface variables generally limit SO_2 deposition below about 70% RH on limestone and below at least 95% RH on marble. Aerodynamic variables generally limit deposition at higher relative humidity or when the surface is wet.

Key words: Dry deposition, sulfur dioxide, chamber, wind tunnel, corrosion, limestone, marble, humidity, moisture.

1. Introduction

Dry deposition of SO_2 is dependent upon environmental factors and characteristics of the receiving surface (e.g., McMahon and Denison, 1979; Sehmel, 1980; Lipfert, 1989; Sherwood *et al.*, 1990a,b). Important environmental factors include SO_2 concentration, wind speed, temperature, orientation, and size and shape of the surface. Important surface factors include wetness, composition, roughness, and area. Temporal changes in deposition rates reflect these variables, particularly wind speed and surface wetness. Well-controlled laboratory studies capable of isolating the effects of these variables are needed to obtain a quantitative understanding of the dry deposition process. This report describes some results of a laboratory study designed to quantify the relationship between relative humidity, aerodynamic conditions, SO_2 concentration, and SO_2 deposition to limestone and marble.

SO_2 deposition to materials has been shown to be enhanced by high relative humidity and surface moisture (Gilardi, 1966; Spedding, 1969; Johansson *et al.*, 1988; Johnson *et al.*, 1990; Spiker *et al.*, 1992a,b; Mangio, 1991). Water on exposed surfaces accelerates the deposition of SO_2 because of its solubility. Sulfur deposition on structures is strongly influenced by differences in surface temperature and wetness, caused by differences in radiative cooling at night and insolation during the day. The amount of moisture present on the surface and the chemical reactivity of the surface, determined to a great extent by its alkalinity, appear to be the two most important factors affecting uptake of SO_2 on buildings.

The dry deposition or uptake of SO_2 on stone and other building materials can vary by one to two orders of magnitude. In order to properly compare the wide range in SO_2 dry deposition rates reported for various building materials, it is necessary to account for differences in aerodynamic conditions in the different studies. Dry deposition involves turbulent transport of the pollutant to the near-surface boundary layer, followed by convection and diffusion to the surface, and ending with chemical or physical capture by the surface. Dry deposition rates are generally reported as a dry deposition velocity, v_d, defined as the flux to a surface, F, in grams per square centimeter per second (g cm^{-2} s^{-1}), normalized by the airborne pollutant concentration [C], in grams per cubic centimeter (g cm^{-2} s^{-1}),

$$v_d = F / [C] \quad (cm/s) \qquad (1)$$

The dry deposition process is conveniently represented in terms of a resistance analog of mass transfer (Slinn et al., 1978), where v_d is inversely proportional to the series resultant of the aerodynamic resistance (r_a), the boundary-layer resistance (r_b), and the surface uptake resistance (r_c), such that

$$v_d = (r_a + r_b + r_c)^{-1} \qquad (2)$$

Any one of the mass transfer resistances can be rate limiting, depending on the circumstances. The aerodynamic resistance is a function of wind speed and air turbulence. The transfer of a gas across the surface boundary layer depends on turbulence and molecular diffusivity. The surface uptake resistance r_c is a function of the surface chemistry, surface roughness and porosity, and surface moisture. Generally speaking, an increase in wind speed decreases the aerodynamic and boundary-layer resistances, r_a and r_b, and an increase in surface moisture decreases the surface resistance, r_c. The magnitude of r_c is what distinguishes the rate of dry deposition for different materials. Materials with some acid buffering capacity, such as sea water, wet zinc, or wet carbonate minerals, tend to have low surface resistance and a relatively high SO_2 deposition velocity (Sherwood et al., 1990a,b).

2. Experimental Methods

An experimental chamber was designed and constructed to directly determine the surface resistance component of SO_2 deposition to material surfaces over a range of typical environmental conditions (Spiker et al., 1992b). Care was taken to insure that the environmental chamber and sample holder were aerodynamically well characterized and uniform in the sample region. Key variables related to gas turbulent transfer and uptake include wind speed, turbulence characteristics, ambient air temperature and humidity, surface material, surface texture, surface temperature and wetness, and pollutant gas species and concentration. The environmental chamber, essentially a recirculating flow wind tunnel, automatically controlled these over ranges representative of ambient values typical of natural exposure.

The present study utilized samples of fresh Salem limestone and Shelburne marble. These stones have been used in many buildings and monuments and were selected as test stones for the field exposure program of the U.S. National Acid Precipitation Assessment Program (NAPAP)(Baedecker et al., 1992). Details relating to the characteristics of the stones can be found in McGee (1989). The total porosity was about 0.75 percent for the marble and about 18.6 percent for the limestone. The exposure surface of the stone was finished similar to stone used for exteriors of buildings: an 80-grit ground surface on the marble, and a smooth planar finish on the limestone.

Cylindrical samples of the finished limestone and marble, 3.8 cm in diameter and about 1 cm thick, were washed and dried, then pre-conditioned in a sealed container for at least 16 hours at approximately the same relative humidity to be used in the test. After exposure in the chamber, samples were leached with water to remove the absorbed sulfur species and the leachate was analyzed by ion chromatography (Spiker et al., 1992b).

Spiker et al. (1992b) found that the rate of SO_2 deposition to limestone and marble was initially high during the first few hours of exposure, which they attributed to SO_2 dissolution in the surface moisture on the stones. The SO_2 deposition gradually slowed and approached a constant rate within about 4 hours; subsequent deposition essentially balanced the reaction with the stone. Therefore, in the present study, the SO_2 deposition during the first 4 hours of exposure was not used to calculate the deposition velocity, in order to better estimate the long-term average rate of reaction with the stone. Replicate samples were removed from the chamber after the first 4 hours of exposure, and the amount of SO_2 deposited on these samples was subtracted from the amount deposited on the remaining replicate samples at the end of the exposure period, which usually lasted for 8 to 18 hours. The exposure times were corrected accordingly.

The SO_2 dry deposition velocity v_d was calculated directly from the flux of SO_2 deposited and the SO_2 concentration in the chamber air, according to Equation (1). The reproducibility, sample to sample within a single run, was approximately ± 5%. The surface resistance r_c was determined by using Equation (2) and assuming the aerodynamic and boundary-layer resistances $r_a + r_b = 128.2U^{-0.8}$, where U is the wind speed in centimeters per second (Spiker et al., 1992b). The $r_a + r_b$ values were determined empirically in the chamber utilizing wet carbonate-coated filter paper, with near-zero surface resistance, as the receptor surface.

3. Results

Figure 1 shows the relationship between relative humidity and SO_2 dry deposition for samples of fresh Salem limestone and Shelburne marble. With increasing relative humidity, limestone showed a rapid increase in deposition and a decrease in surface resistance r_c. There was much less SO_2 deposition on the marble. Based on the results of this study, the r_c of clean, smooth limestone ranges from about 13 s cm^{-1} at 50% RH to about 0.4 s cm^{-1} at 90% RH, and the r_c of clean, smooth marble ranges from about 75 s cm^{-1} at 70% RH to about 7 s cm^{-1} at 90% RH. The r_c of the marble was approximately twenty five times greater than that of the limestone at 80% RH. There was little deposition below 20% RH on clean limestone and below 60% RH on clean marble.

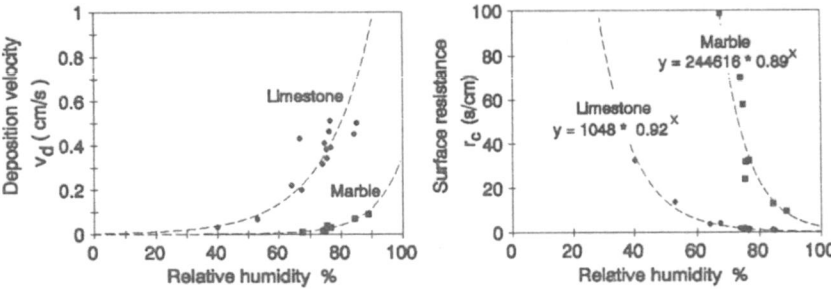

Fig. 1. The SO_2 dry deposition velocity, v_d, and the surface resistance r_c plotted versus the percent relative humidity, for fresh Salem limestone and Shelburne marble. The SO_2 concentration in the chamber air was 50 ppb SO_2 and the temperature was 26° C. Wind speed ranged from about 2.5 to 4.8 m s^{-1}, which accounts for the greater apparent variability of the deposition velocity values compared to the resistance values.

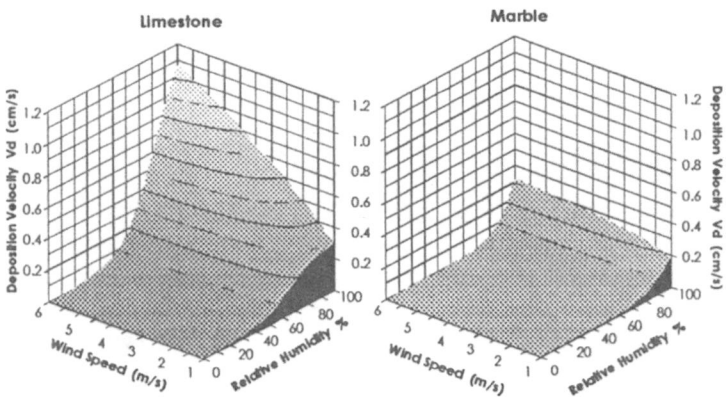

Fig. 2. The SO_2 dry deposition velocity, v_d, plotted versus the percent relative humidity and the wind speed in the experimental chamber, for fresh Salem limestone and Shelburne marble. The SO_2 concentration in the chamber air was 50 ppb SO_2 and the temperature was 26° C.

The relationship of deposition velocity to relative humidity and wind speed in our experimental chamber is summarized graphically in Figure 2. The surface resistance r_c, which is independent of wind speed, decreases exponentially at high relative humidity (Figure 1). On the other hand, the aerodynamic resistances ($r_a + r_b$), which are independent of relative humidity, decrease exponentially at high wind speed, ($r_a + r_b$) = $128.2U^{-0.8}$ (Spiker et al., 1992b). Therefore, the deposition velocity v_d, which is an inverse function of the total resistance ($r_a + r_b + r_c$), is significantly affected by both relative humidity and windspeed, and is highest when both wind speed and relative humidity are greatest (Figure 2).

The relative importance of wind speed versus relative humidity in our experimental chamber is shown in plots of the resistance values versus relative humidity (Figure 3). Although the total resistance to deposition decreases with increasing relative humidity, the relative importance of $r_a + r_b$ increases, especially at low wind speeds when $r_a + r_b$

is greatest. In our chamber, $r_a + r_b$ varies from about 3.2 - 0.9 s cm^{-1} over the wind speed range of 1 - 5 m s^{-1}. Figure 3 shows the increase in $r_a + r_b$, with decreasing wind speed. In our chamber, the surface resistance r_c severely limits deposition to the limestone below about 70% RH, while the aerodynamic $r_a + r_b$ values become rate limiting at higher relative humidity. In contrast, the r_c for marble dominates up to at least 95% RH. However, under other experimental conditions or under natural ambient conditions, the relative importance of wind speed versus relative humidity could differ significantly from these relations depending on a number of factors, including differences in orientation, size and shape of the surface, and differences in temperature and wetness, possibly caused by radiative cooling at night and insolation during the day.

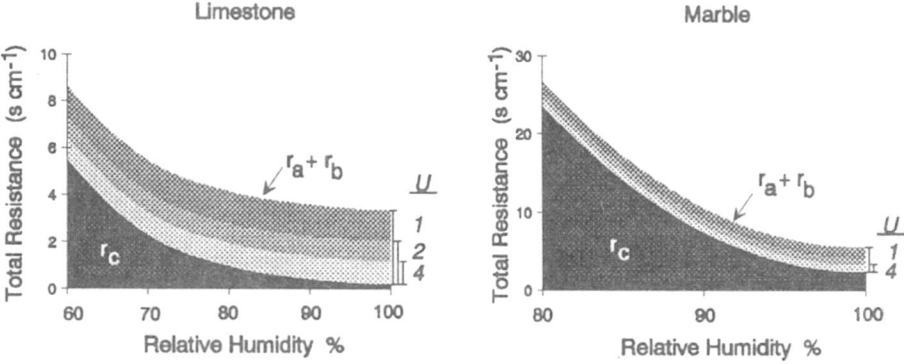

Fig. 3. The resistance to SO_2 dry deposition ($r_a + r_b$, aerodynamic and boundary-layer resistance; r_c, surface resistance) plotted versus the percent relative humidity in the experimental chamber, for fresh Salem limestone and Shelburne marble. The SO_2 concentration in the chamber air was 50 ppb SO_2, the temperature was 26° C. Values of $r_a + r_b$ are 3.22, 1.85, and 1.06 s cm^{-1}, respectively, corresponding to wind speeds of U = 1, 2, and 4 m s^{-1}.

4. Discussion

The large difference in deposition between the limestone and marble probably reflects limestone's greater porosity and surface roughness, which increases its effective surface area and the volume of adsorbed surface water (Sherwood *et al.*, 1990b). Water adsorption analysis indicates that the increase in SO_2 deposition with increasing relative humidity may correspond to a gradual coalescence of the adsorbed moisture film on the stone, suggesting the rate of SO_2 deposition is controlled by the wetted surface area of the stone (Spiker *et al.*, 1992a).

Comparison with other studies is difficult because most typically report only the overall SO_2 deposition velocity, with no separate determination of r_c. There are few published estimates of r_c for any building materials. However, it appears the v_d and r_c values observed here are similar to those reported elsewhere, indicating these results may be generally applicable. Lipfert (1989) gave a compilation from the literature of SO_2 dry deposition velocity data for various materials, including his estimates of the surface resistance values in several cases. Lipfert estimated the r_c of limestone to be in the range 2.1 - 3.9 s cm^{-1} at about 85% RH; this estimate is based on the results of

Braun and Wilson (1970). Lipfert (1989) also listed similar values for dry limestone and sandstone at 2 - 2.5 s cm^{-1}, based on the results of Gilardi (1966). Lipfert (1989) estimated the r_c of wet zinc as 0.3 s cm^{-1}, based on the results of Edney et al. (1986). Voldner et al. (1986) estimated the average r_c for SO_2 deposition to urban areas to be about 10 s cm^{-1}. They assumed urban areas consisted of a mixture of forests and buildings, presumably including non-carbonate, less reactive materials. This compares with average values of zero s cm^{-1} for water, about 2 s cm^{-1} for grassland, and about 7 s cm^{-1} for forests (Voldner et al., 1986). Wu et al. (1992)(Sherwood et al., 1990a, page 75), reported an average $r_a + r_b$ value of about 3 s cm^{-1} for an equestrian bronze statue in an outdoor park setting.

5. Conclusion

These results show that SO_2 deposition on limestone increases exponentially with increasing relative humidity. Marble behaves similarly, but with a much lower deposition rate. These trends reflect the porosity and effective surface area and adsorbed surface moisture of the stone. SO_2 deposition is driven by a combination of factors, including environmental and aerodynamic conditions, and surface chemistry. Water adsorption and surface wetness is affected by stone type, porosity, surface finish and roughness, the presence of hygroscopic contamination, and differences in radiative cooling at night and insolation during the day. Well controlled laboratory studies capable of isolating the effects of these variables are needed to better determine the dry deposition flux for a variety of material surfaces. Coupled with existing theories of atmospheric mass transfer, these data will help estimate pollutant uptake by stone surfaces in real-world conditions. This information will be useful in improving process-level understanding of the dry deposition process, and in developing sound dose-response relationships.

Acknowledgments

This research was done under the auspices of the National Acid Precipitation Assessment Program. We thank R. J. Pickering and P. A. Baedecker (U.S. Geological Survey) and M. Striegel (U.S. National Park Service) for their support.

References

Baedecker P. A., Reddy, M. M., Reimann K. J., Sciammarella C. A.: 1992, *Atmospheric Environment*, **26b**, 147-158.
Braun, R. C. and Wilson, M. J. G.: 1970, *Atmospheric Environment*, **4**, 371-378.
Edney, E. O., Stiles, D. C., Spence, J. W., Haynie, F. H., Wilson, W. E.: 1986, *Atmospheric Environment*, **20**, 541-548.
Gilardi, E. F.: 1966, *Absorption of atmospheric sulfur-dioxide by clay, brick, and other building materials*, Ph.D dissertation, Rutgers (University Microfilms) 67-8187.
Johansson, L-G., Lindqvist, O., Mangio R. E.: 1988: *Durability Building Material*, **5**, 439-449.
Johnson J. B., Haneef S. J., Hepburn B. J., Hutchinson A. J., Thompson, G. E., Wood G. C.: 1990, *Atmospheric Environment*, **24A**, 2585-2592.
Lipfert, F.W.: 1989, *J. Air Poll. Control Assoc.*, **39**, 446-452.
Mangio, R. E.: 1991, *The influence of various air pollutants on the sulfation of calcareous building materials*, Ph.D dissertation, Gothenburg University, Gothenburg, Sweden.
McGee, E. S.: 1989, *U.S. Geological Survey Bulletin No. 1889*, 25pages.

McMahon T. A. and Denison, P. J.: 1979, *Atmospheric Environment*, **13**, 571-585.
Sehmel, G. A.: 1980, *Atmospheric Environment*, **14**, 983-1011.
Sherwood, S. I., Gatz, D. A., Hosker Jr., R. P., et al.: 1990a, *Processes of deposition to structures*, NAPAP State of Science (SOS/T) Report 20, In: National Acid Precipitation Assessment Program, Acidic Deposition: State of Science and Technology, Vol. III.
Sherwood, S. I., Lipfert, F. W., et al.: 1990b, *Distribution of materials potentially at risk from acidic deposition*, NAPAP SOS/T Report 21, *ibid*.
Slinn, W. G. N., Hasse, L., Hicks, B. B., Hogan, A. W., Lal, D., Liss, P. S., Munnich, K. O., Sehmel, G. A., Vittori, O.: 1978, *Atmospheric Environment*, **12**, 2055-2087.
Spedding, D. J.: 1969, *Atmospheric Environment*, **3**, 683.
Spiker E. C., Comer, V. J., Hosker R. P, Sherwood, S. I.: 1992a, *Dry deposition of SO_2 on limestone and marble: role of humidity*, Proceedings 7th International Congress on Deterioration and Conservation of Stone, Lisbon, June 1992, 397-406.
Spiker E. C., Hosker R. P., Comer, V. J., White, J. R., Werre Jr. R. W., Harmon F. L. Gandy G. D., Sherwood, S. I.: 1992b, *Atmospheric Environment*, **26**, 2885-2892.
Voldner, E. C., Barrie, L. A., Sirois, A.: 1986, *Atmospheric Environment*, **20**, 2101-2123.
Wu, Y. L., Davidson, C. I., Dolske, D. A., Sherwood, S. I.: 1992, *Aerosol Science and Technology*, **16**, 65-81.

A METHODOLOGY FOR THE ECONOMIC ASSESSMENT OF MATERIAL DAMAGE CAUSED BY SO_2 AND NO_x EMISSIONS IN EUROPE

P. MAYERHOFER[1], M. WELTSCHEV[2], A. TRUKENMÜLLER[1], R. FRIEDRICH[1]

[1] *Institute of Energy Economics and the Rational Use of Energy (IER), University of Stuttgart, Heßbrühlstr. 49a, 70565 Stuttgart, Germany.* [2] *Bundesanstalt für Materialprüfung (BAM), Unter den Eichen 87, 12200 Berlin, Germany*

Abstract. Damage to materials causes high economic losses in Europe. A large part of this damage can be attributed to the emissions caused by the energy and the transport sector. In the paper, the procedure for the economic assessment of material damages caused by SO_2 and NO_x emissions in Europe is described. Model and data requirements are outlined, and gaps and uncertainties of the quantification are discussed. Two types of results are presented: First, the marginal (additional) costs of damage to materials caused by an additional power plant are assessed. The analysis covers plants with different technologies. Results for the fossil power plants are in the range of 0.0062 to 0.12 mECU/kWh. In addition, the total economic material damage due to the present air pollution was assessed. It is in the range of 2.9 to 5.3 x 10^9 ECU/year. However, the analysis has many uncertainties. Most noteworthy are the material inventories and partially the damage functions and input data.

1. Introduction

Damage to materials causes high economic losses in Europe. A large part of this damage can be attributed to the emissions caused by the energy and the transport sector. Fossil energy systems emit the pollutants SO_2, NO_x, and particulates which can directly or indirectly (via acid deposition) impair or soil external material surfaces. Pollution related damages to buildings include discolouration, failure of protective coatings, loss of detail in carvings, and structural failure. Damages to natural stone surfaces on buildings of cultural value are especially discussed in the public. However, impacts of acid deposition on materials are not restricted to buildings of cultural value.

Valuation of material impacts is complicated because it is highly dependent on the material and the cultural significance of the object in question. Replacement and maintenance costs are probably the easiest to evaluate. Estimation of effects of cultural value is more difficult because there are no appropriate inventories available, the treatment methods are very diverse and, most important, the knowledge about the relationship between any characteristics of the object and its cultural merit and thus between damage and loss of cultural value is small.

In the following a methodology for the assessment of additional maintenance and repair costs in Europe due to acid deposition is presented. The objectives are to assess the marginal damage (external) costs of emission sources (e.g. power plants) and the total costs of air pollution in Europe. For this purpose the available damage functions are evaluated. Materials inventories have been compiled and inquiries about maintenance criteria and costs have been made.

2. The Impact Pathway Approach

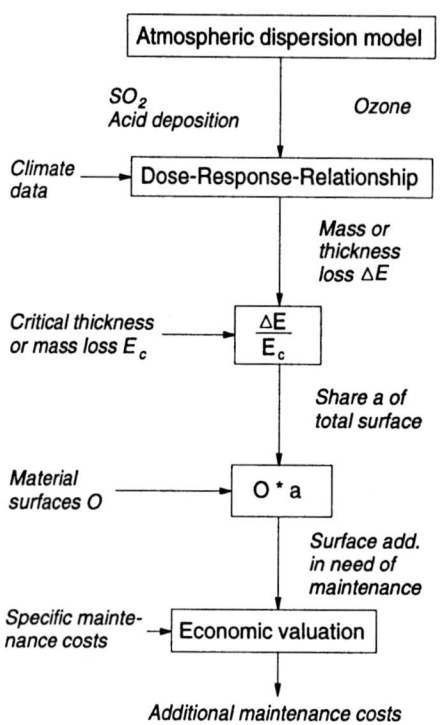

Fig. 1. Quantification approach for additional maintenance and repair costs

In general, economic assessments of material damages rely on regional differentiation studies (e.g. Isecke et al., 1990, Kucera et al., 1993). Maintenance and repair costs in various regions are determined by surveys and compared to the differences in pollution levels between the regions. A prerequisite is that the regions are, apart from the pollution levels, similar to each other. As it is too time-consuming, a Europe-wide application of the method is not possible.

The alternative is the damage-function approach. Corrosion damages are estimated using dose-response functions. Combined with material inventory data, maintenance costs can be determined. This approach has rarely been used in the past but for the assessment of Europe-wide damages it seems to be the most feasible approach. Furthermore it agrees well with the impact pathway concept as developed for other impacts of energy systems (European Commission, 1995).

Figure 1 shows the quantification approach. Input data and models needed are atmospheric dispersion calculations, dose-response relationships, climate data, material inventories, maintenance criteria, and specific costs.

3. Atmospheric dispersion modelling

Atmospheric pollutants from fossil fuel cycles are mainly emitted from the tall stacks of the power plants. They are transported by wind and diluted by atmospheric turbulence until they are deposited on the ground by either turbulent diffusion (dry deposition) or precipitation (wet deposition). Once emitted from the stack, some of the so-called primary pollutants take part in chemical reactions in the atmosphere to form secondary pollutants like sulfuric acid (H_2SO_4) and nitric acid (HNO_3).

The Harwell trajectory model which was used to estimate the concentration and deposition of acid species on a regional scale was originally developed by Derwent and Nodop (1986) and extended by Derwent et al. (1988). The model is a receptor-orientated Lagrangian plume model employing an air parcel with a constant mixing height of 800 m moving with a representative wind speed. The authors have developed a new implementation of the Harwell model which considers receptor specific meteorological input data. The model is called Windrose Trajectory Model (WTM). Base line emissions of NO_x, SO_2, and NH_3 for Europe were taken from the 1990 EMEP inventory.

4. Dose-Response Relationships

All materials exposed to the outdoor environment are subject to degradation caused by natural weathering processes and, more important nowadays, by air pollutants. Acid-forming and oxidizing pollutants are of concern, for inorganic materials especially acid-forming pollutants like SO_2 and NO_x.

The lime components of calcareous stones (limestone, sandstone, marble) are slightly soluble in water (natural weathering). In the presence of SO_2 a much faster chemical attack occurs if either liquid water or water vapour are available. Rain acidity also increases the solubility rate (European Commission, 1995). New empirical evidence shows that in general the effect of dry deposition of SO_2 is more important than that of wet acid deposition (Kucera, 1994).

Atmospheric corrosion of metals is generally an electrochemical process. It occurs only when the surface is wet. Thus, the corrosion rate of metals is determined by an interaction of different climatic parameters. The most important of these are humidity, precipitation, temperature and levels of atmospheric pollutants. Among the atmospheric pollutants, SO_2 causes most damage with the exception of coastal regions, where chlorides play a significant role, too (European Commission, 1995).

Air pollutants reduce the effectiveness of paint by retarding drying, by increasing the erosion rate of the protective coating, and by penetrating the paint film. All processes expose the surface to attack. When the paint surface shows flaws or scratches SO_2 may be important in the degeneration of paint systems (European Commission, 1995).

A multitude of corrosion damage functions have been theoretically or empirically derived in the past (Lipfert, 1987), (Lipfert, 1989), (Butlin et al., 1992a), (Butlin et al., 1992b), (Kucera et al., 1995). For the choice of dose-response relationships, transferability, experimental technique, exposure times, and agreement with available knowledge on lifetime data have been taken into account. At the moment the International Cooperative Programme (ICP) of the UN-ECE constitutes the best source for European damage functions (Kucera et al., 1995). E.g. for zinc the following damage function is used:

$$\Delta E = \frac{\Delta M}{\rho} = \frac{1}{\rho} \times \left[1.56 \times TOW \times \Delta SO_2 + 0.12 \times \Delta H^+\right]$$

where
- ΔE additional annual thickness loss in μm
- ΔM additional annual mass loss in g/m²
- ρ density in kg/dm³
- ΔSO_2 SO_2 concentration increment in μg/m³
- ΔH^+ wet acid deposition increment in meq/m²/year
- TOW fraction of time with relative humidity > 80 % and temperature > 0 °C

For comparison damage functions derived by Lipfert for zinc (Lipfert, 1987) and natural stone (Lipfert, 1989) and a stone function derived by Butlin (Butlin et al., 1992a) were applied as well. The zinc relationship was derived in a meta-analysis of results from multi-national test programs. The Lipfert stone relationship is based on theoretical considerations and has been compared to empirical results. The Butlin stone function is based on first results of the National Materials Exposure Programme (NMEP) of the UK.

The zinc relationships were applied to galvanised steel, too. For mortar the sandstone functions were used as proposed by Short (1994). Similarly the function was applied to rendering, too. The basis for this estimation is the assumption that calcite is the cementing agent in mortar and rendering that binds the sand aggregate together.

As ICP does not yet provide an applicable damage function for paint, a damage function derived by Haynie (1986) for carbonate paints was used as suggested by Short (1994). As this function was derived from exposures of unflawed paint and of paint in inert substrates and as different paints systems are used nowadays, there is only low confidence in the results of this function.

There are no relationships available for soiling.

5. Material Surface Inventories, Maintenance Criteria and Other Data

The material surfaces of buildings and constructions cannot be found in any statistics. For its derivation two steps are necessary. First, the material surfaces of representative buildings (building identikits) have to be assessed. Secondly, these identikits have to be spatially extrapolated based on building or population distributions.

In the ECOTEC study of 1986, building identikits for Birmingham (UK), Dortmund and Köln (Germany) were compiled (ECOTEC, 1986), (Hoos *et al.*, 1987). These have been extrapolated to western Europe based on population data with the exception of Germany for which building statistics are available. Furthermore, the extensive inventory of galvanised steel in the UK derived for (European Commission, 1995) has been used. For Stockholm (Sweden), Sarpsborg (Norway), and Prague (Czech Republic) statistically based inventories of outdoor material surfaces were compiled by Kucera *et al.* (1993). These inventories were extrapolated to northern and eastern Europe based on population data. Unfortunately there are no specific building identikits known for southern Europe.

The dose-response functions yield thickness or mass losses. It is assumed that there is a critical value which induces maintenance or repair measures. These values can be derived from lifetime and pollution data using damage functions. Published information and expert assessment were used as far as available. Based on the data given in Kucera *et al.* (1993) and Short (1994) critical thickness losses between 25 and 200 µm were derived for the different galvanised steel categories in Europe. Critical thicknesses for paint range between 20 to 250 µm. For natural stone and rendering the lifetime data would suggest a critical thickness of about 70 µm. But for this material a critical thickness of 3 to 5 mm is used, as the reason for the short lifetime is certainly more soiling than corrosion only. Large ranges are used for the critical values to cover the differences in individual behaviour.

The specific costs for maintenance and repair were taken from different sources. Country-specific values were used as far as possible. For Germany inquiries were made and for the UK data from ECOTEC were taken. Specific costs for western Europe were derived from the British and German ones. For eastern and northern Europe the costs given in Kucera *et al.* (1993) were used.

Table I

Power plant characteristics and quantified material damage costs

Power plant	COAL1 1989	COAL2 2005	OIL1 1989	OIL2 2005	GAS1 1989	GAS2 2005
Power Plant Characteristics						
Plant type	Pulverised coal, FGD, DENOX, dedusting	Combined cycle, coal gasification	Gas-turbine	Combined cycle, FGD	Gas-turbine	Combined Cycle
Net efficiency	43 %	48.5 %	31.1 %	47.0 %	33.2 %	57.6 %
SO_2 emissions [g/MWh]	570	150	1088	798	0	0
NO_x emissions [g/MWh]	570	300	822	798	577	208
Material damages in ECU/MWh						
Natural stone	0.0022	0.00068	0.0044	0.0035	0.00037	0.00014
Zinc and galvanised steel	0.039	0.012	0.071	0.050	0.012	0.0036
Mortar and Rendering	0.0084	0.0029	0.017	0.012	0.0029	0.0010
Paint	0.014	0.0045	0.024	0.018	0.0042	0.0015
Total	0.063	0.020	0.12	0.083	0.019	0.0062

6. Results and Discussion

Results were calculated for six hypothetical fossil power plants, all located in Lauffen near Stuttgart, Germany. The technologies of these power plants were based on the IKARUS database (Wehovsky *et al.*, 1993). Present technologies (reference year 1989) as well as future technologies (reference year 2005) were used for coal, oil, and natural gas.

Some characteristics of these power plants and the material damages quantified for them are summarised in Table I. They are in the range of 0.0062 to 0.12 ECU/MWh. The COAL1 power plant without FGD and DENOX would cause 0.37 ECU/MWh damage costs, 0.31 ECU/MWh more than COAL1. As there are several unquantified impacts (e.g. soiling, cultural value loss) and the galvanised steel inventory for countries other than the UK is incomplete, these results are only a lower boundary of the total economic costs due to the emissions of the power plants. Galvanised steel is the most affected material. The results of the Lipfert function for zinc are about one order of magnitude lower than those of the ICP function. For natural stone the Butlin function yields higher and the Lipfert function lower results than the ICP function. The results differ by factors from 5 to 10. The uncertainty of all results is high. The reasons are above others the material inventories and partially the damage functions and input data.

The total economic material damage in Europe was calculated by comparison between the present air pollution as calculated by WTM and SO_2 concentrations and acid depositions as valid for the moment in the north of Ireland. These costs are in the range of 2.9 to 5.3 x 10^9 ECU/year. Such results could be used for the attempt to correct the present calculations of the gross national product for environmental quality (green accounting).

7. Conclusions

First attempts were made to estimate marginal and overall costs of material damages in Europe. The damage costs for six exemplary fossil power plants range from 0.0062 to 0.12 ECU/MWh. The overall costs in Europe are in the range of 2.9 to 5.3 x 10^9 ECU/year. The range of uncertainty is very high. The results constitute only a lower boundary of the total damage costs as there are several material damages not yet quantified. As the quantification of Europe-wide damages is not only important for the concept of external costs but also for green accounting, research efforts should be intensified in the direction of the remaining uncertainties and gaps of knowledge.

Acknowledgements

This work was supported by the Stiftung Energieforschung Baden-Württemberg, the European Commission, and the Friedrich-Flick-Stiftung.

References

Butlin, R.N., Coote, A.T., Devenish, M., et al.: 1992a, *Atmospheric Environment* **26B**, 189-198
Butlin, R.N., Coote, A.T., Devenish, M., et al.: 1992b, *Atmospheric Environment* **26B**, 199-206
Derwent, R.G., Nodop, K.: 1986, *Nature* **324**, 356-358.
Derwent, R.G., Dollard, G.J., Metcalfe, S.E.: 1988, *Q.J.R. Meteorol. Soc.* **114**, 1127-1152.
ECOTEC: 1986, *Identification and Assessment of Materials Damage to Buildings and Historic Monuments by Air Pollution*, Report to the UK Department of the Environment.
European Commission, DGXII, Science, Research and Development, JOULE: 1995, *Externalities of Fuel Cycles, ExternE project, Report No. 3, Coal and Lignite Fuel Cycles*, (in press).
Haynie, F.H.: 1986, *Atmospheric Acid Deposition Damage due to Paints*, US-EPA Report EPA/600/M-85/019.
Hoos, D., Jansen, R., Kehl, J., et al.: 1987, *Gebäudeschäden durch Luftverunreinigungen - Entwurf eines Erhebungsmodells und Zusammenfassung von Projektergebnissen*, Institut für Umweltschutz, Universität Dortmund.
Isecke, B., Weltschev, M., Heinz, I.: 1990, *Volkswirtschaftliche Verluste durch umweltverschmutzungsbedingte Materialschäden in der Bundesrepublik Deutschland*, UBA-Forschungsbericht 90-101 03 110, Bundesanstalt für Materialforschung und -prüfung (BAM), Berlin.
Kucera, V., Henriksen, J., Knotkova, D., Sjöström, Ch.: 1993, *Model for Calculations of Corrosion Cost Caused by Air Pollution and Its Application in Three Cities*, Report No. 084, Swedish Corrosion Institute, Roslagsvägen.
Kucera, V., Tidblad, J., Henriksen, J., et al.: 1994, *Statistical analysis of 4 year materials exposure and acceptable deterioration and pollution levels*, Report No. 18, Swedish Corrosion Institute, Stockholm.
Lipfert, F.W.: 1987, *Materials Performance* **26**, 12-19.
Lipfert, F.W.: 1989, *Atmospheric Environment* **23**, 415-429.
Short, N.R.: 1994, *External Costs of Fuel Cycles - Impact on Materials*, Report for Natural Environmental Research Council & CEC DGXII, Aston Materials Services Ltd., Birmingham.
Wehovsky, P., Leidemann, W., Lezuo, A.: 1993, *Daten: Umwandlungssektor, IKARUS-Teilprojekt 4*, Siemens AG - Energieerzeugung KWU, Erlangen.

MAPPING OF CRITICAL LOADS AND LEVELS FOR POLLUTION DAMAGE TO BUILDING MATERIALS IN THE UNITED KINGDOM

R.N.BUTLIN, T.J.S.YATES, AND B.CHAKRABARTI

Building Research Establishment, Garston, Watford, UK, WD2 7JR

Abstract: The United Kingdom National Materials Exposure Programme was initiated in 1986 to study the effects of acid deposition on building materials. The output data in the form of empirical dose-response equations (described elsewhere) have been incorporated into a geographical information system (GIS). In addition, data for the stock at risk of building materials has also been used. The dose-response relations indicate a dominance of dry deposition of sulphur dioxide in the decay process. Critical level/load maps have been determined for a number of materials. General pollution and meteorological data sets are also included in the mapping process. Maps give 'exceedence squares' on a 20 km square grid basis, indicating the unprotected areas or those still at risk for a given scenario for SO_2 reduction in the context of the UNECE protocol for sulphur. In order to derive maps of areas sensitive to pollutants in the future a model, HARM 7.2, is used for the prediction of distribution of emissions of pollutants in the UK. A series of maps has now been produced for different materials at 70% and 80% scenarios for the reduction of SO_2. Studies of the sensitivity of the exceedence maps to the accuracy or variation of the components in the dose-response equations have been undertaken. Results from the mapping programme and the sensitivity analysis are presented together with discussion of the concept of critical loads of materials.

Key words: Critical loads, Sulphur, dose-response functions, Damage functions, acid deposition.

1. Introduction

The United Kingdom has taken a leading role in developing the science of the critical loads approach. This approach is intended to improve the efficiency and effectiveness with which the problems of the control of acid rain can be addressed on a national and European basis. The concept of critical loads is already being applied to the biological environment and it can be demonstrated that the sensitivity of freshwater systems and vegetation to the effects of air pollution and acidic rain varies from place to place depending on the local geology and other factors.

The critical load for a particular receptor-pollutant combination is defined as the highest deposition load that the receptor can withstand without sustaining long-term damage (Bull 1991). The critical load can be considered as a threshold deposition below which significant damage does not occur. There is, however, a difference between the response of building materials and the natural environment to pollutant input. Building materials deteriorate under natural conditions, that is in the absence of pollutants, while natural systems can tolerate certain conditions without any damage.

Different methods that can be used to define critical levels or loads and methods for mapping these were discussed at a workshop run under the auspices of the UNECE in March 1993. The results of the workshop were included in the UN/ECE Draft Mapping Manual (UN/ECE 1993).

2. Determining critical loads for buildings

The research programme that is being undertaken for the UK Department of the Environment includes the integration of the results from the National Materials Exposure Programme (NMEP) into the critical loads/levels framework and consequent mapping of predicted degradation. The mapping of the UK is based on a 20km grid and uses data for 1987-89 for SO_2, rain acidity and rainfall and projections of SO_2 concentrations for a number of future emission scenarios. This data is provided by participants in the UK Critical Loads Advisory Group (CLAG). It is possible to use this approach to identify areas where the degradation of materials is predicted to be unacceptable and to examine the effect of future reductions in emissions of acid gases.

Deterioration rates in mm year^{-1} can be calculated using dose-response functions. Functions have been derived from field and laboratory research programmes in the UK and elsewhere. In general, these functions describe the degradation of materials in rain-washed conditions but some are available that predict the deposition of sulphate in rain sheltered conditions. At present, dose-response functions are available for a number of calcareous stones and for ferrous and non-ferrous metals. Those used in this study are described in Yates *et al.*, (1993) and Butlin *et al* (1995). It is important to quantify the relative sensitivity of other building materials, in particular concrete, mortar and bricks. There is uncertainty in extrapolating deterioration rates determined in laboratory and from the use of small coupons in site exposure programmes to large buildings. It is also important to consider the interaction of pollutants, for example synergistic or catalytic effects, and the possible modification of critical levels by other factors, for example rainfall or temperature.

The range of values over which the dose-response function is considered valid must be considered. At present, there is no evidence that pollution episodes are important and it is long-term (annual mean) values that are important. NO_x is not currently included in the dose-response functions as there is only limited evidence that it is important for calcareous stones or metals. It is also not possible to include any expression for smoke or particles that may result have catalytic effects or result in soiling. It may also be necessary to include a factor for the expected lifetime of a building materials or component in the dose-response function in order to identify those components that will undergo significant deterioration within the time-scale being considered. It is appropriate to determine a Background Deterioration Rate because materials will deteriorate in pollution free climates. In the UK this rate has been defined as the rate of deterioration of a material or component when subjected to the background load. In general, the background rate will be equal to the rate of deterioration that will occur in the absence of pollutants and can be calculated using the dose-response functions with SO_2 concentration equal to zero. It can also be measured directly in areas remote from pollution sources. This requires data on rainfall (mm year^{-1}) and rainfall acidity (H^+) to be available. The absence of a threshold means that it is necessary to define an "acceptable" level of damage. This could be defined in terms of an absolute or relative degradation rate. Appropriate criteria might be in terms of an "acceptable" lifetime or where the pollution-enhanced degradation did not exceed a defined multiple of the natural (i.e. background degradation). This can be expressed as:

Acceptable Deterioration Rate = [n] x the Background Deterioration Rate

The value of n is to be determined by further UN/ECE research but is likely to be in the range 1.2 - 1.5. If required, this concept can also be used to define target levels or loads, that is a pollutant concentration or deposition which results in a rate or extent of degradation that is considered acceptable, and which must be attained by a specified date.

In order to determine those areas or regions with the largest building stock at risk (in absolute, heritage or economic terms) some form of map of building stock is required. A number of approaches can be used ranging from determining the total building stock-at-risk by survey of individual buildings to the mapping of urban and suburban buildings from satellite land use maps. However, in all of these cases some estimate of the materials and their quantities in each building needs to be made.

3. Applying Critical Loads Mapping to the United Kingdom

At the UN/ECE workshop on mapping critical loads for buildings and building materials held in Bath in March 1993 (UN/ECE 1993), it was recognised that mapping can be undertaken at three levels. These are: L1 basic exceedence rates, L2 exceedance rates with stock of material and economic costs, and L3 inclusion of more materials, synergistic effects and climate modifying. It is also possible to use this approach to examine the effect of future reductions in emissions of acid gases and to assess the effect of using different "acceptable levels of damage". This report describes the results of applying Level I to the UK. It includes calculation of SO_2 concentrations that will result in the deterioration rate being reduced to an acceptable rate. The mapping has also included the integration of satellite data on the areas classified as urban and suburban. The mapping has been carried out using the datasets currently available as part of the Department of the Environment research on critical loads and levels.

3.1 CRITICAL LOADS FOR BUILDING MATERIALS IN THE UK

The second series of maps produced as part of this programme shows the areas where the predicted deterioration rate, expressed as the ratio of total deterioration/background deterioration, for a specific material is greater than a nominated value (n=1.2, 1.5 or 2.0). The maps show the areas at risk under the same three scenarios used in the first series. The areas identified as being 'unprotected' have then been classified by their critical SO_2 concentration, that is the concentration of SO_2 (in mg m^{-3}) predicted to give a deterioration rate equal to the nominated value (N) x background, an example is given in Figure 1. The final map shows 'unprotected' areas classified by the percentage of the square that is classified as urban/suburban.

The spatial distribution of unprotected' areas for the different materials and the effect of assuming different values of N can clearly be seen in this series of maps. In addition, the effect of different future emission scenarios can also be seen. The extent of areas 'unprotected', both the total area of the squares and the area classified as urban/suburban, under the different scenarios are summarised in Tables 1 and 2. Tables 3 and 4 illustrate the effect of assuming different values of N on two of the six materials for which results are available.

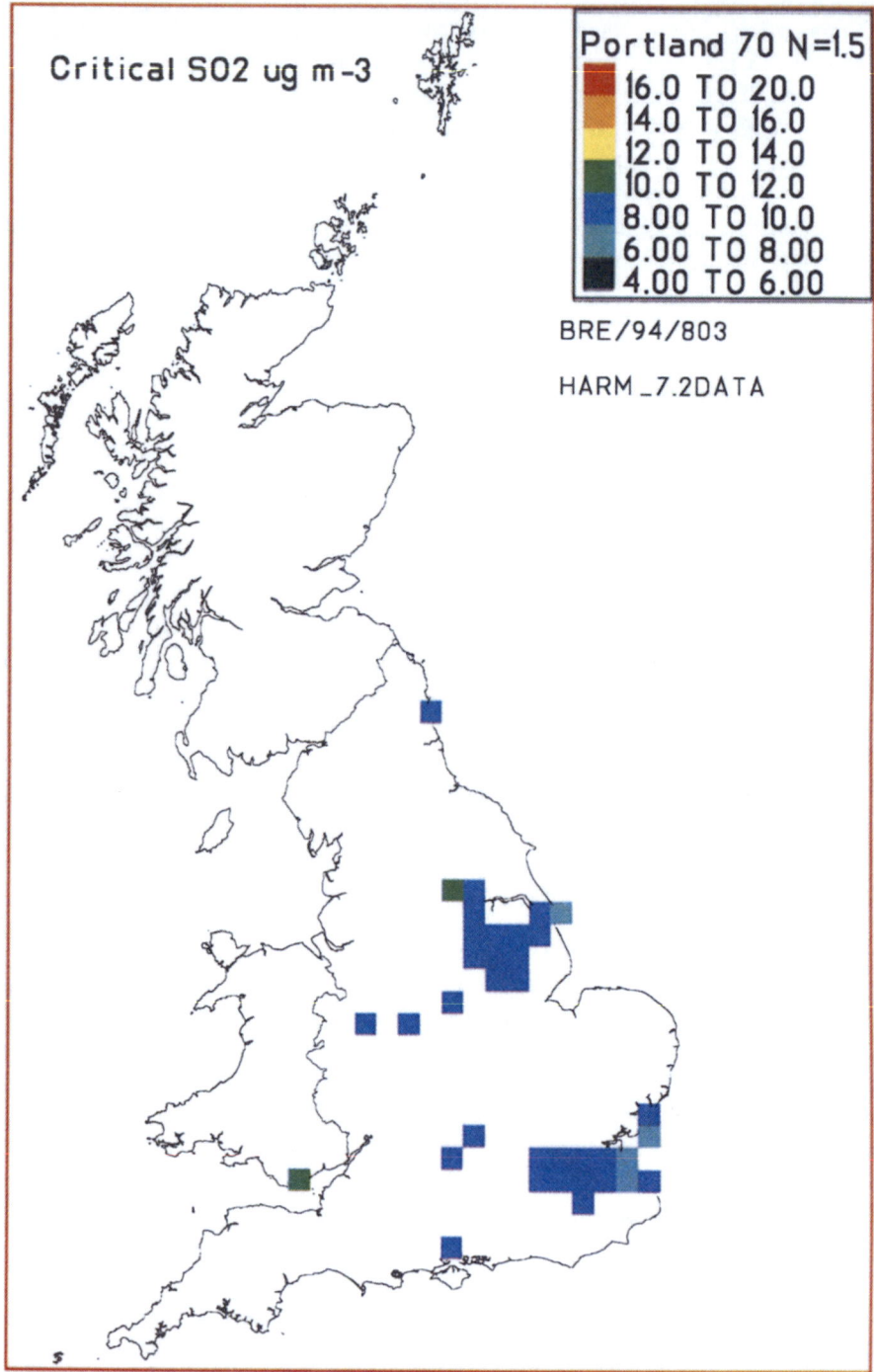

Figure 1. Map of the UK showing areas where the predicted deterioration rate for Portland limestone based on 1987 data is greater than 1.5 x background. These areas have then been classified by their critical sulphur dioxide concentration (mg m^{-3}). See text for further details.

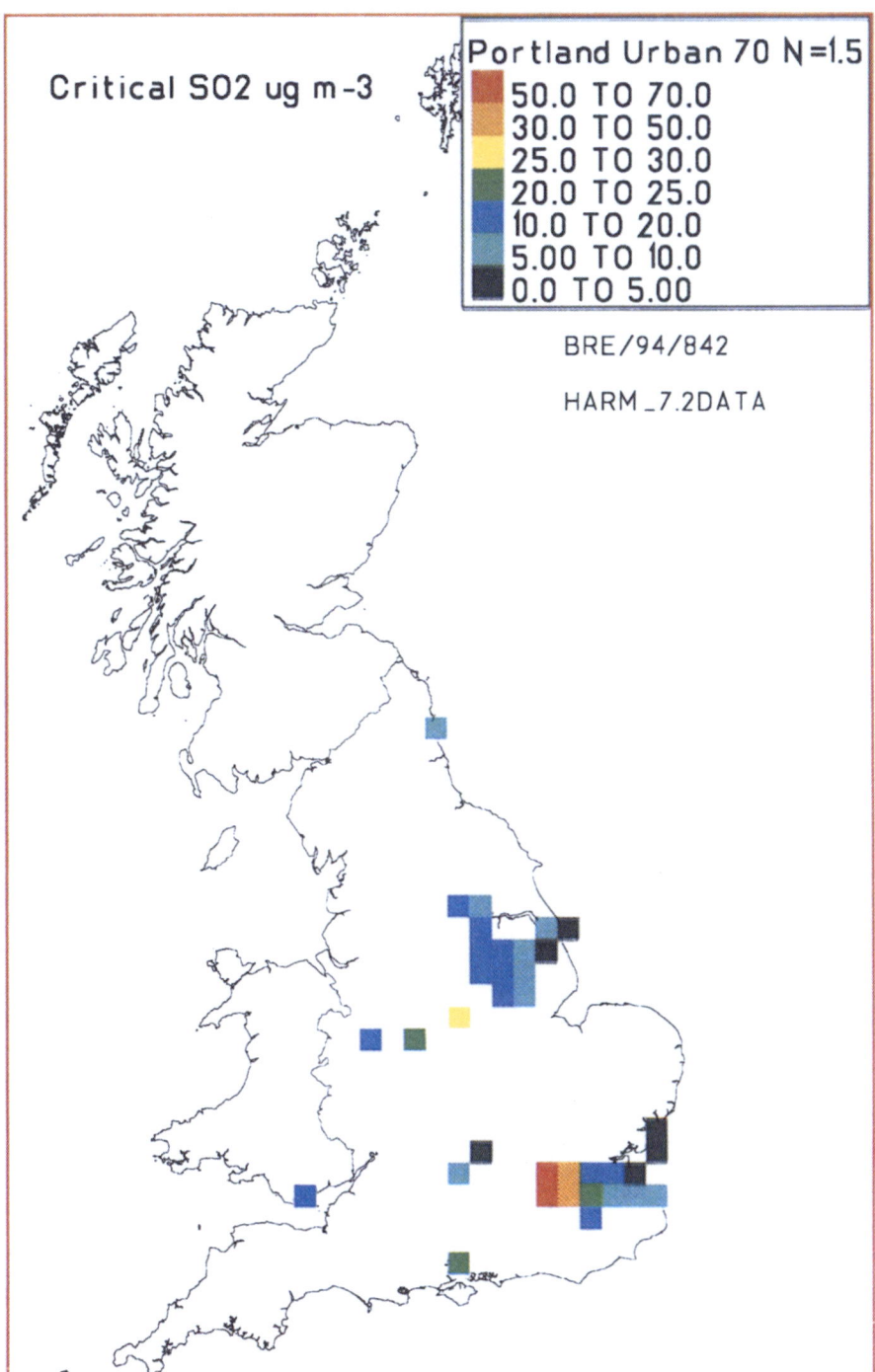

Figure .2 Map of the UK showing areas where the predicted deterioration rate for Portland limestone assuming a 70% reduction in emissions is greater than 1.5 x background. These areas have then been classified by the percentage or the square described as uran/suburban.

Table 1
Number of squares unprotected at N = 1.5 Figures in brackets are total area unprotected in $10^3 km^2$. Note: total number of squares = 782 (312.8).

Material	1987 (Present)	70% Reduction	80% Reduction
Portland	268 (107.2)	36 (14.4)	5 (2.0)
Mansfield	401 (160.4)	267 (106.8)	184 (76.3)
Mild Steel	130 (52.0)	2 (0.8)	1 (0.4)
Galvanised	110 (44.0)	2 (0.8)	0 (0.0)
Copper	12 (4.8)	0 (0.0)	0 (0.0)
Aluminium	32 (12.8)	0 (0.0)	0 (0.0)

Table 2
Total area classified as urban/suburban in $10^3 km^2$ and unprotected at N = 1.5.
Note: total area classified as urban/suburban = 15.75 x $10^3 km^2$

Material	1987 (Present)	70% Reduction	80% Reduction
Portland	12.36	2.22	0.24
Mansfield	14.23	11.84	8.52
Mild Steel	8.50	0.09	0.06
Galvanised	7.79	0.09	0.00
Copper	1.07	0.00	0.00
Aluminium	2.83	0.00	0.00

Table 3
Number of squares unprotected at N = 1.5. Figures in brackets are total area unprotected in $10^3 km^2$

Material/N	1987 (Present)	70% Reduction	80% Reduction
Portland N=1.2	452 (180.8)	302 (120.8)	242 (96.8)
Portland N=1.5	268 (107.2)	36 (14.4)	5 (2.0)
Portland N=2.0	88 (35.2)	1 (0.4)	0 (0.0)
Steel N=1.2	381 (152.4)	274 (107.6)	167 (66.8)
Steel N=1.5	130 (52.0)	2 (0.8)	1 (0.4)
Steel N=2.0	60 (24.0)	0 (0.0)	0 (0.0)

Table 4
Total area classified as urban/suburban in $10^3 km^2$ and unprotected at N = 1.5

Material/N	1987 (Present)	70% Reduction	80% Reduction
Portland N=1.2	14.44	12.88	10.80
Portland N=1.5	12.36	2.22	0.04
Portland N=2.0	6.84	0.06	0.00
Steel N=1.2	13.98	12.24	8.16
Steel N=1.5	8.50	0.09	0.06
Steel N=2.0	5.25	0.00	0.00

3.2 VALUING THE BENEFITS OF REDUCED EMISSIONS

The dose-response functions and the distribution of the stock-at-risk can be used to estimate the financial benefits associated with different scenarios for the reduction of SO2 emissions. These benefits will occur during the period of reduction (possibly relative to the baseline scenario) and also beyond the time when the acceptable level is reached but the full cycle of repair has not been completed. Full details of the methodology being applied to estimate the benefits associated with reduction in SO_2 emissions can be found in the report prepared by ECOTEC Research and Consulting Limited for the United Kingdom Department of the Environment (Report - Critical Levels and the Spatial Distribution of the Stock at Risk, December 1993). It must also be noted that any economic assessment must not be restricted to buildings in the UK but include consideration of all aspects of the environment on the UK and Europe. Overall, it seems unlikely that there will be many occasions when economics alone will be able to give all of the necessary guidance and some combination of scientific and economic assessment will be required.

4. Conclusions

- The mapping of exceedances on a 20km square grid based on dose-response functions derived from filed data enables the prediction of the likely significance of future abatement policies to be assessed for building materials.
- The mapping has been undertaken with the concept of critical loads for buildings using a background deterioration rate which is the deterioration rate at zero sulphur dioxide concentration.
- The economic benefit of reduced emissions for buildings can be estimated from knowledge of the building stock and dose-response functions.

References

Bull, K. 1991, *Critical Loads Maps for the United Kingdom.* Paper produced by the Critical Loads Advisory Group (CLAG) for the UK Department of the Environment.
Butlin, R.N., Yates, T.J.S., Murray, M. and Ashall, G.: forthcoming, *Water, Air and Soil Pollut.*, Porc. Gotenberg Conference.
Ecotec. 1993, *Critical levels and the spatial distribution of the stock at risk.* Report for UK Dept of Environment.
United Nations Economic Commission for Europe (UN/ECE): 1993, Convention on Long-Range Transboundary Air Pollution Draft Manual on methodologies and criteria for Mapping Critical Levels/Loads.
Yates, T.J.S., Butlin, R.N. and Coote, A.T.: 1993, *Durability of Building Materials and Components*, 6, 579- 588.

DETERIORATION OF COPPER AND BRONZE CAUSED BY ACIDIFYING AIR POLLUTANTS

A. REISENER, B. STÖCKLE AND R. SNETHLAGE
*Bavarian State Conservation Office, P.O. Box 10 02 03,
80 076 Munich, Germany*

Abstract. The 'ICP on effects on materials' was launched in 1985, within the framework of the Convention on Long-range Transboundary Air Pollution with the aim to find an approach for clearing the main gaps in the knowledge about material degradation caused by environmental impacts. The project was organized as a staggered 8 year's material exposure programme accompanied by an extensive measuring programme for environmental parameters. During the evaluation of the 4 years data set it took shape that the most efficient parameters for quantifying the material degradation are the mass loss and especially for copper the corrosion volumes. Bronze already reacts extremely sensitive on low SO_2 concentrations. The starting corrosion rate for copper is suprisingly high. A strong impact of chloride on the formation of pin holes for sheltered copper could be detected without showing high mass loss. At simultanous presence of ozone the corrosive action is catalysed by it's oxidation power and leads to severe mass loss. Based upon the 4 years data set for most of the materials preliminary dose-response-functions have been prepared. For the first time a synergetic effect of SO_2 and ozone has been demonstrated in a field exposure.

Key words: copper, bronze, field exposure, material degradation, 3d roughness measurements, dose-response-functions, ozone, sulfur dioxide

1. Introduction

That acid rain and air pollution are not only a problem of our decade shows the fact that the first experimental report dealing with atmospheric corrosion versus particular metals goes back to the year 1924 (Vernon, 1924). In the following decades different approaches concerning atmospheric corrosion considering all kinds of aspects have been made and it became clear that atmospheric corrosion varies quite considerably within geographical areas and with local conditions (Evans, 1948; Bartok, 1976; Wranglén, 1985). In order to assess the rate of material corrosion and the role of pollutants in accelerating this process a field measuring programme was carried out within the framework of the Convention on Long-range Transboundary Air Pollution. This study is part of this extensive field exposure programme within the United Nations Economic Commission for Europe (UN/ECE). Until now no study has been conducted to quantify the effects of acidifying air pollutants in such an extensive range. Therefore the study aims to evaluate the response of environmental pollution on material decay by conventional methods and new experimental approaches (contactless 3d surface scanning methods) accompanied from statistical treatment of the data sets received after each exposure period.

2. Experimental

The project was organized as a staggered 8 year's material exposure programme, which was started in September 1987. A Task Force is organizing the programme with Sweden as lead country and the Swedish Corrosion Institute as Main research centre. Subcentres in the UK, Czech Republic, Norway and Germany have been appointed, each responsible for their own group of materials. All samples are exposed under standardized conditions at 39 test sites in 14 European countries, including 3 test sites in the United States and Canada. The triplet samples were mounted to identical racks, where the specimen are subject to open and sheltered corrosion attack (UN/ECE (2), 1993). Gaseous pollutants, precipitation and climatic parameters are measured at or nearby each test site. For characterizing the corrosion environment the parameters chosen correspond to those resulting from the EMEP (European Monitoring and Evaluation Programme) programme. The collected data were used to build up a unique data base within the 14 membership countries (UN/ECE (10), 1993). The materials exposed are composed of structural metals, stone materials, paint coatings, electric contact materials and in extension of the programme glass and polymers. The german contribution to this programme consists of the investigation and interpretation of the corrosion action on copper (DIN 1787) and bronze (DIN 1705). For detailed information concerning the cleaning procedure of the metal samples after ISO/DIS 8407 to determine weight change and mass loss see (UN/ECE (5), 1991). The standard analytical techniques used were X-ray analysis, eddy current measurements and colour measurements. For selected samples additional SEM/EDS investigations were performed.

3. Data Analysis & Results

During the evaluation of the 4 years data set it took shape that the most efficient parameters for quantifying the material degradation are the mass loss and the corrosion volumes. Whereas the mass loss $[g/m^2]$ shows the absolute dimension of corrosive action the corrosion volumes $[mm^3]$ specify additional structural information. Besides the chemical treatment of the samples as pickling for determining the mass loss and identifying the corrosion products on the metal surface, the linking of the quantitative corrosion data to the environmental data by statistical treatment is an important tool for evaluation. The correlation analysis (Pearson correlation) serves as a starting point for regression analysis, indicating the variables of interest. On the basis of 2d-roughness measurements a strong impact of chloride on the formation of pin holes for sheltered copper could be detected without showing high mass loss. A simultanous presence of ozone the corrosive action is catalysed and leads to severe mass loss. This central impact for copper corrosion could be confirmed first by correlation analysis (R_z–Cl: $R = 0.54$). Most frequent a multiple linear regression is performed and the backward procedure is applied for selecting the variables used in the equation. Afterwards the residuals are examined to check the validity of the model. Based upon the 4 years data set for most of the materials preliminary dose-response-functions have been prepared. The equations shown here have not the best explain variability but are self-consistent from the chemical and physical point of view (UN/ECE (15), 1993).

TABLE 1. Dose-response-functions of copper and bronze

Copper		
Unsheltered Exposure		
$ML(g/m^2) = 3.2 + 0.015\ [SO_2]\ [O_3] + 0.35\ [O_3]$	$N = 38$	$R^2 = 0.55$
$V_2(mm^3) = 0.0562 + 0.0003\ [SO_2]\ [O_3] + 0.0005\ [O_3]$	$N = 35$	$R^2 = 0.62$
Sheltered Exposure		
$ML(g/m^2) = -5.4 + 0.008\ [SO_2]\ [O_3] + 0.42\ [O_3]$	$N = 39$	$R^2 = 0.32$
$V_2(mm^3) = -0.144 + 0.0003\ [SO_2]\ [O_3] + 0.004\ [O_3]$	$N = 36$	$R^2 = 0.51$
Bronze		
Unsheltered Exposure		
$ML(g/m^2) = 11.8 + 0.047\ Tow\ [SO_2]\ [O_3]$	$N = 34$	$R^2 = 0.59$
Sheltered Exposure		
$ML(g/m^2) = 5.3 + 0.024\ Tow\ [SO_2]\ [O_3]$	$N = 34$	$R^2 = 0.52$

$[O_3],[SO_2]:\mu g/m^3$ Tow:time of wetness (RH>80%, T>0°C), time fraction

4. Discussion

Towards the questions of prime importance the effects of wet and dry deposition have to be stressed, as both cause corrosive actions. Although the average corrosion level of the bronze test panels is lower compared to copper, bronze already reacts extremely sensitive against SO_2 even below 10 $\mu g/m^3$ (see figure 1).

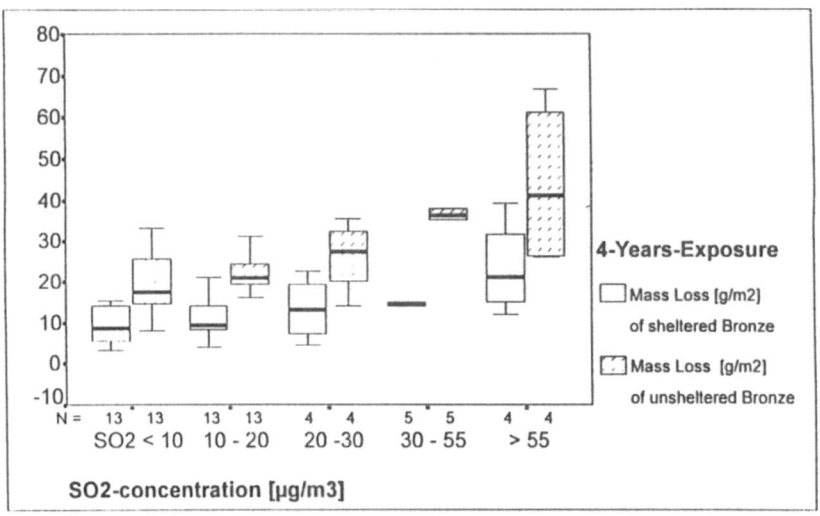

Figure 1.   Corrosion of bronze: Comparison of wet and dry deposition effects

The very first corrosion product formed on bronze surface is anglesite which cannot be detected anymore at higher corrosion levels when other products like cuprite and the sulfates prosnjakite and brochantite are formed (UN/ECE (5), 1991). The mass loss and colour changes are obviously caused by interaction of several environmental parameters. In this connection the photocatalytical processes of ozone and NO_2 definitely act as key reactions (UN/ECE (11), 1993). The starting corrosion rate for copper at rural test sites with moderate SO_2 levels but relatively high ozone concentrations is suprisingly high. The oxidic layer formed during this process retards the corrosion rate.

Figure 2 shows quite impressively that the surface topology of a test panel exposed at the rural test site Waldhof and the one exposed at the czech industrial test site Kopisty is quite different after 4 years of unsheltered exposure (Reisener et al., 1995). A 3d roughness scan of the pickled surfaces of the samples is shown additionally in the lower section of the figure. By the method used for the determination of the corrosion volumes by roughness measurements, the volume enclosing the whole range of corrosive action is splitted into two parts: top and bottom volume. The volume enclosing the grown up corrosion products (top volume V_1 [mm^3]) in Kopisty is twice as high as in Waldhof (see figure 2). Scanning the pickled metal surface by using contactless 3d scanning methods (Rodenstock RM 300: using ISO 4287/1) gives more detailed information of the material decay. At the rural test sites a lot of very tiny pin holes have been formed, at the industrial test site comparingly large areas of the former surface are already missing.

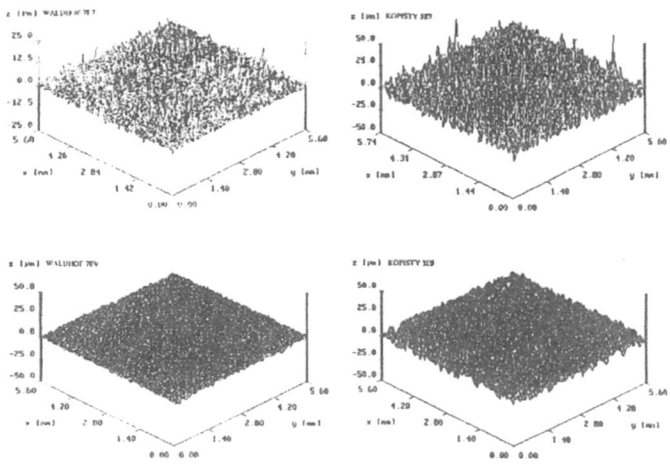

Figure 2. Surface topology of copper samples before and after pickling

The corrosion volume V_2 (bottom volume [mm^3]) that corresponds to the pin hole corrosion rises exponentially with higher SO_2 levels, simultanously the density of the corrosion layer decrease as no longer the oxide cuprite is formed but the sulfate prosnjakite which — by dehydration — changes to brochantite (see figure 3).

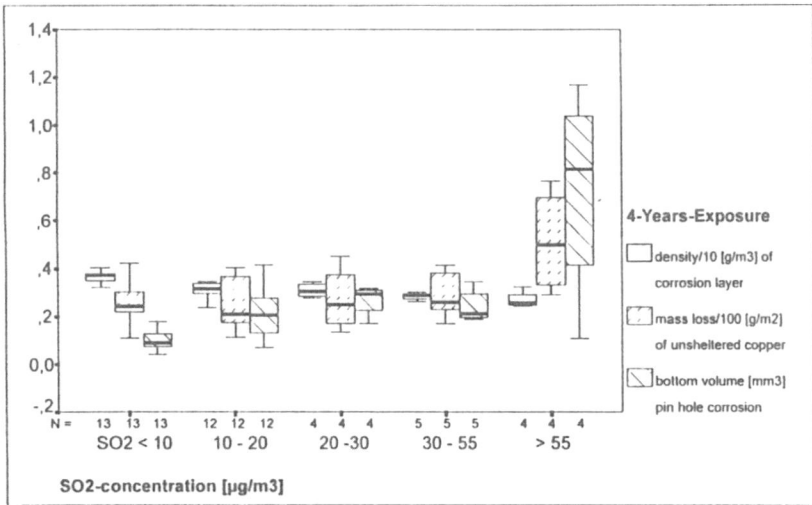

Figure 3. Copper corrosion and density of the corrosion layer at unsheltered exposure

5. Conclusion

Among the acidifying air pollutants besides SO_2, ozone, NO_2 and chloride are regarded as the main sources of corrosive action for copper and bronze. For the first time a synergetic effect of SO_2 and ozone has been demonstrated in a field exposure. The role of NO_2 is not yet been clarified, but will be topic of future work.

Acknowledgments

The German Contribution was financed and supported by the Umweltbundesamt, Berlin. Special thanks to Mrs. Ruth Müller, technician, for her excellent cooperation.

References

Bartok, K. (1976) *Protection Against Atmospheric Corrosion*. Wiley, London.
Evans, U. (1948) *An Introduction to Metallic Corrosion*. Arnold, London.
Reisener, A., Müller, R., Mach, M., Snethlage, R. (1995) *Korrosionsprozesse an Metalloberflächen in natürlicher Umgebung*. Umweltforschungsplan des Bundesministers des Inneren, Forschungsabschlußbericht 108 07 020/07, BLfD, München.

UN/ECE International Co-Operative Programme On Effects On Materials, Including Historic And Cultural Monuments:
Report No. 2 (1993) *Description of Test Sites*. Swedish Corrosion Institute, Stockholm.
Report No.5 (1991) *Corrosion attack on copper and cast bronze. Evaluation after 1 and 2 years of exposure*. Bavarian State Conservation Office, Munich.
Report No.10 (1993) *Environmental Data Report September 1990 to August 1991*. Norwegian Institute For Air Research, Lillestrøm.
Report No.11 (1993) *Corrosion attack on copper and cast bronze. Evaluation after 4 years of exposure*. Bavarian State Conservation Office, Munich.
Report No.15 (1993) *Corrosion attack on electric contact materials. Evaluation after 4 years of exposure*. Swedish Corrosion Institute, Stockholm.

Vernon, W.H.J. (1924) *First (Experimental) Report to the Atmospheric Corrosion Research Committee (of the British Non-Ferrous Metals Research Association*. Trans. Farad. Soc. **19**, 839-900.
Wranglén, G. (1985) *Corrosion and Protection of Metals*. Chapman & Hall, New York.

REACTIONS OF GASES ON CALCAREOUS STONES UNDER DRY CONDITIONS IN FIELD AND LABORATORY STUDIES

J.F. HENRIKSEN

Norwegian Institute for Air Research
P.O. Box 100, N-2007 Kjeller, Norway

Abstract. Dry deposition of gases plays an important role for the deterioration of stone materials and a better understanding of the processes involved will improve our ability to maintain stone monuments and buildings. As a part of an EU-project an investigation with four calcareous stone types have been exposed outdoor at two test sites in Norway for two years. The exposure has been carried out in sheltered position and the amount of reaction products and the penetration depth of SO_2 into the stones was determined as soluble sulphate after half a year and after one and two years. Even if most of the sulphate was found in the upper 0.3 mm of the stone, there was an increase in the sulphate content in stone even down to the center of the stone sample. In laboratory tests with SO_2, NO_2 and changing relative humidity the synergistic effect of NO_2 and the importance of the relative humidity was investigated. The uptake rates were calculated from the laboratory studies by analyzing the gas concentrations before and after the exposure chamber. By calculating the deposition velocity from the field study by using the amount of sulphate found in the stones together with the average outdoor concentration of SO_2 at the test sites, the values were a magnitude higher than in the laboratory test, highest at the industrial paper mill sites with high concentrations both of SO_2 and some hypochlorite and lower in urban atmosphere with fairly low values of SO_2 and high values of NO_2.

Key words: Stone deterioration, dry deposition, air pollution, calcareous stones, field test, laboratory test.

1. Introduction

An extremely important part of our cultural heritage is monuments and buildings of calcareous stone material, and the deterioration that takes place on these monuments and buildings is of great concern to the society. The deterioration of stone material is a complex process, involving many climatic and pollution factors (e.g. Amoroso and Fassina, 1983). An important part of this complex process is the effect of the dry deposition of gases on calcareous materials and the reaction that takes place on the surface. As part of an EEC-STEP project STEP-CT90-0108 NILU has carried out a field and laboratory study of the dry deposition of SO_2 on four different calcareous Mediterranean stone types.

2. Measuring programme

The climate chamber test. The chamber used was a chamber produced by Chalmers Technical University. The temperature control is better than ±0,2° C and had a humidity control from 0% to 100% RH. By permeation tubes it is possible to introduce 1-3 different gases into the chamber at the same time (Henriksen, 1994).

The stone used, Pentelic marble, Carrara marble, Vicenza limestone and Pietra Serena calcareous sandstone have been tested in two different tests. One test studied the humidity impact on the deposition rate of SO_2 in combination with NO_2 for the four stone types. The procedure was to expose one stone sample at the time for 24 hours at each relative humidity selected. The sample (5x5x1 cm) was exposed vertically, standing on a porcelain screen. The amount deposited on the surface was calculated from the concentration differences for SO_2 and NO_2 measured at the inlet and at the outlet of the chamber. The values measured after one day of exposure was taken as the steady state value for the deposition at each humidity level.

All the tests were carried out at 22° C. The flow rate through the chamber was 2 l/min. which gave a windspeed less than 1 mm/s. The concentration at the inlet was in the range of 350-400 µg/m³ for both SO_2 and NO_2. The tests carried out started at a humidity level of 90% RH, went down to 10% RH and up to 90% RH again.

The other test carried out was a test to compare the SO_2 sensitivity of the four stones with other stone and building materials. This SO_2 sensitivity test is offered to the stone industry in Norway as part of a testing programme for property description of the stone materials by The Norwegian Research Centre for Natural Stone. The difference between the concentration measured before the test and after 3 days of exposure gives a value for the SO_2-susceptibility of the stone material. The SO_2-concentrations used were in the range of 350-400 µg/m³.

Field tests. The stones (5x5x1cm) were exposed in vertical position in a sheltered position under a flat roof to prevent the impact of rain events at two of the ECE-ICP materials test sites in Norway. The exposure time selected was first ½ year, second ½ year, 1 year and 2 years. The exposure started in July 1992 and ended in June 1994.

After the exposure the sulphur concentration in the stone was determined in thin layers from the surface. The layers analyzed for SO_4^{2-} were 0-300 µm, 300-600 µm, 600-1200 µm, 1200-2500 µm and 2500-5000 µm. The samples were taken as powder from the stone by rotating hard metal end mill with a diameter of 16 mm (Henriksen, 1994).

3. Results

Laboratory studies. The SO_2 concentration in the test chamber was controlled with a continuous monitoring instrument during all tests and showed that SO_2 concentration dropped in the chamber when calcareous stones were introduced and that a new steady state concentration was observed in less than 1/2 hour. The NO_2 concentration was controlled by taking frequent samples for analysis at our laboratory. No detectable changes were observed in any of the tests. The calculated deposition velocity (Vg) of SO_2 for the four stone types at different relative humidity is shown in Figure 1. A linear regression fits the results for Vicenza limestone, Pietra Serena calcareous sandstone and Pentelic marble, while a semi-log correlation fits Carrara marble better. The equation for the best lines are also given in Figure 1. Both limestone and the calcareous sandstone absorb SO_2 at all humidities tested (10-90% RH), while the marbles seem to be inactive at low humidities ≤20% RH. The effect of NO_2 to the deposition velocity was observed by comparing the results from 90% RH with results from the SO_2 sensitivity test. In Table I the results from the sensitivity test for several stones are shown.

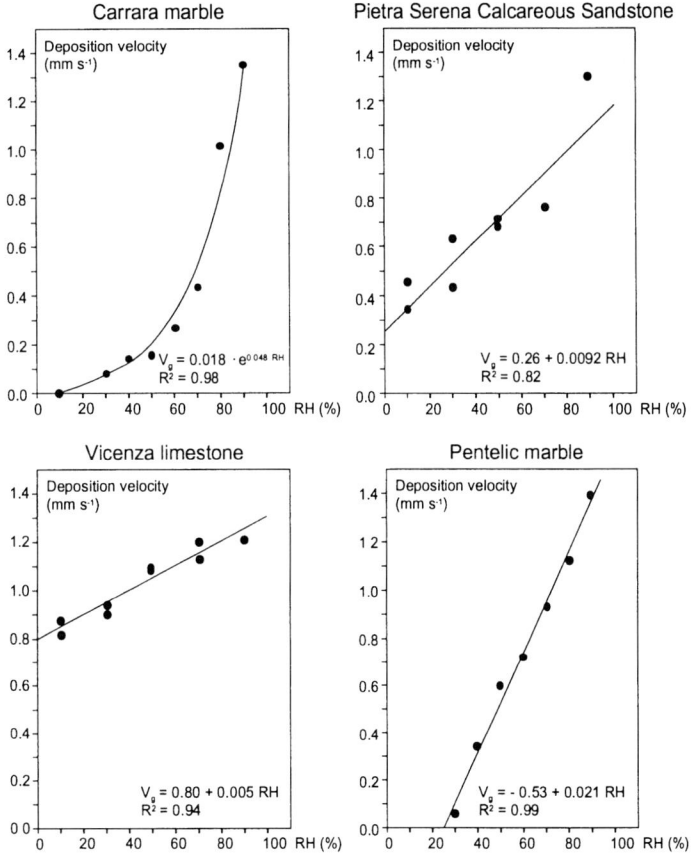

Figure 1: Deposition velocity for four calcareous stones at different Relative Humidities in an atmosphere with SO_2 and NO_2.

The field tests were carried out at one industrial test site, Borregaard in south-east Norway, where the annual gas concentrations were SO_2 = 28,3 µg/m³ and NO_2 = 17,2 µg/m³ and at the urban site in Oslo with SO_2 = 6,4 µg/m³ and NO_2 = 51,9 µg/m³. The SO_2 attack on the stone sample exposed in the field was analyzed by dissolving the soluble amount of SO_4^{2-} in the stone.

TABLE I

The results of the SO$_2$ sensitivity test for different stones at 90% RH, 22° C and 350-400 μg SO$_2$/m^3.

Materials	Deposition velocity mm s^{-1}	% change with 400 μg NO$_2$ present
Calcareous rendering	1,12	
Vicenza limestone	1,00	+25%
Pietra Serena sandstone	0,93	+17%
Pentelic marble	0,69	+97%
Carrara marble	0,58	+133%
Granite red and grey	0,38	
Soap stones	0,07-0,40	
Monzonite green	0,27	
Syenite red	0,27	
Monzonite blue	0,09-0,18	
Sodium glass	~0	

As seen from Table II most of the SO$_2$ had reacted in the upper crust layer. The results for Pietra Serena sandstone were quite different. The SO$_4^{2-}$ concentration in the stone itself is much higher than the other stone types, and this leads to a higher uncertainty in the analysis. Up to one year exposure increased amount of SO$_4^{2-}$ in the stone was only observed in the layers down to 0.6 mm. However, after two years SO$_4^{2-}$ increase was measured down to 5.0 mm. At Borregaard test site the SO$_4^{2-}$ concentration measured in the Pietra Serena sandstone the second year in the upper 0-0.3 mm was lower than after one year.

TABLE II

Distribution of soluble SO$_4^{2-}$ in the stone types after 2 years of exposure.

Material	Exposure site	% of the total found in layers from the surface after 2 years				
		0-0.3 mm	0.3-0.6 mm	0.6-1.2 mm	1.2-2.5 mm	2.5-5.0 mm
Pentelic marble	Oslo	80.0	7.0	5.0	5.0	3.0
	Borregaard	84.5	10.5	2.0	1.5	1.5
Carrara marble	Oslo	81.0	9.0	5.0	3.0	2.0
	Borregaard	73.0	8.0	8.5	6.5	4.0
Vicenza limestone	Oslo	62.0	7.0	7.0	11.0	13.0
	Borregaard	81.0	6.5	5.5	4.0	3.0
Pietra Serena	Oslo	47.5	6.5	10	19.5	16.5
	Borregaard	32.0	13	13	19	23

4. Discussion

The influence of the relative humidity on the deposition velocity was strong for all the stone material tested. Even so, the pattern was quite different. The strongest effect was observed on marbles. At low humidities, 10-20%, no detectable adsorption took place and even if the best fitted curve for Carrara marble was a semi-log curve and for Pentelic linear, the effect at high humidity was comparable. Both Vicenza limestone and Pietra Serena sandstone seems to absorb SO_2 in combination with NO_2 at all relative humidities. The slope of the lines was lower than for marble. By comparing the deposition velocities with and without NO_2 present, a drastic increase in the deposition velocity was observed on the marbles, approximately 100% increase at 90% RH, while for Vicenza limestone 25% and for Pietra Serena sandstone 17% higher were observed. The deposition velocity for the four different stones was fairly close at 90% RH. The deposition velocities calculated from the amount of SO_4^{2-} which could be desolved from a sample of the stone were comparable with the calculated one from the concentration measurements in the laboratory test.

However, the deposition velocity calculated from the dissolved amount of SO_4^{2-} for the samples exposed in the field test gave much higher velocities as shown in Table III.

The large deposition velocity observed at Borregaard is comparable to earlier results reported in polluted atmosphere, Lipfert (1989) and Richardson (1960). In a laboratory study Mangio and Johansson (1989) has shown that the oxidizing effect of ozone will give a deposition velocity in the same range as for Oslo.

TABLE III

Calculated deposition velocities for two years field exposure of calcareous stones.

	Deposition velocity mm s^{-1}	
	Borregaard	Oslo
Pentelic marble	8.0	3.1
Carrara marble	4.4	3.7
Vicenza limestone	8.0	6.0
Pietra Serena sandstone	not calculated	not calculated

The even higher results from Borregaard can probably be explained by an additional oxidizing agent, hypochlorite, since it is used in the paper bleaching plant. In field tests at Borregaard with other materials, similar results are observed (Kucera, 1992). In the STEP project deposition velocity measurements were also carried out by De Santis (1994) with a specially designed flat denuder. The results with the denuder in a mixed SO_2 and NO_2 atmosphere at 90% were in good agreement with the result presented.

Vicenza limestone seems to be the most sensitive stone for reaction with SO_2 in the field test. However, this effect is probably connected to a larger area exposed, since the stone is very porous and the available area for surface reaction large. Pentelic marble is more sensitive to SO_2 than Carrara marble, particularly in polluted atmosphere. Pentelic marble also seems to react more at the surface than Carrara marble. The Pietra Serena sandstone results are more complicated, and the spread in the background sulphate concentration makes the interpretation more uncertain. Even the laboratory test gave a larger spread in the results than the other stones, and the value for the correlation coefficient is lower than for the other regressions.

The laboratory test for SO_2 sensitivity of different stones and other porous materials seems to give reliable results, ranging calcareous rendering as the most sensitive material and sodium glass as the least one. Together with physical tests and morphological studies, the SO_2-sensitivity test can give useful information for selection of stone materials for construction purposes.

Acknowledgements

The author will express his gratitude to the EU-STEP research programme and to the Norwegian Institute for Air Research for the financial support to this study.

References

Amoroso, G.G. and Fassina, V.: 1983, *Stone Decay and Conservation.* Elsevier, Amsterdam, 135-155.
Henriksen, J.F.: 1994, Proceedings of the 3rd International Symposium, 189-194. Venezia 1994.
Katsanos, N.A., De Santis, F., Fassina, V., Henriksen, J.F. and Johansson, L.G.: 1994, *Final Report..* Contract No. STEP-CT90-0108 (TSTS).
Kucera, V.: 1992, Proceedings 12th Scandinavian Corrosion Congress & EUROCORR 92., Vol. 1, 9-24. Espoo 1992.
Lipfert, F.W.: 1989, *Atmos. Environ.* **23,** No. 2, 415-429.
Mangio, R. and Johansson, L.G.: 1989, 11th Scandinavian Corrosion Congress NKM 11, Stavanger June 19-20, paper F-45.
Richardson, E.G. (ed.): 1960: *Aerodynamic Capture of Particles. Oxford Pergamon Press,* 63-88.

RESPONSE OF POROUS BUILDING STONES TO ACID DEPOSITION

C. M. GROSSI[1]; M. MURRAY[2] and R.N. BUTLIN[2].
[1] *Departamento de Geología, Universidad de Oviedo, 33005 Oviedo (Spain).* [2] *Building Research Establishment, Garston, Watford WD2-7JR (UK)*

Abstract. This work investigates the response of porous carbonate building stones to acid deposition during a short-term exposure period and the characteristics that influence their reactivity and/or durability. Several carbonate porous stones used in Spanish and English monuments were exposed to English urban and suburban environments. In each location they were both exposed to and sheltered from rainfall. Monthly analyses were carried out in order to investigate any possible sign of reaction. In addition, some physical properties of the stones relating to transfer of moisture were determined. Results indicate that the reactivity of these stones is relatively high, significant signs of reaction were detected within only a few months of exposure. Under the same environmental conditions, the response and reactivity of porous carbonate stones are determined by their petrophysical characteristics.

1. Introduction

Limestones and dolomitic limestones are very often used as building stones. However these carbonate stones are very sensitive to environmental impact. As a result, in recent times exhaustive research into the response of carbonate building stones to the environment has been carried out (Butlin *et al*, 1985). The weathering of these stones is an interaction of physical, chemical and biological processes. Chemical degradation of carbonate building stones is due to dry deposition of acid gases and aerosols, dissolution by acid species in rainwater and dissolution by unpolluted water (Butlin *et a.l*, 1993; Livingstone, 1992; Cooke and Gibbs, 1993).

This paper continues this line of research and presents some results of the collaborative work between the University of Oviedo (Spain) and the Building Research Establishment (UK) on the interaction between atmospheric pollution and carbonate building materials. The aim of this paper is to assess the response of several carbonate porous stones to English urban and suburban environments during a short-term exposure period and to determine the physical characteristics of these stones that influence this response.

2. Experimental

2.1. MATERIALS

The selected materials were four carbonate stones frequently used as building materials in Spanish and English monuments: Laspra dolomite (Spain), Hontoria limestone

(Spain), Portland limestone (UK) and one of the varieties of Bath stone: Combe Down limestone (UK).

Laspra is a Eocene micrite mainly composed of dolomite. The main component of the other stones is calcite. Portland and Combe Down are Jurassic oolite limestones and Hontoria is a bioclastic limestone from the Cretaceous.

2.2. EXPOSURE

The selected materials have been exposed for one year in two different environments: a) Central London (exterior of Westminster Abbey), urban environment; b) Garston-Hertfordshire (exposure site at the Building Research Establishment), suburban environment.

At each of these sites, pairs of freely rotating carousels of stone tablets (5 cm x 5cm x 1cm) were installed following the methodology used for many years by researchers at the Building Research Establishment (BRE) and others (Butlin *et al.*, 1985; Jaynes and Cooke, 1987). One of the carousels was sheltered from rainfall and exposed only to dry deposition. The other was exposed to wet and dry deposition. Table I shows the pollution conditions at both sites (average exposure period).

TABLE I
Environmental Conditions at exposure sites

Site	SO_2 ($\mu g/m^3$)	Smoke ($\mu g/m^3$)	NO_x ($\mu g/m^3$)
London	11.7	18.7	61.33
Garston[1]	21	6.5	42.1

[1] Estimated from St. Albans data

2.3. ANALYSIS

In order to investigate any possible sign of reaction, monthly analyses were undertaken including analysis of soluble salts by ion chromatography, surface composition by means of Fourier Transform Infra-Red (FT-IR) spectroscopy, and measurements of reflectance and weight variation. Simultaneously, porosity and moisture transfer properties of stones were determined.

3. Results

Significant signs of reaction were detected only a few months after exposure both in samples exposed to and sheltered from rainfall in London and Garston. Notable amounts of sulphate were detected in sheltered positions, and an increase in sulphate was also found in exposed tablets. Some nitrates and chlorides were also present in sheltered stones. Oxalates were also detected in most of the samples. Ammonium salts were found in exposed samples of Laspra dolomitic limestone.

The surface recession measured by weight loss in exposed samples, was higher in the first months of exposure and then tended to stabilise.

3.1. SHELTERED SAMPLES

The concentration of anions and cations was determined by ion-chromatography in successive drillings of 0.5 mm from the surface to 2 mm depth. Table II shows the correlation between anions and cations in terms of the correlation coefficient r.

TABLE II
Correlation Table Cations/anions in sheltered samples

	Ca^{++}	Mg^{++}	Na^+
$SO_4^=$	0.97		
NO_3^-		0.77	0.93
Cl^-	0.70		

This table shows that sulphates correlate strongly with calcium, and nitrates with sodium. An ion balance showed that sulphates were present as calcium sulphate. Fourier Transform infra-red spectroscopy (FT-IR) showed that the calcium sulphate was in the form of gypsum ($CaSO_4.2H_2O$), which could be detected on the surfaces of sheltered stones by after three months of exposure. This agrees with results of other authors (Girardet and Connor, 1992).

3.2. UNSHELTERED SAMPLES

The surface recession of the stones was estimated from the weight loss in samples exposed to rainfall, assuming the material is lost evenly across the whole sample (Butlin et al, 1993) (Table III).

TABLE III
Surface recession of unsheltered samples

	≈ Surface recession (μm/year)	
	Garston	Westminster
Laspra	41.1	47.6
Hontoria	11.8	11.4
Portland	12.0	13.9
Combe Down	40.8	42.7

3.3. PHYSICAL PROPERTIES

Some physical properties that relate to transfer of moisture were determined in order to assess their relationship with the response of carbonate stones to the environment (Table IV).

TABLE IV

Physical properties relating to moisture transfer of the different carbonate stones

	Open porosity (%)[1]	% pore radius<1μm [1]	Specific surface area[1] (m²/g)	Ψ_w (%)	K_v (g.m^{-1}.h^{-1}.mmHg^{-1})	Ψ_H (mg/g) a	b
Laspra	≈ 31	94.4	4.03	12	1.3x10^{-3}	12.5	113.3
Hontoria	≈ 20	29.5	0.26	6	8.5x10^{-4}	0.05	0.24
Portland	≈ 20	51.6	1.21	7.5	9.1x10^{-4}	0.38	2.8
Combe Down	≈ 28	70.7	1.38	9	9.0x10^{-4}	0.81	3.6

[1] Determined by means Hg-porosimetry; Ψ_w = free water absorption by total immersion; Kv = Vapour permeability; Ψ_H = Hygroscopicity (water vapour adsorption) (T = 23°C) (a = 60%; b = saturated)

3. 4. DISCUSSION

Although there is not enough climatic data for a reliable relationship to be determined, a correlation table has been made relating the weathering data from the exposure sites to the physical properties of the stones. Results are summarised on Table V.

TABLE V

Correlation Table weathering data / physical properties of the stones

	Surface recession	[SO$_4^=$]	[NO$_3$]	R(s)	Kv	Ψ_H	Ψ_w
Specific surface area	*		**[2]		**[2]	**[2]	**
Open porosity	**[1]				*	*	**[1]
Pollutants	-	*		**			
[Ca^{++}]	*		**	*			

R(s) = % Reflectance variation (sheltered samples) Pollutants: [SO$_2$], [NO$_x$] and [smoke]
(1) related to porosity (**) correlated (r>0.8)
(2) related to specific surface area (*) weakly correlated (r=0.6-0.8)

3.4.1. Surface recession

The three mechanisms responsible for the stone loss are dry deposition, natural dissolution and acid dissolution. The contribution of each of these will depend on the environmental conditions. Butlin et al., (1993) indicates that acid rain can contribute 2.6% of weight loss in calcareous stones, unpolluted rain 61.6%, and dry SO$_2$ deposition 35.8%. Results of NAPAP (1990) show that 'rainwater' can account for up to 70% of erosion. Livingstone (1992) pointed out that natural dissolution may be important even in rural areas with significant acid rain. For the stone samples in this project, the importance of natural dissolution is reflected by the strong correlation of surface recession to open porosity (Table V). This is because the open pore structure allows more water into the sample and so allows more carbonate to dissolve. The surface recession was found to be higher in the first months, this is due to loss of carbonate cement or stone matrix (Cooke and Gibbs,1993).

The theoretical maximum weight loss due to acid rain was estimated following the model of Webb et al., 1992, and assuming 1 m precipitation per year and a pH=4.2 (Lipfert, 1989) (Table VI). The dry deposition was calculated from the amount of gypsum in sheltered samples, this was converted into an equivalent calcite or dolomite weight. The resultant balance corresponded to natural solution being the most important factor in carbonate stone decay.

TABLE VI
Percentage of predicted carbonate losses after 1 year exposure

	≈ g of carbonate loss by dry deposition		≈ Dry deposition (%)		≈ Acid rain (%)		≈ Natural solution (%)	
	West	Garston	West	Garston	West	Garston	West	Garston
Laspra	0.13	0.04	19.0	8.4	3.3	3.8	77.7	87.8
Hontoria	0.04	0.02	26.0	12.8	6.5	6.3	67.5	80.9
Portland	0.04	0.02	20.7	11.8	5.3	6.2	74.0	82.0
Combe Down	0.05	0.03	8.3	5.0	1.9	2.0	89.8	93.0

TABLE VII
Sulphate concentration in sheltered samples (µg/g)

	Garston 4 months	Garston 12 months	Westminster 4 months	Westminster 12 months
Laspra	1078	3680	2731	11262
Hontoria	675	2910	1656	5553
Portland	934	3103	1657	5042
Combe Down[1]	3340	4306	1163	8170

[1] The background concentration in clean Combe Down samples is very variable.

3.4.2. Sulphate concentration

The presence of sulphate is caused by dry deposition of gaseous SO_2 and aerosols to the stone, thus the concentration of sulphates in sheltered samples (Table VII) should be related to the pollutant concentrations in the atmosphere (Table V). However, there is not enough data to determine this. The deposition of SO_2 on material surfaces will be affected by physical characteristics of the stone, such as surface area. Laboratory experiments have shown some relationship between sulphate deposition on dry stones and their specific surface area (Grossi et al, 1994). Other factors will also influence the deposition of SO_2. According to several authors (Spedding, 1969; Johansson et al, 1988; Spiker et al, 1992; Goturk et al, 1993) the uptake of SO_2 depends on the humidity. The specific surface area controls the uptake of moisture by the stone (Tables IV and V), and Spiker et al (1992) pointed out that SO_2 deposition may be controlled by the wetted surface area of the stone. The presence of some salts will alter this. $NaNO_3$ and $CaCl_2$, both found in the stone will absorb moisture at humidities over 75% and 35% respectively.

3.4.3. Other reaction products

Nitrates were found in sheltered samples, present as sodium nitrate and magnesium nitrate. Their concentration was related to specific surface area, water vapour adsorption and water vapour permeability. NO_x can transform to NO_3^- in the water adsorbed on the surface of the stone. However, the process is slow, the larger surface area could allow more nitrate to be absorbed into surface moisture layers. Finally, biological deterioration can be seen in the stones, mainly as oxalates. This was more evident in Garston than in London.

4. Conclusions

Degradation of carbonate stones is a result of the combination of environmental factors and stone characteristics. Carbonate stones are very reactive to rainwater, SO_2 deposition and biological agents. Gypsum and oxalates can be found in the surfaces of some stones after only a few months of exposure.

The physical characteristics of carbonate stones relating to moisture transfer will affect their response to the environment. Porosity and specific surface area are relevant properties conditioning this response. A study which includes a greater number of well characterised carbonate stones exposed to the same environment needs to be done in order to determine quantitative relationships between porosity, specific surface area and stone durability.

Acknowledgements

This study was supported by the "Ministerio de Eucación y Ciencia" (Spain). The authors wish also to acknowledge Dr. R.M. Esbert ("Universidad de Oviedo" -Spain).

References

Butlin, R.N. , Cooke, R.U., Jaynes, S.M. and Sharp: 1985, V^{th} *International Congress on Deterioration and Conservation of Stone,* Press Polytechniques Romandes, Lausanne, 537-546.

Butlin, R.N., Yates, T.J.S., Coote, A.T, Lloyd, G.O. and Massey, S.W. :1993: *The first phase of the National Exposure Programme, 1987-1991, Report. CR 253/93* Building Res. Establishment, Watford, 70 pages.

Cooke, R.U. and Gibbs, G.B.: 1993, *Crumbling Heritage? Studies on Stone weathering in polluted atmospheres,* National Power plc, Swindon, 68 pages.

Girardet, F. and Connor, M.:1992, *Analusis Magazine,* **20**, n.5, M26-M32.

Goturk, H., Volkan, M. and Kahveci, S.: 1993, *Conserv. of Stone and Other Materials,* **1**, E&FN SPON, 83-90

Grossi, C.M., Esbert; R.M.; Lewry, A.J. and Butlin, R.N.: 1994 , *Seventh International Congress International Association of Engineering Geology,* A.A. Balkema, Rotterdam, 3573-3582.

Jaynes, S.M. and Cooke, R.U: 1987, *Atmos. Environ.* **21**, 1601-1622

Johansson, L.G., Lindqvist, O and Mangio, R.E.: 1988, *Durability of Building Materials,* **5**, 439-449.

Lipfert, F.W.: 1989: *Atmos. Environ..* **23**, 415-429.

Livingstone, R.A. 1992, 7th *Int. Cong. on Deterioration and Conservation of Stone,* LNEC, Lisbon, 375-386.

NAPAP, 1991: *Effects of acidic deposition on materials. National Acid Precipitation Assessment Program,* Office of the Director of Research, Washington., D.C, 280 pages.

Spedding, D.J.: 1969: *Atmos. Environ.,* **3**, 683.

Spiker, E.C., Comer, V.J., Hosker, R.P. and Sherwood, S.I.: 1992, 7th *Int. Cong. on Deterioration and Conservation of Stone,* Labóratorio Nacional de Engenharia Civil , Lisbon, 641-650.

Webb, A.H, Bawden, R.J., Busby, A.K and Hopkins, J.N. : 1992, *Atmos. Eviron.,* **26B**, 165-181.

Key words: building stones; porous carbonate stones; acid deposition; urban environment; stone reactivity; porosity; petrophysics; physical properties.

EFFECT OF ACID RAIN ON SANDSTONE: THE ROYAL PALACE AND THE RIDDARHOLM CHURCH, STOCKHOLM

A. G. NORD and K. TRONNER

Conservation Institute of National Antiquities, P. O. Box 5405, S-114 84 Stockholm, Sweden

Abstract. The deterioration of two kinds of sandstone is discussed for two 18:th century buildings in central Stockholm: the Royal Palace, and the Royal Carolean Burial Chapel (Karolinska gravkoret) annexed to the mediaeval Riddarholm church. The facades of calcitic Gotland sandstone show many signs of serious decay, such as gypsum formation, pulverized surface, exfoliation, discolouration, and salt efflorescence. The socles are built of the more resistant quartzitic Roslagen sandstone, displaying some discolouration, cracks, and slight exfoliation. In total about 300 samples have been analysed. The surface concentration of sulphur is highest at ground level and at rain-sheltered positions. Chemical and sulphur isotope data indicate that the stone decay to a large part may be attributed to anthropogenic sources like acid deposition and car traffic.

Key words: Air pollution, acid deposition, sandstone deterioration, gypsum formation, sulphur isotopes.

1. Introduction

It is well known that air pollution aggravates the weathering of stone (e.g. Fassina, 1988; Del Monte, 1991). One of the least resistant rocks used for facades and sculptural decorations in Sweden is calcitic Gotland sandstone. Two Stockholm buildings of great cultural value are discussed here: the Royal Palace, and the Riddarholm church. The 18:th century Palace has facades of Gotland sandstone and a socle of quartzitic Roslagen sandstone. The mediaeval Riddarholm church is built of bricks. However, the attached 18:th century Royal Carolean Burial Chapel (Karolinska gravkoret) has facades of calcitic Gotland sandstone and a quartzitic sandstone socle. The calcitic sandstone displays serious signs of decay (gypsum formation, exfoliation, discolouration, salt efflorescence, etc). From time to time, extensive restoration work has been carried out.

The buildings are situated in the Stockholm Old Town, close to heavy traffic and parking lots (cf. Figure 1), where exhaust gases and particles are emitted. Also, various pollutants are transported to the area with the winds. The main purpose of this study was to find if there is any correlation between observed damage and anthropogenic sources like acid deposition and other pollutants. The air pollution situation in central Stockholm has been very bad for many years, with average sulphur dioxide concentrations around 200 $\mu g/m^3$. The situation is now much improved, and the sulphur dioxide concentration seldom exceeds 10 $\mu g/m^3$. The climate is also harmful to stone. In the winter there is often frost and snow, with a high relative humidity (80-90%), and a temperature in the interval from 0 to -10°C. It is often windy in the Old Town due to its position on two small islands. The annual precipitation (rain and snow) is around 550 mm.

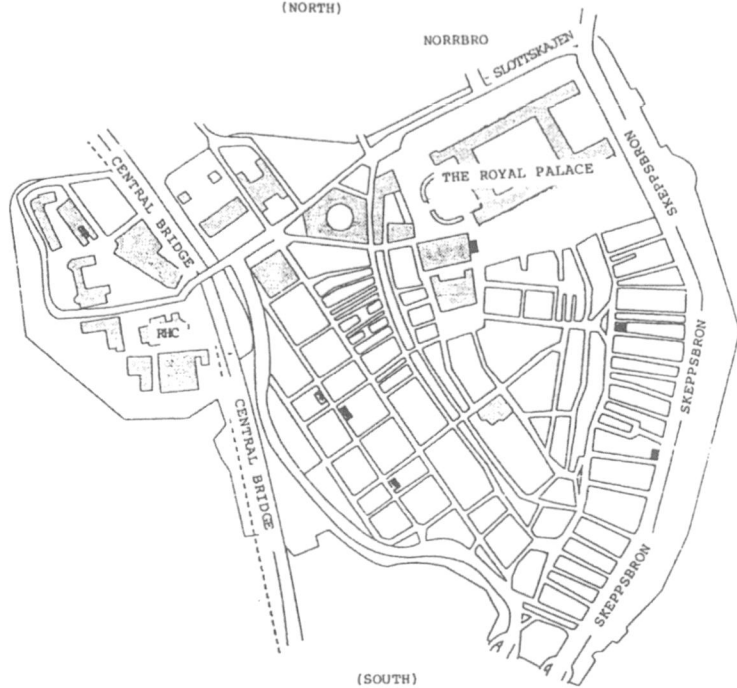

Fig. 1. Map of the Old Town in Stockholm. RHC denotes the Riddarholm church, with the Royal Carolean Burial Chapel to the NE. Heavy traffic passes along the Central Bridge (100 000 vehicles per day) and Norrbro-Slottskajen-Skeppsbron (50 000 vehicles per day).

2. Materials and Methods

The grey calcitic Gotland sandstone is composed of about 60 wt% quartz grains (10-100 µm), cemented together by 7-10 wt% calcite, and with lower amounts of clay minerals, micas, feldspar minerals, pyrite, and glauconite. The porosity is unusually large, around 15 % by volume, which facilitates the penetration of acid rain into the stone. Calcite may thus be chemically transformed into gypsum, in which case the molar volume will be increased by a factor of 2, causing internal strain, cracks, and exfoliation. Gypsum is partly soluble in rain water, but also calcite may be dissolved according to the chemical reaction $CaCO_3(s) + H_2O(l) + CO_2(g,aq) \longrightarrow Ca^{2+}(aq) + 2HCO_3^-(aq)$. There is also dry deposition of gaseous pollutants and particles from traffic and other sources. All these effects in combination with the climatic factors make the facades dirty and cause the stone surface to weather and fall off as loose powder or flakes. The socles of the buildings are constructed of quartz-cemented Roslagen sandstone, mainly consisting of quartz with some potassium feldspar giving a pink colour to fresh undamaged stone. This stone is fairly resistant and only shows discolouration and slight signs of cracks or exfoliation. Moreover, the mediaeval brick is in fairly good condition.

About 250 samples of Gotland sandstone (loose or scratched powder, flakes) were taken from the facades. Samples were also obtained of Roslagen sandstone, marble, mortar, salt efflorescence, and brick (from the Riddarholm church). All samples have been analysed with a JEOL scanning electron microscope equipped with a LINK/EDS unit for energy-dispersive X-ray microanalysis (SEM/EDS). For some selected samples other analytical instruments have also been utilized like ICP, HPLC, XRF, a carbon analysing device, and a mass spectrometer for determination of stable isotopes.

3. Results: The Royal Palace

Sampling and inspection of the Gotland sandstone facades prior to the latest restauration showed that the inner courtyard (cf. Figure 1) was in a slightly better condition than the outer facades, of which the northern one was most deteriorated. The surface often appears to be undamaged although with a thin gypsum-rich layer. Typical damages are pulverization, sometimes to a depth of 50 mm, exfoliation, and a dirty surface. Serious deterioration is illustrated in Figure 2.

Fig. 2. Severe deterioration of calcitic Gotland sandstone at the Royal Palace.

The chemistry is here focused on calcium and sulphur, in undamaged Gotland sandstone 3-4 wt% calcium respectively <0.1 wt% sulphur. Gypsum is ubiquitous, but rain-washed parts differ from sections protected from rain. At a rain-washed, flat stone surface, both calcite and gypsum have been partially dissolved, thus lowering the calcium concentration.

The S/Ca weight ratio is close to 0.8 which is the value for pure gypsum, $CaSO_4 \cdot 2H_2O$. In powderized sections or exfoliations, the calcium surface concentration is usually in the range 3-8 wt%, and the S/Ca weight ratio is generally around 0.5 - 0.8. Accordingly, there is a significant enrichment of gypsum at the surface. By HPLC measurements traces of nitrate (< 0.2 wt%) were found, probably from air pollution. At rain-sheltered sections gypsum-rich crusts were often observed, with 10-23 wt% Ca and S/Ca around 0.80. Sulphur is found several mm inside the porous stone, and the complete amount of "gypsum sulphur" at the Royal Palace may be estimated to 3-4 tons. For further details, cf. Nord (1995). Eleven gypsum-rich samples have been analysed for the $^{34}S/^{32}S$ isotope ratio. The $\delta^{34}S$ values are around +5 per mille relative to the international Canyon Diablo troilite standard (CDT), thus indicating a substantial origin from acid deposition (cf. Longinelli and Bartelloni, 1978; Buzek and Srámek, 1985; Pye and Schiavon, 1989).

In the latest restauration works, old, damaged stone blocks were substituted by fresh stone. We have obtained samples from Gotland sandstone blocks replaced 1-3 years ago. Sampling and analysis were undertaken with the purpose to determine how much sulphur has been deposited since the replacement. The chemical analysis showed that the outermost 1/2 mm now contained 0.9-3.6 wt% sulphur (average: 2.6 wt%), i.e. the deposition of sulphur is still significant. Finally, sixteen samples of quartzitic Roslagen sandstone were analysed. The surface sulphur concentration was 0.2-3.7 wt%, presumably from joint mortar transformed into gypsum. Black discolourations contained soot, grime and iron oxyhydroxides.

4. The Riddarholm Church with the Royal Carolean Burial Chapel

The mediaeval brick and the quartzitic sandstone socle show comparatively few signs of deterioration. The sulphur surface concentration was usually 1-3 wt% due to enrichment of gypsum, iron sulphates, and particles of car-tyre rubber, asphalt, etc. An earlier GC-MS analysis of a thin black layer from the socle has revealed the presence of more than 100 constituents, most of them organic (cf. Nord and Ericsson, 1993; Nord et al., 1994).

The most obvious damages are gypsum formation and salt efflorescence on the Gotland sandstone facades of the Royal Carolean Burial Chapel. We have obtained 15 drilled cores and 24 samples of powdered calcitic Gotland sandstone at various heights. Chemical analysis indicates that gypsum has been formed all over the surface down to a depth of 3-5 mm. Detailed studies reveal that the concentration of gypsum is higher close to the ground, and at rain-sheltered positions, with maximum sulphur values around 18-19 wt% (i.e. almost pure gypsum). The most common salt efflorescence on the facades, gypsum excepted, consists of thenardite (sodium sulphate). The origin of the sodium may be sea water, road salt, cleaning agents, or sodium-rich minerals in the sandstone.

From HPLC measurements of water-soluble anions from samples soaked in distilled water, it is clear that sulphate is dominantt, with significantly lower concentrations for chloride (0.1-0.4 wt%) and nitrate (<0.1 wt%). Carbonates, if formed on outside walls, are likely to have been rapidly transformed by acid deposition into sulphates. In the crypt of the Burial Chapel, though, salt crystals of alkali carbonates as well as alkali sulphates were frequently occurring. The building is situated directly on a rock in Lake Mälaren, and a partial contribution of marine sulphur is therefore probable. The $\delta^{34}S$ value of seawater sulphate is around +20 per mille (rel. CDT) all over the world (Thode et al., 1961). Stable isotope studies have been undertaken (cf. Löfvendahl et al., 1995; and this work). The $\delta^{13}C$ values obtained for carbon from black discolourations are in the region from -9.7 to -25.6 per mille relative to the PDB isotope standard, the latter value close to the average of fossil fuels (cf. Faure, 1977). The $\delta^{34}S$ values for gypsum and thenardite at the facades or from the crypt are somewhat more uniform, spread between +6 and +14 per mille relative to CDT. This indicates air pollution ("acid rain") as the major source of sulphur, but sulphur is also likely to originate from pyrite inclusions or marine salt (cf. Rösch and Schwarz, 1993).

5. Discussion

The chemical analysis has shown that gypsum has been formed all over the calcitic Gotland sandstone facades, concentrated in pulverized areas or exfoliations, while thick gypsum crusts are less frequent. Exfoliation in the form of thin flakes or contour scaling seems to be the first stage in complete pulverization of the surface. The calcium concentration at an apparently undamaged surface has often slightly decreased as compared to fresh stone because calcite has partially been transformed into gypsum, some of which has been dissolved and rinsed away by rain-water. In case a hard, thin gypsum crust has been formed, the calcium surface concentration has instead increased. Also, areas rich in loose powder and/or flakes usually display an increase in calcium contents. Discolouration by soot and grime is often observed on the sandstone facades. The quartz-cemented Roslagen sandstone, though, is harder, less porous and chemically more resistant than the calcitic sandstone. Typical damages are grey or black discolourations from soot, grime, and iron oxide hydroxides; cracks, and slight exfoliation.

The large deposition of sulphur at the Gotland sandstone facades, for the Royal Palace amounting to several tons, indicates a substantial origin from atmospheric pollution. Stable isotope studies have been undertaken to help settle this matter. For the Royal Palace the observed isotope data indicate acid deposition to be the major source of sulphur, while the results of the Royal Carolean Burial Chapel suggest additional sulphur sources. An increased surface concentration of nitrate may be attributed to acid deposition and NO_x pollutants from car traffic. The traffic is also responsible for deposition of particles of soot, car-tyre rubber, asphalt, heavy metals, and other organic constituents. The latter effects are most obvious as thin black layers on rough parts of the sandstone surface, on protruding sculptural decorations, or at sections protected from rain. It is indeed difficult to deduce, from an observed damage, which was the main cause of deterioration. When considering pollutants in general, emissions from local heating, industries, power plants etc must also be considered. However, the present study has shown that acid deposition is responsible for a great part of the deterioration of the calcitic

Gotland sandstone, in combination with effects of weather and wind. The decay of the quartzitic Roslagen sandstone and the bricks of the Riddarholm church is less noticeable, and more difficult to correlate to effects caused by acid deposition. Further studies are in progress to elucidate these matters.

Acknowledgements

We are very grateful to Adrian Boyce (SURRC, East Kilbride, Scotland) and Kjell Billström (Swedish Museum of Natural History, Stockholm) for their kind help with the stable isotope analysis, and to Anders Säfström for valuable technical assistance. Göran Åberg and Harald Sundlin are cordially thanked for their help with sampling.

References

Buzek, F., Srámek, J.: 1985, *Studies in Conservation* **30**, 171-176.
Del Monte, M.: 1991, *Analysis* **20**, M20-M23.
Fassina, V.: 1988, *Durability of Building Materials* **5**, 317-358.
Faure, G.: 1977, *Principles of Isotope Geology*. J. Wiley and Sons, New York.
Longinelli, A., Bartelloni, M.: 1978, *Water, Air and Soil Pollution* **10**, 335-341.
Löfvendahl, R., Nord, A.G., Thorssander, P.: 1995, to be published.
Nord, A.G.: 1995, *Geol. Fören. Stockholm Förhandl.* **117**, 43-48.
Nord, A.G., Ericsson, T.: 1993, *Studies in Conservation* **38**, 25-38.
Nord, A.G., Svärdh, A., Tronner, K.: 1994, *Atmospheric Environment* **28**, 2615-2622.
Pye, K., Schiavon, N.: 1989: *Nature* **342**, 663-664.
Rösch, H., Schwarz, H.J.: 1993: *Studies in Conservation* **38**, 224-230.
Thode, H. G., Monster, J., Dunford, H. B.: 1961, *Geochim. Cosmochim. Acta* **25**, 159-174.

CORROSION OF ARCHAEOLOGICAL BRONZE ARTEFACTS IN ACIDIC SOIL

K. TRONNER[1], A. G. NORD[1] and G. Ch. BORG[2]

[1] *Conservation Institute of National Antiquities, P.O.Box 5405, S-114 84 Stockholm, Sweden*

[2] *Department of Geology, Chalmers University of Technology, S-412 96 Göteborg, Sweden*

Abstract. Corrosion of bronze in soil is a well-known phenomenon. In particular, archaeological artefacts which may remain in the soil for thousands of years are subject to severe corrosion. However, bronze objects excavated 50-100 years ago seem to be less corroded than those found today. Therefore, recent pollution of the soil is suspected to accelerate the corrosion. An interdisciplinary project has been started in Sweden to search for correlations between the degree of bronze corrosion, corrosion products, general archaeological and environmental conditions, and parameters characterizing the soil chemically.

From three archaeological sites near Stockholm (Birka, Fresta, and Valsta), 33 bronze artefacts and related samples of soil have been investigated. All corrosion products and the metal core (if any) were analysed by SEM/EDS and XRD. Metal oxides, carbonates, sulphates, chlorides and phosphates have been identified. Each soil sample has been geologically classified, and a number of chemical analyses have been undertaken: pH in water and KCl, resistivity, loss on ignition, exchangeable acidity, chloride, phosphate, sulphur contents, acid-soluble cations extracted in two different ways, etc. About 8000 data have been compiled in an EXCEL data base. A statistical evaluation including multivariate modelling and analysis utilizing the SIMCA-S system, has been undertaken. The results so far obtained are only tentative but suggest that high concentrations of soot, sulphur or phosphate in the soil may have accelerated the corrosion of the investigated bronze objects. The influence of low pH values, though, is less clear.

Key words: Archaeological artefacts, bronze corrosion, soil chemistry.

1. Introduction

Archaeologists in Sweden and other countries have observed that artefacts from ancient remains when excavated today in general seem to be relatively more decayed than those found in the same region 50-100 years ago. In particular this is true for areas with substantial acidic deposition (cf. Scharff, 1993; Fjaestad and Ullén, 1995). It was reasonable to assume that acidic deposition and other pollutants have accelerated the decay or corrosion of the artefacts during the last decades. However, corrosion of metal objects in soil is a complex process, and it is certainly not easy to distinguish between effects of anthropogenic pollutants and other factors like metal composition, archaeological context, soil chemistry and characteristics, bacteria and other microorganisms, vegetation, climate, land use, geography, etc. In order to increase the knowledge of these matters, and help to settle appropriate countermeasures, an interdisciplinay project was started at the Central Board of National Antiquities in Sweden by Mrs Gunnel Werner (deceased) and Dr Agneta Lagerlöf. Many colleagues are now involved within various fields: archaeology, chemistry, geology, corrosion science, conservation, statistics, etc. (cf. Mattsson, 1993, Nord et al.,

1994; Borg et al., 1995). Classical works in this field have been published for instance by Logan (1942), Geilmann (1956), Arrhenius (1967), Booth et al. (1967), and Tylecote (1979).

In the first step we have restricted the material to bronze artefacts (including samples of soil) from iron age remains in the Stockholm area. After some preliminary investigations (Werner, 1992) we have now investigated three excavations: theViking age town of *Birka* on a small island in Lake Mälaren (28 km W Stockholm), a flat burial ground at *Fresta* 25 km NW Stockholm, and a hillside burial area at *Valsta* (Norrsunda) 33 km north of Stockholm. Archaeological, geographical and other comprehensive data have been obtained for the three sites. For each object a "degree of deterioration" has been determined. Metal remains, corrosion products and soil samples have been chemically analysed. All data have been compiled in an EXCEL data base and processed by the statistical SIMCA-S system. The results, still tentative, are summarized in the last section, which also includes guidelines for further efforts.

2. Materials and methods

Bronze artefacts. In total 33 bronze artefacts have been investigated, namely 7 from Birka, 13 from Fresta, and 13 from Valsta. The "degree of deterioration", a crucial parameter, was determined by an experienced conservator by means of visual inspection including an estimation of the extension and depth of the corrosion, and (whenever possible) a comparison of the volume between metal core and original size, supplemented by X-ray photographs. A tiny fragment of the metal core, if any was left, was used for a semi-quantitative analysis by SEM/EDS. The complex corrosion products, on the average two types per object, were analysed by SEM/EDS and XRD. Altogether, there are about 32 chemical data per bronze artefact.

Soil samples. In total there are about 200 samples of soil, taken close to the objects and, for reference, some distance away. A number of variables have been determined from chemical analysis: $pH(H_2O)$, $pH(KCl)$, humidity, loss on ignition, electric conductivity, exchangeable acidity, concentration of chloride, sulphur and phosphate, and an approximate total composition from XRF. In addition, certain metal ions extracted with 1-M ammonium acetate (buffered with acetic acid to pH = 4.8) respectively a concentrated solution of nitric and hydrochloric acid (3:1) have been analysed with a computer-controlled Spectra-Flame-Modula plasma spectrograph. The classification scheme developed by the Geological Survey of Sweden (SGU, Uppsala) has been simplified and adopted for computer processing in order to better accord with the statistical analysis and the specific archaeological application. For some selected soil samples a grain size distribution curve was also obtained. There are 37 analytical data for each sample of soil.

Comprehensive parameters. These comprise variables characterizing the archaeological context like type of remains (grave, site, village), occurrence of soot, other objects found near the bronze artefact, etc. Altogether there are around 33 archaeological parameters, which are used to classify each object. In addition every site has been coded (by 46 parameters) in terms of geology, geography, vegetation, land use, neighbouring local polluting sources, expected amount of atmospheric acid deposition, etc.

3. Some selected results

A survey of the result tables reveals some interesting differences between the three sites. The metal cores contain copper and tin, usually also some zinc and lead, and often traces of other elements. The bronze objects from Fresta seem to be somewhat better preserved than objects from the two other sites. Generally, the most common corrosion products are cuprite, malachite and amorphous hydrated tin dioxide. Other constituents are basic copper chlorides such as atacamite or paratacamite, tenorite (CuO), or copper sulphates like brochantite, antlerite, posnjakite or bonattite, sometimes smithsonite (zinc carbonate) or cerussite (lead carbonate). Sulphates as corrosion products were only found in Birka and Valsta. Copper chloride occurring as patches of corrosion resembling the so called "bronze disease" (e.g. Mattsson, 1992) were rather common on the Fresta objects.

The Birka soil is typically anthropogenic, with noteworthy high concentrations of soot, organic material, calcium, phosphate, sulphur, copper, zinc, lead, etc, and with a low acidity (pH around 8). It is significant that copper sulphates are frequently occurring on corroded Birka bronzes. The soil samples from Fresta and Valsta somewhat resemble each other, with a sandy or moraine type of soil, and lower pH values. The concentration of chloride is significantly higher in soil from Fresta, probably due to winter-time salting of a nearby road with heavy traffic. This agrees with the observation that atacamite, $Cu_2Cl(OH)_3$, is a frequent corrosion product in Fresta. A three-dimensional display of the geographical distribution of chlorides at Fresta is shown in Figure 1. Few bronze objects were found in soil with the highest concentrations of chloride; however, this may be due to the fact that any copper-rich objects may have corroded away completely.

The complete material now comprises more than 8000 numerical data. The statistical evaluation was undertaken at Umetri AB (Umeå, Sweden) with multivariate analysis by means of the SIMCA-S system in two steps: principal component analysis (PCA), and partial least-squares projections to latent structures (PLS). In the first step all soil parameters were analysed by PCA, indicating that the Birka soil is clearly separated from the two other sites (cf. Figure 2). A further separation of Fresta and Valsta was then carried out, which showed that the concentrations of calcium, aluminium, chloride and phosphate had the greatest influence on the separation. For these reasons, the continued regression analyses to link the bronze objects to the soil samples had to be carried out for each site separately. This procedure also had the advantage that a large number of the "comprehensive parameters" could be omitted in the analysis.

The ultimate aim of the study, however, is to correlate the degree of bronze deterioration to all other variables. Some results will be briefly discussed. Whatever site is considered, most regression coefficients (rc) turn out to be close to zero. In fact, half of the coefficients are in the region $-0.025 < rc < 0.025$ and should be regarded as insignificant. It is clear that there is a significant positive correlation between the degree of deterioration

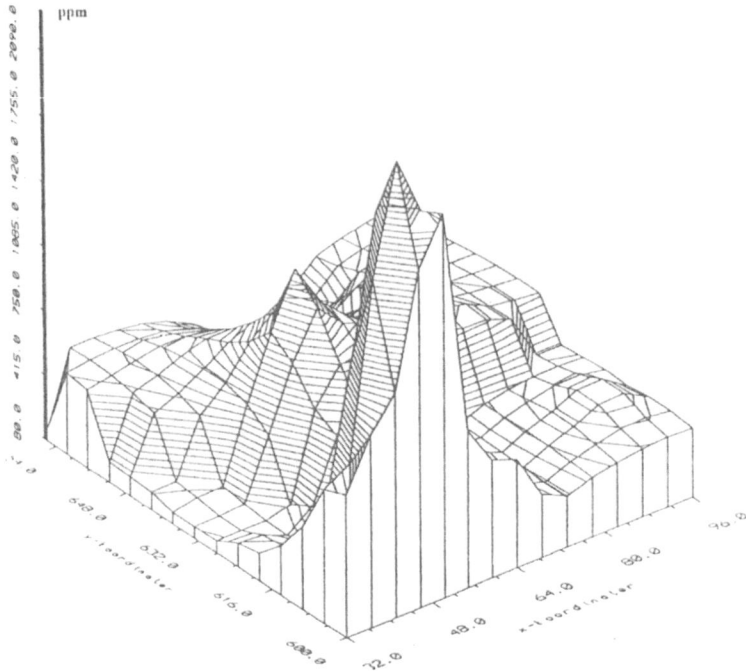

Fig. 1. Concentration of chlorine (ppm) for the soil samples from Fresta.

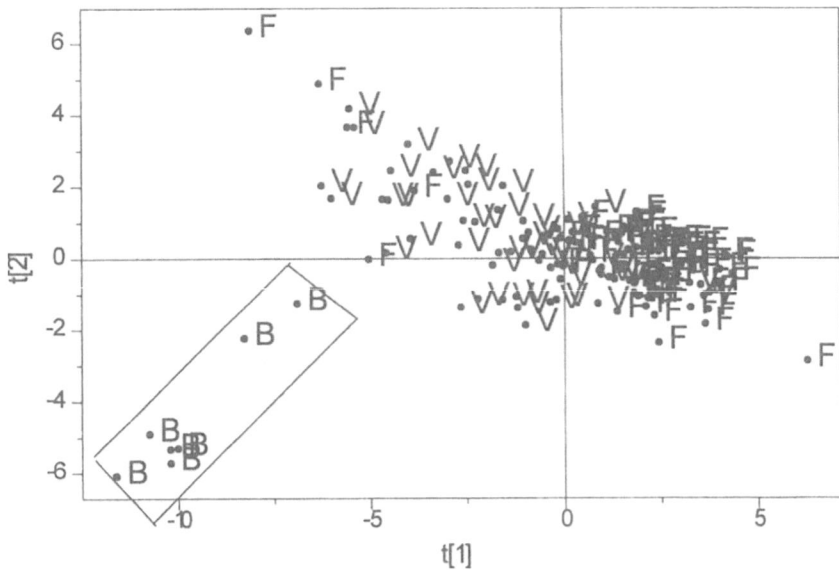

Fig. 2. The two first principal components for all soil samples plotted in a two-dimensional diagram. It is evident that Birka (B) differs significantly from Fresta (F) and Valsta (V).

and "corrosion depth", and a significant negative correlation for "remaining metal core". These two correlations are evident and relevant, since they confirm the reliability of the statistical analysis. Conclusions on the effect of pH values can not be drawn until objects from more acidification sensitive soil on the Swedish west coast have been included in the study. For the same reason, the dependence of soil type on the bronze corrosion cannot be definitely settled with the limited data present today, although a fine-grained soil seems to better preserve the objects. Significant positive correlations were in some cases observed for variables such as concentration of sulphur or phosphate in the soil, and certain parameters to describe the archaeological context, e.g. the occurrence of soot in the soil, or the presence of burnt bone close to the bronze objects. The respective regression coefficients (rc) were around 0.10-0.15. On the whole, though, so far only some tendencies but no definite answer to the question which factors most strongly affect the bronze objects in soil have been found.

4. Conclusions

The basic result of the present study is that so far only indications have been found as regards the most crucial parameters which accelerate the corrosion of archaeological bronze objects in soil. It must be remembered that the present corrosion problem is indeed very complex, which is evident from the literature (cf. Camitz and Vinka, 1989; Sederholm et al., 1992; Årebäck, 1994). In order to increase the reliability of the statistical analysis, additional data must be collected and included in the present data base. Some modifications in the investigation should be made:

-Incorporate other materials in the study, such as bone and iron.
-It is imperative that additional data from archaeological sites at other parts of the country, with different environmental and geological conditions as compared to the three investigated sites, be added to the study.
-The investigation of bronze artefacts excavated 50-100 years ago must be extended.
-The authors plan to establish co-operation with international groups interested in these problems.
-Practical experiments with bronze objects buried in soil.

With these modifications in mind the project will, hopefully, come to more definite conclusions. Still, the present conclusions can refute conventional ideas like those stating that archaeological bronze artefacts are protected in a soil rich in soot or phorphorus, or with a high pH value. The results may help decide for which type of archaeological site, and in which region, an excavation must be undertaken without delay in order to prevent further serious deterioration of the artefacts. They are also helpful for prognostication of expected conservation measures.

Acknowledgements

We are indeed grateful to many of our colleagues for valuable discussions and ideas: Ulf Lindborg, Einar Mattsson, Agneta Lagerlöf, Monika Fjaestad, and Inga Ullén. We also appreciate the assiduous and skilful technical assistance of Ingemar Österling, Per-Olof Ekström and Katalin Holényi.

References

Arrhenius, O.: 1967, *Geological Survey of Sweden (SGU) Series C*, **6 1**, Report C-626, 1-39.
Booth, G.H., Cooper, A.W., Cooper, P.M., Wakerley, D.S.: 1967, *Brit. Corrosion J.* **2**, 104-118.
Borg, G., Jonsson, L., Lagerlöf, A., Mattsson, E., Ullén, I., Werner, G.: 1995, *Konserveringstekniska Studier*, **9** , Central Board of National Antiquities, Stockholm (in Swedish with English summary). pp 1-190.
Camitz, G., Vinka, T.G.: 1989, *Proc. 2:nd Internat. Confer. Corrosion.*, CEOCOR, Napoli, pp 77-90.
Fjaestad, M., Ullén, I.: 1995, To be published.
Geilmann, W.: 1956, *Angew. Chemie* **68**, 201-211.
Logan, K.H.: 1942, *National Bureau of Standards, Washington*, **28**, RP1460, 379-400.
Mattsson, E.: 1992, *Konserveringstekniska Studier*, **5**, 1-27. Central Board of National Antiquities, Stockholm.
Mattsson, E.: 1993, *8:th International Restorer Seminar, Sárespatak 1993*, 223-238.
Nord, A.G., Tronner, K., Holényi, K., Österling, I., Fjaestad, M., Svärdh, A.: 1994, *Report RIK-A-1602:3074,3079*, pp 1-141, Conservation Institute of National Antiquities, Stockholm (in Swedish).
Scharff, W.: 1993, *Gefährdung archäologischer Funde durch immissionsbedingte Bodenversauerung*. Report, Landesdenkmalamt Baden-Wuerttemberg.
Sederholm, B., Svensson, T., Vinka, T.G.: 1992, *Byggforskningsrådet* Report R7:1992, Stockholm, Sweden.
Tylecote, R.F.: 1979, *J. Archaeolcg. Sci.* **6**, 345-368.
Werner, G.: 1992, personal communication.
Årebäck M.: 1994, Thesis, (Report A77), Inst. of Geology, Chalmers Univrsity of Technology, Göteborg.

The Conference Sad Song

The Acid Rain Effects Web

1. I heard that acidity damaged a lake
   Make no mistake, it damaged a lake
   Perhaps it will die

2. I heard air pollution damaged a tree
   About how it does it , we still don't agree
   Perhaps it will die

3. I heard that ozone damaged a crop
   It grew straight up and then started to flop
   Perhaps it will die

4. I heard acid rain damaged a church
   And still as we measured and still as we search
   It started to die

5. I heard aluminum damaged a root
   And as we can't see it, we don't give a hoot
   Perhaps it will die

6. I heard that ammonia damaged a heath
   The heather has gone, there's just grass beneath
   I think that it died

7. I heard photoxidants damaged a child
   And Rough he has asthma, it's only quite mild
   But perhaps he will die

8. I heard of a world that worshipped the car
   They tell me this worship is going too far
   Perhaps it will die

9. I heard energy use continues to soar
   Till it all gets used up and we can't find any more
   Perhaps we will die

10. So I'm back where I started, where the story began
    A tale of the world and the actions of man
    And tell me, who will die?

Tune: I know an old woman that swallowed a fly

The Conference Happy Song

The Acid Rain Scientists Web

1   T'was on the Monday morning, KingCarl Gustav came to call
He opened up the conference in front of one and all.
He identified the problems and his speech touched on one spot
That we all must work together if we wish to learn a lot.

Chorus   And it all makes work for the scientist to do.

2   So we started on emissions and our knowledge is quite high
About what comes from man and nature and gets into the sky
But once it has reacted, the problem is quite vexed
So we call in the deposition scientists to find what happens next

Chorus   And it all makes work for the scientist to do.

3   Now deposition scientists are quite a special breed
With fluxes and with scavenging, they give us what we need
But once the pollution comes to earth, interest goes from their mind
And we then call in the effects team to tell us what we'll find

Chorus   And it all makes work for the scientist to do.

4   Now the effects of air pollution are both complex and widespread
From acid lakes and damaged crops to heaths and trees so dead
From crumbling cultural heritage to acidifying soil
These dedicated scientists, they labour and they toil

Chorus   And it all makes work for the scientist to do.

5   But once they've got the answers on impacts great and small
We've got to use this data to the benefit of all
We've argued and debated and tried so many modes
That it came as a relief when we agreed on critical loads

Chorus   And it all makes work for the scientist to do.

6   Now the critical loads and levels gang, they help to fill the gaps
By providing lots and lots of lovely multicoloured maps
But once they have produced them, the rot quickly then sets in
As their results are swallowed up as the modollers move in

Chorus   And it all makes work for the scientist to do.

7   Now modellers are a curious breed with acronyms galore,
From PROFILE up to MAGIC and RAINS for evermore
But once their models crank away and out results do come
Then the argument increases and the air begins to hum

Chorus   And it all makes work for the scientist to do.

8   But rather like a marriage, for better or worse I Fear,
The analyst still strove to make a "silk purse from a sows ear"
They translate all the science for the policy men to see
Who make these last decisions that affect both you and me

Chorus   And it all makes work for the scientist to do.

9   So the wheel has gone full circle and its Friday closing day
We've had a week of science and what more can I say
We've still got lots of problems to solve on acid rain
Se see you all in Japan in five years time to do it all again!

Chorus   And it all makes work for the scientist to do.

Tune: The Gas Man Cometh with apologies to Michael Flanders and Donald Swan

SPRINGER NATURE

GPSR Compliance

The European Union's (EU) General Product Safety Regulation (GPSR) is a set of rules that requires consumer products to be safe and our obligations to ensure this.

If you have any concerns about our products, you can contact us on ProductSafety@springernature.com

In case Publisher is established outside the EU, the EU authorized representative is:

Springer Nature Customer Service Center GmbH
Europaplatz 3
69115 Heidelberg, Germany

The manufacturer's authorised representative in the EU is Springer Nature Customer Service Centre GmbH, Europaplatz 3, 69115 Heidelberg, Germany. If you have any concerns regarding our products, please contact ProductSafety@springernature.com

Printed and bound by CPI Group (UK) Ltd, Croydon, CR0 4YY

25/03/2026
02078236-0001